Library of Congress Cataloging-in-Publication Data

Nowell, Lisa H.
    Pesticides in stream sediment and aquatic biota : distribution, trends, and governing factors / Lisa H. Nowell, Paul D. Capel, Peter D. Dileanis.
          p.   cm. — (Pesticides in the hydrologic system ; v.4)
    Includes bibliographical references and index.
    ISBN 1-56670-469-3
    1. Pesticides—Environmental aspects—United States. 2. Organochlorine compounds—Environmental aspects—United States. 3. Water—Pollution—United States. 4. Contaminated sediments—United States. 5. Aquatic organisms—Effect of water pollution on—United States. I. Capel, Paul D. II. Dileanis, Peter D. III. Title. IV. Series.
    TD427.P35 N68 1999
    628.1'6842—dc21   99-052684
                                                                                                                                    CIP

No claim to original U.S. Government works
International Standard Book Number 1-56670-469-3
Library of Congress Card Number 99-052684
Printed 1999 in the United States of America 1 2 3 4 5 6 7 8 9 0
Printed on acid-free paper

INTRODUCTION TO THE SERIES

Pesticides in the Hydrologic System is a series of comprehensive reviews and analyses of our current knowledge and understanding of pesticides in the water resources of the United States and of the principal factors that influence contamination and transport. The series is presented according to major components of the hydrologic system—the atmosphere, surface water, bed sediments and aquatic organisms, and ground water. Each volume:

- summarizes previous review efforts;
- presents a comprehensive tabulation, review, and analysis of studies that have measured pesticides and their transformation products in the environment;
- maps locations of studies reviewed, with cross references to original publications;
- analyzes national and regional patterns of pesticide occurrence in relation to such factors as the use of pesticides and their chemical characteristics;
- summarizes processes that govern the sources, transport, and fate of pesticides in each component of the hydrologic system;
- synthesizes findings from studies reviewed to address key questions about pesticides in the hydrologic system, such as:

    How do agricultural and urban areas compare?

    What are the effects of agricultural management practices?

    What is the influence of climate and other natural factors?

    How do the chemical and physical properties of a pesticide influence its behavior in the hydrologic system?

    How have past study designs and methods affected our present understanding?

    Are water-quality criteria for human health or aquatic life being exceeded?

    Are long-term trends evident in pesticide concentrations in the hydrologic system?

This series is unique in its focus on review and interpretation of reported direct measurements of pesticides in the environment. Each volume characterizes hundreds of studies conducted during the past four decades. Detailed summary tables include such features as spatial and temporal domain studied, target analytes, detection limits, and compounds detected for each study reviewed.

Pesticides in the Hydrologic System is designed for use by a wide range of readers in the environmental sciences. The analysis of national and regional patterns of pesticide occurrence, and their relation to use and other factors that influence pesticides in the hydrologic system, provides a synthesis of current knowledge for scientists, engineers, managers, and policy makers at all levels of government, in industry and agriculture, and in other organizations. The interpretive analyses and summaries are designed to facilitate comparisons of past findings to current and future findings. Data of a specific nature can be located for any particular area of the country. For educational needs, teachers and students can readily identify example data sets that meet their requirements. Through its focus on the United States, the series covers a large portion of the global database on pesticides in the hydrologic system and international readers will find much that applies to other areas of the world. Overall, the goal of the series is to provide readers from a broad range of backgrounds in the environmental sciences with a synthesis of the factual data and interpretive findings on pesticides in the hydrologic system.

The series has been developed as part of the National Water Quality Assessment Program of the U.S. Geological Survey, Department of Interior. Assessment of pesticides in the nation's water resources is one of the top priorities for the Program, which began in 1991. This comprehensive national review of existing information serves as the basis for design and interpretation of studies of pesticides in major hydrologic systems of the United States now being conducted as part of the National Water Quality Assessment.

Series Editor

Robert J. Gilliom
U. S. Geological Survey

PREFACE

Residues of pesticides, especially the organochlorine pesticides, in bed sediment and aquatic biota have been an environmental concern since the 1960s. Because of their toxicity and persistence, the majority of organochlorine pesticides (including DDT) were banned in the United States during the 1970s. Yet, more than 20 years later, residues of DDT and other organochlorine pesticides continue to be detected in air, rain, soil, surface water, bed sediment, and aquatic and terrestrial biota throughout the world. Moreover, recent research suggests that low levels of some organochlorine pesticides have the potential to affect the development, reproduction, and behavior of fish and wildlife, and possibly of humans as well.

The primary goal of this book is to assess the current understanding of the occurrence and behavior of pesticides in bed sediment and aquatic biota—the two compartments of the hydrologic system in which organochlorine pesticides are likely to reach their highest levels. This book has two objectives. Much of the book concerns organochlorine pesticides—evaluation of their environmental fate, their distribution throughout United States rivers and streams, the extent to which residues have declined since most of these pesticides were banned, and the potential biological significance of the remaining residues. This coverage is a natural consequence of the historical importance of these compounds and their tendency to accumulate in sediment and biota. A second objective of this book—and an important one, despite there being relatively little information on this topic in the existing literature—is an assessment of the potential for currently used pesticides to accumulate in bed sediment and aquatic biota of hydrologic systems.

Previous reviews of pesticides in bed sediment or aquatic biota provide fairly thorough treatment of the occurrence, distribution, and trends of many organochlorine pesticides in the Great Lakes region and in coastal and estuarine areas of the United States. However, existing reviews do not provide the same perspective for bed sediment and aquatic biota in United States rivers and streams. To accomplish this, we have compiled the results of most published studies that measured pesticides in bed sediment or aquatic biota, or both, in rivers and streams in the United States. These studies include monitoring studies, which range from local to national in scale, as well as field experiments designed to assess the environmental fate of pesticides in hydrologic systems. The initial literature search covered reports published up to 1993, but many articles and reports published after 1993 were included as they became available. For all the studies reviewed, concise summaries of study sites, target analytes, and results are provided in a series of tables (at the back of the book).

There were good technical arguments for combining the review of pesticides in sediment and aquatic biota into a single book, despite the large volume of literature in each of these two areas. Because of their physical and chemical properties, the same chemicals tend to accumulate in both media. Also, a number of studies measured pesticides in both media at the same time, so that separating these media into two separate books would require duplication of effort for both the authors and the readers.

This book was made possible by the National Water Quality Assessment Program of the U.S. Geological Survey. The authors wish to express their thanks and appreciation for the suggestions, reviews, and assistance provided by many individuals in the development of this book. We are indebted to Steven Larson (U.S. Geological Survey) for his assistance in conducting bibliographic searches, for providing references and other materials, and for valuable discussions. We also wish to thank Loreen Kleinschmidt (Toxicology Documentation Center at the University of California, Davis) for her support in conducting literature searches, obtaining references, and assisting in many other ways during the research and writing phase of this book. Thanks also go to William Fitzpatrick and Joyce Calipto (formerly undergraduates at the University of California, Davis) for obtaining many references and entering them in a bibliographic database, to Jean Lucas (U.S. Geological Survey) for providing copies of references and other materials, and Gail Thelin (U.S. Geological Survey) for Geographic Information System support. To the many individuals, too numerous to mention by name, who sent us copies of their papers and reports, or provided lists of references, we are very grateful. We also wish to thank several individuals who assisted us in obtaining electronic data: Larry Shelton and Kathy Shay (U.S. Geological Survey data), Thomas O'Connor (National Oceanic and Atmospheric Administration data), and Peter Lowe and L. Rod DeWeese (U.S. Fish and Wildlife Service data). Thanks also go to Robert Gilliom, Jack Barbash, and Michael Majewski (U.S. Geological Survey) for helpful discussions of various topics covered in this book. Special thanks are due to Steven Goodbred (U.S. Geological Survey) for his technical review of portions of the book, helpful discussions of various topics, providing references on endocrine disruption, and assistance in summarizing some of the monitoring studies reviewed in this book. We are indebted to Herman Feltz (U.S. Geological Survey) and Gregory Foster (George Mason University) for providing timely, thorough, and helpful reviews of the manuscript. Their excellent suggestions greatly improved the quality of this book. We also wish to thank the authors and publishers who gave permission to reproduce various figures or tables from their publications.

Several employees of the data, cartography, and publications sections of the U.S. Geological Survey contributed to the production of this work. We are indebted to Naomi Nakagaki and Thomas Haltom for producing the maps presented in this book. Our grateful thanks go to Susan Davis, Yvonne Gobert, and Glenn Schwegmann for their considerable efforts in the preparation and editing of text, tables, illustrations, and references, and in producing a high-quality camera-ready work. Finally, we are indebted to our technical editor, Thomas Sklarsky, for his thorough, painstaking work in editing the manuscript, and his masterful job of organizing the production of the finished, camera-ready work.

Lisa H. Nowell
Paul D. Capel
Peter D. Dileanis

EDITOR'S NOTE

This work was prepared by the U.S. Geological Survey (USGS). Although it has been edited for commercial publication, some of the style and usage incorporated is based on the USGS's publication guidelines (i.e., *Suggestions to Authors*, 7th edition, 1991). For example, references with more than two authors are cited in the text as "Smith and others (19xx)," rather than "Smith, et al. (19xx)," and some common-use compound modifiers are hyphenated. For units of measure, the international system of units is used for most based and derived units except for the reporting of pesticide use, which is commonly expressed in English units (e.g., pound[s] active ingredient), and the concentration, usually expressed in the metric equivalent (e.g., micrograms per liter). In addition to the standard use of italics, identification of new terms when first used, or of technical terms when first defined, are also denoted by italic type.

Every attempt has been made to design figures and tables as "stand-alone," without the need for repeated cross reference to the text for interpretation of illustrations or tabular data. Some exceptions have been made, however, because of the complexity or breadth of the figure or table. As an aid in comparison, the same shading patterns are shown in the Explanation of all pesticide usage maps, though each pattern may not necessarily apply to every map. In some cases, a figure is shown just before its mention in the text to avoid continuity with unrelated figures or to promote effective layout. The U.S. Department of Agriculture's 10 farm production regions (Figure 3.13 and Section 3.3) are capitalized to denote proper names of specific geographical areas.

Some of the longer tables are located at the end of the chapter or in the appendixes to maintain less disruption of text. In Tables 2.1–2.3 (Appendixes A, B, and C), the analyte names are reported as in the original reference, and so, multiple common names for the same pesticide (e.g., DCPA and dacthal) occur in these tables. In Table 2.2, geographic and personal names are spelled in the same way as in the original reference. In addition to the complete spelling of a name, its abbreviation may have been used in the large tables (e.g., California and Calif.; River and R.) when necessary to economize cell space.

As an organizational aid to the author and reader, chapter headings, figures, and tables are identified in chapter-numbered sequence. The list of abbreviations and acronyms in the front of the book do not include chemical names, which are listed in Appendix D, or symbols and functions in mathematical equations, which are defined when used. Rather than creating new abbreviations for common terms, some of the abbreviations listed have multiple representations (e.g., "na" for *not applicable* and *not analyzed*), though the abbreviations are defined more precisely when used.

With the exception of the index, this work was edited, illustrated, and produced as camera-ready copy by the USGS's Pesticide National Synthesis publications team, Sacramento, California.

CONTENTS

CHAPTER 4

LIST OF FIGURES

LIST OF TABLES

Note: Pages out of sequence indicate that some tables have been placed at the end of the book.

CONVERSION FACTORS

Multiply	By	To obtain
centimeter (cm)	0.3937	inch (in.)
cubic meter (m^3)	35.31	cubic foot (ft^3)
gram (g)	0.03527	ounce, avoirdupois (oz)
hectare (ha)	2.469	acre
kilogram (kg)	2.205	pound, avoirdupois (lb)
kilometer (km)	0.6214	mile (mi)
liter (L)	0.2642	gallon (gal)
meter (m)	3.281	foot (ft)
square kilometer (km^2)	0.3861	square mile (mi^2)
square meter (m^2)	10.76	square foot (ft^2)

Multiply	By	To obtain
acre	0.405	hectare (ha)
cubic foot (ft^3)	0.02832	cubic meter (m^3)
foot (ft)	0.3048	meter (m)
gallon (gal)	3.7854	liter (L)
inch (in.)	2.54	centimeter (cm)
mile (mi)	1.6093	kilometer (km)
ounce, avoirdupois (oz)	28.350	gram (g)
pound, avoirdupois (lb)	0.45359	kilogram (kg)
square foot (ft^2)	0.09290	square meter (m^2)
square mile (mi^2)	2.5900	square kilometer (km^2)

Multiply	By	To obtain
atmosphere (atm)	1.01325×10^5	pascal (Pa)
calorie (cal)	4.1868	joule (J)
joule (J)	0.2388	calorie (cal)
pascal (Pa)	9.869×10^{-6}	atmosphere (atm)
pascal (Pa)	1.4507×10^{-4}	pounds per square inch (psi)
pounds per square inch (psi)	6.8947×10^3	pascal (Pa)

Temperature is given in degrees Celsius (°C), which can be converted to degrees Fahrenheit (°F) by the following equation:

$$°F = 1.8(°C) + 32$$

ABBREVIATIONS AND ACRONYMS

Note: Clarification or additional information is provided in parentheses. Abbreviations for chemical compounds are included in Appendix E.

Common Abbreviations

α, alpha
β, beta
δ, delta
γ, gamma
μ, mu (micro prefix)
μg, microgram
μm, micrometer
μmol, micromole
cm, centimeter
d, day
ft, foot (feet)
g, gram
h, hour
ha, hectare
in., inch
J, joule
kg, kilogram
kg_{oc}, kilogram of sediment organic carbon
kj, kilojoule
km, kilometer
L, liter, lake
lb, pound
lb a.i., pound(s) active ingredient
m, meter
mg, milligram
mm, millimeter
mol, mole
mv, millivolt
ng, nanogram
nm, nanometer
oz, ounce
pg, picogram
Pa, pascal
s, second

Government, Private Agencies, and Legislation

BofCF, Bureau of Commercial Fisheries
CDFA, California Department of Food and Agriculture
CSWRCB, California State Water Resources Control Board
FAO–WHO, Food and Agriculture Organization (United Nations)–World Health Organization
FDA, Food and Drug Administration
FFDCA, Federal Food Drug and Cosmetic Act
FIFRA, Federal Insecticide, Fungicide, and Rodenticide Act
FQPA, Food Quality Protection Act
FWS, Fish and Wildlife Service
NAS/NAE, National Academy of Sciences and National Academy of Engineering
NOAA, National Oceanic and Atmospheric Administration
NYDH, New York Department of Health
RCRA, Resource Conservation and Recovery Act
U.S., United States
USDA, U.S. Department of Agriculture
USEPA, U.S. Environmental Protection Agency
USGS, U.S. Geological Survey

Monitoring Programs and Surveys

BEST, Biomonitoring of Environmental Status and Trends
NASQAN, National Stream Quality Accounting Network
NAWQA, National Water Quality Assessment (Program)
NCBP, National Contaminant Biomonitoring Program
NMPFF, National Monitoring Program for Food and Feed
NS&T, National Status and Trends (Program)
NSCRF, National Study of Chemical Residues in Fish
NYSDEC, New York State Department of Conservation
NUPAS, National Urban Pesticide Applicator Survey
NURP, Nationwide Urban Runoff Program
PMN, Pesticide Monitoring Network
TSMP, Toxic Substances Monitoring Program (California)

Miscellaneous Abbreviations and Acronyms

A, air
AB, aquatic biota
ADP, adenosine diphosphate
aER, alligator estrogen receptor
AET, apparent effects threshold
AET-H, apparent effects threshold–high
AET-L, apparent effects threshold–low
AGRICOLA, a bibliographic database of the National Agricultural Library (USDA)

AL, action level
ATP, adenosine triphosphate
BAF, bioaccumulation factor
BCF, bioconcentration factor
BMF, biomagnification factor
BSAF, biota–sediment accumulation factor
Bt, *Bacillus thuringiensis* var. *Kurstaki*
C, Celsius, channel
C_{GM}, geometric mean concentration
C_H, high concentration (one standard deviation above the mean)
C_{max}, the maximum concentration reported in a study
COA, co-occurrence analysis
CP, chlorophenoxy (chlorophenol)
CPG, Compliance Policy Guide (FDA)
DL, detection limit
DOM, dissolved organic matter
$E_2/11KT$ ratio, ratio of 17β estradiol to 11-ketotestosterone
E_a, activation energy
EC_{50}, median effective concentration
EDL, elevated data level
E_H, redox potential
EP, equilibrium partitioning, extraction procedure
ERL, effects range–low
ERM, effects range–medium
F, f test statistic in analysis of variance
f, fugacity
FCL, Four County Landfill
FCV, final chronic value
f_{oc}, fraction of organic carbon (of a particle)
FTC, fish tissue concentration
FY, fiscal year
GC/MS, gas chromatography with mass spectrometric detection
GIT, gastrointestinal tract
GW, ground water
i, insufficient guidelines to determine a Tier 1–2 boundary value
ISQG, interim sediment quality guideline (Canadian)
K_D, solid-water distribution coefficient
kg_{oc}, kilograms of organic carbon in sediment
K_{oc}, organic carbon normalized sediment-water distribution coefficient
K_{ow}, n-octanol-water partition coefficient
L, lower screening value
LC, Lincoln City
LC_{50}, median lethal concentration
LD_{50}, median lethal dose
MIA, Miami International Airport

MFO, mixed function oxidase

misc, miscellaneous

na, not available, not analyzed

$NADH_2$, reduced nicotinamide-adenine-dinucleotide

NADPH, nicotinamide-adenine dinucleotide phosphate in its reduced form

nd, not detected

NOAEL, no observed adverse effects level

NoL, Niagara-on-the-Lake

nr, not reported

ns, not significant

NTIS, National Technical Information Service

NWR, National Wildlife Refuge

O, overbank

OC, organochlorine

OCDD, octachlorodibenzo-p-dioxin

OP, organophosphate

p, significance level

PAH, polynuclear aromatic hydrocarbon

PBPK, physiologically based pharmacokinetic (model)

PCDD, polychlorinated dibenzo-p-dioxin

PCDF, polychlorinated dibenzofuran

PEL, probable effect level

ppb, part per billion

ppm, part per million

ppt, part per trillion

$q_1{}^*$, cancer potency factor

QSAR, quantitative structure-activity relations

R, rain, river

r, correlation coefficient

RfD, reference dose

r.m., river mile

S, water solubility

SCV, secondary chronic value

SEC, steric effect coefficient

SLC, screening level concentration

sp., species

SPMD, semipermeable membrane sampling device

SQAL, sediment-quality advisory level

SQC, sediment-quality criterion

SQC_{oc}, sediment-quality criterion (organic carbon-normalized)

SSB, spiked sediment bioassay

STORET, STOrage and RETrieval System (USEPA)

STP, sewage treatment plant

SVOC, semivolatile organic compound

SW, surface water

t, total

$t_{1/2}$, half-life

TCB, trichlorobenzene

TCDD-EQ (or TEQ), 2,3,7,8-TCDD equivalents

TEF, toxic equivalency factor

TEL, threshold effect level

TEQ (or TCDD-EQ), 2,3,7,8-TCDD equivalents

TOC, total organic carbon

U, upper screening value

VOC, volatile organic compound

WC, Washington City

WMA, wildlife management area

WQC, water quality criterion

wt, weight

yr, year

>>, much greater than

<<, much less than

PESTICIDES IN STREAM SEDIMENT AND AQUATIC BIOTA

Distribution, Trends, and Governing Factors

Lisa H. Nowell, Paul D. Capel, and Peter D. Dileanis

ABSTRACT

More than 400 monitoring studies and 140 review articles published from 1960 to 1993 were reviewed to assess pesticide contamination in bed sediment and aquatic biota of United States rivers and streams. The studies reviewed included reports from five major national programs that monitored pesticides in river or estuarine sediment, fish, or mollusks. The individual studies reviewed differed substantially in objectives, study design, and analytical methods. Forty-one pesticides or their metabolites were detected in sediment and 68 in aquatic biota, representing 44 percent of those pesticides targeted in sediment and 64 percent of those targeted in biota. Most of the pesticides detected were organochlorine insecticides or their metabolites. This reflects the hydrophobicity and persistence of this class of pesticides, as well as bias in the target analyte list. In the studies reviewed, pesticides other than organochlorine insecticides (in bed sediment and aquatic biota) were rarely analyzed. Of these other pesticides, a number of moderately hydrophobic, moderately persistent ones were detected, although at lower frequencies than the more persistent organochlorine insecticides. These "moderate" pesticides have water solubilities of 0.05 to 1 milligram per liter and estimated soil half-lives of 30 to 150 days. Examples of these moderate pesticides include the insecticides chlorpyrifos, carbofuran, and diazinon; and the herbicides 2,4-D, dacthal, oxadiazon, and trifluralin.

Pesticide detection frequencies and concentrations reflect a combination of historical agricultural use, water solubility, environmental persistence, and analytical detection limits. In some studies, a few compounds were associated with land uses other than agriculture. These compounds include chlordane (urban and industrial use), mirex (red fire ant control and manufacturing inputs), and lindane and α-HCH (Superfund and industrial sites).

Existing data are adequate to assess nationwide trends for a number of organochlorine insecticides in freshwater fish, but not in bed sediment. For example, residues of DDT, chlordane, dieldrin, endrin, lindane, and α-HCH in freshwater fish declined nationally during the 1970s, then appeared to level off during the early 1980s. High concentrations have persisted in some local areas. Toxaphene residues in fish declined nationally during the early 1980s. The available data on currently used pesticides are not sufficient for assessment of trends in either bed sediment or aquatic biota.

CHAPTER 1

Introduction

About 1 billion pounds of pesticides currently are used each year in the United States to control many different types of weeds, insects, and other pests in a wide variety of agricultural and nonagricultural applications. The total quantity used, and the number of different chemicals applied, increased substantially from the early 1960s, when the first reliable records were established, to around 1980, then appeared to decrease or level off. For example, national use of pesticides in agriculture grew from about 370 million pounds of active ingredient in 1964 to a peak of 840 million pounds in 1979; since 1982, usage has fluctuated between 660 and 800 million pounds (Aspelin, 1997). Increased use of pesticides has resulted in benefits such as increased crop production, lower-cost maintenance, and control of public health hazards. At the same time, however, concerns and public awareness about the potential adverse effects of pesticides on the environment and human health also have grown.

Pesticide monitoring activities in the United States began in the mid-1960s in response to a directive from President John F. Kennedy in 1963 to implement recommendations made by the President's Science Advisory Committee that federal agencies develop a network to monitor pesticide residues in air, water, soil, fish, wildlife, and humans (Bennett, 1967). During the late 1960s and the 1970s, several federal agencies began the following monitoring programs: (1) the U.S. Environmental Protection Agency (USEPA) monitored pesticides in humans (adipose tissue, blood serum, and urine), soils (agricultural and urban), raw agricultural crops, surface water (including bed sediment), estuarine fish and shellfish, and ambient air in suburban locales; (2) the Food and Drug Administration analyzed pesticides in processed, ready-to-eat foods, in raw foods, and in animal feeds; (3) the U.S. Department of Agriculture monitored pesticides in meat and poultry; and (4) the U.S. Department of the Interior monitored pesticides in water, sediment, fish, and migratory and nonmigratory birds (Carey and Kutz, 1985). National monitoring activities, many of which continued into the 1980s and some into the 1990s, have been supplemented by many hundreds of state and local monitoring studies.

In many respects, the greatest potential for unintended adverse effects of pesticides is through contamination of the hydrologic system, which supports aquatic life and related food chains and is used for recreation, drinking water, and many other purposes. Water is one of the primary media in which pesticides are transported from targeted application sites to other parts of the environment. Thus, there is potential for movement into and through all components of the hydrologic cycle (illustrated in Figure 1.1). This is demonstrated by the repeated detection of

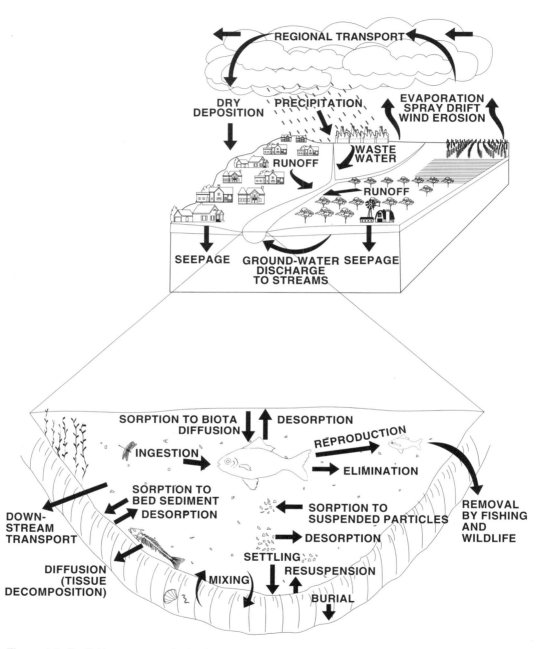

Figure 1.1. Pesticide movement in the hydrologic cycle. Inset: pesticide movement to, from, and within sediment and aquatic biotic phases of the hydrologic system. Adapted from Majewski and Capel (1995).

pesticides in United States ground waters (as in U.S. Environmental Protection Agency, 1990a; Barbash and Resek, 1996); freshwater lakes, reservoirs, rivers and streams (as in Gilliom and others, 1985; Allan and Ball, 1990; Pereira and Rostad, 1990; Schmitt and others, 1990; Larson and others, 1997); marine and estuarine systems (as in National Oceanic and Atmospheric Administration, 1987, 1988); and precipitation (as in Nations and Hallberg, 1992; Majewski and Capel, 1995). The detection of DDT and its transformation products in fish and mammals from the Arctic (Cade and others, 1968; Addison and Smith, 1974) and the Antarctic (George and Frear, 1966; Sladen and others, 1966; Peterle, 1969) demonstrates the global scale of pesticide distribution.

Pesticide contamination of bed sediment and aquatic biota is of long-standing concern. The pesticides of greatest concern in these media have been the *organochlorine insecticides*. These compounds contain predominantly carbon, hydrogen, and chlorine. As a class, the organochlorine insecticides tend to be *hydrophobic* (having little or no affinity for water), and are characterized by low water solubility, high octanol-water partition coefficients, and low-to-moderate volatility. Organochlorine insecticides are typically present at very low levels in the water column itself and are commonly associated with suspended particles. Many organochlorine insecticides also tend to be persistent because they are resistant to abiotic or microbial degradation in water or sediment, and to metabolism by aquatic and terrestrial organisms. Hydrophobic organochlorine insecticides may accumulate to substantial levels in bed sediment and aquatic biota via various phase-transfer and transport processes, as illustrated in the enlarged view of a cross-section of the stream shown in Figure 1.1. Therefore, these compounds may be detectable in bed sediment or aquatic biota even when concentrations in the water column are too low to be detected using conventional sampling and analytical methods. As a result, contaminant research and monitoring efforts frequently rely on sediment and aquatic biota as sampling media when targeting hydrophobic organic compounds.

The organochlorine insecticides were heavily used in agriculture, subterranean termite control, and malaria control programs from the mid-1940s to the mid-1960s. The persistence of the organochlorine insecticides, their tendency to accumulate in soil, sediment, and biota, and their impacts on wildlife brought this class of compounds into disfavor and eventually resulted in the restriction or suspension of most of them in the United States during the 1970s and early 1980s. Only four organochlorine insecticides were still used in agriculture in the United States in 1988: dicofol, endosulfan, lindane, and methoxychlor (Gianessi and Puffer, 1992b). However, DDT, toxaphene, and other organochlorine insecticides continue to be used elsewhere in the world, especially in developing countries in the southern hemisphere and in tropical areas (Tanabe and others, 1982).

Despite the use restrictions of organochlorine insecticides in the United States and Canada, these compounds continue to be detected in sediment and biological tissue samples collected throughout the United States today. Several national and multistate monitoring programs have measured organochlorine insecticides in aquatic biota and, to a lesser extent, in bed sediment. There also have been hundreds of smaller-scale studies that measured organochlorine insecticides in bed sediment and aquatic biota, and hundreds of other studies that focused on the processes by which such compounds are sorbed or bioaccumulated. Structurally similar compounds, such as polychlorinated biphenyls (PCB), chlorinated dibenzo-*p*-dioxins, and chlorinated dibenzofurans, also have been the focus of extensive research and monitoring efforts. These compounds also contain predominantly carbon, hydrogen, and chlorine, and are sometimes grouped with organochlorine insecticides under the label *organochlorine compounds.* Although these compounds are primarily industrial in origin, they resemble the organochlorine insecticides in physical and chemical properties, environmental fate processes, and tendency to accumulate in sediment and biota.

1.1 PURPOSE

The purpose of this book is to review the current understanding of pesticide contamination of bed sediment and aquatic biota in rivers and streams of the United States, with an emphasis on integration of information from studies conducted across a wide range of spatial and temporal scales. Existing information is used to evaluate, to the extent possible, the occurrence, geographic distribution, and trends of pesticide detections in bed sediment and aquatic biota in rivers and streams throughout the United States, factors that affect pesticide concentrations in these media, and potential effects of pesticides in bed sediment and aquatic biota on water quality. This book is one of a series of reviews of current knowledge of pesticide contamination of the hydrologic system that are being done as part of the Pesticide National Synthesis Project of the U.S. Geological Survey's (USGS) National Water Quality Assessment (NAWQA) Program. Other reviews in this series focus on pesticide contamination in the atmosphere (Majewski and Capel, 1995), in ground water (Barbash and Resek, 1996), and in surface water and suspended sediment (Larson and others, 1997). These national topical reviews are intended to complement field investigations conducted in more than 50 NAWQA study units across the United States. The NAWQA study units are major hydrologic basins, typically 10,000 to 30,000 mi^2 (Hirsch and others, 1988; Gilliom and others, 1995).

1.2 PREVIOUS REVIEWS

Previous reviews of existing information on one or more specific aspects of pesticides in sediment and aquatic biota in the United States have been published. A number of these review articles and books are listed in Table 1.1 (located at end of chapter), along with a brief description of their scope. Many are process-oriented reviews, focusing on bioconcentration or bioaccumulation (e.g., Johnson, 1973; Hamelink and Spacie, 1977; Kenaga and Goring, 1980; Spacie and Hamelink, 1985; Farrington, 1989; Barron, 1990; Landrum and others, 1992), sediment sorption and transport processes (e.g., Pierce and others, 1971; Lick, 1984; Elzerman and Coates, 1987), or sediment bioavailability (e.g., Neff, 1984; Knezovich and others, 1987). Many are reviews of monitoring efforts, but each is in some way limited in scope. Some monitoring reviews focus on a specific region of the United States, such as the Great Lakes (e.g., Baumann and Whittle, 1988; Alan and Ball, 1990), California coastal waters (e.g., Brown and others, 1983), specific estuaries such as Tampa Bay (Long and others, 1991) or Boston Harbor (MacDonald, 1991), or specific river basins such as the Illinois River (Steffeck and Striegl, 1989) or Kansas River (Jordan and Stamer, 1991) basins. Other monitoring reviews are confined to a specific pesticide or pesticide class, such as chlordane (Shigenaka, 1990), mirex (Kaiser, 1978; Lum and others, 1987), or chlorophenoxy acid herbicides (Norris, 1981), as noted in the scope of each review (see Table 1.1). Some reviews focus on specific pesticides in specific hydrologic systems, such as toxaphene in the Great Lakes (Sullivan and Armstrong, 1985), or kepone in the James River (Huggett and Bender, 1980; Huggett, 1989; Nichols, 1990); and a few focus on specific types of organisms, such as invertebrates (Kerr and Vass, 1973; Rosenberg, 1975) or zooplankton (Day, 1990). Several reviews summarize the goals, procedures, or findings of long-term national monitoring programs, such as the National Oceanic and Atmospheric Administration's (NOAA) National Status and Trends Program for Marine Environmental Quality (Shigenaka, 1990; O'Connor, 1991; O'Connor and Ehler, 1991) and the Fish and Wildlife Service's (FWS) National Contaminant Biomonitoring Program (NCBP) (O'Shea and others, 1979; Ludke and Schmitt, 1980). Among the most comprehensive reviews of their types are those by Shigenaka (1990), which reviews chlordane in the marine and estuarine environments of the United States; by Mearns and others (1988), which reviews PCB and organochlorine insecticide contamination of fish and shellfish in the marine and estuarine environments of the United States; and by Allan and Ball (1990), which reviews organochlorine contamination in water and sediment of the Great Lakes system. Table 1.1 lists the number of references cited in each review, which is a general indicator of the scope or technical detail of the review.

Perhaps because of the sheer abundance of data on hydrophobic pesticides and structurally related compounds in bed sediment and, especially, aquatic biota, the body of work in these areas has not been reviewed comprehensively. Together, the existing reviews in Table 1.1 provide a relatively good understanding of factors that affect the sources, transport, and fate of pesticides in sediment and aquatic biota. Existing reviews also provide fairly thorough treatment of the occurrence, distribution, and trends of many pesticides in sediment and aquatic biota in the Great Lakes area and in coastal and estuarine areas of the United States. Existing reviews do not, however, provide the same perspective for United States rivers and streams. The most recent publication on organochlorine contamination from the NCBP summarizes results of fish sampling from 1976 to 1984 (Schmitt and others, 1990). Because of the size and quality of this data set, this article alone provides much information on the geographic distribution and trends in organochlorine insecticide contamination of fish from major United States rivers. This NCBP data set serves as a foundation on which the present review will build, by considering bed sediment in conjunction with biological tissue analyses, and by including results of hundreds of other studies, both large and local in scale.

1.3 APPROACH

The primary emphasis of this book is on studies that assessed pesticide contamination in bed sediment or aquatic biota in freshwater rivers and streams in the United States at spatial scales ranging from individual stream reaches to the entire nation. The goal of the review process was to locate all studies within this scope that have been published in an accessible format, including journal articles, federal and state reports, and university reports. Additional studies were selectively reviewed to help explain a particular mechanism or phenomenon. These include studies from outside the United States (mainly Canada); studies on lakes, coastal or estuarine areas in the United States; studies that measured PCBs or chlorinated dioxins and furans (which are structurally related to organochlorine insecticides); and process-oriented field experiments.

An emphasis was placed on studies that measured pesticides other than the organochlorine insecticides in bed sediment or aquatic biota. In evaluating studies, original data generally were not reanalyzed.

The studies reviewed in this book were assembled through the combined use of bibliographic databases, references from reviewed reports, and pesticide files at the Toxicology Documentation Center at the University of California, Davis. The electronic databases used were AGRICOLA (U.S. Department of Agriculture), ChemAbstracts (American Chemical Society), Current Contents, MedLine, NTIS (U.S. Department of Commerce, National Technical Information Service), and Selected Water Resource Abstracts (USGS). In addition, published bibliographies on pesticides in the environment from the Pollution Abstracts database (1970–1989) were searched manually. Although all of the reports and papers identified from these databases were obtained, if possible, and evaluated, other studies undoubtedly exist in the literature that were not identified in the bibliographic searches. Although many reports from studies conducted by state and local agencies are included, many of the unpublished reports could not be obtained for this book. Other reports were out of print and could not be obtained at the time of this review. Therefore, the book reflects primarily the information available in the open scientific literature as of 1993.

Electronic databases were searched by linking one or more compound or compound class terms (such as DDT, chlordane, pesticide, herbicide, insecticide, or chlorinated hydrocarbon) with one or more appropriate media terms (such as sediment, fish, or mollusk). Citations that were excluded from the retrievals included publications focusing on water treatment, analytical method development, toxicology, pesticide application methods, clean-up of contaminated sites, computer model descriptions, and legislative or policy analyses.

The studies were then evaluated and are presented in four main sections. In Chapter 2, all studies reviewed are tabulated along with selected study features such as location, study period, media sampled, number of sites and samples, target analytes, detection frequencies, and analytical detection limits. This tabulation serves as an overview of (and reference to) the studies reviewed and provides the basis for an initial characterization of the nature, degree, and emphasis of study effort that has accumulated.

In Chapter 3, a national perspective on the occurrence, geographic distribution, and trends of pesticides in bed sediment and aquatic biota is developed from the results of the monitoring studies reviewed, especially the major national programs. This overview defines the status of existing information on geographic and long-term temporal distributions of different pesticides in sediment, fish, and shellfish from rivers and estuaries in the United States.

In Chapter 4, the primary factors that affect pesticide concentrations in bed sediment and aquatic biota are reviewed. This provides a basis for understanding the observed patterns in occurrence and distribution and for addressing specific key topics in more detail.

Finally, in Chapters 5 and 6, the reviewed studies are used to address key topics related to the occurrence of pesticides in bed sediment and aquatic biota in United States rivers. These topics represent basic points that should be understood in order to evaluate the sources, degree, and potential significance of pesticide contamination of these media. The answers vary in their completeness, reflecting the strengths and weaknesses of existing information.

Table 1.1. Summary of review articles and books on pesticides in bed sediment and aquatic biota in the United States or on processes governing sorption and bioaccumulation

[Reference: Citations for reviews are listed in the References column. Study Type: Process, review of processes governing sorption and bioaccumulation. Monitoring, review of results from monitoring studies that measured pesticides in bed sediment or aquatic biota. Location(s), locations of studies reviewed; for monitoring studies that were reviewed, this column identifies the states, bodies of water, and countries in which the reviewed studies were conducted; laboratory or artificial ecosystem studies that were reviewed are also identified. Abbreviations and symbols: E_H, redox potential; FWS, Fish and Wildlife Service; K_{ow}, n-octanol-water partitioning coefficient; na, not applicable; NOAA, National Oceanic and Atmospheric Administration; QSAR, quantitative structure-activity relations; R, river; USEPA, U.S. Environmental Protection Agency; USGS, U.S. Geological Survey; "x" indicates that the review addresses this topic; blank cell indicates that the review does not address this topic. Abbreviations of pesticides are defined in the Glossary in Appendix D.]

| Study No. | Reference | Study Type | | Location(s) | Sampling Media | | Number of References Cited | Summary Description |
		Process	Monitoring		Sediment	Biota		
1	Butler, 1966		x	Florida, South Carolina, Gulf of Mexico		x	5	Reviews laboratory studies on acute and chronic toxicity of pesticides to estuarine mollusks; pesticide residues in estuarine fish collected in fish kills.
2	Woodwell, 1967	x	x	Not specified		x	0	Nontechnical review of the fate of DDT in the environment, especially in relation to meteorological and biological cycles.
3	Johnson, 1968		x	Includes Arizona, California, Idaho, Montana, Washington; the Yellowstone R., Mississippi R.; laboratory studies	x	x	157	Review of pesticide toxicity to fish, including mechanisms of action and factors affecting toxicity and pesticide effects on and in aquatic food chains. Also reviews pesticide residues in surface water, suspended and surficial sediment, and aquatic biota. Discusses fish residues in detail by species and tissue type. Data reviewed includes both fresh water and saltwater.
4	Ware and Roan, 1970	x		na		x	84	Review of the interactions between pesticides and aquatic microorganisms, including sources of pesticides to the aquatic environment, pesticide toxicity, bioaccumulation, and metabolism. Includes organochlorine and organophosphate insecticides and herbicides.

Table 1.1. Summary of review articles and books on pesticides in bed sediment and aquatic biota in the United States or on processes governing sorption and bioaccumulation—*Continued*

Study No.	Reference	Study Type		Location(s)	Sampling Media		Number of References Cited	Summary Description
		Process	Monitoring		Sediment	Biota		
5	Pierce and others, 1971	x		na	x		71	Review of literature pertaining to pesticide adsorption in soils and sediments. Includes the effects of temperature, light, microorganisms, and soil type, and discusses theories of adsorption.
6	Cox, 1972	x	x	Pacific Ocean, Sargasso Sea		x	38	Review of DDT residues in marine phytoplankton. Also reviews toxic effects of DDT on marine phytoplankton.
7	Johnson and Ball, 1972	x	x	Great Lakes	x	x	41	Review of organic pesticide pollution in the Great Lakes, especially DDT in Lake Michigan.
8	Kenaga, 1972	x	x	United States		x	63	Early review of DDT and metabolite residues in various environmental compartments, including bioaccumulation factors. Discusses body weight, dietary intake, and biological and environmental factors affecting bioaccumulation. Also discusses the relation between bioaccumulation and physical/chemical properties of a pesticide. Note: terms (e.g., bioconcentration) are defined differently from current usage.
9	Mackenthun and Keup, 1972	x	x	Tennessee, Michigan, New York, Canada; laboratory studies		x	153	Review of 1970 publications concerning water pollution. Studies reviewed include reports on the following: pesticides in water, plankton, and mollusks in lakes and rivers; pesticide bioaccumulation; and pesticide toxicity.
10	Walsh, 1972	x	x	Estuaries worldwide, laboratory studies		x	159	Review of laboratory studies of the bioaccumulation and toxicity of insecticides and herbicides. Also includes field observations of pesticide residues in estuarine systems worldwide and field studies where residues are measured after application of a known quantity of a test pesticide.
11	Weber, 1972	x		na	x		406	Review of the interactions between organic pesticides and particulates in soil and aquatic systems. Includes characteristics of particulate matter, properties of ionic and nonionic pesticides (by class).

Table 1.1. Summary of review articles and books on pesticides in bed sediment and aquatic biota in the United States or on processes governing sorption and bioaccumulation—*Continued*

Study No.	Reference	Study Type		Location(s)	Sampling Media		Number of References Cited	Summary Description
		Process	Monitoring		Sediment	Biota		
12	Butler, 1973a		x	United States estuaries		x	5	Review of the results of National Estuarine Monitoring Program analyses of pesticide residues in estuarine mollusks, 1965–1972. Includes analysis of trends and seasonal variation.
13	Johnson, 1973	x	x	United States		x	159	Review of factors affecting pesticide residues in fish. Discusses uptake, persistence, degradation, and elimination of pesticide residues. Also discusses sources of, and hazards due to, pesticide residues in fish. Includes residue levels in whole fish across the United States.
14	Kerr and Vass, 1973	x	x	Worldwide		x	165	Review of chronic pesticide residues in aquatic invertebrates from diverse habitats. Marine and freshwater species are included. Also reviews sources of contaminant residues, bioaccumulation pathways, and factors influencing residue levels, and presents a model of pesticide bioaccumulation. Also reviews contaminant effects on invertebrate species and communities.
15	McDonald, 1973		x	Mississippi R. (Iowa)		x	3	Brief review of water quality in the Mississippi R. bordering Iowa, including pesticides detected in fish.
16	Pionke and Chesters, 1973	x		na	x		150	Review of pesticide–sediment–water interactions occurring within a watershed. Discusses pesticide persistence/degradation in soils, pathways of transport to aquatic systems, and losses to the atmosphere, to ground water, and in surface runoff. Includes limnological characteristics affecting spatial distribution of sediment and associated contaminants, the sorption process, and persistence/degradation in the aquatic system.
17	Butler, 1974		x	United States estuaries		x	20	Review of factors affecting persistent organochlorine residues in estuarine mollusks, such as the effects of species, age, season, tissue type, and time.
18	Butler, 1975		x	United States estuaries		x	2	Review of the USEPA National Estuarine Monitoring Program, including summary of contaminant residues in mollusks from 1965 to 1972.

Table 1.1. Summary of review articles and books on pesticides in bed sediment and aquatic biota in the United States or on processes governing sorption and bioaccumulation—*Continued*

Study No.	Reference	Study Type		Location(s)	Sampling Media		Number of References Cited	Summary Description
		Process	Monitoring		Sediment	Biota		
19	Harris and Miles, 1975		x	Great Lakes and tributaries	x	x	93	Review of pesticide residues in soils, crops, and aquatic ecosystems in the Great Lakes region. Includes residues in water, bottom sediment, and biota of tributaries to the Great Lakes. Discusses organochlorine, organophosphate, and carbamate insecticides; and fungicides, herbicides, and other pesticides.
20	Hurlbert, 1975	x	x	Not reported	x	x	198	Review of secondary effects of pesticides on aquatic ecosystems. Includes table of pesticide levels in water, suspended sediment, mud, and vegetation following pesticide treatment. Discusses effect of exposure route (e.g., diet).
21	Rosenberg, 1975	x	x	Various (includes Mississippi, Atlantic coast; United Kingdom)		x	70	Review of literature reporting contaminant residues in invertebrate communities to determine whether there is evidence for trophic level effects. Includes field studies and laboratory studies where feeding was allowed to occur. Discusses the role of habitat and feeding behavior.
22	U.S. Environmental Protection Agency, 1975	x	x	United States	x	x	406	Review of the scientific and economic aspects of the USEPA decision to ban DDT. Includes USEPA's findings on the effects of DDT on fish and wildlife (177 references cited), its effects on humans (85 references), residues in the environment and humans (112 references), and economic impacts of the ban (32 references). Summarizes residues in soil, sediment, fish, crops and other human foods. Discusses persistence in soil and the aquatic environment, aerial drift from application sites, and soil erosion.
23	Bevenue, 1976	x	x	United States, Canada, Europe, Asia Australia		x	215	Review of bioaccumulation of DDT. Includes sources of DDT contamination (soil, microorganisms, air, rain, water, and sediment). Discusses food chain magnification, dietary uptake, effects on birds, and human exposure. Note: terms (e.g., bioconcentration) are defined differently from current usage.

Table 1.1. Summary of review articles and books on pesticides in bed sediment and aquatic biota in the United States or on processes governing sorption and bioaccumulation—Continued

Study No.	Reference	Study Type		Location(s)	Sampling Media		Number of References Cited	Summary Description
		Process	Monitoring		Sediment	Biota		
24	Grant, 1976	x	x	California, Missouri R., Mississippi R., unspecified		x	49	Review of the toxicity, uptake and bioaccumulation, and residues of endrin in freshwater. The principal focus is on toxicity, with only four references cited pertaining to residues in biota, suspended material, or water.
25	Terry and Hughes, 1976		x	Mississippi, Pennsylvania, Great Lakes, Canada. South Africa	x	x	141	Review of 1975 publications on pollution effects on surface and ground waters. Selected studies measured pesticides or PCBs in water, sediment, fish, and(or) soil. Data reviewed include lakes, rivers, and coastal areas.
26	Faust, 1977	x		na	x		88	Review of transport and transformation mechanisms of organic contaminants in aquatic environments.
27	Hamelink and Spacie, 1977	x	x	Various; laboratory studies		x	57	Reviews laboratory and field studies pertaining to bioaccumulation of hydrophobic pesticides by fish. Also reviews partition and kinetic models of bioaccumulation, considering growth and bioenergetics.
28	Li, 1977		x	United States estuaries (and some rivers)	x	x	65	Review of pesticide residues in water, sediment, mollusks, and seabirds from United States rivers and estuaries. Includes discussion of routes of pesticide entry into aquatic environments and reports of fish kills by pollution source. Briefly reviews the effects of pesticides on estuarine organisms. Focuses on agricultural use of pesticides.
29	Metcalf, 1977	x	x	Lakes Michigan and Cayuga, New York; laboratory studies		x	42	Review of bioaccumulation and biotransformation of contaminants in aquatic systems. Includes field measurements of bioaccumulation factors. Note: terms (e.g., bioconcentration) are defined differently from current usage.
30	Teal, 1977	x		na		x	36	Presents evidence for food web magnification in aquatic systems. Discusses selective uptake, uptake from water via the gills, and hydrocarbon storage and depuration.

Table 1.1. Summary of review articles and books on pesticides in bed sediment and aquatic biota in the United States or on processes governing sorption and bioaccumulation—Continued

Study No.	Reference	Study Type		Location(s)	Sampling Media		Number of References Cited	Summary Description
		Process	Monitoring		Sediment	Biota		
31	Whittle and others, 1977		x	Worldwide		x	137	Review of contaminant residues in marine animals worldwide. Includes organochlorine pesticides. Describes metabolic and environmental degradation.
32	Johnson, 1978		x	United States, especially Great Lakes		x	18	Brief, general review of the occurrence of toxic organic residues in fish. Topics covered include multiple residues and effects on higher trophic levels.
33	Kaiser, 1978		x	Great Lakes, Niagara R., St. Lawrence R., Atlantic coast, (Georgia, North Carolina, South Carolina)		x	26	Review of the use of mirex and the history of its detection as an environmental contaminant in the United States and Canada. Includes residues of mirex and photomirex in fish and bird eggs. Briefly discusses the history of kepone contamination due to manufacturing inputs in Virginia.
34	Phillips, 1978	x		na		x	187	Review of the use of biological indicator organisms to monitor aquatic pollution by organochlorine compounds. Includes biological factors (e.g., body lipid, age, size, weight, sex, and sexual cycle of the organism sampled) and environmental factors (e.g., salinity, temperature) affecting residue levels.
35	Strachan and Glass, 1978		x	Lake Superior	x	x	44	Review of existing data on organochlorine contaminant residues in water, surficial sediment, and fish of Lake Superior.
36	Macek and others, 1979	x		na		x	29	Laboratory study of aqueous and dietary uptake of the organophosphate insecticide, leptophos, and two other organic compounds by bluegills. Also reviews other studies on bioaccumulation of pesticides (DDT and kepone) in aquatic food chains.
37	O'Shea and others, 1979		x	United States nationwide		x	0	Nontechnical review of the goals, general approach, and findings (through 1977) of the FWS National Contaminant Biomonitoring Program. Includes fish and waterfowl.
38	Eisenreich and others, 1980	x	x	Lake Superior	x		76	Review of sources and pathways of PCBs and DDT in Lake Superior, especially dynamics in surficial sediment.

Table 1.1. Summary of review articles and books on pesticides in bed sediment and aquatic biota in the United States or on processes governing sorption and bioaccumulation—*Continued*

Study No.	Reference	Study Type Process	Study Type Monitoring	Location(s)	Sampling Media Sediment	Sampling Media Biota	Number of References Cited	Summary Description
39	Huggett and Bender, 1980	x	x	James R. (Virginia)	x	x	16	Review of kepone residues and toxic effects in the James R. (Virginia), adjacent to a former manufacturing facility.
40	Kenaga and Goring, 1980	x		na	x	x	215	Compilation of bioconcentration factors in fish and sorption coefficients for 170 chemicals. Also analyzes relations among these parameters, water solubility, and *n*-octanol–water partition coefficient.
41	Ludke and Schmitt, 1980		x	United States		x	0	Review of monitoring efforts and procedures used in the National Pesticide Monitoring Program. Includes summary of results (1967–1977) from freshwater fish monitoring by the FWS: compounds detected, geographic distribution, and temporal trends.
42	National Research Council, 1980	x		na		x	11*	*References on halogenated hydrocarbons only. Review of the use of bivalves as biomonitors for environmental contaminants, as in global Mussel Watch Program. Addresses compositing strategy, chemical analysis, target analytes, and lipid normalization.
43	Phillips, 1980	x		na		x		Review of biological indicator organisms to monitor aquatic pollution by organochlorine and other contaminants. Includes effects of biological factors such as body lipid, age, size, weight, sex, and sexual cycle of the organism sampled on residue levels.
44	Norris, 1981	x	x	Stream examples include California, Florida, New Hampshire, North Carolina, Oklahoma, Oregon; laboratory and artificial ecosystem studies	x		210	Reviews use of phenoxy herbicides in forests; their environmental fate in forests, including residues in air, vegetation, forest floor, and soil; their fate and persistence in forest streams; and bioaccumulation. Also reviews fate of TCDD in forests, including photodegradation, residues in vegetation, soil, and water, and bioaccumulation. Data reviewed included lab and field studies. Few examples were available on bioaccumulation in aquatic systems.

Table 1.1. Summary of review articles and books on pesticides in bed sediment and aquatic biota in the United States or on processes governing sorption and bioaccumulation—*Continued*

Study No.	Reference	Study Type		Location(s)	Sampling Media		Number of References Cited	Summary Description
		Process	Monitoring		Sediment	Biota		
45	Bahner and Oglesby, 1982	x		na		x	11	Review of models for predicting bioaccumulation and toxicity of kepone.
46	Freitag and Klein, 1982	x		na		x	20	Review of laboratory and field experiments investigating bioaccumulation of pesticide residues in fish. Includes metabolism and elimination studies.
47	Reish and others, 1982		x	Rhode Island, California, Virginia, Washington, Italy, Denmark, Mediterranean Sea, North Pacific and Indian Oceans	x	x	392	Review of 1980–1981 publications pertaining to marine and estuarine pollution. Includes papers on biomonitoring, bioaccumulation, toxicity, and residues (in water, suspended particulates, bottom sediment, and biota) of pesticides.
48	Brown and others, 1983	x	x	Southern California coast	x	x	54	Review of contaminants in sediments and marine organisms in California coastal waters. Includes residues in sediment and biota, changes in community assemblages, food web magnification studies, and enzyme assays to determine whether detoxification capacity is exceeded for contaminant classes.
49	Burns, 1983	x		na	x	x	56	Review of quantitative bases for describing transport, transfer, and degradation processes of chemicals in aquatic systems. Includes hydrodynamic transport, ionization, sorption, particle transport, interactions with benthic organisms, bioaccumulation, volatilization, photolysis, oxidation, hydrolysis, and microbial transformation. Provides aquatic fate codes used in several computer models.
50	Kauss and others, 1983		x	Great Lakes and tributaries		x	11	Review of PCB residues in water, suspended and bottom sediments, and biota of the Great Lakes and tributaries. Also includes some data on DDT and mirex in fish.

Table 1.1. Summary of review articles and books on pesticides in bed sediment and aquatic biota in the United States or on processes governing sorption and bioaccumulation—*Continued*

Study No.	Reference	Study Type		Location(s)	Sampling Media		Number of References Cited	Summary Description
		Process	Monitoring		Sediment	Biota		
51	Wennekens, 1983	x		Western Alaska		x	30	Review of FWS Resource Contaminant Assessment Program objectives and resources to be protected in Western Alaska. Reference is an annual progress report containing several unrelated chapters pertaining to different aspects of the program. Reviews types of local contaminants by source: urban (including household pesticides), atmospheric, service industries, and military facilities. Also reviews their impacts on wildlife. Reviews local uses of herbicides and their toxicity. Analyzes local impacts of mining. Reviews use of bivalves to monitor selected contaminants in coastal waters.
52	Adams, 1984	x		na	x	x	16	Review of bioavailability of lipophilic organic chemicals as it relates to feeding habits of freshwater benthic invertebrates, route of exposure, and the impact of sediment organic carbon. The fugacity model is discussed.
53	Biddinger and Gloss, 1984	x		na		x	212	Review of the literature concerning trophic transfer and biomagnification of 11 metals and 13 organic chemicals (all priority pollutants). Also reviews theoretical information pertaining to bioconcentration, biomagnification, and bioavailability.
54	Eidt and others, 1984		x	Examples include California, Great Lakes, Canada		x	69	General review of risks and benefits of pesticide use in forestry and agriculture, pesticide toxicity, techniques for minimizing adverse effects on aquatic systems, and regulation. Includes case histories of pollution problems.
55	Kay, 1984	x		Worldwide; laboratory studies		x	127	Review of literature on biomagnification of contaminants within aquatic ecosystems. Contaminants with potential to biomagnify in aquatic food chains are identified. Includes laboratory and field studies, and marine and freshwater systems.
56	Lick, 1984	x		na	x		5	Review of sediment transport processes in aquatic systems, including flocculation, entrainment, deposition, wave action, and currents.

Table 1.1. Summary of review articles and books on pesticides in bed sediment and aquatic biota in the United States or on processes governing sorption and bioaccumulation—*Continued*

Study No.	Reference	Study Type		Location(s)	Sampling Media		Number of References Cited	Summary Description
		Process	Monitoring		Sediment	Biota		
57	Neff, 1984	x		na	x	x	27	Review of bioaccumulation factors from sediment to biota for nonpolar organic micropollutants. Focuses on PAHs and PCBs. Includes laboratory and field studies, and freshwater and marine biota.
58	Oliver, 1984	x	x	Niagara R. and Lake Ontario	x	x	32	Review of processes responsible for losses of chlorobenzenes (including HCB) from water of the Niagara R. and Lake Ontario: volatilization, sedimentation, and discharge.
59	Segar and Davis, 1984		x	New York Bight; worldwide	x	x	314	Comparison of contamination in the New York Bight with that in coastal marine waters worldwide, especially in relation to population, industrialization, and hydrographic characteristics. Contaminants include trace metals, chlorinated insecticides, and PCBs.
60	Strachan and Edwards, 1984	x	x	Lake Ontario and tributaries	x	x	54	Review of organic contaminants in Lake Ontario and its tributaries. Includes sources, loadings, and residues in water, suspended sediment, surficial sediment, and biota. Includes several organochlorine pesticides and dioxins.
61	U.S. Environmental Protection Agency, 1984		x	United States		x	124	Position document describing USEPA's regulatory actions concerning dicofol and its determination that uses of dicofol contaminated with DDT compounds will result in unreasonable adverse effects on nontarget wildlife. Presents risk assessment for dicofol. Includes summary of DDT residues in fish and birds nationwide and in sensitive areas (California, Arizona, Texas, Florida)
62	Carey and Kutz, 1985		x	United States	x		15	Summary of results from national pesticide monitoring programs in the United States in the 1970s. Environmental media discussed include drinking water, soil, selected crops, air, surface water, sediment, and human urine.
63	Czuczwa and Hites, 1985	x	x	Worldwide	x		15	Review of the fate of dioxins and dibenzofurans in air, soil, water, and aquatic sediments.

Table 1.1. Summary of review articles and books on pesticides in bed sediment and aquatic biota in the United States or on processes governing sorption and bioaccumulation—*Continued*

Study No.	Reference	Study Type		Location(s)	Sampling Media		Number of References Cited	Summary Description
		Process	Monitoring		Sediment	Biota		
64	Hallett, 1985	x		Great Lakes and selected tributaries	x	x	30	Review of sources, loadings, and fate of dioxins in the Great Lakes ecosystem, including Niagara R., and Cayuga and Black creeks. Also includes history of monitoring and remediation efforts.
65	Kreis and Rice, 1985		x	Lake Huron	x	x	126*	* Sum of references at end of all 8 chapters. Review of organic residues in the atmosphere, water, fish, algae, sediments, and herring gull eggs from Lake Huron.
66	Lathrop and Davis, 1985	x	x	Includes Great Lakes, Washington, Soviet Union, Argentina; laboratory studies	x		235	Review of 1984 publications concerning aquatic sediments. Studies mentioned include reports on PCBs, dioxins and PAHs in freshwater and marine sediments, and on pesticide fate processes.
67	Mosher, 1985	x		Great Lakes		x	0	Commentary/review on soil erosion from farmland and related impacts, such as effects of nutrients and pesticides on aquatic systems. Briefly addresses pesticide toxicity, bioaccumulation of pesticides, and fish consumption advisories for the Great Lakes.
68	Spacie and Hamelink, 1985	x		na		x	126	Review of the bioaccumulation process, including uptake from water, special transport, adsorption, uptake from food, bioavailability in water, and elimination. Also reviews kinetic models of bioaccumulation (1- and 2-compartment bioconcentration models and food chain transfer model) and methods for estimating bioconcentration/ bioaccumulation.
69	Sullivan and Armstrong, 1985	x	x	Great Lakes		x	61	Review of the status of toxaphene in the Great Lakes, including residues in water and fish. Discusses sources of toxaphene to the Great Lakes. Also reviews physical properties of toxaphene, its analytical methodology, environmental fate, and toxicity to aquatic life and wildlife.

Table 1.1. Summary of review articles and books on pesticides in bed sediment and aquatic biota in the United States or on processes governing sorption and bioaccumulation—*Continued*

Study No.	Reference	Study Type		Location(s)	Sampling Media		Number of References Cited	Summary Description
		Process	Monitoring		Sediment	Biota		
70	Esser, 1986	x		na		x	71	Review of factors affecting bioaccumulation, including correlations between bioaccumulation and various physical/chemical properties.
71	Grassle and others, 1986		x	Deep oceans worldwide		x	119*	* Number of references from chapters on body burdens (Ch. IIIA and IIIB). Review of existing data on body burdens of organic chemicals and trace metals for deep ocean organisms and other open ocean biota. Also reviews relative sensitivities of deep sea animals.
72	Hallett and Brooksbank, 1986		x	Lake Ontario and Niagara R.	x	x	35	Review of TCDD residues in the Lake Ontario ecosystem, including temporal trends, sources, and loadings. Data reviewed are for surficial sediment, fish, and herring gulls in Lake Ontario, and water and suspended sediment in the Niagara R. Also summarizes residues in human milk and adipose tissue.
73	Lathrop and Davis, 1986	x	x	Includes New York, Alaska, Great Lakes, Mediterranean and North Seas; laboratory studies	x	x	305	Review of 1985 publications concerning aquatic sediments. Studies mentioned include reports on pesticides, PCBs or other organics in river, lake, and ocean sediments, and on pesticide fate processes.
74	Rossman, 1986		x	Lake Huron	x	x	39	Review of the results of 1980 monitoring program in Lake Huron. Includes summary of existing data, including organic contaminants in plankton, filamentous algae, fish, herring gull eggs, water, sediment, and atmosphere.
75	Charles and Hites, 1987		x	Great Lakes, Pettaquamscutt R. (Rhode Island), Hudson R. (New York), California coast	x		71	Review of approaches and techniques used to conduct dated sediment core measurements. Includes examples of reconstructed histories, especially from the Great Lakes.

Table 1.1. Summary of review articles and books on pesticides in bed sediment and aquatic biota in the United States or on processes governing sorption and bioaccumulation—Continued

Study No.	Reference	Study Type		Location(s)	Sampling Media		Number of References Cited	Summary Description
		Process	Moni-toring		Sedi-ment	Biota		
76	Davis and others, 1987	x	x	Includes California, Maine, Wisconsin, Great Lakes, Canada Switzerland; laboratory studies	x		200	Review of 1986 publications concerning aquatic sediments. Studies mentioned include reports on pesticides, PCBs or PAHs in lake, estuarine, and wetland sediments, and on pesticide fate processes.
77	Eadie and Robbins, 1987	x		Great Lakes	x		76	Review of the role of particulate matter on contaminant movement in the Great Lakes: lake particulate matter, particle-contaminant interactions, bioturbation and deposition-resuspension processes, and geochronology in sediment cores. Also presents model of contaminant fate in sediment.
78	Elzermann and Coates, 1987	x		na	x		137	Text-like review of the sorption process for hydrophobic organic compounds onto sediments. Also reviews empirical, conceptual, and mechanistic phases of model development to predict environmental fate of hydrophobic compounds. Equilibrium exchange and kinetic models are discussed.
79	Farrington and others, 1987		x	United States, United Kingdom		x	25	Overview of the rationale for, and limitations of, using bivalves as indicators of coastal environmental quality. Includes summary of results from Mussel Watch programs in United Kingdom and United States.
80	Knezovich and others, 1987	x		na	x	x	91	Review of processes and mechanisms through which organic contaminants are mobilized from sediments to biota, and factors affecting bioavailability of sediment-associated contaminants. Also discusses effects of contaminated sediments on aquatic biota.
81	Logan, 1987		x	Lake Erie	x	x	49	Review of sources and loadings of major pollutants to Lake Erie. Includes summary of pesticide residues in lake water, sediments, and biota.

Table 1.1. Summary of review articles and books on pesticides in bed sediment and aquatic biota in the United States or on processes governing sorption and bioaccumulation—*Continued*

Study No.	Reference	Study Type		Location(s)	Sampling Media		Number of References Cited	Summary Description
		Process	Monitoring		Sediment	Biota		
82	Lum and others, 1987		x	St. Lawrence R. (New York, Canada)	x	x	36	Analysis of relative importance of suspended particulates in Lake Ontario versus eels migrating down the St. Lawrence R. to the long-range transport of mirex from Lake Ontario to the St. Lawrence R. estuary. Includes mass balance for mirex and sediment budget for Lake Ontario.
83	Moriarty and Walker, 1987	x		na		x	16	Review of several approaches to investigating accumulation of pollutants along food chains. Discusses the influence of organism size and metabolism.
84	Oliver, 1987	x	x	Niagara R. and Lake Ontario	x	x	58	Review of the fate of chlorobenzenes in the Niagara R., the river plume, and Lake Ontario. Includes monitoring results and discussion of sediment-water partitioning, sedimentation, volatilization, photodegradation, biodegradation, and bioaccumulation.
85	Allan, 1988	x	x	St. Lawrence R. (Canada, New York)	x	x	28	Review of toxic metal and organic chemical sources and fate in the Lawrence R. and its upper estuary. Includes mirex.
86	Baumann and Whittle, 1988		x	Great Lakes	x	x	37	Review of the status and trends in residues of total DDT, PCBs, PCDD, PCDF, and PAHs in fish from the Great Lakes. Some data were reviewed on residues in sediment, birds, plankton and other aquatic biota from the Great Lakes.
87	Connell, 1988	x		na		x	94	Review of the process of bioaccumulation of organic chemicals by aquatic organisms, including bioconcentration and biomagnification mechanisms, QSAR for bioaccumulation, and kinetics.
88	D'Itri, 1988		x	Great Lakes and tributaries		x	79	Review of literature reporting contaminant residues in Great Lakes fish from 1966 to 1986, and analysis of trends.

Table 1.1. Summary of review articles and books on pesticides in bed sediment and aquatic biota in the United States or on processes governing sorption and bioaccumulation—*Continued*

Study No.	Reference	Study Type		Location(s)	Sampling Media		Number of References Cited	Summary Description
		Process	Monitoring		Sediment	Biota		
89	Davis and Denbow, 1988	x	x	Includes Great Lakes, Washington, Florida, Japan; laboratory studies	x		219	Review of 1987 publications concerning aquatic sediments. Studies mentioned include reports on pesticides, PCBs or other organics in river, estuarine, and ricefield sediments, pesticide fate processes, and effects of bioturbation on contaminant fate or toxicity.
90	Eadie and others, 1988	x		Great Lakes	x	x	49	Review of contaminant behavior in sediment that affects benthic invertebrates: sediment mixing (bioturbation); bioaccumulation and food web transport; and chemical stresses on, and alteration of, community structure.
91	Hellawell, 1988	x		na		x	106	Review of the behavior of toxic pollutants in aquatic systems and their environmental effects.
92	Kennicutt and others, 1988		x	Gulf of Mexico	x	x	129	Review of data on distribution of organic contaminants in sediments, water, and biota from the Gulf of Mexico.
93	Long and others, 1988		x	San Francisco Bay (California)	x	x	104	Review of status and trends in concentrations of selected toxicants in San Francisco Bay and estimation of the potential for biological effects. Synthesizes existing information from many surveys and monitoring programs. Includes DDT and metabolite residues in sediment, fish, crustaceans, and bivalve mollusks from four basins in the San Francisco Bay system, including some peripheral waterways. Emphasizes relations between contaminant residues and biological effects in the field.
94	McCain and others, 1988		x	Near urban centers in California and Washington	x	x	32	Review of contaminant residues in surficial sediments and biota and associated fish pathologies in coastal waters near San Diego, Los Angeles, San Francisco, and Seattle–Tacoma. Includes total DDT residues.

Table 1.1. Summary of review articles and books on pesticides in bed sediment and aquatic biota in the United States or on processes governing sorption and bioaccumulation—*Continued*

Study No.	Reference	Study Type		Location(s)	Sampling Media		Number of References Cited	Summary Description
		Process	Monitoring		Sediment	Biota		
95	Mearns and others, 1988		x	United States estuaries and coastal areas		x	128	Review of data from 217 national and regional surveys measuring PCBs and chlorinated pesticides in United States marine and estuarine fish and shellfish (1940–1985). National data were analyzed for geographic and long-term trends in residues. A 15–20 year history of pesticide and PCB contamination was developed for commercial fish stocks, e.g., menhaden, striped bass.
96	Nalepa and Landrum, 1988	x	x	Great Lakes		x	125	Review of interactions between benthic organisms and contaminants in the Great Lakes. Includes field assessments of contaminant effects on benthic communities, and laboratory bioassays. Discusses the role of bioturbation in contaminant cycling, bioaccumulation of organic contaminants, and trophic transfer. Includes pesticide residues in Great Lakes benthos.
97	Norstrom and Muir, 1988	x	x	Arctic and sub-Arctic marine environments		x	75	Review of existing data on organochlorine compounds in Arctic and sub-Arctic biota. Discusses main transport and bioaccumulation mechanisms of organochlorines in the Arctic marine environment. Includes apparent bioconcentration factors between trophic levels. Compares results to those in biota from other oceans, seas, and Great Lakes.
98	O'Connor and Huggett, 1988		x	North Atlantic coast of the United States (Maine to Virginia)	x	x	94	Review of contaminant levels (1975–1983) in sediment and biota from the North Atlantic coast, including Chesapeake Bay. Includes tributaries to the Chesapeake Bay and New York Bight. Also reviews field observations of biological and biochemical effects of pollutants, including fish disease, developmental abnormalities, and effects on habitat and community structure.
99	Phillips and Spies, 1988		x	San Francisco Bay and Delta and its upstream catchment (California)	x	x	50	Review of data on chlorinated contaminants in sediments and biota of the San Francisco Bay and Delta and its upstream catchments. Includes sediment data from 39 streams that flow into the Bay and fish data from the Central Valley catchment waters.

Table 1.1. Summary of review articles and books on pesticides in bed sediment and aquatic biota in the United States or on processes governing sorption and bioaccumulation—*Continued*

Study No.	Reference	Study Type		Location(s)	Sampling Media		Number of References Cited	Summary Description
		Process	Monitoring		Sediment	Biota		
100	Smith and others, 1988	x		na	x	x	79	Review of the partitioning behavior of nonionic organic compounds in aquatic systems: sorption to soil/sediment organic matter, and bioconcentration to lipid reservoirs. Also addresses effect of dissolved organic matter on contaminant solubility.
101	Thomas and others, 1988	x	x	Lake Ontario and tributaries, Niagara R., St. Lawrence R.	x	x	97	Review of major contaminant issues in Lake Ontario, especially sources, loadings, and fate of selected contaminants. Includes mirex and photomirex residues in water, suspended sediment, surficial sediment, and biota in Lake Ontario and its tributaries (including the Niagara R.) and the St. Lawrence R. Includes DDT residues in biota from the lake. Also addresses incidences of fish pathology and reproductive failure associated with pollutants.
102	Allan, 1989	x	x	Great Lakes	x	x	56	Review of the major processes and factors controlling the fate of toxic organic chemicals in the Great Lakes. Focuses on interaction between compartments: land–water, air–water, sediment–water, nutrient–water (including biota and food webs).
103	Andreasen, 1989		x	United States		x	0	Review of environmental monitoring programs of the U.S. Fish and Wildlife Service. Includes summaries of trends from fish and bird analyses in the National Contaminant Biomonitoring Program and examples of monitoring efforts in national wildlife refuges.
104	Bradbury and Coats, 1989	x				x	49	Review of pyrethroid toxicokinetics (absorption, biotransformation, distribution, and elimination) and toxicodynamics (biochemical/physiological effects) in fish as critical factors influencing species selectivity.
105	Buchman, 1989		x	Estuaries in Oregon	x	x	102	Review of geographic and temporal trends in trace contaminants in Oregon estuaries. Synthesizes existing information from many surveys and monitoring programs. Also summarizes data on biological effects of pollutants. Includes organochlorine pesticide residues in water, sediment, fish, crustaceans, mollusks, and seabirds. Also includes limited data on organophosphate insecticides in fish.

Table 1.1. Summary of review articles and books on pesticides in bed sediment and aquatic biota in the United States or on processes governing sorption and bioaccumulation—*Continued*

Study No.	Reference	Study Type		Location(s)	Sampling Media		Number of References Cited	Summary Description
		Process	Moni-toring		Sedi-ment	Biota		
106	Farrington, 1989	x		na		x	75	Review of the bioaccumulation process, including physical/chemical properties and their role in bioavailability, uptake and elimination kinetics, and dietary sources of organic pollutants. Examples are provided for bivalve mollusks and amphipods.
107	Huggett, 1989		x	James R. (Virginia)	x	x	12	Review of the contamination of the James R. with kepone, including the effects of remediation efforts.
108	Ramade, 1989	x	x	Examples only, worldwide		x	112	General discussion of contaminant transport in the atmosphere, transfer to water, and uptake by aquatic biota. Includes mathematical description of magnification in food chains, some examples of organochlorine concentrations in water and biota of various trophic levels. General discussion of contaminant effects on populations, communities, and ecosystems.
109	Sergeant and Onuska, 1989	x	x	Nationwide (fish); Georgia, Florida, unspecified (sediment)	x	x	203	Review of the manufacture, use, analysis, environmental fate, and mobility of toxaphene. Includes residues in precipitation, air, water, soil, sediment, fish, and crops. Also includes laboratory and microcosm studies of bioaccumulation and toxicity to aquatic biota, and laboratory and field studies of environmental fate and persistence. Focus of review is on analytical methods.
110	Spooner and others, 1989	x	x	Kansas, Montana, Texas, Utah; India	x	x	243	Review of 1988 publications dealing with nonpoint source pollution. Pesticides were a topic of major concern; several studies mentioned analyzed pesticides in ground water, drinking water, surface water, sediment, and biota.
111	Steffeck and Striegl, 1989		x	Upper Illinois R. basin (Illinois, Indiana, Wisconsin)		x	247	Review of aquatic biology studies relating to stream-water quality in the Upper Illinois R. Basin. Includes residues of pesticides in aquatic biota. Also includes ecological surveys, evaluations of organism health, and toxicity studies conducted in river or sediment. Report also describes the environmental setting of the basin.

Table 1.1. Summary of review articles and books on pesticides in bed sediment and aquatic biota in the United States or on processes governing sorption and bioaccumulation—*Continued*

Study No.	Reference	Study Type — Process	Study Type — Monitoring	Location(s)	Sampling Media — Sediment	Sampling Media — Biota	Number of References Cited	Summary Description
112	Allan and Ball, 1990		x	Great Lakes	x		188	Monograph reviewing data on selected pesticides, PCBs, and other contaminants in water, suspended sediment, and bottom sediment of the Great Lakes and connecting channels. Includes brief discussion of toxic chemical sources and fate. Interlake and intralake comparisons, nearshore and offshore comparisons, and trends are discussed. Attached bibliography (805 references) includes toxic contaminants in biota also.
113	Barron, 1990	x		na		x	87	Review of the hydrophobicity model of bioconcentration, including the effect of species, body size, lipid content, and environmental conditions that affect uptake and elimination (e.g., temperature). Also presents an alternative physiological model of bioconcentration that considers distributional processes (e.g., blood flow, membrane permeability) and biotransformation.
114	Bero and Gibbs, 1990	x	x	Hudson Estuary (New York)	x		37	Reviews contaminant loading and mechanisms of sediment transport in the Hudson Estuary.
115	Day, 1990		x	Worldwide		x	75	Review of the literature on pesticide residues in freshwater and marine zooplankton. Includes microcosm and mesocosm studies, modeling residues in zooplankton, and relation between residues and toxicity.
116	Fowler, 1990		x	Worldwide	x	x	265	Review of DDT and other contaminant residues in marine waters, sediments, and biota.
117	Huckle and Millburn, 1990	x		na		x	276	Text-like review of metabolism, bioconcentration, and toxicity of pesticides in fish. Includes review of the fate of various pesticide classes in fish and of the factors affecting the bioconcentration factor in fish.
118	Nichols, 1990	x	x	James R. estuary (Virginia)	x	x	51	Review of field studies of kepone in the James R. estuary and hydrodynamic model analyses addressing contaminant fate. Includes kepone residues in water, suspended sediment, bed sediment, plankton, mollusks, and fish. Focus is on kepone fate.

Table 1.1. Summary of review articles and books on pesticides in bed sediment and aquatic biota in the United States or on processes governing sorption and bioaccumulation—*Continued*

Study No.	Reference	Study Type		Location(s)	Sampling Media		Number of References Cited	Summary Description
		Process	Moni-toring		Sedi-ment	Biota		
119	Opperhuizen and Sijm, 1990	x		na		x	40	Review of the literature on bioconcentration and biomagnification of chlorinated dioxins and furans in fish. Discusses effect of biotransformation and the effect of membrane permeation on bioconcentration factors in relation to degree of chlorination.
120	Shigenaka, 1990	x	x	United States and worldwide	x	x	262	Summarizes results for marine environmental residues (surficial sediment, fish, bivalves) of chlordane from NOAA's National Status and Trends Program. Also reviews existing data on chlordane residues from other monitoring studies (United States and worldwide). Provides general information on manufacture, metabolites, and worldwide use of chlordane. Also reviews environmental fate, regulatory status, and toxicity of chlordane.
121	Boyer and Chapra, 1991	x	x	United States, Europe, South America	x	x	225	Review of 1990 publications concerning fate of environmental pollutants. Studies mentioned include reports on the fate of pesticides, chlorinated dioxins, PCBs or other organic compounds; the distribution of organochlorine compounds in water, sediment, and biota (in Europe, South America, and the United States), and bioaccumulation studies with pesticides.
122	Di Toro and others, 1991	x		na	x		81	Review presenting the technical basis for establishing sediment-quality criteria using equilibrium partitioning. Includes review of equilibrium partitioning theory, sediment toxicity and bioaccumulation data (especially relation with pore water concentration), sorption of nonionic chemicals, and applicability of water-quality criteria as the effects levels for benthic organisms. Proposed sediment-quality criteria are based on chemical concentration in sediment organic carbon.
123	Gobas and Russell, 1991	x		na		x	19	Review of chemical sorption and membrane permeation processes that control the bioavailability of organochlorines in natural waters. Includes simple models to estimate bioavailability of organic chemicals in natural waters.

Table 1.1. Summary of review articles and books on pesticides in bed sediment and aquatic biota in the United States or on processes governing sorption and bioaccumulation—*Continued*

Study No.	Reference	Study Type — Process	Study Type — Monitoring	Location(s)	Sampling Media — Sediment	Sampling Media — Biota	Number of References Cited	Summary Description
124	Jordan and Stamer, 1991		x	Lower Kansas R. (Kansas, Nebraska)	x	x	162	Review of existing water-quality data for the lower Kansas R. basin. Includes pesticide residues in surface water, sediments, and fish (1964–1986).
125	Kauss, 1991		x	St. Mary's R. (Michigan, Canada)	x	x		Review of water and habitat quality of St. Mary's R. Includes organic contamination of water, surficial sediments, and biota.
126	Loganathan and Kannan, 1991		x	Japan; global		x	22	Discussion of temporal trends in organochlorine concentrations on a global scale. Presents general schematic of temporal trends in relation to source, life span, and physical/chemical properties. Based on residues in fish, human adipose tissue, and striped dolphin samples from Japan, used to represent point source, nonpoint source, and remote areas of contamination, respectively.
127	Long and others, 1991		x	Tampa Bay (Florida)	x	x	80	Review of status and trends in concentrations of selected toxicants in Tampa Bay (Florida) and estimation of the potential for biological effects. Synthesizes existing information from many surveys and monitoring programs. Includes organochlorine pesticide residues in sediment and bivalve mollusks from Tampa Bay, including some tributaries.
128	MacDonald, 1991		x	Boston Harbor (Massachusetts)	x	x	61	Review of status and trends in concentrations of selected toxicants in Boston Harbor and estimation of the potential for biological effects. Synthesizes existing information from many surveys and monitoring programs. Reports total DDT residues in sediment, fish, crustaceans, and bivalve mollusks. Includes the inner and outer harbors and lower reaches of many tributaries.
129	O'Connor, 1991		x	Coastal and estuarine United States	x	x	12	Review of the results of NOAA's National Status and Trends monitoring of coastal and estuarine sediments (1984–1987) and mollusks (1986–1988) for selected chlorinated contaminants.

Table 1.1. Summary of review articles and books on pesticides in bed sediment and aquatic biota in the United States or on processes governing sorption and bioaccumulation—*Continued*

Study No.	Reference	Study Type		Location(s)	Sampling Media		Number of References Cited	Summary Description
		Process	Monitoring		Sediment	Biota		
130	O'Connor and Ehler, 1991		x	Coastal and estuarine United States	x	x	21	Review of contaminant data in fine-grained sediments from NOAA's National Status and Trends sites in the coastal and estuarine United States. Sediment data are used to describe spatial distribution of contaminants and mollusk data to describe temporal trends. Sediment concentrations are compared with sediment toxicity thresholds from the literature.
131	Adams and others, 1992	x		United States	x		76	Review of the status of sediment assessment in the United States. Compares several approaches to assessing sediment quality, including equilibrium partitioning and apparent effects threshold approaches.
132	Chapra and Boyer, 1992	x	x	United States, Canada, Baltic Sea	x	x	207	Review of 1991 publications concerning fate of environmental pollutants. Studies mentioned include reports on pesticide fate and bioaccumulation, and on organic contaminant residues in soil, water, groundwater, fish, precipitation, and air. Several studies derived models of pesticide fate or bioaccumulation.
133	Landrum and others, 1992	x		na		x	126	Review of kinetic models available for predicting toxicant accumulation for nonsteady-state exposures and multiple uptake routes. Includes water only and sediment exposures. Compares kinetic and steady-state models.
134	Rowan and Rasmussen, 1992	x	x	Great Lakes		x	88	Review of published data on DDT and PCB residues in the Great Lakes ecosystem, and multiple regression analysis of the relation between residues in fish with those in sediment and water (among basins). Residues in water and sediment can explain between-basin variability in fish residues only when basin-specific ecological attributes (fish lipid content, trophic level, and trophic structure of the food chain in the basin) are considered.
135	Wenning and others, 1992	x	x	Newark Bay and New York Bight		x	85	Review of TCDD isomer residues in biota from Newark Bay and the New York Bight. Focus is on sources and residues of 1,2,8,9-TCDD.

Table 1.1. Summary of review articles and books on pesticides in bed sediment and aquatic biota in the United States or on processes governing sorption and bioaccumulation—*Continued*

Study No.	Reference	Study Type		Location(s)	Sampling Media		Number of References Cited	Summary Description
		Process	Monitoring		Sediment	Biota		
136	Wolfe and Macalady, 1992	x		na	x		29	Review of abiotic transformation of organic contaminants in groundwater and sediments. Discusses system variables that affect transformation (such as E_H, pH) and the effect of sorption.
137	Anonymous, 1993		x	Commercial (United States)		x	0	Nontechnical guide for consumers on contamination and mislabeling in commercial fish, including contamination with bacteria, metals, PCBs, and pesticides.
138	Fletcher and McKay, 1993	x	x	Includes Massachusetts, Michigan, Minnesota, Missouri, New York, Great Lakes, Europe, Japan	x	x	91	Review of polychlorinated dibenzo-*p*-dioxins and dibenzofurans in the aquatic environment, including sources, partitioning between water and sediment, sediment transport processes, and bioaccumulation, including dietary uptake and food chain studies. Laboratory and field studies were reviewed.
139	Rinella and others, 1993		x	Yakima R. basin (Washington)	x	x	8	Nontechnical review reporting results from USGS's National Water Quality Assessment Program studies in the Yakima R. Basin (Washington). Also discusses results of prior monitoring studies in the basin, the history of irrigation and DDT use in the basin, and hazardous characteristics of DDT.
140	Smith and others, 1993	x	x	United States rivers nationwide		x	53	Review of stream water quality in the United States during the 1980s. Summarizes results from the USGS monitoring of chemical, physical, and sanitary indicators nationwide, FWS monitoring of contaminants in whole fish tissues nationwide, and a USGS study of herbicide occurrence in streams draining agricultural areas in the midwest. Also includes land use information.

CHAPTER 2

Characteristics of Studies Reviewed

The studies that are reviewed in this book have investigated pesticide occurrence in bed sediment or in one or more species of aquatic biota. The emphasis is on studies that investigated pesticides in bed sediment or aquatic biota in rivers and streams in the United States. The publications reviewed are summarized in Tables 2.1, 2.2, and 2.3 (located at the end of the book), respectively, according to three main categories: national and multistate monitoring studies, state and local monitoring studies, and process and matrix distribution studies. Within each of Tables 2.1, 2.2, and 2.3, individual studies are listed by publication date, and then alphabetically. Review papers (Table 1.1), laboratory studies, and selected papers on pesticides in lakes or in foreign countries are cited in the text as needed, but they are not included in Tables 2.1, 2.2, and 2.3.

National and multistate monitoring studies (Table 2.1) are occurrence surveys for one or more classes of pesticides at several (or many) sites in multiple states. Table 2.1 includes published national and multistate studies of United States estuaries, as well as rivers. That is because these studies individually contribute a lot to our understanding of the national distribution of pesticide contaminants in sediment and aquatic biota. Results from some national programs were reported in a series of publications, either sequential reports that each present 1–4 years of national data (such as reports from the Fish and Wildlife Service's [FWS] National Contaminant Biomonitoring Program [NCBP]), or regional reports that divided the national data set geographically (such as reports from the National Oceanic and Atmospheric Administration's [NOAA] National Benthic Surveillance Project). For programs with multiple publications, Table 2.1 contains a data summary for each individual report published, and these report summaries are grouped by program. Under each program title, individual publications are ordered first by publication date, then by author. For some national programs, more than one federal agency participated in sediment or biota data collection; in these cases, both agencies are named in the program title in Table 2.1, for example, the U.S. Geological Survey's (USGS)–U.S. Environmental Protection Agency's (USEPA) Pesticide Monitoring Network. To the extent possible, each report in Table 2.1 represents a unique data set. Where sequential papers were published from a given program, later papers sometimes included a summary of data from earlier papers; in these cases, only the new (previously unpublished) data are summarized in Table 2.1. Occasionally, two publications contained identical data sets; in these cases, the two publications are listed together under the most recent publication date and authors. The location of sampling

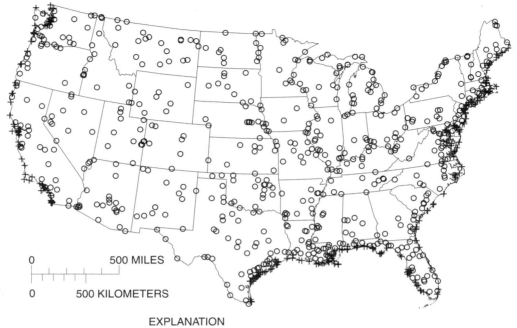

EXPLANATION
o Pesticide Monitoring Network, 1975–1980 (USGS–USEPA)
+ National Status and Trends Program, 1984–1989 (NOAA)

Figure 2.1. Site locations for major national programs that sampled bed sediment. Abbreviations: NOAA, National Oceanic and Atmospheric Administration; USEPA, U.S. Environmental Protection Agency; USGS, U.S. Geological Survey

sites from the major national programs for bed sediment and aquatic biota are presented in Figures 2.1 and 2.2, respectively.

State and local monitoring studies (Table 2.2) are occurrence and distribution surveys for specific compounds or compound classes, usually at multiple sites within a specific area. They include publications from statewide monitoring programs, one-time reconnaissance (screening) surveys for specific areas or basins, and longer-term monitoring efforts at a limited number of sites. Single basin studies with sites in more than one state were considered local (rather than regional) studies, and are included in Table 2.2. The only state and local studies reviewed were those that sampled United States rivers, although some of these studies included estuarine, coastal, or lake sites in addition to river sites. State- and local-scale studies of United States estuaries and coastal areas are not included in Table 2.2, unless these studies also included some riverine sites, because of the sheer volume of such studies and the focus of this book on United States rivers and streams. For studies that had riverine sites as well as estuarine or coastal sites, data for all sites are summarized in Table 2.2 (not just data for river sites). The geographic distribution of the state and local monitoring studies reviewed is shown in Figure 2.3.

Process and matrix distribution studies (Table 2.3) generally measured the distribution of one or more pesticides among various hydrologic compartments, including bed sediment and

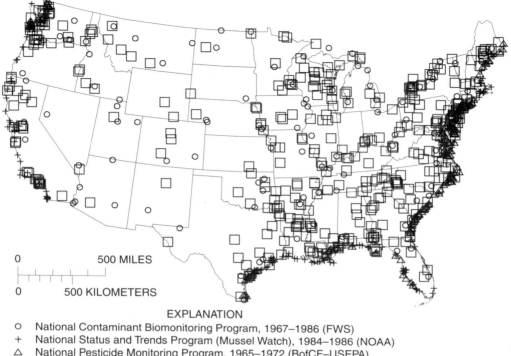

EXPLANATION

○ National Contaminant Biomonitoring Program, 1967–1986 (FWS)
+ National Status and Trends Program (Mussel Watch), 1984–1986 (NOAA)
△ National Pesticide Monitoring Program, 1965–1972 (BofCF–USEPA)
□ National Study of Chemical Residues in Fish, 1986–1987 (USEPA)

Figure 2.2. Site locations for major national programs that sampled aquatic biota. Abbreviations: BofCF, Bureau of Commercial Fisheries; FWS, Fish and Wildlife Service; NOAA, National Oceanic and Atmospheric Administration; USEPA, U.S. Environmental Protection Agency.

aquatic biota, to investigate environmental fate processes or to determine dissipation rates. These studies tended to fall into two categories: (1) dissipation studies in which a specified quantity of a given pesticide was applied to a treatment area and monitored for its movement away from the site of application or into various environmental compartments, and (2) transplantation studies in which nonresident biota or semipermeable membrane devices (SPMD) were deployed in a study area to measure pesticide bioconcentration. Most field experiments investigated the environmental fate and persistence of a specific pesticide, and involved relatively specialized sampling at one or several sites for several days, weeks, or months. The geographic distribution of the process and matrix distribution studies reviewed is shown in Figure 2.4.

2.1 GENERAL DESIGN FEATURES

Table 2.4 summarizes selected characteristics of the studies reviewed. Several national or multistate monitoring studies generated multiple reports (such as the NCBP and the NOAA's

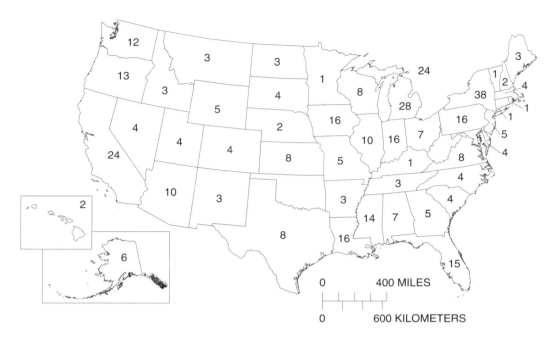

Figure 2.3. Geographic distribution of state and local monitoring studies that were reviewed. These studies are listed in Table 2.2. The number in each state refers to the number of reports reviewed from state and local monitoring studies conducted within the state. No number in a state indicates that no reports of this type from this state were reviewed. Ontario, Canada, had 24 reviewed studies.

National Status and Trends [NS&T] Program for Marine Environmental Quality). Therefore, although 49 reports from national and multistate monitoring studies are listed in Table 2.1, these reports represent only 25 separate studies. In computing study characteristics for Table 2.4, each national and multistate study was counted only once. There were far more state and local monitoring studies (318) than process and matrix distribution studies (47) or national and multistate monitoring studies (25).

National and multistate monitoring studies had predictably more sites and samples than either state and local monitoring studies or process studies (as shown in Table 2.4). However, the maximum numbers of sites for state and local studies (229 and 303 sites for sediment and biota, respectively) were almost as great as those for national and multistate studies (285 and 388 sites for sediment and biota, respectively), indicating that some state and local studies collected samples at a large number of sites. Note that the state and local study category contains some statewide studies, some of which covered a large geographic area (such as California). The study duration also was longer for national and multistate studies than it was for the other study types. State and local studies and process studies ranged in duration from 1 month to about 11 years, and had a median duration of just under 1 year. National and multistate studies ranged from 3 months to 24 years, with the Food and Drug Administration's (FDA) National Monitoring

Figure 2.4. Geographic distribution of process and matrix distribution studies that were reviewed. These studies are listed in Table 2.3. The number in each state refers to the number of reports reviewed from process and matrix distribution studies conducted within the state. No number in a state indicates that no reports of this type from this state were reviewed. Ontario, Canada, had four reviewed studies.

Program for Food and Feed (NMPFF) responsible for the maximum value, and with a median duration of 2 years.

Study design ranged from monitoring a single pesticide at a single site to national studies of multiple pesticide classes. There was little consistency in site selection strategy, sediment collection methods, species of organisms sampled, tissue type analyzed, or analytical detection limits. In contrast, there was great consistency in target analytes: the majority of all three types of studies focused on organochlorine insecticides (as shown in Table 2.4). This was particularly true of monitoring studies (from national to local in scale) as opposed to process studies, regardless of the decade of sampling, as will be discussed in more detail in the next section. Even among studies that measured organochlorine insecticides, there were considerable differences in analytical methods (such as use of packed-column versus capillary column gas chromatography, or quantitation of technical chlordane versus individual chlordane and nonachlor isomers), analytical detection limits, and data reporting and normalization. This is particularly important when comparing older studies to more recent ones, since improvements in analytical methodologies over time have tended to remove interferences, improve precision, and decrease method detection limits. A few studies did not report some characteristics of study design, such as analytes that were targeted but not detected, analytical method detection limits, species or tissue type sampled, number of sampling sites or samples, sampling site location, or even sampling dates.

Table 2.4. General study characteristics condensed from Tables 2.1, 2.2, and 2.3

[Abbreviations: CP, chlorophenoxy acid; OC, organochlorine; OP, organophosphate]

Study Characteristics	Study Type		
	National and Multistate Monitoring Studies (from Table 2.1)	State and Local Monitoring Studies (from Table 2.2)	Process and Matrix Distribution Studies (from Table 2.3)
Number of studies:	25 (49 reports)	318	47
Number of sites sampled:			
— **Sediment**			
Range	17–285	1–229	1–43
Median	117	8	5.5
— **Biota**			
Range	9–388	1–303	1–156
Median	57	6	4.5
Number of samples taken:			
— **Sediment**			
Range	117–1,426	1–771	1–108
Median	783	14	22
— **Biota**			
Range	26–13,262	1–1,310	1–644
Median	85	30	68
Study duration (years):			
Range	0.25–24	0.08–11	0.08–10
Median	2	0.8	0.9
Matrix sampled (number of studies sampling each matrix):			
Bed sediment	6	219	26
Aquatic biota (any)	23	202	39
Fish	20	173	25
Mollusks	8	32	12
Other invertebrates	4	40	19
Plants	0	8	8
Algae	0	7	2
Amphibians	0	8	5
Other aquatic biota	0	21	4
Pesticide class (number of studies analyzing each class):			
OC insecticides	25	312	29
OP insecticides	6	52	6
Carbamate insecticides	0	8	1
Other insecticides	0	2	1
CP acid herbicides	1	37	5
Triazine herbicides	1	10	2
Other herbicides	4	40	8
Fungicides	4	24	3

2.2 GEOGRAPHIC DISTRIBUTION

Figures 2.1 and 2.2 show the location of sampling sites for major national programs that measured pesticides in sediment and aquatic biota, respectively. The distribution of study effort, by state, is shown for state and local monitoring studies in Figure 2.3, and for process and matrix distribution studies in Figure 2.4. Figures 2.1 and 2.2 show that national studies have provided good geographic coverage of the United States, particularly for aquatic biota sampling. National estuarine and coastal monitoring programs have provided extensive coverage along the Atlantic, Pacific, and Gulf coasts; freshwater monitoring programs have provided fair coverage in inland states. Monitoring efforts by the state and local studies reviewed (which sampled rivers and streams, rather than lakes or coastal areas) have been heavier in the Great Lakes area, along the Mississippi River, and on the west coast, especially in California (see Figure 2.3). The geographic distribution of process and matrix distribution studies that were reviewed is highly uneven, with no studies in many states, especially in the Great Plains and Rocky Mountain areas (see Figure 2.4). Again, there appears to have been greater process-type study efforts in the Great Lakes area, along the Mississippi River, and in a few other states, especially Florida and California.

2.3 TEMPORAL DISTRIBUTION

Two national studies spanned almost the entire duration of time encompassed by this report. The FDA's NMPFF has measured pesticide residues in fish and shellfish samples from 1963 to the present. However, these samples are fish and shellfish products in commerce in the United States, and are not necessarily representative of the water resources of the United States. Moreover, published FDA reports did not provide information on sampling location, species of organism, tissue type, or type of hydrological system (lake, river, or marine system) sampled. The FDA results are useful for an assessment of human exposure (discussed in Section 6.2), but do not contribute much to our understanding of pesticides in the hydrologic system and will not be discussed further in this context. In contrast, the FWS's NCBP, which sampled whole fish nationwide from 1967 to 1986, provides a unique data set for assessing the distribution and trends in pesticide residues across the United States.

For monitoring studies in general, there appeared to be an increase in study effort per decade. This is illustrated both in Figure 2.5, which shows the total study duration (in years) by decade of sampling and by pesticide class, and in Figure 2.6, which shows the percentage of studies conducted in each decade by study type. The NMPFF was not included in compiling data for these figures for the reasons given above. In Figure 2.6, national and multistate monitoring studies and state and local monitoring studies show similar patterns, with the number of studies conducted at these scales increasing with each decade. Between 5–15 percent of monitoring studies at both scales were conducted during the 1960s, 30–35 percent during the 1970s, and about 50 percent during the 1980s. In contrast, the distribution of study effort by decade for process and matrix distribution studies shows that most of these studies were conducted during the 1970s, with fewer studies conducted during the 1980s than during the 1960s. Many process and matrix distribution studies were field experiments designed to assess the environmental fate,

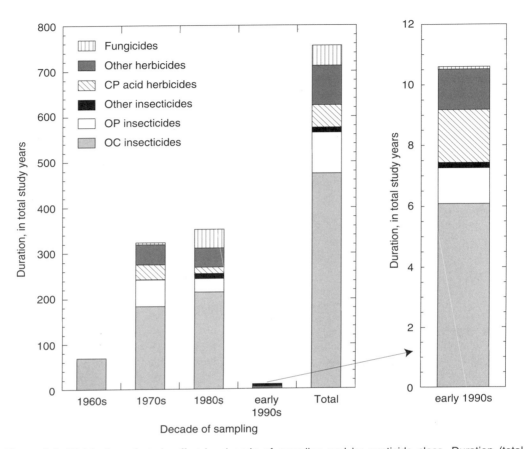

Figure 2.5. Distribution of study effort by decade of sampling and by pesticide class. Duration (total study years) of study effort was calculated from all the monitoring studies listed in Tables 2.1 and 2.2. Each study was assigned to the most appropriate decade of sampling. Inset: enlargement of data for the early 1990s (1990–1994). Abbreviations: CP, chlorophenoxy; OC, organochlorine; OP, organophosphate.

distribution, and persistence of a target pesticide after application of a known quantity of that pesticide, sometimes to meet requirements for pesticide registration. Although there are exceptions (such as studies that transplanted mollusks or that used SPMDs), many process studies were exploratory in nature. The observed temporal patterns in monitoring studies versus process studies suggests that a decrease during the 1980s in field experiments designed to assess the general environmental fate and persistence of pesticides was accompanied by an increase in monitoring studies that focused on analytes known to be present in sediment and aquatic biota (namely, the organochlorine insecticides and a few other, moderately hydrophobic pesticides).

Of the national monitoring programs listed in Table 2.1, only the NS&T Program (besides the NMPFF) has continued into the 1990s. Many state monitoring programs are also continuing during the 1990s. Although the NCBP has been discontinued, additional national-scale sampling

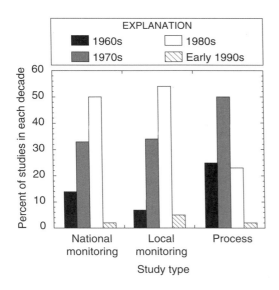

Figure 2.6. Temporal distribution of studies reviewed, by the decade of sampling and by the study type. Study type: National monitoring—national and multistate monitoring studies; Local monitoring—state and local monitoring studies; Process—process and matrix distribution studies. Graph is based on all the studies listed in Tables 2.1–2.3.

of pesticides in fish during the 1990s will be done under the National Water Quality Assessment (NAWQA) Program (which also is analyzing bed sediment) and the USGS's Biomonitoring of Environmental Status and Trends (BEST) Program (formerly a program of the FWS).

2.4 SAMPLING MATRICES

The scope of this review is limited to studies that investigated pesticide occurrence in bed sediment and aquatic biota. A companion book (Larson and others, 1997) reviewed studies on pesticides in water-column matrices, including whole (unfiltered) water, filtered water, suspended solids (biotic or abiotic particles separated from water by filtration or centrifugation), colloidal or dissolved organic carbon, and the surface microlayer. Because a number of studies measured one or more water-column matrices as well as bed sediment or biota, there is understandably some overlap between studies in Larson and others (1997) and in this review.

Specific matrices included in this review are bed sediment, fish, mollusks, other invertebrates, plants, algae, amphibians, and other aquatic biota. The number of studies that investigated pesticide occurrence in each of these matrices is listed in Table 2.4. Many studies also measured pesticide residues in other media (such as water, suspended sediment, mammals, or birds). This information is not included in Table 2.4, but is noted under the listing for each individual study in Tables 2.1–2.3.

Table 2.4 shows that more national and multistate monitoring studies measured pesticides in aquatic biota (23 of 25 studies) than in sediment (6 of 25 studies). The focus of monitoring efforts on the state and local scale was much more evenly divided, with slightly more studies measuring residues in bed sediment (219 studies) than in aquatic biota (202 studies). More process and matrix distribution studies measured pesticide residues in aquatic biota (39 studies) than in bed sediment (26 studies).

Of aquatic biota investigated, more studies of all three types measured residues in fish than in other aquatic organisms. Other invertebrates and mollusks placed a distant second and third, respectively, in state and local monitoring studies and in process studies. In national and multistate monitoring studies, mollusks placed second relative to fish, and other invertebrates a distant third. As shown in Table 2.4, a number of state and local monitoring studies and process studies measured pesticide residues in plants, algae, amphibians, or other aquatic biota, whereas no national or multistate monitoring studies did. This is particularly true for process studies, since this group includes matrix distribution studies specifically designed to determine the distribution of the target pesticide in various environmental compartments. For studies that measured pesticide residues in fish, there was little consistency in the type of tissue sampled: whole fish, fillets or muscle tissue, liver, or other organs.

Information on species sampled is provided for each study reviewed, when available (see Tables 2.1, 2.2, and 2.3). Species generally are referred to by common name. The aquatic species named in this report are listed in Appendix D.

2.5 TARGET ANALYTES

Chemical and common names of pesticide analytes discussed in this book are listed in Appendix E. As noted in Section 2.2, all three types of studies tended to focus on organochlorine insecticides: 100 percent of national and multistate monitoring studies, 98 percent of state and local monitoring studies, and 62 percent of process and matrix distribution studies analyzed for organochlorine insecticides. Proportionately more process studies than monitoring studies focused on pesticides other than the organochlorine insecticides. Such studies established the extent to which certain pesticides were expected to be found in various environmental media, including sediment and biota. In contrast, monitoring studies targeted compounds known to be found in sediment and biota.

The pesticide analytes investigated in the studies listed in Tables 2.1, 2.2, and 2.3 are classified into six pesticide groups: organochlorine insecticides, organophosphate insecticides, other insecticides, chlorophenoxy acid herbicides, other herbicides, and fungicides. The distribution of monitoring effort devoted to these six pesticide groups as a function of sampling decade is displayed in Figure 2.5. A few studies that analyzed for pesticides also analyzed for residues of tetrachlorodibenzo-p-dioxin (TCDD), usually the 2,3,7,8-isomer, which is a contaminant in some pesticide formulations, as well as a byproduct of incineration and some chlorination processes. For such studies, either TCDD or 2,3,7,8-TCDD (as appropriate for a given study) is listed as an analyte in Tables 2.1–2.3, although this information is not included in Table 2.4 or Figure 2.5. In compiling the data for Figure 2.5, all the monitoring studies that were reviewed (i.e., those listed in Tables 2.1 and 2.2) were included if they reported the pesticide analytes and sampling dates. As stated previously, the NMPFF was not included in Figure 2.5 because this program sampled fish and shellfish in interstate commerce; also, the published reports from this program listed the pesticide analytes for all raw agricultural commodities, but not fish and shellfish alone (Duggan and others, 1971; Food and Drug Administration, 1980, 1981, 1988, 1989a, 1990a, 1991, 1992; Yess and others, 1991a,b). For Figure 2.5, each state or local study was assigned to a single decade on the basis of the decade with the most years in which samples were collected. For national studies published as a series of reports, the data from each report was assigned to the appropriate decade. The number of years of study effort per decade was computed by summing the durations of studies that sampled during the appropriate decade, regardless of whether the studies looked at pesticides in sediment, aquatic biota, or both. The number of analytes, sampling sites, and samples were not taken into account in Figure 2.5; nonetheless, this figure gives a general indication of monitoring efforts in bed sediment and aquatic biota over the last few decades.

As shown in Figure 2.5, all monitoring studies that sampled during the 1960s targeted the organochlorine insecticides exclusively. The organochlorine insecticides have accounted for the majority of overall study effort from the 1960s through the early 1990s, even though the use of most of these compounds in the United States has been restricted since the 1970s. This is because these compounds are both hydrophobic and persistent in the environment; they therefore tend to be associated with organic materials in soil and sediment, and to bioaccumulate in living organisms (as discussed in Chapter 4). Also, organochlorine compounds continue to be used in other countries, and have the potential for long-range atmospheric transport and for contamination of migratory wildlife spending parts of their lives in those countries.

During the 1970s, 57 percent of total study effort (defined as the duration of study in years devoted to each class of pesticides, summed for all classes) focused on the organochlorine insecticides, with 43 devoted to pesticides in other compound classes. About 18 percent of total study effort was devoted to organophosphate insecticides; for the most part, this represented bed sediment sampling. Selected organophosphate insecticides (especially chlorpyrifos and diazinon) were occasionally analyzed in aquatic biota, also. Chlorophenoxy acid herbicides (2,4-D, 2,4,5-T, and silvex) and other herbicides (such as dacthal and trifluralin) constituted 10 and 14 percent, respectively, of total study effort during the 1970s.

During the 1980s, the study effort increased overall, with the relative proportion devoted to pesticides other than organochlorine insecticides (44 percent) remaining about the same as during the 1970s. Fungicides (primarily pentachlorophenol and its metabolite, pentachloroanisole) became more important target analytes during the 1980s, making up 11 percent of study effort. Study effort devoted to organophosphate insecticides and chlorophenoxy acid herbicides declined to 8 and 4 percent, respectively, during the 1980s. Because detection frequencies for herbicides, fungicides, and organophosphate insecticides never reached the levels in sediment or aquatic biota that have been observed for organochlorine insecticides, it is not surprising that they have not been targeted with the same consistency in sediment and biota monitoring studies.

At the time this review was conducted, relatively few data were available for the 1990s. For the most part, the studies reviewed were limited to those published as of the end of 1992, although a few studies published in 1993 were included. The total duration of study effort during the 1990s is too brief to discern the distribution of study effort by analyte at the scale used in Figure 2.5. However, enlargement of the 1990s part of Figure 2.5 (see inset) shows a similar distribution of study effort among target analyte classes as in the two previous decades.

2.6 ANALYTICAL DETECTION LIMITS

The *detection limit* values listed in Tables 2.1 and 2.2 are those concentrations below which data were reported as nondetections in the studies reviewed. Depending on the study, the detection limit may be a *method detection limit* (the lowest concentration that can be detected by the analytical method in an ideal matrix following specified test procedures) or a *reporting limit* (a threshold concentration that is higher than the method detection limit, but below which data are not reported, often due to matrix interferences). Many of the individual studies reviewed did not specify which kind of detection limit was used.

A major problem in comparing results from different studies is that of unknown or variable detection limits. In some cases, analytical detection limits were inferred from the reported data when less-than values were given. In other studies, a range of detection limits was reported for a single analyte in a given matrix, reflecting variable performance by different laboratories, changes in analytical methodologies, or the presence of interferences in some samples. In some studies, detection limits were provided for some analytes or for some matrices only. Unfortunately, in a number of studies, detection limits were not reported for analytes that were not detected. This confounds data interpretation, since a zero detection frequency means one thing when the lower limit of detection is 50 micrograms per kilogram (µg/kg), and quite another when the lower limit of detection is 1 µg/kg. In other studies, detection limits were not provided for analytes that were detected in all samples, a less significant omission. Where detection limits were provided, they ranged up to three orders of magnitude for a given analyte among different studies.

Analytical detection limits were reported for over 80 percent of national and multistate monitoring studies (32 of 39 reports, excluding reports from the NMPFF). Four of the seven exceptions were reports from the NS&T program (Benthic Surveillance Project) that focused on the incidence of toxicopathic fish disease in relation to chemical contaminants. By comparison, detection limits were reported for only 65 percent of state and local monitoring studies. It is encouraging to note that detection limits were reported more often in studies with later publication dates. In fact, the percentage of studies that provided detection limits for at least some analytes, by decade of publication, was 40 percent for studies published during the 1960s, 45 percent for the 1970s, 67 percent for the 1980s, and 78 percent for the early 1990s.

The analytical detection limit for a pesticide in any matrix is partially determined by the volume or mass of the sample. In general, increasing the sample mass will result in a lowering of detection limits, provided that sampling and extraction efficiencies remain the same. Certain changes in analytical methodology, such as a change from packed-column to capillary-column gas chromatography, will markedly lower method detection limits. In general, improvements in analytical methods for the organochlorine compounds have lowered method detection limits substantially over the last 30 years.

Because detection limits can influence the results, and ultimately the interpretation of a study, they must be considered in any comparison among studies. This is especially true when comparing detection frequencies from different studies. For example, the Bureau of Commercial Fisheries and the USEPA monitored pesticide residues in mollusks in coastal estuaries throughout the United States from 1965 to 1972 (Butler, 1973b). A subset of sites was resampled in 1977, and detection frequencies from the two sampling periods compared (Butler and others, 1978). The authors concluded that the extent of pesticide contamination in these estuaries had decreased over time. These conclusions were based on the dramatic decrease in detection frequencies observed in the later study relative to the earlier one, together with site-specific decreases in average total DDT concentrations in 1977 compared with the previous 12 months of sampling (1972 or earlier for different estuaries). Specifically, Butler (1973b) reported that total

DDT was detected in 64 percent of the 1965–1972 mollusk samples; dieldrin was detected in 15 percent of samples; and toxaphene, endrin, and mirex were each detected in 0–2 percent of samples. In contrast, total DDT was detected in only 8 percent of 1977 mollusk samples, and other organochlorine insecticides were not detected at all (Butler and others, 1978). However, it is possible that a change in detection limits may have contributed to the apparent decrease in detection frequencies that was observed between the 1965–1972 and 1977 studies. Butler (1973b) reported the number of 1965–1972 samples with residues greater than 5 µg/kg, whereas Butler and others (1978) used a detection limit of 10 µg/kg for total DDT in 1977. This means that concentrations between 5 and 10 µg/kg would have been reported as detections during 1965–1972, but not in 1977. Because the mean concentration of total DDT in mollusks also decreased between the 1965–72 and 1977 studies (Butler, 1973b), it is likely that total DDT in estuarine mollusks did decline during this time. However, the evidence for such a decline would be stronger if detection frequencies above 10 µg/kg had been computed for the 1965–1972 data for direct comparison with data from 1977 at a common detection limit.

CHAPTER 3

National Distribution and Trends

This chapter provides an overview of pesticide occurrence, geographic distribution, and trends in bed sediment and aquatic biota in the nation's rivers and estuaries, synthesized from existing studies and review articles and books as described in Chapter 2. General patterns of pesticide occurrence in bed sediment and aquatic biota (Section 3.1) are discussed, followed by national patterns of pesticide use (Section 3.2). Then, the geographic distribution (Section 3.3) and long-term trends (Section 3.4) of individual pesticides in bed sediment and aquatic biota are evaluated in relation to pesticide use. As is clear from the discussion in this chapter, pesticides in aquatic biota from United States rivers have been studied more extensively than in bed sediment, especially at the national scale.

3.1 PESTICIDE OCCURRENCE

Existing studies can be used as the basis for a preliminary assessment of pesticide occurrence in sediment and aquatic biota in the United States. Process and matrix distribution studies generally investigate the environmental fate and persistence of a single pesticide applied in known quantities or measure pesticide residues in artificial media (such as semipermeable membrane devices). Because these types of studies do not address ambient conditions, they do not provide much information on the occurrence and distribution of pesticides in bed sediment and aquatic biota in the nation's rivers. Monitoring studies (regardless of scale) assess ambient conditions, but the results depend largely on study design characteristics.

To assess pesticide occurrence, monitoring data from all national, multistate, state, and local studies were combined and aggregate detection frequencies were calculated. These aggregate detection frequencies provide an indication of how often individual pesticides have been detected in monitoring studies in the United States. Computing detection frequencies from the combined data set offers two benefits: (1) the combined data set is bigger and more extensive over space and time than that provided by any individual study; and (2) any biases in design of individual studies may be averaged out. On the other hand, these calculations are simplistic in that for a given analyte, the bias caused by quantity of data and differences among the designs of the studies combined is unknown. In contrast, detection frequencies can be used from a single study, in which case the results can be considered in the context of study design and sampling location.

Thus, aggregate detection frequencies presented below (Section 3.1.1) are followed by a discussion of the results of the individual national programs that monitored pesticides in sediment or aquatic biota (Section 3.1.2). Each national program comprises a data set large enough to provide nationwide perspective on pesticide occurrence and distribution in bed sediment or aquatic biota in United States rivers. Although state and local monitoring studies also provide considerable information on the occurrence and distribution of pesticides, the results of these studies are difficult to compare because of the large variability in site selection strategy, sampling methods, time of sampling, analytical methods, and detection limits. For each national program discussed in Section 3.1.2, important study characteristics and highlights of the results provide a synopsis of that program's contribution to our understanding of pesticide occurrence in bed sediment and aquatic biota in United States rivers on a national scale. Finally, comparison of results from the major national programs (Section 3.1.3) provides an overview of pesticide occurrence in sediment and aquatic biota in United States rivers and streams.

3.1.1 AGGREGATE DETECTION FREQUENCIES OF PESTICIDES

To examine general patterns of pesticide occurrence in sediment and aquatic biota, results from national and multistate monitoring studies (listed in Table 2.1) were combined with those from state and local monitoring studies (listed in Table 2.2). Two types of aggregate detection frequencies were calculated for individual pesticides: the percentage of total sites where a pesticide was detected in at least one sample, and the percentage of total samples with detectable residues of that pesticide. Because many of the pesticides targeted in sediment and aquatic biota have not been used in agriculture in the United States since the 1970s, it is possible that detection frequencies have declined over time. Therefore, aggregate detection frequencies were calculated for each decade of sampling.

As noted above, these calculations are necessarily simplistic because they do not consider differences in analytical detection limits, sample volumes, or quantitation methods, all of which may affect the sensitivity of the analytical method, and therefore the probability of detection. The individual monitoring studies that were combined had different study designs and collected data for different durations of time. For site detection frequencies, each study is weighted according to the number of sites, not the number of samples or years of sampling. For example, a single Fish and Wildlife Service's (FWS) National Contaminant Biomonitoring Program (NCBP) site, which may have been sampled 10–12 times between 1967 and 1986, would be weighted equally with a site from a local study that was sampled only once. For the sample detection frequencies, however, each study is weighted according to the total number of samples collected. To continue with the above example, the NCBP study, which collected hundreds of samples, would be weighted more heavily than a local study that collected only a few samples.

For some multicomponent residues, different studies reported data for different analytes. For example, some studies reported chlordane results as total chlordane, whereas others reported results for individual components of technical chlordane. Because these data could not be combined, aggregate detection frequencies were calculated for each individual analyte as reported. Therefore, *cis*-chlordane represents the detection frequency only for studies that reported that analyte and does not include studies that reported the detection frequency only for total chlordane. If a study reported data for individual chlordane components (such as *cis*-chlordane, *trans*-nonachlor) as well as for total chlordane, data from this study were included in the calculations for all of these analytes.

Target analytes for bed sediment and aquatic biota are listed in Tables 3.1 and 3.2, respectively, along with the total number of sites and samples in the combined data set and the corresponding aggregate detection frequencies, by decade of sampling. Data in Table 3.2 are combined for whole fish, fish muscle, and mollusk tissue. Sample detection frequencies for each of these three types of tissue are presented separately in Table 3.3, also by sampling decade. For Tables 3.1–3.3, each study was assigned to the most appropriate decade based on the years of sampling. For long-term national studies that reported data in a series of sequential reports, data from each report were assigned to the appropriate decade. Aggregate detection frequencies for individual pesticide analytes that were analyzed at 15 or more sites in the combined data set are shown graphically in Figures 3.1 (bed sediment) and 3.2 (aquatic biota). The aggregate detection frequencies in Figures 3.1 and 3.2 do not take into account sampling year, the percentage of samples in which a given analyte was detected, or the number of years in which it was detected. Because detection frequencies in Tables 3.1, 3.2, and 3.3 are presented for each decade of sampling, they represent only studies that reported the sampling year. However, the calculations for the corresponding Figures 3.1 and 3.2 also included data from studies that did not report the sampling year.

The national, multistate, state, and local monitoring studies, taken together, show that a large number of pesticide analytes have been detected in sediment and aquatic biota at some time over the last 35 years. The term "pesticide analytes" encompasses pesticides, individual components of technical mixtures, and pesticide transformation products. Altogether, 109 pesticide analytes were measured in sediment, and 129 pesticide analytes were measured in the most commonly sampled types of aquatic biota tissues (whole fish, fish muscle, or mollusk tissue). Some studies did not report the data necessary to calculate aggregate detection frequencies (i.e., the total number of sites, the number of sites with detections, the total number of samples, and number of samples with detections). However, for studies that did report sufficient data, 41 of 93 pesticide analytes (44 percent) were detected in bed sediment in at least one study, and 68 of 106 pesticide analytes (64 percent) in aquatic biota.

Most of the pesticides detected are organochlorine insecticides or their transformation products, despite the fact that most of the organochlorine insecticides were banned or severely restricted during the 1970s. The prevalence of organochlorine insecticides in sediment and biota samples across the United States reflects both the extreme hydrophobicity and persistence of these compounds and the bias in the target analyte list (discussed below). The most commonly detected compounds in both sediment and aquatic biota were DDT and its metabolites, chlordane compounds, and dieldrin.

A few compounds in pesticide classes other than the organochlorine insecticides had fairly high detection frequencies in sediment or biota. In sediment, the herbicide diuron was detected at 100 percent of sites at which it was targeted, although diuron was targeted at only 15 sites nationwide. Additional pesticides from other classes were detected at 10–30 percent of sites: the organophosphate insecticide zytron; the herbicides ametryn, dacthal, 2,4-DB, and dicamba; and the wood preservative pentachlorophenol. All of these compounds except ametryn contain two or more chlorines. Trifluralin, which contains fluorine, was detected at 8 percent of sites. Many of

Table 3.1. Total number of sites and samples, and corresponding aggregate detection frequencies (in percent) of pesticides in bed sediment from United States rivers, calculated by combining data from the monitoring studies in Tables 2.1 and 2.2

[Data include some estuarine sites and samples for some national studies. Results are listed by decade of sampling. Blank cell indicates that no samples were collected in studies from the noted decade. Abbreviations: nr, not reported; PCNB, pentachloronitrobenzene]

Target Analytes	Total Number of Sites Sampled for Sediment				Percentage of Sites with Detectable Residues in Sediment				Total Number of Sediment Samples				Percentage of Sediment Samples with Detectable Residues			
	1960s	1970s	1980s	1990s	1960s	1970s	1980s	1990s	1960s	1970s	1980s	1990s	1960s	1970s	1980s	1990s
Acephate			3				0				3				0	
Alachlor		nr	nr	15		nr	nr	0		28	nr	15		0	nr	0
Aldicarb			3				0				3				0	
Aldrin	70	792	839	83	9	5	5	7	555	2,346	1,505	84	6	2	6	7
Ametryn		15				20				38				16		
Atrazine		141	12	15		0	0	0		413	14	15		0	0	0
Azinphosmethyl		nr	3			nr	0			nr	3			nr	0	
Butylate				15				0				15				0
Carbaryl		nr	3			nr	0			nr	3			nr	0	
Carbofuran		nr	15	15		nr	0	0		28	17	15		0	0	0
Carbofuran, 3-hydroxy			nr				nr				nr				nr	
Carbophenothion		nr	nr			nr	nr			nr	nr			nr	nr	
Chlorbenside			nr				nr				nr				nr	
Chlordane, α-		25	184	46		56	36	0		25	672	46		56	53	0
Chlordane, γ-		25	225	46		48	28	0		52	275	46		35	19	0
Chlordane, total	58	818	894	35	12	41	41	77	551	2,736	2,328	36	5	20	43	75
Chlordene			nr				nr				nr				nr	
Chlordene, hydroxy-	nr				nr				nr				nr			
Chlordene, α-			4	nr			75	nr			4	nr			75	nr
Chlordene, γ-			4	nr			25	nr			4	nr			25	nr
Chlorobenzilate			28				0				28				0	
Chlorpyrifos			5	15			20	15			5	15			20	0
Coumaphos			3				0				3				0	
Cyanazine		nr		15		nr		0		nr		15		nr		0
D, 2,4- (or ester)		247	102	64		2	3	9		636	125	64		1	2	9
Dacthal (DCPA)		2	58	nr		50	16	nr		72	63	nr		1	16	nr
DB, 2,4-				20				15				20				15
DBP			nr				nr				nr				nr	

Table 3.1. Total number of sites and samples, and corresponding aggregate detection frequencies (in percent) of pesticides in bed sediment from United States rivers, calculated by combining data from the monitoring studies in Tables 2.1 and 2.2—*Continued*

Target Analytes	Total Number of Sites Sampled for Sediment				Percentage of Sites with Detectable Residues in Sediment				Total Number of Sediment Samples				Percentage of Sediment Samples with Detectable Residues			
	1960s	1970s	1980s	1990s	1960s	1970s	1980s	1990s	1960s	1970s	1980s	1990s	1960s	1970s	1980s	1990s
DDD, *o,p'-*	54	21	255	46	28	76	36	0	530	34	768	46	6	56	39	0
DDD, *p,p'-* or total	76	884	892	145	53	57	44	41	578	2,597	1,607	146	18	37	46	41
DDE, *o,p'-*	76	21	255	46	46	81	35	0	578	35	768	46		60	20	0
DDE, *p,p'-* or total	76	880	903	145	46	54	49	38	578	2,900	1,618	146	12	35	54	38
DDMS, *p,p'-*			11	nr			0	nr			11	nr			0	nr
DDMU, *p,p'-* or total	54	nr	16	nr		nr	88	nr	530	nr	11	nr		nr	82	nr
DDT, *o,p'-*	54	106	326	46	4	57	34	0	530	144	825	46	1	54	25	0
DDT, *p,p'-* or (*o,p'+p,p'-*)	88	882	862	145	39	40	26	27	578	2,879	1,653	146	7	26	34	27
DDT, total	nr	142	647	14	nr	79	72	100	nr	243	1,689	14	nr	77	75	100
DEF		nr				nr				nr				nr		
Demeton			3			nr	0			nr	3			nr	0	
Diazinon		584	118	nr		1	9	nr		1,648	123	nr		1	10	nr
Dicamba		20	9			20	0			44	9			9	0	
Dichlorvos			3				0				3				0	
Dicofol		nr	2	nr		nr	0	nr		nr	2	nr		nr	0	nr
Dieldrin	82	850	1,272	125	26	48	37	20	584	3,105	3,105	126	8	27	37	20
Dimethoate			3				0				3				0	
Disulfoton		nr	3			nr	0			nr	3			nr	0	
Diuron		15				100				38				97		
DP, 2,4-		49	38	20		2	0	0		75	55	20		1	0	0
Endosulfan, total		127	315	35		2	6	9		254	363	36		4	5	8
Endosulfan I		23	260	64		30	8	11		23	288	64		30	9	11
Endosulfan II		23	282	64		26	11	3		23	296	64		26	10	3
Endosulfan sulfate			186	64			9	22			198	64			8	22
Endosulfan sulfate I			nr				nr				nr				nr	
Endothal			19				0				19				0	
Endrin	68	798	827	145	6	9	4	3	570	2,798	1,047	146	4	3	2	3
Endrin aldehyde		19	79	64		0	0	14		19	81	64		0	0	14
Endrin ketone	54		63		2		0		530		72		2		0	
EPN			3				0				3				0	

Table 3.1. Total number of sites and samples, and corresponding aggregate detection frequencies (in percent) of pesticides in bed sediment from United States rivers, calculated by combining data from the monitoring studies in Tables 2.1 and 2.2—*Continued*

Target Analytes	Total Number of Sites Sampled for Sediment				Percentage of Sites with Detectable Residues in Sediment				Total Number of Sediment Samples				Percentage of Sediment Samples with Detectable Residues			
	1960s	1970s	1980s	1990s	1960s	1970s	1980s	1990s	1960s	1970s	1980s	1990s	1960s	1970s	1980s	1990s
Ethion		493	113			0	0			1,419	118			0	0	
Ethoprop			3	15			0	0			3	15			0	0
Famphur			3				0				3				0	
Fenthion			3				0				3				0	
Fenvalerate			nr				nr				20				0	
Fonofos				15				0				15				0
Glyphosate	nr		19		nr		0				19				0	
HCH, α-		42	479	46		14	10	0		42	646	46		14	2	0
HCH, β		42	359	46		0	10	0		69	527	46		29	4	0
HCH, δ-			196	110			0	1			249	110			0	1
HCH, total	54	10	49		13	0	10	nr	530	20	46	nr	3	0	11	nr
Heptachlor	64	636	847	103	3	4	5	19	549	1,732	1,547	104	2	2	3	19
Heptachlor epoxide	72	768	904	145	7	12	5	6	557	2,516	1,569	146	1	5	4	5
Heptachlor, total		nr	nr			nr	nr			nr	nr			nr	nr	
Hexachlor			8				0				9				0	
Hexachlorobenzene	54	65	845	46	17	57	46	0	530	59	2,526	46	6	53	37	0
Isodrin			28				0				28				0	
Isopropalin			nr				nr				nr				nr	
Kepone		nr	28			nr	0			nr	28			nr	0	
Lindane	22	798	1,065	81	0	3	17	0	48	2,753	2,823	82	0	1	19	0
Linuron		22				0				79				0		
Malathion		575	116			0	0			1,642	121			0	0	
Methamidophos			3				0				3				0	
Methiocarb			3				0				3				0	
Methomyl			3				0				3				0	
Methoxychlor	nr	323	647	79	nr	6	6	1	nr	1,261	730	80	nr	1	6	1
Methyl parathion		567	116	nr		0	0	nr		1,606	141	nr		0	2	nr
Methyl trithion		474	113			0	0			1,397	118			0	0	
Metolachlor				15				0				15				0
Metribuzin				15				0				15				0
Mevinphos			3				0				3				0	

Table 3.1. Total number of sites and samples, and corresponding aggregate detection frequencies (in percent) of pesticides in bed sediment from United States rivers, calculated by combining data from the monitoring studies in Tables 2.1 and 2.2—*Continued*

Target Analytes	Total Number of Sites Sampled for Sediment				Percentage of Sites with Detectable Residues in Sediment				Total Number of Sediment Samples				Percentage of Sediment Samples with Detectable Residues			
	1960s	1970s	1980s	1990s	1960s	1970s	1980s	1990s	1960s	1970s	1980s	1990s	1960s	1970s	1980s	1990s
Mirex	237	298	847	81	35	7	16	0	241	381	2,580	82	36	6	18	0
Molinate																
Monocrotophos			3				0				3				0	
Nitrofen																
Nonachlor, *cis-*		nr	117	46		nr	3	0		nr	163	46		nr	2	0
Nonachlor, *trans-*		nr	17	46		nr	12	0		nr	0	46		nr	nr	0
Nonachlor, total			134	nr			19	nr			623	nr			52	nr
Oxadiazon			3				100	nr			3	nr			100	nr
Oxamyl			3				0				3				0	
Oxychlordane	nr	nr	177	46	nr	nr	4	0	nr	nr	323	46	nr	nr	2	0
Parathion		569	299	nr		2	3	nr		1,605	450	nr		1	2	nr
PCNB																
Pentachloroanisole		nr				nr				nr				nr		
Pentachlorophenol		19	146	12		0	21	33		19	189	12		0	16	33
Permethrin							nr				20				15	
Perthane		107	264	35		0	1	0		201	300	36		0	1	0
Phorate		nr	3	15		nr	0	0		nr	3	15		nr	0	0
Photomirex		nr				nr				nr				nr		
Picloram		20	9			5	0			44	9			2	0	
Silvex		253	48	20		3	2	0		702	67	20		1	1	0
T, 2,4,5-		239	45	20		1	4	10		647	61	20		0	3	10
Terbufos			3	15			0	0			3	15			0	0
Tetrachlorvinphos																
Tetradifon			36				3				37				3	
Thiobencarb																
Toxaphene	54	682	565	65	11	4	2	0	530	2,197	694	66	2	1	3	0
Trichlorfon			3				0				3	35			0	3
Trifluralin		nr	48	35		nr	13	3		nr	52			nr	12	
Trithion		455	113			0	0			1,351	118			0	0	
Zytron (Xytron)			48				29				52				31	

Table 3.2. Total number of sites and samples, and corresponding aggregate detection frequencies (in percent) of pesticides in aquatic biota from United States rivers, calculated by combining data from the monitoring studies in Tables 2.1 and 2.2

[Data are for whole fish, fish muscle, and mollusk tissue, combined. Data include some estuarine sites and samples for some national studies. Results are listed by decade of sampling. Blank cell indicates that no samples were collected in the noted decade by the studies reviewed. Abbreviations: nr, not reported; PCNB, pentachloronitrobenzene]

Target Analytes	Total Number of Sites Sampled for Aquatic Biota				Percentage of Sites with Detectable Residues in Aquatic Biota				Total Number of Aquatic Biota Samples				Percentage of Aquatic Biota Samples with Detectable Residues			
	1960s	1970s	1980s	1990s	1960s	1970s	1980s	1990s	1960s	1970s	1980s	1990s	1960s	1970s	1980s	1990s
Acephate			3				0				3				0	
Alachlor		5	2	nr		0	0	nr		215	20	nr		0	0	nr
Aldicarb			3				0				3				0	
Aldrin	306	154	413	65	11	3	6	0	12,146	2,047	2,073	70	1	1	3	0
Ametryn		nr				nr				nr				nr		
Atrazine		3	2	nr		0	0	nr		209	20	nr		0	0	nr
Azinphosmethyl		110	5			0	0			1,702	97			0	0	
Butylate				nr				nr				nr				nr
Carbaryl		nr	3			nr	0			24	3			13	0	
Carbofuran		nr	5	nr		nr	20	nr		36	98	nr		0	2	nr
Carbofuran, 3-hydroxy-			2				100				69				6	
Carbophenothion		110	nr			2	nr			1,702	nr			0	nr	
Chlorbenside			nr				nr				nr				nr	
Chlordane, α-		235	877	108		74	63	52		744	2,709	150		79	58	57
Chlordane, γ-		227	835	103		78	54	43		744	2,263	141		59	43	51
Chlordane, total	109	329	1,075	276	78	46	81	40	1,575	2,734	3,836	71	10	24	69	45
Chlordene		28	31			7	0			28	39			7	0	
Chlordene, hydroxy-	nr		31		nr		0		75		39		17		0	
Chlordene, α-			nr	61			nr	3			nr	66			nr	3
Chlordene, γ-			nr	61			nr	5			nr	66			nr	5
Chlorobenzilate			nr				nr				nr				nr	
Chlorpyrifos			571	61			21	15			1,226	66			15	17
Coumaphos			3				0				3				0	
Cyanazine		3	nr	nr		0		nr		173	nr	nr		0		nr
D, 2,4- (or ester)		2	3	nr		0	0	nr		30	14	nr		7	0	nr
Dacthal (DCPA)		123	431	61		33	67	23		450	1,484	66		18	26	24

Table 3.2. Total number of sites and samples, and corresponding aggregate detection frequencies (in percent) of pesticides in aquatic biota from United States rivers, calculated by combining data from the monitoring studies in Tables 2.1 and 2.2—*Continued*

Target Analytes	Total Number of Sites Sampled for Aquatic Biota				Percentage of Sites with Detectable Residues in Aquatic Biota				Total Number of Aquatic Biota Samples				Percentage of Aquatic Biota Samples with Detectable Residues			
	1960s	1970s	1980s	1990s	1960s	1970s	1980s	1990s	1960s	1970s	1980s	1990s	1960s	1970s	1980s	1990s
DB, 2,4-				nr				nr				nr				nr
DBP			5				60				6				67	
DCA		nr				nr				nr				nr		
DDD, o,p'-	nr	32	259	108	nr	13	28	24	nr	69	1,202	150	nr	16	51	31
DDD, p,p'- or total	264	392	647	114	95	60	66	59	3,334	3,591	2,607	151	51	77	70	63
DDE, o,p'-		4	254	108		75	22	21		52	1,190	150		58	27	29
DDE, p,p'- or total	277	479	1,174	121	97	77	92	86	4,198	3,988	3,185	167	68	85	90	86
DDMS, p,p'-			nr	61			nr	0			nr	66			nr	0
DDMU, p,p'- or total		28	9	61		4	100	18		28	nr	66		4	nr	17
DDT, o,p'-	23	80	262	101	96	23	29	15	285	182	1,171	127	2	37	36	13
DDT, p,p'- or (o,p'+p,p')-	394	375	629	114	95	66	55	39	4,759	3,654	2,786	156	57	58	57	38
DDT, total	412	543	1,159	83	98	78	96	89	9,064	6,078	4,002	105	67	80	95	90
DEF		110				0				1,702				0		
Demeton		110	3			0	0			1,702	3			0	0	
Diazinon		111	203	61		0	2	3		1,703	646	66		0	1	3
Dicamba		nr	nr			nr	nr			nr	nr			nr	nr	
Dichlorvos			3				0				3				0	
Dicofol		2	595	61		0	12	0		6	1,363	66		0	8	0
Dieldrin	454	540	1,784	108	75	66	58	39	12,774	5,955	6,378	145	24	50	57	46
Dimethoate			3				0				3				0	
Disulfoton		2	3			0	0			6	3			0	0	
Diuron		nr				nr				nr				nr		
DMDT methoxychlor			nr				nr				nr				nr	
DP, 2,4-			nr	nr			nr	nr			nr	nr			nr	nr
Endosulfan, total		112	210	61		0	13	25		1,717	766	66		0	12	24
Endosulfan I		nr	41	68		nr	61	31		nr	122	89		nr	24	37
Endosulfan II		nr	43	45		nr	56	2		nr	157	49		nr	25	4
Endosulfan sulfate			18	45			0	4			103	49			1	6
Endosulfan sulfate I			14				0				28				0	

Table 3.2. Total number of sites and samples, and corresponding aggregate detection frequencies (in percent) of pesticides in aquatic biota from United States rivers, calculated by combining data from the monitoring studies in Tables 2.1 and 2.2—Continued

Target Analytes	Total Number of Sites Sampled for Aquatic Biota				Percentage of Sites with Detectable Residues in Aquatic Biota				Total Number of Aquatic Biota Samples				Percentage of Aquatic Biota Samples with Detectable Residues			
	1960s	1970s	1980s	1990s	1960s	1970s	1980s	1990s	1960s	1970s	1980s	1990s	1960s	1970s	1980s	1990s
Endotetrasulfuron I			16				0				21				0	
Endothal			nr				nr				nr				nr	
Endrin	294	321	1,300	108	13	41	12	9	12,125	3,331	3,628	150	2	21	10	12
Endrin aldehyde		6	30	nr		0	23	nr		6	112	nr		0	15	nr
Endrin ketone	nr		16		nr		0		nr		21		nr		0	
EPN			3				0				3				0	
Ethion		111				2				1,703				1	nr	
Ethoprop			3	nr			nr	nr			3	nr			0	nr
Famphur			3				0				3				0	
Fenthion			3				0				3				0	
Fenvalerate			2				0				109				2	
Fonofos				nr				nr				nr				nr
HCH, α-	50	137	1,029	101	98	85	42	7	147	657	3,139	127	90	51	27	8
HCH, β-		9	406	101		0	7	5		24	1,468	127		0	10	6
HCH, δ-			356	96			4	0			1,244	118			2	0
HCH, total	nr	92	243	61	nr	16	23	8	2,150	150	725	66	8	45	14	9
Heptachlor	295	216	841	72	11	6	9	6	12,200	2,072	2,626	93	1	1	8	8
Heptachlor epoxide	307	127	1,056	108	17	26	22	19	12,214	1,984	3,212	150	3	6	14	27
Heptachlor, total		177	365			46	21			2,726	841			9	11	
Hexachlor			5				60				14				64	
Hexachlorobenzene	nr	288	1,509	101		53	38	15		912	4,535	127		26	25	14
Hexachlorobutadiene			7				57				31				19	
Isodrin	nr				nr				nr				nr			
Isophorone			nr				nr				nr				nr	
Isopropalin			362				4				560				3	
Kepone		6	63			67	10			nr	85				7	
Lindane	15	300	1,395	108	7	13	37	19	46	2,743	4,533	150	2	3	32	27
Linuron			nr				nr				nr				nr	

Table 3.2. Total number of sites and samples, and corresponding aggregate detection frequencies (in percent) of pesticides in aquatic biota from United States rivers, calculated by combining data from the monitoring studies in Tables 2.1 and 2.2—*Continued*

Target Analytes	Total Number of Sites Sampled for Aquatic Biota				Percentage of Sites with Detectable Residues in Aquatic Biota				Total Number of Aquatic Biota Samples				Percentage of Aquatic Biota Samples with Detectable Residues			
	1960s	1970s	1980s	1990s	1960s	1970s	1980s	1990s	1960s	1970s	1980s	1990s	1960s	1970s	1980s	1990s
Malathion		110	3			0	0			1,738	3			0	0	
Methamidophos			3				0				3				0	
Methiocarb			3				0				3				0	
Methomyl			3				0				3				0	
Methoxychlor	182	135	666	68	0	1	8	1	8,113	1,781	1,853	89	0	0	4	1
Methyl parathion		21	3	61		5	0	2		1,524	93	66		0	16	2
Methyl trithion		nr	nr			nr	nr			nr	nr			nr	nr	
Metolachlor			2	nr			0	nr			20	nr			0	nr
Metribuzin				nr				nr				nr				nr
Mevinphos			3				0				3				0	
Mirex	180	313	1,288	40	5	12	33	0	8,095	2,304	4,508	61	0	18	21	0
Molinate			2				100				2				0	
Monocrotophos			3				0				3				0	
Nitrofen			362				3				560				2	
Nonachlor, *cis-*		261	793	96		66	37	28		770	2,142	118		47	35	25
Nonachlor, *trans-*		235	881	108		69	71	57		744	2,734	145		60	61	62
Nonachlor, total			38				87				105				58	
Octachlor epoxide			9				33				27				22	
Oxadiazon			3	61			100	11			5	66			100	12
Oxamyl			3				0				3				0	
Oxychlordane	nr	181	800	108	nr	30	30	25	nr	369	2,128	150	nr	25	27	35
Parathion		119	182	61		3	2	0		1,715	589	66		0	1	0
PCNB			362				1				560				1	
Pentachloroanisole		54	506			35	60			54	1,235			35	37	
Pentachlorophenol		34	52	nr		3	40	nr		34	383	nr		3	13	nr
Permethrin			nr				nr				90				7	
Perthane		6	368	nr		0	1	nr		19	650	nr		0	1	nr
Phorate		110	3	nr		0	0	nr		1,702	3	nr		0	0	nr
Phosdrin			nr				nr				nr				nr	

Table 3.2. Total number of sites and samples, and corresponding aggregate detection frequencies (in percent) of pesticides in aquatic biota from United States rivers, calculated by combining data from the monitoring studies in Tables 2.1 and 2.2—*Continued*

Target Analytes	Total Number of Sites Sampled for Aquatic Biota				Percentage of Sites with Detectable Residues in Aquatic Biota				Total Number of Aquatic Biota Samples				Percentage of Aquatic Biota Samples with Detectable Residues			
	1960s	1970s	1980s	1990s	1960s	1970s	1980s	1990s	1960s	1970s	1980s	1990s	1960s	1970s	1980s	1990s
Photomirex		37	41			16	37			45	125			22	67	
Picloram		nr	nr			nr	nr			nr	nr			nr	nr	
Silvex		nr	3	nr		nr	0	nr		nr	14	nr		nr	0	nr
T, 2,4,5-		nr	3	nr		nr	0	nr		36	14	nr		0	0	nr
Terbufos			3	nr			0	nr			3	nr			0	nr
Tetrachlorvinphos			nr				nr				nr				nr	
Tetradifon			15				7				51				2	
Thiobencarb			2				100				nr				nr	
Toxaphene	285	277	644	94	6	43	32	19	10,997	4,232	2,732	116	2	13	32	17
Trichlorfon			3	nr			0	nr			3	nr			0	nr
Trifluralin		110	417			0	16			1,702	673			0	15	
Trithion		nr	nr			nr	nr			nr	nr			nr	nr	
Zytron (Xytron)			14				64				28				39	

these detected compounds are intermediate in hydrophobicity (as discussed in Section 5.4). A few other organophosphate insecticides (chlorpyrifos, diazinon, and parathion), the herbicide picloram, the acaricide tetradifon, and some chlorophenoxy acid herbicides (2,4-D, silvex, 2,4,5-T, and 2,4-DP) were detected at 1–5 percent of sites targeted. In general, of these pesticides (other than organochlorine insecticides) analyzed in sediment, most were targeted at relatively few sites (15–868 sites) nationwide, and many came from one or a few studies (see Bias From Selection of Target Analytes, below).

Many of the same pesticides from other chemical classes (i.e., other than the organochlorine insecticides) that were found in bed sediment were also found in aquatic biota. Of these compounds, detection frequencies tended to be higher in aquatic biota than in bed sediment. The organophosphate insecticide zytron, the herbicide dacthal, and pentachloroanisole (a metabolite of pentachlorophenol) were detected in biota at over 50 percent of sites at which they were targeted. Pentachlorophenol, the insecticide chlorpyrifos, and the herbicides oxadiazon and trifluralin were detected in biota at 15–25 percent of sites targeted. The acaricide tetradifon, the organophosphate insecticides methyl parathion, parathion, diazinon, carbophenothion, and ethion, the herbicides isopropalin and nitrofen, and the fungicide pentachloronitrobenzene were detected at 1–10 percent of sites. For compounds from pesticide classes other than the organochlorine insecticides, most of those detected contained one or more chlorine (or fluorine) constituents, and were intermediate in hydrophobicity (see Section 5.4). As in bed sediment, pesticides from other chemical classes tended to be analyzed in aquatic biota at fewer sites (14–632 sites) compared with the organochlorine insecticides.

Bias From Selection of Target Analytes

It should be noted that the results in Tables 3.1, 3.2, and 3.3 are for the target analytes reported in the scientific literature reviewed. The absence of a pesticide analyte from these lists, and from Figures 3.1 and 3.2, does not necessarily mean that that analyte was not present in bed sediment and aquatic biota, only that previous monitoring studies did not look for this analyte in these media. Even for those compounds with zero or few detections, these results do not necessarily imply absence from bed sediment and aquatic biota throughout the United States. This is especially true for compounds that were analyzed at only a few sites or in relatively few samples.

In general, the fewer the data available for a given pesticide analyte, the riskier the assumption that the results are not biased from some aspect of study design, such as site selection near urban areas or point sources. If all available data for a given pesticide came from one or two studies, then the aggregate detection frequencies are especially susceptible to study design bias. If a pesticide was analyzed at only a few sites nationwide, localized sampling may result in a geographic bias as well.

In the monitoring studies reviewed, 42 organochlorine insecticides (including components of technical mixtures and pesticide transformation products) were targeted in sediment, and 47 organochlorine insecticides were targeted in aquatic biota. Of pesticides in other classes, 58 were analyzed in sediment and 53 in aquatic biota. However, pesticides in other chemical classes were analyzed in sediment and aquatic biota in fewer studies (see Table 2.4), at fewer sites, and in fewer samples (see Tables 3.1 and 3.2) compared with organochlorine insecticides.

Table 3.3. Total number of sites and corresponding aggregate detection frequencies (in percent) of pesticides in whole fish, fish muscle, and mollusk samples from United States rivers, calculated by combining data from the monitoring studies in Tables 2.1 and 2.2

[Data include some estuarine sites and samples for some national studies. Results are listed by decade of sampling. Blank cell indicates that no samples were collected in studies from the noted decade. Abbreviations: nr, not reported; PCNB, pentachloronitrobenzene]

Target Analytes	Total Number of Whole Fish Samples				Percentage of Whole Fish Samples with Detectable Residues				Total Number of Fish Muscle Samples			
	1960s	1970s	1980s	1990s	1960s	1970s	1980s	1990s	1960s	1970s	1980s	1990s
Acephate			3				0				nr	
Alachlor		191	8	nr		0	0	nr		nr	12	nr
Aldicarb			3				0				nr	
Aldrin	943	1,552	308	nr	8	0	1	nr	2,260	99	320	nr
Ametryn		nr				nr				nr		
Atrazine		191	8	nr		0	0	nr		nr	12	nr
Azinphosmethyl		1,524	97			0	0			nr	nr	
Butylate				nr				nr				nr
Carbaryl		24	3			13	0			nr	nr	
Carbofuran		18	98	nr		0	2	nr		nr	nr	nr
Carbofuran, 3-hydroxy-			69				6				nr	
Carbophenothion		1,524	nr			0	nr			nr	nr	
Chlorbenside			nr				nr				nr	
Chlordane, α-		742	1,237	51		79	55	88		2	113	9
Chlordane, γ-		742	1,173	51		59	46	76		2	113	9
Chlordane, total	736	2,374	824	nr	20	25	74	nr	9	57	297	5
Chlordene		28	39			7	0			nr	nr	
Chlordene, hydroxy-	75		39		17		0		nr		nr	
Chlordene, α-			nr	nr			nr	nr			nr	nr
Chlordene, γ-			nr	nr			nr	nr			nr	nr
Chlorobenzilate			nr				nr				nr	
Chlorpyrifos			3	nr			0	nr			36	nr
Coumaphos			3				0				nr	
Cyanazine		173		nr		0		nr		nr		nr
D, 2,4- (or ester)		24	nr	nr		8	nr	nr		nr	nr	nr
Dacthal (DCPA)		350	773	nr		21	28	nr		96	63	nr
DB, 2,4-				nr				nr				nr
DBP			nr				nr				nr	
DCA		nr				nr				nr		
DDD, o,p'-	nr	28	401	51	nr	4	19	45	nr	41	81	9
DDD, p,p'- or total	1,006	3,048	1,514	57	76	82	70	91	2,305	202	200	4
DDE, o,p'-		nr	401	51		nr	13	45		52	69	9
DDE, p,p'- or total	1,006	3,164	1,534	57	93	96	90	96	2,305	164	200	9
DDMS, p,p'-			nr	nr			nr	nr			nr	nr
DDMU, p,p'- or total		28	nr	nr		4	nr	nr		nr	nr	nr
DDT, o,p'-	184	61	421	28	1	16	12	11	101	61	85	9
DDT, p,p'- or (o,p'+p,p')-	1,567	3,110	1,394	57	81	62	48	67	2,305	172	187	9
DDT, total	785	5,273	1,052	39	98	81	95	95	6	391	325	nr
DEF		1,524								nr		

Table 3.3. Total number of sites and corresponding aggregate detection frequencies (in percent) of pesticides in whole fish, fish muscle, and mollusk samples from United States rivers, calculated by combining data from the monitoring studies in Tables 2.1 and 2.2—*Continued*

Target Analytes	Percentage of Fish Muscle Samples with Detectable Residues				Total Number of Mollusk Samples				Percentage of Mollusk Samples with Detectable Residues			
	1960s	1970s	1980s	1990s	1960s	1970s	1980s	1990s	1960s	1970s	1980s	1990s
Acephate		nr				nr				nr		
Alachlor		nr	0	nr		nr	nr	nr		nr	nr	nr
Aldicarb		nr				nr				nr		
Aldrin	2	0	0	nr	8,943	193	328	nr	0	0	10	nr
Ametryn		nr				nr				nr		
Atrazine		nr	0	nr		nr	nr	nr		nr	nr	nr
Azinphosmethyl		nr	nr		178	nr				0	nr	
Butylate				nr				nr				nr
Carbaryl		nr	nr		nr					nr		
Carbofuran		nr	nr	nr		nr	nr	nr		nr	nr	nr
Carbofuran, 3-hydroxy-		nr				nr				nr		
Carbophenothion		nr	nr		178	nr				0	nr	
Chlorbenside		nr				nr				nr		
Chlordane, α-		100	63	44		nr	331	nr		nr	89	nr
Chlordane, γ-		100	58	33		nr	75	nr		nr	33	nr
Chlordane, total	33	67	45	0	830	241	1,398	nr	1	5	98	nr
Chlordene		nr	nr			nr	nr			nr	nr	
Chlordene, hydroxy-	nr		nr		nr		nr		nr		nr	
Chlordene, α-			nr	nr			nr	nr			nr	nr
Chlordene, γ-			nr	nr			nr	nr			nr	nr
Chlorobenzilate		nr				nr				nr		
Chlorpyrifos		8	nr			nr	nr			nr	nr	
Coumaphos		nr				nr				nr		
Cyanazine		nr		nr	nr			nr		nr		nr
D, 2,4- (or ester)		nr	nr	nr		nr	nr	nr		nr	nr	nr
Dacthal (DCPA)		9	40	nr		nr	nr	nr		nr	nr	nr
DB, 2,4-				nr				nr				nr
DBP		nr				nr				nr		
DCA		nr				nr				nr		
DDD, o,p'-	nr	24	4	0	nr	nr	578	nr	nr	nr	74	nr
DDD, p,p'- or total	40	83	61	100	848	272	645	nr	25	27	80	nr
DDE, o,p'-		58	26	0		nr	578	nr		nr	40	nr
DDE, p,p'- or total	67	97	93	100	848	282	644	11	40	32	86	27
DDMS, p,p'-			nr	nr			nr	nr			nr	nr
DDMU, p,p'- or total		nr	nr	nr		nr	nr	nr		nr	nr	nr
DDT, o,p'-	5	62	24	0	nr	60	537	nr	nr	33	44	nr
DDT, p,p'- or (o,p'+p,p')-	53	51	49	0	848	272	582	nr	21	19	55	nr
DDT, total	100	97	100	nr	8,095	232	1,586	nr	63	22	98	nr
DEF		nr			178					0		

Table 3.3. Total number of sites and corresponding aggregate detection frequencies (in percent) of pesticides in whole fish, fish muscle, and mollusk samples from United States rivers, calculated by combining data from the monitoring studies in Tables 2.1 and 2.2—*Continued*

Target Analytes	Total Number of Whole Fish Samples				Percentage of Whole Fish Samples with Detectable Residues				Total Number of Fish Muscle Samples			
	1960s	1970s	1980s	1990s	1960s	1970s	1980s	1990s	1960s	1970s	1980s	1990s
Demeton		1,524	3			0	0			nr	nr	
Diazinon		1,524	3	nr		0	0	nr	nr	36	nr	nr
Dicamba		nr	nr			nr	nr			nr	nr	
Dichlorvos			3				0				nr	
Dicofol		nr	142	nr		nr	25	nr		nr	8	nr
Dieldrin	1,535	4,978	1,699	51	66	53	49	73	2,291	571	259	4
Dimethoate			3				0				nr	
Disulfoton		nr	3			nr	0			nr	nr	
Diuron		nr				nr				nr		
DMDT methoxy-chlor			nr				nr				nr	
DP, 2,4-		nr	nr	nr		nr	nr	nr		nr	nr	nr
Endosulfan, total		1,544	26	nr		0.06	15	nr		4	105	nr
Endosulfan I		nr	64	23		nr	28	74		nr	26	nr
Endosulfan II		nr	68	nr		nr	32	nr		nr	26	nr
Endosulfan sulfate			44	nr			0	nr			45	nr
Endosulfan sulfate I			28				0				nr	
Endotetrasulfuron I			21				0				nr	
Endothal			nr				nr				nr	
Endrin	931	2,895	1,405	51	7	23	14	33	2,251	287	158	9
Endrin aldehyde		nr	34	nr		nr	0	nr			45	nr
Endrin ketone	nr		21		nr		0		nr		nr	
EPN			3				0			·	nr	
Ethion		1,524	nr			1	nr			nr	nr	
Ethoprop			3	nr			0	nr			nr	nr
Famphur			3				0				nr	
Fenthion			3				0				nr	
Fenvalerate			19				0				nr	
Fonofos			nr				nr					nr
Glyphosate			nr				nr				nr	
HCH, α-	147	638	1,126	28	90	52	32	0	nr	13	208	9
HCH, β-		18	365	28		0	9	0		nr	103	9
HCH, δ-			247	28			3	0			78	9
HCH, total	nr	146	246	nr	nr	46	3	nr	2,150	4	20	nr
Heptachlor	1,006	1,584	303	23	8	1	15	30	2,251	217	317	nr
Heptachlor epoxide	1,006	1,784	724	51	21	4	18	75	2,260	135	354	9
Heptachlor, total		2,726	636			9	15			nr	nr	
Hexachlor		14				64				nr		
Hexachlorobenzene		837	1,120	28		25	16	14		15	207	9
Hexachlorobutadiene		nr				nr				nr		
Isodrin	nr	nr			nr	nr			nr	nr		
Isophorone		nr				nr				nr		
Isopropalin		nr				nr				nr		
Kepone		nr	80			nr	8			nr	5	

Table 3.3. Total number of sites and corresponding aggregate detection frequencies (in percent) of pesticides in whole fish, fish muscle, and mollusk samples from United States rivers, calculated by combining data from the monitoring studies in Tables 2.1 and 2.2—*Continued*

Target Analytes	Percentage of Fish Muscle Samples with Detectable Residues				Total Number of Mollusk Samples				Percentage of Mollusk Samples with Detectable Residues			
	1960s	1970s	1980s	1990s	1960s	1970s	1980s	1990s	1960s	1970s	1980s	1990s
Demeton		nr	nr		178	nr			0	nr		
Diazinon		nr	17	nr	178	nr	nr		0	nr	nr	
Dicamba		nr	nr			nr	nr			nr	nr	
Dichlorvos		nr				nr				nr		
Dicofol		nr	0	nr		nr	14	nr		nr	7	nr
Dieldrin	28	49	56	25	8,943	275	1,787	nr	15	31	87	nr
Dimethoate		nr				nr				nr		
Disulfoton		nr	nr			nr	nr			nr	nr	
Diuron		nr				nr				nr		
DMDT methoxy-chlor		nr				nr				nr		
DP, 2,4-		nr	nr	nr		nr	nr	nr		nr	nr	nr
Endosulfan, total		0	3	nr	187	2	nr		1	0	nr	
Endosulfan I		nr	0	nr		nr	21	nr		nr	48	nr
Endosulfan II		nr	0	nr		nr	21	nr		nr	48	nr
Endosulfan sulfate			0	nr			3	nr			0	nr
Endosulfan sulfate I		nr				nr				nr		
Endotetrasulfuron I		nr				nr				nr		
Endothal		nr				nr				nr		
Endrin	5	9	25	0	8,943	69	82	nr	1	6	0	nr
Endrin aldehyde			0	nr			2	nr			0	nr
Endrin ketone	nr		nr		nr		nr		nr		nr	
EPN		nr				nr				nr		
Ethion		nr	nr		178				0			
Ethoprop			nr	nr			nr	nr			nr	nr
Famphur		nr				nr				nr		
Fenthion		nr				nr				nr		
Fenvalerate		nr				nr				nr		
Fonofos			nr					nr				nr
Glyphosate		nr				nr				nr		
HCH, α-	nr	0	40	0	nr	6	97	nr	nr	0	52	nr
HCH, β-		nr	0	0		6	39	nr		0	26	nr
HCH, δ-			5	0			28	nr			0	nr
HCH, total	8	0	5	nr	830	nr	nr	nr	3	nr	nr	nr
Heptachlor	1	2	6	nr	8,943	193	379	nr	0	0	26	nr
Heptachlor epoxide	3	5	13	11	8,943	15	396	nr	0	53	72	nr
Heptachlor, total		nr	nr			nr	nr			nr	nr	
Hexachlor		nr				nr				nr		
Hexachlorobenzene		13	16	0		nr	1,751	nr		nr	22	nr
Hexachloro-butadiene		nr					31				19	
Isodrin	nr		nr		nr		nr		nr		nr	
Isophorone		nr				nr				nr		
Isopropalin			nr				nr				nr	
Kepone		nr	0				10				0	

Table 3.3. Total number of sites and corresponding aggregate detection frequencies (in percent) of pesticides in whole fish, fish muscle, and mollusk samples from United States rivers, calculated by combining data from the monitoring studies in Tables 2.1 and 2.2 —*Continued*

Target Analytes	Total Number of Whole Fish Samples				Percentage of Whole Fish Samples with Detectable Residues				Total Number of Fish Muscle Samples			
	1960s	1970s	1980s	1990s	1960s	1970s	1980s	1990s	1960s	1970s	1980s	1990s
Lindane	23	2,249	602	51	4	3	14	47	nr	215	265	9
Linuron		nr				nr				nr		
Malathion		1,542	3			0	0			nr	nr	
Methamidophos			3				0				nr	
Methiocarb			3				0				nr	
Methomyl			3				0				nr	
Methoxychlor	nr	1,524	182	23	nr	0	16	4	nr	29	166	nr
Methyl parathion		1,524	3	nr		0	0	nr		nr	nr	nr
Methyl trithion		nr	nr			nr	nr			nr	nr	
Metolachlor			8	nr			0	nr			12	nr
Metribuzin				nr				nr				nr
Mevinphos			3				0				nr	
Mirex	nr	1,969	1,247	28	nr	20	8	0	nr	142	182	9
Molinate			nr				nr				nr	
Monocrotophos			3				0				nr	
Nitrofen			nr				nr				nr	
Nonachlor, *cis-*		768	1,104	28		47	45	4		2	113	9
Nonachlor, *trans-*		742	1,256	51		60	55	86		2	113	4
Nonachlor, total			97				59				8	
Octachlor epoxide			nr				nr				27	
Oxadiazon			5	nr			100	nr			nr	nr
Oxamyl			3				0				nr	
Oxychlordane	nr	369	1,171	51	nr	25	27	71	nr	nr	20	9
Parathion		1,524	3	nr		0	0	nr		13	nr	nr
PCNB			nr				nr				nr	
Pentachloroanisole		54	675			35	22			nr	nr	
Pentachlorophenol		28	51	nr		4	41	nr		nr	48	nr
Permethrin			nr				nr				nr	
Perthane		6	8	nr		0	0	nr		4	82	nr
Phorate		1,524	3	nr		0	0	nr		nr	nr	nr
Phosdrin			nr				nr				nr	
Photomirex		28	18			4	56			17	39	
Picloram		nr	nr			nr	nr			nr	nr	
Silvex		nr	nr	nr		nr	nr	nr		nr	nr	nr
T, 2,4,5-		18	nr	nr		0	nr	nr		nr	nr	nr
Terbufos			3	nr			0	nr			nr	nr
Tetrachlorvinphos			nr				nr				nr	
Tetradifon			40				3				nr	
Thiobencarb			nr				nr				nr	
Toxaphene	738	3,908	1,388	28	1	12	47	25	2,164	8	377	9
Trichlorfon			3	nr			0	nr			nr	nr
Trifluralin		1,524	50			0	56			nr	63	
Trithion			nr			nr	nr			nr	nr	
Zytron (Xytron)			28				39				nr	

Table 3.3. Total number of sites and corresponding aggregate detection frequencies (in percent) of pesticides in whole fish, fish muscle, and mollusk samples from United States rivers, calculated by combining data from the monitoring studies in Tables 2.1 and 2.2—*Continued*

Target Analytes	Percentage of Fish Muscle Samples with Detectable Residues				Total Number of Mollusk Samples				Percentage of Mollusk Samples with Detectable Residues			
	1960s	1970s	1980s	1990s	1960s	1970s	1980s	1990s	1960s	1970s	1980s	1990s
Lindane	nr	0	6	0	18	193	1,762	nr	0	0	69	nr
Linuron		nr				nr				nr		
Malathion		nr	nr			178	nr			0	nr	
Methamidophos			nr				nr				nr	
Methiocarb			nr				nr				nr	
Methomyl			nr				nr				nr	
Methoxychlor	nr	7	0	nr	8,113	203	31	nr	0	0	32	nr
Methyl parathion		nr	nr	nr		nr	nr	nr		nr	nr	nr
Methyl trithion		nr	nr			nr	nr			nr	nr	
Metolachlor		0		nr			nr	nr			nr	nr
Metribuzin			nr					nr				nr
Mevinphos			nr				nr				nr	
Mirex	nr	12	7	0	8,095	193	1,802		0	0	32	
Molinate			nr				nr				nr	
Monocrotophos			nr				nr				nr	
Nitrofen			nr				nr				nr	
Nonachlor, *cis-*		100	33	0		nr	40	nr		nr	0	nr
Nonachlor, *trans-*		50	65	75		nr	331	nr		nr	1	nr
Nonachlor, total			50				nr				nr	
Octachlor epoxide			22				nr				nr	
Oxadiazon			nr	nr			nr	nr			nr	nr
Oxamyl			nr				nr				nr	
Oxychlordane	nr	nr	0	11	nr	nr	29	nr	nr	nr	0	nr
Parathion		8	nr	nr		178	nr			0	nr	
PCNB			nr				nr				nr	
Pentachloroanisole		nr	nr			nr	nr			nr	nr	
Pentachlorophenol		nr	33	nr			8	nr			63	nr
Permethrin			nr				nr				nr	
Perthane		0	0	nr		9	nr	nr		0	nr	nr
Phorate		nr	nr	nr		178	nr	nr		0	nr	nr
Phosdrin			nr				nr				nr	
Photomirex		53	15			nr	nr			nr	nr	
Picloram		nr	nr			nr	nr			nr	nr	
Silvex		nr	nr	nr		nr	nr	nr		nr	nr	nr
T, 2,4,5-		nr	nr	nr		nr	nr	nr		nr	nr	nr
Terbufos		nr	nr			nr				nr		
Tetrachlorvinphos			nr				nr				nr	
Tetradifon			nr				nr				nr	
Thiobencarb			nr				nr				nr	
Toxaphene	2	38	16	0	8,095	263	53	nr	2	21	0	nr
Trichlorfon			nr	nr			nr	nr			nr	nr
Trifluralin		nr	2			178	nr			0	nr	
Trithion		nr	nr			nr	nr			nr	nr	
Zytron (Xytron)			nr				nr				nr	

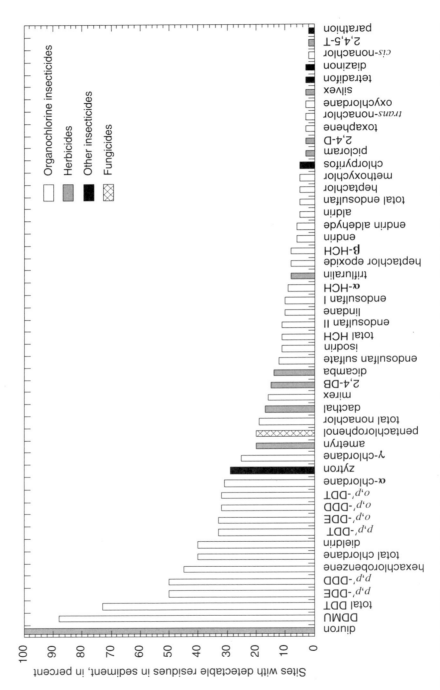

Figure 3.1. Pesticides detected in bed sediment, shown by the percentage of sites with detectable residues at one or more sites at any time for individual pesticide analytes. Shading indicates the type or class of pesticide. Data are combined from all the monitoring studies listed in Tables 2.1 and 2.2. Only those pesticide analytes that were analyzed at 15 or more sites nationally are included.

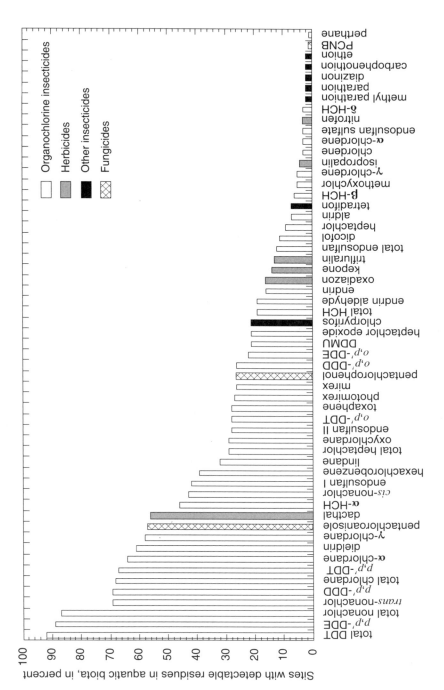

Figure 3.2. Pesticides detected in aquatic biota, shown by the percentage of sites with detectable residues at one or more sites at any time for individual pesticide analytes. Shading indicates the type or class of pesticide. Data are combined for whole fish, fish muscle, and shellfish from all the monitoring studies listed in Tables 2.1 and 2.2. Only those pesticide analytes that were analyzed at 15 or more sites nationally are included.

In sediment studies, the median number of sites was 373 for organochlorine insecticides and 19 for pesticides in other chemical classes. For organochlorine insecticides in sediment, 31 percent of analytes were targeted at more than 1,000 sites and only 10 percent at fewer than 15 sites. In contrast, no pesticides from other chemical classes were targeted at over 1,000 sites, but 34 percent were targeted fewer than 15 sites; because the sample size is small, there is a strong likelihood of bias from study design or localized sampling, and these aggregate detection frequencies may not be representative of United States rivers and streams.

In aquatic biota studies, there also was a paucity of studies that targeted pesticides other than the organochlorine insecticides. The median number of sites was 542 for organochlorine insecticides in aquatic biota, and 5 for pesticides in other chemical classes. For organochlorine insecticides in aquatic biota, 43 percent of analytes were targeted at over 1,000 sites, and only 2 percent at fewer than 15 sites. By contrast, no pesticides from other chemical classes were targeted at over 1,000 sites, but 62 percent were targeted at fewer than 15 sites. Detection frequencies in aquatic biota for a few compounds (chlorpyrifos, dacthal, pentachloroanisole, and trifluralin) were based on data from more than 500 sites nationwide. Even so, these detection frequencies are not necessarily representative of freshwater rivers in the United States. The majority of sites for three of these four compounds (all except dacthal) were sampled as part of the U.S. Environmental Protection Agency's (USEPA) National Study of Chemical Residues in Fish (NSCRF); 80 percent of the 400 sites sampled in this study were located near potential point or nonpoint sources, including many industrial and urban sites. These results are better considered in the context of study design and will be discussed below (Sections 3.1.2 and 3.3).

Comparison of Bed Sediment and Aquatic Biota

The number of pesticides detected in aquatic biota is slightly greater than the number detected in bed sediment. This reflects to some extent the fact that fewer pesticides have been targeted for analysis in sediment than in biota. However, a higher percentage of the pesticides looked for have been found in aquatic biota (64 percent) than in bed sediment (44 percent). This comparison also may be biased by differences in the types of analytes targeted in sediment versus biota (for example, organophosphate insecticides were more frequently targeted in sediment studies). Aggregate detection frequencies for most organochlorines were higher in aquatic biota than in bed sediment (as shown in Tables 3.1 and 3.2, Figures 3.1 and 3.2). Moreover, direct comparison between sediment and biota samples collected from the same sites as part of the same study supports this. In samples from the National Oceanic and Atmospheric Administration's (NOAA) National Status and Trends (NS&T) Program for Marine Environmental Quality, both detection frequencies and concentrations (dry weight) of most hydrophobic organic contaminants were higher in estuarine fish tissues than in associated sediment (e.g., see Zdanowicz and Gadbois, 1990).

3.1.2 PESTICIDE OCCURRENCE IN MAJOR NATIONAL MONITORING PROGRAMS

Six major national programs have monitored pesticides in bed sediment or aquatic biota throughout the United States. These programs are discussed in this section in the order in which sample collection began. The design features of these programs are summarized in Table 3.4. Of

Table 3.4. Design features of major national programs that measured pesticide residues in bed sediment or aquatic biota

[Abbreviations and symbols: PAH, polynuclear aromatic hydrocarbon; PCB, polychlorinated biphenyl; μm, micrometer]

Study	Study Period	Media Sampled	Objectives and Study Design
Food and Drug Administration's National Monitoring Program for Food and Feed	1963–present	Freshwater and estuarine fish (edible) and shellfish	To locate foods containing unsafe levels of pesticides and to maintain surveillance of identities and levels of pesticides in foods. Raw agricultural products were sampled at unspecified harvesting and distribution centers nationwide.
Bureau of Commercial Fisheries–U.S. Environmental Protection Agency's National Pesticide Monitoring Program	1965–1971, 1977	Estuarine bivalve mollusks	To determine extent of organochlorine pollution of estuaries throughout the United States. Composite mollusk samples were collected in estuaries in 15 coastal states.
U.S. Fish and Wildlife Service's National Contaminant Biomonitoring Program	1967–1986	Freshwater fish (whole)	To document temporal and geographic trends in contaminant concentrations that may threaten fish and wildlife resources. Target analytes included organochlorine insecticides, PCBs, and trace metals. Composite fish samples were collected in rivers and lakes nationwide.
U.S. Geological Survey–U.S. Environmental Protection Agency's Pesticide Monitoring Network	1975–1979	Riverine bed sediment (surficial)	To assess levels of pesticides in runoff and bottom sediment and to identify problem areas. Target analytes were organochlorine and organophosphate insecticides, triazine and chlorophenoxy herbicides. Sites were on large United States rivers. Sediment samples were composites from cross-sectional transects.
U.S. National Oceanic and Atmospheric Administration's National Status and Trends Program—Mussel Watch and Benthic Surveillance Projects	1984–present	Coastal and estuarine benthic fish (livers, stomach contents), bivalve mollusks, bed sediment (surficial)	To determine the current status and to detect trends in the environmental quality of the nation's coastal and estuarine areas. Target analytes included organochlorine insecticides, PCBs, PAHs, and trace elements. Composite samples were collected in estuaries and coastal areas. Sites were representative, not near point sources. Sediments were collected from depositional zones and were >20% fines (<63 μm).
U.S. Environmental Protection Agency's National Study of Chemical Residues in Fish	1986–1987	Freshwater and a few estuarine fish (game fish fillets, whole bottom feeders)	To determine the prevalence of selected chemicals in fish, to identify sources, and to estimate human health risks. Analytes included organochlorine insecticides, PCBs, selected industrial compounds, and chlorinated dioxins and furans. Samples were collected near point and nonpoint sources, at large river and reference sites.

these six national programs, two monitored pesticides in bed sediment: one in major rivers and one in coastal and estuarine areas. Five national programs measured pesticides in aquatic biota: two targeted marine and estuarine biota, and two targeted primarily freshwater fish (as shown in Table 3.4). The fifth program, the Food and Drug Administration's (FDA) National Monitoring Program for Food and Feed (NMPFF), did not report sources of fish and shellfish samples, but these sources probably included both freshwater and marine systems.

The six major national programs in Table 3.4 differ in their scope, objectives, and study design. These differences affect the study results and need to be considered in integrating the results of these studies into a comprehensive picture of pesticides in bed sediment and aquatic biota in the United States. Comparison of results within and among these programs is complicated by many factors, including differences in species collected, type of tissue analyzed (fish fillet, liver, or whole body), sampling season, site selection strategy, and year and duration of sampling.

All national programs except the NMPFF share one common design feature: they targeted predominantly or exclusively hydrophobic, persistent contaminants that are expected to sorb to sediment and to bioaccumulate, such as the organochlorine insecticides. Individual reports documenting contaminant results from these national programs are included in Table 2.1.

The FDA's National Monitoring Program for Food and Feed

Under the NMPFF, samples of domestically produced and imported foods were analyzed for hundreds of pesticides from 1963 to the present. The target analyte list included currently used pesticides as well as organochlorine insecticides. As noted previously, it is not clear whether all types of food and feed samples were analyzed for all pesticides on the target analyte list. Fish and shellfish samples analyzed under this program represent products in interstate commerce in the United States, rather than the water resources in the United States. FDA data generally were combined for all domestic fish and shellfish samples. Moreover, published FDA reports did not provide information on sampling location, species of organism, tissue type, or even type of hydrological system (lake, river, marine system) sampled. Therefore, FDA results are useful to an assessment of human exposure (discussed in Section 6.2.1), but do not contribute much to our understanding of pesticides in the hydrologic system and will not be discussed further in this chapter.

The Bureau of Commercial Fisheries–USEPA's National Pesticide Monitoring Program

The Bureau of Commercial Fisheries (BofCF), and later the USEPA, monitored residues of organochlorine insecticides and polychlorinated biphenyls (PCB) in estuarine biota from 1965 to 1977 to determine the extent of pesticide pollution of estuaries throughout the United States. Mollusks were monitored monthly at 180 sites from 1965 to 1972 (Butler, 1973b), and again at a subset of 87 sites in 1977 to determine trends since the previous sampling (Butler and others, 1978). From 1972 to 1976, estuarine fish were targeted for sampling because fish were considered to store synthetic compounds longer than mollusks (Butler and Schutzmann, 1978).

Fish were sampled once or twice a year in 144 estuaries. The 1965–1972 mollusk samples were analyzed for organochlorine compounds. For analyses of the 1972–1976 fish and 1977 mollusk samples, the target list was expanded to include organophosphate insecticides as well.

Total DDT was the most frequently detected pesticide in mollusks and fish during all sampling periods (see data summaries in Table 2.1, listed under the BofCF–USEPA's National Pesticide Monitoring Program). Dieldrin, toxaphene, endrin, and mirex were detected in some mollusk samples collected during 1965–1972 (Butler, 1973b). The only detectable residues in 1977 mollusk samples were for total DDT (Butler and others, 1978). Organochlorine detection frequencies in mollusks appeared to decrease between the 1965–1972 survey and the 1977 resampling. However, this apparent decrease in detection frequency may be partly a function of changing detection limits (see discussion in Section 2.6). The highest total DDT residues (>1,000 micrograms per kilogram [µg/kg]) in mollusks were collected during 1965–1972 from drainage basins with heavy agricultural development in California, Florida, and Texas (Butler, 1973b). Estuarine fish samples contained dieldrin, chlordane, toxaphene, heptachlor epoxide, methyl parathion, ethion, carbophenothion, and parathion as well as total DDT (Butler and Schutzmann, 1978). The fish samples from Delaware, Florida, and New York contained 1,000–4,000 µg/kg total DDT— levels greater than those measured in mollusks collected from the same estuaries between 1965 and 1972.

The FWS's National Contaminant Biomonitoring Program

The FWS analyzed organochlorine compounds and trace elements in whole freshwater fish nationwide every 1–3 years from 1967 to 1986 as part of the NCBP, formerly part of the interagency National Pesticide Monitoring Program. The program objective was to document temporal and geographic trends in concentrations of environmental contaminants that may threaten fish and wildlife resources. Stations were located in major river systems throughout the United States, including Alaska and Hawaii, and in the Great Lakes. Over time, the program was expanded from 50 sites to 112 sites nationwide. Because of the program focus on potential threats to fish and wildlife resources, the NCBP measured whole-body residues in fish (Schmitt and others, 1981).

Results of the NCBP from 1967 to 1984 were published in a series of reports (Henderson and others, 1969, 1971; Schmitt and others, 1981, 1983, 1985, 1990). The most recent of these reports summarized the results of data collected from 1976 to 1984. Data collected prior to 1976 were less reliable because of limitations in some of the early analytical methods used (Henderson and others, 1971; Schmitt and others, 1981) and discrepancies among the multiple laboratories that analyzed NCBP samples (Henderson and others, 1969, 1971; Schmitt and others, 1981, 1983). Nonetheless, the earlier reports contain useful information for selected compounds. The 1986 NCBP data were obtained directly from the FWS (U.S. Fish and Wildlife Service, 1992).

Total DDT, chlordane, and dieldrin were detected consistently at NCBP sites throughout the program (see data summaries in Table 2.1, listed under the FWS's National Contaminant Biomonitoring Program). Total DDT was detected at over 97 percent of sites, and *cis*-chlordane, *trans*-nonachlor, and dieldrin at over 70 percent of sites every year. Other organochlorine compounds were detected at an intermediate number of sites: toxaphene (59–88 percent of sites in different years), α-HCH (47–90 percent), endrin (22–47 percent), heptachlor epoxide (38–55 percent), methoxychlor (32 percent), and lindane (8–31 percent). Pentachloroanisole, a metabolite of the wood preservative pentachlorophenol, was detected at low levels at 24–30 percent of sites. The herbicide dacthal was detected at 28–45 percent of sites since its analysis was begun in 1978. Because α-HCH, lindane, methoxychlor, and pentachloroanisole are shorter-lived than most organochlorine insecticides, their detection in environmental samples probably indicates recent inputs (Schmitt and others, 1985, 1990). A few sites were contaminated with unusually high residues of aldrin, endrin, or hexachlorobenzene (>50 µg/kg), or an unusually high proportion of o,p'-DDT and homologues (>20 percent of total DDT); these residues were probably a result of contamination from chemical manufacturing, or chemical storage, facilities (Schmitt and others, 1983, 1990). Some Great Lakes sites were among the most contaminated sites for many compounds, including total DDT, chlordane, dieldrin, α-HCH, total heptachlor, methoxychlor, mirex, and toxaphene. High levels of selected pesticides were detected in fish from agricultural sites (Schmitt and others, 1990). Examples include dacthal (>70 µg/kg, in the lower Rio Grande and lower Snake rivers), total DDT (>1,000 µg/kg, in the Yazoo, Colorado, and Rio Grande rivers), dieldrin (>200 µg/kg, in major rivers draining the Corn Belt), and toxaphene (>4,500 µg/kg, in the Cotton Belt). Urban and suburban areas may now constitute the primary source of chlordane to aquatic ecosystems (Schmitt and others, 1983, 1990). The distributions of α-HCH and toxaphene have been attributed in part to atmospheric transport (Schmitt and others, 1983, 1985).

The USGS–USEPA's Pesticide Monitoring Network

The first national effort to monitor pesticides in bed sediment was the Pesticide Monitoring Network (PMN), operated by the U.S. Geological Survey (USGS) and USEPA as part of the interagency National Pesticide Monitoring Program (Gilliom and others, 1985). The USGS collected whole water and bed sediment samples at 160–180 sites on major rivers throughout the United States, and the USEPA analyzed them for pesticides. The objective was to assess levels of pesticides in runoff and bed sediment, and to identify problem areas. Bed sediment data for 1975 to 1980 were analyzed by Gilliom and others (1985). Surficial bed sediment samples were collected along a cross-section of the river and composited, then analyzed unsieved. Target analytes in bed sediment included organochlorine and organophosphate insecticides, and a few chlorophenoxy acid and triazine herbicides. Pesticide concentrations in bed sediment were reported on a dry weight basis.

Organochlorine insecticides were detected at more sites and in a higher percentage of samples in bed sediment than in river water (Gilliom and others, 1985). DDE was the compound detected in bed sediment at the most sites (see data summaries in Table 2.1, listed under the USGS–USEPA's Pesticide Monitoring Network). Detection frequencies for an individual organochlorine insecticide were found to reflect a combination of its degree of use on farms (by agricultural region), and its water solubility, environmental persistence, and analytical detection limit. Two organophosphate insecticides (ethion and diazinon) and three chlorophenoxy acid herbicides (2,4-D, 2,4,5-T, and silvex) also were detected in sediment, but their detection frequencies in sediment were lower than in water. Atrazine was not detected in sediment at all, although it was the most frequently detected pesticide in water of any compound analyzed (Gilliom and others, 1985).

The NOAA's National Status and Trends Program

The NS&T Program consists of two complementary projects: the National Mussel Watch Project and the National Benthic Surveillance Project. The National Mussel Watch Project has analyzed bivalve mollusks and associated surficial sediment at about 150 coastal and estuarine sites in the United States since 1986. The National Benthic Surveillance Project has analyzed benthic fish (livers and stomach contents) and associated surficial sediment from about 50 coastal and estuarine sites around the United States since 1984. The overall program objectives are to determine the current status and to detect trends in the environmental quality of the nation's coastal and estuarine areas (Lauenstein and others, 1993). Residues in sediment were used to determine geographic distributions of contaminants (discussed in Section 3.3) because contaminant concentrations in biological tissues may vary with species, age, sex, size, and other factors (National Oceanic and Atmospheric Administration, 1989). Residues in fish and mollusks were used to monitor temporal trends in contamination (Section 3.4) because contaminant levels in biota change relatively rapidly in response to surroundings (National Oceanic and Atmospheric Administration, 1988). The National Benthic Surveillance Project also monitors the association of chemical contaminants with fish disease.

NS&T sites were selected to be representative of their surroundings, so that small-scale areas of contamination and known point source discharges were avoided (National Oceanic and Atmospheric Administration, 1989). Forty-five percent of NS&T sites were within 20 km of population centers with more than 100,000 people. All sites were subtidal (never exposed at lowest tides). Target analytes included organochlorine insecticides, PCBs, polynuclear aromatic hydrocarbons (PAH), and trace elements.

Sediment was collected from depositional zones as close as possible to the corresponding biota sampling site (Lauenstein and others, 1993). All sediment samples were composites of the top 1–3 cm of three grabs or cores. Sediment data were normalized by dividing the raw concentration of contaminant in a composite by the weight fraction of sediment particles that were less than 63 μm in diameter. This assumed that no contaminants were associated with sand-sized particles, that the presence of sand merely diluted the concentration of contaminants (National Oceanic and Atmospheric Administration, 1991). The National Oceanic and Atmospheric Administration (1988, 1991) chose not to normalize contaminant concentrations in sediment by total organic carbon (TOC) content because TOC, like trace contaminants, was high near urban areas. This indicated that TOC was influenced by human activity and was itself behaving as a contaminant. Sediment data from the National Mussel Watch Project and the National Benthic Surveillance Project were analyzed together and published in two NS&T progress reports. The first of these (National Oceanic and Atmospheric Administration, 1988) was superseded by the second (National Oceanic and Atmospheric Administration, 1991), which presented sediment data from 1984 to 1989.

Mollusk and fish tissue data were reported on a dry weight basis (this makes comparison with other studies difficult, since contaminant residues in biota generally are reported on a wet weight basis). Two species of mussels and two species of oysters were collected in the National Mussel Watch Project. Contaminant data for bivalve mollusks were published in two technical memorandums; the first of these (National Oceanic and Atmospheric Administration, 1987) was superseded by the second (National Oceanic and Atmospheric Administration, 1989), which covers mollusk data from 1986 to 1988. For estuarine fish, a series of regional reports was published discussing fish contamination at National Benthic Surveillance Project sites. Seven fish species were collected from 31 sites on the Pacific coast from 1984 to 1986 (Varanasi and others, 1988, 1989; Myers and others, 1993), five species from 20 sites on the Atlantic coast from 1984 to 1986 (Zdanowicz and Gadbois, 1990; Johnson and others, 1992a, 1993), and two species from 16 sites on the Atlantic and Gulf coasts from 1984 to 1985 (Hanson and others, 1989).

Results for sediment, mollusks, and fish livers are summarized in Table 2.1 (see individual reports listed under the NOAA's National Status and Trends Program). Total DDT was the most commonly detected pesticide in sediment. Total chlordane, dieldrin, hexachlorobenzene, lindane, and mirex also were detected in sediment (total chlordane was defined as the sum of *cis*-chlordane, *trans*-nonachlor, heptachlor, and heptachlor epoxide). Most contaminants occurred together and their concentrations (as well as TOC) were related to human population levels (National Oceanic and Atmospheric Administration, 1991). An exception was total DDT, which was not correlated with human population levels on a national scale, but was highly associated with southern California. Because 45 percent of NS&T sites were near urban areas, the concentrations measured probably overestimated the extent of contamination in typical United States coastal sediment, but may grossly underestimate concentrations found at hot spots, such as near point source discharges (National Oceanic and Atmospheric Administration, 1991).

In mollusks, total DDT and total chlordane were detected in 98 percent of samples nationally (1986–1988), followed by dieldrin (91 percent), lindane (70 percent), mirex (30 percent), and hexachlorobenzene (23 percent). The highest levels of total DDT, total chlordane, dieldrin, and lindane in mollusks were found in urban areas (National Oceanic and Atmospheric Administration, 1989; O'Connor, 1992). West coast sites had the highest average total DDT levels in bivalves during 1986–1988, followed by east coast sites, with the lowest average levels at Gulf coast sites (Sericano and others, 1990b).

In estuarine fish livers, the primary contaminants detected were total DDT, dieldrin, and total chlordane. Again, organochlorine concentrations tended to be greater at urban sites than at nonurban sites (Varanasi and others, 1989; Hanson and others, 1989; Zdanowicz and Gadbois, 1990). The incidence of certain fish diseases also was higher near urban locations (Hanson and others, 1989). Generally, concentrations of organic contaminants in the southeast were average-to-low relative to other parts of the United States (Hanson and others, 1989).

The USEPA's National Study of Chemical Residues in Fish

In 1983, USEPA initiated the National Dioxin Study, a 2-year nationwide investigation of 2,3,7,8-tetrachlorodibenzo-*p*-dioxin (TCDD) contamination in soil, water, sediment, air, and fish. As an outgrowth of this study, USEPA conducted a one-time nationwide survey of contaminant residues in fish, titled the NSCRF (formerly titled the National Bioaccumulation Study). The study objectives were to determine the prevalence of selected bioaccumulative chemicals in fish, to identify sources of these contaminants, and to estimate human health risks. Fish were collected during 1986–1987 at almost 400 sites nationwide. Most sites were on rivers and lakes; a few were estuarine or coastal sites. About 80 percent of total sites were located near potential point and nonpoint sources (called "targeted sites"), 10 percent were in areas expected to be relatively free of contamination to provide background concentrations, and 10 percent were collocated with a subset of USGS National Stream Quality Accounting Network (NASQAN) sites to provide geographic coverage. Two composite samples were collected from each site: a representative bottom-feeding fish, which was analyzed whole, and a representative game fish, which was analyzed as a fillet (skin off). Target analytes included chlorinated dibenzo-*p*-dioxins and dibenzofurans, PCBs, organochlorine insecticides, and selected other pesticides. Results from the NSCRF were published in U.S. Environmental Protection Agency (1992a,b).

The most frequently detected contaminant in the NSCRF was *p,p'*-DDE, which was detected in fish at 99 percent of sites (see data summary in Table 2.1 under the USEPA's National Study of Chemical Residues in Fish). Other pesticide analytes detected at over 50 percent of sites were *trans*-nonachlor, pentachloroanisole, *cis*- and *trans*-chlordane, dieldrin, and α-HCH. Hexachlorobenzene, lindane, mirex, oxychlordane, and chlorpyrifos were detected at between 25–50 percent of sites; heptachlor epoxide, endrin, and trifluralin at 10–20 percent of sites; and dicofol, heptachlor, methoxychlor, isopropalin, and nitrofen were detected at fewer than 8 percent of sites. Mean or median residues of *p,p'*-DDE, chlorpyrifos, trifluralin, and dicofol were highest at agricultural sites. Mean or median residues of total chlordane, dieldrin, total nonachlor, hexachlorobenzene, HCH isomers, and isopropalin were highest at Superfund sites, sites in industrial and urban areas, and sites near refineries and other industry. The median concentrations of pentachloroanisole (a metabolite of the wood preservative pentachlorophenol) and 2,3,7,8-TCDD (a byproduct of paper and pulp mill bleaching processes that use chlorine) were highest at sites near paper mills. Whole-body residues in bottom feeders were higher than residues in game fish fillets for some pesticides (chlordane compounds, HCH isomers), but not others (DDE and dieldrin) (U.S. Environmental Protection Agency, 1992a,b).

3.1.3 COMPARISONS OF MAJOR NATIONAL PROGRAMS

The differences in design features of the various national programs (see Table 3.4) make comparisons among these studies problematic. Nonetheless, comparison of detection frequencies for analytes in common indicates different results from different studies, and suggests factors that may be responsible. Figures 3.3 and 3.4 compare detection frequencies from the principal studies that measured pesticide residues in bed sediment or aquatic biota during the 1970s and 1980s, respectively. Detection limits from each study are provided, since these may substantially affect detection frequencies.

During the 1970s, one study measured pesticides in bed sediment from major rivers (the PMN, Gilliom and others, 1985) and one in whole fish from major rivers and lakes (the NCBP, Schmitt and others, 1983). Comparison between these studies (Figure 3.3) suggests that detection frequencies for organochlorine compounds may be lower in bed sediment (Gilliom and others, 1985) than in whole fish (Schmitt and others, 1983). This is not due to differences in detection limits, since detection limits are lower for sediment than fish. Although there was some overlap in site selection between the two studies (about 10–15 percent of sites from the two studies were identical or nearby on the same rivers), for the most part, sediment and fish samples were not collected at the same sites or at the same times. Also, sediment concentrations may have been higher if sediment samples had been collected from depositional zones rather than along a cross-section of the river. Nonetheless, this comparison suggests that the occurrence of residues in fish was higher than in sediment from major rivers during the late 1970s. This is consistent with results of the NS&T Program (National Benthic Surveillance Project), in which paired benthic fish and sediment samples were collected from each coastal or estuarine site. Both

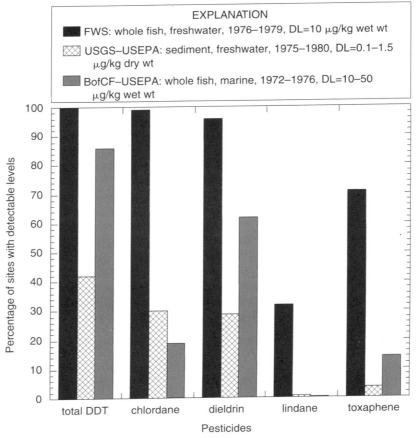

Figure 3.3. Pesticide occurrence in major national programs that sampled bed sediment or aquatic biota during the 1970s. Percentage of sites with detectable levels are determined for each program listed in the Explanation, which also specifies the sampling medium, water type, period of sampling, and detection limit. Data are from Schmitt and others, 1983 (for FWS data); Gilliom and others, 1985 (for USGS–USEPA data); and Butler and Schutzmann, 1978 (for BofCF–USEPA data). Abbreviations: BofCF, Bureau of Commercial Fisheries; DL, detection limit; FWS, Fish and Wildlife Service; kg, kilogram; USEPA, U.S. Environmental Protection Agency; USGS, U.S. Geological Survey; µg, microgram; wt, weight.

detection frequencies and dry weight concentrations of most organochlorine insecticides were higher in benthic fish tissues (liver and stomach contents) than in associated sediment at NS&T sites (Zdanowicz and Gadbois, 1990). The exceptions were two parent compounds that are metabolized fairly rapidly in fish tissue: heptachlor and aldrin (Schnoor, 1981; U.S. Environmental Protection Agency, 1992b). In both cases, the parent compounds were more prevalent in sediment than in fish tissue samples; however, residues of the metabolites (heptachlor epoxide and dieldrin, respectively) fit the expected pattern and were higher in fish tissues than in sediment (Zdanowicz and Gadbois, 1990).

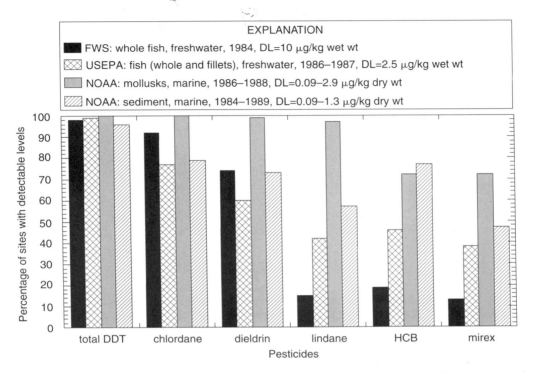

Figure 3.4. Pesticide occurrence in major national programs that sampled bed sediment or aquatic biota during the 1980s. Percentage of sites with detectable levels are determined for each program listed in the Explanation, which also specifies the sampling medium, water type, period of sampling, and detection limit. Data are from Schmitt and others, 1990 (for FWS data); U.S. Environmental Protection Agency, 1992a,b (for USEPA data); National Oceanic and Atmospheric Administration, 1989 (for NOAA mollusk data); National Oceanic and Atmospheric Administration, 1991 (for NOAA sediment data). Abbreviations: DL, detection limit; FWS, Fish and Wildlife Service; kg, kilogram; HCB, hexachlorobenzene; NOAA, National Oceanic and Atmospheric Administration; USEPA, U.S. Environmental Protection Agency; μg, microgram; wt, weight.

Figure 3.4 compares detection frequencies from four studies that were conducted during the 1980s: one study that measured pesticides in sediment (the NS&T Program) and three studies that measured pesticides in aquatic biota (the NCBP and the NSCRF and NS&T Programs). This comparison indicates the following: (1) Total DDT was detected at over 95 percent of sites in all four studies. (2) For the other pesticides in common (chlordane, dieldrin, lindane, hexachlorobenzene, and mirex), detection frequencies were higher in estuarine mollusks sampled by NOAA than in fish sampled by either the FWS or the USEPA. This may be caused, at least in part, by the lower detection limit in the NOAA study. Also, FWS and USEPA fish samples were from freshwater systems, whereas NOAA mollusk samples were collected from estuarine and coastal areas. For chlordane, however, the results shown in Figure 3.4 for the four different studies are not directly comparable because of differences in which components of total chlordane are represented. The chlordane data in Figure 3.4 for the FWS and USEPA studies represent *trans*-nonachlor only, whereas the data for NOAA represent the sum of *cis*-chlordane, *trans*-nonachlor, heptachlor, and heptachlor epoxide. (3) Detection frequencies for lindane, hexachlorobenzene, and mirex were higher in fish sampled during 1986–1987 by the U.S. Environmental Protection Agency (1992a) than in fish sampled in 1984 by the FWS (Schmitt and others, 1990), whereas detection frequencies of chlordane and dieldrin were comparable or slightly lower. The detection limit used by USEPA was five-fold lower than that used by the FWS, which should increase USEPA detection frequencies relative to those of the FWS. Also, the two studies had different criteria for site selection, with the USEPA study targeting 80 percent of its sampling sites near potential point and nonpoint sources. This included more industrial sites than in the NCBP. (4) Detection frequencies for most organochlorine compounds were higher in estuarine mollusks than in estuarine sediment collected by NOAA, except for hexachlorobenzene (National Oceanic and Atmospheric Administration, 1989). The mollusk sites were a subset of the sediment sites. Because hexachlorobenzene is not significantly biotransformed (U.S. Environmental Protection Agency, 1992b), it is not clear why its detection frequency should be lower in mollusks than in associated sediment.

3.2 NATIONAL PESTICIDE USE

National pesticide use estimates are available from a number of sources, including the USEPA (Aspelin, 1997), United States Department of Agriculture (USDA) (Eichers and others, 1968, 1970, 1978; Andrilenas, 1974), and Resources for the Future (Gianessi and Puffer, 1991, 1992a, 1992b). Unfortunately, information is not generated in a consistent manner on a yearly basis or for all segments of the pesticide industry. In general, more quantitative information is available for agricultural use (discussed in Section 3.2.1) than for residential, industrial, and other nonagricultural uses of pesticides (discussed in Section 3.2.2). Trends in total pesticide use, including quantitative estimates by chemical class, are discussed in Section 3.2.3.

In comparing quantitative estimates of pesticide use from various sources, it is important to consider which pesticide chemicals are included for a given estimate. For example, the term "herbicides" may or may not include plant growth regulators, and "insecticides" may or may not include miticides. In the following discussions, terminology from USEPA (Aspelin, 1997) is used. The term "pesticides" refers to any agent used to kill or control undesired insects, weeds, rodents, fungi, bacteria, or other organisms. "Conventional pesticides" are developed and produced primarily for use as pesticides. These include herbicides, insecticides, fungicides, acaricides, plant growth regulators, fumigants, nematicides, rodenticides, molluscicides, insect regulators, piscicides, and bird pesticides. "Other pesticides" include chemicals that are produced mostly for other purposes, but that are sometimes used as pesticides, such as sulfur or petroleum. These terms do not include wood preservatives, disinfectants, or chlorine and hypochlorite used for water treatment, although these chemicals may be considered pesticides in the broadest sense.

3.2.1 AGRICULTURAL USE

According to USEPA estimates (Aspelin, 1997), the quantity of conventional pesticides used in agriculture increased from about 370 million pounds of active ingredient (lb a.i.) in 1964 to a peak of about 840 million lb a.i. in 1979 (as shown in Figure 3.5). Since 1982, it has

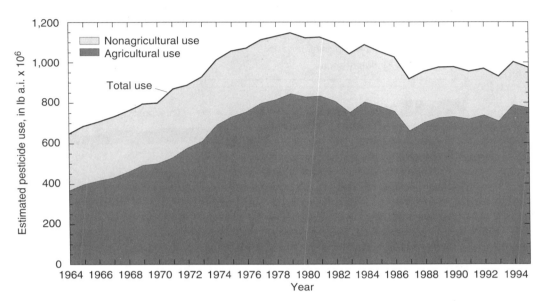

Figure 3.5. Estimated total use of conventional pesticides, showing agricultural and nonagricultural use, in the United States from 1964 to 1995. Graph is based on U.S. Environmental Protection Agency data from Aspelin (1997).

fluctuated between 660 and 800 million lb a.i. per year. Pesticide use in agriculture can vary considerably, depending on factors such as weather, pest outbreaks, crop acreage, and crop prices (Aspelin, 1997). The percentage of total pesticide use that was used in agriculture increased slowly but steadily from 57 percent in 1964 to 79 percent in 1995 (Aspelin, 1997). The quantity of conventional pesticides used in agriculture showed noticeable dips in 1983 and in 1987, and a pronounced peak in 1994 (as shown in Figures 3.5 and 3.6). Driven principally by changes in agricultural use, the total quantity of conventional pesticides used in all market sectors also showed this same pattern (see Figure 3.6). The 1983 dip appears to be caused by a decrease in the annual agricultural use of herbicides (Figure 3.7), whereas the 1987 dip was caused by decreases in annual agricultural use of herbicides (Figure 3.7), insecticides (Figure 3.8), and fungicides (Figure 3.9). The 1994 peak was primarily a function of an increased annual use of herbicides in agriculture (Figure 3.7), which occurred because of increased crop acreage and unusual pest control problems associated with major flooding and unseasonable weather in the Midwest and western United States (Aspelin, 1997).

Quantitative estimates of agricultural use for individual pesticides (in pounds active ingredient) are available from several sources. Table 3.5 lists national use estimates for organic pesticides used in agricultural and other settings in the United States. Agricultural use estimates for selected pesticides in 1964, 1966, 1971, and 1976 are from the USDA (Eichers and others, 1968, 1970, 1978; Andrilenas, 1974). More recent (1988) estimates of agricultural pesticide use are from Resources for the Future (Gianessi and Puffer, 1991, 1992a, 1992b). Table 3.5 also lists, for the 25 most commonly used conventional pesticides in United States crop production, USEPA estimates of agricultural use in 1995 (Aspelin, 1997). In comparing the use estimates in

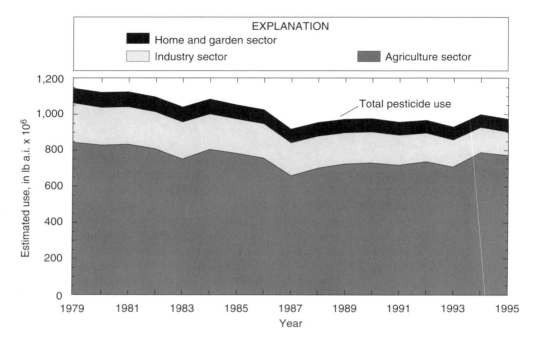

Figure 3.6. Estimated annual use of conventional pesticides in the United States by market sector from 1979 to 1995. Home and garden sector: homeowner applications (both indoor and outdoor) to homes (both single- and multiple-unit housing), lawns, and gardens. Industry sector: applications by business owners and commercial applicators to industrial, commercial, and government facilities, buildings, sites, and land; and by commercial applicators to homes, lawns, and gardens. Agriculture sector: applications by owners and commercial applicators to farms and facilities involved in production of raw agricultural commodities, principally food, fiber, and tobacco. Graph is based on U.S. Environmental Protection Agency data from Aspelin (1997).

Table 3.5 that come from different sources, it is important to note that there are differences in how agricultural pesticide use is defined. For example, the 1976 use estimates from USDA are for applications to major crops only: corn, cotton, wheat, sorghum, rice, other grains, soybeans, tobacco, peanuts, alfalfa, other hay and forage, and pasture and rangeland (Andrilenas, 1974). In contrast, USDA estimates for 1964–1971 apply to all crops, pasture and rangeland, livestock, and other uses by farmers (Eichers and others, 1968, 1970, 1978). The 1988 use estimates from Resources for the Future (Gianessi and Puffer, 1991, 1992a, 1992b) are for applications to cropland only, excluding postharvest use, greenhouse use, and ornamental use. The 1995 USEPA estimates are for applications by owner-operators and commercial applicators to farms and facilities involved in production of raw agricultural commodities, principally food, fiber, and tobacco; they include noncrop and postharvest uses, as well as field and crop applications (Aspelin, 1997).

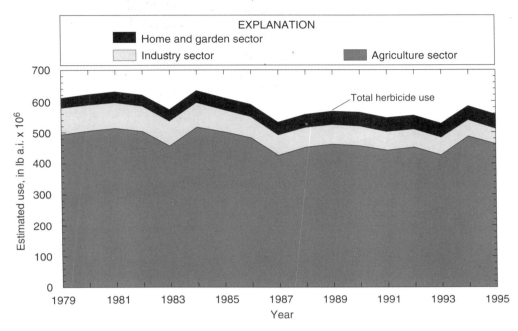

Figure 3.7. Estimated annual herbicide use in the United States by market sector from 1979 to 1995. Home and garden sector: homeowner applications (both indoor and outdoor) to homes (both single- and multiple-unit housing), lawns, and gardens. Industry sector: applications by business owners and commercial applicators to industrial, commercial, and government facilities, buildings, sites, and land; and by commercial applicators to homes, lawns, and gardens. Agriculture sector: applications by owners and commercial applicators to farms and facilities involved in production of raw agricultural commodities, principally food, fiber, and tobacco. Graph is based on U.S. Environmental Protection Agency data from Aspelin (1997).

3.2.2 NONAGRICULTURAL USES

Nonagricultural uses of pesticides include use in lawn and garden care, control of nuisance insects (indoor and outdoor), subterranean termite control, landscape maintenance and rights-of-way, control of public-health pests, industrial settings, forestry, roadways and rights-of-way, and direct application to aquatic systems. Nonagricultural uses of pesticides were reviewed in more detail by Larson and others (1997). There is relatively little information available on types and quantities of individual pesticides used in various nonagricultural applications, at least on a national scale. Some quantitative information is available on national pesticide use in and around homes and gardens; in subterranean termite control; on industrial, commercial, and government buildings and land; and in forestry. These nonagricultural applications are discussed below.

Overall, nonagricultural use makes up about 20–25 percent of total conventional pesticide use (Aspelin, 1997), as shown in Figures 3.5 and 3.6. USEPA divided nonagricultural use into two market sectors: in and around homes and gardens, and on industrial, commercial, and

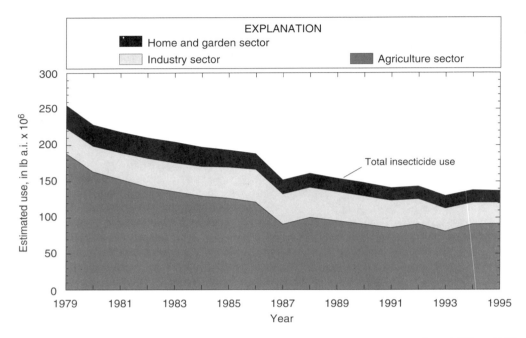

Figure 3.8. Estimated annual insecticide use in the United States by market sector from 1979 to 1995. Home and garden sector: homeowner applications (both indoor and outdoor) to homes (both single- and multiple-unit housing), lawns, and gardens. Industry sector: applications by business owners and commercial applicators to industrial, commercial, and government facilities, buildings, sites, and land; and by commercial applicators to homes, lawns, and gardens. Agriculture sector: applications by owners and commercial applicators to farms and facilities involved in production of raw agricultural commodities, principally food, fiber, and tobacco. Graph is based on U.S. Environmental Protection Agency data from Aspelin (1997).

government buildings and land (hereinafter, referred to simply as the industry sector). Figure 3.10 shows how the total use of individual types of pesticides (such as herbicides, insecticides) breaks down by market sector. Agricultural use is the dominant use for all types of pesticides. For the nine most commonly used pesticides in each sector, USEPA reported the approximate quantities (in pounds active ingredient) used in 1995 in the home and garden and in the industry sectors of the market (Aspelin, 1997). These 1995 use estimates are included in Table 3.5.

Home and Garden

Home and garden use of pesticides consists of applications to homes, lawns, and gardens by homeowners and by professional pest control firms. The consumer and professional applicator markets for pesticides were each estimated at $1.1 billion in sales, at the manufacturer's level, in 1991. This compares with $4.9 billion in sales in the agricultural market (Hodge, 1993). On an active ingredient basis, in 1981, about 85 million lb a.i. were applied to homes, lawns, and gardens by homeowners (Aspelin, 1997), compared with 47 million lb a.i. applied to lawns,

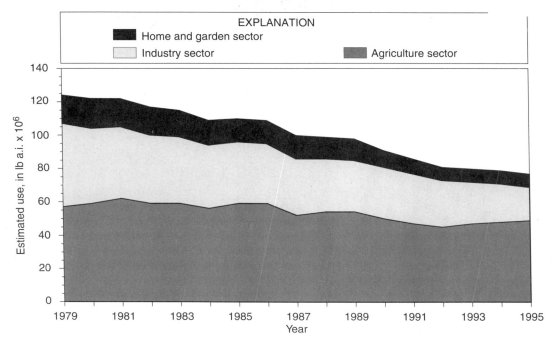

Figure 3.9. Estimated annual fungicide use in the United States by market sector from 1979 to 1995. Home and garden sector: homeowner applications (both indoor and outdoor) to homes (both single- and multiple-unit housing), lawns, and gardens. Industry sector: applications by business owners and commercial applicators to industrial, commercial, and government facilities, buildings, sites, and land; and by commercial applicators to homes, lawns, and gardens. Agriculture sector: applications by owners and commercial applicators to farms and facilities involved in production of raw agricultural commodities, principally food, fiber, and tobacco. Graph is based on U.S. Environmental Protection Agency data from Aspelin (1997).

trees, and structures by commercial applicators (Immerman and Drummond, 1984). Agricultural use of conventional pesticides in the same year (1981) was much larger (831 million lb a.i.) (Aspelin, 1997). Because quantitative data on home and garden use of pesticides by homeowners, and by commercial applicators, come from different sources, these two segments of the home and garden pesticide use market are discussed separately below.

USEPA estimates of home and garden use of pesticides during the period 1979–1995 (Aspelin, 1997) are represented graphically in Figures 3.6–3.9. As defined by USEPA, the home and garden sector consists of homeowner applications (both indoor and outdoor) to homes (both single- and multiple-unit housing), lawns, and gardens (Aspelin, 1997). Therefore the data in Figures 3.6–3.9 do not include applications by commercial applicators to homes, lawns, and gardens, which USEPA defined as part of its industry sector (discussed in the following section). As shown in Figure 3.6, the volume of home and garden pesticide use by homeowners has declined very gradually in recent years, from 85 million lb a.i. in 1979 to 74 million lb a.i. in 1995 (Aspelin, 1997). In 1995, the quantity applied in this home and garden sector constituted

Table 3.5. Estimated pesticide use in agricultural, home and garden, and industry settings in the United States, and detections in ground water, surface water, rain, air, sediment, and aquatic biota

[Agricultural use for 1964–1988 is in millions of pounds (active ingredient). Home and garden use for 1990 is in thousands of products that households had on hand, and thousands of outdoor applications. Use in 1995 (agricultural, home and garden, and industry) is an estimated range in millions of pounds (active ingredient) for the largest conventional pesticides only. Industry use consists of applications to industrial, commercial, and government buildings and land. Pesticides are listed in descending order of agricultural use in 1988, then home and garden use in 1990 (products), except for organochlorine insecticides, which are listed in parentheses (see footnote 2 and Glossary). Agricultural use data are from Eichers and others (1968) (data for 1964); Eichers and others (1970) (data for 1966); Andrilenas (1974) (data for 1971); Eichers and others (1978) (data for 1976); Gianessi and Puffer (1991, 1992a,b) (data for 1988), and Aspelin (1997) (data for 1995). Home and garden use data are from Whitmore and others (1992) (data for 1990) and Aspelin (1997) (data for 1995). Industry use data (for 1995) are from Aspelin (1997). Data sources are Barbash and Resek (1996) (for ground water); Larson and others (1997) (for surface water); Majewski and Capel (1995) (for rain and air); and Tables 2.1 and 2.2 (for bed sediment and aquatic biota). Abbreviations and symbols: A, air; AB, aquatic biota; BS, bed sediment; GW, ground water; R, rain; SW, surface water; "•" indicates that the pesticide was detected at least once in the sampling medium specified; lb a.i., pounds active ingredient; na, data not available; PCNB, pentachloronitrobenzene; blank cell indicates no data]

Compound	Agricultural Use — Crops, Livestock, and Other Uses (lb a.i. × 10⁶) 1964	1966	1971	Agricultural Use — Major Crops[1] 1976	1988	Agricultural Use — Crops 1995	Home and Garden Use — Products (×1,000) 1990	Outdoor Applications (×1,000) 1990	Estimated Range (lb a.i. ×10⁶) 1995	Industry Use (lb a.i. ×10⁶) 1995	GW	SW	R	A	BS	AB
ORGANOCHLORINE INSECTICIDES																
Toxaphene	38.91	34.61	37.46	33.10			74				•	•		•	•	•
DDT	33.54	27.00	14.32				202				•	•	•	•	•	•
Aldrin	11.15	14.76	7.93	0.87							•	•	•	•	•	•
DDD (TDE)	3.39	2.90	0.24	0.22							•	•		•	•	•
Strobane	2.72	2.02	0.22													
Endrin	2.17	0.75	1.43	0.56								•		•	•	•
Methoxychlor	1.43	2.58	3.01	3.81	0.11		3,564	3,692				•	•	•	•	•
Heptachlor	1.31	1.54	1.21	1.62			72	177			•	•	•	•	•	•
Lindane	1.01	0.70	0.65	0.18	0.07		1,638	1,355			•	•	•	•	•	•
Dieldrin	0.94	0.72	0.33								•	•	•	•	•	•

Table 3.5. Estimated pesticide use in agricultural, home and garden, and industry settings in the United States, and detections in ground water, surface water, rain, air, sediment, and aquatic biota—*Continued*

Agricultural Use values are in lb a.i. × 10⁶ ($\times 10^6$). Home and Garden Use: Products (×1,000), Outdoor Applications (×1,000), Estimated Range (lb a.i. × 10⁶), Industry Use (lb a.i. × 10⁶). Pesticide Detected In: GW, SW, R, A, BS, AB.

Compound	1964	1966	1971	1976	Major Crops[1] 1976	Crops 1988	Crops 1995	Products (×1,000) 1990	Outdoor Applications (×1,000) 1990	Est. Range 1995	Industry Use 1995	GW	SW	R	A	BS	AB
Chlordane	0.55	0.53	1.89	1.42				1,156	478			•	•	•	•	•	•
BHC (HCH)	0.35																
Endosulfan		0.79	0.88	0.81		1.99		111	561			•	•	•	•	•	
Dicofol (Kelthane)	1.18	0.89	0.45	0.63		1.72		4,587	4,179								•
Paradichlorobenzene								1,098	538	30–35							
Others	0.44	0.35	0.29	0.08													
Total organochlorine insecticide use:	99.07	90.13	70.32	43.07		3.89		12,502	10,980								
Percent of total insecticide use:	66.6	63.0	42.2	30.4		3.6		3.8	1.8								

ORGANOPHOSPHATE INSECTICIDES

Compound	1964	1966	1971	1976	Major Crops[1] 1976	Crops 1988	Crops 1995	Products (×1,000) 1990	Outdoor Applications (×1,000) 1990	Est. Range 1995	Industry Use 1995	GW	SW	R	A	BS	AB
Chlorpyrifos				9.99		16.73	9–13	16,652	41,900	2–4	9–13	•	•	•	•	•	•
Methyl parathion		8.00	27.56	22.79		8.13						•	•	•	•	•	•
Terbufos				2.49		7.22	6–9					•					
Phorate	1.27	0.33	4.18	6.32		4.78						•	•	•	•		
Fonofos (Dyfonate)				5.01		4.04									•		
Malathion	4.77	5.22	3.60	2.80		3.19		9,551	16,597	2–3		•	•	•	•	•	
Disulfoton	0.89	1.95	4.08	5.50		3.06		2,364	6,464			•			•		

Table 3.5. Estimated pesticide use in agricultural, home and garden, and industry settings in the United States, and detections in ground water, surface water, rain, air, sediment, and aquatic biota—*Continued*

Compound	Agricultural Use (lb a.i. × 10^6)						Home and Garden Use				Pesticide Detected In:					
	Crops, Livestock, and Other Uses			Major Crops[1]	Crops	Crops	Products (×1,000)	Outdoor Applications (×1,000)	Estimated Range (lb a.i. × 10^6)	Industry Use (lb a.i. × 10^6)	GW	SW	R	A	BS	AB
	1964	1966	1971	1976	1988	1995	1990	1990	1995	1995						
Acephate					2.97		4,940	19,167								
Dimethoate				0.58	2.96		301	132			•		•			
Parathion	6.43	8.45	9.48	6.56	2.85	4–7										
Azinphosmethyl	2.27	1.47	2.65	0.32	2.48		37	348			•			•	•	•
Diazinon	2.31	5.61	3.17	1.64	1.71		15,703	56,758	2–4	3–4	•	•	•	•	•	•
Ethoprop				1.15	1.64											
Ethion		2.01	2.33		1.25		39									•
Profenofos					1.22											
Methamidophos					1.14						•					
Phosmet				0.52	1.06		371	173			•					
Dicrotophos (Bidrin)	0.28	1.86	0.81	0.27	0.96											
Sulprofos					0.87						•					
Fenamiphos					0.76											
Mevinphos					0.46											
Methidathion					0.40									•		
Oxydemeton-methyl					0.37		1,032	670				•	•			
Naled					0.22		158									
Trichlorfon	1.08	1.06	0.62	0.63	0.04		41				•					
Ronnel (Fenchlorphos)	0.76	0.39	0.48	0.48			38									
Dichlorvos (DDVP)	0.26	0.91	2.43	0.97			8,953	13,043								

Table 3.5. Estimated pesticide use in agricultural, home and garden, and industry settings in the United States, and detections in ground water, surface water, rain, air, sediment, and aquatic biota—*Continued*

Compound	Crops, Livestock, and Other Uses (lb a.i. × 10⁶)			Major Crops[1] (lb a.i. × 10⁶)	Crops (lb a.i. × 10⁶)	Crops (lb a.i. × 10⁶)	Products (×1,000)	Outdoor Applications (×1,000)	Estimated Range (lb a.i. × 10⁶)	Industry Use (lb a.i. × 10⁶)	GW	SW	R	A	BS	AB
	1964	1966	1971	1976	1988	1995	1990	1990	1995	1995						
Coumaphos				0.52												
Famphur				0.49												
Crufomate				0.12												
Fenthion				0.44												
Crotoxyphos				0.22			40									
Dioxathion				0.10												
Tetrachlorvinphos				0.23			2,423									
Fensulfothion (Dasanit)				0.74												
EPN				6.25												
Monocrotophos				1.92												
Phosalone							118	93								
Isofenphos							100					•				
Chlorfenvinphos							105									
Others	3.60	2.71	9.32	0.41												
Total organophosphate insecticide use:	33.89	39.97	70.71	69.47	70.50		62,966	155,345								
Percent of total insecticide use:	22.8	27.9	42.4	49.1	64.8		19	26								

Table 3.5. Estimated pesticide use in agricultural, home and garden, and industry settings in the United States, and detections in ground water, surface water, rain, air, sediment, and aquatic biota—*Continued*

| Compound | Agricultural Use (lb a.i. × 10⁶) | | | | | | Home and Garden Use | | | | Pesticide Detected In: | | | | | |
| | Crops, Livestock, and Other Uses | | | Major Crops[1] | Crops | Crops | Products (×1,000) | Outdoor Applications (×1,000) | Estimated Range (lb a.i. × 10⁶) | Industry Use (lb a.i. × 10⁶) | | | | | | |
	1964	1966	1971	1976	1988	1995	1990	1990	1995	1995	GW	SW	R	A	BS	AB
CARBAMATE INSECTICIDES																
Carbaryl	14.95	12.39	17.84	9.33	7.62		18,437	31,735	1–3		•	•		•		•
Carbofuran			2.85	11.61	7.06						•	•	•			•
Propargite					3.79							•				
Aldicarb				0.59	3.57						•					
Methomyl			1.08	2.49	2.95		346	629			•					
Propoxur							21,484	53,594			•					
Total carbamate insecticide use:	14.95	12.39	21.77	24.02	24.99		40,267	85,958								
Percent of total insecticide use:	10.1	8.7	13.1	17.0	23.0		12.4	14.1								
OTHER INSECTICIDES																
Cryolite					2.97											
Thiodicarb					1.71											
Permethrin (mixed)					1.12		7,397	18,461						•		
Oxamyl					0.73											
Fenbutatin oxide					0.56		248	2,147			•					
Formetanate HCl					0.41							•				
Esfenvalerate					0.29											
Tefluthrin					0.20											

Table 3.5. Estimated pesticide use in agricultural, home and garden, and industry settings in the United States, and detections in ground water, surface water, rain, air, sediment, and aquatic biota—*Continued*

Compound	Crops, Livestock, and Other Uses				Major Crops[1]	Crops	Products (×1,000)	Outdoor Applications (×1,000)	Estimated Range (lb a.i. × 10^6)	Industry Use (lb a.i. × 10^6)	GW	SW	R	A	BS	AB
	1964	1966	1971	1976	1988	1995	1990	1990	1995	1995						
Cypermethrin					0.19		198									
Trimethacarb					0.13											
Lambdacyhalothrin					0.11											
Cyfluthrin					0.11		1,139	5,654								
Oxythioquinox					0.10											
Amitraz					0.08											
Fenvalerate					0.07		3,192	2,937			•	•				
Metaldehyde					0.04		5,144	27,094								
Tralomethrin					0.04											
Diflubenzuron					0.04											
Bifenthrin					0.03											
Abamectin					0.01											
Cyromazine					0.01											
Piperonyl butoxide[2]							41,729	58,991								
Pyrethrins							34,609	39,289								
MGK-264[2]							27,558	13,249								
Diethyltoluamide (Deet)							21,544	14,134	5–7			•				
Allethrin, total							18,543	52,277								
Tetramethrin							12,962	31,464								
Resmethrin, total							12,506	34,576								

Table 3.5. Estimated pesticide use in agricultural, home and garden, and industry settings in the United States, and detections in ground water, surface water, rain, air, sediment, and aquatic biota—*Continued*

Compound	Agricultural Use — Crops, Livestock, and Other Uses (lb a.i. × 10^6)				Major Crops[1] (lb a.i. × 10^6)	Crops	Home and Garden Use — Products (×1,000)	Outdoor Applications (×1,000)	Estimated Range (lb a.i. × 10^6)	Industry Use (lb a.i. × 10^6)	Pesticide Detected In: GW	SW	R	A	BS	AB
	1964	1966	1971	1976	1988	1995	1990	1990	1995	1995						
Sumithrin							8,089	31,856								
Rotenone							3,997	4,510								
Methoprene							2,709	1,999								
Hydramethylnon							2,389	10,485								
Allethrin							1,388	2,005								
Warfarin							1,145	545								
Brodifacoum							906	296								
Bendiocarb							697	1,778								
Tricosene							307	453								
Dienochlor							265	1,591								
Methiocarb							237	386			●					
Diphacinone							89									
Pindone							81									
Bromadiolone							70	177								
Fenoxycarb							54									
Cythioate							37									
Chlordimeform				4.49												
Bufencarb (Bux)		0.04	3.61													
Others	0.78	0.50	0.37	0.46	0.39											
Total other insecticide use:	0.78	0.54	3.98	4.94	9.34		209,229	356,354								

Table 3.5. Estimated pesticide use in agricultural, home and garden, and industry settings in the United States, and detections in ground water, surface water, rain, air, sediment, and aquatic biota—*Continued*

Compound	Agricultural Use						Home and Garden Use			Industry Use (lb a.i. × 10⁶)	Pesticide Detected In:					
	Crops, Livestock, and Other Uses (lb a.i. × 10⁶)				Major Crops[1]	Crops	Products (×1,000)	Outdoor Applications (×1,000)	Estimated Range (lb a.i. × 10⁶)		GW	SW	R	A	BS	AB
	1964	1966	1971	1976	1988	1995	1990	1990	1995	1995						
Percent of total insecticide use:	0.5	0.4	2.4	3.5	8.6		64.4	58.5								
TOTAL INSECTICIDE USE:	148.69	143.03	166.77	141.50	108.71		324,964	608,637								

TRIAZINE HERBICIDES

Compound	1964	1966	1971	1976	1988	1995	Products (×1,000) 1990	Outdoor Applications (×1,000) 1990	Estimated Range 1995	Industry Use 1995	GW	SW	R	A	BS	AB
Atrazine	10.90	23.52	57.45	90.34	64.24	68–73	134	488			•	•	•	•		
Cyanazine				10.57	22.89	24–29					•	•	•			
Metribuzin				5.21	4.82						•	•	•			
Propazine		0.58	3.17	3.89	4.02						•	•	•			
Simazine	0.32	0.19	1.74	2.53	3.96		172				•	•	•	•		
Prometryn				0.66	1.81							•				
Terbutryn				0.84	1.11						•					
Prometon				0.20			1,244	1,281			•	•				
Others			1.45	0.03	1.02											
Total triazine use:	11.31	24.29	63.80	114.26	103.87		1,550	1,769	11.31							
Percent of total herbicide use:	15.4	22.0	28.6	30.6	22.7		3.3	1.4								

Table 3.5. Estimated pesticide use in agricultural, home and garden, and industry settings in the United States, and detections in ground water, surface water, rain, air, sediment, and aquatic biota—*Continued*

Compound	Agricultural Use (lb a.i. × 10^6)						Home and Garden Use				Pesticide Detected In:					
	Crops, Livestock, and Other Uses			Major Crops[1]	Crops	Crops	Products (×1,000)	Outdoor Applications (×1,000)	Estimated Range (lb a.i. ×10^6)	Industry Use (lb a.i. ×10^6)						
	1964	1966	1971	1976	1988	1995	1990	1990	1995	1995	GW	SW	R	A	BS	AB
AMIDE HERBICIDES																
Acetochlor						22–27						•		•		
Alachlor		·14.75	88.55	55.19	19–24						•	•	•	•		
Metolachlor				49.71	59–64						•	•	•			
Propanil	3.85	2.59	6.66	6.85	7.52	6–10					•	•	•			
Propachlor		2.27	23.73	11.02	3.99						•	•	•	•		
Diphenamid				0.56	0.93											
Napropamide					0.70											
Naptalam	1.06	1.00	3.33	4.27	0.66											
CDAA	3.67	4.90														
Others	na		0.79	0.08	0.25											
Total amide use:	8.58	10.76	49.27	111.32	118.96		0	0								
Percent of total herbicide use:	11.7	9.7	22.1	29.8	26.0		0	0								
CARBAMATE HERBICIDES																
EPTC	[3]0.63	3.14	4.41	8.60	37.19	9–13	125				•	•	•			
Butylate			5.92	24.41	19.11						•	•	•			
Molinate	[3]0.63			1.16	4.41						•	•	•			
Triallate				0.86	3.51									•		•

Table 3.5. Estimated pesticide use in agricultural, home and garden, and industry settings in the United States, and detections in ground water, surface water, rain, air, sediment, and aquatic biota—*Continued*

Compound	Agricultural Use (lb a.i. × 10^6)						Home and Garden Use				Pesticide Detected In:					
	Crops, Livestock, and Other Uses			Major Crops[1]	Crops	Crops	Products (×1,000)	Outdoor Applications (×1,000)	Estimated Range (lb a.i. × 10^6)	Industry Use (lb a.i. × 10^6)	GW	SW	R	A	BS	AB
	1964	1966	1971	1976	1988	1995	1990	1990	1995	1995						
Thiobencarb					1.36						•	•		•		•
Cycloate					1.18						•					
Vernolate	[3]0.63		3.74		0.86						•					
Pebulate		0.15	1.06	0.33	0.65											
Propham				0.15	0.32											
Chlorpropham	0.81	1.15		0.78	0.26											
Barban				0.26	0.05											
Others	0.16	0.80		1.48	0.14											
Total carbamate herbicide use:	1.61	5.24	15.13	38.02	69.02		125	0								
Percent of total herbicide use:	2.2	4.7	6.8	10.2	15.1		0.3	0								

CHLOROPHENOXY ACID HERBICIDES

Compound	1964	1966	1971	1976	1988	1995	Products 1990	Outdoor 1990	Est. Range 1995	Industry 1995	GW	SW	R	A	BS	AB
2,4-D	34.45	40.14	34.61	38.39	33.10	31–36	14,324	44,054	7–9	10–13	•	•		•	•	
MCPA	1.52	1.67	3.30		4.34		119	121			•					
Diclofop					1.45								•			
2,4-DB					1.37						•					
2,4,5-T	1.66	0.76	1.53				84				•			•	•	
Silvex, total							341	88			•	•		•	•	
Others	0.76	1.49	0.61	3.38	0.04						•	•		•	•	

Table 3.5. Estimated pesticide use in agricultural, home and garden, and industry settings in the United States, and detections in ground water, surface water, rain, air, sediment, and aquatic biota—Continued

Compound	Agricultural Use (lb a.i. × 10^6)						Home and Garden Use				Pesticide Detected In:					
	Crops, Livestock, and Other Uses			Major Crops[1]	Crops	Crops	Products (×1,000)	Outdoor Applications (×1,000)	Estimated Range (lb a.i. × 10^6)	Industry Use (lb a.i. × 10^6)	GW	SW	R	A	BS	AB
	1964	1966	1971	1976	1988	1995	1990	1990	1995	1995						
Total chlorophenoxy acid herbicide use:	38.39	44.07	40.05	41.77	40.30	23–28	14,868	44.263								
Percent of total herbicide use:	52.2	39.9	18.0	11.2	8.8		31.2	34.5								

OTHER HERBICIDES

Compound	1964	1966	1971	1976	1988	1995	Products 1990	Outdoor 1990	Est. Range 1995	Industry Use 1995	GW	SW	R	A	BS	AB
Trifluralin	[4]0.74	5.23	11.43	28.33	27.12	23–28	483	547			•	•		•		•
Pendimethalin				0.14	12.52	23–28	158	338			•	•	•	•		
Glyphosate				0.12	11.60	25–30	8,110	25,618	5–7	8–11	•	•	•			
Dicamba	0.02	0.22	0.43	3.56	11.24	6–10	3,636	6,431	3–5		•	•	•	•	•	
Bentazon				3.82	8.21	4–8					•	•				
MSMA				1.98	5.07	4–8	240	267		3–4			•			
Ethalfluralin					3.52											
Paraquat				0.65	3.03		201	79			•					
Chloramben (Amiben)	1.21	3.77	9.56	4.46	3.02						•	•				
Picloram				0.42	2.93						•	•		•		
Clomazone					2.72											
Bromoxynil					2.63								•			
Linuron	0.22	1.43	1.80	8.35	2.62		733	368	1–3			•		•		
Fluometuron			3.33	5.28	2.44	5–9						•	•			
Dacthal (DCPA)					2.22						•	•		•		•

Table 3.5. Estimated pesticide use in agricultural, home and garden, and industry settings in the United States, and detections in ground water, surface water, rain, air, sediment, and aquatic biota—*Continued*

Compound	Agricultural Use (lb a.i. × 10⁶)						Home and Garden Use				Pesticide Detected In:					
	Crops, Livestock, and Other Uses			Major Crops[1]	Crops	Crops	Products (×1,000)	Outdoor Applications (×1,000)	Estimated Range (lb a.i. ×10⁶)	Industry Use (lb a.i. ×10⁶)	GW	SW	R	A	BS	AB
	1964	1966	1971	1976	1988	1995	1990	1990	1995	1995						
Diuron	1.12	1.62	1.23	0.94	1.99					2–4	•				•	
Norflurazon					1.77							•				
DSMA				1.46	1.71		38	38								
Acifluorfen					1.48		1,845	7,081			•					
Oryzalin				0.36	1.43		117	1,766			•					
Benefin (Benfluralin)	[4]0.74			0.88	1.17		79	81	1–3							
Bromacil					1.16						•					
Asulam					1.09											
Imazaquin					1.07											
Sethoxydim					0.79											
Fluazifop					0.73											
Bensulide					0.63		36	73								
Profluralin				0.26	0.62											
Tebuthiuron					0.61											
Oxyfluorfen					0.60		234	643			•					
Diethatyl ethyl					0.50											
Dalapon	2.06	0.04	1.04		0.45						•					
Nitralin		0.01	2.71													
Fluorodifen			1.33													
Noruron (Norea)		0.24	1.32													
MCPP (Mecoprop)					0.03		12,692	35,644	3–5							

Table 3.5. Estimated pesticide use in agricultural, home and garden, and industry settings in the United States, and detections in ground water, surface water, rain, air, sediment, and aquatic biota—*Continued*

Compound	Agricultural Use (lb a.i. × 10⁶) Crops, Livestock, and Other Uses 1964	1966	1971	1976	Major Crops[1] 1988	Crops 1995	Home and Garden Use Products (×1,000) 1990	Outdoor Applications (×1,000) 1990	Estimated Range (lb a.i. × 10⁶) 1995	Industry Use (lb a.i. × 10⁶) 1995	Pesticide Detected In: GW	SW	R	A	BS	AB
Triclopyr, total					0.07		683	823			•					
Chlorflurenol-methyl							501	1,067								
Fluazifop-butyl							421	426			•					
Diquat					0.17		211	635								
Dichlobenil					0.06		167	124								
Amitrole	0.30						163	40								
Metam-sodium						49–54	128	74								
Endothal (Endothall)					0.20		124	43								
Sodium thiocyanate							82									
Monuron	0.24															
TCA	0.39															
2,3,6-TBA	2.22	2.97														
Chlorobromuron				0.21												
Dinitramine				0.60												
Others	5.21	10.54	20.45	6.04	5.64											
Total other herbicide use:	13.73	26.07	54.63	67.86	124.82		31,082	82,206								
Percent of total herbicide use:	18.7	23.6	24.5	18.2	27.3		65.3	64.1								
TOTAL HERBICIDE USE:	73.60	110.42	222.87	373.23	456.97		47,625	128,238								

Table 3.5. Estimated pesticide use in agricultural, home and garden, and industry settings in the United States, and detections in ground water, surface water, rain, air, sediment, and aquatic biota—*Continued*

Compound	Agricultural Use						Home and Garden Use				Pesticide Detected In:					
	Crops, Livestock, and Other Uses		Major Crops[1]	Crops	Crops		Products (×1,000)	Outdoor Applications (×1,000)	Estimated Range (lb a.i. × 10^6)	Industry Use (lb a.i. × 10^6)						
	(lb a.i. × 10^6)															
	1964	1966	1971	1976	1988	1995	1990	1990	1995	1995	GW	SW	R	A	BS	AB
FUNGICIDES																
Chlorothalonil				4.43	9.93	8–12	1,399	2,602			•					
Mancozeb					8.66	6–9	113					•				
Captan		6.87	6.49	0.10	3.71		3,067	4,682			•					
Maneb		4.44	3.88	0.10	3.59		345	878								
Ziram	[5]3.80				1.89		81									
Benomyl				1.72	1.34		684	3,704			•					•
PCNB				4.23	0.80		64	64								
Iprodione					0.74											
Fosetyl-Al					0.69											
Metiram					0.64											
Metalaxyl					0.64		41									
Thiophanate methyl					0.53											
Triphenyl tin hydroxide					0.42											
Ferbam	[5]3.80	2.95	1.40	0.07	0.34		82									
DCNA					0.29											
Dodine	0.90	[6]1.145	[7]1.19		0.28											
Propiconazole					0.27											

Table 3.5. Estimated pesticide use in agricultural, home and garden, and industry settings in the United States, and detections in ground water, surface water, rain, air, sediment, and aquatic biota—*Continued*

Compound	Agricultural Use (lb a.i. × 10^6)						Home and Garden Use			Industry Use (lb a.i. × 10^6)	Pesticide Detected In:					
	Crops, Livestock, and Other Uses			Major Crops[1]	Crops	Crops	Products (×1,000)	Outdoor Applications (×1,000)	Esti-mated Range (lb a.i. × 10^6)							
	1964	1966	1971	1976	1988	1995	1990	1990	1995	1995	GW	SW	R	A	BS	AB
Thiram					0.24		397	249								
Triademefon					0.15											
Anilazine					0.14		106	71								
Thiabendazole					0.14											
Myclobutanil					0.12											
Etridiazole					0.10											
Vinclozolin					0.10											
Streptomycin					0.09											
Triforine					0.08		3,150	14,286								
Fenarimol					0.06											
Oxytetracycline					0.04											
Carboxin					0.02											
Dinocap			[7]1.19		0.01		703	796								
Folpet							3,314	4,347								
Limonene							819									
Zineb		6.90	1.97				684	1,413								
Pentachlorophenol	0.37						576	89			•	•			•	•
Dextrin							79									
Dichlorophene							43									
Dichlone	1.04						40									
Thymol							37	166								

Table 3.5. Estimated pesticide use in agricultural, home and garden, and industry settings in the United States, and detections in ground water, surface water, rain, air, sediment, and aquatic biota—*Continued*

Compound	Crops, Livestock, and Other Uses				Major Crops[1]	Crops	Products (×1,000)	Outdoor Applications (×1,000)	Estimated Range (lb a.i. × 10^6)	Industry Use (lb a.i. × 10^6)	GW	SW	R	A	BS	AB
	1964	1966	1971	1976	1988	1995	1990	1990	1995	1995						
Difolatan				0.17												
Dithane M45				0.20												
Karathane	0.08	[6]1.14														
Other fungicides	12.06	3.33	10.81	0.02												
Total organic fungicide use:	23.93	25.64	25.74	11.04	36.04		15,824	33,347								

MITICIDES

Compound	1964	1966	1971	1976	1988	1995	1990	1990	1995	1995	GW	SW	R	A	BS	AB
Formetante HCl				0.07												
Omite				0.30											•	•
Chlorobenzilate	0.24	0.47	0.81													
Aramite	0.18	0.10	0.02													
Tetradifon	0.59	0.37														
Other miticides	0.91	0.68	0.33	0.01												
Total miticide use:	1.92	1.61	1.16	0.38												

FUMIGANTS

Compound	1964	1966	1971	1976	1988	1995	1990	1990	1995	1995	GW	SW	R	A	BS	AB
Chloropicrin				1.28												
D-D mixture	[8]14.57	13.96	7.02	1.24							•	•				
Dibromochloro- propane		3.91	3.60	2.91												

Table 3.5. Estimated pesticide use in agricultural, home and garden, and industry settings in the United States, and detections in ground water, surface water, rain, air, sediment, and aquatic biota—*Continued*

Compound	Agricultural Use: Crops, Livestock, and Other Uses (lb a.i. × 10⁶)				Major Crops[1]		Home and Garden Use: Products (×1,000)	Outdoor Applications (×1,000)	Estimated Range (lb a.i. × 10⁶)	Industry Use (lb a.i. × 10⁶)	Pesticide Detected In:					
	1964	1966	1971	1976	Crops 1988	Crops 1995	1990	1990	1995	1995	GW	SW	R	A	BS	AB
Ethylene dibromide	2.71			1.66							•	•				
Methyl bromide	4.61	5.71	5.92	6.58		39–46				6–8						
Telone (1,3-D)	[8]14.57	0.39	6.95	1.47												
Sodium methyldithio-carbamate	0.65															
Other fumigants	1.89	12.78	5.47	0.01												
Total fumigant use:	24.42	36.75	28.96	15.16												

OTHER MISCELLANEOUS PESTICIDES

	1964	1966	1971	1976												
DEF+Folex	3.39	4.23	5.05	3.39												
Maleic hydrazide	2.56	3.29	4.22	3.22												
T-148				3.05												

[1] Major crops are corn, cotton, wheat, sorghum, rice, other grains, soybeans, tobacco, peanuts, alfalfa, other hay and forage, and pasture and rangeland.
[2] Pesticide synergist.
[3] Sum of EPTC, molinate, and vernolate.
[4] Sum of trifluralin and benefin.
[5] Sum of ziram and ferbam.
[6] Sum of dodine and karathane.
[7] Sum of dodine and dinocap.
[8] Sum of D-D mixture and Telone.

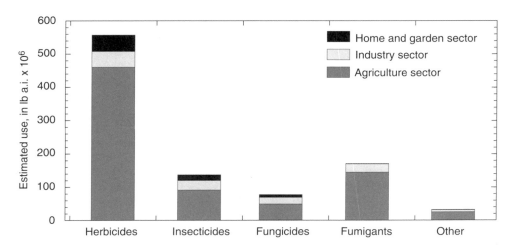

Figure 3.10. Estimated use of conventional pesticides in the United States in 1995, by market sector and pesticide type. Home and garden sector: homeowner applications (both indoor and outdoor) to homes (both single- and multiple-unit housing), lawns, and gardens. Industry sector: applications by business owners and commercial applicators to industrial, commercial, and government facilities, buildings, sites, and land; and by commercial applicators to homes, lawns, and gardens. Agriculture sector: applications by owners and commercial applicators to farms and facilities involved in production of raw agricultural commodities, principally food, fiber, and tobacco. Graph is based on U.S. Environmental Protection Agency data from Aspelin (1997).

about 8 percent of total conventional pesticide use. The decline in pesticide use in USEPA's home and garden sector during 1979–1995 is caused by decreases in the home and garden use of both insecticides (shown in Figure 3.8) and fungicides (Figure 3.9) during this time period. The quantity of herbicides used in the home and garden sector actually increased between 1979 and 1995 (shown in Figure 3.7).

Table 3.5 lists the number of pesticide products present, and the number of outdoor applications of pesticide active ingredients made by homeowners in and around homes and gardens in 1990. These data are from the USEPA's National Home and Garden Survey (Whitmore and others, 1992). This survey was not designed to collect quantitative pesticide use information, but only the products each interviewed household had on hand that contained a particular pesticide and the number of times they applied each of them. This survey did not include use of insecticides for subterranean termite control (Aspelin and others, 1992). The principal pesticides used in and around homes and gardens in 1990 were not necessarily the same ones that were commonly used in agriculture in 1988 (as shown in Table 3.5). Home and garden use (1990) of herbicides was dominated by chlorophenoxy acid herbicides and other (miscellaneous) herbicides, whereas agricultural use (1988) was dominated by herbicides in the triazine, amide, and carbamate classes. For insecticides, home and garden use (1990) was dominated by pyrethroid insecticides, pyrethroid synergists, the insect repellant N,N-diethyl-m-toluamide, and other compounds in the other (miscellaneous) insecticide category. In contrast, the other insecticide category constituted less than 10 percent of agricultural insecticide use (1988).

Nonagricultural use estimates (in pounds active ingredient) for individual pesticides are available only for the highest-use conventional pesticides in the United States in 1995 (Aspelin, 1997). These also are provided in Table 3.5. The conventional pesticides that were most commonly applied by homeowners to homes and gardens were the herbicides 2,4-D, glyphosate, dicamba, MCPP, benefin, and dacthal; and the insecticides diazinon, chlorpyrifos, and carbaryl. In addition, paradichlorobenzene and naphthalene were used in moth control, and N,N-diethyl-m-toluamide as an insect repellant. Five of the top nine herbicides and insecticides in the home and garden sector (MCPP, benefin, dacthal, carbaryl, and diazinon) were different from the top 20 conventional pesticides used in agriculture.

Pesticide use by professional pest control firms in lawn care, tree care, and treatment of structures was estimated in a 1981 survey, the National Urban Pesticide Applicator Survey (NUPAS). As mentioned previously, total use by professional applicators in 1981 was estimated as 47 million lb a.i. (Immerman and Drummond, 1984). Treatment of structures with insecticides, much of which may have been indoors, accounted for more than 50 percent of the total amount of pesticides used by professional applicators in 1981. This NUPAS estimate, therefore, overlaps with estimates of pesticide application for subterranean termite control, discussed below. The NUPAS survey identified 1,073 pesticide products containing 338 different active ingredients. The data from this survey (not shown) are somewhat out of date, but this is the most recent compilation available. More recent estimates of pesticide sales (1991) indicate that herbicides now constitute about 50 percent and insecticides about 30 percent of the professional market, while the consumer market is still dominated (approximately 75 percent) by insecticide sales (Hodge, 1993).

Industrial, Commercial, and Government Buildings and Land

As defined by USEPA, this sector consists of applications by business owners and operators and commercial applicators to industry, commercial, and government facilities, buildings, sites, and land, and by commercial applicators to homes, lawns, and gardens. In the remainder of this chapter, this sector is referred to as the "industry sector." The industry sector presumably includes, but is not limited to, subterranean termite control. In 1995, about 13 percent of total conventional pesticide use was in the industry sector (Aspelin, 1997). The quantities of herbicides (Figure 3.7) and fungicides (Figure 3.9) used in the industry sector declined between 1979 and 1995, as did the total conventional pesticides used in this sector (Figure 3.6), whereas the quantity of insecticides used (Figure 3.8) remained fairly constant. The individual pesticides that were most commonly applied in the industry sector were the herbicides 2,4-D, glyphosate, MSMA, and diuron; the fumigant methyl bromide; the fungicide copper sulfate; and the insecticides chlorpyrifos, diazinon, and malathion. Five of these top nine pesticides used in the industry sector (MSMA, diuron, copper sulfate, diazinon, and malathion) were different from the top 20 conventional pesticides used in agriculture.

Subterranean Termite Control

USEPA estimates of pesticides used in subterranean termite control are available for 1980 (U.S. Environmental Protection Agency, 1983) and 1985 (Esworthy, 1987). In 1980, about 10 million lb a.i. chlordane, 1–2 million lb a.i. heptachlor, and a few thousand lb a.i. chlorpyrifos were used in termite control (U.S. Environmental Protection Agency, 1983). The total quantity of pesticides used in termite control in 1985 (about 5–6 million lb a.i.) represents a reduction by about half from that used in 1980 (Esworthy, 1987). The primary insecticide applied in 1985 for subterranean termite control was chlordane (3.0–3.5 million lb a.i.), which constituted slightly less than 60 percent of the total. Other insecticides used for subterranean termite control in 1985 included heptachlor (0.75–1.0 million lb a.i.), aldrin (1.0–1.5 million lb a.i.), and chlorpyrifos (<1.0 million lb a.i.) (Esworthy, 1987). Aldrin use was expected to decline because products were being formulated from existing supplies of technical aldrin, since the manufacture and sale of aldrin was discontinued in 1974 (U.S. Environmental Protection Agency, 1983; Esworthy, 1987). The use of chlorpyrifos was expected to increase, as were the market shares of more recently registered termiticides (Esworthy, 1987): endosulfan and isofenphos (both registered in 1983) and synthetic pyrethroids (registered in 1984). Since 1988, the commercial use of existing stocks of chlordane products for subterranean termite control was prohibited (U.S. Environmental Protection Agency, 1990b).

The total quantity of organochlorine insecticides used in termite control in 1985 (4–6 million lb a.i.) is very small relative to the total quantity of all insecticides applied in agriculture in 1985 (225 million lb a.i., Aspelin and others, 1992) or even in 1988 (170 million lb a.i., Gianessi and Puffer, 1992b). However, it exceeds the total quantity of organochlorine insecticides applied in agriculture in 1988 (3.9 million lb a.i., Gianessi and Puffer, 1992b; see Table 3.5). Although current data on use of insecticides for subterranean termite control are not available, the existing data (1985 and earlier) suggest that urban and residential settings may have been a source of some hydrophobic insecticides (especially chlordane, heptachlor, and chlorpyrifos) to hydrologic systems in some areas. According to USEPA estimates, chlorpyrifos was the top insecticide applied in the industry sector in 1995 (Aspelin, 1997); some (unknown) fraction of the 9–13 million lb a.i. chlorpyrifos that were applied in 1995 by pest control firms, commercial applicators, and business owners and operators was probably for termite control.

Forestry

In their review of pesticides in surface waters in the United States, Larson and others (1997) estimated the total use of pesticides on forested land in the United States by extrapolating from use data for national forest land published by the U.S. Forest Service. The U.S. Forest Service, which administers 191 million acres of the approximately 800 million acres of forested land in the United States, has collected its own pesticide use data since the 1970s (U.S. Forest Service, 1978, 1985, 1989, 1990, 1991, 1992, 1993, 1994). Comparison of U.S. Forest Service use data with additional forestry use data available for two states (California and Virginia) suggests that pesticide use on national forest land is representative of use on the remainder of forested land in the United States (Larson and others, 1997). The quantity of pesticides applied in forestry is a small fraction of that applied in agriculture, as is the acreage treated. The total mass of herbicides, insecticides, and fungicides applied to national forest land from 1977 to 1993 is shown in Figure 3.11. Although these values represent only about 20 percent of United States forested land, they can be used to indicate general trends in pesticide use in forestry (Larson and others, 1997).

Insecticides are used on forested land primarily to control outbreaks of specific pests in localized areas, and are not a part of routine silvicultural practice (Larson and others, 1997). As

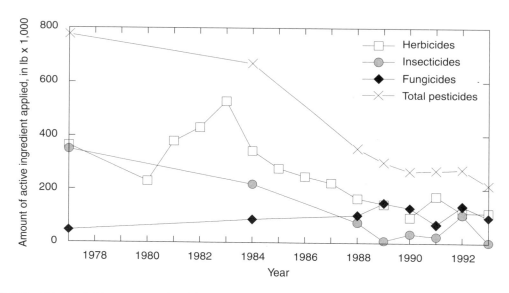

Figure 3.11. Pesticide use on national forest land, 1977–1993, in pounds active ingredient. Insecticide amounts do not include use of *Bacillus thuringiensis* var. *kurstaki* (Bt). From Larson and others (1997).

in agriculture, there has been a dramatic shift in the types of insecticides used in forestry since the 1960s. DDT and other organochlorine insecticides, which were used during the 1950s and 1960s, were largely replaced by organophosphate and carbamate insecticides during the 1970s and 1980s. The apparent decline in insecticide use since the mid-1980s is probably due to replacement of chemical insecticides by the bacterial agent, *Bacillus thuringiensis* var. *kurstaki* (Bt), since quantities of Bt used are not included in Figure 3.11 (Larson and others, 1997). Herbicide use in forestry also has declined since the early 1980s. This is partially due to a 1984 ban on aerial application of herbicides in national forests and the increased costs of preparing required environmental impact statements (Wehr and others, 1992). As a result, this decline may not have occurred in forested land outside the national forests (Larson and others, 1997). The type of herbicides used also have changed over time, with 2,4-D, picloram, and hexazinone being the highest use compounds during the 1970s–1980s, and triclopyr now the highest use herbicide (Larson and others, 1997). Fungicide use on national forest land since 1977 has remained fairly constant or increased slightly. Total pesticide use on national forest land has declined dramatically since the mid-1980s due to declining use of both herbicides and insecticides (Figure 3.11).

3.2.3 TRENDS AND CONCLUSIONS

Total use of conventional pesticides grew from about 650 million lb a.i. in 1964 to over 1,100 million lb a.i. in 1979, then remained fairly constant or decreased, reaching about 970 million lb a.i. in 1995 (Figure 3.5). Substantial changes (>5 percent) in total use of conventional pesticides occurred in 1983 and 1987 (both decreases) and in 1994 (an increase). Herbicides constitute the largest share of total conventional pesticides used (Figure 3.10). The total annual use of herbicides (Figure 3.7) remained fairly constant between 1979 and 1995, showing the same minor dips and peaks observed in the annual use of total conventional pesticides (Figure 3.6). Total annual insecticide use in the United States declined between 1979 and 1995, mostly because of declining agricultural use, with the sharpest drop occurring in 1987 (Figure 3.8). Total annual use of fungicides (Figure 3.9) also declined from 1979 to 1995, especially in the nonagricultural sectors.

Table 3.5 summarizes quantitative information on use of individual pesticides and pesticide classes in agriculture and in selected nonagricultural applications. The following important points regarding pesticide use emerge from Table 3.5:

1. A comparison of the top 50 pesticides used in agriculture and the home and garden sector indicates that there is only about a 20 percent overlap in the pesticides used. In other words, the principal pesticides used in agriculture tend to differ from those used in and around homes and gardens.

2. There was a pronounced change in the types of insecticides used between the 1960s and the 1980s as organochlorine insecticides were replaced by organophosphate, carbamate, and other insecticides. Agricultural use of organochlorine insecticides decreased markedly since the mid-1960s, both in absolute mass and as a percentage of total insecticide used. Most organochlorine insecticides were banned or severely restricted during the early 1970s. Only four organochlorine insecticides were used in agriculture in 1988: endosulfan, dicofol, methoxychlor, and lindane. A number of organochlorine insecticides were used in and around the home and garden in 1990, although they accounted for fewer than 4 percent of products and 2 percent of outdoor applications. Large quantities of paradichlorobenzene were used by homeowners for moth control in 1995.

3. The total quantity of organophosphate insecticides used in agriculture increased from 1964 to 1971, then remained fairly constant in 1976 and 1988. This contrasts to total insecticide use, which declined during this time. As a result, organophosphate use as a percentage of total insecticide use increased from 1964 to 1988. Organophosphate insecticides constituted 65 percent of total agricultural insecticide use in 1988 (based on pounds active ingredient) and 26 percent of home and garden insecticide use in 1990 (based on number of outdoor applications). A few organophosphate insecticides (chlorpyrifos, malathion, acephate, diazinon, and dichlorvos) are commonly applied in nonagricultural settings by business owners and operators, commercial applicators, and pest control firms.

4. The total quantity of carbamate insecticides used increased from 1966 to 1971, then remained fairly constant in 1976 and 1988. The percentage of total pesticide use comprised by carbamate insecticides increased from 10 percent in 1964 to 23 percent in 1988. Carbamate insecticides constituted about 14 percent of home and garden insecticide use in 1990 (based on number of outdoor applications). Carbaryl and propoxur were two of the most commonly applied insecticides in the home and garden sector in 1995.

5. Most insecticides used around homes and gardens in 1990 (64 percent of products and 59 percent of outdoor applications) fall into the "other insecticides" category. Home and garden insecticide use was dominated by pyrethroid insecticides (pyrethrins, allethrin, tetramethrin, resmethrin, sumithrin, and permethrin) and pyrethroid synergists (piperonyl butoxide and MGK-264). In contrast, although the use of other insecticides in agriculture has increased since the 1960s, they still account for less than 10 percent of total insecticide use in agriculture. The insecticide in this category that had the largest agricultural use in 1988 was cryolite (Gianessi and Puffer, 1992b), an inorganic fluorine compound (Meister Publishing Company, 1996) that was used primarily on grapes in California (82 percent of total insecticide use in pounds active ingredient). Except for the insect repellent N,N-diethyl-m-toluamide, no insecticides in the other insecticides category were among the top insecticides used in 1995 for any of the agricultural, home and garden, or industry sectors (Aspelin, 1997).

6. Nonagricultural use estimates on a mass basis (pounds active ingredient) are available for only selected insecticides in 1995 (Aspelin, 1997). Comparison of these data with recent agricultural use estimates (1988 or 1995) indicates that, for at least some insecticides, nonagricultural applications appear to constitute a significant fraction of the total quantity of insecticide applied annually. For chlorpyrifos and diazinon (organophosphate insecticides), the quantities applied in nonagricultural settings in 1995 equaled or exceeded those applied in agricultural settings in 1988 or 1995. For malathion (another organophosphate insecticide), use by the industry sector in 1995 was about 80 percent of that used by the agriculture sector in 1988. For carbaryl (a carbamate insecticide), home and garden use in 1995 was about 25 percent of the quantity used in agriculture in 1988.

7. Agricultural use of herbicides has increased markedly since the 1960s. Most herbicide classes have contributed to this increase, except the chlorophenoxy acid herbicides. The quantity of chlorophenoxy acid herbicides used in agriculture remained fairly constant (about 40 million lb a.i.) from 1964 to 1988, but their use as a percentage of total herbicide use in agriculture decreased from over 50 percent in 1964 to less than 9 percent in 1988.

8. On the basis of the number of outdoor applications in 1990, over 98 percent of herbicides applied around homes and garden are classified as either chlorophenoxy acid herbicides (35 percent) or other herbicides (64 percent). This is in contrast to agricultural use, where these classes represent 9 and 27 percent, respectively, of total herbicides applied to cropland in 1988 on a mass (pounds active ingredient) basis. Because the 1990 data on pesticide use around homes and gardens are in numbers of products and of outdoor applications, they cannot be compared directly with agricultural use estimates (pounds active ingredient). However, nonagricultural use estimates on a mass basis are available for selected herbicides in 1995 (Aspelin, 1997). Comparison of these data with recent agricultural use estimates (1988 or 1995) indicates that, for at least some herbicides, nonagricultural applications constitute a significant fraction of the total quantity of that herbicide

applied annually. For dacthal, diuron, benefin, and MCPP, the quantities used in the home and garden or the industry sectors equaled or exceeded the quantities used recently (1988 or 1995) in agriculture. Even for 2,4-D and glyphosate, which were used heavily in agriculture in 1995, the quantities used in nonagricultural sectors were 50 percent or greater of the quantities used in agriculture.

9. Triazine, amide, and carbamate herbicides constituted 23, 26, and 15 percent, respectively, of total herbicides used in agriculture in 1988, but less than 2 percent each of outdoor applications around the home and garden in 1990.

10. The USDA estimate of agricultural fungicide use in 1976 appears to be markedly lower than the USDA estimates for 1964–1971. However, this is because the 1976 use estimate applies to major crops only (Andrilenas, 1974), whereas the use estimates for 1964–1971 apply to all crops, pasture and rangeland, livestock, and other uses by farmers (Eichers and others, 1968, 1970, 1978). Between 1971 and 1976, organic fungicide use on major crops increased by about 55 percent, and total fungicide use on all crops increased by about 9 percent (Andrilenas, 1974). Estimates of fungicide use on cropland from the Resources for the Future (about 36 million lb a.i. in 1988) are higher than USDA estimates for crops, livestock, and other uses by farmers (about 25 million lb a.i. during 1964–1971). This may reflect differences in how estimates were calculated. According to USEPA estimates (Aspelin, 1997), the total use of fungicides decreased steadily from 124 million lb a.i. in 1970 to 77 million lb a.i. in 1995. Agricultural use of fungicides declined slightly during this time, constituting about 6–8 percent of total fungicide use.

3.3 GEOGRAPHIC DISTRIBUTION IN RELATION TO USE

The occurrence and geographic distribution of a pesticide in a hydrologic system are influenced by sources of that pesticide to the hydrologic system, characteristics of the hydrologic system (such as discharge, slope, pH, organic carbon content of the water and sediment), and the physical and chemical properties of the specific pesticide (such as vapor pressure and water solubility) that affect its environmental fate. Moreover, detection of pesticides in any environmental medium will be biased by various aspects of study design. Examples include which compounds were looked for in samples (compounds not analyzed in samples will never be detected), how study sites were selected (such as agricultural versus urban areas), how samples were collected (such as sediment from depositional areas versus cross-sectional samples, and different species of biota), when samples were collected (such as seasonal sampling on the basis of seasonal pesticide use, low versus high stream flow, or life cycle stage of the organism sampled), and analytical methods (such as accuracy and precision of recovery, and analytical detection limits).

As a first approximation, the likelihood of detecting pesticides in sediment or aquatic biota in a certain location may be expected to relate directly to the quantities of pesticides applied in the drainage area. In the following analysis, pesticide detection frequencies are compared with available pesticide use information. Ideally, detection frequencies and use levels would be compared on a drainage-basin scale. Because consistent basin-scale pesticide use information is not available nationwide, comparisons can be done only on a national or regional scale.

In the following discussion, pesticide detection data are compared with national and regional estimates of pesticide use in agriculture. The focus of these comparisons is on agricultural use for two reasons. The first is data availability. Quantitative estimates (in pounds active ingredient) of national use in agriculture are available for hundreds of individual pesticides. This is not the case for home and garden use or other nonagricultural uses of pesticides. Second, agricultural use constitutes the largest portion of total pesticide use (Figure 3.7), total insecticide use (Figure 3.8), total herbicide use (Figure 3.9), and total fungicide use (Figure 3.10) in the United States. The following analysis has two stages. First, national agricultural use estimates are compared with aggregate detection frequencies for individual pesticides calculated as described in Section 3.1.1 (using the combined data from all monitoring studies listed in Tables 2.1 and 2.2). Separate comparisons are made for bed sediment and aquatic biota. Second, a more detailed examination is made for selected pesticides employing regional agricultural use estimates and concentration data from the most appropriate national studies for bed sediment and aquatic biota.

In the following analysis, a distinction is made between historically used organochlorine insecticides (discussed in Section 3.3.1) and currently used pesticides (discussed in Section 3.3.2). This is done for practical reasons. As implied by the term "historically used," organochlorine insecticides were used extensively on cropland during the 1970s or earlier, but are no longer used in agriculture in the United States. Therefore, the occurrence of these compounds cannot be compared with current information on agricultural pesticide use in the United States. However, the persistence of these restricted organochlorine insecticides suggests that it may be interesting to compare current occurrence data for these compounds with use information from the period of heaviest use in agriculture (mid-1960s). For the restricted organochlorine insecticides, therefore, 1966 pesticide use estimates from the USDA (Eichers and others, 1970) are employed in the analysis below. The two national programs providing the best long-term national data on organochlorine insecticides in bed sediment and aquatic biota from United States rivers are the PMN for bed sediment (1975–1980) and the NCBP for whole fish (1967–1986).

In this book, "currently used pesticides" refer to pesticides that either or both (1) were used on cropland in the United States in quantities of greater than 500,000 lb a.i. in 1988 (Gianessi and Puffer, 1991, 1992a,b), and (2) are listed in Meister Publishing Company (1996) as used in crop production. These include only four organochlorine insecticides: dicofol, endosulfan, lindane, and methoxychlor. For currently used pesticides, the following analysis employs the most recent agricultural use estimates available, which are 1988 estimates from Resources for the Future (Gianessi and Puffer, 1991, 1992a,b). Two national studies targeted a substantial number of currently used pesticides in United States rivers: the PMN (1975–1980) for bed sediment and the NSCRF (1986–1987) for fish.

Those pesticides that were detected in bed sediment or aquatic biota, as well as in rain, air, ground water, and surface water, are indicated in Table 3.5. Tables 3.6A and 3.6B show, for sediment and aquatic biota respectively, the total number of sites analyzed for a given pesticide in all monitoring studies (from national to local scale) combined; the percentage of those sites with detectable levels; and recent use estimates for the agriculture sector (1988) and the home and garden sector (1990). As discussed in Section 3.1.1, the list of compounds detected is biased by the target analyte list. It is clear that the focus of most sediment and biota monitoring studies was on the organochlorine insecticides, particularly those no longer used in agriculture in the United States (Tables 2.4, 3.5, and 3.6). Relatively few studies on pesticides in bed sediment or aquatic biota analyzed for herbicides or for insecticides other than the organochlorines (Tables 2.4 and 3.6).

3.3.1 HISTORICALLY USED ORGANOCHLORINE INSECTICIDES

Use of the organochlorine insecticides in agriculture in the United States began during the 1940s and increased to peak levels during the 1950s–1960s. Most were banned or severely restricted during the 1970s because of their persistence, tendency to bioaccumulate, and potential effects on wildlife and human health (see Sections 6.1 and 6.2, respectively). This resulted in a shift from use of organochlorine compounds to other classes of insecticides (as discussed in Section 3.2.3). Although most organochlorine insecticides are no longer used in agriculture in the United States today, these compounds continue to be the focus of sediment and biota monitoring efforts because of their extreme persistence. In general, organochlorine compounds are hydrophobic, with low water solubilities and high n-octanol-water partition coefficients (K_{ow}), and resistant to environmental (abiotic or biologically mediated) degradation (see Chapter 4).

In Table 3.6, organochlorine insecticides are grouped separately from other pesticide classes. The list of organochlorine insecticides includes a number of transformation products and components of technical mixtures. This list includes the more hydrophobic organochlorine insecticides that are no longer used in agriculture in the United States (such as DDT), as well as the four less-hydrophobic insecticides that are still used in agriculture in the United States today (dicofol, endosulfan, lindane, and methoxychlor). Also included with the organochlorine insecticides is the fungicide hexachlorobenzene, which is also a common industrial by-product. Hexachlorobenzene contains only carbon and chlorine and is extremely hydrophobic. Because its physical and chemical properties and its environmental fate are similar to the most hydrophobic organochlorine insecticides, hexachlorobenzene is grouped with them in Table 3.6.

Organochlorine insecticides generally were targeted at more sites and detected at higher frequencies than pesticides in other classes (Figures 3.1 and 3.2, and Table 3.6). This was particularly true for the historically used organochlorine insecticides, such as DDT, chlordane, dieldrin, endrin, and heptachlor. Of these organochlorine insecticides, chlordane and heptachlor were applied outdoors around homes and gardens in 1990 (Whitmore and others, 1992). Although reportedly not applied outdoors in 1990, products containing DDT and toxaphene were on hand in some households (Table 3.5). As noted in Section 3.2.2 (Home and Garden), insecticide use for subterranean termite control was not included in the USEPA's National Home and Garden Survey (Aspelin and others, 1992), so this use is not reflected in the 1990 home and garden use estimates (Tables 3.5, 3.6) from Whitmore and others (1992). Chlordane, heptachlor, aldrin, and chlorpyrifos were the primary insecticides used for subterranean termite control in 1985. Use of endosulfan, isofenphos, and permethrin products as termiticides was expected to increase from their zero-to-low 1985 market share (Esworthy, 1987).

Table 3.6. Analysis of pesticides in bed sediment (3.6A) and analysis of pesticides in aquatic biota (3.6B). Total number of sites sampled, aggregate site detection frequencies (in percent), and pesticide use estimates both on crops in 1988 (in million pounds active ingredient) and in and around homes and gardens in 1990 (in thousands of products)

[Organochlorine insecticides are listed separately from pesticides in other classes. Within each group, pesticide analytes are listed in descending order of total number of sites sampled. Total number of sites and aggregate detection frequencies were calculated for all monitoring studies listed in Tables 2.1 and 2.2, combined. For pesticide components and pesticide transformation products, use estimates are listed in parentheses and refer to the technical mixture and the parent compound, respectively (see Glossary). Pesticide use data are from Gianessi and Puffer (1991, 1992a,b) (for 1988 crop use); and Whitmore and others (1992) (for 1990 home and garden use). Abbreviations: AA-H, amino acid herbicide; Amide-H, acid amide herbicide; Amine-H, amine herbicide; BA-H, chlorobenzoic acid herbicide; Carb-I, carbamate insecticide; CP-H, chlorophenoxy acid herbicide; Dinitr-H, dinitroaniline herbicide; Fung, fungicide; IR, insect repellent; lb a.i., pounds active ingredient; M, metabolite; OC-Fung, organochlorine fungicide; OC-I, organochlorine insecticide; OC-Misc, miscellaneous organochlorine contaminant (pesticide by-product); OP-I, organophosphate insecticide; Other-H, other herbicide; Other-I, other insecticide; PCNB, pentachloronitrobenzene; Pyreth-I, pyrethroid insecticide; Thcarb-H, thiocarbamate herbicide; Triaz-H, triazine herbicide; Urea-H, urea herbicide]

3.6A: BED SEDIMENT

Target Analytes	Class	Total Number of Sites	Percentage of Sites with Detections	Use on Crops in 1988 (lb a.i. × 1,000)	Home and Garden Use in 1990, Outdoor Applications (× 1,000)
ORGANOCHLORINE INSECTICIDES					
Dieldrin	OC-I	2,329	40		
DDE, *p,p'*- or total	OC-I(M)	2,004	50		
DDD, *p,p'*- or total	OC-I(M)	1,997	50		
DDT, *p,p'*- or (*o,p'+p,p'*)-	OC-I	1,977	33		
Lindane	OC-I	1,966	10	66	1,355
Heptachlor epoxide	OC-I(M)	1,889	8		(177)
Endrin	OC-I	1,838	6		
Chlordane, total	OC-I	1,805	40		478
Aldrin	OC-I	1,784	5		
Heptachlor	OC-I	1,650	5		177
Mirex	OC-I	1,463	16		
Toxaphene	OC-I	1,366	3		
Methoxychlor	OC-I	1,049	5	109	3,692
Hexachlorobenzene	OC-Fung	956	45		
DDT, total	OC-I	803	73		
HCH, α-	OC-I	567	9		
DDT, *o,p'*-	OC-I	532	32		
Endosulfan, total	OC-I	477	5	1,992	561
HCH, β-	OC-I	447	8		
Perthane	OC-I	406	1		
DDD, *o,p'*-	OC-I(M)	376	32		
Endosulfan II	OC-I	369	11	(1,992)	(561)
Endosulfan I	OC-I	347	10	(1,992)	(561)
DDE, *o,p'*-	OC-I(M)	322	33		
HCH, δ-	OC-I	306	0		

Table 3.6. Analysis of pesticides in bed sediment (3.6A) and analysis of pesticides in aquatic biota (3.6B). Total number of sites sampled, aggregate site detection frequencies (in percent), and pesticide use estimates both on crops in 1988 (in million pounds active ingredient) and in and around homes and gardens in 1990 (in thousands of products)—*Continued*

3.6A: BED SEDIMENT

Target Analytes	Class	Total Number of Sites	Percentage of Sites with Detections	Use on Crops in 1988 (lb a.i. × 1,000)	Home and Garden Use in 1990, Outdoor Applications (× 1,000)
Chlordane, *trans-*	OC-I	296	25		(478)
Chlordane, *cis-*	OC-I	255	31		(478)
Endosulfan sulfate	OC-I(M)	250	12	(1,992)	(561)
Oxychlordane	OC-I(M)	223	3		(478)
Nonachlor, *cis-*	OC-I	163	2		(478)
Endrin aldehyde	OC-I(M)	162	6		
Nonachlor, total	OC-I	134	19		(478)
Endrin ketone	OC-I(M)	117	1		
HCH, total	OC-I	113	11		
Isodrin	Other-I	82	11		
Nonachlor, *trans-*	OC-I	63	3		(478)
Kepone	OC-I	28	0		
DDMU, *p,p'-* or total	OC-I(M)	16	88		
DDMS, *p,p'-*	OC-I(M)	11	0		
Chlordene, α-	OC-I	4	75		(478)
Chlordene, γ-	OC-I	4	25		(478)
Dicofol	OC-I	2	0	1,718	4,179

PESTICIDES IN OTHER CLASSES

Target Analytes	Class	Total Number of Sites	Percentage of Sites with Detections	Use on Crops in 1988 (lb a.i. × 1,000)	Home and Garden Use in 1990, Outdoor Applications (× 1,000)
Parathion	OP-I	868	2	2,848	
Diazinon	OP-I	702	3	1,710	56,758
Malathion	OP-I	691	0	3,188	16,597
Methyl parathion	OP-I	683	0	8,131	
Ethion	OP-I	606	0	1,249	
Methyl trithion	OP-I	587	0		
Trithion	OP-I	568	0		
2,4-D (or ester)	CP-H	413	3	33,096	44,054
Silvex	CP-H	321	3		88
2,4,5-T	CP-H	304	2		
Pentachlorophenol	Fung	177	20		89
Atrazine	Triaz-H	168	0	64,236	488
2,4-DP	CP-H	107	1		
Trifluralin	Dinitr-H	83	8	27,119	547
Dacthal (DCPA)	BA-H	60	17	2,219	368
Zytron (Xytron)	OP-I	48	29		
Tetradifon	Other-I	36	3		
Carbofuran	Carb-I	30	0	7,057	
Dicamba	BA-H	29	14	11,240	3,636

Table 3.6. Analysis of pesticides in bed sediment (3.6A) and analysis of pesticides in aquatic biota (3.6B). Total number of sites sampled, aggregate site detection frequencies (in percent), and pesticide use estimates both on crops in 1988 (in million pounds active ingredient) and in and around homes and gardens in 1990 (in thousands of products)—*Continued*

3.6A: BED SEDIMENT

Target Analytes	Class	Total Number of Sites	Percentage of Sites with Detections	Use on Crops in 1988 (lb a.i. × 1,000)	Home and Garden Use in 1990, Outdoor Applications (× 1,000)
Picloram	Amine-H	29	3	2,932	
Chlorobenzilate	Other-I	28	0		
Linuron	Urea-H	22	0	2,623	
Chlorpyrifos	OP-I	20	5	16,725	41,900
2,4-DB	CP-H	20	15	1,368	
Endothal	Other-H	19	0	199	43
Glyphosate	AA-H	19	0	11,595	25,618
Ethoprop	OP-I	18	0	1,636	
Phorate	OP-I	18	0	4,782	
Terbufos	OP-I	18	0	7,218	
Alachlor	Amide-H	15	0	55,187	
Ametryn	Triaz-H	15	20		
Butylate	Thcarb-H	15	0	19,107	
Cyanazine	Triaz-H	15	0	22,894	
Diuron	Urea-H	15	100	1,986	
Fonofos	OP-I	15	0		
Metolachlor	Amide-H	15	0	49,713	
Metribuzin	Triaz-H	15	0	4,822	
Acephate	OP-I	3	0	2,965	19,167
Aldicarb	Carb-I	3	0	3,573	
Azinphosmethyl	OP-I	3	0	2,477	348
Carbaryl	Carb-I	3	0	7,622	31,735
Coumaphos	OP-I	3	0		
Demeton	OP-I	3	0		
Dichlorvos	OP-I	3	0		13,043
Dimethoate	OP-I	3	0	2,960	132
Disulfoton	OP-I	3	0	3,058	6,464
EPN	OP-I	3	0		
Famphur	OP-I	3	0		
Fenthion	OP-I	3	0		
Methamidophos	OP-I	3	0	1,135	
Methiocarb	Carb-I	3	0		
Methomyl	Carb-I	3	0	2,952	629
Mevinphos	OP-I	3	0	463	
Monocrotophos	OP-I	3	0		
Oxadiazon	Other-H	3	100		
Oxamyl	Carb-I	3	0	726	
Trichlorfon	OP-I	3	0	36	

Table 3.6. Analysis of pesticides in bed sediment (3.6A) and analysis of pesticides in aquatic biota (3.6B). Total number of sites sampled, aggregate site detection frequencies (in percent), and pesticide use estimates both on crops in 1988 (in million pounds active ingredient) and in and around homes and gardens in 1990 (in thousands of products)—*Continued*

3.6B: AQUATIC BIOTA

Target Analytes	Class	Total Number of Sites	Percentage of Sites with Detections	Use on Crops in 1988 (lb a.i. × 1,000)	Home and Garden Use in 1990, Outdoor Applications (× 1,000)
ORGANOCHLORINE PESTICIDES					
Dieldrin	OC-I	2,886	61		
DDT, total	OC-I	2,197	92		
DDE, *p,p'*- or total	OC-I(M)	2,051	89		
Endrin	OC-I	2,023	16		
Hexachlorobenzene	OC-I	1,898	39		
Mirex	OC-I	1,821	26		
Lindane	OC-I	1,818	32	66	1,355
Chlordane, total	OC-I	1,789	68		478
Heptachlor epoxide	OC-I(M)	1,598	21		
DDT, *p,p'*- or (*o,p'+p,p'*)-	OC-I	1,512	67		
Heptachlor	OC-I	1,424	9		177
DDD, *p,p'*- or total	OC-I(M)	1,417	69		
HCH, α-	OC-I	1,317	46		
Toxaphene	OC-I	1,300	28		
Nonachlor, *trans-*	OC-I	1,224	69		(478)
Chlordane, *cis-*	OC-I	1,220	64		(478)
Chlordane, *trans-*	OC-I	1,165	58		(478)
Nonachlor, *cis-*	OC-I	1,150	43		(478)
Oxychlordane	OC-I(M)	1,089	29		(478)
Methoxychlor	OC-I	1,051	5	109	3,692
Aldrin	OC-I	938	7		
Dicofol	OC-I	658	11	1,718	4,179
Heptachlor, total	OC-I	542	29		(177)
HCH, β-	OC-I	516	6		
DDT, *o,p'*-	OC-I	466	28		
HCH, δ-	OC-I	452	3		
Endosulfan, total	OC-I	403	13	1,992	561
DDD, *o,p'*-	OC-I(M)	399	26		
HCH, total	OC-I	396	19		
Perthane	OC-I	374	1		
DDE, *o,p'*-	OC-I(M)	366	22		
Endosulfan I	OC-I	109	42	(1,992)	(561)
DDMU, *p,p'*- or total	OC-I(M)	98	21		
Endosulfan II	OC-I	88	28	(1,992)	(561)

Table 3.6. Analysis of pesticides in bed sediment (3.6A) and analysis of pesticides in aquatic biota (3.6B). Total number of sites sampled, aggregate site detection frequencies (in percent), and pesticide use estimates both on crops in 1988 (in million pounds active ingredient) and in and around homes and gardens in 1990 (in thousands of products)—*Continued*

3.6B: AQUATIC BIOTA

Target Analytes	Class	Total Number of Sites	Percentage of Sites with Detections	Use on Crops in 1988 (lb a.i. × 1,000)	Home and Garden Use in 1990, Outdoor Applications (× 1,000)
Photomirex	OC-I(M)	78	27		
Kepone	OC-I	69	14		
Endosulfan sulfate	OC-I(M)	63	3	(1,992)	(561)
Chlordene, α-	OC-I	61	3		(478)
Chlordene, γ-	OC-I	61	5		(478)
DDMS, p,p'-	OC-I(M)	61	0		
Chlordene	OC-I	59	3		(478)
Nonachlor, total	OC-I	38	87		(478)
Endrin aldehyde	OC-I(M)	36	19		
Hydroxychlordene	OC-I(M)	31	0		(478)
Endrin ketone	OC-I(M)	16	0		
Endosulfan sulfate I	OC-I(M)	14	0	(1,992)	(561)

PESTICIDES IN OTHER CLASSES

Target Analytes	Class	Total Number of Sites	Percentage of Sites with Detections	Use on Crops in 1988 (lb a.i. × 1,000)	Home and Garden Use in 1990, Outdoor Applications (× 1,000)
Chlorpyrifos	OP-I	632	21	16,725	41,900
Dacthal (DCPA)	BA-H	615	56	2,219	368
Pentachloroanisole	Fung(M)	560	57		(89)
Trifluralin	Dinitr-H	527	13	27,119	547
Diazinon	OP-I	375	2	1,710	56,758
Isopropalin	Dinitr-H	362	4		
Nitrofen	Other-H	362	3		
Parathion	OP-I	362	2	2,848	
PCNB	Fung	362	1		
Azinphosmethyl	OP-I	115	0	2,477	348
Demeton	OP-I	113	0		
Malathion	OP-I	113	0	3,188	16,597
Phorate	OP-I	113	0	4,782	
Ethion	OP-I	111	2	1,249	
Carbophenothion	OP-I	110	2		
DEF	OP-Misc	110	0		
Pentachlorophenol	Fung	86	26		89
Methyl parathion	OP-I	85	2	8,131	
Oxadiazon	Other-H	64	16		
Tetradifon	Other-I	15	7		
Zytron (Xytron)	OP-I	14	64		
Alachlor	Amide-H	7	0	55,187	
Atrazine	Triaz-H	5	0	64,236	488
Carbofuran	Carb-I	5	20	7,057	

Table 3.6. Analysis of pesticides in bed sediment (3.6A) and analysis of pesticides in aquatic biota (3.6B). Total number of sites sampled, aggregate site detection frequencies (in percent), and pesticide use estimates both on crops in 1988 (in million pounds active ingredient) and in and around homes and gardens in 1990 (in thousands of products)—*Continued*

3.6B: AQUATIC BIOTA

Target Analytes	Class	Total Number of Sites	Percentage of Sites with Detections	Use on Crops in 1988 (lb a.i. × 1,000)	Home and Garden Use in 1990, Outdoor Applications (× 1,000)
2,4-D (or ester)	CP-H	5	0	33,096	44,054
DBP	IR	5	60		
Disulfoton	OP-I	5	0	3,058	6,464
Acephate	OP-I	3	0	2,965	19,167
Aldicarb	Carb-I	3	0	3,573	
Carbaryl	Carb-I	3	0	7,622	31,735
Coumaphos	OP-I	3	0		
Cyanazine	Triaz-H	3	0	22,894	
Dichlorvos	OP-I	3	0		13,043
Dimethoate	OP-I	3	0	2,960	132
EPN	OP-I	3	0		
Ethoprop	OP-I	3	0	1,636	
Famphur	OP-I	3	0		
Fenthion	OP-I	3	0	.	
Methamidophos	OP-I	3	0	1,135	
Methiocarb	Carb-I	3	0		
Methomyl	Carb-I	3	0	2,952	629
Mevinphos	OP-I	3	0	463	
Monocrotophos	OP-I	3	0		
Oxamyl	Carb-I	3	0	726	
Silvex	CP-H	3	0		88
2,4,5-T	CP-H	3	0		
Terbufos	OP-I	3	0	7,218	
Trichlorfon	OP-I	3	0	36	
3-Hydroxycarbofuran	Carb-I(M)	2	100	(7,057)	
Fenvalerate	Pyreth-I	2	0	73	2,937
Metolachlor	Amide-H	2	0	49,713	
Molinate	Thcarb-H	2	100	4,408	
Thiobencarb	Thcarb-H	2	100	1,359	

Agricultural use estimates (Gianessi and Puffer, 1992b) indicate that four organochlorine insecticides were used on cropland in the United States in quantities of greater than 500,000 lb a.i. in 1988: endosulfan, dicofol, methoxychlor, and lindane (Table 3.5). These four compounds also were applied around homes and gardens in 1990 (Whitmore and others, 1992). For two of these compounds (dicofol and endosulfan), relatively little sampling effort has gone into measuring these compounds in monitoring studies (Table 3.6). Not coincidentally, the four organochlorine insecticides still used in agriculture are less persistent than DDT and the other restricted organochlorine insecticides. Their aggregate detection frequencies in sediment and aquatic biota tend to be low relative to the restricted organochlorine compounds (Table 3.6). Endosulfan, dicofol, methoxychlor, and lindane will be discussed in more detail in the section on currently used pesticides (Section 3.3.2).

Figure 3.12 shows the results of the first and most simplistic stage in the analysis of national pesticide occurrence in relation to use. The aggregate site detection frequencies for individual organochlorine insecticides in sediment (Figure 3.12A) and aquatic biota (Figure 3.12B) are shown as a function of national use in agriculture in 1966 (Eichers and others, 1970). For transformation products (such as heptachlor epoxide and dieldrin), aggregate site detection frequencies are plotted against use of the appropriate parent pesticides (such as heptachlor and

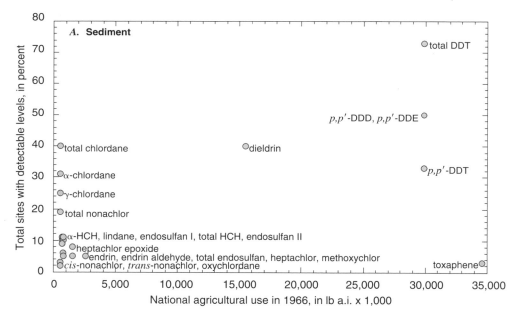

Figure 3.12. The relation between aggregate site detection frequency and 1966 national agricultural use, shown for organochlorine insecticides and transformation products detected in (*A*) sediment and (*B*) aquatic biota. The percentage of sites with detectable levels is for all monitoring studies (listed in Tables 2.1 and 2.2) combined. Data for aquatic biota consist of whole fish, fish muscle, and shellfish, combined. Only pesticides targeted at more than 15 sites were included. Agricultural use data (1966) are U.S. Department of Agriculture estimates from Eichers and others (1970). Abbreviation: lb a.i., pounds active ingredient.

aldrin plus dieldrin, respectively). For multicomponent pesticides, detection frequencies for the total product and for all of its individual components are plotted separately against use of the technical mixture. For example, some studies reported data for individual components of chlordane, and other studies reported data for total chlordane only. These data could not be combined, so separate detection frequencies were calculated for each analyte as reported, and these detection frequencies are all plotted versus technical chlordane use in Figure 3.12. Therefore, Figure 3.12 contains separate points corresponding to total chlordane, *cis*- and *trans*-chlordane, *cis*- and *trans*-nonachlor, and (in aquatic biota) oxychlordane. The aggregate detection frequencies for total chlordane were not based on the same set of studies as the detection frequencies for individual components of chlordane; each reflects only those studies reporting data in that way. There was an exception to this rule: studies reporting data for DDT (unspecified), or for the sum of *o,p'*- plus *p,p'*-DDT, were grouped with those reporting data for *p,p'*-DDT, which is the primary component of DDT. DDD and DDE were handled the same way. As noted in Section 3.1.1 the aggregate site detection frequency is the percent of sites with one or more detections based on data combined from all national and multistate (Table 2.1) and state and local (Table 2.2) monitoring studies. It represents a crude composite detection frequency from the 1960s to the 1990s, and is subject to an unknown bias because of the different designs of the individual studies that make up the composite for a given analyte.

Figure 3.12 does not indicate any apparent correlation between the percent of sites with detectable levels of organochlorine insecticides and their 1966 national use levels in agriculture. A closer look suggests factors that may be obscuring any relation between these parameters.

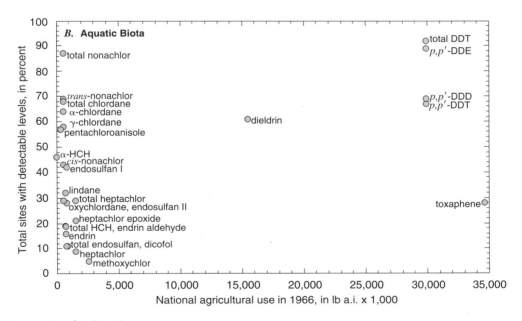

Figure 3.12. *Continued.*

Note that the patterns revealed for bed sediment and aquatic biota shown in Figures 3.12 *A* and 3.12*B*, respectively, are very similar. When a few compounds are excluded (the chlordane compounds, toxaphene, and, in biota, pentachloroanisole), there appears to be a positive relation between national agricultural use in 1966 and aggregate detection frequency in both bed sediment and aquatic biota.

For each pesticide analyte that does not fit this relation, there are special circumstances that may obscure the relation between detection frequency and agricultural use. The detection frequencies for chlordane compounds and pentachloroanisole were higher than expected on the basis of 1966 agricultural use alone, whereas that for toxaphene was much lower. Chlordane has been used in nonagricultural settings (such as in and around homes and gardens, and in buildings for subterranean termite control), so it is not surprising that the aggregate detection frequencies for chlordane components exceed predictions from agricultural use alone. Pentachloroanisole is a metabolite of pentachlorophenol and its detection frequency is plotted against the 1966 agricultural use of phenols (because a 1966 use estimate for pentachlorophenol alone was not available). Despite this overestimate of pentachlorophenol use, pentachloroanisole was detected in biota at a frequency higher than would be expected from agricultural use alone, probably because there are other sources of pentachlorophenol. Pentachlorophenol was widely used as a wood preservative and a slimicide by the forest products industries (National Research Council of Canada, 1974), and is itself a metabolite of hexachlorobenzene, which is both a fungicide and an industrial contaminant. The aggregate detection frequencies for toxaphene in sediment and aquatic biota are much lower than expected on the basis of national agricultural use in 1966. In fact, because national agricultural use of toxaphene remained high through the mid-1970s (Table 3.5), its low detection frequencies in sediment and biota appear particularly surprising. This may reflect both the use patterns and analytical characteristics of toxaphene. About 80–90 percent of the toxaphene used in agriculture from the mid-1960s to the early 1970s was applied to cotton (see discussion of toxaphene later in this section). Most of the sampling sites used in this analysis are probably not in cotton-growing areas, which constitute a relatively small part of the United States and are concentrated in the southernmost states. Moreover, toxaphene is a complex mixture of over 200 different compounds, which complicates its analysis in environmental samples. Because different components undergo physical weathering, chemical degradation, and metabolism in the environment at different rates, the analytical fingerprint for toxaphene in environmental samples is often considerably different from that in analytical standards (Bidleman and others, 1988). There is potential for packed-column gas chromatography methods (used in earlier monitoring studies) to underestimate residues of weathered toxaphene in environmental samples (Ribick and others, 1982). Moreover, the analytical detection limit for toxaphene is generally higher, so that low-level residues are less likely to be detected, than for single component organochlorine insecticides.

To improve the chances of seeing a relation between pesticide occurrence and pesticide use, the variability in both data sets can be reduced. First, pesticide occurrence data can be taken from a single national program at a time. Second, pesticide use estimates can be improved by using USDA pesticide use estimates by farm production region (Eichers and others, 1970). USDA divided the continental United States into the following 10 farm production regions, shown in Figure 3.13: Appalachian, Corn Belt, Delta States, Lake States, Mountain, Northeast, Northern Plains, Pacific, Southeast, and Southern Plains. These USDA farm production regions are used in the discussion throughout the remainder of this chapter. In this second stage of analysis of the relation between pesticide occurrence and use that follows, results are presented first for bed sediment, then for aquatic biota.

Bed Sediment

To assess the national occurrence and distribution of organochlorine insecticides in bed sediment, only one national program provides long-term national data for United States rivers—the PMN, which sampled river bed sediment during the 1970s. The NS&T Program has published long-term national data on contaminants in surficial sediment in United States estuaries and near-shore coastal areas for the period 1984–1989. The geographic distributions of aldrin and dieldrin (Figure 3.14) and total DDT (Figure 3.15) from these studies are discussed below in relation to pesticide use by USDA farm production region. Data treatment for Figures 3.14 and 3.15 is described in detail in the following section.

Aldrin and Dieldrin

Environmental residues of dieldrin may result from introduction of either dieldrin or aldrin. Aldrin is a soil insecticide, most uses of which were suspended in 1974.   Aldrin is rapidly transformed into dieldrin in the environment (U.S. Environmental Protection Agency, 1980a). Dieldrin, the epoxide of aldrin, was applied directly to control soil-dwelling insects, public health insects, and termites (U.S. Environmental Protection Agency, 1992b). Both aldrin and

Figure 3.13. The U.S. Department of Agriculture's farm production regions. Redrawn from Eichers and others (1970).

dieldrin were used on corn and citrus. Agricultural use of aldrin during the 1960s exceeded use of dieldrin by more than ten-fold (Eichers and others, 1970). The use of both compounds decreased rapidly after 1966, primarily because of increased insect resistance and the development and availability of substitute chemicals (U.S. Environmental Protection Agency, 1980a). By 1972, all uses of aldrin and dieldrin were banned, except for subsurface termite control, dipping of nonfood roots and tops, and moth proofing in a closed system by manufacturing processes (U.S. Environmental Protection Agency, 1980a); these remaining uses were voluntarily cancelled by industry (U.S. Environmental Protection Agency, 1992b). As indicated by its physical and chemical properties, which are summarized in Table 3.7, dieldrin has a fairly low volatility (as evidenced by a moderately low Henry's law constant), has low water solubility, sorbs strongly to soil (high K_{oc}, which is the n-organic carbon-normalized distribution coefficient), and has a strong tendency to bioaccumulate (high K_{ow} and bioconcentration factor). Dieldrin hydrolyzes very slowly at neutral pH, with a half-life of about 10.5 years (U.S. Environmental Protection Agency, 1992b). It also photolyzes in water, with a half-life of about 2 months. Dieldrin is very slowly transformed by soil microbes under aerobic or anaerobic conditions (U.S. Environmental Protection Agency, 1992b), with a field dissipation half-life of about 3 years (Agricultural Research Service, 1997; see Table 3.7).

The geographic distribution of aldrin plus dieldrin (also called "total dieldrin") in river bed sediment at USGS–USEPA's PMN sites is shown in Figure 3.14. The individual PMN sites are represented as circles, each shaded according to the median concentration of total dieldrin at that site during 1972–1982. This distribution is shown in relation to USDA estimates of 1966 aldrin plus dieldrin use by farm production region (Eichers and others, 1970). The total dieldrin occurrence data used for this analysis were obtained from the U.S. Geological Survey (1996) for the period 1972–1982. Most sites were sampled at least five times during this period. Over 85 percent of these sites had median total dieldrin concentrations less than the detection limit (<0.30 µg/kg). Therefore, sites in Figure 3.14 are shaded to correspond to concentration categories that are based on 50th and 90th percentiles of all nonzero site median concentrations. The first (lowest) concentration category in Figure 3.14 contains sites with median concentrations below detection (<0.30 µg/kg). The second category contains sites with median concentrations between the detection limit (0.30 µg/kg) and the 50th percentile value (1.08 µg/kg) for all nonzero site median concentrations. The third category contains sites with median concentrations between the 50th and 90th percentile values (1.09–3.62 µg/kg) for all nonzero site median concentrations. The fourth (highest) category contains sites with maximum concentrations greater than the 90th percentile value (>3.62 µg/kg) for all nonzero site median concentrations.

Regional 1966 agricultural use levels in Figure 3.14 were determined from USDA estimates of aldrin plus dieldrin use in 10 farm production regions of the United States (Eichers and others, 1970). Use data for the sum of aldrin plus dieldrin are employed because either may be the source of dieldrin residues in biota. The quantity of aldrin plus dieldrin (lb a.i.) used in each region was normalized by the number of square miles of agricultural land within the region (U.S. Geological Survey, 1970). This was done because some regions have less agricultural acreage than others, so that the total pesticides applied in those regions tend to be concentrated in relatively smaller land areas. The normalized 1966 agricultural use levels for the 10 USDA farm production regions were divided into quartiles, and the agricultural land within each region was shaded to correspond to the appropriate use quartile. The first (lowest) category contains regions with normalized use levels that are less than the 25th percentile value. Regions in the second and

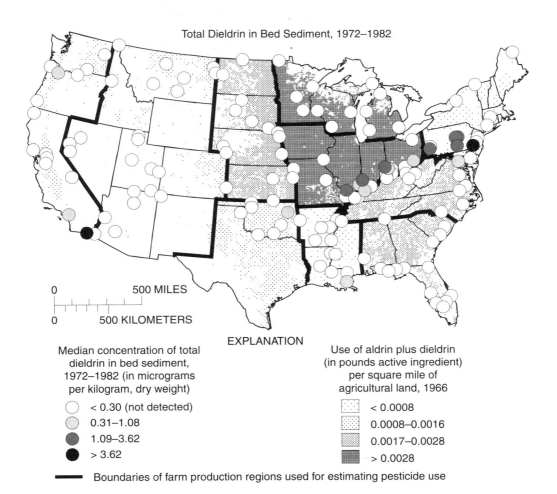

Total Dieldrin in Bed Sediment, 1972–1982

0   500 MILES

0   500 KILOMETERS

EXPLANATION

Median concentration of total
dieldrin in bed sediment,
1972–1982 (in micrograms
per kilogram, dry weight)

○   < 0.30 (not detected)

◔   0.31–1.08

◑   1.09–3.62

●   > 3.62

Use of aldrin plus dieldrin
(in pounds active ingredient)
per square mile of
agricultural land, 1966

  < 0.0008

  0.0008–0.0016

  0.0017–0.0028

  > 0.0028

━━━ Boundaries of farm production regions used for estimating pesticide use

Figure 3.14. The geographic distribution of total dieldrin in bed sediment of major United States rivers (1972–1982) from USGS–USEPA's Pesticide Monitoring Network, shown in relation to 1966 agricultural use of aldrin plus dieldrin by farm production region (in pounds active ingredient per square mile of agricultural land). Total dieldrin is the sum of residues of aldrin and dieldrin. Each site (circle) is shaded to represent the median concentration of total dieldrin, in micrograms per kilogram dry weight, at that site. Total dieldrin concentration categories are based on 50th and 90th percentiles of all nonzero site median concentrations as follows: <DL; DL–50th percentile; >50th–90th percentile; >90th percentile. Total dieldrin concentration data (1972–1982) are from U.S. Geological Survey (1996). Use data are U.S. Department of Agriculture estimates of 1966 agricultural use of aldrin plus dieldrin by farm production region (Eichers and others, 1970), normalized by the number of square miles of agricultural land within the region (U.S. Geological Survey, 1970). Agricultural land within each region is shaded to correspond to the appropriate use quartile. Abbreviations: DL, detection limit; USEPA, U.S. Environmental Protection Agency; USGS, U.S. Geological Survey.

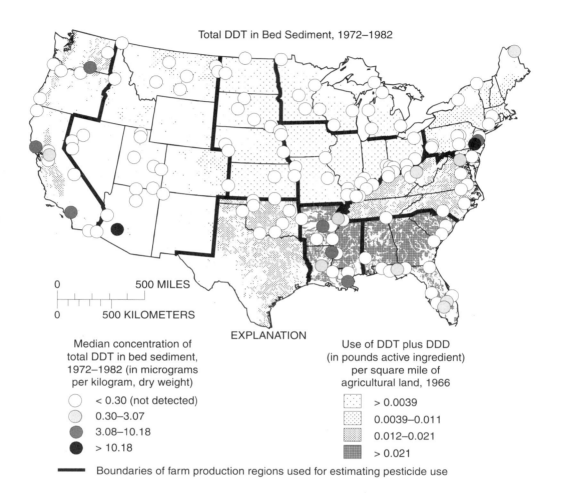

Figure 3.15. The geographic distribution of total DDT in bed sediment of major United States rivers (1972–1982) from USGS–USEPA's Pesticide Monitoring Network, shown in relation to 1966 agricultural use of DDT plus DDD by farm production region (in pounds active ingredient per square mile of agricultural land). Total DDT is the sum of residues of o,p'-DDT, o,p'-DDE, o,p'-DDD, p,p'-DDT, p,p'-DDE, and p,p'-DDD. Each site (circle) is shaded to represent the median concentration of total DDT, in micrograms per kilogram dry weight, at that site. Total DDT concentration categories are based on 50th and 90th percentiles of all nonzero site median concentrations as follows: < DL; DL–50th percentile; >50th–90th percentile; >90th percentile. Total DDT concentration data (1972–1982) are from U.S. Geological Survey (1996). Use data are U.S. Department of Agriculture estimates of 1966 agricultural use of DDT plus DDD by farm production region (Eichers and others, 1970), normalized by the number of square miles of agricultural land within the region (U.S. Geological Survey, 1970). Agricultural land within each region is shaded to correspond to the appropriate use quartile. Abbreviations: DL, detection limit; USEPA, U.S. Environmental Protection Agency; USGS, U.S. Geological Survey.

third categories have normalized use levels between the 25th and 50th percentile values, and the 50th and 75th percentile values, respectively. Regions in the fourth (highest) category have normalized use levels greater than the 75th percentile value.

The highest agricultural use of aldrin plus dieldrin in 1966 occurred in the Corn Belt and Lake States regions (Figure 3.14). Several sites along the southern border of the Corn Belt Region had high median total dieldrin concentrations (>1.08 µg/kg). However, median concentrations of total dieldrin were higher than expected at several sites in the Northeast and Pacific regions, and lower than expected in the Lake States Region. Similar conclusions were drawn by Gilliom and others (1985), who evaluated pesticide concentrations in bed sediment at PMN sites on major rivers (1975–1980) in relation to USDA estimates of pesticide use by farm production region (for 1971). Gilliom and others (1985) noted that dieldrin was detected relatively often in the Northeast and Pacific regions, despite relatively low regional use on farms in these regions. These authors suggested that higher than expected detection in the Northeast Region may have been due to intensive nonfarm use or incidental release from chemical manufacturing plants in urban and industrial areas.

Total DDT

DDT was first synthesized in 1874 and used as an insecticide in 1939 (U.S. Environmental Protection Agency, 1992b). It was widely used in the United States on a variety of crops and for control of insect-borne diseases (U.S. Environmental Protection Agency, 1992). Its use in the United States peaked at 80 million lb a.i. in 1959, decreasing steadily to less than 12 million lb a.i. by 1972 (U.S. Environmental Protection Agency, 1975, 1980d). DDD (also called TDE) also was applied as an insecticide for several years, but its agricultural use was only about 10 percent or less of that of DDT (Table 3.5). USEPA cancelled all products containing DDD in 1971, and all remaining crop uses of DDT in 1972 (U.S. Environmental Protection Agency, 1975, 1980c). DDT biodegrades very slowly in the environment, with the rate and products of degradation depending on environmental conditions. The half-life of DDT in soil has been estimated at about 15 years (Mischke and others, 1985; Agricultural Research Service, 1997). Its principal metabolites are DDE and DDD, which also are environmentally persistent. DDT is fairly volatile (high Henry's law constant), has a very low water solubility, sorbs strongly to soil (high K_{oc}), and has a strong tendency to bioaccumulate (high K_{ow} and bioconcentration factor), as shown in Table 3.7. The metabolites DDE and DDD are less volatile than DDT, but have comparably low water solubilities, strong tendencies to sorb to soil and to bioaccumulate, and long soil half-lives (Table 3.7).

Environmental concentrations are frequently calculated for total DDT, which consists of the sum of *o,p'* and *p,p'* isomers of DDT, DDD, and DDE. Environmental residues of total DDT may be derived from use of DDT, DDD, or dicofol, which may contain up to 0.1 percent of DDT and related compounds as impurities (U.S. Environmental Protection Agency, 1992b). In a study conducted in California, it was found that dicofol usage probably was not a significant source of environmental residues of total DDT (Mischke and others, 1985).

Figure 3.15 shows the geographic distribution of total DDT in river bed sediment at PMN sites during 1972–1982. The individual PMN sites are represented as circles, each shaded according to the median concentration of total DDT at that site. This distribution is shown in relation to USDA estimates of agricultural use of DDT plus DDD in 1966, by farm production

Table 3.7. Selected physical and chemical properties of pesticide analytes that were targeted in bed sediment and aquatic biota studies

[Citations are listed in a key at the end of the table. Citations that reported the same value for a given property are listed as a group (separated by commas). For citations that reported different values for a given property, the citations, and their respective property values, are separated by a slash (/). All values cited from Agricultural Research Service (1997) are the preferred values (as designated therein), if available, followed by the range of values (in parentheses) reported therein. Some values in the table are rounded to three significant figures. Abbreviations and symbols: BCF, bioconcentration factor; C, Celsius; K_{oc}, soil organic carbon partition coefficient; K_{ow}, n-octanol-water partition coefficient; mg/L, milligram per liter; nr, not reported; Pa-m^3/mol, Pascal cubic meter per mole; PCNB, pentachloronitrobenzene; S, water solubility; Temp, temperature; —, no data found]

Compound	S (mg/L)	Temp (°C)	Citation (see Key)	log K_{ow}	Citation (see Key)	log K_{oc}	Citation (see Key)
Aldrin	0.027	27	([22])	—	—	4.24	([22])
Atrazine	33	22	([22])	2.68	([22])	2.17	([22])
Carbofuran	350	25	([22])	1.41	([22])	1.66	([22])
Carbophenothion	0.63	20	([22])	4.75	([22])	4.70	([22])
Chlordane	0.06	25	([22])	6	([22])	4.78	([22])
Chlorpyrifos	1	25	([22])	5.0 (4.7–5.3)	([22])	4.00 (3.78–4.17)	([22])
D, 2,4- (acid)	23,200 (pH 7)	25	([22])	−0.75	([22])	1.68	([22])
Dacthal	0.5	25	([22])	3.83/4.87	([10])/([9])	3.75 (3.60–3.81)	([22])
DDD, o,p'-	0.1	25	([22])	5.06–6.22	([22])	5.36	([22])
DDD, p,p'-	0.05	25	([22])	5.06–6.22	([22])	5.38	([9])
DDE, o,p'-	0.0013	nr	([22])	5.69–6.96	([22])	5.58	([22])
DDE, p,p'-	0.065	24	([22])	5.69–6.96	([22])	5.95	([22])
DDT, o,p'-	—	—	—	5.98/6.00	([1])/([8])	5.63	([22])
DDT, p,p'-	0.0077	20	([22])	5.98/6.00	([1])/([8])	5.63	([22])
Diazinon	60	22	([22])	3.3	([22])	3.18	([22])
Dicofol	0.8	20	([22])	4.28	([22])	3.78	([22])

Table 3.7. Selected physical and chemical properties of pesticide analytes that were targeted in bed sediment and aquatic biota studies—*Continued*

Compound	BCF	Citation (see Key)	Henry's Law Constant (Pa-m³/mol)	Temp (°C)	Citation (see Key)	Soil Half-Life (days)	Citation (see Key)
Aldrin	28	[13]	91.2	20	[22]	365	[22]
Atrazine	—	—	2.48×10^{-4}	25	[22]	173(13–402)	[22]
Carbofuran	—	—	5.20×10^{-5}	25	[22]	41(17–90)	[22]
Carbophenothion	—	—	0.58	nr	[22]	30	[22]
Chlordane	14,000	[12]	9.51	25	[22]	365	[22]
Chlorpyrifos	450–470	[12]	0.743	25	[22]	43(4–139)	[22]
D, 2,4- (acid)	—	—	1.8×10^{-7} (pH 7)	25	[22]	14	[22]
Dacthal	—	—	0.219	25	[22]	50(14–100)	[22]
DDD, *o,p'-*	53,600	[15]	—	—	—	730–5,690	[22]
DDD, *p,p'-*	53,600	[15]	0.900	nr	[22]	730–5,690	[22]
DDE, *o,p'-*	53,600–180,000	[12]	1.02	25	[22]	730–5,690	[22]
DDE, *p,p'-*	53,600–180,000	[12]	1.02	25	[22]	730–5,690	[22]
DDT, *o,p'-*	53,600	[15]	394	25	[6]	2,390	[22]
DDT, *p,p'-*	53,600	[15]	394	25	[6]	110–5,480	[22]
Diazinon	—	—	0.0720	25	[22]	7(2.8–13)	[22]
Dicofol	10,000–>25,000	[12]	0.0245	—	[22]	57(45–68)	[22]

Table 3.7. Selected physical and chemical properties of pesticide analytes that were targeted in bed sediment and aquatic biota studies—*Continued*

Compound	S (mg/L)	Temp (°C)	Citation (see Key)	log K_{ow}	Citation (see Key)	log K_{oc}	Citation (see Key)
Dieldrin	0.14	20	[22]	3.69–6.2	[22]	4.08	[7, 22]
Endosulfan	0.32	22	[5, 11, 22]	3.13	[22]	4.09	[5, 10, 22]
Endothal	1.00×10^5	20	[22]	−0.87/−0.89	[3]/[9]	1.9	[22]
Endrin	0.24	25	[22]	3.21–5.34	[22]	4	[22]
Ethion	1.1	20	[3]	5.07/5.1	[3]/[1]	4.19	[3, 8, 10]
Glyphosate	1.20×10^4	25	[4, 9, 11, 22]	−1.6	[22]	3.32	[22]
HCH, α-	1.63	25	[6]	3.8/3.81	[3]/[1, 6]	3.28	[9]
Heptachlor	0.056	25–29	[22]	4.4–5.5	[22]	4.38	[7, 9, 22]
Heptachlor epoxide	0.275/0.35	25	[6, 9]	3.65	[6]	3.89	[3]
Hexachlorobenzene	0.04	20	[1, 9, 22]	3.93–6.42	[22]	4.7	[22]
Isopropalin	0.1	25	[22]	5.74	[12]	4	[22]
Kepone	7.6	24	[9]	4.07	[9]	4.74	[9]
Lindane	7	20	[22]	3.24/3.61	[6]/[3]	3.13	[22]
Malathion	130	nr	[22]	2.7	[22]	3.08	[22]
Methoxychlor, *p,p'*-	0.1	25	[10, 11, 22]	4.68	[6, 8]	4.88	[22]
Methyl parathion	55	20	[22]	2.86	[3]	3.8	[22]
Mirex	7×10^{-5}	22	[22]	6.9	[1]	6	[22]
Nitrofen	1	nr	[22]	—	—	4	[22]
Nonachlor	0.06	nr	[10]	5.66	[10]	4.86	[10]
Oxadiazon	0.7	25	[22]	4.7	[22]	3.52 (3.20–3.72)	[22]
Oxychlordane	200	nr	[10]	2.6	[10]	2.48	[10]

Table 3.7. Selected physical and chemical properties of pesticide analytes that were targeted in bed sediment and aquatic biota studies—*Continued*

Compound	BCF	Citation (see Key)	Henry's Law Constant (Pa-m^3/mol)	Temp (°C)	Citation (see Key)	Soil Half-Life (days)	Citation (see Key)
Dieldrin	4,670	(^{12})	0.0650	20	(^{22})	1,000	(^{22})
Endosulfan	270	(^{16})	1.09	25	(^{22})	60(4–200)	(^{22})
Endothal	—	—	—	—	—	2.8	(^{22})
Endrin	3,970/ 1,480–13,000	(^{17})/(^{12})	0.148	25	(^{22})	4,300	(^{22})
Ethion	—	—	0.0384	25	(^{9})	2.18	(^{5})
Glyphosate	—	—	<1.4×10^{-7}	25	(^{22})	37(2–174)	(^{22})
HCH, α-	130	(^{12})	1.07	25	(^{3})	2–19	(^{2})
Heptachlor	11,200/ 1,000–21,000	(^{18})/(^{12})	353	25	(^{22})	250	(^{22})
Heptachlor epoxide	11,200/ 850–14,400	(^{21})/(^{12})	3.2	25	($^{6,\ 9}$)	4.7–79	(^{2})
Hexachlorobenzene	8,690/7,800– 22,000	(^{14})/(^{12})	10.32	20	(^{22})	1,000	(^{22})
Isopropalin	25,000	(^{12})	3.71	25	(^{22})	100	(^{22})
Kepone	—	—	3,150	24–25	(^{9})	—	—
Lindane	130/130– 1,400	(^{19})/(^{12})	0.183	20	(^{22})	423(100– 1,420)	(^{22})
Malathion	—	—	0.00114	nr	(^{22})	9(0.2–25)	(^{22})
Methoxychlor, *p,p'*-	190–8,300	(^{12})	1.60	25	(^{3})	128(7–210)	(^{22})
Methyl parathion	—	—	9.57×10^{-4}	20	(^{22})	10(1–30)	(^{22})
Mirex	2,600–41,000	(^{12})	2.23	25	(^{12})	3,000	(^{22})
Nitrofen	16,000	(^{12})	0.314	25	(^{12})	15	(^{22})
Nonachlor	22,000	(^{10})	0.0203	25	(^{10})	—	—
Oxadiazon	—	—	0.0072	25	(^{22})	75(30–180)	(^{22})
Oxychlordane	19	(^{10})	0.0304	25	(^{10})	—	—

Table 3.7. Selected physical and chemical properties of pesticide analytes that were targeted in bed sediment and aquatic biota studies—*Continued*

Compound	S (mg/L)	Temp (°C)	Citation (see Key)	log K_{ow}	Citation (see Key)	log K_{oc}	Citation (see Key)
Parathion	11	20	([22])	3.83	([22])	3.88	([22])
PCNB	0.032/0.44	25	([3])/([11])	4.64/5.45	([3, 10])/([12])	3.7/3.9/4.15	([5, 10])/ ([3])/ ([12])
Pentachloroanisole	0.2	20	([10])	5.66	([10, 12])	4.62	([10, 12])
Pentachlorophenol	14 (pH 5)	20	([9])	3.32–5.86	([22])	4.3	([22])
Perthane	0.1	nr	([10])	7.14	([10])	6.04	([10])
Phorate	22	25	([22])	3.92	([22])	3.02	([22])
Silvex	140	25	([3, 6])	2.44/3.41	([6])/([3])	3.41	([3]),([8])
T, 2,4,5- (acid)	150	25	([22])	2	([22])	1.9	([7]),([22])
Toxaphene	3	nr	([22])	3.3	([1, 6])	5	([5, 10, 22])
Trifluralin	2	25	([22])	5.07	([9, 10, 11, 22])	3.86 (3.08–4.14)	([22])

[1]Suntio and others (1988).

[2]Howard and others (1991).

[3]Howard (1991).

[4]Tomlin (1994).

[5]Wauchope and others (1992).

[6]U.S. Environmental Protection Agency (1992e).

[7]Jury and others (1987).

[8]Kenaga (1980b).

[9]Montgomery (1993).

[10]U.S. Environmental Protection Agency (1991).

[11]Milne (1995).

Table 3.7. Selected physical and chemical properties of pesticide analytes that were targeted in bed sediment and aquatic biota studies—*Continued*

Compound	BCF	Citation (see Key)	Henry's Law Constant (Pa-m³/mol)	Temp (°C)	Citation (see Key)	Soil Half-Life (days)	Citation (see Key)
Parathion	—	—	0.0240	nr	[22]	14	[22]
PCNB	6.3–79	[12]	10	nr	[10]	30–100	[2]
Pentachloroanisole	10,000	[12]	811	25	[12]	—	—
Pentachlorophenol	770–1,000	[12]	0.430	20	[22]	48	[22]
Perthane	660,000	[10]	6.180	25	[10]	—	—
Phorate	—	—	1.01	25	[22]	37(2–173)	[22]
Silvex	—	—	0.00133	nr	[3]	—	—
T, 2,4,5- (acid)	—	—	1.20×10^{-6}	20	[22]	30	[22]
Toxaphene	13,100	[20]	0.0730	20	[22]	9	[22]
Trifluralin	1,800–6,000	[12]	1.53	nr	[22]	81(15–132)	[22]

[12]U.S. Environmental Protection Agency (1992b).

[13]U.S. Environmental Protection Agency (1980a).

[14]U.S. Environmental Protection Agency (1980c).

[15]U.S. Environmental Protection Agency (1980d).

[16]U.S. Environmental Protection Agency (1980e).

[17]U.S. Environmental Protection Agency (1980f).

[18]U.S. Environmental Protection Agency (1980g).

[19]U.S. Environmental Protection Agency (1980h).

[20]U.S. Environmental Protection Agency (1980i).

[21]U.S. Environmental Protection Agency (1993c).

[22]Agricultural Research Service (1997).

region (Eichers and others, 1970). Both use and occurrence data were obtained and treated as previously described for total dieldrin (Figure 3.14). In summary, PMN sites are shaded to correspond to concentration categories that are based on 50th and 90th percentiles of all nonzero site median concentrations only. Therefore, the first (lowest) concentration category in Figure 3.15 contains sites with median concentrations below detection (<0.30 µg/kg). The second category contains sites with median concentrations between the detection limit and the 50th percentile value (0.30–3.07 µg/kg) for all nonzero site median concentrations. The third category contains sites with median concentrations between the 50th and 90th percentile values (3.08–10.18 µg/kg), and the fourth (highest) category contains sites with maximum concentrations greater than the 90th percentile value (>10.18 µg/kg) for all nonzero site median concentrations. Agricultural use data (1966) were summed for DDT and DDD because either may be a source of environmental residues of total DDT in biota. The use level for each USDA farm production region was normalized by acreage of cropland within the region, and agricultural land within each region is shaded to correspond to the appropriate use quartile.

The highest median concentrations of total DDT in river bed sediment at PMN sites (1972–1982) were observed at sites along the lower Mississippi River and at scattered sites in the Pacific, Mountain, and Northeast regions (Figure 3.15). These same sites also had high detection frequencies (not shown) of total DDT, but many additional sites scattered through several regions did also. The high median total DDT concentrations observed at sites in the Delta States and Southeast regions are consistent with the high agricultural use of DDT plus DDD (1966) in these regions. In the Pacific Region, where several sites had fairly high median concentrations (>3.07 µg/kg) of total DDT, agricultural use of DDT plus DDD was also moderately high (>0.011 lb a.i. per square mile of agricultural land). However, agricultural use of DDT plus DDD was fairly low in the Northeast and Mountain regions, where one or more sites had high median concentrations of total DDT (Figure 3.15). Gilliom and others (1985) also noted higher than expected detection frequencies for DDT, DDE, and DDD at PMN sites in the Northeast Region, and attributed this to intensive nonfarm use or to incidental release from chemical manufacturing plants in urban and industrial areas. Some sites in the Mountain and Pacific regions that had high median sediment concentrations of total DDT (Figure 3.15) were located near urban areas (Phoenix, Arizona; Los Angeles and San Francisco, California).

In estuarine bed sediment sampled by NOAA under the NS&T Program, total DDT was the most frequently detected organochlorine insecticide. National Oceanic and Atmospheric Administration (1991) reported mean total DDT concentrations in fine-grained (<63 µm) sediment at individual NS&T sites. These concentrations were determined by dividing the raw concentration in each composite sediment sample by the fraction (by weight) of sediment that was less than 63 µm in diameter. This is equivalent to assuming that no contaminants are associated with larger, sand-sized particles (National Oceanic and Atmospheric Administration, 1991). The distribution among sites of mean contaminant concentrations was highly skewed towards low values, so that, when plotted as logarithms, the means were approximately normally distributed (National Oceanic and Atmospheric Administration, 1991). National Oceanic and Atmospheric Administration (1991) defined "high concentration" as the mean plus one standard deviation of the lognormal distribution. For total DDT, a high concentration was greater than 37 µg/kg dry weight in fine-grained sediment. Because 45 percent of NS&T sites were near urban areas, the geometric mean and high concentrations probably overestimated the extent of contamination typical of United States coastal sediments. On the other hand, hot spots (such as near point source discharges) may have contaminant concentrations in sediment that far exceed the high concentrations calculated from the NS&T data set (National Oceanic and Atmospheric Administration, 1991).

Figure 3.16 shows the distribution of mean total DDT concentrations in fine-grained (<63 µm) sediment at NS&T sites during 1984–1989 (National Oceanic and Atmospheric Administration, 1991) in relation to agricultural use of DDT plus DDD in 1966, by farm production region (Eichers and others, 1970). Enlarged views of the west, southeast, and northeast map areas are also shown. NS&T sites in Figure 3.16 are shaded according to concentration categories. The first (lowest) category contains sites with mean concentrations below detection limits, which were variable, but within the range of 0.01–0.2 µg/kg. The second category contains sites with total DDT concentrations between the detection limit and the mean of the lognormal distribution (<6.6 µg/kg), and the third category contains sites between this mean (6.6 µg/kg) and the mean plus one standard deviation of the lognormal distribution (37 µg/kg). The fourth (highest) category contains sites with mean total DDT concentrations greater than 37 µg/kg, which is the high concentration as defined by National Oceanic and Atmospheric Administration (1991). Note that the detection limits were low (0.01–0.2 µg/kg dry weight) and there were very few NS&T sites with no detectable total DDT residues in sediment.

The agricultural use data (USDA regional estimates for 1966) and use level categories in Figure 3.16 are identical to those described previously for Figure 3.15. Figure 3.16 shows no apparent relation between agricultural use of DDT plus DDD in 1966 and detection of total DDT in estuarine and coastal bed sediment during 1984–1989. In particular, high total DDT concentrations were observed in areas of low agricultural use (such as New York and Massachusetts) or in areas with little agricultural land (such as southern California). Fewer high concentrations were detected on the Gulf coasts of Louisiana and Mississippi despite high agricultural use of DDT plus DDD in the Delta States Region (shown in the southeast map area).

National Oceanic and Atmospheric Administration (1991) reported that, according to factor analysis of mean concentrations, most contaminants (as well as TOC in sediment) occurred together, and their concentrations were related to human population levels. Total DDT was not correlated with human population levels on a national scale because it was too highly associated with southern California. Total DDT residues in southern California were higher than elsewhere in the country, even compared with other highly populated areas (National Oceanic and Atmospheric Administration, 1991). Figure 3.17 shows the distribution of mean total DDT concentrations in fine-grained sediment at NS&T sites during 1984–1989 (National Oceanic and Atmospheric Administration, 1991) in relation to population density in people per square mile in 1990, by county (U.S. Department of Commerce, 1995). Enlarged views of the west, southeast, and northeast map areas are also shown. The total DDT concentration data and concentration categories for NS&T sites in Figure 3.17 are identical to those in Figure 3.16. Therefore, the darkest shaded sites in Figure 3.17 have high mean total DDT concentrations as defined by National Oceanic and Atmospheric Administration (1991), described above. Population density classes used in Figure 3.17 are from Hitt (1994). The two lowest classes designate nonurban land uses: rural (<130 people per square mile), and low density residential (131–1,000 people per square mile). The three highest classes are considered urban: medium density residential (1,001–5,180 people per square mile), high density residential (5,181–13,000 people per square mile), and very high density residential (>13,000 people per square mile). Figure 3.17 illustrates

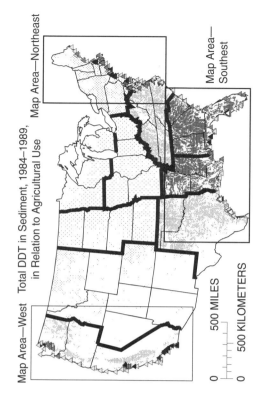

Map Area—Northeast

Map Area—Southest

Total DDT in Sediment, 1984–1989, in Relation to Agricultural Use

Map Area—West

500 MILES

500 KILOMETERS

0

0

Figure 3.16. The geographic distribution of total DDT in sediment at coastal and estuarine sites (1984–1989) from NOAA's National Status and Trends Program, shown in relation to 1966 agricultural use of DDT plus DDD by farm production region (in pounds active ingredient per square mile of agricultural land). Enlarged views of west, southeast, and northeast map areas are shown. Total DDT is the sum of residues of o,p'-DDT, o,p'-DDE, o,p'-DDD, p,p'-DDT, p,p'-DDE, and p,p'-DDD. Each site (triangle) is shaded to represent the mean total DDT concentration in sediment, in micrograms per kilogram of fine-grained (<63 µm) sediment dry weight, at that site. Total DDT concentration categories are based on the geometric mean concentration (C_{GM}) and high concentration (C_H), which is one standard deviation above the mean), as follows: <DL; DL–<C_{GM}; C_{GM}–C_H; >C_H. Total DDT concentration data are from National Oceanic and Atmospheric Administration (1991). Use data are U.S. Department of Agriculture estimates of 1966 agricultural use of DDT plus DDD by farm production region (Eichers and others, 1970), normalized by the number of square miles of agricultural land within the region (U.S. Geological Survey, 1970). Agricultural land within each region is shaded to correspond to the appropriate use quartile. Abbreviations: DL, detection limit; NOAA, National Oceanic and Atmospheric Administration; µm, micrometer.

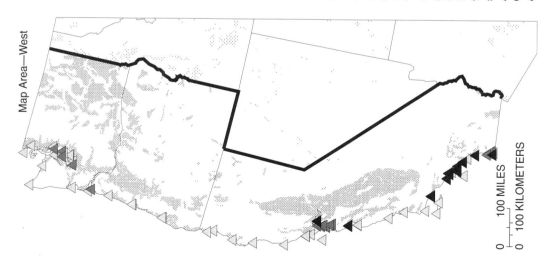

Map Area—West

100 MILES

100 KILOMETERS

0

0

Table 3.8. Aldrin plus dieldrin use on corn and toxaphene use on cotton, 1964–1976

[Pesticide use is in thousands of pounds active ingredient and in percent of total use on crops. Data are U.S. Department of Agriculture estimates from Eichers and others (1968, 1970, 1978) and Andrilenas (1974). Abbreviations: lb a.i., pounds active ingredient]

Year of Application	Aldrin plus Dieldrin Use on Crops[1]			Toxaphene Use on Crops[1]		
	Total Use (lb a.i. × 1,000)	Use on Corn (lb a.i. × 1,000)	Percentage of Total that Was Used on Corn	Total Use (lb a.i. × 1,000)	Use on Cotton (lb a.i. × 1,000)	Percentage of Total that Was Used on Cotton
1964	12,046	10,779	89	34,189	26,915	79
1966	15,455	14,296	93	30,924	27,345	88
1971	8,228	7,759	94	32,867	28,112	86
1976	[2]865	850	98	[2]30,721	26,289	86

[1]Unless otherwise indicated, total use applies to all crops, pasture and rangeland, livestock, and other uses by farmers.
[2]Total use applies only to major field crops, hay, and pasture and rangeland.

coasts do not appear to correspond either with high agricultural use or with highly contaminated inland sites from the NCBP. One exception is the presence of contaminated sites (both inland NCBP sites and coastal and estuarine NS&T sites) on or near the Mississippi River (Figure 3.21, southeast map area). Because NOAA reported concentrations of organochlorine contaminants in mollusks on a dry weight basis, and the FWS reported concentrations in fish on a wet weight basis, absolute concentrations between the two studies cannot be directly compared. High dieldrin residues in estuarine mollusks in the Northeast Region (Figure 3.21) appear to be associated with urban areas (urban areas correspond to the three highest population density classes in Figure 3.17).

Chlordane

Technical chlordane is a mixture of more than 140 components derived from cyclopentadiene and hexachlorocyclopentadiene. The most abundant individual components in technical chlordane are *cis*- and *trans*-chlordane. A typical composition of technical chlordane is about 24 percent *trans*-chlordane, 19 percent *cis*-chlordane, 10 percent heptachlor, 21.5 percent chlordene isomers, and 7 percent nonachlor, with the remaining 18.5 percent consisting of other structurally related chlorinated compounds (U.S. Environmental Protection Agency, 1980b). The term "chlordane" usually refers to technical chlordane or total chlordane, although the latter has been defined as different combinations of components by different authors. Chlordane was widely used in the United States as an insecticide on corn, grapes, strawberries, and other crops; as a home and garden insecticide; and for termite control (U.S. Environmental Protection Agency, 1980b). All uses were cancelled in 1978 except dipping of roots and tips of nonfood plants (which was cancelled in 1987) and subsurface termite control, although some uses were phased out gradually through 1983 (U.S. Environmental Protection Agency, 1992b) and no commercial sale, distribution, or use was permitted after 1988 (U.S. Environmental Protection

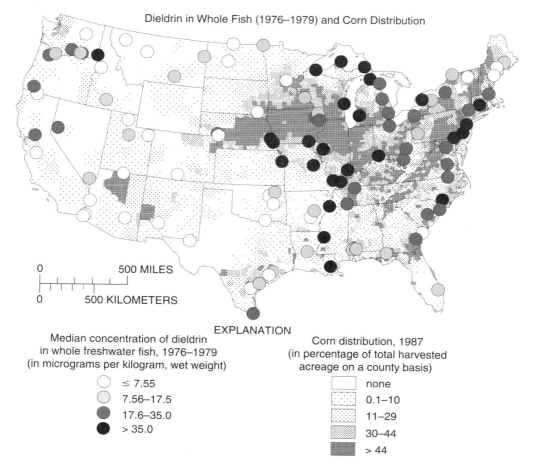

Figure 3.20. The geographic distribution of dieldrin in whole freshwater fish (1976–1979) from FWS's National Contaminant Biomonitoring Program, shown in relation to 1987 corn production (in percentage of total harvested acreage on a county basis). Each site (circle) is shaded to correspond to the median dieldrin concentration in whole fish (in micrograms per kilogram wet weight) at that site during 1976–1979. Dieldrin concentration categories are based on quartiles for data from the period 1976–1979, so that dieldrin concentration data and categories are identical to those in Figure 3.18. Dieldrin concentration data (1976–1979) are from U.S. Fish and Wildlife Service (1992). Corn production data (1987) are from the U.S. Department of Commerce (1995). Abbreviation: FWS, Fish and Wildlife Service.

Agency, 1990b, 1992b). Chlordane is moderately volatile (moderately high Henry's law constant), has low water solubility, sorbs moderately to soil (moderate K_{oc}), and tends to bioaccumulate (high K_{ow} and bioconcentration factor), as shown in Table 3.7. Different chlordane components vary in their persistence. For example, *cis*-chlordane can hydrolyze under alkaline conditions, but *trans*-chlordane does not (U.S. Environmental Protection Agency, 1992b). *trans*-Nonachlor is the most persistent component of technical chlordane (Schmitt and others, 1990).

Figures 3.22 and 3.23 show median total chlordane concentrations in whole freshwater fish at NCBP sites during 1976–1979 and in 1986, respectively, in relation to 1966 agricultural use of chlordane by farm production region. In the NCBP, total chlordane consists of the sum of *cis*-chlordane, *trans*-chlordane, *cis*-nonachlor, *trans*-nonachlor, and oxychlordane. Treatment of both use data and concentration data in Figures 3.22 and 3.23 is comparable with that described for aldrin plus dieldrin in the previous section (Figures 3.18 and 3.19).

Figure 3.22 shows a fairly strong association between estimated chlordane use in 1966 and median total chlordane concentration in fish during 1976–1979. Higher than expected total chlordane concentrations occurred in the Northeast Region, and at scattered sites near the Texas coast and in the Pacific and Mountain regions. Chlordane was used extensively on corn (Eichers and others, 1970; Andrilenas, 1974). The 1987 U.S. Census of Agriculture shows substantial corn acreage (greater than 20,000 acres per county) in the vicinity of sites with higher than expected concentrations in California, Washington, Texas, and in the northeastern United States (not shown). Heavily populated areas, such as in the Northeast and on the Texas coast (Figure 3.17), also may have received chlordane inputs from nonfarm uses such as subterranean termite control. The quantity of chlordane used in termite control (10 million lb in 1980; U.S. Environmental Protection Agency, 1983) was actually greater than that used in agriculture (0.5– 2 million lb per year during 1964–1976; Eichers and others, 1968, 1970, 1978; Andrilenas, 1974).

Comparison of Figures 3.22 and 3.23 indicates some decrease in median total chlordane concentrations in the western and southeastern United States, and in Texas between the period 1976–1979 and 1986, as indicated by the reduction in the number of darker shaded sites in Figure 3.23 relative to Figure 3.22. However, areas of high total chlordane contamination have persisted in the Lake States and Northeast regions. The relative abundance of various chlordane components in NCBP fish samples also changed over time, with *cis*-chlordane the most abundant component before 1980 and *trans*-nonachlor from 1980 to 1986 (Schmitt and others, 1990; U.S. Fish and Wildlife Service, 1992). As previously noted, *cis*-chlordane is one of the most abundant components of technical chlordane, but *trans*-nonachlor is the most persistent. On the basis of the low overall concentrations of chlordane components, and the increasing predominance of *trans*-nonachlor in NCBP fish samples, Schmitt and others (1990) suggested that the rate of chlordane influx to aquatic ecosystems was reduced after the mid-1970s, when agricultural use of chlordane was terminated. Urban and suburban sources may now constitute the primary source to aquatic ecosystems (Bevenue and others, 1972; Tanita and others, 1976; Truhlar and Reed, 1976; Arruda and others, 1987; Schmitt and others, 1990).

Figure 3.24 compares the geographic distribution of median total chlordane concentrations in whole freshwater fish at NCBP sites with its distribution in estuarine mollusks at NS&T (specifically, the National Mussel Watch Project of NS&T) sites. Data from both programs are for 1986. Enlarged views of the west, southeast, and northeast map areas are shown. In Figure 3.24, NCBP sites are represented by circles shaded to represent the concentration quartile (where quartiles are based on 1986 data) that corresponds to the median concentration at that site (Although the 1986 median concentrations for individual NCBP sites in Figures 3.23 and 3.24 are the same, the sites are shaded according to different quartiles in the two figures. Concentration quartiles for NCBP sites in Figure 3.23 were based on 1976–1979 data to emphasize changes in concentration that occurred between the period 1976–1979 and 1986.

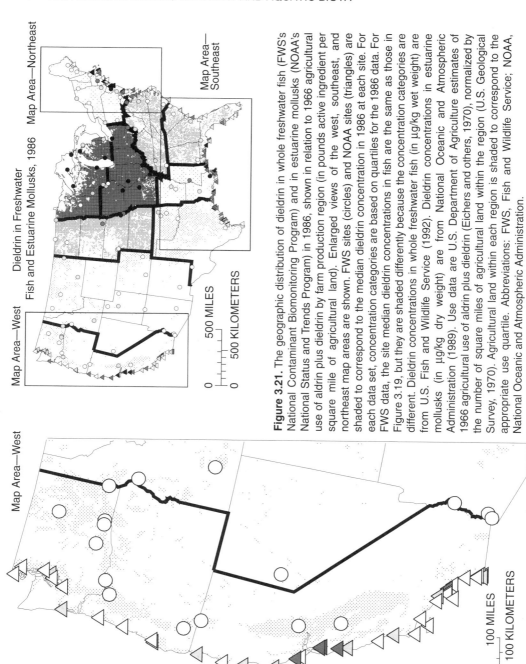

Figure 3.21. The geographic distribution of dieldrin in whole freshwater fish (FWS's National Contaminant Biomonitoring Program) and in estuarine mollusks (NOAA's National Status and Trends Program) in 1986, shown in relation to 1966 agricultural use of aldrin plus dieldrin by farm production region (in pounds active ingredient per square mile of agricultural land). Enlarged views of the west, southeast, and northeast map areas are shown. FWS sites (circles) and NOAA sites (triangles) are shaded to correspond to the median dieldrin concentration in 1986 at each site. For FWS data, the site median dieldrin concentrations in fish are the same as those in Figure 3.19, but they are shaded differently because the concentration categories are different. Dieldrin concentrations in whole freshwater fish (in µg/kg wet weight) are from U.S. Fish and Wildlife Service (1992). Dieldrin concentrations in estuarine mollusks (in µg/kg dry weight) are from National Oceanic and Atmospheric Administration (1989). Use data are U.S. Department of Agriculture estimates of 1966 agricultural use of aldrin plus dieldrin (Eichers and others, 1970), normalized by the number of square miles of agricultural land within the region (U.S. Geological Survey, 1970). Agricultural land within each region is shaded to correspond to the appropriate use quartile. Abbreviations: FWS, Fish and Wildlife Service; NOAA, National Oceanic and Atmospheric Administration.

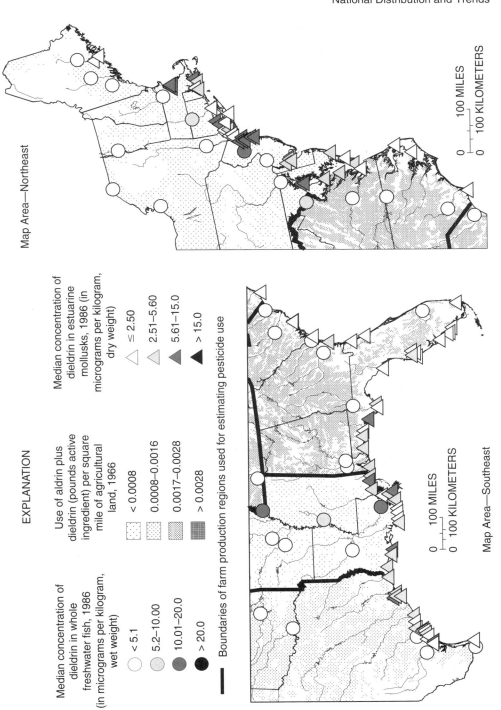

EXPLANATION

Median concentration of dieldrin in whole freshwater fish, 1986 (in micrograms per kilogram, wet weight)

○ < 5.1
◐ 5.2–10.00
◑ 10.01–20.0
● > 20.0

Use of aldrin plus dieldrin (pounds active ingredient) per square mile of agricultural land, 1966

< 0.0008
0.0008–0.0016
0.0017–0.0028
> 0.0028

Median concentration of dieldrin in estuarine mollusks, 1986 (in micrograms per kilogram, dry weight)

△ ≤ 2.50
◁ 2.51–5.60
◀ 5.61–15.0
▲ > 15.0

—— Boundaries of farm production regions used for estimating pesticide use

Map Area—Northeast

Map Area—Southeast

0   100 MILES
0   100 KILOMETERS

Figure 3.21. *Continued.*

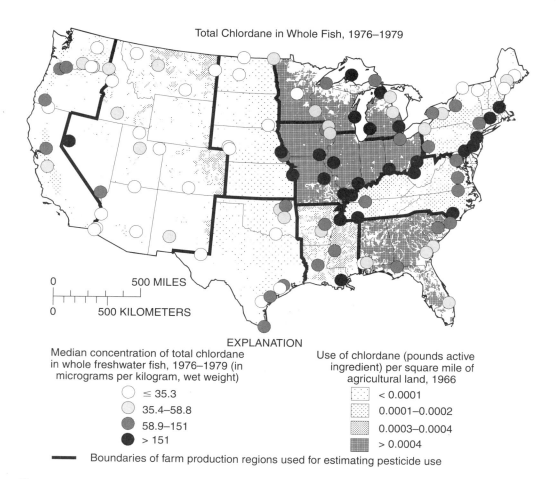

Figure 3.22. The geographic distribution of total chlordane in whole freshwater fish (1976–1979) from FWS's National Contaminant Biomonitoring Program, shown in relation to 1966 agricultural use of chlordane by farm production region (in pounds active ingredient per square mile of agricultural land). Total chlordane is the sum of residues of *cis*- and *trans*-chlordane, *cis*- and *trans*-nonachlor, and oxychlordane. Each site (circle) is shaded to correspond to the median total chlordane concentration (in micrograms per kilogram wet weight) at that site during 1976–1979. Total chlordane concentration categories are based on quartiles for 1976–1979 data. Total chlordane concentration data (1976–1979) are from U.S. Fish and Wildlife Service (1992). Use data are U.S. Department of Agriculture estimates of 1966 agricultural use of chlordane by farm production region (Eichers and others, 1970), normalized by the number of square miles of agricultural land within the region (U.S. Geological Survey, 1970). Agricultural land within each region is shaded to correspond to the appropriate use quartile. Abbreviations: FWS, Fish and Wildlife Service

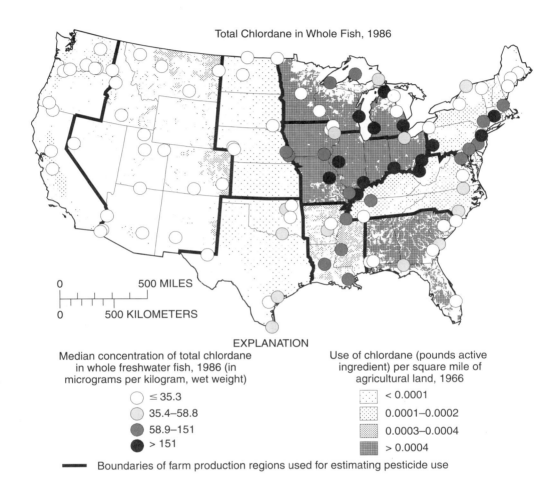

Figure 3.23. The geographic distribution of total chlordane in whole freshwater fish (1986) from FWS's National Contaminant Biomonitoring Program, shown in relation to 1966 agricultural use of chlordane by farm production region (in pounds active ingredient per square mile of agricultural land). Total chlordane is the sum of residues of *cis*- and *trans*-chlordane, *cis*- and *trans*-nonachlor, and oxychlordane. Each site (circle) is shaded to correspond to the median total chlordane concentration in whole fish (in micrograms per kilogram wet weight) at that site in 1986. Total chlordane concentration categories are based on quartiles for data from the period 1976–1979, so the categories are the same as those in Figure 3.22. Total chlordane concentration data (1986) are from U.S. Fish and Wildlife Service (1992). Use data are U.S. Department of Agriculture estimates of 1966 agricultural use of chlordane by farm production region (Eichers and others, 1970), normalized by the number of square miles of agricultural land within the region (U.S. Geological Survey, 1970). Agricultural land within each region is shaded to correspond to the appropriate use quartile. Abbreviations: FWS, Fish and Wildlife Service.

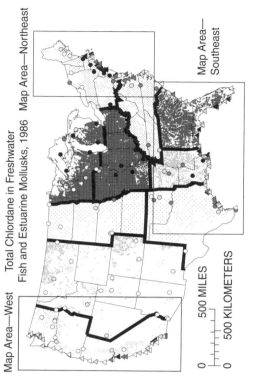

Map Area—West   Total Chlordane in Freshwater
Fish and Estuarine Mollusks, 1986   Map Area—Northeast

Map Area—
Southeast

Map Area—West

Figure 3.24. The geographic distribution of total chlordane in whole freshwater fish (FWS's National Contaminant Biomonitoring Program) and in estuarine mollusks (NOAA's National Status and Trends Program) in 1986, shown in relation to 1966 agricultural use of chlordane by farm production region (in pounds active ingredient per square mile of agricultural land). Enlarged views of the west, southeast, and northeast map areas are shown. FWS sites (circles) and NOAA sites (triangles) are shaded to correspond to the median total chlordane concentration in 1986 at each site. For each data set, concentration categories are based on quartiles for the 1986 data. For FWS data, the site median total chlordane concentrations in fish are the same as those in Figure 3.23, but they are shaded differently because the concentration categories are different. Total chlordane concentrations in whole freshwater fish (in μg/kg wet weight) are from U.S. Fish and Wildlife Service (1992). Total chlordane concentrations in estuarine mollusks (in μg/kg dry weight) are from National Oceanic and Atmospheric Administration (1989). For FWS sites, total chlordane is the sum of *cis-* and *trans*-chlordane, *cis-* and *trans*-nonachlor, and oxychlordane. For NOAA sites, total chlordane is the sum of *cis*-chlordane, *trans*-nonachlor, heptachlor, and heptachlor epoxide. Use data are U.S. Department of Agriculture estimates of 1966 agricultural use of chlordane by farm production region (Eichers and others, 1970), normalized by the number of square miles of agricultural land within the region (U.S. Geological Survey, 1970). Agricultural land within each region is shaded to correspond to the appropriate use quartiles. Abbreviations: FWS, Fish and Wildlife Service; NOAA, National Oceanic and Atmospheric Administration.

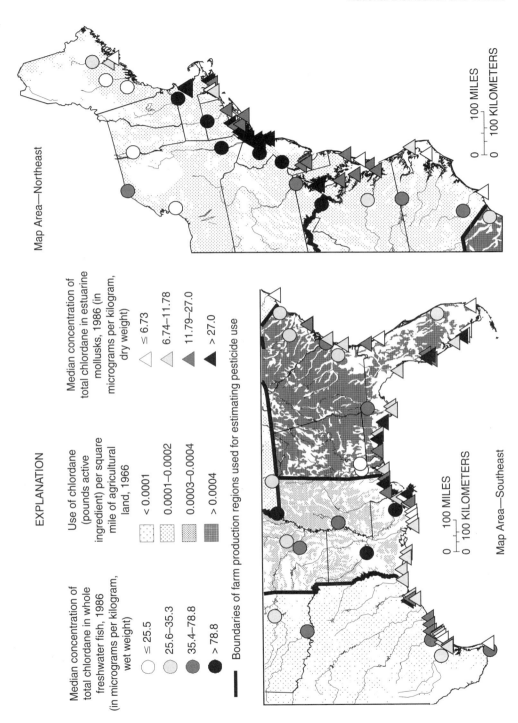

Map Area—Northeast

EXPLANATION

Median concentration of
total chlordane in estuarine
mollusks, 1986 (in
micrograms per kilogram,
dry weight)

△ ≤ 6.73

◁ 6.74–11.78

◀ 11.79–27.0

◀ > 27.0

Use of chlordane
(pounds active
ingredient) per square
mile of agricultural
land, 1966

< 0.0001

0.0001–0.0002

0.0003–0.0004

> 0.0004

Boundaries of farm production regions used for estimating pesticide use

Median concentration of
total chlordane in whole
freshwater fish, 1986
(in micrograms per kilogram,
wet weight)

○ ≤ 25.5

25.6–35.3

35.4–78.8

> 78.8

Map Area—Southeast

0   100 MILES

0   100 KILOMETERS

Figure 3.24. *Continued.*

Concentration quartiles for NCBP sites in Figure 3.24 were based on 1986 data for relative comparison with NS&T data from 1986.). In Figure 3.24, NS&T sites are represented by triangles shaded to represent the concentration quartile (where quartiles are based on 1986 data) that corresponds to the median concentration at that site.

Figure 3.24 shows no apparent relation between mollusk contamination at NS&T sites in 1986 and agricultural use in 1966. As shown in the northeast map area, aquatic biota at both inland NCBP sites (whole fish) and coastal and estuarine NS&T sites (mollusks) are contaminated along the northeast Atlantic coast. This is a highly urban area, as shown in Figure 3.17. Urban areas along the Pacific and Gulf coasts (also shown in Figure 3.17) also had high total chlordane concentrations in estuarine mollusks (Figure 3.24). The southeast map area (Figure 3.24) shows some contamination both at inland sites along the Mississippi River (NCBP fish) and at coastal and estuarine sites near its outflow (NS&T mollusks). The west map area shows total chlordane contamination in mollusks at scattered NS&T sites on the Pacific coast; however, few inland sites (NCBP fish) were sampled in this area. Absolute concentrations of total chlordane in freshwater fish at NCBP river sites cannot be directly compared with those in estuarine mollusks at NS&T sites for two reasons. First, as noted previously for dieldrin, the NCBP reported wet weight concentrations in fish, and the NS&T Program reported dry weight concentrations in mollusks. Second, in the NS&T Program, total chlordane residues consisted of the sum of *cis*-chlordane, *trans*-nonachlor, heptachlor and heptachlor epoxide, whereas the NCBP measured the sum of *cis*-chlordane, *trans*-chlordane, *cis*-nonachlor, *trans*-nonachlor, and oxychlordane.

Total DDT

As discussed in the previous section on Bed Sediment, total DDT consists of the sum of *o,p′* and *p,p′* isomers of DDT and its metabolites, DDD and DDE. Environmental residues of total DDT may result from the insecticidal use of DDT or DDD. These insecticides were used extensively on a variety of crops and in control of insect-borne diseases during the 1940s–1960s. DDT and metabolites have low water solubilities, strong tendencies to sorb to soil and to bioaccumulate, and long half-lives in soil; DDT also is fairly volatile (see Table 3.7).

Figure 3.25 shows the relation between median concentration of total DDT in whole freshwater fish at individual NCBP sites during 1976–1979 and USDA estimates of agricultural use of DDT plus DDD in 1966, by farm production region. Treatment of both use data and concentration data in Figure 3.25 is comparable with that previously described for aldrin plus dieldrin (Figure 3.18).

There appears to be some positive association between median total DDT concentration in whole fish during 1976–1979 and agricultural use in 1966, except that higher than expected concentrations occurred in the Great Lakes area and along the northeast Atlantic coast. The same hypothesis suggested by Gilliom and others (1985) to explain higher than expected detection of DDT in river bed sediment in the northeast (intensive nonfarm use or incidental release from chemical manufacturing plants in urban and industrial areas) may also apply to total DDT residues in fish from the sites in the Northeast and Lakes States regions. Atmospheric deposition of DDT, which has been quantified for the Great Lakes (Majewski and Capel, 1995), constitutes another potential source. Strachan and Eisenreich (1990) estimated that atmospheric deposition contributed between about 22–31 percent (for Lakes Erie and Ontario) to more than 97 percent (for Lakes Superior, Huron, and Michigan) of total inputs to the Great Lakes, depending on the magnitude of other inputs (such as agriculture) in the basin. High total DDT residues observed at scattered sites in the Pacific states may reflect localized use on orchards.

Figure 3.26 shows median total DDT concentrations in whole fish at NCBP sites a decade later (1986), again in relation to 1966 agricultural use of DDT plus DDD by farm production region. Total DDT concentration quartiles based on the 1976–1979 data (from Figure 3.25) were retained in Figure 3.26, so that concentration changes occurring from the period 1976–1979 to 1986 can be identified by comparing Figures 3.25 and 3.26. Figure 3.26 contains fewer dark shaded circles than Figure 3.25, indicating that site median total DDT concentrations in 1986 generally were lower than during 1976–1979. However, some areas of high total DDT contamination did persist, especially in the Lakes States and Delta States regions, and in southern Texas. One site in Idaho showed an increase in median total DDT residues between the period 1976–1979 and 1986. However, this probably reflects variability in the individual samples collected, rather than an increase in contamination at this site, since the range in total DDT concentration in the individual samples was 190–1,100 µg/kg in 1976 and 90–480 µg/kg in 1986.

In Figure 3.27, the geographic distribution of total DDT in whole freshwater fish at NCBP sites (circles) is compared with the distribution in estuarine mollusks at NS&T (National Mussel Watch Project) sites (triangles). Data for both studies are site median concentrations in 1986 (The site median concentrations at individual NCBP sites in Figure 3.27 are the same as in Figure 3.26, but they are shaded according to different quartiles in the two figures. As explained previously for aldrin and dieldrin, concentration quartiles in Figure 3.26 were based on 1976–1979 data, and those in Figure 3.27 on 1986 data.). Enlarged views of the west, southeast, and northeast map areas also are shown in Figure 3.27.

The distribution of median concentrations of total DDT in mollusks at NS&T sites in 1986 shows clustering of high concentrations on the northeast Atlantic, southwest Pacific, and Gulf coasts (Figure 3.27). Figure 3.27 shows no apparent relation between mollusk contamination in 1986 and agricultural use in 1966. Although agricultural use of DDT plus DDD in 1966 was highest in the Southeast and Delta States regions, mollusk contamination at NS&T sites appears to be somewhat less on the southeast Atlantic coast than on the Gulf, Pacific, and northeast Atlantic coasts (Figure 3.27). This also was true of benthic fish sampled as part of the NS&T Program (not shown): concentrations of organic contaminants in benthic fish from the southeast were average-to-low relative to other parts of the United States (Hanson and others, 1989). As noted previously, regional differences in fish and mollusk species collected by the NS&T Program may contribute to apparent regional differences in contaminant residues. The west, southeast, and northeast map areas (Figure 3.27) show that contaminated coastal or estuarine sites (NS&T) are sometimes, but not always, associated with contaminated inland sites (NCBP). For both mollusks (Figure 3.27) and sediment (Figure 3.17) from the NS&T Program, a cluster of highly contaminated sites in the Los Angeles area is not located near farmland. This contamination is attributed to offshore urban outfalls (Stephenson and others, 1995).

Heptachlor and Heptachlor Epoxide

Heptachlor was widely used to control fire ants in the southeast and as a soil insecticide on corn and other food crops until the mid-1970s. All uses except dipping of roots and tops of nonfood plants and subsurface termite control were cancelled by 1983. Heptachlor is no longer

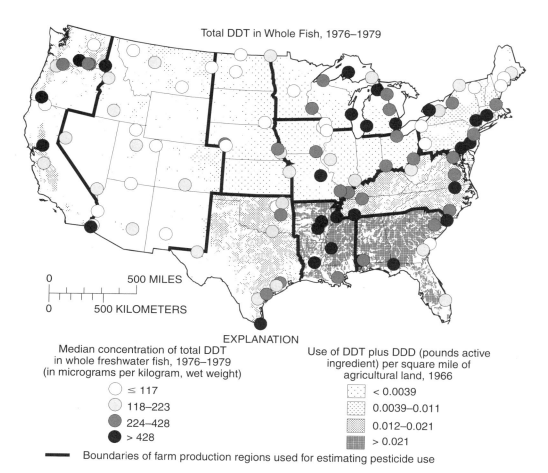

Figure 3.25. The geographic distribution of total DDT in whole freshwater fish (1976–1979) from FWS's National Contaminant Biomonitoring Program, shown in relation to 1966 agricultural use of DDT plus DDD by farm production region (in pounds active ingredient per square mile of agricultural land). Total DDT is the sum of residues of o,p'-DDT, o,p'-DDE, o,p'-DDD, p,p'-DDT, p,p'-DDE, and p,p'-DDD. Each site (circle) is shaded to correspond to the median total DDT concentration in whole fish (in micrograms per kilogram wet weight) at that site during 1976–1979. Total DDT concentration categories are based on quartiles for 1976–1979 data. Total DDT concentration data (1976–1979) are from U.S. Fish and Wildlife Service (1992). Use data are U.S. Department of Agriculture estimates of 1966 agricultural use of DDT plus DDD by farm production region (Eichers and others, 1970), normalized by the number of square miles of agricultural land within the region (U.S. Geological Survey, 1970). Agricultural land within each region is shaded to correspond to the appropriate use quartile. Abbreviation: FWS, Fish and Wildlife Service.

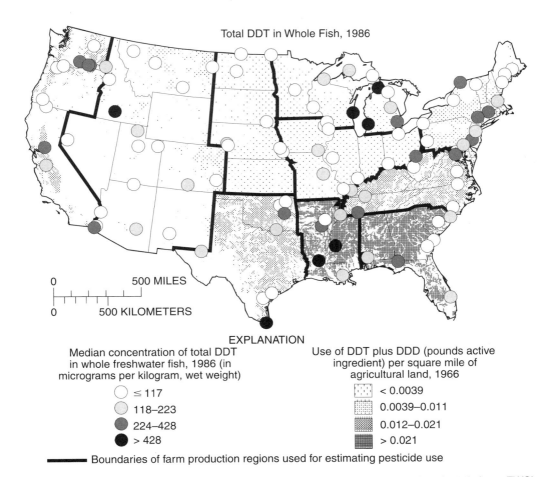

Total DDT in Whole Fish, 1986

EXPLANATION

Median concentration of total DDT in whole freshwater fish, 1986 (in micrograms per kilogram, wet weight)

○ ≤ 117
◔ 118–223
◕ 224–428
● > 428

Use of DDT plus DDD (pounds active ingredient) per square mile of agricultural land, 1966

< 0.0039
0.0039–0.011
0.012–0.021
> 0.021

━━━ Boundaries of farm production regions used for estimating pesticide use

Figure 3.26. The geographic distribution of total DDT in whole freshwater fish (1986) from FWS's National Contaminant Biomonitoring Program, shown in relation to 1966 agricultural use of DDT plus DDD by farm production region (in pounds active ingredient per square mile of agricultural land). Total DDT is the sum of residues of o,p'-DDT, o,p'-DDE, o,p'-DDD, p,p'-DDT, p,p'-DDE, and p,p'-DDD. Each site (circle) is shaded to correspond to the median total DDT concentration in whole fish (in micrograms per kilogram wet weight) at that site in 1986. Total DDT concentration categories are based on quartiles for data from the period 1976–1979, so the categories are the same as those in Figure 3.25. Total DDT concentration data (1986) are from U.S. Fish and Wildlife Service (1992). Use data are U.S. Department of Agriculture estimates of 1966 agricultural use of DDT plus DDD by farm production region (Eichers and others, 1970), normalized by the number of square miles of agricultural land within the region (U.S. Geological Survey, 1970). Agricultural land within each region is shaded to correspond to the appropriate use quartile. Abbreviations: FWS, Fish and Wildlife Service.

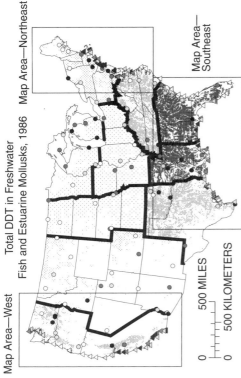

Map Area—Northeast

Total DDT in Freshwater
Fish and Estuarine Mollusks, 1986   Map Area—Northeast

Map Area—West

Map Area—
Southeast

0   500 MILES

0   500 KILOMETERS

Figure 3.27. The geographic distribution of total DDT in whole freshwater fish (FWS's National Contaminant Biomonitoring Program) and in estuarine mollusks (NOAA's National Status and Trends Program) in 1986, shown in relation to agricultural use of DDT plus DDD in 1966 by farm production region (in pounds active ingredient per square mile of agricultural land). Enlarged views of the west, southeast, and northeast map areas are shown. FWS sites (circles) and NOAA sites (triangles) are shaded to correspond to the median total DDT concentration in 1986 at each site. For each data set, concentration categories are based on quartiles for the 1986 data. For FWS data, the site median total DDT concentrations in fish are the same as those in Figure 3.26, but they are shaded differently because the concentration boundaries are different. Total DDT concentrations in freshwater whole fish (in µg/kg wet weight) are from U.S. Fish and Wildlife Service (1992). Total DDT concentrations in estuarine mollusks (in µg/kg dry weight) are from National Oceanic and Atmospheric Administration (1989). For both studies, total DDT is the sum of residues of o,p'-DDT, o,p'-DDE, o,p'-DDD, p,p'-DDT, p,p'-DDE, and p,p'-DDD. Use data are U.S. Department of Agriculture estimates of 1966 agricultural use of DDT plus DDD by farm production region (Eichers and others, 1970), normalized by the number of square miles of agricultural land within the region (U.S. Geological Survey, 1970). Agricultural land within each region is shaded to correspond to the appropriate use quartile. Abbreviations: FWS, Fish and Wildlife Service; NOAA, National Oceanic and Atmospheric Administration.

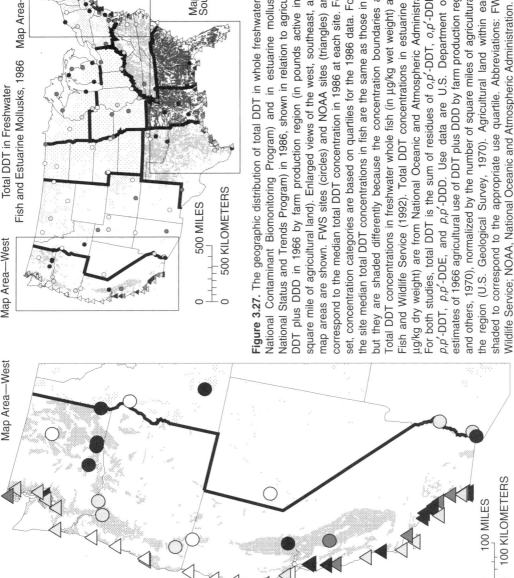

Map Area—West

0   100 MILES

0   100 KILOMETERS

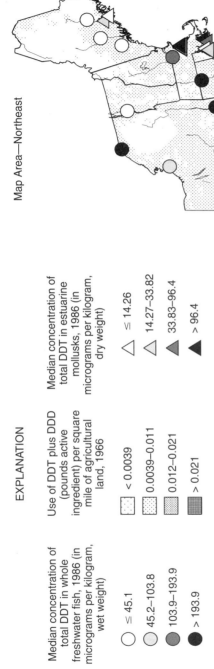

Map Area—Northeast

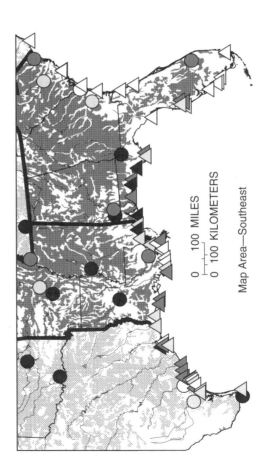

EXPLANATION

Median concentration of total DDT in whole freshwater fish, 1986 (in micrograms per kilogram, wet weight)

- ○ ≤ 45.1
- ◔ 45.2–103.8
- ◑ 103.9–193.9
- ● > 193.9

Use of DDT plus DDD (pounds active ingredient) per square mile of agricultural land, 1966

- ▫ < 0.0039
- ▨ 0.0039–0.011
- ▩ 0.012–0.021
- ▦ > 0.021

Median concentration of total DDT in estuarine mollusks, 1986 (in micrograms per kilogram, dry weight)

- △ ≤ 14.26
- ◮ 14.27–33.82
- ◭ 33.83–96.4
- ▲ > 96.4

— Boundaries of farm production regions used for estimating pesticide use

Map Area—Southeast

Figure 3.27. *Continued.*

sold in the United States, but remaining stock was used in some states by commercial exterminators for termite control (U.S. Environmental Protection Agency, 1992b). Heptachlor also is a minor component (10 percent or less) of technical chlordane (U.S. Environmental Protection Agency, 1980b). Heptachlor is moderately volatile (high Henry's law constant), has low water solubility, sorbs strongly to soil or sediment (high K_{oc}), and tends to bioaccumulate (high K_{ow} and bioconcentration factor), as shown in Table 3.7. In water, heptachlor undergoes both hydrolysis (to 1-hydroxychlordene with a half-life of 1–3 days) and photolysis. Heptachlor is biotransformed to heptachlor epoxide, but at a slower rate than abiotic hydrolysis (U.S. Environmental Protection Agency, 1992b).

Heptachlor epoxide is a metabolite of heptachlor that is produced by plants, animals, and microorganisms. It also is a byproduct in the production of heptachlor and chlordane (U.S. Environmental Protection Agency, 1992b). It is slightly more water soluble than heptachlor and slightly less bioaccumulative (Table 3.7). However, heptachlor epoxide is very persistent, with a half-life in soil of several years. It is stable to hydrolysis, photolysis, oxidation or biotransformation (U.S. Environmental Protection Agency, 1992b). "Total heptachlor" refers to the sum of heptachlor and heptachlor epoxide residues.

Data from the studies reviewed indicate that heptachlor itself makes up relatively little of the total heptachlor detected in biological tissues. For example, unmetabolized heptachlor was rarely detected in NCBP fish samples after 1977 (Schmitt and others, 1990). Heptachlor was detected in fish at 2 percent of sites sampled as part of the NSCRF during 1986–1987, compared with heptachlor epoxide at 16 percent of sites. During the same years (1986–1987), heptachlor and heptachlor epoxide residues were detected in estuarine mollusks collected from the Gulf of Mexico coast: heptachlor was detected in 23–37 percent and heptachlor epoxide in 86–98 percent of mollusk samples from various sites (Sericano and others, 1990a). Estuarine fish data from the NS&T (specifically, the National Benthic Surveillance Project of NS&T) suggest that environmental residues of heptachlor comprise a source of heptachlor epoxide in fish. Whereas heptachlor residues at Atlantic NBSP sites were higher in sediment than in benthic fish tissues from the same site, heptachlor epoxide residues were higher in fish liver than in associated sediment (Zdanowicz and Gadbois, 1990).

The NCBP analyzed heptachlor and heptachlor epoxide in whole freshwater fish from 1967 to 1986. The maximum heptachlor epoxide concentration detected decreased from over 8,000 µg/kg during 1967–1968, to about 1,000 µg/kg during 1970–1974, to less than 300 µg/kg during 1980–1981. The geographic distribution of total heptachlor at NCBP sites was not mapped because median concentrations of total heptachlor generally were at or below the detection limit (10 µg/kg) after 1974. Most detections of total heptachlor at NCBP sites were found in areas of high heptachlor use (in the Corn Belt, Lake States, and Delta States regions). However, there were some detections in the Northeast Region and at a few sites in the Mountain and Pacific regions, where heptachlor was not used in agriculture during the 1960s (Eichers and others, 1970). These residues may have been due to the use of heptachlor for termite control in urban areas, or to the use of technical chlordane, which contains about 10 percent heptachlor, and which was applied to cropland in the Northeast and Pacific regions (Figure 3.22). Both heptachlor and chlordane were used extensively on corn (Eichers and others, 1970; Andrilenas, 1974), so it is not surprising that residues of heptachlor and chlordane components in NCBP fish samples are positively correlated (Schmitt and others, 1983).

Mirex

Mirex was used in the southeastern United States to control red fire ants from 1962 to 1975, and in Hawaii to control the pineapple mealybug (U.S. Environmental Protection Agency, 1992b). Mirex also was widely used as a flame retardant and polymerizing agent in plastics (Schmitt and others, 1990). All registered uses were cancelled as of 1977, and the sale, distribution, and use of existing stocks were prohibited after 1978 (U.S. Environmental Protection Agency, 1992b). On the basis of its physical and chemical properties, mirex is expected to have fairly low volatility (low Henry's law constant) from water or soil (U.S. Environmental Protection Agency, 1992b). Mirex has very low water solubility, a strong tendency to sorb to sediment or soil (high K_{oc}), and a strong tendency to bioaccumulate (high K_{ow} and bioconcentration factor), as shown in Table 3.7. Mirex is stable to thermal degradation (abiotic degradation in the absence of sunlight) and to biodegradation, but there is some evidence that it is transformed to kepone in soil. Mirex undergoes slow photolysis to photomirex (U.S. Environmental Protection Agency, 1992b).

Mirex was analyzed in NCBP samples beginning in 1980. The maximum mirex concentration detected rose from 210 µg/kg during 1980–1981 to 440 µg/kg during 1984–1985, then declined to 60 µg/kg in 1986. The geometric mean, median, and 90th-percentile values for the site median mirex concentrations all were below the detection limit (10 µg/kg). Most detections were concentrated in the Great Lakes area (especially Lake Ontario and the St. Lawrence River) and in the southeastern United States. The manufacturing and industrial use of mirex was known to result in the contamination of Lake Ontario (Kaiser, 1978). The geographic distribution of mirex residues in NCBP whole fish samples (not shown) suggests that contamination was limited to areas treated for control of red fire ants (the southeastern United States) or to manufacturing inputs (the Great Lakes). Results from the BofCF–USEPA's National Pesticide Monitoring Program estuarine mollusk surveys (1965–1972) are consistent with this; mirex was detected in only 2 percent of samples, all of which were from South Carolina (Butler, 1973b).

Toxaphene

Toxaphene is a complex insecticidal mixture of chlorinated camphenes and bornanes that is typically 67–69 percent chlorine. Toxaphene was used on corn, cotton, fruit, small grains, vegetables and soybeans (Worthing and Walker, 1987). Toxaphene was the most heavily used insecticide throughout the 1970s (see Table 3.5), having replaced many of the agricultural applications of DDT (U.S. Environmental Protection Agency, 1980d). Most uses of toxaphene in the United States were cancelled in 1982 (U.S. Environmental Protection Agency, 1990b). Some components are more susceptible to environmental degradation and biodegradation than others, so that the component distribution of environmental residues change over time (Ribick and others, 1982). Toxaphene is moderately volatile and has low water solubility (Majewski and Capel, 1995); it sorbs moderately to soil (moderate K_{oc}) and has some tendency to bioaccumulate (moderately high K_{ow} and bioconcentration factor), as shown in Table 3.7.

Figure 3.28 shows the relation between the toxaphene concentration in whole fish at NCBP sites during 1976–1979 and USDA estimates of agricultural use of toxaphene in 1966, by farm production region. Treatment of toxaphene use data in Figure 3.28 is comparable with that previously described for aldrin plus dieldrin (Figure 3.18). NCBP sites are shown as circles in

Figure 3.28, with each circle shaded to represent the concentration category corresponding to the median concentration at that site. However, concentration categories for toxaphene were not based on quartiles for 1976–1979 data because the median concentrations during these years (during all years) were less than the detection limit. Therefore, concentration categories were based instead on the 50th and 90th percentiles of all nonzero site median concentrations measured in any sampling year from 1972 to 1986. Therefore, the first (lowest) toxaphene concentration category consists of nondetections (<100 µg/kg), the second category consists of concentrations between the detection limit and the 50th percentile of all detections during 1972–1986 (100–300 µg/kg), the third category consists of concentrations between the 50th and 90th percentiles of all detections during 1972–1986 (301–1,600 µg/kg), and the fourth (highest) category consists of concentrations greater than the 90th percentile of all detections during 1972–1986 (>1,600 µg/kg). This means that the darkest shaded sites in Figure 3.28 had median toxaphene concentrations during 1976–1979 in the top 10 percent of all nonzero site median concentrations determined at any time from 1972 to 1986. These highest toxaphene concentrations in 1976–1979 were located in the Lake States, Delta States, Southeast, and lower Corn Belt regions.

Figure 3.29 shows median toxaphene concentrations in whole fish at NCBP sites a decade later (in 1986), in relation to agricultural use of toxaphene in 1966, by farm production region. Exactly as in Figure 3.28, the toxaphene concentration categories were based on all nonzero site median concentrations occurring during 1972–1986. Therefore, concentration changes that occurred between the period 1976–1979 and 1986 can be identified by comparing Figures 3.28 and 3.29. There was a dramatic reduction in the number of dark-shaded sites in Figure 3.29 relative to Figure 3.28, indicating that toxaphene concentrations decreased substantially between the period 1976–1979 and 1986, with residues remaining high at a few sites in the Lake States and Delta States regions.

The geographical distribution of toxaphene residues does not appear to correlate consistently with agricultural use of toxaphene in 1966 (Figure 3.28). It is true that toxaphene concentrations during 1976–1979 were high in the Delta States and Southeast regions, which had the highest toxaphene use in 1966. However, higher than expected toxaphene residues were observed in several low-use areas, especially in the Lake States Region. Another point to note is that a few sites with undetectable residues during 1976–1979 had low, but measurable, residues detected in 1986 (such as along the northeast Atlantic coast). These anomalies suggest it may be interesting to further investigate the distribution of toxaphene detections over time, and to attempt to relate this distribution to toxaphene use patterns and other factors.

Because about 80 percent or more of the toxaphene used in agriculture from the mid-1960s to the mid-1970s was applied to cotton (as shown in Table 3.8), it seems likely that the geographic distribution of toxaphene residues may be associated with cotton production areas. Figures 3.30–3.34 show the geographic distribution of toxaphene residues at NCBP sites during 1972–1974, 1976–1979, 1980–1981, 1984, and 1986, respectively, compared to cotton production acreage on a county basis in 1987 (U.S. Department of Commerce, 1995). As in Figures 3.28 and 3.29, toxaphene concentration categories are based on the percentiles of toxaphene detections only (during 1972–1986). Because site median concentrations in Figures 3.30–3.34 are shaded according to a single set of concentration categories, it is easy to identify sites at which median toxaphene concentrations changed over time.

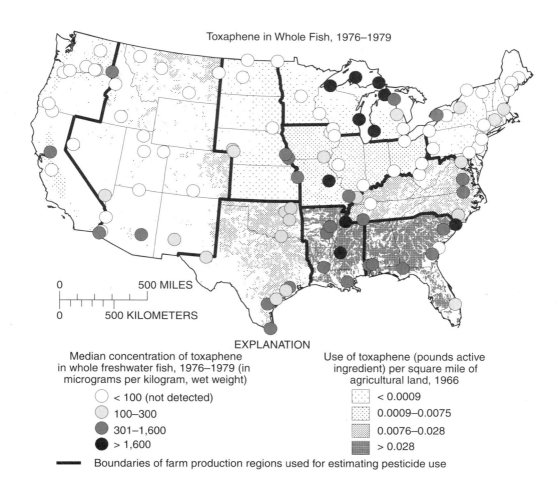

Figure 3.28. The geographic distribution of toxaphene in whole freshwater fish (1976–1979) from FWS's National Contaminant Biomonitoring Program, shown in relation to 1966 agricultural use of toxaphene by farm production region (in pounds active ingredient per square mile of agricultural land). Each site (circle) is shaded to correspond to the median toxaphene concentration in whole fish (in micrograms per kilogram wet weight) at that site during 1976–1979. Toxaphene concentration categories are based on quartiles for all nonzero site median concentrations measured in any sampling year during 1972–1986. Toxaphene concentration data (1976–1979) are from U.S. Fish and Wildlife Service (1992). Use data are U.S. Department of Agriculture estimates of 1966 agricultural use of toxaphene by farm production region (Eichers and others, 1970), normalized by the number of square miles of agricultural land within the region (U.S. Geological Survey, 1970). Agricultural land within each region is shaded to correspond to the appropriate use quartile. Abbreviations: FWS, Fish and Wildlife Service.

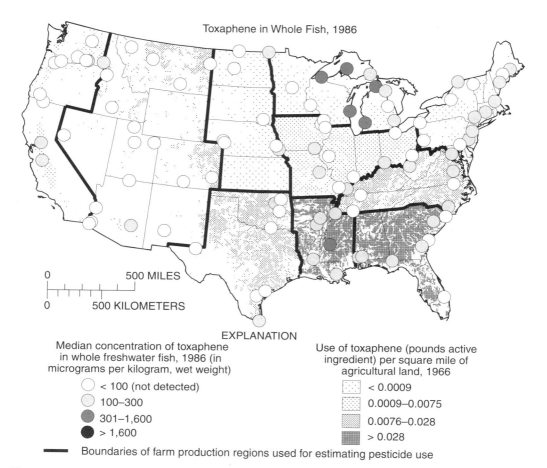

Figure 3.29. The geographic distribution of toxaphene in whole freshwater fish (1986) from FWS's National Contaminant Biomonitoring Program, shown in relation to 1966 agricultural use of toxaphene by farm production region (in pounds active ingredient per square mile of agricultural land). Each site (circle) is shaded to correspond to the median toxaphene concentration in whole fish (in micrograms per kilogram wet weight) at that site in 1986. Toxaphene concentration categories are based on quartiles for all nonzero site median concentrations measured in any sampling year during 1972–1986, so the categories are the same as those in Figure 3.28. Toxaphene concentration data (1986) are from U.S. Fish and Wildlife Service (1992). Use data are U.S. Department of Agriculture estimates of 1966 agricultural use of toxaphene by farm production region (Eichers and others, 1970), normalized by the number of square miles of agricultural land within the region (U.S. Geological Survey, 1970). Agricultural land within each region is shaded to correspond to the appropriate use quartile. Abbreviations: FWS, Fish and Wildlife Service.

Comparison of Figures 3.30–3.34 indicates the following. First, toxaphene was detected at only a few sites during 1972–1974; however, the concentrations detected were very high, and the sites with detectable residues were located in cotton-growing areas in the southeastern United States. No toxaphene was detected in cotton-growing areas elsewhere in the country. Second, by the period 1976–1979, high toxaphene residues were more widespread and showed, at best, only a partial association with cotton production areas. Some cotton-growing areas in the southwestern United States showed toxaphene detections. However, unexpectedly high toxaphene residues were observed at several Great Lakes sites, along the Missouri and upper Mississippi rivers and in the Appalachian states. Third, several of the sites with the highest toxaphene concentrations during 1976–1979 showed somewhat lower concentrations during 1980–1981; examples include several sites in the Great Lakes area. During the same time (from the periods 1976–1979 to 1980–1981), however, concentrations at numerous sites across the United States actually increased, especially in Indiana, Ohio, and in the northeastern United States. Fourth, toxaphene residues decreased at many sites across the United States from the period 1980–1981 to 1986, including sites with high-level and low-level contamination.

These temporal and geographic distribution patterns may relate to a number of factors: (1) atmospheric deposition of toxaphene, (2) changes in toxaphene use patterns in the 1970s and 1980s, and (3) improvements in the analytical methods for detection of toxaphene. First, regional atmospheric transport of toxaphene may explain the high residues of toxaphene found in fish at NCBP sites in the Great Lakes (Schmitt and others, 1990). At first, these high concentrations appear inconsistent with the low use of toxaphene in the Great Lakes states (Eichers and others, 1968, 1970, 1978; Andrilenas, 1974). Furthermore, there are no toxaphene manufacturing facilities located in the Great Lakes area and no evidence of accidental spillage (Rice and others, 1986). Nonetheless, toxaphene has been detected in air, rain, snow, surface water, and soil in the Great Lakes drainage basin (Bidleman and Olney, 1974; Eisenreich and others, 1981; Rapaport and others, 1985; Rapaport and Eisenreich, 1986, 1988; Rice and others, 1986; Bidleman and others, 1988; McConnell and others, 1993). Regional atmospheric transport of toxaphene into the Great Lakes from higher-use areas in the south has been fairly well-established (Majewski and Capel, 1995). Toxaphene was extensively used in the southern United States (Eichers and others, 1968, 1970, 1978; Andrilenas, 1974) until its cancellation, and it is still used in other countries around the world, including Mexico, the Czech Republic, Poland, and Hungary (Majewski and Capel, 1995). Current atmospheric concentrations are believed to be due in part to volatilization of persistent soil residues (Bidleman and other, 1988, 1989). Toxaphene then is transported into the Great Lakes area by southerly winds from the Gulf of Mexico that flow in a northeasterly direction (Majewski and Capel, 1995). In fact, a toxaphene concentration gradient that decreases from south to north along this air transportation corridor has been measured (Rice and others, 1986).

Second, changes in toxaphene use patterns during the 1970s–1980s may contribute to the increase in low-level contamination at sites in various parts of the United States. Schmitt and others (1985, 1990) noticed that the peak maximum toxaphene residues observed in fish samples from the NCBP occurred during the early 1970s, whereas median residues and detection frequencies continued to increase until the early 1980s. These authors hypothesized that these patterns may be related to toxaphene use patterns. As noted previously, toxaphene was principally used on cotton from 1964 to 1976 (Table 3.8). After 1976, toxaphene use on cotton declined substantially. Total toxaphene use on major crops in 1976 was 26 million lb a.i., of

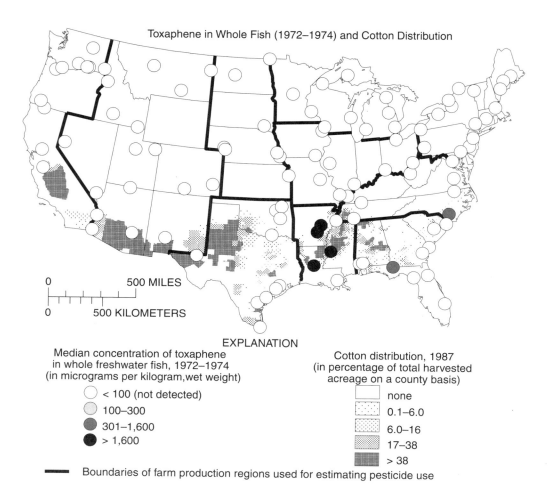

Figure 3.30. The geographic distribution of toxaphene in whole freshwater fish (1972–1974) from FWS's National Contaminant Biomonitoring Program, shown in relation to 1987 cotton production (in percentage of total harvested acreage on a county basis). Each site (circle) is shaded to correspond to the median toxaphene concentration in whole fish (in micrograms per kilogram wet weight) at that site during 1972–1974. Toxaphene concentration categories are based on quartiles for all nonzero site median concentrations measured in any sampling year during 1972–1986, so the categories are the same as those in Figures 3.28 and 3.29. Toxaphene concentration data (1972–1974) are from U.S. Fish and Wildlife Service (1992). Cotton production data (1987) are from U.S. Department of Commerce (1995). Abbreviation: FWS, Fish and Wildlife Service.

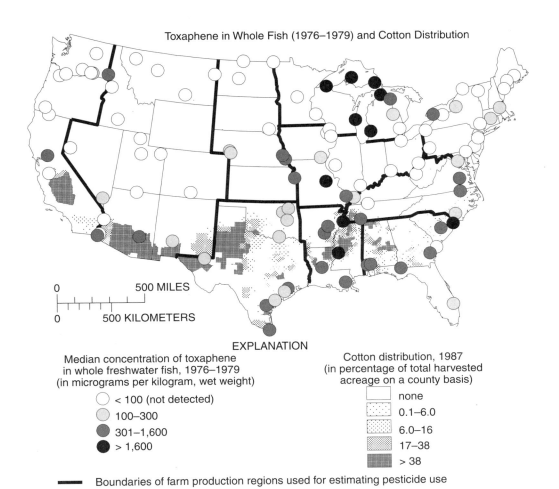

Figure 3.31. The geographic distribution of toxaphene in whole freshwater fish (1976–1979) from FWS's National Contaminant Biomonitoring Program, shown in relation to 1987 cotton production (in percentage of total harvested acreage on a county basis). Each site (circle) is shaded to correspond to the median toxaphene concentration in whole fish (in micrograms per kilogram wet weight) at that site during 1976–1979. Toxaphene concentration categories are based on quartiles for all nonzero site median concentrations measured in any sampling year during 1972–1986, so the categories are the same as those in Figures 3.28–3.30. Site median toxaphene concentration data are identical to those in Figure 3.28. Toxaphene concentration data (1976–1979) are from U.S. Fish and Wildlife Service (1992). Cotton production data (1987) are from U.S. Department of Commerce (1995). Abbreviation: FWS, Fish and Wildlife Service.

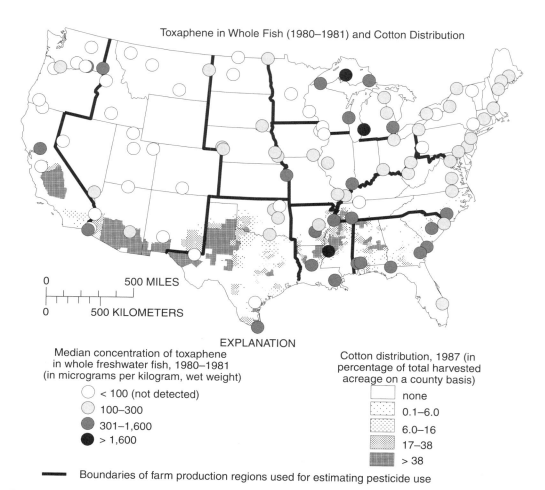

Figure 3.32. The geographic distribution of toxaphene in whole freshwater fish (1980–1981) from FWS's National Contaminant Biomonitoring Program, shown in relation to 1987 cotton production (in percentage of total harvested acreage on a county basis). Each site (circle) is shaded to correspond to the median toxaphene concentration in whole fish (in micrograms per kilogram wet weight) at that site during 1980–1981. Toxaphene concentration categories are based on quartiles for all nonzero site median concentrations measured in any sampling year during 1972–1986, so the categories are the same as those in Figures 3.28–3.31. Toxaphene concentration data (1980–1981) are from U.S. Fish and Wildlife Service (1992). Cotton production data (1987) are from U.S. Department of Commerce (1995). Abbreviation: FWS, Fish and Wildlife Service.

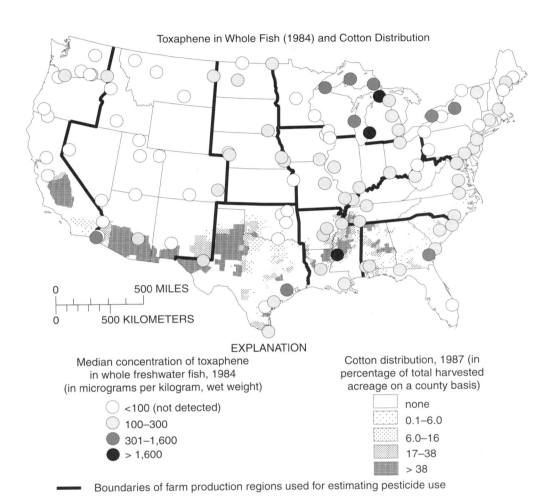

Toxaphene in Whole Fish (1984) and Cotton Distribution

EXPLANATION

Median concentration of toxaphene
in whole freshwater fish, 1984
(in micrograms per kilogram, wet weight)

○ <100 (not detected)
◦ 100–300
● 301–1,600
● > 1,600

Cotton distribution, 1987 (in
percentage of total harvested
acreage on a county basis)

☐ none
▨ 0.1–6.0
▨ 6.0–16
▨ 17–38
▨ > 38

━━━ Boundaries of farm production regions used for estimating pesticide use

Figure 3.33. The geographic distribution of toxaphene in whole freshwater fish (1984) from FWS's National Contaminant Biomonitoring Program, shown in relation to 1987 cotton production (in percentage of total harvested acreage on a county basis). Each site (circle) is shaded to correspond to the median toxaphene concentration in whole fish (in micrograms per kilogram wet weight) at that site in 1984. Toxaphene concentration categories are based on quartiles for all nonzero site median concentrations measured in any sampling year during 1972–1986, so the categories are the same as those in Figures 3.28–3.32. Toxaphene concentration data (1984) are from U.S. Fish and Wildlife Service (1992). Cotton production data (1987) are from U.S. Department of Commerce (1995). Abbreviation: FWS, Fish and Wildlife Service.

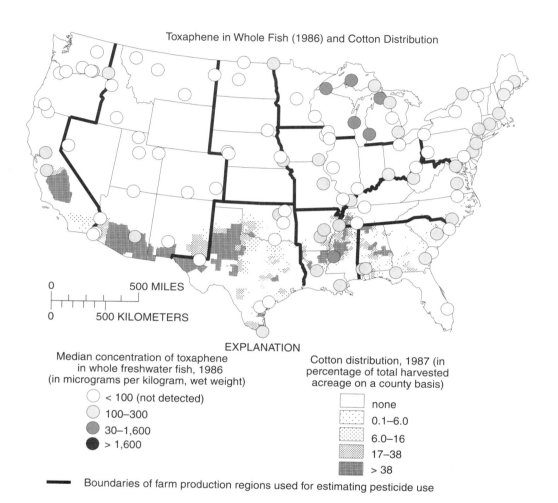

Figure 3.34. The geographic distribution of toxaphene in whole freshwater fish (1986) from FWS's National Contaminant Biomonitoring Program, shown in relation to 1987 cotton production (in percentage of total harvested acreage on a county basis). Each site (circle) is shaded to correspond to the median toxaphene concentration in whole fish (in micrograms per kilogram wet weight) at that site in 1986. Toxaphene concentration categories are based on quartiles for all nonzero site median concentrations measured in any sampling year during 1972–1986, so the categories are the same as those in Figures 3.28–3.33. Toxaphene concentration data (1986) are from U.S. Fish and Wildlife Service (1992). Cotton production data (1987) are from U.S. Department of Commerce (1995). Abbreviation: FWS, Fish and Wildlife Service.

which 86 percent (or 22 million lb) was used on cotton (Eichers and others, 1978; see Table 3.8). By 1982, total toxaphene use was 15 million lb a.i., of which only 15 percent (or 2.3 million lb) was used on cotton (U.S. Environmental Protection Agency, 1982). Most uses of toxaphene were cancelled in 1982 (U.S. Environmental Protection Agency, 1990b). However, prior to its cancellation, toxaphene became more widely used on crops other than cotton, such as wheat, sorghum, sunflowers, soybeans, and peanuts (Schmitt and others, 1985). Schmitt and others (1985) suggested that the peak maximum residues observed during the early 1970s may have been associated with cotton-growing areas. Figure 3.30 supports this. As use on cotton decreased, the maximum residues observed decreased accordingly (Figures 3.31–3.33). At the same time, increases were observed in median toxaphene residues (compare Figures 3.31 and 3.32) and detection frequencies (not shown) at many NCBP sites during 1980–1981; these increases may have been caused by the increasing use of toxaphene on other crops (Schmitt and others, 1985). Soybean acreage, for example, was extensive in the Corn Belt (Figure 3.35). Sorghum production may explain some low to moderate detections in the Southern Plains Region (Figure 3.36).

A third factor that probably contributes to the temporal and geographic distributions of toxaphene residues in fish at NCBP sites is analytical method changes that were made during the early 1980s. The apparent increase in low-level residues of toxaphene in fish at NCBP sites across the country between the periods 1976–1979 and 1980–1981 (Figures 3.31 and 3.32) may be partly an artifact of a change from packed-column to capillary-column gas chromatography that was implemented in 1980. Although capillary-column gas chromatography permits a much lower detection limit than the packed-column method, Schmitt and others (1985) imposed the same lower limit of detection on data generated using the capillary method as existed for earlier data generated using the packed-column method (100 µg/kg wet weight). This was necessary to permit comparison of detection frequencies among sample-collection years. Even with this detection limit correction, however, the two methods may not yield comparable results for environmental samples containing toxaphene concentrations near the detection limit (Schmitt and others, 1990) because toxaphene is a complex mixture whose chromatogram contains multiple peaks. Because of interferences that occur early in the chromatogram for toxaphene when analyzed by packed-column chromatography, only the second portion of the chromatogram was used for quantitation using the packed-column method. In contrast, the capillary method permits quantitation over the whole range of toxaphene component peaks (Ribick and others, 1982). This is important because the distribution of toxaphene components in environmental samples changes over time, with real fish samples showing a reduction relative to standard solutions in some of the late-eluting peaks (Ribick and others, 1982; Schmitt and others, 1990). Therefore, the capillary technique was more likely to detect lower concentrations in real fish samples than the packed-column method, which relies exclusively on the late-eluting peaks for quantitation (Schmitt and others, 1990). This complicates the interpretation of trends in toxaphene residues between the 1970s and 1980s. For a site that showed detectable toxaphene residues for the first time during 1980–1981, the apparent increase in toxaphene concentration may either represent a real increase in toxaphene residues at that site, or indicate merely that the improved analytical method was able to detect the low concentrations of toxaphene already present at the site.

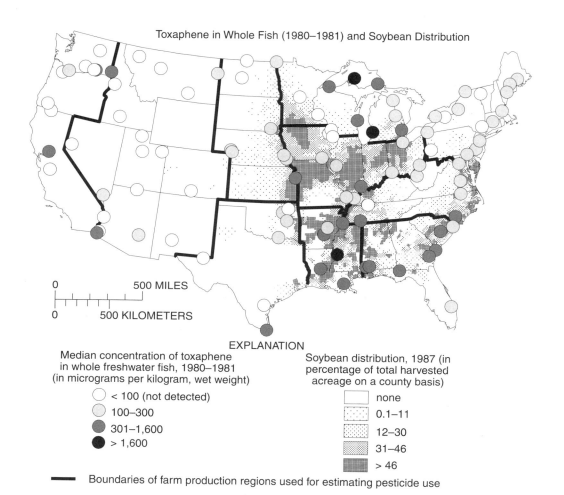

Figure 3.35. The geographic distribution of toxaphene in whole freshwater fish (1980–1981) from FWS's National Contaminant Biomonitoring Program, shown in relation to 1987 soybean production (in percentage of total harvested acreage on a county basis). Each site (circle) is shaded to correspond to the median toxaphene concentration in whole fish (in micrograms per kilogram wet weight) at that site during 1980–1981. Toxaphene concentration categories are based on quartiles for all nonzero site median concentrations measured in any sampling year during 1972–1986, so the categories are the same as those in Figures 3.28–3.34. Site median toxaphene concentration data are identical to those in Figure 3.32. Toxaphene concentration data (1980–1981) are from U.S. Fish and Wildlife Service (1992). Soybean production data (1987) are from U.S. Department of Commerce (1995). Abbreviation: FWS, Fish and Wildlife Service.

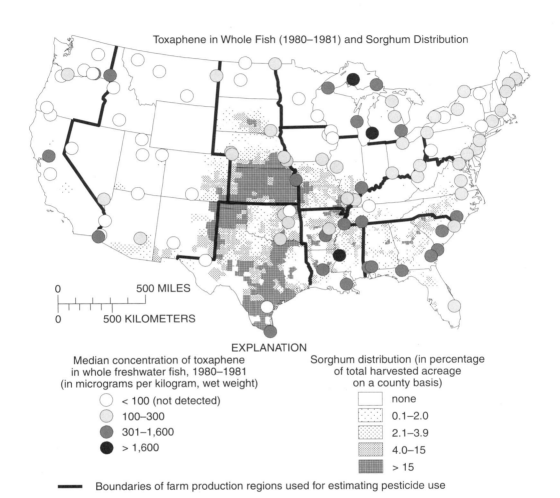

Toxaphene in Whole Fish (1980–1981) and Sorghum Distribution

EXPLANATION

Median concentration of toxaphene
in whole freshwater fish, 1980–1981
(in micrograms per kilogram, wet weight)

○ < 100 (not detected)
◔ 100–300
◉ 301–1,600
● > 1,600

Sorghum distribution (in percentage
of total harvested acreage
on a county basis)

□ none
0.1–2.0
2.1–3.9
4.0–15
> 15

━━━ Boundaries of farm production regions used for estimating pesticide use

Figure 3.36. The geographic distribution of toxaphene in whole freshwater fish (1980–1981) from the Fish and Wildlife Service's National Contaminant Biomonitoring Program, shown in relation to 1987 sorghum production (in percentage of total harvested acreage on a county basis). Each site (circle) is shaded to correspond to the median toxaphene concentration in whole fish (in micrograms per kilogram wet weight) at that site during 1980–1981. Toxaphene concentration categories are based on quartiles for all nonzero site median concentrations measured in any sampling year during 1972–1986, so the categories are the same as those in Figures 3.28–3.35. Site median toxaphene concentration data also are identical to those in Figures 3.32 and 3.35. Toxaphene concentration data (1980–1981) are from U.S. Fish and Wildlife Service (1992). Sorghum production data (1987) are from U.S. Department of Commerce (1995).

3.3.2 CURRENTLY USED PESTICIDES

This section examines detection frequencies for currently used pesticides in relation to estimates of agricultural use in 1988 (if available). The discussion below also includes a few superseded pesticides that were discontinued recently and that are not organochlorine insecticides. One hundred pesticides (56 herbicides, 32 insecticides, and 12 fungicides) were used in agriculture in quantities of greater than 500,000 lb a.i. in 1988 (see Table 3.5). Four of these high agricultural-use pesticides are organochlorine insecticides: dicofol, endosulfan, lindane, and methoxychlor. These compounds are somewhat less hydrophobic and less persistent than DDT and the other organochlorine insecticides that were cancelled during the 1970s. Although all four of these currently used organochlorine insecticides have been targeted in sediment and aquatic biota monitoring studies, two of the four (dicofol and endosulfan) were analyzed at far fewer sites than the more hydrophobic restricted-use organochlorine insecticides (such as DDT and dieldrin), as shown in Table 3.6. These four currently used organochlorine insecticides are included in the analysis that follows.

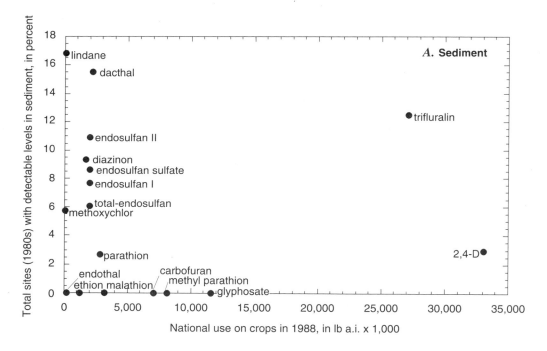

Figure 3.37. The relation between the aggregate site detection frequency during the 1980s and the 1988 national use on crops, shown for currently used pesticides detected in (*A*) sediment and (*B*) aquatic biota. The percentage of sites with detectable levels is for all monitoring studies (listed in Tables 2.1 and 2.2) that sampled primarily during the 1980s, combined. Data for aquatic biota consist of whole fish, fish muscle, and shellfish, combined. Only pesticides targeted at more than 15 sites were included. Use data (1988) are Resources for the Future estimates of national pesticide use on crops (Gianessi and Puffer, 1991, 1992a,b). Abbreviations: lb a.i., pounds active ingredient.

Of the remaining 96 high agricultural-use pesticides, fewer than 40 percent were targeted in either sediment (37 pesticide analytes, listed in Table 3.6) or aquatic biota (32 pesticide analytes, listed in Table 3.6). They include representatives of many pesticide classes, including organophosphate insecticides, carbamate insecticides, chlorophenoxy acid herbicides, triazine herbicides, acid amide herbicides, and fungicides. A few are pesticide metabolites. Of the high agricultural-use pesticides targeted in sediment (37 pesticides) or aquatic biota (32 pesticides), about one-third were detected at least once for both sediment and aquatic biota. The list of pesticides targeted in monitoring studies also includes an additional 20 pesticides (for bed sediment) or 21 pesticides (for aquatic biota) that are neither organochlorine insecticides nor among the 100 highest agricultural-use compounds in 1988 (see Table 3.6). These additional pesticides include some superseded pesticides (such as the chlorophenoxy acid herbicide, 2,4,5-T, and the organophosphate insecticide, zytron). However, most of these additional pesticides are still registered for use in agriculture, although 1988 agricultural use estimates are not available. Nine of the additional 20 pesticides were detected in sediment, and 10 of the additional 21 pesticides were detected in aquatic biota.

Figure 3.37 shows the aggregate site detection frequency for currently used pesticides in bed sediment (Figure 3.37A) and aquatic biota (Figure 3.37B) during the 1980s in relation to national use in agriculture in 1988 (Gianessi and Puffer, 1991, 1992a, 1992b). Only pesticides with available 1988 agricultural use data (and their metabolites) are included. For purposes of this analysis, aquatic biota is defined as whole fish, fish muscle, or shellfish. Calculation of

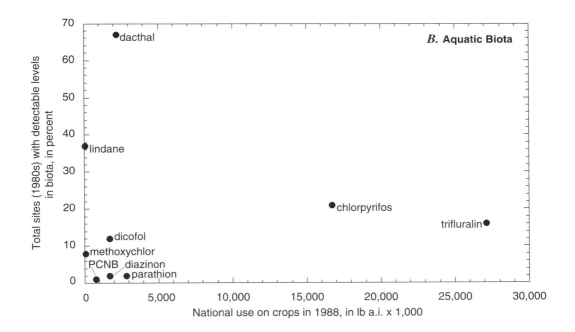

Figure 3.37. *Continued.*

aggregate site detection frequencies is described in Section 3.1.1. The aggregate site detection frequency for the 1980s is the percent of sites with one or more detections calculated by combining data from all monitoring studies (listed in Tables 2.1 and 2.2) that sampled primarily during the 1980s. Thus, the aggregate site detection frequencies in Figure 3.37 represent crude composite detection frequencies for the 1980s. Separate aggregate site detection frequencies were calculated for individual pesticide analytes, and the detection frequency for each pesticide analyte was plotted against the 1988 agricultural use of the parent compound or technical mixture. For example, aggregate detection frequencies for total endosulfan, endosulfan I, and endosulfan II are plotted against the use level for endosulfan in Figure 3.37A.

Of the 100 top agricultural-use pesticides during 1988 (Table 3.5), 14 were targeted in bed sediment at a total of 15 or more sites during the 1980s (see total number of sites sampled for sediment, in Table 3.1). These include three of the four organochlorine insecticides used in agriculture in 1988 (lindane, endosulfan, and methoxychlor). The other 11 high-use pesticides targeted at 15 sites or more during the 1980s were from the following pesticide classes: organophosphate insecticides (diazinon, parathion, ethion, malathion, and methyl parathion), carbamate insecticides (carbofuran), and herbicides (dacthal, endothall, glyphosate, 2,4-D, and trifluralin). Figure 3.37A does not indicate any relation between the aggregate site detection frequencies for these pesticides in sediment during the 1980s and national use on crops in 1988 (Gianessi and Puffer, 1991, 1992a, 1992b). The national agricultural use estimates for the three pesticides detected at the most sites during the 1980s were very wide-ranging: 0.066 million lb a.i. for lindane (detected at 17 percent of sites), 2.2 million lb a.i. for dacthal (16 percent of sites), and 27 million lb a.i. for trifluralin (13 percent of sites). Furthermore, the pesticide with the highest use in 1988 (2,4-D), relative to the other pesticides in Figure 3.37A, was detected at only 3 percent of sites. The five most frequently detected pesticides represented several pesticide classes: organochlorine insecticides (lindane, endosulfan II), herbicides (dacthal, trifluralin), and organophosphate insecticides (diazinon). The probable influence of physical and chemical properties can be seen in Figure 3.37A. The currently used organochlorine insecticides (lindane, endosulfan, and methoxychlor) were all detected in sediment at higher frequencies than most other pesticides, especially considering their relative 1988 agricultural use estimates. Although these three organochlorines are moderately hydrophobic and moderately persistent relative to DDT and most other organochlorine insecticides, they tend to be more persistent than other pesticide classes (see Table 3.7). The organophosphate insecticides, for example, undergo hydrolysis in aquatic environments. The single organophosphate insecticide detected at a much higher frequency than the others (diazinon) is more environmentally persistent than other organophosphate insecticides (Gilliom and others, 1985). Glyphosate had a relatively high use in agriculture (in 1988), but it has a short half-life in soil (Table 3.7) and was not found in river bed sediment during the 1980s. In general, it is important to remember that the aggregate site detection frequencies plotted in Figure 3.37A are composites of data from a variable number of studies with different study designs and analytical methods. This may introduce an unknown bias into the results.

Nine pesticides that were used on crops in 1988 were targeted in aquatic biota (whole fish, fish fillet, or shellfish samples) at 15 sites or more during the 1980s (see Table 3.2, column on total number of sites sampled for aquatic biota). These pesticides include three of the four organochlorine compounds used on crops in 1988 (dicofol, lindane, and methoxychlor). The other six high-use pesticides targeted at 15 sites or more during the 1980s are organophosphate insecticides (chlorpyrifos, diazinon, and parathion); herbicides (dacthal and trifluralin); or fungicides (pentachloronitrobenzene). Figure 3.37*B* does not indicate any clear relation between detection frequency in aquatic biota during the 1980s and national use on cropland in 1988, even when the three organochlorine insecticides are excluded. The two highest agricultural-use compounds, trifluralin and chlorpyrifos, were detected at a substantial number of sites (16 percent and 21 percent, respectively). The pesticide detected at the most sites (67 percent) was dacthal, which had low-to-moderate use on crops in 1988. Dacthal has fairly comparable water solubility (S of 0.5 mg/L) and log K_{ow} value (3.8–4.9) to those of chlorpyrifos (S of 1 mg/L and log K_{ow} of 4.7–5.3) and trifluralin (S of 2 mg/L and log K_{ow} of 5.1). Chlorpyrifos undergoes hydrolysis with half-lives ranging from 35 to 78 days, depending on the pH. On the basis of its national agricultural use estimates (1988) and its physical and chemical properties, dacthal would be expected to have a lower detection frequency than trifluralin and chlorpyrifos. The observed discrepancy may be due to bias introduced by aggregating data from studies with different sampling designs, as discussed in more detail below.

　　To improve the chances of seeing a relation between pesticide occurrence and pesticide use, the variability in both data sets can be reduced. First, pesticide occurrence data can be taken from a single national program at a time. Second, pesticide use data can be improved by using 1988 pesticide use estimates on a county basis (Gianessi and Puffer, 1991, 1992b). This constitutes the next stage in the analysis, described separately for bed sediment and aquatic biota.

Bed Sediment

　　The only national data set for pesticides in bed sediment that includes pesticides other than the organochlorine insecticides was the USGS–USEPA's PMN (1975–1980). In addition to organochlorine compounds, this study measured several herbicides and organophosphate insecticides in bed sediment of United States rivers (Gilliom and others, 1985). Ethion and diazinon were the only organophosphate compounds detected in sediment (at 1.2 percent and 0.6 percent of sites, respectively). Ethion has a lower water solubility (1.1 mg/L) than most organophosphate compounds (see Table 3.7). Diazinon has a low detection limit and is more environmentally persistent (Gilliom and others, 1985). Three chlorophenoxy acid herbicides were analyzed (2,4-D, 2,4,5-T, and silvex); all three were detected in sediment, but at 1 percent (or fewer) of sites. The triazine herbicide atrazine was not detected in any sediment samples, although it was the most frequently detected pesticide in water of any compound analyzed (Gilliom and others, 1985). For diazinon and 2,4-D, the percent of sites with detectable residues was higher in water than in bed sediment. For ethion, 2,4,5-T, and silvex, detection frequencies (percent of sites or samples with detections) in sediment were the same as or higher than in water, but all were less than 2.5 percent.

　　Because these bed sediment data were collected during 1975–1980, they do not correspond in time with recent agricultural pesticide use estimates (1988). Gilliom and others (1985) compared pesticide detection frequencies for the period 1975–1980 with pesticide use estimates at the time (USDA estimates of agricultural pesticide use in 1971). However, there were too few detections of pesticides other than the organochlorines in sediment to permit association with pesticide use (Gilliom and others, 1985).

Aquatic Biota

The largest portion of the combined data on currently used pesticides in aquatic biota during the 1980s came from a single study, the USEPA's NSCRF, which sampled during 1986–1987 (U.S. Environmental Protection Agency, 1992a,b). Both chlorpyrifos and trifluralin were analytes in this study (362 sites), but dacthal was not. This suggests that differences in study design (such as site selection strategy) may have influenced the aggregate site detection frequency of dacthal relative to those of the NSCRF analytes. The largest sources of data for dacthal in aquatic biota came from two programs: the NCBP (113 sites) and California's Toxic Substances Monitoring Program (194 sites). The NSCRF probably targeted more industrial and urban sources (about 60 percent of sites) and fewer agricultural sources (about 9 percent of sites) than either the NCBP or California studies. For the NSCRF, this suggests that detection frequencies of pesticides that are used at high rates in agriculture might be higher for the subset of agricultural sites than for the study as a whole. This is discussed below.

Results from the NSCRF and other large-scale monitoring studies that detected currently used pesticides in aquatic biota are considered below in relation to available use data and to physical and chemical properties. The following currently used pesticides were most frequently detected in aquatic biota: dicofol, lindane, methoxychlor, endosulfan (organochlorine insecticides), chlorpyrifos (organophosphate insecticide), and dacthal and trifluralin (herbicides).

Dicofol, Lindane, and Methoxychlor

Dicofol, lindane, and methoxychlor are discussed together because these three organochlorine insecticides were target analytes in fish samples from the NSCRF. Dicofol is an insecticide and acaricide used on cotton, citrus, apples, ornamental plants, turf, and other minor crops. Lindane (also called γ-HCH) is one of several isomers of technical HCH. All products containing technical HCH were voluntarily cancelled by industry, and the sale, distribution, and use of existing stocks of HCH isomers other than γ-HCH are prohibited in the United States (U.S. Environmental Protection Agency, 1990b). Lindane is used as an insecticide or acaricide on a variety of fruits and vegetables, on tobacco, in forestry, in commercial warehouses and feed storage areas, on farm animal premises, on wooden structures, and by homeowners for domestic outdoor and indoor uses (U.S. Environmental Protection Agency, 1992b). Application is permitted only by a certified applicator. Its primary uses in 1982 were for seed treatment, livestock, and hardwood lumber (U.S. Environmental Protection Agency, 1992b). Methoxychlor is registered for insect control in foliar treatment, dormant application, soil or seed treatment, and postharvest application of many crops. Its primary uses are to control houseflies, black flies, and mosquitoes in areas of human habitation (U.S. Environmental Protection Agency, 1992b).

As noted previously, these three compounds tend to be less hydrophobic and less persistent than DDT and most other organochlorine insecticides. Lindane has a higher water solubility (7 mg/L) than most organochlorine insecticides (see Table 3.7). Both dicofol and lindane undergo alkaline-catalyzed hydrolysis, with half-lives at pH 9 of 26 minutes and 14 days, respectively (U.S. Environmental Protection Agency, 1992b). Dicofol is essentially nonvolatile from water or soil (U.S. Environmental Protection Agency, 1992b; low Henry's law constant), and has a fairly strong tendency to sorb to soil (high K_{oc}) and to bioaccumulate (high K_{ow} and bioconcentration factor), as shown in Table 3.7. Lindane volatilizes relatively slowly from water and faster from soil; it is moderately sorptive to soil and sediment (fairly high K_{oc}), bioaccumulative to some extent (moderate bioconcentration factor), and biodegrades under anaerobic conditions (U.S. Environmental Protection Agency, 1992b). Methoxychlor has a low water solubility, is fairly volatile (intermediate Henry's law constant), sorbs strongly to soil (high K_{oc}), and is bioaccumulative (high K_{ow} and fairly high bioconcentration factor), as shown in Table 3.7. Methoxychlor undergoes direct and indirect photolysis (with a half-life of 4.5 months to 5 hours) and hydrolysis (with a half-life of about 1 year at 27 °C), independent of pH (U.S. Environmental Protection Agency, 1992b). Methoxychlor is more susceptible to biotransformation than its structural analogue, DDT, because it has the more labile methoxy substituents where DDT has chlorine substituents.

In Figure 3.37B, which combines aquatic biota data from all the monitoring studies reviewed, dicofol, methoxychlor, and lindane had higher aggregate detection frequencies in biota during the 1980s than might be expected on the basis of agricultural use (in 1988) alone. This was particularly pronounced for lindane. The same relation held when only data from the NSCRF (1986–1987) were compared with 1988 agricultural use estimates (data not shown). All three compounds are also used in residential settings (Table 3.5), so environmental inputs are not from agriculture alone. Moreover, as noted previously, dicofol, methoxychlor, and lindane tend to be more hydrophobic and more persistent than other currently used pesticides.

Endosulfan

The fourth organochlorine insecticide used in agriculture in 1988, and in residential settings in 1990, was endosulfan (Table 3.5). Endosulfan is used as an insecticide and acaricide on a variety of crops, including citrus, small fruits, coffee, forage crops, and grains, vegetables, and tobacco (Meister Publishing Company, 1996). Endosulfan is a technical mixture of two isomers: endosulfan I (α-endosulfan) and endosulfan II (β-endosulfan). Endosulfan has a low water solubility (0.32 mg/L), sorbs fairly strongly to soil (fairly high K_{oc}), and bioaccumulates to some extent (moderate K_{ow} and bioconcentration factor). Endosulfan degrades in soil with a half-life of 30–70 days (Tomlin, 1994). Its principal breakdown product, endosulfan sulfate, is persistent in the environment (Rasmussen and Blethrow, 1990; Tomlin, 1994).

Very few studies targeted any endosulfan compounds (endosulfan I, endosulfan II, endosulfan sulfate, or total endosulfan) in aquatic biota. The only national program that did so was the BofCF–USEPA's National Pesticide Monitoring Program, which analyzed endosulfan in whole juvenile (yearling) estuarine fish collected during 1972–1976 (Butler and Schutzmann, 1978). Endosulfan was not detected in any samples in this study, although the detection limit was fairly high (20 μg/kg wet weight).

Endosulfan was detected in a few local studies in California during the early 1980s (Mearns and others, 1988). Examples include marine mollusks and fish from sloughs and bays in coastal Monterey County and fish from the nearby Salinas River (Ali and others, 1984), and mussels from additional sites in California (Stephenson and others, 1986). Increasing concentrations in mussels collected between 1979 and 1981 were observed at 3 of 11 California sites (Stephenson and others, 1986). In California's Toxic Substances Monitoring Program, fish and invertebrates were monitored from over 200 water bodies throughout the state (Rasmussen and Blethrow, 1990). The highest concentrations of endosulfan detected in the state during

1978–1987 were from two intensively farmed areas, the Watsonville–Salinas area and the Colorado River Basin. Detection frequencies in these two areas were about twice the statewide average. Endosulfan also was detected in fish from an Oregon estuary (Claeys and others, 1975). Data from these local studies suggest that endosulfan contamination may occur in rivers and estuaries near agricultural drainages (Mearns and others, 1988). Endosulfan is very toxic to fish, and in fact is used as a piscicide (Greve and Wit, 1971; Capel and others, 1988).

Chlorpyrifos

Chlorpyrifos is used in agriculture to control soil and foliar pests on cotton, peanuts, and sorghum, and to control root-infesting and boring insects on a variety of fruits, nuts, vegetables, and field crops. It also has been used to control ticks on cattle and sheep; in subterranean termite control (Esworthy, 1987); and around homes and gardens (Whitmore and others, 1992) to control household insects such as ants, cockroaches, fleas, and mosquitoes (U.S. Environmental Protection Agency, 1992b). Chlorpyrifos has moderate water solubility, sorbs fairly strongly to soil (fairly high K_{oc}), and has some tendency to bioaccumulate (a high K_{ow} and moderate bioconcentration factor), as shown in Table 3.7. It undergoes hydrolysis at acid, neutral, and (particularly) alkaline pH, and photolyzes in both air and soil.

In the NSCRF, the U.S. Environmental Protection Agency (1992a) examined detection frequencies and concentrations of various contaminants as a function of land use. The highest mean chlorpyrifos concentration for all site categories was for sites in agricultural areas. Also, of the five sites with the highest chlorpyrifos concentrations, three were in agricultural areas and two were in urban and industrial areas. In Figure 3.38, the geographic distribution of chlorpyrifos residues in fish at NSCRF sites is mapped in relation to its use on crops in 1988 (in lb a.i. per square mile of agricultural land in the region). The use data shown in Figure 3.38 were obtained from county-based estimates of chlorpyrifos use on crops in 1988 (Gianessi and Puffer, 1992b). Regional use estimates were calculated by summing use estimates for all counties within each of the 10 USDA farm production regions. The regional use estimates then were normalized by dividing by the number of square miles of agricultural land within each region (U.S. Geological Survey, 1970). The normalized 1988 agricultural use levels for the ten regions were divided into quartiles and the agricultural land within each region was shaded to correspond to the appropriate use quartile.

Figure 3.38 shows that the highest concentrations of chlorpyrifos were observed in or near the Corn Belt Region, which received fairly high chlorpyrifos use (between the 50th and 75th use percentiles). A few other sites also had high chlorpyrifos residues, such as sites in the Imperial Valley in California (which was in the highest use quartile) and near San Antonio, Texas (which was in the lowest use quartile). A number of sites in the highest chlorpyrifos use quartile (Pacific and Mountain regions) did not contain detectable chlorpyrifos residues in fish. Low-level detections of chlorpyrifos occurred in regions in all four use quartiles. Thus, Figure 3.38 does not show a clear-cut association between chlorpyrifos detections in NSCRF fish samples (1986–1987) and regional estimates of chlorpyrifos use in agriculture in 1988.

California's Toxic Substances Monitoring Program analyzed for chlorpyrifos in fish and invertebrates from 200 water bodies throughout California during 1978–1987. As for endosulfan (noted previously), concentrations of chlorpyrifos were highest in two agricultural areas: the

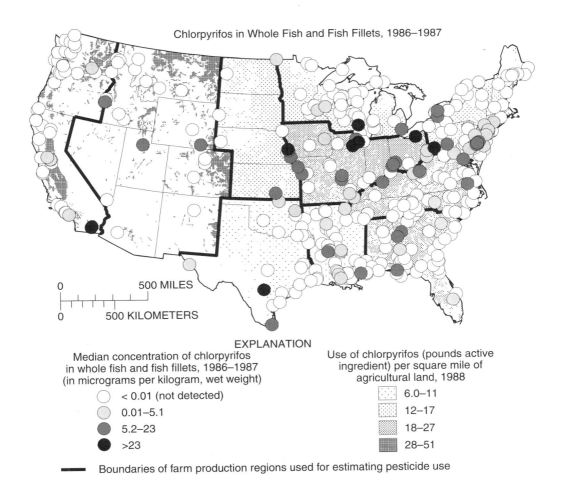

Figure 3.38. The geographic distribution of chlorpyrifos in fish (1986–1987) from USEPA's National Study for Chemical Residues in Fish, shown in relation to 1988 agricultural use of chlorpyrifos by farm production region (in pounds active ingredient per square mile of agricultural land). Each site (circle) is shaded to correspond to the median chlorpyrifos concentration in fish (in micrograms per kilogram wet weight) at that site during 1986–1987. Chlorpyrifos concentration categories are based on quartiles for all nonzero site median concentrations. Chlorpyrifos concentration data (1986–1987) are from U.S. Environmental Protection Agency (1992b). Regional use data (1988) are based on Resources for the Future estimates of national chlorpyrifos use on crops by county (Gianessi and Puffer, 1991) and were calculated by summing use estimates for all counties within each of the 10 U.S. Department of Agriculture farm production regions (Eichers and others, 1970), then normalizing this sum by the number of square miles of agricultural land within the region (U.S. Geological Survey, 1970). Agricultural land within each region is shaded to correspond to the appropriate use quartile. Abbreviations: lb a.i., pounds active ingredient; USEPA, U.S. Environmental Protection Agency.

Watsonville–Salinas area and the Colorado River Basin (Rasmussen and Blethrow, 1990). Its detection frequencies in these areas were about twice the statewide average.

Dacthal

Dacthal (also called DCPA) is a widely used broad-spectrum herbicide registered for use on ornamental plants, turf, and certain vegetable and field crops (Schmitt and others, 1990). It has low-to-moderate water solubility (0.5 mg/L), sorbs fairly strongly to soil (fairly high K_{oc}), and is moderately hydrophobic (intermediate K_{ow}), as shown in Table 3.7.

Dacthal was analyzed in two large-scale studies: the NCBP (113 sites) and California's Toxic Substances Monitoring Program (194 sites). In both studies, the highest concentrations occurred in areas that were intensively farmed. The highest residues at NCBP sites in 1984 were from the lower Rio Grande River in Texas (Schmitt and others, 1990). In California's Toxic Substances Monitoring Program, the highest dacthal concentrations in the state (1978–1987) were from two heavily farmed areas: the Watsonville–Salinas area and the lower Colorado River Basin (Rasmussen and Blethrow, 1990). As was observed for endosulfan and chlorpyrifos, the detection frequencies of dacthal in these two areas were about twice the statewide average.

Trifluralin

Trifluralin is used to control broadleaf weeds and annual grasses in a variety of agricultural crops, especially on cotton and soybeans. Nonagricultural uses include the control of grasses in rights-of-way, at outdoor industrial sites, and around homes and gardens (U.S. Environmental Protection Agency, 1992b; Whitmore and others, 1992). Trifluralin has moderate water solubility (2 mg/L), sorbs fairly strongly to soil (fairly high K_{oc}), and tends to bioaccumulate (high K_{ow} and fairly high bioconcentration factor), as shown in Table 3.7.

Trifluralin was detected in fish at 11 percent of NSCRF sites (U.S. Environmental Protection Agency, 1992a). The highest mean concentrations of trifluralin were for sites in agricultural areas and for sites located on major rivers (at NASQAN sampling stations). In Figure 3.39, the geographic distribution of trifluralin residues in fish at NSCRF sites is mapped in relation to its use on crops in 1988. These use data were prepared and mapped as described previously for chlorpyrifos (Figure 3.38). As shown in Figure 3.39, most of the sites with detectable trifluralin residues in fish were located in the regions of high trifluralin use: the Corn Belt, Northern Plains, and Delta States regions. A few additional sites had high trifluralin concentrations; these sites were located in agricultural areas in regions of intermediate trifluralin use: the Lakes States, Mountain, and Pacific regions. Thus, there is a fairly strong association between trifluralin concentration in fish at NSCRF sites and regional estimates of trifluralin use. U.S. Environmental Protection Agency (1992a) reported that most detections of trifluralin were in states that had the highest trifluralin use (Arkansas, Illinois, Iowa, Minnesota, Missouri, North Dakota, South Carolina, Tennessee, and Texas) or were near pesticide manufacturing plants.

Other Currently Used Pesticides

For other currently used pesticides targeted in the NSCRF, land use comparisons indicated primarily urban or industrial sources (U.S. Environmental Protection Agency, 1992a,b). The

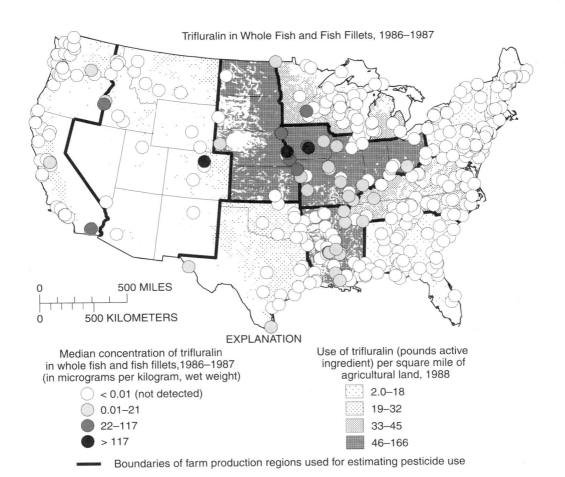

Trifluralin in Whole Fish and Fish Fillets, 1986–1987

0 ⊢⊢⊢⊢⊢ 500 MILES

0 ⊢⊢⊢⊢⊢ 500 KILOMETERS

EXPLANATION

Median concentration of trifluralin
in whole fish and fish fillets,1986–1987
(in micrograms per kilogram, wet weight)

○ < 0.01 (not detected)
◯ 0.01–21
● 22–117
● > 117

Use of trifluralin (pounds active
ingredient) per square mile of
agricultural land, 1988

2.0–18
19–32
33–45
46–166

━━━ Boundaries of farm production regions used for estimating pesticide use

Figure 3.39. The geographic distribution of trifluralin in fish (1986–1987) from USEPA's National Study for Chemical Residues in Fish, shown in relation to 1988 agricultural use of trifluralin by farm production region (in pounds active ingredient per square mile of agricultural land). Each site is shaded to correspond to the median trifluralin concentration in fish (in micrograms per kilogram wet weight) at that site during 1986–1987. Trifluralin concentration categories are based on quartiles for all nonzero site median concentrations. Trifluralin concentration data (1986–1987) are from U.S. Environmental Protection Agency (1992b). Regional use data (1988) are based on Resources for the Future estimates of national trifluralin use on crops by county (Gianessi and Puffer, 1991) and were calculated by summing use estimates for all counties within each of the 10 U.S. Department of Agriculture farm production regions (Eichers and others, 1970), then normalizing this sum by the number of square miles of agricultural land within the region (U.S. Geological Survey, 1970). Agricultural land within each region is shaded to correspond to the appropriate use quartile. Abbreviations: USEPA, U.S. Environmental Protection Agency.

herbicide isopropalin, which was discontinued in 1993 (Meister Publishing Company, 1996), had the highest mean concentrations in fish from industrial and urban sites. Pentachloroanisole (a metabolite of the wood preservative pentachlorophenol) had its highest mean and median concentrations near paper mills that do not use chlorine in the bleaching process, followed by paper mills that do use chlorine. The fungicide pentachloronitrobenzene was detected at only four sites; the two highest concentrations were near chemical manufacturing plants or Superfund sites. The herbicide nitrofen was detected at only five sites in the NSCRF, all of which were in agricultural areas or near Superfund sites or chemical manufacturing facilities. Nitrofen has not been used in the United States since 1984 (U.S. Environmental Protection Agency, 1992b), so it is not included in Figure 3.37. Prior to 1984, it was used to control weeds on vegetables.

Some state and local studies noted associations between the occurrence of currently used pesticides in aquatic biota or bed sediment and agricultural or other land uses (also see Section 5.1). In California's Toxic Substances Monitoring Program, which monitored contaminants in fish and invertebrates, the highest residues of several currently used pesticides were from two intensively-farmed areas: the Watsonville–Salinas area and the Colorado River Basin in California (Rasmussen and Blethrow, 1990). Moreover, detection frequencies for these pesticides in fish from these two areas were about twice the statewide average. These pesticides included the organophosphate insecticides diazinon and parathion, as well as the pesticides discussed previously (chlorpyrifos, dacthal, and endosulfan).

In one local study, high levels of the herbicide oxadiazon were measured in bed sediment, fish, and transplanted mollusks in the San Diego Creek drainage, which enters Newport Bay in southern California (Crane and Younghans-Haug, 1992). Residues of oxadiazon reached 87 µg/kg dry weight in creek sediment, 2,200 µg/kg wet weight in native fish, and 2,400 µg/kg dry weight in mollusks transplanted to the creek. Mussels transplanted into the Newport Bay also contained oxadiazon residues (up to 83 µg/kg dry weight) that decreased in concentration with increasing distance from San Diego Creek. The use of oxadiazon in California has increased steadily from 1976 and doubled between 1986 and 1987. Its primary uses in California were maintenance of landscape and rights-of-way (Crane and Younghans-Haug, 1992). Crane and Younghans-Haug (1992) suggested that the source of oxadiazon in San Diego Creek and tributaries was from its use by parks, golf courses, and nurseries in the drainage basin. In field studies in Japan, oxadiazon was detected in common carp from Lake Kojima 2–9 months after it was applied locally (Imanaka and others, 1981). Subsequent studies showed that oxadiazon was present in lake water (up to about 6 µg/L), as well as in carp (up to 1,800 µg/kg in edible tissues). In fact, oxadiazon concentrations in carp and in water followed very similar patterns during 1980–1982, with residue levels in carp reflecting the concentrations in the water, almost without any time lag (Imanaka and others, 1985). In laboratory experiments, the biological half-life of oxadiazon in willow shiner (*Gnathopogon caerulescens*) was 2.3 days (Tsuda and others, 1990). Measured bioconcentration factors for oxadiazon ranged from 200 to 400 for edible tissues of carp from Lake Kojima (Imanaka and others, 1985) and were about 1,200 for whole-body willow shiner in the laboratory (Tsuda and others, 1990). These studies suggest the following: (1) oxadiazon may be present in hydrologic systems receiving drainage from areas of continual or repeated use; (2) if present in the water, oxadiazon is likely to be present at substantial levels in aquatic biota; and (3) oxadiazon residues in aquatic biota would be expected to decline if oxadiazon use were discontinued and concentrations in the water column declined.

3.4 LONG-TERM TRENDS

Data from existing monitoring studies and reviews are adequate to assess national trends for several organochlorine insecticides in aquatic biota. Data for sediment are more limited. The monitoring studies reviewed vary in study duration from less than 1 month to 11 years for state and local studies, and from 3 months to 20 years (excluding FDA's NMPFF) for national and multistate studies (Table 2.4). Determination of nationwide trends from these studies is made difficult by differences among these studies, such as study design, sampling procedures, analytical methods, and data reporting. Therefore, the discussion below emphasizes those studies that were longer in duration, especially national and multistate studies (Table 2.1), each of which used uniform procedures over a large geographic area. The discussion focuses on freshwater studies, but results of available marine or estuarine studies are mentioned for comparative purposes. Trends are discussed first for the historically used organochlorine insecticides that were banned or severely restricted during the 1970s (Section 3.4.1) and then for currently used pesticides (Section 3.4.2). Trends data are available for two of the less persistent organochlorine insecticides still used in agriculture in 1988 (lindane and methoxychlor), which are discussed in Section 3.4.2.

3.4.1 HISTORICALLY USED ORGANOCHLORINE INSECTICIDES

Because most sediment or aquatic biota monitoring studies targeted organochlorine compounds, considerable data are available to assess trends for a number of organochlorine insecticides. Chlordane compounds, DDT and metabolites, and dieldrin tend to be the most frequently detected organochlorine insecticides.

Bed Sediment

Organochlorine insecticides were measured in bed sediment of major United States rivers as part of the PMN. Trends during that time (1975–1980) were either nondetectable or predominately downward, both nationally and in all regions except the southeastern United States (Gilliom and others, 1985). Bed sediment data were adequate to assess trends for 123 chemical–site combinations, of which 36 were statistically significant. Seven of these 36 significant trends were upward and 29 were downward. Significant trends tended to occur at only a few sites. Of the seven upward trends nationwide, five were at a single site in the southeastern United States. Of 29 downward trends nationwide, 18 occurred at 6 sites, and the remaining 11 at 11 different sites. In general, trends were discernible most often for those chemicals that were most frequently detected: chlordane, DDD, DDE, DDT, and dieldrin.

A more recent national assessment (1990s) of organochlorine pesticides in bed sediment from United States rivers will be obtained from ongoing sampling by the National Water Quality Assessment (NAWQA) Program. The design of the NAWQA Program calls for sampling river bed sediment throughout 55–60 surface-water basins nationwide on a rotational basis, which began in 1991.

Surficial sediment from coastal and estuarine areas of the United States has been sampled nationwide since 1984 as part of the NS&T Program. As of this writing, data have been published only for 1984–1986, which are not yet sufficient to provide information on long-term trends in organochlorine concentrations in coastal and estuarine sediment. The NS&T Program, however, relies on residues in fish and mollusks, rather than on sediment (which were used to determine spatial distribution), to monitor temporal trends in contamination because contaminant levels in biota change relatively rapidly in response to surroundings (National Oceanic and Atmospheric Administration, 1988).

Aquatic Biota

The NCBP provides information on long-term trends for several organochlorine insecticides in whole freshwater fish (1967–1986). The discussion below draws upon data from the NCBP, supplemented by selected long-term local monitoring studies, and on review papers. Trends in freshwater biota are compared, where data are available, with trends in estuarine fish and shellfish contamination. Information on trends in estuarine biota comes from the review by Mearns and others (1988) and from two national long-term studies that measured organochlorine insecticides in estuarine biota: the BofCF—USEPA's National Pesticide Monitoring Program (1965–1972, 1977) and the NS&T Program (1984–present). Results are presented below for individual pesticides.

In the future, updated information on pesticide residues (1990s) in aquatic biota from United States rivers and estuaries will come from several ongoing studies. The NAWQA Program has sampled pesticide residues in whole fish and mollusks from United States rivers nationwide beginning in 1991. The Biological Resources Division of the USGS is conducting several pilot studies as part of a joint effort by two programs started by the FWS: the NCBP and the Biomonitoring of Environmental Status and Trends (BEST) Program. One of these pilot studies covers the Mississippi River Basin. In the NCBP–BEST pilot studies, various contaminants in whole fish (including pesticides) are being measured in association with several biomarkers of fish health. The NS&T Program continues to sample estuarine mollusks and benthic fish liver, as well as associated sediment.

Aldrin and Dieldrin

Temporal trends in dieldrin residues in whole fish at NCBP sites from 1969 to 1986 are shown in Figure 3.40. For each sampling year, a box plot shows the distribution of concentrations as follows. The bottom and top of the box represent the 25th and 75th percentile concentrations, respectively; the bar across the box represents the median concentration; and the bottom and top of the lines extending below and above the box represent the 10th and 90th percentiles, respectively. In plotting these percentiles in Figure 3.40, nondetections are shown as one half of the detection limit (10 µg/kg). There is considerable year-to-year variability in dieldrin concentrations at the more contaminated NCBP sites, as shown by variations among years in the 75th and 90th percentiles (Figure 3.40). Nonetheless, residues at more contaminated sites (those above the median) during the 1980s are clearly lower than those during the early 1970s. Median dieldrin concentrations remained at 10–20 µg/kg wet weight from 1969 to 1986 (except during 1973–1974, when the median concentration was below the detection limit). Use of aldrin and dieldrin was heaviest in the Corn Belt and Great Lakes regions (Figure 3.14). Schmitt and colleagues reported that dieldrin residues in fish from the Great Lakes and major rivers of the Midwest did not change appreciably between the period 1978–1979 and 1984

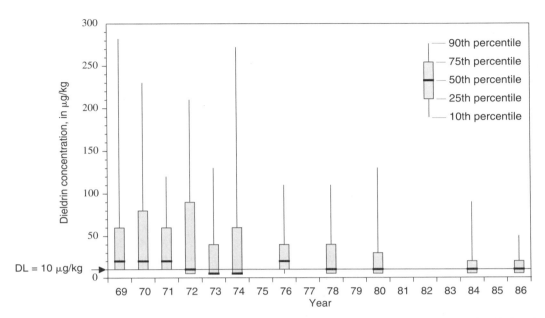

Figure 3.40. Temporal trends in dieldrin concentrations in whole fish sampled by the FWS's National Contaminant Biomonitoring Program from 1969 to 1986. Concentrations are in μg/kg wet weight. Data are from U.S. Fish and Wildlife Service (1992). Abbreviations: DL, detection limit; FWS, Fish and Wildlife Service; μg/kg, micrograms per kilogram.

(Schmitt and others, 1990), and concluded that dieldrin was still being carried into receiving waters from fields in the Midwest, even though no aldrin had been used in agriculture since 1974 (Schmitt and others, 1985). Declining dieldrin residues probably indicate that aldrin inputs to hydrologic systems also have declined since the early 1970s. Aldrin residues in whole fish at NCBP sites were rarely seen after 1977 (Schmitt and others, 1983).

Data for coastal and estuarine biota tell a similar story. There is evidence for declining dieldrin residues in bivalve mollusks, fish liver, and whole fish at some marine sites, but it remains to be determined whether there is a nationwide trend for coastal and estuarine areas, especially in fish (Mearns and others, 1988). Dieldrin was detected in 15 percent of all mollusk samples collected between 1965 and 1972 as part of the National Pesticide Monitoring Program (Butler, 1973b), but it was not detected in any 1977 samples (Butler and others, 1978). This suggests a declining trend in dieldrin contamination in the estuaries sampled. However, as discussed in Section 2.6, the increase in detection limit that occurred between the period 1965–1972 (5 μg/kg) and 1977 (10 μg/kg) may have contributed to the apparent decrease in detection frequency during this time. Dieldrin was detected in whole juvenile (yearling) estuarine fish at 17 percent of sites during 1972–1976 (Butler and Schutzmann, 1978) and in estuarine fish livers at 58 percent of NS&T sites in 1984 (Mearns and others, 1988), but these data were insufficient to evaluate trends.

Aldrin was not detected in bivalve mollusks (1965–1972, 1977) or whole juvenile estuarine fish (1972–1976) from the National Pesticide Monitoring Program (Butler, 1973a,b; Butler and others, 1978; Butler and Schutzmann, 1978). Aldrin was at, or below, the detection limit of 1 µg/kg wet weight in estuarine fish livers (1984) collected at NS&T sites (Mearns and others, 1988). However, decreasing dieldrin contamination in estuarine biota probably indicates that aldrin contamination also has decreased (Mearns and others, 1988).

Chlordane

Temporal trends in total chlordane concentrations in NCBP whole freshwater fish samples (1976–1986) are shown in Figure 3.41. Total chlordane is defined as the sum of *cis-* and *trans-*chlordane, *cis-* and *trans-*nonachlor, and oxychlordane (except during 1976–1977, when oxychlordane was not determined). Earlier data were either not comparable because of analytical method differences (1967–1969) or not reported by the FWS because of analytical difficulties (1970–1974). The median concentration of total chlordane increased from the period 1976–1977 to 1978–1979 (some part of the apparent increase may have been due to quantitation of oxychlordane during 1978–1979), and afterwards decreased slightly or stayed the same (Figure 3.41). The 75th and 90th percentile concentrations (see Figure 3.41) and the maximum concentrations (not shown) observed followed a similar pattern, except that there was an increase in the maximum concentration between the period 1980–1981 and 1984 (Schmitt and others, 1990). Overall, total chlordane concentrations in 1984 and 1986 appear to have been lower than

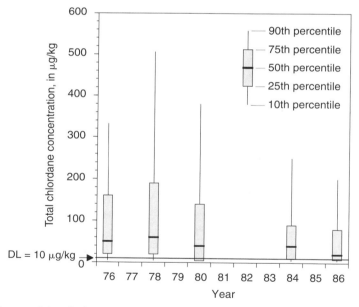

Figure 3.41. Temporal trends in total chlordane concentrations in whole fish sampled by the FWS's National Contaminant Biomonitoring Program from 1976 to 1986. Concentrations are in µg/kg wet weight. Total chlordane is the sum of *cis-* and *trans-*chlordane, *cis-* and *trans-*nonachlor, and oxychlordane. Data are from U.S. Fish and Wildlife Service (1992). Abbreviations: DL, detection limit; FWS, Fish and Wildlife Service; µg/kg, micrograms per kilogram.

in previous years, although residues may have leveled off during the 1980s. The proportional composition of total chlordane residues in fish at NCBP sites also changed between the period 1976–1977 and 1986, with *cis*-chlordane being the most abundant component prior to 1980, and *trans*-nonachlor (the most persistent component of technical chlordane) being the most abundant component since 1980 (Schmitt and others, 1990; U.S. Fish and Wildlife Service, 1992). This suggests that chlordane residues in aquatic biota are weathering over time.

For estuarine biota, it is difficult to determine trends in chlordane residues because either chlordane compounds were not determined, or concentrations were below detection limits (Mearns and others, 1988). Moreover, comparison among studies is difficult because chlordane concentrations were reported differently (such as technical chlordane versus different combinations of chlordane components) by different studies. Several studies determined chlordane residues in bivalve mollusks. The results of the following five studies, considered together, are inconclusive regarding trends in chlordane residues: (1) Chlordane (constituents not specified) was not detected in any coastal or estuarine bivalve mollusk samples collected nationwide during 1965–1972 or in 1977 as part of the National Pesticide Monitoring Program, which had a detection limit of 5–10 μg/kg wet weight (Butler, 1973b; Butler and others, 1978). (2) In contrast, chlordane was detected in 98 percent of coastal and estuarine bivalve mollusk samples collected a decade later (1986–1988) as part of the NS&T Program (National Oceanic and Atmospheric Administration, 1989). However, detection limits were lower in the NS&T Program (0.19–1.0 μg/kg dry weight) than in the National Pesticide Monitoring Program (5–10 μg/kg wet weight). Although a period of 3 years is too short to permit testing for trends, monotonic increases in total chlordane concentrations in bivalve mollusks from 1986 to 1988 were observed at three sites, and decreases were observed at six sites, nationwide (National Oceanic and Atmospheric Administration, 1989). In the NS&T Program, chlordane was defined as the sum of *cis*-chlordane, *trans*-nonachlor, heptachlor, and heptachlor epoxide. (3) Chlordane (constituents not specified) was detected in mollusks and finfish from the Chesapeake Bay and its Maryland tributaries from 1976 to 1980 (Eisenberg and Topping, 1984, 1985). Pooled means and medians, by species, suggest a possible decrease in chlordane residues over the study period for oysters, but not finfish or clams. However, the apparent trend for oysters was not supported by site-specific data (Shigenaka, 1990). (4) Another Chesapeake Bay watershed study conducted from 1981 to 1985 showed no temporal trends in chlordane residues in oysters (Murphy, 1990). (5) The California State Mussel Watch Program has measured eight constituents of technical chlordane (*cis*- and *trans*-chlordane, *cis*- and *trans*-nonachlor, α- and γ-chlordene, oxychlordane, and heptachlor) in mollusks from 1977 to the present. Temporal trends in summed concentrations of five chlordane compounds (*cis*-chlordane, *trans*-chlordane, oxychlordane, *trans*-nonachlor, and heptachlor) in mussels (*Mytilus californianus*) for seven sites sampled more than five times between 1979 and 1988 are described in Shigenaka (1990). These results, shown in Figure 3.42, suggest a declining chlordane trend at most of those sites. From the same California State Mussel Watch Program, total chlordane residues (the sum of all eight constituents listed above, minus heptachlor) in *M. californianus* declined significantly at about half of the (47 total) sites that were sampled frequently from 1977 to 1992 (Stephenson and others, 1995). Declines in total chlordane residues were observed for several urban sites (including sites in San Francisco Bay, Los Angeles Harbor, and San Diego Bay) and for two of three sites located near agricultural fields (Stephenson and others, 1995).

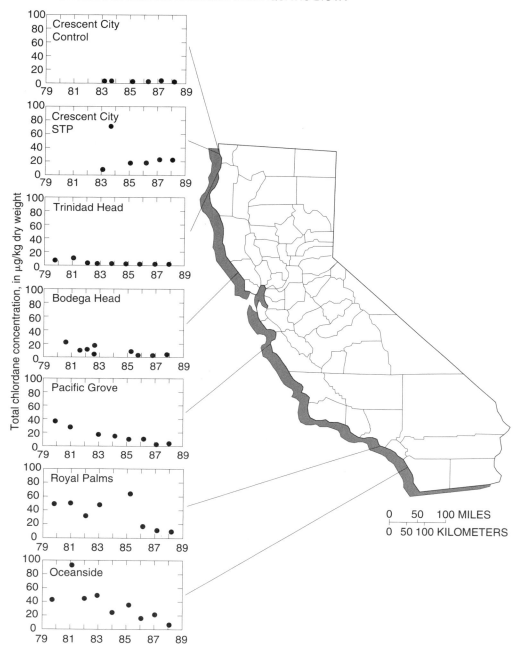

Figure 3.42. Temporal trends in total chlordane concentrations in mussels (*Mytilus californianus*) at seven estuarine and coastal sites sampled by the California Mussel Watch from 1979 to 1988. Total chlordane is the sum of *cis-* and *trans*-chlordane, *cis-* and *trans*-nonachlor, α- and γ-chlordene, oxychlordane, and heptachlor. Concentrations are in μg/kg dry weight. Redrawn from Shigenaka (1990). Abbreviations: STP, Sewage Treatment Plant; μg/kg, micrograms per kilogram.

Total DDT

Total DDT concentration data for whole freshwater fish from the NCBP were published for the period 1967–1984, although data collected prior to 1976 are less reliable because of analytical limitations (Schmitt and others, 1990). Nationally, both median and 90th-percentile total DDT concentrations in whole fish declined from 1969 to 1986, except in 1974, as shown in Figure 3.43. The declines were greatest during the early 1970s, with residues during the late 1970s and early 1980s showing a slower decline, or even an apparent plateau. Schmitt and others (1990) noted the same pattern for geometric mean total DDT concentrations in whole fish, with declines observed from 1976 to 1981, but not in 1984. Of all the components of total DDT (*o,p'* and *p,p'* isomers of DDT, DDD, and DDE), only *p,p'*-DDT showed a significant decline in mean concentration between the period 1980–1981 and 1984. At the same time, the proportion of total DDT consisting of DDE rose slightly from 70 percent during 1974–1979 to 73 percent in 1984 (Schmitt and others, 1990). The detection frequency for total DDT remained high (greater than 97 percent of sites) in all years sampled (Table 2.1). About 4 percent of 1986 samples had concentrations that exceeded 1,000 µg/kg, indicating that fairly contaminated areas appear to be persisting. These results indicate low influx and slow but continued weathering of DDT in the

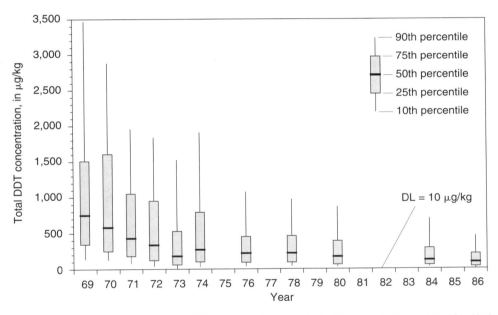

Figure 3.43. Temporal trends in total DDT concentrations in whole fish sampled by the FWS's National Contaminant Biomonitoring Program from 1969 to 1986. Total DDT is the sum of residues of *o,p'*-DDT, *o,p'*-DDE, *o,p'*-DDD, *p,p'*-DDT, *p,p'*-DDE, and *p,p'*-DDD. Concentrations are in µg/kg wet weight. Data are from U.S. Fish and Wildlife Service (1992). Abbreviations: DL, detection limit; FWS, Fish and Wildlife Service; µg/kg, micrograms per kilogram.

environment (Schmitt and others, 1990). Long-term fish data from a study in Lake Michigan show a similar pattern. The maximum total DDT in Lake Michigan bloater chubs declined rapidly from 9,900 µg/kg in 1969 to the mid–1970s, then appeared to level off at about 1,000 µg/kg during the early 1980s (Eadie and Robbins, 1987).

While residues of DDT itself have declined nationwide since its use was discontinued, there is evidence that DDE residues may have stabilized (U.S. Environmental Protection Agency, 1984, 1986c), at least in some areas. Long-term monitoring efforts in areas of past high DDT use, such as the Yakima River Basin in Washington (Rinella and others, 1993) and the agricultural areas of California (Rasmussen and Blethrow, 1990), do not indicate any decline in mean levels of total DDT in fish since the cancellation of DDT. The apparent plateau in total DDT levels may be the result of continued inputs of total DDT residues as contaminated soils are carried into receiving waters.

Evidence for total DDT trends in estuarine and marine biota comes from monitoring programs that sampled bivalve mollusks, whole fish, fish muscle, and fish liver. Existing studies have shown that, nationally, DDT residues have declined substantially in the last two decades, especially during the 1970s. Benthic fish liver data from 1984 suggest that this decline continued into the 1980s (Mearns and others, 1988). This national decline in total DDT residues in aquatic biota is illustrated in Table 3.9, which is adapted from Mearns and others (1988). DDT residues appear to have persisted in three coastal areas: southern California, upper Delaware Bay, and southern Laguna Madre, Texas (Mearns and others, 1988). Long-term bivalve mollusk data (*M.Californianus*) from the California coast indicates that total DDT concentrations declined at about half of the sites frequently sampled between 1977 and 1992 (Stephenson and others, 1995).

Endrin

Detection frequencies and concentrations for endrin in NCBP whole freshwater fish samples were typically lower than for most other organochlorine insecticides because this compound is relatively short-lived in the environment and was used on relatively few crops (Schmitt and others, 1985). Endrin concentrations were at, or below, the detection limits from 1970 to 1974. Prior to 1976, changes in percent occurrence of endrin probably reflected varying detection limits (Schmitt and others, 1981). Endrin concentrations (see data summaries in Table 2.1 for Schmitt and others, 1983, 1985, 1990) were highest during 1976–1977, then declined. After 1977, both geometric mean (Schmitt and others, 1990) and median concentrations (Table 2.1) were below the detection limit (10 µg/kg). However, neither the maximum concentration nor the percent of sites with detectable residues declined consistently from 1978 to 1986, indicating a possible leveling off in endrin residues during the early 1980s.

In estuarine bivalve mollusks collected as part of the National Pesticide Monitoring Program, endrin was detected in less than 2 percent of mollusk samples during 1965–1972. No endrin was detected when a subset of sites was resampled in 1977. These data are insufficient to assess trends at marine and estuarine sites.

α-*HCH*

α-HCH was a component of technical lindane prior to 1977. Because the HCH isomers are shorter-lived than most organochlorine insecticides, their detection in environmental samples indicates recent inputs (Schmitt and others, 1990). In national studies, α-HCH was determined only in whole fish from freshwater systems, as part of the NCBP. Data are available for 1969 and 1970 (after which α-HCH was dropped from the target analyte list), and for 1976 (when it was put back on the list) through 1986. When analysis was resumed during 1976–1977, the percentage of sites with detectable levels had dropped to 76 percent (from 99 percent in 1970), and the median concentration had decreased to 10 µg/kg (from 20 µg/kg in 1970). α-HCH occurrence decreased gradually from the period 1976–1977 to 1986, with only 7 percent of sites having detectable residues in 1986. After 1977, α-HCH concentrations were very close to the detection limit (10 µg/kg). Except for one station probably contaminated by an abandoned chemical dump site, Schmitt and others (1983) attributed the widespread low-level distribution of α-HCH residues to atmospheric transport.

Table 3.9. Trends in residues of DDT and metabolites in aquatic biota: Median residues from several national surveys

[Adapted from Mearns and others (1988). Abbreviations and symbols: BofCF, Bureau of Commercial Fisheries; FWS, Fish and Wildlife Service; NOAA, National Oceanic and Atmospheric Administration; USEPA, U.S. Environmental Protection Agency; µg/kg, microgram per kilogram; —, no data]

Organism	Type of Hydrologic System	Total DDT or DDE (µg/kg wet weight) by Sampling Period[1]			
		1965–1972	1972–1975	1976–1977	1984–1986
Bivalve mollusks	Marine	[2]24	—	[3]10, [4]1	[5]3
Fish—whole, juvenile	Marine	—	[6]14	—	—
Fish—muscle	Marine	—	[7]110	[8]12	—
Fish—liver	Marine	—	—	[8]220	[9]54
Fish—whole	Freshwater	[10]430–750	[11]180–335	[12]220	[13]90–120

[1]Data for marine systems are from Mearns and others (1988); data for freshwater systems are from Fish and Wildlife Service (1992).
[2]Median of 180 site means for total DDT, from BofCF–USEPA's National Pesticide Monitoring Program, 1965–1972.
[3]Median of 89 site means for DDE, from BofCF–USEPA's National Pesticide Monitoring Program, 1977.
[4]Median of 80 site means or site values for DDE, from USEPA's Mussel Watch, 1976–1978.
[5]Median of 145 site means for total DDT, NOAA's National Status and Trends Program (Mussel Watch Project), 1986.
[6]Median of 144 site means for total DDT, BofCF–USEPA's National Pesticide Monitoring Program, 1972–1976.
[7]Median of area or site means for total DDT, from area-weighted compilation of data from 31 areas sampled by Stout and coworkers, 1972–1975.
[8]Median of 19 site means for total DDT, from NOAA–USEPA's Cooperative Estuarine Monitoring Program and additional sampling by NOAA, 1976–1977.
[9]Median of 42 site medians for total DDT, from NOAA's National Status and Trends Program (Benthic Surveillance Project), 1984.
[10]Range in annual median for total DDT, from FWS's National Contaminant Biomonitoring Program, 1969–1971.
[11]Range in annual median for total DDT, from FWS's National Contaminant Biomonitoring Program, 1972–1974.
[12]Range in annual median for total DDT, from FWS's National Contaminant Biomonitoring Program, 1976–1977.
[13]Range in annual median for total DDT, from FWS's National Contaminant Biomonitoring Program, 1984–1986.

Heptachlor and Heptachlor Epoxide

As previously reported, heptachlor is rapidly metabolized to heptachlor epoxide by fish and other organisms. Environmental residues of heptachlor or its epoxide may derive from insecticidal use of either heptachlor or technical chlordane, of which heptachlor is a minor component (less than 10 percent).

The NCBP reported data for total heptachlor (the sum of heptachlor and heptachlor epoxide) residue in whole freshwater fish. Data analyzed prior to 1976 had been subject to analytical interference (Schmitt and others, 1983). Since 1976, the median concentration of total heptachlor has been below detection (less than 10 µg/kg; see Table 2.1). The geometric mean concentration of total heptachlor (10–20 µg/kg) did not change substantially from the period 1976–1977 to 1984 (Schmitt and others, 1990). The maximum total heptachlor concentration increased from 780 µg/kg during 1976–1977 to 1,170 µg/kg during 1978–1979, but has been below 300 µg/kg since the period 1980–1981. The percentage of sites with detectable residues has not declined consistently, varying between 24 and 38 percent for all sampling years. Despite the decline in maximum concentrations, low but stable geometric mean concentrations and detection frequencies suggest continued inputs to hydrologic systems.

Insufficient data are available to evaluate trends in total heptachlor residues at estuarine and coastal sites. Heptachlor and heptachlor epoxide were below detection limits in estuarine mollusks (1965–1972 and 1977) and in estuarine juvenile fish (1972–1976) sampled as part of the National Pesticide Monitoring Program (Butler, 1973b; Butler and others, 1978; Butler and Schutzmann, 1978). Heptachlor and heptachlor epoxide concentrations were near or below detection limits in estuarine fish liver (1984) sampled as part of the NS&T Program (Mearns and others, 1988).

Hexachlorobenzene

Data from the NCBP indicate that hexachlorobenzene residues in whole freshwater fish appeared to decline nationwide from 1976 to 1986. The percentage of sites with detectable residues decreased gradually from 45 percent during 1976–1977 to 7 percent in 1986 (Table 2.1). Geometric mean concentrations of hexachlorobenzene also declined from the period 1976–1977 to 1984 (Schmitt and others, 1990). The maximum hexachlorobenzene concentration varied from year to year, but the highest concentration (700 µg/kg) was observed in 1977.

Data on hexachlorobenzene residues in estuarine biota are insufficient to assess trends. In 1984, hexachlorobenzene was detected in about 60 percent of fish liver samples from NS&T sites, but most detections were at concentrations less than 10 µg/kg wet weight (Mearns and others, 1988).

Toxaphene

At NCBP sites, toxaphene was monitored in whole freshwater fish from 1967 to 1986, although results were not considered quantitative until 1972 (Schmitt and others, 1981). Interpretation of toxaphene trends is confounded by the analytical method change from packed-column to capillary-column gas chromatography in 1980 (Schmitt and others, 1990), as explained in Section 3.3.1 (Toxaphene). However, data collected from 1980–1981 onward were not influenced by any change in analytical method. Trends in toxaphene residues at NCBP sites from 1972 to 1986 are summarized in Figure 3.44. From 1972 to 1974, fewer than 8 percent of sites had detectable toxaphene residues, but at those sites, the concentrations tended to be quite high. Maximum concentrations of toxaphene during 1972–1974 ranged from 30,000 to 50,000 μg/kg (Table 2.1). Maximum toxaphene residues peaked in 1974, whereas geometric mean residues of toxaphene continued to increase until the period 1976–1977 and then declined (Schmitt and others, 1990). Similarly, the 90th-percentile concentration peaked during 1976–1977, whereas the median concentration peaked during 1980–1981 (as shown in Figure 3.44). Between the period 1980–1981 and 1986, toxaphene residues appeared to decline nationally on the basis of observed decreases in detection frequency, maximum, median, and geometric mean concentrations. These unusual patterns may be the result of three factors, as discussed in Section 3.3.1: (1) atmospheric deposition of toxaphene, (2) decreasing use of toxaphene in cotton-growing areas accompanied by small but increasing use on other crops during the late 1970s, and (3) the change from packed-column to capillary-column gas chromatography implemented in 1980. In summary, toxaphene residues in NCBP fish samples have declined nationally since about 1980. Residues at the most contaminated sites peaked earlier, around 1974.

California's Toxic Substance Monitoring Program detected toxaphene, which was widely used in that state, in 22 percent of fish sampled statewide from 1978 to 1987 (Rasmussen and Blethrow, 1990). All detections were within five of nine geographical regions within the state: the Central Coast, Los Angeles, Santa Ana, Central Valley, and Colorado River Basin regions.

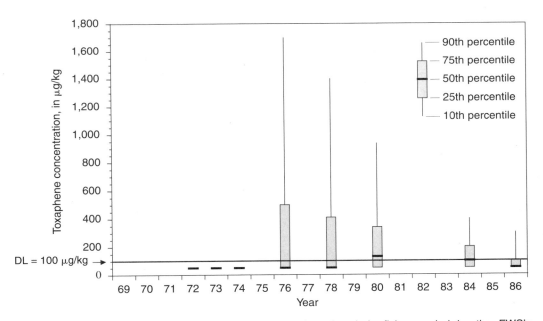

Figure 3.44. Temporal trends in toxaphene concentrations in whole fish sampled by the FWS's National Contaminant Biomonitoring Program from 1972 to 1986. Concentrations are in μg/kg wet weight. Data are from U.S. Fish and Wildlife Service (1992). Abbreviations: DL, detection limit; FWS, Fish and Wildlife Service; μg/kg, micrograms per kilogram.

Consistently high residues were found in the San Joaquin River. There was no evidence to suggest a decrease in the detection frequency of toxaphene statewide from 1978 to 1987, even though the use of toxaphene in California decreased from over one million lb a.i. in 1978 to less than 1,000 lb a.i. in 1987 (Rasmussen and Blethrow, 1990).

Results of coastal and estuarine studies suggest that toxaphene residues in biota have declined nationally, but that concentrations may remain high at selected hot spots, such as in Georgia, Texas, and California (Mearns and others, 1988). In national studies, toxaphene was monitored in estuarine mollusks and whole juvenile fish as part of the National Pesticide Monitoring Program. In estuarine mollusks, toxaphene was detected (greater than 250 µg/kg) at only 9 percent of sites sampled during 1965–1972. The contaminated sites were located in Georgia, Texas, and California. Concentrations at these sites were high; for example, the maximum concentration in oysters was 54,000 µg/kg (Butler, 1973b). However, toxaphene was not detected (less than 50 µg/kg) when selected sites were resampled in 1977. The resampled estuaries included at least some sites in Georgia and Texas where toxaphene had been detected in the previous 12-month sampling period (Butler and others, 1978), indicating that toxaphene residues in estuarine mollusks had declined at those sites. In whole juvenile estuarine fish sampled during 1972–1976, toxaphene was detected in less than 1 percent of samples nationwide. All detections (greater than 50 µg/kg) were in Louisiana, Mississippi, and Texas (Butler and Schutzmann, 1978). In addition, several local studies reviewed by Mearns and others (1988) measured high toxaphene concentrations (200–23,000 µg/kg wet weight) in estuarine biota. Examples include mussels from San Francisco Bay and freshwater bivalves from Salinas River–Elkhorn Slough and Mugu Lagoon sampled during 1984–1986 (Stephenson and others, 1986); estuarine fish from the San Francisco Bay and Delta, California, sampled during 1978–1980 (Mearns and others, 1988); estuarine fish from the Arroyo Colorado and Rio Grande drainages in Texas sampled during 1974–1979 (White and others, 1983); and mollusks, crustaceans, and fish sampled during 1974 from seven estuaries near Brunswick, Georgia, that were contaminated with toxaphene from industrial wastes (Reimold and Durant, 1972, 1974).

3.4.2 CURRENTLY USED PESTICIDES

Relatively few currently used pesticides were analyzed in bed sediment or aquatic biota by national monitoring programs. These few include two of the less persistent organochlorine insecticides that were used in agriculture in the United States during 1988 (lindane and methoxychlor), as well as a few pesticides in other chemical classes.

Bed Sediment

The only national data on currently used pesticides in bed sediment are for rivers sampled as part of the PMN during the late 1970s. There are no national data available for estuarine sediment, since NS&T sediment sampling efforts for pesticides were limited to organochlorine insecticides. The PMN determined seven organophosphate insecticides, three chlorophenoxy acid herbicides, and one triazine herbicide in bed sediment of major United States rivers during 1975–1980. Of these, only two organophosphate insecticides (ethion and diazinon) and three chlorophenoxy acid herbicides (2,4-D, 2,4,5-T, and silvex) were detected in bed sediment.

Detection frequencies in sediment were all low (less than 2 percent of sites, and less than 1 percent of samples). These detections were too few to permit testing of trends. Two moderately persistent organochlorine insecticides (lindane and methoxychlor) used in agriculture in 1988 also were analyzed in sediment. Both of these compounds were detected at less than 1 percent of sites and in 0.1 percent of samples. These detection frequencies were much lower than those seen in the same samples for the more persistent organochlorines, such as DDT, chlordane, and dieldrin (detected at 20–30 percent of sites), and even aldrin (detected at 3 percent of sites). No significant trends were found for lindane or methoxychlor (Gilliom and others, 1985).

Aquatic Biota

For aquatic biota in rivers, long-term national data for currently used pesticides are available only for two pesticides (lindane and methoxychlor) from one national study (the NCBP). The NCBP also targeted one additional currently used pesticide (dacthal) and one pesticide metabolite (pentachloroanisole). However, these analytes were added to the target analyte list fairly recently, so that long-term data from this program on these compounds are not available. Several currently used pesticides were analyzed as part of the NSCRF, but this was a one-time survey (1986–1987), and so it did not provide information on long-term trends. The available information on long-term trends in residues of lindane, methoxychlor, dacthal, and pentachloroanisole is discussed in the following sections.

Lindane

As noted previously, the detection of lindane in environmental samples indicates recent inputs because lindane (γ-HCH) is shorter-lived than most organochlorine insecticides (Schmitt and others, 1990). In NCBP whole freshwater fish samples, the occurrence of lindane decreased in frequency and magnitude between the period 1976–1977 and 1978–1979, then remained fairly constant through 1984 (Schmitt and others, 1990). Residues were very close to the detection limit (10 µg/kg) after 1977.

Data for estuarine and coastal areas are insufficient to evaluate trends in lindane contamination. Lindane was not detected in any mollusk samples collected during 1965–1972 or in 1977 as part of the National Pesticide Monitoring Program (Butler, 1973b; Butler and others, 1978). In estuarine fish livers, lindane was detected at 47 percent of sites sampled in the NS&T Program during 1984 (Mearns and others, 1988). However, the detection limit in the NS&T Program was lower than in the earlier National Pesticide Monitoring Program.

Methoxychlor

There are insufficient data to assess national trends in methoxychlor in either freshwater or estuarine biota. Methoxychlor was a target analyte in the NCBP only during 1980–1981. It was detected at low concentrations (less than 190 µg/kg) in whole freshwater fish at 32 percent of sites (Schmitt and others, 1985).

Methoxychlor was not detected in measurable concentrations in any estuarine mollusk samples collected during 1965–1972 or in 1977 as part of the National Pesticide Monitoring Program (Butler, 1973b; Butler and others, 1978). A few local studies have reported detections of methoxychlor in California and in the Chesapeake Bay (Stephenson and others, 1986; Mearns and others, 1988).

Dacthal

Dacthal was determined in whole freshwater fish as part of the NCBP from 1978–1979 to 1986 (Table 2.1). The limited data available do not indicate a national decline in dacthal contamination at these freshwater sites. Geometric mean residues were at the detection limit (10 μg/kg) during 1978–1979 and have been below detection since then (Schmitt and others, 1990). Median concentrations also were below detection during every sampling year (Table 2.1). The maximum concentration was substantially higher during 1978–1979 (1,200 μg/kg) than in later sampling years (180–450 μg/kg). However, the frequency of dacthal detection increased from the period 1978–1979 (detected at 33 percent of sites and in 19 percent of samples) to 1986 (detected at 60 percent of sites and in 32 percent of samples).

Pentachloroanisole

Pentachloroanisole, a metabolite of the wood preservative pentachlorophenol, was determined in NCBP whole freshwater fish samples beginning in 1980. The percentage of NCBP sites with detectable pentachloroanisole residues rose slightly from the period 1980–1981 to 1984 (24 to 30 percent), as did the percentage of samples with detectable residues (17 to 20 percent) and the maximum residues observed (70 to 100 μg/kg wet weight). In 1986, site and sample detection frequencies and maximum concentrations all returned to approximately the same levels observed during 1980–1981. Geometric mean (Schmitt and others, 1985, 1990) and median (Table 2.1) pentachloroanisole concentrations were below detection during each sampling year. In the NSCRF, pentachloroanisole was detected in 64 percent of whole bottom fish and game fish fillet samples (1986–1987). These studies provide insufficient data to evaluate trends, but they suggest that additional sampling for pentachloroanisole in rivers and estuaries may be warranted.

CHAPTER 4

Governing Processes

An understanding of the occurrence and distribution of pesticides in bed sediment and aquatic biota requires consideration of pesticide sources, transport processes, and mechanisms of transformation and removal from bed sediment and aquatic biota. In general, the movement of a pesticide from the point of application to bed sediment and aquatic biota is controlled first by processes that deliver the pesticide to the stream, and then by processes that deliver the pesticide from the water column to the bed sediment or biota. Once in the bed sediment or aquatic biota, environmental processes continue to act upon the pesticide. The environmental processes that govern the behavior and fate of a pesticide in surface water can be classified into three types: (1) phase transfer processes, which control its movement among environmental compartments, such as water, biota, suspended sediment, bed sediment, and the atmosphere; (2) transport processes, which move it away from its initial point of introduction to the environment and throughout the surface-water system; and (3) transformation processes, which change its chemical structure. In general, the short-term behavior and long-term fate of a pesticide in surface-water systems are controlled by the physical, chemical, and biological properties of the pesticide (which in turn are determined by its chemical structure) and by the environmental conditions in the hydrologic system.

The following chapter is an overview of the sources of pesticides to bed sediment and aquatic biota, and processes underlying the short-term behavior and long-term fate of pesticides in bed sediment and aquatic biota. Each section of this chapter addresses a subject that has been studied and written about extensively. An exhaustive review of this literature is not attempted here. Instead, the objective of this chapter is to present the major concepts and terminology that will assist some readers with the remainder of the book. For more detailed or mathematical treatment of these topics, appropriate review articles and books are cited in the text.

4.1 PESTICIDE SOURCES

Sources of pesticides to bed sediment and aquatic biota are summarized in the following sections. First, sources to surface water systems (Section 4.1.1) are discussed, followed by pathways by which pesticides reach bed sediment (Section 4.1.2) and aquatic biota (Section 4.1.3).

4.1.1 SOURCES TO SURFACE WATER SYSTEMS

Pesticides are purposely introduced into the environment for numerous reasons and by various methods. Pesticides are commonly used in agriculture, forestry, transportation (for roadside weed control), urban and suburban areas (for control of pests in homes, buildings, gardens, and lawns), lakes and streams (for control of aquatic flora and fauna), and various industries. A fraction of the pesticides used for all of these purposes is transported from the site of application and enters the broader environment, where it is perceived as an environmental contaminant, rather than as a useful chemical. In addition to pesticide applications, pesticides may enter the environment in waste streams (such as from manufacturing or formulation plants) or in accidental discharge (such as spills or leakage). Some of the pesticide contamination may enter surface waters directly, whereas some fraction first enters the atmosphere, soil, or ground water with the potential for eventual transport to surface water. Various sources of pesticides to surface waters, and the processes involved in transport of pesticides from each source to surface waters, are reviewed in detail in Larson and others (1997). The remainder of Section 4.1.1 draws from this review and summarizes information on sources of pesticides to surface water systems. This subject is a necessary precursor to discussion of factors that affect movement of pesticides within surface water systems to bed sediment and aquatic biota

Agriculture

Agricultural use of pesticides constitutes the major source of most pesticides to surface waters (Larson and others, 1997). In 1995, agricultural use accounted for almost 80 percent of total pesticide use in the United States (Aspelin, 1997). The pesticides used vary tremendously in chemical structure, application rate, and their potential for movement to surface waters. Pesticide input from an agricultural field to surface water generally occurs through runoff or drainage induced by rain or irrigation. However, a given pesticide also may enter the atmosphere (in vapor form) or it may enter ground water, the vadose zone, or surface water (in dissolved form). Runoff of pesticides from agricultural fields can occur via overland flow, interflow (water that enters the shallow subsurface and then returns to the soil surface), and flow through tile-drainage networks. Generally, the water is routed to drainage ditches or natural topographic drains and ultimately to a surface water system (Larson and others, 1997).

The potential for movement of pesticides from agricultural areas to surface waters is a function of environmental conditions, agricultural management practices, and properties of the pesticide itself. Leonard (1990) summarized the following four dominant factors that affect pesticide transport in runoff: (1) climate, including the duration, amount, and intensity of rainfall; the timing of rainfall with respect to pesticide application; and rainwater temperature; (2) soil characteristics, including soil texture and organic matter content, surface crusting and compaction, water content prior to rainfall, slope and topography of the field, and degree of soil aggregation and stability; (3) the physical and chemical properties of the pesticide itself, such as water solubility, acid or base properties, ionic properties, sorption properties, and persistence; and (4) agricultural management practices, which include pesticide formulation, application rate, application placement (soil surface, soil incorporation, or foliar), erosion control practices, plant residue management, use of vegetative buffer strips, and irrigation practices.

Pesticides removed from a treated agricultural area in runoff constitute only a small percentage of the total application of the compound. In their reviews of the literature, Wauchope

(1978) and Leonard (1990) concluded from field plot studies that normal runoff losses are approximately 1 percent of application for foliar-applied organochlorine insecticides, 2 to 5 percent of application for pesticides formulated as wettable powders, and less than 0.3 percent of application for the remaining pesticides. Nonetheless, runoff from agricultural fields is the major source of the pesticide load to most surface waters that are in proximity to, or downstream from, agricultural areas (Larson and others, 1997).

Forestry

Pesticide use in forested land is small relative to agricultural use, both in terms of the mass of pesticide applied and the acreage involved (Larson and others, 1997). Nonetheless, pesticide application in forests is important for several reasons. Forested lands are often relatively pristine and highly valued for their aesthetic and recreational uses; they serve as a habitat for wildlife, and support a number of important fisheries. Many of the national parks and wilderness areas in the United States border forested land that may be treated with pesticides. Also, the headwaters of most of the nation's major river systems are in forested areas (Larson and others, 1997).

Certain characteristics of the forest environment affect the movement of pesticides to surface waters. Forested land in the United States generally has higher slope and receives more precipitation than agricultural land. Forests also tend to have relatively shallow soils with high infiltration capacity, low pH, and high organic carbon content (Larson and others, 1997). The presence of year-round vegetation in forests is another important difference from agricultural land (Norris, 1981; Norris and others, 1982). Pesticides applied in forests may reach surface waters by several different routes. Direct input to streams or lakes can result from aerial application of liquid or pellet formulations, or from spray drift from ground spraying. Pellets or liquids that are aerially applied to dry ephemeral stream channels have a high probability of entering surface waters, especially if rainfall occurs shortly after application. Pesticides may move in overland flow (surface runoff) or in subsurface (downslope) flow to surface waters (Larson and others, 1997).

Urban and Suburban Areas

Pesticides are introduced into the urban environment in a variety of ways (Larson and others, 1997). Herbicides, insecticides, and fungicides are applied to lawns and gardens, golf courses, cemeteries, and some parks. Insecticides are applied to building foundations and the surrounding soil to control termites or other destructive insects. In some parts of the United States, insecticides are applied to control mosquitoes for both public health and aesthetic purposes. Some lakes and reservoirs in urban areas are treated with herbicides for control of algae or undesirable weeds (such as Eurasian watermilfoil). Control of specific insect pests for agricultural purposes (such as medfly control in California) has included aerial spraying of insecticides in urban areas (Larson and others, 1997).

Processes affecting the movement of pesticides to surface waters in urban areas are the same as in agricultural areas, but there are some important differences between the two environments that can affect this movement (Larson and others, 1997). Urban areas have large expanses of impermeable surfaces, such as concrete and asphalt roads and sidewalks, over which pesticides can be easily transported by runoff water, with little or no loss from sorption. Pesticides applied in urban areas may reach impervious surfaces via spray drift, direct aerial application, or runoff from gardens, lawns, and other turf areas. Once a pesticide reaches these impervious surfaces, there is a relatively high probability that it will be transported to surface water bodies, compared with a pesticide applied to an agricultural field. In studies with turf plots, it has been found that very little runoff of water occurs from well-maintained grass, even with large amounts of precipitation (Harrison and others, 1993). So, at least for applications to lawns, the limiting step may be the reaching of impervious surfaces (Larson and others, 1997). Storm sewer systems also provide a direct pathway for movement of pesticides to lakes or rivers. Effluents from sewage treatment plants often flow directly into rivers. These effluents may contain pesticides, particularly in urban areas in which storm sewers and sanitary sewers are combined, although some pesticides may be removed from water during treatment at the plant (Larson and others, 1997).

Industrial Waste and Accidental Discharge

All manufactured pesticides have the potential to be released into the environment as part of an industrial waste stream. Several studies have attributed the presence of certain pesticides in surface waters to manufacturing waste disposal (Larson and others, 1997). During the 1970s, the occurrence of the insecticide mirex in the water, bed sediment, and biota of Lake Ontario was attributed either to inputs from the manufacturing waste stream or to disposal of unused chemical from a secondary industrial user (Kaiser, 1974; Scrudato and DelPrete, 1982). The presence of mirex in Lake Ontario could not be explained on the basis of any legitimate agricultural use of the compound, since it had a narrow registration and was primarily used to control fire ants in the southeastern region of the United States. Oliver and Nicol (1984) observed constant low-level inputs of organochlorine compounds to the Niagara River over a 2-year time period, presumably from waste disposal sites; they also observed a number of erratic concentration spikes, probably indicating direct discharges from industrial sources.

Residues of pesticides or pesticide components in fish collected at a few sites from the Fish and Wildlife Service's (FWS) National Contaminant Biomonitoring Program (NCBP) were attributed to contamination from industrial spills, manufacturing, or chemical storage (Schmitt and others, 1990). Examples are high o,p'-DDT (suggesting a source other than insecticidally applied DDT) at sites near previously identified manufacturing or chemical storage point sources, and high *trans*-chlordane concentrations (relative to other chlordane components) at one site on the Mississippi River near a pesticide manufacturing site. Fish from the latter site also contained residues of endrin (Schmitt and others, 1990). Fish from the Mississippi River were reported to contain precursors to cyclodiene insecticides (Yurawecz and Roach, 1978).

There also are reports of pesticide inputs from manufacturing facilities in other countries, which may be looked at as examples of potential problems in the United States (Larson and others, 1997). One example is the accidental discharge of the insecticide endosulfan into the Rhine River in 1969, where it accumulated in bed sediment (Greve and Wit, 1971). In 1986, about 1.5 metric tons of pesticides were discharged into the Rhine River because of a fire at a chemical manufacturing facility in Switzerland (Capel and others, 1988). Wherever there is manufacturing or storage of pesticides near surface water bodies, the possibility of direct inputs of waste and of spill discharges exists (Larson and others, 1997). Such discharges are sometimes a significant acute local environmental problem; their significance as a source of frequent or long-term low-level inputs to surface water systems is unknown.

Atmospheric Deposition

Another route of pesticide movement to surface water is through atmospheric deposition. Many pesticides have been observed in various atmospheric matrices (air, aerosols, rain, snow, and fog). In their review of pesticides in the atmosphere, Majewski and Capel (1995) reported that 63 pesticides and pesticide transformation products have been identified in atmospheric matrices. The extent to which a pesticide enters the atmosphere is a function of its physical and chemical properties, the application method, and the pesticide formulation. In general, those pesticides that are more volatile (as indicated by a high Henry's law constant) and those that are applied aerially have a greater chance of entering the atmosphere (Majewski and Capel, 1995). Many of the pesticides observed in bed sediment and tissues (such as the organochlorine insecticides) are also routinely observed in the atmosphere.

Pesticides enter the atmosphere through a variety of processes both during and after application. If a pesticide is applied aerially or by ground-based spraying, some fraction of the pesticide is likely to remain in the atmosphere (Larson and others, 1997). Pesticides sorbed to soil can enter the atmosphere through volatilization (also called *vapor desorption*) or through wind erosion of soil particles. Other routes of entry to the atmosphere include air-water exchange from surface waters into the air and direct release by plant surfaces. Once in the atmosphere, a pesticide can be transported by wind currents, undergo photochemical and hydrolytic degradation, and be deposited to aquatic and terrestrial surfaces (Larson and others, 1997). The movement of pesticides from the atmosphere to surface waters can occur by either wet deposition (which involves precipitation) or dry deposition (which does not). Pesticides in the vapor phase may partition into water droplets or sorb onto atmospheric particles. Pesticides associated with atmospheric particles may undergo gravitational settling or be scavenged by water droplets. Both dry and wet deposition of pesticides may occur directly to surface water systems, or indirectly after deposition to soil or plant surfaces and subsequent runoff to surface waters (Majewski and Capel, 1995).

Although atmospheric deposition of pesticides occurs globally (discussed further in Section 5.1), the relative importance of atmospheric inputs of pesticides to a particular surface water body is directly dependent on the magnitude of the other sources of pesticides to that water body (Larson and others, 1997). An example is the Great Lakes. Strachan and Eisenreich (1990) estimated that atmospheric deposition comprises more than 97 percent of the total DDT burden in Lakes Superior, Huron, and Michigan, but only 22 percent in Lake Erie and 31 percent in Lake Ontario, both of whose lake basins are more heavily agricultural.

4.1.2 SOURCES TO BED SEDIMENT

The Water Column

Pesticides may be in the dissolved phase or in association with soil particles when they enter the surface water column. Regardless of how it enters the water, a pesticide will redistribute itself between the water and aquatic (suspended) particles in the water column. The distribution of a pesticide between the water (aqueous phase) and aquatic particles (sorbed phase) is controlled by phase-transfer processes that include nonspecific sorption and specific charged interactions. The particle-associated pesticide then can be deposited to bed sediment.

For the most part, the presence of pesticides in the bed sediment of streams is due to sedimentation of pesticide-contaminated particles and subsequent accumulation of those particles in long-term depositional areas. These phase-transfer and transport processes are discussed in Section 4.2. For more technical detail, see reviews by Karickhoff (1984), Lick (1984), Eadie and Robbins (1987), Elzerman and Coates (1987), Smith and others (1988), Lyman and others (1990), Schwarzenbach and others (1993), and Chiou (1998).

Aquatic Biota

Contaminated biota constitute a source of pesticides to bed sediment when (1) dead and living biotic particles (such as algae and detritus) containing pesticides settle to the sediment–water interface; (2) higher organisms die and settle to bed sediment, where they decompose; and (3) excretions (such as fecal material) containing pesticide contaminants are released in, or settle to, bed sediment.

Ground Water

A description of the movement of pesticides between ground water and surface water (through bed sediment) is provided in a companion review on pesticide occurrence in ground waters of the United States (Barbash and Resek, 1996). In rivers adjacent to alluvial aquifers, the water and pesticides in the water can move from the stream through the bed sediment to the aquifer during periods of high flow (Squillace and others, 1993). During periods of low flow in the river, pesticide movement tends to be reversed, and the input of pesticides from the alluvial aquifer to the river becomes significant and perhaps dominant (Barbash and Resek, 1996). For example, Squillace and others (1993) observed a seasonal pattern for herbicide movement to and from the Cedar River in Iowa. Herbicides moved from the river to the alluvial aquifer during the spring (high flow period), then from the alluvial aquifer to the river during the autumn and winter.

Pesticides in an alluvial aquifer can move through the bed sediment to enter an adjacent river. Nonetheless, pesticide inputs to rivers from ground water probably do not constitute a significant source of the pesticides that are found in bed sediment. A pesticide must have certain physical, chemical, and biological characteristics for ground water inputs to be a significant source of that pesticide to surface waters (Barbash and Resek, 1996). The pesticide must be relatively water soluble and have low affinity for solid surfaces. These characteristics allow the pesticide to enter and readily move through the ground water system. The pesticide must also have relatively slow transformation rates, with a residence time in ground water of at least a few months. Given these constraints, relatively few pesticides have a strong potential to be delivered in appreciable quantities from ground water to surface waters. The most common example in the mid-continental United States is atrazine (Squillace and Thurman, 1992; Squillace and others, 1993; Schottler and others, 1994). Some of these same physical and chemical properties (in particular, high water solubility and low sorption tendency) work against accumulation of a pesticide in bed sediment. Atrazine was a target analyte in bed sediment in relatively few monitoring studies (listed in Tables 2.1 and 2.2). However, of the 442 total sediment samples that were collected from a total of 168 study sites and analyzed for atrazine in the monitoring studies reviewed, atrazine was not detected in a single sample.

4.1.3 SOURCES TO AQUATIC BIOTA

The Water Column

A pesticide that enters the surface water column will redistribute itself between the water and carbon-rich compartments (such as sediment and biota) in the water column. Uptake by biota occurs via (1) direct uptake from water, (2) ingestion of contaminated food or other suspended particles, or (3) sorption directly from sediment. Their relative importance appears to depend on the concentration of the pesticide in water, the species of organisms in the food web, and the physical and chemical properties of the pesticide (see Section 5.2.5). Direct uptake of pesticides from water is controlled by the phase-transfer processes of sorption and diffusion (see Section 4.3.1). Sorption may constitute the first step in bioaccumulation and is particularly important for microorganisms. Diffusion occurs through the semipermeable membranes in an organism, such as the gills, the lining of the mouth, and the gastrointestinal tract. For small organisms, diffusion may occur through the integument. A number of reviews discuss uptake of contaminants from the water column; for example, Biddinger and Gloss (1984), Spacie and Hamelink (1985), Connell (1988), and Gobas and Russell (1991).

Bed Sediment

A pesticide that reaches the sediment–water interface will distribute itself among available compartments: sorbed phase (sediment), pore water, and overlying water. Sediment-sorbed contaminants may be transferred to biota via three probable pathways: (1) pore water, (2) ingested sediment (organic and inorganic), and (3) direct contact with sediment particles (Knezovich and others, 1987). The relative importance of these pathways is a subject of some controversy (see Section 5.2.5, Uptake of Sediment-Sorbed Chemicals), but appears to depend on the type of sediment, species of organism, and physical and chemical properties of the contaminant. The process of diffusion is important for direct uptake from pore water and after ingestion of sediment. The sorption and desorption processes are important for uptake via direct contact with sediment. These phase-transfer processes are described in Section 4.3.1. Additional detail is provided in reviews by Adams (1984) and Knezovich and others (1987).

4.2 BEHAVIOR AND FATE OF PESTICIDES IN BED SEDIMENT

Once a pesticide reaches bed sediment, it can undergo a number of processes that will determine its short-term behavior and long-term fate. Generally, a hydrophobic pesticide will arrive at the sediment–water interface in association with some type of particulate matter. In the particle-rich environment of bed sediment, where the concentration of solids is on the order of 10^6 mg/L (compared with 10^1 to 10^3 mg/L in the water column), sorptive processes are critical to the overall behavior and fate of a pesticide. Once a pesticide is delivered to the sediment–water interface, some fraction of that pesticide will undergo desorption into pore water or overlying water as the pesticide attempts to reequilibrate between water and sediment in its new environment. In fact, the particle with which it was associated when it arrived at the sediment–water interface is likely to be changed by microbiological activity and by the sedimentation process. In the sedimentation process, particulate organic constituents are fractionated, and the more polar, water-soluble components are removed to form dissolved organic matter and colloids (Chiou, 1998). Therefore, much of the particulate organic carbon that is delivered to the sediment–water interface is decomposed and reintroduced back into the water column as either dissolved organic carbon or mineralized carbonate species (Cole, 1983; Chiou, 1998). The sorption–desorption cycle of the pesticide continues throughout its lifetime in the bed sediment as its environment continues to change. The physical location of sediment particles, together with their associated pesticides, also is likely change. After initial deposition, the particle may be buried with fresh sediment, transported downstream as part of the bed load, resuspended and reintroduced back into the water column, or ingested by aquatic biota near or in the bed sediment. Moreover, the pesticide itself may be chemically or biologically transformed into other organic (and inorganic) compounds. Altogether, the interaction of multiple processes acting on a bed sediment particle may result in a number of possible fates for that particle and any associated pesticides. Individual phase-transfer, transport, and transformation processes that affect pesticides in bed sediment are described in Sections 4.2.1, 4.2.2, and 4.2.3, respectively.

4.2.1 PHASE-TRANSFER PROCESSES

Phase-transfer processes operating on contaminants in bed sediment and nearby water (pore water and overlying water) include sorption and desorption, and various charged interactions. Additional detail is provided in reviews by Weber (1972), Karickhoff (1984), Eadie and Robbins (1987), Elzerman and Coates (1987), Smith and others (1988), Lyman and others (1990), Schwarzenbach and others (1993), Chiou (1998), and Rathbun (1998).

Sorption and Desorption

Sorption has been defined as any accumulation of a dissolved organic chemical by solid particles (Voice and Weber, 1983). There is some disagreement in the literature as to the relative importance of *adsorption* (the formation of a physical or chemical bond between the chemical and the solid surface) and *partitioning* (the distribution of the chemical between the water and organic matter associated with the solid) (Rathbun, 1998). However, for nonionic organic chemicals in aqueous systems, sorption to particulates is expected to be predominantly a partitioning process (Chiou and others, 1979; Smith and others, 1988; Rathbun, 1998). A simple, mass-based, solid–water distribution coefficient (K_D) can be used to quantify the extent to which a pesticide is distributed between the aqueous and particulate phases. This is often defined as the concentration of the pesticide sorbed to sediment (C_s, in units of pesticide mass per dry weight sediment mass) divided by the concentration of the pesticide in the water (C_w, in units of pesticide mass per volume):

$$K_D = C_s / C_w \qquad (4.1)$$

The fraction of organic carbon in a particle is very important in determining the extent of sorption that many pesticides will undergo in that specific environment. In many cases, nonionic pesticides are sorbed primarily to the organic fraction of the sediment (Karickhoff and others, 1979; Wolfe and others, 1990). For this reason, an organic carbon normalized distribution coefficient, K_{oc}, has been defined and used extensively in the literature. K_{oc} is defined as:

$$K_{oc} = K_D/f_{oc} \qquad (4.2)$$

where f_{oc} is the fraction of organic carbon of the particle. K_{oc} is often used as a generic measure of the degree to which a particular pesticide will sorb in natural waters and in soil–water systems (Kenaga and Goring, 1980; Howard, 1991; Wauchope and others, 1992). In fact, normalizing the sorption coefficient with the fraction of organic carbon removes much, but not all, of the variability observed for sorption of a given compound on different soils and sediments (Kenaga and Goring, 1980; Lyman, 1990; Rathbun, 1998).

The extent of sorption of a nonionic organic compound to an aquatic particle is determined principally by the chemical structure of the compound and the nature of the particle. Sorption also may be influenced by certain environmental conditions, such as temperature, pH, and ionic strength. Organic compounds interact with water to varying degrees depending on their chemical structures. The most important structural attributes of organic compounds are their molecular size and the number and types of functional groups. A hydrophobic pesticide displays a stronger affinity for carbon-containing environmental phases (such as sediment particles or biota) than for water. As a result, the extent of sorption of many nonionic organic pesticides is inversely related to its subcooled water solubility and directly related to K_{ow}, the n-octanol-water partition coefficient (a measure of a chemical's hydrophobicity). A number of structure-activity relations have been proposed for the sorption of organic chemicals in water as a function of either water solubility or K_{ow}. These relations are based on linear regression of data from laboratory experiments using a single type of aquatic particle, so that the only variable is the structure of the chemical (see reviews by Kenaga and Goring, 1980; Karickhoff, 1984; Lyman, 1990; Schwarzenbach and others, 1993; Chiou, 1998).

The second important variable in determining the extent of sorption of hydrophobic organic compounds to aquatic particles is the composition of the particles. The most important attribute of the particles is their organic carbon content (Karickhoff and others, 1979; Schwarzenbach and others, 1993). Particle size and surface area also have been shown to be important. In many natural waters, the organic carbon content of the bed sediment is inversely related to particle size because the natural organic carbon accumulates on the finer-grained sediment. Surface-water particles contain variable amounts of organic carbon, ranging from highly organic particulates, which consist primarily of phytoplankton and bacteria, to primarily inorganic sediment with a thin organic coating (Allan, 1989). For particles deep in bed sediment or in water columns with very low productivity, the fraction of organic carbon (f_{oc}) may be less than 1 percent, but for particles that are (or very recently were) living, the f_{oc} may be on the order of 40 percent. The organic carbon in aquatic particles provides a thermodynamically favorable environment, compared with the water itself, for hydrophobic compounds. The larger the f_{oc} of an aquatic particle, the more favorable an environment it provides to a hydrophobic compound. Studies have shown that in natural waters, pesticides with low water solubility tend to follow the particulate organic carbon in the hydrologic system (Eisenreich and others, 1981; Halfon, 1987). Structure–activity relations for sorption as a function of water solubility (or K_{ow}) have been modified to include the effects of sediment organic carbon (Schwarzenbach and Westfall, 1981).

Environmental conditions that may influence sorption include temperature, pH, and ionic strength. Because adsorption is an exothermic process, values of K_D (or K_{oc}) should decrease with increasing temperature (Lyman, 1990). However, studies on the effect of temperature on sorption have yielded somewhat contradictory results. Examples are provided in Chiou and others (1979) and in reviews by ten Hulscher and Cornelissen (1996) and Rathbun (1998). The various results found in the existing literature probably occur because sorption consists of both adsorption and partitioning to a degree that depends on both the structure of the hydrophobic compound and the nature of the particle (Mingelgrin and Gerstl, 1996; Rathbun, 1998).

In the pH range that is considered normal for surface waters (pH 5–9), the sorption of nonionic organic compounds is not affected very much by pH (Lyman, 1990). Similarly, changes in ionic strength or salinity have little effect on the sorption of nonionic organic compounds, although they can significantly affect the sorption of charged or ionizable molecules (Pionke and Chesters, 1973; Rathbun, 1998). However, large changes in salinity (such as by a factor of two) can modify the structure of the colloids or particulates to which a compound is sorbed, causing a shift in K_{oc} and a corresponding change in the extent of sorption (Means and Wijayaratne, 1982).

Equations 4.1 and 4.2 are equilibrium expressions in which C_s and C_w refer to concentrations at equilibrium of a given compound in the sorbed phase and in water, respectively. It is commonly assumed that equilibrium is rapid and reversible. However, some experimental studies show that the attainment of equilibrium may be slow or that sorption may proceed in stages, with rapid sorption occurring during the first few minutes, followed by slow continued sorption over a period of days to weeks (Kenaga and Goring, 1980; Rathbun, 1998). Moreover, there is laboratory evidence that desorption coefficients may be smaller than sorption coefficients and that desorption may have two distinct stages (slow and fast), with the slow stage reflecting the desorption of more strongly bound molecules (Rathbun, 1998). Kinetic models of the sorption–desorption process (Wu and Gschwend, 1986) have been developed and are discussed in more detail in Schwarzenbach and others (1993) and Rathbun (1998).

Charged Interactions

Pesticides or pesticide transformation products that are ionizable or strongly polarizable have other mechanisms of interaction with aquatic particles besides the hydrophobic sorption discussed in the previous section. Ionizable compounds, which exist as anions or cations in water or bed sediment in certain pH and salinity regions, can undergo specific interactions with charged groups on the particle surface. These types of sorptive interactions do not follow the simple water solubility–K_{oc} relations described above. Other chemical–solid interactions include ion exchange (discussed below), hydrogen bonding, surface complexation (charge transfer), and covalent bonding to organic materials on the particle surface (chemisorption).

For compounds that may occur as more than one species in solution, the sorptive interactions of each of the species must be considered (Schwarzenbach and others, 1993). For weak acids (such as the herbicides 2,4-D and picloram), the neutral form of the molecule sorbs more strongly to the organic carbon in a sediment particle than the corresponding anion does (Lyman, 1990). The reason for this strong sorption is that the anion has stronger interactions with the water and weaker interactions with the particulate organic matter. Thus, environmental pH will affect both water solubility and the degree of sorption. If the solution pH is so high that the anionic species is much more abundant than the neutral species, then sorption of the anion cannot be neglected (Schwarzenbach and others, 1993). For organic bases, the dissociated form (cation) may sorb strongly to particles carrying a net negative charge (Lyman, 1990). Increasing salinity can significantly lower the sorption coefficient of bases that are in the cationic form. This may be caused by displacement of the cations from the particle, or by the lower activity of the compound as the ionic strength of the solution increases (Lyman, 1990).

For cationic pesticides such as paraquat and diquat, the most important type of interaction is *ion exchange* (in which cations added to a solid–water mixture replace some of the counter-ions, such as sodium ion, that surround the negatively charged surface of the aquatic particle). Because these cationic pesticides have high water solubilities (for example, 620,000 mg/L for paraquat), they might be expected to have minimal sorptive interactions in natural waters. However, because these pesticides can undergo specific ionic interactions, their sorption tendency is high. A nonionic organic compound with a water solubility comparable to that of paraquat would be expected to have a very small sorption coefficient ($K_{oc} \approx 2.8$, from Kenaga and Goring, 1980). However, paraquat actually has a measured K_{oc} value of 1,000,000 (Wauchope and others, 1992) because of its capability to undergo ion exchange.

Additional detail on ion exchange and other charged interactions is provided in Schwarzenbach and others (1993). The overall effect of these charged interactions is that certain hydrophilic pesticides will be strongly associated with aquatic particles and, therefore, may accumulate in bed sediment.

4.2.2 TRANSPORT PROCESSES

Transport processes acting on sediment and associated pesticides include settling (sedimentation), deposition, burial, resuspension, bioturbation, and downstream transport. Additional detail or mathematical treatment is provided in reviews by Pionke and Chesters (1973), Lick (1984), and Eadie and Robbins (1987).

Settling, Deposition, and Burial

Pesticides that are sorbed to suspended particulates are transported to the sediment–water interface as these particulates sink to the bottom of the lake, pond, or river. Particle settling times are a function of particle size and density, and the temperature, depth, and turbulence of the water body. In stagnant water, the velocity at which particles settle (v_s, in cm/s) varies with the square of the particle radius (r, in cm), as follows:

$$v_s = 0.22gr^2(d_p/d_w-1)/\,\eta_k \qquad (4.3)$$

where d_p is particle density (in g/cm^3), d_w is the density of water (in g/cm^3), η_k is water's kinematic viscosity (absolute viscosity divided by the density at the same temperature, in units of cm^2/s), and g is gravitational acceleration (in units of cm/s^2). For an average particle with density 2.6 g/cm^3, equation 4.3 becomes:

$$v_s = 34,400r^2 \qquad (4.4)$$

assuming d_w is 1.0 g/cm^3, η_k is 0.0101 cm^2/s at 20 °C, and g is 980 cm/s^2. So, for example, a sand particle with a radius of 100 μm would fall through 1 m of stagnant water at 20 °C in 29 seconds, while a clay particle with a radius of 0.5 μm would take 13.5 days to fall the same distance (Crosby, 1998). Even in deep systems like the Great Lakes, particle settling times are generally less than 1 year (Eadie and Robbins, 1987).

Equations 4.3 and 4.4 indicate that small particles (silts, clays, and colloids) are more likely to remain suspended than larger particles, especially in rivers where *advection* (transport with the current) and *turbulence* (rapid mixing) can affect particle settling. In rivers, these processes result in a sorting of bed sediment particles by size as the water moves downstream (Leopold and others, 1964). At least in the ideal case, bed sediment particle size tends to decrease downstream. The rate of decrease is relatively rapid in the headwater reaches, and much slower farther downstream, indicating that materials become fairly well sorted after a relatively short distance of transport. Of course, tributary entrances complicate the picture by introducing sediment of different origin and history into the main river (Leopold and others, 1964). Also, there are spatial variations in bed sediment particle size even within a given reach of the stream. For example, fine-grained sediment particles tend to accumulate and remain in depositional areas within the stream. Such depositional areas tend to occur in the low-energy zones of surface water systems, such as the deepest areas of lakes and reservoirs, the shallow back-water areas of streams, and behind dams in reservoirs or obstacles in streams.

Because long-term depositional areas contain mostly fine-grained sediment (silt and clay), which tend to be high in particulate organic carbon, this sediment may contain relatively high concentrations of sorbed hydrophobic contaminants. As fresh sediment continues to accumulate, a sorbed pesticide associated with previous accumulations will be slowly buried. While a pesticide resides in the upper few centimeters of bed sediment, it may be reintroduced into the water column either in its dissolved state (by desorption, followed by chemical diffusion) or in association with particles (via physical or biological mixing). Once a pesticide has been buried to a depth beyond this zone of mixing, at least in lakes, it probably has been removed from most possibilities of reentering into the water column (Eisenreich and others, 1989). In rivers prone to flooding or high seasonal flows, depositional areas may be scoured and sediment resuspended and carried downstream.

Resuspension, Bioturbation, and Downstream Transport

Pesticides that remain in the upper few centimeters of the bed sediment may comprise a long-term source of contamination to the water column through a variety of processes. Resuspension of bottom materials, driven by energy inputs into the system (such as strong wind-induced currents, lake turnover, large releases from reservoirs, and storms resulting in large water discharges), can erode the long-term sedimentation areas and move particle-associated pesticides into the water column. Also, the movement and feeding activities of benthic organisms in the upper layers of sediment (up to about 20 cm) mixes recently deposited sediment with older material. This process, referred to as *bioturbation*, moves buried particles and their associated pesticides back toward the sediment–water interface where they can more easily be resuspended. The depth of mixing depends on the organism. For example, in the Great Lakes (Eadie and Robbins, 1987), the mixed-layer zone at the sediment surface varies from a few centimeters (for the amphipod *Pontoporeia hoya*) to about 20 cm (for unionid clams). Thus, the depth of sediment affected by bioturbation tends to be shallower than that affected by physical erosion.

In rivers, pesticides associated with suspended sediment tend to be transported downstream. Transport of sediment may occur as suspended load (for fine-grained sediment) or as bed load (for coarser particles). Fine-grained suspended sediment may be redeposited and resuspended in and out of bed sediment until it reaches a long-term depositional area. Even these depositional areas may be scoured in rivers that are subject to high seasonal flows or flooding, causing the bed sediment to be resuspended and carried downstream.

Bed Sediment as a Long-Term Sink for Pesticides in Surface Water Systems

The processes of deposition and burial create a long-term environmental sink for pesticides that are hydrophobic and *recalcitrant* (resistant to degradation). The depositional zones of rivers make up some fraction of this sink, although many rivers are prone to catastrophic flooding that may mix even relatively deep sediment into the water column. From a global perspective, the ultimate sink is the bottom sediment of the world's oceans. Once the pesticide becomes buried in this sediment, it will probably remain resident there until it is transformed, regardless of the time span.

The presence of persistent, hydrophobic pesticides in bed sediment, and their consequent reintroduction into the water column, contribute to measurable concentrations in the water and biota of many surface water systems. Although some of the present organochlorine contamination of surface waters can be attributed to atmospheric deposition and fresh additions of historically contaminated soil particles, a large fraction can be attributed to the release of in-place pesticides in many surface water systems (Baker and others, 1985; Gilliom and Clifton, 1990). Because most in-place pesticides are widely distributed and present at low concentrations (milligram per kilogram in sediment), remediation usually is not practical. Bed sediment will continue to be a source of persistent, hydrophobic pesticides to surface waters, albeit at a slow and diminishing rate over time.

4.2.3 TRANSFORMATION PROCESSES

The transformation of a pesticide results in changes in its chemical structure—one or more new compounds are produced, while the original pesticide disappears. The new compounds formed can be organic or inorganic, and nonionic or ionic. From an environmental-effects point of view, the ideal fate for a pesticide is its complete *mineralization* (degradation to inorganic species such as water, carbon dioxide, and chloride ions). A pesticide may degrade via multiple pathways, sometimes simultaneously, with the relative importance of each pathway depending on environmental conditions. The initial transformation products may be subsequently trans-formed, so that a large number of transformation products can be formed. Some of these inter-mediate compounds may be recalcitrant (resistant to further degradation), and some may have pesticidal properties. Most organochlorine pesticides undergo fairly slow transformation and, when they do degrade, the products formed are long-lived and can accumulate in bed sediment. An example is the transformation of DDT to DDE and DDD.

Transformations may be mediated by chemical or biological means. In bed sediment, *abiotic* reactions (chemically induced, rather than mediated by any living organism) can occur, of which the most important are hydrolysis and oxidation–reduction reactions. Photochemical degradation is a transformation reaction induced by the energy from sunlight; it is important only in the water column or near the sediment–water interface in those hydrologic systems shallow enough for sunlight to penetrate. For contaminants in bed sediment, biotransformation (biologically mediated reactions, discussed below) is probably the most important transformation mechanism. Microorganisms can induce pesticides to undergo both hydrolysis and oxidation–reduction reactions. The transformation products from hydrolysis may be the same whether the reactions were biologically mediated or abiotic, although these products may be formed at different rates. The same is true for oxidation–reduction reactions.

Reactions generally take place in several integrated steps that, together, represent a reaction mechanism:

$$A + B \rightleftharpoons [AB] \rightarrow C + D \tag{4.5}$$

[AB] represents a transition state whose formation generally is the rate-limiting step in the reaction and requires the most energy. For abiotic reactions, this *activation energy* (E_a) is supplied by heat or light.

Most environmental reactions are more complex than simple unimolecular reactions. However, because the concentrations of common environmental reactants (such as water, hydronium ion, hydroxide anion, and dissolved organic matter [DOM]) tend to be so large that their concentrations remain virtually constant, many environmental reactions appear to follow *first order kinetics* (where the rate of degradation of a contaminant depends only on the concentration of that contaminant). In these cases, transformation reactions can be grouped together in a kinetic expression to describe the disappearance of a pesticide with an aggregated, *pseudo-first order* reaction rate constant. The reaction is called pseudo-first order because the reaction appears to be first order, but the rate constant actually incorporates the (unchanging) concentrations of one or more common environmental reactants. This is the approach generally used in interpreting or predicting the fate of pesticides in bed sediment. The rate of contaminant disappearance can be expressed as:

$$\ln (C / C_o) = -k_1 t \tag{4.6}$$

where C is the contaminant concentration at any given time (t), C_o is the initial concentration of the contaminant, and k_1 (in units of time^{-1}) is the pseudo-first order rate constant. One characteristic of pseudo-first order degradation is that the time required for half of the contaminant concentration to degrade, referred to as the *half-life* (or $t_{1/2}$), remains constant.

$$t_{1/2} = \ln2/k_1 = 0.693/k_1 \tag{4.7}$$

Reaction rates also depend on temperature. For example, the hydrolysis of esters (which have E_a values between 40 and 80 kJ/mol) proceeds 2–3 times faster at 30 °C than at 20 °C (Schwarzenbach and others, 1993).

Hydrolysis, oxidation and reduction, photochemical degradation, and biotransformation reactions in bed sediment are described briefly below. Additional detail is provided in reviews of abiotic transformations by Mabey and Mill (1978), Macalady and others (1986), and Wolfe and others (1990); reviews of biotransformation by Alexander (1981), Bollag and Liu (1990), and Scow (1990); and general reviews by Tinsley (1979), Vogel and others (1987), Crosby (1998), and Larson and Weber (1994).

Hydrolysis

Hydrolysis is the reaction (biologically mediated or abiotic) of a pesticide with water, usually resulting in the cleavage of the molecule into smaller, more water-soluble portions and in the formation of new C–OH or C–H bonds. Hydrolysis is an example of a larger class of reactions called *nucleophilic substitution* reactions, in which a *nucleophile* (an electron-rich species with an unshared pair of electrons) attacks an *electrophilic* (electron-deficient) atom in the molecule. In general, hydrolysis requires a hydrolyzable functional group such as an aliphatic halogen, epoxide, ester, or amide group. Although hydrolysis can proceed via a number of mechanisms (see Mabey and Mill, 1978; Schwarzenbach and others, 1993), the most common reaction mechanisms are nucleophilic substitution and nucleophilic addition–elimination, examples of which are shown in equations 4.8 and 4.9, respectively:

$$R\text{---}CH_2X \xrightarrow{\ H_2O,\ OH^-\ } R\text{---}CH_2OH + HX \tag{4.8}$$

$$\begin{array}{c} H \\ | \quad | \\ \text{---}C\text{---}C\text{---} \\ | \quad | \\ \quad X \end{array} \xrightarrow{\ H_2O,\ OH^-\ } \begin{array}{c} | \quad | \\ \text{---}C=C\text{---} \end{array} + HX \tag{4.9}$$

Pesticides that can undergo hydrolysis in bed sediment on environmentally relevant time scales (half-lives of days to years) include alkyl halides, aliphatic and aromatic esters, carbamates, and phosphoric acid esters (Vogel and others, 1987). Some pesticides, such as dichlorvos, undergo hydrolysis at rates too fast (half-lives of minutes) to be found at significant concentrations in surface waters. Other pesticides, such as DDT and chlordane, undergo hydrolysis at rates that are too slow (half-lives of years to decades) to warrant consideration of this transformation process. Others, such as pentachlorophenol and benefin, contain no hydrolyzable functional groups (Howard, 1991).

The hydrolysis rate of a given organic compound is dependent on the characteristics of the solution, such as pH, temperature, and the presence of certain metal ions or other catalyzing substances (Armstrong and Chesters, 1968; Mabey and Mill, 1978; Burkhard and Guth, 1981; Wolfe and others, 1990). The most important environmental factor affecting the rate of hydrolysis is pH. Hydrolysis reactions can be a result of attack by the water molecule (H_2O), the hydronium ion (H_3O^+), or the hydroxide ion (OH^-). These reactions are referred to as "neutral," "acid-catalyzed," and "base-catalyzed hydrolysis," respectively. At low pH, reactions are dominated by acid-catalyzed hydrolysis, whereas at high pH, reactions are dominated by base-catalyzed hydrolysis. At intermediate pH values, both neutral and acid-catalyzed reactions (or both neutral and base-catalyzed reactions) can be important to the overall rate of hydrolysis. Temperature is also an important factor. There is generally a two- to four-fold increase in the reaction rate for a temperature increase of 10 °C (Mabey and Mill, 1978).

Hydrolysis reactions can occur in moist soil or sediment, as well as in water. Surface-catalyzed (or surface-enhanced) hydrolysis has been reported for several pesticides. For example, the hydrolysis of chloro-*s*-triazines has been proposed to occur through the interaction of the sorbed pesticide with carboxyl groups of particulate organic matter (Armstrong and Konrad, 1974). Surface-enhanced hydrolysis of organophosphate esters also has been reported. Possible mechanisms include interaction with particle-bound metal ions or with strongly polarized water molecules that are associated with exchangeable cations (Wolfe and others, 1990). Hydrolysis also can be catalyzed by the acidic sites of clay minerals, or by free esterases present in soil (Crosby, 1998).

Oxidation and Reduction

Oxidation–reduction reactions are biologically mediated, chemical, or photochemical reactions that involve a transfer of electrons. The process requires two chemical species to react as a couple: one chemical undergoes *oxidation* (losing one or more electrons) while the other undergoes *reduction* (gaining one or more electrons). Because oxidation and reduction reactions are always coupled, they are often referred to as "redox reactions." Whether an organic contaminant will undergo redox reactions in the environment depends on its chemical structure (such as the presence of oxidizable or reducable functional groups) and on the redox conditions of the hydrologic environment. The redox condition is quantified by the *redox potential* (E_H), which is the electrical potential difference between a water sample and the hydrogen electrode. The presence of dissolved molecular oxygen (a strong oxidizing agent and electron acceptor) controls the redox condition of the water or sediment. For natural waters in equilibrium with air, the redox conditions are strongly oxidizing (E_H of about 750 mV) (Larson and Weber, 1994). For a typical water-saturated sediment, the E_H within the top 2.5 cm depth is about 300 mV (Crosby, 1998). The rates of redox reactions in the environment are dependent on the pH and the magnitude of the redox potential. For example, the reduction half-life of the organophosphate insecticide parathion in strongly reducing environments (low redox potential) is on the order of minutes (Macalady and others, 1986).

Redox conditions both influence and are influenced by microbial activity. As long as molecular oxygen is present, microorganisms undergo aerobic respiration, which results in the oxidation of organic compounds and the consumption of O_2. Once all the oxygen has been consumed, conditions become anaerobic, and microorganisms utilize alternative electron acceptors such as NO_3^- and MnO_2. When NO_3^- and MnO_2 are no longer present, and the only available electron acceptors have low reduction potentials, there is a pronounced decrease in the redox potential (Schwarzenbach and others, 1993). Thus, microbial activity influences the availability of oxygen and other oxidizing and reducing agents (and therefore the redox conditions). Conversely, the redox conditions of the water or sediment will largely determine the kinds of organisms that live and function in that environment. Some organisms can live only in well oxygenated, strongly oxidizing environments; others need reducing conditions, and still others can adapt themselves to a range of redox conditions. The mediation of redox reactions by biota is important over the complete range of E_H found in natural waters. Schwarzenbach and others (1993) contains a detailed discussion of redox reactions of organic chemicals in natural waters.

Oxidation reactions in surface waters tend to be biologically or photochemically induced (Larson and others, 1997). Microorganisms are the main cause of oxidations in soil (Crosby, 1998) or sediment. Oxidation also can be catalyzed by extracellular oxidative enzymes, probably originating from dead organisms, and by transition metals such as copper and manganese (Crosby, 1998). Biologically mediated redox reactions are discussed later in this section (Biotransformation).

Reductions in bed sediment generally take place in pore water under anaerobic conditions (Crosby, 1998). Pesticides may be reduced microbially or sometimes, when organic or inorganic reducing agents are present, abiotically. These reducing agents include certain transition metals (iron, nickel, cobalt, chromium), extracellular enzymes, iron porphyrins, and chlorophylls (Vogel and others, 1987). Abiotic reduction appears to be important for many pesticides in soil because the degradation rates of several pesticides, such as amiben, amitrole, atrazine, and pronamid, have been shown to be the same in both sterilized and unsterilized soil (Crosby, 1998).

Photochemical Reactions

The photochemical degradation of pesticides is caused by the absorption of energy from sunlight. Three kinds of photochemical reactions may occur: direct, sensitized, and indirect photochemical reactions. In *direct photochemical reactions*, the pesticide itself absorbs energy from sunlight, raising the molecule to an excited state. A small fraction of the electronically excited molecules releases the excess energy in a chemical reaction, such as bond cleavage, dimerization, oxidation, hydrolysis, or rearrangement (most molecules release the excess energy as heat or light, or by energy transfer in collisions with other molecules). In the environment, direct photodegradation will occur only if the pesticide absorbs light at wavelengths present in solar radiation (>290 nm).

In sensitized and indirect photochemical reactions, sunlight is absorbed by a molecule other than the pesticide of interest. *Sensitized photochemical reactions* occur when a light-absorbing chemical transfers energy to an acceptor (such as a pesticide), which then photodegrades. Photosensitizers present in natural waters include riboflavin, chlorophyllin, and tryptophan (Tsao and Eto, 1994). In *indirect photochemical reactions*, sunlight excites a photon absorber (such as nitrate or dissolved organic matter) in natural waters, which in turn reacts with dissolved oxygen to form highly reactive oxidants. These highly reactive species then attack the pesticide. In natural waters, the two most important indirect photochemical reactions probably involve singlet oxygen and nitrate-induced photooxidation (Hoigne and others, 1989). Singlet oxygen is a selective oxidant that is generated from humic substances in natural waters and soils (Zepp and others, 1981; Gohre and Miller, 1986; Miller and Donaldson, 1994). Nitrate-induced photooxidation generates hydroxyl radical, which is a relatively nonselective oxidant that can react with all organic molecules (as well as other molecules present, including water, DOM, and dissolved oxygen) (Hoigne and others, 1989).

Organic compounds that are sorbed to surfaces also can undergo photodegradation (Crosby, 1994). Sorption affects the vibrational and rotational properties of molecules (Miller and Donaldson, 1994). Therefore, a pesticide that is sorbed to soil or sediment may have a different absorption spectrum at sunlight wavelengths (Crosby, 1994) and form different photoproducts (Miller and Donaldson, 1994) than when the same pesticide is dissolved in water. In practice, it can be difficult to distinguish photochemical reactions occurring on bed sediment from those occurring on adjacent suspended particulates or in the water column (Crosby, 1994). However, photodegradation of organic contaminants sorbed to suspended particles has been reported in the laboratory. Examples are photoreduction of DDE and 3,4-dichloroaniline (Miller and Zepp, 1979a), photooxidation of polynuclear aromatic hydrocarbons (PAH) (Andelman and Suess, 1971), and indirect photolysis of substituted anilines and phosphorothioates on the surface of suspended algae (Zepp and Schlotzhauer, 1983).

In natural waters, the photochemical reactions that a given pesticide will undergo depend on the structure of the pesticide and on specific environmental conditions. Natural waters are rich both in chemical reagents, such as oxygen, halides, nitrate and nitrite, transition metal ions, and DOM, and in hydroxide or hydronium ions (depending on pH). A particular pesticide may undergo multiple reactions, and the resulting photoproducts may themselves continue to degrade. Phototransformation products are often similar to those derived from other abiotic reactions or from biotransformation.

The rate of a photochemical reaction is directly related to the sunlight radiation intensity at the wavelengths absorbed by the chemical. In natural waters, sunlight radiation intensity attenuates with increasing depth because of absorption and scattering (Zepp and Cline, 1977). The zone of sunlight penetration is called the *photic zone*. Increased stream velocity may bring deeper water, as well as suspended particles with sorbed contaminants, up into the photic zone. However, very high turbidity may decrease light intensity such that photochemical reactions do not occur. In hydrologic systems, photodegradation will tend to be important only in the water column, on suspended particles, and at the sediment–water interface where the water is shallow enough for sunlight to penetrate.

Biotransformation

The *biotransformation* of a contaminant refers to any change in its chemical structure that is mediated by living organisms using enzymes. Although abiotic transformation reactions can cause structural changes in an organic chemical, biotransformation is the only transformation process able to completely mineralize a pesticide (Alexander, 1981). Of particular importance is the capability of microorganisms to cleave the aromatic ring (Tinsley, 1979). Biotransformation that results in degradation of the pesticide molecule is also called *biodegradation*, although the latter term sometimes refers to degradation processes in which the pesticide serves as a substrate for growth (e.g., Bollag and Liu, 1990). All naturally produced organic compounds can be biotransformed, though for some chemicals this is a slow process. On the other hand, some synthetically produced organic chemicals, including many pesticides, have structures that are unfamiliar to microorganisms; therefore, microorganisms may not have the enzymes needed to biotransform these compounds. This is the primary reason why some pesticides and pesticide transformation products, such as DDE, hexachlorobenzene, and mirex, are recalcitrant (and thus, long lived) in the environment.

However, even recalcitrant compounds are biotransformed, albeit very slowly, probably due to a process called cometabolism (Bollag and Liu, 1990). In *cometabolism*, the microorganisms use other substrates (carbon sources) for growth and energy, and the unfamiliar synthetic compound enters into the process and is transformed. The microorganisms derive no particular benefit from the biotransformation of this compound. In addition to biodegradation and cometabolism, microbial transformation of pesticides also may result in polymerization or conjugation (Bollag and Liu, 1990). In *polymerization*, pesticide molecules are linked together with other pesticides, pesticide intermediates, or other synthetic or naturally occurring organic compounds to form larger polymers. Polymerization may play a role in the incorporation of pesticides and other synthetic organic compounds into particulate organic matter (Bollag and Liu, 1990). In *conjugation*, pesticides or pesticide intermediates are linked with other compounds resulting in methylated, acetylated, or alkylated compounds, glycosides, or amino acid conjugates (Bollag and Liu, 1990). Conjugation will be discussed in more detail in Section 4.3.3.

Biotransformation of a synthetic organic compound may lead to accumulation of intermediate products that have decreased, comparable, or increased toxicity relative to the parent compound. In some cases, such intermediates may inhibit both microbial growth and further metabolism (Bollag and Liu, 1990).

The rate of biotransformation of a pesticide depends on its chemical structure, the environmental conditions, and the microorganisms that are present. The structure of the organic chemical determines the types of enzymes needed to bring about its transformation. The concentration of the chemical also can affect its rate of transformation. At high concentrations a chemical may be toxic to microorganisms; at very low concentrations it can be overlooked by the organisms as a potential substrate. The environmental conditions (such as temperature, pH, moisture, oxygen availability, salinity, and concentration of other substrates) determine the species and viability of the microorganisms present. Finally, the microorganisms themselves control the rate of biotransformation, depending on their species composition, spatial distribution, population density and viability, previous exposure to the compound of interest, and enzymatic content and activity (Scow, 1990). In general the microbiological community is extremely viable in bed sediment. Indeed, this is where much of the decomposition of natural organic matter takes place in surface waters.

4.3 BEHAVIOR AND FATE OF PESTICIDES IN AQUATIC BIOTA

This section discusses the short-term behavior and long-term fate of pesticides that are taken up by aquatic biota. These are processes by which pesticides enter the organism, are transported through the organism, are biotransformed, and leave the organism. Also included are processes by which pesticide-contaminated organisms move within (or even outside) the hydrologic system, carrying the associated pesticides with them. Although most of this discussion focuses on macrobiota (especially fish and invertebrates), much of it applies to microorganisms as well. Some special considerations that apply specifically to microbial accumulation are also discussed.

Bioaccumulation of a pesticide by an aquatic organism may begin with sorption to the surface of the organism. Generally, a pesticide will enter the organism by diffusion, either at the gills, at the lining of the mouth, in the gastrointestinal tract, or through the body surface (such as skin or integument). Once a pesticide enters the organism, its bioaccumulation is facilitated by blood flow (perfusion) from the gills and by digestion of food from the gastrointestinal tract, each of which transports a pesticide away from the site of diffusion. Diffusion also is important in the distribution of a pesticide among various tissues and organs within an organism. Once in an organism, a pesticide may be biotransformed or excreted, or both. The distribution of a pesticide among tissues and the extent to which the pesticide is biotransformed by an organism are variable, depending on both the pesticide and the organism. The physical location of a pesticide within the biota compartment may change as a pesticide-contaminated organism is transported downstream or migrates, or as the pesticide passes from one organism to another, such as during reproduction or predation. If a pesticide-contaminated organism dies, the pesticide may be released to water or to particulate phases during settling or tissue decomposition.

Pesticide accumulation by aquatic biota varies as a function of the pesticide, the organism, and environmental conditions (discussed in detail in Section 5.2.5). Chemical characteristics that influence transfer of pesticides across cell membranes include lipid and water solubility, degree of ionization, chemical stability, molecular size, shape, and structure (Esser and Moser, 1982). Concentrations of a given pesticide in biota have been shown to vary with the species of organism, sex, age, body size or weight, surface-to-volume ratio, life stage or reproductive state, lipid content, trophic level, vertical distribution, physical condition, tissue or organ analyzed, migration pattern, and the season in which samples were collected. Some of these biological factors are seasonal, many are interdependent (Huckle and Millburn, 1990), and the processes responsible for some of these differences are not fully understood. Differences in metabolic capabilities, growth rates, and reproductive strategies probably play a role, since the concentration of a hydrophobic pesticide in an organism will be decreased by growth dilution, biotransformation, and reproduction (Sijm and others, 1992). Environmental factors that affect pesticide accumulation include water temperature, pH, salinity, hardness, food availability, the types and concentrations of aquatic particles, and the presence of other contaminants (Huckle and Millburn, 1990).

In the literature on contaminant bioaccumulation, a distinction is made between the terms bioconcentration, bioaccumulation, and biomagnification (Brungs and Mount, 1978). In *bioconcentration*, a contaminant enters an aquatic organism directly from water via gill or epithelial tissue. In *bioaccumulation*, a contaminant may enter an aquatic organism through the gills, epithelial tissue, dietary intake, or other sources. Use of this term does not imply any particular route of exposure. It is commonly used when referring to field measurements of contaminant residues in biota, where the routes of exposure are unknown. *Biomagnification* is the process whereby the tissue concentrations of a contaminant increase as it passes up the food chain through two or more trophic levels.

Three related terms are commonly used to describe the degree of contaminant bioaccumulation by aquatic biota in laboratory studies or in field populations. The *bioconcentration factor* (BCF) is the ratio of contaminant concentrations in biota and the surrounding water, as shown in equation 4.10. At long exposure times (equilibrium), the BCF also equals the ratio of uptake and elimination constants, k_1/k_2 (Mackay, 1982).

$$\text{BCF} = C_b/C_w = k_1/k_2 \qquad\qquad (4.10)$$

where C_b is the concentration in biota (in units of pesticide mass per tissue mass) and C_w is the concentration in the surrounding water (in units of pesticide mass per volume). The uptake (k_1) and elimination (k_2) constants refer to diffusive uptake and release only. The BCF is used to describe results of laboratory studies in which an organism is exposed to a pesticide in the surrounding water.

The *bioaccumulation factor* (also called the BAF) is analogous to the BCF, but applies to field measurements or to laboratory measurements with multiple exposure routes. The BAF is the ratio of contaminant concentration measured in biota in the field (or under multiple exposure conditions) to the concentration measured in the surrounding water. The kinetic expression for the BAF (see Section 5.2.1) is considerably more complicated than that shown in equation 4.10 for the BCF. Occasionally, the term *biomagnification factor* (BMF) is used in the literature to refer to the ratio of contaminant concentration in biota to that in the surrounding water when the biota was exposed via contaminated food. These terms are used in the remainder of this book.

Individual phase-transfer, transport, and transformation processes involved in bioaccumulation are described in Sections 4.3.1, 4.3.2, and 4.3.3, respectively. Many of the factors that affect bioaccumulation (such as the K_{ow} of the chemical, and temperature) can be explained in terms of their effects on these processes. Other factors, especially the biological ones, are not fully understood because of the structural complexity of living organisms and the role of physiology in bioaccumulation. Biological, chemical, and environmental factors that affect bioaccumulation are discussed in more detail in Section 5.2.5. Also discussed in Section 5.2 are theories and models of pesticide accumulation by aquatic biota (Sections 5.2.1–5.2.3), and uptake mechanisms and elimination rates (Section 5.2.5).

4.3.1 PHASE-TRANSFER PROCESSES

Phase-transfer processes involve the movement of a pesticide from one environmental matrix to another. The important phase transfers by which hydrophobic pesticides enter aquatic biota or are removed from aquatic biota include: transfer from water to biota (via diffusion or sorption); transfer from sediment to biota (via sorption or ingestion followed by diffusion); transfer from biota to water (via diffusion or other elimination processes); and transfer from biota to particulate matter (via elimination in feces or during tissue decomposition after the death of the organism). The next two sections describe fundamental processes underlying these phase transfers: passive diffusion and sorption–desorption. The last two sections in Section 4.3.1 discuss phase transfer of pesticides during elimination and tissue decomposition. Additional detail is provided in Section 5.2 and in reviews by Kenaga (1972), Metcalf (1977), Spacie and Hamelink (1985), Knezovich and others (1987), Connell (1988), Day (1990), and Huckle and Millburn (1990). Although the focus of this book is on pesticides, some of the studies cited below investigated the behavior and fate of structurally similar compounds, especially polychlorinated biphenyls (PCB) and tetrachlorodibenzo-*p*-dioxins (TCDD).

Passive Diffusion

Passive diffusion occurs through semipermeable membranes such as the gill, the lining of the mouth, and the gastrointestinal tract, when there is a concentration gradient across the membrane (Spacie and Hamelink, 1985). The lipid bilayer of biological membranes permits diffusion of nonionic lipophilic organic molecules (including hydrophobic pesticides), but little diffusion of water, ions, and polar molecules. Water, small ions, and small molecules (with molecular weights less than 100 daltons) can diffuse through proteinaceous pores in the membrane.

The gill membranes are particularly susceptible to diffusion by organic contaminants such as pesticides because they are thin (2–4 µm) and represent 2 to 10 times the surface area of the body. Furthermore, blood flow transports chemicals away from the site of diffusion, thus maintaining a strong concentration gradient. Measured uptake rates are greater for living organisms than for dead ones because of the role of blood circulation in contaminant transport. Physical factors that affect the rate of diffusion include surface area, membrane thickness, respiration rate, and the size and lipophilicity of the contaminant (Spacie and Hamelink, 1985). Rate constants for uptake at the gills are fairly constant for a wide range of chemicals, except that a decrease in uptake rate constant was observed for chemicals with molecular weight above about 300 daltons (Spacie and Hamelink, 1982). It appears that large molecules are not efficiently transferred across gill membranes, though fish can assimilate them from food (Niimi and Oliver, 1983; Skea and others, 1981). Dietary uptake rates of organochlorine compounds in goldfish increased when these compounds were administered in low-lipid food, indicating that intestinal absorption was due to passive diffusion, rather than to lipid coassimilation (Gobas and others, 1993a). Passive diffusion of pesticides through the body surface has been reported for some invertebrates. Examples include diffusion of DDT or metabolites through the cuticle of midge larvae (Derr and Zabik, 1974), and the integument of daphnids (Crosby and Tucker, 1971) and dragonfly nymphs (Wilkes and Weiss, 1971).

Typically, transport across biological membranes requires that the contaminant be present in a dissolved form (Spacie and Hamelink, 1985). Therefore, factors and processes that reduce the amount of contaminant in solution also reduce the rate of uptake. These include sorption (to particles, colloids, and humic materials) and ionization (Spacie and Hamelink, 1985; Farrington, 1989; Day, 1990).

Passive elimination of neutral organic contaminants can occur across the gills and skin by the same diffusion processes involved in uptake. Nonpolar organic contaminants that are not readily transformed (such as organochlorine insecticides) are eliminated from fish primarily through the gills at rates that are an inverse function of their lipophilicity (Maren and others, 1968).

Sorption and Desorption

More is known about sorption to microorganisms, because of their relative structural simplicity, than about sorption to fish and other aquatic macrobiota. However, studies of sorption by microorganisms (such as algae and bacteria) can help us understand the role of sorption in bioaccumulation by higher organisms. Sorption to the surface of an organism is important primarily as the initial step in bioaccumulation. Surface sorption is especially important for microorganisms because of their high surface-area/volume ratio. Sorption has been reported for PCBs and toxaphene in bacteria (Paris and others, 1977, 1978), and for DDT, PCBs, and TCDDs in algae (Sodergren, 1968, Hamelink and others, 1971; Cox, 1970b, 1971, 1972; Swackhamer and Skoglund, 1991). Because sorption is a physical process, it is equally effective with dead and living organisms (Tinsley, 1979; Swackhamer and Skoglund, 1991). The uptake of pesticides and other contaminants by microorganisms is not limited to surface adsorption. Sodergren (1968) reported that DDT first sorbed to the surface of *Chlorella* cells, then passed across the cell membranes. Swackhamer and Skoglund (1991) observed rapid uptake of PCB congeners by phytoplankton in Green Bay, Wisconsin, under low growth conditions, with 40 to 90 percent of uptake occurring within the first 24 hours. Uptake then appeared to level off, then increased slowly and steadily through the end of the experiment (up to 20 days). These results were interpreted as rapid surface sorption followed by slower transfer into lipids in the cell matrix.

The direct sorption of pesticides to the body wall or exoskeleton, or sorption through the integument, also has been reported for higher organisms (Knezovich and others, 1987). Examples include pesticide sorption to crustacean chitin and chitosan (Davar and Wightman, 1981; Kemp and Wightman, 1981) and annelid cuticle (Lord and others, 1980).

Elimination

In vertebrates, polar pesticides and metabolites of nonpolar pesticides usually are eliminated in bile via the liver or in urine via the kidney. The relative proportion depends on the molecular weight of the pesticide (or metabolite) and on the presence of polar functional groups. Chemicals of moderate to high molecular weight (greater than about 400 daltons) tend to be eliminated in bile, whereas compounds of molecular weight less than 400 daltons are eliminated predominantly in urine. Metabolites formed in the liver of vertebrates are transported to the gallbladder. They are then discharged with bile into the small intestine, where they may be eliminated with the feces or reabsorbed into the intestine and returned to the blood. Bile is probably the major route of excretion for metabolites of a number of lipophilic pesticides and other contaminants (Spacie and Hamelink, 1985; Huckle and Millburn, 1990).

Renal excretion (elimination in urine) of a pesticide by fish depends on the pesticide's molecular structure and on its physical and chemical properties. Most organic contaminants in the blood are small enough (molecular weight greater than 600 daltons) to be removed by glomerular filtration in the kidney. However, nonpolar compounds in the glomerular filtrate may be reabsorbed by passive diffusion across renal (kidney) membranes (Huckle and Millburn, 1990). Organic acids and bases may be actively secreted into urine (Spacie and Hamelink, 1985). When winter flounder (*Pleuronectes americanus*) was dosed with DDT and its polar metabolite DDA, DDA was eliminated in urine 250 times more quickly than DDT (Pritchard and others, 1977). Plasma binding was comparable for both chemicals. DDA was actively accumulated by the renal organic acid transport system, whereas DDT was not actively transported into urine, and probably was reabsorbed from tubular fluids.

Different routes of elimination in aquatic organisms can be affected differently by biological factors (such as tissue damage) or environmental factors (such as temperature, water chemistry, presence of competing chemicals, and previous exposure to toxicants). This contributes to the variability in contaminant residues observed in natural populations.

Tissue Decomposition

Dead and living biotic particles (such as algae and detritus) that contain pesticides may settle to the sediment–water interface. Also, higher organisms may die and settle to bed sediment, where they decompose. Pesticide contaminants in these biotic tissues may be released into other environmental compartments via diffusion to water or pore water. Diffusive release from superficial layers of fish tissue (Gakstatter and Weiss, 1967) was shown in experiments with mosquitofish, *Gambusia affinis* (Ferguson and others, 1966), in which the rate of release of endrin from dead fish was similar to that from live fish (live fish were obtained from a ditch that drained cotton fields and were resistant to endrin toxicity). Diffusive release may occur during settling, at the sediment–water interface, or in bed sediment. The diffusion process would be facilitated as the tissues were being decomposed. The primary site of release during tissue decomposition is likely to be surficial bed sediment, which contains organic matter in various stages of decomposition. Bacterial populations and metabolic activity increase by several orders of magnitude from the overlying water to surface sediment and decrease rapidly with increasing depth in the sediment (Wetzel, 1983). The decomposition of contaminated fish is believed to be a primary source of mirex contamination to some tributaries of Lake Ontario (Low, 1983; Lewis and Makarewicz, 1988). Pacific salmon (*Oncorhynchus* sp.) migrate from Lake Ontario (which has mirex-contaminated bed sediment) upstream into these tributaries to spawn. After spawning, the salmon normally die in the tributaries. Stream resident fish may take up mirex directly, by ingesting salmon eggs or portions of the dead carcasses, or indirectly, after release of mirex from the salmon carcasses to water or sediment (Lewis and Makarewicz, 1988).

4.3.2 TRANSPORT PROCESSES

Aquatic biota, and any pesticide contaminants they carry, are physically transported within a hydrologic system by a number of processes. Consumption by predators, migration or downstream movement, and reproduction are discussed in the following sections.

Consumption by Predators

Consumption of pesticide-contaminated organisms by predators may result in physical transport of the pesticide within a hydrologic system. If the primary source of hydrophobic pesticides to an aquatic system is slow release from bed sediment, then it is plausible that uptake by benthic organisms, followed by predation by bottom-feeding fish, may be a significant route of exposure for pelagic fish to these pesticides (Farrington, 1989). Consumption of fish by humans may constitute a significant mechanism of removal of hydrophobic pesticides from a hydrologic system (Humphrey, 1987; Lewis and Makarewicz, 1988).

Migration or Downstream Movement

Hydrophobic contaminant residues have been detected in fish downstream of known point sources (Schmitt and others, 1990). However, it is difficult to determine whether these fish were exposed upstream and then traveled downstream, or whether they were exposed downstream because of the downstream movement of contaminated water and sediment.

For some hydrologic systems, migratory species containing hydrophobic pesticide residues probably contribute to the transport of these pesticides outside the hydrologic system in which residues originated. Movement of contaminated eels (*Anguilla rostrata*) has been postulated as a mechanism for downstream transport of mirex from Lake Ontario to the upper St. Lawrence estuary (Lum and others, 1987) and of DDT and other organochlorine compounds from a eutrophic estuary in Scandinavia to the Sargasso Sea (Larsson and others, 1991). Pacific salmon migrating into tributaries upstream from Lake Ontario to spawn are believed to be a source of mirex contamination to these tributaries, which have no known mirex sources (Low, 1983; Lewis and Makarewicz, 1988). Resident (nonmigrating) fish species in these tributaries contained mirex, whereas resident fish from a dammed tributary (not accessible to migrating fish from Lake

Ontario) did not (Lewis and Makarewicz, 1988). This study was discussed previously in Section 4.3.1 (Tissue Decomposition).

In artificial stream channels, grazers (snails) and shredders (caddisflies) enhanced downstream transport of a PCB congener, relative to control streams, probably by ingestion of particles and subsequent defecation, dislodging fragments of periphyton, and processing of leaves (Sallenave and others, 1994).

Reproduction

Female fish deposit fats into maturing eggs as the spawning season approaches. Because organochlorine contaminants are often associated with fats, this results in a transfer of contaminants from parent to offspring. A sequential concentration of organochlorine pesticides was observed in ovarian tissues of winter flounder as the female approached spawning condition. Adult males also show changes in feeding and fat metabolism that may lower pesticide concentrations in muscle tissue as spawning season approaches (Smith and Cole, 1970). Reproduction can be a significant route of hydrophobic contaminant elimination. For example, in flounder (*Platichthys flesus*) and Atlantic cod (*Gadus morhua*) from Norway, production of eggs during the reproductive period (spring) resulted in loss of substantial PCBs in females (Marthinsen and others, 1991). In juvenile (sexually immature) guppies (*Poecilia reticulata*) that were fed a PCB-contaminated diet, second generation guppies contained PCB concentrations that were comparable with those of their parents (Sijm and others, 1992).

4.3.3 BIOTRANSFORMATION PROCESSES

Several authors have reviewed biotransformation of pesticides and other organic contaminants in fish and other aquatic organisms (e.g., Metcalf, 1977; Farrington, 1989; Huckle and Millburn, 1990). The following summary draws from these reviews as well as from the primary literature.

Biotransformation (defined in Section 4.2.3) is often referred to as "metabolism," especially in higher organisms. Fish, crustaceans, and polychaetes have active enzyme systems that can transform pesticides, PCBs, PAHs, and similar contaminants. In general, rates of contaminant biotransformation in fish are slower than in many mammalian species (James and others, 1977; Franklin and others, 1980). Typically, hydrophobic compounds are converted to more water-soluble compounds that can be excreted (or conjugated with a biochemical and then excreted), rather than stored in tissue lipids. Most biotransformation reactions of organic contaminants occur in the liver, although some reactions also take place in the kidney, plasma, intestine, intestinal microflora, or brain (Fukami and others, 1969; Glickman and Lech, 1981; Edwards and others, 1987b). Species differences may be important here. For example, in rainbow trout (*Oncorhynchus mykiss*), oxidative activity in the kidney is as important as in the liver (Pesonen and others, 1985). For fish, biotransformation products (also called "metabolites") of contaminants generally are eliminated in urine via the kidney, in bile via the liver, or by diffusion through the gills into the surrounding water.

Parent compounds with certain functional groups (such as alkyl side chains or methoxy groups) will be biotransformed more readily, thus making the parent compound less persistent and less likely to bioaccumulate. For example, methoxychlor is structurally identical to DDT, except that the aromatic chlorine substituents in DDT are replaced by methoxy groups in methoxychlor. The methoxy groups undergo O-dealkylation, which makes methoxychlor much less bioaccumulative relative to DDT. Measured BCF values for methoxychlor in fish ranged from 185 in flow-through tests to 1,550 in static tests; the analogous BCF values for DDT in fish were 61,600 and 84,500, respectively (Kenaga, 1980a).

A given pesticide may be biotransformed by multiple pathways involving multiple enzymes. Among the most important enzymes involved in biotransformation are the microsomal oxidases (mixed-function oxidases), especially cytochrome P-450 monooxygenases (Metcalf, 1977). Metabolism does not necessarily mean detoxication, since metabolites may have comparable or greater toxicity to the organism than the parent compound. The phosphorothioate esters (organophosphate compounds containing a P=S bond) provide a classic example. The thiono group (P=S) is biotransformed by oxidative desulfuration to the corresponding, and more toxic, phosphate group (P=O), resulting in an increase in toxicity (Klaassen and others, 1996). When a parent compound is biotransformed to a more toxic metabolite, this process is called *metabolic activation*. An example is the oxidation of parathion to paraoxon.

Biotransformation reactions in fish can be classified as either Phase I or Phase II reactions (Williams and Millburn, 1975). *Phase I reactions*, summarized in Table 4.1, insert functional groups such as hydroxyl, amine, or carboxylic acid groups into the pesticide by oxidation, reduction, or hydrolysis reactions. The addition of this functional group generally increases the biochemical reactivity of the pesticide. In *phase II reactions*, the pesticide (or its metabolite from phase I, which now contains a reactive functional group) reacts with a highly polar or ionic species (such as sulphate or an amino acid) to form a product that has a much higher water solubility than the original pesticide, and can be excreted. Phase II reactions, also called *conjugation* reactions (see Section 4.2.3), for pesticides in fish are summarized in Table 4.2.

Both biological and environmental factors influence the biotransformation of pesticides and other contaminants in fish. Important biological parameters include species, strain, sex, age, and reproductive state. For example, male fish have higher liver cytochrome P-450 content and monooxygenase activity than do females of the same species (Stegeman and Chevion, 1980; Koivusaari and others, 1981). Also, monooxygenase activity is influenced by sex hormones (Forlin and Hansson, 1982) and, therefore, varies as a function of reproductive state (Huckle and Millburn, 1990). Important environmental factors that may affect biotransformation include water temperature, pH, previous exposure to contaminants, and the degree of water oxygenation. For example, adult rainbow trout in northern regions adapt to the seasonal temperature changes of their environment with a temperature compensation pattern that affects phase I, but not phase II, biotransformation reactions (Koivusaari and others, 1981; Hanninen and others, 1984; Andersson and Koivusaari, 1986; Huckle and Millburn, 1990). Prior exposure to certain contaminants (including organochlorine insecticides) causes *enzyme induction* (increased enzyme activity), which can increase biotransformation rates. Exposure to polluted water has been shown to induce hepatic (liver) monooxygenase activities in fish (Kezic and others, 1983; Kleinow and others, 1987; Melancon and others, 1987; Huckle and Millburn, 1990). Prior exposure of fish populations to environmental contaminants that induce monooxygenase activity has been proposed as an explanation for observed differences in biotransformation among strains (Pedersen and others, 1976).

Table 4.1. Phase I biotransformation reactions of pesticides in fish

[Adapted from Huckle and Millburn (1990) with permission of the publisher. Copyright 1990 John Wiley & Sons, Ltd.]

Reaction Type	Pesticide (class)	Fish Species[1]
Oxidation:		
Hydroxylation	Cypermethrin (pyrethroid insecticide)	RT
	Rotenone (piscicide and insecticide)	BG, C
	Endrin (organochlorine insecticide)	BG
	Carbaryl (carbamate insecticide)	BG, CC, GF, KG, YP
Epoxidation	Aldrin (organochlorine insecticide)	YBC, BG, GS, MF, SH
O-Dealkylation	Chlorpyrifos (organophosphate insecticide)	GF
N-Dealkylation	Aminocarb (carbamate insecticide)	RT
P=S group to P=O	Diazinon (organophosphate insecticide)	BG, CC
Oxidative dehydrochlorination	Dieldrin (organochlorine insecticide)	BG
Reduction:		
Nitro-reduction	Parathion (organophosphate insecticide)	A, BB, LMB, BG, PS, SC, SU, WF
Reductive dechlorination	DDT (organochlorine insecticide)	GF
Hydrolysis:		
Hydrolysis of carboxylic esters	Malathion (organophosphate insecticide)	P
Hydrolysis of phosphoric esters	Fenitrothion (organophosphate insecticide)	M, K

[1]These fish include examples of freshwater, euryhaline, and marine species:

A = Alewife	M = Striped mullet
BB = Black bullhead	MF = Mosquito fish (*Gambusia affinis*)
BG = Bluegill	P = Pinfish
C = Carp	PS = Pumpkinseed
CC = Channel catfish	RT = Rainbow trout
G = Guppy (*Poecilia reticulata*)	SC = Shorthorn sculpin
GF = Goldfish	SH = Golden shiner
GS = Green sunfish	SU = White sucker
K = Killifish	WF = Winter flounder
KG = Kissing gourami	YBC = Yellow bullhead catfish
LMB = Largemouth bass	YP = Yellow perch

Oxidation

Oxidation is the most important type of phase I reaction. Oxidation reactions (Table 4.1) are generally catalyzed by cytochrome P-450 monooxygenases, which are in the liver microsomal fraction. These oxidations require nicotinamide-adenine dinucleotide phosphate in its reduced form (NADPH) and molecular oxygen (Huckle and Millburn, 1990). The electrons from NADPH are transferred to cytochrome P-450 by another microsomal enzyme. The P-450 then inserts one atom of the oxygen into the contaminant, and the second oxygen atom is reduced to produce water (Huckle and Millburn, 1990). Animals have multiple cytochrome

Table 4.2. Phase II biotransformation (conjugation) reactions of pesticides in fish

[Adapted from Huckle and Millburn (1990) with permission of the publisher. Copyright 1990 John Wiley & Sons, Ltd.]

Reaction Type	Pesticide (Type)	Fish Species[1]
Glucuronidation	3-Trifluoromethyl-4-nitrophenol (Selective sea lamprey larvicide)	RT
	[2]4′-Hydroxycypermethrin and Cl$_2$CA	RT
	[3]5,6-Dihydro-5,6-dihydroxycarbaryl	BG, CC, GF, KG, YP
	[3]1-Naphthol	RT
Sulphation	[4]3-(4-Hydroxyphenoxy)benzoic acid	BM, RT
Glutathione conjugation and mercapturic acid biosynthesis	Molinate (Herbicide)	C
Taurine conjugation	2,4-D (Herbicide)	D, SF, WF
	2,4,5-T (Herbicide)	D, SF
	[5]DDA	SF
	[6]3-Phenoxybenzoic acid	BM, RT
Acetylation	[7]Ethyl 3-aminobenzoate (MS 222)	D, RT

[1]These fish include examples of freshwater, euryhaline and marine species:

BG = bluegill	KG = Kissing gourami
BM = Armed bullhead minnow	RT = Rainbow trout
C = Carp	SF = Southern flounder
CC = Channel catfish	WF = Winter flounder
D = Spiny dogfish	YP = Yellow perch
GF = Goldfish	

[2]4′-Hydroxycypermethrin and 3-(2,2-dichlorovinyl)-2,2-dimethylcyclopropanecarboxylic acid (Cl$_2$CA) are the Phase I metabolites of the pyrethroid insecticide cypermethrin resulting from oxidation and hydrolysis, respectively.

[3]Phase I metabolite of the carbamate insecticide carbaryl.

[4]Hydroxylated derivative of 3-phenoxybenzoic acid, a metabolite of many pyrethroid insecticides.

[5]DDA is 1,1-bis-(4-chlorophenyl)acetic acid, a metabolite of the organochlorine insecticide DDT.

[6]A metabolite of many pyrethroid insecticides.

[7]MS 222 is not a pesticide, but is a widely used fish anesthetic.

P-450 isoenzymes, each with a slightly different substrate specificity. For example, five hepatic cytochrome P-450 isoenzymes were isolated from rainbow trout (Williams and Buhler, 1984) and four from Atlantic cod (Goksoyr, 1985).

The importance of microsomal oxidases in fish has been demonstrated in the laboratory using the enzyme inhibitor piperonyl butoxide. In green sunfish exposed to a series of pesticides, piperonyl butoxide increased methoxychlor accumulation by 15 times and shifted the degradation pathway from O-dealkylation (an oxidation pathway) to dehydrochlorination (a reduction pathway). Aldrin accumulation increased by 17 times and dieldrin formation was reduced. Trifluralin accumulation increased 45-fold and the rate of N-dealkylation decreased (Reinbold and Metcalf, 1976).

Species differences in oxidative capability were illustrated using ^{14}C-labeled aldrin in a 3-day model ecosystem (Lu, 1973), as shown in Figure 4.1. For the various species in the model ecosystem, the extent of oxidation of aldrin to dieldrin occurred in the following order: *Oedogonium* (green alga) and *Daphnia* (zooplankton) were less than *Culex* (mosquito) which was less than *Physa* (snail) which was less than *Gambusia* (mosquitofish). Other metabolites (hydroxylation and conjugation products) were relatively minor.

Reduction

Many biological reductions are mediated by cytochrome P-450 monooxygenase. However, here the contaminant (rather than oxygen) accepts the electrons from NADPH. Because this process competes with the normal cytochrome P-450 oxidations, it is strongly inhibited by molecular oxygen (Crosby, 1994). Reduction of pesticides was observed in several species of fish

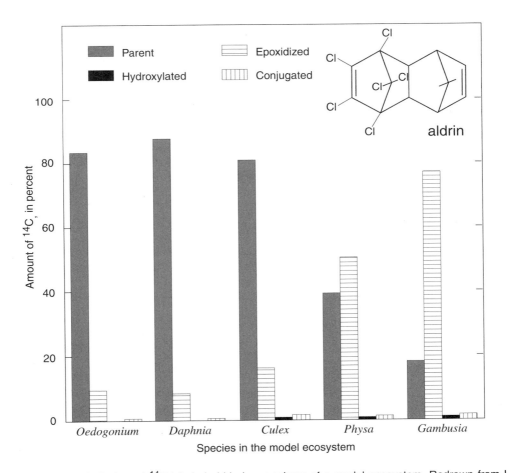

Figure 4.1. Metabolic fate of ^{14}C-labeled aldrin in organisms of a model ecosystem. Redrawn from Lu and Metcalf (1975).

(listed in Table 4.1). Examples include nitro-reduction of parathion (Hitchcock and Murphy, 1967) and reductive dechlorination of DDT to DDD (Young and others, 1971).

Hydrolysis

Phase I hydrolysis occurs for carboxylic esters (such as malathion), phosphoric esters (such as fenitrothion), and carbamates (such as carbaryl). In rainbow trout (Statham and others, 1975), hydrolysis was the major route of biotransformation of carbaryl compared to hydroxylation (which is an oxidation reaction). However, rainbow trout were only able to hydrolyze pyrethroids rather slowly (Edwards and others, 1987b), which probably contributes to the toxicity of pyrethroids to trout and other salmonids.

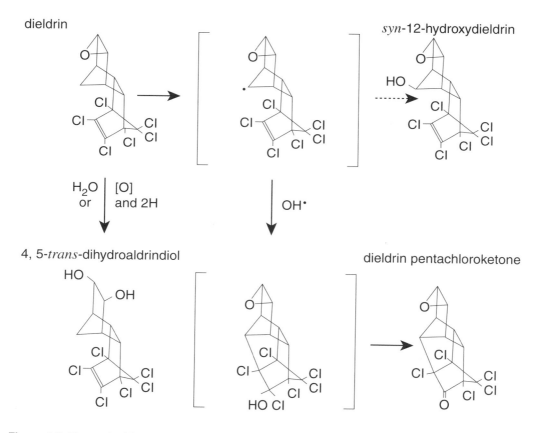

Figure 4.2. The major biotransformation pathways of dieldrin in bluegills. The solid arrows indicate the major routes, and dashed arrows the minor routes, of biotransformation. Brackets indicate transient intermediates. Redrawn from Huckle and Millburn (1990) with permission of the publisher. Copyright 1990 John Wiley & Sons, Ltd.

Conjugation

Phase II conjugation of organic contaminants in fish includes glucuronidation, sulphation, and glutathione conjugation (Table 4.2). The enzymes involved are located in the liver: UDP glucuronyl-transferase, sulphotransferases, and glutathione S-transferases, respectively. Although a number of amino acids are used by mammals to conjugate carboxylic acids derived from pesticides, taurine derivatives are the only amino acid conjugates identified in fish (Huckle and Millburn, 1990). The enzyme responsible (acyl-CoA: amino acid N-acyltransferase) is located in kidney and liver, but its activity is higher in the kidney for most fish species studied (James, 1978).

Examples of Biotransformation Reactions

Biotransformation of a pesticide may occur via multiple pathways simultaneously. This is illustrated in Figures 4.2–4.5, which show the major metabolic pathways in fish for selected pesticides: dieldrin, DDT (organochlorine insecticides), fenitrothion (organophosphate insecticide), and carbaryl (carbamate insecticide).

Organochlorine insecticides are biotransformed relatively slowly. The biotransformation of cyclodienes (aldrin, chlordane, and dieldrin) in fish is complex. For example, eight radiolabeled metabolites were separated following exposure of bluegills (*Lepomis macrochirus*) to ^{14}C-labeled dieldrin (Sudershan and Khan, 1979, 1980a–c, 1981) (see Figure 4.2). DDT is biotransformed by dehydrochlorination to the relatively stable DDE or by reductive

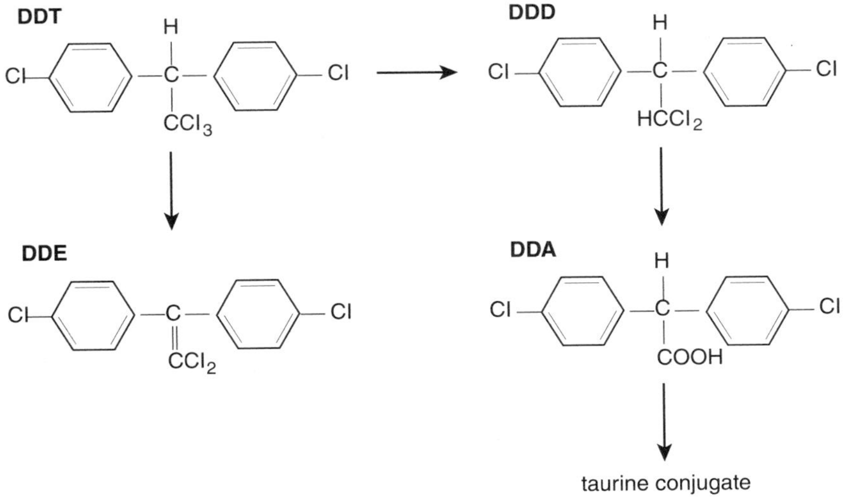

Figure 4.3. The major biotransformation pathways of DDT in fish. Redrawn from Huckle and Millburn (1990) with permission of the publisher. Copyright 1990 John Wiley & Sons, Ltd.

dechlorination to DDD, as shown in Figure 4.3. DDD, in turn, breaks down via several intermediates to DDA, which may be conjugated and excreted (James, 1978, 1987).

Many organophosphate insecticides contain the P=S moiety, which is transformed in fish (as in mammals) by oxidative desulfuration to the corresponding P=O compound. Because these oxon derivatives are more potent cholinesterase inhibitors than the parent compounds, this constitutes metabolic activation. Detoxication then occurs by transformation of the active phosphoric acid triester into the inactive diester, which can occur via several enzymatic mechanisms simultaneously (Hutson and Roberts, 1985), as shown for fenitrothion in Figure 4.4. N-methylcarbamate insecticides undergo oxidation, hydrolysis, and N-methyl oxidation, as shown for carbaryl in Figure 4.5. Hydrolysis appears to be the major biotransformation pathway, often followed by conjugation (Huckle and Millburn, 1990).

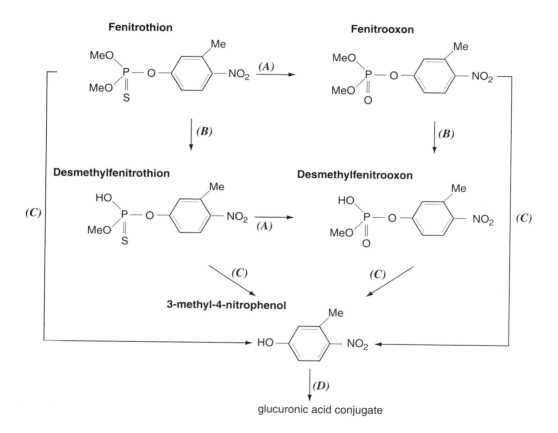

Figure 4.4. The biotransformation pathways of the organophosphate insecticide fenitrothion in fish. Reaction types: (*A*) oxidative desulfuration, (*B*) demethylation, (*C*) hydrolysis, and (*D*) conjugation. Redrawn from Huckle and Millburn (1990) with permission of the publisher. Copyright 1990 John Wiley & Sons, Ltd.

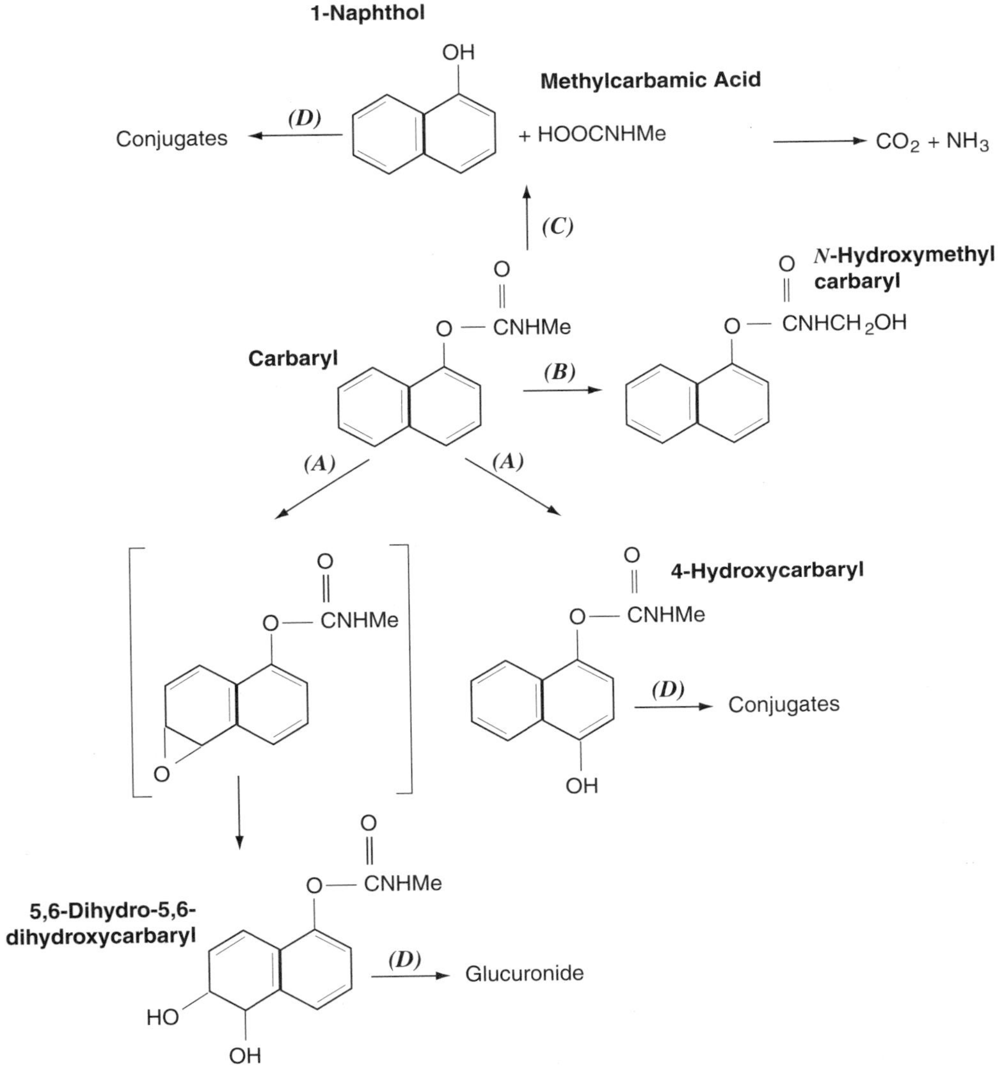

Figure 4.5. The biotransformation pathways of the *N*-methylcarbamate insecticide carbaryl in fish. Reaction types: (*A*) oxidation, (*B*) N-methyl oxidation, (*C*) hydrolysis, and (*D*) conjugation. Brackets indicate transient intermediates. Redrawn from Huckle and Millburn (1990) with permission of the publisher. Copyright 1990 John Wiley & Sons, Ltd.

Analysis Of Key Topics—Sources, Behavior, And Transport

The preceding overviews of national distribution and trends of pesticides in bed sediment and aquatic biota, and of governing factors that affect their concentrations in these media, leaves many specific questions unanswered. The next two chapters draw on information in the literature reviewed to discuss, in detail, several important topics related to pesticides in bed sediment and aquatic biota. Each key topic falls into one of two categories: (1) sources, behavior, and transport (Chapter 5), or (2) environmental significance (Chapter 6).

5.1 EFFECT OF LAND USE ON PESTICIDE CONTAMINATION

The terrestrial environment has a strong influence on the water quality of adjacent hydrologic systems. Both natural and anthropogenic characteristics of the terrestrial environment are important. For example, concentrations of major chemical constituents (such as sulfate, calcium, and pH) in a hydrologic system are influenced by geology, and the concentration of suspended sediment is influenced by soil characteristics, topography, and land cover. Land use activities, such as row crop agriculture, pasture, forestry, industry, and urbanization, also can affect adjacent water bodies. Any pesticide associated with a land use can potentially find its way to the hydrologic system and, if the pesticide has persistent and hydrophobic properties (see Section 5.4), it will tend to accumulate in bed sediment and aquatic biota. The following section addresses the observed link between land use and the detection of pesticides in bed sediment and aquatic biota. Four types of land use will be discussed: agriculture, forestry, urban areas and industry, and remote or undeveloped areas. In many cases, forested areas also could be described as remote or undeveloped areas. The critical distinction here, however, is that many forested areas have been managed with the use of pesticides whereas remote and undeveloped areas have not.

5.1.1 AGRICULTURE

By far the largest use of most pesticides, both presently and historically, has been in agriculture (Aspelin and others, 1992; Aspelin 1994). The soils of many agricultural areas still contain residues of hydrophobic, persistent pesticides that were applied during the 1970s or earlier. This was documented in 1970 for 35 states, mostly east of the Mississippi River (Crockett and others, 1974), in 1985 in California (Mischke and others, 1985), and during 1988–1989 in Washington (Rinella and others, 1993). In the California study (Mischke and others, 1985), only fields with known previous DDT use were targeted. This study obtained 99 soil samples from fields in 32 counties. Every sample analyzed contained residues of total DDT (the sum of DDT and its transformation products). The investigators compared the concentrations of the parent DDT with the concentrations of its transformation products (DDD and DDE) and found that the ratio of the parent DDT to total DDT was 0.49. That is, 49 percent of the total DDT remaining in the soils at least 13 years after use still existed as the parent compound. In the U.S. Environmental Protections Agency's (USEPA) National Study of Chemical Residues in Fish (NSCRF), which measured fish contaminants at sites in different land-use categories (such as agricultural sites, industrial and urban sites, paper mills using chlorine, other paper mills, and Superfund sites), sites in agricultural areas had the highest mean and median concentrations of p,p'-DDE in fish, as well as four of the top five individual fish sample concentrations. Agricultural sites also had the second highest mean concentration of dieldrin in fish (second to Superfund sites), as well as two of the top five individual sample concentrations. Soils containing residues of DDT and similar recalcitrant pesticides from past agricultural use constitute a reservoir for these pesticides today; they have been, and will continue to be, a source of these compounds to hydrologic systems, thus leading to contamination of surface water, bed sediment, and aquatic biota.

Those pesticides currently used in agriculture (Table 3.5) are not as persistent as the restricted organochlorine compounds. As discussed in Section 3.3, some moderately hydrophobic, moderately persistent pesticides have been detected in bed sediment and aquatic biota, although at lower detection frequencies than the more persistent organochlorine compounds. It is probable that additional pesticides with moderate water solubilities and persistence may be found in bed sediment or aquatic biota if they are targeted in these media (see Section 5.4), especially in high use areas. A few moderately hydrophobic, moderately persistent compounds were analyzed in fish by the NSCRF (U.S. Environmental Protection Agency, 1992a): dicofol, lindane, α-HCH, and methoxychlor (organochlorine insecticides or insecticide components); chlorpyrifos (organophosphate insecticide); and trifluralin, isopropalin, and nitrofen (herbicides). Of these compounds, several were found in association with agricultural areas. Agricultural sites had the highest mean and maximum concentrations of dicofol and chlorpyrifos, and they had the highest mean concentration of trifluralin, in fish. Moreover, sites with the highest trifluralin residues in fish were in states with the highest agricultural use of trifluralin (Arkansas, Illinois, Iowa, Minnesota, Missouri, North Dakota, South Carolina, Tennessee, and Texas). In California's Toxic Substance Monitoring Program, which monitored pesticides in fish and invertebrates from over 200 water bodies throughout the state, the highest concentrations of several currently used pesticides in fish during 1978–1987 were from two intensively farmed areas (Rasmussen and Blethrow, 1990). These pesticides are the insecticides chlorpyrifos, diazinon, endosulfan, and parathion, and the herbicide dacthal. The highest residues of dacthal in whole fish analyzed by the Fish and Wildlife Service's (FWS) National Contaminant Biomonitoring Program (NCBP) also occurred in intensively farmed areas (Schmitt and others, 1990).

5.1.2 FORESTRY

A number of studies monitored one or more pesticides in forest streams or lakes after known application. Most of these studies were at sites in the forests of the southeastern United States (e.g., Yule and Tomlin, 1970; Neary and others, 1983; Bush and others, 1986; Neary and Michael, 1989), northwestern United States (e.g., Sears and Meehan, 1971; Moore and others, 1974), or Canada (e.g., Kingsbury and Kreutzweiser, 1987; Sundaram, 1987; Feng and others, 1990; Kreutzweiser and Wood, 1991; Sundaram and others, 1991). The majority of these studies can be described as field experiments, in which a known amount of a certain pesticide was applied to a section of a watershed, with subsequent sampling of water, bed sediment, or aquatic biota for a period of weeks to years. These studies are considered process and matrix distribution studies and are described in Table 2.3 (if they were conducted in United States streams and they sampled bed sediment or aquatic biota). Few, if any, studies have reported on the ambient concentrations of pesticides in bed sediment or aquatic biota after routine use of pesticides, so little is known about the long-term presence of pesticides in streams from forest applications. On the basis of the reported field experiments and information on pesticide use in forestry, a few conclusions can be drawn.

The choice of chemicals used for forest applications has changed over time (Freed, 1984). Before the mid-1940s, the only pesticides that were used were inorganic compounds. Organic pesticides were introduced after World War II. Aerial spraying of pesticides began following the availability of suitable airplanes. The chlorophenoxy acid herbicides, 2,4-D and 2,4,5-T, and the organochlorine insecticide, DDT, were the first of the organic pesticides to be widely used. In subsequent years, a wide variety of herbicides and insecticides were introduced into forestry use. Most of the major classes of herbicides were represented, including triazines, ureas, uracils, and chlorophenoxy acids. The insecticides used included most of the organochlorine compounds and numerous organophosphate, carbamate, and pyrethroid compounds. Since the 1980s, the use of chemical pesticides in forestry has declined (Larson and others, 1997). The chemical insecticides have largely been replaced by biological pesticides. The current use of pesticides in forestry (Section 3.2.2) and forestry as a source of pesticides in surface water systems (Section 4.1.1) were previously discussed. The potential impacts on water quality are covered in more detail in Larson and others (1997).

The pesticides used in forestry since the 1940s may have caused some environmental impact at the time of application. However, many of these pesticides do not persist long in forest soils or streams, so are unlikely to have lasting or long-term effects on stream biota after a period of time (days, months, or years, depending on the chemical) has elapsed since application. The exceptions are pesticides that are hydrophobic and recalcitrant (and thus long-lived), such as the organochlorine insecticides. Because of their physical and chemical properties (see Section 5.4), organochlorine insecticides may persist in bed sediment and aquatic biota of forest streams, and forest soils containing organochlorine insecticide residues may be washed into the stream for many years after the period of application. Also, as with pesticides applied in agricultural areas (see Section 3.3.2), there is potential for detection of moderately hydrophobic, moderately persistent silvicultural pesticides in bed sediment or biota, especially in high-use areas.

Triclopyr is now the highest-use herbicide on national forest land. The next most commonly used herbicides in national forests in 1992 were 2,4-D, hexazinone, glyphosate, and picloram (Larson and others, 1997). Except for *Bacillus thuringiensis* var. *Kurstaki* (Bt), carbaryl was the highest-use insecticide in national forests in 1992 (Larson and others, 1997). Of these compounds, only 2,4-D was targeted in sediment or aquatic biota at more than 30 (total) sites in all the monitoring studies reviewed (Tables 3.1 and 3.2). When data from all monitoring studies were combined, 2,4-D was detected in 1 percent of (825 total) sediment samples and in 5 percent of (44 total) biota samples. Of the other recently used pesticides, picloram was detected bed sediment in 2 percent of (53) samples; detection data were not reported for biota. Carbaryl was detected in aquatic biota (11 percent of 27 samples), but not in bed sediment (only 3 samples analyzed). Glyphosate was not detected in any of 19 total bed sediment samples; data for biota were not reported. Triclopyr and hexazinone were not targeted in bed sediment or aquatic biota in any of the monitoring studies reporting detection data.

Five process and matrix distribution studies (or field experiments) in forest streams provide some indication of the behavior of organochlorine insecticides, pyrethroid insecticides, and other selected pesticides following application in forestry. In one study in New Brunswick, Canada (Yule and Tomlin, 1970), DDT and its transformation products, DDE and DDD, were studied in water and bed sediment of a stream after application to nearby forests for the control of Spruce budworm. One motivation for this study was that fish-kills occurred following the use of DDT in forests in this area. The stream had high concentrations of DDT in the surface of the water column immediately after application, but these subsided to the background concentration (about 0.7 μg/L) after a few hours. The deeper stream water (12–18 in. below the surface) did not show the same immediate DDT concentration spike; however, DDT levels there were relatively consistent for 2 years following the application. Twelve months after application, every bed sediment sample (18 total) collected from the vicinity of the site of application to the mouth of the river, about 50 mi downstream, had measurable concentrations of total DDT. The average bed sediment concentration was about 12 percent of the forest soil concentration on a dry weight basis. There was a trend of decreasing concentration downstream, and also a change in the ratio of DDT/total DDT. As the distance from the point of application increased, the transformation products constituted a greater percentage of the total DDT, indicating in-stream transformation. Unfortunately, no time series data were presented for the bed sediment. The authors suggested that DDT persists in forest soils, predominantly as the parent compound, and that the long-term transport to streams is through runoff of soil particles. The presence of DDT components in the bed sediment throughout the river system 1 year after application, and the presence of DDT components in the water 2 years after application, suggest that there is long-term storage of DDT in the forest soil and in the bed sediment of the river system, and that the soil and bed sediment constitute a constant source of contaminant to the river water.

Prior to its cancellation in the early 1980s, endrin was used in forestry as a coating on aerially applied tree seeds to protect them from seed-eating rodents. One study (Moore and others, 1974) examined the presence of this compound in the water and aquatic biota of two Oregon watersheds after seeding. The actual amount of endrin applied to the watersheds was estimated to be 2.5 to 10 grams a.i. per hectare. Endrin was observed consistently in the stream water for about 9 days (maximum concentration was about 12 ng/L), then was nondetectable until a high flow period about 21 days after application. At this time, it was detected in the water again. This second period of detection suggests that the endrin was stored either in the forest soils or in the bed sediment of the stream and then released with higher streamflow. Fish (coho salmon [*Onchorhynchus kisutch*] and sculpins [family Cottidae]) and various unidentified aquatic insects were analyzed for endrin. Because of sample contamination, the results are somewhat ambiguous. The authors did conclude that endrin was present in all biotic samples obtained within days after application. Samples collected 12 and 30 months after the application of endrin did not contain detectable traces of endrin. Bed sediment was not collected during this study.

A third example is the study of permethrin in Canadian streams (Kreutzweiser and Wood, 1991; Sundaram, 1991). Permethrin, a synthetic pyrethroid, is known not only for its high insecticidal activity and its ability to control lepidopterous defoliators, but also for its high acute toxicity to fish and strong sorption tendencies. Kreutzweiser and Wood (1991) examined the presence of permethrin in a forest stream after aerial application. They detected the compound in water, bed sediment, and fish. The concentration in water declined with time and distance from application. Permethrin was seldom seen in the bed sediment of the stream (only 8 percent of the samples). Atlantic salmon (*Salmo salar*), brook trout (*Salvelinus fontinalis*), and slimy sculpin (*Cottus cognatus*) were analyzed, and permethrin was detected in about half of the samples during the first 28 days after application. The fish were sampled again 69 to 73 days after application and no traces of permethrin were detected. Sundaram (1991) studied the behavior of permethrin by adding it directly into a forest stream. He found that it was not detected in the stream water near the site of application after 5 hours and that it was seldom detected in the bed sediment of the system, probably because of the low sediment organic carbon content. Sundaram (1991) did detect permethrin in aquatic plants (water arum, *Calla palustris*), stream detritus, caged crayfish (*Orconectes propinquus*), and caged brook trout collected during the study (up to 7 to 14 hours after application). No permethrin was detectable in caged stoneflies (*Acroneuria abnormis*) throughout the study duration (14 hours). Permethrin also was detected in invertebrate drift collected 280–1,700 m downstream of the application point. The longer-term presence of permethrin in this system was not studied.

In a fourth example, 2,4-D was sprayed on clearcut forested lands in Alaska (Sears and Meehan, 1971). The results of this study show potential for at least initial accumulation in biota. Residues of 2,4-D were detected in river water samples (up to 200 μg/L), and in a single composite sample of coho salmon fry (500 μg/kg), collected 3 days after spraying. Unfortunately, later samples were not taken, so no information is provided on dissipation rates.

Finally, a dissipation study of the organophosphate pesticide chlorpyrifos-methyl was conducted in a forest stream in New Brunswick, Canada (Szeto and Sundarum, 1981). The results of this study indicate that there is potential for initial accumulation in stream bed sediment and aquatic biota, but that residues are unlikely to persist. After aerial application, chlorpyrifos-methyl residues persisted in balsam fir foliage and forest litter for the duration of the experiment (125 days). Residues in bed sediment (10–180 μg/kg dry weight) persisted for at least 10 days; at the next sampling time (105 days post-application), residues in sediment were nondetectable (less than 1 μg/kg wet weight). In stream water, chlorpyrifos-methyl dissipated rapidly within the first 24 hours after application, and it was not detectable in water (less than 0.02 μg/L) after four days. Residues of up to 46 μg/kg chlorpyrifos-methyl were detected in fish (slimy sculpin and brook trout); only trace levels (less than 3 μg/kg wet weight) were detected after 9 days, and chlorpyrifos-methyl was nondetectable (less than 1.5 μg/kg wet weight) after 47 days. Concentrations in brook trout were consistently higher than in slimy sculpin sampled at the same time.

The results from these limited studies suggest that the behavior of pesticides in forested streams are in agreement with their behavior in agricultural streams. DDT and its transformation products appear to have the longest residual time in the bed sediment. Endrin, permethrin, and chlorpyrifos-methyl, although persisting for days to months in the bed sediment or biota, gradually dissipated. Carbaryl, 2,4-D, and picloram are moderate in water solubility, but would be expected to degrade in the environment eventually. Moderately hydrophobic, moderately persistent pesticides may be expected to be found in some bed sediment or biota samples, especially in areas of high or repeated use.

5.1.3 URBAN AREAS AND INDUSTRY

Another source of pesticides to surface water systems, and thus to bed sediment and aquatic biota, is from urban areas. Pesticides are applied to control pests for public health or aesthetic reasons in and around homes, yards, gardens, public parks, urban forests, golf courses, and public and commercial buildings (Buhler and others, 1973; Racke, 1993). The available data suggest that the patterns of urban pesticide use have changed during the past few decades, much as has pesticide use in agriculture and forestry. Many of the high use organochlorine insecticides have been banned and replaced by organophosphate, carbamate, and pyrethroid insecticides. The use of herbicides in and around homes and gardens has increased, whereas herbicide applications to industry, commercial, and government buildings and land have decreased (Aspelin, 1997).

The major pesticides used in and around homes and gardens in 1990 are listed in Table 3.5. An examination of Table 3.5 shows that most of the organochlorine pesticides that are commonly observed in bed sediment (Figure 3.1) and aquatic biota (Figure 3.2) are no longer used in urban areas, with the exception of dicofol, chlordane, heptachlor, lindane, and methoxychlor. The commercial use of existing stocks of chlordane in urban environments was banned in 1988, and homeowner use of existing stocks is likely to have declined since then also. Although the kind of data in Table 3.5 does not exist for the time period of the 1950s through mid-1970s, it is known that many of the organochlorine insecticides had significant urban uses, including aldrin, chlordane, DDT, dieldrin, endosulfan, heptachlor, and lindane (Meister Publishing Company, 1970). In 1970, lindane was used predominantly in the urban environment; there was also considerable urban use of chlordane (Meister Publishing Company, 1970). It seems that endrin was the exception, with little or no urban use. Of the moderately hydrophobic, moderately persistent pesticides that have been observed, when targeted, in sediment or aquatic biota, several are used in and around the home and garden (Table 3.5). These include chlorpyrifos, diazinon, carbaryl, permethrin, and 2,4-D.

A number of local-scale studies have monitored pesticides in the sediment or aquatic biota of urban areas. Mattraw (1975) examined the occurrence and distribution of dieldrin and DDT components in the bed sediment of southern Florida. The study area included the urbanized areas on the Atlantic coast (such as Miami and Fort Lauderdale), the Everglades water conservation area and two nearby agricultural areas. Mattraw reported the data as concentration frequency plots, shown in Figures 5.1 and 5.2. In the case of DDD (Figure 5.1), urban areas had a mean concentration and a general distribution between those of the two agricultural areas, and well above those of the undeveloped area. In the case of dieldrin (Figure 5.2), the urban areas had a mean bed sediment concentration and a general concentration distribution greater than all other land use activities. In another example, Kauss (1983) measured 15 different organochlorine insecticides and transformation products in the Niagara River below Buffalo, New York. This is an area with many large chemical production facilities. It is thought that some of the chemicals in the sediment of this river are due either to transport from Lake Erie (the source of water for the Niagara River) or to localized inputs. One example of a potential localized input is disposal of 1,700 metric tons of endosulfan at disposal sites in the area. In another urban area study, Thompson (1984) reported DDE, DDE, DDT, dieldrin, heptachlor, methoxychlor, silvex, and 2,4-D in the sediment of the Jordan River in Salt Lake City, Utah. Pariso and others (1984) reported that DDT and chlordane were observed in bed sediment and in various species of fish collected from the Milwaukee Harbor and Green Bay urban areas of Wisconsin, during a study of contaminants in the rivers draining into Lake Michigan. Lau and others (1989) reported the presence of *trans*-chlordane, DDE, DDD, and DDT in the suspended sediment of the St. Clair and Detroit rivers on the Michigan and Ontario border. Fuhrer (1989) reported the presence of chlordane, DDD, DDE, DDT, and dieldrin in the bed sediment of the Portland, Oregon harbor. Capel and Eisenreich (1990) reported concentrations of α-HCH, DDE, DDD, and DDT in the bed sediment and tissues of mayfly (*Hexagenia*) larvae from the harbor in Lake Superior at Duluth, Minnesota. Crane and Younghaus-Hans (1992) detected oxadiazon residues in fish (red shiner, *Cyprinella lutrensis*) and bed sediment from San Diego Creek, California. Oxadiazon was also detected in transplanted clams (*Corbicula fluminea*) in the San Diego Creek and in transplanted mussels (*Mytilis californianus*) in the receiving estuary, Newport Bay. Oxadiazon is widely used in landscape and rights-of-way maintenance in California, and the high residues observed in this study were attributed to its use on golf courses upstream of the study area.

Although there have been numerous local-scale studies, there has been no systematic large-scale study of pesticides in the bed sediment or aquatic biota of urban freshwater hydrologic systems. The National Oceanic and Atmospheric Administration's (NOAA) National Status and Trends (NS&T) Program targeted coastal and estuarine sites near urban population centers, and found a correlation between most organic contaminants in bottom sediment and human population levels (National Oceanic and Atmospheric Administration, 1991). However, there is no comparable nationwide study of pesticides in bed sediment or aquatic biota from rivers in urban areas. In the U.S. Geological Survey (USGS)–USEPA's Pesticide Monitoring Network (PMN), which sampled bed sediment from major United States rivers, only 10 of about 180 sites sampled between 1975–1980 were in urban areas (Gilliom and others, 1985). Nonetheless, two of these urban sites (Philadelphia, Pennsylvania, and Trenton, New Jersey) were among the 10 sites with the highest frequency of pesticide detection. The only national-scale study of pesticides in rivers near urban centers was the USEPA's Nationwide Urban Runoff Program (NURP), which analyzed water samples for pesticides in urban areas nationwide during 1980–1983 (Cole and others, 1983, 1984). The NURP samples were analyzed for the priority pollutants, which include 20 organochlorine insecticides or transformation products, at 61 residential and commercial sites across the United States. Of these 20 organochlorine insecticides, 13 were observed in at least one water sample. The most frequently observed organochlorine insecticides were α-HCH (in 20 percent of samples), endosulfan I (in 19 percent), pentachlorophenol (in 19 percent), chlordane (in 17 percent), and lindane (in 15 percent). During this time period, all of these chemicals were still in active use in urban areas. Because of the hydrophobicity of these compounds, their detection in the water column suggests that they also would have been present at detectable levels in bed sediment and aquatic biota in these urban environments.

Although many monitoring studies have reported the frequent detection of organochlorine pesticides in bed sediment, aquatic biota, and water in urban areas, the actual sources of these

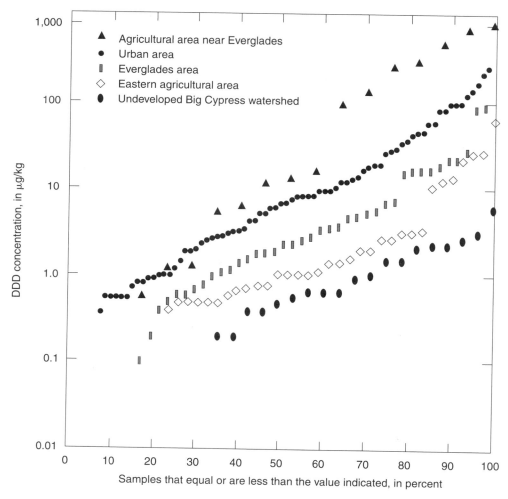

Figure 5.1. Concentration frequency plot for DDD in bed sediment from agricultural, urban, and undeveloped areas in southern Florida (1968–1972). Redrawn from Mattraw (1975) with permission of the author.

pesticide residues are not completely known. Since the organochlorine insecticides had both extensive urban and agricultural uses, their presence in urban areas could have been derived from either source, since many urban areas are located downstream from agricultural areas. Conversely, some rivers flowing through agricultural areas may be located downstream of urban areas. Examples are the Mississippi River below Minneapolis and St. Paul, Minnesota, and below St. Louis, Missouri. In such cases, residues may derive from urban, as well as from agricultural, origin. It is reasonable to suppose that most pesticides currently in bed sediment and aquatic biota in urban areas are derived from both agricultural and urban uses, although the relative contribution of each of the two sources probably varies by location and compound.

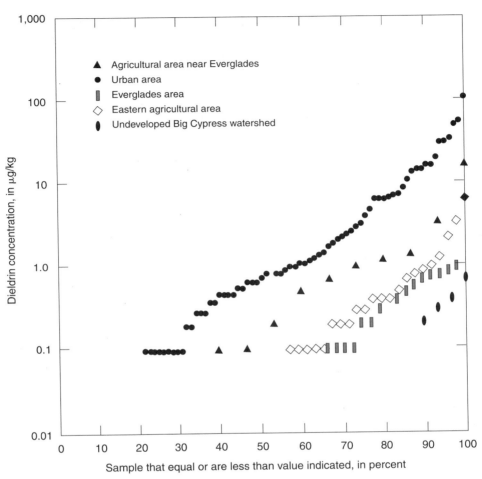

Figure 5.2. Concentration frequency plot for dieldrin in bed sediment from agricultural, urban, and undeveloped areas in southern Florida (1968–1972). Redrawn from Mattraw (1975) with permission of the author.

5.1.4 REMOTE OR UNDEVELOPED AREAS

Pesticides, particularly the organochlorine insecticides, are often observed in bed sediment and aquatic biota in remote areas of the United States and of the rest of the world. Their presence in remote or undeveloped areas is seldom due to local use, but rather to atmospheric transport and deposition. Majewski and Capel (1995) have reviewed the presence and movement of pesticides in the atmosphere and the deposition processes involved in their delivery to remote areas. For some pesticides, particularly the organochlorine insecticides, regional atmospheric transport is common and serves as a mechanism to disperse them throughout the world, particularly toward the polar regions.

Pesticides are introduced into the atmosphere either by volatilization or wind erosion. Once they are in the atmosphere, they can either be deposited locally (in the range of tens of kilometers) or move into the upper troposphere and stratosphere for more widespread regional, or possibly global, distribution. Once in the upper atmosphere, the global wind circulation patterns control their long-range transport. The general global longitudinal circulation is a form of thermal convection driven by the difference in solar heating between equatorial and polar regions. Over the long-term, upper air masses tend to be carried poleward and descend into the subtropics, subpolar, or polar regions. These air masses are then carried back toward the tropics in the lower atmosphere (Levy, 1990). Once in the atmosphere, the residence time of a pesticide depends on how efficiently it is removed by either deposition or chemical transformation. Atmospheric deposition processes can be classified into two categories: those involving precipitation (wet deposition) and those not involving precipitation (dry deposition). The effectiveness of a particular removal process depends on the physical and chemical properties of the pesticide, the meteorological conditions, and the terrestrial or aquatic surface to which deposition is occurring. Risebrough (1990) described the airborne movement of pesticides from their point of application as a global gas-chromatographic system where pesticide molecules move many times between the vapor-soil-water-vegetation phases, maintaining an equilibrium of chemical potential between these phases. That is, after a pesticide is deposited from the atmosphere to a terrestrial or aquatic surface, it can reenter the atmosphere and be transported and redeposited downwind repeatedly until it is chemically transformed or globally distributed.

Virtually all studies of pesticides in remote areas have been conducted on remote lakes and oceans, rather than on rivers and streams. A few examples of these studies will be presented to illustrate the global nature of atmospheric deposition. One of the earliest reports that attributed the presence of DDT in a remote surface water body to atmospheric deposition was a study by Swain (1978) conducted in the national park in Isle Royale, Michigan. Although this island is in Lake Superior and is removed hundreds of kilometers from agricultural uses of DDT, DDT was found in the water, sediment, and fish (lake trout, *Salvelinus namaycush*, and lake whitefish, *Coregonus clupeaformis*) of Siskitwit Lake on Isle Royale. The probable explanation for this contamination was through atmospheric deposition. Organochlorine contamination of air, snow, water, and aquatic biota in the Arctic has been extensively studied (Hargrave and others, 1988; Patton and others, 1989; Bidleman and others, 1990; Gregor, 1990; Muir and others, 1990) and also is attributed to atmospheric transport. All of the common organochlorine insecticides have been observed in Arctic studies, but the two most prevalent were α-HCH and lindane. These are the two organochlorine insecticides with the highest vapor pressures and their abundance supports the idea of the global gas-chromatographic effect of pesticides being transported to the polar regions described above. Although organochlorine concentrations in the Arctic water are low, these contaminants bioaccumulate in aquatic biota and appear to be magnified in aquatic and terrestrial food webs, reaching quite elevated levels in the Arctic mammals. Addison and Zinck (1986) found that the DDT concentration in the Arctic ringed seal (*Phoca hispida*) did not decrease significantly between 1969 and 1981, while the concentration of polychlorinated biphenyls (PCB) did decline. They attributed this to continued atmospheric deposition of DDT from its use in areas of eastern Europe during this time, compared with declining global PCB use.

5.2 PESTICIDE UPTAKE AND ACCUMULATION BY AQUATIC BIOTA

Historically, there has been controversy in the literature as to the mechanisms of contaminant uptake and bioaccumulation by aquatic biota. Probably the strongest controversy concerns whether biomagnification occurs in aquatic systems. In common usage, "biomagnification" may refer either to a process or to an effect (or phenomenon). The biomagnification process, in which the tissue concentrations of a contaminant increase as it passes up the food chain through two or more trophic levels, results in the phenomenon of biomagnification, in which organisms at higher trophic levels are observed to possess higher contaminant levels than their prey. An alternative school of thought holds that contaminant accumulation by aquatic organisms can be described using equilibrium partitioning theory, in which contaminant concentrations in water, blood, and tissue lipids approach equilibrium and the concentrations in these phases are related by partition coefficients. Regardless of the mechanism of uptake, the contaminant partitions into and out of these phases according to its relative solubility. Until fairly recently, biomagnification and equilibrium partitioning theories were considered mutually exclusive, since the phenomenon of biomagnification appeared to violate thermodynamic conditions of equilibrium. However, recent equilibrium partitioning models have attempted to incorporate and explain the biomagnification process (see Section 5.2.5).

The relative importance of contaminant uptake from the diet and from water via partitioning has also been debated in the literature. Although dietary uptake has been associated with biomagnification and uptake by partitioning with equilibrium partitioning theory, this is not a true dichotomy. Uptake of a contaminant by aquatic organisms can occur via partitioning of the contaminant from water, pore water, or sediment; and via ingestion of contaminated food or sediment. These are not mutually exclusive mechanisms of uptake, and indeed it is frequently assumed that bioaccumulation in the field results from multiple routes of uptake. Dietary uptake is not inconsistent with equilibrium partitioning theory; the critical issue in equilibrium partitioning theory is that, however a contaminant enters an organism, it will partition within the organism or be eliminated from the organism according to its relative solubility in these compartments, or phases.

Hundreds of laboratory and field studies have been performed that attempt to elucidate uptake mechanisms or to test the theories of biomagnification or equilibrium partitioning. Although each theory has been the dominant one at some time in the past, the extensive discussion and effort put into experimentation, field monitoring, and modeling during the past three decades have begun to achieve some resolution between the two schools of thought. In summary, the route of uptake (diet versus partitioning) and the mechanism of bioaccumulation (biomagnification versus equilibrium partitioning) in aquatic systems appear to depend on the characteristics of the chemical (such as hydrophobicity, and molecular weight and structure), on the organisms involved (such as species, age, body size, reproductive state, lipid content, and metabolic capability), and on environmental factors (such as temperature).

In the remainder of this section, some terminology and simple models of bioaccumulation are defined (Section 5.2.1). Next, the two theories of biomagnification (Section 5.2.2) and equilibrium partitioning (Section 5.2.3) are described, and then some laboratory and field studies that attempted to test these theories are examined (Section 5.2.4). Finally, an emerging resolution between competing mechanisms of bioaccumulation is presented by describing the current understanding of the processes of uptake and elimination; the biological, chemical, and environmental factors that affect contaminant accumulation; and examples of some different types of bioaccumulation models (Section 5.2.5).

5.2.1 BIOACCUMULATION TERMINOLOGY AND SIMPLE MODELS

In the early literature, individual authors defined their own terminology to describe uptake and accumulation by biota. Because the same terms were not used consistently by different authors, this added confusion to the already complex subject under investigation. Today, conventional definitions of several terms exist that have facilitated organized discussion of contaminant uptake mechanisms. These terms were introduced in Section 4.3, and are described in more detail below. In general, contaminant accumulation can be viewed as a function of competing processes of uptake and elimination.

Bioconcentration refers to chemical residue obtained directly from water via gill or epithelial tissue (Brungs and Mount, 1978). The bioconcentration process is viewed as a balance between two kinetic processes, uptake and elimination, as quantified by pseudo-first-order rate constants k_1 and k_2, respectively.

$$dC_b/dt = k_1 C_w - k_2 C_b \qquad (5.1)$$

where C_b is the concentration in biota (in units of pesticide mass per tissue mass) and C_w is the concentration in the surrounding water (in units of pesticide mass per volume). The elimination constant, k_2, refers to diffusive release only. This simple model assumes that there is no contaminant uptake from food, no metabolism, no excretion, and no growth dilution. The *bioconcentration factor* (BCF) is defined as the ratio of a contaminant concentration in biota to its concentration in the surrounding medium (water). At long exposure times (equilibrium), the BCF also equals the ratio of the uptake constant (k_1) to the elimination constant (k_2) (Mackay, 1982).

$$BCF = C_b/C_w = k_1/k_2 \qquad (5.2)$$

The BCF can be measured in the laboratory in either of two ways. First, using the steady-state approach, biota (usually fish) are exposed to an aqueous solution of the target contaminant for a fixed length of time. The BCF then is calculated as the ratio of the concentration measured in fish to the concentration measured in water at the end of the experiment. Second, using the kinetic approach, uptake and elimination rate constants are measured in separate experiments and the BCF is calculated as the ratio k_1/k_2. At equilibrium, the two methods should give the same results. For extremely hydrophobic contaminants that require a long time to reach equilibrium, the kinetic approach permits estimation of the BCF over a shorter exposure time.

Bioaccumulation is the process whereby a chemical enters an aquatic organism through the gills, epithelial tissue, dietary intake, and other sources (Brungs and Mount, 1978). Use of this term does not imply any particular route of exposure. It is commonly used when referring to field measurements of contaminant residues in biota, where the routes of exposure are unknown. Bioaccumulation, like bioconcentration, is viewed as a balance between processes of uptake and elimination, except that in a bioaccumulation model, multiple routes of uptake and elimination are possible. The kinetics of a bioaccumulation model can be described as:

$$dC_b/dt = k_1 C_w + \alpha\beta C_f - k_2 C_b - k_e C_b - k_m C_b - Rk_r C_b - GC_b \qquad (5.3)$$

This specific model considers uptake via water and food, as well as elimination via excretion from gills and feces, biotransformation, reproduction, and growth (Gobas and others, 1989b; Sijm and others, 1992). The terms are defined as follows: C_b is the concentration in biota; C_w is the concentration in water; C_f is the concentration in food; k_1 and k_2 are diffusion-controlled constants for uptake and elimination, respectively; α is the absorption efficiency of a chemical from food, which varies from 0 to 1; β is the food consumption rate; k_e is the rate constant for elimination in feces; k_m is the biotransformation rate constant; k_r is the zero-order reproduction rate; R is a trigger value that is either 0 or 1 (depending on whether reproduction takes place or not); and G is the growth dilution factor. The *bioaccumulation factor* (BAF) is analogous to the BCF, but applies to field measurements or to laboratory measurements with multiple exposure routes. The BAF is the ratio of contaminant concentration measured in biota in the field (or under multiple exposure conditions) to the concentration measured in the surrounding water. At steady state, chemical fluxes into and out of the fish are equal, so the quantity (dC_b/dt) equals zero. Therefore:

$$\text{BAF} = C_b/C_w = (k_1 + \alpha\beta \text{BCF}_f)/(k_2 + k_e + k_m + Rk_r + G) \qquad (5.4)$$

where BCF_f is the bioconcentration factor of the food (i.e., the ratio of C_f/C_w).

Biomagnification is the process whereby the tissue concentrations of a chemical increase as it passes up the food chain through two or more trophic levels (Brungs and Mount, 1978). Biomagnification is also called the "food chain effect." Occasionally, the term *biomagnification factor* (BMF) is used in the literature to refer to the ratio of contaminant concentration in biota to that in the surrounding water when the biota was exposed via contaminated food.

5.2.2 BIOMAGNIFICATION

In theory, biomagnification begins with ingestion by a predator of a lower trophic level organism whose tissues contain contaminant residues. This theory was supported initially by field observations and later by food chain models (see Section 5.2.4). These include many observations of increasing contaminant residues at higher trophic levels, as well as higher residues of metabolites in predators than prey. Also, field-measured BAFs often were higher than BCFs measured in the laboratory during water-only exposures, indicating that partitioning from water did not adequately account for residues bioaccumulated by aquatic organisms in natural systems.

The available evidence suggests that biomagnification may occur under conditions of low water concentration for compounds of high lipophilicity, high persistence, and low water solubility (Biddinger and Gloss, 1984). Biomagnification is most likely to occur for chemicals with log n-octanol-water-partition coefficient (K_{ow}) values greater than 5 or 6 (Connell, 1988; Gobas and others, 1993b) and for top predators with long lifetimes. Dietary intake and biomagnification are very important for air-breathing vertebrates (see Section 5.2.5, subsection on Uptake Processes).

The mechanism by which biomagnification operates is not completely understood. As previously noted, this subject has been controversial, since biomagnification appeared to be inconsistent with thermodynamic conditions (see Section 5.2.5, subsection on Dietary Uptake and Biomagnification). During the 1960s, the hypothesis prevailed that bioaccumulation in aquatic systems was controlled by mass transfer through the food chain (Rudd, 1964; Hunt, 1966; Woodwell, 1967; Woodwell and others, 1967; Harrison and others, 1970). This was based on the observation that hydrophobic chemical concentrations increased with increasing trophic levels in aquatic systems (Hunt and Bischoff, 1960; Woodwell, 1967; Woodwell and others, 1967) and by analogy to terrestrial species, for which food was usually the dominant route of uptake (Moriarty and Walker, 1987). This food chain effect was traditionally explained by the loss of biomass in the food chain due to respiration and excretion as biomass is transferred from one trophic level to the next (Woodwell, 1967). This assumes that, for each step in the food chain, more chemical residues are retained than energy or body mass (Hamelink and Spacie, 1977).

Subsequently, it was pointed out (Hamelink and Spacie, 1977) that this mechanism must take growth efficiency and dietary uptake efficiency into account. The growth efficiency of fish is about 8 percent: thus dietary uptake efficiencies should exceed this value for any increase in contamination to occur (Connell, 1988). Dietary uptake efficiencies reported for some organochlorine compounds in fish ranged from 9 to 68 percent and tend to decline with increasing concentration, reaching a steady state (Hamelink and Spacie, 1977). Moreover, some observations of food chain effects can be explained by lipid-based partitioning (Section 5.2.3, Relation Between Contaminant Residues and Trophic Levels). The original mass-transfer mechanism is now considered unlikely to account for steadily increasing contaminant concentration with increasing trophic level (Connell, 1988). More recently, it was proposed that food digestion and absorption from the gastrointestinal tract, accompanied by inflow of more contaminated food, increase the concentration of the chemical in the gastrointestinal tract relative to that in the original food (Connolly and Pedersen, 1988; Gobas and others, 1988, 1993b; also see Section 5.2.5, subsection on Uptake Processes).

5.2.3 EQUILIBRIUM PARTITIONING THEORY

The hypothesis of food chain transfer was first questioned around 1971 (Hamelink and others, 1971; Woodwell and others, 1971). Hamelink and others (1971) instead proposed that organisms continuously exchange pesticide residues with the surrounding water, in theory reaching a chemical equilibrium with their environment. As an approximation, the organism was viewed as a pool of lipophilic material, and contaminant accumulation was proposed to be controlled by sorption to body surfaces and partitioning into lipids from water. This equilibrium partitioning hypothesis prevailed for almost 15 years (e.g., National Research Council, 1979, 1985; Levin and others, 1985). Recently, some equilibrium partitioning models have attempted to incorporate dietary intake and to explain the phenomenon of biomagnification (Section 5.2.5).

The equilibrium partitioning theory holds that, at equilibrium, the thermodynamic activity of a chemical will be the same in all phases of the system (Hamelink and others, 1971). The organism is considered to be a single, uniform compartment, with the solubility of the chemical in the organism controlled by the chemical's solubility in lipid. The rate of uptake is controlled by the concentration gradient between the organism and the surrounding water. This simple model assumes the following: uptake and elimination show pseudo-first-order kinetics; uptake is limited only by diffusion; the BCF is controlled by the hydrophobicity of the chemical and the lipid content of the fish; and there is negligible growth or metabolism. This theory was supported by laboratory experiments that demonstrated that experimentally determined values for BCF were directly correlated with the K_{ow}, and inversely correlated with water solubility (Section 5.2.4). n-Octanol is a convenient surrogate for the lipid phase (Mackay, 1982), and the K_{ow} is a useful estimate of the degree of hydrophobicity (Farrington, 1989). Laboratory experiments that show correlations between BCF and chemical properties do not prove the equilibrium partitioning theory, but they are consistent with it. On the other hand, instances where BCF fails to correlate with these chemical properties may indicate limitations in the equilibrium partitioning model.

One key bioaccumulation model (also see Section 5.2.5, subsection on Bioaccumulation Models) is the *fugacity model* developed by Mackay (1982). This is a simple equilibrium partitioning model that views an organism as an inanimate volume consisting of multiple phases of differing chemical composition. A chemical diffuses between the organism and water because of a concentration gradient. The rate of uptake can be expressed using Fick's law, which holds that sorption of a lipid-soluble chemical through an integument is generally pseudo-first-order, with the rate of sorption proportional to the surface area and concentration of the diffusing chemical, and inversely proportional to the thickness of the integument. When the two phases (organism and water) are not in equilibrium, the concentration gradient determines which direction the chemical will diffuse to reach equilibrium. This situation can be described using fugacity concepts and terminology. In general, *fugacity* is a thermodynamic measure of the escaping tendency of a chemical from a phase, and is equivalent to chemical activity or potential. Fugacity has units of pressure and is proportional to concentration in the phase. Mass diffuses from high to low fugacity under nonsteady-state conditions. When the escaping tendencies of a chemical from two phases are equal, the phases are in equilibrium. According to the fugacity model, contaminant uptake by the organism is determined by the chemical fugacity differential between the organism and the surrounding medium (water). At low concentrations (such as those that commonly occur in the environment), fugacity is related to concentration as follows:

$$C = (Z)(f) \tag{5.5}$$

where C is concentration (in units of mole per cubic meter, or mol/m^3), f is fugacity (Pascal), and the proportionality constant Z is the fugacity capacity ($mol/m^3/Pascal$). The fugacity capacity depends on the temperature, the pressure, the chemical, and the environmental medium; it quantifies the capacity of each phase for fugacity. For biota, actual uptake may be a combination of uptake from the surrounding medium (water) and from food, which also may be at or approaching equilibrium with the surrounding water.

Some of the predictions of the fugacity model have been tested using field data (Section 5.2.4). For example, in its simplest form, the fugacity model predicts that the animal/water fugacity ratio will be 1 at equilibrium and that the concentration of a contaminant in the lipids of all animals must be equal, regardless of trophic position. This condition is termed *equifugacity*. Only under nonequilibrium conditions may the fugacity ratio deviate from 1.

5.2.4 EVIDENCE FROM LABORATORY AND FIELD STUDIES

Some key laboratory and field studies that attempted to test the validity of the biomagnification or equilibrium partitioning theories are discussed in this section. Some studies looked for evidence of biomagnification in the field, and other studies attempted to test predictions of equilibrium partitioning theory. These studies have helped to elucidate the biological, environmental, and chemical factors that affect bioaccumulation.

Evidence of Biomagnification in the Field

Three types of evidence of biomagnification in the field will be discussed: (1) correlations between contaminant concentrations and trophic levels in aquatic biota, (2) comparison of laboratory BCFs that are based on water exposure only with field-measured BAFs, and (3) development and validation of food chain models.

Effect of Trophic Level on Contaminant Concentrations

There are many examples of field studies in which contaminant concentrations in aquatic biota were observed to increase with increasing trophic levels. In a Long Island (New York) salt marsh, DDT residues in marine organisms increased with increasing organism size and increasing trophic level (Woodwell and others, 1967). Total DDT residues ranged over three orders of magnitude, from 40 µg/kg wet weight in plankton to 2,070 µg/kg in a carnivorous fish (the Atlantic needlefish, *Strongylura marina*) to 75,500 µg/kg in ring-billed gulls (*Larus delawarensis*), as shown in Table 5.1. In later examples, accumulation was found to be directly related to position in the food chain for the following: chlordane, total DDT, and dieldrin in zooplankton, forage fish, and predator fish in the Great Lakes (Whittle and Fitzsimons, 1983); total DDT in amphipods (*Pontoporeia affinis*), various fish species, and ducks from Lake Michigan (Ware and Roan, 1970); DDT in krill, benthic fish, and Weddell seals (*Leptonychotes weddelli*) in the Antarctic Ocean (Hidaka and others, 1983); kepone in the James River food chain (Connolly and Tonelli, 1985); PCBs in the lake trout food chain in Lake Michigan (Thomann and Connolly, 1984); PCBs in the yellow perch (*Perca flavescens*) food chain in the Ottawa River (Norstrom and others, 1976); organochlorine compounds in micro- and macro-zooplankton off the Northumberland coast (Robinson and others, 1967); pesticides and PCBs in periphyton, green algae, macrophytes, snails, and various fish in the Schuylkill River, Pennsylvania (Barker, 1984); and hexachlorobenzene and PCBs in white bass (*Morone chrysops*) from Lake Erie (Russell and others, 1995). Tanabe and others (1984) reported increasing concentrations from zooplankton to squid for total DDT and PCBs, but not for total HCH (which is less hydrophobic and has a lower K_{ow}). In examining field data on residues in benthic animals from

Table 5.1. Residues of total DDT in samples from the Carmans River Estuary, Long Island, New York

[Residues are in mg/kg wet weight of the whole organism, unless otherwise indicated. Proportions of DDD, DDE, and DDT are expressed as a percentage of total DDT. Letters in parentheses indicate replicate samples in original reference as follows: there were three common tern replicates (a–c), two green heron replicates (a–b), six herring gull replicates (a–f), and 2 least tern replicates (a–b). Abbreviations and symbols: mg/kg, milligrams per kilogram; —, no data. Reproduced from Woodwell and others (1967) with permission of the publisher. Copyright 1967 American Association for the Advancement of Science]

Sample	DDT Residues (mg/kg)	Percentage of Residue as:		
		DDT	DDE	DDD
Water[1]	[1]0.00005	—	—	—
Plankton, mostly zooplankton	0.040	25	75	Trace
Cladophora gracilis	0.083	56	28	16
Shrimp[2]	0.16	16	58	26
Opsanus tau, oyster toadfish (immature)[2]	0.17	None	100	Trace
Menidia menidia, Atlantic silverside[2]	0.23	17	48	35
Crickets[2]	0.23	62	19	19
Nassarius obsoletus, mud snail[2]	0.26	18	39	43
Gasterosteus aculeatus, threespine stickleback[2]	0.26	24	51	25
Anguilla rostrata, American eel (immature)[2]	0.28	29	43	28
Flying insects, mostly Diptera[2]	0.30	16	44	40
Spartina patens, shoots	0.33	58	26	16
Mercenaria mercenaria, hard clam[2]	0.42	71	17	12
Cyprinodon variegatus, sheepshead minnow[2]	0.94	12	20	68
Anas rubripes, black duck	1.07	43	46	11
Fundulus heteroclitus, mummichog[2]	1.24	58	18	24
Paralichthys dentatus, summer flounder[3]	1.28	28	44	28
Esox niger, chain pickerel	1.33	34	26	40
Larus argentatus, herring gull, brain (d)	1.48	24	61	15
Strongylura marina, Atlantic needlefish	2.07	21	28	51
Spartina patens, roots	2.80	31	57	12
Sterna hirundo, common tern (a)	3.15	17	67	16
Sterna hirundo, common tern (b)	3.42	21	58	21
Butorides virescens, green heron (a) (immature, found dead)	3.51	20	57	23
Larus argentatus, herring gull (immature) (a)	3.52	18	73	9
Butorides virescens, green heron (b)	3.57	8	70	22
Larus argentatus, herring gull, brain[4] (e)	4.56	22	67	11
Sterna albifrons, least tern (a)	4.75	14	71	15
Sterna hirundo, common tern (c)	5.17	17	55	28

Table 5.1. Residues of total DDT in samples from the Carmans River Estuary, Long Island, New York—*Continued*

Sample	DDT Residues (mg/kg)	Percentage of Residue as:		
		DDT	DDE	DDD
Larus argentatus, herring gull (immature) (b)	5.43	18	71	11
Larus argentatus, herring gull (immature) (c)	5.53	25	62	13
Sterna albifrons, least tern (b)	6.40	17	68	15
Sterna hirundo, common tern (five abandoned eggs)	7.13	23	50	27
Larus argentatus, herring gull (d)	7.53	19	70	11
Larus argentatus, herring gull[4] (e)	9.60	22	71	7
Pandion haliaetus, osprey (one abandoned egg)[5]	13.8	15	64	21
Larus argentatus, herring gull (f)	18.5	30	56	14
Mergus serrator, red-breasted merganser (1964)[3]	22.8	28	65	7
Phalacrocorax auritus, double-crested cormorant (immature)	26.4	12	75	13
Larus delawarensis, ring-billed gull (immature)	75.5	15	71	14

[1] In units of milligrams per liter.
[2] Composite sample of more than one individual.
[3] From Captree Island, New York, 20 miles (32 kilometers) west-southwest of study area.
[4] Found moribund and emaciated, north shore of Long Island, New York.
[5] From Gardiners Island, Long Island, New York.

the Great Lakes, Bierman (1990) observed that body burdens of various organic chemicals were significantly higher for carp than for all other organisms, and higher than predicted by equilibrium partitioning theory. However, body burdens in forage fish were not significantly different from those in benthic macroinvertebrates.

Moreover, several authors reported that the relative concentration ratios of pesticides to pesticide metabolites in fish varied with respect to trophic levels. Fish at higher trophic levels contained a higher percentage of pesticide metabolites (DDE, DDD, heptachlor epoxide, dieldrin) than fish at lower trophic levels (Hannon and others, 1970; Johnson, 1973). Organisms at lower trophic levels had proportionally more DDT (parent compound) residues, relative to organisms at higher trophic levels (Woodwell and others, 1967; Johnson, 1973).

In some studies, no clear relation between hydrophobic contaminant residues and trophic level was observed. Examples include the following: dieldrin in aquatic invertebrates in the Rocky River, South Carolina (Wallace and Brady, 1971); PCBs in cod (*Gadus morhua*) (livers and fillets) and prey organisms from the western Baltic Sea (Schneider, 1982); and organo-chlorine residues in amphipods and other stream animals from Swedish streams (Sodergren and others, 1972). The lack in finding any food chain effects has been attributed to the complexity of food chains in the communities sampled (Schneider, 1982), differences in metabolic capability, habitat conditions, seasonal effects, or subtle differences in feeding strategy (Wallace and Brady, 1971).

Hamelink and others (1971) conducted mesocosm studies investigating the behavior of DDT in food chains. Fish rapidly accumulated total DDT after *p,p'*-DDT was added to the water, and there was no difference in residues between complete food chains (algae, invertebrates, fish), or broken food chains (algae, fish; or algae, invertebrates). These authors observed a stepwise increase in residue levels between trophic levels, whether or not food chains were broken or complete. They questioned the biomagnification theory and proposed that the uptake mechanism involved sorption and partitioning into body lipids. One factor complicating interpretation of these results is that the broken food chains were fed, while those in the complete food chains were not, even though the food supply was inadequate to maintain the fish in prime condition.

Biddinger and Gloss (1984) reviewed field, laboratory, and artificial ecosystem studies that assessed bioconcentration, dietary uptake, and potential biomagnification of organic contaminants. They concluded that food chain biomagnification was not well substantiated in the literature at that time, but that it was most likely to occur under conditions of low water concentration for compounds of high lipophilicity, low water solubility, and high persistence. They also pointed out that most cases of high residues that occurred in organisms of high trophic levels had not been shown to be the result of trophic transfer; rather, factors such as age, size, sex, season, lipid content, and physical condition may have been involved. This does not disprove the theory of biomagnification, but merely illustrates the difficulty in deducing cause and effect from field studies.

The observed progression in residue levels with trophic level was explained by some authors as an artifact that organisms at higher trophic levels have greater lipid pools than those at lower trophic levels (Hamelink and others, 1971; Clayton and others, 1977; Goerke and others, 1979; Ellgehausen and others, 1980). This suggests that lipid normalization of residues would reduce or eliminate any trophic level effect observed for wet-weight residues, which was the case in a few studies. When concentrations were lipid-normalized, mean PCB concentrations for marine zooplankton were similar to those for marine fish (such as herring and salmon—species not specified) (Clayton and others, 1977). PCB levels per weight of extractable lipids in cod (*G. morhua*) and prey organisms from the western Baltic Sea were more uniform than wet weight residues, indicating the important role of lipids in PCB bioaccumulation (Schneider, 1982). Lipid content and composition have been suggested as one basis for seasonal effects in contaminant accumulation (discussed in Section 5.3.5), as well as for differences among species and tissue types (discussed in Section 5.2.4, subsection on Lipid Normalization).

Other explanations have been offered as the basis for the trophic level effects commonly observed in the field. Biddinger and Gloss (1984) noted that increases in contaminants with trophic level have occurred only for a few extremely hydrophobic contaminants (such as DDT and PCBs), and these increases generally were less than an order of magnitude over the whole aquatic food chain. The apparent trophic level effects observed in field surveys have been attributed to nonequilibrium conditions that exist in the field; because the direct uptake (via partitioning) of extremely hydrophobic compounds is slow, feeding may provide significant exposure to these compounds for high trophic levels (Connolly and Pedersen, 1988). Also, because population turnover rates are more rapid at lower and intermediate trophic levels than at high trophic levels, it has been suggested that apparent biomagnification may be an artifact of the period of exposure of different trophic levels (Grzenda and others, 1970).

There remain some observations of trophic level effects in the field that have not been explained by differences in lipid content or other factors. For example, Crossland and others (1987) monitored distribution of 2,5,4'-trichlorobiphenyl in ponds stocked with grass carp

(*Ctenopharyngodon idella*) and rainbow trout (*Oncorhynchus mykiss*). By eight days after exposure began, the trout had significantly higher residues than carp on a lipid-weight basis. The stomach contents of the fish were examined to determine what foods were consumed. The stomach contents of all of the grass carp contained aquatic vegetation and no invertebrates, whereas those of all of the trout contained zooplankton, snails, arthropods, and no aquatic vegetation. The higher accumulation of 2,5,4'-trichlorobiphenyl by trout could not be explained in terms of differences in lipid content, growth rates, or metabolic rates. Crossland and others (1987) suggested that accumulation via the food chain was responsible. In another example, lipid-based BCF values did not explain the high PCB concentrations observed at upper trophic levels in Lake Michigan (Thomann and Connolly, 1984). Also, lipid-normalized PCB residues in four invertebrate species and one fish species (sole, *Solea solea*) from the Wadden Sea were correlated with trophic level (Goerke and others, 1979).

Bioaccumulation Factors

For extremely hydrophobic contaminants, BAF values measured in field surveys (where biota may be exposed to contaminants via multiple routes, such as water, food, and sediment) are commonly higher than BCF measurements made in the laboratory on the basis of aqueous exposure only. For example, this has been observed for DDT (Biddinger and Gloss, 1984), hexachlorobenzene (Oliver and Niimi, 1983), mirex (Oliver and Niimi, 1985), and PCBs (Oliver and Niimi, 1985; Porte and Albaiges, 1994). Some authors have concluded that uptake from water alone underestimates residues of these contaminants in aquatic biota, indicating that these residues are partly derived from dietary uptake (Biddinger and Gloss, 1984; Oliver and Niimi, 1985; Porte and Albaiges, 1994). In contrast, for hydrophobic contaminants with short half-lives in fish, laboratory-derived BCFs were comparable with field-measured BAFs. Examples include lindane, α-HCH, 1,2,4-trichlorobenzene, and 1,2,3,4-tetrachlorobenzene in rainbow trout from Lake Ontario (Oliver and Niimi, 1983, 1985). For these compounds, direct uptake from water can account for residues observed in field surveys.

The observation that field-based BAF measurements are higher than laboratory-derived BCF values for some contaminants does not by itself indicate food chain transfer. Because organisms in the field are exposed over a lifetime, they may be closer to equilibrium than in short-term laboratory experiments (Connolly and Pedersen, 1988). This is particularly relevant for extremely hydrophobic compounds, since the time to achieve equilibrium increases with increasing K_{ow} (Veith and others, 1979a; Hawker and Connell, 1985). This is illustrated in Figure 5.3 for chlorobenzenes in rainbow trout (Oliver and Niimi, 1983). The higher the degree of chlorination, the longer it was required for the systems to equilibrate, and thus, the higher the BCF value. Hexachlorobenzene did not reach equilibrium within the duration of the experiments (about 120 days). Oliver and Niimi (1985) subsequently reported that *p,p'*-DDE, *cis*- and *trans*-chlordane, several PCB congeners (18, 40, 52, 101, 155), octachlorostyrene, and mirex did not reach equilibrium in 96-day laboratory experiments to measure BCF values.

For contaminants that take a long time to reach equilibrium, the field-based BAF can be compared with the BCF at theoretical equilibrium, which is estimated using the kinetic approach. As described in Section 5.2.1, this entails measuring the uptake and elimination rate constants in separate experiments, then calculating the BCF as the ratio of the rate constants (Equation 5.2). In contrast, the BCF that is measured using the steady-state approach (in which the BCF is calculated as the ratio of the measured concentrations of the contaminant in fish and in water at the end of the exposure time) is likely to underestimate the BCF value for compounds that require a long time to reach equilibrium. In tests with rainbow trout, kinetic and steady-state BCF values were similar for compounds (such as lindane) with short half-lives in fish (less than about 30 days). For compounds with longer half-lives in fish, the two BCF values did not agree; the steady-state approach underestimated the BCF because steady state probably had not yet been reached for chemicals that were eliminated slowly by fish (Oliver and Niimi, 1985). These laboratory-derived BCFs were compared with field BAFs for rainbow trout from Lake Ontario (Oliver and Niimi, 1985). Predicted BAF values (based on laboratory BCFs) for lindane, α-HCH, 1,2,4-trichlorobenzene, and 1,2,3,4-tetrachlorobenzene were close to the observed BAF values. These compounds all have short half-lives in fish. For compounds (such as PCBs) with longer half-lives in fish, both steady-state and kinetic BCFs underestimated field BAF values by a factor of 3 to 220. Because residues in Lake Ontario fish were higher than what could result from bioconcentration from water, Oliver and Niimi (1985) concluded that dietary uptake was the major source of contamination for some compounds.

Field Modeling

A bioaccumulation model incorporating dietary intake and biomagnification was developed by Norstrom and others (1976) for PCBs in yellow perch; this model included such factors as dietary efficiency, contaminant concentration in food, and caloric requirements for growth and respiration. This type of model was later expanded to apply to entire food chains (e.g., Thomann, 1981, 1989; Thomann and Connolly, 1984). Aquatic food chain models have predicted high residues of hydrophobic contaminants in top predators (e.g., Weininger, 1978; Thomann, 1981; Biddinger and Gloss, 1984) and the importance of dietary sources (Thomann and Connolly, 1984; Thomann, 1989).

The food chain model developed by Thomann (1989) contains four trophic levels (above phytoplankton), and assumes steady-state conditions and uptake from water and food. For

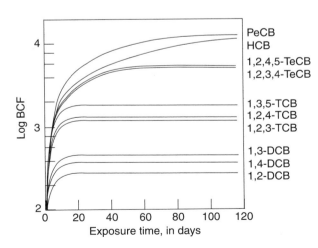

Figure 5.3. The logarithm of the bioconcentration factor (log BCF) for 10 chlorobenzenes measured in rainbow trout as a function of exposure time (days). Abbreviations: DCB, dichlorobenzene; HCB, hexachlorobenzene; PeCB, pentachlorobenzene; TeCB, tetrachlorobenzene; TCB, trichlorobenzene. Redrawn from Oliver and Niimi (1983) with permission of the publisher. Copyright 1983 American Chemical Society.

compounds with log K_{ow} values between 3.5–6.5, BAF values predicted by the model were found to approximate the values observed. For compounds with log K_{ow} values greater than 6.5, field BAF values predicted by the model were higher than those observed in the field, with the magnitude of the difference depending on assumed values for certain parameters of the model (the chemical assimilation efficiency, the BCF for phytoplankton, and the predator growth rate). According to the model, food chain accumulation becomes significant for compounds with log K_{ow} values above 5.0. At a log K_{ow} of 6.5, accumulation in the top predator was attributed almost entirely to the food chain.

Other field models have shown the importance of the food chain in fish contaminated with DDT or PCBs. For example, PCB concentrations in lake trout from a wide range of lakes in Ontario, Canada, were determined by the number of pelagic trophic levels (length of the food chain), fish lipid content, and distance from urban–industrial centers (Rasmussen and others, 1990). Empirical models of variability in fish contamination between lakes of the Great Lakes showed that concentrations of PCBs and DDT in water and sediment could explain variability in fish contamination between basins only when basin-specific ecological attributes were included (Rowan and Rasmussen, 1992). The most important factors were fish lipid content, fish trophic level, and the trophic structure of the food chain. Multiple regressions of these variables explained 59 percent (DDT) to 72 percent (PCBs) of the variation in contaminant concentrations of 25 species of Great Lakes fish.

Testing Predictions of Equilibrium Partitioning Theory

The next four groups of studies attempted to test predictions of equilibrium partitioning theory. These studies assessed (1) correlations between measured BCFs and chemical properties, (2) fish/sediment ratios, (3) the effect of trophic level on fugacity, and (4) the effect of lipid normalization on data variability.

Correlation Between Bioconcentration Factor and Chemical Properties

The equilibrium partitioning theory of uptake (Hamelink and others, 1971) was supported by many laboratory experiments demonstrating that experimentally determined values for BCF were directly correlated with K_{ow}, the *n*-octanol-water partition coefficient (Neely and others, 1974; Sugiura and others, 1979; Veith and others, 1979a; Kenaga, 1980a,b; Kenaga and Goring, 1980; Mackay, 1982; Shaw and Connell, 1984) and inversely correlated with water solubility (Kapoor and others, 1973; Chiou and others, 1977; Kenaga, 1980a,b; Kenaga and Goring, 1980; Mackay and others, 1980; Bruggeman and others, 1981). As noted above, equilibrium partitioning theory views an aquatic organism as a pool of lipophilic material and chemical accumulation as primarily a lipid–water partitioning process. *n*-Octanol is a convenient surrogate for lipids, and the K_{ow} is a useful estimate of the degree of hydrophobicity (Farrington, 1989). In an early example, Neely and others (1974) demonstrated a linear relation between log BCF (measured in the muscle of rainbow trout) and log K_{ow}:

$$\log BCF = (0.542)(\log K_{ow}) + 0.124 \tag{5.6}$$

This regression line is one of three such regression lines shown in Figure 5.4. In the study by Neely and others (1974), BCF values were measured using the kinetic approach (i.e., as the ratio of uptake and elimination rate constants). This approach permits measurement of BCF even if equilibrium is not reached before the end of the experiment. Metcalf and colleagues demonstrated that bioconcentration of organic chemicals by mosquitofish (*Gambusia affinis*) in artificial ecosystems was directly related to K_{ow} (Lu and Metcalf, 1975), as shown in Figure 5.5, and inversely related to water solubility (Metcalf, 1977), as shown in Figure 5.6.

Several authors have noted that some points on a plot of log BCF versus log K_{ow} depart from the general linear correlation (e.g., Mackay, 1982; Oliver and Niimi, 1983; Shaw and Connell, 1984). Proposed explanations for this have included metabolism, food input, differential uptake and elimination (O'Connor and Pizza, 1987), lipid type and content (Chiou, 1985), and inaccurate K_{ow} measurements (Oliver and Niimi, 1985). Mackay (1982) plotted log BCF and log K_{ow} for 50 compounds originally compiled by Veith and others (1979a) as well as additional values from the literature. After deleting values suspected to be in error (such as values attributed to error in calculated K_{ow} or to potential ionization), Mackay (1982) found that BCF and K_{ow} were directly proportional and developed the following relation:

$$\text{BCF} = f_{\text{lipid}} \times K_{ow} = 0.048 \times K_{ow} \tag{5.7}$$

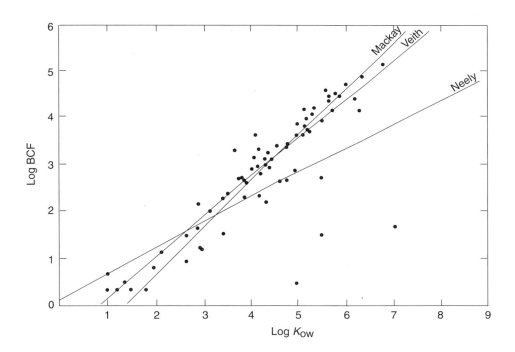

Figure 5.4. Correlations between the logarithm of the bioconcentration factor (log BCF) and the logarithm of the *n*-octanol-water partition coefficient (log K_{ow}) from the work of Mackay (1982), and of Neely and others (1974) and Veith and others (1979a) as cited in Mackay (1982). Redrawn from Mackay (1982) with permission of the publisher. Copyright 1982 American Chemical Society.

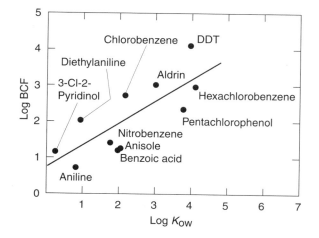

Figure 5.5. The relation between the logarithm of the bioconcentration factor (log BCF) in mosquitofish (*Gambusia affinis*) and the logarithm of the *n*-octanol-water partition coefficient (log K_{ow}) for various organic chemicals in laboratory model ecosystem studies. The regression line was computed by the method of least squares. Redrawn from Lu and Metcalf (1975).

where f_{lipid} is the fraction of tissue composed of lipid. This relation is plotted in Figure 5.4, as are the previous correlations determined by Neely and others (1974) and Veith and others (1979a). However, Mackay (1982) noted that this relation failed for extremely hydrophobic compounds (log K_{ow} greater than 6), compounds with BCF values of less than 10, and compounds that were metabolized with half-lives less than, or equivalent to, the uptake time. Oliver and Niimi (1985) observed that the relation between BCF and K_{ow} values may fail for large, high molecular weight compounds. Thus, there appears to be an optimum K_{ow} range for bioconcentration to occur (log K_{ow} between 2 and 6). This observation is consistent with measurements of direct uptake of several organic chemicals across the gills of rainbow trout made using an in vivo fish model (McKim and others, 1985). In this study, uptake efficiencies were calculated by measuring the concentration in inspired and expired water of trout exposed to each chemical. Uptake efficiencies for compounds with very low K_{ow} values (log K_{ow} less than 0.9) were found to be low and unrelated to log K_{ow}; from log K_{ow} of 0.9–2.8, uptake efficiencies were positively correlated with log K_{ow}; from log K_{ow} of 2.8–6.2, uptake efficiency was constant; and at log K_{ow} greater than 6.2, uptake efficiency appeared to be inversely correlated with log K_{ow} (shown in Figure5.7). These results are discussed in more detail in Section 5.2.5 (subsection on Uptake Processes).

The observed correlations between measured BCF values and chemical properties (such as K_{ow} and water solubility) do not prove equilibrium partitioning theory, although they are consistent with it. On the other hand, the failure of these correlations observed for contaminants with high K_{ow} values suggests that uptake from water may not be the only (or even the most important) mechanism of uptake for extremely hydrophobic or high molecular weight compounds. This suggests that dietary uptake and biomagnification may be important for these compounds.

Fish/Sediment Concentration Ratios

The equilibrium partitioning model postulates that concentrations of a contaminant in fish and sediment will be in equilibrium through their respective equilibria with the water. This is

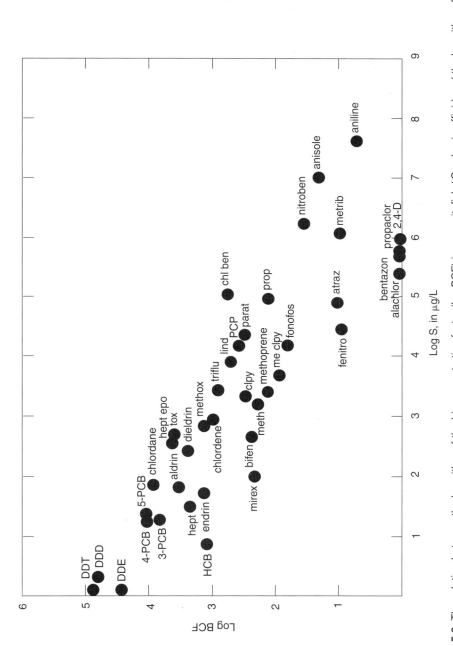

Figure 5.6. The relation between the logarithm of the bioconcentration factor (log BCF) in mosquitofish (*Gambusia affinis*) and the logarithm of the water solubility (log S, μg/L) for various organic chemicals in laboratory model ecosystem studies. Abbreviations: atraz, atrazine; bifen, bifenthrin; chl ben, chlorobenzene; clpy, chlorpyrifos; fenitro, fenitrothion; HCB, hexachlorobenzene; hept, heptachlor; hept epo, heptachlor epoxide; lind, lindane; me clpy, chlorpyrifos-methyl; meth, 2,2-bis-(4-methylphenyl)-1,1,1-trichloroethane; metrib, metribuzin; μg/L, microgram per liter; nitroben, nitrobenzene; parat, parathion; 3-PCB, 2,5,2'-trichlorobiphenyl; 4-PCB, 2,5,2',5'-tetrachlorobiphenyl; 5-PCB, 2,4,5,2',5'-pentachlorobiphenyl; PCP, pentachlorophenol; prop, propoxur; tox, toxaphene; triflu, trifluralin. Redrawn from Metcalf (1977) with permission of the publisher. Copyright 1977 John Wiley & Sons, Inc. Data are from Lu and Metcalf (1975).

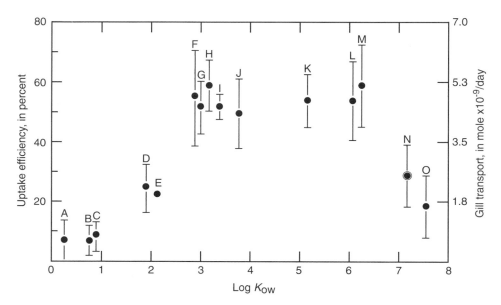

Figure 5.7. The gill uptake efficiency (percent) and 24-hour gill transport (mole × 10^{-9} per day), measured using an in vivo fish (rainbow trout) mode, in relation to the logarithm of the *n*-octanol-water partition coefficient (log K_{ow}) for 15 organic chemicals. Each point represents the mean of four trout (except *p*-cresol, which is a single trout), while the bars around each point correspond to the standard deviation. Data for fenvalerate (circled) are from Bradbury (1983). Abbreviations: A, ethyl formate; B, ethyl acetate; C, *n*-butanol; D, nitrobenzene; E, *p*-cresol; F, chlorobenzene; G, 2,4-dichlorophenol; H, decanol; I, pentachlorophenol; J, 2,4,5-trichlorophenol; K, dodecanol; L, 2,5,2′,5′-tetrachlorobiphenyl; M, hexachlorobenzene; N, fenvalerate; O, mirex. Adapted from McKim and others (1985) with permission of the publisher and the author. Copyright 1985 Academic Press, Inc.

true regardless of the extent to which contaminants sorbed to particles are available for uptake by fish (Connor, 1984a). Several authors have calculated fish/sediment concentration ratios for use as tools in predicting bioaccumulation. These include the fish/sediment concentration (Connor, 1984a), partitioning factor (Lake and others, 1987), preference factor (McElroy and Means, 1988), accumulation factor (Lake and others, 1990; Ferraro and others, 1990, 1991), biota–sediment factor (DiToro and others, 1991), bioavailability index (Muir and others, 1992), and biota–sediment accumulation factor (BSAF) (Boese and others, 1995, 1996). These ratios also can be used to test predictions of the equilibrium partitioning model.

In most of these ratios, chemical concentrations in biota and sediment are normalized by biotic lipid content and sediment organic carbon content, respectively. For example:

$$\text{BSAF} = [(C_b / f_{lipid}) / (C_s / f_{oc})] \tag{5.8}$$

where BSAF is the biota–sediment accumulation factor, C_b is the chemical concentration (in micrograms per kilogram) in aquatic biota, C_s is the chemical concentration in total sediment (in micrograms per kilogram), f_{lipid} is the fraction of the organism that is composed of lipid, and f_{oc} is the fraction of total sediment that consists of organic carbon (Boese and others, 1995). This model assumes equilibrium or steady-state conditions in the aquatic biota and the sediment, no chemical transformation, no phase transfer resistance, and chemical partitioning primarily between the organic material in the sediment and the lipid pool in the biota (Ferraro and others, 1991; Boese and others, 1995).

In theory, the BSAF value should equal 1 if tissue lipids and sediment organic carbon are equivalent distributional phases; tissue lipids and sediment organic carbon from all biota and all sediments are the same; the phases are in equilibrium; and there is no contaminant metabolism or degradation. Also, a semiempirical theoretical BSAF value of about 2 can be derived that allows for differences in contaminant affinity for tissue lipids and sediment organic carbon as distributional phases (McFarland, 1984). This semiempirical theoretical BSAF value was derived using the following empirical relations relating contaminant hydrophobicity (K_{ow}) with BCF_{lipid} in fish (Köneman and van Leeuwen, 1980) and with the organic carbon normalized sediment–water distribution coefficient, K_{oc} (Karickhoff, 1981):

$$\log BCF_{lipid} = 0.980 \log K_{ow} - 0.063 \qquad (5.9)$$

$$\log K_{oc} = 0.989 \log K_{ow} - 0.346 \qquad (5.10)$$

where BCF_{lipid} is the BCF with the concentration in biota expressed on a lipid basis (i.e., $(C_b/f_{lipid})/C_w$). Subtracting Equation 5.10 from Equation 5.9, and assuming the K_{ow} terms to be equal and canceling them, results in the following equation:

$$BSAF = (BCF_{lipid}/K_{oc}) = 10^{0.283} = 1.9 \qquad (5.11)$$

Note that the exact value of the semiempirical theoretical BSAF value will depend on the empirical relations used.

Table 5.2 lists BSAF values were measured in some laboratory and field studies. The compounds listed in Table 5.2 are pesticides or certain structurally similar organochlorine compounds (namely, tetrachlorodibenzo-p-dioxins [TCDD], total PCBs [or technical PCB mixtures], and hexachlorinated PCB congeners). For these compounds, measured BSAF values for organochlorine contaminants ranged from 0.2 to 2.8 in several laboratory studies and from 0.3 to 5.9 in several field studies (Table 5.2). Most of these studies measured BSAF values close to the range of theoretical BSAF values of 1–2. A few studies measured BSAF values for some organisms or contaminants that were considerably less than 1 (e.g., Kuehl and others, 1987; McElroy and Means, 1988) or considerably higher than 2 (e.g., Lake and others, 1987; Lake and others, 1990). Measured BSAF values sometimes were found to vary with regard to species, sediment type, and contaminant concentration (Boese and others, 1995). At least in laboratory studies, the considerable variability in BSAF values measured for a given chemical may be due in part to a failure to achieve equilibrium conditions during laboratory exposures (McFarland and others, 1994).

Table 5.2. Biota–sediment accumulation factors measured in some laboratory and field studies

[Taxon: scientific name included only if common name is general. Abbreviations: BSAF, biota–sediment accumulation factor; HCB, hexachlorobenzene]

Study	Taxon	Study Type (location)	Chemical	Mean BSAF
Boese and others, 1996	Clam (*Macoma nasuta*)	Laboratory	PCB-153 HCB	2.8 2.0
Boese and others, 1995	Clam (*M. nasuta*)	Laboratory	PCB congeners HCB	[1,2]2.8 [2]2.5
McFarland and others, 1994	Fish (Japanese medaka)	Laboratory	PCB-52	2.3
Pruell and others, 1993	Clam (*M. nasuta*)	Laboratory	PCB-153 2,3,7,8-TCDD	[3]1.8 [3]0.9
	Polychaete (*Nephtys incisa*)	Laboratory	PCB-153 2,3,7,8-TCDD	[3]1.4 [3]0.5
	Grass shrimp	Laboratory	PCB-153 2,3,7,8-TCDD	[3]2.1 [3]0.7
Muir and others, 1992	Mussel (*Anodonta grandis*) Fish (white sucker)[5]	Mesocosm[4] Mesocosm[4]	1,3,6,8-TCDD 1,3,6,8-TCDD	1.7 0.8
Ankley and others, 1992b	Oligochaete (*Lumbriculus variegatus*)	Laboratory	Hexachloro-PCB Total PCB	1 0.8
	Oligochaetes (predominantly *L. hoffmeisteri* and *L. cervix*)	Field (Wisconsin)	Hexachloro-PCB Total PCB	1.4 0.9
	Fish (fathead minnow)	Laboratory	Hexachloro-PCB Total PCB	0.5 0.3
	Fish (black bullhead)	Field (Wisconsin)	Hexachloro-PCB Total PCB	3.4 1.9
Ferraro and others, 1991	Clam (*M. nasuta*)	Laboratory	PCB congeners PCB-153	[6]0.2–2.1 [7]0.6–0.8
Ferraro and others, 1990	Clam (*M. nasuta*)	Laboratory	*p,p'*-DDE *p,p'*-DDD Aroclor 1254	[7]0.7–2.8 [7]0.52–1.0 [7]0.5–1.8
Lake and others, 1990	Clam (*Mercenaria mercenaria*) and polychaete (*N. incisa*)	Field (New York, Massachusetts, and Rhode Island)	Aroclor 1254 PCB-153	[8]1.6 [8]4.6
McElroy and Means, 1988	Bivalve (*Yoldia limatula*)	Laboratory	Hexachloro-PCB	[9]0.9,1.7
	Polychaete (*N. incisa*)	Laboratory	Hexachloro-PCB	[9]0.2, 0.4
Kuehl and others, 1987	Fish (common carp)	Field (Wisconsin)	2,3,7,8-TCDD	0.3

Table 5.2. Biota–sediment accumulation factors measured in some laboratory and field studies—*Continued*

Study	Taxon	Study Type (location)	Chemical	Mean BSAF
Lake and others, 1987	Polychaete (*N. incisa*)	Field (Rhode Island)	*trans*-Chlordane	5.9
			cis-Chlordane	4.2
			p,p'-DDD	4.2
			Aroclor 1254	4
	Bivalve (*Y. limatula*)	Field (Rhode Island)	*trans*-Chlordane	4.5
			cis-Chlordane	4
			p,p'-DDD	4
			Aroclor 1254	3.3

[1]Mean value for all PCB congeners.
[2]Measured in fine sediment.
[3]Estimated by eye from figure in reference.
[4]Exposed for 10–24 days.
[5]Carcass (minus gills and gastrointestinal tract).
[6]Range of mean values measured for all PCB congeners and all treatments, 0–2 cm depth.
[7]Range of mean values for all treatments, 0–2 cm depth.
[8]Mean of data for all species and stations.
[9]Mean values for sediments with high and low contamination, respectively.

As was seen for correlation between BCF and chemical properties, discussed previously, these results do not prove equilibrium partitioning theory, but they do tend to support its general applicability. When deviations from predicted values or results do occur, they sometimes indicate violations of assumptions or suggest limitations in the applicability of equilibrium partitioning theory. For example, Boese and others (1995) investigated the effect of sediment organic carbon and K_{ow} on BSAFs. These authors exposed marine clams (*Macoma nasuta*) to sediment spiked with PCB congeners or hexachlorobenzene using standard sediment bioaccumulation test procedures. Steady-state tissue concentrations were attained within 28–42 days for 12 of the 14 compounds tested. The exceptions were trichlorobiphenyl (which had rapid uptake but variable data, indicating possible metabolism by clams) and octachlorobiphenyl (which reached steady state before 28 days but had very low tissue uptake). Boese and others (1995) found the following. First, BSAFs and BAFs were lower for highly chlorinated (more chlorine substituents) and highly hydrophobic (log K_{ow} ≥7) PCB congeners. Second, BAF values and tissue residues were higher in sediment with the lowest f_{oc}. This is consistent with both equilibrium partitioning and selective feeding behavior. In the equilibrium partitioning model, neutral organic contaminants partition between sediment organic carbon and tissue lipids; thus a high f_{oc} in sediment reduces contaminant bioavailability. Because a constant amount of contaminant was added to each sediment type in this experiment, the low-carbon sediments had a higher contaminant concentration per unit carbon. Deposit feeders such as *M. nasuta* selectively ingest particles with finer grain size and higher f_{oc} than in the aggregate sediment; thus the clams feeding on the low-carbon sediment had a higher ingested dose than clams in high-carbon sediment. Third, BSAFs were less variable than BAFs, suggesting that BSAFs may be better predictors of bioaccumulation potential than BAFs. Fourth, the BSAFs determined were higher than those predicted by equilibrium partitioning, indicating that a simple equilibrium partitioning model does not account for all the uptake of highly hydrophobic compounds into sediment-ingesting organisms. Therefore, for highly hydrophobic compounds, BSAF values may under-estimate bioavailability, especially when the sediment tested has low contaminant concentrations and low sediment organic carbon content (Boese and others, 1996). In subsequent work with *M. nasuta*, Boese and others (1996) suggested that chemical partitioning also occurs in the digestive tract; gut BSAF values determined on the basis of ingested sediment and fecal organic carbon were consistently smaller and less variable than traditional BSAFs (that were calculated on the basis of the surrounding sediment).

To test the equilibrium partitioning postulate that fish and sediment will be in equilibrium through their respective equilibria with the water, Connor (1984a) computed fish/sediment concentration ratios from the literature using field studies that reported contaminant concentrations in sediment and in fish living over that sediment. He found that (for a given contaminant), the fish/sediment concentration ratio was dependent on the (water) residence time in the hydrologic system. Ratios were lower for poorly flushed estuarine areas than for lakes, and lower still for well-flushed estuarine areas. Connor (1984a) concluded that surficial sediment was not in equilibrium with fish lipid pools, i.e., that areas flushed by uncontaminated waters would have lower water concentrations and, therefore, lower fish concentrations. From the data, he estimated that the fish/sediment concentration ratio may reach a plateau at flushing times (water residence times) greater than 100 days. He also found that the fish/sediment concentration ratio depended on the metabolic capacity of the organism for that contaminant. This results in differences between chemicals (for example, fish/sediment ratios for chlorinated hydrocarbons were about three orders of magnitude higher than for aromatic hydrocarbons with the same K_{ow} values) and between organisms (for example, ratios for different phyletic groups were inversely related to their mixed-function oxidase activity).

Muir and others (1992) measured a related fish/sediment ratio, the bioavailability index, for organisms in outdoor mesocosms (enclosures in two lakes) exposed to polychlorinated dibenzo-*p*-dioxin (PCDD) and polychlorinated dibenzofurans (PCDF). These authors found that the highest bioavailability indexes were achieved by organisms that ingest or filter particles at the sediment–water interface (mussels [*Anodonta grandis*], chironomid larvae, and mayfly [*Hexagenia*] nymphs) and by those that feed on benthic organisms (crayfish, *Orconectes virilis*, and suckers, *Catostomus* species). They also observed accumulation of octachlorodibenzo-*p*-dioxin (OCDD) in fish and invertebrates, although direct uptake of this congener from water had been shown to be limited by steric effects in laboratory studies. These observations suggest the importance of the food chain pathway for bioaccumulation from contaminated sediment to fish (Muir and others, 1992).

Tracey and Hansen (1996) conducted a statistical analysis of BSAF values from the published literature. The data included field and laboratory measurements of BSAF values for 27 species in freshwater and marine systems. BSAF values were similar among species and among habitat types. Data also were analyzed by chemical class, with the tested chemicals classified as either pesticides, PCBs, or polynuclear aromatic hydrocarbons (PAH). The median BSAF for all species pooled by chemical class was lower for PAHs (0.29) than for PCBs (2.10) or pesticides (2.69). However, there were no dramatic changes in K_{ow}-specific median BSAFs over a wide range of log K_{ow} values (2.8–7.0 for pesticides)—in other words, there was no evidence that highly hydrophobic compounds have higher BSAF values than compounds with low K_{ow} values. However, variability in the data set was high. In fact, within-species variability was 1.5–2 times greater than the between-species variability. This indicates that factors that contribute to within-species variability (such as experimental design and site-specific conditions) are more important contributors to overall BSAF variation than factors such as chemical characteristics (within the class), feeding type, and species (Tracey and Hansen, 1996).

Effect of Trophic Level on Fugacity

In its simplest form, the fugacity model of bioaccumulation predicts that the animal/water fugacity ratio will be 1 at equilibrium and that the concentration of a contaminant in the lipids of all animals must be equal, regardless of trophic position. This condition is called *equifugacity*. Only under nonequilibrium conditions may the fugacity ratio deviate from 1. This prediction has been tested using data from field studies.

In aquatic food chains, animal/water fugacity ratios have been observed to increase with increasing trophic level and, for higher trophic levels, to exceed the fugacity in water (Clark and others, 1988; Connolly and Pedersen, 1988; Oliver and Niimi, 1988). Examples include PCB in Lake Michigan deep-water fish (Figure 5.8), four pesticides in fish from Coralville Reservoir, Iowa (Figure 5.9), and organochlorine compounds in herring gulls (*Larus argentatus*), alewife (*Alosa pseudoharengus*), smelt (family Osmeridae), and bed sediment from Lake Ontario (Figure 5.10). All of the organochlorine compounds for which fugacities are shown in Figure 5.10 have high K_{ow} values (5–7). Animal/water fugacity ratios also appear to vary with the K_{ow} of the chemical. In rainbow trout in Lake Ontario, the animal/water fugacity ratio increased from about 1 (for chemicals with log K_{ow} values in the range of 3–4) to 10–100 (for chemicals with log K_{ow} values of about 6), as shown in Figure 5.11. In laboratory water-only exposures, animal/water fugacity ratios were substantially less than 1 for high K_{ow} chemicals, probably because of nonequilibrium conditions. The finding that the chemical activity (fugacity) of extremely hydrophobic contaminants in biota can be greater than that in water suggests movement of these contaminants (from water to prey to predator) against a thermodynamic gradient, from low to high fugacity (Gobas and others, 1993b). Neither nonequilibrium conditions nor differences in growth rates could account for the magnitude of the difference in fugacity ratios observed (Connolly and Pedersen, 1988). As discussed in Section 5.2.5, this has been attributed to food chain transfer (Connolly and Pedersen, 1988).

Lipid Normalization

The equilibrium partitioning model (Mackay, 1982) predicts that differences in lipid content may explain to some extent the high variability observed in contaminant residues among different species, different individuals of the same species, or different organs and tissues in a single organism. If so, normalizing contaminant residues by lipid content would be expected to reduce variability in the data set.

Existing studies do not provide a consistent view of the usefulness of lipid normalization. Normalization of wet weight concentrations by lipid content has been reported to reduce variability between species to about a five-fold difference (Smith and others, 1988). Examples include PCBs in cod (*G. morhua*) livers and fillets and prey organisms from the western Baltic Sea (Schneider, 1982); DDT and dieldrin in 28 species of fish from the five Great Lakes (Reinert, 1970); and PCBs in 4 invertebrate species and 1 fish species (*S. solea*) from a single sampling

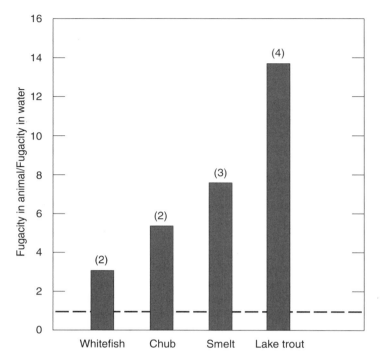

Figure 5.8. The ratio of (fugacity in animal)/(fugacity in water) for PCBs in deep-water fish of Lake Michigan. Numbers in parentheses indicate trophic level: 2, omnivore; 3, lower carnivore; 4, top carnivore. There are no herbivores (trophic level 1) among the fish shown. The dotted line indicates a fugacity ratio of 1. Redrawn from Connolly and Pedersen (1988) with permission of the publisher. Copyright 1988 American Chemical Society.

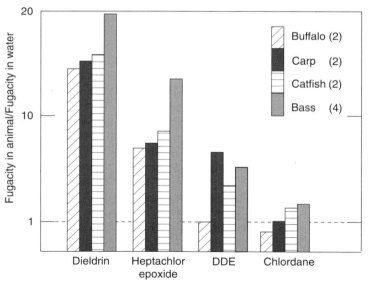

Figure 5.9. The ratio of (fugacity in animal)/(fugacity in water) for four pesticides in four fish species from Coralville Reservoir in Iowa. Numbers in parentheses indicate trophic level: 2, omnivore; 4, top carnivore. There are no herbivores (trophic level 1) or lower carnivores (trophic level 3) among the fish shown. The dotted line indicates a fugacity ratio of 1. Redrawn from Connolly and Pederson (1988) with permission of the publisher. Copyright 1988 American Chemical Society.

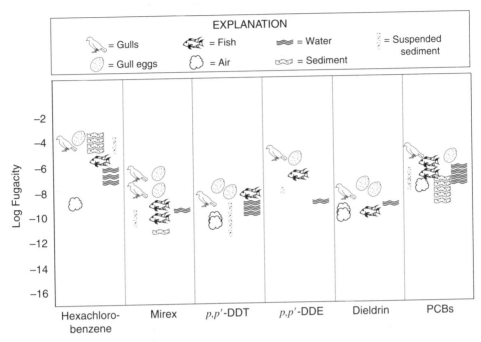

Figure 5.10. The logarithm of the fugacity (Pa) of organochlorine compounds in various media from the Lake Ontario region. Fugacities are calculated from concentrations reported in the literature. Fugacity values for fish apply to alewife and smelt. A single fish symbol indicates that the fugacities reported for these two species are approximately the same. For symbols other than fish, multiple symbols delineate the range in fugacity values reported for that compound and sampling medium. Abbreviation: Pa, Pascal. Redrawn from Clark and others (1988) with permission of the publisher. Copyright 1988 American Chemical Society.

area in the Wadden Sea (Goerke and others, 1979). In laboratory tests with 1-year-old brook trout exposed to methoxychlor, the fish consumed significantly more food, so that both the lipid content and the methoxychlor concentration increased as the daily food intake increased (Oladimeji and Leduc, 1975). Oladimeji and Leduc suggested that fish exposed to pesticides adapt to their surroundings by consuming more food, thus increasing their lipid content, which protects them from the toxic effects of lipophilic pesticides. Some field studies are consistent with this interpretation. For example, both lipid content (chloroform–methanol extractable) and DDT residues were much higher in fish (bream and perch—species not specified) from the polluted Danube Delta than from the Dniepr-Bug Estuary (Maslova, 1981).

In contrast, lipid normalization did not reduce variability in residue levels in a number of other studies. Examples include dieldrin and DDT residues in common carp (*Cyprinus carpio*) muscle tissue from the Des Moines River in Iowa (Hubert and Ricci, 1981); and endosulfan residues in livers of catfish (*Tandanus tandanus*), bony bream (*Nematolusa erebi*), and carp (*C. carpio*) collected from cotton-growing areas in Australia (Nowak and Julli, 1991). In fact, Huckins and others (1988) reported that lipid-normalization of total PCB residues in fish samples (various species) increased the variability by five-fold, instead of reducing it. In long-term data

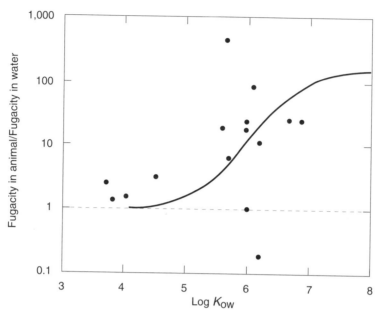

Figure 5.11. The ratio of (fugacity in animal)/(fugacity in water) for Lake Ontario rainbow trout in relation to the logarithm of the *n*-octanol-water partition coefficient (log K_{ow}) for 15 organochlorine compounds. The solid curve indicates the fugacity ratio predicted for large fish (top carnivores) in a four-level food chain. The dotted line indicates a fugacity ratio of 1. The compounds plotted are: *cis*-chlordane; *trans*-chlordane; *p,p'*-DDE; α-HCH; 2,4,5,2',4',5'-hexachlorobiphenyl; lindane; mirex; octachlorostyrene; 2,4,5,2',5'-pentachlorobiphenyl; 2,3,4,5,6-pentachlorotoluene; 1,2,3,4-tetrachlorobenzene; 2,3,2',3'-tetra-chlorobiphenyl; 2,5,2',5'-tetrachlorobiphenyl; 1,2,4-trichlorobenzene; 2,5,2'-trichlorobiphenyl. Redrawn from Connolly and Pedersen (1988) with permission of the publisher. Copyright 1988 American Chemical Society.

on organochlorine contaminants in whole fish from the NCBP, lipid normalization did not improve precision or explain interspecies differences among samples collected at the same site (Schmitt and others, 1990). In another example, the percentage of fat in whole-body yellow perch and white bass from Lake Erie generally increased late in the year, but there were no seasonal changes observed in DDT, PCB, or dieldrin residues (Kelso and Frank, 1974). Kelso and Frank suggested that metabolism of storage lipid (as opposed to structural lipid, which varies little) did not necessarily affect organochlorine residues. In this case, normal seasonal changes in lipid content would not affect organochlorine concentrations; then, if organochlorine concentrations were normalized on a lipid basis, the variability might actually increase (relative to that in wet weight concentrations).

In some studies, lipid normalization reduced variability for some species, some contaminants, or some tissues, but not others. For example, Wharfe and van den Broek (1978) measured organochlorine residues in macroinvertebrates and fish from the Lower Medway Estuary in Kent, England. Significant correlations were found between hexane-extractable lipid and dieldrin levels in muscle tissues of eel (*Anguilla anguilla*), whiting (*Merlangius merlangus*), flounder (*Platichthys flesus*), and plaice (*Pleuronectes platessa*), and in whole-body gobies

(*Pomatoschistus minutus*) and sprats (*Sprattus sprattus*). Although liver tissues had a higher lipid content than muscle, no correlation was found between lipid level and dieldrin residues in livers. No correlation was observed between DDT or PCB residues and lipid content.

Lipid normalization of data is usually accomplished by dividing the wet weight concentration by the lipid concentration to form a ratio. This approach is appropriate when there is an isometric (direct proportional) relation between contaminant concentration and lipid concentration, but it may lead to erroneous conclusions or mask other important factors if it is used when contaminant concentrations do not vary in direct proportion to lipid content (Hebert and Keenleyside, 1995). For example, contaminant distribution may vary with lipid composition (Schneider, 1982; Kammann and others, 1990). Hebert and Keenleyside (1995) recommended an alternative normalization approach on the basis of analysis of covariance, which removed the variation associated with lipid from contaminant concentrations. Another factor complicating interpretation of lipid-normalized data is that biotic lipid contents measured may vary considerably, depending on the extraction solvent used (de Boer, 1988; Randall and others, 1991).

5.2.5 CONVERGING THEORIES OF BIOACCUMULATION

The equilibrium partitioning and biomagnification theories of bioaccumulation have recently been resolved to some degree. This is due to the extensive effort put into laboratory experimentation, field monitoring, and modeling during the past three decades. An exhaustive review of this subject is beyond the scope of this book. However, the remainder of this section provides a summary of the current understanding of bioaccumulation on the basis of representative studies in the literature. There is some overlap in content with study results discussed in Section 5.2.4, which were presented by type of study or analysis, and which focused on evidence pertaining to the equilibrium partitioning and biomagnification theories of bioaccumulation. In the following summary, however, information is synthesized by topic; i.e., results of all of these studies, taken together, are evaluated to determine our current understanding of the following topics: contaminant uptake; contaminant elimination; biological, chemical, and environmental factors affecting contaminant accumulation; and bioaccumulation models. This summary owes much to previous review articles and books on bioaccumulation (cited below), especially the following: Cox (1972), Johnson (1973), Murty (1986a,b), Day (1990), Gobas and Russell (1991), and Swackhamer and Skoglund (1991) on bioaccumulation by various types of organisms; Kenaga (1972), Metcalf (1977), Macek and others (1979), Shaw and Connell (1984), Spacie and Hamelink (1985), Knezovich and others (1987), Connell (1988), and Huckle and Millburn (1990) on factors affecting bioaccumulation; and Clark and others (1988), Barron (1990), and Landrum and others (1992) on bioaccumulation models.

Uptake Processes

Uptake by an organism can occur via partitioning (diffusion through surface membranes) from the surrounding medium, or via ingestion of contaminated food or particles. Food is usually the dominant route of uptake of contaminants for terrestrial species (Moriarty, 1985; Moriarty and Walker, 1987). In fish, bioconcentration (partitioning) processes may predominate for some contaminants (Ellgehausen and others, 1980; Bruggeman and others, 1981). The relative importance of uptake via food or water depends on the conditions of exposure, duration, dose level, and the individual fish (Huckle and Millburn, 1990).

Partitioning From Water

Many studies demonstrated direct uptake of hydrophobic contaminants from water by aquatic organisms on aqueous exposure in the laboratory or in artificial ecosystems. Such studies used a variety of test methods and test species. A few examples are uptake of toxaphene by microorganisms, estuarine shrimp and fish (Sergeant and Onuska, 1989); uptake of DDT, dieldrin, and lindane by bluegills and goldfish (*Carassius auratus*) (Gakstatter and Weiss, 1967); uptake of dieldrin by channel catfish (*Ictalurus punctatus*) (Bulkley and others, 1974); and uptake of DDT by mosquitofish (Murphy, 1971). McKim and others (1985) measured direct uptake of *xenobiotics* (synthetic chemicals) across the gills of rainbow trout using an in vivo fish model, in which uptake efficiency was calculated by measuring the concentration in the inspired and expired water of trout exposed to each chemical.

The mechanism of bioconcentration (partitioning) involves transfer from water to the gills or body surface, then to the circulatory fluid (blood), followed either by metabolism and excretion of products or by storage in body lipids. With small organisms, diffusion may be the principal transfer mechanism. When a lipophilic molecule dissolves in water, an envelope of water molecules forms around it, structured by the polarity of the water molecule. When this envelope meets a lipophilic phase, the orderly water shell disintegrates and the molecule dissolves in the nonpolar phase (Connell, 1988). The change in entropy (as well as enthalpy) provides a driving force for bioconcentration (Connell, 1988 [based on Hansch, 1969]).

Hamelink and others (1971) proposed that chlorinated hydrocarbon uptake by aquatic biota occurred primarily through transfer from water to blood through the gills, and then from the blood to lipids. Murphy and Murphy (1971) later confirmed that the gills were the primary site of transfer in fish. In this case, the rate of chemical uptake may be influenced by factors that affect the rate of water movement across the gills, such as the ventilation rate. Murphy and Murphy (1971) reported a linear correlation between oxygen and DDT uptake in mosquitofish; both declined when the temperature was reduced. Neely (1979) and Norstrom and others (1976) developed equations for uptake by fish, showing that the efficiency of transfer across the gill membrane was a function (probably nonlinear) of log K_{ow}. Physiological factors such as body size, growth rate, physical activity, and physiological state (such as age and spawning) all affect the metabolic rate, which governs oxygen requirements supplied by water moving over the gills. The gill area/body weight ratio changes with fish size, suggesting that small fish may accumulate hydrophobic chemicals more rapidly (Murphy and Murphy, 1971).

Unicellular organisms probably take up hydrophobic organic chemicals by sorption onto the cell surface, followed by diffusion into the cell (Kerr and Vass, 1973). Uptake of lipophilic compounds by zooplankton (Harding and Vass, 1979; Southward and others, 1978; Addison, 1976) and the dragonfly (*Tetragoneuria* sp.) nymph (Wilkes and Weiss, 1971) also may occur directly from water through the outer body surface. In these cases, the organic chemicals appear to be following the uptake route of oxygen. For benthic infauna, Shaw and Connell (1987) suggested that polychaetes bioconcentrate PCBs from the interstitial water. Courtney and

Langston (1978) reported uptake of PCBs from water and sediment by polychaetes in laboratory aquaria.

Uptake of Sediment-Sorbed Chemicals

Contaminants in sediment may be transferred to biota via three probable pathways: (1) interstitial water; (2) ingested sediment (organic and inorganic); and (3) direct body contact with sediment particles. The relative contributions appear to depend on the type of sediment, class of chemical, and species of organism (Knezovich and others, 1987). In cases where organisms living in or on sediment contained higher residues than pelagic organisms in the same system, it was thought that contaminants were desorbed from sediment and then taken up from water (Kobylinski and Livingston, 1975; Roesijadi and others, 1978). More recently, direct uptake of sediment-sorbed contaminants has been recognized. For example, dietary uptake of PCBs by a benthic fish (spot, *Leiostomus xanthurus*) was demonstrated by Rubinstein and others (1984) in a multiphase experiment in which fish were exposed to PCB-contaminated sediment and worms. A few cases of direct sorption of contaminants to the body wall or exoskeleton of an organism or absorption through its integument have been reported (Knezovich and others, 1987). Examples include pesticide sorption to crustacean chitin and chitosan (Davar and Wightman, 1981; Kemp and Wightman, 1981), and sorption through annelid cuticle (Lord and others, 1980).

The relative contributions of sediment and water pathways to chemical uptake are not well understood (Knezovich and others, 1987). Some authors have concluded that water exposure is more important (Roesijadi and others, 1978; Adams, 1984), but others suggest that pathways other than water must be involved (Lynch and Johnson, 1982; Landrum and Scavia, 1983). Interstitial water (Adams and others, 1985) or direct sediment contact (Fowler and others, 1978) may play an important role for infaunal organisms. Chemicals sorbed to suspended particles may be taken up by deposit- or filter-feeding organisms (Langston, 1978). Ingestion of suspended-sediment sorbed contaminants is probably important for suspension-feeding fish. Recent studies with suspension-feeding fish indicate that the gill rakers in these fish do not act as filters to trap food particles, as previously thought. Instead, the gill rakers appear to contribute to a flow pattern that directs particle-laden water to the mucus-covered roof of the oral cavity, where particles are retained (Sanderson and others, 1991; Cheer and others, 1993; Sanderson and Cheer, 1993).

Dietary Uptake and Biomagnification

Contaminant accumulation from exposure in the diet has been shown in laboratory studies for some hydrophobic contaminants, but not others. Examples of dietary accumulation include guppies (*Lebistes reticulatus*) that were fed DDT-contaminated daphnids (*Daphnia magna*) (Reinbold and others, 1971); goldfish that were fed a DDT- or dieldrin-contaminated diet (Grzenda and others, 1970, 1971); coho salmon (Gruger and others, 1975) and juvenile (sexually immature) guppies (*Poecilia reticulata*) (Sijm and others, 1992) that were fed a PCB-contaminated diet; bluegills fed kepone- or mirex-contaminated daphnids (Skaar and others, 1981); and striped bass (*Morone saxatilis*) administered PCBs by gavage (Pizza and O'Connor, 1983). On the other hand, methoxychlor, which is metabolized by the guppy (*L. reticulatus*), did not bioaccumulate appreciably when guppies were fed contaminated *D. magna* at the same rate as resulted in DDT accumulation (Reinbold and others, 1971). Uptake of endrin in channel catfish exposed to endrin in diet was proportional to the dose in the food, reaching steady-state levels after about 200 days (Grant, 1976).

Dietary uptake also has been shown for sediment-dwelling organisms. Uptake of PCBs by a benthic fish (spot, *L. xanthurus*) was demonstrated by Rubinstein and others (1984) in a multiphase experiment in which the fish were exposed to PCB-contaminated sediment and polychaete worms (sandworms, *Nereis virens*). Note that, for filter-feeders, uptake from sediment is not readily distinguishable from uptake from the diet.

In some aquatic food chains, contaminant concentrations (and fugacities) in organisms were observed to increase with increasing trophic level and, for higher trophic levels, exceeded the fugacity in water (Clark and others, 1988; Connolly and Pedersen, 1988), as discussed in Section 5.2.4. This is believed to be due to biomagnification, or dietary accumulation (Gobas and others, 1993b). Despite observations of apparent biomagnification in the field, the mechanism by which it might operate has not been well understood (Connell, 1988). The finding that the chemical activity (fugacity) of hydrophobic contaminants in biota can be greater than that in water suggested that contaminants appeared to be moving (from water to prey to predator) against a thermodynamic gradient, from low to high fugacity (Gobas and others, 1993b). The initial mechanism proposed (that biomagnification was the result of the loss of food substances due to respiration, whereas resistant contaminants were retained by the organism) is now considered unlikely (Connell, 1988). More recently, it has been proposed that food digestion and absorption from the gastrointestinal tract, accompanied by inflow of more contaminated food, increase the concentration (and also the fugacity) of the chemical in the gastrointestinal tract relative to that in the original food (Connolly and Pedersen, 1988; Gobas and others, 1988, 1993b). This creates a fugacity gradient (or *fugacity pump*) that drives the passive diffusion of chemical from the gastrointestinal tract into the organism, raising the fugacity of the predator over that of the prey (the consumed food) (Gobas and others, 1993b). Assuming that the net transfer of chemical from the prey to the predator exceeds the rate of mass transfer between the predator and water phase, the predator will maintain a fugacity higher than that of water. The fugacity pump model is discussed in more detail later in Section 5.2.5 (Bioaccumulation Models). Consumption by an animal still higher in the food chain would result in a further increase in fugacity, or biomagnification (Connolly and Pedersen, 1988).

The fugacity pump theory of biomagnification is a modified fugacity (partitioning) theory that subscribes to, and attempts to explain, observations of biomagnification from a thermodynamic perspective. As such, it represents a convergence of previously competing theories of partitioning and biomagnification. Like all models, fugacity and other partitioning models are oversimplifications of a complicated biological system (organism). Other models, such as physiological models or physiologically based pharmacokinetic models, incorporate bioenergetics and physiological processes, such as blood flow and metabolism (Barron, 1990). These models are discussed in more detail in Section 5.2.5 (Bioaccumulation Models).

Factors Affecting Route of Uptake

A number of laboratory and artificial ecosystem studies attempted to assess the relative importance of contaminant uptake from water versus food. Most of these studies observed greater accumulation from water than from food (e.g., Reinert, 1967; Chadwick and Brocksen,

1969; Reinert, 1972; Moore and others, 1977; Macek and others, 1979; Hansen, 1980; Day and Kaushik, 1987). A few laboratory or microcosm studies did report greater uptake from contaminated food than from water for at least some analytes (Macek and Korn, 1970; Metcalf and others, 1973; Reinert and others, 1974a; Day, 1990). Reinert and others (1974a) reported that DDT and dieldrin residues taken up by lake trout from food and water appeared to be additive. Other studies reported that natural food and water sources did not appear to be additive (Lenon, 1968; Chadwick and Brocksen, 1969; Reinert, 1972).

At least three factors appeared to be important: the identity (characteristics) of the contaminant, the relative dose, and the organism tested. Dietary contributions to the total body burden in fish reported for different pesticides vary considerably. For example, dietary contributions to the total body burden in fish were estimated to comprise 0.1 percent for kepone (Bahner and others, 1977), 1.2 percent for the organophosphate insecticide leptophos (Macek and others, 1979), 3 percent for lindane (Hamelink and others, 1977), 10 percent for endrin (Grant, 1976), about 50 percent for dieldrin (Reinert and others, 1974a); 27–62 percent for DDT (Jarvinen and others, 1977); and 50–70 percent for DDE (Hamelink and others, 1977). In studies reviewed by Macek and others (1979), only DDT seems to have significant contribution from dietary sources (Table 5.3). Such differences may be attributed to differences in chemical characteristics such as hydrophobicity, molecular weight and structure (see Section 5.2.5, subsection on Chemical Characteristics that Affect Bioaccumulation). In a key study by McKim and others (1985), direct uptake of organic chemicals across the gills of rainbow trout was measured in vivo. Uptake efficiency through the gill varied with K_{ow}; specifically, there appeared to be a range of log K_{ow} values (about 2.8 to 6) for optimum chemical uptake by fish (also see Section 5.2.5, subsection on Solubility). Chemicals with very low K_{ow} values may be too insoluble in fat to pass biological membranes, while those with very high log K_{ow} values may bind to lipid membranes and can not pass into blood or cellular fluid (Hansch and Clayton, 1973). Macek and others (1979) suggest that elimination rates (which are listed for selected

Table 5.3. The relative importance of dietary sources to bioaccumulation of various chemicals in studies reviewed in Macek and others (1979)

[Abbreviations and symbols: BAF, bioaccumulation factor; DEHP, bis(2-ethylhexyl)phthalate; TCB, 1,2,4-trichlorobenzene. Reproduced from Macek and others (1979) with permission of the publisher. Copyright 1979 ASTM]

Chemical	Species	BAF[1]	Coefficient of Variation (in Percent)[2]	Dietary Contribution (in Percent)[3]
DEHP	Bluegill	112	17	14
TCB	Bluegill	182	47	6
Leptophos	Bluegill	773	17	1.2
Endrin	Fathead minnow	6,800	29	10
Kepone	Sheepshead minnow	7,400	([4])	<0.1
DDT	Fathead minnow	133,000	47	27–62

[1]Estimated BAF at equilibrium.
[2]Coefficient of variation associated with the estimated body burden at equilibrium.
[3]Percent of total body residue contributed by dietary sources during bioaccumulation.
[4]Insufficient data for calculation.

chemicals in Table 5.4) are the most useful data for identifying chemicals for which dietary sources may be important (such as DDT, as shown in Table 5.3). Elimination rates also tend to be related to hydrophobicity (see Section 5.2.5, Elimination Processes).

In comparing contaminant uptake from aqueous sources with that from dietary sources, the relative concentration or dose is a critical aspect of study design. Macek and others (1979) pointed out the importance of exposing the food organisms to the same aqueous concentration as the consumer organisms. In tests with daphnids (*D. magna*), the percentage of PAH taken up by ingestion of suspended yeast cells was 25–50 percent for benzo(*a*)pyrene and 1.3–15 percent for anthracene, depending on the concentration of particulates. The relative contribution of food ingestion to total accumulation was found to be a function of the fraction sorbed to particles and may be important for highly hydrophobic chemicals (McCarthy, 1983). The importance of relative dose in determining the predominant route of contaminant uptake can be seen from Equation 5.3, repeated below:

$$dC_b/dt = k_1 C_w + \alpha\beta C_f - k_2 C_b - k_e C_b - k_m C_b - Rk_r C_b - GC_b \qquad (5.3)$$

In Equation 5.3, the first term ($k_1 C_w$) represents uptake from water and the second term ($\alpha\beta C_f$) uptake from food. When C_w is very high, the uptake rate from water is large and may dominate. In the environment, C_w for hydrophobic organochlorine contaminants tends to be very low; thus, uptake from water is relatively low. In contrast, C_f may be very large in the environment, especially for organisms near the top of the food chain. Uptake from water in some laboratory exposures may be biased if artificially high concentrations of contaminants in water (relative to ambient conditions) are used.

Table 5.4. Time required by biota to eliminate 50 percent of the body burden ($t_{1/2}$, in days) during depuration in uncontaminated flowing water

[Data are from studies reviewed in Macek and others (1979). Abbreviations: DEHP, bis(2-ethylhexyl)phthalate; TCB, trichlorobenzene; $t_{1/2}$, half-life. Reproduced from Macek and others (1979) with permission of the publisher. Copyright 1979 ASTM]

Chemical	Species	Half-life $t_{1/2}$ (days)
DEHP	Bluegill	<3
1,2,4-TCB	Bluegill	<7
Leptophos	Bluegill	<10
Endrin	Catfish	<10
Kepone	Sheepshead minnow	<28
Kepone	Oysters	<4
Aroclor 1254	Grass shrimp	<14
Aroclor 1254	Spot	<42
Aroclor 1254	Fathead minnow	<42
DDT	Rainbow trout	>160
DDT	Lake trout	>125

The predominant route of uptake also appears to depend on the organism involved. At one extreme are air-breathing vertebrates such as sea birds, seals, and whales. These organisms have no external surface for rapid exchange, such as gills in fish (Clayton and others, 1977); thus, contaminant accumulation must be by biomagnification (Connell, 1988). For example, seals showed relatively greater bioaccumulation of PCBs relative to zooplankton and fish, even after lipid normalization (Clayton and others, 1977).

At the other extreme are autotrophic organisms that draw their food from dissolved components in water; for such organisms, bioconcentration must be the uptake mechanism (Baughman and Paris, 1981). For lower trophic levels, uptake by adsorption and diffusion is believed to be more important than dietary uptake (Hall and others, 1986). Initial uptake by phytoplankton is probably accomplished by (rapid) surface sorption, followed by slower partitioning to lipids (Peters and O'Connor, 1982; Swackhamer and Skoglund, 1991). Examples include uptake of PCBs by marine zooplankton in Puget Sound, where the lipid fraction of the biota accounted for 74–79 percent of the variability (Clayton and others, 1977); and accumulation of DDT and dieldrin by both living and killed (by autoclaving) mycelia of the yeast, *Streptomyces* (Chacko and Lockwood, 1967). DDT uptake by the unicellular green alga, *Chlorella*, was rapid and quantitative, and the incorporated residues were not released into clean medium (Sodergren, 1968). Uptake of DDT by the alga, *Dunaliella salina*, in the laboratory was proportional to cell density, indicating that saturation had taken place (Cox, 1970b). However, saturation did not occur in natural populations of marine phytoplankton. Linear concentration-dependent uptake of PCBs by zooplankton was observed in outdoor artificial ecosystems, which is consistent with an equilibrium partitioning uptake mechanism (Larsson, 1986). Suspended particulates and colloids may affect uptake of contaminants by zooplankton by diminishing the quantity of dissolved contaminants in the water. For example, PCB accumulation by filter-feeding plankton in highly contaminated water may be higher than predicted from partition coefficients (Peters and O'Connor, 1982). PCB levels in amphipod crustaceans (*Gammarus* sp.) and microzooplankton were correlated with suspended particulate concentrations (O'Connor and others, 1980).

For intermediate organisms, both mechanisms described above probably occur, with their relative importance depending on the organism, the contaminant, and various other factors (see Section 5.2.5, subsection on Biological Factors that Affect Accumulation). Bioconcentration is considered to be more important (Connell, 1988) by many scientists. Chemical elimination processes and rates appear to be important in determining the potential for biomagnification (see Section 5.2.5, Elimination Processes). Biomagnification is most likely to occur with organisms that have long lifetimes, such as top predators (Connell, 1988).

In summary, from a review of artificial ecosystem, field, and laboratory studies that assess bioconcentration, dietary uptake, and potential biomagnification of organic chemical contaminants, it appears that biomagnification may occur under conditions of low water concentration for compounds of high lipophilicity, high persistence, and low water solubility (Biddinger and Gloss, 1984). Biomagnification is most likely to occur for chemicals with log K_{ow} values greater than 5 or 6 (Connell, 1988; Gobas and others, 1993b) and for top predators with long lifetimes. Dietary intake and biomagnification are extremely important for air-breathing vertebrates.

Elimination Processes

Contaminant elimination from an aquatic organism occurs by biotransformation (metabolism); by excretion via the gills, skin, urine or feces; or by growth dilution. Organism growth, although not strictly an elimination process, causes the contaminant concentration in an organism to decrease as the body mass of the organism increases. Therefore, growth often is grouped together with true elimination processes, such as metabolism or depuration via the gills.

The rate of elimination is a critical factor in determining whether or not a chemical will accumulate in an organism. For example, in small fish (such as guppies, goldfish, fathead minnows [*Pimephales promelas*], and juvenile rainbow trout), the BCFs for PCDDs with up to 5–6 chlorines were low relative to other chlorinated aromatic hydrocarbons (Figure 5.12). The same was true for BMFs (defined as the ratio of the contaminant concentration in fish to that in water when the fish were exposed via contaminated food), as shown in Figure 5.13. The low BCF and BMF values for these PCDDs were not due to slow uptake rates, since these PCDDs

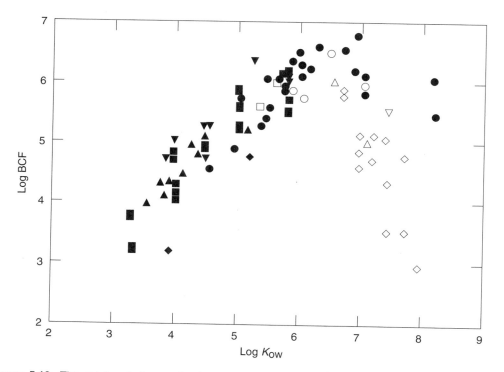

Figure 5.12. The relation between the logarithm of the bioconcentration factor (log BCF) and the logarithm of the *n*-octanol-water partition coefficient (log K_{OW}) for the following compounds: polychlorinated dibenzofurans (△), polychlorinated dibenzo-*p*-dioxins (◇), polybrominated biphenyls (○), polychlorinated naphthalenes (▼), polychlorinated anisoles (▲), polybrominated benzenes (◆), polychlorinated biphenyls (●), and polychlorinated benzenes (□); not defined in original reference (■ and ▽). Adapted from Opperhuizen and Sijm (1990) with permission of the publisher. Copyright 1990 Society of Environmental Toxicology and Chemistry (SETAC).

have uptake rate constants comparable to those of other compounds (as shown in Figure 5.14). However, PCDDs with fewer than 5–6 chlorines have high elimination rate constants with respect to K_{ow} (Figure 5.15). Therefore, their low BCF and BMF values were due to rapid elimination, probably by biotransformation (Opperhuizen and Sijm, 1990).

Chemical elimination rate tends to be related to hydrophobicity. This was demonstrated in the in vivo fish experiment conducted by McKim and others (1985), in which transport rates of organic chemicals were measured across the gills of rainbow trout exposed to a given chemical for 1–6 hours. Chemicals with log K_{ow} values less than 2.8 were eliminated rapidly within 4 hours. Chemicals with log K_{ow} values greater than 2.8 were not eliminated from the gills within the dosing period. The slope of log BCF versus log K_{ow} must be controlled by excretion rates because BCF was directly related to log K_{ow} for chemicals with log K_{ow} values of 2–6 (Neely and others, 1974; Veith and others, 1979a), but uptake rates were the same for chemicals in this range (McKim and others, 1985). This is supported by other observations of the inverse relation between elimination rate and log K_{ow} (for example, Figure 5.16). Elimination rate constants were independent of K_{ow} for chemicals with low-to-moderate hydrophobicity (Gobas and others, 1986), but they decreased with increasing K_{ow} for more hydrophobic chemicals (Spacie and Hamelink, 1982; Gobas and others, 1986; Gobas and others, 1989b; Lydy and others, 1992). This difference in elimination rates may result from differences in their relative affinities for internal tissues, such as fish lipid compartments (Gobas and others, 1986), and the rate of

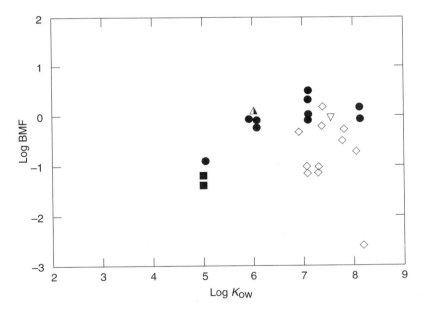

Figure 5.13. The relation between the logarithm of the biomagnification factor (log BMF) and the logarithm of the n-octanol-water partition coefficient (log K_{ow}) for the following compounds: polychlorinated dibenzo-p-dioxins ($\diamond$), polychlorinated biphenyls ($\bullet$), and mirex ($\triangle$); not defined in original reference ($\blacksquare$ and $\triangledown$). Adapted from Opperhuizen and Sijm (1990) with permission of the publisher. Copyright 1990 Society of Environmental Toxicology and Chemistry (SETAC).

transport through the body from storage sites to site of excretion (Gakstatter and Weiss, 1967). A hydrophobic chemical like DDT would be excreted very slowly because of the difficulty in transporting DDT to the site of excretion (Gakstatter and Weiss, 1967). According to one model of bioconcentration in fish, elimination is controlled by membrane diffusion for less hydrophobic compounds and diffusion through aqueous layers for more hydrophobic chemicals (Gobas and others, 1986). For extremely hydrophobic chemicals, then, elimination in feces may be more important than diffusive release. In Figure 5.16, the solid line shows the theoretical relation between the logarithm of the rate constant for diffusive release (log k_2) and log K_{ow} (Gobas and others, 1989b). Experimental data relating the logarithm of the elimination rate (log k_T) to log K_{ow} deviate from this theoretical relation for chemicals with very low K_{ow} values (log K_{ow} less than 4) or very high K_{ow} values (log K_{ow} greater than 7). Gobas and others (1989b) attributed these deviations to two factors: (1) metabolism for the lower K_{ow} chemicals, and (2) elimination into feces for the higher K_{ow} chemicals. The dashed line in Figure 5.16 represents the expected relation between the rate constant for elimination in feces (log k_e) and log K_{ow} (this relation was derived from the observed relation between the uptake efficiency of organic substances from food and K_{ow}). In Figure 5.16, the data points for chemicals with log K_{ow} values greater than 7

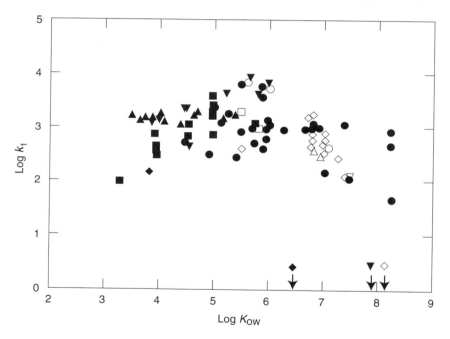

Figure 5.14. The relation between the logarithm of the uptake rate coefficient (log k_1) and the logarithm of the *n*-octanol-water partition coefficient (log K_{ow}) for the following compounds: polychlorinated dibenzofurans (△), polychlorinated dibenzo-*p*-dioxins (◇), polybrominated biphenyls (○), polychlorinated naphthalenes (▼), polychlorinated anisoles (▲), polybrominated benzenes (◆), polychlorinated biphenyls (●), and polychlorinated benzenes (□); not defined in original reference (■ and ▽).Adapted from Opperhuizen and Sijm (1990) with permission of the publisher. Copyright 1990 Society of Environmental Toxicology and Chemistry (SETAC).

appear to fit this dashed line. These findings are consistent with the suggestion by Macek and others (1979) that elimination rates are the most useful data for identifying chemicals for which dietary sources may be important.

The fugacity pump model of biomagnification (Gobas and others, 1993b), mentioned previously (Section 5.2.5, Uptake) offers an explanation for the importance of elimination rates and chemical hydrophobicity in determining biomagnification potential. If the combined rate of contaminant metabolism, elimination through the gills, skin, or urine, and growth dilution is small relative to the rate of contaminant elimination in feces, then the high chemical fugacity in the organism is maintained and biomagnification may be evident. If the combined rate of metabolism, elimination through the gills, skin, or urine, and growth dilution is high relative to the fecal elimination rate, then the high fugacity of the organism is not maintained, and there should be no evidence of biomagnification (Gobas and others, 1993b). Gobas and others suggested that this is why chemicals with log K_{ow} values less than 6 do not biomagnify, even if they are nonmetabolizable. For nonmetabolizable compounds with log K_{ow} values greater than 6, the rates of elimination via the gills, skin, urine, or metabolism are too low to reduce the high fugacity in the organism.

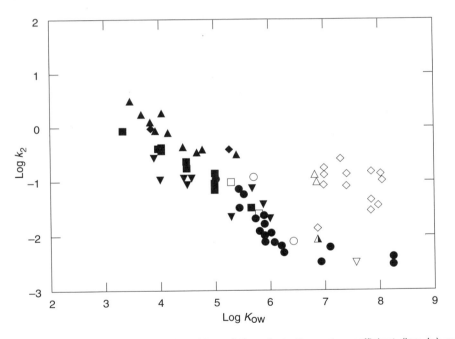

Figure 5.15. The relation between the logarithm of the elimination rate coefficient (log k_2) and the logarithm of the n-octanol-water partition coefficient (log K_{ow}) for the following compounds: polychlorinated dibenzofurans (△), polychlorinated dibenzo-p-dioxins (◇), polybrominated biphenyls (○), polychlorinated naphthalenes (▼), polychlorinated anisoles (▲), polybrominated benzenes (◆), polychlorinated biphenyls (●), polychlorinated benzenes (□), and mirex (▲); not defined in original reference (■ and ▽). Adapted from Opperhuizen and Sijm (1990) with permission of the publisher. Copyright 1990 Society of Environmental Toxicology and Chemistry (SETAC).

Elimination rates also have been shown to be dependent on the initial concentration in the fish, a characteristic of second-order kinetics (Ellgehausen and others, 1980). Elimination rates appeared to decrease over time, especially for lipophilic compounds, probably because of binding in deeper compartments (Huckle and Millburn, 1990).

Biological Factors that Affect Bioaccumulation

Table 5.5 lists biological factors reported to affect contaminant accumulation by aquatic biota. For each of these factors, results from some relevant laboratory and field studies are summarized in the following sections.

Body Length, Body Weight, and Age

Many field observations suggest that concentrations of hydrophobic contaminants in aquatic biota are related to body weight, body length, and age (Kenaga, 1972; Matsumura, 1977; Ellgehausen and others, 1980; Biddinger and Gloss, 1984; Huckle and Millburn, 1990). Because these variables tend to be covariant, they will be discussed together. Body length (shown in

Table 5.5. Factors reported in the literature to affect contaminant accumulation by aquatic biota

[Abbreviations: DOM, dissolved organic matter; K_{ow}, n-octanol-water partition coefficient]

Biological Characteristics	Physical and Chemical Properties	Environmental Conditions
Age	Molecular weight	Temperature
Body weight and length	Molecular size and shape (cross-section, parachor)	pH
Lipid content	Molecular structure (steric effect coefficient)	Salinity
Gill ventilation volume	Degree of ionization	Oxygen concentration
Epithelial characteristics (surface area, thickness, permeability)	Water solubility	Particulate concentration
Blood Flow	Lipid solubility (K_{ow})	DOM concentrations
Metabolism	Chemical stability	Water quality
Growth rate or efficiency	Volatility	Season
Reproductive state		
Species		
Tissue Analyzed		
Sex		
Enzyme induction		
Feeding habits		
Vertical distribution		

Figure 5.17) and age (shown in Figure 5.18) also tend to be positively correlated with lipid content (Bulkley and others, 1976; Insalaco and others, 1982).

In several studies, organochlorine residues in fish samples have been observed to increase with an increase in body size or age. Note that most of these studies also have shown a positive correlation with fat or oil content. For example, DDT residues in whole-body northern pike (*Esox lucius*) in the Richelieu River (Canada) increased with increasing age, especially when expressed on a wet weight basis (Boileau and others, 1979). In lake trout from Lakes Michigan and Superior, and in coho salmon from Lake Michigan, DDT residues increased with the length of the fish and with oil content (Reinert and Bergman, 1974). In channel catfish from the Des Moines River (Iowa), dieldrin levels in dorsal muscle tissue were related to body length (as shown in Figure 5.19) and age (as shown in Figure 5.20). In lake trout from Lake Michigan and Lake Superior and in walleye (*Stizostedion vitreum*) from Lake Erie, DDT and dieldrin levels increased with the size of the fish (Reinert, 1970). Frank and others (1974) measured DDT and dieldrin residues in whole-body samples of 40 species of fish from rivers and lakes in Canada. Within any single species, DDT and dieldrin residues, as well as extractable lipids, increased with increasing body size and weight. In fish (various species) from Iowa streams, dieldrin residues increased with increasing fish length (Morris and Johnson, 1971).

Laboratory tests with channel catfish showed a similar positive relation between dieldrin accumulation and body size (Bulkley and others, 1974). Larger fish (35–40 cm) accumulated

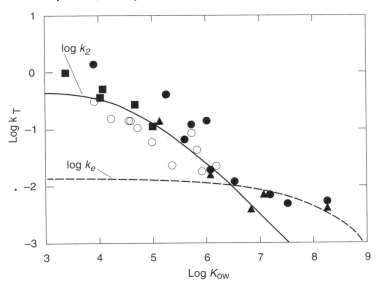

Figure 5.16. The relation between the logarithm of the overall elimination rate constant (log k_T) in the guppy, and the logarithm of the n-octanol-water partition coefficient (log K_{ow}), for selected halogenated compounds. Theoretical relations are shown for the logarithm of k_2 (solid line) and the logarithm of k_e (dotted line), where k_2 is the rate constant for diffusive release and k_e is the rate constant for elimination in feces. Data shown are from Gobas and others (1989b) (●); Bruggeman and others (1984) (▲); Opperhuizen and others (1985) (○); and Könemann and van Leeuwen (1980) (■). Adapted from Gobas and others (1989b) with permission of the publisher. Copyright 1989 Society of Environmental Toxicology and Chemistry (SETAC).

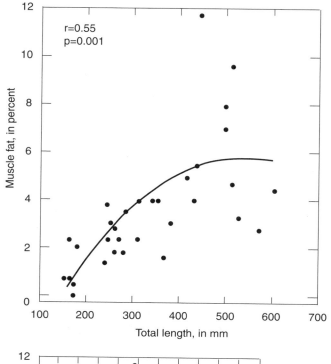

Figure 5.17. The relation between percentage of fat in dorsal muscle tissue and total body length (mm) for channel catfish from the Des Moines River. The fitted curve relation is percent of muscle fat = − 4.65 + 0.0374 (total length) − 0.000034 (total length)^2. Abbreviations: mm, millimeter; p, significance level; r, correlation coefficient. Redrawn from Bulkley and others (1976) with permission of the publisher. Copyright 1976 American Fisheries Society.

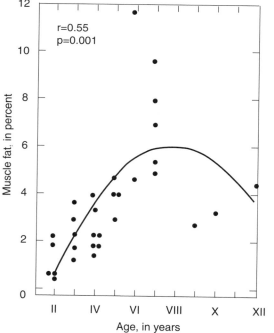

Figure 5.18. The relation between percentage of fat in dorsal muscle tissue and age (years) for channel catfish from the Des Moines River. The fitted curve relation is percent of muscle fat = − 3.66 + 2.4133(age) − 0.150422(age)^2. Abbreviations: p, significance level; r, correlation coefficient. Redrawn from Bulkley and others (1976) with permission of the publisher. Copyright 1976 American Fisheries Society.

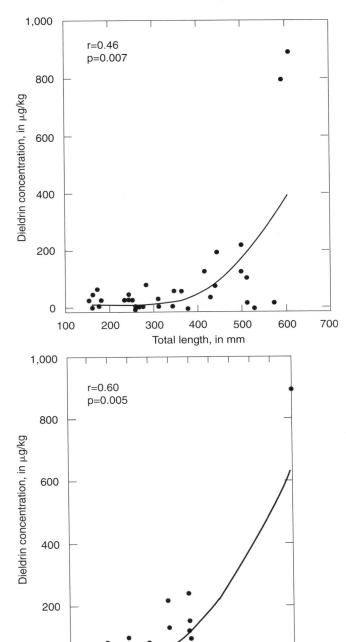

Figure 5.19. The relation between dieldrin concentration in dorsal muscle tissue (µg/kg) and total body length (mm) for channel catfish from the Des Moines River. The fitted curve relation is dieldrin concentration (µg/kg) = 355.39 − 2.4396 (total length) + 0.004257(total length)^2. Abbreviations: mm, millimeter; p, significance level; r, correlation coefficient; µg/kg, microgram per kilogram. Redrawn from Buckley and others (1976) with permission of the publisher. Copyright 1976 American Fisheries Society.

Figure 5.20. The relation between dieldrin concentration in dorsal muscle tissue (µg/kg) and age (years) for channel catfish from the Des Moines River. The fitted curve relation is dieldrin concentration (µg/kg) = 186.09 − 80.3199(age) + 9.998156 (age)^2. Abbreviations: p, significance level; r, correlation coefficient; µg/kg, microgram per kilogram. Redrawn from Buckley and others (1976) with permission of the publisher. Copyright 1976 American Fisheries Society.

dieldrin from both food and water faster than did smaller fish (15–23 cm). Larger fish also reached significantly higher residues after 28 days of uptake, for all three exposure regimes (food only, food plus water, and water only). The 30–35 cm group eliminated dieldrin significantly more slowly than the two smaller classes (10–15 and 17–25 cm) within the 14-day depuration period. Bulkley and colleagues suggest that the larger fish may have a greater capacity for accumulation and retention because of higher fat content and that smaller fish may be able to excrete dieldrin faster because of a higher metabolic rate.

Other field studies showed mixed results. For example, total DDT and PCB residues increased with increasing body length in round whitefish (*Prosopium cylindraceum*) fillets from Lake Huron (Miller and Jude, 1984). However, residues of dieldrin and chlordane were not consistently related to size, although these compounds were detected at low levels in this study. Dieldrin residues were highest in intermediate size fish (34–42 cm) and declined in longer fish. Chlordane residues increased with an increase in size in males, but not in females. In a second study, mirex was measured in edible tissues of chinook salmon (*Onchorhynchus tshawytscha*) and coho salmon from Lake Ontario (Insalaco and others, 1982). The mirex concentration increased significantly with an increase in body weight in some tissues (fillet, belly flap, and dorsal loin) but not in others (skin, red muscle, or caudal peduncle). In tissues where the increase was significant, Insalaco and others (1982) attribute it to an increase in tissue body fat with age and the tendency for hydrophobic organic compounds to accumulate in fish with higher lipid content. For the other tissues, Insalaco and others (1982) suggest that the small sample size and variability in the data prevented weight from being a significant variable in the multivariate analysis. Simple linear regression of skin data (or red muscle or caudal peduncle) versus weight showed a small but steady increase in residues with weight. In a third study, dieldrin and total DDT residues were monitored in muscle tissue of common carp from the Des Moines River, Iowa (Hubert and Ricci, 1981). The mean dieldrin concentrations (wet weight) were significantly related to age in June, but not in September. When dieldrin concentrations were expressed on a lipid basis, no age effect was observed in either June or September. Total DDT levels in 4-year olds were higher than in 3-year olds.

Not all field and laboratory studies observed increasing contaminant concentrations with increasing body size or age. For example, TCDD residues in livers of largemouth bass (*Micropterus salmoides*) and brown bullhead (*Ameiurus nebulosus*) sampled downstream of a bleach-kraft paper mill on the St. Johns River (in Florida) were not related to fish age (Schell and others, 1993). In another example, DDT and PCB levels were measured in perch (*Perca fluviatilis*) and roach (*Leuciscus rutilus*) from a cooling-water recipient area near a nuclear power plant (Edgren and others, 1981). No correlation was found between lipid-weight concentrations and body size of an individual when each month and sampling area was tested individually. Because data were reported on a lipid basis, any lipid-related body size effect would not be apparent. In some other studies (e.g., Hubert and Ricci, 1981) size- or age-related effects were smaller or nonexistent when contaminant concentrations were expressed on a lipid-weight basis. In laboratory tests, Murphy (1971) exposed mosquitofish of different sizes to DDT (0.041 ng/L) in water for 48 hours. Higher (wet weight) concentrations were taken up by smaller fish (as shown in Figure 5.21).

Several explanations for the observed age- and body size-related effects have been suggested: larger, older fish have greater lipid content (Bulkley and others, 1974; Insalaco and others, 1982); elimination rates are faster in smaller fish (Bulkley and others, 1974); elimination

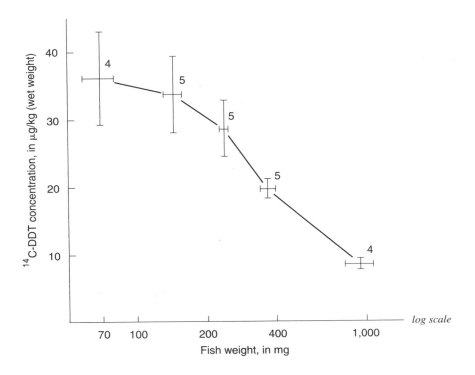

Figure 5.21. The relation between ^{14}C-DDT concentration in mosquitofish (µg/kg wet weight) and fish body weight (mg). Concentrations were measured after 48 hours in water with an initial concentration of 0.041 ng/L. Numbers on graph (e.g., "4" or "5") indicate the number of fish in each weight class. Abbreviations: mg, milligram; ng/L, nanogram per liter; µg/kg, microgram per kilogram. Redrawn from Murphy (1971) with permission of the publisher and the author. Copyright 1971 Springer-Verlag New York,

rates are slower than uptake rates, so that the length of time required to reach steady state will be long and may exceed the lifetime of the organism (Metcalf, 1977); larger fish may take longer to reach steady-state concentrations (Moriarty and Walker, 1987); and larger fish have a smaller surface-to-volume ratio (Matsumura, 1977; Ellgehausen and others, 1980; Biddinger and Gloss, 1984).

Lipid Content

The simple equilibrium partitioning model assumes that bioaccumulation of hydrophobic organic compounds is a partitioning process (bioconcentration) between the body lipid of an organism and the surrounding water (Mackay, 1982). Accordingly, the total contaminant concentrations in individuals within a population would be a function of their respective lipid contents, assuming equivalent physiological characteristics and exposure. This model is an oversimplification; clearly, organisms are not composed entirely of lipids, and all lipids are not identical. Nonetheless, lipids have been shown to play an important role in bioaccumulation of hydrophobic contaminants by aquatic biota. Contaminant levels in individual fish have been reported to correspond with their total lipid content (Goldbach and others, 1976; Roberts and others, 1977). Moreover, differences in lipid content may explain interspecies differences in contaminant concentrations or the distribution of a contaminant among tissues of an organism (Connell, 1988).

Many studies have examined species differences in contaminant residues in relation to differences in lipid content. These include both laboratory and field studies, and the results are inconsistent. In laboratory tests with HCH isomers and chlorinated biphenyls in four species of fish (golden orf, *Leuciscus idus*; guppy, *P. reticulata*; common carp; and brown trout, *Salmo trutta*), BCF values were found to be proportional to the lipid content of the fish (Suguira and others, 1979). Interspecific differences in several field studies also have been attributed to lipid content; for example, DDT and metabolites in mackerel (*Scomber scombrus*) and brook trout from Canadian waters (Duffy and O'Connell, 1968); DDT and dieldrin in herring (*Clupea harengus*) and cod (*G. morhua*) from Swedish waters (Jensen and others, 1969); DDT and PCBs in yellow perch, white perch (*Morone americana*), coho salmon, brown trout, and lake trout from Lake Ontario (Armstrong and Sloan, 1980); and pesticides and PCBs in the American eel and other fish from the Schuylkill River, Pennsylvania (Barker, 1984). On the other hand, no relation between residues and lipid content was observed in three fish species from Lake Erie (Kelso and Frank, 1974). This was attributed to small sample size, high individual variability, and low levels of the residues (Kelso and Frank, 1974). In composite whole fish samples from the NCBP, differences in contaminant concentrations between species collected from a given site could not be explained solely on the basis of lipid content (Schmitt and others, 1981, 1983). In some studies, the results were compound-dependent. For example, lipid content could explain interspecific differences in residues of DDT, DDE, and PCBs, but not lindane, DDD, aldrin, and dieldrin, in 12 species of fish from Lake Paijanne in Finland (Hattula and others, 1978).

A number of studies have reported that hydrophobic contaminant residues were greatest in adipose tissue or in other tissues with high lipid content. Examples include hydrocarbons in mussels (Connell, 1978); PCB in rainbow trout (Lieb and others, 1974); DDT in fat of various fish species (Meeks, 1968; Maslova, 1981; Ziliukiene, 1989) and in blubber of Weddell seals (Hidaka and others, 1983); and TCDD in rainbow trout (Hektoen and others, 1992). In some studies, the relation with lipid content held for residues in all tissue types (or organs) sampled. For example, in bowfin (*Amia calva*), largemouth bass, brown bullhead, and blue crab (*Callinectes sapidus*) downstream of bleach-kraft paper mill on the St. Johns River in Florida, TCDD residues were highly correlated with lipid content, regardless of the species or tissue sampled (Schell and others, 1993). Also, residues of total DDT in organs and tissues of Weddell seals depended roughly on lipid contents (Hidaka and others, 1983). Preferential accumulation and storage of residues in adipose tissues of organisms was attributed to lipid solubility and fat utilization by the organism (Kenaga, 1972). However, other studies have reported no correlation between organochlorine residues and tissue lipid content. For example, in laboratory studies with goldfish fed a diet contaminated with DDT or dieldrin, there was no correlation between the pesticide concentration in individual tissues and the tissue lipid content (Grzenda and others, 1970, 1971). In laboratory studies with cod (*G. morhua*), and rainbow trout exposed to radiolabeled 2,3,7,8-TCDD, lipid content could explain interspecific differences in radioactivity in the liver, but not in the brain (Hektoen and others, 1992).

If bioaccumulation is a function of lipid solubility, as predicted by equilibrium partitioning theory, contaminant concentrations would be expected to be less variable when normalized by lipid content. As discussed in Section 5.2.4 (Lipid Normalization), existing studies do not provide a consistent view of the usefulness of lipid normalization. Normalizing residues to lipid content has been reported to reduce variability between species to about a five-fold difference (Smith and others, 1988). In other studies, including the long-term national data set from the NCBP (Schmitt and others, 1981, 1983), lipid normalization did not reduce variability in residue levels.

In evaluating the contribution of lipid content to differences in contaminant concentrations between species, between individual organisms, or among different tissues within an organism, a number of factors complicate both data collection and data interpretation. First, lipid content changes during growth and development and it tends to be covariant with age and body size. In goldfish, for example, the organochlorine concentrations in the ovary increased by a factor of six or more during maturation, accompanied by an increase in lipid content (Grzenda and others, 1971). An increase in DDT residues in ovarian and liver tissues was observed with an increase in age in landlocked salmon (*Salmo salar*) in Maine (Anderson and Everhart, 1966). Age and fat content were found to be interdependent in the same species; residue levels decreased with age in fish with high fat content and remained constant with age in fish with low fat content (Anderson and Fenderson, 1970).

Second, both lipid content and contaminant residues change seasonally (also see Section 5.3.5). For example, Ziliukiene (1989) measured organochlorine pesticides in various tissues of vimba (*Vimba vimba* L.) from Lithuania. Total DDT residues in internal fat were higher than in liver, muscle, or gonad tissues throughout the year. At any point in time, either liver or muscle ranked second. In those months when muscle was second highest, it generally had higher fat content as well. During months in which the lipid content of muscle and liver was similar, total DDT residues were higher in muscle during winter and higher in liver during summer and autumn (the period of active feeding). Ziliukiene (1989) concluded that contaminants that entered fish by way of food, accumulated first in the liver, and later migrated with the blood to lipid-rich tissues and organs. Lack of correlation, or even inverse correlation, between lipid content and contaminant levels could result from mobilization of fat during the period immediately prior to spawning (Murty, 1986a).

Third, contaminant distribution appears to vary with lipid composition. Schneider (1982) measured PCB concentrations in spawning female and spawning male cod (*G. morhua*) from the western Baltic Sea during 1978–1979. The distribution of PCBs among tissues (fillets, gonads, and livers) was not proportional to extractable lipid content. However, the solubility of PCBs was much less in phospholipids than in other lipids tested (fats, free fatty acids, and cholesterol and its esters). Mean PCB concentrations were proportional to the contents of these other lipids in the various tissues. Schneider (1982) noted that brain lipids consist primarily of phospholipids, which may explain low concentrations of hydrophobic organics in brain tissues observed by Holden (1962) and others. Lipid composition also was shown to be an important factor influencing the tissue distribution of PCBs in female dabs (*Limanda limanda* L.) from the German Bight (Kammann and others, 1990). Individual tissues showed a distinct PCB congener pattern that was independent of sampling site or season. Trichlorobiphenyls and tetrachlorobiphenyls were high in the kidneys and gallbladder, which had a high percentage of polar (methanol-soluble) lipids. In contrast, pentachlorobiphenyls and hexachlorobiphenyls were predominant in the liver, ovaries, gills, and stomach, which were high in neutral (*n*-hexane-soluble) lipids. The PCB congener pattern in the stomach reflected the food composition of the dab. The PCB pattern in the ovaries reflected transfer of liver lipids. The congener pattern of the gills was not comparable to that in water. The authors concluded that lipid composition was more important for determining congener distribution than equilibration with sea water was. Because hydrophobic contaminants initially pass through the gill, lipids of the gill may be more important in controlling uptake than lipids in the rest of the fish (Henderson and Tocher, 1987; Barron, 1990). These studies suggest that differences in lipid composition may account in part for differences in contaminant residues between species or among tissues.

A fourth potentially important factor that is related to lipid composition is the method used to extract the lipid. In laboratory tests with various aquatic biota tissues, lipid concentrations in replicate samples varied by a factor of 3.5 when different extraction solvents were used (Randall and others, 1991). The tissues tested were from bluefish (*Pomatomus saltatrix*) flesh and liver, sandworms (*N. virens*), and blue mussel (*Mytilus edulis*). De Boer (1988) showed the importance of extraction solvent in determining both lipid content and contaminant residues measured. De Boer found that fish contain about 5 g/kg of bound lipids, which consist chiefly of phospholipids. The remaining lipids (free lipids) consist of neutral lipids. In tests with five fish species, nonpolar solvents extracted free lipids, but not (or only partially) bound lipids. Therefore, the lipid contents measured depended on the extraction solvent used. Chlorobiphenyls were observed to accumulate in the bound lipids as well as in depot fats, so that the extraction solvent used also affected the quantity of contaminant residues measured (de Boer, 1988).

In summary, considerable evidence suggests that lipids play an important role in bioaccumulation of contaminants by aquatic biota. In some studies, differences in lipid content can account for some part of the differences observed between species, between individual organisms, or among tissues within an organism. However, the relation between lipid content and contaminant levels frequently is not as straightforward as predicted by simple equilibrium partitioning model. (This is not surprising, since these are clearly and intentionally simplistic models.) A number of complicating factors may obscure any relation between lipid content and contaminant levels, such as lipid composition, seasonality in both lipid content and residue levels, and the analytical method used to extract lipids and contaminant residues. Moreover, additional variables, such as body size, age, and reproductive state, that affect contaminant bioaccumulation may be covariant with lipid content.

Gill Ventilation Volume and Other Gill Characteristics

Gill ventilation volume is often identified as a physiological variable that affects contaminant uptake (Thomann, 1989; Erickson and McKim, 1990; Hayton and Barron, 1990). In the bioconcentration process, uptake of chlorinated hydrocarbon contaminants by aquatic biota occurs via transfer from the water through the gills to the blood, and then from blood to lipids (Hamelink and others, 1971; Murphy and Murphy, 1971). The rate of chemical uptake may be affected by factors affecting rate of water movement across gills, which is regulated by the ventilation rate.

Other related physiological characteristics of the gill are also important. Gill design (such as its large surface area and high rate of blood perfusion) maximizes the rate of exchange of solutes between water and blood. The gill consists of rows of filaments that maintain a flow of water countercurrent to blood flow in the lamellae (Nilsson, 1986). Lamellae are covered with mucus film and the water immediately above this film is stagnant. Hayton and Barron (1990) describe the pathway that an organic chemical takes as consisting of: (1) transport by water flow to near the gill epithelium, (2) diffusion across the aqueous stagnant layer to the epithelial surface, (3) diffusion across the epithelium to the blood, and (4) distribution through the organism by the blood. In this model, there are four potential barriers to uptake: water flow across gill, the aqueous stagnant layer, the gill epithelium, and blood flow through the gill (Hayton and Barron, 1990). The rate-limiting barrier is determined by the physical and chemical properties of the chemical (such as K_{ow}, molecular weight, and presence of polar functional groups, all of which affect the diffusion coefficients in the epithelium and in water, the blood–water distribution coefficient, and the epithelium–water distribution coefficient), the physiology of the organism (such as ventilation volume, epithelial surface area, epithelial permeability, epithelial thickness, and effective blood flow past the gill), and environmental conditions (such as temperature). Epithelial resistance is likely to be rate-limiting for compounds with high water solubility or high molecular weight, or for compounds that contain polar functional groups. When epithelial resistance declines, blood flow is generally limiting. For strongly lipophilic compounds, uptake is controlled by the rate of blood flow to storage tissues (which are poorly diffused), rather than by the rate of uptake by the gill. This means that ventilation volume, epithelial surface area, and epithelial thickness may affect uptake of chemicals of low-to-moderate hydrophobicity (Hayton and Barron, 1990).

Blood Flow

Blood flow or blood circulation time has been suggested as an important factor affecting the accumulation of hydrophobic contaminants (Kenaga, 1972; Erickson and McKim, 1990; Hayton and Barron, 1990). Blood flow and blood–tissue partitioning will determine the rate of transfer during uptake and elimination (Andersen, 1981), especially for highly hydrophobic contaminants (Hayton and Barron, 1990). Blood flow is different to various tissues, making kinetically distinct tissue compartments. Slow blood flow may prevent rapid distribution of a chemical into lipid storage compartments, which may be responsible for the time it takes for many hydrophobic chemicals to reach steady-state (Barron and others, 1990a; Hayton and Barron, 1990).

Metabolism

Metabolism, or biotransformation, is an important route of elimination for many chemicals. The types of metabolic reactions by which chemicals are biotransformed are described in Section 4.3.3. As mentioned there, some synthetic organic chemicals, including many pesticides, have structures that are totally unfamiliar to microorganisms and other biota, which may not have the enzymes needed for degradation of these compounds. This is the primary reason why many chlorinated pesticides (such as mirex and DDE) and structurally similar industrial compounds (such as hexachlorobenzene and some PCB congeners) are recalcitrant in the environment.

When hydrophobic compounds are metabolized, they typically are converted to more water-soluble compounds that can be excreted or conjugated with a biochemical and then excreted, rather than stored in tissue lipids (National Research Council, 1979; Stegeman, 1981; Varanasi and others, 1985). Metabolites of contaminants generally are eliminated in the urine, via bile into feces, or by diffusion through the gills into the surrounding water. The rate of contaminant metabolism is influenced by species (Johnson, 1973; Nagel, 1983) and strain (Pedersen and others, 1976). Contaminant metabolism also may vary for different tissues, lengths of exposure, and habitats or geographic regions (Johnson, 1973).

Contaminant metabolism is also an important factor in toxicity. There appears to be an inverse correlation between the susceptibility of a taxonomic group and pesticide concentrations in their tissues (Johnson, 1973). However, metabolites are not always less toxic than the parent compounds (National Research Council, 1979; Stegeman, 1981). For example, DDE has comparable toxicity to DDT, but paraoxon (the oxidation product of parathion) is more toxic than parathion (Klaassen and others, 1996).

When biotransformation occurs, it increases the rate of elimination of hydrophobic contaminants, as shown by several studies. In one study, juvenile guppies (*P. reticulata*) were fed PCB-contaminated food for 30 weeks, then elimination was measured for 2 years. A kinetic life-cycle biomagnification model was developed, taking into account growth, reproduction, and biotransformation. Uptake efficiencies were low in juveniles, then increased with age. Growth, biotransformation, and reproduction decreased the concentration of hydrophobic contaminants. The higher chlorinated biphenyls were not biotransformed, so that growth dilution was the only important elimination process. Biotransformation was the most important elimination process for PCB congeners that had chlorine in the *ortho* positions and that were unsubstituted at *meta* and *para* positions (Sijm and others, 1992).

In a study by Opperhuizen and Sijm (1990) that was discussed in Section 5.2.5 (Elimination Processes), PCDDs and PCDFs had lower BCFs (Figure 5.12) and higher elimination constants (Figure 5.15) in fish than other chlorinated aromatic hydrocarbons with comparable K_{ow} values. The low BCF values for PCDDs with up to 5–6 chlorines were attributed to rapid elimination by biotransformation (Opperhuizen and Sijm, 1990). Biotransformation of PCDDs and PCDFs in fish has been shown in a number of studies (e.g., Muir and others, 1985, 1986; Mehrle and others, 1988; Sijm and Opperhuizen, 1988). Moreover, a few studies have identified polar metabolites (Muir and others, 1986; Kleeman and others, 1988; Sijm and Opperhuizen, 1988).

The inhibition of metabolism increases bioaccumulation of a variety of chemicals (Stehly and Hayton, 1989; Opperhuizen and Sijm, 1990). Compounds that are readily metabolized (such as pyridine ester) will have kinetically determined BCF values far lower than those predicted on basis of hydrophobicity alone (Barron, 1990). This means that, for compounds that are metabolized to successively increasing degrees, the partition coefficient becomes less reliable as an indicator of bioaccumulation (Moriarty and Walker, 1987).

Growth Rate

Growth can be a principal route of contaminant elimination, especially for compounds that are not biotransformed. Even when the total quantity of contaminant in an organism remains unchanged, organism growth results in dilution so that the contaminant concentration in tissue decreases. After termination of exposure, organochlorine residue levels in coho salmon decreased in proportion to growth; growth may be more important than excretion (Willford and others, 1969). In juvenile guppies (*P. reticulata*) fed a PCB-contaminated diet (Sijm and others, 1992), growth, biotransformation, and reproduction were observed to decrease the concentration of PCB congeners to varying degrees. For the higher chlorinated biphenyls, growth dilution was the only important elimination process. In goldfish, the dieldrin residue content of the ovaries increased with the state of maturity, although the residue content of the testes was not related to the state of maturity (Grzenda and others, 1970).

In phytoplankton from Green Bay, Wisconsin, growth rate was an important factor affecting PCB congener bioaccumulation (Swackhamer and Skoglund, 1991). Measured BAFs were higher for phytoplankton under low-growth conditions. The effect of rapid growth was a decrease in BAF resulting from a gradual dilution of the biomass or an increase in excretion of extracellular products because of increased cellular metabolism, or both. In winter, when growth rate was low, field populations of phytoplankton from Green Bay, Wisconsin, showed a strong relation between log BAF and log K_{ow}. This was similar to results observed in the laboratory under low-growth conditions. In contrast, the relation between log BAF and log K_{ow} was very weak during the summer, when phytoplankton growth rate was rapid, suggesting a lack of equilibrium. In fact, the rate of PCB uptake during the summer was similar to the rate of phytoplankton growth. This means that, because PCB congeners reach equilibrium very slowly, it is unlikely that equilibrium would be reached in phytoplankton during periods of normal growth.

A number of bioaccumulation models incorporate terms for growth rate or efficiency (e.g., Norstrom and others, 1976; Spacie and Hamelink, 1982; Thomann, 1989). Even a simple equilibrium partitioning model should consider growth rate under field conditions (Thomann, 1989). This is particularly important for extremely hydrophobic chemicals, such as DDT and PCBs (Hamelink and Waybrant, 1976), because of the time required to achieve steady-state (Branson and others, 1975). Chemical equilibrium (equifugacity) between the fish and water cannot be reached if the chemical is metabolized by the organism or eliminated in fecal matter, or if the organism undergoes significant growth during the time necessary to reach equilibrium (Gobas and others, 1993b). Bioaccumulation models can be highly sensitive to the relative values assumed for the assimilation efficiency and loss terms (due to depuration and growth), especially at higher trophic levels (Spacie and Hamelink, 1985).

Reproductive State

Reproduction can decrease the concentration of hydrophobic contaminants. In juvenile guppies (*P. reticulata*) fed a PCB-contaminated diet (Sijm and others, 1992), reproduction was a significant route of hydrophobic contaminant elimination. Second generation guppies contained PCB concentrations that were comparable to those of their parents. Reproductive state also can affect hydrophobic contaminant elimination in other ways. Sex hormones influence mono-oxygenase activity in rainbow trout (Forlin and Hansson, 1982), so that enzyme activities (and metabolism) will be subject to seasonal variations because of reproductive status (Huckle and Millburn, 1990).

Seasonal changes in contaminant residues were sometimes associated with the reproductive cycle (Langston, 1978; National Research Council, 1980; Phillips, 1980; Huckle and

Millburn, 1990; also see Section 5.3.5). For example, in flounder (*P. flesus*) and cod (*G. morhua*) from Norway, production of eggs during the reproductive period (spring) resulted in loss of substantial PCBs in females (Marthinsen and others, 1991). Higher PCB residues in female coho salmon spawners also has been reported (Armstrong and Sloan, 1980).

In a number of studies, feeding activity or lipid content, or both, also changed during spawning, so that seasonal changes in contaminant levels were associated with more than one variable. For example, Porte and Albaiges (1994) measured hydrocarbons and PCB congeners in marine mussels (*Mytilus galloprovincialis*), crabs (*Macropipus tuberculatus*), red mullet (*Mullus barbatus* and *Mullus surmuletus*), and tuna (*Thunnus thynnus*) on the Catalan coast of the Mediterranean. Mussels and red mullet had maximum residues in spring, and a decline in summer. These changes were associated with spawning, during which feeding activity and lipid content were reduced.

In another example, PCB and hexachlorobenzene residues were measured in female dabs from the German Bight as a function of organ (or tissue) and lipid content (Kammann and others, 1990). During the spawning season, lipids were mobilized and used for anabolism of the eggs. Both total lipid and neutral lipid in the liver decreased from December, the beginning of spawning season, to April. Total PCB and hexachlorobenzene levels in the liver decreased from December to February, then increased slightly. Contaminant levels in the ovaries increased dramatically from January to April. The PCB pattern in the ovaries (predominantly pentachloro-biphenyls and hexachlorobiphenyls) reflected transfer of lipids from the liver.

Seasonal variations also were reported both in organochlorine concentrations and in lipid content of whelks (*Buccinum undatum*) from the German Bight (Knickmeyer and Steinhart, 1990). These authors attributed these changes to a combination of the reproductive cycle and feeding behavior. Whelks begin spawning in November–December, at which time their feeding activity declines sharply. They resume feeding 4–5 months later, after egg laying. Growth of the ovary and oviduct coincided with low feeding activity and with a decrease in the size of the digestive gland. Both lipid reserves and protein from the large foot muscle were mobilized and used for gonad maturation. Lipid content was higher in females than in males from the end of November to the end of January. Males showed an increase in lipids in February and females an increase in April and May. Males had higher PCB and DDE residues from the end of November to the end of February. Females had increased PCB, hexachlorobenzene, and DDE contamination in April, corresponding to the maximal feeding rate observed in April. Residues of lindane (the least lipophilic of the compounds investigated) were not correlated to lipid content. Total PCB and DDE residues in males were inversely correlated with lipid content. Hexachlorobenzene residues were positively correlated with lipid content.

Such studies suggest that organisms taken prior to egg laying or spawning may have high lipid content, while those collected afterwards may show a decrease in both lipid content and contaminant residues (Philips, 1980; Insalaco and others, 1982). Therefore, a few weeks difference in sampling during a spawning run may greatly influence the results (Armstrong and Sloan, 1980).

Species

Interspecific differences among contaminant residues have been widely reported (e.g., Matsumura, 1977; Ellgehausen and others, 1980; Biddinger and Gloss, 1984; Huckle and

Millburn, 1990). Interspecific differences have been attributed to feeding habitat (e.g., in the Town River, insecticide residues were higher in bottom-feeders than in predators [Mick and McDonald, 1971]); lipid content (e.g., DDT and metabolites in mackerel, or *S. scombrus*, and brook trout [Duffy and O'Connell, 1968]); and DDT and dieldrin in herring and cod [Jensen and others, 1969, 1972]; metabolic capability (e.g., dieldrin in aquatic invertebrates [Wallace and Brady, 1971]); and trophic level (e.g., PCBs and pesticides in various fish from the Schuylkill River, Pennsylvania [Barker, 1984]; and PCBs in invertebrates and sole (*S. solea*) from the Wadden Sea [Goerke and others, 1979]). The role of lipid content in interspecific differences was discussed previously (Section 5.2.5, subsection on Lipid Content). In many cases, there was no clearcut or single explanation for the observed differences (e.g., DDTs and PCBs in various mussels and benthic organisms from the Northern Adriatic Sea and Limski Canal in Yugoslavia [Najdek and Bazulic, 1988]). Observed interspecific differences may be at least partly attributable to a combination of covariant factors, such as body size (Davies and Dobbs, 1984) and ambient temperature (Barron, 1990).

Temperature may contribute to apparent differences in contaminant accumulation by warm water and cold water fish species. Water temperature affects respiration, caloric intake (Spacie and Hamelink, 1985), and metabolic enzyme activity (Huckle and Millburn, 1990; Mayer and others, 1970). Trout generally have a lower BCF than fathead minnows (Veith and others, 1979a), as well as a lower lipid content (Henderson and Tocher, 1987). However, the exposure temperature for trout, which can affect bioaccumulation, also is lower (Barron, 1990). Higher temperatures tend to increase rates of uptake and elimination (Barron and others, 1987; Barron, 1990). For example, the BCF of a PCB mixture (Aroclor 1254) in green sunfish (*Lepomis cyanellus*) increased from about 6,000 to about 50,000 when the temperature was increased from 5° to 15 °C; similarly the BCF in rainbow trout increased from 7,000 to 10,000 (Veith and others, 1979a). For additional information on the effects of temperature on biotransformation and bioaccumulation, see Section 4.3.3 and Section 5.2.5 (subsection on Environmental Conditions that Affect Bioaccumulation).

Tissue

Many studies investigated the distribution of hydrophobic contaminants among tissues. The role of lipid content in tissue distribution of contaminant residues was discussed previously (Section 5.2.5, subsection on Lipid Content). In summary of that discussion, hydrophobic contaminant residues are generally, but not always, greater in tissues with high lipid content. Examples include hydrocarbons in mussels (Connell, 1978); PCB in rainbow trout (Lieb and others, 1974); DDT in blubber of Weddell seals (Hidaka and others, 1983); DDT in various fish species (Meeks, 1968; Maslova, 1981; Ziliukiene, 1989); and TCDD in rainbow trout (Hektoen and others, 1992). The following factors may obscure or contribute to the relation between tissue contamination and lipid content: (1) lipid content changes during growth and development and it tends to be covariant with age and body size; (2) many aquatic organisms undergo seasonal variations in lipid content, principally in response to reproductive cycles; (3) contaminant distribution appears to vary with lipid composition; and (4) measurements of both contaminant residues and lipid content are affected by the solvent extraction method used.

In addition to lipid content, lipid composition can be important in controlling the distribution of hydrophobic contaminants among tissues. This is illustrated by a study of PCB congeners in female dabs from the German Bight (Kammann and others, 1990). Individual tissues showed a distinct PCB congener pattern, independent of sampling site or season. Trichlorobiphenyls and tetrachlorobiphenyls were high in the kidneys and gallbladder, which have a high percentage of polar (methanol-soluble) lipids. Pentachlorobiphenyls and hexachlorobiphenyls were predominant in the liver, ovaries, gills, and stomach, which are high in neutral (*n*-hexane-soluble) lipids. The pattern of PCB congeners in the stomach reflected the food composition of the dab.

Hydrophobic contaminant residues often tend to be high in the liver, which plays a role in residue elimination (Johnson, 1973; also see Section 4.3.3). In summary, the liver is an important site of contaminant biotransformation (Huckle and Millburn, 1990). Metabolites formed in the liver are transported to the gallbladder, then discharged with bile into the small intestine, where they may be eliminated with the feces (Spacie and Hamelink, 1985). Bile is probably the primary route of excretion for metabolites of a number of lipophilic compounds (Huckle and Millburn, 1990). High concentrations of hydrophobic contaminants were noted in the liver in several studies: TCDD in cod (Hektoen and others, 1992); total DDT and dieldrin in common carp, white sucker (*Catostomus commersoni*), redhorse sucker (*Moxostoma* sp.), northern pike, walleye, and smallmouth bass (*Micropterus dolomieu*) from the Grand River in Ontario, Canada (Kelso and others, 1970); organochlorine compounds in various macroinvertebrates and fish from the Lower Medway Estuary in Kent, United Kingdom (Wharfe and van den Broek, 1978); and PCBs in cod (*G. morhua*) from the western Baltic Sea (Schneider, 1982).

Several studies reported that hydrophobic contaminant residues were particularly high in the gonads of aquatic biota. Examples include kepone in female blue crab from the James River and the lower Chesapeake Bay (Roberts, 1981); and TCDD in bowfin, largemouth bass, brown bullhead, and blue crab collected downstream of a paper mill on the St. Johns River in Florida (Schell and others, 1993). In some studies, seasonal variations in contamination of the ovaries were associated with spawning season. Examples include DDT, DDE, and heptachlor epoxide in winter flounder (*Pleuronectes americanus*) from the Weweantic River Estuary in Massachusetts (Smith and Cole, 1970); and PCB congeners in female dabs from the German Bight (Kammann and others, 1990). Also, nonlipid tissues in some organisms (such as the carapace of crustaceans) may be storage sites for hydrophobic contaminants (Barron and others, 1988).

Sex

Sex has been identified as a parameter affecting contaminant residues in fish (Matsumura, 1977; Ellgehausen and others, 1980; Biddinger and Gloss, 1984; Huckle and Millburn, 1990). Sex differences could be caused by different lipid distributions; different reproductive cycles and seasonal fluctuations; different enzyme activities, and therefore metabolic capabilities; and differential response to inducing agents. In general, male fish have higher cytochrome P-450 content and monooxygenase activity than females of the same species (Huckle and Millburn, 1990). The effects of inducing agents also vary depending on age and sex (e.g., Forlin, 1980; Goksoyr and others, 1987). Moreover, sex hormones also influence monooxygenase activity in rainbow trout (Forlin and Hansson, 1982) so that enzyme activities will be subject to seasonal variations due to reproductive status (Huckle and Millburn, 1990).

Sex differences in hydrophobic contaminant residues have been detected in some studies, but not in others. For example, organochlorine pesticides in round whitefish fillets from Lake

Huron were higher for males than females (Miller and Jude, 1984). In blue crabs from the James River and lower Chesapeake Bay, males contained more kepone in backfin muscle than females (Roberts, 1981). On the other hand, no significant differences were observed with respect to sex for pesticides and PCBs in chub (*Leuciscus cephalus*) and bleak (*Alburnus alburnus*) from the River Po in Italy (Galassi and others, 1981), or in mosquitofish from a rice field in the Ebro River Delta in Spain (Porte and others, 1992). In both cases, samples were composites of fish muscle, grouped by sex and size. Conflicting reports as to sex differences in contaminant levels may result because of the large number of interrelated variables that may affect residue levels, such as age, size, lipid content, and reproductive state.

Other Biological Factors

A number of other biological factors have been suggested as affecting residue levels in fish. These include enzyme induction and increased metabolic ability in response to previous exposure to DDT or related compounds (Kenaga, 1972); feeding habits and fat storage metabolism (Kenaga, 1972); and vertical distribution in the water column (Biddinger and Gloss, 1984).

Interaction Among Biological Factors

Some of the biological factors discussed in the previous sections directly affect contaminant uptake or elimination rates, or both. Examples include gill ventilation volume, blood flow, and growth rate. Other factors, such as species, tissue type, sex, age, body size or weight, and reproductive state, were identified empirically from laboratory tests and field monitoring studies. Of these empirical factors, some are seasonal and many are interdependent (Huckle and Millburn, 1990). It is unclear whether some of the observed differences relating to these empirical factors may be due to the influence of underlying physiological characteristics or environmental conditions (Davies and Dobbs, 1984), many of which are also interdependent. For example, different species have characteristic lipid contents, tissue lipid distributions (e.g., Hektoen and others, 1992), and metabolic capabilities (Nagel, 1983). Also, the metabolic rate in each fish species depends on temperature, salinity, feeding rate, and physical stress (Farmer and Beamish, 1969; Niimi and Beamish, 1974). The body weight of the fish influences its oxygen consumption (and therefore metabolic activity), the rate of pesticide uptake, and lipid content. Physiological changes (such as growth and spawning), seasonal factors (such as temperature and food availability), and habitat selection all contribute to the bioenergetics of the fish and contaminant bioaccumulation (Huckle and Millburn, 1990).

Chemical Characteristics that Affect Bioaccumulation

The bioaccumulation potential of an organic chemical is affected by its chemical characteristics, as listed in Table 5.5: molecular size, shape, and structure; physical and chemical properties such as its solubility in water and lipids; and its resistance to degradation, both abiotic and biologically mediated (see Table 5.5).

Molecular Size, Shape, and Structure

The transfer of an organic chemical across a membrane is influenced by its molecular size, shape, and structure, including polarity, presence of functional groups, and degree of ionization (Esser and Moser, 1982; Shaw and Connell, 1984; Opperhuizen and others, 1985; Esser, 1986; Huckle and Millburn, 1990). A decrease in uptake rate constant has been reported for compounds with high molecular weights (greater than 300 daltons); large cross-sectional areas (greater than 0.95 nm); or large parachors (greater than 500), which is the molar volume corrected for surface tension (Oliver and Niimi, 1985; Opperhuizen and others, 1985). It appears that large molecules are not efficiently transferred across gill membranes, though fish can assimilate them from food (Skea and others, 1981; Niimi and Oliver, 1983).

The permeability of biological membranes is inversely proportional to molecular weight. For fish, compounds with molecular weights greater than about 600 daltons were not taken up from water (Zitko and Hutzinger, 1976). As noted previously, bioaccumulative chemicals tend to have molecular weights less than about 300 daltons, with a decline in capacity at higher values (Brookes and others, 1986; Connell, 1988; Oliver and Niimi, 1985; Opperhuizen and others, 1985). Molecular size-related parameters are predictive of bioconcentration for several classes of organic chemicals (Connell and Schuurmann, 1988; Schuurmann and Klein, 1988). Some PCDD and PCDF congeners (such as OCDD and octachlorodibenzofuran) were taken up at very slow rates or efficiencies (if at all) by several fish species after dietary or aqueous exposure (Opperhuizen and Sijm, 1990; Muir and others, 1985, 1986; Muir and Yarechewski, 1988). For PCDD and PCDF congeners with an effective cross-section greater than 0.95 nm, BCF values actually decreased with increasing hydrophobicity. BCF values for polychlorinated biphenyls and naphthalenes also were observed to depend on molecular size, rather than on hydrophobicity (Opperhuizen and others, 1985). These observations have been attributed to decreased permeability with an increase in molecular size (Zitko, 1980; Opperhuizen and others, 1985; Connell, 1988; Opperhuizen and Sijm, 1990).

Opperhuizen and Sijm (1990) proposed that these slow uptake rates and efficiencies probably were a function of the morphology and physiology of fish membranes, rather than just the physical and chemical properties of the chemical. Opperhuizen and others (1985) suggested that the phospholipid bilayer of the gill epithelium can restrict the uptake of hydrophobic molecules with long chain length or large cross-sectional area. Transfer from the gill epithelium into the blood also may be limited by slow solvation of large hydrophobic molecules (Plant and others, 1983). The determining factor appears to be molecular size (volume or cross-sectional area) rather than molecular weight (Opperhuizen and others, 1985; Barron, 1990).

Tulp and Hutzinger (1978) suggested that steric configuration and molecular size may influence uptake, and they proposed that molar volume, molar refractivity, and the parachor are parameters that reflect this. BCF values were found to be linearly correlated with parachor (Tulp and Hutzinger, 1978). Tulp and Hutzinger found that the parachor places too much emphasis on the number of chlorine substituents and too little on their position. Matsuo (1980a,b) developed sum-i and sum-o parameters that quantitatively described the superimposition of a compound and fish tissue. Matsuo (1979, 1980a,b) found that a nonplanar configuration can prevent a molecule from efficient interaction with fish tissue, as was observed for hexachlorobenzene and some PCB congeners. Shaw and Connell (1980, 1982) developed a steric effect coefficient (SEC), calculated from empirical coefficients corresponding to particular substitution patterns, for PCBs. The SEC was linearly related to uptake within groups of PCB congeners, but the product of SEC and log K_{ow} was a still better predictor of bioaccumulation than SEC alone. The SEC was related to the planarity of the PCB congeners as measured by their chromatographic adsorption behavior on carbon (Shaw and Connell, 1984). The uptake of PCB molecules was hindered by three structural features: (1) three or four *ortho* chlorines, (2) three or four chlorines adjacent to each other on a phenyl ring, and (3) chlorine substituents in the 3- and 5- positions. Increasing *ortho* substitution increased the angle of twist between biphenyl rings, reducing planarity. The maximum bioaccumulation was observed for PCB congeners with five to seven chlorines. Those with fewer chlorines had unfavorable lipophilicity (low K_{ow} values), and those with more chlorines had unfavorable stereochemistry and sorption characteristics (Shaw and Connell, 1984).

The degree of ionization of a chemical also affects bioaccumulation. For an ionizable chemical, the ionized form has greater polarity, higher water solubility, and a lower K_{ow} than the nonionized (neutral) form. The K_{ow} may be affected by the charge on the molecule, the pH of the solution, the pK_a or pK_b value of the compound, and the presence of neutral salts (Connell, 1988). In general, nonionized forms of organic chemicals more readily penetrate biological membranes than do ionized forms (Saarikoski and others, 1986). Therefore, the uptake of organic acids and bases is influenced by the environmental pH (Huckle and Millburn, 1990).

Solubility

The bioaccumulation potential of a chemical is affected by its solubility in both water and lipid. A chemical that is readily soluble in lipid (*lipophilic*) also tends to have a low solubility in water (*hydrophobic*); thus the terms "lipophilic" and "hydrophobic" are often used synonymously. A high K_{ow} value indicates lipophilicity (hydrophobicity).

A chemical's water solubility controls its extent of sorption, which affects the extent of exposure of an aquatic organism to the chemical in its dissolved form. Water solubility tends to be inversely related to lipid solubility, to K_{ow}, and (as shown in Figure 5.6) to BCF (Mackay, 1982; Moriarty and Walker, 1987; Connell, 1988; Huckle and Millburn, 1990; also see Section 5.2.4, subsection on Correlation between Bioconcentration Factor and Chemical Properties). Bioaccumulative chemicals tend to have water solubilities between 0.002 and 0.02 mol/m^3, with declining capacity at values that are either below or above this range (Connell, 1988).

In aqueous exposure of bluegills and goldfish to DDT, dieldrin, and lindane, both sorption from water and elimination from fish were related to water solubility (Gakstatter and Weiss, 1967). The order of decreasing water solubility (lindane greater than dieldrin; dieldrin greater than DDT) corresponded to the order of increasing uptake from water and decreasing elimination from fish. Differences in elimination rates may result from differences in the relative affinities of these pesticides for internal tissues and in their rate of transport through the body from storage sites to the site of excretion. For example, DDT excretion would be the slowest because it is the most difficult to transport to the site of excretion (Gakstatter and Weiss, 1967).

Bioaccumulation tends to increase with increasing lipophilicity, as reflected in the *n*-octanol-water partition coefficient, K_{ow}. Because biological membranes consist of a phospholipid bilayer, the extent of organic chemical penetration through membranes is a function of lipid solubility (Huckle and Millburn, 1990). Both degree of accumulation (Neely and others, 1974; Hamelink and Spacie, 1977) and toxicity (Yang and Sun, 1977) in fish have been observed to increase with an increase in the log K_{ow} (Huckle and Millburn, 1990). Actually, there appears to be an optimum K_{ow} (log K_{ow} of 2–6) for bioaccumulation of chemicals in fish (Connell, 1988). This is a function of effects on both uptake and elimination processes.

The logarithm of the uptake rate constant for the guppy (*P. reticulata*) increased linearly with an increase in K_{ow} for chemicals with log K_{ow} values less than 6 (Gobas and others, 1989b), as shown in Figure 5.22. However, for chemicals with a log K_{ow} greater than 6, bioaccumulation declined as the K_{ow} continued to increase (Hawker and Connell, 1986; Gobas and others, 1989b). A similar pattern was observed by McKim and others (1985), who measured direct uptake and elimination of organic chemicals across the gills of rainbow trout using an in vivo fish model (also see Section 5.2.5, subsection on Factors Affecting Route of Uptake). These authors found that uptake efficiencies for chemicals with very low K_{ow} values (log K_{ow} less than 0.9) were low and unrelated to K_{ow} (Figure 5.7). For chemicals with log K_{ow} values between 0.9 and 2.8, uptake efficiencies were positively correlated with log K_{ow}. For chemicals with log K_{ow} values between 2.8 and 6.2, uptake efficiency was constant. At log K_{ow} greater than 6.2, uptake efficiency appeared to be inversely correlated with log K_{ow}. Hansch and Clayton (1973) hypothesized, on the basis of review of bioassay data, that chemicals with very low K_{ow} values were too insoluble in fat to pass through biological membranes, while those with very high log K_{ow} values tended to bind to lipid membranes and could not pass into blood or cellular fluid.

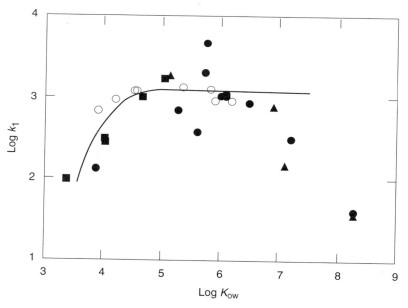

Figure 5.22. The relation between the logarithm of the uptake rate constant (log k_1) in the guppy, and the logarithm of the *n*-octanol-water partition coefficient (log K_{ow}), for selected halogenated compounds. The solid line represents the theoretical relation for uptake by diffusive release (k_1). Data shown are from Gobas and others (1989b) (●); Bruggeman and others (1984) (▲); Opperhuizen and others (1985)(○); and Könemann and van Leeuwen (1980) (■). Adapted from Gobas and others (1989b) with permission of the publisher. Copyright 1989 Society of Environmental Toxicology and Chemistry (SETAC).

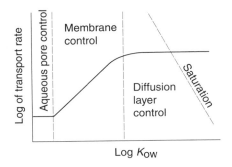

Figure 5.23. The relation between the logarithm of the transport rate and the logarithm of the *n*-octanol-water partition coefficient (log K_{ow}) when pore transport is not negligible. Redrawn from McKim and others (1985) with permission of the publisher and the author. Copyright 1985 Academic Press, Inc. The original figure in McKim and others (1985) was modified from Yalkowsky and Morozowich (1980).

This is consistent with a model describing flux of a chemical across a biological membrane bounded by one or more aqueous compartments (Flynn and Yalkowsky, 1972; Yalkowsky and Morozowich, 1980; McKim and others, 1985). Chemicals with very low K_{ow} values would readily pass through the aqueous diffusion layers, but would be transported very slowly across the lipid membrane. For chemicals with molecular weights less than 100 daltons, molecules may be transported through aqueous pores in the membrane (Figure 5.23). For these chemicals, flux is not related to log K_{ow} because movement is limited by molecular size. As K_{ow} increases, the chemicals pass more readily through the lipid membrane and flux across the membrane is related to log K_{ow}. As K_{ow} continues to increase, flux becomes diffusion-layer controlled; high K_{ow} chemicals are not hindered by the membrane, and there is again no relationship between flux and K_{ow}. At very high K_{ow} values, the water solubility of the chemical limits transport. Additional factors, such as molecular volumes, molecular weight, chemical self-association (McKim and others, 1985), or reduced availability of the chemical due to sorption or interaction with organic or other matter in the water phase (Gobas and others, 1989b), also may reduce membrane transport at log K_{ow} values greater than 7.

Elimination rates also are influenced by K_{ow}. From the same in vivo model of McKim and others (1985), chemicals with log K_{ow} values less than 3 were eliminated (through the gills into the water) rapidly within 4 hours. Chemicals with log K_{ow} values greater than 3 were not eliminated from the gills within the 4-hour dosing period. Chemicals with higher K_{ow} may require metabolism to increase their polarity sufficiently that excretion via any route can take place. As noted above, uptake rates as measured using this in vivo model were the same for chemicals with log K_{ow} values from 2.8 to 6.2. Therefore, in the region of log K_{ow} values from 2.8 to 6.2, the slope of the log BCF versus log K_{ow} relation was controlled by excretion rates, with excretion rates slower for chemicals with higher log K_{ow} (also see Section 5.2.5, Elimination Processes). An inverse relation between elimination rate and log K_{ow} (e.g., Figure 5.16) also was observed by Spacie and Hamelink (1982) and Gobas and others (1989b). According to the model of bioconcentration in fish developed by Gobas and others (1986), this relation is explained by different rates of chemical release from fish lipid compartments. Elimination is controlled by the lipid membrane for less hydrophobic compounds, and by diffusion through aqueous diffusion layers for more hydrophobic compounds. Hayton and Barron (1990) proposed that for strongly lipophilic compounds, uptake is controlled by the rate of blood flow to storage tissues (poorly diffused), rather than by the rate of uptake by the gill. Hydrophobicity is the principal determinant of bioconcentration only when other steps in the accumulation process are not rate-limiting.

Potentially rate-limiting steps include ventilation volume, membrane permeability, and blood flow (Hayton and Barron, 1990; Erickson and McKim, 1990). For extremely hydrophobic chemicals (log BCF greater than 6.5 or 7), elimination into the feces becomes more important than elimination across the gills (Gobas and others, 1989b). This is illustrated in Figure 5.16, which shows experimental data relating the logarithm of the elimination rate (log k_T) for the guppy (*P. reticulata*) to log K_{ow}. For chemicals with very high K_{ow} values (log K_{ow} greater than 7), experimental data are consistent with the expected relation between K_{ow} and the logarithm of the rate constant for elimination in feces (k_e). This expected relation (the dashed line in Figure 5.16) was derived from the observed relation between K_{ow} and the uptake efficiency of organic substances from food. Gobas and others (1989b) pointed out that the ratio of the chemical concentration in fish to that in water (BAF) represents a fish–water partition coefficient (equal to k_1/k_2, the BCF as defined in Equation 5.2) only when elimination into the water is the predominant route of elimination by fish. The fish–water partition coefficient, or BCF (where uptake and elimination occur by diffusion), is expected to show a strong relation with hydrophobicity, or K_{ow}. However, if chemical elimination in feces becomes significant, this will lower the measured BAF (equal to $k_1/[k_2 + k_e]$ assuming negligible growth or metabolism; see Equation 5.4) below the BCF. Thus, BAF values for extremely hydrophobic chemicals, as discussed earlier, are lowered by chemical elimination in feces (Gobas and others, 1989b), as well as by decreased uptake rates.

A modified fugacity model developed by Gobas and others (1993b) emphasizes the importance of elimination rates in biomagnification potential. When an organism ingests a hydrophobic chemical contaminant, its fugacity is high relative to the surrounding water; this fugacity can be reduced by the processes of chemical transformation, gill elimination, and growth dilution in the organism. If the combined rate of these processes is slow relative to chemical elimination in feces, then the high fugacity in the organism is maintained and biomagnification occurs. In contrast, if the combined rate of metabolism and elimination is high relative to the fecal elimination rate, then the high fugacity of the organism is not maintained, and there should be no evidence of biomagnification. Gobas and others (1993b) suggest that this is why chemicals with log K_{ow} values less than 6 do not biomagnify. For nonmetabolizable compounds with log K_{ow} values greater than 6, the rates of elimination via the gills, skin, urine, metabolism, or growth dilution are too slow to reduce the high fugacity in the organism.

Chemical Stability

Bioaccumulative chemicals tend to be resistant to both abiotic degradation and biologically mediated degradation (metabolism). They tend to have a high proportion of C–C (aliphatic or aromatic), C–H, and C–Cl bonds, relative to unsaturated bonds. These bond types confer stability on the parent compound, as reflected in the environmental persistence of organochlorine insecticides (Connell, 1988). Bioaccumulation is not likely to occur if elimination via metabolism is rapid.

Sorption, volatilization, and degradation reactions (such as hydrolysis, oxidation, and photolysis) are competing fate processes to bioaccumulation and tend to reduce the aqueous concentration of a contaminant, thus reducing the aqueous exposure of aquatic biota. Hydrophobic chemicals may require a relatively long period of time to bioaccumulate, but recalcitrant chemicals are available for long periods of time (Hawker and Connell, 1986). Many bioaccumulative chemicals also tend to sorb to soil or sediment. A chemical that is resistant to microbial degradation, or degrades very slowly, may accumulate in sediment. Thus, the sediment may act as both a sink and source of that chemical in the hydrologic system.

Environmental Conditions that Affect Bioaccumulation

Because bioaccumulation is the net result of competing processes of uptake and elimination, factors that affect the rate of one or both processes will affect the degree of bioaccumulation. Environmental factors that influence the transfer of organic contaminants across chemical membranes include water temperature, pH, salinity, degree of oxygenation, exposure concentration and duration, ambient concentrations of dissolved organic matter (DOM) and particulates, and water quality (Table 5.5).

Temperature

Temperature tends to increase rates of both uptake and elimination (Barron and others, 1987; Barron, 1990). For example, the BCF of a PCB in green sunfish increased from 6,000 to 50,000 when temperature was increased from 5° to 15 °C; similarly, the BCF in rainbow trout increased from 7,400 to 10,000 (Veith and others, 1979a). Murphy and Murphy (1971) reported a linear correlation between the uptake of oxygen and DDT in mosquitofish; both uptake rates declined when temperature was reduced. Temperature also influences rates of biotransformation (discussed in Section 4.3.3). Generally, BCF values increase with an increase in temperature (Reinert and others, 1974b; Edgren and others, 1979). This may be due to a temperature-dependent increase in chemical uptake (relative to elimination) rates, although this seems to be species-dependent (Veith and others, 1979a).

pH and Salinity

pH has little effect on the bioconcentration of nonionic compounds, but can influence the uptake rate of organic acids and bases (Huckle and Millburn, 1990). In general, nonionized forms more readily penetrate biological membranes than do ionized forms (Saarikoski and others, 1986). The ionized form of a chemical will have greater polarity, higher water solubility, and a lower K_{ow}. The K_{ow} may be affected by the charge on the molecule, the pH of the solution, the pK_a or pK_b value of the compound, and the presence of neutral salts (Connell, 1988). The pH also may affect biotransformation rates. For example, lowering the pH from 6.7 to 3 significantly decreased aryl hydrocarbon hydroxylase activity in lavaret (*Coregonus lavaretus*) and splake, a hybrid offspring of two trout species (Laitinen and others, 1982). pH also affects environmental degradation rates (such as hydrolysis and photolysis) of some pesticides (see Section 4.2.3), thus affecting the concentration and duration of exposure.

High salinity reduced uptake rates of organochlorine compounds in fish (Murphy, 1970; Kenaga, 1972; Tulp and others, 1979), but appeared to increase uptake of the organophosphate insecticide fenitrothion in blue crab (Johnston and Corbett, 1986). Salinity also affected the metabolic fate of fenitrothion in blue crab, with more fenitrooxon (a toxic metabolite) formed at high salt concentrations (Johnston and Corbett, 1986). In fish, BCF values for a given chemical tend to be lower in salt water species than in freshwater species (Davies and Dobbs, 1984).

Dissolved Organic Matter and Particulate Concentrations

Two routes of exposure by which organisms bioaccumulate hydrophobic organic contaminants are (1) transport across bioavailable membranes exposed to the water phase and (2) direct ingestion of contaminated food and other particles (Suffet and others, 1994). Particulate and dissolved material within the water and sediment environments influence the rate and magnitude of chemical transport across biological membranes from the water phase, and thus affect the bioavailability of the chemical (Barron, 1990). The hydrophobic molecule must be in solution (each molecule with a hydration shell) to cross the absorbing epithelium (Landrum and others, 1985). Experiments have shown that uptake of nonionic organic compounds by biota may be reduced in the presence of DOM or humic-like material (e.g., Boehm and Quinn, 1976; McCarthy and others, 1985). Sorption, entrapment, or sequestering of nonionic organic contaminants on or within particles also reduces their bioavailability (Farrington, 1989). Binding to particulates and DOM, including highly reversible binding, reduce accumulation by reducing the unbound, bioavailable fraction (Landrum and others, 1987; Black and McCarthy, 1988). However, in a study of hydrophobic compounds in the Great Lakes, binding to DOM was generally 5 percent or less (Eadie and others, 1990). For highly hydrophobic compounds, the rate of desorption from particulates may determine the rate of uptake (Opperhuizen and Stokkel, 1988). The exact role that sorption plays in controlling bioavailability via ingestion is poorly understood (Suffet and others, 1994). However, the contribution from bound fractions may become important for certain organisms via ingestion. Catabolic reactions in the gut may render ingested particle or DOM-associated contaminants bioavailable (Suffet and others, 1994).

Oxygen Concentration

The degree of water oxygenation can influence gill absorption of contaminants (McKim and Goeden, 1982). In general, oxygen consumption influences metabolic activity (Huckle and Millburn, 1990). Oxygen concentration has been proposed to affect uptake rates by influencing the ventilation volume of water passing over the gills (Neely, 1979). However, in a study by Opperhuizen and Schrap (1987), the rate of PCB diffusion within the guppy had a greater effect on uptake and elimination than did the ventilation volume of water passing over the gills. Endrin uptake by the gills of brook trout (12 °C) increased with decreasing oxygen at high oxygen concentrations, but flattened out at low oxygen concentrations (McKim, 1994). The author postulated that, at the highest oxygen concentration, uptake appeared to be limited primarily by water flow. As the oxygen concentration declined, the water flow rate increased to maintain oxygen consumption and provided more endrin for adsorption. However, as water flow continued to increase, diffusion became more important in controlling uptake.

Other Environmental Conditions

Water quality may affect the degree of contaminant bioaccumulation. Prior exposure to chemical contaminants can reduce bioaccumulation if enzyme induction has occurred, resulting in increased metabolism and elimination (Binder and others, 1984). When fish are exposed to multiple chemicals, uptake of one chemical may be influenced by another (Mayer and others, 1970). The presence of emulsifiers also may affect the uptake of a contaminant from water (Kenaga, 1972).

Season of the year influenced contaminant residues in some studies (e.g., Kellogg and Bulkley, 1976; Kelso and others, 1970; Knickmeyer and Steinhart, 1990), but not in other studies (e.g., Kelso and Frank, 1974; Morris and others, 1972). Observed seasonality may be related to environmental conditions (such as seasonal changes in water temperature; Section 5.3.4), biological factors (such as the reproductive cycle and seasonal variations in lipid content; Section 5.3.5), as well as to seasonal changes in aqueous concentrations or inputs of pesticides (Sections 5.3.1 and 5.3.2).

Bioaccumulation Models

Bioaccumulation models have contributed to the development of our understanding of bioaccumulation processes, and have been useful in exposure and hazard assessments of new and existing chemicals. Various types of models used to describe bioaccumulation are briefly described, and some key examples given. The emphasis of the discussion will be on the contribution of bioaccumulation models to our understanding of the mechanisms involved. This discussion draws on reviews by Spacie and Hamelink (1982, 1985), Connell (1988), Farrington (1989), Barron (1990), Day (1990), and Landrum and others (1992), as well as on the primary literature. Some of these reviews, especially Spacie and Hamelink (1982, 1985), Barron (1990), and Landrum and others (1992) present individual models, including the underlying mathematics, in more detail. Use of bioaccumulation models in exposure and hazard assessments is discussed by Landrum and others (1992).

Correlations have been well documented between some measure of bioaccumulation in fish (such as BCF, BMF, lipid- or wet-weight concentration) and physical and chemical properties. For example, BCF and BMF tend to be positively correlated with K_{ow} (e.g., Neely and others, 1974; Metcalf, 1977; Veith and others, 1979a; Mackay, 1982), parachor (Tulp and Hutzinger, 1978; Briggs, 1981), molecular connectivity index (Sabljic, 1987), molecular weight (Kanazawa, 1982), and soil sorption coefficient (Kenaga and Goring, 1980). BCF and BMF are inversely correlated with water solubility (e.g., Metcalf, 1977; Kenaga and Goring, 1980; Moriarty and Walker, 1987). For example, see Figures 5.4–5.6, 5.12, and 5.13 and Section 5.2.5, Chemical Characteristics that Affect Bioaccumulation. Incorporation of a steric effect coefficient improved the correlation between uptake and K_{ow} for PCBs (Shaw and Connell, 1984).

Such correlations are useful, but have limitations. They do not allow prediction of tissue residues of metabolizable chemicals, and they do not account for the many independent factors affecting uptake and elimination rates (Spacie and Hamelink, 1982). Kinetic models permit better prediction of residue dynamics, and can take into account factors affecting uptake and elimination, thereby contributing further to our understanding of the bioaccumulation process. The following discussion distinguishes between simple kinetic models used to describe bioconcentration at equilibrium or steady state and more complex kinetic models that can be used to predict nonsteady-state, nonequilibrium contaminant concentrations from fluctuating exposures or multiple exposure routes. Following Landrum and others (1992), two types of complex kinetic models will be discussed: compartment-based kinetic models, which describe chemical movement between compartments, and physiologically based kinetic models, which describe the kinetics and dynamics of chemical accumulation in relation to physiological processes.

Steady-State or Equilibrium Partitioning Models

Equilibrium models of bioaccumulation were introduced in Section 5.2.3. Early models of this kind used simple kinetics (uptake and elimination rates) to estimate steady-state accumulation from aqueous exposures (e.g., Branson and others, 1975; Neely, 1979). Such correlations have been successfully used to predict contaminant distributions among ecosystem compartments at equilibrium (Landrum and others, 1992). The well-recognized correlation between BCF and K_{ow} derives from separate correlations for uptake (k_1) and elimination (k_2) rate constants (Spacie and Hamelink, 1982). This was expressed in Equation 5.2 (with the terms defined in Section 5.2.1).

$$\text{BCF} = C_b/C_w = k_1/k_2 \tag{5.2}$$

According to this model, sometimes called the *hydrophobicity model*, bioconcentration consists of partitioning between the organism and the surrounding water with no physiological barriers to impede accumulation. The organism is considered to be a single, uniform compartment, with the solubility of the chemical in the organism controlled by the chemical's solubility in lipid. The rate of uptake is controlled by the concentration gradient between the organism and the surrounding water. This simple model assumes that uptake and elimination show pseudo-first-order kinetics, uptake is limited only by diffusion, the BCF is controlled by the hydrophobicity of the chemical and the lipid content of the fish, and there is negligible growth or metabolism. Regression models (such as shown in Figure 5.4) assume that *n*-octanol is a good surrogate for the lipid phase in organisms, which appears to be a valid assumption (Mackay, 1982).

One variation of the steady-state model is the simplest fugacity model, which was also described briefly in Section 5.2.3. Initially proposed as an equilibrium model (Mackay, 1979, 1982), the organism can be viewed as an inanimate volume of material approaching thermodynamic equilibrium with the surrounding medium (water). Fugacity is the thermodynamically driven escaping tendency of a chemical from a particular phase. It has units of pressure, and at low concentrations (such as usually occur in the environment) it is proportional to concentration. A difference in fugacity provides a driving force for passive chemical diffusion from high to low fugacity. When the escaping tendencies of a chemical from two phases are equal, the phases are in equilibrium. When they are not equal, mass diffuses from high to low fugacity.

The fugacity concept was introduced into equilibrium models and steady-state models of bioconcentration in fish by Mackay and coworkers (Mackay, 1979; Mackay and Paterson, 1981). At chemical equilibrium, chemical fugacities in the fish and water are equal. Equilibrium cannot be reached if the chemical is metabolized by the fish or eliminated in fecal matter, or if the fish undergoes significant growth during the time necessary to reach equilibrium (Gobas and others, 1993b). At steady state, chemical fluxes into and out of the fish are equal. Steady-state concentrations are driven by metabolic processes (such as biotransformation) as well as by thermodynamics. Metabolic processes may result in accumulations very different from those at thermodynamic equilibrium.

A steady-state fugacity model successfully described the uptake and elimination of dichlorinated biphenyls, trichlorinated biphenyls, and tetrachlorinated biphenyls (all PCBs) in goldfish

(Mackay and Hughes, 1984). It is important to note that the PCB congeners that were tested had fast enough fish-to-water elimination rates that they did not show biomagnification (Gobas and others, 1993b). Steady-state models also have been used to describe bioconcentration by zooplankton, though these models have rarely been verified (Day, 1990).

Steady-state fugacity models also have been developed for bioaccumulation of sediment-sorbed neutral organic chemicals (e.g., Lake and others, 1987; Ferraro and others, 1990, 1991). Concentrations in tissue and sediment were normalized by animal lipid and sediment organic carbon, respectively, and a BSAF calculated as in Equation 5.8 (introduced in Section 5.2.4, subsection on Fish/Sediment Concentration Ratios):

$$\text{BSAF} = [(C_b/f_{\text{lipid}})/(C_s/f_{\text{oc}})] \tag{5.8}$$

where C_b is the chemical concentration (µg/kg) in aquatic biota, C_s is the chemical concentration in total sediment (µg/kg), f_{lipid} is the fraction of animal lipid, and f_{oc} is the fraction of sediment organic carbon. This model was tested in the laboratory with clams (*Macoma nasuta*) exposed to contaminated field sediment containing multiple pollutants at widely varying concentrations (Ferraro and others, 1990). Contaminants included *p,p'*-DDD, *p,p'*-DDE, some PCB congeners, and some PAHs. Except for sediment with low organic carbon and low contaminant concentrations, the BSAF was a conservative predictor of bioaccumulation potential in that its use tended to slightly overestimate the contaminant concentration in clam tissue. This model assumes steady state, no transformation, and no phase resistance. If these model assumptions are not met, the value of C_b, would be reduced and experimentally derived BSAFs would underestimate the actual bioaccumulation potential (Ferraro and others, 1990). BSAFs for some neutral compounds were found to differ among species (Foster and others, 1987; Rubinstein and others, 1987; Clarke and others, 1988; McElroy and Means, 1988). Also, the BSAF value obtained may depend on the lipid extraction methods used (Ferraro and others, 1990), since lipid concentrations may vary by a factor of 3 to 4 when different extraction solvents are used (Randall and others, 1991).

Equilibrium assumptions are violated in many laboratory and most field situations. Even steady-state assumptions may not be valid in the field (where, for example, source concentrations may fluctuate). Many steady-state models allow for chemical uptake from water only, although this is not necessarily the only uptake route in field situations. Subsequently, the fugacity concept was used in more complex kinetic models that incorporated physiological factors, allowed for dietary uptake, or applied to nonequilibrium conditions (e.g., Gobas and others, 1989b; Gobas and others, 1993a) or nonsteady-state conditions (Mackay and Paterson, 1991). The fugacity concept also has been used in pharmacokinetic models of contaminant accumulation (e.g., Clark and others, 1990; Clark and Mackay, 1991).

Compartment-Based Kinetic Models

Compartment models are mathematical descriptions of chemical flux between compartments representing an organism (or parts of an organism) and the surrounding media. A *compartment* is defined as that quantity of chemical that behaves as if it were in a homogeneous, well-mixed container and crosses the compartment boundary with a single uptake or elimination rate constant (Landrum and others, 1992). A compartment may or may not represent a physical entity. As the number of compartments in the model increases, the model tends to become more complex. The reader should be warned that terminology, units, and symbols are not used consistently in the literature (Landrum and others, 1992). Even terms that describe compartment models may be used differently. A two-compartment model may consist of a water compartment and a single organism compartment (e.g., Landrum and others, 1992). Alternatively, it may indicate that the organism is represented by two compartments (e.g., Spacie and Hamelink, 1985; Barron, 1990). This book will use the latter convention, counting only the biological compartments.

Using this naming convention, in a one-compartment bioconcentration model, the organism is considered to consist of a single (lipid) phase. The organism is exposed to a chemical in the water only, and chemical uptake is proportional to the concentration in water. Elimination follows pseudo-first-order kinetics. A growth term may or may not be incorporated. The steady-state models discussed above can be considered simple one-compartment bioconcentration models. Landrum and others (1992) described three types of compartment-based kinetic models: pseudo-first-order rate coefficient models (e.g., Spacie and Hamelink, 1985), fugacity models (e.g., Gobas and Mackay, 1987), and clearance volume models (e.g., Barron and others, 1990b). These three types of kinetic models are mathematically equivalent for exposures with a single uptake route (Landrum and others, 1992). Landrum and others (1992) included mathematical derivations for these three types of one-compartment model.

In multiple-compartment bioconcentration models, the organism is assumed to have more than one compartment. As an example of a simple two-compartment model, chemical uptake and elimination may occur only through a fast compartment, whereas chemical residues accumulate in a slow compartment (representing lipid storage) as a function of its concentration in the fast compartment. Metabolites may be eliminated at different rates from that of the parent chemical (Spacie and Hamelink, 1985). An example is the fugacity model derived by Gobas and others (1989b) that explained the BCFs for low- and high-K_{ow} chemicals on the basis of transfer between fish, water, and feces. In another example, a three-compartment model was used for bivalve mollusks, in which the organism was considered as consisting of gills, circulatory fluid, and lipid storage (Stegeman and Teal, 1973). Initial uptake across gills was rapid, transfer to circulatory fluid occurred more slowly, and transfer to lipid was much slower. According to this model, the reservoir of contaminants in the lipid pool became larger over time until it approached its maximum capacity or equilibrium. If the mollusk was transferred to clean water, the process was reversed, except that contaminants present in the lipids may be released slowly. The multicompartment model explains mathematically why bivalve mollusks transplanted from contaminated areas to clean areas do not rapidly release their contaminant load (Boehm and Quinn, 1977).

Some bioaccumulation models incorporate a term for assimilation of a chemical by ingestion into a one- or two-compartment model with growth (e.g., Thomann, 1981; Gobas and others, 1993b). Contribution from food will be insignificant if the loss term is large (i.e., for chemicals with short depuration half-lives). For substances with long depuration half-lives, food intake becomes important. These models assume that food and water sources are additive, and that elimination of food- and water-derived residues are equivalent and follow pseudo-first-order kinetics. In practice, the uptake, assimilation, and elimination rates are affected by the metabolic rate of the organism. For example, water temperature affects respiration and caloric intake (Spacie and Hamelink, 1985).

As discussed in Section 5.2.4 (subsection on Effect of Trophic Level on Fugacity), fugacity calculations from field studies have been observed to deviate from equifugacity (Clark and others, 1988; Connolly and Pedersen, 1988). Modified fugacity models have been developed that attribute these deviations to dietary accumulation (Gobas and others, 1993a,b). Conceptually, these models are important by resolving the lipid-partitioning mechanism with previously conflicting observations of biomagnification. Gobas and others (1993a) tested two possible mechanisms for biomagnification: (1) intestinal absorption of the chemical occurs by passive diffusion across a fugacity gradient, so that the fugacity is increased in the gastrointestinal tract (relative to that of the food consumed and that of the fish) by the processes of food digestion and absorption; and (2) biomagnification occurs in the tissues (rather than in the gastrointestinal tract) as lipids are converted into energy, leaving behind a higher concentration (and fugacity) of the chemical contaminant. In the latter case, chemical uptake from the gastrointestinal tract is due to lipid coassimilation, in which the chemical is transported across the gastrointestinal tract in association with dietary lipids. This was tested by measuring dietary uptake rates in goldfish of chlorinated benzenes and PCBs administered in diets with different lipid contents. The dietary uptake efficiency from low-lipid food was found to be higher than from high-lipid food, which is consistent with diffusion-controlled uptake rate. If uptake had occurred via lipid coassimilation, uptake efficiencies should have been higher when chemicals were administered in high-lipid diets (Gobas and others, 1993a).

In these biomagnification models, then, food digestion and absorption were proposed to raise the fugacity of food in the gastrointestinal tract above the fugacity of the consumed food, or prey (Connolly and Pederson, 1988; Gobas and others, 1988, 1993a,b). This is illustrated in Figure 5.24, which (along with the following explanation) is from Gobas and others (1993b). The gastrointestinal tract is viewed as a continuously mixed compartment that receives a constant flow of food and releases a constant flow of fecal matter. The fugacity of feces is assumed to be equal to that of the gastrointestinal tract. Initially (as shown in Figure 5.24A), the chemical fugacity of the fish (f_B) is equal to that in the water (f_W) and in the diet, or prey, of the fish (f_D). Also, the fugacity of the gastrointestinal tract (f_G) is equal to that of the diet (f_D) (i.e., $f_B = f_W = f_D = f_G = 1$). In Figure 5.24B, food is absorbed from the gastrointestinal tract, but initially no chemical is absorbed because f_B equals f_G. At the same time, the inflow of contaminated food (G_D) to replace the absorbed food increases the concentration of the chemical in the gastrointestinal tract (and also the fugacity, f_G) relative to that in the original food. Now, f_G is higher than f_D and f_B (Figure 5.24B). In Figure 5.24C, food digestion in the gastrointestinal tract alters the composition of the consumed food in the gastrointestinal tract. For example, the hydrolysis of lipids and absorption of the reaction products decrease the capability of the gastrointestinal tract to dissolve hydrophobic organic contaminants, which lowers the fugacity capacity of the gastrointestinal tract (Z_G) for the chemical below the fugacity capacity of the original food (Z_D). This also raises the f_G, since fugacity is inversely proportional to fugacity capacity. Digestion and absorption of food, therefore, act as a fugacity pump to raise the fugacity in the gastrointestinal tract over that in the consumed food and in the fish (f_G greater than f_D, f_B). As shown in Figure 5.24D, this fugacity gradient drives the passive diffusion of chemical from the gastrointestinal tract into the organism, raising the fugacity of the predator (f_B) over that in the prey (f_D) (i.e., biomagnification). If the combined rate of chemical transformation, gill elimination, and growth dilution is slow relative to the rate of chemical elimination in feces, then the high chemical fugacity in the organism is maintained and biomagnification will occur, as

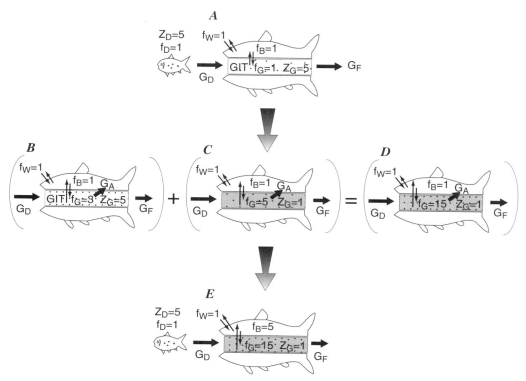

Figure 5.24. Conceptual diagram of the proposed mechanism of organic chemical biomagnification in fish, illustrating the increase of the chemical fugacity in the gastrointestinal tract (GIT) caused by food digestion and absorption. (*A*) equifugacity between predator (f_B), its GIT (f_G), and its food or prey (f_D); (*B*) f_G increases as food is absorbed from GIT; (*C*) food digestion lowers Z_G, thus raising f_G above f_D and f_B; (*D*) this fugacity gradient drives the diffusion of the chemical from the GIT into the organism, raising the fugacity of the predator (f_B) over that in the prey (f_D); (*E*) biomagnification occurs when the high f_B is maintained, which requires that the combined rate of chemical transformation, gill elimination, and growth dilution be slow relative to G_F. Dots in GIT represent chemical concentration, and gray shading represents changes in the fugacity capacity of the GIT contents. Z_B, Z_D, and Z_G are the fugacity capacity (mol/m^3 × Pa) of the fish, diet, and GIT, respectively; f_B, f_D, f_G, and f_W are the fugacity (Pa) of the fish, diet, GIT, and water, respectively; G_D, G_F, and G_A are the rates (m^3/day) of food intake, fecal release, and food absorption from the GIT, respectively. Abbreviations: m^3, cubic meter; mol, mole; Pa, Pascal. Redrawn from Gobas and others (1993b) with the permission of the publisher. Copyright 1993 American Chemical Society.

shown in Figure 5.24*E*. If this combined rate of elimination is fast relative to the fecal elimination rate, then the high fugacity of the organism is not maintained, and the chemical should not biomagnify (Gobas and others, 1993b).

Gobas and others (1993b) suggested that nonmetabolizable compounds with log K_{ow} values greater than 6 are observed to biomagnify because the rates of elimination via the gills, skin, urine, or metabolism are too slow to reduce the high fugacity in the organism. This fugacity pump operates each time one organism is consumed by another, causing the fugacity to increase with increasing trophic level. In contrast, chemicals with log K_{ow} values less than 6 are eliminated via the gills, skin, urine, or metabolism faster than they are eliminated in feces, thus reducing the fugacity in the organism, so that biomagnification is not evident.

Fugacity models have been applied to fish (e.g., Gobas and Mackay, 1987; Gobas and others, 1989b), benthic organisms (e.g., Bierman, 1990), aquatic insects (e.g., Gobas and others, 1989a), and aquatic macrophytes (e.g., Gobas and others, 1991). Some fugacity models have considered dietary uptake (e.g., Clark and Mackay, 1991) or gill uptake efficiency (e.g., Gobas and Mackay, 1987). These are more complex versions of the hydrophobicity model. However, most compartment models tend to assume negligible metabolism or a lack of steric hindrance in the gill, or both, and generally do not consider the role of blood flow in controlling uptake, distribution, and elimination (Barron, 1990). The permeability of the plasma membrane in the gill and other tissues is a barrier to absorption (diffusion) and transfer via blood to lipoidal tissues (Hamelink and others, 1971). Hansch (1969) considered chemical uptake as a random walk across a series of aqueous and lipid compartments, each acting as a potential barrier to movement. Absorption of the chemical across the gill can be viewed as a series of steps, each of which can govern the uptake rate and extent of accumulation (Hayton and Barron, 1990). Simple partitioning (hydrophobic) models predict that chemical transfer into an organism should increase with increasing K_{ow}, following a log–log relation. In practice, uptake of chemicals of increasing hydrophobicity (log K_{ow}) follows a sigmoidal relation, and the rate-limiting step appears to vary with log K_{ow} (Spacie and Hamelink, 1982; McKim and others, 1985; Erickson and McKim, 1990).

Physiologically Based Kinetic Models

Hydrophobicity is the principal determinant of bioconcentration only when other steps in accumulation process are not rate-limiting (Barron, 1990). Potentially rate-limiting steps include ventilation volume, membrane permeability, and blood flow (Erickson and McKim, 1990; Hayton and Barron, 1990). Physiological models consider the importance of physiological processes and metabolism in chemical accumulation. Two kinds of physiologically based kinetic models will be discussed: pharmacokinetic models and bioenergetic models.

Pharmacokinetic Models. Physiologically based pharmacokinetic (PBPK) models describe the accumulation and distribution of a chemical among tissues within an organism (Landrum and others, 1992). The organism is mathematically characterized using multiple compartments, each consisting of organs or tissues that are kinetically equivalent. In PBPK models, a series of compartments is generally arranged as parallel shunts between arterial and venous blood flows, using physiological parameters (Barron, 1990). A schematic representation of a PBPK model is shown in Figure 5.25. Development of a PBPK model requires extensive data, such as tissue volumes, blood flow rates, partition coefficients between tissue compartments, and biotransformation rates for different compartments. Differential mass balance equations then are written to describe accumulation, elimination, and metabolism of the chemical (Landrum and others, 1992).

PBPK models were originally developed to describe the metabolism of pharmaceuticals in mammals and were then extended to apply to other animals, including aquatic biota. For

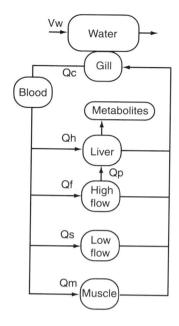

Figure 5.25. Schematic representation of a physiologically based pharmacokinetic model for a chemical absorbed through the gill and metabolized in the liver. Q is the blood flow to tissues; Vw is the ventilation volume; Qc is the cardiac output; Qh is the liver blood flow; Qp is the hepatic portal flow; Qf is the flow to highly perfused tissues; Qs is the flow to slowly perfused tissues; Qm is the muscle blood flow. Redrawn from Barron (1990) with the permission of the publisher. Copyright 1990 American Chemical Society.

example, PBPK models were developed for phenol red in dogfish shark, *Squalus acanthias* (Bungay and others, 1976), and methotrexate in the sting rays, *Dasyatidae sabrina* and *D. sayi* (Zaharko and others, 1972). More recently, they have been applied to bioaccumulation of environmental contaminants in fish and invertebrates. Examples include triclopyr in coho salmon (Barron and others, 1990a); pentachlorophenol in goldfish (Stehly and Hayton, 1989); HCH, 2,3,7,8-TCDD, and various halogenated aromatic compounds in rainbow trout and other fish (Barber and others, 1988); PCBs in salmonids (Barber and others, 1991); pentachlorophenol in rainbow trout (McKim and others, 1986); lindane in mollusks (Thybaud and Caquet, 1991); and various organic contaminants in rainbow trout (Erickson and McKim, 1990) and in amphipods (Landrum and Stubblefield, 1991).

There are several advantages of PBPK models (Landrum and others, 1992). They focus on the mechanisms occurring within the organism, not simply on rate processes. Compartments have real physiological meaning. Also, models have the potential to be scaled from one species or size to another by inserting the appropriate physiological processes or substituting appropriate parameter values. On the other hand, PBPK models assume that a particular mechanism (such as gill ventilation) is rate-limiting, sometimes without empirical evidence (Landrum and others, 1992). In this case, PBPK models may not be any more realistic than the compartment model approach used by rate-clearance kinetic models, but instead represent merely a different approach (Landrum and others, 1992). Also, data requirements are extensive (Hamelink and Spacie, 1977), and some data may not be available because analyzing tissue volumes, or taking blood samples from small fish or invertebrates, is difficult (Landrum and others, 1992).

Bioenergetic Models. Bioenergetic models describe chemical accumulation and loss in terms of the energy requirements of the organism. The organism generally is treated as a single compartment (Landrum and others, 1992). These models predict chemical uptake as a function of the flux of water across its gills and the fluxes of food and sediment through its gut. Uptake is assumed to be proportional to flux for each source (Landrum and others, 1992). Biological and chemical parameters incorporated into such models may include growth rates, respiration rates, fluxes of water and food, chemical assimilation efficiencies from water and from food, and the BCF for the chemical.

Bioenergetic models have been used to predict chemical residues in fish (e.g., Norstrom and others, 1976; Jensen and others, 1982; Connolly and Tonelli, 1985; Borgmann and Whittle, 1992) and to model transport through a food chain (Thomann and Connolly, 1984; Thomann, 1989). Food chain models, discussed in Section 5.2.4 (subsection on Field Modeling), are expanded versions of a bioaccumulation model that incorporate transfer through an aquatic food chain with multiple trophic levels (Spacie and Hamelink, 1985). A four-level example would be phytoplankton, zooplankton, small fish, and large fish. Bioaccumulation at each trophic level can be estimated from the bioconcentration factor at each level and food chain transfer coefficients for the consumers. Such models are highly sensitive to the relative values of assimilation efficiency and loss due to depuration and growth, especially at higher trophic levels (Spacie and Hamelink, 1985).

Bioenergetic models typically are used to predict the uptake of chemicals under field conditions or to determine the importance of uptake routes (Landrum and others, 1992). Several bioenergetic models have shown the importance of biomagnification for certain hydrophobic compounds. For example, accumulation was directly related to trophic level (on the basis of wet weight concentrations) for PCB in Ottawa River yellow perch (Norstrom and others, 1976), PCBs in the Lake Michigan lake trout food chain (Thomann and Connolly, 1984), and kepone in the James River food chain (Connolly and Tonelli, 1985). According to the food chain model for PCBs in the Lake Michigan lake trout food chain, 99 percent of the PCB in an adult trout was taken up from food. Furthermore, the model predicted that, after reduction in water column PCB levels, it would take 5 years for the higher PCB residues to be eliminated from lake trout. The dissolved PCB concentration would have to be 0.5–2.5 ng/L to reduce PCB residues in all age classes of lake trout below 5 mg/kg (wet weight) in edible trout tissue, which is the level used to determine suitability of fish for human consumption (Thomann and Connolly, 1984).

Effect of Environmental and Physiological Conditions

Biological and chemical parameters incorporated into a given model will vary, but may include uptake rate, growth rate, biotransformation rates in various tissues, blood flow rates to various tissues, respiration rate, tissue volumes, fluxes of water and food, chemical assimilation efficiency from food, and partition coefficients between tissues. Kinetic models are sensitive to factors that affect rate constants and other parameter values used in the models. Some parameters are likely to change as the physiology of the fish changes, as it does during growth, maturity, spawning, and senescence (Hamelink and Spacie, 1977). For example, growth and reproductive state may affect lipid content, which in turn may affect uptake and elimination rates and partition coefficients between tissues. Environmental conditions may either affect the physiology of the organism or change the chemistry and kinetics (uptake and elimination rates) of the contaminant (Landrum and others, 1992). For example, temperature may affect uptake, biotransformation, and elimination rates (Huckle and Millburn, 1990). Changes in pH may alter environmental degradation rates (which affect ambient exposure levels) and the degree of ionization (which affects the K_{ow} and bioavailability), which in turn may affect uptake rates. The concentrations of particulates and DOM also may affect the amount of a chemical that is bioavailable and its rate of uptake (Landrum and others, 1992).

5.3 SEASONAL CHANGES IN PESTICIDE RESIDUES

Because bed sediment and aquatic biota may act as long-term sinks for hydrophobic contaminants in a hydrologic system, pesticide residues in bed sediment and aquatic biota generally are considered to be more stable temporally than in the water column. Nonetheless, seasonal events may affect the accumulation of pesticide residues in sediment and aquatic biota. They do this by influencing the movement of pesticides to the hydrologic system and the fate and degradation of pesticides within the system.

The primary sources of a pesticide to surface waters are seasonal application of the pesticide, and (if residues are present in soils from application in previous years) transport of pesticide-contaminated soil to surface waters. Leonard (1990) summarized four factors affecting pesticide transport in surface runoff: (1) climate, including duration, amount and intensity of rainfall, and timing of rainfall after pesticide application; (2) soil characteristics, including soil texture and organic carbon content, water content of the soil prior to rainfall, slope and topography of the field, and degree of soil aggregation and stability; (3) the physical and chemical properties of the pesticide; and (4) agricultural management practices, including pesticide formulation, application rate, erosion control practices, and irrigation practices. For the organochlorine insecticides, most of which have very low water solubilities (<<1 mg/L) and are no longer used in agriculture in the United States, transport of contaminated soil is the most important source of insecticides to the hydrologic system (Larson and others, 1997). Thus, events that contribute to soil erosion and surface water runoff (such as weather-driven and streamflow-related events, and certain agricultural management practices) are likely to affect the introduction of pesticides into a hydrologic system and their subsequent accumulation in bed sediment and tissues. Other seasonal parameters may affect contaminant accumulation in sediment and aquatic biota. Environmental conditions such as temperature and salinity may affect sorption to sediment, pesticide degradation, and uptake by aquatic biota discussed previously in Sections 4.2.1, 4.2.3, and 5.2.5, respectively. The concentration of pesticides in biota are also affected by seasonal biological or physiological factors such as lipid content, reproductive state, and enzyme activity.

A number of monitoring studies sampled bed sediment or biological tissues several times per year (often monthly) and noted short-term temporal variations in pesticide residues. Far more studies investigated seasonal variations in contaminant residues in biota than in bed sediment. In many of these studies, the authors were able to offer explanations for the observed fluctuations by pinpointing correlations in time with specific conditions or events; these studies are specified below in discussions of individual anthropogenic, environmental, or biological factors involved. Other studies noted seasonal fluctuations, but did not offer explanations, or made only general suggestions as to factors or processes that may be operating without providing details or making specific correlations (e.g., Kolipinski and others, 1971; Edgren and others, 1981; Miller and Jude, 1984; Nettleton and others, 1990). Still other studies with comparable sampling schedules observed little or no seasonal variation in pesticide residues in sediment or biota (e.g., Morris and others, 1972; Kelso and Frank, 1974; Najdek and Bazulik, 1988). In a few studies, seasonal variations observed during one year were not detected or were different the following year (e.g., Hubert and Ricci, 1981; Nowak, 1990). In some studies, seasonal patterns were different for different species (e.g., Kellogg and Bulkley, 1976; Galassi and others, 1981), different age groups (e.g., Smith and Cole, 1970; Bulkley and others, 1976; Boileau and others, 1979), different target compounds (e.g., Smith and Cole, 1970; Miller and Jude, 1984; Nettleton and others, 1990), or different sampling sites (Kolipinski and others, 1971; Reinert and Bergman, 1974; Marchand and others, 1976). Possible explanations usually were offered. In general, it is likely that multiple factors are interacting in any given situation. Although circumstantial evidence may strongly suggest the importance of one or more factors in a given situation, actual causes of observed seasonal fluctuations often cannot be conclusively established in the field. However, the occurrence of seasonal fluctuations (whatever the cause) points out the importance of sampling in periods that are well-defined ecologically and hydrologically. Residue monitoring programs should consider seasonality in sampling design and data interpretation.

In the discussion that follows, those studies that looked for short-term temporal variations in pesticide residues in sediment and aquatic biota are examined in relation to the following potential explanatory variables: seasonal variations in pesticide use, weather-driven and streamflow-related events, agricultural management practices, environmental conditions, and biological factors.

5.3.1 PESTICIDE USE

Agricultural Sources

Pesticides in agriculture generally are applied during particular seasons to protect crops from pests at a particular stage of growth. Herbicides are often used just before or after the emergence of the crop, whereas insecticides tend to be used later in the year. In temperate areas with well-defined growing seasons, a large peak in pesticide delivery commonly occurs with the first runoff-inducing rainfall or irrigation event after pesticide application. The rest of the year, pesticide input to aquatic systems will be lower (Phillips, 1980). Hydrophobic organic pesticides like the organochlorine compounds generally will be carried away from application sites in association with particles. In equatorial areas with multiple cropping and harvesting patterns, multiple peaks may occur, each after a pesticide application (Phillips, 1980). The actual timing of use of any particular chemical is dependent on the region of the country, the crop, and the local weather conditions. Pesticides are commonly applied in the Corn Belt during April–June for many herbicides and May–August for many insecticides. In California, insecticides are commonly applied to orchards as dormant spray in February, to alfalfa in March–April, and to rice fields in May–June. The exact timing in any particular year depends on temperature and rainfall.

This seasonal use of pesticides in agriculture often results in the seasonal input of pesticides to surface waters, causing significant seasonal concentration changes of pesticides in the water column. Kellogg and Bulkley (1976) observed this for dieldrin in the Des Moines River in Iowa during 1971–1973. They observed increased levels of dieldrin in both the dissolved and particulate phases starting soon after application, which occurs in late April through early June. The maximum concentrations were reached in June and early July of each year, then declined rapidly through late summer and autumn. This same pattern has been documented for many different pesticides in surface waters in various regions of the country (Larson and others, 1997). Bradshaw and others (1972) reported that pesticides entered Utah Lake in Utah in three surges during one year (1970–1971), with surges of individual pesticides (aldrin, heptachlor, and HCH) corresponding to times and locations of applications in the drainages. These short term fluctuations in the water have the potential to add both to the burden of the bed sediment through sorption followed by particle sedimentation, and to aquatic biota via bioaccumulation.

Seasonal peaks in pesticide levels measured in bed sediment or aquatic biota have been attributed to seasonal pesticide applications by several authors. A few generalizations are possible. The most clearcut examples tend to be in rivers or estuaries receiving runoff directly from agricultural land. In estuaries and coastal waters that are farther from the site of application, seasonal changes due to pesticide use may be obscured by other factors. Seasonal patterns in estuaries and coastal waters may correlate with freshwater inputs and (for residues in aquatic biota) physiological changes in the biota. For persistent compounds such as DDT, application may result in an initial peak followed by a fluctuating or steady decline (Phillips, 1980). Discussed below are examples of studies that measured seasonal patterns in association with agricultural pesticide application. Currently used pesticides are discussed first, followed by studies that measured organochlorine residues prior to their restricted use in the early 1970s.

Few studies have investigated seasonal patterns of currently used pesticides in bed sediment or aquatic biota. Permethrin and methyl parathion were detected in wetland sediment from the Moon Lake watershed in Mississippi during the spray season (1983–1984), but fenvalerate was not (Cooper, 1991). All three compounds were detected sporadically in water and in various species of fish. Nowak and Julli (1991) measured endosulfan residues monthly in livers of wild fish collected from cotton-growing areas in Australia (1987–1989). Endosulfan was applied to cotton during the summer (November–February). Endosulfan residues were detected in catfish (*T. tandanus*), bony bream (*N. erebi*), and carp (*C. carpio*) from cotton-growing areas during all seasons sampled. Residues in fish from cotton-growing areas were significantly higher than in control fish from non-cotton-growing areas. For all species, endosulfan residues were highest during the summer of 1988 and lowest during the winter of 1988. Residues during the summers of 1987 and 1989 were intermediate. Endosulfan residues were significantly higher during the summer of 1988 than during the other seasons when fish were collected. Nowak and Julli (1991) attributed this to heavy rainfall during the summer of 1988. In another example, organophosphate residues were analyzed in mosquitofish collected from rice fields in the Ebro Delta in Spain. Residues in muscle tissue peaked in August–September, corresponding to organophosphate insecticide use in the rice fields (Porte and others, 1992).

Most studies of pesticides in sediment and aquatic biota (including the seasonal investigations) have targeted organochlorine insecticides, and most of the pesticides detected in bed sediment and aquatic biota are organochlorine insecticides that are no longer used in agriculture in the United States. For these pesticides, current agricultural use has not been a relevant factor in the seasonality of residues detected in sediment and aquatic biota since the 1970s. However, several studies conducted during the 1960s and early 1970s assessed the seasonal influx of organochlorine insecticides to surface waters in relation to their seasonal application. For example, total DDT and dieldrin levels in sediment from Big Creek and the Erieau Drainage Ditch in Ontario (1970) were highest in May, which was attributed to local application for cutworm control (Miles and Harris, 1971). Correlation of seasonal peaks in organochlorines in aquatic biota with local agricultural use were seen for DDT and metabolites, aldrin, and methoxychlor in mussels in a Michigan River (Bedford and others, 1968); for endrin in biota of Tule Lake, California (Godsil and Johnson, 1968); for dieldrin and DDT in various species of fish from the Grand River, Ontario (Kelso and others, 1970); and for DDT in salmonids from Alaska (Reed, 1966), Pennsylvania (Cole and others, 1967), and New Brunswick, Canada (Sprague and others, 1971).

Seasonal changes in dieldrin concentrations measured both in bed sediment and in a variety of aquatic biota from the Des Moines River (1972–1973) were attributed largely to seasonal water column concentration changes that were driven by the seasonal application of dieldrin (Kellogg and Bulkley, 1976). This example will be described in some detail. Dieldrin concentrations in bed sediment peaked in July (maximum concentration 2,400 µg/kg dry weight), then declined to below detection (less than 100 µg/kg) in September and November. In mayfly larvae (*Potamanthus* sp.) in 1972, the mean concentration of dieldrin increased by almost a factor of three between April–May (44 µg/kg wet weight) and June (115 µg/kg), and then declined again by late July (48 µg/kg) before emergence of the adult mayflies. The seasonal changes in mayfly larvae concentrations were not as drastic in 1973 as in 1972, but the maximum water column concentrations were also lower in 1973 compared with 1972. Dieldrin concentrations measured in eight other types of aquatic insects in this river also showed seasonal concentration differences, but not as pronounced as in the mayflies. In most insect genera, there appeared to be a second maximum concentration during late October. In contrast, crayfish (*Orconectes rusticus*) exhibited little seasonal variation in dieldrin residues. Four types of small fish also showed some seasonal variations in dieldrin residues, with the maximum concentration occurring in June, the minimum concentration in August, and concentrations increasing through November. The largest concentration changes were exhibited by the spotfin shiner (*Cyprinella spiloptera*), followed by sand shiner (*Notropis stramineus*), bluntnose minnow (*Pimephales rotatus*), and carpsucker (*Carpiodes* sp.). Finally, the study examined dieldrin in the muscle of various sizes of channel catfish. There, seasonal trends were less evident, but the authors demonstrated statistical seasonal variations in the channel catfish in all but one length group. The larger fish, which had a longer life cycle, seemed to be somewhat less affected by short term variations in water column concentration. The authors attributed the secondary concentration maxima for many species in October–November to seasonal physiological factors in the organisms, such as fat content, metabolic rate, reproductive activity, and activity of detoxifying enzymes.

Butler and others (1972) observed that seasonal profiles of organochlorines in estuarine biota were related to availability in runoff waters, which was a function of local agricultural use. For example, total DDT in estuarine oysters (*Crassostrea virginica*) was correlated with DDT use in the catchments of each estuary (Figure 5.26). Moreover, the timing of peak residues in oysters was correlated with pesticide application periods and with rainfall. One catchment (the Laguna Madre, [A] in Figure 5.26) had three applications of DDT per year and showed three peaks in total DDT residues. The other two catchments in Figure 5.26 were treated only once per year and showed a single peak in total DDT residues (Butler and others, 1972).

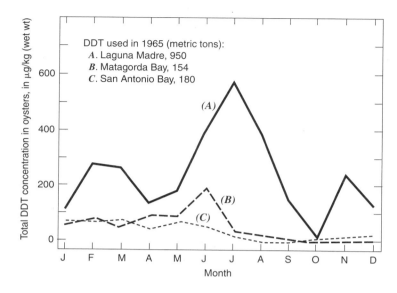

Figure 5.26. Seasonal variations in total DDT concentrations (μg/kg wet weight) in estuarine oysters (*Crassostrea virginica*) from three catchment basins in relation to DDT use (metric tons) in 1985. DDT was applied to catchment *A* three times per year and to catchments *B* and *C* only once per year. Abbreviations: μg/kg, microgram per kilogram; wt, weight. Redrawn from Butler and others (1972) with the permission of the publisher. Copyright 1972 Food and Agriculture Organization of the United Nations.

Hansen and Wilson (1970) measured total DDT residues in fish from the Escambia and Pensacola Bays, Florida. These authors observed a well-defined seasonal peak in total DDT residues in fish (various species) from the lower estuary that corresponded to local applications in adjacent land masses, whereas residues in fish from the upper estuary were lower and more stable year-round. Consistent seasonal peaks in organochlorine residues at different sites in an estuary have been used to indicate a source in the upper reaches of the estuary. Examples include dieldrin in hard-shelled clams (*Mercenaria mercenaria*) (Check and Canario, 1972), malathion in brown and white shrimp (*Penaeus aztecus* and *P. setiferus*, respectively) (Conte and Parker, 1975), and organochlorine insecticides in various epibenthic organisms (Livingston and others, 1978). The effect of local agricultural use on seasonality may be obscured by other factors (such as environmental conditions or physiological factors). Alternatively, other factors (such as lipid content) may coincide with local application periods (Phillips, 1980), thus exaggerating the correlation between local use and seasonal pesticide residues.

In more recent studies of organochlorine residues in bed sediment and aquatic biota, seasonal changes are not driven by local agricultural use patterns. Such studies have been used to elucidate effects of other types of factors (such as irrigation or reproductive state) that affect seasonality of contaminant residues and are discussed in Sections 5.3.2–5.3.5.

Industrial Sources

Less information on application periods and use levels is available for industrial and urban pesticide uses than for agricultural uses. For industrial sources of pesticides, seasonal profiles may be more erratic (Butler, 1969a; Foehrenback, 1972). Available concentrations of organochlorines from industrial sources probably depend on both industrial discharge rates and seasonal runoff (Phillips, 1978). Clegg (1974) reported different seasonal profiles in total DDT in oysters (*Crassostrea commercialis*) from two sites in Moreton Bay, Australia, one exposed to industrial sources and the second to agricultural runoff from a rural area. Seasonal fluctuations in methoxychlor residues in caged and wild fish (rainbow trout, white sucker, finscalesucker [*Catostomus catostomus*], and flathead chub [*Platygobio gracilis*]) in a river in Alberta (Canada) were correlated with its use as a black fly larvicide (Lockhart and others, 1977).

PCBs, which share many physical and chemical properties and have similar environmental fate processes to the organochlorine insecticides, are exclusively industrial or urban in origin. Industrial discharges of PCBs are not necessarily erratic (Phillips, 1978). For example, seasonal patterns in PCB residues observed in oysters from Escambia Bay in Florida, were fairly smooth; PCB discharge rates appeared fairly constant, and the seasonal PCB variation in oysters was attributed to runoff rates and oyster physiology (Duke and others, 1970).

5.3.2 WEATHER-DRIVEN AND STREAMFLOW-RELATED EVENTS

Weather-driven seasonal events, such as precipitation and snowmelt with subsequent runoff, are similar to irrigation practices in terms of their effect on pesticide delivery to, and transport within, surface water systems. Reservoir water release or flood events may dramatically increase water levels and stream discharge. These processes all tend to transport soil and associated contaminants into aquatic systems as well as increase the velocity and turbulence of streamflow. Strong winds and increased stream discharge can have a strong impact on the pesticides that are already present in the bed sediment by resuspending and transporting the sediment.

In general, weather-driven and streamflow-related events may be seasonal (such as spring snowmelt and runoff, release of irrigation water, or storms) or episodic (such as floods or reservoir water release for flood control). As an example, in much of the eastern portion of the United States, a significant number of storms occur during the spring months (April–June). These storms are often accompanied by strong winds and heavy rainfall. The rainfall runs off into surface water carrying eroded soil particles and increases the stream discharge. The increased water discharge increases the stream velocity, which in turn can resuspend and transport bed sediment within the stream. The maximum stream discharge and maximum concentration of suspended particles for many rivers and streams are often related. The release of irrigation water can have the same effects as rainfall in terms of soil erosion, river discharge, and bed sediment resuspension. The overall effect is a short-term, and seasonally predictable, redistribution of pesticides that are attached to particulates in the surface water system. This gives rise to temporary increases in the concentrations of these chemicals in the water column that can act as a source to bed sediment and biota.

Because the most persistent and hydrophobic organochlorine insecticides are no longer used in agriculture in the United States, their distribution in the environment is no longer affected by local seasonal use (i.e.,current agricultural use). Nonetheless, residues from past use of these insecticides are still present in many agricultural soils (e.g., Mischke and others, 1985; Johnson and others, 1988b) and in the upper layers of sediment (e.g., Gilliom and others, 1985; Agee, 1986; see Tables 2.1, 2.2). As described above, weather-driven and streamflow-related events can cause erosion of contaminated soil into streams and redistribute persistent organochlorine insecticides within streams.

Irrigation

The Yakima River at Kiona in Washington was monitored for total DDT many times each year from 1968–1980 (Johnson and others, 1988b). The Yakima River Basin experienced heavy use of DDT from the 1940s to 1972, when it was banned. Between 1968 and 1973, the detection frequency of DDT in the water was much greater during the months of crop irrigation (April–September) than during the months with no irrigation (October–March). After 1973, total DDT was seldom detected in the river water. However, in an intensive sampling period during 1985, Johnson and others (1988b) showed that total DDT levels were measurable in whole water samples from the Yakima River and its tributaries during the irrigation season, but were below the detection limit after irrigation ceased. These data suggest that irrigation has caused DDT-contaminated soil to run off into the drainage system and ultimately into the Yakima River. During this same general time period, the concentrations of total DDT in whole fish were monitored by the FWS at a site immediately upriver from Kiona. Individual samples were composites of 3–5 whole fish. Multiple species were sampled. On a long-term scale, the fish concentrations, just like the water concentrations, declined significantly after the ban on DDT use. In the short-term, there were substantial differences between concentrations measured in different samples collected during the same year and between samples collected during consecutive years. For example, the concentrations ranged from about 500 to 2,200 µg/kg in four samples during 1973 and from about 2,000 to 4,200 µg/kg in four samples during 1974. The sources of this sample variability are not known, but it could be due in part to species differences or to differences in irrigation runoff conditions.

In a more recent study of the Yakima River, Rinella and others (1993) measured total DDT residues in soil, water, bed sediment, and fish. Their findings agreed with the earlier studies. Over 15 years after DDT was banned, there was still a considerable reservoir of DDT in the soils of the Yakima River Basin. Throughout the basin, the highest concentrations in the river water occurred during the irrigation and heavy rainfall periods. The soil concentration of total DDT was about four times greater than its concentrations in either the suspended sediment or the bed sediment in the river. A very strong correlation was observed between the concentration of total DDT in whole water and the suspended sediment concentration. Rinella and others (1993) suggest that the seasonally elevated concentrations of total DDT in the water came from a combination of fresh soil running off into the river and bed sediment being resuspended because of increased water discharge.

Seasonal changes in pesticide residues in fish from rice fields in the Ebro Delta, Spain, (1985–1988) also have been attributed in part to irrigation practices (Ruiz and Llorente, 1991; Porte and others, 1992). The delta is flooded by diverting water from the Ebro River through irrigation channels. These channels are open from April to December. After the channels are closed, the rice fields dry up, gradually leaving fish in small ponds in which food and pollutants are concentrated (January–March). On a wet weight basis, DDT and PCB residues in eels and common carp peaked in April (1985), corresponding to the opening of irrigation channels at the beginning of the rice cycle (Ruiz and Llorente, 1991). The inflow of water stirred up the sediment, thus increasing the availability of hydrophobic contaminants to the biota. A second peak in both organochlorine residues and lipid content in carp and eels occurred during December. This was attributed to the closing of irrigation channels and decreasing water level, concentrating both food and pollutants. Subsequent studies with mosquitofish (1987–1988) showed high PCB and organophosphate residues in February, which also were attributed to the closing of irrigation channels, decreasing water levels, and consequent exposure of fish to more concentrated levels of pollutants (Porte and others, 1992). These Ebro Delta rice field studies also showed a correlation between organochlorine residues and the reproductive cycle (Section 5.3.5).

Precipitation and Storm Events

One example of seasonal changes in bed sediment is a study of wind driven inputs from the bed sediment to the water column in Lake Superior (Baker and others, 1985). In the spring there are frequently storms accompanied by strong winds in this area. The winds can, at times, create seiches in Lake Superior, which can disturb the bed sediment. During a June sampling event, after a storm, elevated concentrations of suspended sediment, PCBs, and p,p'-DDE were observed. The authors suggested that all three parameters were consistent with a short-term bed sediment source that was introduced into the water from the energy of the seiche. They estimated that 2 to 7 mm of bottom sediment had to have been resuspended in order to sustain the concentrations of suspended sediment, PCBs, and p,p'-DDE that were observed. By the next sampling trip two weeks later, the concentrations of these three parameters had returned to the concentrations that were usually observed in this part of Lake Superior.

Several authors have attributed peak residues of organochlorines in fish to seasonal rains washing contaminated soil into the aquatic system. Examples include DDT and dieldrin in fish (largemouth bass and other species) from an Indiana reservoir (Vanderford and Hamelink, 1977) and endosulfan residues in catfish (*T. tandanus*) from cotton-growing areas in Australia (Nowak and Julli, 1991). In the Sado Estuary, Portugal, winter runoff was associated with a sharp increase in DDT residues (lipid weight) in oysters (*Crassostrea angulata*), but not PCB residues (Castro and others, 1990).

Winter or Spring Snowmelt and Runoff

Several studies have reported that maximum organochlorine residues in aquatic biota were associated with heavy surface runoff because of midwinter thaw or spring snowmelt. For example, organochlorine residues in amphipods from Swedish streams increased from February to April, which coincided with snowmelt, surface runoff into streams, and high stream discharge (Sodergren and others, 1972). Olsson and others (1978) reported that seasonal variation in PCB concentrations in roach (in Sweden) was associated with the spring flood.

Maximum DDT residues in juvenile winter flounder from the Weweantic River Estuary in Massachusetts (1966–1967) occurred during the late spring (Smith and Cole, 1970). They were not associated with any known uses in the drainage basin, and the authors suggested they were the result of spring runoff containing contaminated soils or sediment. The maximum DDE residues occurred about six months later, probably because of environmental degradation of DDT. Maximum exposure to heptachlor occurred during the winter, coinciding with midwinter thaw and substantial surface runoff; maximum residues of the metabolite, heptachlor epoxide, occurred two months later. No seasonal pattern was observed for dieldrin residues, but dieldrin was detected throughout the year at low levels.

Reservoir Water Release

Controlled water release may cause short-term temporal changes in contaminant residues in bed sediment. For example, Wang and others (1979) measured PCBs and DDT and metabolites in bed sediment of the St. Lucie River and its estuary in central Florida. The St. Lucie is a controlled river that drains Lake Okeechobee. The lake receives runoff from agricultural areas in Florida and is known to be contaminated with a variety of agricultural contaminants. In three sampling trips, spaced about one week apart, increased bed sediment concentrations of DDT, DDE, and DDD components in both the river and the estuary were observed after water was released from the lake to the estuary, as compared with the bed sediment concentrations before the water was released. This suggested that the lake sediment was disturbed with the water release, and the suspended sediment with its associated contaminants was later deposited in the river and estuary over a short-term period. Periodic release of water from a reservoir is an example of a seasonal event that does not directly depend on weather, but can influence the concentration of hydrophobic pesticides both in the water column and in bed sediment.

Water Inflow to Estuaries

Seasonal peaks in estuarine organisms have been attributed to the river discharge into an estuary. Examples include dieldrin in the marsh clam from a Texas estuary (Petrocelli and others, 1975); and DDT in oysters from estuaries in Alabama (Casper and others, 1969), Florida (Butler, 1966), and Texas (Butler, 1969b; Petrocelli and others, 1975).

Livingston and others (1978) measured pesticide residues in various species of fish above and below the Jim Woodruff Dam (Lake Seminole and the Apalachicola River, respectively, in Florida) and in the Apalachicola Estuary from 1972–1974. Residues in Lake Seminole and the Apalachicola River did not vary with time or with site location (above versus below the dam). However, in the Apalachicola River Estuary, total DDT and PCB residues in biota were highest during winter and early spring and coincided with river flooding.

5.3.3 AGRICULTURAL MANAGEMENT PRACTICES

Irrigation (discussed in Section 5.3.2 as a streamflow-related event) is one of the most important agricultural management practices affecting pesticide accumulation in sediment and aquatic biota. In addition, a number of agricultural management practices may affect the transport of pesticides to surface waters. Because hydrophobic organochlorine insecticides in surface water runoff and drainage exist predominantly in association with soil particles, reduced soil erosion is expected to reduce the quantity of these insecticides that are transported to surface waters. For example, tillage practices in which the mechanical manipulation of the soil is either diminished (*reduced-till management*) or eliminated completely (*no-till management*) may reduce agrichemical inputs to surface waters by reducing soil erosion. Some agricultural practices that reduce pesticide transport to surface waters may tend to increase pesticide infiltration into ground water, although this is not always the case. For example, practices such as limiting the land area to which pesticides are applied, or increasing the interval between multiple applications, may decrease pesticide transport to both surface waters and ground water. On the other hand, practices such as reduced- or no-till management or increasing the depth to which pesticides are incorporated into soil, which tend to reduce pesticide transport to surface waters, may have an inverse effect on pesticide infiltration to the subsurface (Barbash and Resek, 1996). However, for the hydrophobic organochlorine insecticides, which have high K_{oc} values, interactions with soil organic matter will retard movement into the subsurface (Barbash and Resek, 1996).

Other than irrigation, the effect of agricultural management practices on pesticide accumulation in sediment or aquatic biota was addressed by very few of the monitoring and process-oriented studies reviewed in this book (Tables 2.1–2.3). An exception is Agee (1986), who investigated the source of DDT residues in sediment from the Blanco Drain in the Salinas Valley (California). This study noted that agricultural fields containing DDT-contaminated soils on the east leg of the drain were plowed over the edge and into the drain. East leg sediments contained the fingerprint of soil-based DDT (similar to technical DDT), whereas lower portions of the drain contained DDD and DDT ratios that were more characteristic of sediment (weathered DDT). The authors concluded that this practice may be a major source of DDT to the drain, although other erosion events also may have contributed to the DDT found in the drain.

5.3.4 ENVIRONMENTAL CONDITIONS

Ambient Water Concentrations

A pesticide in water will redistribute itself between the water and other (carbon-rich) compartments in the river, such as suspended particles and aquatic biota. This distribution among compartments is controlled by phase-transfer processes, including sorption to particles and passive diffusion through biological membranes (Sections 4.2.1 and 4.3.1). Although equilibrium or steady-state concentrations may not be reached in lotic systems (such as rivers), levels of hydrophobic pesticides in bed sediment and aquatic biota would be expected to track concentrations in water, although probably with some time lag. Contaminant levels in biota are expected to change relatively rapidly in response to surroundings (National Oceanic and Atmospheric Administration, 1988). However, highly hydrophobic contaminants may have slow elimination rates possibly because of slow release from deeper compartments (Gobas and others, 1986; Huckle and Millburn, 1990) or difficulty in transporting the contaminant from storage sites to the site of excretion (Gakstatter and Weiss, 1967). The relation between contaminant concentrations in bed sediment and water depends on the particle settling times, which is influenced by factors such as particle size and density, water temperature, and the depth of the water body (Eadie and Robbins, 1987). However, a pesticide reaching the sediment–water interface is expected to distribute itself among available compartments: sorbed phase (sediment), pore water, and overlying water.

Several studies noted seasonal changes in pesticide residues in sediment or biota that appeared to be related to pesticide concentrations in the water column. The relations observed were sometimes, but not always, straightforward. In rice fields in the Ebro Delta in Spain, the increase in organochlorine residues that occurred in winter for common carp, eels (Ruiz and Llorente, 1991), and mosquitofish (Porte and others, 1992) was attributed to an increase in contaminant concentrations in ambient water. After the closing of irrigation channels in December, the rice fields dried up, gradually exposing the fishes to an increase in ambient organochlorine contaminant concentrations in water. In another example, Kellogg and Bulkley (1976) attributed seasonal changes in dieldrin residues in bed sediment and fish tissues (various species) from the Des Moines River to seasonal water column concentration changes driven by seasonal application of dieldrin. Also, seasonal fluctuations in dieldrin residues in mayfly (*Potamanthus* sp.) larvae were stronger in 1972 than those in 1973; maximum water concentrations in 1972 also were higher than those in 1973.

In contrast, dieldrin residues in catfish (flesh) from Iowa rivers (measured during 1968–1971) were not lower during the dry season (winter), despite the fact that water concentrations of total DDT and dieldrin showed seasonal fluctuations, with maximum water concentrations occurring following the wet period of the year. However, the highest dieldrin residues in catfish flesh were found in rivers with the highest concentrations of dieldrin in whole water and with the highest silt loads (Morris and others, 1972). Morris and others (1972) suggest that bed sediment provides a constant supply of contaminant to the hydrologic system.

The importance of ambient water concentrations was demonstrated by Larsson (1986) in large outdoor model ecosystems in Sweden. PCBs were added to the model ecosystems in treated sediment. PCBs were transported into water according to a seasonal cycle, with higher levels in summer and lower levels in winter. PCB loss from sediment to water was greater during the first year than it was during the second year, possibly because of lower bioturbation the second year (the number of chironomid larvae was reduced 10-fold) or because allochthonous materials (such as leaves) built up on sediment, thus reducing the transport rate. PCB uptake by zooplankton (*D. magna*) followed a positive linear relation with concentration in water (Figure 5.27). This is consistent with an equilibrium partitioning mechanism, but, as pointed out by Larsson, the life span of the daphnids is only 20–30 days (at 25 °C). Therefore, the decreasing concentrations in zooplankton that occurred in the autumn–winter of 1983 and the summer of 1984 are probably a result of different generations of daphnids being exposed to different concentrations of PCBs in the water column. In the winter, the filtration rate of daphnids decreases (at low temperature), thus decreasing PCB uptake from food; this factor may contribute to the lower levels of PCBs observed in daphnids during the winter. In a separate model ecosystem containing PCB-contaminated sediment, PCBs were taken up by bottom-feeding eels (*A. anguilla*) and planktivorous rudd (*Scardinius erythrophthalamus*); fish in an untreated reference system contained much lower PCB concentrations. Rudd, exposed to high levels after the first summer's maximum water concentrations, maintained those high levels during second summer when the water concentration was lower, indicating that elimination was slow.

Temperature

Temperature changes may influence pesticide residues in aquatic biota in a number of ways, some relating to changes in ambient concentrations (i.e., exposure) and some to changes in the organisms themselves. Temperature may affect the degradation, either abiotic or biologically mediated, of a pesticide in a hydrologic system, which in turn affects the exposure of organisms therein (Bulkley and others, 1974). Deep lakes in temperate regions may undergo seasonal temperature changes large enough to cause thermal stratification. In stratified lakes,

Seasonal fluctuations in organochlorine residues in zooplankton from the North Adriatic Sea were related to species composition (as noted above) and to food source or trophic level. Maximum organochlorine (DDT, PCBs, and HCH) residues occurred during summer (July to early September), at which time the samples were predominantly cladocerans (carnivorous zooplankton) relative to herbivorous copepods. As previously noted, a linear correlation between PCB and cladocerans was found (Cattani and others, 1981).

Interestingly, the studies reviewed in this section (also see Tables 2.1, 2.2, and 2.3) indicate that both pelagic organisms (such as phytoplankton, zooplankton, and predator fish) and sediment-dwelling organisms (including bivalve mollusks and bottom-feeding fish) often show seasonal fluctuations in organochlorine residues. In some cases, seasonal changes for bivalves have been attributed to changes in contaminant availability. The studies reviewed provide no evidence that organisms living and feeding in bed sediment are shielded from seasonal variations in contaminant exposure.

Migration

Migration can have a direct, pronounced effect on the exposure of biota to hydrophobic contaminant residues, as the migrating individuals may migrate into or out of a contaminated habitat. Other effects of migration (such as mobilization of lipid stores) are indirect and may be difficult to see empirically. The different seasonal patterns in organochlorine residues observed for juvenile and adult winter flounder in the Weweantic River Estuary in Massachusetts (1966–1967) were attributed to migration (Smith and Cole, 1970). Migratory adults enter the estuary in late October, spawn from midwinter to early spring, and leave in May; in contrast, juveniles stay in the estuary year round. Juveniles had maximum residues of DDT in late spring and maximum heptachlor residues in winter (associated with spring runoff and midwinter thaw, respectively). The degradation products DDE and heptachlor epoxide peaked six and two months later, respectively. Adult flounder from the estuary had lower DDE residues than juveniles (because they were absent from the estuary in late spring and summer), but comparable heptachlor levels (because they were present during the maximum exposure period).

In a few studies, migration of contaminated fish was implicated as a source of contaminants to other parts of the hydrologic system. Migration in association with spawning would be a seasonal event. For example, Pacific salmon (*Oncorhynchus* sp.) migrating from Lake Ontario to spawn are believed to be the primary source of mirex contamination to tributaries with no known mirex sources (Low, 1983; Lewis and Makarewicz, 1988), as discussed in Section 4.3.1 (subsection on Tissue Decomposition). In another example, total DDT residues were measured in water, sediment, and biota of the Tennessee River system (Water and Air Research, Inc., 1980). The source of these residues was a manufacturing plant that discharged waste into a tributary system to the Tennessee River from 1947–1970. Sediment analyses showed the tributary system to be the major source of total DDT to the river. Downstream of the tributary system, 80 percent of channel catfish had residues greater than 5 ppm, suggesting that contamination resulted from in situ conditions rather than from migration. However, in largemouth bass collected upstream, total DDT residues were more variable and suggested migration as the source. The indirect effects of migration on organochlorine concentrations are illustrated by PCB concentrations in tuna (*T. thynnus*) from the Mediterranean coast of Spain (Porte and Albaiges, 1994). In 1986, PCB concentrations in July were two-fold higher in muscle than in liver, whereas in October, liver levels were 22 times higher than in muscle tissue. Porte and Albaiges suggest this was due to the postspawning (which occurs in May–June) migration within the Mediterranean in which lipid reserves were mobilized.

5.3.6 SUMMARY

Bed Sediment

Relatively few studies have attempted to measure seasonal changes in contaminant residues in sediment. However, when these studies are considered in relation to seasonal and episodic events, it suggests that there are seasonal perturbations of the bed sediment that cause changes in the concentrations of hydrophobic organic chemicals. For a river basin that contains soils contaminated with historically used pesticides, a yearly scenario of surficial bed sediment concentrations might be as follows. Total DDT will be used as an example. In the period of low flow, the water column concentration will be at a minimum because of a lack of total DDT input into the water column. Whatever total DDT is present will be distributed between the dissolved phase, particulate phase, and the biota. During this period of low flow, some of the total DDT associated with particles or recently deceased biota will settle to the sediment–water interface. Some of this material will be returned quickly to the water column; some will be transferred to the bed sediment. Through this process, the total DDT concentration in surficial bed sediment will slowly increase over time. When irrigation or the rainy season starts, there will be a sudden influx of fresh soil and its associated total DDT into the hydrologic system, beginning with the first runoff events. This may substantially increase the water column concentrations of total DDT and suspended solids, but it may only slightly increase the water discharge. As the irrigation or rains continue, the water discharge and velocity of the river will increase. Increases in discharge, velocity, and turbulence are likely to resuspend some of the bed sediment. At this time, the total DDT concentration in surficial bed sediment will be at a minimum. The water column will be at or near its highest concentration, receiving both soil and bed sediment inputs at this time. As the irrigation or rains subside, the water discharge and velocity will decrease, and the suspended particles will settle out of the water to the depositional zones of bed sediment. Total DDT concentrations in bed sediment will increase as the stream velocity dissipates. The cycle may repeat itself, depending on the local schedule of each irrigation or rainy season. Reservoir release, which tends to occur on an episodic rather than seasonal basis, would have similar effects on soil erosion, streamflow velocity and sediment disturbance as irrigation or rainfall. The deeper bed sediment that is protected (by virtue of being deep) from resuspension is not affected by the normal seasonal cycle, but it may be moved by infrequent catastrophic events that cause extremely high stream velocity (greater than 2.5 ft/s), such as the flooding of the upper Mississippi River in 1993.

Aquatic Biota

A typical scenario of seasonal changes in pesticide residues of aquatic biota is difficult to develop, because seasonal variations are dependent on the species of organism, including its body size, habitat and trophic level, reproductive cycle, and life span. In general, organisms are able to respond to changes in concentrations of pesticides in their environment relatively quickly in terms of bioaccumulation. Smaller organisms are likely to respond more quickly than larger organisms. This is true of phytoplankton, for which contaminant uptake occurs via sorption and subsequent partitioning. It has been shown that the kinetics of uptake of PCBs into algae are of the same order of magnitude as the kinetics of growth. For higher organisms, a few generalizations can be made. Seasonal variations in pesticide residues have been observed for biota from both saltwater and freshwater systems, although variations are more pronounced in estuarine than offshore waters. Both pelagic organisms (such as phytoplankton, zooplankton, and predator fish) and sediment-dwelling organisms (including bivalve mollusks and bottom-feeding fish) have been observed to show seasonal fluctuations in contaminant residues. Interestingly, the studies reviewed provide no evidence that organisms living or feeding in bed sediment are shielded from seasonal variations in contaminant exposure. Most of the factors affecting seasonality in aquatic biota appear to act by influencing either or both (1) organochlorine availability (determined by a combination of pesticide application, industrial discharge rates, streamflow, runoff, etc.) and (2) physiological changes (especially related to lipid content and the reproductive cycle). Seasonal fluctuations in lipid content generally are associated with the reproductive cycle, in that lipid content tends to increase as the spawning period approaches and decrease during and after spawning. Contaminant availability does not necessrily correspond with reproductive stage and high lipid content, so that multiple peaks may be observed in seasonal profiles. Differences in the timing of seasonal maxima and minima may occur among different species at the same location and among organisms of the same species from different locations; the latter may be caused by differences in contaminant availability (because of differences in runoff or local pesticide application) or by differences in the timing of the reproductive cycle and associated lipid changes (because of differences in temperature, salinity, or other factors).

The multiple factors affecting seasonality are not necessarily synchronized. It is perhaps surprising that seasonal fluctuations have been observed as consistently as they have. Only four of the studies reviewed in this section found no evidence of seasonal fluctuation in tissue residues, contrasted to over 30 studies that did. Given the number of factors involved and the interaction between them, it may be difficult to attribute seasonal profiles observed in the field to individual factors or combinations of factors. The occurrence of seasonal fluctuations, whatever the cause, points out the importance of sampling in periods that are well-defined ecologically and hydrologically. Residue monitoring programs should consider seasonality in sampling design and data interpretation.

5.4 PHYSICAL AND CHEMICAL PROPERTIES OF PESTICIDES IN SEDIMENT AND AQUATIC BIOTA

One significant area of research in environmental science over the past few decades has been in environmental chemodynamics, that is, the study of the relations between various physical and chemical properties of an organic chemical and its behavior, transport, and fate in the environment (e.g., Mackay, 1979; Tinsley, 1979; Stumm and others, 1983). The basis for this is that a similar chemical property (such as low water solubility) for different chemicals will lead to similar behavior (such as strong association with aquatic particles). Chemodynamic relations provide both a fundamental interpretative basis for a chemical's behavior as observed in various environmental matrices and a predictive tool for new chemicals that are or will be introduced into the environment. Agencies that are responsible for regulating pesticides or other toxic chemicals in the environment have relied on both aspects of chemodynamics for existing and new chemicals.

In this section, physical and chemical properties that control the tendency of a pesticide to accumulate in bed sediment or aquatic biota are discussed (Section 5.4.1). Then, detection frequencies for individual pesticides in bed sediment and aquatic biota (aggregated from the monitoring studies reviewed) are plotted as a function of selected physical and chemical properties (Section 5.4.2). Finally, the results are used as a predictive tool to identify pesticides that were not extensively analyzed in bed sediment or aquatic biota by the studies reviewed, but that may be present in these media (Section 5.4.3).

5.4.1 PROPERTIES THAT CONTROL ACCUMULATION IN SEDIMENT AND AQUATIC BIOTA

For pesticides in bed sediment and aquatic biota, there are two categories of physical and chemical properties that govern their behavior and allow for the interpretive and predictive roles of chemodynamics. First are those characteristics of a pesticide that promote its association with sediment or biota (such as low water solubility and high K_{ow}). Second, a pesticide must be environmentally persistent for it to accumulate to substantial levels over time in bed sediment or aquatic biota.

Hydrophobicity

The phase-transfer processes of sorption–desorption and partitioning control the movement of nonionic pesticides into bed sediment (Section 4.2.1) and aquatic biota (Section 4.3.1). Several interrelated physical and chemical properties show a relation with the sorption of nonionic organic chemicals to sediment from water and also with their bioconcentration in aquatic biota. The extent of sorption or bioconcentration tends to be inversely related to water solubility and directly related to K_{ow}, which is a measure of lipophilic as well as hydrophobic character. Other relevant properties include degree of ionization, molecular size, shape, and structure.

Water solubility is thought to be a key chemical property that affects the extent of both sorption and bioconcentration. The water solubility of an organic chemical is governed by two thermodynamic interactions between the chemical and the water—enthalpy and entropy. The enthalpic aspects are from the electrostatic interactions between the polar covalent water and dipoles or charged portions of the organic molecule. Entropic effects are largely driven by the size and shape of the organic molecule. In order to accommodate (solubilize) an organic molecule, the solvent (water) must open a cavity large enough for the molecule to fit. This causes an increase in the ordering of the water around the solubilized molecule and increases the entropy. The combination of these two factors, electrostatic interactions and cavity formation, determine the solubility of an organic molecule. Generally, smaller molecules or molecules with polarized or charged functional groups have greater water solubilities (Capel, 1993). As an example, DDT is often observed in bed sediment and tissues. The size of the DDT molecule is large (two aromatic rings and five chlorines with a total molecular weight of 354.5 daltons). Also, there are no strongly polarized or charged functional groups (the chlorine portions of the DDT have minimal electrostatic interactions with the water). These two factors combine to give DDT an extremely low water solubility, on the order of 1 μg/L.

For a hydrophobic molecule, the processes of sorption to particles and bioconcentration by aquatic biota are driven by the removal of the molecule from the water (which yields a decrease in entropy), as well as by interactions of the molecule with the particle or biota (which yield a gain in enthalpy). Since the sorptive and bioconcentrative interactions of hydrophobic chemicals can be described on a thermodynamic basis, their behavior in hydrologic systems can be, and has been, described by chemodynamic (structure–activity) relations.

Data are widely available for two properties reflecting hydrophobicity: water solubility and K_{ow} (Table 3.7). In the following analysis (Section 5.4.2), these properties are used to develop chemodynamic relations for pesticides in bed sediment and aquatic biota.

Persistence

The second property affecting the presence of a pesticide in bed sediment or aquatic biota is that of environmental persistence. Both bed sediment and aquatic biota tend to be long-term accumulators of chemicals. This means that for a chemical to be found frequently in measurable quantities in bed sediment and tissues, it generally must be recalcitrant and, therefore, long-lived. Organochlorine insecticides, most of which have not been used in agriculture since the 1970s, continue to be detected in bed sediment and aquatic biota during the 1980s and 1990s (Tables 2.1 and 2.2), suggesting strong environmental persistence. Field monitoring studies (e.g., Mischke and others, 1985) indicate that DDT half-lives in soil are on the order of 15 years or longer (Section 5.5).

Transformation of a pesticide can occur in many environmental compartments. For most pesticides that are fairly soluble in water, most transformation reactions occur in the soil to which they were applied, although some transformation processes may occur in the water column after the pesticide enters a hydrologic system (Larson and others, 1997). This difference probably is due to the greater concentration of microorganisms in soils (and therefore the greater extent of biologically mediated degradation), as well as the presence of trace elements that may catalyze abiotic degradation reactions. For hydrophobic pesticides that are commonly detected in bed sediment and aquatic biota, however, transformation reactions in soil may be very slow and the rate of degradation may be accelerated once the pesticide enters a hydrologic system, typically in association with soil particles (Agee, 1986; also see Section 5.5.3). As noted in Section 4.2.3, the microbial community in bed sediment is extremely viable, and this is where much of the decomposition of natural organic matter takes place in surface waters.

The environmental transformation processes for pesticides were discussed earlier for bed sediment (Section 4.2.3), aquatic biota (Section 4.3.3), and the water column (Larson and others, 1997). Pesticide transformation may occur by a combination of chemical reactions (hydrolysis, oxidation, and reduction), photochemical reactions, and biologically mediated processes. In general, there is more information available for the reaction rates and mechanisms of pesticide transformation in soils than there is for surface waters or bed sediment. However, the transformation processes (chemical, photochemical, and microbiological) that a pesticide undergoes in the water column or in bed sediment are the same as, or similar to, the transformation processes that occur in the soil to which it was applied, although these reactions generally have different rates in different compartments. Therefore, to a large extent, information that exists on the relative persistence of pesticides in soils can reasonably be extrapolated to persistence in surface waters or bed sediment, where there have been fewer studies of persistence. Soil half-life (Table 3.7) has been used in Section 5.4.2 to develop chemodynamic relations for pesticides in bed sediment and aquatic biota. Although laboratory and field-plot estimates of soil half-lives for a given chemical may vary considerably, depending on test conditions (e.g., see Howard and others, 1991), soil half-life is still the best measure of persistence that is available for a large number of pesticides.

For residues in aquatic biota, pesticide metabolism also must be considered (Section 4.3.3). A pesticide may be transformed by higher organisms at a faster rate than in the water column or bed sediment via chemical or microbiological processes. For example, two organochlorine pesticides (aldrin and heptachlor) are rapidly metabolized by fish and other organisms (Schmitt and others, 1983, 1990). Detection of these parent compounds in biota probably indicates recent exposure (Schmitt and others, 1983) and might be expected to occur at higher frequencies in sediment than in biota. This is borne out by comparison of overall site detection frequencies for both compounds in sediment versus in biota, as aggregated from all national, multistate, state, and local studies (Tables 3.1 and 3.2). During the 1990s, aldrin was detected in sediment at 7 percent of the total sites sampled and in biota at 0 percent of sites; heptachlor was detected in sediment at 19 percent of the total sites sampled and in biota at 6 percent of sites. This may be circumstantial, since for the most part, sediment and biota were not sampled at the same sites. However, a stronger case can be made by considering individual studies that analyzed both parent compounds and metabolites. Aldrin and heptachlor were rarely observed in whole freshwater fish samples from the NCBP after 1977, although their metabolites (dieldrin and heptachlor epoxide, respectively) continued to be detected in some samples through 1986 (Schmitt and others, 1983, 1990). At NS&T (Benthic Surveillance Project) sites, paired sediment and benthic fish samples were collected at each site. Detection frequencies and concentrations of most organochlorine pesticides tended to be higher in benthic fish tissues (liver and stomach contents) than in sediment, except for aldrin and heptachlor. Both aldrin and heptachlor were more prevalent in sediment than in fish tissue samples, but residues of their metabolites (dieldrin and heptachlor epoxide, respectively) fit the expected pattern and were higher in fish tissues than in sediment (Zdanowicz and Gadbois, 1990). Metabolism is also important for PAHs, which are another class of hydrophobic organic chemical pollutants. PAHs are frequently detected in bed sediment (e.g., National Oceanic and Atmospheric Administration, 1991), but not in fish because PAHs are readily metabolized by fish (Crawford and Luoma, 1994). In mollusks, which metabolize PAHs slowly, PAH residues may be detected (Crawford and Luoma, 1993). Thus, for hydrophobic contaminants that are metabolized significantly by aquatic biota, metabolism may limit the success of chemodynamic relations that use soil half-life as a measure of persistence.

5.4.2 CHEMODYNAMIC RELATIONS

The presence or absence of a given pesticide in bed sediment and aquatic biota should be predictable from chemodynamic relations. In this section, we will look for relations between detection frequency (aggregated from all the monitoring studies reviewed, which are listed in Tables 2.1and 2.2) and measures of hydrophobicity and persistence. Two different measures of hydrophobicity are used: water solubility and K_{ow}. Soil half-life is used as a measure of persistence. The values used for water solubility, K_{ow}, and soil half-life are shown in Table 3.7. For consistency, property values in Table 3.7 were obtained (if available) from the Pesticide Properties Database of the U.S. Department of Agriculture (Agricultural Research Service, 1997). For properties with multiple values or a range of values given in Table 3.7, the mean value was used for chemodynamic relations.

Water Solubility and Soil Half-Life

Figure 5.30 shows the logarithm of water solubility plotted against the logarithm of soil half-life for pesticides analyzed in sediment or aquatic biota by the monitoring studies reviewed. In order to minimize bias from site selection in individual studies, only pesticides targeted at 50 sites or more were included in Figure 5.30. Figure 5.30 distinguishes between pesticides in three groups: the rarely detected group (empty circles, ○), which consists of pesticides detected at fewer than 5 percent of sites sampled for sediment and biota; the intermediate group (half-filled squares, ◩), which consists of pesticides detected at 5–20 percent of sites sampled for sediment and 5–60 percent of sites sampled for biota; and the commonly detected group (filled squares, ■), which consists of pesticides detected at greater than 20 percent of sites in both sediment and biota. Many pesticides in the commonly detected group were detected at over 50 percent of biota sites. The second, intermediate group was defined this way because several compounds were detected at over 20 percent of sites in biota, but at between 5 and 20 percent of sites in sediment. All pesticides fit in one of these three groups except for toxaphene, which was detected at over 20 percent of sites for biota, but at less than 5 percent for sediment; toxaphene was classed with the intermediate group (half-filled squares, ◩). A few pesticides in Figure 5.30 were targeted at sufficient sites in only one medium (bed sediment or aquatic biota only), and these are categorized by the detection frequency in the medium sampled.

Figure 5.30 indicates that the expected chemodynamic relations are supported by field observations; that is, the pesticides with low water solubility (S) and long soil half-life ($t_{1/2}$) tend to be observed in bed sediment and aquatic biota, whereas the pesticides with high water solubility and short soil half-life generally are not. For the following discussion, Figure 5.30 is divided into six blocks, A–F. Most of the commonly detected pesticides (filled squares, ■) were highly hydrophobic (S less than 1 mg/L) and recalcitrant (soil $t_{1/2}$ over 300 days); these values correspond to block A in Figure 5.30. The single exception (pentachlorophenol), located in block E, had a higher detection frequency than expected from its water solubility (S = 14 mg/L at pH 5, 20 °C) and soil half-life (soil $t_{1/2}$ = 48 days). Pentachlorophenol was detected at 26 percent of sites in biota and at 20 percent of sites in sediment. These detection frequencies are lower than those observed for the more hydrophobic organochlorine compounds in the commonly detected group, such as p,p'-DDE (89 percent in biota and 50 percent in sediment) and dieldrin (61 percent in biota and 40 percent in sediment). However, pentachlorophenol is a weak acid (negative log of acid dissociation constant pK_a = 4.74), so it will be partly dissociated at environmental pH values. Thus, its distribution between acid and phenolate forms, and therefore its water solubility and persistence, will depend on the local (environmental) pH.

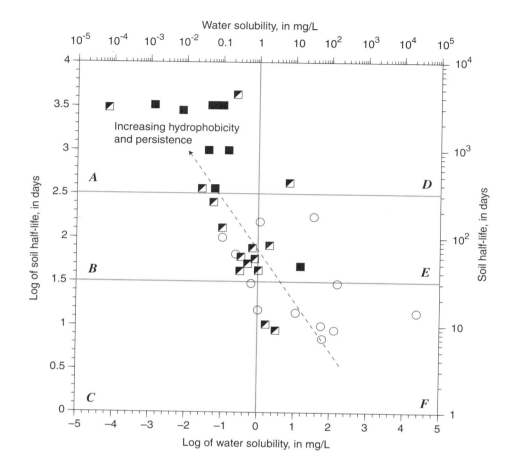

EXPLANATION

○ Rarely detected—detected at <5 percent of sites in sediment and biota

◪ Intermediate—detected at 5–20 percent of sites in sediment and 5–60 percent in biota

■ Commonly detected—detected at >20 percent of sites in sediment and biota

Figure 5.30. The relation between the logarithm of soil half-life (days) and the logarithm of the water solubility (log S, in mg/L) for pesticides analyzed in sediment and aquatic biota by the monitoring studies reviewed. Only pesticides targeted at 50 sites or more were included. Pesticides were classified into one of three groups: rarely detected—pesticides detected at <5 percent of total sites for sediment and biota (○); intermediate—pesticides detected at 5–20 percent of total sites sampled for sediment and 5–60 percent of sites sampled for biota (◪); and commonly detected—pesticides detected at >20 percent of total sites in both sediment and biota (■). Graph is divided into six blocks (A–F) as follows: Water solubility is <1 mg/L for blocks A, B, and C, and >1 mg/L for blocks D, E, and F. Soil half-life is >300 days for blocks A and D, 30–300 days for blocks B and E, and <30 days for blocks C and F. Dashed line shows direction of increasing hydrophobicity and persistence. Abbreviation: mg/L, milligram per liter.

At the other extreme, most of the pesticides in the rarely detected group (empty circles, ○) tended to be more water soluble (blocks E and F) and less persistent (blocks C and F) than those in the intermediate and commonly detected groups. However, four pesticides in the rarely detected group were in (or very near) block B; that is, they had water solubilities (S ≤ 1 mg/L) and soil half-lives (soil $t_{1/2}$ ≥ 30 days) comparable with compounds in the intermediate and commonly detected groups. These four pesticides are isopropalin, pentachloronitrobenzene (PCNB), ethion, and trithion. All four compounds were detected in biota at 1–4 percent of sites; only ethion and trithion were analyzed in sediment at 50 or more sites, and neither was detected. Trithion (an organophosphate insecticide) and isopropalin (a nitrogen-containing herbicide) were discontinued in 1987 and 1993, respectively (Meister Publishing Company, 1996).

Of pesticides in the intermediate group (half-filled squares, ◩), only three (mirex, endrin, and aldrin) are in block A, i.e., they have water solubilities and soil half-lives comparable with DDT and the other organochlorines in the commonly detected group. Their relatively lower detection frequencies may be partly a function of metabolism (aldrin) or limited use in agriculture (mirex and endrin). As noted above (Section 5.4.1), aldrin (detected at 7 percent of sites in biota and 5 percent in sediment) is readily metabolized to dieldrin. Mirex (detected at 26 percent of sites in biota, but only 16 percent in sediment) was applied almost exclusively in the southeastern United States. Endrin (detected at 16 percent of sites in biota and 6 percent in sediment) was used in agriculture in quantities that were only 5–20 percent of the quantities used for toxaphene, DDT, and aldrin plus dieldrin during the 1960s and 1970s (see Table 3.5). Moreover, endrin is slightly more water soluble than DDT (Table 3.7).

Two pesticides in the intermediate group (half-filled squares, ◩) are located in block F of Figure 5.30, i.e., they have higher detection frequencies than expected on the basis of their water solubility and persistence. These compounds are toxaphene and α-HCH. Recall that toxaphene was classified in the intermediate group (detected at 5–20 percent of sites) even though it was detected at less than 5 percent of sediment sites, on the basis of its detection at 28 percent of biota sites.

However, the majority of pesticides in the intermediate group (half-filled squares, ◩) were moderate in hydrophobicity and persistence. A cluster of nine such pesticides appears in (or very near) block B in Figure 5.30. These intermediate compounds have water solubilities between 0.05 and 1 mg/L and soil half-lives between 30 and 300 days. The cluster consists of chlorpyrifos, dacthal, dicofol, endosulfan I, heptachlor, heptachlor epoxide, methoxychlor, oxadiazon, and trifluralin. Three of these compounds (chlorpyrifos, trifluralin, and dacthal) are high use pesticides in agriculture today (see Table 3.5). The presence of these compounds in sediment and aquatic biota suggests that other pesticides of intermediate hydrophobicity and persistence also might be found, if looked for, in these sampling media.

In Figure 5.31, a fourth group of pesticides has been plotted by water solubility and soil half-life, superimposed on the data from Figure 5.30. This fourth group (plus symbols in Figure 5.31) is the rarely analyzed group. It consists of pesticides that were (1) analyzed in sediment and aquatic biota at fewer than 50 sites and (2) in the top 500 pesticides used in agriculture in 1988 (Gianessi and Puffer, 1991, 1992a,b) or used in homes and gardens in 1990 (Whitmore and others, 1992), except that, for simplicity, pesticides with water solubilities greater than 10^5 mg/L were excluded. Because most pesticides in this rarely analyzed group have fairly high water solubility (blocks D, E, and F) and fairly short soil half-lives (blocks C and F), they are not likely to accumulate in bed sediment and aquatic biota. None of the pesticides in the rarely analyzed group fall into block A (where DDT and the other highly hydrophobic and persistent pesticides are located), indicating that all pesticides that (1) are highly hydrophobic and persistent and (2) were in the top 500 pesticides used in agriculture in 1988 or were used around homes and gardens in 1990, have been analyzed at a total of 50 sites or more by existing monitoring studies. However, several pesticides in the rarely analyzed group fall in block B, so have water solubilities and soil half-lives that are comparable with many pesticides in the intermediate group. This indicates some potential for these pesticides to accumulate in sediment and aquatic biota, as discussed further in Section 5.4.3.

n-Octanol-Water Partition Coefficient and Soil Half-Life

Figure 5.32 shows the logarithm of K_{ow} plotted against the logarithm of soil half-life, with pesticides grouped according to detection frequency in sediment and biota as described for Figure 5.30. As was done for Figure 5.30, Figure 5.32 is divided into six blocks, with block A designating the region of greatest hydrophobicity and persistence. As expected, pesticides in the commonly detected group (filled squares, ■) tend to fall in block A (i.e., they have log K_{ow} greater than 3 and soil half-life greater than 300 days). The single exception (as also occurred in Figure 5.30) was again pentachlorophenol (in block B), which was detected at greater frequencies than expected from its soil half-life. As discussed above, pentachlorophenol is a weak acid (pK$_a$ = 4.74) and will be partially ionized at environmental pHs, so its hydrophobicity and persistence will vary with local (environmental) pH. At the other extreme, the rarely detected pesticides (open circles, ○) tend to have log K_{ow} values less than 3 (blocks E and F) and soil half-lives less than 30 days (blocks C and F). Finally, the majority of intermediate pesticides (half-filled squares, ◪) are in block B and have log K_{ow} values of 3–5 and soil half-lives of 30–300 days. The exceptions are aldrin, mirex, and endrin in block A (which have lower detection frequencies than expected) and toxaphene and α-HCH in block C (which have higher detection frequencies than expected). These exceptions are identical to those in Figure 5.30, since their soil half-life values set them apart from the other compounds in their respective groups.

Figure 5.33 contains a fourth group of pesticides plotted by K_{ow} and soil half-life, superimposed on the data from Figure 5.32. This fourth group (plus symbols in Figure 5.33) is the rarely analyzed group defined as in Figure 5.31. The same patterns hold here as were discussed for Figure 5.31. Most pesticides in the rarely analyzed group have log K_{ow} values less than 3 (blocks D, E, and F) or soil half-lives less than 30 days (blocks C and F), so are not likely to accumulate in bed sediment and aquatic biota. However, some of the pesticides in the rarely analyzed group fall in block B, so have water solubilities and soil half-lives that are comparable with many pesticides in the intermediate group. This again indicates some potential for accumulation in sediment and aquatic biota (discussed in Section 5.4.3).

5.4.3 PREDICTIONS FROM CHEMODYNAMIC RELATIONS

The data shown in Figures 5.31 and 5.33 can be used to identify pesticides that were analyzed at fewer than 50 total sites in sediment or biota by the monitoring studies reviewed, but

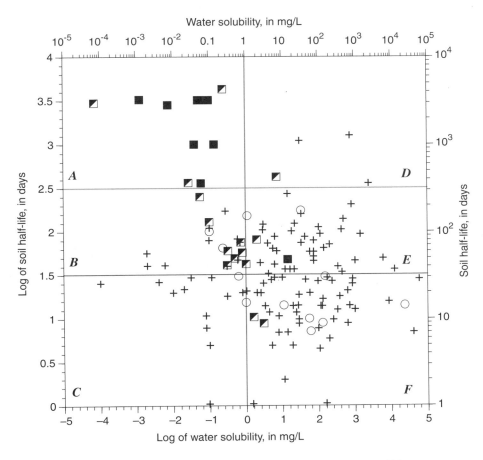

○   Rarely detected—detected at <5 percent of sites in sediment and biota

▟   Intermediate—detected at 5–20 percent of sites in sediment and 5–60 percent in biota

◼   Commonly detected—detected at >20 percent of sites in sediment and biota

+   Rarely analyzed—analyzed at <50 sites in sediment and biota

Figure 5.31. The relation between the logarithm of soil half-life (days) and the logarithm of the water solubility (log S, in mg/L) for pesticides analyzed in sediment and aquatic biota by the monitoring studies reviewed, compared with pesticides that were rarely analyzed in these media. Only pesticides targeted at 50 sites or more were included. Pesticides were classified into one of four groups: rarely detected—pesticides detected at <5 percent of total sites for sediment and biota (○); intermediate—pesticides detected at 5–20 percent of total sites sampled for sediment and 5–60 percent of sites sampled for biota (▟); commonly detected—pesticides detected at >20 percent of total sites in both sediment and biota (◼); and rarely analyzed—pesticides that were analyzed at fewer than 50 total sites (+). Graph is divided into six blocks (A–F) as follows: Water solubility is <1 mg/L for blocks A, B, and C, and >1 mg/L for blocks D, E, and F. Soil half-life is >300 days for blocks A and D, 30–300 days for blocks B and E, and <30 days for blocks C and F. Abbreviation: mg/L, milligram per liter.

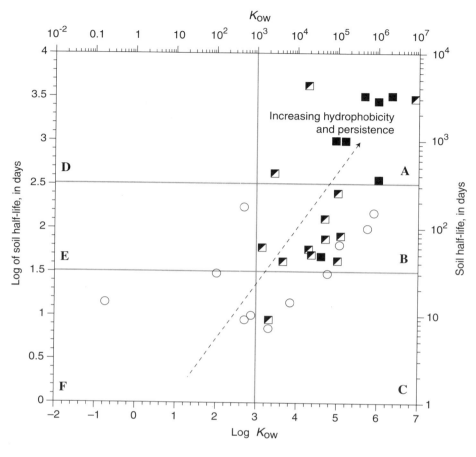

EXPLANATION

○ Rarely detected—detected at <5 percent of sites in sediment and biota

▨ Intermediate—detected at 5–20 percent of sites in sediment and 5–60 percent in biota

■ Commonly detected—detected at >20 percent of sites in sediment and biota

Figure 5.32. The relation between the logarithm of soil half-life (days) and the logarithm of the n-octanol-water partition coefficient (log K_{ow}) for pesticides detected in sediment and aquatic biota by the monitoring studies reviewed. Only pesticides targeted at 50 sites or more were included. Pesticides were classified into one of three groups: rarely detected—pesticides detected at <5 percent of total sites for sediment and biota (○); intermediate—pesticides detected at 5–20 percent of total sites sampled for sediment and 5–60 percent of sites sampled for biota (▨); and commonly detected—pesticides detected at >20 percent of total sites in both sediment and biota (■). Graph is divided into six blocks (A–F) as follows: log K_{ow} is >3 for blocks A, B, and C, and <3 for blocks D, E, and F. Soil half-life is >300 days for blocks A and D, 30–300 days for blocks B and E, and <30 days for blocks C and F. Dashed line shows direction of increasing hydrophobicity and persistence.

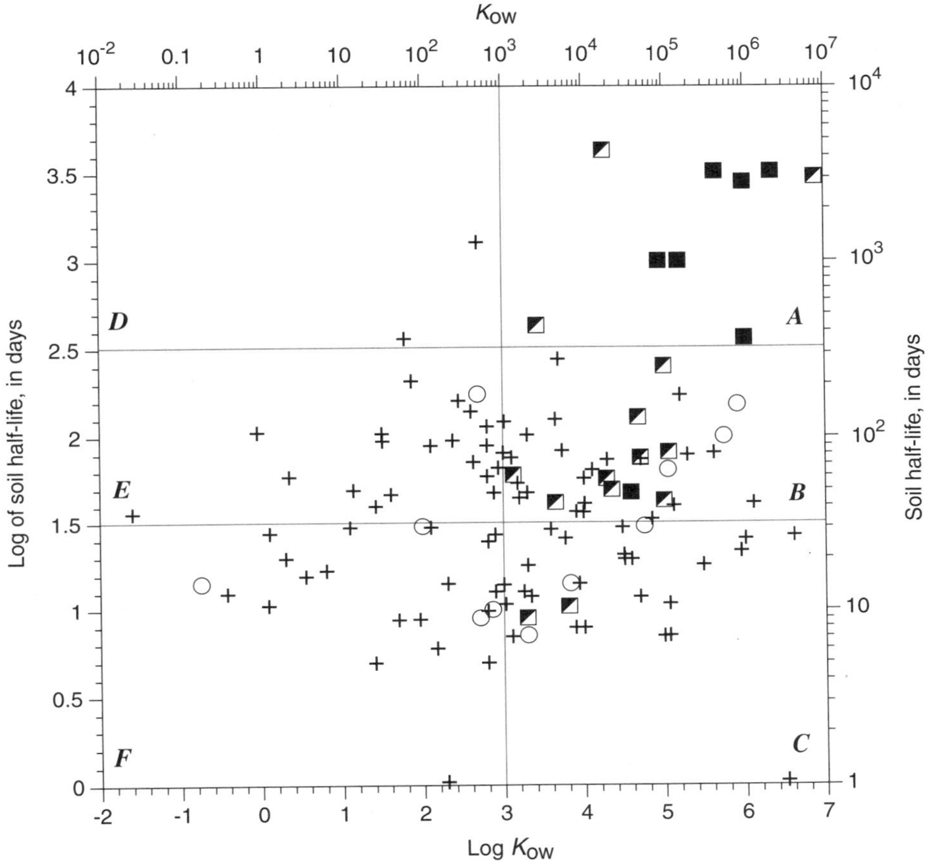

○ Rarely detected—detected at <5 percent of sites in sediment and biota

◩ Intermediate—detected at 5–20 percent of sites in sediment and 5–60 percent in biota

■ Commonly detected—detected at >20 percent of sites in sediment and biota

+ Rarely analyzed—analyzed at <50 sites in sediment and biota

Figure 5.33. The relation between the logarithm of soil half-life (days) and the logarithm of the *n*-octanol-water partition coefficient (log K_{ow}) for pesticides analyzed in sediment and aquatic biota by the monitoring studies reviewed, compared with pesticides that were rarely analyzed in these media. Only pesticides targeted at 50 sites or more were included. Pesticides were classified into one of four groups: rarely detected—pesticides detected at <5 percent of total sites for sediment and biota (○); intermediate—pesticides detected at 5–20 percent of total sites sampled for sediment and 5–60 percent of sites sampled for biota (◩); commonly detected—pesticides detected at >20 percent of total sites in both sediment and biota (■); and rarely analyzed—pesticides that were analyzed at fewer than 50 total sites (+). Graph is divided into six blocks (*A–F*) as follows: log K_{ow} is >3 for blocks *A*, *B*, and *C*, and <3 for blocks *D*, *E*, and *F*. Soil half-life is >300 days for blocks *A* and *D*, 30–300 days for blocks *B* and *E*, and <30 days for blocks *C* and *F*.

that have potential for being detected in bed sediment and aquatic biota, if they are analyzed in these media. Table 5.6 lists currently used pesticides that meet two selection criteria: (1) the water solubility is less than 1 mg/L or the log K_{ow} is greater than 3, or both, and (2) the soil half-life is greater than 30 days. Table 5.6 was generated by applying these selection criteria to all pesticides listed in Table 3.5 as having been used in agriculture in 1988 or around homes and gardens in 1990. On the basis of their hydrophobicity and persistence (Sections 5.4.1 and 5.4.2), all the pesticides in Table 5.6 have some potential for accumulation in sediment and biota.

Pesticides in Table 5.6 are listed in order of estimated agricultural use in 1988 (Gianessi and Puffer, 1990, 1992a,b). Several pesticides in Table 5.6 were used on crops in quantities that exceeded 1 million lb a.i. in 1988: chlorpyrifos, phorate, propargite, endosulfan, dicofol, and permethrin (insecticides); and trifluralin, pendimethalin, ethalfluralin, triallate, dacthal, and benfluralin (herbicides). Pesticides in Table 5.6 that were used around homes and gardens in 1990 (Whitmore and others, 1992) are endosulfan, dicofol, methoxychlor, and lindane (organochlorine insecticides); chlorpyrifos (organophosphate insecticide); permethrin and fenvalerate (pyrethroid insecticides); trifluralin, pendimethalin, dacthal, benfluralin, bensulide (herbicides); and dichlone, PCNB, and pentachlorophenol (fungicides or wood preservatives). For penta-chlorophenol, there may be other (industrial) sources to hydrologic systems besides its use as a wood preservative. The herbicide oxadiazon also is included in Table 5.6, although it was not among the top 500 pesticides used in agriculture in 1988. Although national use data for oxadiazon are not available, oxadiazon had fairly high (and increasing) use in California during the 1980s for maintenance of landscape, rights of way, structural pest control, ornamentals, turf, and roses (Rasmussen and Blethrow, 1991).

Table 5.6 also lists aggregate detection frequencies in sediment and biota, as well as the total number of sediment and biota samples analyzed by all of the monitoring studies reviewed. Detection frequencies greater than 10 percent are shaded for emphasis. Of the 22 pesticides listed in Table 5.6, 8 pesticides were not analyzed in bed sediment or aquatic biota by any of the monitoring studies reviewed. These eight pesticides are propargite and esfenvalerate (insecticides); pendimethalin, ethalfluralin, triallate, benfluralin, and bensulide (herbicides); and dichlone (fungicide). Pendimethalin had the third highest agricultural use in 1988.

Of the 14 pesticides in Table 5.6 that were analyzed by one or more monitoring studies, 13 pesticides were analyzed in 50 or more total samples of bed sediment or aquatic biota (or both) when data from all monitoring studies were combined, and 1 pesticide (fenthion) was analyzed at a total of only three sites (in both bed sediment and aquatic biota). For the 13 pesticides that were analyzed in 50 or more samples of bed sediment or aquatic biota, 8 were detected in over 10 percent of the total samples analyzed in sediment or aquatic biota, or both. These eight pesticides are chlorpyrifos, endosulfan, permethrin, and lindane (insecticides); trifluralin, dacthal, and oxadiazon (herbicides); and pentachlorophenol (wood preservative). Of these, trifluralin and chlorpyrifos each had the highest agricultural use levels in 1988, and chlorpyrifos and permethrin each had greater than 15 million outdoor applications around homes and gardens in 1990. Of the 13 pesticides that were analyzed in 50 or more samples, only 1 pesticide (phorate, an organophosphate insecticide) was not detected in any of the samples analyzed (18 bed sediment and 1,705 aquatic biota samples). It should be noted, however, that the aggregate detection frequencies shown for pesticides listed in Table 5.6 may not be representative of conditions elsewhere, especially for compounds that were analyzed in relatively few samples.

In general, moderately hydrophobic, moderately persistent pesticides would be expected to occur in sediment and biota at lower frequencies than DDT and the other highly hydrophobic, highly persistent organochlorine insecticides that are commonly found in sediment and aquatic biota (Figures 3.1, 3.2, 5.30, 5.32; Tables 3.1 and 3.2). However, results from one local study in California suggest that residues of such intermediate compounds may reach fairly high concentrations in fish and sediment under some conditions. In this study (Crane and Younghans-Haug, 1992), residues of oxadiazon were detected in 100 percent of samples collected from San Diego Creek, at concentrations reaching 87 µg/kg dry weight in bed sediment, 2,200 µg/kg wet weight in fish, and 2,400 µg/kg dry weight in transplanted mollusks. Oxadiazon was detected in samples from this study over multiple years (1988–1990). Starting in 1988, oxadiazon was added to the target analyte list of California's Toxic Substances Monitoring Program (TSMP), which analyzes toxic substances in freshwater and estuarine fish from California waters statewide (Rasmussen and Blethrow, 1991). Oxadiazon was detected from about 50 percent of TSMP sites sampled for whole fish and 2–8 percent of TSMP sites sampled for fish fillets during both 1988–1989 and 1990 (Rasmussen and Blethrow, 1991; Rasmussen, 1992). Although the highest concentrations by far were detected in the same watershed (San Diego Creek, Riverside County) described by Crane and Younghans-Haug (1992), oxadiazon also was detected in freshwater fish at other sites in both southern California (San Diego and Los Angeles counties) and northern California (Santa Clara and Solano counties).

Several studies of oxadiazon in Japan, where this compound is applied as a rice herbicide, provide some insight into the observations made in California. In fish and water collected from the Tenjin River (Tsuda and others, 1991), the mean BAF values for oxadiazon were 1,180 in pale chub (*Zacco platypus*) and 1,372 in ayo sweetfish (*Plecoglossus altivelis*). In laboratory studies with willow shiner (Tsuda and others, 1990), the mean BCF for oxadiazon was 1,226. However, oxadiazon was shown to be rapidly excreted from willow shiner (a half-life of 2.3 days) in the same study. In fact, oxadiazon residue levels in fish and shellfish from Lake Kojima, Japan, have been shown to reflect levels in the water, almost without any time lag (Imanaka and others, 1985). Oxadiazon residues were detected in both fish and water nine months after application to rice in Japan (Imanaka and others, 1981), probably because oxadiazon is relatively stable in soil and relatively resistant to degradation in river water (Imanaka and others, 1985). These studies from Japan suggest that the high oxadiazon residues observed in sediment, fish, and transplanted mollusks in California probably were accompanied by high concentrations in water. This in turn suggests that there may have been repeated or continual inputs to the hydrologic system during the sampling period. Thus, there may be potential for actively used pesticides that are intermediate in hydrophobicity and persistence to reach fairly high concentrations and detection frequencies in sediment and aquatic biota, if sampled during or shortly after a period of use.

In summary, most of the pesticides in high use during the 1990s are more water soluble and have shorter environmental residence times than the organochlorine insecticides that are the most common pesticides found in bed sediment and aquatic biota. Nonetheless, there is potential for currently used pesticides that are moderately hydrophobic and moderately persistent to occur in sediment and aquatic biota. Chemodynamic relations indicate that compounds with (1) a water solubility less than 1 mg/L or a log K_{ow} greater than 3, and (2) a soil half-life greater than 30 days, have some potential to accumulate in sediment and aquatic biota. Currently used pesticides that meet these selection criteria are listed in Table 5.6: the organochlorine

Table 5.6. Pesticides predicted to be potential contaminants in bed sediment and aquatic biota on the basis of hydrophobicity and persistence, with estimated pesticide use, the percentage of total samples with detectable concentrations, and the total number of samples

[Detection frequencies greater than 10 percent are shaded for emphasis. Data on pesticide properties are mean values from Table 3.7. Data on pesticide use are from Gianessi and Puffer (1991, 1992a,b) (1988 crop use); and Whitmore and others (1992) (1990 home and garden use). Pesticide class: BA-H, benzoic acid herbicide; Dinitro-H, dinitroaniline herbicide; Fung, fungicide or wood preservative; OC-I, organochlorine insecticide; OP-I, organophosphate insecticide; Other-H, other herbicide; Other-I, other insecticide; Pyreth-I, pyrethroid insecticide; Thcarb-H, thiocarbamate herbicide. Abbreviations and symbols: K_{ow}, n-octanol–water partition coefficient; lb a.i., pounds active ingredient; mg/L, milligram per liter; na, data not available; PCNB, pentachloronitrobenzene; S, water solubility; —, no data]

Currently Used Pesticides that Meet the Selection Criteria[1]	Pesticide Class	S (mg/L)	log K_{ow}	Soil Half-life (days)	Agricultural Use in 1988 (lb a.i.× 10[6])	Home and Garden Use (1990)		Bed Sediment		Aquatic Biota	
						Products (× 1,000)	Outdoor Applications (× 1,000)	Percentage of Samples with Detections	Total Number of Samples Analyzed[2]	Percentage of Samples with Detections	Total Number of Samples Analyzed[2]
Trifluralin	Dinitr-H	2	5.1	81	27.12	483	547	12	52	4	2,375
Chlorpyrifos	OP-I	1	5.0	43	16.73	16,652	41,900	5	20	15	1,292
Pendimethalin	Dinitr-H	0.275	5.2	174	12.52	158	338	—	0	—	0
Phorate	OP-I	22	3.9	37	4.78	—	—	0	18	0	1,705
Propargite	Other-I	0.6	3.7	84	3.79	—	—	—	0	—	0
Ethalfluralin	Dinitr-H	0.3	5.1	41	3.52	—	—	—	0	—	0
Triallate	Thcarb-H	4	4.3	74	3.51	—	—	—	0	—	0
Dacthal	BA-H	0.5	4.4	50	2.22	733	368	10	135	24	2,006
Endosulfan	OC-I	0.32	3.1	60	1.99	111	561	[3]6–13	[3]375–653	[3]4–29	[3]206–2,563
Dicofol	OC-I	0.8	4.3	57	1.72	4,587	4,179	0	2	8	1,435
Benfluralin	Dinitr-H	0.1	5.3	80	1.17	79	81	—	0	—	0
Permethrin	Pyreth-I	0.006	6.1	42	1.12	7,397	18,461	15	20	7	90
PCNB	Fung	0.24	5.1	65	0.80	64	64	—	0	1	560
Bensulide	Other-H	5.6	4.1	64	0.63	36	73	—	0	—	0
Esfenvalerate	Pyreth-I	0.002	4.0	42	0.29	—	—	—	0	—	0
Methoxychlor	OC-I	0.1	4.7	128	0.10	3,564	3,692	3	2,071	1	11,836
Fenvalerate	Pyreth-I	0.002	4.0	57	0.07	3,192	2,937	0	20	2	109

Table 5.6. Pesticides predicted to be potential contaminants in bed sediment and aquatic biota on the basis of hydrophobicity and persistence, with estimated pesticide use, the percentage of total samples with detectable concentrations, and the total number of samples—*Continued*

Currently Used Pesticides that Meet the Selection Criteria[1]	Pesticide Class	S (mg/L)	log K_{ow}	Soil Half-life (days)	Agricultural Use in 1988 (lb a.i. × 10⁶)	Home and Garden Use (1990)		Bed Sediment		Aquatic Biota	
						Products (× 1,000)	Outdoor Applications (× 1,000)	Percentage of Samples with Detections	Total Number of Samples Analyzed[2]	Percentage of Samples with Detections	Total Number of Samples Analyzed[2]
Lindane	OC-I	7	3.4	423	0.07	1,368	1,355	10	5,706	21	7,472
Fenthion	OP-I	4.2	4.8	34	na[4]	—	—	0	3	0	3
Oxadiazon	Other-H	0.7	4.7	75	na	—	—	100	3	18	71
Pentachlorophenol	Fung	14 (pH 5)	4.6	48	na	576	89	14	220	12	417
Dichlone	Fung	0.1	5.6	82	na	40	—	—	0	—	0

[1]Selection criteria: (1) S < 1.0 mg/L or log K_{ow} > 3, and (2) soil half-life > 30 days.
[2]Aggregated from all monitoring studies reviewed (from Tables 2.1 and 2.2).
[3]Range shown is for total endosulfan, endosulfan I, and endosulfan II.
[4]Although not in the top 500 pesticides used in 1988, agricultural use of fenthion in 1976 was 440,000 lb a.i. (Eichers and others, 1978).

insecticides dicofol, endosulfan, lindane, and methoxychlor; the organophosphate insecticides chlorpyrifos, fenthion, and phorate; the pyrethroid insecticides esfenvalerate, fenvalerate, and permethrin, the other insecticide propargite; the dinitroaniline herbicides benfluralin, ethalfluralin, pendimethalin, and trifluralin; the other herbicides bensulide, dacthal, oxadiazon, and triallate; and fungicides and wood preservatives dichlone, PCNB, and pentachlorophenol. If analyzed in bed sediment and aquatic biota, such moderately hydrophobic, moderately persistent pesticides may be found, although at lower detection frequencies than the highly hydrophobic and persistent organochlorine insecticides, such as DDT. However, currently used pesticides that are intermediate in water solubility and persistence may reach fairly high concentrations in sediment or aquatic biota in areas of high or repeated use. The probability of detection would likely increase if study designs considered the location and time of application.

5.5 COMPOSITION OF TOTAL DDT AS AN INDICATOR OF DDT SOURCES AND PERIOD OF USE

The composition of total DDT residues detected in bed sediment or aquatic biota from a given hydrologic system may provide information on the likely source of the residues and the length of time the residues have been in that system. These inferences can be drawn from (1) information on the composition of DDT from various sources and (2) knowledge of the environmental fate and persistence of the various components of technical DDT and their metabolites. However, incorrect inferences can be, and have been, drawn in the past on the basis of a partial understanding of DDT environmental fate and persistence. The discussion below will assess the extent to which current knowledge of DDT in the environment permits inference—on the basis of measured total DDT residues in sediment or aquatic biota—about the length of time the DDT has been in the environment and the probable source of those residues. Three types of inferences will be considered: (1) whether the $(o,p'\text{-DDT} + p,p'\text{-DDT})$/total DDT ratio indicates the length of time the DDT has been in the environment (i.e., historic use versus recent application); (2) whether the $(o,p'\text{-DDD} + p,p'\text{-DDD})$/total DDT ratio indicates that the source of residues was insecticidal use of DDD; and (3) whether the ratio of $(o,p'\text{-DDT} + o,p'\text{-DDE} + o,p'\text{-DDD})$/total DDT indicates either the source of DDT (insecticide use versus discharge or leakage from a chemical manufacturing or storage facility) or the length of time it has been in the environment. These three types of inferences are discussed in Sections 5.5.3–5.5.5, respectively, after background information is presented on sources of environmental residues of DDT (Section 5.5.1) and the environmental fate of DDT (Section 5.5.2).

For convenience, the following conventions are used in this section:
- Total DDT refers to the sum of $o,p'\text{-DDT}$, $p,p'\text{-DDT}$, $o,p'\text{-DDD}$, $p,p'\text{-DDD}$, $o,p'\text{-DDE}$, and $p,p'\text{-DDE}$;
- DDT, or DDT isomers, refers to the sum of $o,p'\text{-DDT}$ and $p,p'\text{-DDT}$;
- DDE, or DDE isomers, refers to the sum of $o,p'\text{-DDE}$ and $p,p'\text{-DDE}$;
- DDD, or DDD isomers, refers to the sum of $o,p'\text{-DDD}$ and $p,p'\text{-DDD}$;
- $o,p'\text{-DDX}$ refers to the sum of $o,p'\text{-DDT}$, $o,p'\text{-DDD}$, and $o,p'\text{-DDE}$; and
- $p,p'\text{-DDX}$ refers to the sum of $p,p'\text{-DDT}$, $p,p'\text{-DDD}$, and $p,p'\text{-DDE}$.

5.5.1 POSSIBLE SOURCES OF DDT RESIDUES IN THE ENVIRONMENT

Residues of total DDT can be derived from insecticidal uses of DDT, DDD, or dicofol. Agricultural and commercial use of DDT was widespread in the United States after 1945. DDT usage in the United States peaked at about 80 million lb a.i. in 1959, then decreased to under 12 million lb a.i. by the early 1970s. Of the total amount used during 1970–1972, about 80 percent was used on cotton; most of the remainder was applied to soybeans and peanuts. DDT also was used on other vegetables, and on greenhouse and nursery plants (U.S. Environmental Protection Agency, 1975). Besides its use in agriculture, DDT was also used in urban and suburban settings against residential pests and for municipal control of disease vectors and nuisance insects (U.S. Environmental Protection Agency, 1975; Mischke and others, 1985). DDD also was used as an insecticide (common name, TDE) in agriculture (Mischke and others, 1985). Dicofol, a miticide marketed under the trade name Kelthane, is manufactured from DDE and may contain up to 0.1 percent of DDT and related compounds as impurities (Mischke and others, 1985).

In addition, total DDT residues in the environment can derive from industrial discharge, spills, or leakage (such as from chemical manufacturing, storage, or disposal sites). The insecticide DDT was applied as technical DDT, which contained two DDT isomers: about 80 percent p,p'-DDT and about 20 percent o,p'-DDT (Worthing and Walker, 1987). The manufacture of technical DDT entailed enrichment in the active ingredient, p,p'-DDT. Manufacturing waste, therefore, tended to have a higher proportion of o,p'-DDT than technical DDT.

5.5.2 ENVIRONMENTAL FATE OF DDT

DDT is recalcitrant in the environment; i.e., resistant to degradation. However, DDT does biodegrade slowly under both aerobic and anaerobic conditions, and it is metabolized by higher organisms. However, its principal degradation products (see Figure 4.3) also tend to be recalcitrant. DDD and DDE are the most common degradation products formed in soil or sediment, as well as the most common metabolites formed in higher organisms. When DDT breaks down, the relative amounts of DDD and DDE formed depend on environmental conditions. DDD is formed via reductive determination of DDT under anaerobic conditions. Anaerobic incubation of DDT in soil containing unidentified microorganisms resulted primarily in DDD, with traces of DDE, DDA, dicofol, p-chlorobenzoic acid, 4,4'-dichlorobenzophenone, and 4,4-dichlorodiphenyl-methane. No metabolites were observed in sterilized soil, indicating that degradation was microbially mediated (Guenzi and Beard, 1967). DDD breaks down over time to DDMU, DDMS, and other products (Agee, 1986). DDE is more stable than DDD. In fish (Figure 4.3), the primary metabolite of DDT is DDE. Reductive dechlorination to DDD occurs in some species, possibly by gut microflora, but DDD is typically a minor component of total DDT residues in fish (Schmitt and others, 1990).

The half-life of DDT has been measured in various soils under various conditions in the laboratory. Laboratory and field-plot estimates of soil half-lives ranged from 2 to 16 years (Lichtenstein and others, 1971; Ware and others, 1971, 1974, 1978; Howard and others, 1991). As will be discussed below, more recent field monitoring studies (e.g., Mischke and others, 1985) indicate that DDT half-lives in soil can be on the order of 15 years or longer.

Although isomer designations are frequently omitted in discussion of metabolites, isomeric structure is conserved during degradation or metabolism. p,p'-DDT breaks down to p,p'-DDE and p,p'-DDD, and o,p'-DDT breaks down to the analogous o,p'-metabolites. Although generally thought to be less stable than p,p'-DDT, o,p'-DDT has been found to have comparable stability to p,p'-DDT in at least some soils (Agee, 1986).

5.5.3 THE DDT/TOTAL DDT RATIO AS AN INDICATOR OF TIME IN THE ENVIRONMENT

Because DDT is expected to degrade to its fairly stable products DDD and DDE in soil, the DDT/total DDT ratio has been used as an indicator of the length of time that the DDT residues have been in the environment (White and Krynitsky, 1986). Soil residues with an unexpectedly high ratio of DDT/total DDT, and with stable or increasing residues of DDE in fish and wildlife subsequent to the 1972 ban, have prompted concern about recent DDT use (Clark and Krynitsky, 1983; Agee, 1986; White and Krynitsky, 1986; King and others, 1992).

The prevalence of total DDT residues in California produce, soil, sediment, and fish over ten years after DDT was banned, led the California Department of Food and Agriculture (CDFA) and the California State Water Resources Control Board (CSWRCB) to investigate possible sources of contamination, including recent illegal use of DDT; use of dicofol or other pesticides containing DDT as a contaminant; leaking waste dumps; and long-lived residues from previous legal applications of DDT. The CSWRCB conducted a case study of sediment and fish from the Salinas River and the Blanco Drain and its adjacent soils (Agee, 1986). The CDFA monitored DDT residues in soil from agricultural areas where DDT has historically been used (Mischke and others, 1985). These monitoring efforts looked at DDT isomers and breakdown products, rather than just total DDT.

Case Study: Salinas River and Blanco Drain

DDT had been found at moderate-to-high levels in the Salinas River and Moss Landing watershed for many years prior to the study. Fish from the Salinas River had unusually high ratios of DDT/total DDT. Fish and sediment samples indicated that one source of this material was the Blanco Drain, which empties into the Salinas River. DDT isomers constituted about 25 percent of total DDT in sediment and fish from the mouth of the Blanco Drain (Agee, 1986).

The CSWRCB analyzed sediment from Blanco Drain and soil from adjacent fields to characterize contamination in the Blanco Drain and to identify possible sources of DDT to the drain. Soils from fields adjoining the west and east legs of the drain contained up to 5,000 and 3,000 µg/kg total DDT, respectively. Except for one site, soils from fields adjoining all parts of the drain contained 66–80 percent DDT isomers (average 72 percent) and 1.4–5.6 percent DDD isomers (average 3.5 percent). Moreover, most soil samples contained 15–20 percent o,p'-DDT as a percentage of the sum of DDT isomers, which is very close to the original formulation of technical DDT. o,p'-DDT, originally thought to be less stable than the p,p' isomer, was found to be more stable in soils adjacent to the drain than had been previously thought (Agee, 1986).

Bottom sediment in the west and east legs of the drain contained 800–6,200 and 200–1,700 µg/kg total DDT, respectively. These levels were 200–400 times higher than those in the Salinas River above the Blanco Drain outfall. The ratio of DDT/total DDT was highly variable; expressed as a percentage, this ratio was 29–80 percent on the east leg and 19–45 percent on the west leg of the drain. Sediment samples from the Blanco Drain fit into one of two categories:

(1) over 60 percent DDT isomers and less than 10 percent DDD, which probably were recently deposited soils; and (2) less than 45 percent DDT isomers and over 20 percent DDD. The former were generally characteristic of soil, and the latter of stream sediment, in the area (Agee, 1986).

The CSWRCB also monitored DDT and metabolite residues in biota from the Blanco Drain. Fish (suckers, *Catostomus* sp.) from the outlet of the drain contained just under 1,000 µg/kg total DDT. Freshwater clams (*Corbicula* sp.) transplanted to the drain accumulated up to 3,800 µg/kg total DDT. The ratio of DDT/total DDT varied from 19–38 percent (Agee, 1986). Degradation of DDT in fish tends to be relatively rapid compared to that in soil or sediment, with half-lives on the order of 1 to 2 months (Grzenda and others, 1970; Agee, 1986).

From these studies, the CSWRCB concluded the following. First, the uniform distribution of total DDT throughout the Blanco Drain area soils (low variability, similarity in composition, absence of hot spots) was consistent with a pattern of residues from past use and not with new illegal usage or leaky depositories of DDT. Second, similarities between total DDT levels and composition in Blanco Drain sediment and in soil from adjacent fields suggested that soil erosion from the fields was the source of total DDT in the drain. Third, both *o,p'* and *p,p'* isomers of DDT have long lifetimes in Salinas clay soils. Salinas area soils comprise a reservoir of DDT that is being released into the aquatic environment through soil erosion. Fourth, fields on the east leg of the drain are plowed over the edge and into the drain. East leg sediment contains the fingerprint of soil-based DDT. This practice may be a major source of DDT to the drain. Fifth, DDT breaks down slowly in drain sediment. When it reaches the Salinas River, it consists of about 20 percent DDT, 35 percent DDD, and 45 percent DDE. Sixth, extremely high accumulation of total DDT by transplanted clams (filter feeders) suggested that much of the total DDT is transported via fine suspended materials (Agee, 1986).

Case Study: California Soil Monitoring Survey

CDFA analyzed 99 soil samples from 32 counties in California (Mischke and others, 1985). Samples were collected from individual fields with a known history of DDT application and from fields in areas known to have widespread and repeated applications of DDT. Total DDT residues were present in soil wherever DDT was used legally in the past; all 99 samples from 32 counties contained total DDT residues. The ratio of DDT/total DDT varied from 0 to 100 percent, but was 49 percent on average. Moreover, the soils contained an average of 19 percent *o,p'*-DDT as a percentage of the sum of DDT isomers, which was close to that of technical DDT (about 20 percent). This confirmed the findings of the Blanco Drain study (Agee, 1986), which showed that *o,p'*-DDT has a comparable lifetime in soil to *p,p'*-DDT. These results indicate that DDT appeared to be more stable in California soils than had been previously thought.

Conclusions

First, high total DDT residues and a high ratio of DDT/total DDT in soil, sediment, or biota samples do not necessarily indicate recent use of DDT. On the basis of laboratory estimates, the lifetime of DDT in soils appears to be longer than had been previously thought as indicated by (1) the presence of total DDT residues in soil wherever DDT was used legally in the past in a statewide California soil survey conducted in 1985; (2) the unexpectedly high ratio of

DDT/total DDT in California soils (mean of 49 percent in the same survey); and (3) the presence, in most soil samples, of 15–20 percent *o,p'*-DDT (as a percentage of technical DDT), which is very close to the original formulation of technical DDT.

Second, a high ratio of DDT/total DDT in sediment or aquatic biota appears to indicate that the residues recently entered the hydrologic system (Aguillar, 1984; Agee, 1986), such as by erosion of DDT-contaminated soil (Johnson and others, 1988). DDT appears to break down more quickly once it enters a river or other waterway. In the Blanco Drain study (Agee, 1986), sediment samples could be categorized as either (1) recently deposited soil with a similar composition to soils adjacent to the drain (typically greater than 60 percent DDT and less than 10 percent DDD) or (2) typical of local stream sediment (containing less than 45 percent DDT and greater than 20 percent DDD). By the time sediment from the drain reached the Salinas River, it had broken down to about 20 percent DDT, 35 percent DDD, and 45 percent DDE.

Therefore, the ratio of DDT/total DDT appears to be an effective tool in identifying of recently mobilized DDT in rivers and other waterways (Agee, 1986). The CSWRCB (Agee, 1986) has used a ratio of DDT/total DDT exceeding 10 percent as one criterion to indicate the *relative freshness* of DDT in environmental samples (discussed further in Section 5.5.5). The ratio of DDT/total DDT was also used to interpret fish data from the NCBP. During 1980–1981, nine sites had samples that exceeded 500 µg/kg *p,p'*-DDT or that contained an unusually high ratio of DDT/total DDT (Schmitt and others, 1985); this was true for only two sites in 1984 (Schmitt and others, 1990), which indicated continued weathering of DDT in the environment and suggested a decline during this time in erosion of DDT-contaminated soils.

5.5.4 THE DDD/TOTAL DDT RATIO AS AN INDICATOR OF DDD USE

An unexpectedly high ratio of DDD/total DDT in aquatic biota may indicate past use of the insecticide DDD (TDE), once marketed under the tradename Rothane, for control of insects on fruit and vegetables. This is not true for residues in sediment, since DDD is a common breakdown product under anaerobic conditions. However, in fish the primary metabolite of DDT is DDE; although reductive dechlorination to DDD occurs in some species, possibly by gut microflora, DDD is typically a minor component of total DDT in fish (Schmitt and others, 1990). The inference that residues derive partly from DDD use may not be appropriate if the fish or shellfish sampled is living in sediment that contains high DDD residues. When available, it is helpful to consider concurrent data on DDT and metabolite residues in the surrounding sediment and historical data on fish residues from that site. For example, three samples from a single NCBP site in Hawaii (in 1984) contained an abnormally high proportion (79–83 percent) of DDD (Schmitt and others, 1990). Previous data from this site contained substantial total DDT, but mostly as *p,p'*-DDE and DDT. Schmitt and others (1990) suggested that there may have been recent direct inputs of the insecticide DDD.

5.5.5 THE *o,p'*-DDX/TOTAL DDT RATIO AS AN INDICATOR OF INDUSTRIAL ORIGIN OR TIME IN THE HYDROLOGIC SYSTEM

As previously mentioned, technical DDT (applied for insecticidal use) contained about 80 percent *p,p'*-DDT and 20 percent *o,p'*-DDT. Deviations from this ratio in fish samples have been used to draw inferences regarding the length of time the DDT has been in the hydrologic system or the origin of the DDT (insecticidal use versus industrial inputs). The relative stability of o,p' and p,p' isomers is a critical factor here. o,p'-DDT was once thought to be less stable in the environment than p,p'-DDT (Anderson and others, 1982), although field studies have shown that o,p'-DDT appears to be as stable as p,p'-DDT in at least some soils (Lichtenstein and others, 1971; Mischke and others, 1985; Agee, 1986). Both isomers of DDT break down more rapidly in fish than in soil or sediment (Agee, 1986).

The CSWRCB devised several criteria to indicate the relative freshness of DDT in the environment: (1) p,p'-DDT constitutes at least 15 percent of p,p'- DDX; (2) DDT constitutes at least 10 percent of total DDT; (3) o,p'-DDT constitutes at least 8 percent of DDT; and (4) total DDT in fish exceeds 1,000 µg/kg wet weight (Agee, 1986). The second and third criteria have proved to be the most useful of the four (Agee, 1986). Using the third criterion, fish with o,p'-DDT greater than 8 percent of DDT would suggest recent mobilization of DDT into the hydrologic system.

An unusually high ratio o,p'-DDT/total DDT in fish samples (greater than 20–30 percent) has been used to indicate industrial (noninsecticidal) sources of DDT. Waste products from the manufacture of technical DDT were enriched in o,p'-DDT relative to p,p'-DDT, which was the active ingredient. In the NCBP, a few sites had fish samples with higher than expected proportions of o,p' homologs (greater than 20 percent of total DDT), which suggested a source other than insecticidally applied DDT, such as a chemical manufacturing or storage facility (Schmitt and others, 1990). There was a known industrial source nearby for several of these sites: the Arkansas River at Pine Bluff, Arkansas (near the Pine Bluff Arsenal); the Tombigbee River, Alabama (near a pesticide manufacturing facility); and the Tennessee River, Tennessee (downstream of the former Red Stone Arsenal, now the Wheeler National Wildlife Refuge).

Because o,p'-DDT can be used as an indicator of two very different events (recent mobilization into the hydrologic system or industrial origin), some caution is warranted in drawing any inferences on the basis of high o,p'-DDT levels in fish samples. However, note that the CSWRCB uses an 8 percent ratio of o,p'-DDT/DDT as an indicator of relative freshness of DDT residues in the environment. Assuming that the ratios of o,p'-DDD/DDD and o,p'-DDE/DDE are similar, this would correspond to a 8 percent ratio of o,p'-DDX/total DDT. Given that technical DDT contains about 20 percent of o,p'-DDT/DDT, a ratio of o,p'-DDX/total DDT below 8 percent implies some preferential degradation of o,p' isomers over p,p' isomers. In contrast, the NCBP looked for a much higher ratio of o,p' isomers/total DDT (greater than 20–30 percent) in fish samples to indicate industrial origin. If a ratio of 20–30 percent is exceeded, this means that the fish sample is enriched in o,p' isomers relative to technical DDT (15–20 percent); especially where known industrial sources are nearby, it seems reasonable to postulate industrial origin in such cases. In cases where the measured ratios of o,p'-DDT/DDT or o,p'-DDX/total DDT are above 8 percent but less than 20 percent, there is some ambiguity. It may be worthwhile to check whether such a case meets other criteria, such as whether the DDT/total DDT ratio exceeds 10 percent, that tend to indicate recent mobilization of DDT-contaminated soils into a hydrologic system.

CHAPTER 6

Analysis Of Key Topics—Environmental Significance

Because hydrophobic pesticides tend to associate with particulates and organic matter, they may be found in bed sediment and aquatic biota even when they are not detectable in water samples from the same hydrologic system. Thus, their detection in bed sediment and aquatic biota serves as an indicator that these compounds are present as contaminants in the hydrologic system. Clearly, the monitoring studies reviewed (Tables 2.1 and 2.2) indicate that many pesticides are frequently found in these media in rivers and streams of the United States. Yet, what is the significance of these residues to biota that live in or are dependent on the hydrologic systems? As noted in Chapter 4, organisms in the water column and in sediment may be exposed to pesticides in the dissolved form, or in association with colloids or particulates, or by ingestion of contaminated food or particulates. Pesticide-contaminated fish may be consumed by both wildlife and humans, thus reintroducing the pesticide into the terrestrial environment. These exposures thus result in the potential for adverse effects on ecosystems and human health. The potential effects of hydrophobic pesticides on aquatic organisms and fish-eating wildlife, and on human health, are discussed in Sections 6.1 and 6.2, respectively.

6.1 EFFECTS OF PESTICIDE CONTAMINANTS ON AQUATIC ORGANISMS AND FISH-EATING WILDLIFE

A review of the toxicity of organochlorine and other hydrophobic contaminants to fish, other aquatic organisms, and wildlife is beyond the scope of this book. The literature in this area is voluminous (e.g., U.S. Environmental Protection Agency, 1975; Eisler, 1985, 1990; Eisler and Jacknow, 1985; Murty, 1986a,b). For example, DDT has been shown to adversely affect a variety of organisms, from phytoplankton to fish-eating birds and mammals. In a report supporting its decision to ban DDT, U.S. Environmental Protection Agency (USEPA) reviewed the adverse effects of DDT on fish and wildlife known at that time (U.S. Environmental Protection Agency, 1975). This review documented effects on phytoplankton (reduced photosynthesis and growth rates), aquatic invertebrates (acute toxicity at microgram per liter levels, and reproductive failure and other sublethal effects at nanogram per liter levels), and fish (acute toxicity at microgram per liter levels, reproductive impairment, and sublethal effects including abnormal utilization of amino acids, inhibition of thyroid activity, and interference with temperature regime selection).

Exposure of fish to DDT may compound stress due to thermal pollution by affecting temperature selection and increasing oxygen consumption. Secondary effects of DDT on higher trophic levels (such as starvation following acute kills of prey organisms) have been observed. The disruptive effects of DDT on birds (mortality, eggshell thinning, abnormal courtship and reproductive behavior, and reproductive failure) have been studied extensively. U.S. Environmental Protection Agency (1975) concludes that a clear-cut time relation existed between eggshell production, concomitant reproductive failures, and DDT use in the United States. The report also concludes that DDT had contributed to the reproductive impairment of a number of fish species in natural waters. Fish-eating mammals also were shown to accumulate high levels of DDT and metabolites from their diets. More recent studies have shown that DDT and metabolites can have estrogenic effects on animals (e.g., Denison and others, 1981; Fry and Toone, 1981; Fry and others, 1987; Bustos and others, 1988; Guillette and others, 1994).

Most other organochlorine compounds also are very toxic to aquatic organisms. Mayer and Ellersieck (1986) compiled acute toxicity data for xenobiotics from tests performed at the Columbia National Fisheries Laboratory (then part of the U.S. Fish and Wildlife Service [FWS]) and its field laboratories between 1965–1984. These tests covered 410 chemicals and 66 species of aquatic organisms under a variety of test conditions. Acute toxicity data from this report (Mayer and Ellersieck, 1986) for all pesticides and transformation products detected in aquatic biota are summarized in Table 6.1. The range of acute toxicity values (48-hour EC_{50} or 96-hour LC_{50}) are presented in Table 6.1 for the principal species tested (daphnids, rainbow trout, fathead minnow, and bluegill), as well as for stoneflies and other species that were sensitive to specific chemicals. In Table 6.1, no attempt was made to describe test conditions (such as static versus flow-through, pH, hardness, and temperature) or other test characteristics (such as life stage of the organism and pesticide formulation) that may affect toxicity. Chlordane, DDD, dieldrin, endrin, heptachlor, methoxychlor, and toxaphene were acutely toxic to aquatic organisms, including daphnids and rainbow trout, at the low microgram per liter level (Table 6.1). Endosulfan also was toxic to amphipod crustaceans (*Gammarus* species) and several species of fish (including rainbow trout and channel catfish) in the microgram per liter range (Mayer and Ellersieck, 1986). Although aquatic organisms are fairly resistant to mirex in short-term toxicity tests (Eisler, 1985; Mayer and Ellersieck, 1986), delayed toxicity to mirex has been observed for fish and crustaceans (Eisler, 1985). Many organochlorine pesticides besides DDT can cause impaired reproduction and other sublethal effects at low concentrations. Examples include toxaphene (Eisler and Jacknow, 1985), mirex (Eisler, 1985), chlordane, and dieldrin (Johnson, 1973; Thomas and Colborn, 1992).

The toxicity of pesticide transformation products to aquatic organisms was reviewed by Day (1991). Transformation products may have lesser, greater, or comparable toxicity relative to the parent compound, depending on the pesticide, type of organisms tested, and even test conditions. The organochlorine insecticides tend to be persistent in the environment, but they do degrade (albeit slowly) under natural conditions to transformation products that are more stable than the parent compound (Sections 4.2.3 and 4.3.3). Some of these transformation products appear to be less toxic than the parent compound (e.g., DDD versus DDT), whereas others appear to have comparable or even greater toxicity (e.g., dieldrin versus aldrin), as shown in Table 6.1. Some metabolites of organophosphate, carbamate, and pyrethroid insecticides may be more toxic than the parent compound to some organisms. As discussed in Section 4.3.3, when a parent compound is biotransformed to a more toxic metabolite, the process is referred to as "metabolic activation." The oxidation of the organophosphate insecticide parathion to paraoxon is a classic example of metabolic activation. For many herbicides, transformation products tend to be less toxic than the parent compound (Day, 1991).

A number of guidelines have been developed that indicate threshold concentrations above which pesticide levels may adversely affect aquatic organisms or wildlife. In the discussion below, these guidelines will be used to assess the potential for biological effects in United States rivers on the basis of the maximum pesticide concentrations reported by the monitoring studies reviewed in this book. The fraction of studies in which these guidelines are exceeded provides some measure of the potential for biological effects in study areas where pesticide residues were detected. As used in this book, the term "guidelines" refers to threshold values that have no regulatory status but are issued in an advisory capacity. The issuing agency may use a different term to describe a given set of guidelines (such as criteria, advisories, guidance, or recommendations). For the most part, the guidelines used in this book were established on the basis of acute and chronic toxicity to aquatic organisms and wildlife. Except for some sediment guidelines (which also include field-based measures of biological effects), most of the guidelines used were based on the results of single-species toxicity tests conducted in the laboratory. Some additional toxicity-related issues (effects of chemical mixtures and potential endocrine-disrupting effects of pesticides) are briefly discussed in Section 6.1.4.

Exposure of an aquatic organism to a pesticide in a hydrologic system may occur via physical contact with the pesticide in the water column, bed sediment, or sediment pore water, and by ingestion of contaminated water, food, or particulates. In the following discussion, potential effects are addressed separately for organisms exposed via the water column (Section 6.1.1), benthic organisms exposed via sediment (Section 6.1.2), and wildlife that consume contaminated aquatic biota (Section 6.1.3).

6.1.1 TOXICITY TO ORGANISMS IN THE WATER COLUMN

USEPA water-quality criteria for the protection of aquatic organisms were designed to protect aquatic life from adverse effects of toxic pollutants in hydrologic systems (Nowell and Resek, 1994). These criteria are expressed on a water concentration basis. However, they can be used in conjunction with fish bioconcentration factors (BCF) to estimate pesticide concentrations in fish tissue that may be associated with potential biological effects. Additional information on potential biological effects of pesticide residues in fish can be gained by looking at fish kills and the incidence of fish diseases in association with chemical contaminant residues. Selected studies relating to these topics are discussed below.

USEPA's Water-Quality Criteria for Protection of Aquatic Organisms

Ambient water-quality criteria were issued by USEPA for *priority pollutants* (pollutants designated as toxic under the Clean Water Act), in accordance with USEPA's mandate under Section 304(a) of that Act. This list of priority pollutants includes a number of pesticides, including most of the organochlorine insecticides (Code of Federal Regulations, Volume 40, Part 423, Appendix A). Water-quality criteria for the protection of aquatic organisms (also called "aquatic life criteria") for most of the priority pollutant pesticides were issued in 1980 (U.S. Environmental Protection Agency, 1980a–h). However, aquatic life criteria for a few pesticides

Table 6.1. Acute aquatic toxicity of pesticides detected in aquatic biota

[Pesticides listed are those that were detected in aquatic biota by one or more monitoring studies listed in Tables 2.1 and 2.2. The range of acute toxicity values (48-h EC_{50} or 96-h LC_{50}), in micrograms per liter, are listed. Total number of toxicity tests: the total number of toxicity tests, for all test species, listed in Mayer and Ellersieck (1986). Number of tests: the number of toxicity tests listed for a given species. Other most sensitive species: If the most sensitive test species is some other species than daphnids, stoneflies, rainbow trout, fathead minnow, or bluegill, then this species is specified and the number of tests and range in toxicity values are listed. Blank cell indicates that no other species tested were more sensitive than the species listed in previous columns of the table (scientific names included only if common name is general). Abbreviations and symbols: EC_{50}, median effective concentration (usually refers to immobilization); h, hour; LC_{50}, median lethal concentration; PCA, pentachloroanisole; PCNB, pentachloronitrobenzene; µg/L, microgram per liter; —, no data. Compiled from Mayer and Ellersieck (1986)]

Pesticides Detected in Aquatic Biota (from Tables 2.1, 2.2)	Total Number of Toxicity Tests	Daphnids[1]		Stoneflies[2]		Rainbow Trout	
		Number of Tests	48-Hour EC_{50} (µg/L)	Number of Tests	96-Hour LC_{50} (µg/L)	Number of Tests	96-Hour LC_{50} (µg/L)
Aldrin	26	3	23–32	1	1.3	6	2.6–14.3
Carbaryl	136	5	5.6–11	1	5.6	20	<320–3,500
Carbofuran	15	0	—	0	—	2	380–600
Chlordane	56	3	20–29	1	15	23	2.9–59
Chlorpyrifos	23	0	—	2	0.57–10	4	<1–51
D, 2,4-	21	0	—	0	—	1	110,000
Dacthal (DCPA)	9	4	27,000–>10^5	0	—	0	—
DDD	21	5	3.2–9.1	1	380	2	70
DDE	3	0	—	0	—	1	32
DDT	85	4	0.36–4.7	4	1.2–7	7	4.1–11.4
Diazinon	10	3	0.8–1.8	1	25	1	90
Dicofol	8	0	—	1	650	1	(^{3})
Dieldrin	37	3	190–250	3	0.5–0.58	6	1.2–2.3
Endosulfan	10	0	—	1	2.3	4	1.1–2.9
Endrin	53	6	4.2–74	3	0.076–0.54	6	0.74–2.4
Fenvalerate	29	1	2.1	0	—	11	0.32–1.7
Heptachlor	31	3	42–80	3	0.9–2.8	6	7.0–43
Heptachlor Epoxide	2	0	—	0	—	1	20
Hexachloro-benzene	12	0	—	0	—	0	—
Kepone	14	1	260	0	—	2	29–30
Lindane	45	3	460–880	2	1.0–4.5	5	18–41
Methoxychlor	146	3	0.78–5.6	3	1.4–25	10	11–62
Methyl parathion	31	3	0.14–0.37	0	—	2	2,750–3,700
Mirex	13	3	>100–>1,000	0	—	2	>10^5
Nitrofen	1	0	—	0	—	0	—

Table 6.1. Acute aquatic toxicity of pesticides detected in aquatic biota—*Continued*

Pesticides Detected in Aquatic Biota (from Tables 2.1, 2.2)	Fathead Minnow		Bluegill		Other Most Sensitive Species		
	Number of Tests	96-Hour LC_{50} (μg/L)	Number of Tests	96-Hour LC_{50} (μg/L)	Species	Number of Tests	96-Hour LC_{50} (μg/L)
Aldrin	1	8.2	6	5.6–12			
Carbaryl	3	7,700–14,600	14	1,800–39,000	*Isogenus* sp.	6	2.8–12
Carbofuran	3	872–1,990	2	88–240			
Chlordane	2	56–115	3	57–128	Channel catfish	5	0.8–230
Chlorpyrifos	0	—	5	1.7–4.2	*Gammarus lacustris*	1	0.11
D, 2,4-	1	133,000	1	180,000	Cutthroat trout	12	37,000–172,000
Dacthal (DCPA)	0	—	1	>100	*G. pseudolimnaeus*	2	26.2–>100
DDD	1	4,400	1	42	*G. fasciatus*	2	0.6–0.9
DDE	0	—	1	240			
DDT	3	9.9–13.2	7	1.6–8.6	*Orconectes nais*	7	0.18–100
Diazinon	0	—	1	168	*G. fasciatus*	1	0.2
Dicofol	0	—	1	520	Cutthroat trout	2	53–158
Dieldrin	1	3.8	8	3.1–18			
Endosulfan	1	1.5	1	1.2			
Endrin	2	0.24–1.8	7	0.19–0.73			
Fenvalerate	2	2.15–2.35	11	0.42–1.35	*G. pseudolimnaeus*	1	0.032
Heptachlor	1	23	3	13–17	*O. nais*	1	0.5
Heptachlor Epoxide	0	—	1	5.3			
Hexachloro-benzene	1	>10,000	2	>1,000–12,000	Channel catfish	3	7,000–>10^6
Kepone	2	340–420	1	72			
Lindane	4	67–87	6	25–68			
Methoxychlor	1	31	5	25–79			
Methyl parathion	3	7,200–9,960	3	1,000–4,380			
Mirex	1	>10^5	2	>10^5			
Nitrofen	0	—	0	—	*G. fasciatus*	1	3,100

Table 6.1. Acute aquatic toxicity of pesticides detected in aquatic biota—*Continued*

Pesticides Detected in Aquatic Biota (from Tables 2.1, 2.2)	Total Number of Toxicity Tests	Daphnids[1]		Stoneflies[2]		Rainbow Trout	
		Number of Tests	48-Hour EC_{50} ($\mu g/L$)	Number of Tests	96-Hour LC_{50} ($\mu g/L$)	Number of Tests	96-Hour LC_{50} ($\mu g/L$)
Oxadiazon	0	—	—	—	—	—	—
Oxychlordane	0	—	—	—	—	—	—
Parathion	47	4	0.37–0.6	3	1.5–5.4	4	780–1,430
PCA	0	—	—	—	—	—	—
PCNB	0	—	—	—	—	—	—
Pentachloro-phenol	15	2	240–410	0	—	4	34–121
Permethrin	29	1	1.26	0	—	10	2.9–8.2
Photomirex	0	—	—	—	—	—	—
Strobane	0	—	—	—	—	—	—
Tetradifon	5	0	—	0	—	2	1,200–1,350
Toxaphene	89	4	10–19	3	1.3–2.3	7	1.8–12
Trifluralin	62	3	560–900	1	2,800	25	22–1,600
Xytron	0	—	—	—	—	—	—

[1]*Daphnia magna, D. pulex,* or *Simocephalus serrulatus.*
[2]*Claasenia sabulosa, Pteronarcys californica,* or *Pteronarcella badia.*
[3]Not measured.

(toxaphene, chlorpyrifos, pentachlorophenol, and parathion) were issued or revised in 1986 (U.S. Environmental Protection Agency, 1986b–e). USEPA aquatic-life criteria are national numerical criteria designed to prevent unacceptable long-term and short-term effects on aquatic organisms in rivers, streams, lakes, reservoirs, oceans, and estuaries. Separate criteria were determined for freshwater and saltwater organisms and for acute (short-term) and chronic (long-term) exposures. Criteria were established on the basis of toxicity tests with at least one species of aquatic animal in at least eight families (Stephan and others, 1985; U.S. Environmental Protection Agency, 1986f). Criteria values were established to protect 95 percent of the genera tested (Stephan and others, 1985), which suggests that effects on fewer than 5 percent of organisms would not be unacceptable. Also, it is assumed that an aquatic ecosystem can recover as long as the average concentration over a prescribed period (1 hour for acute criteria and 4 days for chronic criteria) does not exceed the applicable criterion more than once every 3 years on average. USEPA water-quality criteria for pesticides are expressed as pesticide concentrations in whole water, so they cannot be directly compared with pesticide residues in sediment or aquatic biota (Nowell and Resek, 1994).

In their review of pesticides in surface waters of the United States, Larson and others (1997) compared ambient pesticide concentrations in surface waters (from national, multistate, state, and local monitoring studies) with USEPA water-quality criteria. Every one of the organochlorine insecticides targeted in the surface water studies reviewed by Larson and others

Table 6.1. Acute aquatic toxicity of pesticides detected in aquatic biota—*Continued*

Pesticides Detected in Aquatic Biota (from Tables 2.1, 2.2)	Fathead Minnow		Bluegill		Other Most Sensitive Species		
	Number of Tests	96-Hour LC_{50} (µg/L)	Number of Tests	96-Hour LC_{50} (µg/L)	Species	Number of Tests	96-Hour LC_{50} (µg/L)
Oxadiazon	—	—	—	—			
Oxychlordane	—	—	—	—			
Parathion	2	2,090–2,350	7	18–400	*O. nais*	2	0.04–15
PCA	—	—	—	—			
PCNB	—	—	—	—			
Pentachloro-phenol	1	205	2	32–215	Chinook salmon	3	31–68
Permethrin	2	5.7	10	4.5–13	*G. pseudolimnaeus*	1	0.17
Photomirex	—	—	—	—			
Strobane	—	—	—	—			
Tetradifon	0	—	1	880	*G. fasciatus*	1	111
Toxaphene	10	5.6–23	13	2.4–18	Channel catfish	26	0.82–13.1
Trifluralin	3	105–160	17	8.4–400			
Xytron	—	—	—	—			

(1997) exceeded the applicable aquatic-life criterion (if one had been established) in one or more surface water studies. The number of studies in which criteria were exceeded was probably an underestimate, because detection limits were frequently higher than the USEPA chronic criteria (Larson and others, 1997).

However, many contaminant monitoring studies do not analyze for organochlorine insecticides in water, but instead rely on bed sediment and aquatic biota sampling to determine whether these hydrophobic contaminants are present in the hydrologic system. Both national trends (Section 3.4.1) and local trends (Larson and others, 1997) in organochlorine contamination are more easily seen in fish or sediment than water. Also, because analytical detection limits in water are commonly at or above chronic water-quality criteria for the organochlorine pesticides (Larson and others, 1997), detection of organochlorine compounds in surface waters at biologically significant levels may require special monitoring techniques such as large-volume samplers. For organochlorine compounds, the use of water-column concentrations to assess their aquatic toxicity is complicated by the fact that a significant fraction of these hydrophobic compounds in whole water may be associated with dissolved organic carbon or suspended particles in the water, thereby reducing their bioavailability to organisms in the water column (Landrum and others, 1985; Knezovich and others, 1987). Existing USEPA water-quality criteria for pesticides in surface water were derived on the basis of total concentrations of contaminant in water, rather than on the dissolved or bioavailable fractions; at the time the criteria were developed, there was no attempt to make such a distinction (Nowell and Resek, 1994).

Detection of organochlorine insecticides for aquatic biota or bed sediment in a hydrologic system indicates the presence of these contaminants in that system. However, it is difficult to assess the biological significance of this contamination to fish and other aquatic organisms in the system. No guidelines exist that relate contaminant concentrations in fish tissues, for example, to toxicity to the fish. In theory, chronic aquatic-life criteria (which indicate a threshold concentration in water above which chronic toxicity may occur) can be used to estimate contaminant concentrations in fish tissues that may be associated with biological effects. This extrapolation would require that equilibrium between the pesticide concentrations in fish and in ambient water be assumed. At equilibrium, the fish tissue concentration (FTC) corresponding to the chronic aquatic-life criterion would be calculated as follows:

$$FTC = WQC_{chronic} \times (BCF) \qquad (6.1)$$

where the FTC is in units of μg/kg fish (wet weight), $WQC_{chronic}$ is the USEPA chronic water-quality criterion for the protection of aquatic organisms (in μg/L), and BCF is the fish bioconcentration factor (in L/kg fish).

Pesticide concentrations in fish and the surrounding water are unlikely to be in thermodynamic equilibrium in the environment, given factors such as the mobility, metabolism, and growth of fish and the dynamic nature of water flow, pesticide input, and pesticide degradation. Maintenance of steady-state concentrations (discussed in Section 5.2.1) would be a more realistic simplifying assumption. Despite the crudeness of the FTC approach, it gives some indication of which pesticide concentrations in fish tissue may be associated with potential biological effects.

There are two potential problems with the FTC approach described in equation 6.1 that need to be addressed. First, this extrapolation is very sensitive to the BCF values used. BCF values can be calculated or determined experimentally, and they frequently vary depending on the species and test conditions (Howard, 1991; U.S. Environmental Protection Agency, 1992b). Also, laboratory-measured BCF values for organochlorine compounds are frequently lower than field-measured bioaccumulation factors (BAF) (Section 5.2.4, see subsection on Bioaccumulation Factors). The higher the BCF value used, the higher the corresponding FTC, and the less likely that the FTC will be exceeded by measured concentrations in fish.

Second, the chronic aquatic-life criterion must be a threshold concentration for chronic toxicity to organisms in the water column. Although this sounds self-evident, this is not necessarily the case for all priority pollutants. For some priority pollutants (including DDT and chlordane), the chronic aquatic-life criterion is actually lower than the threshold for chronic toxicity. This happens because the chronic aquatic-life criterion is designed to protect aquatic organisms and their uses. Specifically, the USEPA guidelines for deriving water-quality criteria for the protection of aquatic organisms (Stephan and others, 1985) dictate that the chronic aquatic-life criterion be the lowest of three values: the final chronic value (which measures chronic toxicity to aquatic animals), the final plant value (which measures toxicity to aquatic plants), and the final residue value (which protects fish-eating wildlife and the marketability of fish). The rationale for this is that, if ambient concentrations do not exceed the lowest of these three values, then all of the corresponding uses will be protected. For several organochlorine insecticides (including DDT, chlordane, and dieldrin), the final residue value (generally the Food and Drug Administration [FDA] action level; see Section 6.2.3) was the lowest of the three values, so it was selected as the chronic aquatic-life criterion. For these organochlorine insecticides, the final chronic value (rather than the chronic aquatic life criterion itself) should be used to estimate the FTC.

In Table 6.2, FTC values were extrapolated from final chronic values (μg/L) for several pesticides: chlordane, chlorpyrifos, dieldrin, endosulfan, endrin, hexachlorobenzene, lindane, and toxaphene. With one exception, all final chronic values in Table 6.2 are from water-quality criteria documents (U.S. Environmental Protection Agency, 1980b,c,e,f,h; 1986b,e). The exception is that of dieldrin, which is from the sediment quality criteria document (U.S. Environmental Protection Agency, 1993a), in which acute and chronic toxicity data were updated from the original water-quality criteria document for aldrin and dieldrin (U.S. Environmental Protection Agency, 1980a), and the final chronic value revised. For consistency, BCF values (normalized to 3 percent lipids, which is the weighted average for consumed fish) were taken from the same references (U.S. Environmental Protection Agency, 1980a,b,c,e,f,h,i) when available. Although these BCF values are not the most recent estimates, they were geometric means of values available at the time and they form a consistent data set. The BCF value for chlorpyrifos was taken from U.S. Environmental Protection Agency (1992b). Several pesticides that have been detected in bed sediment and aquatic biota are missing from Table 6.2 because data were insufficient to compute FTC values: DDT, heptachlor, heptachlor epoxide, pentachlorophenol, and parathion. For DDT, heptachlor, and heptachlor epoxide, there were insufficient aquatic toxicity data to determine final chronic values at the time the water-quality criteria were developed (U.S. Environmental Protection Agency, 1980d,g). For pentachloro-phenol, the freshwater final chronic value in the water-quality criteria document is pH dependent (U.S. Environmental Protection Agency, 1986d). For parathion, no BCF value was available.

Pesticides in Whole Fish—Analysis of Potential Fish Toxicity

The FTC values described in the preceding section were compared with maximum concentrations of pesticides in whole fish that were reported by individual monitoring studies (Tables 2.1 and 2.2). Even when only recently published (1984–1994) monitoring studies were considered, the FTCs of several of these pesticides were exceeded by the maximum concentrations in whole fish reported by one or more studies (Table 6.2): 11 percent of studies for dieldrin (7 of the 64 studies measuring dieldrin in whole fish), 25 percent of studies for total endosulfan (1 of 4 studies), 14 percent of studies for endrin (2 of 14 studies), 17 percent of studies for lindane (6 of 35 studies), and 100 percent of studies for toxaphene (15 of 15 studies). No studies reported maximum concentrations that exceeded the extrapolated FTCs for total chlordane (out of 18 studies that measured total chlordane in whole fish), chlorpyrifos (out of 2 studies), or hexachlorobenzene (out of 44 studies). For studies in which the FTC was exceeded, this indicates potential for toxicity to fish at the most contaminated site in each study. Because only maximum concentrations have been compared with FTCs, Table 6.2 does not indicate what fraction of sites or samples exceeded the FTC in each study.

Table 6.2. Potential chronic toxicity to aquatic biota in studies that monitored pesticides in whole fish

[Development of FTC guidelines: The USEPA final chronic value (μg/L) and BCF at 3 percent lipid (L/kg fish) are multiplied to give the fish tissue concentration (μg/kg fish). Monitoring studies that reported concentrations in whole fish: Data are based on monitoring studies in Tables 2.1 and 2.2 that (1) were published during 1984–1994 and (2) reported concentrations in whole fish. The total number of studies, the range in maximum concentrations reported in these studies, the number of studies for which the maximum concentration exceeded the applicable FTC, and the percentage of studies for which the maximum concentration exceeded the applicable FTC. Abbreviations: BCF, bioconcentration factor; C_{max}, the maximum concentration reported in a study; FTC, fish tissue concentration; nd, not detected; kg, kilogram; L, liter; μg, microgram]

Pesticide	Development of FTC guidelines			Recently Published (1984–1994) Monitoring Studies—Whole Fish			
	Final Chronic Value (μg/L)	BCF at 3 Percent Lipid (L/kg Fish)	FTC (μg/kg Fish)	Total Number of Studies	Range in C_{max} (μg/kg)	Number of Studies with C_{max} that Exceeded FTC	Percentage of Studies with C_{max} that Exceeded FTC
Chlordane	0.17	14,100	2,400	18	nd–870	0	0
Chlorpyrifos	0.041	470	19	2	nd	0	0
Dieldrin	0.0625	4,670	292	64	nd–5,680	7	11
Endosulfan	0.056	270	15	4	nd–170	1	25
Endrin	0.061	3,970	242	14	nd–2,060	2	14
Hexachlorobenzene	3.68	8,690	32,000	45	nd–27,000	0	0
Lindane	0.08	130	10	35	nd–120	6	17
Toxaphene	0.013	13,100	170	15	nd–280,330	15	100

Fish Kills Attributed to Pesticides

A number of fish kills that occurred during the 1950s and 1960s were attributed to organochlorine insecticides (Madhun and Freed, 1990). For example, it was estimated that 10 to 15 million fish were killed during 1960–1963 in the Mississippi and Atchafalaya rivers and associated bayous in Louisiana. The organochlorine insecticide endrin was singled out as a major cause of the mortality (Mount and Pudnicki, 1966). Forest spraying with DDT to control insects such as spruce budworm and black-headed budworm caused fish kills in the Yellowstone River system (Cope, 1961), the Miramachi River in New Brunswick, Canada (Kerwill and Edwards, 1967), and the forests of British Columbia (Crouter and Vernon, 1959) and Maine (Warner and Fenderson, 1962). Thousands of fish were killed following DDT treatment of a tidal ditch in Florida (Crocker and Wilson, 1965). Over 400,000 fish were killed in a total of 48 fish kills that occurred in California during 1965–1969 (Hunt and Linn, 1970). Most of the kills were attributed to organochlorine insecticides, although organophosphate insecticides and pentachlorophenol were implicated in a few of the fish kills. In addition, the accidental discharge or leakage of pesticides has resulted in a number of severe fish kills (Madhun and Freed, 1990). Examples include the flushing of endrin and the fungicide nabam from a potato sprayer into the Mill River on Prince Edward's Island, Canada, in 1962 (Saunders, 1969) and the accidental discharge of pesticides into the Rhine River in 1987 (Capel and others, 1988).

A National Oceanic and Atmospheric Administration (NOAA) report evaluated over 3,600 reported fish kills that occurred between 1980 and 1989 in rivers, streams, and estuaries in 22 states (Lowe and others, 1991). Of these fish kills, 41 percent were attributed to low dissolved oxygen levels, 10 percent were of unknown cause, and 4 percent were attributed to pesticides. About 2.2 million fish were killed in the pesticide-driven events, which constituted only 0.5 percent of the total fish killed in all 3,600 events. In contrast, a report on fish kills in coastal waters of South Carolina (including estuaries and tidally influenced rivers, lagoons, and harbors) identified pesticides as the cause of 19 percent of the 259 total kills that occurred from 1978 to 1988 (Trim and Marcus, 1990). A seasonal pattern was noted, with fish kills attributed to anthropogenic causes occurring mostly in early to midspring and in early to midautumn. Of the fish kills attributed to pesticides, 18 were reported to be from agricultural use, 19 from herbicides used to control aquatic and terrestrial weeds, and 12 from insecticides used for insect vector control. As pointed out by Larson and others (1997), Trim and Marcus (1990) did not report how they were able to attribute fish kill events to pesticides used in specific applications.

High chemical pollutant loads were identified as a possible factor that has contributed to the decline in the striped bass (*Morone saxatilis*) population in the Sacramento–San Joaquin Delta (California) since the mid–1970s (Cashman and others, 1992). Other factors that may have contributed to the decline include reduced adult stock (thus producing fewer eggs), reduced food production in the upper delta, and loss of larval fish into water diversion projects. A seasonal die-off of adult striped bass was noted during the summer, which corresponds in time to the discharge of several herbicides used in the cultivation of rice. Dying striped bass were found to have unusually high concentrations of various organic contaminants in their livers compared to apparently healthy fish from the delta or from the Pacific Ocean. Chemical residues that were detected included pollutants from industrial (such as aliphatic hydrocarbons and esters), agricultural (such as rice herbicides), and urban (such as dialkyl phthalates and petroleum-based compounds) sources. The variability in liver contaminant concentrations was high, but the number and quantities of contaminants found suggest that chemical contamination may be partially responsible (perhaps acting as multiple stressors) for the summer die-off and decline in the striped bass population (Cashman and others, 1992).

Fish Diseases Associated with Chemical Residues

The NOAA's National Status and Trends (NS&T) Program (Benthic Surveillance Project) monitors the association of chemical contaminants in benthic fish, from coastal and estuarine areas, with fish disease (Section 3.1.2). The highest incidence of fish pathology (fin erosion, liver and kidney lesions) was found to occur at the most contaminated sites (Varanasi and others, 1988, 1989), although the types of disease observed varied among sites (McCain and others, 1988). Toxicopathic liver lesions occurred in primarily urban areas (Myers and others, 1993, 1994). An association between heavy chemical contamination, incidence of fish disease, and urban land use was observed in all three regions (Northeast, Southeast, and Pacific) of the United States (McCain and others, 1988; Hanson and others, 1989; Johnson and others, 1992a; Myers and others, 1993, 1994).

DDTs, polychlorinated biphenyls (PCB), and aromatic hydrocarbons were found to be risk factors for five types of hepatic (liver) lesions in three species of fish, whereas chlordane and dieldrin were risk factors for two types of hepatic lesions in one or two species of fish (Myers and others, 1993, 1994). Three types of renal (kidney) lesions were irregularly associated with exposure to DDTs, chlordane, dieldrin, PCBs, aromatic hydrocarbons, and some metals (Myers and others, 1993). These contaminants tended to be covariant in sediment and fish tissues, so that benthic fish were exposed to a mixture of contaminants. Therefore, it was not possible to quantify the relative contributions of individual risk factors (Myers and others, 1994). Hepatic lesions also were associated with several biological risk factors: age, sex, and season (Johnson and others, 1992b, 1993). Nonetheless, these results together suggest that organic contaminant residues in fish may contribute to the incidence of disease in fish from the coastal and estuarine United States.

In addition, a few field studies have found reproductive impairment associated with high concentrations of chemical contaminants (Slooff and DeZwart, 1983; Stott and others, 1983; Johnson and others, 1992b). Life cycle tests with chemical stressors have shown that reproduction can be impaired at concentrations that do not affect embryonic development, hatching, or growth (Folmar, 1993). Reproductive hormones and vitellogenin may be suppressed in fish exposed to xenobiotic chemicals in the field or laboratory (Folmar, 1993).

6.1.2 TOXICITY TO BENTHIC ORGANISMS

To evaluate potential adverse effects of sediment contaminants on benthic organisms in the monitoring studies reviewed (Tables 2.1 and 2.2), the maximum concentrations of individual pesticides in sediment that were reported in these studies were compared with an array of available sediment guidelines. These guidelines consist of reference values above which sediment contaminant concentrations have some potential for adverse effects on benthic organisms. Sediment guidelines have been determined using a number of approaches, which are discussed in more detail below. The following sediment guidelines were used in this analysis: (1) sediment background levels for organochlorine compounds in Lakes Huron and Superior

(Persaud and others, 1993); (2) proposed USEPA sediment quality criteria for protection of freshwater benthic organisms (U.S. Environmental Protection Agency, 1993a,b); (3) USEPA sediment quality advisory levels (U.S. Environmental Protection Agency, 1997a); (4) Canadian interim sediment quality guidelines for freshwater organisms (Environment Canada, 1995; Canadian Council of Ministers of the Environment, 1998); (5) effects range–median and effects range–low values for marine and estuarine sediment quality (Long and Morgan, 1991; Long and others, 1995); (6) probable effect levels and threshold effect levels developed for coastal Florida (MacDonald, 1994); and (7) apparent effects thresholds developed for Puget Sound, Washington (Barrick and others, 1988).

These different guidelines were developed using multiple approaches. Each approach has its uncertainties and limitations (U.S. Environmental Protection Agency, 1997a), and no single type of sediment guideline is generally accepted in the scientific literature (Persaud and others, 1993). Furthermore, for most individual pesticides, there are inconsistencies among the different guideline values derived by different methods for the same pesticide. Rather than choose one type of guideline over another, the analysis below follows a procedure developed by the U.S. Environmental Protection Agency (1997a), which used multiple sediment quality assessment methods to evaluate sediment contaminant data in USEPA's National Sediment Inventory. According to this U.S. Environmental Protection Agency (1997a) procedure, the available sediment guidelines for a given individual contaminant are used to classify sites into categories on the basis of the *probability of adverse effects* on aquatic life at those sites. In the analysis below, the U.S. Environmental Protection Agency (1997a) procedure was modified in that individual studies are classified into categories (because site-specific data were not always available) on the basis of the probability of adverse effects on aquatic life at the most contaminated site in each study. This procedure is described in more detail below.

The effective use of guidelines in water-quality assessment requires an understanding of how the guidelines were derived (Nowell and Resek, 1994) so that their applicability and limitations will be understood. Therefore, the analysis below of potential sediment toxicity in the monitoring studies reviewed (Tables 2.1 and 2.2) is preceded by some background information. First, various approaches to assessing sediment quality are described and the specific guidelines used in the analysis below defined, including their derivations and underlying assumptions. Next, more detail is provided on the modified USEPA procedure used to classify studies according to the probability of adverse effects on aquatic life. Finally, the sediment monitoring studies reviewed in this book (Tables 2.1 and 2.2) are analyzed for potential effects on aquatic organisms at the most contaminated site in each study.

Approaches to Assessing Sediment Quality

Sediment guidelines have been determined using a number of causal or empirical correlation methods (U.S. Environmental Protection Agency, 1997a). The following discussion distinguishes between four general approaches, or types of methods: sediment background, equilibrium partitioning, empirical biological effects correlation, and sediment toxicity.

Sediment Background Approach

The sediment background approach uses reference or background concentrations in pristine areas as a standard for comparison with measured concentrations in sediment.

Background concentrations are usually specific to a region or geographic area. This approach may be especially useful for determining where enrichment in naturally occurring contaminants, such as metals, has occurred. For synthetic contaminants such as pesticides, contamination in pristine areas may be caused by atmospheric contamination (Majewski and Capel, 1995). The main disadvantage of this method is that it has no biological effects basis. Measured concentrations that exceed background concentrations may indicate that local sources exist or have existed in the past, but do not necessarily indicate that biological effects have occurred.

Equilibrium Partitioning Approach

In the equilibrium partitioning approach, an equilibrium partition coefficient (K_{oc}) is used to calculate the chemical concentration in sediment that ensures that the concentration in pore water does not exceed a threshold aqueous concentration expected to cause toxic effects on aquatic organisms (Di Toro and others, 1991). The threshold aqueous concentration used is typically the USEPA chronic water-quality criterion for protection of aquatic life. This approach assumes that chemical concentrations in pore water and sediment organic carbon are at equilibrium and that the toxicity of the chemical to benthic organisms is equivalent to its toxicity to water-column species. This model explains two observations arising from sediment toxicity testing. First, in toxicity tests using different sediments but the same chemical and test organism, there was no relation between toxicity and chemical concentration (dry weight) in sediment. However, if the chemical concentration in sediment was expressed on an organic carbon basis (or, except for highly hydrophobic compounds, if the chemical concentration in pore water was used), then biological effects took place at similar concentrations (within a factor of 2) for different sediments (Di Toro and others, 1991). Second, the biological-effects concentration expressed on a pore water basis is similar to the biological effects concentration determined in toxicity tests with water-only exposures. This approach differs from the biological effects correlation approach, which is empirically based, in that it postulates a theoretical causal relation between chemical bioavailability and chemical toxicity in different sediments (U.S. Environmental Protection Agency, 1997a).

Of the guidelines used in the analysis below, two were developed based on the equilibrium-partitioning approach: proposed USEPA sediment quality criteria (U.S. Environmental Protection Agency, 1993a,b) and USEPA sediment quality advisory levels (U.S. Environmental Protection Agency, 1997a). Also, some biological effects-based guidelines (for example, the effects range values from Long and Morgan, 1991; Long and others, 1995) were based on an array of studies that included equilibrium partitioning studies.

Biological Effects Correlation Approach

The biological effects correlation approach consists of matching sediment chemistry measurements with biological effects measurements to relate the incidence of biological effects to the concentration of an individual contaminant at a particular site. These data sets are used to identify a level of concern for contaminant concentrations on the basis of the probability of observing adverse effects on benthic organisms. This approach is empirically based and does not indicate a direct cause and effect relation between chemical contamination and biological effects. It assumes that the contaminant measured is responsible for the effects observed, although field sediment samples typically contain mixtures of chemical contaminants. This approach also assumes that the influence of the chemical contaminant is greater than the influence of environmental conditions (Long and Morgan, 1991).

Several of the guidelines used in the analysis below were developed using a biological effects correlation approach: the apparent effects thresholds (Barrick and others, 1988); Florida's probable effect levels and threshold effect levels (MacDonald, 1994); the effects range–medians and effects range–lows (Long and Morgan, 1991; Long and others, 1995); and Canada's interim sediment quality guidelines (Environment Canada, 1995; Canadian Council of Ministers of the Environment, 1998).

Sediment Toxicity Approach

The sediment toxicity approach encompasses acute and chronic toxicity tests with potentially contaminated field-collected sediment (such as dredged material) or sediment elutriates, and spiked sediment bioassays in which organisms are exposed to pristine sediment spiked in the laboratory with a known amount of a test chemical. An uncertainty factor is generally applied to the toxicity level for the most sensitive species. One disadvantage of this approach is that the toxicity of a given chemical to a given species of organism may vary for different sediments (Long and Morgan, 1991; U.S. Environmental Protection Agency, 1997a). Also, there is a lack of standardized techniques for spiking sediment, and measured toxicity levels may be affected by the technique used to spike the sediment (Persaud and others, 1993). Whereas acute sediment toxicity tests (with field-collected sediment samples) are widely accepted by the scientific and regulatory communities, more work is required on chronic toxicity testing (U.S. Environmental Protection Agency, 1997a) before these tests are accepted to the same degree.

Data from sediment toxicity tests were used in determining some of the biological effects-based guidelines used in the analysis below; for example, effects range–medians and effects range–lows (Long and Morgan, 1991; Long and others, 1995).

Definitions of Sediment Quality Guidelines

The sediment quality guidelines used in this book are described briefly below. Each type of guideline is based on one or more of the approaches just described. Numerical guideline values for individual pesticides in sediment are listed in Table 6.3 (background levels) and Table 6.4 (all other guidelines).

Sediment Background Levels (Lakes Huron and Superior)

Background concentrations have been reported for organochlorine concentrations in surficial sediment from Lake Huron and Lake Superior (Persaud and others, 1993). These values are listed in Table 6.3. Measured concentrations exceeding these background levels do not give any information regarding potential biological effects. However, it is interesting to compare these background levels with biological effects-based guidelines to see what probability of adverse effects may occur at background-level contamination. Also, because contamination of pristine areas with synthetic contaminants such as pesticides may be due to atmospheric

Table 6.3. Background levels for pesticides in bed sediment from Lakes Huron and Superior

[Abbreviations and symbols: kg, kilogram; wt, weight; µg, microgram. Reproduced from Persaud and others (1993) with permission of the publisher. Copyright 1993 Queen's Printer for Ontario]

Target Analyte	Background levels (µg/kg dry wt)
Aldrin	1
Chlordane	1
DDD, *p,p'*-	2
DDE, *p,p'*-	3
DDT, (*o,p'* + *p,p'*)-	5
DDT, total	10
Dieldrin	1
Endrin	1
HCB	1
HCH, α-	1
HCH, β-	1
Heptachlor	1
Heptachlor epoxide	1
Lindane	1
Mirex	1

contamination (Majewski and Capel, 1995), measured concentrations at other sites that exceed background levels probably indicate local inputs.

USEPA's Sediment Quality Criteria

Under Section 304(a) of the Clean Water Act, USEPA is developing sediment quality criteria for the protection of benthic organisms for some pollutants designated as toxic. Thus far, USEPA has proposed sediment quality criteria (SQC) for only two pesticides: dieldrin and endrin. According to U.S. Environmental Protection Agency (1993a,b, 1994a), benthic organisms should be acceptably protected in sediment containing dieldrin or endrin concentrations at or below the criteria, except possibly where a locally important species is very sensitive or where sediment organic carbon is less that 0.2 percent. Proposed USEPA SQCs for these two compounds are based on the equilibrium partitioning approach, described above. Specifically,

$$SQC_{oc} = (FCV) \times (K_{oc}) \times (10^{-3} kg_{oc}/g_{oc}) \qquad (6.2)$$

where SQC_{oc} is expressed on a sediment organic carbon basis (in microgram per gram of sediment organic carbon, or $\mu g/g_{oc}$), FCV is the final chronic value (in micrograms per liter), K_{oc} is the organic-carbon normalized distribution coefficient (in liter per kilogram of sediment organic carbon, or L/kg$_{oc}$) (Section 4.2.1), and the last term is added to make the units cancel appropriately. The K_{oc} is calculated from the n-octanol-water partitioning coefficient (K_{ow}). The K_{oc} value is presumed to be independent of sediment type, so that sediment quality criteria (SQC$_{oc}$) also will be independent of sediment type. SQC$_{oc}$ values are based on chronic toxicity data and K_{ow} data that have been judged to be high quality after extensive peer review. Separate SQC were derived for freshwater and marine aquatic life. The proposed USEPA SQC values for dieldrin and endrin listed in Table 6.4 are for freshwater.

Because SQC$_{oc}$ values are expressed on a sediment organic-carbon basis (micrograms of pesticide per gram of sediment organic carbon), measured contaminant concentrations in total sediment must be converted accordingly before comparisons with USEPA criteria are made. This conversion consists of dividing the measured pesticide concentration in total sediment, dry weight (microgram of pesticide per kilogram of total sediment), by the organic carbon content in the sediment (gram organic carbon per kilogram of total sediment).

USEPA's Sediment Quality Advisory Levels

USEPA sediment quality advisory levels (SQAL) are interim guidelines for selected contaminants for which USEPA has not yet developed SQCs (U.S. Environmental Protection Agency, 1997a). The U.S. Environmental Protection Agency (1997a) provided SQAL values for δ-HCH, lindane, dieldrin, endosulfan (I, II, and total), endrin, malathion, methoxychlor, and toxaphene (Table 6.4). SQALs are derived the same way as SQCs (on the basis of the equilibrium partitioning approach), except that SQALs may be calculated with less extensive data than SQCs (U.S. Environmental Protection Agency, 1997a).

$$SQAL_{oc} = (FCV\ or\ SCV)(K_{oc})(10^{-3}kg_{oc}/g_{oc}) \qquad (6.3)$$

where SQAL$_{oc}$ is expressed on a sediment organic carbon basis (in microgram per gram of sediment organic carbon, or μg/g$_{oc}$); FCV is the final chronic value (in micrograms per liter); SCV is a secondary chronic value for aquatic toxicity (also in micrograms per liter), used when no FCV is available; K_{oc} is the organic carbon-water partitioning coefficient (in liter per kilogram of sediment organic carbon, or L/kg$_{oc}$); and the last term is added to make the units cancel appropriately. As with SQCs, K_{oc} values are calculated from K_{ow} values. The best available chronic toxicity values and K_{ow} values are used, selected according to a hierarchy described by U.S. Environmental Protection Agency (1997a). The USEPA SQALs used in the analysis below are listed in Table 6.4.

Apparent Effects Thresholds

Apparent effects thresholds (AET) were derived using a statistically based method that attempts to relate individual sediment contaminant concentrations with observed biological effects. For a given chemical, studies are compiled that measured a statistically significant difference (for some biological indicator) relative to appropriate reference conditions. The AET for this data set is defined as the sediment concentration above which there is always a statistically significant difference for the biological indicator measured (Tetra Tech Inc., 1986; Barrick and others, 1988). Biological indicators can be either field-measured effects (such as

Table 6.4. Sediment-quality guidelines and boundary values for pesticides in bed sediment

[For guidelines on a sediment-organic carbon basis, default values at 1 percent total organic carbon are shown in italics. Each guideline is designated (in bold) as an upper screening value (**U**) or a lower screening value (**L**). Abbreviations and symbols: AET-H, apparent effects threshold-high; AET-L, apparent effects threshold-low; ERL, effects range–low; ERM, effects range–median; i, insufficient guidelines to determine a Tier 1–2 boundary value; ISQG, interim sediment-quality guideline; μg/kg, microgram per kilogram; μg/kg$_{oc}$, microgram per kilograms of organic carbon in sediment; NOAA, National Oceanic and Atmospheric Administration; NS&T, National Status and Trends; PEL, probable effect level; SQAL, sediment-quality advisory level; SQC, sediment-quality criteria; SV, screening value; TEL, threshold effect level; TOC, total organic carbon in sediment; USEPA, U.S. Environmental Protection Agency; wt, weight; —, no guideline available]

Screening Value:	U				L	U
Target Analytes	USEPA SQC[1] (μg/kg$_{oc}$)	USEPA SQC[1] at 1% TOC (μg/kg$_{oc}$)	USEPA SQAL[2] (μg/kg$_{oc}$)	USEPA SQAL[2] at 1% TOC (μg/kg)	Canadian Interim Sediment Quality Guidelines, Freshwater[3]	
					ISQG (μg/kg dry wt)	PEL (μg/kg dry wt)
Chlordane[7]	—	—	—	—	4.50	8.87
Chlordane, cis-	—	—	—	—	—	—
Chlordane, trans-	—	—	—	—	—	—
DDD, p,p'-	—	—	—	—	[8]3.54	[8]8.51
DDE, p,p'-	—	—	—	—	[8]1.42	[8]6.75
DDT, p,p'-	—	—	—	—	[8,10]1.19	[8,10]4.77
DDT, total	—	—	—	—	—	—
Diazinon	—	—	19	0.19	—	—
Dieldrin	11,000	110	—	—	2.85	6.67
Endosulfan[7]	—	—	540	5.4	—	—
Endosulfan I	—	—	290	2.9	—	—
Endosulfan II	—	—	1,400	14	—	—
Endrin	4,200	42	—	—	2.67	62.4
HCH[7]	—	—	370	3.7	—	—
HCH, α-	—	—	—	—	—	—
HCH, β-	—	—	—	—	—	—
HCH, δ-	—	—	13,000	130	—	—
Heptachlor epoxide	—	—	—	—	0.60	2.74
Hexachlorobenzene	—	—	—	—	—	—
Lindane	—	—	370	3.7	0.94	1.38
Malathion	—	—	67	0.67	—	—
Methoxychlor	—	—	1,900	19	—	—
Nonachlor, cis-	—	—	—	—	—	—

Table 6.4. Sediment-quality guidelines and boundary values for pesticides in bed sediment—*Continued*

Screening Value:	L	U	L	U	L	U	Bound-ary Value Tier 2–3	Bound-ary Value Tier 1–2
	NOAA NS&T[4]		Florida Department of Environmental Conservation[5]		Puget Sound (Barrick and others)[6]		Lowest Lower SV	2nd Lowest Upper SV
Target Analytes	ERL (µg/kg dry wt)	ERM (µg/kg dry wt)	TEL (µg/kg dry wt)	PEL (µg/kg dry wt)	AET-L (µg/kg dry wt)	AET-H (µg/kg dry wt)	(µg/kg dry wt)	(µg/kg dry wt)
Chlordane[7]	0.5	6	2.26	4.79	—	—	0.5	6
Chlordane, *cis-*	—	—	2.26	4.79	—	—	2.26	i
Chlordane, *trans-*	—	—	2.26	4.79	—	—	2.26	i
DDD, *p,p'-*	2	20	1.22	7.81	16	43	1.22	8.51
DDE, *p,p'-*	[9]2.2	[9]27	2.07	374	9	15	1.42	15
DDT, *p,p'-*	1	7	1.19	4.77	34	34	1	7
DDT, total	[9]1.58	[9]46.1	3.89	51.7	9	15	1.58	46.1
Diazinon	—	—	—	—	—	—	0.19	i
Dieldrin	0.02	8	0.715	4.3	—	—	0.02	6.67
Endosulfan[7]	—	—	—	—	—	—	5.4	i
Endosulfan I	—	—	—	—	—	—	2.9	i
Endosulfan II	—	—	—	—	—	—	14	i
Endrin	0.02	45	—	—	—	—	0.02	45
HCH[7]	—	—	0.32	0.99	—	—	0.32	3.7
HCH, α-	—	—	0.32	0.99	—	—	0.32	i
HCH, β-	—	—	0.32	0.99	—	—	0.32	i
HCH, δ-	—	—	0.32	0.99	—	—	0.32	130
Heptachlor epoxide	—	—	—	—	—	—	0.6	i
Hexachlorobenzene	—	—	—	—	22	230	22	i
Lindane	—	—	0.32	0.99	—	—	0.32	1.38
Malathion	—	—	—	—	—	—	0.67	i
Methoxychlor	—	—	—	—	—	—	19	i
Nonachlor, *cis-*	—	—	2.26	4.79	—	—	2.26	i

Table 6.4. Sediment-quality guidelines and boundary values for pesticides in bed sediment—*Continue*

Screening Value:	U				L	U
					Canadian Interim Sediment Quality Guidelines, Freshwater[3]	
Target Analytes	USEPA SQC[1] ($\mu g/kg_{oc}$)	*USEPA SQC[1] at 1% TOC ($\mu g/kg_{oc}$)*	USEPA SQAL[2] ($\mu g/kg_{oc}$)	*USEPA SQAL[2] at 1% TOC ($\mu g/kg$)*	ISQG ($\mu g/kg$ dry wt)	PEL ($\mu g/kg$ dry wt)
Nonachlor, *trans-*	—	—	—	—	—	—
Pentachlorophenol	—	—	—	—	—	—
Toxaphene	—	—	10,000	*100*	[11]1.5	—

[1]U.S. Environmental Protection Agency (1993a,b).
[2]U.S. Environmental Protection Agency (1997a).
[3]Canadian Council of Ministers of the Environment (1998).
[4]Long and Morgan (1991) unless noted otherwise.
[5]MacDonald (1994) and U.S. Environmental Protection Agency (1997a).

changes in benthic community structure) or effects measured in the laboratory (such as sediment toxicity tests performed with field-collected sediment samples). The AET method is very sensitive to the species and effects endpoints that are selected. If the data used consist of multiple species and endpoints, the least sensitive species will predominate and other more sensitive species may not be protected (Persaud and others, 1993). The AET method assumes that effects observed above the AET are due to the target chemical, whereas effects observed at concentrations below the AET are due to some other chemical. Because this method determines the concentration above which biological effects always occur, AET-based guidelines may be under-protective, in that effects may be observed at lower concentrations in some sediment samples (Persaud and others, 1993).

The AET guideline values used in the analysis below (Table 6.4) are based on data for sediment from Puget Sound, Washington (Barrick and others, 1988). Following the procedure of U.S. Environmental Protection Agency (1997a), two threshold AET-based guidelines are used in this analysis: The AET-L is the lowest sediment concentration for which any particular biological indicator showed an effect (i.e., the lowest of the AET values for the different indicators available for the target chemical). The AET-H is the sediment concentration corresponding to the highest of the available AET values for a given target chemical. For example, the AET-L for total DDT was based on benthic infaunal abundance and the AET-H on amphipod toxicity tests. AETs were determined for individual isomers of DDT and metabolites, plus hexachlorobenzene and pentachlorophenol.

Effects Range Values for Aquatic Sediment

Long and coworkers (Long and Morgan, 1991; Long and others, 1995) compiled and evaluated data from the literature on contaminant concentrations in sediment and associated

Table 6.4. Sediment-quality guidelines and boundary values for pesticides in bed sediment—*Continued*

Screening Value	L	U	L	U	L	U	Bound-ary Value Tier 2–3	Bound-ary Value Tier 1–2
	NOAA NS&T[4]		Florida Department of Environmental Conservation[5]		Puget Sound (Barrick and others)[6]		Lowest Lower SV (μg/kg dry wt)	2nd Lowest Upper SV (μg/kg dry wt)
Target Analytes	ERL (μg/kg dry wt)	ERM (μg/kg dry wt)	TEL (μg/kg dry wt)	PEL (μg/kg dry wt)	AET-L (μg/kg dry wt)	AET-H (μg/kg dry wt)		
Nonachlor, *trans*-	—	—	2.26	4.79	—	—	2.26	i
Pentachlorophenol	—	—	—	—	360	690	360	i
Toxaphene	—	—	—	—	—	—	1.5	i

[6]Barrick and others (1988) and U.S. Environmental Protection Agency (1997a).
[7]Total or unspecified.
[8]Applies to the sum of *o,p'* and *p,p'* isomers.
[9]Long and others (1995).
[10]Provisional; marine guideline adopted.
[11]Provisional; calculated from the Canadian water-quality guideline.

biological effects to identify informal guidelines for use in evaluating potential biological effects of contaminant residues in sediment measured by the NS&T program. The resulting guidelines are the effects range–median (ERM) and effects range–low (ERL) values. Their derivation is described in detail in Long and Morgan (1991).

Long and coworkers (Long and Morgan, 1991; Long and others, 1995) screened studies using a variety of biological approaches, including equilibrium partitioning, spiked sediment bioassays, and several biological effects correlation methods (including AETs). For a given chemical, any individual study was included if both biological and sediment chemistry data were reported, there was a discernible gradient in contamination for the chemical among samples, methods were adequately documented, and biological and chemical data were from the same locations (Long and Morgan, 1991). For each study, the contaminant concentration observed (or predicted) to be associated with biological effects was determined. Studies were compiled for each contaminant, and those concentrations observed or predicted by the different studies to be associated with biological effects were listed in ascending order. The lower 10th and median percentiles were identified as informal guidelines for predicting biological effects. The ERL, defined as the lower 10th percentile concentration of the available data in which effects were detected, approximates the concentrations at which adverse effects were first observed. The ERM, defined as the median concentration of the available data in which effects were detected, approximates the concentrations at or above which adverse effects were often observed. An example (dieldrin) is shown in Table 6.5.

By using multiple approaches to determine ERL and ERM values, the influence of any single data point in setting these guidelines was minimized. Long and Morgan (1991) referred to this as "establishing the preponderance of evidence." For each chemical considered, the accuracy of the ERL and ERM guidelines was limited by the quantity and consistency of the available data. For most pesticides, Long and Morgan (1991) evaluated the degree of confidence in the accuracy of the ERL and ERM guidelines as low or moderate, due to a lack of consistency among data from various approaches or a lack of data from multiple approaches. Effects range values for chlordane, p,p'-DDT, p,p'-DDD, dieldrin, and endrin (Table 6.4) were based on both freshwater and saltwater studies, but the majority were estuarine or coastal studies (Long and Morgan, 1991). Effects range values for p,p'-DDE and total DDT (Table 6.4) are revised values from Long and others (1995) that are based on an expanded data set that excluded freshwater studies (i.e., estuarine and coastal studies only). ERM and ERL values are expressed on a sediment dry weight basis.

Florida's Probable Effect Levels and Threshold Effect Levels

Similar preponderance of evidence, biological effects-based guidelines were developed for the Florida Department of Environmental Protection for application to sediment off the Florida coast (MacDonald, 1994). Again, data from studies using multiple approaches were compiled for a given chemical. For each chemical, two data sets were compiled: studies where concentrations of that chemical were associated with biological effects (the effects data), and those where concentrations of that chemical were associated with no observed effects (the no-effects data).

Table 6.5. Effects range–low (ERL) and effects range–median (ERM) values for dieldrin and the 14 concentrations, arranged in ascending order, that were used to determine these values

[Endpoint: type of approach or test used to measure toxicity. The 10th and 50th percentile concentrations, which correspond to the ERL and ERM, respectively, are shaded. Abbreviations: AET, apparent effects threshold; COA, co-occurrence analysis; EP, equilibrium partitioning; LC_{50}, median lethal concentration; SLC, screening level concentration; SSB, spiked sediment bioassay; TOC, total organic carbon in sediment. Reproduced from Long and Morgan (1991) with permission of the author]

Concentrations (in microgram per kilogram)	Endpoint
0.01	EP 99 percentile chronic marine
0.02	**ERL**
0.02	EP 95 percentile chronic marine
0.21	Freshwater SLC at 1 percent TOC
4.1	SSB LC_{50} for *Crangon septemspinosa*
6.6	AET, San Francisco Bay, California
6.6	AET, San Francisco Bay, California
7.4	Benthos COA, Kishwaukee River, Illinois
8.0	**ERM**
8.2	Bioassay COA, San Francisco Bay, California
10.3	Bioassay COA, San Francisco Bay, California
11.9	EP freshwater lethal threshold
16.0	Benthos COA, DuPage River, Illinois
57.7	EP interim marine criteria
199.0	EP interim freshwater criteria
13,000.0	SSB LC_{50} for *Neanthes virens*

Studies in each data set were listed in ascending order, by chemical concentration in sediment. The threshold effect level (TEL) was calculated as the geometric mean of the lower 15th percentile from the effects data and the 50th percentile concentration from the no-effects data. The TEL represents the concentration below which toxic effects rarely occurred. The probable effect level (PEL) was calculated as the geometric mean of the 50th percentile concentration from the effects data and the 85th percentile concentration from the no-effects data. Toxic effects usually or frequently occurred at concentrations above the PEL.

Florida's TELs and PELs were derived for seven pesticides or pesticide groups: components of technical chlordane, p,p'-DDE, p,p'-DDD, p,p'-DDT, total DDT, dieldrin, and HCH isomers. The capabilities of TELs and PELs to predict the toxicity of sediment accurately were evaluated on the basis of independent sets of field data from Florida and the Gulf of Mexico, and found to be about 85 percent. TELs and PELs (Table 6.4) are expressed on a sediment dry weight basis.

Canada's Interim Sediment Quality Guidelines

Canada is in the process of developing sediment quality guidelines that are based on a spiked-sediment toxicity test approach and on a modified version of the approach used in the NS&T program described above (Canadian Council of Ministers of the Environment, 1995, 1998; Environment Canada, 1995). In the meantime, Canada has published interim sediment quality guidelines that are based on the modified NS&T approach alone (Environment Canada, 1995; Canadian Council of Ministers of the Environment, 1998). This approach is comparable to that used by the Florida Department of Environmental Protection (MacDonald, 1994) in that it involves compiling chemical and biological data into an effects data set and a no-effects data set for each chemical; listing the studies in each data set in ascending order by chemical concentration in sediment; and determining a TEL and PEL. The TEL and PEL are defined exactly as described in the preceding subsection on the Florida guidelines. Then, the Canadian interim sediment quality guideline (ISQG) is defined as equivalent to the TEL (Canadian Council of Ministers of the Environment, 1998). Unlike Florida, which developed marine guidelines only, Canada developed TELs and PELs for both freshwater and marine sediment by maintaining separate data sets for freshwater and marine studies (Environment Canada, 1995). A minimum of 20 entries in each data set was required to calculate TEL and PEL values. The entries included data from equilibrium partitioning studies, guidelines from other jurisdictions, spiked-sediment toxicity tests, and field studies from throughout North America. Canada's TEL (ISQG) defines a concentration below which adverse effects are rarely anticipated, and above which adverse effects are occasionally anticipated; the PEL defines a concentration above which adverse effects are frequently anticipated (Environment Canada, 1995). In studies that were used to validate the Canadian interim TELs and PELs, it was found that (1) at concentrations below the TEL, 20 percent or fewer studies showed adverse biological effects, and (2) for most pesticides at concentrations above the PEL, 50 percent or more studies showed adverse biological effects. For two pesticides (p,p'-DDE and lindane), only 47 and 49 percent of studies, respectively, showed adverse biological effects at concentrations that exceeded the freshwater PEL, thus indicating less confidence in these PEL values (relative to PELs for other pesticides) as indicators of the probable effect concentration (Environment Canada, 1995). The Canadian ISQGs and PELs for freshwater sediment are listed in Table 6.4. Canadian guidelines are available for chlordane, DDD, DDE, DDT, dieldrin, endrin, heptachlor epoxide, lindane, and toxaphene.

USEPA's Procedure for Classifying Sites by Probability of Adverse Effects

As mentioned previously, USEPA developed a procedure that uses all available sediment quality guidelines for a given contaminant to estimate the probability of adverse effects on aquatic life at a site, on the basis of measured contaminant concentrations in sediment at that site. This procedure was intended as part of a screening-level analysis, which designates potentially contaminated sites that can be noted for further study. This procedure was used by USEPA to evaluate data in the National Sediment Inventory (U.S. Environmental Protection Agency, 1997a). USEPA's procedure permits sediment quality assessment despite the considerable inconsistency among the different sediment guidelines available for a given chemical (as shown in Table 6.4).

The first step in the USEPA procedure was to assemble the available sediment quality guidelines for an individual sediment contaminant. Note that the guidelines assembled and used by U.S. Environmental Protection Agency (1997a) to analyze data in the National Sediment Inventory included both freshwater guidelines (such as USEPA's proposed SQC) and saltwater guidelines (such as ERMs and ERLs from Long and others, 1995). Next, individual guidelines were labeled as either *lower screening values*, which indicate a threshold concentration above which one begins to see adverse effects on some benthic organisms (such as the ERL, AET-L, and TEL), or *upper screening values*, above which more frequent or severe biological effects may be expected (such as the ERM, AET-H, and PEL).

Individual sites were then classified into tiers on the basis of the probability of adverse effects on aquatic life at those sites, as follows:

Tier 1: high probability of adverse effects—sites for which any sediment chemistry measurement exceeded either (1) the USEPA's proposed SQC (for dieldrin and endrin only) or (2) at least two of the upper screening values (for all other pesticides and for dieldrin and endrin at sites with no sediment organic carbon data).

Tier 2: intermediate probability of adverse effects—sites for which any sediment chemistry measurement exceeded any single one of the lower screening values.

Tier 3: no indication of adverse effects—sites for which the sediment chemistry measurements were below all available screening values.

Modified Procedure for Classifying Studies by Probability of Adverse Effects

In the analysis below (see Section 6.1.1, subsection on Pesticides in Bed Sediment—Analysis of Potential Adverse Effects on Benthic Organisms) of potential effects of sediment contaminants in the monitoring studies reviewed, three modifications to the USEPA procedure were made, mostly because of data availability. First, each study (rather than each site) was classified according to the probability of adverse effects at the most contaminated site within the study, by comparing the available sediment quality guidelines with the maximum concentration reported in each study. Second, because sediment organic carbon data were not available for all monitoring studies, the maximum measured concentrations of dieldrin and endrin could not be compared directly with USEPA's proposed SQCs (which are expressed on an organic-carbon basis) for all studies. Therefore, for all pesticides, studies were classified into Tier 1 if two upper screening values were exceeded (This is consistent with the USEPA procedure for dieldrin and endrin at sites where sediment organic carbon data are not available). Third, as described below, some additional sediment quality guidelines were added to those used by U.S. Environmental Protection Agency (1997a).

For a given chemical, the following guidelines were considered to be upper screening values in the analysis below: the USEPA's proposed SQC using a default value of 1 percent sediment organic carbon, the USEPA SQAL using a default value of 1 percent sediment organic carbon, the AET-H, the ERM, Florida's PEL, and Canada's freshwater PEL. These are identical to the set of upper screening values used in U.S. Environmental Protection Agency (1997a), except that Canada's freshwater PEL (which was not included in USEPA's analysis) has been added. Also, some additional ERM values from Long and Morgan (1991) were used in the analysis below for a few pesticides that did not have ERM and ERL values listed in Long and others (1995). The addition of the Canadian PEL values, and the extra ERM values, increased the number of pesticides for which at least two upper screening values were available. Inclusion of the Canadian guidelines also increased the number of screening values determined specifically for freshwater. Note that USEPA's probability of adverse effects procedure was developed at the Second National Sediment Inventory Workshop in April 1994 (U.S. Environmental Protection Agency, 1997a), which preceded publication of the Canadian interim guidelines in draft form (Environment Canada, 1995).

In the analysis below, the following lower screening values are used: the AET-L, the ERL, Florida's TEL, and Canada's freshwater ISQG. All except Canada's freshwater ISQG were also used by U.S. Environmental Protection Agency (1997a) to analyze data in the National Sediment Inventory. Also, some extra ERL values from Long and Morgan (1991) were added for a few pesticides that did not have ERL values listed in Long and others (1995).

As the next step in the analysis, the sediment guidelines available for a given pesticide were compared with the maximum sediment concentrations reported for each monitoring study that analyzed that pesticide in sediment. Individual studies were categorized according to the probability of adverse effects at the most contaminated site (where the maximum concentration was detected) within the study, as follows:

Tier 1 studies: high probability of adverse effects at the most contaminated site— studies in which the maximum sediment concentration exceeded at least two of the applicable upper screening values for a given pesticide. The second lowest upper screening value was called the "Tier 1–2 boundary value," since it serves as the boundary between Tier 1 and Tier 2.

Tier 2 studies: intermediate probability of adverse effects at the most contaminated site—studies in which the maximum sediment concentration exceeded the lowest of the applicable lower screening values for a given pesticide. The lowest of the lower screening values was called the "Tier 2–3 boundary value," since it serves as the boundary between Tier 2 and Tier 3.

Tier 3 studies: no indication of adverse effects at any sites—studies in which the maximum sediment concentration was less than the lowest of the applicable screening values (i.e., the Tier 2–3 boundary value). This category includes studies in which the target pesticide was not detected at any sites.

Note that this analysis, like the USEPA procedure it is based on, requires some consistency among upper screening values to put studies in Tier 1 (the high probability category), in that two upper screening values must be exceeded for this classification. Designation of a study as a Tier 2 study (the intermediate probability category) is more conservative (overprotective), in that only one lower screening value (the lowest and most sensitive) needs to be exceeded for this classification. For some pesticides, there was only one upper screening value available; for these compounds, no Tier 1 classification was possible. Therefore, for these compounds, only a Tier 2–3 boundary value could be determined; measured concentrations exceeding this boundary value indicated some potential for adverse effects on aquatic life, but it is not possible to differentiate between intermediate and higher probabilities of adverse effects.

Pesticides in Bed Sediment—Comparison with Background Levels

Background levels from Lakes Huron and Superior (Table 6.3) were compared with reported maximum concentrations from individual monitoring studies reviewed in this book (see Tables 2.1 and 2.2). The background levels present at these reference sites are most likely due to atmospheric contamination. These background levels were exceeded at the most contaminated site in over 50 percent of studies reviewed for the following pesticides: chlordane, p,p'-DDE, total DDT, and dieldrin. Sediment concentrations of DDD, DDT, and hexachlorobenzene exceeded background levels at the most contaminated sites in over 20 percent of all studies reviewed. This suggests that local sources existed at, or upstream of, these sites, but provides no information on potential biological effects at these sites.

It is interesting to note that background levels are the same as or higher than some lower screening values for total chlordane, DDD, DDE, DDT, total DDT, dieldrin, and endrin. This indicates that an intermediate probability of adverse effects on aquatic life may be associated with background levels found in Lakes Huron and Superior.

Pesticides in Bed Sediment—Analysis of Potential Adverse Effects on Benthic Organisms

The maximum concentrations reported by individual monitoring studies in Tables 2.1 and 2.2 were compared with applicable sediment guidelines. These sediment guidelines, as well as Tier 1–2 and Tier 2–3 boundary values, are listed in Table 6.4 for individual pesticides and pesticide transformation products.

In the monitoring studies reviewed, pesticide concentrations in sediment generally were reported on a dry weight basis. As a result, the concentration data summarized in Tables 2.1 and 2.2 (median concentration and concentration ranges) generally are presented in units of micrograms per kilogram of total sediment, dry weight. Although most sediment guidelines are expressed on a total sediment basis (dry weight), two guidelines are expressed on a sediment organic-carbon basis: USEPA's proposed SQCs and USEPA's SQALs. For these USEPA guidelines, Table 6.4 lists guideline values on a sediment organic carbon basis (in micrograms per kilogram of sediment organic carbon) as well as default values that assume 1 percent total organic carbon (TOC) in sediment (which converts them to units of micrograms per kilogram of total sediment). Incidentally, the default TOC value of 1 percent is very close to the mean TOC content (1.2 percent) of marine and estuarine sediment in the biological effects database compiled by Long and others (1995). The 1 percent default value also was used by U.S. Environmental Protection Agency (1997a) in its evaluation of data from sites in the National Sediment Inventory that had no TOC data. For any sediment sample with a TOC greater than 1 percent, the pesticide guideline values will be proportionally higher than the 1 percent TOC default values listed in Table 6.4.

In comparing maximum concentrations from monitoring studies with the available guidelines, it is important to remember three points. First, if the Tier 1–2 boundary value for a given pesticide is exceeded by a study, this indicates only the following: the site with the highest concentration of that pesticide in that study exceeded that boundary value; therefore, that site has a high probability of adverse effects on aquatic life. Classification of a study into Tier 1 does not provide any information on the number of sites in that study that have a high probability of adverse effects on aquatic life. The same is true for studies classified in Tier 2 (intermediate probability of adverse effects on aquatic life). Second, the monitoring studies reviewed span a long period of time, with publication dates from 1960 to 1994. Where guidelines are exceeded, it is worthwhile to consider the study date, since concentrations may be expected to have declined since the 1970s, when most organochlorine insecticides were banned or severely restricted. When only studies published during the last decade (1984–1994) were considered, pronounced reductions in the maximum reported concentrations were apparent for chlordane, total DDT, endosulfan, mirex, and toxaphene (Table 6.6). This suggests that fewer studies may exceed sediment quality guidelines for these compounds, when only recently published studies are considered. Third, some of the monitoring studies reviewed did not report the maximum concentration detected for each analyte, so these studies are not represented in the analysis, tables, and figures in this section.

To examine the effect of study date, one ideally would compare studies conducted at different times or in different decades. Because many studies spanned a considerable period of time, and sampling dates were inconsistently reported, it was not always possible to identify the sampling date corresponding to the maximum concentration (or even the sampling date for the whole study). Instead, the publication date was used in the analysis below as an indicator of the relative sampling date. It is recognized that this is approximate at best, since the time elapsed from sampling to publication in any given study is variable. Nonetheless, when recently published (1984–1994) studies were compared with all studies published (1960–1994), a decline in the extent to which sediment quality guidelines were exceeded was observed for many pesticides (see below). This comparison is not made to establish trends, for three reasons. First, publication date is not a perfect surrogate for sampling date. Second, the distributions of these studies are overlapping, since the second group (recently published studies) are included among the first group (all studies reviewed). Moreover, for many compounds, the majority of studies were recently published, in which case these studies dominate the results for all studies reviewed. Third, even if recently published studies were compared with studies published prior to 1984, the two groups of studies did not sample at the same site locations. A better understanding of trends can be obtained from national programs that monitored residues at the same sites over time (Section 3.4). The results for recently published (1984–1994) studies are provided to show whether there is potential for adverse effects on benthic organisms at the most contaminated site when older studies (which may have sampled before many of the organochlorine insecticides were banned) are excluded.

Table 6.6. Selected results of studies that monitored pesticides in bed sediment

[Results are presented for all monitoring studies (published 1960–1994), and for recent monitoring studies (published 1984–1994) studies that are listed in Tables 2.1 and 2.2. Concentration range in detection limits: "nr" indicates that one or more studies did not report detection limits. Abbreviations: nd, not detected; nr, not reported; µg/kg, microgram per kilogram]

Target Analytes	All Monitoring Studies (Published 1960–1994)				Recent Monitoring Studies (Published 1984–1994)			
	Concentration Range (µg/kg dry weight)		Total Number of Studies that Reported Data	Percentage of Studies with Detectable Residues in at Least One Sample	Concentration Range (µg/kg dry weight)		Total Number of Studies that Reported Data	Percentage of Studies with Detectable Residues in at Least One Sample
	In Detection Limits	In Maximum Concentrations			In Detection Limits	In Maximum Concentrations		
Aldrin	nr, 0.05–480	nd–1,065	119	28	nr, 0.1–480	nd–1,065	70	24
Chlordane[1]	nr, 0.01–4,800	nd–1,000	111	74	nr, 0.1–4,800	nd–510	62	69
Chlordane, cis-	nr, 0.5–100	nd–293	25	24	nr, 0.5–100	nd–90	22	14
Chlordane, trans-	nr, 0.5–50	nd–149	28	32	nr, 0.5–50	nd–149	25	24
DDD[1]	nr, 0.01–38	nd–5,100	95	91	nr, 0.01–10	nd–260	46	91
DDD, o,p'-	nr, 0.1–50	nd–1,312	29	28	nr, 0.1–50	nd–1,312	27	22
DDD, p,p'-	nr, 0.05–960	nd–5,820	50	52	nr, 0.1–960	nd–5,820	44	48
DDE[1]	nr, 0.01–24	nd–10,000	101	88	nr, 0.01–20	nd–430	48	90
DDE, o,p'-	nr, 0.1–50	nd–292	29	21	nr, 0.1–50	nd–292	27	15
DDE, p,p'-	nr, 0.1–960	nd–1,870	52	60	nr, 0.1–960	nd–1,870	45	53
DDT[2]	nr, 0.01–10	nd–6,480	96	80	nr, 0.01–10	nd–2,280	45	76
DDT, o,p'-	nr, 0.1–50	nd–807	35	37	nr, 0.1–50	nd–807	29	28
DDT, p,p'-	nr, 0.1–960	nd–3,752	51	53	nr, 0.1–960	nd–3,752	43	47
DDT, total	nr, 0.01–50	nd–30,200,000	44	80	nr, 0.1–50	nd–4,443,500	25	68
Diazinon	nr, 0.1–100	nd–13	42	19	nr, 0.1–50	nd–0.2	16	13
Dieldrin	nr, 0.01–100	nd–440	163	65	nr, 0.01–100	nd–440	97	54
Endosulfan[1]	nr, 0.1–100	nd–4,530	50	26	nr, 0.1–10	nd–339	34	21
Endosulfan I	nr, 0.1–480	nd–96	14	50	nr, 0.1–480	nd–96	13	46
Endosulfan II	nr, 0.1–960	nd–140	14	50	nr, 0.1–960	nd–140	13	46

Table 6.6. Selected results of studies that monitored pesticides in bed sediment—*Continued*

Target Analytes	All Monitoring Studies (Published 1960–1994)				Recent Monitoring Studies (Published 1984–1994)			
	Concentration Range (µg/kg dry weight)		Total Number of Studies that Reported Data	Percentage of Studies with Detectable Residues in at Least One Sample	Concentration Range (µg/kg dry weight)		Total Number of Studies that Reported Data	Percentage of Studies with Detectable Residues in at Least One Sample
	In Detection Limits	In Maximum Concentrations			In Detection Limits	In Maximum Concentrations		
Endrin	nr, 0.01–100	nd–120	131	25	nr, 0.1–100	nd–43	80	20
HCH[1]	nr, 0.1–100	nd–20	10	30	nr, 0.1–100	nd–20	9	22
HCH, α-	0.03–480	nd–110	44	20	nr, 0.1–480	nd–65	39	13
HCH, β-	nr, 0.1–480	nd–2,800	33	15	nr, 0.1–480	nd–2,800	30	17
HCH, δ-	nr, 0.1–480	nd–0.8	25	4	nr, 0.1–480	nd–0.8	25	4
Heptachlor	nr, 0.01–480	nd–16.6	119	21	nr, 0.1–480	nd–12	72	22
Heptachlor epoxide	nr, 0.01–480	nd–106	123	32	nr, 0.01–480	nd–106	78	28
Hexachlorobenzene	nr, 0.03–400	nd–7,500,000	47	32	nr, 0.1–400	nd–7,500,000	41	24
Lindane	nr, 0.05–480	nd–221	128	19	nr, 0.1–480	nd–65	75	16
Malathion	nr, 0.1–100	0	41	0	nr, 0.1–50	0	14	0
Methoxychlor[3]	nr, 0.1–4,800	nd–366	70	20	nr, 0.1–4,800	nd–366	50	20
Mirex	nr, 0.01–100	nd–1,834	85	13	nr, 0.01–50	nd–310	64	9
Nonachlor, *cis*-	nr, 0.5–100	nd–5.5	22	9	nr, 0.5–100	nd–5.5	22	9
Nonachlor, *trans*-	nr, 0.5–100	nd–16	21	10	nr, 0.5–100	nd–16	21	10
Pentachlorophenol	nr, 0.5–1,200	nd–770	13	31	nr, 0.1–1,200	nd–770	8	25
Toxaphene	nr, 0.05–9,600	nd–1,858,000	100	12	nr, 0.1–9,600	nd–2,800	65	11

[1] Total or unspecified.
[2] $(o,p' + p,p')$- or unspecified.
[3] p,p'- or unspecified.

Table 6.6 compares the range in maximum pesticide concentrations for all monitoring studies reviewed (published during 1960–1994) with that for recently published (1984–1994) studies only. Table 6.6 also lists the range in detection limits reported in the studies reviewed and the percentage of studies with detectable residues of the target pesticide in one or more bed sediment samples. Only pesticide analytes for which one or more sediment guidelines are available are included in Table 6.6. These data were compiled from all national, multistate, state, and local monitoring studies (Tables 2.1 and 2.2) that reported maximum pesticide concentrations in bed sediment.

Table 6.7 lists the percent of studies in each of Tier 1 (high probability of adverse effects), Tier 2 (intermediate probability of effects), and Tier 3 (no indication of effects) for all monitoring studies reviewed (i.e., published during 1960–1994) and for recently published (1984–1994) studies. The fraction of monitoring studies that exceeded each applicable guideline is shown graphically for a few pesticides (see Figures 6.1–6.5) to illustrate the range in the different guideline values that may exist for a given chemical and the way in which boundary values were determined. The following chemicals are plotted in Figures 6.1–6.5: dieldrin, total chlordane, and total DDT (the most commonly detected pesticides or pesticide groups); p,p'-DDE (the most abundant component of total DDT); and diazinon (as an example of a pesticide with insufficient guidelines to determine a Tier 1–2 boundary value). The results for these and other pesticides are discussed individually below.

Aldrin and Dieldrin

There were no sediment quality guidelines for aldrin, except a background level (from Lakes Huron and Superior) of 1 μg/kg dry weight (Table 6.3). This background level was exceeded by maximum concentrations in 16 percent of all monitoring studies reviewed. This indicates potential local sources of aldrin at the most contaminated site in those studies. Aldrin was detected in sediment in only 28 percent of all studies (Table 6.6), probably because aldrin is converted to dieldrin relatively quickly in the environment (Schnoor, 1981).

Dieldrin was detected in more studies than aldrin (in 65 percent of all monitoring studies reviewed, compared with 28 percent for aldrin), as shown in Table 6.6. Figure 6.1 shows the cumulative frequency distribution of the maximum concentrations reported by the studies reviewed for dieldrin, overlaid by the applicable sediment quality guideline values. Individual guideline values are shown as vertical lines on the graph, each line corresponding to the appropriate dieldrin concentration. At the point where the cumulative frequency intersects a given vertical line, the corresponding Y-value indicates what percentage of studies have maximum concentrations that exceed that guideline concentration. Vertical lines representing individual guidelines are designated with either a "U" for upper screening values or an "L" for lower screening values (the only exception is the vertical line representing the background level, which is not a biological effects-based guideline, so is not considered either an upper or a lower screening value). Tier 1–2 and Tier 2–3 boundary values are shown as bold vertical lines.

Upper screening values are consistently higher than lower screening values for dieldrin (this is not the case for all pesticides, as seen below). Lower screening values range from 0.02 to 2.85 μg/kg dry weight, and upper screening values from 4.3 to 110 μg/kg dry weight. Note that the SQC for dieldrin (using 1 percent TOC) is considerably higher than the other upper screening values. Of all the dieldrin monitoring studies reviewed (published during 1960–1994), 33 percent have maximum concentrations that exceeded the Tier 1–2 boundary value, so have been classified as Tier 1 studies (i.e., there is a high probability of adverse effects at one or more sites in the study). Of all monitoring studies, 32 percent have maximum concentrations exceeding the Tier 2–3 boundary value, but less than the Tier 1–2 boundary value, and so, have been classified as Tier 2 studies (i.e., there is a intermediate probability of adverse effects at one or more sites in the study). The remaining 35 percent of studies were classified in Tier 3 (i.e., there is no indication of adverse effects at any sites in the study). All of the Tier 3 studies had no detectable dieldrin because the Tier 2–3 boundary value (the ERL, at 0.02 µg/kg) was below the lowest of the maximum dieldrin concentrations (0.1 µg/kg) reported by the studies reviewed (Figure 6.1).

When only recently published studies (1984–1994) were considered, there was a modest drop in the percent of studies in Tiers 1 and 2, and a corresponding increase in the percent of Tier 3 studies (Table 6.7, Figure 6.1). Note in Table 6.6 that there was no difference in the maximum dieldrin concentration range for all monitoring studies (published during 1960–1994) compared with studies recently published during 1984–1994 because the study reporting the highest maximum dieldrin concentration (440 µg/kg dry weight) was published after 1983.

Chlordane

The percentage of monitoring studies in Tiers 1–3 was calculated for total chlordane (Table 6.4), rather than for its major individual components (*cis*- and *trans*-chlordane, and *cis*- and *trans*-nonachlor). This is because (1) there are a limited number of guidelines available for individual components of chlordane, and (2) the guidelines available for these chlordane components actually apply to total chlordane. Therefore, comparison of these guidelines with the concentration of only one component of total chlordane would tend to underestimate the probability of adverse effects for that study.

Total chlordane was detected in 74 percent of all the monitoring studies in which it was targeted. Note that this includes studies that reported results for chlordane (unspecified), technical chlordane, and total chlordane. Figure 6.2 shows the cumulative frequency distribution of the maximum total chlordane concentrations reported by the studies reviewed, overlaid by the applicable sediment quality guideline values. There is fair agreement among upper screening values (4.8–8.9 µg/kg dry weight) and among lower screening values (0.5–4.5 µg/kg dry weight) for total chlordane in sediment (Table 6.4). The boundary values are listed in Table 6.4, and the percent of studies with maximum concentrations in each of Tiers 1, 2, and 3 are shown in Table 6.7. Figure 6.2 illustrates this information graphically. Fifty-five percent of all monitoring studies reviewed have a high probability of adverse effects at one or more sites in the study (Tier 1 studies), 19 percent have an intermediate probability of adverse effects at one or more sites in the study (Tier 2 studies), and the remaining 26 percent have no indication of adverse effects at any sites (Tier 3 studies). All Tier 3 studies had no detectable chlordane at any sites because the lowest screening value (the ERL at 0.5 µg/kg) was below the lowest of the maximum total chlordane concentrations (1 µg/kg) reported by any of the studies reviewed (Figure 6.2).

When only recently published (1984–1994) studies were considered, the highest maximum concentration reported by any of the studies reviewed was 510 µg/kg (Table 6.6), indicating that the highest total chlordane concentration in sediment reported by all the monitoring studies reviewed (1,000 µg/kg) was from a study published prior to 1984. When only recently published (1984–1994) studies were considered, the percentage of Tier 1 studies decreased (from 55 to 42

Table 6.7. Percentage of studies in *probability of adverse effects* classes, Tiers 1, 2, and 3, for pesticides in bed sediment

[Results are presented for all monitoring studies reviewed (published 1960–1994) and for recently published (1984–1994) studies. Tier 1: "i" indicates that there were insufficient guidelines to make a Tier 1 designation, so the corresponding Tier 2 studies have an intermediate or higher probability of adverse effects at the most contaminated site in each study. Abbreviation: i, insufficient guidelines to determine a Tier 1–2 boundary value]

Target Analytes	All Monitoring Studies (Published 1960–1994) Percentage of Studies in:			Recent Monitoring Studies (Published 1984–1994) Percentage of Studies in:		
	Tier 1	Tier 2	Tier 3	Tier 1	Tier 2	Tier 3
Chlordane[1]	55	19	26	42	27	31
DDD, *p,p'*-	46	4	50	43	2	55
DDE, *p,p'*-	44	12	44	40	11	49
DDT[2]	23	40	38	16	33	51
DDT, *p,p'*-	49	4	47	44	2	53
DDT, total	55	23	23	52	12	36
Diazinon	i	17	83	i	6	94
Dieldrin	33	32	35	29	25	46
Endosulfan[1]	i	10	90	i	9	91
Endosulfan I	i	50	50	i	46	54
Endosulfan II	i	29	71	i	23	77
Endrin	2	24	75	0	20	80
HCH[1]	10	20	70	11	11	78
HCH, α-	i	18	82	i	11	89
HCH, β-	i	15	85	i	17	83
HCH, δ-	0	4	96	0	4	96
Heptachlor epoxide	i	23	77	i	21	79
Hexachlorobenzene	i	21	79	i	17	83
Lindane	9	6	85	7	5	88
Malathion	i	0	100	i	0	100
Methoxychlor	i	7	93	i	8	92

Table 6.7. Percentage of studies in *probability of adverse effects* classes, Tiers 1, 2, and 3, for pesticides in bed sediment—*Continued*

Target Analytes	All Monitoring Studies (Published 1960–1994)			Recent Monitoring Studies (Published 1984–1994)		
	Percentage of Studies in:			Percentage of Studies in:		
	Tier 1	Tier 2	Tier 3	Tier 1	Tier 2	Tier 3
Nonachlor, *cis-*	[1]	5	95	[1]	5	95
Nonachlor, *trans-*	[1]	5	95	[1]	5	95
Pentachlorophenol	[1]	15	85	[1]	25	75
Toxaphene	[1]	12	88	[1]	11	89
Probability of adverse effects on aquatic life at the most contaminated site in each study:	High	Intermediate	No indication of adverse effects	High	Intermediate	No indication of adverse effects

[1]Total or unspecified.
[2](o,p' + p,p')-, or unspecified.

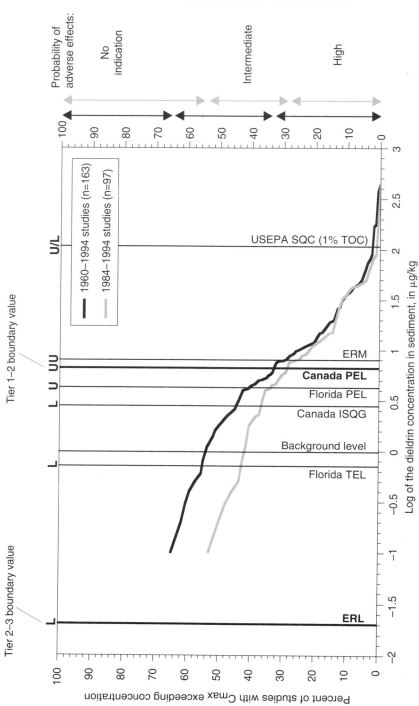

Figure 6.1. Cumulative frequency distribution of the maximum concentrations of dieldrin in sediment (μg/kg dry weight) reported by the monitoring studies reviewed, shown for all studies (published during 1960–1994) and recently published (1984–1994) studies. Guidelines for dieldrin in sediment are shown as light and heavy vertical lines. "Tier 1–2 boundary value" and "Tier 2–3 boundary value" indicate boundary concentrations separating Tier 1 from Tier 2 studies, and Tier 2 from Tier 3 studies, respectively; these boundary values are shown as heavy vertical lines. Abbreviations: C_{max}, maximum concentration in the study; ERL, effects range–low; ERM, effects range–median; ISQG, interim sediment quality guideline; **L**, lower screening value; n, number of studies; PEL, probable effect level; SQC, sediment-quality criterion; TEL, threshold effect level; TOC, total organic carbon (in sediment); **U**, upper screening value; **U/L**, USEPA considers this guideline to be both an upper and a lower screening value; USEPA, U.S. Environmental Protection Agency; μg/kg, microgram per kilogram.

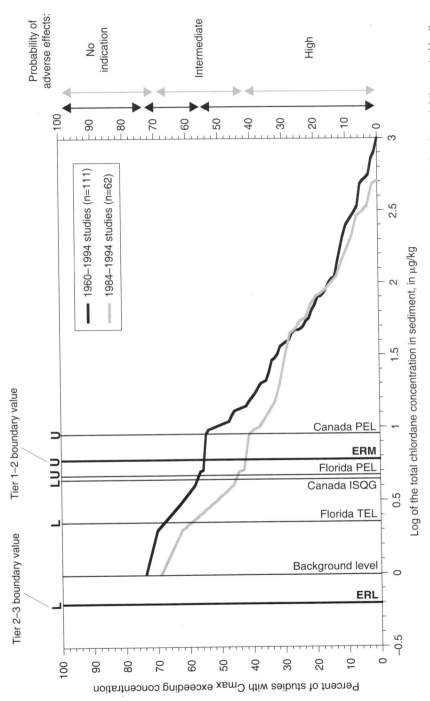

Figure 6.2. Cumulative frequency distribution of the maximum concentrations of total chlordane in sediment (μg/kg dry weight) reported by the monitoring studies reviewed, shown for all studies (published during 1960–1994) and recently published (1984–1994) studies. Studies that measured chlordane (unspecified), technical chlordane, or total chlordane are included. Guidelines for chlordane in sediment are shown as light and heavy vertical lines. "Tier 1–2 boundary value" and "Tier 2–3 boundary value" indicate boundary concentrations separating Tier 1 from Tier 2 studies, and Tier 2 from Tier 3 studies, respectively; these boundary values are shown as heavy vertical lines. Abbreviations: C_{max}, maximum concentration in the study; ERL, effects range–low; ERM, effects range–median; ISQG, interim sediment quality guideline; **L**, lower screening value; n, number of studies; PEL, probable effect level; TEL, threshold effect level; **U**, upper screening value; μg/kg, microgram per kilogram.

percent), the percentage of Tier 2 studies increased (from 19 to 27 percent), and the percentage of Tier 3 studies increased (from 26 to 31 percent), as shown in Table 6.7 and Figure 6.2.

DDT and Metabolites

Comparison of reported concentrations with sediment guidelines for the DDT family is complicated by the fact that different studies reported these compounds differently. Some studies analyzed for individual o,p' and p,p' isomers of DDD, DDE, and DDT; other studies analyzed for DDD, DDE, and DDT (isomers unspecified); some of both types of studies also reported data for total DDT; and still other studies reported data for total DDT only. Moreover, some sediment guidelines were established for individual isomers (such as p,p'-DDD) and others for total DDT. In Table 6.7, separate comparisons with applicable guidelines are made for reported maximum concentrations of p,p'-DDD, p,p'-DDE, p,p'-DDT, DDT (the sum of o,p' and p,p' isomers), and total DDT. These are the most commonly detected DDT compounds for which sediment quality guidelines are available. For example, a single study that reported data for p,p'-DDD, p,p'-DDE, p,p'-DDT, and total DDT would be included in the calculations for each of these compounds. Note that studies that reported data for DDT (unspecified) were grouped with those that measured o,p'- plus p,p'-DDT.

The cumulative frequency distribution of p,p'-DDE is shown in Figure 6.3. The lower screening values for p,p'-DDE are fairly close together (1.4–9 µg/kg), but there is a large spread in the upper screening values (6.8–374 µg/kg) for this compound. Moreover, there is some overlap between upper and lower screening values, with the AET-L (a lower screening value) higher than Canada's PEL (an upper screening value). Using the USEPA-based procedure for classifying studies, the probability of adverse effects at one or more sites is high for 44 percent of studies (Tier 1 studies); the probability is intermediate for another 12 percent of studies (Tier 2 studies), and there is no indication of adverse effects at any sites for the remaining 44 percent of studies (Tier 3 studies), as shown in Table 6.7. When only recently published (1984–1994) studies were considered, there was only a small decrease in the number of studies that exceeded Tier 1–2 and Tier 2–3 boundary values: 40 percent of recently published studies were in Tier 1, 11 percent in Tier 2, and 49 percent in Tier 3 (Table 6.7). There was no change in the maximum concentration observed because the highest p,p'-DDE concentration reported (1,870 µg/kg dry weight) was from a study published after 1983.

The results for p,p' isomers of DDT and DDD were fairly similar to those for p,p'-DDE (Table 6.7). Between 46–49 percent of studies were classified as Tier 1 studies (high probability of adverse effects at the most contaminated site), 4 percent of studies as Tier 2 studies (intermediate probability of adverse effects at the most contaminated site), and 47–50 percent as Tier 3 studies (no indication of adverse effects at any sites in the study). When only recently published (1984–1994) studies were considered, there was only a slight reduction in the percent of studies that exceeded boundary values for p,p'-DDT and p,p'-DDD (Table 6.7). Note that there was no change in maximum concentrations of p,p' isomers of DDT or DDD, when only recently published (1984–1994) studies were considered because the highest maximum concentrations of these compounds came from studies published after 1983.

The percentage of studies with maximum concentrations in Tiers 1–3 are quite different for DDT (23 percent in Tier 1, 40 percent in Tier 2, and 38 percent in Tier 3), than for p,p'-DDT (49 percent in Tier 1, 4 percent in Tier 2, and 47 percent in Tier 3), as shown in Table 6.7. Two

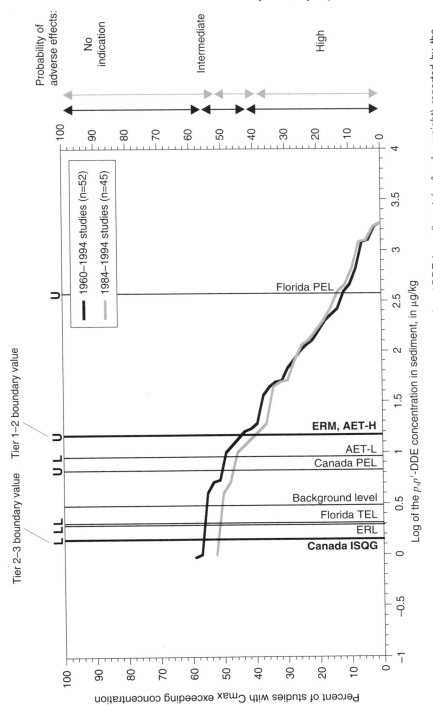

Figure 6.3. Cumulative frequency distribution of the maximum concentrations of *p,p'*-DDE in sediment (μg/kg dry weight) reported by the monitoring studies reviewed, shown for all studies (published during 1960–1994) and recently published (1984–1994) studies. Guidelines for *p,p'*-DDE in sediment are shown as light and heavy vertical lines. "Tier 1–2 boundary value" and "Tier 2–3 boundary value" indicate boundary concentrations separating Tier 1 from Tier 2 studies, and Tier 2 from Tier 3 studies, respectively; these boundary values are shown as heavy vertical lines. Abbreviations: AET-H, apparent effects threshold–high; AET-L, apparent effects threshold–low; C$_{max}$, maximum concentration in the study; ERL, effects range–low; ERM, effects range–median; ISQG, interim sediment quality guideline; L, lower screening value; n, number of studies; PEL, probable effect level; TEL, threshold effect level; U, upper screening value; μg/kg, microgram per kilogram.

factors probably contribute to this difference. First, the Tier 1–2 boundary value is lower for p,p'-DDT (7 µg/kg) than DDT (34 µg/kg), as shown in Table 6.4. Logically, we may expect these two pesticide analytes to be equivalent in toxicity. However, in practice, the boundary values are determined on the basis of different sets of empirical data. Second, the cumulative frequency distributions of the maximum concentrations for these two analytes represent two different groups of studies. A higher fraction of all monitoring studies that measured p,p'-DDT were recently published (84 percent of all monitoring studies were published during 1984–1994) than the monitoring studies that measured DDT (47 percent). Recall that the latter include studies that reported data for DDT (unspecified) as well as for the sum of o,p'- and p,p'-DDT. As analytical methodology improved over time, more studies analyzed for individual isomers of DDT and metabolites. The higher percent of Tier 3 studies for p,p'-DDT (47 percent) than for DDT (38 percent) may be due in part to declining residues of DDT in sediment since the 1972 ban. When only recently published (1984–1994) studies were considered, the percent of studies in Tier 3 is very similar for the two analytes (51 percent for p,p'-DDT and 53 percent for DDT).

For total DDT, the cumulative frequency distribution of maximum concentrations is shown in Figure 6.4. The lower screening values (1.58–9 µg/kg) and the upper screening values (15–52 µg/kg) for total DDT show good agreement. The USEPA-based procedure permits classification of studies into "probability of adverse effects" categories on the basis of the maximum concentration of total DDT in each study. The probability is high for adverse effects at one or more sites for 55 percent of all studies reviewed (Tier 1 studies); the probability is intermediate for another 23 percent of studies (Tier 2 studies); and there is no indication of adverse effects at any sites for 23 percent of studies (Tier 3 studies). When only recently published (1984–1994) studies were considered, the maximum concentration observed decreased from 30,200 mg/kg (all studies) to 4,443.5 mg/kg total DDT (recently published studies). When only recently published (1984–1994) studies were considered, there was little change in the percentage of Tier 1 studies (from 55 to 52 percent), but the percentage of Tier 2 studies decreased (from 23 to 12 percent), and the percentage of Tier 3 studies increased (from 23 to 36 percent).

To summarize for the DDT family, the probability of adverse effects at the most contaminated site appears to be high for about 23–55 percent of all monitoring studies reviewed; there is no indication of adverse effects at any sites for another 23–50 percent of studies; and the remainder (4–27 percent) of studies have an intermediate probability of adverse effects at the most contaminated site, as shown in Table 6.7. When only recently published studies (1984–1994) were considered, there was some decrease in the percentage of Tier 1 studies and a corresponding increase in the percentage of Tier 3 studies. When only recently published studies were considered, the decrease in Tier 1 studies and the increase in Tier 3 studies were greater for DDT than for p,p'-DDT because a high proportion of studies that targeted p,p'-DDT were recently published (84 percent) compared with those that targeted DDT (47 percent).

Other Pesticides

Of the remaining pesticide analytes that have sediment quality guidelines, only four analytes had sufficient guidelines (see Table 6.4) to determine a Tier 1–2 boundary value (which requires at least two upper screening values): endrin, HCH, δ-HCH, and lindane. For endrin, 2 percent of all monitoring studies (published during 1960–1994) were in Tier 1 (high probability of adverse effects at one or more sites in the study), 24 percent were in Tier 2 (intermediate

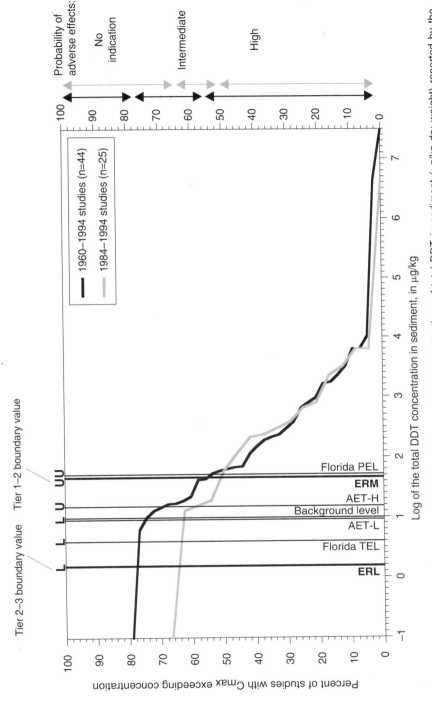

Figure 6.4. Cumulative frequency distribution of the maximum concentrations of total DDT in sediment (μg/kg dry weight) reported by the monitoring studies reviewed, shown for all studies (published during 1960–1994) and recently published (1984–1994) studies. Total DDT is the sum of residues of o,p'-DDT, o,p'-DDE, o,p'-DDD, p,p'-DDT, p,p'-DDE, and p,p'-DDD. Guidelines for total DDT in sediment are shown as light and heavy vertical lines. "Tier 1–2 boundary value" and "Tier 2–3 boundary value" indicate boundary concentrations separating Tier 1 from Tier 2 studies, and Tier 2 from Tier 3 studies, respectively; these boundary values are shown as heavy vertical lines. Abbreviations: AET-H, apparent effects threshold–high; AET-L, apparent effects threshold–low; C_{max}, maximum concentration in the study; ERL, effects range–low; ERM, effects range–median; n, number of studies; PEL, probable effect level; TEL, threshold effect level; **U**, upper screening value; μg/kg, microgram per kilogram.

probability of adverse effects at one or more sites), and 75 percent in Tier 3 (no indication of adverse effects at any sites), as shown in Table 6.7. HCH (which includes technical HCH and total HCH measurements) and lindane had a higher percentage of all studies in Tier 1 (9–10 percent) than δ-HCH (0 percent). HCH also had 20 percent of studies in Tier 2, compared with only 4–6 percent for lindane and δ-HCH. When only recently published (1984–1994) studies were considered, the percentage of studies in Tier 3 increased for these four pesticide analytes. The percentage of studies in Tier 1 actually increased slightly (from 10 to 11 percent) for HCH. This increase occurred because there are only 10 total studies that measured HCH (total or technical); 9 of these studies were published recently (after 1983), including the single study in Tier 1.

The remaining pesticides in Table 6.7 do not have sufficient sediment quality guidelines available to determine a Tier 1–2 boundary value (see Table 6.4): diazinon, the endosulfan compounds, α- and β-HCH, heptachlor epoxide, hexachlorobenzene, malathion, methoxychlor, pentachlorophenol, and toxaphene. For these analytes, therefore, we can only distinguish between Tier 2 studies (with an intermediate or higher probability of adverse effects at one or more sites in the study) and Tier 3 studies (with no indication of adverse effects at any sites). This is illustrated for diazinon in Figure 6.5, which shows the cumulative frequency distribution of the maximum diazinon concentrations in all monitoring studies (published during 1960–1994) that were reviewed. For diazinon, 17 percent of all studies were Tier 2 studies and the remaining 83 percent were Tier 3 studies. Recently published studies were not plotted in Figure 6.5 because only two studies reported maximum diazinon concentrations above the detection limit. Of the pesticide analytes with no Tier 1–2 boundary value, endosulfan I has the highest percentage of Tier 2 studies (50 percent). Three analytes have 20–30 percent of studies in Tier 2 (endosulfan II, heptachlor epoxide, and hexachlorobenzene), followed by several analytes with 10–20 percent of studies in Tier 2 (diazinon, endosulfan [total or technical], α- and β-HCH, pentachlorophenol, and toxaphene). Methoxychlor has 6 percent of studies in Tier 2. For malathion, which was not detected at any sites in any of the monitoring studies that analyzed for this compound (41 total studies), there is no indication of adverse effects at any sites in these studies (100 percent Tier 3 studies). When only recently published (1984–1994) studies were considered, most analytes showed a slight decrease in the percentage of Tier 2 studies and a corresponding increase in the percentage of Tier 3 studies. The percentages for malathion (100 percent Tier 3 studies) and toxaphene (94 percent Tier 3 studies) remained the same. When only recently published studies were considered, three pesticide analytes showed slight increases in the percentage of Tier 2 studies: β-HCH, methoxychlor, and pentachlorophenol. These increases occurred because the majority of studies for these three analytes were recently published (after 1983), including all or most of the Tier 1 studies.

It is important to note that there are relatively few (less than 15) total studies for several pesticides: endosulfan I and II, HCH, δ-HCH, pentachlorophenol, and toxaphene. Because of the small sample sizes, the results for these pesticide analytes may not be at all representative of conditions in United States rivers and streams.

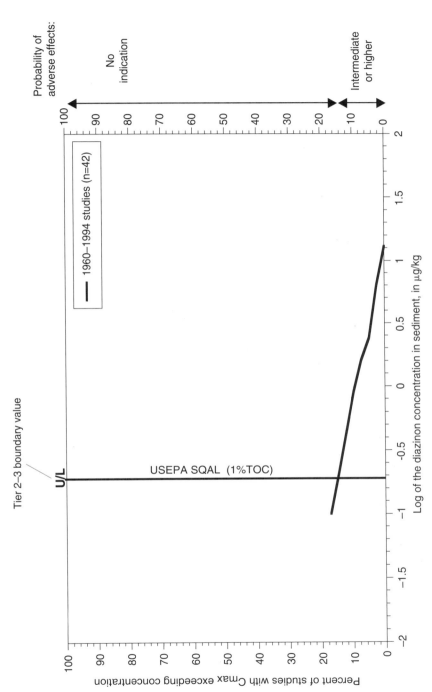

Figure 6.5. Cumulative frequency distribution of the maximum concentrations of diazinon in sediment (µg/kg dry weight) reported by the monitoring studies reviewed, shown for all studies (published during 1960–1994). The only available guideline for diazinon in sediment is shown as a heavy vertical line. There are insufficient guidelines to determine a Tier 1–2 boundary value. "Tier 2–3 boundary value" indicates the boundary concentration separating Tier 2 from Tier 3 studies. Abbreviations: C_{max}, maximum concentration in the study; n, number of studies; SQAL, sediment-quality advisory level; TOC, total organic carbon (in sediment); **U/L**, USEPA considers this guideline to be both an upper and a lower screening value; USEPA, U.S. Environmental Protection Agency; µg/kg, microgram per kilogram.

Summary and Conclusions

The maximum concentrations reported in the monitoring studies reviewed were compared with available sediment guidelines for individual pesticide analytes, and each study was classified according to the probability of adverse effects at the most contaminated site (or sites) in that study. The procedure used to classify studies was based on methodology developed by the U.S. Environmental Protection Agency (1997a) and used to classify individual sites in the National Sediment Inventory. Table 6.7 lists the pesticide analytes for which sediment guidelines are available, as well as the percentage of studies in each of three tiers: Tier 1 studies have a high probability of adverse effects at one or more sites in each study; Tier 2 studies have an intermediate probability of adverse effects at one or more sites in each study; and Tier 3 studies have no indication of adverse effects at any sites in the study.

The percentage of studies in Tiers 1, 2, and 3 for individual pesticide analytes is shown graphically in Figure 6.6. Analytes are listed in order of increasing percentage of Tier 3 studies

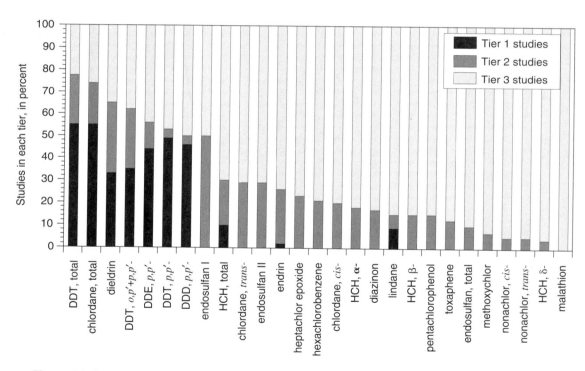

Figure 6.6. Percentage of all monitoring studies (published 1960–1994) in *probability of adverse effects* categories. For Tier 1 studies, there is a high probability of adverse effects at the most contaminated site, or sites, in each study; for Tier 2 studies, there is an intermediate probability of adverse effects at the most contaminated site, or sites, in each study; for Tier 3 studies, there is no indication of adverse effects at any sites in each study. Compounds are listed in order of increasing percentage of Tier 3 studies

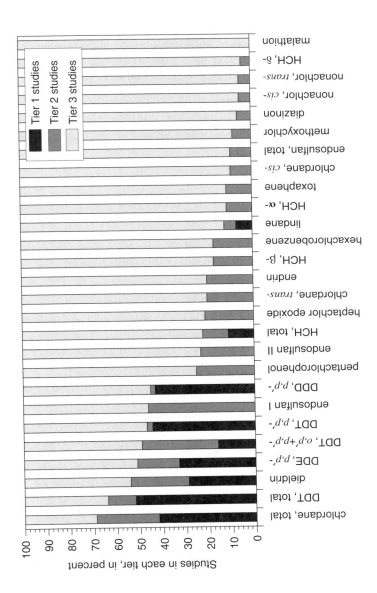

Figure 6.7. Percentage of recently published (1984–1994) monitoring studies in *probability of adverse effects* categories. For Tier 1 studies, there is a high probability of adverse effects at the most contaminated site, or sites, in each study; for Tier 2 studies, there is an intermediate probability of adverse effects at the most contaminated site, or sites, in each study; for Tier 3 studies, there is no indication of adverse effects at any sites in each study. Compounds are listed in order of increasing percentage of Tier 3 studies.

(i.e., from the highest to the lowest percentage of studies with an intermediate-to-high probability of adverse effects). These analytes fall naturally into two groups: For the DDT compounds, total chlordane, dieldrin, and endosulfan I, this analysis indicates an intermediate-to-high probability of adverse effects (Tiers 1 and 2) occurring at the most contaminated site or sites in a substantial percentage (50 percent or more) of the monitoring studies reviewed. It is important to note that this analysis provides no information on the number or proportion of sites within each study that has a high or intermediate probability of adverse effects on aquatic life. Analytes in the second group all have fewer than 50 percent of studies in Tiers 1 and 2 (Figure 6.6).

When only recently published (1984–1994) studies are considered, pesticide analytes again fall into two groups. As illustrated in Figure 6.7, the DDT compounds, total chlordane, dieldrin, and endosulfan I all have more than 45 percent of studies in Tiers 1 and 2, compared with fewer than 25 percent of studies for the other pesticide analytes.

6.1.3 TOXICITY TO WILDLIFE

Potential toxicity to wildlife will be estimated by comparing maximum pesticide concentrations from the monitoring studies reviewed with available guidelines for whole fish. Because wildlife generally consume the entire fish, guidelines for protection of wildlife apply to residues in whole fish. All national, multistate, state, and local monitoring studies (Tables 2.1 and 2.2) that reported pesticide residues in whole fish are included in these comparisons.

Guidelines for Pesticides in Whole Fish

Reported pesticide residues in whole fish are compared with two types of guidelines for the protection of wildlife: maximum recommended tissue concentrations from the National Academy of Sciences and National Academy of Engineering (NAS/NAE) and fish-flesh criteria from the New York State Department of Environmental Conservation (NYSDEC). Guideline values for whole fish are shown in Table 6.8.

NAS/NAE's Guidelines for Protection of Fish-Eating Wildlife

In 1973, the NAS/NAE issued maximum recommended concentrations in water for protection of freshwater aquatic life (which preceded the USEPA ambient water-quality criteria for the protection of aquatic organisms). However, because certain organochlorine pesticides persist and accumulate in aquatic organisms, these maximum recommended concentrations in water were not considered protective of predators. The National Academy of Sciences and National Academy of Engineering (1973) reasoned that, because of trophic accumulation, birds and mammals that occupy higher trophic levels in the food web may acquire body burdens that are lethal or that have significant sublethal effects on reproductive capacity, even though the concentration in water remains very low. Therefore, the NAS/NAE made recommendations for maximum tissue concentrations of selected pesticides for protection of predators. These guidelines are over 20 years old and were considered preliminary even at the time. Moreover, there is little information in the NAS/NAE report (National Academy of Sciences and National Academy of Engineering, 1973) concerning the technical basis for these guidelines. Nonetheless, they remain the only national guidelines available for the protection of fish-eating predators.

New York's Fish Flesh Criteria for Piscivorous Wildlife

The NYSDEC has developed fish flesh criteria for the protection of *piscivorous* (fish-eating) wildlife (Newell and others, 1987). This effort focused on organochlorine chemicals that were found in spottail shiners from the Niagara River. Because few feeding studies with wildlife were available, the NYSDEC used the same extensive laboratory animal toxicology database that is used to derive criteria for the protection of human health. Instead of extrapolating from laboratory animals to humans, the NYSDEC extrapolated from laboratory animals to wildlife. From all target species, the bird and mammal with the greatest ratios of daily food consumption to body weight were selected to derive the wildlife criteria. Because several birds consume about 20 percent of their body weight per day, a generic bird was selected with a body weight of 1 kg and a food consumption rate of 0.2 kg/d. To represent fish-eating mammals, the mink was selected, which has an average body weight of 1 kg and a food consumption rate of 0.15 kg/d. Various uncertainty factors were applied on the basis of the type and quantity of the toxicity data available. For five chemicals for which wildlife feeding studies were available, the criteria that were based on laboratory animal data were compared with criteria derived directly from the wildlife feeding studies. For four of the five chemicals, the criteria values that were based on laboratory animal data were somewhat lower than the values derived directly from wildlife feeding studies. The criteria based on laboratory animal data were therefore judged protective of target wildlife species.

Fish flesh criteria were derived both for noncarcinogenic endpoints of toxicity and for carcinogenic effects. The carcinogenicity-based criteria values were based on an acceptable risk of 1 in 100 for protection of wildlife populations, although this assumption could not be fully justified at the time (Newell and others, 1987). For potential human carcinogens, USEPA considers risk levels of 10^{-6} to 10^{-4} to be protective of public health (Nowell and Resek, 1994). However, because it is unclear what level of cancer risk to wildlife would be acceptable, only the noncarcinogenicity-based criteria are used in the analysis below.

Pesticides in Whole Fish—Analysis of Potential Adverse Effects on Fish-Eating Wildlife

Guidelines for the protection of fish-eating wildlife from the National Academy of Sciences and National Academy of Engineering (1973) and the NYSDEC (Newell and others, 1987) are shown in Table 6.8. Note that the NAS/NAE guidelines apply to the combined residues of a number of pesticides. Table 6.8 includes NYSDEC fish-flesh criteria for both noncancer and cancer effects, although the latter are not used in the analysis below.

For all monitoring studies that measured pesticide concentrations in whole fish, the maximum concentrations reported were compared with the applicable NAS/NAE and NYSDEC guidelines (Table 6.9). Thus, measured concentrations that exceed a guideline by a given study indicates the potential for effects on fish-eating wildlife only at the most contaminated site in the study. This comparison provides no information on the number of sites in each study where

Table 6.8. Standards and guidelines for pesticides in aquatic biota

[Whole fish: Guidelines are for protection of fish-eating wildlife. Fish muscle: Standards and guidelines are for protection of human health. Shellfish: Standards are for protection of human health. NAS/NAE guidelines are from National Academy of Sciences and National Academy of Engineering (1973). NYSDEC guidelines are from Newell and others (1987). FDA action levels for fish and shellfish are from Food and Drug Administration (1989b). USEPA tolerances are from the Code of Federal Regulations, volume 40, Part 180 (with the subpart specified in appropriate footnotes). USEPA screening values are from U.S. Environmental Protection Agency (1995a). Abbreviations and symbols: FDA, Food and Drug Administration; NAS/NAE, National Academy of Sciences and National Academy of Engineering; NYSDEC, New York State Department of Environmental Conservation; USEPA, U.S. Environmental Protection Agency; q_1*, cancer potency factor; µg/kg, microgram per kilogram; —, no standard or guideline available]

Target Analytes	Whole Fish			Fish Muscle			Shellfish
	NAS/NAE Maximum Recommended Concentrations (µg/kg)[1]	NYSDEC Fish Flesh Criteria for Piscivorous Wildlife (µg/kg)		FDA Action Level (µg/kg)	USEPA Tolerance (µg/kg)	USEPA Guidance for Use in Fish Advisories: Screening Value[1] (µg/kg)[1]	FDA Action Level (µg/kg)
		Noncancer Effects[1]	Cancer Effects (1:100)[2]				
Aldrin	[3]100	[4]120	[4]22	[4]300	—	—	[4]300
Chlordane	[3]100	[5]500	[5]370	[6]300	—	[7,8]80	[6]300
Chlordane, *cis-*	[3]100	[5]500	[5]370	[6]300	—	[7,8]80	[6]300
Chlordane, *trans-*	[3]100	[5]500	[5]370	[6]300	—	[7,8]80	[6]300
Chlordene	—	—	—	[6]300	—	—	[6]300
Chlorpyrifos	—	—	—	—	—	30,000	—
D, 2,4-	—	—	—	—	[9]1,000	—	—
DDD	[10]1,000	[10]200	[10]266	[11]5,000	—	[7,12]300	[11]5,000
DDD, *o,p'-*	[10]1,000	[10]200	[10]266	[11]5,000	—	[7,12]300	[11]5,000
DDD, *p,p'-*	[10]1,000	[10]200	[10]266	[11]5,000	—	[7,12]300	[11]5,000
DDE	[10]1,000	[10]200	[10]266	[11]5,000	—	[7,12]300	[11]5,000
DDE, *o,p'-*	[10]1,000	[10]200	[10]266	[11]5,000	—	[7,12]300	[11]5,000
DDE, *p,p'-*	[10]1,000	[10]200	[10]266	[11]5,000	—	[7,12]300	[11]5,000
DDT	[10]1,000	[10]200	[10]266	[11]5,000	—	[7,12]300	[11]5,000

Table 6.8. Standards and guidelines for pesticides in aquatic biota—*Continued*

Target Analytes	Whole Fish				Fish Muscle		Shellfish
	NAS/NAE Maximum Recommended Concentrations (µg/kg)[1]	NYSDEC Fish Flesh Criteria for Piscivorous Wildlife (µg/kg)		FDA Action Level (µg/kg)	USEPA Tolerance (µg/kg)	USEPA Guidance for Use in Fish Advisories: Screening Value[1] (µg/kg)[1]	FDA Action Level (µg/kg)
		Noncancer Effects[1]	Cancer Effects (1:100)[2]				
DDT, *o,p'*-	[10]1,000	[10]200	[10]266	[11]5,000	—	[7,12]300	[11]5,000
DDT, *p,p'*-	[10]1,000	[10]200	[10]266	[11]5,000	—	[7,12]300	[11]5,000
DDT, total	[10]1,000	[10]200	[10]266	[11]5,000	—	[7,12]300	[11]5,000
Diazinon	—	—	—	—	—	900	—
Dicofol	—	—	—	—	—	10,000	—
Dieldrin	[3]100	[4]120	[4]22	[4]300	—	[7]7	[4]300
Diquat	—	—	—	—	[13]100	—	—
Disulfoton	—	—	—	—	—	500	—
Endosulfan	[3]100	—	—	—	—	60,000	—
Endosulfan I	[3]100	—	—	—	—	[14]60,000	—
Endosulfan II	[3]100	—	—	—	—	[14]60,000	—
Endrin	[3]100	25	—	300	—	3,000	300
Ethion	—	—	—	—	—	5,000	—
Fluridone	—	—	—	—	[15]500	—	—
Glyphosate	—	—	—	—	[16]250	—	—

Table 6.8. Standards and guidelines for pesticides in aquatic biota—*Continued*

Target Analytes	Whole Fish			Fish Muscle			Shellfish
	NAS/NAE Maximum Recommended Concentrations (µg/kg)[1]	NYSDEC Fish Flesh Criteria for Piscivorous Wildlife (µg/kg)		FDA Action Level (µg/kg)	USEPA Tolerance (µg/kg)	USEPA Guidance for Use in Fish Advisories: Screening Value[1] (µg/kg)[1]	FDA Action Level (µg/kg)
		Noncancer Effects[1]	Cancer Effects (1:100)[2]				
HCH, total	[3]100	100	510	—	—	—	—
Heptachlor	[3]100	[17]200	[17]210	[17]300	—	—	[17]300
Heptachlor epoxide	[3]100	[17]200	[17]210	[17]300	—	[7]10	[17]300
Hexachlorobenzene	—	330	200	—	—	[7]70	—
Lindane	[3]100	100	510	—	—	[7]80	—
Mirex	—	330	330	100	—	[7,18]60	100
Nonachlor, *cis-*	—	[5]500	370	[6]300	—	[7,8]80	[6]300
Nonachlor, *trans-*	—	[5]500	370	[6]300	—	[7,8]80	[6]300
Simazine	—	—	—	—	[19]12,000	—	—
Terbufos	—	—	—	—	—	1,000	—
Toxaphene	[3]100	—	—	5,000	—	[7]100	5,000

[1] Based on chronic toxicity, unless otherwise specified.
[2] Based on 1 in 100 cancer risk.
[3] Applies to total residues of aldrin, chlordane, dieldrin, endosulfan, endrin, HCH, heptachlor, heptachlor epoxide, lindane, and toxaphene, either singly or in combination.
[4] Applies to aldrin plus dieldrin.
[5] Applies to technical (or total) chlordane.
[6] Applies to sum of *cis-* and *trans-*chlordane, *cis-* and *trans-*nonachlor, oxychlordane, α-, β-, and δ-chlordene, and chlordene.
[7] Based on 1 in 100,000 cancer risk.
[8] Applies to sum of *cis-* and *trans-*chlordane, *cis-* and *trans-*nonachlor, and oxychlordane.

[9]Applies to use of dimethylamine salt for water hyacinth control in slow-moving or quiescent water bodies federal, state, or local public agencies (Part 180.142(f)); or to use of dimethylamine salt or its butoxyethanol ester for Eurasian watermilfoil control in Tennessee Valley Authority programs (Part 180.142(i)).

[10]Applies to total DDT residues, including DDD and DDE.

[11]Applies to residues of DDT, DDD and DDE, either singly or in combination.

[12]Applies to sum of o,p' and p,p' isomers of DDT, DDD and DDE.

[13]Calculated as the cation. Applies to use of the dibromide salt to slow-moving or quiescent water bodies in programs of federal or state public agencies, or to ponds, lakes, and drainage ditches with little or no outflow of water and which are totally under the control of the user (Part 180.226(b)).

[14]Applies to endosulfan I plus endosulfan II.

[15]For the combined residues (free and bound) of fluridone (Part 180.420(a)).

[16]For the combined residues of glyphosate and its metabolite, aminomethyl-phosphonic acid. Applies to use of glyphosate isopropylamine salt and glyphosate monoammonium salt (Part 180.364(a)).

[17]Applies to heptachlor plus heptachlor epoxide.

[18]Calculated using the equation provided in U.S. Environmental Protection Agency (1995a) and the q_1* of 1.8 $(mg/kg/day)^{-1}$, which is from U.S. Environmental Protection Agency (1992b).

[19]For the combined residues of simazine and its metabolites, 2-amino-4-chloro-6-ethylamino-s-triazine and 2,4-diamino-6-chloro-s-triazine (Part 180.213(a)).

Table 6.9. Selected results of studies that monitored pesticides in whole fish and the percentage of those studies that exceeded guidelines for the protection of fish-eating wildlife

[Results are presented for all monitoring studies (published 1960–1994) and recently published (1984–1994) monitoring studies that are listed in Tables 2.1 and 2.2. All concentrations are wet weight. NAS/NAE guideline, maximum recommended concentration for protection of fish-eating wildlife. NYSDEC guideline: New York fish flesh criterion for protection of piscivorous wildlife. Abbreviations and symbols: C_{max}, maximum concentration in the study; NAS/NAE, National Academy of Sciences and National Academy of Engineering; NYSDEC, New York State Department of Environmental Conservation; na, data not available; nd, not detected; nr, one or more studies did not report this information; µg/kg, microgram per kilogram; —, no guideline available]

Target Analytes	Whole Fish: All Monitoring Studies (Published 1960–1994)					
	Concentration Range (µg/kg wet weight)		Total Number of Studies	Percentage of Studies with Detectable Residues	Percentage of Studies with C_{max} that Exceeded NAS/NAE Guideline[1]	Percentage of Studies with C_{max} that Exceeded NYSDEC Guideline[2]
	In Detection Limits	In Maximum Concentrations (C_{max})				
Aldrin	nr, 0.01–50	nd–12	32	19	0	0
Acephate	50	nd	1	0	—	—
Alachlor	nr, 10–100	nd	2	0	—	—
Aldicarb	50	nd	1	0	—	—
Atrazine	10–8,000	nd	3	0	—	—
Azinphos-methyl	50–300	nd	2	0	—	—
Carbaryl	5–50	nd–50	1	0	—	—
Carbofuran	nr, 20–200	nd–560	1	100	—	—
Carbofuran, 3-hydroxy-	20–200	1,490	1	100	—	—
Carbophenothion	50	nd	1	0	—	—
Chlordane[1]	nr, 0.1–100	nd–870	27	63	33	4
Chlordane, cis-	nr, 0.01–100	nd–1,090	44	61	20	5
Chlordane, trans-	nr, 0.01–50	nd–970	42	62	12	7
Chlorpyrifos	nr, 10	nd	2	0	—	—
Coumaphos	nr	nd	1	0	—	—
Cyanazine	10	nd	1	0	—	—
D, 2,4-	1	6	1	100	—	—
Dacthal (DCPA)	nr, 2–10	nd–13,400	14	79	—	—
DDD[1]	nr, 0.1–50	nd–12,500	19	84	26	26
DDD, o,p'-	nr, 0.01–50	nd–420	34	29	0	9
DDD, p,p'-	nr, 0.01–100	nd–31,000	61	77	10	31
DDE[1]	nr, 0.1–100	nd–31,500	21	90	33	62
DDE, o,p'-	nr, 0.01–50	nd–360	35	23	0	3
DDE, p,p'-	nr, 0.01–100	nd–140,000	61	89	28	43
DDT[2]	nr, 0.1–50	nd–6,750	22	77	27	50
DDT, o,p'-	nr, 0.01–50	nd–720	37	35	0	5

Table 6.9. Selected results of studies that monitored pesticides in whole fish and the percentage of those studies that exceeded guidelines for the protection of fish-eating wildlife—*Continued*

Target Analytes	Whole Fish: Recent Monitoring Studies (Published 1984–1994)					
	Concentration Range (µg/kg wet weight)		Total Number of Studies	Percentage of Studies with Detectable Residues	Percentage of Studies with C_{max} that Exceeded NAS/NAE Guideline[1]	Percentage of Studies with C_{max} that Exceeded NYSDEC Guideline[2]
	In Detection Limits	In Maximum Concentrations (C_{max})				
Aldrin	nr, 0.01–50	nd–12	25	16	0	0
Acephate	nr	nd	1	0	—	—
Alachlor	na	na	0	—	—	—
Aldicarb	nr	nd	1	0	—	—
Atrazine	300–8,000	nd	1	0	—	—
Azinphos-methyl	50–300	nd	2	0	—	—
Carbaryl	50	nd	1	0	—	—
Carbofuran	nr, 20–200	nd–560	2	50	—	—
Carbofuran, 3-hydroxy-	20–200	1,490	1	100	—	—
Carbophenothion	50	nd	1	0	—	—
Chlordane[1]	nr, 0.1–100	nd–870	18	67	33	6
Chlordane, *cis*-	nr, 0.01–106	nd–1,090	43	60	21	5
Chlordane, *trans*-	nr, 0.01–50	nd–970	41	61	12	7
Chlorpyrifos	nr, 10	nd	2	0	—	—
Coumaphos	nr	nd	1	0	—	—
Cyanazine	na	na	0	—	—	—
D, 2,4-	na	na	0	—	—	—
Dacthal (DCPA)	nr, 2–12	nd–2,300	12	75	—	—
DDD[1]	nr, 0.1–50	nd–1,290	9	67	11	11
DDD, *o,p'*-	nr, 0.01–50	nd–420	34	29	0	9
DDD, *p,p'*-	nr, 0.01–100	nd–19,000	54	74	9	28
DDE[1]	nr, 0.1–50	nd–1,460	9	78	22	56
DDE, *o,p'*-	nr, 0.01–50	nd–360	35	23	0	3
DDE, *p,p'*-	nr, 0.01–100	nd–31,000	53	87	28	43
DDT[2]	nr, 0.1–50	nd–1,500	10	60	10	20
DDT, *o,p'*-	nr, 0.01–50	nd–720	35	31	0	6

Table 6.9. Selected results of studies that monitored pesticides in whole fish and the percentage of those studies that exceeded guidelines for the protection of fish-eating wildlife—*Continued*

Target Analytes	Concentration Range (µg/kg wet weight)		Total Number of Studies	Percentage of Studies with Detectable Residues	Percentage of Studies with C_{max} that Exceeded NAS/NAE Guideline[1]	Percentage of Studies with C_{max} that Exceeded NYSDEC Guideline[2]
	In Detection Limits	In Maximum Concentrations (C_{max})				
DDT, *p,p'*-	nr, 0.01–100	nd–4,600	55	58	7	20
DDT, total	nr, 0.1–100	nd–28,880	34	100	50	82
Demeton	50	nd	1	0	—	—
Diazinon	nr, 100	nd	2	0	—	—
Dichlorvos	nr	nd	1	0	—	—
Dicofol	nr, 1–10	nd–560	10	40	—	—
Dieldrin	nr, 0.01–100	nd–12,500	87	74	31	30
Dimethoate	41–50	nd	1	0	—	—
Disulfoton	nr	nd	1	0	—	—
Endosulfan[1]	nr, 0.1–20	nd–170	6	67	17	—
Endosulfan I	nr, 0.02–10	nd–285	8	38	13	—
Endosulfan II	nr, 2–10	nd–40	7	29	0	—
Endosulfan sulfate	nr, 2–200	nd–20	6	17	—	—
Endrin	nr, 0.01–100	nd–2,060	65	28	5	9
Endrin aldehyde	nr, 200	nd	2	0	—	—
Endrin ketone	nr	nd	1	0	—	—
EPN	200	nd	1	0	—	—
Ethoprop	nr	nd	1	0	—	—
Famphur	nr	nd	1	0	—	—
Fensulfothion	nr	nd	1	0	—	—
Fenthion	nr	nd	1	0	—	—
Fenvalerate	nr	nd–11	2	50	—	—
HCH[1]	nr, 0.1–100	nd–170	12	50	17	17
HCH, α-	nr, 0.01–100	nd–610	34	35	—	—
HCH, β-	nr, 0.01–100	nd–900	31	23	—	—
HCH, δ-	nr, 0.01–100	nd–410	25	12	—	—
Heptachlor	nr, 0.01–50	nd–600	34	35	6	6
Heptachlor epoxide	nr, 0.01–104	nd–480	64	50	5	2
Hexachlorobenzene	nr, 0.01–100	nd–27,000	51	45	—	14
Kepone	nr, 10–50	nd–2,800	6	50	—	—
Lindane	nr, 0.01–100	nd–120	42	24	2	2
Malathion	nr, 1–50	nd	3	0	—	—
Methamidophos	nr	nd	1	0	—	—

Table 6.9. Selected results of studies that monitored pesticides in whole fish and the percentage of those studies that exceeded guidelines for the protection of fish-eating wildlife—*Continued*

Target Analytes	Whole Fish: Recent Monitoring Studies (Published 1984–1994)					
	Concentration Range (µg/kg wet weight)		Total Number of Studies	Percentage of Studies with Detectable Residues	Percentage of Studies with C_{max} that Exceeded NAS/NAE Guideline[1]	Percentage of Studies with C_{max} that Exceeded NYSDEC Guideline[2]
	In Detection Limits	In Maximum Concentrations (C_{max})				
DDT, *p,p'*-	nr, 0.01–100	nd–4,500	48	54	6	17
DDT, total	nr, 0.1–100	nd–23,490	22	100	45	77
Demeton	nr	nd	1	0	—	—
Diazinon	nr, 100	nd	2	0	—	—
Dichlorvos	nr	nd	1	0	—	—
Dicofol	nr, 1–10	nd–560	10	40	—	—
Dieldrin	nr, 0.01–100	nd–5,680	64	66	27	25
Dimethoate	41–50	nd	2	0	—	—
Disulfoton	nr	nd	1	0	—	—
Endosulfan[1]	nr, 0.1–20	nd–170	4	50	25	—
Endosulfan I	nr, 0.02–10	nd–10	7	29	0	—
Endosulfan II	nr, 2–10	nd–40	7	29	0	—
Endosulfan sulfate	nr, 2–200	nd–20	6	17	—	—
Endrin	nr, 0.01–100	nd–2,060	56	25	5	9
Endrin aldehyde	nr, 200	nd	2	0	—	—
Endrin ketone	nr	nd	1	0	—	—
EPN	200	nd	1	0	—	—
Ethoprop	nr	nd	1	0	—	—
Famphur	nr	nd	1	0	—	—
Fensulfothion	nr	nd	1	0	—	—
Fenthion	nr	nd	1	0	—	—
Fenvalerate	nr	nd–11	2	50	—	—
HCH[1]	nr, 0.1–100	nd–160	10	40	10	10
HCH, α-	nr, 0.01–100	nd–610	34	35	—	—
HCH, β-	nr, 0.01–100	nd–900	31	16	—	—
HCH, δ-	nr, 0.01–100	nd–410	25	12	—	—
Heptachlor	nr, 0.01–50	nd–600	27	41	7	7
Heptachlor epoxide	nr, 0.01–104	nd–480	54	41	6	2
Hexachlorobenzene	nr, 0.01–100	nd–27,000	45	38	—	11
Kepone	nr, 10–50	nd–150	5	40	—	—
Lindane	nr, 0.01–100	nd–120	35	23	3	3
Malathion	nr, 50	nd	2	0	—	—
Methamidophos	nr	nd	1	0	—	—

Table 6.9. Selected results of studies that monitored pesticides in whole fish and the percentage of those studies that exceeded guidelines for the protection of fish-eating wildlife—*Continued*

Target Analytes	Whole Fish: All Monitoring Studies (Published 1960–1994)					
	Concentration Range (µg/kg wet weight)		Total Number of Studies	Percentage of Studies with Detectable Residues	Percentage of Studies with C_{max} that Exceeded NAS/NAE Guideline[1]	Percentage of Studies with C_{max} that Exceeded NYSDEC Guideline[2]
	In Detection Limits	In Maximum Concentrations (C_{max})				
Methiocarb	nr	nd	1	0	—	—
Methomyl	nr	nd	1	0	—	—
Methoxychlor[3]	nr, 0.1–100	nd–130	15	47	—	—
Methyl parathion	nr, 50	nd–60	3	33	—	—
Metolachlor	nr	nd	1	0	—	—
Mevinphos	nr	nd	1	0	—	—
Mirex	nr, 0.01–100	nd–1,810	46	28	—	11
Monocrotophos	nr	nd	1	0	—	—
Nonachlor, *cis-*	nr, 0.01–100	nd–156	38	29	—	—
Nonachlor, *trans-*	nr, 0.01–107	nd–1,550	45	64	—	—
Oxadiazon	nr	2,200	1	100	—	—
Oxamyl	nr	nd	1	0	—	—
Oxychlordane	nr, 0.01–105	nd–640	50	34	—	—
Parathion	30–50	nd	2	0	—	—
Pentachloroanisole	nr	33–160	2	100	—	—
Pentachlorophenol	nr, 0.1–3,000	nd–4,520	2	50	—	—
Permethrin	nr	0.53	1	100	—	—
Perthane	0.01–1	nd	3	0	—	—
Phorate	50	nd	1	0	—	—
Phosdrin	20	nd	1	0	—	—
Photomirex	nr	196–400	2	100	—	—
T, 2,4,5-	0.2	nd	1	0	—	—
Terbufos	50	nd	1	0	—	—
Tetrachlorvinphos	20	nd	1	0	—	—
Tetradifon	nr, 1–10	nd–2	5	20	—	—
Toxaphene	nr, 0.1–2,000	nd–280,330	51	37	37	—
Trichlorfon	50–80	nd	2	0	—	—
Trifluralin	nr, 2	7–126	2	100	—	—

[1]Total or unspecified.
[2]$(o,p' + p,p')$- or unspecified.
[3]p,p'- or total.

Table 6.9. Selected results of studies that monitored pesticides in whole fish and the percentage of those studies that exceeded guidelines for the protection of fish-eating wildlife—*Continued*

Target Analytes	Whole Fish: Recent Monitoring Studies (Published 1984–1994)					
	Concentration Range (μg/kg wet weight)		Total Number of Studies	Percentage of Studies with Detectable Residues	Percentage of Studies with C_{max} that Exceeded NAS/NAE Guideline[1]	Percentage of Studies with C_{max} that Exceeded NYSDEC Guideline[2]
	In Detection Limits	In Maximum Concentrations (C_{max})				
Methiocarb	nr	nd	1	0	—	—
Methomyl	nr	nd	1	0	—	—
Methoxychlor[3]	nr, 0.1–100	nd–130	15	47	—	—
Methyl parathion	nr, 50	nd–16	3	33	—	—
Metolachlor	nr	nd	1	0	—	—
Mevinphos	nr	nd	1	0	—	—
Mirex	nr, 0.01–100	nd–1,400	44	25	—	9
Monocrotophos	nr	nd	1	0	—	—
Nonachlor, *cis-*	nr, 0.01–100	nd–156	37	27	—	—
Nonachlor, *trans-*	nr, 0.01–107	nd–1,550	44	64	—	—
Oxadiazon	nr	2,200	1	100	—	—
Oxamyl	nr	nd	1	0	—	—
Oxychlordane	nr, 0.01–105	nd–640	49	35	—	—
Parathion	30–50	nd	2	0	—	—
Pentachloroanisole	nr	33–160	2	100	—	—
Pentachlorophenol	0.1–3,000	nd–4,520	2	50	—	—
Permethrin	nr	0.53	1	100	—	—
Perthane	0.01–1	nd	3	0	—	—
Phorate	50	nd	1	0	—	—
Phosdrin	20	nd	1	0	—	—
Photomirex	nr	196–400	2	100	—	—
T, 2,4,5-	na	na	0	—	—	—
Terbufos	50	nd	1	0	—	—
Tetrachlorvinphos	20	nd	1	0	—	—
Tetradifon	nr, 1–10	nd–2	5	20	—	—
Toxaphene	nr, 0.1–2,000	nd–280,330	46	33	33	—
Trichlorfon	50–80	nd	2	0	—	—
Trifluralin	nr, 2	7–126	2	100	—	—

measured concentrations exceeded the applicable guidelines. The NAS/NAE and NYSDEC guidelines for DDT compounds actually apply to total DDT (Table 6.8). Therefore, comparison of these guidelines with the concentration of any single DDT compound or isomer (such as DDE or p,p'-DDE) probably underestimates the percentage of studies that exceed the guideline. The same is true for the NYSDEC guidelines for the chlordane compounds (*cis*- and *trans*-chlordane, *cis*- and *trans*-nonachlor), which apply to total chlordane. Another example is the NAS/NAE guideline of 100 µg/kg, which actually applies to the combined residues of a number of pesticides: aldrin, chlordane, dieldrin, endosulfan, endrin, HCH, heptachlor, heptachlor epoxide, lindane, and toxaphene (Table 6.8). Therefore, for these individual compounds, the percentage of studies that exceeded the NAS/NAE guideline (listed in Table 6.9) may be an underestimate.

Table 6.9 also shows the effect of study date on the percentage of studies with maximum concentrations that exceeded the NAS/NAE or NYSDEC guidelines. The results for recently published (1984–1994) studies are provided to show whether there is potential for adverse effects on fish-eating wildlife at the most contaminated site when older studies (which may have sampled before many organochlorine insecticides were banned) are excluded. As discussed in Section 6.1.2, comparison of all monitoring studies reviewed (published 1960–1994) with recently published (1984–1994) studies is not a reliable indicator of trends for three reasons: (1) Publication date is not a perfect surrogate for sampling date; (2) The distributions of these studies are overlapping, since the second group (recently published studies) are included among the first group (all studies reviewed); (3) The two groups of studies did not sample at the same site locations. A better understanding of trends can be obtained from national programs that monitored residues at the same sites over time (Section 3.4).

For those pesticide analytes with a substantial percentage of studies that exceeded NAS/NAE or NYSDEC guidelines, Figures 6.8–6.11 show the cumulative frequency distribution of the maximum concentrations reported by monitoring studies (from Tables 2.1 and 2.2) that analyzed that pesticide analyte in whole fish. These pesticide analytes are dieldrin (Figure 6.8), total chlordane (Figure 6.9), total DDT (Figure 6.10), and toxaphene (Figure 6.11). Each figure compares the cumulative frequency distribution for all monitoring studies reviewed (published during 1960–1994) with that for recently published (1984–1994) studies. The applicable guidelines are shown as vertical lines at the appropriate pesticide concentration.

Aldrin and Dieldrin

Aldrin was detected in whole fish in 19 percent of all monitoring studies reviewed, but the maximum concentrations reported did not exceed the applicable NAS/NAE and NYSDEC guidelines in any of the studies reviewed (Table 6.9). Aldrin is metabolized by fish to dieldrin (U.S. Environmental Protection Agency, 1980a).

Dieldrin was detected in whole fish at one or more sites by 74 percent of all monitoring studies. As shown in Figure 6.8 and Table 6.9, 31 percent of these studies had maximum dieldrin concentrations that exceeded the NAS/NAE guideline for combined residues of dieldrin plus several other pesticides (aldrin, chlordane, endosulfan, endrin, HCH, heptachlor, heptachlor epoxide, lindane, and toxaphene). Thirty percent of all studies had maximum dieldrin concentrations that exceeded the NYSDEC guideline for aldrin plus dieldrin. When only recently published studies were considered, the NAS/NAE and NYSDEC guidelines were exceeded by 27 and 25 percent of studies, respectively. Recently published studies also had a lower percent of

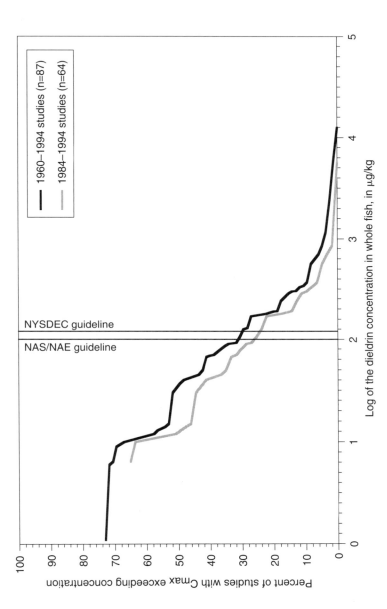

Figure 6.8. Cumulative frequency distribution of the maximum concentrations of dieldrin in whole fish (μg/kg wet weight) reported by the monitoring studies reviewed, shown for all studies (published during 1960–1994) and recently published (1984–1994) studies. Guidelines for dieldrin in whole fish (for protection of fish-eating wildlife) are shown as vertical lines. Abbreviations: C_{max}, maximum concentration in the study; n, number of samples; NAS/NAE, National Academy of Sciences and National Academy of Engineering; NYSDEC, New York State Department of Environmental Conservation; μg/kg, microgram per kilogram.

studies with detectable residues (66 percent) and a smaller range of maximum concentrations (Table 6.9).

Chlordane

Figure 6.9 shows the cumulative frequency distributions for maximum concentrations of total chlordane measured in whole fish by the monitoring studies reviewed. Total chlordane was detected in whole fish at one or more sites in 63 percent of all monitoring studies reviewed (Table 6.9). Thirty-three percent of all studies reviewed had maximum concentrations of total chlordane that exceeded NAS/NAE guideline for combined residues of several pesticides, including chlordane (Figure 6.9, Table 6.9). Relatively few studies (4 percent) had maximum concentrations that exceeded the NYSDEC guideline for total chlordane. The results were similar when only recently published (1984–1994) studies were considered: 67 percent of recently published studies had detectable chlordane residues, 33 percent had maximum concentrations that exceeded NAS/NAE guidelines, and 6 percent had maximum concentrations that exceeded NYSDEC guidelines (Table 6.9). Two-thirds of all studies reviewed were published after 1983 (Table 6.9).

DDT and Metabolites

Figure 6.10 shows the cumulative frequency distributions for maximum concentrations of total DDT measured in whole fish by the monitoring studies reviewed. The NAS/NAE and NYSDEC guidelines for total DDT were exceeded by maximum concentrations of total DDT in 50 and 82 percent of studies, respectively. When only recently published (1984–1994) studies were considered, these guidelines were exceeded for 45 and 77 percent of studies, respectively. A reduction in the maximum concentration range also occurred (Table 6.9), which indicated that the highest maximum total DDT concentration came from a study published prior to 1984. All the monitoring studies that targeted total DDT in whole fish (100 percent) detected total DDT at one or more sites. Note that recent studies generally calculated total DDT concentration by summing the concentrations of individual o,p' and p,p' isomers of DDT, DDD, and DDE, whereas older studies frequently calculated total DDT residues from concentrations of DDT, DDD, and DDE.

For individual components of total DDT (DDD, DDE, DDT, and individual o,p' and p,p' isomers of DDD, DDE, and DDT), Table 6.9 lists the percentages of studies with maximum concentrations that exceeded NAS/NAE and NYSDEC guidelines. However, these percentages probably underestimate potential effects on fish-eating wildlife in the studies reviewed because the guidelines apply to total DDT residues.

Toxaphene

Only one guideline for protection of fish-eating wildlife (from NAS/NAE) is available for toxaphene. As shown in Figure 6.11, all studies in which toxaphene was detected had maximum concentrations that exceeded this guideline. This represented 37 percent of all monitoring studies reviewed (published during 1960–1994), and 33 percent of recently published (1984–1994)

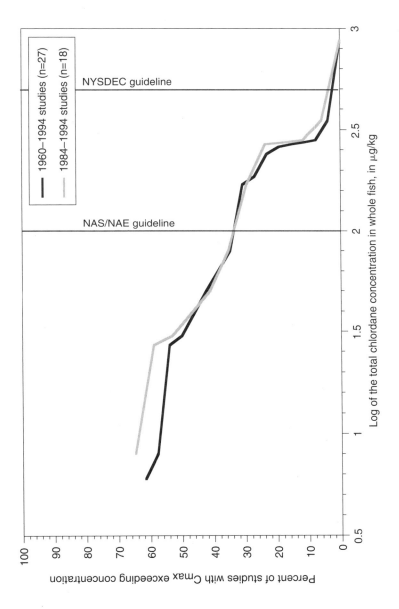

Figure 6.9. Cumulative frequency distribution of the maximum concentrations of total chlordane in whole fish (μg/kg wet weight) reported by the monitoring studies reviewed, shown for all studies (published during 1960–1994) and recently published (1984–1994) studies. Studies that measured chlordane (unspecified) and total chlordane are included. Guidelines for chlordane in whole fish (for protection of fish-eating wildlife) are shown as vertical lines. Abbreviations: C_{max}, maximum concentration in the study; n, number of samples; NAS/NAE, National Academy of Sciences and National Academy of Engineering; NYSDEC, New York State Department of Environmental Conservation; μg/kg, microgram per kilogram.

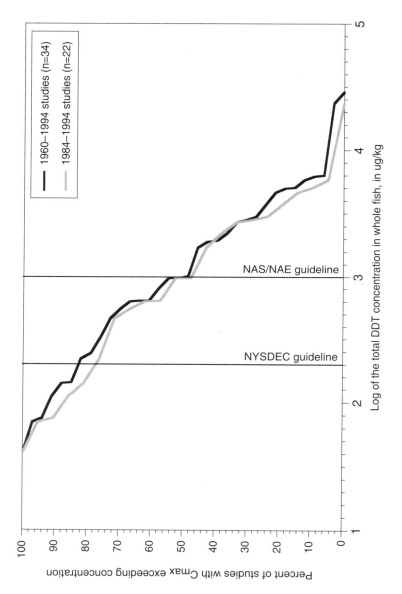

Figure 6.10. Cumulative frequency distribution of the maximum concentrations of total DDT in whole fish (μg/kg wet weight) reported by the monitoring studies reviewed, shown for all studies (published during 1960–1994) and recently published (1984–1994) studies. Guidelines for total DDT in whole fish (for protection of fish-eating wildlife) are shown as vertical lines. Abbreviations: C_{max}, maximum concentration in the study; n, number of samples; NAS/NAE, National Academy of Sciences and National Academy of Engineering; NYSDEC, New York State Department of Environmental Conservation; μg/kg, microgram per kilogram.

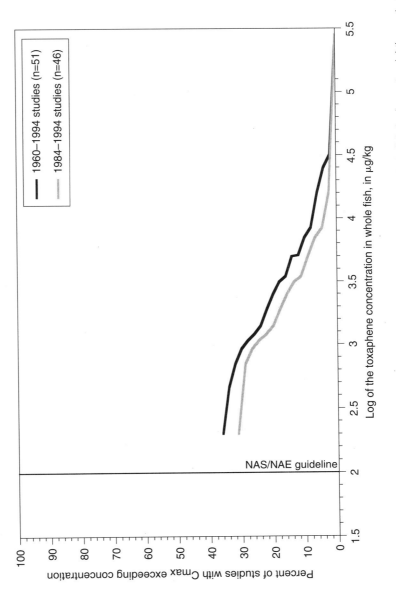

Figure 6.11. Cumulative frequency distribution of the maximum concentrations of toxaphene in whole fish (µg/kg wet weight) reported by the monitoring studies reviewed, shown for all studies (published during 1960–1994) and recently published (1984–1994) studies. Guideline for toxaphene in whole fish (for protection of fish-eating wildlife) is shown as a vertical line. Abbreviations: C_{max}, maximum concentration in the study; n, number of samples; NAS/NAE, National Academy of Sciences and National Academy of Engineering; µg/kg, microgram per kilogram.

studies. As noted previously, this may underestimate potential effects on fish-eating wildlife, since the NAS/NAE guideline applies to the summed concentrations of chlordane compounds, dieldrin, endosulfan, endrin, lindane and other HCH isomers, heptachlor epoxide, and toxaphene.

Other Pesticides

For the remaining pesticides with NAS/NAE or NYSDEC guidelines for protection of fish-eating wildlife, the applicable guidelines were exceeded by the maximum concentrations in whole fish in fewer than 20 percent of all monitoring studies reviewed (Table 6.9). When only recently published (1984–1994) studies were considered, the applicable NAS/NAE guidelines were exceeded by maximum concentrations of total endosulfan (in one of only four studies, or 25 percent), endrin (5 percent of studies), total HCH (10 percent), heptachlor (7 percent), heptachlor epoxide (6 percent), and lindane (3 percent). These results probably underestimate the percent of studies with detectable residues since the applicable guideline actually applies to the combined residues of all these pesticides plus chlordane, dieldrin, endosulfan, and toxaphene. Applicable NYSDEC guidelines were exceeded by maximum concentrations of endrin (in 9 percent of recently published studies), total HCH (10 percent), heptachlor (7 percent), heptachlor epoxide (2 percent), hexachlorobenzene (11 percent), lindane (3 percent), and mirex (9 percent).

Summary

The comparisons in Table 6.9 indicate that the NAS/NAE and NYSDEC guidelines for protection of fish-eating wildlife have been exceeded by pesticide concentrations at the most contaminated site (or sites) in a substantial fraction of the studies reviewed. This is true even in recently published (1984–1994) studies for several pesticides. The NAS/NAE guidelines were most often exceeded for total DDT compounds (45 percent of recently published studies), total chlordane (33 percent), toxaphene (33 percent), and dieldrin (27 percent). NYSDEC fish flesh criteria for noncancer effects were most often exceeded for total DDT (77 percent of recently published studies) and dieldrin (25 percent). These comparisons indicate the percentage of studies with maximum concentrations that exceeded the applicable guidelines, but the comparisons provide no information about the number of sites within each study that may have exceeded these guidelines. Because the NAS/NAE guidelines apply to the combined residues of a number of pesticides, some values shown in Table 6.9 for the percentage of studies with maximum concentrations that exceeded the NAS/NAE guidelines are probably underestimates.

6.1.4 OTHER TOXICITY CONCERNS

The existing guidelines used in the preceding analysis were developed considering acute and chronic toxicity to aquatic organisms and (to the extent data were available) to wildlife. However, such guidelines generally do not consider certain more complex issues related to toxicity, such as the effect of chemical mixtures on toxicity and the potential for endocrine-disrupting effects on the development, reproduction, and behavior of fish and wildlife populations.

Effects of Chemical Mixtures

Several studies noted co-occurrence of organochlorine contaminants in sediment (e.g., National Oceanic Atmospheric Administration, 1991) or aquatic biota (e.g., Schmitt and others, 1983). Organisms at contaminated sites are likely to be exposed to a mixture of contaminants. The potential for toxicological interactions among chemical contaminants exists. Possible interactions are generally categorized as one of the following: (1) *antagonistic*, where the mixture is less toxic than the chemicals considered individually; (2) *synergistic*, where the mixture is more toxic than the chemicals considered individually; or (3) *additive*, where the mixture has toxicity equivalent to the sum of the toxicities of the individual chemicals in the mixture.

Chemicals within a particular class (such as organochlorine insecticides) may have similar mechanisms of toxicity and produce similar effects. For example, most organochlorine compounds induce the mixed function oxidase (enzyme) system, which is involved in many Phase I biotransformation reactions of xenobiotics (as discussed in Section 4.3.3). This enzyme system is present in both fish and mammals. Prior or simultaneous exposure to chemicals that induce the same enzyme system may affect toxicity. In some cases, toxicity may be increased, while in others toxicity may be decreased because of detoxication. Exposure to multiple organochlorine contaminants has been shown to affect contaminant uptake and accumulation, as well as toxicity. For example, simultaneous exposure of mosquitofish (*Gambusia affinis*) to DDT and endrin reduced DDT accumulation in the gall bladder (Denison and others, 1985). While DDT residues in the gall bladder were not affected by posttreatment with endrin, they were increased by posttreatment with aldrin or dieldrin. Mixtures of DDT and mirex showed more than additive effects on fish survival and reproduction (Koenig, 1977). A mixture of DDT and methyl parathion appeared to increase acute toxicity to mosquitofish synergistically (Ferguson and Bingham, 1966).

In some cases, the presence of a pesticide and its transformation products may have synergistic, additive, or antagonistic effects, depending on the test system and endpoint of toxicity (Day, 1991). For example, Stratton (1984) exposed the blue-green alga *Anacystis inaequalis* to atrazine and two transformation products (desisopropylatrazine and desethylatrazine). The photosynthetic response showed an antagonistic response, whereas culture growth showed an additive response. In another example, fathead minnows (*Pimephales promelas*) exposed to malathion and two hydrolysis products (diethyl fumarate and dimethyl phosphorodithioic acid), showed a pronounced synergistic increase in toxicity (Bender, 1969).

Multiple contaminants have been proposed to explain adverse health effects observed for fish-eating birds from the Great Lakes area (Giesy and others, 1994a). Polychlorinated diaromatic hydrocarbons are believed to have a common mechanism of action involving a common aromatic hydrocarbon receptor. This suggests that quantitative structure-activity relations may be used to integrate the effects of multiple contaminants of this type. This group of compounds includes the DDT family, PCB congeners, polychlorinated dibenzo-*p*-dioxins (PCDD), and polychlorinated dibenzofurans (PCDF). Because 2,3,7,8-tetrachlorodibenzo-*p*-dioxin (TCDD) is the most toxic compound in this group, the toxic potency of other compounds are compared with it using *toxic equivalency factors* (TEF). TEFs can be based on various endpoints (such as lethality, deformities, or enzyme induction). The toxic potency of a mixture of compounds is then expressed as 2,3,7,8-TCDD equivalents (TCDD-EQ, also called TEQ). Using an additive model, the TCDD-EQ concentration of a mixture can be calculated from

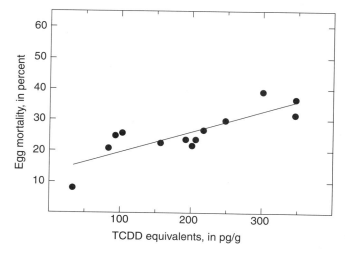

Figure 6.12. Egg mortality (percent) in double-crested cormorants from 12 colonies in the Great Lakes and one reference colony (Manitoba, Canada) as a function of the concentration (pg/g) of H4IIE bioassay-derived TCDD equivalents. The regression line is Y = 0.067(X) + 13.1 (r^2 = 0.703, p = 0.0003). Abbreviations: p, significance level; pg/g, picogram per gram; r, correlation coefficient; TCDD, tetrachlorodibenzo-p-dioxin. Adapted from Tillitt and others (1992) with permission of the publisher. Copyright 1992 Society of Environmental Toxicology and Chemistry (SETAC).

instrumentally derived concentrations of individual compounds in the mixture and their TEFs. An alternative to the additive model involves use of an in vitro bioassay that integrates the potency of mixtures of compounds with varying toxicities, as well as interactions between them. One such bioassay, the *H4IIE bioassay* (Giesy and others, 1994a), determines the potency for induction of specific mixed-function oxidase enzymes (under the control of CYP1A locus in H4IIE cells). Using the H4IIE bioassay to measure the concentrations of TCDD-EQ, Tillitt and others (1992) demonstrated that egg mortality (percent) in double-crested cormorants (*Phalacrocorax auritus*) from 13 colonies was correlated with concentrations of TCDD-EQ (Figure 6.12). Twelve of these colonies were from the Great Lakes, and the thirteenth (which had only 8 percent egg mortality, the lower left-most point in Figure 6.12) was a reference site from Manitoba, Canada. In this study, the total concentration of TCDD-EQ was a better predictor of egg mortality rates (r^2=0.703) than was the total concentration of PCBs (r^2=0.319). Correlations also have been observed between concentrations of TCDD-EQ and rates of deformities in cormorant and Caspian tern chicks (Giesy and others, 1994a). Planar PCBs, PCDDs, and PCDFs were believed to account for the greatest proportion of the TCDD-EQ predicted by the additive model. However, there were TCDD-EQs that could not be accounted for by concentrations of individual congeners of PCBs, PCDDs, and PCDFs. This may be a result of the presence of other compounds not quantified experimentally or to possible interactions between compounds present (Giesy and others, 1994a).

Potential Endocrine-Disrupting Effects of Pesticides

A number of studies have raised concerns about potential endocrine-disrupting effects of chemicals (including several pesticides) on wildlife and other animals (Colborn and Clement, 1992a). This is a controversial topic in both scientific and political arenas, and it has been frequently discussed in the popular literature (e.g., Begley and Glick, 1994; Lemonick, 1994) as well as in more technical publications. The discussion below is not a comprehensive review of the rapidly growing literature on this subject, but attempts to summarize some key findings regarding potential effects on fish and wildlife. This discussion relies on previous review articles and books by Colborn and others (1993), Hileman (1994), Raloff (1994a,b,c), and U.S. Environmental Protection Agency (1997b), as well as on primary publications. Potential endocrine-disrupting effects of pesticides on human health are discussed in Section 6.2.

For purposes of the discussion below, an *environmental endocrine disruptor* (or hormone disruptor) is defined as an exogenous agent that interferes with the synthesis, secretion, transport, binding, action, or elimination of natural hormones in the body that are responsible for the maintenance of homeostasis, reproduction, development, or behavior (U.S. Environmental Protection Agency, 1997b). A large number of chemicals have been shown to have some kind of hormonal function in animals, including birds, fish, reptiles, and mammals (Colborn and Clement, 1992a; U.S. Environmental Protection Agency, 1997b). These include both synthetic chemicals (xenobiotics) and naturally occurring ones. A lot of research and public attention has focused on *xenoestrogens* (xenobiotics with estrogenic activity). However, estrogen is only one part of an extremely complicated, integrated endocrine system. Endocrine-disrupting chemicals are believed to act in several ways: by antagonizing the effects of normal endogenous hormones, mimicking the effects of endogenous hormones such as estrogens and androgens, altering the pattern of synthesis and metabolism of endogenous hormones, and altering the binding characteristics of receptor sites on cell surfaces (Colborn and Clement, 1992a; McKinney and Waller, 1994). Steroid sex hormones (estrogens and androgens) regulate sexual differentiation and development of gametes, and coordinate various cellular metabolic activities. Therefore, exposure to endocrine-disrupting chemicals may cause abnormal sexual development and impaired reproduction, probably via multiple pathways. For example, DDE acts as both an estrogen mimic (or xenoestrogen, which binds to estrogen receptors, triggering a cascade of biochemical events) and as an androgen blocker (or antiandrogen, which ties up the androgen receptor, preventing true androgens from binding there) (Raloff, 1994b).

A number of pesticides have been shown to be endocrine disruptors. These include several persistent organochlorine insecticides and metabolites (DDT, DDE, dicofol, dieldrin, heptachlor, hexachlorobenzene, kepone, lindane and other HCH isomers, methoxychlor, mirex, and toxaphene), synthetic pyrethroids, triazine herbicides, and ethylene bis-dithiocarbamate fungicides. Other chemicals with estrogenic activity include some PCB congeners, 2,3,7,8-TCDD and other dioxins, 2,3,7,8-TCDF and other furans, some alkylphenols (breakdown products of alkylphenol polyethoxylates, which are commonly used in detergents and other household products), bisphenol A (a breakdown product of polycarbonate plastics), some phthalate esters, and synthetic estrogens such as diethylstilbestrol (Colborn and Clement, 1992a; Raloff, 1994a; U.S. Environmental Protection Agency, 1997b). Naturally occurring chemicals with estrogenic activity include the endogenous steroidal estrogens (estradiol), *phytoestrogens* (naturally occurring plant estrogens, such as coumestrol), some mycotoxins (fungal toxins, such as zearalenone), and some metals (Colborn and Clement, 1992a; U.S. Environmental Protection Agency, 1997b; Safe and Gaido, 1998).

In general, xenoestrogens tend to have much weaker estrogenic activity than the steroid hormone estradiol. This is illustrated in Table 6.10, from Vonier and others (1996), who investigated the capability of various xenobiotics to bind the estrogen receptor (aER) in the American alligator (*Alligator mississippiensis*). In competitive binding experiments, several pesticides inhibited binding of 17β-estradiol to the aER, but their potencies were much weaker

Table 6.10. Inhibitor concentrations necessary for 50 percent inhibition of [³H]17β-estradiol binding to aER by environmental chemicals

[Data are representative of at least three independent experiments with three replicates. Abbreviations and symbols: aER, alligator estrogen receptor; DDOH, 2,2–bis(p-chlorophenyl)-ethanol; IC_{50}, concentration that inhibits 50 percent of [³H]17β-estradiol binding to aER; ns, not significant; PCB, polychlorinated biphenyl; μmol/L, micromole per liter; >, greater than. From Vonier and others (1996)]

Chemical	aER Binding IC_{50} (μmol/L)
17β-Estradiol	0.0078
o,p'-DDD	2.26
o,p'-DDT	9.1
DDOH	11.1
o,p'-DDE	37.25
Dicofol	45.6
p,p'-DDT	>50[1]
p,p'-DDD	>50[1]
p,p'-DDE	>50[1]
p,p'-DDA	ns
Methoxychlor	ns
Nonachlor, trans-	10.6
Cyanazine	19
Atrazine	20.7
Alachlor	27.5
Kepone	34
Aroclor 1242 (PCB)	37.2
Nonachlor, cis-	40
Endosulfan I	>50[1]
Endosulfan sulfate	>50[1]
Dieldrin	ns
Endosulfan II	ns
Toxaphene	ns
2,4-D	ns

[1]Compounds that inhibited [3H]17β-estradiol but were insoluble at concentrations necessary to achieve 50 percent inhibition.

(by a factor ranging from 300 to 3,000) than 17β-estradiol itself (Table 6.10). However, dose, body burden, timing, and duration of exposure at critical periods of life are important considerations for assessing adverse effects of an endocrine disruptor (U.S. Environmental Protection Agency, 1997b). Another important factor is the potential for interactive effects of multiple chemicals. When competitive binding assays were conducted with chemical mixtures, a mixture of DDT metabolites showed an additive decrease in 17β-estradiol binding to the aER; some combinations of pesticides (at concentrations equivalent to those seen in alligator eggs) decreased 17β-estradiol binding in a greater than additive manner (Vonier and others, 1996).

Several field studies have shown an association between exposure to endocrine-disrupting chemicals in the environment and various adverse effects on animals (U.S. Environmental Protection Agency, 1997b). These effects include thyroid dysfunction in birds and fish; decreased reproductive success in birds, fish, shellfish, reptiles, and mammals; metabolic abnormalities in birds, fish, and mammals; behavioral abnormalities in birds; demasculinization and feminization of male fish, birds, reptiles, and mammals; defeminization and masculinization of female fish, gastropods, and birds; and compromised immune systems in birds and mammals (Colborn and Clement, 1992a; Colborn and others, 1993; U.S. Environmental Protection Agency, 1997b). These effects have been observed in areas where multiple synthetic chemicals were present. In some cases, reproductive impairment in field populations has been attributed to specific chemical contaminants. Examples include DDT in western gulls (*Larus occidentalis*) in southern California (Fry and Toone, 1981); DDE in American alligators from Lake Apopka, Florida (Guillette and others, 1994); polynuclear aromatic hydrocarbons (PAH) in marine polychaetes (Fries and Lee, 1984); PCBs, PCDDs, and PCDFs in fish-eating birds from the Great Lakes (Giesy and others, 1994a); and PCBs in common seals from the Wadden Sea (Reijnders, 1986). In some cases, however, the action of other chemicals could not be ruled out. Effects on steroid hormones in field populations exposed to contaminants have been shown for English sole (*Pleuronectes vetulus*) exposed to PAHs (Sol and others, 1995); lake whitefish (*Coregonus clupeaformis*) and white sucker (*Catostomus commersoni*) (Munkittrick and others, 1992a,b, 1994) and mosquitofish (Fox, 1992) exposed to pulp and paper mill effluents; rainbow trout (*Oncorhynchus mykiss*) and common carp (*Cyprinus carpio*) exposed to sewage treatment effluent (Purdom and others, 1994); and common carp exposed to organochlorine contaminants, PAHs, phenols, and phthalates in Las Vegas Wash (Bevans and others, 1996). Multiple chemicals clearly were present in pulp and paper mill and sewage treatment effluents, and the chemicals responsible for estrogenic activity were not conclusively established. Studies with laboratory animals have reproduced some of the abnormalities observed in field populations, for example, the effects of DDT and other pesticides on California gulls (*Larus californicus*) and western gulls (Fry and Toone, 1981), and the work of Guillette and colleagues with DDE and alligator eggs (Gross and Guillette, 1995; Guillette and others, 1994; Hileman, 1994; Raloff, 1994b).

Residues of most organochlorine pesticide residues in sediment and aquatic biota have declined nationally since the early 1970s, although high concentrations have persisted in some local areas (Section 3.4.1). A legitimate question is whether the remaining low-level residues are sufficient to disrupt the development and reproduction in some fish and wildlife species. There is strong evidence that endocrine disrupting chemicals have caused abnormal development and reproduction in some fish and wildlife species (U.S. Environmental Protection Agency, 1997b), at least in the more contaminated areas of the United States. Examples include demasculinization and reduced hatching success in American alligators from Lake Apopka, Florida (Guillette and others, 1994); thyroid abnormalities and poor egg survival in salmon (*Oncorhynchus* sp.) (Leatherland, 1992); decreased hatching success and increased developmental abnormalities in snapping turtles (*Chelydra serpentina*) (Bishop and others, 1991); and reproductive loss and early mortality in fish-eating birds from the Great Lakes (Kubiak and others, 1989; Colborn and others, 1993; Giesy and others, 1994a). It is more difficult to assess potential effects on fish and wildlife populations living in less contaminated areas. The USEPA recently reviewed the scientific literature on endocrine disruption and concluded that it was not clear at present whether the adverse effects of environmental endocrine disruptors on aquatic life and wildlife that have been seen at various sites were confined to localized areas or were representative of more widespread conditions (U.S. Environmental Protection Agency, 1997b).

Thus, the extent to which populations in the environment are affected by xenobiotic endocrine disruptors remains unclear. First, it is difficult to determine cause and effect in field populations. Some adverse effects of endocrine disruptors occur on exposure to very low levels and may not be observed until long after the period of exposure. In some cases, effects may not be observed until subsequent generations mature (U.S. Environmental Protection Agency, 1997b). Second, as discussed above, animals in the environment generally are exposed to multiple contaminants (Colborn and Clement, 1992b). Third, wildlife also is exposed to an abundance of plant estrogens in the environment. Estrogenic substances have been found in more than three hundred plants from at least sixteen families (Setchell, 1985; Hughes, 1988). The plants containing hormone mimics include many foods (such as rice, wheat, potatoes, carrots, sunflower seeds and oil, and apples) consumed by humans and foods (such as rye grass and clover) consumed by other animals. Naturally occurring hormone mimics also have been observed to reduce fertility and disrupt development in mammals (e.g., Bennetts and others, 1946; Whitten and Naftolin, 1992; Whitten and others, 1992, 1993). One fundamental difference between naturally occurring plant estrogens and xenoestrogens is that plant estrogens are rapidly metabolized and excreted (Hughes, 1988), while many synthetic chemicals (including organochlorine pesticides) are resistant to degradation and accumulate in the body.

A few recent or ongoing large-scale studies have looked for direct evidence of endocrine disruption in field populations at sites throughout the United States that have varying degrees of environmental contamination. The USGS conducted a reconnaissance study to determine whether endocrine systems of fish are being disrupted on a widespread scale throughout streams in the United States (Goodbred and others, 1995, 1996, 1997; Smith and Sorensen, 1995, 1997). Common carp were sampled from streams believed to represent a range of contaminant conditions commonly occurring in United States streams. A total of 25 sites were sampled from 11 major river basins across the United States. Several bioindicators, including levels of three steroid hormones (17β-estradiol, 11-ketotestosterone, and testosterone), vitellogenin, and histopathology of gonads, were measured. The ratio of 17β–estradiol to 11–ketotestosterone (E_2/11-KT ratio) appears to be a sensitive marker of abnormal sex steroid hormone levels (Bevans and others, 1996). The E_2/11-KT ratio in blood plasma of carp were negatively correlated with time-weighted mean concentrations of total pesticides in water, at 11 sites that had adequate data on pesticides in water (Goodbred and others, 1996, 1997). Organochlorine pesticides in fish tissue also showed a negative correlation with male 17β-estradiol (Goodbred and others, 1997). Similar results were seen in largemouth bass (*Micropterus salmoides*) from a site in Florida that has been contaminated with organochlorine pesticides (Gross and others, 1995) and in laboratory studies with lindane and α-HCH in catfish (Singh and Singh, 1987). In a regional analysis of sites in the Mississippi River Basin, E_2/11-KT ratios were significantly lower at contaminated sites relative to a reference site (Goodbred and others, 1996). Vitellogenin is a phospholipoprotein (an egg yolk precursor) that is normally found in female egg-laying vertebrates, but not in males. Vitellogenin is naturally induced in females in response to estrogen (typically 17β-estradiol); when found in male fish, it indicates exposure to environmental estrogens (U.S. Environmental Protection Agency, 1997b). Vitellogenin was detected in one or more male carp at more than half of the study sites, although concentrations were well below those in female carp (Goodbred and others, 1997). The results of the national reconnaissance study suggested variable endocrine status of carp at study sites, some of which may be related to contaminants (Goodbred and others, 1995, 1996, 1997).

In an ongoing study conducted by the USGS (Biomonitoring of Environmental Status and Trends), selected NCBP sites in the Mississippi River Basin have been revisited, and both contaminant residues and an array of bioindicators in fish were measured (Steven Goodbred, U.S. Geological Survey, personal commun., 1998). Such studies will continue to elucidate the extent to which endocrine disruption occurs in fish from streams across the United States and its degree of correlation with pesticide (and other synthetic chemical) contamination.

Endocrine-disrupting effects of chemicals have not been evaluated in standard toxicological studies, nor have they been considered in establishing most standards and guidelines for protection of human health and wildlife. The U.S. Environmental Protection Agency (1997b) is establishing a Federal advisory working group to develop a screening and testing strategy for new and existing chemicals that may act as endocrine disruptors. The existing guidelines for protection of aquatic life or wildlife (discussed in Sections 6.1.1–6.1.3) generally indicate thresholds for chronic toxicity. Life cycle tests with chemical stressors have shown that reproduction can be impaired at concentrations that do not affect embryonic development, hatching, or growth (Folmar, 1993). This suggests that effects on development and reproduction in fish and wildlife may occur at concentrations below existing guidelines. Therefore, the percentage of studies with maximum concentrations that exceed various guidelines for protection of aquatic life or wildlife (Tables 6.2 and 6.9) may underestimate potential adverse effects on fish and fish-eating wildlife because of potential effects on development and reproduction in exposed populations.

6.1.5 SUMMARY

Detection of hydrophobic pesticides in bed sediment and aquatic biota serves as an indicator that these compounds are present as contaminants in the hydrologic system. It is more difficult to assess the biological significance of these pesticide residues to aquatic life or wildlife feeding within the hydrologic system. Organochlorine pesticides such as DDT have been shown to adversely affect a variety of organisms in laboratory tests, including phytoplankton, aquatic invertebrates, fish, birds, and mammals. In the field, the disruptive effects of DDT on reproduction of fish and birds have been demonstrated, and fish-eating mammals have been shown to accumulate high levels of DDT and metabolites from their diets. Pesticide contamination in the environment also has been associated with fish kills and with fish diseases. However, most studies that reported pesticide concentration data in bed sediment or biota do not investigate the incidence of fish pathology or other adverse effects on biota.

In the monitoring studies reviewed (Tables 2.1 and 2.2), hydrophobic pesticide residues were frequently detected in bed sediment and aquatic biota. Preliminary assessment of potential effects of these residues on aquatic organisms and fish-eating wildlife living in or near the hydrologic systems being sampled was made by comparing measured pesticide residue levels with existing guidelines designed for protection of aquatic life or wildlife. Because of data availability limitations, each study was represented in this analysis by the maximum concentration of each pesticide reported in the study and by its date of publication. Despite the limitations of this kind of analysis, some important points emerge. Even when only recently published (1984–1994) studies were considered, the maximum concentrations reported in a large number of studies exceeded applicable guidelines for protection of aquatic life or wildlife (Tables 6.7 and 6.9).

For pesticides in bed sediment, measured concentrations were compared with various sediment quality guidelines. Using a "probability of adverse effects" method, studies were classified according to the probability of adverse effects on aquatic life at the most contaminated site reported in the study. This method was based on the procedure used by USEPA to evaluate sites in the National Sediment Inventory. Sediment guideline values indicating a high probability of adverse effects on aquatic life were exceeded by maximum concentrations in 25–50 percent of recently published (1984–1994) studies for several pesticides: total chlordane, total DDT, and dieldrin (Table 6.7, Figure 6.6). This indicates potential effects on benthic organisms at the most contaminated site for 25–50 percent of recently published studies.

For pesticides measured in whole fish, it is appropriate to assess potential impacts on both fish and fish-eating wildlife. No guidelines exist that relate contaminant concentrations in fish tissues directly to fish toxicity. In theory, corresponding fish tissue concentrations can be extrapolated from chronic water-quality criteria for the protection of aquatic organisms by assuming equilibrium between the pesticide concentration in fish and that in the ambient water. Because of the simplifying assumptions necessary and the considerable uncertainty in the data required for the extrapolation (BCF and chronic aquatic toxicity data), this approach provides only a crude estimate of pesticide concentrations in fish that may be associated with potential toxicity to aquatic life. For the following pesticides, the extrapolated fish tissue concentrations were exceeded by the maximum concentrations reported in some recently published (1984–1994) studies that monitored residues in whole fish (Table 6.2): dieldrin (11 percent of recently published studies), total endosulfan (1 of 4 studies, or 25 percent), endrin (14 percent of studies), lindane (17 percent), and toxaphene (100 percent). This indicates potential toxicity to aquatic life from these pesticides at the most contaminated site in at least 10–20 percent of recently published studies.

Two sets of guidelines for pesticides in whole fish are designed to protect fish-eating wildlife (Table 6.8). Even when only recently published (1984–1994) monitoring studies were considered, 45 percent of studies had maximum concentrations that exceeded the NAS/NAE maximum recommended tissue concentration for total DDT, and about 30 percent of studies for total chlordane, dieldrin, and toxaphene (Table 6.9). About 75 and 25 percent of recently published studies had maximum concentrations that exceeded New York fish flesh criteria for total DDT and dieldrin, respectively. This analysis indicates potential toxicity to fish-eating wildlife at the most contaminated site in 25–75 percent of recently published studies.

For both sediment and whole fish, therefore, a substantial number of recently published (1984–1994) studies reported maximum concentrations of some pesticides that exceeded applicable guidelines for protection of aquatic organisms or wildlife. The maximum concentrations of certain organochlorine compounds measured by individual studies appear to be sufficient to adversely affect benthic organisms (in over 45 percent of studies that targeted these pesticides in bed sediment), fish (in 10–20 percent of studies sampling whole fish), and fish-eating wildlife (in 25–75 percent of studies sampling whole fish). The pesticides most frequently present at levels that may cause toxicity are DDT and metabolites, dieldrin, and chlordane. These comparisons indicate what percentage of studies may have adverse effects at the most contaminated site in each study, but not what proportion of sites in each study may be affected.

Most of the guidelines used in the above analysis were based on the results of single-species, single-chemical toxicity tests conducted in the laboratory. However, such guidelines generally do not consider certain more complex issues related to toxicity, such as the effect of chemical mixtures on toxicity and the potential for endocrine-disrupting effects on development, reproduction, and behavior of fish and wildlife populations. As a result, the preceding analysis may underestimate potential adverse effects of contaminants in sediment and aquatic biota on fish, other aquatic biota, and fish-eating wildlife.

6.2 EFFECTS OF PESTICIDE CONTAMINANTS IN AQUATIC BIOTA ON HUMAN HEALTH

This section discusses potential human health effects resulting from consumption of fish and shellfish contaminated with pesticides. For the most part, this discussion focuses on chronic toxicity and carcinogenicity, although some additional information (such as on reproductive or developmental effects) is presented when available. First, some key studies that document the presence of pesticide residues in human tissues or assess potential sources of human exposure (including the consumption of fish) are presented. Second, the general toxicological characteristics of the pesticides found in aquatic biota are described. Third, the potential for human health effects of pesticide levels found in the monitoring studies reviewed (Tables 2.1 and 2.2) is evaluated by comparing pesticide concentrations measured in these studies with standards and guidelines designed for the protection of human health. Finally, the following selected toxicity-related issues are discussed: exposure to multiple contaminants, effects of fish preparation and cooking on residue levels, cancer risks resulting from consumption of carcinogen-contaminated fish relative to other sources of carcinogens, and potential endocrine-disrupting effects of pesticides detected in aquatic biota.

6.2.1 HUMAN EXPOSURE TO PESTICIDES

Human exposure to pesticides has been documented by analytical measurements of pesticides and their transformation products in human tissues, as well as in various foods and drinking water.

Pesticide Residues in Human Tissues

Organochlorine compounds, including pesticides, PCBs, and PCDDs and PCDFs, are commonly detected in samples of human breast milk, adipose tissue, reproductive tissues, blood, and urine. The literature on this topic is extensive. The following discussion is not intended to be a comprehensive review, but to show the variety of studies in the literature and to present some key findings. This discussion draws upon review articles and books by Ackerman (1980), Geyer and others (1986), Jensen and Slorach (1991), and Thomas and Colborn (1992), and on selected other studies.

Breast Milk

Many studies have measured total DDT and other organochlorine residues in human breast milk in countries throughout the world. Examples include Japan (Doguchi, 1973), Canada (Mes and Malcom, 1992), and India (Nair and Pillai, 1992). Typical concentrations of total DDT in breast milk are about 30 µg/kg in whole milk, or 1,000 µg/kg in milk fat. In developing countries where DDT is still used, total DDT concentrations may be ten to a hundred times higher (Jensen and Slorach, 1991). For example, DDT levels in breast milk from Vietnam (11,000–12,000 µg/kg fat) were much higher than from the United States and Germany (Schecter and others, 1989). Although the details of DDT production and use in Vietnam are unknown, Kannan and others (1992) reported that DDT probably was still being used in Vietnam to combat malaria and to a lesser extent for agricultural purposes. In a study conducted in Central Java, total DDT levels in breast milk samples were related to the length of time since the last DDT spraying. In three villages that were sprayed 2, 8, and 24 years ago, the median total DDT residues in human breast milk were 39,000, 7,000, and 3,400 µg/kg fat, respectively (Noegrohati and others, 1992).

Pesticide levels in breast milk fat worldwide were assessed from the scientific literature by Thomas and Colborn (1992). In this review, average levels for organochlorine insecticides in human breast milk fat were reported as follows: chlordane (80 µg/kg fat); total DDT (1,000 µg/kg fat); dieldrin (50 µg/kg fat); total heptachlor (50 µg/kg fat); hexachlorobenzene (100 µg/kg fat); β-HCH (1,000 µg/kg fat); and lindane (50 µg/kg fat). The highest levels of organochlorine pesticides in human breast milk found worldwide were from the following countries: for dieldrin, Australia, Iraq, and Uruguay; for heptachlor and heptachlor epoxide, Spain and Italy; for chlordane, Mexico and Iraq; for total DDT, Guatemala; for β-HCH, Chile and China; and for lindane, Italy. β-HCH is the most persistent component of technical HCH, and often constitutes about 90 percent of the total HCH residues in breast milk (Thomas and Colborn, 1992). For total PCBs, which have similar physical and chemical properties to the organochlorine insecticides, the average level in human breast milk fat is about 1,000 µg/kg fat. In breast milk from Innuit women from Hudson Bay (1987–1988), PCB levels averaged 3,600 µg/kg fat, reflecting mothers' diets of marine fish and mammals and freshwater fish (Dewailly and others, 1989).

In the United States, a national study conducted by USEPA during 1975–1976 found that dieldrin was detected in over 80 percent, oxychlordane in 74 percent, and heptachlor epoxide in 63 percent, of breast milk samples (U.S. Environmental Protection Agency, 1976). This study did not analyze for DDT and metabolites. There were regional differences in fat-adjusted contaminant levels in breast milk that could not be explained on the basis of differences among regions in the proportion of urban and rural residents. Milk in women from the southeastern United States had the highest mean levels (fat-adjusted) of dieldrin, heptachlor epoxide, and oxychlordane, and women from the northwestern United States had the lowest. In another nationwide study (Suta, 1977), mirex was not detected in breast milk but kepone was (in 3 percent of samples). These kepone detections occurred for mothers' milk from North Carolina, Alabama, and Georgia (in 3.9, 7.7, and 2.6 percent of mothers in those states, respectively). The highest localized concentrations of kepone probably were due to discharges from manufacturing (Suta, 1977). Chlordane and its isomers were found more frequently in breast milk from the United States than from most other parts of the world (Thomas and Colborn, 1992). The concentration of total DDT in breast milk in the United States did not change substantially from 1970 to 1982, although concentrations have decreased in several European countries over the last twenty years (Calabrese, 1982; Jensen and Slorach, 1991).

As shown for dieldrin by Ackerman (1980), organochlorine pesticides are transferred to the fetus during pregnancy, and the child receives additional amounts after birth in breast milk or dairy products. Lactation probably is a significant route of organochlorine excretion (Ackerman, 1980), as indicated by the significant relation found between women with low levels of dieldrin and oxychlordane in their breast milk and women who have nursed several children (U.S. Environmental Protection Agency, 1976). This was shown also in a Canadian study in which hexachlorobenzene concentrations in breast milk declined significantly throughout 98 days of lactation (Mes and others, 1984). In a study of Michigan children at age 4, breast feeding was believed to be the primary source of exposure to DDT, polybrominated biphenyls (PBB), and PCBs (Jacobson and others, 1989). Blood serum levels of DDT in these children were significantly correlated with duration of nursing, having older mothers, and residing in a rural area. The mothers in the study were exposed by eating contaminated fish or other foods (Thomas and Colborn, 1992).

Adipose Tissue

In a review of 13 studies from the United States, Canada, England, and Holland, Ackerman (1980) reported that dieldrin levels in human adipose tissue remained fairly constant (150–220 µg/kg wet weight) from 1963 to 1976. This is shown in Table 6.11, which lists mean

Table 6.11. Dieldrin residues in human adipose tissue, 1963–1976

[—, data not reported. From Ackerman (1980)]

Year	Number of Persons Sampled	Percentage Positive	Mean Concentration (µg/kg wet weight)	Sampling Location
1963	—	—	150±20	United States
1964	—	—	160	Canada
1966	47	—	210	England
1965–1967	146	—	220	United States
1968	11	100	170	Netherlands
1970	200	100	200	United States
1970	1,412	97	180	United States
1971	1,615	99	220	United States
1972	1,913	98	180	United States
1973	1,094	99	180	United States
1974	898	99	150	United States
1975	779	96	170	United States
1976	682	94	150	United States

dieldrin levels in human adipose tissue from 1963 to 1976. In another literature review (Geyer and others, 1986), mean dieldrin concentrations in adipose tissue of nonoccupationally exposed humans from different countries ranged from 20 to 450 μg/kg wet weight. For the United States, the mean dieldrin concentration was 100 μg/kg wet weight.

In studies reviewed by Thomas and Colborn (1992), mean p,p'-DDE levels in human adipose tissue were 2,400 μg/kg fat in Japan (during 1986–1988), 6,300 μg/kg fat in Spain (1985–1987), 3,200 μg/kg wet weight in Canada (1984), and 3,900 μg/kg wet weight in Texas (1979–1980). These levels are fairly consistent with the mean concentrations of total DDT in human adipose tissue from various countries reported by Geyer and others (1986): 1,730 to 5,110 μg/kg wet weight. For the United States, the mean total DDT concentration reported by Geyer and others (1986) was 3,230 μg/kg wet weight. In other studies, total DDT residues were reported to range from 0 to 26,000 μg/kg wet weight in Australia during 1985–1986 (Ahmad and others, 1988), and from 170 to 9,100 μg/kg wet weight in India during 1988–1989 (Nair and Pillai, 1992).

Geyer and others (1986) reported the following mean concentrations of HCH isomers in human adipose tissue from different countries: α-HCH (6.7–20 μg/kg wet weight); β-HCH (180–630 μg/kg wet weight); and lindane (7–57 μg/kg wet weight). As noted with reference to human breast milk, β-HCH is the most persistent component of technical HCH (Thomas and Colborn, 1992). In studies reviewed by Thomas and Colborn (1992), mean β-HCH levels were 84 μg/kg wet weight in Canada (in 1984), 840 μg/kg fat in Japan (1986–1988), and 3,100 μg/kg fat in Spain (1985–1987).

Geyer and others (1986) reported that the mean hexachlorobenzene concentration in human adipose tissue from different countries ranged from 35 to 1,900 μg/kg wet weight. For the United States, the mean hexachlorobenzene concentration was 35 μg/kg wet weight. Hexachlorobenzene was detected in 98 percent of human adipose tissue samples collected nationwide (U.S. Environmental Protection Agency, 1986g). From 1975 to 1985, hexachlorobenzene levels in adipose tissue stayed fairly constant or increased (U.S. Environmental Protection Agency, 1986g). An increasing trend in hexachlorobenzene levels in adipose tissue was also observed in more recent surveys of the United States, United Kingdom, and Canada (Jensen and Slorach, 1991). The primary source of hexachlorobenzene in environmental samples today is probably as an industrial byproduct, rather than from its use as a fungicide. Hexachlorobenzene has been detected in human adipose and breast milk samples in India, where it was never used as a fungicide; moreover, its highest concentrations were found in samples from industrial areas (Nair and Pillai, 1989).

Other pesticides detected in human adipose tissue include mirex and pentachlorophenol. In studies reviewed by Geyer and others (1986), mean concentrations of pentachlorophenol in human adipose tissue from various countries ranged from 15 to 86 μg/kg wet weight (Geyer and others, 1986). Residues of mirex in human adipose tissue in the United States ranged from trace levels to 1,300 μg/kg (Suta, 1977).

Geyer and others (1986) reported that populations of different countries had very similar human *bioconcentration factors* (human BCFs, which is defined as the ratio of the concentration of a pesticide in human tissue to that in the diet). These authors grouped organochlorine contaminants according to their bioconcentration potential in humans, as follows: Organochlorine compounds with high bioconcentration potential (human BCF >100) were DDT, hexachlorobenzene, β-HCH, PCBs, and TCDD; compounds with medium bioconcentration potential (human BCF between 10 and 100) were dieldrin, α-HCH, and lindane; and compounds with low bioconcentration potential (human BCF <10) were δ-HCH and pentachlorophenol.

Blood

Pesticides also have been detected in human blood samples. As will be discussed below (Sources of Human Exposure to Pesticides), organochlorine residues in human blood serum have been shown to be related to consumption of contaminated fish (Kuwabara and others, 1979; Humphrey 1987; Gossett and others, 1989).

In a number of studies reviewed by Thomas and Colborn (1992), several organochlorine compounds were detected in 90–100 percent of blood samples taken during the 1980s: *p,p'*-DDE and hexachlorobenzene in Norway (during 1981–1982); *p,p'*-DDE, *p,p'*-DDT, β-HCH, lindane, and PCBs in Yugoslavia (1985–1986); *p,p'*-DDE, hexachlorobenzene, and PCBs in Yugoslavia (1989); *p,p'*-DDE and β-HCH in Pakistan (1987); PCBs in India (1986–1987); *p,p'*-DDE, *p,p'*-DDT, dieldrin, hexachlorobenzene, β-HCH, and lindane, in Spain (1985–1987); *p,p'*-DDE and heptachlor epoxide in Israel (1984–1985); chlordane in Japan (1984–1985), and *p,p'*-DDE in Texas (1982–1983). Mean residues in these studies were in the low microgram per kilogram range: for example, 6–16 μg/kg *p,p'*-DDE and <1–18 μg/kg β-HCH. The highest residues found in these studies were 43 μg/kg wet weight *p,p'*-DDE (from Israel) and 31 μg/kg wet weight β-HCH (from Yugoslavia). High concentrations of total DDT concentrations in human blood (up to about 12,000 μg/L) were reported for Tokyo in 1971 (Doguchi, 1973) and for Delhi during 1988–1989 (Nair and Pillai, 1992); total HCH residues in these studies were 9 μg/L in Tokyo in 1971 (Doguchi, 1973), but 430–2,800 μg/L in Delhi during 1988–1989 (Nair and Pillai, 1992). In the United States, kepone was detected in venous blood from people (19 percent of those sampled) who lived within 1 mile of a kepone manufacturing plant (in Hopewell, Virginia), but who did not work at the plant (Suta, 1977).

Other Human Tissues

Selected pesticides and metabolites have been detected in human urine, which was monitored throughout the United States from 1976 to 1980 (Carey and Kutz, 1985). The results are summarized in Table 6.12. The compounds detected include metabolites of several organophosphate insecticides (such as chlorpyrifos) and several carbamate insecticides (including propoxur, which is primarily used in suburban and urban areas, rather than in agriculture, as shown in Table 3.5). Also detected was the herbicide dicamba, which was among the top10 herbicides used in both the home and garden and the agricultural sectors in 1995 (see Table 3.5), and the wood preservative pentachlorophenol, which also is a metabolite of hexachlorobenzene and lindane.

In another study, pentachlorophenol was analyzed in various tissues from seven deceased male subjects (from Oregon) that ranged in age from 22 to 75 years old (Wagner and others, 1991). Pentachlorophenol was detected in all human tissues (from all seven subjects) that were sampled: liver, kidney, prostate, testes, and adipose tissue. The highest concentrations (on a lipid-weight basis) were found in the testes, followed by the kidney, prostate, and liver. These

Table 6.12. Frequency of pesticides and metabolites detected in urine from persons 12–74 years old, United States, 1976–1980

[Results are based on the analysis of 6,990 samples collected from the general population via the Health and Nutrition Examination Survey II, National Center for Health Statistics. Adapted from Carey and Kutz (1985) with permission of the publisher and the author. Copyright 1985 Kluwer Academic Publishers]

Pesticide Origin	Chemical Detected	Frequency of Detection (in Percent)
Pentachlorophenol, lindane, and hexachlorobenzene	Pentachlorophenol	71.6
Any organophosphate insecticide containing these phosphate or thiophosphate molecules	Dimethylphosphate (DMP)	10.5
	Diethylphosphate (DEP)	6.2
	Dimethyl phosphorothionate (DMTP)	5.7
	Diethyl phosphorothionate (DETP)	5.3
Chlorpyrifos	3,5,6-Trichloro-2-pyridinol	5.8
Propoxur	Isopropoxyphenol	3.6
2,4,5-Trichlorophenol (disinfectant use); or as a metabolite of certain insecticides	2,4,5-Trichlorophenol	3.4
Carbofuran	Carbofuranphenol	3.1
	3-Ketocarbofuran	3.3
Methyl and ethyl parathion	p-Nitrophenol	2.4
Dicamba	Dicamba	1.4
Carbaryl and naphthalene	α-Naphthol	1.4
Malathion	α-Monocarboxylic acid	1.1
	Dicarboxylic acid	0.5

residues were 40-fold higher than those in adipose tissue. Residues of nonachloro-2-phenoxyphenol (NCPP), which is an impurity in technical pentachlorophenol, were detected in all tissues sampled, although three individual subjects did not have detectable NCPP in all tissues. The sums of all pentachlorophenol and NCPP residues in all tissues were positively correlated with the subject age at death (r=0.81), which suggests that pentachlorophenol and NCPP concentrations increased with time (Wagner and others, 1991). Pentachlorophenol is readily absorbed into the blood after ingestion, then combines with plasma proteins and is transported throughout body. Pentachlorophenol is eliminated fairly rapidly (with an elimination half-life of about 20 days), primarily in urine (as the glucuronic acid conjugate). The high concentrations of pentachlorophenol observed in the kidney and liver may result from these organs being the principal site of excretion and metabolism, respectively (Wagner and others, 1991).

Summary and Conclusions

In general, concentrations of organochlorine compounds are about the same in milk, adipose tissue and muscle tissue when calculated on an extractable fat basis. However, concentrations in blood will vary depending on mobilization from fat and recent intake of these compounds. Mobilization of contaminants stored in adipose tissue occurs during lactation and starvation, and as a result of disease. In general, people in countries still using DDT tend to have higher DDT residues in their tissues than those in countries where DDT has been banned or restricted. People in industrialized countries tend to have higher residues of PCBs, PCDDs, and PCDFs in their tissues than those in developing nations do.

Concentrations of some organochlorine concentrations in human tissues do not appear to be declining, although most organochlorine pesticides were banned or severely restricted in the United States during the 1970s. Total DDT levels in human breast milk did not decline appreciably in the United States between 1975 and 1985, although concentrations have decreased in several European countries over the last twenty years. Dieldrin residues in human adipose tissue remained fairly constant from 1963 to 1976. Residues of hexachlorobenzene in adipose tissue have remained constant or increased from the 1970s to the 1980s in the United States, United Kingdom, and Canada. Although hexachlorobenzene was used to some extent as a fungicide, its main source in the environment today is probably as an industrial contaminant. Hexachlorobenzene is formed as a byproduct in the production of various other chlorinated compounds, in the incineration of municipal waste, and in the chlorination of industrial process water (U.S. Environmental Protection Agency, 1992b).

Sources of Human Exposure to Pesticides

Several studies of human exposure to pesticides indicate that food consumption is the principal mode of intake of hydrophobic pesticide contaminants. Davies (1990) estimated food and other sources of human exposure to organochlorine pesticides and PCBs in the Great Lakes basin. Previously reported concentrations in air, drinking water, and food were multiplied by estimates of air inhaled, water ingested, and food consumed, respectively. The average amount of food consumed was estimated from food purchase data (for 1982). Contaminant concentrations were from local monitoring studies in Toronto and southern Ontario, Canada, conducted during the early 1980s. Specific monitoring data used included ambient air concentrations in the Great Lakes basin (in 1981); mean concentrations in Toronto drinking water (during 1978–1984); data for fish from the Toronto Harbor (in 1980); residues in cows' milk (in 1983), beef and pork fat (in 1981), and chicken and egg fat (during 1981–1982) in Ontario. Davies (1990) concluded that, on average, fish accounted for 26 percent of exposure to PCBs and several organochlorine pesticides (total DDT, aldrin plus dieldrin, heptachlor, hexachlorobenzene, α-HCH, β-HCH, and lindane); 30 percent was from beef and pork, 29 percent from fruit and vegetables, 8 percent from cows' milk, and 7 percent from chicken and eggs. Total food accounted for 89 percent of exposure, with 7 percent from drinking water and 4 percent from air.

Several authors have estimated the contributions of various food groups to human exposure from food. For example, Geyer and others (1986) estimated that food was the source of more than 90 percent of PCBs, TCDD, DDT, hexachlorobenzene, dieldrin, and HCH isomer residues occurring in the nonoccupationally exposed populations of industrialized countries, including the United States. On the basis of market basket studies in the United States, the United Kingdom, and Germany, the main foods responsible for this human exposure were fish, dairy products, and animal fat (Geyer and others, 1986). Kannan and others (1992) reported that the primary source of DDT in the human diet was from consumption of fish and shellfish. Cereals and vegetables were primary sources of PCBs and HCHs. The primary human exposures to mirex and kepone were estimated to be from consumption of seafood from contaminated areas (Suta, 1977). In a review of dieldrin studies, Ackerman (1980) concluded that the major source of dieldrin in adipose tissue was probably food. On the basis of a total diet study (in 1974), dieldrin was detectable mainly in meat, fish and poultry (100 percent positive) and dairy products (97 percent). Less important food sources included garden fruits (60 percent positive), potatoes (30 percent), leafy vegetables (10 percent), and fruits (7 percent).

Some key studies have demonstrated that consumption of fish results in human exposure to organochlorine compounds. Most data are for PCBs, which are structurally similar to organochlorine pesticides and have similar physical and chemical properties. Humphrey (1976, 1983a,b, 1987) found a relation between fish consumption and blood serum levels of organo-chlorine contaminants in Michigan sport fishers. Blood serum levels of PCB in those who regularly ate Lake Michigan fish were compared with levels in those who did not. PCB blood serum levels briefly increased after each fish meal and were correlated with annual fish consumption and with the number of years fish had been consumed. PCB levels in Lake Michigan fish-eaters were up to 30-fold higher than levels in those who did not eat fish. PCB, DDT, and DDE were detectable in nearly every Lake Michigan fish-eater who was tested.

In another study, PCB residues in the blood of two human subjects were monitored before and after consumption of commercial fish—cutlass fish and yellowtail—obtained at a market in Japan (Kuwabara and others, 1979); the fish contained 900 and 400 µg/kg total PCB, respec-tively. Total PCB levels in the blood increased steadily after consumption of the fish, reaching a maximum concentration 3–5 hours after consumption and returning to preingestion levels by the next morning. The PCB congener pattern in blood was different from that in the fish, and the pattern of PCB in the blood did not change after ingestion. The congeners that were observed in blood included a few PCB congeners with three and four chlorine substituents and many different congeners with five to eight chlorines.

Blood serum levels of DDTs and PCBs were quantified in humans who had consumed sportfish from the Los Angeles area at least three times per week for 3 years and compared with controls who had consumed little or no fish (Gossett and others, 1989). DDT was manufactured within the Los Angeles county from 1947 to 1980, and the local coastal waters have been contaminated by ocean disposal of waste sludges and by the Los Angeles County Sanitation District ocean outfall. Sport fishers consuming local fish were found to have blood serum levels of DDT and metabolites that were significantly higher than controls. The estimated number of lifetime sportfish meals was significantly correlated with DDE blood serum levels. Total PCB levels in blood serum were not correlated with fish consumption, possibly because of multiple sources.

PCBs have been implicated in a couple of studies demonstrating short-term health effects in infants whose mothers consumed Great Lakes fish (Swain, 1991). Women who consumed as little as two or three meals per month of Lake Michigan fish had shorter gestation times and gave birth to infants who weighed less and had smaller head circumferences than mothers in the same communities who did not consume fish (Fein and others, 1984). PCBs are the predominant contaminants in fish from Lake Michigan (Hileman, 1993). The mothers' annual PCB-in-fish consumption rate was determined from interviews and measured residues in different fish species. Maternal fish consumption was associated with alterations in motor maturity, amount of startle, lability of state, and number of abnormally weak reflexes (Jacobson and others, 1984).

Early childhood visual recognition memory deficits were related to intrauterine exposure to PCB, but not to gestational age, birth size, or postpartum PCB exposure from breast-feeding (Jacobson and others, 1985). Another study of mothers and infants from Wisconsin (1980–1981) examined the relation between maternal consumption of fish ·from Lake Michigan and the Sheboygan River, maternal PCB levels during pregnancy, and health effects in infants (Smith, 1984). Mothers in the study ate Lake Michigan or Sheboygan River fish at least twice per month for 3 years. Maternal serum PCB level during pregnancy was positively associated with the number and type of infectious illnesses (colds, earaches, flu) suffered by the infant during the first 4 months.

Dietary Intake of Pesticides

The FDA Total Diet Study has estimated the dietary intake of pesticides from foods in the United States periodically since 1961 (Food and Drug Administration, 1988). Foods were purchased from supermarkets, prepared ready to eat, composited by commodity groups (such as dairy products, cereal, and fish and shellfish), and analyzed for pesticides. Prior to 1982, dietary intake estimates were made for three subpopulations: 6-month infants, 2-year toddlers, and 15- to 20-year-old males (Yess and others, 1991a). Major changes were made to the FDA Total Diet Study in 1982. The diet basis was updated using information from two new national surveys (the U.S. Department of Agriculture's Nationwide Food Survey of 1977–1978 and the National Center for Health Statistics' Health and Nutrition Examination Survey II of 1976–1980). Also, intake estimates were made for eight subpopulations: 6- to 11-month infants, 2-year toddlers, 14- to 16-year-old males and females, 25- to 30-year-old males and females, and 60- to 65-year-old males and females. Foods were collected from four geographical regions, then prepared, and composited (Yess and others, 1991b).

The number of pesticides detected and the percentage of samples containing the most frequently detected pesticides in the FDA Total Diet Survey (1978–1991) are provided in Table 6.13 (pesticides are listed in the order of detection frequency in 1991). In general, detection frequencies (see Table 6.13) and dietary intake estimates for the organochlorine pesticides have declined since the mid–1960s (Yess and others, 1991b). One exception is endosulfan, which was detected in 5–8 percent of samples each year, without appearing to decline over time (Table 6.13). Detection frequencies for the chlorinated fungicides hexachlorobenzene and PCNB also declined during this time. For several insecticides in other classes (malathion, chlorpyrifos-methyl, methamidophos, acephate, and chlorpropham) and the fungicide dicloran, the percent occurrence has stayed about the same over time. The percent occurrence values for two organophosphate insecticides, diazinon and chlorpyrifos, appeared to peak around 1987, then decline (Table 6.13).

Table 6.14 lists dietary intake estimates for selected pesticides in 1991 (Food and Drug Administration, 1992), along with corresponding human health guideline values. Pesticides listed are those detected in aquatic biota in the monitoring studies reviewed (Tables 2.1 and 2.2) and for which pesticide intake estimates are available. The human health guidelines used are the USEPA reference dose (U.S. Environmental Protection Agency, 1996b, 1997c) and the United Nations' Food and Agriculture Organization–World Health Organization (FAO–WHO) acceptable daily intake (Food and Drug Administration, 1992). Both types of guidelines indicate the highest estimated daily exposure that is expected to have no deleterious health effects in

Table 6.13. Percentage of samples that contained the most frequently detected pesticides, and the number of pesticides detected, in the FDA Total Diet Study from 1978 to 1991

[Pesticides are listed in order of frequency of detection in 1991. Data were compiled from Food and Drug Administration (1988, 1989a, 1990a, 1992); and Yess and others (1991a,b). Abbreviations and symbols: Carb-I, carbamate insecticide; FDA, Food and Drug Administration; Fung, fungicide, wood preservative, or seed protectant; na, not applicable; OC-I, organochlorine insecticide; OP-I, organophosphate insecticide; —, no data]

Pesticide	Pesticide Class	Percentage of Samples with Detectable Residues, in:					
		1978–1982	1983–1986	1987	1988	1989	1991
Malathion	OP-I	18	22	23	21	20	18
DDT, total	OC-I	22	24	22	20	13	10
Chlorpyrifos-methyl	OP-I	—	—	5	10	10	10
Dieldrin	OC-I	24	14	12	10	7	8
Endosulfan	OC-I	7	5	7	8	5	7
Methamidophos	OP-I	—	4	5	5	6	6
Chlorpyrifos	OP-I	3	8	12	11	8	5
Dicloran	Fung	7	6	4	6	4	5
Diazinon	OP-I	11	10	21	15	9	4
Acephate	OP-I	—	—	3	3	3	4
Heptachlor, total	OC-I	12	7	6	5	3	3
Chlorpropham	Carb-I	8	4	3	3	5	3
Hexachlorobenzene	Fung[1]	18	9	10	7	5	2
Lindane	OC-I	11	5	6	4	4	2
HCH, α- + β-	OC-I	23	12	6	4	3	1
PCNB	Fung	9	2	3	3	3	1
Chlordane, total	OC-I	7	6	5	4	—	1
Pentachlorophenol	Fung[1]	7	13	—	—	—	—
Total number of pesticides detected	na	42	62	53	55	53	51

[1]Also an industrial contaminant.

humans over an individual's lifetime, under certain exposure conditions. These guidelines generally are based on a *no observed adverse effect level* (NOAEL) determined from toxicity tests with animals and incorporate uncertainty factors (depending on the quantity and quality of data used) to extrapolate (1) from animals to humans and (2) from relatively short-term test conditions to a lifetime exposure (Nowell and Resek, 1994). The USEPA reference dose considers chronic toxicity, but not carcinogenicity.

Dietary intake estimates for all pesticides were below FAO–WHO acceptable daily intake and USEPA reference dose values. In fact, for all pesticides except one, the dietary intake estimates were less than 1 percent of the applicable FAO–WHO and USEPA guidelines (Food

Table 6.14. Pesticide intake (μg/kg body weight/d) in total diet analyses for three age–sex groups in 1991 and corresponding FAO–WHO acceptable daily intake and USEPA reference dose values

[Pesticides listed in column 1 are those detected in aquatic biota in the monitoring studies reviewed (as listed in Tables 2.1 and 2.2), and for which pesticide intake estimates are available. Pesticide intake data and FAO–WHO acceptable daily intake values are from Food and Drug Administration (1992). USEPA reference dose values are from U.S. Environmental Protection Agency (1996b), except those that are designated in footnotes as being from USEPA's Office of Pesticide Programs, which are from U.S. Environmental Protection Agency (1997c). Abbreviations: FAO–WHO, Food and Agriculture Organization and World Health Organization; FDA, Food and Drug Administration; USEPA, U.S. Environmental Protection Agency; μg/kg, microgram per kilogram; weight/d, weight per day; —, no guideline available. Adapted from Food and Drug Administration (1992) with the permission of the publisher. Copyright 1993 AOAC International]

Pesticides Detected in Aquatic Biota in Monitoring Studies (from Tables 2.1, 2.2)	Pesticide Intake in Total Diet Analyses in 1991 (μg/kg body weight/d)			FAO–WHO Acceptable Daily Intake (μg/kg body weight/d)	USEPA Reference Dose (μg/kg body weight/d)
	6–11 month-old toddlers	14–15 year-old males	60–65 year-old females		
Carbaryl	0.1801	0.09	0.0811	10	100
Carbofuran	0.0002	0.0001	0.0004	[1]10	5
Chlordane, total	0.0001	0.0003	0.0002	0.5	0.06
Chlorpyrifos	0.0082	0.0034	0.0024	10	3
Dacthal (DCPA)	0.0002	0.0001	0.0002	—	[2,3]10
DDT, total	0.0095	0.0056	0.0043	[1]20	[4]0.5
Diazinon	0.0049	0.0022	0.0022	2	0.09
Dicofol, total	0.0218	0.0077	0.0235	25	[3]1.2
Dieldrin	0.0027	0.0021	0.0021	[1]0.1	0.05
Endosulfan, total	0.0173	0.0158	0.0242	[1]6	[2,3]6
Endrin	<0.0001	<0.0001	<0.0001	0.2	0.3
Heptachlor, total	0.0003	0.0003	0.0002	[1]0.5	[4]0.5
Hexachlorobenzene	0.0003	0.0004	0.0002	—	0.8
Lindane	0.0004	0.0008	0.0003	10	0.3
Methoxychlor, p,p'-	0.0006	0.0007	0.0002	100	5
Methyl parathion	0.0007	0.0001	0.0001	20	0.25
Parathion	0.0097	0.0016	0.0042	5	[3]0.33
Pentachlorophenol	0.0016	0.0004	0.0008	—	30
Permethrin, total	0.0251	0.0338	0.0495	50	50
Toxaphene	0.0033	0.0059	0.0024	—	[3]0.25

[1]Includes other related compounds.
[2]USEPA reference dose is under review.
[3]Value listed is from USEPA's Office of Pesticide Programs.
[4]Parent compound only.

and Drug Administration, 1988). The single exception was dieldrin, for which the dietary intake estimates constituted up to 5 percent of USEPA reference dose values (Table 6.14).

6.2.2 POTENTIAL TOXICITY TO HUMANS

A comprehensive review of the human health effects of pesticides that are commonly or occasionally found in aquatic biota is beyond the scope of this book. However, the general toxicological characteristics of pesticides found in aquatic biota in the monitoring studies reviewed (Tables 2.1 and 2.2) are discussed briefly, by pesticide class, in this section.

Several endpoints of toxicity are typically distinguished: for example, acute toxicity, chronic toxicity, developmental toxicity, carcinogenicity, and interactive effects. *Acute toxicity* and *chronic toxicity* refer to effects associated with short- and long-term exposure, respectively. In general, chronic toxicity has more bearing on potential health effects associated with consumption of contaminated fish (U.S. Environmental Protection Agency, 1994b). Chronic toxicity may encompass immunological and reproductive system effects. *Developmental toxicity* refers to adverse effects on the developing organism that may result from exposure prior to conception, during prenatal development, or postnatally up to the time of sexual maturation (U.S. Environmental Protection Agency, 1986h). Because the maternal barrier is not as effective at preventing harm to the fetus as once thought, chemical exposures that affect cell replication and developmental processes may lead to serious birth defects, stillbirths, miscarriages, developmental delays, and other adverse effects (U.S. Environmental Protection Agency, 1994b). Developmental toxicity can be demonstrated in laboratory studies, but it is difficult to establish cause and effect in epidemiological studies because of confounding factors such as variability in human exposure and (for some effects) the time lag between exposure and effects. Potential for human carcinogenicity is reflected in cancer potency factors determined by applying the linearized multistage model to results of animal tests (U.S. Environmental Protection Agency, 1989). Interactive effects may include additive, synergistic, and antagonistic effects (as discussed with reference to aquatic toxicity in Section 6.1.4).

Table 6.15 lists the USEPA reference dose, along with additional characteristics of mammalian toxicity, for selected pesticide analytes. The compounds listed are those detected in aquatic biota in the monitoring studies reviewed (Tables 2.1 and 2.2). Note that data are not available for several pesticide transformation products (marked with an "M" in column 2, "Pesticide Class"). The *USEPA reference dose*, defined as the estimated daily exposure that is without appreciable risk of deleterious health effects in the human population over an individual's lifetime (70 years), is a measure of chronic (noncarcinogenic) toxicity (U.S. Environmental Protection Agency, 1988). The reference dose generally does not consider developmental effects (U.S. Environmental Protection Agency, 1994b).

Two USEPA guidelines in Table 6.15 relate to potential carcinogenicity: cancer group and cancer potency factor. Each pesticide is classified into one of five cancer groups according to their carcinogenic potential (U.S. Environmental Protection Agency, 1986i) as follows:

Group A: Human carcinogen—compounds for which there is sufficient evidence in epidemiologic studies to support causal association between exposure and cancer.

Table 6.15. Mammalian toxicity characteristics of pesticide analytes that were detected in aquatic biota in the monitoring studies reviewed

[Pesticide class: Carb-I, carbamate insecticide; CP-H, chlorophenoxy acid herbicide; Dinitr-H, dinitroaniline herbicide; Fung, fungicide, wood preservative, or seed protectant; (M), metabolite; OC-I, organochlorine insecticide; OP-I, organophosphate insecticide; Other-H, other herbicide; Other-I, other insecticide. Cancer group: If two values are listed, separated by a slash, these represent cancer group assignments from USEPA's Office of Water and Office of Pesticide Programs, respectively. Cancer groups are: B2, probable human carcinogen (and known animal carcinogen); C, possible human carcinogen; D, not classified; and E, no evidence of carcinogenicity for humans. MFO Induction column and Developmental or Reproductive Effects column: "•" indicates that there is evidence for this type of effect; blank cell indicates that this effect has not been observed or that no information is available on this type of effect. RfD values are from U.S. Environmental Protection Agency (1996b, 1997c) and Nowell and Resek (1994). Cancer group and cancer potency factor data are from U.S. Environmental Protection Agency (1992b,g, 1994b, 1996b, 1997c). Data on MFO induction are from U.S. Environmental Protection Agency (1994b). Data on developmental and reproductive effects are from U.S. Environmental Protection Agency (1992b, 1994b). Abbreviations and symbols: MFO, mixed function oxidase; mg/kg/d, milligrams per kilogram of body weight per day; PCNB, pentachloronitrobenzene; RfD, reference dose; USEPA, U.S. Environmental Protection Agency; —, no guideline available]

Pesticides Detected in Aquatic Biota (from Tables 2.1, 2.2)	Pesticide Class	USEPA Guidelines			MFO Induction	Developmental or Reproductive Effects
		RfD (mg/kg/d)	Cancer Group	Cancer Potency Factor $(mg/kg/d)^{-1}$		
Aldrin	OC-I	0.00003	B2	17		
Carbaryl	Carb-I	0.1	D/C	—		
Carbofuran	Carb-I	0.005	E	—		
Carbofuran, 3-hydroxy-	Carb-I(M)	—	—	—		•
Chlordane	OC-I	0.00006	B2	1.3	•	
Chlordene, α-	OC-I	—	—	—		
Chlordene, 1-hydroxy	OC-I(M)	—	—	—		
Chlorpyrifos	OP-I	0.003	D/E	—		•
D, 2,4-	CP-H	0.01	D	—		
Dacthal (DCPA)	Other-H	[1,2]0.01	D	—		
DDD	OC-I(M)	—	B2	0.24		
DDE	OC-I(M)	—	B2	0.34		
DDMU	OCOI(M)	—	—	—		
DDT	OC-I	0.0005	B2	0.34	•	•

Table 6.15. Mammalian toxicity characteristics of pesticide analytes that were detected in aquatic biota in the monitoring studies reviewed—Continued

Pesticides Detected in Aquatic Biota (from Tables 2.1, 2.2)	Pesticide Class	USEPA Guidelines			MFO Induction	Developmental or Reproductive Effects
		RfD (mg/kg/d)	Cancer Group	Cancer Potency Factor $(mg/kg/d)^{-1}$		
Diazinon	OP-I	0.00009	E	—		•
Dicofol	OC-I	[2]0.0012	C	0.44	•	•
Dieldrin	OC-I	0.00005	B2	16	•	•
Endosulfan	OC-I	[1,2]0.006	E	—	•	•
Endosulfan sulfate	OC-I(M)	—	—	—		
Endrin	OC-I	0.0003	D	—	•	•
Endrin aldehyde	OC-I(M)	—	—	—		
Fenvalerate	Other-I	0.025	—	—		
Heptachlor	OC-I	0.0005	B2	4.5		
Heptachlor epoxide	OC-I(M)	0.00001	B2	9.1	•	•
Hexachlorobenzene	Fung	0.0008	B2	1.6	•	•
Kepone	OC-I	—	—	—		
Lindane	OC-I	0.0003	C/B2 or C	1.3	•	•
Methoxychlor	OC-I	0.005	D	—		
Methyl parathion	OP-I	0.00025	D	—		
Mirex	OC-I	0.0002	B2	1.8	•	
Nitrofen	Other-H	—	[3]B2	—		•
Oxadiazon	Other-H	0.005	C	—		•
Oxychlordane	OC-I(M)	—	—	—		
Parathion	OP-I	[2]0.00033	C	—		

Table 6.15. Mammalian toxicity characteristics of pesticide analytes that were detected in aquatic biota in the monitoring studies reviewed—Continued

Pesticides Detected in Aquatic Biota (from Tables 2.1, 2.2)	Pesticide Class	USEPA Guidelines			MFO Induction	Developmental or Reproductive Effects
		RfD (mg/kg/d)	Cancer Group	Cancer Potency Factor $(mg/kg/d)^{-1}$		
PCNB	Fung	0.003	C	—		•
Pentachloroanisole	Fung-I(M)	—	—	—		
Pentachlorophenol	Fung	0.03	[4]B2	0.12		•
Permethrin	Other-I	0.05	C	0.018		
Photomirex	OC-I(M)	—	—	—		
Strobane	OC-I	—	—	—		
Tetradifon	Other-I	—	—	—		
Toxaphene	OC-I	[2]0.00025	B2	1.1	•	•
Trifluralin	Dinitr-H	0.0075	C	0.0077		
Zytron (Xytron)	OP-I	—	—	—		

[1]USEPA reference dose is under review.
[2]Value listed is for USEPA's Office of Pesticide Programs.
[3]Probable classification. This compound was canceled before USEPA's carcinogenicity guidelines (USEPA, 1986h) were developed.
[4]For technical pentachlorophenol.

Group B: Probable human carcinogen—compounds for which there is limited evidence in epidemiologic studies (group B1) or sufficient evidence in animal studies (group B2) to support causal association between exposure and cancer.

Group C: Possible human carcinogen—compounds for which there is limited or equivocal evidence in animal studies and inadequate or no data in humans to support causal association between exposure and cancer.

Group D: Not classified—compounds for which there is inadequate or no human or animal evidence of carcinogenicity.

Group E: No evidence of carcinogenicity for humans—compounds for which there is no evidence of carcinogenicity in at least two adequate animal tests in different species or in adequate epidemiological and animal studies.

For chemicals in cancer groups A and B (known or probable carcinogens), USEPA computes a *cancer potency factor* (also called the "q_1*"), which is a quantitative measure of potential carcinogenicity and has units of $(mg/kg/d)^{-1}$ (Nowell and Resek, 1994).

A few pesticides have been assigned to different cancer groups by two different programs within USEPA (the Office of Water and the Office of Pesticide Programs). For these pesticides, both cancer groups are listed in Table 6.15 (for example, "D/E" indicates that the Office of Water has assigned the pesticide to cancer group D, and the Office of Pesticide Programs to cancer group E). Cancer potency factors (q_1* values) are listed in Table 6.15 for B2 and some C carcinogens. Of the pesticides detected in aquatic biota in the monitoring studies reviewed, those reported to induce microsomal enzymes or to have developmental or reproductive toxicity are designated in Table 6.15.

Organochlorine Insecticides

Organochlorine compounds are readily absorbed via the digestive system and often accumulate in lipid-rich tissues, including brain, adipose tissue, liver, and human milk. Effects of organochlorine insecticides include neurological effects (resulting from interference with axonic transmission of brain impulses) and induction of the hepatic microsomal enzyme system (U.S. Environmental Protection Agency, 1994b; Klaassen and others, 1996). The sites and mechanisms of toxic action differ for the three classes of organochlorine insecticides: dichlorodiphenylethane compounds (such as DDT), chlorinated cyclodiene compounds (such as dieldrin), and the chlorinated benzenes and cyclohexanes (such as hexachlorobenzene and lindane, respectively). For example, DDT inhibits several functions in neurons, reducing the rate at which the nerve membrane is repolarized following stimulation; the neurons therefore are very sensitive to complete depolarization by small stimuli. Insecticides in the chlorinated cyclodiene, benzene, and cyclohexane groups appear to affect the central nervous system, rather than the perpipheral (sensory) neurons.

Many organochlorine insecticides have endocrine-disrupting effects in a variety of animals (Colborn and Clement, 1992a) and interfere directly or indirectly with fertility and reproduction (Klaassen and others, 1996). From Table 6.15, it is clear that the organochlorine pesticides found in aquatic biota in the monitoring studies reviewed have moderate to high chronic toxicity; they tend to be associated with developmental or reproductive effects in animal studies; they are potent enzyme inducers; and most (with three exceptions) are probable human carcinogens (cancer group B2). The exceptions are endrin and methoxychlor (classified in group D), and endosulfan (classified in the group E).

Organophosphate Insecticides

The organophosphate insecticides are absorbed efficiently via the digestive system (U.S. Environmental Protection Agency, 1994b). The acute toxicity of organophosphate insecticides tends to be much greater than that of the organochlorine insecticides. The toxicity of individual organophosphate compounds depends on the rate at which they are metabolized in the body (U.S. Environmental Protection Agency, 1994b). Organophosphate esters may be attacked by enzymes simultaneously at a number of points in the molecule (Klaassen and others, 1996). One such reaction (oxidation of a phosophorothioate ester to its oxygen analog) results in a substantial increase in toxicity. An example is the oxidation of parathion to paraoxon. Further metabolism, principally by hydrolysis, results in detoxication in mammals. Insects are frequently deficient in the hydrolyze enzymes (which catalyze hydrolysis), so are more sensitive than mammals to these agents (Klaassen and others, 1996).

Organophosphate insecticides inhibit *acetylcholinesterase* activity in nerve tissue. This is the enzyme responsible for the destruction and termination of the biological activity of the neurotransmittor, acetylcholine (Klaassen and others, 1996). The reaction between an organophosphate compound and the active site in acetylcholinesterase results in formation of an intermediate complex, which then hydrolyzes to form a stable, phosphorylated, and fairly unreactive (inhibited) enzyme. The loss of enzyme function results in accumulation of acetylcholine at the nerve endings of cholinergic nerves, causing continual stimulation of electrical activity. Toxic signs and symptoms include those resulting from stimulation of muscarinic receptors in the parasympathetic autonomic nervous system (in smooth muscles, the heart, and exocrine glands), stimulation and subsequent blockade of nicotinic receptors (including autonomic ganglia and the junctions between nerves and muscles), and effects on the central nervous system (Klaassen and others, 1996). A few organophosphate compounds have caused delayed neurotoxicity. Dephosphorylation of the phosphorylated enzyme (to produce free enzyme) is extremely slow, so that inhibition by organophosphate insecticides is only very slowly reversible. With some organophosphate insecticides, the enzyme is irreversibly inhibited and toxicity will persist until sufficient new enzyme is synthesized in 20–30 days to destroy the excess neurotransmittor. In an organophosphate compound with the structural formula, XO-P(O)(Z)-OY, the substituents (X, Y and Z) determine the strength of binding to the enzyme active site and the rate of deposphosphorylation to produce free enzyme (Klaassen and others, 1996).

Many organophosphate compounds are classified in the cancer group D (that is, there is insufficient data to determine potential carcinogenicity). Chlorpyrifos and diazinon, which are the two organophosphate insecticides detected in aquatic biota in the monitoring studies reviewed (Tables 2.1 and 2.2), were classified as D/E and E, respectively (Table 6.15). Chlorpyrifos was fetotoxic in several species; there was insufficient data to assess potential for sublethal developmental effects (U.S. Environmental Protection Agency, 1994b). Diazinon showed developmental effects in multiple studies, including teratogenicity, abnormal growth, and severe behavioral abnormalities (U.S. Environmental Protection Agency, 1994b).

Carbamate Insecticides

Like the organophosphate insecticides, the mode of action of the carbamate insecticides is inhibition of acetylcholinesterase (Klaassen and others, 1996). However, the action of the carbamates is shorter in duration and milder in intensity than that of the organophosphate insecticides. This is because decarbamylation of the enzyme occurs more rapidly than does dephosphorylation. Therefore, the carbamates generally are reversible inhibitors of cholinesterase. Also, the carbamates are rapidly metabolized in vivo to intermediates that lack biological activity (Klaassen and others, 1996).

Two carbamate insecticides were detected in bed sediment or aquatic biota monitoring studies: carbaryl and carbofuran. These compounds were classified by USEPA in cancer groups C/D and E, respectively (Table 6.15).

Pyrethroid Insecticides

Synthetic pyrethroids are widely used in agricultural, household, and veterinary insecticide applications (Eisler, 1992). They are structural analogues of natural pyrethrins of botanical origin. In general, they are more stable (to photochemical, chemical, and microbial degradation), less toxic to mammals, and more toxic to insects than natural pyrethrins (Eisler, 1992). Pyrethroids induce repetitive activity in the nervous system by acting on the sodium channel in the nerve membrane (Eisler, 1992). They are not known to cause carcinogenic, mutagenic, or teratogenic effects. Two distinct types of pyrethroids have different behavioral, neurophysiological, chemical, and biochemical profiles. Type II pyrethroids contain an α-cyano group (such as fenvalerate) and are more neurotoxic than Type I pyrethroids (such as permethrin), which lack the α-cyano group (Eisler, 1992).

Very few studies targeted synthetic pyrethroids in bed sediment or aquatic biota. Two synthetic pyrethroids were detected in these media: fenvalerate and permethrin. Permethrin has been classified in cancer group C (possible human carcinogen). Fenvalerate is a potential tumor promoter (Eisler, 1992).

Herbicides

The herbicides in use today in the United States fall into a number of chemical classes and vary in their mechanisms of toxicity. By definition, all herbicides are capable of killing or severely injuring plants. The mechanisms of action of several different herbicide classes are summarized in Table 6.16 (adapted from Klaassen and others, 1996 [based on Jager, 1983]). With a few exceptions, most herbicides have relatively low toxicity to mammals, with oral LD_{50} (median lethal dose) values for rats on the order of 100 to 10,000 mg/kg. Some of these chemicals may be mutagenic, carcinogenic, or teratogenic (Klaassen and others, 1996).

Only a few herbicides have been detected in bed sediments and aquatic biota in the monitoring studies reviewed (Tables 2.1 and 2.2). These herbicides fall into the following chemical classes: chlorophenoxy acids (2,4-D), chlorobenzoic acids (dacthal), diphenyl ethers (nitrofen), oxadiazol compounds (oxadiazon), and dinitroanilines (trifluralin). Nitrofen was canceled before USEPA's carcinogenicity guidelines were developed, but probably would have a B2 classification (U.S. Environmental Protection Agency, 1992b; Nowell and Resek, 1994). The others have been classified in either cancer group C (oxadiazon and trifluralin), C/D (dacthal) or

Table 6.16. Mechanisms of action of several different herbicide classes

[Abbreviations: ADP, adenosine diphosphate; ATP, adenosine triphosphate; $NADH_2$, reduced nicotinamide-adenine-dinucleotide. From Klaassen and others (1996), with minor corrections based on the original data source, Jager (1983), with the permission of the publishers. Copyright 1996 McGraw-Hill Companies. Copyright 1983 John Wiley & Sons, Inc.]

Mechanisms	Herbicide Classes
Inhibition of photosynthesis by disruption of light reactions and blockade of electron transport	Ureas, 1,3,5-triazines, 1,4-triazinones, uracils, pyridazinones, 4-hydroxybenzonitriles, N-arylcarbamates, acylanilides (some)
Inhibition of respiration by blockade of electron transport from $NADH_2$ to O_2 or blocking the coupling of electron transfer in the formation of ATP from ADP	Dinitrophenols, halophenols, nitrofen
Growth stimulants, "auxins"	Aryloxyalkylcarboxylic acids, benzoic acids
Inhibitors of cell and nucleus division	Alkyl N-arylcarbamates, dinitroanilines
Inhibition of protein synthesis	Chloroacetamides
Inhibition of carotenoid synthesis, protective pigments in chloroplasts to prevent chlorophyll from being destroyed by oxidative reactions	O-substituted diphenyl ethers
Inhibition of lipid synthesis	S-alkyl dialkylcarbamodithioates, aliphatic chlorocarboxylic acids (some)
Unknown mechanisms	Dichlobenil, chlorthiamid, bentazone, diphenamid, benzoylpropethyl

D (2,4-D). Reproductive effects in laboratory tests have been observed for nitrofen (U.S. Environmental Protection Agency, 1992b).

Fungicides

Hexachlorobenzene and pentachloronitrobenzene (fungicides) and pentachlorophenol (a wood preservative) were detected in aquatic biota in the monitoring studies reviewed (Tables 2.1 and 2.2). Note that contamination with hexachlorobenzene and pentachlorophenol also may derive from industrial sources. In general, these compounds were targeted in very few studies. These compounds are characterized by fairly low toxicity to mammals (acute oral LD_{50} values for rats of over 1,000 mg/kg). Hexachlorobenzene has been classified in cancer group B2 (probable carcinogen), and has caused both teratogenic and fetotoxic effects in laboratory tests. Pentachloronitrobenzene has been classified in cancer group C (possible human carcinogen), and pentachlorophenol in cancer group B2 (probable carcinogen). Cleft palates were observed in offspring of mice exposed orally to pentachloronitrobenzene and pentachlorophenol was feto-toxic in laboratory tests with rats (U.S. Environmental Protection Agency, 1992b).

6.2.3 STANDARDS AND GUIDELINES FOR PESTICIDES IN EDIBLE FISH AND SHELLFISH

Various human health standards and guidelines (numerical values that indicate a threshold for potential effects on human health) exist for pesticides in fish and shellfish. These standards and guidelines will be described below and compared with pesticide concentrations measured in the monitoring studies reviewed in Section 6.2.4.

The terms "standards" and "guidelines" are used in different ways by different authors and thus require careful definition. In this book, "standards" refer to threshold values that are legally enforceable by agencies of the United States government, whereas "guidelines" refer to threshold values that have no regulatory status, but are issued in an advisory capacity. To determine potential effects on human health, two types of standards (USEPA tolerances and FDA action levels) and one set of guidelines (USEPA screening values, which were issued as part of USEPA's guidance for use in setting fish advisories) are used. These standards and guidelines all apply to edible fish and shellfish tissue.

These standards and guidelines were developed using risk assessment and risk management techniques. *Risk assessment* is a scientifically based procedure used to estimate the probability of adverse health effects from a specific source under specific exposure conditions. *Risk manage-ment* is the process of integrating risk assessment data with social, economic, and political information to decide how to reduce or eliminate the potential risks in question (Reinert and others, 1991).

USEPA's Tolerances

The Federal Food Drug and Cosmetic Act (FFDCA) authorizes USEPA to set tolerances for pesticides in raw agricultural commodities. A *tolerance* is the maximum amount of a pesti-cide residue that can be legally present in or on a raw agricultural commodity (Section 408 of the FFDCA). Tolerances are enforceable standards. Under the FFDCA, a tolerance or tolerance exemption is required when USEPA grants registration under the Federal Insecticide, Fungicide, and Rodenticide Act (FIFRA) for the use of a pesticide in food or feed production in the United States. Thus, USEPA tolerances exist only for currently registered pesticides.

The tolerance provisions of both the FFDCA and FIFRA were amended by the Food Quality Protection Act (FQPA) of 1996. All existing tolerances must be reassessed within 10 years, and a new safety standard must be met. The new safety standard is "a reasonable certainty that no harm will result from aggregate exposure to the pesticide chemical residue" (Section 408(a)(4) of the FFDCA). Thus, USEPA is required to consider aggregate exposure to all non-occupational sources of exposure, including drinking water, residential, and dietary exposures, when setting tolerances. The FFDCA, as amended by the FQPA, also requires a specific determination that tolerances are safe for infants and children, and that various other factors (such as cumulative and aggregate exposure to chemicals with a common mechanism of toxicity) be considered in setting tolerances (U.S. Environmental Protection Agency, 1998).

Table 6.8 lists the existing tolerances for pesticides in fish and shellfish (as of July 1, 1997, as listed in the Code of Federal Regulations, Volume 40, Part 180). Of the pesticides that were analyzed in edible fish tissue in the monitoring studies in Tables 2.1 and 2.2, only one (2,4-D) has a USEPA tolerance.

FDA's Action Levels

An FDA *action level* is an enforceable regulatory limit for unavoidable pesticide residues in or on a food or animal feed. Its purpose is to protect the general public from contaminants in fish shipped under interstate commerce (Reinert and others, 1991). FDA action levels exist only for pesticides that do not have USEPA tolerances, but that may be present as unavoidable residues in fish or shellfish. Such pesticides include a number of environmentally persistent organochlorine insecticides that are no longer registered for use in the United States: total chlordane, total DDT, total dieldrin, endrin, total heptachlor, mirex, and toxaphene. FDA action levels for pesticides in edible fish and shellfish are listed in Table 6.8.

Action levels are published in FDA's Compliance Policy Guide (CPG) 7141.01, Attachment B (Food and Drug Administration, 1989b). When residues are at or above this action level, FDA can, at the agency's discretion, take legal action to remove the adulterated food from interstate commerce. Food and Drug Administration (1990b) published a policy statement explaining that agency's use of action levels to regulate pesticides in food or feed. The following is an abbreviated summary (Nowell and Resek, 1994) of that policy statement:

> The FFDCA provides that, in the absence of a USEPA tolerance, any amount of a pesticide residue in a food or feed is unsafe and renders the food or feed adulterated. However, FDA recognized that food or feed may contain residues of certain pesticides for which no tolerances existed (because the pesticides had been canceled by USEPA) but that persist in the environment (for example, the organochlorine insecticides). FDA found that in such cases, the level of pesticide was frequently so low that it was not of any regulatory or public health significance and that pursuing enforcement action would provide little or no benefit to the public and would not be the most prudent way to expend agency resources. FDA therefore adopted the concept of action levels to define amounts of a particular pesticide that FDA regarded as rendering a food or feed adulterated. These action levels represent an exercise of FDA's discretion under Section 306 of the FFDCA to refrain from initiating enforcement action for minor violations and to decide when and how to enforce the FFDCA. However, FDA action levels provide guidance only on how FDA may exercise its enforcement discretion. FDA may decide that an enforcement action is not warranted even though the residue level in a food or feed exceeds the action level, or it may decide that an action is warranted against a food or feed that contains a pesticide residue level that is below the applicable action level.

An action level is established by FDA but is based on USEPA's recommendation. The action levels recommended by USEPA were derived primarily from FDA's monitoring data, which provided an indication of the extent to which residues of a particular pesticide cannot be avoided by good agricultural practice or by current good manufacturing practice (U.S. Environmental Protection Agency, 1992c). These recommended action levels also considered the existing analytical detection limits (Food and Drug Administration, 1990b) and the economic impacts likely to result on the commercial fishing industry (Reinert and others, 1991; U.S. Environmental Protection Agency, 1992c), as well as the health risks to the general public. Thus, the FDA action levels incorporate risk management, as well as risk assessment, considerations (Reinert and others, 1991).

USEPA's Guidance for Use in Fish Advisories

U.S. Environmental Protection Agency (1994b, 1995a) developed guidance for state, local, regional, and tribal environmental health officials who are responsible for issuing fish advisories for chemically contaminated noncommercial fish. For individual target analytes, U.S. Environmental Protection Agency (1995a) determined *screening values*, defined as concentrations of analytes in fish and shellfish that are of potential health concern and that are used as benchmarks against which levels of contamination in similar tissue collected from the ambient environment can be compared. Screening values were derived using a risk assessment method, assuming average values for consumer body weight, meal size, and meal frequency. This risk assessment method is provided in U.S. Environmental Protection Agency (1995a), along with procedures for calculating modified screening values that reflect local conditions, populations consuming higher-than-average quantities of fish (for example, sport and subsistence fishers), or sensitive subpopulations (for example, children and pregnant women). U.S. Environmental Protection Agency (1994b) also has published, for high-priority chemical fish contaminants, tables of monthly fish consumption limits that are based on potential chronic and cancer health effects. The numbers in these tables are based on the same risk assessment method, but the format of the tables allows a user to determine the maximum number of meals that should be consumed per month for a given level of contamination (depending on local conditions) and for a variety of exposure levels. For example, the number of fish meals allowed per month will vary for an adult versus a child, for different meal sizes (3–16 ounces), and (for potential carcinogens) for different acceptable cancer risk levels.

USEPA screening values (U.S. Environmental Protection Agency, 1995a) and the associated monthly fish consumption limits (U.S. Environmental Protection Agency, 1994b) were calculated using health-based numbers (a dose that corresponds to a specific health endpoint) and making certain assumptions regarding exposure. For noncarcinogens, the health-based number is the reference dose, which, as discussed in Section 6.2.2, is the estimated daily exposure that is without appreciable risk of deleterious health effects in the human population over a 70-year lifetime (U.S. Environmental Protection Agency, 1988). The reference dose is based on chronic toxicity. Developmental effects were not considered in developing screening values because USEPA had not yet developed reference dose values that were based specifically on developmental effects for most analytes (U.S. Environmental Protection Agency, 1994b). However, U.S. Environmental Protection Agency (1994b) included data summaries on developmental toxicity of individual analytes as part of its guidance for use in setting fish advisories (this information is incorporated into Table 6.15).

For potential carcinogens, separate screening values were calculated for noncarcinogenic effects (chronic toxicity) and carcinogenic effects. The screening value for noncarcinogenic effects is calculated from the reference dose, as just described for noncarcinogens. For carcinogenic effects, the health-based number used is the cancer potency factor (the q_1^*, which is the upper 95th percentile confidence limit providing a low-dose estimate of cancer risk using a linearized multistage model). Because this model is conservative, the resulting cancer risk is an upper limit estimate (U.S. Environmental Protection Agency, 1989). An essential component of the screening value calculation for carcinogenic effects is the *maximum acceptable cancer risk level*, sometimes called simply the "cancer risk level." The model of carcinogenicity used by USEPA is a nonthreshold model (i.e., in theory, even one molecule of a carcinogen has a finite risk of causing cancer). According to this model, there is no safe level of a carcinogen. USEPA instead calculates the concentration (screening value) that would result, under specified exposure conditions, in a specified risk of excess cancer (e.g., 10^{-5}, or 1 extra case of cancer in 100,000 people). The choice of maximum acceptable cancer risk level is a risk management decision. Cancer risk levels generally are between 10^{-4} and 10^{-7} (as discussed in Section 6.2.5; see Relative Cancer Risks). The importance of the choice of cancer risk level is illustrated by the following example. The USEPA screening value for dieldrin (a B2 carcinogen) is 7 μg/kg at a maximum acceptable cancer risk level of 10^{-5}, but the screening value becomes 0.7 μg/kg if the cancer risk level is lowered to 10^{-6}. Thus, the USEPA screening values for noncarcinogens are based on risk assessment procedures; the screening values for potential carcinogens are based on both risk assessment and risk management procedures.

The difference between screening values (U.S. Environmental Protection Agency, 1995a) and monthly fish consumption limits (U.S. Environmental Protection Agency, 1994b) lies in the exposure assumptions. The format of the monthly fish consumption limit tables allows the user to see the effect of changing exposure conditions on the maximum allowable meals per month. This format is likely to be useful to health officials in setting fish consumption limits. However, it is not particularly conducive to use in the present analysis, the goal of which is to determine how many of the studies reviewed had maximum concentrations that exceeded USEPA guidance for use in fish advisories. Because a nationally consistent set of guidelines is necessary, the screening values recommended by U.S. Environmental Protection Agency (1995a) were used. These screening values (and therefore, the analysis in Section 6.2.4) apply to the general adult population and assume average values for body weight (70 kg), a mean daily consumption rate of 6.5 g/d of freshwater and estuarine fish over a 70-year lifetime, and that 80 percent of daily exposure comes from consumption of fish and shellfish. For carcinogens, this analysis also assumes an acceptable cancer risk level of 10^{-5}. Note that a mean daily consumption rate of 6.5 g/d is fairly low. This is about 7 ounces per month, assuming 30 days per month. At 4 ounces of fish per meal, this amounts to slightly less than two meals of freshwater and estuarine fish per month. This means that, for a site where measured concentrations in edible fish or shellfish exceed the applicable screening value, there is some potential for adverse effects on human health if individuals consume two meals per month of fish from that site over a 70-year lifetime.

USEPA screening values (U.S. Environmental Protection Agency, 1995a) are provided for most of the same pesticides that have FDA action levels, plus some additional ones: total chlordane, chlorpyrifos, total DDT, diazinon, dicofol, dieldrin, disulfoton, total endosulfan, endrin, ethion, heptachlor epoxide, hexachlorobenzene, lindane, mirex, terbufos, and toxaphene. These screening values are listed in Table 6.8. For potential carcinogens, the screening value listed in Table 6.8 is the lower of that calculated for noncarcinogenic effects and for carcinogenic effects; therefore, the screening value listed is protective of both types of effects. Where applicable, the cancer risk level is specified in a footnote.

Table 6.17. Selected results of studies that monitored pesticides in fish muscle and percentage of those studies that exceeded standards and guidelines for the protection of human health

[Results are presented for all monitoring studies (published 1960–1994), and recently published (1984–1994) studies, listed in Tables 2.1 and 2.2. All concentrations are wet weight. Range in detection limits. Abbreviations and symbols: AL, action level; FDA, Food and Drug Administration; SV, screening value; USEPA, U.S. Environmental Protection Agency; nd, not detected; nr, one or more studies did not report this information; µg/kg, microgram per kilogram; —, no standard or guideline available]

Target Analytes	Fish Muscle: All Monitoring Studies (Published 1960–1994)					
	Concentration Range (µg/kg wet weight)		Total Number of Studies	Percentage of Studies with Detectable Residues	Percentage of Studies with C_{max} that Exceeded FDA AL	Percentage of Studies with C_{max} that Exceeded USEPA SV
	In Detection Limits	In Maximum Concentrations (C_{max})				
Alachlor	nr	nd	1	0	—	—
Aldrin	nr, 0.1–200	nd–280	13	15	0	—
Atrazine	300–8,000	nd	1	0	—	—
Chlordane[1]	nr, 0.1–500	nd–5,200	26	77	46	77
Chlordane, cis-	nr, 5–10	nd–59	9	67	0	0
Chlordane, trans-	nr, 5–10	nd–73	10	70	0	0
D, 2,4-	0.1	nd	1	0	—	—
Dacthal	nr, 1	trace–217	2	100	—	—
DDD[1]	nr, 0.1–100	nd–92,500	11	82	9	45
DDD, o,p'-	nr, 1–100	nd–200	11	27	0	0
DDD, p,p'-	nr, 1–100	nd–1,100	18	89	0	33
DDE[1]	nr, 0.1–100	0.4–90,000	12	100	8	58
DDE, o,p'-	nr, 1–100	nd–300	11	45	0	9
DDE, p,p'-	nr, 1–100	nd–3,620	19	89	0	37
DDT[2]	nr, 1–100	nd–50,000	12	83	8	33
DDT, o,p'-	nr, 1–100	nd–600	12	42	0	8
DDT, p,p'-	nr, 0.1–100	nd–1,700	19	58	0	26
DDT, total	nr, 0.1–200	nd–627,000	17	100	12	94
Dieldrin	nr, 0.1–100	nd–2,100	44	80	14	80
Endosulfan[1]	0.1–40	nd–30	3	33	—	0
Endosulfan I	5	nd	1	0	—	0
Endosulfan II	5	nd	1	0	—	0
Endosulfan sulfate	5–200	nd	2	0	—	—
Endrin	nr, 0.1–100	nd–75	20	30	0	0
Endrin aldehyde	5–200	nd	2	0	—	—
HCH[1]	5–130	5–3,750	2	100	—	—
HCH, α-	nr, 0.1–100	nd–340	15	47	—	—
HCH, β-	nr, 1–100	0–15	8	13	—	—
HCH, δ-	5–100	nd–30	6	17	—	—
Heptachlor	nr, 0.01–100	nd–1,100	12	17	8	—

Table 6.17. Selected results of studies that monitored pesticides in fish muscle and percentage of those studies that exceeded standards and guidelines for the protection of human health—*Continued*

Target Analytes	Fish Muscle: Recent Monitoring Studies (Published 1984–1994)					
	Concentration Range (µg/kg wet weight)		Total Number of Studies	Percentage of Studies with Detectable Residues	Percentage of Studies with C_{max} that Exceeded FDA AL	Percentage of Studies with C_{max} that Exceeded USEPA SV
	In Detection Limits	In Maximum Concentrations (C_{max})				
Alachlor	nr	nd	1	0	—	—
Aldrin	nr, 0.1–10	nd–3	8	13	0	—
Atrazine	300–8,000	nd	1	0	—	—
Chlordane[1]	nr, 0.1–100	nd–2,420	16	75	56	75
Chlordane, *cis-*	nr, 5–10	nd–59	9	67	0	0
Chlordane, *trans-*	nr, 5–10	nd–73	10	70	0	0
D, 2,4-	0.1	nd	1	0	—	—
Dacthal	1	trace	1	100	—	—
DDD[1]	nr, 0.1–20	nd–490	6	67	0	33
DDD, *o,p'-*	nr, 1–40	nd–45	8	25	0	0
DDD, *p,p'-*	nr, 1–40	nd–780	12	83	0	8
DDE[1]	nr, 0.1–10	0.4–3,200	7	100	0	57
DDE, *o,p'-*	nr, 1–20	0–120	8	38	0	0
DDE, *p,p'-*	nr, 1–20	nd–2,400	12	83	0	25
DDT[2]	nr, 1–20	nd–1,070	6	83	0	17
DDT, *o,p'-*	nr, 1–40	0–220	8	38	0	0
DDT, *p,p'-*	nr, 0.1–40	nd–669	13	46	0	8
DDT, total	nr, 0.1–40	nd–19,750	8	100	13	100
Dieldrin	nr, 0.1–50	nd–180	22	73	0	73
Endosulfan 1	0.1–20	nd–30	2	50	—	0
Endosulfan I	5	nd	1	0	—	0
Endosulfan II	5	nd	1	0	—	0
Endosulfan sulfate	5–200	nd	2	0	—	—
Endrin	nr, 0.1–20	nd–75	14	29	0	0
Endrin aldehyde	5–200	nd	2	0	—	—
HCH[1]	5	5	1	100	—	—
HCH, α-	nr, 0.1–100	nd–340	12	42	—	—
HCH, β-	nr, 5–100	nd–15	7	14	—	—
HCH, δ-	5–100	nd–30	6	17	—	—
Heptachlor	nr, 0.1–10	nd	5	0	0	—

Table 6.17. Selected results of studies that monitored pesticides in fish muscle and percentage of those studies that exceeded standards and guidelines for the protection of human health—*Continued*

Target Analytes	Fish Muscle: All Monitoring Studies (Published 1960–1994)					
	Concentration Range (µg/kg wet weight)		Total Number of Studies	Percentage of Studies with Detectable Residues	Percentage of Studies with C_{max} that Exceeded FDA AL	Percentage of Studies with C_{max} that Exceeded USEPA SV
	In Detection Limits	In Maximum Concentrations (C_{max})				
Heptachlor epoxide	nr, 0.1–200	nd–440	20	50	5	50
Hexachlorobenzene	nr, 1–100	nd–280	17	47	—	24
Kepone	nr, 1–10	nd–560	3	67	—	—
Lindane	nr, 0.1–100	nd–110	20	30	—	5
Methoxychlor[3]	0.1–100	nd–100	6	17	—	—
Metolachlor	nd	nd	1	0	—	—
Mirex	nr, 0.1–30	nd–1,142	22	50	45	50
Nonachlor, *cis-*	nr, 5–10	nd–33	9	44	0	0
Nonachlor, *trans-*	nr, 5–10	nd–150	10	90	0	10
Oxychlordane	nr, 10	nd–22	8	38	0	0
Parathion	10	280	1	100	—	—
Pentachloroanisole	nr	60	1	100	—	—
Pentachlorophenol	nr, 1–50	nd–100	3	67	—	—
Perthane	1	nd	1	0	—	—
Photomirex	nr, 5–10	80–490	5	100	—	—
T, 2,4,5-	0.1	nd	1	0	—	—
TP, 2,4,5-	0.1	nd	1	0	—	—
Toxaphene	nr, 10–2,000	nd–6,800	17	24	6	24
Trifluralin	nr, 50	nd–74	2	50	—	—

[1]Total or unspecified.
[2]$(o,p' + p,p')$- or unspecified.
[3]p,p'- or unspecified.

6.2.4 PESTICIDES IN EDIBLE FISH AND SHELLFISH—ANALYSIS OF POTENTIAL ADVERSE EFFECTS ON HUMAN HEALTH

The standards and guidelines discussed in Section 6.2.3 were intended for comparison with contaminant residues in edible fish tissue and shellfish. Different monitoring studies reported pesticide residues in a variety of edible fish tissues, such as fish muscle (or flesh), red muscle, skin-on fillets, or skin-off fillets. All such edible fish tissues were included in comparing results

Table 6.17. Selected results of studies that monitored pesticides in fish muscle and percentage of those studies that exceeded standards and guidelines for the protection of human health—*Continued*

Target Analytes	Fish Muscle: Recent Monitoring Studies (Published 1984–1994)					
	Concentration Range (µg/kg wet weight)		Total Number of Studies	Percentage of Studies with Detectable Residues	Percentage of Studies with C_{max} that Exceeded FDA AL	Percentage of Studies with C_{max} that Exceeded USEPA SV
	In Detection Limits	In Maximum Concentrations (C_{max})				
Heptachlor epoxide	nr, 0.1–100	nd–37	14	50	0	50
Hexachlorobenzene	nr, 1–100	nd–100	13	38	—	15
Kepone	1–10	nd–trace	2	50	—	—
Lindane	nr, 0.1–100	nd–110	14	29	—	7
Methoxychlor[3]	0.1–100	nd	4	0	—	—
Metolachlor	nr	nd	1	0	—	—
Mirex	nr, 0.1–10	nd–1,142	15	40	40	40
Nonachlor, *cis-*	nr, 5–10	nd–33	9	44	0	0
Nonachlor, *trans-*	nr, 5–10	nd–150	10	90	0	10
Oxychlordane	nr, 10	nd–22	8	38	—	—
Parathion	—	—	0	—	—	—
Pentachloroanisole	—	—	0	—	—	—
Pentachlorophenol	1–50	nd–55	2	50	—	—
Perthane	1	nd	1	0	—	—
Photomirex	nr, 5–10	78–160	3	100	—	—
T, 2,4,5-	0.1	nd	1	0	—	—
TP, 2,4,5-	0.1	nd	1	0	—	—
Toxaphene	nr, 10–2,000	nd–1,200	12	8	0	8
Trifluralin	nr, 50	nd–74	2	50	—	—

of the monitoring studies reviewed (Tables 2.1 and 2.2) with the standards and guidelines for edible fish tissue. These standards and guidelines are labeled in Tables 6.8 and 6.17 as applying to "fish muscle."

Standards and guidelines for pesticides in edible fish (fish muscle) and shellfish are listed in Table 6.8. Table 6.17 lists, for individual pesticides that were analyzed in edible fish tissue in the monitoring studies in Tables 2.1 and 2.2, the total number of applicable studies, the range in detection limits for those studies, the range in maximum concentrations reported by those studies, and the percentage of those studies in which the pesticide was detected. Table 6.17 also lists, for pesticides that have human health standards or guidelines, the percentage of those studies in which the maximum concentration exceeded each applicable standard or guideline. This analysis is done for all monitoring studies in Tables 2.1 and 2.2 that targeted edible fish tissue (published during 1960–1994) and for recently published (1984–1994) studies only. Table 6.18 contains the same information and analysis for studies that measured pesticides in shellfish.

The results of this analysis are presented below for individual pesticides that were analyzed in edible fish and shellfish in the monitoring studies in Table 2.1 and 2.2. However, discussion of the comparisons of measured concentrations with applicable standards and guidelines can be simplified by first making two points. First, as mentioned in Section 6.2.3, a USEPA tolerance exists for only one of the pesticides (2,4-D) that were analyzed in edible fish tissue in the monitoring studies reviewed. Technically, this tolerance applies only to certain uses of 2,4-D: (1) use of dimethylamine salt for water hyacinth control in slow-moving or quiescent water bodies by federal, state, or local public agencies, and (2) use of dimethylamine salt or its butoxy-ethanol ester for Eurasian watermilfoil control in Tennessee Valley Authority programs (see Table 6.8). Only a single monitoring study analyzed for 2,4-D in edible fish tissue; 2,4-D was not detected in this study, which had a detection limit (0.1 µg/kg) that was well below the USEPA tolerance for 2,4-D in fish (1,000 µg/kg). Therefore, information on USEPA tolerances is not included in Table 6.17.

Second, it is clear from Table 6.8 that FDA action levels for most organochlorine compounds (all but endrin and mirex) are considerably higher than the corresponding USEPA screening values. The reverse is true for endrin and mirex (the USEPA screening values are higher than the FDA action levels). There are several differences between FDA action levels and USEPA screening values. First, FDA action levels apply to fish or fish products in interstate commerce, whereas USEPA screening values apply to noncommercial fish. Second, FDA action levels are based on a different risk assessment methodology than that used by USEPA in setting guidance for use in fish consumption advisories (Reinert and others, 1991). For example, USEPA uses a more conservative approach when extrapolating from the results of toxicity tests with rodents to estimates of toxicity to humans. Whereas USEPA relies on a ratio of animal-to-human body surface area, FDA uses a ratio of animal-to-human body weight. The USEPA model results in a 6-fold greater risk than the FDA model when data from rats are used and a 14-fold greater risk when data from mice are used (Reinert and others, 1991). Third, FDA action levels are not based on health considerations alone, but also consider factors such as the economic costs of banning a foodstuff (U.S. Environmental Protection Agency, 1992c), analytical detection limits (Food and Drug Administration, 1990b), and the extent to which residues cannot be avoided by good agricultural or current good manufacturing practice (U.S. Environmental Protection Agency, 1992c). Thus, there is not the same direct link between levels of risk and levels of fish consumption that the USEPA risk-based approach provides (Reinert and others, 1991). USEPA screening values were derived using explicitly stated risk-assessment methodology, with specified exposure assumptions and acceptable risk levels. These differences should be considered when interpreting what it means when a given pesticide concentration exceeds, or does not exceed, the applicable FDA action level and USEPA screening value.

For selected pesticides, Figures 6.13–6.15 show the cumulative frequency distribution of the maximum concentrations reported for each pesticide in edible fish tissue from the monitoring studies reviewed. These pesticides are the most commonly detected ones in aquatic biota:

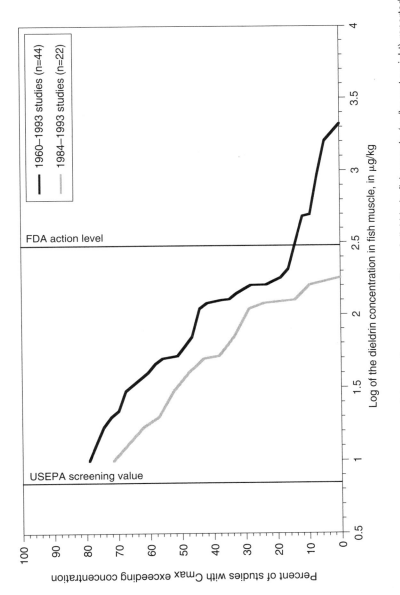

Figure 6.13. Cumulative frequency distribution of the maximum concentrations of dieldrin in fish muscle (µg/kg wet weight) reported by the monitoring studies reviewed, shown for all studies (published during 1960–1994) and recently published (1984–1994) studies. Standards and guidelines for dieldrin in fish muscle (for protection of human health) are shown as vertical lines. Abbreviations: C_{max}, maximum concentration in the study; FDA, Food and Drug Administration; n, number of samples; USEPA, U.S. Environmental Protection Agency; µg/kg, microgram per kilogram.

dieldrin (Figure 6.13), total chlordane (Figure 6.14), and total DDT (Figure 6.15). In Figures 6.13–6.15, the applicable standard and guideline values are shown as vertical lines on the graph, corresponding to the appropriate pesticide concentration. Cumulative frequency distributions for pesticides in shellfish were not plotted because relatively few shellfish studies found pesticide concentrations that exceeded the applicable standards.

Aldrin and Dieldrin

Only two studies detected any aldrin residues in edible fish tissue, and the maximum concentrations detected were below the applicable FDA action level (Table 6.17). For dieldrin, 6 of 44 total (edible fish) monitoring studies (or 14 percent) found dieldrin concentrations greater than the FDA action level for aldrin plus dieldrin. However, when only recently published (1984–1994) studies were considered, none of the 22 studies reported dieldrin levels above the FDA action level (Table 6.17). A total of 12 monitoring studies analyzed dieldrin in shellfish, but none had maximum concentrations in shellfish that exceeded the FDA action level for dieldrin (Table 6.18).

In contrast, 80 percent of all (edible fish) monitoring studies had maximum concentrations that exceeded the USEPA screening value for dieldrin (7 µg/kg), which is much lower than the FDA action level for total dieldrin (100 µg/kg). When only recently published (1984–1994) monitoring studies were considered, the percentage of studies with maximum concentrations that exceeded the USEPA screening value was slightly lower (73 percent). Comparison of the cumulative frequency distributions (Figure 6.13) for recently published (1984–1994) studies with all monitoring studies (published during 1960–1994) indicates that far fewer recently published studies measured dieldrin residues at concentrations greater than about 200 µg/kg. This analysis indicates that there is potential for adverse effects on human health at the most contaminated site in 73 percent of recently published studies, if the fish sampled in these studies were consumed at average levels (about two 4-ounce meals per month) by the general adult population over a 70-year lifetime. The potential adverse health effect expected would be cancer, and the excess cancer risk (at the most contaminated site in the affected studies) may exceed 1 in 100,000 people. Because this analysis deals only with the maximum concentrations reported by each study, it does not provide any information on the number of sites in each study that may exceed applicable standards and guidelines.

Total Chlordane

The FDA action level and the USEPA screening value for total chlordane were exceeded by the maximum concentrations detected for a large percentage of studies reviewed (Table 6.17). This was true even when only recently published (1984–1994) studies were considered: 56 percent had maximum concentrations that exceeded the FDA action level, and 75 percent had maximum concentrations that exceeded the USEPA screening value. In fact, as shown in Figure 6.14, the study date had relatively little effect on the cumulative frequency distribution of maximum concentrations of total chlordane detected in the studies reviewed, except that a higher percentage of recently published studies had maximum concentrations of total chlordane in the range of 150 to 1,000 µg/kg, and earlier monitoring studies (published prior to 1984) accounted

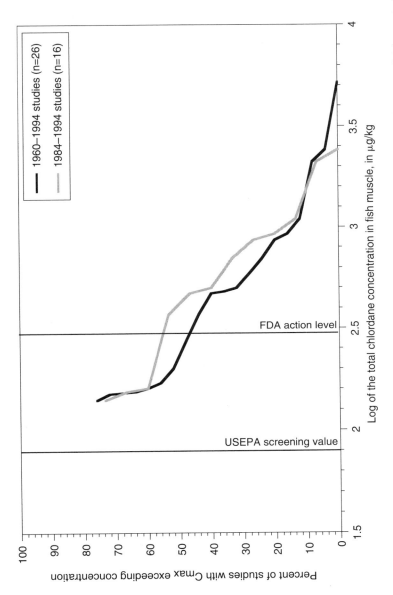

Figure 6.14. Cumulative frequency distribution of the maximum concentrations of chlordane in fish muscle (μg/kg wet weight) reported by the monitoring studies reviewed, shown for all studies (published during 1960–1994) and recently published (1984–1994) studies. Studies that measured chlordane (unspecified) and total chlordane are included. Standards and guidelines for chlordane in fish muscle (for protection of human health) are shown as vertical lines. Abbreviations: C_{max}, maximum concentration in the study; FDA, Food and Drug Administration; n, number of samples; USEPA, U.S. Environmental Protection Agency; μg/kg, microgram per kilogram.

Table 6.18. Selected results of studies that monitored pesticides in shellfish and percentage of those studies that exceeded standards for the protection of human health

[Results are presented for all monitoring studies (published 1960–1994), and recently published (1984–1994) studies, listed in Tables 2.1 and 2.2. All concentrations are wet weight. Abbreviations and symbols: AL, action level; FDA, Food and Drug Administration; nd, not detected; nr, one or more studies did not report this information; μg/kg, microgram per kilogram; —, no standard available]

| Target Analytes | Shellfish: All Monitoring Studies (Published 1960–1994) | | | | |
| | Concentration Range (μg/kg wet weight) | | Total Number of Studies | Percentage of Studies with Detectable Residues | Percentage of Studies with C_{max} that Exceeded FDA AL |
	In Detection Limits	In Maximum Concentrations (C_{max})			
Aldrin	nr, 0.01–20	nd–18	10	20	0
Chlordane[1]	nr, 0.01–100	nd–532	10	60	10
Chlordane, cis-	20–100	nd	2	0	0
Chlordane, trans-	0.01–20	nd–0.3	2	50	0
DDD[1]	nr, 0.01–10	nd–490	5	80	0
DDD, o,p'-	nr, 0.1–40	nd–40	4	25	0
DDD, p,p'-	nr, 0.01–100	nd–490	9	44	0
DDE[1]	nr, 0.01–10	nd–690	5	100	0
DDE, o,p'-	nr, 0.1–20	nd	4	0	0
DDE, p,p'-	nr, 0.01–100	nd–390	9	67	0
DDT[2]	nr, 0.1–10	15.6–1,100	4	100	0
DDT, o,p'-	nr, 0.1–40	nd–1,000	7	43	0
DDT, p,p'-	nr, 0.01–100	nd–540	8	38	0
DDT, total	nr, 0.1–40	nd–1,780	7	71	0
Diazinon	250	nd	1	0	—
Dieldrin	nr, 0.1–100	nd–77	12	42	0
Endosulfan[1]	0.1–20	nd–0.2	2	50	—
Endosulfan I	nr	22.3	1	100	—
Endosulfan II	nr, 0.01	0.75–2.57	2	100	—
Endosulfan sulfate	5–200	nd	2	0	—
Endrin	nr, 0.01–100	nd–10	9	44	0
Endrin aldehyde	nr, 0.01–200	nd–0.69	3	33	—
Ethion	250	nd	1	0	—
Fenchlorphos	250	nd	1	0	—
Guthion	250	nd	1	0	—
HCH, α-	nr, 0.01–100	nd–64	9	56	—
HCH, β-	nr, 0.01–100	nd–19.5	7	29	—
HCH, δ-	0.01–100	nd–1.41	3	33	—
Heptachlor	nr, 0.01–20	nd–37.8	10	20	0
Heptachlor epoxide	nr, 0.1–100	nd–24	10	20	0

Table 6.18. Selected results of studies that monitored pesticides in shellfish and percentage of those studies that exceeded standards for the protection of human health—*Continued*

Target Analytes	Shellfish: Recent Monitoring Studies (Published 1984–1994)				
	Concentration Range (µg/kg wet weight)		Total Number of Studies	Percentage of Studies with Detectable Residues	Percentage of Studies with C_{max} that Exceeded FDA AL
	In Detection Limits	In Maximum Concentrations (C_{max})			
Aldrin	nr, 0.01–20	nd–18	8	25	0
Chlordane[1]	nr, 0.01–100	nd–532	7	57	14
Chlordane, *cis-*	0	20–100	2	0	0
Chlordane, *trans-*	0.01–20	nd–0.3	2	50	0
DDD[1]	nr, 0.01–5	nd–100	4	75	0
DDD, *o,p'-*	nr, 0.1–1	nd–40	4	25	0
DDD, *p,p'-*	nr, 0.01–100	nd–311	7	43	0
DDE[1]	nr, 0.01–5	nd–111	4	100	0
DDE, *o,p'-*	nr, 0.1–20	nd	4	0	0
DDE, *p,p'-*	nr, 0.01–100	nd–40	8	63	0
DDT[2]	nr, 0.1–5	15.6–132	3	100	0
DDT, *o,p'-*	nr, 0.1–40	nd–122	5	60	0
DDT, *p,p'-*	nr, 0.01–100	nd–2.8	6	33	0
DDT, total	nr, 0.1–40	nd–83	4	50	0
Diazinon	250	nd	1	0	—
Dieldrin	nr, 0.1–100	nd–8.2	9	33	0
Endosulfan[1]	0.1–20	nd–0.2	2	50	—
Endosulfan I	nr	22.3	1	100	—
Endosulfan II	nr, 0.01	0.75–2.57	2	100	—
Endosulfan sulfate	200	nd	1	0	—
Endrin	nr, 0.01–100	nd–9.47	7	29	0
Endrin aldehyde	nr, 0.01–200	nd–0.69	3	33	—
Ethion	250	nd	1	0	—
Fenchlorphos	250	nd	1	0	—
Guthion	250	nd	1	0	—
HCH, α-	nr, 0.01–100	nd–64	7	57	—
HCH, β-	nr, 0.01–100	nd–19.5	6	33	—
HCH, δ-	0.01–100	nd–1.41	3	33	—
Heptachlor	nr, 0.01–20	nd–37.8	8	25	0
Heptachlor Epoxide	nr, 0.1–100	nd–24	8	25	0

Table 6.18. Selected results of studies that monitored pesticides in shellfish and percentage of those studies that exceeded standards for the protection of human health—*Continued*

Target Analytes	Shellfish: All Monitoring Studies (Published 1960–1994)				
	Concentration Range (µg/kg wet weight)		Total Number of Studies	Percentage of Studies with Detectable Residues	Percentage of Studies with C_{max} that Exceeded FDA AL
	In Detection Limits	In Maximum Concentrations (C_{max})			
Hexachlorobenzene	0.01–20	nd–20.9	6	50	—
Lindane	nr, 0.01–100	nd–27.3	9	22	—
Malathion	250	nd	1	0	—
Methoxychlor	nr, 0.1–100	nd–4.05	9	11	—
Methyl parathion	250	nd	1	0	—
Mirex	nr, 0.01–20	nd–85.2	9	33	0
Nonachlor, *cis*-	20–100	nd	2	0	0
Nonachlor, *trans*-	20–100	nd	2	0	0
Oxychlordane	20–100	nd	2	0	0
Parathion	250	nd	1	0	—
Pentachlorophenol	nr, 0.1–5	nd–5.1	4	50	—
Perthane	0.1	nd	1	0	—
Strobane	1	58	1	100	—
Toxaphene	nr, 5–2,000	nd–7,680	9	33	22

[1]Unspecified or total.
[2]Unspecified or ($o,p' + p,p'$)-.

for all of the maximum concentrations greater than 2,500 µg/kg. Of 10 shellfish monitoring studies, the FDA action level for total chlordane was exceeded in only 1 study (Table 6.18).

This analysis indicates potential for adverse human health effects at the most contaminated site in 75 percent of recently published (1984–1994) monitoring studies, if the fish sampled by these studies were consumed at average levels (about two 4-ounce meals per month) by the general population over a 70-year lifetime. The USEPA screening value for total chlordane is based on carcinogenicity and assumes a cancer risk level of 10^{-5}. Therefore, the potential adverse health effect expected would be cancer, and the excess cancer risk (at the most contaminated site in the affected studies) may exceed 1 in 100,000 people. There is no information regarding the number of sites in each study that exceeded applicable standard or guideline values.

Total DDT

The FDA action level and USEPA screening value apply to total DDT, rather than to individual DDT compounds. The FDA action level for total DDT (5,000 µg/kg) was exceeded by maximum concentrations in two monitoring studies, one of which was published recently (since 1984) and the other prior to 1984 (Table 6.17). As shown in Figure 6.15, no recently published

Table 6.18. Selected results of studies that monitored pesticides in shellfish and percentage of those studies that exceeded standards for the protection of human health—*Continued*

Target Analytes	Shellfish: Recent Monitoring Studies (Published 1984–1994)				
	Concentration Range (µg/kg wet weight)		Total Number of Studies	Percentage of Studies with Detectable Residues	Percentage of Studies with C_{max} that Exceeded FDA AL
	In Detection Limits	In Maximum Concentrations (C_{max})			
Hexachlorobenzene	0.01–20	nd–20.9	5	40	—
Lindane	nr, 0.01–100	nd–27.3	9	22	—
Malathion	250	nd	1	0	—
Methoxychlor	nr, 0.1–100	nd–4.05	9	11	—
Methyl parathion	250	nd	1	0	—
Mirex	nr, 0.01–20	nd–85.2	9	33	0
Nonachlor, *cis-*	20–100	nd	2	0	0
Nonachlor, *trans-*	20–100	nd	2	0	0
Oxychlordane	20–100	nd	2	0	0
Parathion	250	nd	1	0	—
Pentachlorophenol	nr, 0.1–5	nd–5.1	4	50	—
Perthane	0.1	nd	1	0	—
Strobane	1	58	1	100	—
Toxaphene	nr, 5–2,000	nd–7,680	9	33	22

(1984–1994) studies had maximum concentrations exceeding 20,000 µg/kg. Only seven of the monitoring studies reviewed analyzed total DDT in shellfish; the maximum concentrations from these studies were well below the FDA action level for total DDT (Table 6.18).

As shown in Table 6.8 and in Figure 6.15, the USEPA screening value for total DDT (300 µg/kg) is much lower than the FDA action level. In fact, 94 percent of all 17 monitoring studies reviewed and 100 percent of the eight recently published (1984–1994) studies had maximum total DDT concentrations that exceeded the USEPA screening value (Table 6.17). For the most contaminated site in each recently published study, this indicates the potential for adverse human health effects, if the fish sampled by these studies were consumed at average levels (about two 4-ounce meals per month) by the general population over a 70-year lifetime. Again, the adverse health effects expected would be cancer, and the excess cancer risk (at the most contaminated site in the affected studies) may exceed 1 in 100,000 people. This analysis provides no information on the number of sites in each study that may exceed the screening value or have concentrations sufficient to cause potential effects on human health.

Other Pesticides

Of the remaining pesticide analytes in Tables 6.17 and 6.18, many do not have any applicable human health standards or guidelines for edible fish tissue. Some of these pesticide

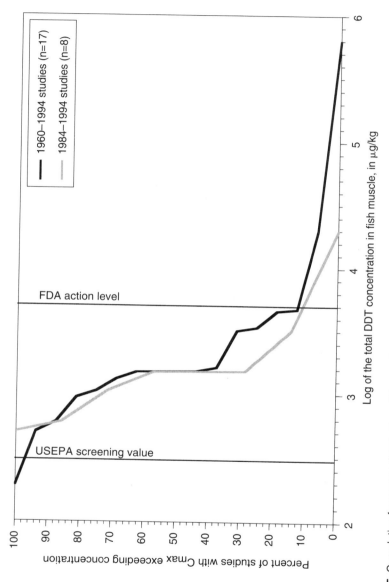

Figure 6.15. Cumulative frequency distribution of the maximum concentrations of total DDT in fish muscle (μg/kg wet weight) reported by the monitoring studies reviewed, shown for all studies (published during 1960–1994) and recently published (1984–1994) studies. Standards and guidelines for total DDT in fish muscle (for protection of human health) are shown as vertical lines. Abbreviations: C_{max}, maximum concentration in the study; FDA, Food and Drug Administration; n, number of samples; USEPA, U.S. Environmental Protection Agency; μg/kg, microgram per kilogram.

analytes (such as dacthal, HCH, α-HCH, kepone, parathion, and photomirex) were detected at fairly high concentrations (over 200 μg/kg) in edible fish tissue in one or more studies. Because the approach used in this analysis relies on comparison of measured concentrations with human health standards and guidelines, no evaluation of potential human health effects is possible for these pesticides. This is a real limitation of this approach.

Those pesticides in Table 6.17 that do have FDA action levels or USEPA screening values are discussed in two groups, depending on their carcinogenic potential. The noncarcinogens (compounds in cancer groups D or E) that have human health standards or guidelines are endosulfan (and its isomers) and endrin. Very few studies analyzed for endosulfan, endosulfan I, endosulfan II, or (the metabolite) endosulfan in edible fish tissue or shellfish. Low levels of the endosulfan compounds were detected in some studies (see Tables 6.17 and 6.18), but the maximum concentrations detected (30 μg/kg endosulfan in edible fish tissue and 22 μg/kg endosulfan I in shellfish) were well below the USEPA screening value for total endosulfan (60,000 μg/kg). Endrin was analyzed in 20 edible-fish studies and 9 shellfish studies, but none of these studies had maximum concentrations that exceeded the FDA action level. The maximum concentrations of endrin detected in edible fish (up to 75 μg/kg) also were well below the USEPA screening value (3,000 μg/kg). Therefore, there is no evidence for potential human health effects due to edible fish contamination with endosulfan or endrin in the monitoring studies reviewed.

Besides dieldrin, chlordane, and total DDT, there are six potential carcinogens in Tables 6.17 and 6.18 that have human health standards or guidelines: heptachlor, heptachlor epoxide, hexachlorobenzene, lindane, mirex, and toxaphene. Four of these pesticide analytes have FDA action levels: heptachlor, heptachlor epoxide, mirex, and toxaphene. For three of the four analytes (heptachlor, heptachlor epoxide, and toxaphene), the applicable FDA action level was exceeded by the maximum concentration in only one edible-fish monitoring study (Table 6.17). For toxaphene, the FDA action level also was exceeded in two of nine shellfish monitoring studies (Table 6.18). Since these studies (for all three analytes) all were published prior to 1984, no recently published (1984–1994) studies that sampled edible fish tissue or shellfish measured heptachlor, heptachlor epoxide, or toxaphene concentrations that exceeded the applicable FDA action levels. Mirex was exceptional in that 45 percent of all edible-fish monitoring studies (published during 1960–1994), and 40 percent of recently published (1984–1994) edible-fish studies, exceeded the FDA action level.

Five of the six potential carcinogens in Table 6.17 (all except heptachlor) have USEPA screening values that were exceeded by maximum concentrations in some percentage of all of the edible-fish monitoring studies reviewed, as follows: lindane (in 5 percent of all studies), hexachlorobenzene (in 24 percent), toxaphene (in 24 percent), mirex (in 50 percent), and heptachlor epoxide (in 50 percent). When only recently published studies were considered, the percentage of studies that exceeded USEPA screening values decreased slightly for some analytes. On the basis of recently published edible-fish studies, this analysis indicates that, under certain exposure conditions, there is potential for adverse human health effects at the most contaminated site in some studies for the following pesticides: lindane (in 7 percent of recently published studies), toxaphene (in 8 percent), hexachlorobenzene (in 15 percent), mirex (in 40 percent), and heptachlor epoxide (in 50 percent). The necessary exposure conditions are that fish sampled by these studies be consumed at average levels (about two 4-ounce meals per month) by the general population over a 70-year lifetime. The adverse health effects expected would be cancer, and the excess cancer risk (at the most contaminated site in the affected studies) may exceed 1 in 100,000 people. This analysis provides no information on the number of sites in each study that may exceed applicable screening values or have concentrations sufficient to cause potential effects on human health.

6.2.5 OTHER TOXICITY-RELATED ISSUES

Effect of Fish Preparation and Cooking on Pesticide Residues

Pesticide residues in raw fish may be reduced by fish preparation and cooking. Pesticide residues are not uniformly distributed among the tissues and organs of the fish, but tend to be concentrated in high-lipid tissues such as skin, liver, and adipose tissue. For example, removal of skin from edible portions of carp fillets significantly reduced both the concentration of PCBs (by 26 percent) and the lipid content (by 30 percent). Skin removal from carp in the largest size class had the least effect on the final concentrations of both PCBs and lipid (Hora, 1981). Removal of the dark lateral line and belly flap in fish before cooking can substantially remove contamination (Zabik and others, 1978).

A number of studies have shown that cooking markedly reduces organochlorine contaminants in fish and meat, although this depends to some extent on the species of organism, the organ or tissue sampled, and cooking method. Although they will not be reviewed here, several studies have demonstrated the effects of various cooking methods (such as baking, frying, stewing, grilling, and microwave cooking) on various meats (e.g., Liska and others, 1967; Ritchey and others, 1967, 1969; McCaskey and others, 1968; Funk and others, 1971; Maul and others, 1971). In studies of organochlorine contaminant residues in fish, about 20–60 percent of part per trillion residues of the contaminant 2,3,7,8-TCDD was lost from carp on charbroiling (Zabik, 1984; Kaczmar and others, 1985). Baking did not significantly reduce concentrations of PCBs, pesticides, or fat in bluefish (*Pomatomus saltatrix*) fillets (Trotter and others, 1989). PCB residues in fillets averaged 2.5 mg/kg both before and after cooking (after oil drippings and skin were discarded). However, the authors attributed this to loss of moisture (and thus, loss of total fillet weight) during cooking because the quantity of PCB in the fillet was reduced after cooking. The principal mode of pesticide removal may be fat rendering (Ritchey and others, 1967, 1969). For PCBs, the effectiveness of removal from fish depended on the species and fat content of the fish (Zabik, 1984). For example, in lake trout (*Salvelinus namaycush*) fillets (averaging 25–29 percent fat), roasting, broiling, or microwave cooking reduced PCB residues by 26–53 percent. In carp fillets (averaging 7.7 percent fat), PCB levels were unaffected by roasting, broiling, poaching, deep-fat frying, or microwave cooking to the same endpoint temperature (Zabik and others, 1982; Zabik, 1984).

Relative Cancer Risks

For probable human carcinogens (in the B2 cancer group), USEPA guidance for use in issuing fish consumption advisories was based on cancer risk. As discussed in Section 6.2.3, this involves selection of a maximum acceptable cancer risk level, which is a risk management decision (U.S. Environmental Protection Agency, 1994b), rather than a component of the risk assessment. For pesticides in the B2 cancer group, the USEPA screening values shown in Table 6.8 are based on carcinogenicity and correspond to an excess cancer risk level of 10^{-5}. In Section 6.2.4, these USEPA screening values for these B2 pesticides were compared with maximum concentrations measured in edible fish tissue by the monitoring studies reviewed. For several B2 pesticides (total chlordane, total DDT, dieldrin, and heptachlor epoxide), 50–100 percent of recently published (1984–1994) monitoring studies had maximum concentrations that would warrant restriction to less than two 4-ounce fish meals per month, based on an acceptable cancer risk of 10^{-5}, if these fish were consumed by the general population. Again, these meal restrictions may be warranted only for the most contaminated site in each study.

What does an excess cancer risk of 10^{-5} mean? It means that the activity in question (in this case, consumption of fish contaminated at the measured level by an 70-kg individual over a 70-year lifetime at a rate of 6.5 grams of fish per day, or about 7 ounces of fish per month) may result in an upper limit estimate of 1 excess cancer for every 100,000 people. USEPA reviews individual state policies on cancer risk levels as part of its water-quality standards oversight function under the Clean Water Act. It is USEPA's policy to accept cancer risk policies from the states in the range of 10^{-7} to 10^{-4} (U.S. Environmental Protection Agency, 1992d). It is useful to consider the risk of excess cancer from consumption of pesticide-contaminated fish in the context of other cancer risks.

The relative cancer risks attributed to both consumption of fish from the Great Lakes by sport fishers and drinking of Great Lakes water were analyzed relative to other sources of carcinogens: namely, other foods, groundwater supplies, and urban air (Bro and others, 1987). Bro and coworkers used levels of sport fish consumption from Humphrey (1976). The cancer potency factors that were used were taken from USEPA and the New York Department of Health (NYDH) databases. It is important to note that USEPA's cancer potency factors for PCBs and DDT at that time were 18- and 10-fold higher than those of NYDH. The different values for PCB were based on the same study of liver tumors in female rats, but on different endpoints: The NYDH used induction of hepatocellular carcinomas, whereas USEPA included neoplastic nodules. For DDT, USEPA used a multigenerational study in which the rate of cancer increased with succeeding generations. The NYDH based its value on studies that used a single generation of animals. Both values are conservative, but the USEPA values are more so (Bro and others, 1987). Note that the USEPA no longer uses the cancer potency value for total DDT that was cited by Bro and others (1987) ($8.4 \ [mg/kg/d]^{-1}$); the more recent cancer potency value used by U.S. Environmental Protection Agency (1996a, 1997c) ($0.34 \ [mg/kg/d]^{-1}$) is even lower than that used by NYDH ($0.88 \ [mg/kg/d]^{-1}$). However, the following analysis by Bro and coworkers nicely illustrates the importance of cancer potency estimates on cancer risk estimates. In their analysis, Bro and others (1987) used 1980–1981 levels of contaminants in raw fillets and assumed that residues in cooked fish were half of those in raw fillets (Cordle and others, 1982). The lifetime risk of cancer from various organochlorine compounds (toxaphene, PCB, dieldrin, DDT plus DDE, chlordane, and others) was estimated for each of seven locations in the Great Lakes.

When the cancer potency factors from USEPA were used, most of the cancer risk associated with consuming fish from all areas except Lake Superior was due to PCBs and DDT plus DDE. The total risk ranged from less than 10^{-3} (1 in 1,000) for Lake Superior to greater than 3×10^{-2} (3 in 100) for southern Lake Michigan. When the NYDH cancer potencies were used, other organochlorine compounds (especially dieldrin) became important also. The risk

Table 6.19. Estimated cancer risks from breathing urban air and consuming various foods and water, including Lake Michigan fish and Niagara River water

[Cancer risk: the number of excess cases of cancer in every 100,000 people, assuming lifetime exposure. All data sources not shown were taken by Bro and others (1987) from Crouch and Wilson (1982). Abbreviations: d, day; L, liter; oz., ounce; USEPA, U.S. Environmental Protection Agency. Reproduced from Bro and others (1987) with the permission of the publisher and the author. Copyright 1987 Springer-Verlag New York, Inc.]

Source[1]	Lifetime Cancer Risk (in 10^5)
Average United States lifetime risk of cancer of all types	25,000
Typical foods	
Four tablespoons peanut butter per day (aflatoxin)	60
One pint milk per day (aflatoxin)	14
8 oz. broiled steak per week (cancer only)	3
One diet soda per day (saccharin)	70
Average United States fish consumption (Connor, 1984b)	33
Great Lakes	
Lake Michigan sport fish	
Median levels of sport fish consumption and USEPA (1980b) potency values	480–3,300
Median levels of sport fish consumption and Kim and Stone (1981) potency values	77–340
Niagara River water: 2 L/d, USEPA (1980b) potency values	0.3
Drinking water	
Average United States ground water: communities >10,000 population; 2 L/d (Connor, 1984b)	1
Urban water supplies: 1976–1977 contaminant levels; 2 L/d (Crouch and others, 1983)	120–5,300
Air	
Average United States urban air: normal breathing (Clement Associates, 1984)	63–560
Average United States urban air: normal breathing (Thomson and others, 1985)	10–100

values were about an order of magnitude less when NYDH values were used for cancer potencies than when USEPA values were used. This shows the importance of considering the assumptions underlying risk estimates (Bro and others, 1987). In both cases, the highest risk was in the southern basin of Lake Michigan. Because the cancer potency values used are conservative, the risk estimates constitute a plausible upper bound to cancer risk (Bro and others, 1987).

Bro and others (1987) estimated that the cancer risk from drinking Niagara River water was three orders of magnitude less than that from consuming Lake Michigan fish. In Table 6.19, the cancer risks from drinking Niagara River water and consuming Lake Michigan fish are compared with other cancer risks. Drinking urban groundwater and breathing urban air may constitute comparable cancer risks to frequent consumption of sport fish from the Great Lakes (Bro and others, 1987).

Connor (1984b) also estimated risks of average consumption of contaminants in United States fish. Connor's risk estimates were lower than those of Bro and others (1987) because the average consumer typically eats less fish than sport fishers. Also, the average consumer eats more ocean fish than Great Lakes fish, and contaminants in ocean fish generally are lower. Connor (1984b) estimated that cancer risks due to fish consumption were around 3×10^{-4} (3 in 10,000), mostly due to DDT and PCBs.

Exposure to Multiple Contaminants

Although humans are exposed to multiple contaminants in the environment, both exposure and risk assessments generally are performed for individual chemicals. As discussed in the context of aquatic toxicity (Section 6.1.4), possible interactions among chemicals may be categorized as any of the following: (1) *antagonistic*, where the mixture is less toxic than the chemicals considered individually; (2) *synergistic*, where the mixture is more toxic than the chemicals considered individually; or (3) *additive*, where the mixture has toxicity equivalent to the sum of the toxicities of the individual chemicals in the mixture.

The organochlorine insecticides may affect the toxicity of other chemicals by inducing hepatic microsomal enzymes. This results in increased rate of metabolism of all other chemicals that are metabolized by this system. This increased metabolism may increase or decrease the toxicity of the other chemical, depending on whether the metabolite is more or less toxic than the parent chemical (U.S. Environmental Protection Agency, 1994b). Prior or simultaneous exposure to organochlorine compounds (including chlordane, DDT, dieldrin, endosulfan, endrin, heptachlor, hexachlorobenzene, lindane, mirex, and toxaphene) have been reported to affect the response to therapeutic drugs, hormones, and other endogenous chemicals.

Pesticides in the same class (such as organophosphate insecticides) often have similar mechanisms of toxicity and produce similar health effects. U.S. Environmental Protection Agency (1986j) recommends the additive approach (that is, summing individual risks for multiple chemicals) for mixtures of carcinogens, or for mixtures of noncarcinogens with the same adverse health endpoints and mechanisms of action. For chemicals that cause dissimilar health effects, risks from these contaminants should be described and assessed separately (U.S. Environmental Protection Agency, 1994b).

Relatively few data are available on the toxic interactions between multiple chemicals (U.S. Environmental Protection Agency, 1994b). Some data on interactive effects can be found in MIXTOX, a USEPA database containing summaries of information on interactive effects from the primary literature (U.S. Environmental Protection Agency, 1992f). Table 6.20 lists specific combinations of pesticides and pesticide byproducts reported to cause synergism or antagonism. The information in Table 6.20 was extracted from the health effects profiles of individual pesticide contaminants found in U.S. Environmental Protection Agency (1994b) as well as from some studies in the primary literature. Clearly, some combinations of pesticides may cause both synergistic or antagonistic effects to at least some animals under some conditions.

U.S. Environmental Protection Agency (1986j) stated that any available data on interactive effects (antagonism and synergism), "must be assessed in terms of both its relevance to subchronic or chronic hazard and its suitability for quantitatively altering the risk assessment." Existing data on interactive effects tend to consist of acute effects interaction and information on mechanisms of action, which are not easily applicable to assessment of chronic health risks of multiple contaminants (U.S. Environmental Protection Agency, 1994b). U.S. Environmental

Table 6.20. Interactive effects reported in the literature for pesticides found in aquatic biota

[Reported synergists: For a given pesticide in column 1, synergism has been reported in the literature with each pesticide listed in column 2. Each line in column 2 represents the results of a separate experiment. Reported antagonists: For a given pesticide in column 1, antagonism has been reported in the literature with each pesticide listed in column 3. Each line in column 3 represents the results of a separate experiment. Compiled from data in U.S. Environmental Protection Agency (1994b)]

Pesticide	Reported Synergists	Reported Antagonists
Chlordane	Aldrin, endrin, methoxychlor Endrin Lindane Malathion, toxaphene	Lindane
Diazinon	Dieldrin	Toxaphene
Dieldrin	Hexachlorobenzene Diazinon	Hexachlorobenzene
Endrin	Chlordane Lindane	Lindane
Hexachlorobenzene	Dieldrin Lindane	Dieldrin Lindane Pentachlorophenol TCDD, 2,3,7,8-
Lindane	Chlordane Hexachlorobenzene Mirex Toxaphene	Chlordane Hexachlorobenzene HCH, α-, β-, δ- Mirex Toxaphene
Mirex	Lindane Toxaphene	Lindane Toxaphene
Toxaphene	Chlordane, malathion Lindane Mirex	Diazinon Lindane Mirex

Protection Agency (1986j, 1994b) has recommended that such data on interactive effects be discussed in relation to long-term risks and interactive effects, but that these data not necessarily be incorporated into a quantitative risk assessment.

Potential Endocrine-Disrupting Effects of Pesticides

Humans have experienced increased incidences of certain developmental, reproductive, and carcinogenic effects during the last 50 years. Examples include increased rates of breast, testicular, and prostate cancers in the United States; an increase in ectopic pregnancies in the United States; an two-fold increase in cryptorchidism (undescended testes) in the United Kingdom; and a possible decrease in sperm production worldwide (Colborn and others, 1993; U.S. Environmental Protection Agency, 1997b). In a recent report, the U.S. Environmental Protection Agency (1997b) examined the working hypothesis (called the *endocrine disruption hypothesis*) that these adverse effects may be caused by environmental chemicals acting to disrupt the endocrine system that regulates these processes.

As discussed in Section 6.1.4, USEPA has defined an environmental endocrine disruptor as an exogenous agent that interferes with the synthesis, secretion, transport, binding, action, or elimination of natural hormones in the body that are responsible for the maintenance of homeostasis, reproduction, development, or behavior (U.S. Environmental Protection Agency, 1997b). Endocrine disruptors include any chemical that adversely affects any aspect of the entire endocrine system, including (but not limited to) environmental estrogens. The endocrine system includes a number of central nervous system–pituitary–target organ feedback pathways. Therefore, there are multiple target organ sites at which an environmental chemical could potentially disrupt endocrine function (U.S. Environmental Protection Agency, 1997b).

Diethylstilbestrol (DES) has been proposed as a model for human exposure to xenoestrogens (Colborn and others, 1993). DES is a synthetic estrogen that was used in the United States from 1948 to 1971 to prevent spontaneous abortions in women. Female offspring of women who took DES showed higher incidence of reproductive organ dysfunction, decreased fertility, immune system disorders, increased incidence of a certain reproductive tract cancer (vaginal clear-cell adenocarcinoma), and periods of depression (Colborn and others, 1993). Male offspring showed increased incidence of abnormalities in the scrotum, cryptorchidism, and malformed urethra, and decreased sperm counts (Hileman, 1994; U.S. Environmental Protection Agency, 1997b). Some effects of DES in humans and in laboratory animals were not noted until maturity. Endocrine-disrupting chemicals may act via multiple mechanisms, some of which may operate only during specific developmental periods (Gray, 1992; Colborn and others, 1993).

The U.S. Environmental Protection Agency (1997b) identified 8 pesticides (out of 63 pesticides screened) that were potential reproductive toxicants in females. Examples include the organochlorine insecticide dicofol (which caused ovarian vacuolation in a multigenerational reproductive study in rats); the organophosphate insecticide oxydemeton-methyl (multiple reproductive organ toxicity in rats); the herbicide molinate (reduced fertility and ovarian histopathology in rats); and the fungicides procymidone and vinclozolin (increases in luteinizing hormone and testosterone following binding to and inhibition of the androgen receptor, ovarian tumors, and sex cord tumors). For many of these pesticides, other endpoints of toxicity (such as cholinesterase inhibition or potential carcinogenicity) occurred at lower doses, so that the reference dose (or risk specific dose) also would be protective of reproductive toxicity (U.S. Environmental Protection Agency, 1997b).

The male reproductive system can be adversely affected by endocrine imbalance (U.S. Environmental Protection Agency, 1997b). Developing males (both prenatally and postnatally) appear to be particularly susceptible. Very early embryos have the potential to develop either a female or a male reproductive system. Normal differentiation and development of the male reproductive tract depends on the action of hormones, especially anti-Mullerian hormone and (the androgen) testosterone. Chemicals with endocrine activity can induce abnormality in the expression of the genome or interference with the action of gene products, as well as acceleration of the rate of cell division. Three classes of chemicals have been shown to influence androgen levels when administered during development (U.S. Environmental Protection Agency, 1997b): antiandrogens (such as DDE, vinclozolin), estrogens (such as DES, butyl benzyl phthalate, and octylphenol), and chemicals that can activate the aryl hydrocarbon receptor (such as 2,3,7,8-TCDD).

A few incidences of association between chemical exposure in humans and reproductive problems or cancer have been demonstrated. For example, some studies reported an association between PCB levels and reduced sperm motility in men with fertility problems (Carlsen and others, 1992) or abnormal spermatogenesis in humans exposed to kepone at a production facility (Cohn and others, 1978). However, at this time, the U.S. Environmental Protection Agency (1997b) considers that the epidemiological evidence for the role of certain environmental chemicals in breast cancer (DDE, PCBs, or PAHs), endometriosis (PCBs, TCDD), and prostate cancer (herbicides or PAHs) is weak. With a few exceptions (such as DES, TCDD, and DDT and its metabolite DDE), epidemiological studies have not demonstrated causal relations between exposure to a specific environmental chemical and an adverse effect on human health operating via an endocrine disruption mechanism (U.S. Environmental Protection Agency, 1997b). In general, however, it is difficult to show cause and effect with epidemiological studies that correlate known levels of chemical exposure with health effects in humans. Such studies are expensive, there is a long lag time between exposure and effects, documentation of disease and exposure are often poor, and there are confounding factors such as exposure to other chemicals (Morris and others, 1984). This is especially true for endocrine disruptors for four reasons: (1) these chemicals may have very different effects on the human embryo and fetus than on the adult, (2) effects are usually seen in the offspring and not the exposed parent, (3) the timing of exposure is critical in determining the effects, and (4) effects of exposure in utero may not be observed until maturity (Colborn and Clement, 1992a).

In addition to pesticides and other synthetic compounds, a number of phytoestrogens and mycotoxins (fungal agents) have estrogenic effects in mammals (Gray, 1992). Estrogenic substances have been found in more than three hundred plants from at least sixteen families (Bradbury and White, 1954; Setchell, 1985). Many foods consumed by humans (including rice, wheat, potatoes, carrots, apples, plums, sunflower seeds and oil, garlic, and coffee) contain hormone mimics. In fact, dietary exposure to xenoestrogens is much lower than to naturally occurring phytoestrogens (Safe, 1995). Plant estrogens have been observed to reduce fertility and disrupt development in laboratory animals (e.g., Bennetts and others, 1946; Whitten and Naftolin, 1992; Whitten and others, 1992, 1993). This suggests that the effects of xenoestrogens on humans may be insignificant compared with those of naturally occurring estrogens. As noted in Section 6.1.4, one fundamental difference between naturally occurring plant estrogens and xenoestrogens is that plant estrogens are rapidly metabolized and excreted (Hughes, 1988), whereas many synthetic chemicals (including organochlorine pesticides) are resistant to degradation and accumulate in the body. Also, some research suggests that xenoestrogens and plant estrogens may affect the metabolism of estradiol in different ways (Davis and Bradlow, 1995).

As pointed out by U.S. Environmental Protection Agency (1997b), the endocrine system is highly regulated in adults, with negative feedback control of hormone concentrations. Moreover, xenobiotics tend to have low ambient concentrations and low binding affinities relative to endogenous hormones (such as estradiol). For example, compared to estradiol, about 1,000 to 10,000 times more of alkylphenols (xenoestrogens) were required to bind 50 percent of the estrogen receptor (White and others, 1994). The U.S. Environmental Protection Agency (1997b) concluded that exposure to a single xenoestrogenic chemical at normal environmental concentrations probably is insufficient to cause an adverse effect in adults. However, it is not clear whether this holds for the fetus and neonate. Another uncertainty involves the effect of chemical mixtures on the endocrine system, including the potential for additive, synergistic, or antagonistic effects.

The U.S. Environmental Protection Agency (1997b) does not consider endocrine disruption to be an adverse effect per se, but rather to be a mode of action that may lead to other outcomes (such as developmental, reproductive, or carcinogenic effects) that are considered in reaching regulatory decisions. Evidence of endocrine disruption can influence priority setting for testing, and the results of such testing could lead to regulatory action if adverse effects are shown to occur (U.S. Environmental Protection Agency, 1997b). In its interim assessment report, the USEPA called for further research and testing on the consequences of endocrine disruption (U.S. Environmental Protection Agency, 1997b).

An important issue is whether existing guidelines and testing protocols are sufficient to detect endocrine-disrupting potential effects in the general population or in sensitive subgroups such as the fetus, children, the infirm, and the elderly (U.S. Environmental Protection Agency, 1997b). Toxicity tests conducted under current USEPA guidelines are not designed specifically to detect endocrine disruption or its mechanisms; they are designed to detect effects on endpoints of reproductive concern that may occur throughout several life stages of the animal regardless of their mechanism of action (U.S. Environmental Protection Agency, 1997b). Specific procedures for identifying better measures of potential endocrine disruption are being developed and incorporated into more recent testing guidelines for development and reproduction (U.S. Environmental Protection Agency, 1995c). These procedures are discussed in USEPA's Reproductive Toxicity Risk Assessment Guidelines (U.S. Environmental Protection Agency, 1996c).

The analysis of the potential human health effects of pesticides in sediment and aquatic biota in Section 6.2.4 did consider potential carcinogenicity, and to some extent, potential reproductive toxicity (because the reference dose is protective of chronic noncancer toxicity, which may include some reproductive endpoints). It is important to note that this analysis does not take potential effects on endocrine disruption into consideration.

6.2.6 SUMMARY

Organochlorine compounds, including pesticides, PCBs, PCDDs, and PCDFs, are commonly detected in samples of human breast milk, adipose tissue, reproductive organ tissue, blood, and urine. Several studies of human exposure to pesticides indicate that food consumption is the principal mode of intake of hydrophobic pesticide contaminants. Fish and shellfish consumption were reported to be a major source of human exposure (relative to other foods) for a number of organochlorine compounds, including DDT, dieldrin, HCH isomers, hexachlorobenzene, kepone, mirex, PCBs, and TCDD. Moreover, organochlorine residues in human blood have been shown to be related to consumption of contaminated fish.

Standards and guidelines for the protection of human health are numerical values that indicate a threshold for potential effects on human health. These can be used to assess potential human health effects from consumption of fish and shellfish contaminated with pesticides. The analysis in Section 6.2.4 compared maximum concentrations of individual pesticides in the monitoring studies reviewed in this book (Tables 2.1 and 2.2) with two sets of national standards and one set of guidelines for pesticides in edible fish or shellfish. The standards are USEPA tolerances for currently registered pesticides (the maximum amount of a pesticide residue that can be legally present in or on a raw agricultural commodity) and FDA action levels for pesticides that are no longer registered for use in agriculture (enforceable regulatory limits for unavoidable pesticide residues in or on food). The guidelines used are USEPA recommended screening values for contaminants in edible fish, published as part of USEPA's guidance to state, local, and tribal health officials for use in setting fish consumption advisories. These USEPA guidelines are based on potential health effects (both chronic toxicity and carcinogenicity), and the risk assessment methodology and underlying assumptions are explicitly stated. FDA action levels for most organochlorine compounds are considerably higher than the corresponding USEPA guidance for use in fish advisories. However, FDA action levels are not based on health considerations alone, but also consider factors such as the economic costs of banning a foodstuff, the extent to which pesticides residues cannot be avoided by good agricultural practice or good manufacturing practice, and analytical detection limits at the time the action levels were established.

Maximum pesticide levels in edible fish tissue (fish muscle) in the monitoring studies reviewed were compared with the applicable standards and guidelines. There is a USEPA tolerance for only one pesticide analyzed in edible fish tissue, and this pesticide (2,4-D) was not detected. FDA action levels or USEPA screening values, or both, exist for most of the organochlorine pesticides analyzed in fish muscle. For pesticides in chemical classes other than the organochlorine insecticides, the few pesticides that were analyzed in edible fish tissue by any of the monitoring studies reviewed (Table 6.17) have no applicable standards or guidelines for protection of human health.

Even when only recently published (1984–1994) studies were considered, a high percentage of studies that monitored pesticides in fish muscle had maximum concentrations that exceeded USEPA screening values for several organochlorine insecticides. Those pesticides that exceeded the USEPA screening values in at least 50 percent of studies were total chlordane, total DDT, dieldrin, and heptachlor epoxide. This suggests that these pesticide residues at the most contaminated site in at least 50 percent of monitoring studies would be high enough to cause adverse human health effects, if the fish from these sites were consumed at average rates by the general adult population over a 70-year lifetime (Table 6.17). No information is provided on the percentage of sites within each study that exceeded the guideline. For these four organochlorine pesticides, the expected adverse health effect would be cancer, and the excess cancer risk (at the most contaminated site in each study) may exceed 1 in 100,000 people. Adverse health risks may be higher for sport or subsistence fishers (who consume more local fish than the average population) or for sensitive subpopulations such as children or pregnant women. It is important to note that the processes of preparing and cooking fish would be likely to reduce pesticide residues in the fish prior to consumption.

FDA action levels were exceeded by the maximum concentrations of several pesticides in edible fish for a substantial percentage of recently published (1984–1994) monitoring studies: total chlordane (in 56 percent of studies), mirex (40 percent), total DDT (13 percent), and toxaphene (6 percent). In recently published (1984–1994) shellfish monitoring studies, maximum pesticide concentrations were below the applicable FDA action levels, except for total chlordane in one study and toxaphene in two studies. For concentrations above the FDA action level, this indicates that, if the contaminated fish or shellfish were in interstate commerce, the pesticide residues would be high enough to warrant enforcement action by the FDA to remove them. No information is available on the number of sites within each study that may have exceeded the FDA action level.

Results from the FDA Total Diet Study indicate that pesticides consumed on average by people in the United States are less than the FAO–WHO acceptable daily intake and the USEPA reference dose, both of which provide a measure of chronic toxicity. The Total Diet Study makes regional assessments (using appropriate diet and residue levels), but does not attempt to assess the potential for local effects. However, the reference dose guidelines do not consider potential carcinogenicity. Most organochlorine pesticides (which are the most common pesticide contaminants in fish) have been classified by USEPA as probable human carcinogens. The total risk of excess cancer from consumption of fish has been estimated as ranging from 3 in 10,000 for the average consumer of fish in the United States up to 30–300 in 10,000 for sport fishers consuming Lake Michigan fish. The excess cancer risk was attributed principally to residues of DDT, PCBs, and dieldrin. Standards and guidelines for protection of human health generally do not consider certain more complex issues related to toxicity, such as the effect of chemical mixtures on toxicity and the potential for endocrine-disrupting effects on development and reproduction. These are important areas of current research. In these respects, the above analysis may underestimate potential adverse effects of contaminants on human health.

CHAPTER 7

Summary and Conclusions

This book reviews more than 400 studies of pesticides in bed sediment and aquatic biota of United States rivers conducted over the last three decades. Considered together, the existing literature provides a basis for at least partial assessment of the extent of pesticide contamination of rivers and estuaries in the United States, although there are areas where our current understanding is incomplete.

The majority of studies reviewed were monitoring studies, ranging in scale from local to national. There was little consistency among the studies reviewed in terms of site selection strategy, sample collection methods, species of organisms sampled, tissue type analyzed, and analytical detection limits. In contrast, there was great consistency in the types of pesticides that were target analytes in bed sediment and aquatic biota studies—the vast majority of studies focused on organochlorine insecticides. Since the 1960s, there has been a decrease in process-type studies (or field experiments) designed to assess the general environmental fate and persistence of individual pesticides, and an increase in monitoring-type studies that focused on analytes known to be present in bed sediment and aquatic biota (as the environmental fate of hydrophobic pesticides became more generally understood).

Results of the hundreds of state and local monitoring studies are difficult to compare because of the large variability in study design (which includes factors such as site selection, sampling methods, and analytical detection limits). However, several national programs have been conducted that constitute individual data sets large enough to provide a nationwide perspective on pesticide occurrence and distribution. Together, these national monitoring studies have provided good geographic coverage of pesticides in bed sediment and aquatic biota of United States rivers and estuaries. Temporally, the national studies have provided the fewest data for freshwater sediment, which was sampled only during the 1970s. The focus of state and local monitoring studies was fairly evenly divided between bed sediment and aquatic biota. Geographically, state and local monitoring efforts have been heaviest in the Great Lakes region, along the Mississippi River, and in western states, especially California. In aquatic biota monitoring studies, pesticide residues were more often measured in fish than in other aquatic organisms, with other invertebrates and mollusks placing a distant second and third. In studies that sampled fish, there was little consistency in the types of tissue analyzed (such as whole fish, fillets, liver, or other organs). No national studies measured pesticide residues in plants, algae, or amphibians, whereas a number of both local monitoring studies and process-type studies did.

Analytical detection limits varied among studies and were not always reported. In general, detection limits tended to be lower and to be reported more frequently in more recently published studies. Changes in analytical methods over time (especially the change from packed-column to capillary column gas chromatography) tended to lower detection limits.

The monitoring studies reviewed show that a large number of pesticides have been detected in bed sediment and aquatic biota at some time over the last 30 years. Forty-one of 93 target analytes (44 percent) were detected in bed sediment in one or more studies, and 68 of 106 analytes (64 percent) were detected in aquatic biota in one or more studies. Most of the target analytes detected were organochlorine insecticides, by-products, or transformation products, despite the fact that most organochlorine insecticides were banned or severely restricted by the mid-1970s. This reflects both the environmental persistence of these compounds and the bias in the target analyte list (which typically was limited to organochlorine compounds). DDT and metabolites, chlordane compounds, and dieldrin were the most commonly detected pesticide analytes in both bed sediment and aquatic biota. Other organochlorine insecticides that sometimes were detected included endosulfan compounds, endrin and metabolites, heptachlor and heptachlor epoxide, mirex, lindane, α- and β-HCH, kepone, methoxychlor, and toxaphene.

Besides the organochlorine insecticides, a few compounds in other pesticide classes were detected in some studies. Most of these pesticides contained chlorine or fluorine substituents and were intermediate in hydrophobicity. Examples in bed sediment included the herbicides ametryn, dacthal, 2,4-DB, dicamba, and diuron; the organophosphate insecticide zytron; and the fungicide/wood preservative pentachlorophenol. In aquatic biota, examples included the herbicides dacthal, oxadiazon, and trifluralin; the organophosphate insecticides zytron and chlorpyrifos; and the fungicide/wood preservative pentachlorophenol and (its metabolite) pentachloroanisole. Of pesticides from other chemical classes that were analyzed in bed sediment or aquatic biota, most were targeted at relatively few sites nationwide, and those sites generally came from one or a few studies. Aggregate detection frequencies (calculated by combining data from the studies reviewed) are not necessarily representative of freshwater resources in the United States. If a given pesticides is absent from the list of pesticides detected in bed sediment or aquatic biota, it does not necessarily mean that the individual pesticide is not present in these sampling media, but that it may not have been looked for in the studies reviewed. Even for those compounds with zero or few detections, the results may have been biased by sampling location or study design.

The monitoring studies reviewed suggest that pesticides were detected more often in aquatic biota than in bed sediment. More pesticides (both the total number and the percentage of those targeted) were detected in aquatic biota than in bed sediment and the cumulative detection frequencies for a given pesticide generally were higher. Moreover, in direct comparison between benthic fish and associated surficial sediment samples collected by the National Oceanic and Atmospheric Administration's (NOAA) National Status and Trends (NS&T) Program (Benthic Surveillance Project), detection frequencies and concentrations (dry weight) of most hydrophobic organic contaminants tended to be higher in estuarine fish liver than in associated sediment. The exceptions were aldrin and heptachlor, both of which are metabolized rapidly in aquatic biota. Their transformation products (dieldrin and heptachlor epoxide, respectively) showed the expected pattern, that is, they were found at higher levels in fish liver samples than in associated sediment.

Table 2.1. Pesticides in bed sediment and aquatic biota from rivers and estuaries in the United States: National and multistate monitoring studies—*Continued*

| Target Analytes | Bed Sediment (µg/kg dry weight) | | | Aquatic Biota (whole fish unless noted) | | | | | Comments |
	DL	Range	Median	% Detection Sites	Samples	DL	Range	Median	
				FDA's National Monitoring Program for Food and Feed					
Aldrin	—	—	—	nr	2*	5	nr	<5*†	* Fish fillets only.
Chlordane	—	—	—	nr	1§	5	nr	<5†§	† Value shown is mean.
DDD	—	—	—	nr	41*	5	nr	130*†	§ Shellfish only. Objec-
DDE	—	—	—	nr	66*	5	nr	490*†	tive: to monitor pesti-
DDT	—	—	—	nr	55*	5	nr	220*†	cide residues in food
Dieldrin	—	—	—	nr	28*	5	nr	20*†	and to assess compli-
Endrin	—	—	—	nr	6*	5	nr	<5*†	ance with USEPA
HCH	—	—	—	nr	8*	5	nr	<5*†	tolerances. Results in
Heptachlor	—	—	—	nr	1*	5	nr	<5*†	this table are for dom-
Heptachlor epoxide	—	—	—	nr	3*	5	nr	<5*†	estic samples from interstate commerce.
Lindane	—	—	—	nr	2*	5	nr	<5*†	Domestic samples may
Toxaphene	—	—	—	nr	2*	5	nr	40*†	be biased by repeated sampling in contami- nated areas. DDT, DDE, DDD, and dieldrin residues were higher in fish than in shellfish.

Table 2.1. Pesticides in bed sediment and aquatic biota from rivers and estuaries in the United States: National and multistate monitoring studies—*Continued*

Reference	Sam-pling Dates	Sampling Locations and Strategy	Sampling Media and Species	Number of Samples	Target Analytes	Bed Sediment % Detection	
						Sites	Samples
Food and Drug Administration, 1980	1974–1976	Raw agricultural products in interstate commerce throughout the United States. Samples are either surveillance (monitoring) or compliance (known contamination).	Fish (fillets), shellfish. Species: not reported.	1,228 (total)*	DDD	—	—
					DDE	—	—
					DDT	—	—
					Dieldrin	—	—
					Endrin	—	—
					HCH	—	—
					Hexachlorobenzene	—	—
Food and Drug Administration, 1981	1976–1977	Raw agricultural products in interstate commerce throughout the United States. Samples are either surveillance (monitoring) or compliance (known contamination).	Fish (fillets), shellfish. Species: not reported.	1,147 (total)*	Chlordane	—	—
					DDD	—	—
					DDE	—	—
					DDT	—	—
					Diazinon	—	—
					Dieldrin	—	—
					Dyfonate	—	—
					Endrin	—	—
					HCH	—	—
					Heptachlor epoxide	—	—
					Hexachlorobenzene	—	—
					Lindane	—	—
					Malathion	—	—
					Methyl parathion	—	—
					Parathion	—	—
					Toxaphene	—	—

Table 2.1. Pesticides in bed sediment and aquatic biota from rivers and estuaries in the United States: National and multistate monitoring studies—*Continued*

Target Analytes	Bed Sediment (µg/kg dry weight)			Aquatic Biota (whole fish unless noted)					Comments
				% Detection		(µg/kg wet weight)			
	DL	Range	Median	Sites	Samples	DL	Range	Median	
DDD	—	—	—	nr	30*	10–100	nr	nr	* For edible fish and shellfish combined—domestic surveillance samples only. Objective: to locate foods containing unsafe levels of pesticides; to monitor identities and levels of pesticides in foods. Detection frequencies were higher in compliance than surveillance samples.
DDE	—	—	—	nr	59*	10–100	nr	nr	
DDT	—	—	—	nr	20*	10–100	nr	nr	
Dieldrin	—	—	—	nr	41*	10–100	nr	nr	
Endrin	—	—	—	nr	<12*	10–100	nr	nr	
HCH	—	—	—	nr	18*	10–100	nr	nr	
Hexachloro-benzene	—	—	—	nr	<10*	10–100	nr	nr	
Chlordane	—	—	—	nr	9*	10–100	nr	nr	* For edible fish and shellfish combined—domestic surveillance samples only. Objectives: to locate foods containing unsafe levels of pesticides; to identify and quantitate pesticides in foods. In fish, detection frequencies for DDT and dieldrin declined after FY72. There was no clear trend for DDE or DDD. The HCH detection frequency increased from FY64–69 (8%) to FY77 (18%). The toxaphene detection frequency has not changed since FY73.
DDD	—	—	—	nr	19*	10–100	nr	nr	
DDE	—	—	—	nr	42*	10–100	nr	nr	
DDT	—	—	—	nr	15*	10–100	nr	nr	
Diazinon	—	—	—	nr	0*	10–100	nr	nr	
Dieldrin	—	—	—	nr	28*	10–100	nr	nr	
Dyfonate	—	—	—	nr	0.6*	10–100	nr	nr	
Endrin	—	—	—	nr	8*	10–100	nr	nr	
HCH	—	—	—	nr	18*	10–100	nr	nr	
Heptachlor epoxide	—	—	—	nr	2*	10–100	nr	nr	
Hexachloro-enzene	—	—	—	nr	11*	10–100	nr	nr	
Lindane	—	—	—	nr	4*	10–100	nr	nr	
Malathion	—	—	—	nr	0*	10–100	nr	nr	
Methyl parathion	—	—	—	nr	0*	10–100	nr	nr	
Parathion	—	—	—	nr	0*	10–100	nr	nr	
Toxaphene	—	—	—	nr	8*	10–100	nr	nr	

Table 2.1. Pesticides in bed sediment and aquatic biota from rivers and estuaries in the United States: National and multistate monitoring studies—*Continued*

Reference	Sampling Dates	Sampling Locations and Strategy	Sampling Media and Species	Number of Samples	Target Analytes	Bed Sediment % Detection	
						Sites	Samples
Food and Drug Administration, 1988	1986–1987	Raw agricultural products in interstate commerce throughout the United States. Samples are either surveillance (monitoring) or compliance (known contamination).	Fish (edible), shellfish. Species: not reported. Other commodities include milk, grains, vegetables, fruits.	499 (total)*	253 pesticides were target analytes. 113 pesticides were detected in domestic and imported commodities (including fish and shellfish).	—	—
Food and Drug Administration, 1989a	1987–1988	Raw agricultural products in interstate commerce throughout the United States. Samples are either surveillance (monitoring) or compliance (known contamination).	Fish (edible), shellfish. Species: not reported. Other commodities include milk, grains, vegetables, and fruits.	447 (total)*	256 pesticides were target analytes. 118 pesticides were detected in domestic and imported commodities (including fish and shellfish).	—	—

Table 2.1. Pesticides in bed sediment and aquatic biota from rivers and estuaries in the United States: National and multistate monitoring studies—*Continued*

Target Analytes	Bed Sediment (µg/kg dry weight)			Aquatic Biota (whole fish unless noted)					Comments
				% Detection		(µg/kg wet weight)			
	DL	Range	Median	Sites	Samples	DL	Range	Median	
253 pesticides were target analytes. 113 pesticides were detected in domestic and imported commodities (including fish and shellfish).	—	—	—	nr	73*†	5–1,000	nr	nr	* For edible fish and shellfish combined—domestic samples. † Detection of any pesticide. Objective: to monitor pesticide residues in foods and to assess compliance with USEPA tolerances. 5% of fish and shellfish samples were violative (exceeded tolerances or contained pesticides with no tolerance).
256 pesticides were target analytes. 118 pesticides were detected in domestic and imported commodities (including fish and shellfish).	—	—	—	nr	72*†	1–10§	nr	nr	* For edible fish and shellfish combined—domestic samples. † Detection of any pesticide. § DL was 5–10 µg/kg for monitoring studies. Objective: to monitor pesticide residues in foods and to assess compliance with USEPA tolerances. No residues were found in over 60% of all domestic and imported samples. About 1% of domestic samples were violative (exceeded tolerances or contained pesticides with no tolerances).

Table 2.1. Pesticides in bed sediment and aquatic biota from rivers and estuaries in the United States: National and multistate monitoring studies—*Continued*

Reference	Sampling Dates	Sampling Locations and Strategy	Sampling Media and Species	Number of Samples	Target Analytes	Bed Sediment % Detection	
						Sites	Samples
Food and Drug Administration, 1990a	1988–1989	Raw agricultural products in interstate commerce throughout the United States. Samples are either surveillance (monitoring) or compliance (known contamination).	Fish (edible), shellfish. Species: not reported. Other commodities include milk, grains, vegetables, and fruits.	414 (total)*	270 pesticides were target analytes. 108 pesticides were detected in domestic and imported commodities (including fish and shellfish).	—	—
Food and Drug Administration, 1991	1989–1990	Raw agricultural products in interstate commerce throughout the United States. Samples are either surveillance (monitoring) or compliance (known contamination).	Fish (edible), shellfish. Species: not reported. Other commodities include milk, grains, vegetables, and fruits.	568 (total)*	268 pesticides were target analytes.	—	—

Table 2.1. Pesticides in bed sediment and aquatic biota from rivers and estuaries in the United States: National and multistate monitoring studies—*Continued*

Target Analytes	Bed Sediment (µg/kg dry weight)			Aquatic Biota (whole fish unless noted) % Detection		(µg/kg wet weight)			Comments
	DL	Range	Median	Sites	Samples	DL	Range	Median	
270 pesticides were target analytes. 108 pesticides were detected in domestic and imported commodities (including fish and shellfish).	—	—	—	nr	65*†	1–10§	nr	nr	* For edible fish and shellfish combined—domestic samples. † Detection of any pesticide. § DL was 5–10 µg/kg for monitoring studies. Objective: to monitor pesticide residues in foods and to assess compliance with USEPA tolerances. About 1% of domestic surveillance samples were violative (exceeded tolerances or contained pesticides with no tolerances).
268 pesticides were target analytes.	—	—	—	nr	68*†	5–1,000	nr	nr	* For edible fish and shellfish combined—domestic samples. † Detection of any pesticide. Objective: to monitor pesticide residues in foods and to assess compliance with USEPA tolerances. About 30% of compliance and <2% of surveillance domestic samples were violative (exceeded tolerances or contained pesticides with no tolerances). 60% of domestic aquaculture products had detectable residues of DDT, HCH, dieldrin, chlordane, or dacthal.

Table 2.1. Pesticides in bed sediment and aquatic biota from rivers and estuaries in the United States: National and multistate monitoring studies—*Continued*

Reference	Sampling Dates	Sampling Locations and Strategy	Sampling Media and Species	Number of Samples	Target Analytes	Bed Sediment % Detection	
						Sites	Samples
Yess and others, 1991a	1978–1982	Raw agricultural products in interstate commerce throughout the United States. Samples are either surveillance (monitoring) or compliance (known contamination).	Fish (edible), shellfish. Species: not reported. Other commodities include milk, grains, vegetables, and fruits.	3,689 (total)*	128 pesticides were detected in domestic and imported commodities (including fish and shellfish).	—	—
Yess and others, 1991b	1983–1986	Raw agricultural products in interstate commerce throughout the United States. Samples are either surveillance (monitoring) or compliance (known contamination).	Fish (edible), shellfish. Species: not reported. Other commodities include milk, grains, vegetables, and fruits.	1,759 (total)*	135 pesticides were detected in domestic and imported commodities (including fish and shellfish).	—	—

Table 2.1. Pesticides in bed sediment and aquatic biota from rivers and estuaries in the United States: National and multistate monitoring studies—*Continued*

Target Analytes	Bed Sediment (µg/kg dry weight)			Aquatic Biota (whole fish unless noted)					Comments
				% Detection		(µg/kg wet weight)			
	DL	Range	Median	Sites	Samples	DL	Range	Median	
128 pesticides were detected in domestic and imported commodities (including fish and shellfish)	—	—	—	nr	76*†	5–1,000	nr	nr	* For edible fish and shellfish combined—domestic samples. † Detection of any pesticide. Objective: to monitor pesticide residues in foods and to assess compliance with USEPA tolerances. 6% of fish and shellfish samples were violative (exceeded tolerances or contained pesticides with no tolerances).
135 pesticides were detected in domestic and imported commodities (including fish and shellfish).	—	—	—	nr	67*†	5–1,000	nr	nr	* For edible fish and shellfish combined—domestic samples. † Detection of any pesticide. Objective: to monitor pesticide residues in foods and to assess compliance with USEPA tolerances. 7% of fish and shellfish samples were violative (exceeded tolerances or contained pesticides with no tolerances).

Table 2.1. Pesticides in bed sediment and aquatic biota from rivers and estuaries in the United States: National and multistate monitoring studies—*Continued*

Reference	Sampling Dates	Sampling Locations and Strategy	Sampling Media and Species	Number of Samples	Target Analytes	Bed Sediment % Detection	
						Sites	Samples
Food and Drug Administration, 1992	1990–1991	Raw agricultural products in interstate commerce throughout the United States. Samples are either surveillance (monitoring) or compliance (known contamination).	Fish (edible), shellfish. Species: not reported. Other commodities include milk, grains, vegetables, and fruits.	531 (total)*	298 pesticides were target analytes. 108 pesticides were detected in domestic and imported commodities (including fish and shellfish).	—	—
BofCF–USEPA's National Pesticide Monitoring Program							
Butler, 1973b	1965–1972	180 coastal estuaries throughout the United States. Samples were composites of 15 or more mature individuals.	Mussels, clams, oysters. Species: 10 species, including *Mytilus edulis*.	8,095 total	Aldrin DDT, total Dieldrin Endrin Heptachlor Heptachlor epoxide Lindane Methoxychlor Mirex Toxaphene	— — — — — — — — — —	— — — — — — — — — —

Table 2.1. Pesticides in bed sediment and aquatic biota from rivers and estuaries in the United States: National and multistate monitoring studies—*Continued*

| Target Analytes | Bed Sediment (µg/kg dry weight) | | | Aquatic Biota (whole fish unless noted) | | | | | Comments |
| | DL | Range | Median | % Detection | | (µg/kg wet weight) | | | |
				Sites	Samples	DL	Range	Median	
298 pesticides were target analytes. 108 pesticides were detected in domestic and imported commodities (including fish and shellfish).	—	—	—	nr	41*†	5–1,000	nr	nr	* For edible fish and shellfish combined—domestic samples. † Detection of any pesticide. Objective: to monitor pesticide residues in foods and to assess compliance with USEPA tolerances. About 19% of compliance and 0.8% of surveillance domestic samples were violative (exceeded tolerances or contained pesticides with no tolerances). 36% of domestic aquaculture products had detectable residues of DDT, dieldrin, HCH, chlordane, dacthal, or chlorpyrifos.

BofCF–USEPA's National Pesticide Monitoring Program—*Continued*

Aldrin	—	—	—	0*	0*	5	nd*	nd*	* Mollusk tissue.
DDT, total	—	—	—	95*	63*	5	nd–5,390*	nr	Objective: to monitor
Dieldrin	—	—	—	54*	15*	5	nd–230*	nr	pesticides in estuarine
Endrin	—	—	—	7*	0.4*	5	nd–32*	nr	mollusks to determine
Heptachlor	—	—	—	0*	0*	5	nd*	nd*	the extent of pollution.
Heptachlor epoxide	—	—	—	0*	0*	5	nd*	nd*	Incidence of total DDT varied among drainages. There was a
Lindane	—	—	—	0*	0*	5	nd*	nd*	decreasing trend in
Methoxychlor	—	—	—	0*	0*	5	nd*	nd*	DDT residues, except
Mirex	—	—	—	5*	0.1*	5	nd–540*	nr	in California, New
Toxaphene	—	—	—	3*	2*	5	nd–54,000*	nr	York, and Virginia. High toxaphene residues in Georgia were due to manufacturing plant effluents.

Table 2.1. Pesticides in bed sediment and aquatic biota from rivers and estuaries in the United States: National and multistate monitoring studies—*Continued*

Reference	Sam-pling Dates	Sampling Locations and Strategy	Sampling Media and Species	Number of Samples	Target Analytes	Bed Sediment % Detection	
						Sites	Samples
Butler and others, 1978	1977	89 sites in coastal estuaries throughout the United States. Two composite samples of 15 bivalves were collected at each station. Site locations in each state were not reported.	Mussels, clams, oysters. Species: 7 species, including *Mytilus edulis*.	178 total	Aldrin	—	—
					Azinphosmethyl	—	—
					Carbophenothion	—	—
					Chlordane	—	—
					DDD	—	—
					DDE	—	—
					DDT	—	—
					DDT, total	—	—
					DEF	—	—
					Demeton	—	—
					Diazinon	—	—
					Dieldrin	—	—
					Endosulfan	—	—
					Ethion	—	—
					Heptachlor	—	—
					Lindane	—	—
					Malathion	—	—
					Methoxychlor	—	—
					Mirex	—	—
					Parathion	—	—
					Phorate	—	—
					Toxaphene	—	—
					Trifluralin	—	—

Table 2.1. Pesticides in bed sediment and aquatic biota from rivers and estuaries in the United States: National and multistate monitoring studies—*Continued*

| Target Analytes | Bed Sediment (µg/kg dry weight) | | | Aquatic Biota (whole fish unless noted) | | | | | Comments |
| | DL | Range | Median | % Detection | | (µg/kg wet weight) | | | |
				Sites	Samples	DL	Range	Median	
Aldrin	—	—	—	0*	0*	10	nd*	nd*	* Mollusk tissue.
Azinphos-methyl	—	—	—	0*	0*	10	nd*	nd*	† Value shown is range of mean values from 87
Carbopheno-thion	—	—	—	0*	0*	50	nd*	nd*	sites. Objective: to monitor pesticide resi-
Chlordane	—	—	—	0*	0*	10	nd*	nd*	dues in some of the
DDD	—	—	—	1*	1*	10	nd–nr*	nd*	same estuaries as
DDE	—	—	—	4*	4*	10	nd–nr*	nd*	Butler (1973b) to
DDT	—	—	—	1*	1*	10	nd–nr*	nd*	determine further
DDT, total	—	—	—	4*	4*	10	nd–122*†	nd*	trends in pollution after
DEF	—	—	—	0*	0*	50	nd*	nd*	a 5–7 year lapse. No
Demeton	—	—	—	0*	0*	10	nd*	nd*	pesticides were
Diazinon	—	—	—	0*	0*	10	nd*	nd*	detected in 85 of 87
Dieldrin	—	—	—	0*	0*	10	nd*	nd*	estuaries. DDT was the
Endosulfan	—	—	—	0*	0*	20	nd*	nd*	only pesticide detected
Ethion	—	—	—	0*	0*	30	nd*	nd*	in the remaining two
Heptachlor	—	—	—	0*	0*	10	nd*	nd*	estuaries: upper
Lindane	—	—	—	0*	0*	10	nd*	nd*	Delaware Bay (New
Malathion	—	—	—	0*	0*	10	nd*	nd*	Jersey, Delaware) and
Methoxychlor	—	—	—	0*	0*	30	nd*	nd*	Muga Lagoon
Mirex	—	—	—	0*	0*	50	nd*	nd*	(California). Ten years
Parathion	—	—	—	0*	0*	10	nd*	nd*	earlier, total DDT
Phorate	—	—	—	0*	0*	10	nd*	nd*	levels at these sites
Toxaphene	—	—	—	0*	0*	50	nd*	nd*	were 3–15 times
Trifluralin	—	—	—	0*	0*	10	nd*	nd*	higher, and other pesticides (e.g., dieldrin, endrin) also were present.

Table 2.1. Pesticides in bed sediment and aquatic biota from rivers and estuaries in the United States: National and multistate monitoring studies—*Continued*

Reference	Sampling Dates	Sampling Locations and Strategy	Sampling Media and Species	Number of Samples	Target Analytes	Bed Sediment % Detection	
						Sites	Samples
Butler and Schutzmann, 1978	1972–1976	144 estuaries in 21 coastal areas, including the Pacific, Gulf, and southeast Atlantic coastal states. Sites were monitored in spring and autumn for 1–4 years. At each site, the same two species (one bottom-feeder and one carnivore) were collected throughout the study period. Samples were composites of 25 fish less than 1 year old. Site locations in each state were not reported.	Fish (whole). Species: 154 species, of which 65 species were collected 3 or more times and 22 species were collected in estuaries of 3 or more states.	1,524	Aldrin	—	—
					Azinphosmethyl	—	—
					Carbophenothion	—	—
					Chlordane	—	—
					DDT, total	—	—
					DEF	—	—
					Demeton	—	—
					Diazinon	—	—
					Dieldrin	—	—
					Endosulfan	—	—
					Ethion	—	—
					Heptachlor	—	—
					Heptachlor epoxide	—	—
					Lindane	—	—
					Malathion	—	—
					Methoxychlor	—	—
					Methyl parathion	—	—
					Mirex	—	—
					Parathion	—	—
					Phorate	—	—
					Toxaphene	—	—
					Trifluralin	—	—

Table 2.1. Pesticides in bed sediment and aquatic biota from rivers and estuaries in the United States: National and multistate monitoring studies—*Continued*

Target Analytes	Bed Sediment (µg/kg dry weight)			Aquatic Biota (whole fish unless noted)					Comments
				% Detection		(µg/kg wet weight)			
	DL	Range	Median	Sites	Samples	DL	Range	Median	
Aldrin	—	—	—	0*	0	10	nd	nd	* Percentage of the 21
Azinphos-methyl	—	—	—	0*	0	10	nd	nd	coastal areas with detectable residues.
Carbopheno-thion	—	—	—	10*	0.2	50	nd–109	nd	Objective: to monitor pesticides in estuarine
Chlordane	—	—	—	19*	2	10	nd–290	nd	fish. DDT was found in
DDT, total	—	—	—	86*	39	10	nd–4,082	nd	some fish that were
DEF	—	—	—	0*	0	50	nd	nd	only a few months old.
Demeton	—	—	—	0	0	10	nd	nd	Samples in California,
Diazinon	—	—	—	0	0	10	nd	nd	New York, Delaware,
Dieldrin	—	—	—	62*	5	10	nd–145	nd	and Florida contained
Endosulfan	—	—	—	0	0	20	nd	nd	1,000–4,000 µg/kg
Ethion	—	—	—	10*	1	30	nd–169	nd	total DDT; this exceeds
Heptachlor	—	—	—	0	0	10	nd	nd	levels in mollusks from
Heptachlor epoxide	—	—	—	5*	0.2	10	nd–15	nd	the same estuaries that were sampled during
Lindane	—	—	—	0	0	10	nd	nd	1965–1972 (Butler,
Malathion	—	—	—	0	0	10	nd	nd	1973b). The mean total
Methoxychlor	—	—	—	0	0	30	nd	nd	DDT concentration did
Methyl parathion	—	—	—	5*	1	10	nd–47	nd	not change from 1972 to 1976 but the DDE/
Mirex	—	—	—	0	0	50	nd	nd	total DDT fraction
Parathion	—	—	—	14*	0.3	10	nd–75	nd	increased and the
Phorate	—	—	—	0	0	10	nd	nd	DDT/total DDT
Toxaphene	—	—	—	14*	0.4	50	nd–504	nd	fraction decreased.
Trifluralin	—	—	—	0	0	10	nd	nd	Most dieldrin detections were in the Chesapeake Bay and the Texas coast.

Table 2.1. Pesticides in bed sediment and aquatic biota from rivers and estuaries in the United States: National and multistate monitoring studies—*Continued*

Reference	Sampling Dates	Sampling Locations and Strategy	Sampling Media and Species	Number of Samples	Target Analytes	Bed Sediment % Detection Sites	Bed Sediment % Detection Samples
Butler and Schutzmann, 1979	1974	(1) 80 sites on New England coast from New Jersey to Nova Scotia (5–200 miles offshore). Samples were composites of 5–10 fish by species. Site locations were not reported.	Fish (liver). Species: 20 species, including spiny dog-fish, ocean pout, yellow-tail flounder, and alewife.	nr (668 total fish in the com-posite sam-ples)	DDT, total	—	—
	1975	(2) Unspecified number of sites in same area. Fish were juveniles only. Site locations were not reported.	Fish (liver). Species: spiny dogfish, goosefish, little skate, and pollock.	14	DDD	—	—
					DDE	—	—
					DDT	—	—
					DDT, total	—	—
	1975	(3) 8 sites in same area. Dogfish were gravid. Similar embryonic stages were analyzed in pooled samples.	Fish (eggs, fetus with yolk sac), mature fetus w/out liver, mature fetal liver. Species: spiny dogfish.	3	DDD	—	—
				3	DDE	—	—
				3	DDT	—	—
				4	DDT, total	—	—

Table 2.1. Pesticides in bed sediment and aquatic biota from rivers and estuaries in the United States: National and multistate monitoring studies—*Continued*

Target Analytes	Bed Sediment (µg/kg dry weight)			Aquatic Biota (whole fish unless noted)					Comments
				% Detection		(µg/kg wet weight)			
	DL	Range	Median	Sites	Samples	DL	Range	Median	
DDT, total	—	—	—	nr	nr	100	20–4,760*	160*†	* Fish liver. † Median of means for 20 species. Objective: to monitor pesticides in estuarine fish. Dogfish had the highest mean total DDT level (4,760 µg/kg). Samples were pooled across sites and were not uniform in size, age of fish.
DDD	—	—	—	nr	57*	10	330–1,600*	330*	* Fish liver. Objective: to sample small fish of species that had the highest residues in 1974. These smaller fish contained low residues. Dogfish had much higher total DDT levels than other species. Dogfish were neonates with empty digestive tracts, so probably acquired DDT before birth.
DDE	—	—	—	nr	100*	10	30–2,460*	900*	
DDT	—	—	—	nr	50*	10	1,120–5,552*	570*	
DDT, total	—	—	—	nr	100*	10	30–10,790*	1,290*	
DDD	—	—	—	nr	100*	10	50–420*	nr	* For all (various) embryonic tissue samples analyzed. Objective: to measure residues in embryonic tissues of spiny dogfish. High concentrations in eggs, fetus with yolk sac, and mature fetal liver indicate that DDT is acquired from mother during embryonic development.
DDE	—	—	—	nr	100*	10	80–940*	nr	
DDT	—	—	—	nr	100*	10	60–630*	nr	
DDT, total	—	—	—	nr	100*	10	190–1,990*	nr	

Table 2.1. Pesticides in bed sediment and aquatic biota from rivers and estuaries in the United States: National and multistate monitoring studies—*Continued*

Reference	Sampling Dates	Sampling Locations and Strategy	Sampling Media and Species	Number of Samples	Target Analytes	Bed Sediment % Detection	
						Sites	Samples
FWS's National Contaminant Biomonitoring Program							
Henderson and others, 1969	1967–1968	50 sites located on major United States rivers and in the Great Lakes. Three composite samples (3–5 fish each) were collected at each site in autumn and spring.	Fish (whole). Species: 62 fish species, including carp, black bass, catfish, and various suckers.	589	Aldrin	—	—
					Chlordane, total	—	—
					DDD, *p,p′-*	—	—
					DDE, *p,p′-*	—	—
					DDT, *p,p′-*	—	—
					DDT, total	—	—
					Dieldrin	—	—
					Endrin	—	—
					Heptachlor	—	—
					Heptachlor epoxide	—	—
					Lindane	—	—
					Toxaphene	—	—
Henderson and others, 1971	1969	50 sites located on major rivers throughout the United States and in the Great Lakes. Three composite samples (3–5 fish each) were collected at each site in autumn.	Fish (whole). Species: 44 species. 35% of samples were carp, channel catfish, and largemouth bass; 18% were various suckers.	147	Aldrin	—	—
					Chlordane, total	—	—
					DDD, *p,p′-*	—	—
					DDE, *p,p′-*	—	—
					DDT, *p,p′-*	—	—
					DDT, total	—	—
					Dieldrin	—	—
					Endrin	—	—
					HCH, α-	—	—
					Heptachlor	—	—
					Heptachlor epoxide	—	—
					Toxaphene	—	—

Table 2.1. Pesticides in bed sediment and aquatic biota from rivers and estuaries in the United States: National and multistate monitoring studies—*Continued*

Target Analytes	Bed Sediment (µg/kg dry weight)			Aquatic Biota (whole fish unless noted)					Comments
				% Detection		(µg/kg wet weight)			
	DL	Range	Median	Sites	Samples	DL	Range	Median	
FWS's National Contaminant Biomonitoring Program—Continued									
Aldrin	—	—	—	58	13	1–10	nd–7,500	—	Objective: to monitor
Chlordane, total	—	—	—	64	22	1–10	nd–7,290	—	pesticide residues in fish nationwide. Cross-
DDD, *p,p'*-	—	—	—	100	94	1–10	nd–2,800	—	checks revealed
DDE, *p,p'*-	—	—	—	100	95	1–10	nd–18,400	—	substantial differences
DDT, *p,p'*-	—	—	—	100	86	1–10	nd–6,670	—	among laboratories
DDT, total	—	—	—	100	99	1–10	nd–45,270	—	used. DDT com-
Dieldrin	—	—	—	98	72	1–10	nd–1,940	—	pounds may be
Endrin	—	—	—	50	10	1–10	nd–1,500	—	overestimated in
Heptachlor	—	—	—	62	12	1–10	nd–8,330	—	samples with high
Heptachlor epoxide	—	—	—	82	32	1–10	nd–8,460	—	PCBs, toxaphene, or chlordane. Fish from
Lindane	—	—	—	68	17	1–10	nd–480	—	widely scattered areas
Toxaphene	—	—	—	12	1	1–10	nd–30	—	have high organo-chlorine residues. There is considerable variation among samples.
Aldrin	—	—	—	0	0	5	nd	nd	Objective: to monitor
Chlordane, total	—	—	—	100	11	5	nd–13,500	nd	pesticide residues in fish nationwide. One
DDD, *p,p'*-	—	—	—	100	100	5	10–10,400	230	laboratory analyzed the
DDE, *p,p'*-	—	—	—	100	100	5	10–42,300	300	samples, with cross-
DDT, *p,p'*-	—	—	—	100	99	5	nd–5,070	150	checks performed by
DDT, total	—	—	—	100	100	5	30–57,770	750	two additional labora-
Dieldrin	—	—	—	96	93	5	nd–1,590	20	tories. DDT and
Endrin	—	—	—	0	0	5	nd	nd	dieldrin occurred in
HCH, α-	—	—	—	98	90	5	nd–4,370	20	almost all fish samples;
Heptachlor	—	—	—	2	2	5	nd–450	nd	residues remained high
Heptachlor epoxide	—	—	—	6	4	5	nd–340	nd	at some stations in 1969. Other organo-
Toxaphene	—	—	—	0	0	5	nd	nd	chlorines occurred in fewer samples and at lower levels than in previous years.

Table 2.1. Pesticides in bed sediment and aquatic biota from rivers and estuaries in the United States: National and multistate monitoring studies—*Continued*

Reference	Sampling Dates	Sampling Locations and Strategy	Sampling Media and Species	Number of Samples	Target Analytes	Bed Sediment % Detection	
						Sites	Samples
Schmitt and others, 1981	1970–1974	109 sites located on major rivers throughout the United States and in the Great Lakes. Three composite samples (3–5 fish each) were collected at each site.	Fish (whole). Species: 71 species. 39% of samples were carp, channel cat-fish, or largemouth bass.	2,106	DDD, *p,p´*- DDE, *p,p´*- DDT, *p,p´*- DDT, total Dieldrin * Endrin HCH† Heptachlor, total Toxaphene	— — — — — — — — —	— — — — — — — — —
Schmitt and others, 1983	1976–1979	112 sites located on major rivers throughout the United States and in the Great Lakes. Three composite samples (3–5 fish each) were collected at each site. Samples were collected at half the stations in autumn of even-numbered years, and the other half in autumn of odd-numbered years.	Fish (whole). Species: 58 species. 45% of samples were carp, channel catfish, largemouth bass, yellow perch, or white sucker.	620	Chlordane, *cis*- Chlordane, total Chlordane, *trans*- Dacthal* DDD, *p,p´*- DDE, *p,p´*- DDT, *p,p´*- DDT, total Dieldrin† Endrin HCH, α- Heptachlor, total Hexachlorobenzene Lindane Nonachlor, *cis*- Nonachlor, *trans*- Oxychlordane Toxaphene	— — — — — — — — — — — — — — — — — —	— — — — — — — — — — — — — — — — — —

Table 2.1. Pesticides in bed sediment and aquatic biota from rivers and estuaries in the United States: National and multistate monitoring studies—*Continued*

Target Analytes	Bed Sediment (µg/kg dry weight)			Aquatic Biota (whole fish unless noted)					Comments
				% Detection		(µg/kg wet weight)			
	DL	Range	Median	Sites	Samples	DL	Range	Median	
DDD, p,p'-	—	—	—	100	81	5	nd–14,000	90	* Sum of aldrin and
DDE, p,p'-	—	—	—	100	95	5	nd–32,000	180	dieldrin.
DDT, p,p'-	—	—	—	98	64	5	nd–11,000	30	† Sum of α and γ
DDT, total	—	—	—	100	97	5	20–48,000	360	isomers. Objective: to
Dieldrin *	—	—	—	98	71	5	nd–9,100	10	monitor pesticide
Endrin	—	—	—	79	24	5	nd–820	nd	residues in fish
HCH†	—	—	—	92	99	5	nd–3,700	20	nationwide. Laboratory
Heptachlor, total	—	—	—	10	1	5	nd–1,000	nd	cross-checks showed that DDT compounds
Toxaphene	—	—	—	30	9	5	nd–51,000	nd	were overestimated during 1970–1971 and that results for toxaphene, HCH, and chlordane were estimates only.
Chlordane, *cis*-	—	—	—	98	85	10	nd–2,530	20	* Analyzed during 1978–1979 only.
Chlordane, total	—	—	—	99	87	10	nd–6,360	50	† Sum of aldrin and dieldrin (for 1976–
Chlordane, *trans*-	—	—	—	89	60	10	nd–540	10	1977 only). Objective: to monitor pesticide
Dacthal*	—	—	—	33	19	10	nd–1,220	nd	residues in fish nation-
DDD, p,p'-	—	—	—	100	91	10	nd–3,010	50	wide. p,p'-DDE resi-
DDE, p,p'-	—	—	—	100	99	10	nd–6,760	140	dues declined, then
DDT, p,p'-	—	—	—	93	57	10	nd–1,990	10	appeared to stabilize by
DDT, total	—	—	—	100	99	10	nd–10,620	220	1976 or 1977. Signi-
Dieldrin†	—	—	—	96	75	10	nd–5,010	10	ficant p,p'-DDT was
Endrin	—	—	—	54	26	10	nd–400	nd	found at only a few
HCH, α-	—	—	—	92	54	10	nd–300	10	stations. However,
Heptachlor, total	—	—	—	70	37	10	nd–1,170	nd	DDT and metabolites remain ubiquitous.
Hexachloro-benzene	—	—	—	47	21	10	nd–700	nd	From 1974 to 1977– 1978, dieldrin residues
Lindane	—	—	—	32	12	10	nd–280	nd	declined nationwide
Nonachlor, *cis*-	—	—	—	78	46	10	nd–710	nd	but not in the Great Lakes. Chlordane was
Nonachlor, *trans*-	—	—	—	96	65	10	nd–2,170	10	the most widespread cyclodiene insecticide.
Oxychlordane	—	—	—	38	28	10	nd–740	nd	
Toxaphene	—	—	—	71	46	100	nd–18,700	nd	

Table 2.1. Pesticides in bed sediment and aquatic biota from rivers and estuaries in the United States: National and multistate monitoring studies—*Continued*

Reference	Sam- pling Dates	Sampling Locations and Strategy	Sampling Media and Species	Number of Samples	Target Analytes	Bed Sediment	
						% Detection	
						Sites	Samples
Schmitt and others, 1985	1980– 1981	107 sites located on major rivers throughout the United States and in the Great Lakes. Three composite samples (3–5 fish each) were collected at each site. Samples were collected at half the stations in the autumn of even-numbered years, and the other half in the autumn of odd-numbered years.	Fish (whole). Species: 53 species. The most frequently collected taxa were carp, white sucker, and largemouth bass.	315	Chlordane, *cis-*	—	—
					Chlordane, total	—	—
					Chlordane, *trans-*	—	—
					Dacthal	—	—
					DDD, *p,p´-*	—	—
					DDE, *p,p´-*	—	—
					DDT, *p,p´-*	—	—
					DDT, total	—	—
					Dieldrin	—	—
					Endrin	—	—
					HCH, α-	—	—
					Heptachlor, total	—	—
					Hexachlorobenzene	—	—
					Lindane	—	—
					Methoxychlor	—	—
					Mirex	—	—
					Nonachlor, *cis-*	—	—
					Nonachlor, *trans-*	—	—
					Oxychlordane	—	—
					PCA	—	—
					Toxaphene	—	—

Table 2.1. Pesticides in bed sediment and aquatic biota from rivers and estuaries in the United States: National and multistate monitoring studies—*Continued*

Target Analytes	Bed Sediment (µg/kg dry weight)			Aquatic Biota (whole fish unless noted)					Comments
				% Detection		(µg/kg wet weight)			
	DL	Range	Median	Sites	Samples	DL	Range	Median	
Chlordane, *cis-*	—	—	—	74	63	10	nd–360	10	Objective: to monitor pesticide residues in
Chlordane, total	—	—	—	88	73	10	nd–1,950	50	fish nationwide. Both mean and maximum
Chlordane, *trans-*	—	—	—	72	56	10	nd–220	10	concentrations of DDT compounds and diel-
Dacthal	—	—	—	28	16	10	nd–400	nd	drin have decreased
DDD, *p,p´-*	—	—	—	97	91	10	nd–3,430	30	since the early 1970s.
DDE, *p,p´-*	—	—	—	100	99	10	nd–2,570	110	Both detection fre-
DDT, *p,p´-*	—	—	—	79	57	10	nd–2,690	10	quency and maximum
DDT, total	—	—	—	100	100	10	20–6,500	170	concentration of toxa-
Dieldrin	—	—	—	75	64	10	nd–720	10	phene have increased
Endrin	—	—	—	22	14	10	nd–300	nd	since the late 1970s.
HCH, α-	—	—	—	54	35	10	nd–40	nd	
Heptachlor, total	—	—	—	38	27	10	nd–270	nd	
Hexachloro-benzene	—	—	—	24	15	10	nd–120	nd	
Lindane	—	—	—	16	6	10	nd–30	nd	
Methoxychlor	—	—	—	32	nr	10	nd–190	nr	
Mirex	—	—	—	18	10	10	nd–210	nd	
Nonachlor, *cis-*	—	—	—	73	59	10	nd–270	10	
Nonachlor, *trans-*	—	—	—	85	71	10	nd–770	10	
Oxychlordane	—	—	—	39	30	10	nd–330	nd	
PCA	—	—	—	24	17	10	nd–70	nd	
Toxaphene	—	—	—	87	73	100	nd–21,000	130	

Table 2.1. Pesticides in bed sediment and aquatic biota from rivers and estuaries in the United States: National and multistate monitoring studies—*Continued*

Reference	Sam-pling Dates	Sampling Locations and Strategy	Sampling Media and Species	Number of Samples	Target Analytes	Bed Sediment % Detection	
						Sites	Samples
Schmitt and others, 1990	1984–1985	112 sites located on major rivers throughout the United States and in the Great Lakes. Three composite samples (3–5 fish each) were collected at each site in the autumn of 1984 or spring of 1985.	Fish (whole). Species: 47 species. The most frequently collected taxa were carp, white sucker, and largemouth bass.	321	Chlordane, *cis-*	—	—
					Chlordane, total	—	—
					Chlordane, *trans-*	—	—
					Dacthal	—	—
					DDD, *p,p´-*	—	—
					DDE, *p,p´-*	—	—
					DDT, *p,p´-*	—	—
					DDT, total	—	—
					Dieldrin	—	—
					Endrin	—	—
					HCH, α-	—	—
					Heptachlor, total	—	—
					Hexachlorobenzene	—	—
					Lindane	—	—
					Mirex	—	—
					Nonachlor, *cis-*	—	—
					Nonachlor, *trans-*	—	—
					Oxychlordane	—	—
					PCA	—	—
					Toxaphene	—	—

Table 2.1. Pesticides in bed sediment and aquatic biota from rivers and estuaries in the United States: National and multistate monitoring studies—*Continued*

Target Analytes	Bed Sediment (µg/kg dry weight)			Aquatic Biota (whole fish unless noted)					Comments
				% Detection		(µg/kg wet weight)			
	DL	Range	Median	Sites	Samples	DL	Range	Median	
Chlordane, *cis-*	—	—	—	85	73	100	nd–660	10	Objective: to monitor pesticide residues in
Chlordane, total	—	—	—	92	79	10–100	30–2,690	50	fish nationwide. Reference discusses trends
Chlordane, *trans-*	—	—	—	69	54	10	nd–350	10	from 1976–1977 to 1984. Mean concentra-
Dacthal	—	—	—	46	27	10	nd–450	nd	tions of DDT, dieldrin,
DDD, *p,p′-*	—	—	—	97	94	10	nd–2,550	20	chlordane, and toxa-
DDE, *p,p′-*	—	—	—	98	97	10	nd–4,740	70	phene decreased, espe-
DDT, *p,p′-*	—	—	—	88	71	10	nd–1,790	10	cially at sites where
DDT, total	—	—	—	98	97	10	20–9,080	120	residues were highest.
Dieldrin	—	—	—	74	61	10	nd–1,390	10	Detection frequencies
Endrin	—	—	—	29	16	10	nd–220	nd	have not declined since
HCH, α-	—	—	—	15	6	10	nd–10	nd	1980–1981, indicating
Heptachlor, total	—	—	—	49	38	10	nd–290	nd	continued input to aquatic systems. The
Hexachloro-benzene	—	—	—	19	13	10	nd–410	nd	relative abundance of components of DDT
Lindane	—	—	—	47	29	10	nd–40	nd	and chlordane mixtures
Mirex	—	—	—	13	7	10	nd–440	nd	indicate continued
Nonachlor, *cis-*	—	—	—	82	66	10	nd–450	10	weathering of residues in the United States.
Nonachlor, *trans-*	—	—	—	89	79	10	nd–1,000	10	Urban uses may consti-tute major source of
Oxychlordane	—	—	—	44	31	10	nd–290	nd	chlordane to aquatic
PCA	—	—	—	30	20	10	nd–100	nd	systems.
Toxaphene	—	—	—	71	58	10	nd–8,200	100	

Table 2.1. Pesticides in bed sediment and aquatic biota from rivers and estuaries in the United States: National and multistate monitoring studies—*Continued*

Reference	Sampling Dates	Sampling Locations and Strategy	Sampling Media and Species	Number of Samples	Target Analytes	Bed Sediment % Detection	
						Sites	Samples
USGS–USEPA's National Pesticide Monitoring Network							
Gilliom and others, 1985	1975–1980	160–180 sites on major rivers throughout the United States (a subset of NASQAN sites). Cross-sectional bed sediment samples (unsieved) were taken twice per year. Water samples were collected quarterly.	Sediment (surficial), water (whole).	1,018 2,950	Aldrin	2.9	0.6
					Atrazine	0	0
					Chlordane	30	9.9
					D, 2,4-	1.4	0.4
					DDD	31	12
					DDE	42	17
					DDT	26	8.5
					Diazinon	1.2	0.2
					Dieldrin	29	12
					Endrin	2.3	0.6
					Ethion	0.6	0.4
					Heptachlor epoxide	5.3	1
					Lindane	0.6	0.1
					Malathion	0	0
					Methoxychlor	0.6	0.1
					Methyl parathion	0	0
					Methyl trithion	0	0
					Parathion	0	0
					Silvex	1.4	0.4
					T, 2,4,5-	0.7	0.2
					Toxaphene	3.5	0.6
					Trithion	0	0

Table 2.1. Pesticides in bed sediment and aquatic biota from rivers and estuaries in the United States: National and multistate monitoring studies—*Continued*

| Target Analytes | Bed Sediment (µg/kg dry weight) | | | Aquatic Biota (whole fish unless noted) | | | | | Comments |
	DL	Range	Median	% Detection Sites	Samples	DL	Range	Median	
				USGS–USEPA's National Pesticide Monitoring Network—Continued					
Aldrin	0.1	nd–12	nd	—	—	—	—	—	Objective: to assess
Atrazine	5	nd	nd	—	—	—	—	—	pesticide levels in
Chlordane	1.5	nd–300	nd	—	—	—	—	—	runoff and bed sedi-
D, 2,4-	5	nd–14.9	nd	—	—	—	—	—	ment of the nation's
DDD	0.5	nd–168	nd	—	—	—	—	—	rivers. Samples were
DDE	0.3	nd–163	nd	—	—	—	—	—	collected as part of the
DDT	0.5	nd–180	nd	—	—	—	—	—	National Pesticide
Diazinon	1	nd–7.1	nd	—	—	—	—	—	Monitoring Network.
Dieldrin	0.3	nd–34	nd	—	—	—	—	—	Detections of organo-
Endrin	0.5	nd–2.9	nd	—	—	—	—	—	chlorines decreased
Ethion	3	nd–191	nd	—	—	—	—	—	erratically, but grad-
Heptachlor epoxide	0.1	nd–1.1	nd	—	—	—	—	—	ually, in both water and bed sediment from
Lindane	0.1	nd–0.3	nd	—	—	—	—	—	1975 to 1980. A few
Malathion	3	nd	nd	—	—	—	—	—	sites had increasing
Methoxychlor	1	nd–1.5	nd	—	—	—	—	—	trends in bed sediment.
Methyl parathion	3	nd	nd	—	—	—	—	—	Detection frequency is a function of use and
Methyl trithion	5	nd	nd	—	—	—	—	—	persistence. Regional use (1971) is correlated
Parathion	2.5	nd	nd	—	—	—	—	—	with detections in bed
Silvex	5	nd–6.3	nd	—	—	—	—	—	sediment for chlor-
T, 2,4,5-	5	nd–9.1	nd	—	—	—	—	—	dane, DDT, endrin,
Toxaphene	3	nd–815	nd	—	—	—	—	—	heptachlor epoxide,
Trithion	5	nd	nd	—	—	—	—	—	and toxaphene.

Table 2.1. Pesticides in bed sediment and aquatic biota from rivers and estuaries in the United States: National and multistate monitoring studies—*Continued*

Reference	Sampling Dates	Sampling Locations and Strategy	Sampling Media and Species	Number of Samples	Target Analytes	Bed Sediment % Detection Sites	Samples
NOAA's National Status and Trends Program							
McCain and others, 1988 Benthic Surveillance Project	nr	8 sites on the Pacific coast: 5 urban and 3 reference (nonurban) sites. Sites were not adjacent to known point sources. 3 composite sediment samples were collected (from 3 different stations) per site. 3 fish liver samples (each a composite of 10 fish) were analyzed. Each species was sampled from at least one urban and one reference site.	Sediment (surficial), fish (liver). Species: barred sand bass, white croaker, starry flounder, and English sole.	24 30	DDT, total	50	nr

Table 2.1. Pesticides in bed sediment and aquatic biota from rivers and estuaries in the United States: National and multistate monitoring studies—*Continued*

| Target Analytes | Bed Sediment (µg/kg dry weight) | | | Aquatic Biota (whole fish unless noted) | | | | | Comments |
| | DL | Range | Median | % Detection | | (µg/kg wet weight) | | | |
				Sites	Samples	DL	Range	Median	
	**NOAA's National Status and Trends Program—Continued**								
DDT, total	nr	nd–620*	nr	100†	nr	nr	30–12,000§	nr	*Range of mean values per site. † Fish liver. § Range of mean levels (dry weight) in fish liver for each species–site combination. Objective: to investigate fish pathology and contaminant residues as indicators of marine pollution near four urban areas. In sediment: total DDT was high near Los Angeles (mean 620 µg/kg) compared with other urban centers (8 µg/kg). No total DDT was found at reference sites. In fish livers: total DDT levels were higher in California (130–12,000 µg/kg) than Oregon and Washington (<1,500 µg/kg). Pollution-related fish diseases appeared highest at the most contaminated sites.

Table 2.1. Pesticides in bed sediment and aquatic biota from rivers and estuaries in the United States: National and multistate monitoring studies—*Continued*

Reference	Sampling Dates	Sampling Locations and Strategy	Sampling Media and Species	Number of Samples	Target Analytes	Bed Sediment % Detection	
						Sites	Samples
Hanson and others, 1989	1984–1985	17 sites on the Southeast and Gulf coasts (North Carolina to Florida, Florida to Texas). Three composite sediment samples were analyzed per site. Fish liver samples were composites of tissue from 10 or more individuals.	Sediment (surficial), fish (liver, stomach contents). Species: Atlantic croaker and spot.	96	Chlordane	18	nr
Benthic Surveillance Project					DDT, total	41	nr
				90	Dieldrin	12	nr
				24	Heptachlor	0	nr
					Heptachlor epoxide	0	nr
					Hexachlorobenzene	6	nr
					Lindane	6	nr
					Mirex	6	nr
					Nonachlor	12	nr

Table 2.1. Pesticides in bed sediment and aquatic biota from rivers and estuaries in the United States: National and multistate monitoring studies—*Continued*

Target Analytes	Bed Sediment			Aquatic Biota (whole fish unless noted)					Comments
	(μg/kg dry weight)			% Detection		(μg/kg wet weight)			
	DL	Range	Median	Sites	Samples	DL	Range	Median	
Chlordane	1	nd–~0.8*	nd*	94†	nr	3	nd– ~32*†	<10*†	* Range and median are
DDT, total	1	nd–7*	nd*	100†	nr	3	2–209*†	54*†	of means for each site–
Dieldrin	1	nd– ~2.2	nd*	81†	nr	3	nd~120*†	<20*†	year combination.
Heptachlor	1	nd	nd	13†	nr	3	nd– ~5*†	nd*†	Approximate values
Heptachlor epoxide	1	nd	nd	50†	nr	3	nd– ~17*†	nd*†	(~) were estimated visually from figures in
Hexachloro-benzene	1	nd– ~25*	nd*	13†	nr	3	nd– ~17*†	nd*†	reference.
Lindane	1	nd– ~0.3*	nd*	19†	nr	3	nd– ~13*†	nd*†	† Fish liver, μg/kg dry weight. Objective: to
Mirex	1	nd– ~0.4*	nd*	50†	nr	3	nd– ~7*†	nd*†	assess environmental
Nonachlor	1	nd– ~0.5*	nd*	100†	nr	3	nd– ~35*†	<10*†	quality of estuaries and
									coastal areas in the southeastern United States using contam-inant residues in fish and fish pathology as indicators. All organo-chlorines were found more often in fish liver than in sediment, and fish concentrations were 10–100 times higher. Levels in stomach contents were intermediate. Contami-nant distributions were related to population density.

Table 2.1. Pesticides in bed sediment and aquatic biota from rivers and estuaries in the United States: National and multistate monitoring studies—*Continued*

Reference	Sampling Dates	Sampling Locations and Strategy	Sampling Media and Species	Number of Samples	Target Analytes	Bed Sediment % Detection	
						Sites	Samples
National Oceanic and Atmospheric Administration, 1989 Mussel Watch Project	1986–1988	177 sites in coastal and estuarine areas throughout the United States. Site selection bias was towards urban areas, but major point sources were avoided. 3 composite samples were collected per site per year in late autumn or winter.	Mollusks. Species: 2 species of mussels and 2 species of oysters.	1,372	Chlordane, total* DDT, total Dieldrin, total Hexachlorobenzene Lindane Mirex	— — — — — —	— — — — — —
Varanasi and others, 1989; Varanasi and others, 1988 Benthic Surveillance Project	1984–1986	31 sites on the Pacific coast (California to Alaska). 22 sites were near urban centers. 5 sites were reference sites. At each site each year, 4 fish samples (3 liver and 1 stomach contents) and 1 sediment sample were analyzed. Fish samples were composites of 10 fish. Sediment samples were composites of 3 grabs per station.	Sediment (surficial), fish (liver, stomach contents). Species: fourhorn sculpin, flathead sole, English sole, starry flounder, white croaker, hornyhead turbot, and barred sand bass.	201 222 71	Aldrin Chlordane, total§ DDT, total Dieldrin Heptachlor, total Hexachlorobenzene Lindane Mirex	nr 35 81 26 nr 61 nr nr	nr nr nr nr nr nr nr nr

Table 2.1. Pesticides in bed sediment and aquatic biota from rivers and estuaries in the United States: National and multistate monitoring studies—*Continued*

| Target Analytes | Bed Sediment (µg/kg dry weight) | | | Aquatic Biota (whole fish unless noted) | | | | | Comments |
	DL	Range	Median	% Detection Sites	Samples	DL	Range (µg/kg wet weight)	Median	
Chlordane, total*	—	—	—	100†	98†	0.2–1.0*	nd–290*§	nr	* Sum of *cis*-chlordane, *trans*-nonachlor, heptachlor, and heptachlor epoxide. † Mollusks, µg/kg dry weight. § Range of annual mean values. Objective: to assess distribution and trends of contaminants in United States estuaries and coastal areas. Mean levels in mussels and sediment were significantly correlated (0.01) for total DDT, total chlordane, and dieldrin.
DDT, total	—	—	—	100†	99†	0.2–1.5*	0.7–1,600*§	nr	
Dieldrin, total	—	—	—	99†	91†	0.2–1.7*	nd–120*§	nr	
Hexachloro-benzene	—	—	—	72†	23†	0.2–2.4*	nd–34*§	nr	
Lindane	—	—	—	97†	70†	0.1–1.7*	nd–41*§	nr	
Mirex	—	—	—	72†	30†	0.2–2.5*	nd–17*§	nr	
Aldrin	nr	nr	nr	0*	0*	20	nd*†	nd*†	* Fish liver, µg/kg dry weight. † Range and median are of mean values per site. § Sum of *cis*-chlordane and *trans*-nonachlor. Objective: to investigate fish pathology in association with contaminant residues in United States estuaries and coastal areas. Results were published as overview (1988) and detailed analysis (1989). DDTs were ≤DLs in sediment from reference sites and most Alaska sites. For each fish species, organochlorine levels were 5–15 times lower in stomach contents than in livers. Fish pathology occurred most often at most contaminated sites.
Chlordane, total§	2	nd–13†	nd†	100*	nr	nr	11–1,200*†	72*†	
DDT, total	5	nd–620†	2.2†	100*	nr	nr	30–21,000*†	970*†	
Dieldrin	1	nd–1.2†	nd†	94*	nr	5	nd–350*†	11*†	
Heptachlor, total	nr	nr	nr	0*	0*	20	nd*†	nd*†	
Hexachloro-benzene	1	nd–3.3†	0.17†	87*	nr	2	nd–140*†	4.3*†	
Lindane	nr	nr	nr	0*	0*	20	nd*†	nd*†	
Mirex	nr	nr	nr	0*	0*	20	nd*†	nd*†	

Table 2.1. Pesticides in bed sediment and aquatic biota from rivers and estuaries in the United States: National and multistate monitoring studies—*Continued*

Reference	Sampling Dates	Sampling Locations and Strategy	Sampling Media and Species	Number of Samples	Target Analytes	Bed Sediment	
						% Detection	
						Sites	Samples
Sericano and others, 1990a Mussel Watch Project	1986–1987	51 sites on the Gulf of Mexico coast (Florida to Texas). Sites were selected to represent broad range of environmental conditions and possible loadings, but point sources were avoided. Oysters were composites of 20 individual oysters per station at 3 stations per site.	Sediment (surficial), mollusks. Species: oysters (*Crassostrea virginica*).	301 147	Aldrin	nr	10–19
					Chlordane, *cis*-	nr	72–77
					DDD, *o,p´*-	nr	42–56
					DDD, p,p´-	nr	66–86
					DDE, *o,p´*-	nr	13–20
					DDE, *p,p´*-	nr	86–96
					DDT, *o,p´*-	nr	30
					DDT, *p,p´*-	nr	44–71
					DDT, total	84	nr
					Dieldrin	nr	43–73
					Heptachlor	nr	17–27
					Heptachlor epoxide	nr	17–22
					Hexachlorobenzene	nr	65–72
					Lindane	nr	19–20
					Mirex	nr	32–48
					non-DDT, total	100	nr
					Nonachlor, *trans*-	nr	70–71

Table 2.1. Pesticides in bed sediment and aquatic biota from rivers and estuaries in the United States: National and multistate monitoring studies—*Continued*

Target Analytes	Bed Sediment (µg/kg dry weight)			Aquatic Biota (whole fish unless noted)					Comments
				% Detection		(µg/kg wet weight)			
	DL	Range	Median	Sites	Samples	DL	Range	Median	
Aldrin	0.02	nd–2.87	nd	nr	9*	0.25	nd–6.66*	nd*	* Mollusks, µg/kg dry weight.
Chlordane, *cis*-	0.02	nd–43.5	0.07†	100*	100*	0.25	0.65–292*	5.83†*	† Value shown is mean of median values for
DDD, *o,p′*-	0.02	nd–319	0.02†	nr	73–89*	0.25	nd–975*	1.66†*	1986 and 1987.
DDD, p,p′-	0.02	nd–2,240	0.25†	nr	91–98*	0.25	nd–1,310*	8.32†*	Objective: to determine
DDE, *o,p′*-	0.02	nd–29.3	nd	nr	24–39*	0.25	nd–64*	nd*	the status and trends of
DDE, *p,p′*-	0.02	nd–195	0.27†	100*	100*	0.25	0.6–1,170*	11.8†*	environmental
DDT, *o,p′*-	0.02	nd–49.5	nd	nr	17–30*	0.25	nd–22.2*	nd*	contaminants in Gulf
DDT, *p,p′*-	0.02	nd–691	0.03†	nr	36–68*	0.25	nd–38.6*	0.38†*	Coast estuaries. The
DDT, total	0.02	nd–3,270	0.88†	100*	nr	0.25	3.0–3,570*	26.3†*	highest organo-
Dieldrin	0.02	nd–9.47	0.06†	nr	95–99*	0.25	nd–52.2*	4.71†*	chlorine levels in
Heptachlor	0.02	nd–1.14	nd	nr	23–37*	0.25	nd–7.04*	nd*	sediment were east of
Heptachlor epoxide	0.02	nd–3.82	nd	nr	86–98*	0.25	nd–27.3*	2.16†*	the Mississippi River, especially in Florida.
Hexachloro-benzene	0.02	nd–3.62	0.05†	nr	14–16*	0.25	nd–4.33*	nd	Organochlorine levels were higher in oysters
Lindane	0.02	nd–1.74	nd	nr	80–82*	0.25	nd–9.06*	0.97†*	than in sediment.
Mirex	0.02	nd–3.58	nd	nr	38–49*	0.25	nd–16.1*	nd*	Average DDT/DDD/
non-DDT, total	0.02	nd–89.4	0.72†	100*	nr	0.25	3.26–623*	22.5†*	DDE proportions were 5/52/44 in sediment
Nonachlor, *trans*-	0.02	nd–31.4	0.07†	100*	99–100*	0.25	nd–289*	4.68†*	and 22/64/14 in oysters. Total DDT contained 80% *p,p′* isomers.

Table 2.1. Pesticides in bed sediment and aquatic biota from rivers and estuaries in the United States: National and multistate monitoring studies—*Continued*

Reference	Sampling Dates	Sampling Locations and Strategy	Sampling Media and Species	Number of Samples	Target Analytes	Bed Sediment % Detection	
						Sites	Samples
Sericano and others, 1990b Mussel Watch Project	1988	63 sites on the Gulf of Mexico coast (Florida to Texas). Oysters were composites of 20 individual oysters per station at 3 stations per site.	Mollusks. Species: oysters (*Crassostrea virginica*).	189	DDD, *o,p´*- DDD, *p,p´*- DDE, *o,p´*- DDE, *p,p´*- DDT, *o,p´*- DDT, *p,p´*- DDT, total	— — — — — — —	— — — — — — —
Zdanowicz and Gadbois, 1990 Benthic Surveillance Project	1984–1986	20 sites on the Northeast coast (Maine to Virginia). 4 fish samples (3 liver, 1 stomach contents) and 3–5 sediment samples were taken per site per year. Fish samples were composites of 10 fish. Sediment samples were composites of 3 grabs per station at 3–5 stations per site.	Sediment (surficial), fish (liver, stomach contents). Species: spot, longhorn sculpin, Atlantic croaker, winter flounder, and windowpane.	162 126 44	Aldrin Chlordane, *cis*- DDD, *o,p´*- DDD, *p,p´*- DDE, *o,p´*- DDE, *p,p´*- DDT, *o,p´*- DDT, *p,p´*- Dieldrin Heptachlor Heptachlor epoxide Hexachlorobenzene Lindane Mirex Nonachlor, *trans*-	85 90 70 85 55 95 65 85 65 65 25 100 95 50 90	39 69 44 65 13 75 19 56 26 19 5 76 57 21 63

Table 2.1. Pesticides in bed sediment and aquatic biota from rivers and estuaries in the United States: National and multistate monitoring studies—*Continued*

Target Analytes	Bed Sediment (µg/kg dry weight)			Aquatic Biota (whole fish unless noted)					Comments
				% Detection		(µg/kg wet weight)			
	DL	Range	Median	Sites	Samples	DL	Range	Median	
DDD, *o,p'*-	—	—	—	nr	92*	0.25	nd–220*	2.6*	* Mollusks, µg/kg dry weight. Objective: to monitor status and trends of DDT compounds in sediment and mollusks on the Gulf Coast. Reference has data for 1986–1988. Because 1986–1987 data are in Sericano and others (1990a), only 1988 data are shown here. The Mississippi River is an important source of total DDT to the Gulf. Comparison with historical data indicated that mean total DDT levels in Gulf Coast oysters peaked in 1968, then declined, although levels at some sites remain high.
DDD, *p,p'*-	—	—	—	nr	99*	0.25	nd–860*	7.8*	
DDE, *o,p'*-	—	—	—	nr	72*	0.25	nd–250*	nd*	
DDE, *p,p'*-	—	—	—	100*	100*	0.25	0.42–370*	15*	
DDT, *o,p'*-	—	—	—	nr	79*	0.25	nd–20*	0.49*	
DDT, *p,p'*-	—	—	—	nr	70*	0.25	nd–42*	nd*	
DDT, total	—	—	—	100*	100*	0.25	2.4–1,400*	32*	
Aldrin	0.02–0.8	nd–10	nr	28*	10*	0.01–6.0*	nd–11*	nd*	* Fish liver, µg/kg dry weight. Objective: to assess status, spatial distribution, and temporal trends in contaminant residues in estuaries and coastal areas in the northeastern United States. Data are presented without interpretation. Data tables in reference do not specify species of each sample.
Chlordane, *cis-*	0.02–0.4	nd–17	nr	100*	95*	0.04–1.5*	nd–498*	nr	
DDD, *o,p'*-	0.04–0.9	nd–26	nd	94*	81*	0.1–2.9*	nd–48*	nr	
DDD, *p,p'*-	0.03–0.7	nd–235	nr	94*	69*	0.2–7.0*	nd–450*	nr	
DDE, *o,p'*-	0.02–1.1	nd–358	nd	28*	7*	0.1–5.3*	nd–36*	nd*	
DDE, *p,p'*-	0.02–0.3	nd–26	nr	100*	99*	0.04	nd–1,845*	nr	
DDT, *o,p'*-	0.03–0.9	nd–19	nd	100*	90*	0.1–2.5*	nd–54*	nr	
DDT, *p,p'*-	0.02–0.7	nd–15	nr	100*	88*	0.1–2.2*	nd–98*	nr	
Dieldrin	0.02–0.8	nd–4.1	nd	100*	93*	0.1–2.0*	nd–794*	nr	
Heptachlor	0.02–0.8	nd–5.8	nd	56*	20*	0.09–2.5*	nd–11*	nd*	
Heptachlor epoxide	0.03–1.0	nd–2.0	nd	100*	80*	0.1–0.9*	nd–40*	nr	
Hexachloro-benzene	0.02–0.2	nd–12	nr	100*	99*	0.4*	nd–39*	nr	
Lindane	0.02–0.7	nd–15	nr	100*	79*	0.05–3.7*	nd–156*	nr	
Mirex	0.02–0.9	nd–6.0	nd	100*	91*	0.1–1.5*	nd–20*	nr	
Nonachlor, *trans*	0.02–0.4	nd–21	nr	100*	99*	0.04	nd–565*	nr	

Table 2.1. Pesticides in bed sediment and aquatic biota from rivers and estuaries in the United States: National and multistate monitoring studies—*Continued*

Reference	Sampling Dates	Sampling Locations and Strategy	Sampling Media and Species	Number of Samples	Target Analytes	Bed Sediment % Detection	
						Sites	Samples
National Oceanic and Atmospheric Administration, 1991; National Oceanic and Atmospheric Administration, 1988 Mussel Watch and Benthic Surveillance Projects	1984–1989	285 sites in coastal and estuarine areas throughout the United States. Site selection was biased towards urban areas, but major point sources were avoided. 3 composite samples were collected (from 3 stations) per site. Only data from sediment samples containing >20% fine-grained material (<63 μm) were used.	Sediment (surficial), adjusted by the weight-fraction less than 63 μm.	1,426	Chlordane, total† DDT, total Dieldrin, total Hexachlorobenzene Lindane Mirex	79 96 73 77 57 47	61 83 48 50 27 22

Table 2.1. Pesticides in bed sediment and aquatic biota from rivers and estuaries in the United States: National and multistate monitoring studies—*Continued*

Target Analytes	Bed Sediment			Aquatic Biota (whole fish unless noted)					Comments
	(µg/kg dry weight)			% Detection		(µg/kg wet weight)			
	DL	Range	Median	Sites	Samples	DL	Range	Median	
Chlordane, total†	0.09–0.6	nd–93†	nr	—	—	—	—	—	* Sum of *cis*-chlordane, *trans*-nonachlor, heptachlor, and heptachlor epoxide.
DDT, total	0.1–1.0	nd–5,900†	6,600§	—	—	—	—	—	† Range of the mean grain-size adjusted concentrations for all sites. Extremely high values were deleted.
Dieldrin, total	0.02–1.3	nd–23†	nr	—	—	—	—	—	
Hexachloro-benzene	0.1–0.9	nd–6.4†	nr	—	—	—	—	—	
Lindane	0.1–1.5	nd–5.7†	nr	—	—	—	—	—	§ Value is geometric mean. Objective: to define geographic distribution of contaminants in sediment from United States estuaries and coastal areas. Results in National Oceanic and Atmospheric Administration (1988) are a subset of those in National Oceanic and Atmospheric Administration (1991), so the latter are shown here. Most of the highest contaminant concentrations were found near urban areas.
Mirex	0.1–1.2	nd–8.5†	nr	—	—	—	—	—	

Table 2.1. Pesticides in bed sediment and aquatic biota from rivers and estuaries in the United States: National and multistate monitoring studies—*Continued*

Reference	Sam-pling Dates	Sampling Locations and Strategy	Sampling Media and Species	Number of Samples	Target Analytes	Bed Sediment % Detection	
						Sites	Samples
O'Connor, 1992 Mussel Watch Project	1990	214 sites nationwide, on the Atlantic (Maine to Florida), Gulf (Florida to Texas), and Pacific (California to Alaska) coasts, and Hawaii: Almost half the sites were in or near urban areas.	Mollusks. Species: 2 species of mussels and 2 species of oysters.	nr	Chlordane, total* DDT, total	— —	— —

Table 2.1. Pesticides in bed sediment and aquatic biota from rivers and estuaries in the United States: National and multistate monitoring studies—*Continued*

| Target Analytes | Bed Sediment (µg/kg dry weight) | | | Aquatic Biota (whole fish unless noted) | | | | | Comments |
| | DL | Range | Median | % Detection | | (µg/kg wet weight) | | | |
				Sites	Samples	DL	Range	Median	
Chlordane, total*	—	—	—	~37†	nr	nr	nd—~120†	14†§	* Sum of *cis*-chlordane, *trans*-nonachlor, heptachlor, and hepta-chlor epoxide.
DDT, total	—	—	—	nr	nr	nr	nr	37†§	† Mollusks, µg/kg dry weight. Approximate values (~) were esti-mated visually from figures in reference.

§ Value is geometric mean. Objective: to assess spatial distribu-tion and trends of con-taminants in sediment and mollusks in United States estuaries and coastal areas. Report contains 1986–1990 data, but all except 1990 data (summarized here) were published elsewhere. Contami-nant levels were high in urban areas. Spear-man correlations were significant between organochlorine con-centrations and popu-lation within 20 km.

Table 2.1. Pesticides in bed sediment and aquatic biota from rivers and estuaries in the United States: National and multistate monitoring studies—*Continued*

Reference	Sam-pling Dates	Sampling Locations and Strategy	Sampling Media and Species	Number of Samples	Target Analytes	Bed Sediment	
						% Detection	
						Sites	Samples
Johnson and others, 1992a; Johnson and others, 1993 Benthic Surveillance Project	1987–1989	22 sites on the Northeast coast (Massachusetts to New Jersey). 14 sites were urban, and 8 sites were nonurban. At each site, 4 composite fish samples were analyzed: 3 liver and 1 stomach contents. 3–5 sediment samples were analyzed per site per year, each a composite of 3 grabs per station.	Sediment (surficial), fish (liver, stomach contents) Species: winter flounder.	nr nr nr	Chlordane, total* DDT, total	nr nr	nr nr

Table 2.1. Pesticides in bed sediment and aquatic biota from rivers and estuaries in the United States: National and multistate monitoring studies—*Continued*

| Target Analytes | Bed Sediment (µg/kg dry weight) | | | Aquatic Biota (whole fish unless noted) | | | | | Comments |
| | DL | Range | Median | % Detection | | (µg/kg wet weight) | | | |
				Sites	Samples	DL	Range	Median	
Chlordane, total*	nr	nr	nr	nr	nr	nr	nr	nr	* Sum of *cis*-chlordane and *trans*-nonachlor.
DDT, total	nr	nr	nr	nr	nr	nr	nr	nr	Objective: to examine relations between hepatic lesions and contaminant concentrations. Contaminant data were not reported. DDTs, chlordanes, and PCBs tended to occur together at urban sites. Levels in fish liver were 2–3 times higher than in stomach contents, which were 10 times higher than levels in sediment. DDTs and chlordanes were significant risk factors for certain fish lesions. Biological risk factors for hepatic lesions were age, sex, and season.

Table 2.1. Pesticides in bed sediment and aquatic biota from rivers and estuaries in the United States: National and multistate monitoring studies—*Continued*

Reference	Sampling Dates	Sampling Locations and Strategy	Sampling Media and Species	Number of Samples	Target Analytes	Bed Sediment	
						% Detection	
						Sites	Samples
Myers and others, 1993 Benthic Surveillance Project	1984–1988	45 sites on the Pacific coast (California to Alaska). 23 sites were in or near urban embayments. 22 sites were nonurban, including 5 reference sites. Sediment and fish liver samples were composites.	Sediment (surficial), fish (liver, stomach contents). Species: flathead sole, English sole, starry flounder, hornyhead turbot, white croaker, and black croaker.	nr nr nr	Chlordane, total* DDT, total Dieldrin Hexachlorobenzene	nr nr nr nr	nr nr nr nr

Table 2.1. Pesticides in bed sediment and aquatic biota from rivers and estuaries in the United States: National and multistate monitoring studies—*Continued*

| Target Analytes | Bed Sediment (µg/kg dry weight) | | | Aquatic Biota (whole fish unless noted) | | | | | Comments |
| | DL | Range | Median | % Detection | | (µg/kg wet weight) | | | |
				Sites	Samples	DL	Range	Median	
Chlordane, total*	nr	nr	nr	nr	nr	nr	nr	nr	* Sum of *cis*-chlordane and *trans*-nonachlor. Objective: to assess the incidence of fish pathology in association with contaminant exposure. Contaminant data are not reported. Toxicopathic hepatic lesions were detected primarily in fish from urban sites. DDTs, chlordanes, dieldrin, PCBs, and aromatic hydrocarbons each were risk factors for some types of hepatic lesions in some fish species. Fish age must be accounted for in evaluating influence of contaminant exposure. DDTs, PCBs, aromatic hydrocarbons, and some metals tended to covary in all media sampled.
DDT, total	nr	nr	nr	nr	nr	nr	nr	nr	
Dieldrin	nr	nr	nr	nr	nr	nr	nr	nr	
Hexachloro-benzene	nr	nr	nr	nr	nr	nr	nr	nr	

Table 2.1. Pesticides in bed sediment and aquatic biota from rivers and estuaries in the United States: National and multistate monitoring studies—*Continued*

Reference	Sampling Dates	Sampling Locations and Strategy	Sampling Media and Species	Number of Samples	Target Analytes	Bed Sediment % Detection	
						Sites	Samples
Myers and others, 1994 Benthic Surveillance Project	1984–1988	27 sites on the Pacific coast (California to Alaska). Sites were depositional areas with multiple contaminant sources. 19 sites were in or near urban areas. 8 sites were nonurban, including 3 reference sites. Composite sediment and fish samples were taken.	Sediment (surficial), fish (liver, stomach contents). Species: English sole, starry flounder, and white croaker.	nr nr nr	Chlordane, total* DDT, total Dieldrin Hexachlorobenzene	nr nr nr nr	nr nr nr nr

Table 2.1. Pesticides in bed sediment and aquatic biota from rivers and estuaries in the United States: National and multistate monitoring studies—*Continued*

Target Analytes	Bed Sediment (µg/kg dry weight)			Aquatic Biota (whole fish unless noted)					Comments
				% Detection		(µg/kg wet weight)			
	DL	Range	Median	Sites	Samples	DL	Range	Median	
Chlordane, total*	nr	1–15†	nr	nr	nr	nr	190–1,700†§	nr	* Sum of *cis*-chlordane and *trans*-nonachlor.
DDT, total	nr	17–670@	nr	nr	nr	nr	1,100–26,000†§	nr	† Range is of maximal mean concentrations (1984–1988) collected in 4 urban areas with highest organic contamination: Puget Sound, San Francisco Bay, Los Angeles, and San Diego.
Dieldrin	nr	0.1–3@	nr	nr	nr	nr	34–300†§	nr	§ Fish liver (dry weight). Objective: to examine relation between toxicopathic hepatic lesions and chemical contaminants in sediment and various fish tissues. Toxicopathic hepatic lesions occurred primarily in urban areas. Some hepatic lesions in some species were associated with DDTs, chlordanes, dieldrin, PCBs, and aromatic hydrocarbons. DDTs, PCBs, aromatic hydrocarbons, and some metals tended to covary in all media sampled.
Hexachloro-benzene	nr	0.1–5@	nr	nr	nr	nr	4–150†§	nr	

Table 2.1. Pesticides in bed sediment and aquatic biota from rivers and estuaries in the United States: National and multistate monitoring studies—*Continued*

Reference	Sampling Dates	Sampling Locations and Strategy	Sampling Media and Species	Number of Samples	Target Analytes	Bed Sediment % Detection	
						Sites	Samples
USEPA's National Study of Chemical Residues in Fish							
U.S. Environmental Protection Agency, 1992a	1986–1987	388 sites in rivers, lakes, estuaries, and coastal areas throughout the United States. 314 sites were targeted (point and nonpoint source pollution), 39 were NASQAN (large river) sites, and 35 were background sites.	Fish (whole bottom-feeders; gamefish fillets) shellfish. Species: 119 species. Most samples belong to 14 species, especially carp and largemouth bass.	560 total	Chlordane, *cis-*	—	—
					Chlordane, *trans-*	—	—
					Chlorpyrifos	—	—
					DDE	—	—
					Dicofol	—	—
					Dieldrin	—	—
					Endrin	—	—
					HCH, α-	—	—
					Heptachlor	—	—
					Heptachlor epoxide	—	—
					Hexachlorobenzene	—	—
					Isopropalin	—	—
					Lindane	—	—
					Methoxychlor	—	—
					Mirex	—	—
					Nitrofen	—	—
					Nonachlor, *cis-*	—	—
					Nonachlor, *trans-*	—	—
					Oxychlordane	—	—
					PCA	—	—
					PCNB	—	—
					Perthane	—	—
					TCDD, 2,3,7,8-	—	—
					Trifluralin	—	—

Table 2.1. Pesticides in bed sediment and aquatic biota from rivers and estuaries in the United States: National and multistate monitoring studies—*Continued*

Target Analytes	Bed Sediment (µg/kg dry weight)			Aquatic Biota (whole fish unless noted)					Comments
				% Detection		(µg/kg wet weight)			
	DL	Range	Median	Sites	Samples	DL	Range	Median	
USEPA's National Study of Chemical Residues in Fish—Continued									
Chlordane, cis-	—	—	—	64*	59*	0.17–2.5	nd–378*	3.7*†	* Data shown are for whole and fillet fish samples (combined). † Median of the highest concentration at each site. Objective: to determine the prevalence of bioaccumulative pollutants in fish and to identify correlations with pollutant sources. Associations with agricultural sites occurred for DDE, chlorpyrifos, dicofol, and methoxychlor. Associations with industrial or urban sites occurred for DDE, chlordane, dieldrin, lindane, and HCB. Mirex was detected primarily in the Southeast and Great Lakes areas. Whole-body bottomfish had higher mean concentrations than gamefish fillets for some analytes (chlordane, HCB, α-HCH, heptachlor epoxide, lindane, nonachlor, and PCA) and about the same for others (chlorpyrifos, dicofol, dieldrin, endrin, mirex, oxychlordane, DDE, and trifluralin)
Chlordane, trans-	—	—	—	61*	54*	0.26–2.5	nd–310*	2.7*†	
Chlorpyrifos	—	—	—	26*	24*	0.34–2.5	nd–344*	nd*†	
DDE	—	—	—	99*	98*	0.38–2.5	nd–14,028*	58*†	
Dicofol	—	—	—	16*	12*	0.26–2.5	nd–74*	nd*†	
Dieldrin	—	—	—	60*	56*	0.28–2.5	nd–450*	4.2*†	
Endrin	—	—	—	11*	8*	0.64–2.5	nd–162*	nd*†	
HCH, α-	—	—	—	55*	44*	0.25–2.5	nd–44*	0.72*†	
Heptachlor	—	—	—	2*	2*	0.28–2.5	nd–76*	nd*†	
Heptachlor epoxide	—	—	—	16*	13*	1.18–2.5	nd–63*	nd*†	
Hexachloro-benzene	—	—	—	46*	38*	0.09–2.5	nd–913*	nd*†	
Isopropalin	—	—	—	4*	3*	2.32–2.5	nd–38*	nd*†	
Lindane	—	—	—	42*	34*	0.29–2.5	nd–83*	nd*†	
Methoxychlor	—	—	—	7*	5*	0.30–2.5	nd–393*	nd*†	
Mirex	—	—	—	38*	31*	0.18–2.5	nd–225*	nd*†	
Nitrofen	—	—	—	3*	2*	0.24–2.5	nd–18*	nd*†	
Nonachlor, cis-	—	—	—	35*	32*	1.58–2.5	nd–127*	nd*†	
Nonachlor, trans-	—	—	—	77*	72*	1.5–2.5	nd–477*	9.2*†	
Oxychlordane	—	—	—	27*	23*	0.46–2.5	nd–243*	nd*†	
PCA	—	—	—	64*	54*	0.07–2.5	nd–647*	0.92*†	
PCNB	—	—	—	1*	1*	0.86–2.5	nd–16*	nd*†	
Perthane	—	—	—	1*	1*	0.29–2.5	nd–5.1*	nd*†	
TCDD, 2,3,7,8-	—	—	—	70*	60*	0.07–7.9	nd–0.204*	0.001*†	
Trifluralin	—	—	—	12*	13*	1.79–2.5	nd–458*	nd*†	

Table 2.1. Pesticides in bed sediment and aquatic biota from rivers and estuaries in the United States: National and multistate monitoring studies—*Continued*

Reference	Sampling Dates	Sampling Locations and Strategy	Sampling Media and Species	Number of Samples	Target Analytes	Bed Sediment % Detection	
						Sites	Samples
Other Studies							
Stout, 1968	1967	9 coastal and estuarine sites in the Pacific Northwest (Oregon, Washington, and British Columbia). Most samples were composites.	Fish (whole, fillets, carcasses— minus fillets), crab. Species include cod, yellowtail rockfish, hake, starry flounder, and English sole.	25 12 4 5	DDD DDE DDT	— — —	— — —
Barthel and others, 1969	1964– 1967	11 sites on the lower Mississippi River and 43 sites on its tributaries (Tennessee, Louisiana, Mississippi, and Arkansas). In 1967, samples were taken only from two contaminated sites. Samples were taken upstream and downstream of potential sources. 12 samples were taken at each site: at 2 water depths and from both banks at each of 3 cross-sections.	Sediment (surficial), water.	548 10	Aldrin	9.3	5.8
					Chlordane	11	5.1
					DDD, *o,p´-* DDD, *p,p´-*	28 44	6.4 13
					DDE, *p,p´-* DDT, *o,p´-* DDT, *p,p´-* Dieldrin	35 3.7 15 19	9.2 1.1 2.1 5.8
					Endrin	5.6	3.8
					Endrin ketone	1.9	1.9
					HCH Heptachlor Heptachlor epoxide	13 3.7 7.4	2.8 2.5 1.3
					Isodrin	17	6.2
					Toxaphene†	11	2.3

Table 2.1. Pesticides in bed sediment and aquatic biota from rivers and estuaries in the United States: National and multistate monitoring studies—*Continued*

| Target Analytes | Bed Sediment (µg/kg dry weight) | | | Aquatic Biota (whole fish unless noted) | | | | | Comments |
	DL	Range	Median	% Detection Sites	Samples	DL	Range	Median	
Other Studies—Continued									
DDD	—	—	—	100	100	nr	trace–244	nr	Objective: to monitor pesticide levels of edible fish in the Pacific Northwest. Yellowtail rockfish fillets contained lower pesticide residues than the remaining carcasses. Higher pesticide residues were detected in yellowtail rockfish from Ilwaco, Washington (Columbia River) than from Hecate Strait, British Columbia, where no major river enters the ocean.
DDE	—	—	—	100	100	nr	38–172	nr	
DDT	—	—	—	100	100	nr	trace–223	nr	
Aldrin	50*	nd–3,000*	nd	—	—				* In mg/kg dry wt. † Includes strobane. Objective: to determine the extent and possible sources of pesticides in streams of the Mississippi River Delta: Reported concentrations were adjusted for percent recovery. Two localized areas of contamination were identified: a manufacturing area near Memphis (Tennessee) and formulating plants in Mississippi. With one exception, DDTs were the only residues originating from nonindustrial (municipal and agricultural) sources.
Chlordane	50*	nd–30,000*	nd	—	—	—	—	—	
DDD, *o,p′-*	50*	nd–10.2*	nd	—	—	—	—	—	
DDD, *p,p′-*	50*	nd–17,400*	nd	—	—	—	—	—	
DDE, *p,p′-*	50*	nd–0.86*	nd	—	—	—	—	—	
DDT, *o,p′-*	50*	nd–0.41*	nd	—	—	—	—	—	
DDT, *p,p′-*	50*	nd–1.62*	nd	—	—	—	—	—	
Dieldrin	50*	nd–9,000*	nd	—	—	—	—	—	
Endrin	50*	nd–12,800*	nd	—	—	—	—	—	
Endrin ketone	50*	nd–3,509*	nd	—	—	—	—	—	
HCH	50*	nd–2.9*	nd	—	—	—	—	—	
Heptachlor	50*	nd–11.1*	nd	—	—	—	—	—	
Heptachlor epoxide	50*	nd–1,090*	nd	—	—	—	—	—	
Isodrin	50*	nd–12,000*	nd	—	—	—	—	—	
Toxaphene†	50*	nd–13*	nd	—	—	—	—	—	

Table 2.1. Pesticides in bed sediment and aquatic biota from rivers and estuaries in the United States: National and multistate monitoring studies—*Continued*

Reference	Sampling Dates	Sampling Locations and Strategy	Sampling Media and Species	Number of Samples	Target Analytes	Bed Sediment % Detection	
						Sites	Samples
Holden, 1970	1967–1968	Samples collected in 11 countries from areas with no history of local pesticide use: Canada, Italy, Finland, Ireland, Netherlands, Norway, Portugal, Spain, Sweden, United Kingdom, and United States. Mussels and fish were sampled before spawning.	Fish (muscle, liver), mollusks, birds. Aquatic species: pike, dogfish, and mussel.	>179 total	DDD DDE DDT Dieldrin	— — — —	— — — —
Giam and others, 1972	1971	17 sites on the Gulf of Mexico coast (Florida, Louisiana, and Texas) and Caribbean Sea. Samples were composites of two or more fish or six invertebrates.	Fish (whole, liver, gonads, gut, muscle) and invertebrates. Species: more than 25 species, including tuna, flounder, red snapper, crab, squid, shark, and shrimp.	20 11, 5 4, 11 13	DDD, *p,p´-* DDE, *p,p´-* DDT, *p,p´-* DDT, total	— — — —	— — — —

Table 2.1. Pesticides in bed sediment and aquatic biota from rivers and estuaries in the United States: National and multistate monitoring studies—*Continued*

Target Analytes	Bed Sediment (µg/kg dry weight)			Aquatic Biota (whole fish unless noted)					Comments
				% Detection		(µg/kg wet weight)			
	DL	Range	Median	Sites	Samples	DL	Range	Median	
DDD	—	—	—	nr	nr	0.1–10	nd–1,500*	nr	* Data shown are for fish and mollusk data, combined. Objective: to measure pesticides in species from areas that are relatively free of local contamination and to compare results from different laboratories. Inter-laboratory comparison of test samples showed coefficient of variation of +25% to +60%. Current levels were between 10–100 µg/kg; levels above 100 µg/kg probably indicated local contamination.
DDE	—	—	—	nr	nr	0.1–10	nd–3,720*	nr	
DDT	—	—	—	nr	nr	0.1–10	nd–6,760*	nr	
Dieldrin	—	—	—	nr	nr	0.1–10	nd–46*	nr	
DDD, *p,p′*-	—	—	—	0	0	0.1–0.3	nd	nd	Objective: to measure concentrations of DDT compounds in biota from the Gulf of Mexico, which receives runoff from 2/3 of the United States and 1/2 of Mexico. Levels generally were higher in samples from coastal areas than from the open ocean. DDE/DDT ratios varied widely. In individual fish, DDT and DDE levels generally were highest in liver, appre-ciable in gonads and digestive tract, and lowest in muscle tissue.
DDE, *p,p′*-	—	—	—	100	100	0.1–0.3	3.3–65	7.7	
DDT, *p,p′*-	—	—	—	100	95	0.1–0.3	nd–141	19	
DDT, total	—	—	—	100	100	0.1–0.3	7–159	38	

Table 2.1. Pesticides in bed sediment and aquatic biota from rivers and estuaries in the United States: National and multistate monitoring studies—*Continued*

Reference	Sampling Dates	Sampling Locations and Strategy	Sampling Media and Species	Number of Samples	Target Analytes	Bed Sediment % Detection	
						Sites	Samples
Holden, 1973	1969–1971	72 sites in 10 countries, including 2 sites in the United States. Sites were in known polluted areas and in areas with no history of local pesticide use. Samples were composites.	Fish (whole, muscle), mollusks, and birds (eggs). Aquatic species: pike, eel, herring, sprat, and mussel (*Mytilus edulis*).	46 (total fish) 35 18	DDD DDE DDT, *p,p´*- Dieldrin	— — — —	— — — —
Johnson and others, 1974	1971–1972	Unspecified number of sites from 17 states throughout the United States. Some samples were from the FWS's NCBP program; others were obtained from various state and national agencies or laboratories.	Fish (whole, eggs, fry, and oil). Species: paddlefish, bigmouth buffalo, coho salmon, channel catfish, carp, white perch, striped bass, and lake trout.	55 18, 2 4	Hexachlorobenzene	—	—

Table 2.1. Pesticides in bed sediment and aquatic biota from rivers and estuaries in the United States: National and multistate monitoring studies—*Continued*

Target Analytes	Bed Sediment (µg/kg dry weight)			Aquatic Biota (whole fish unless noted)					Comments
				% Detection		(µg/kg wet weight)			
	DL	Range	Median	Sites	Samples	DL	Range	Median	
DDD	—	—	—	69*	68*	1–10	nd–510*	16*	* Data are for all fish samples (whole and fillet combined). Objective: to monitor pesticides in fish and wildlife from 10 countries. Concentration ranges from polluted and unpolluted areas overlapped. Levels in herring were lower than in pike or mussels from same areas. Levels in eel were higher than in pike, perhaps because of higher lipid content.
DDE	—	—	—	100*	100*	nr	1–1,100*	117.5*	
DDT, *p,p´-*	—	—	—	69*	73*	5–6	nd–1,000*	22*	
Dieldrin	—	—	—	51*	50*	1–20	nd–950*	1.5*	
Hexachloro-benzene	—	—	—	nr	nr	1	nd–62,000	nd–16,000*	* Range of mean values for 8 species (whole fish only). Objective: to identify an unknown contaminant in NCBP samples and to look for it in other fish sampled in the United States. The unknown was identified as HCB using GC/MS. Excluding one sample (near a chemical storage area), the range was nd–1,000 µg/kg in whole fish, nd–630 µg/kg in fish eggs, and 2–90 µg/kg in fish fry.

Table 2.1. Pesticides in bed sediment and aquatic biota from rivers and estuaries in the United States: National and multistate monitoring studies—*Continued*

Reference	Sampling Dates	Sampling Locations and Strategy	Sampling Media and Species	Number of Samples	Target Analytes	Bed Sediment % Detection	
						Sites	Samples
Markin and others, 1974a	1971	9 areas in the southeastern United States, 7 inside (Mississippi, Alabama, Florida, Georgia, South Carolina, and North Carolina) and 2 outside (Delaware and Maryland) mirex treatment areas. Most samples were composites of individuals from 3 or more sources in the sampling area.	Fish (whole), oysters, squid, clams, crabs, scallops, shrimp, fish meal, and fish oil. Fish species included shad, mullet, red snapper, flounder, and striped bass.	46 4 4, 2 9 1 1, 3 1	DDD DDE DDT DDT, total Mirex	— — — — —	— — — — —
Veith and others, 1979b	1976	57 sites from the lower reaches of major drainages in 18 states throughout the United States. Samples were mixed-species composites.	Fish (whole). Species included shad, sucker, carp, sunfish, crappie, gar, goldfish, catfish, rock bass, perch, buffalo, and eel.	58	Chlordane, *cis-* Chlordane, *trans-* DDT, total Hexachlorobenzene Nonachlor, *cis-* Nonachlor, *trans-*	— — — — — —	— — — — — —

Table 2.1. Pesticides in bed sediment and aquatic biota from rivers and estuaries in the United States: National and multistate monitoring studies—*Continued*

Target Analytes	Bed Sediment (µg/kg dry weight)			Aquatic Biota (whole fish unless noted)					Comments
	DL	Range	Median	% Detection		DL	Range	Median	
				Sites	Samples		(µg/kg wet weight)		
DDD	—	—	—	100	74	1	nd–550	14	Objective: to survey
DDE	—	—	—	100	100	1	4–1,295	50	various types of sea-
DDT	—	—	—	100	93	1	nd–630	35	food and marine pro-
DDT, total	—	—	—	100	100	1	7–2,475	108	ducts to determine the
Mirex	—	—	—	11	11	5	nd–24	nd	extent of mirex con-
									tamination. Mirex was
									found in only 9 of 77
									total samples; all
									detections were in
									samples collected near
									Savannah, Georgia,
									which had extensive
									mirex use. Mirex is not
									as widespread a con-
									taminant of marine
									seafood as DDT and
									PCBs. Earlier reports
									of extensive mirex
									contamination of
									marine seafood may
									have resulted from
									PCB interference.
Chlordane, *cis-*	—	—	—	37	36	nr*	nr*	nr*	* Not quantitated because of poor
Chlordane, *trans-*	—	—	—	37	36	nr*	nr*	nr*	resolution. Objective: to identify major PCB
DDT, total	—	—	—	91	91	50	nd–4,530	610	components and other
Hexachloro-benzene	—	—	—	53	52	5	nd–11,600	6	contaminants in fish from United States
Nonachlor, *cis-*	—	—	—	35	34	nr*	nr*	nr*	watersheds. Other compounds found in
Nonachlor, *trans-*	—	—	—	37	36	nr*	nr*	nr*	one or more samples included PCA, pentachlorophenol, heptachlor, heptachlor epoxide, and chlordene.

Table 2.1. Pesticides in bed sediment and aquatic biota from rivers and estuaries in the United States: National and multistate monitoring studies—*Continued*

Reference	Sam-pling Dates	Sampling Locations and Strategy	Sampling Media and Species	Number of Samples	Target Analytes	Bed Sediment % Detection	
						Sites	Samples
Stout, 1980	1973–1975	9 sites off the southeast coast of the United States (Florida to Texas), the Atlantic Ocean, and Gulf of Mexico. Samples were composited by site and species (10 fish, equal weight from each).	Fish (skin-off fillets). Species: red snapper, red grouper, gag, king mackerel, and Spanish mackerel.	70	DDD, *p,p'*- DDE, *p,p'*- DDT, *p,p'*- DDT, total Dieldrin Endrin	— — — — — —	— — — — — —
Veith and others, 1981	1978	26 sites in 22 major rivers in 6 states near the Great Lakes. Samples were collected near mouths of rivers or confluence of major tributaries. Samples were mixed-species composites.	Fish (whole). Species included freshwater drum, walleye, carp, white sucker, and channel catfish.	26	Chlordane* DDE DDT, total Heptachlor Heptachlor epoxide Hexachlorobenzene Mirex Nonachlor* Oxychlordane PCA	— — — — — — — — — —	— — — — — — — — — —

Table 2.1. Pesticides in bed sediment and aquatic biota from rivers and estuaries in the United States: National and multistate monitoring studies—*Continued*

Target Analytes	Bed Sediment (µg/kg dry weight)			Aquatic Biota (whole fish unless noted)					Comments
				% Detection		(µg/kg wet weight)			
	DL	Range	Median	Sites	Samples	DL	Range	Median	
DDD, *p,p´-*	—	—	—	nr	nr	3	nr	nr	* Data are for fish fillets. Objective: to measure residues in fish from the northwestern Atlantic Ocean and northern Gulf of Mexico. DDT and PCB levels were lowest in red grouper and highest in king and Spanish mackerel. Organochlorine content was related to lipid content, although this was significant for only 3 of 6 species.
DDE, *p,p´-*	—	—	—	nr	nr	3	nr	nr	
DDT, *p,p´-*	—	—	—	nr	nr	3	nr	nr	
DDT, total	—	—	—	100*	nr	3	nd–996*	nr	
Dieldrin	—	—	—	89*	21*	3	nd–26*	nd*	
Endrin	—	—	—	89*	21*	3	nd–27*	nd*	
Chlordane*	—	—	—	38	38	0.5	nd–2,680	nd	* Sum of *cis* and *trans* isomers. Objective: to estimate relative contaminant levels in major watersheds near the Great Lakes. Two intermediate compounds in the manufacture of cyclodiene pesticides were detected in fish from the Wabash River (Indiana). Total DDT residues were <1,000 µg/kg in 81% of samples.
DDE	—	—	—	100	100	nr	10–1,000	230	
DDT, total	—	—	—	100	100	0.5	30–1,660	425	
Heptachlor	—	—	—	23	23	0.5	nd–77	nd	
Heptachlor epoxide	—	—	—	4	4	0.5	nd–5	nd	
Hexachloro-benzene	—	—	—	65	65	0.5	nd–447	8	
Mirex	—	—	—	4	4	0.5	nr	nr	
Nonachlor*	—	—	—	58	58	0.5	nd–3,070	17	
Oxychlordane	—	—	—	12	12	0.5	nd–167	nd	
PCA	—	—	—	58	58	0.5	nr	nr	

Table 2.1. Pesticides in bed sediment and aquatic biota from rivers and estuaries in the United States: National and multistate monitoring studies—*Continued*

Reference	Sam-pling Dates	Sampling Locations and Strategy	Sampling Media and Species	Number of Samples	Target Analytes	Bed Sediment % Detection	
						Sites	Samples
Kuehl and others, 1983	1979	48 sites in major watersheds in 7 states near the Great Lakes. Sites included known problem areas and rivers for which little data were available. Sites were located near the mouths of rivers or confluence of major tributaries. Samples were composites; a few samples contained more than one species.	Fish (whole, fillets). Species included northern pike, carp, white sucker, lake trout, catfish, smallmouth bass, bullhead, walleye, and northern pike.	46 2	Chlordane, *cis-*	—	—
					Chlordane, *trans-*	—	—
					Chlordene	—	—
					DDD, *o,p´-*	—	—
					DDD, *p,p´-*	—	—
					DDE, *o,p´-*	—	—
					DDE, *p,p´-*	—	—
					DDMU, *p,p´-*	—	—
					DDT, *o,p´-*	—	—
					DDT, *o,p´-*	—	—
					Endrin	—	—
					HCH	—	—
					Heptachlor epoxide	—	—
					Hexachlorobenzene	—	—
					Mirex	—	—
					Nonachlor, *cis-*	—	—
					Nonachlor, *trans-*	—	—
					Oxychlordane	—	—
					PCA	—	—
					Pentachlorophenol	—	—
					Photomirex	—	—

Table 2.1. Pesticides in bed sediment and aquatic biota from rivers and estuaries in the United States: National and multistate monitoring studies—*Continued*

| Target Analytes | Bed Sediment (µg/kg dry weight) | | | Aquatic Biota (whole fish unless noted) | | | | | Comments |
| | DL | Range | Median | % Detection | | (µg/kg wet weight) | | | |
				Sites	Samples	DL	Range	Median	
Chlordane, *cis-*	—	—	—	74	74	20	nd–610	28	Objective: to measure contaminants in fish from watersheds of the Great Lakes. The highest chlordane residues were from the Mississippi River (Illinois) and Mad River (Ohio). DDE residues >1,000 µg/kg were found in Lakes Ontario and Michigan, Fox River, Otter Creek (Illinois), Mad River, Muskingum River (Ohio), Milwaukee River, and Pike River (Wisconsin). The highest HCB residues were found in Fields Brook (Ohio). Different types and levels of contaminants were often found in fish from different parts of the same watershed. Unusual industrial products, byproducts, or degradation products were identified.
Chlordane, *trans-*	—	—	—	87	87	20	nd–520	56	
Chlordene	—	—	—	7	7	nr	nr	nr	
DDD, *o,p'-*	—	—	—	4	4	nr	nr	nr	
DDD, *p,p'-*	—	—	—	14	14	nr	nr	nr	
DDE, *o,p'-*	—	—	—	14	14	nr	nr	nr	
DDE, *p,p'-*	—	—	—	93	93	15	nd–5,800	105	
DDMU, *p,p'-*	—	—	—	4	4	nr	nr	nr	
DDT, *o,p'-*	—	—	—	4	4	nr	nr	nr	
DDT, *o,p'-*	—	—	—	25	25	nr	nr	nr	
Endrin	—	—	—	4	4	nr	nr	nr	
HCH	—	—	—	11	11	nr	nr	nr	
Heptachlor epoxide	—	—	—	11	11	nr	nr	nr	
Hexachloro-benzene	—	—	—	65	65	5	nd–1,500	5.5	
Mirex	—	—	—	4	4	nr	nr	nr	
Nonachlor, *cis-*	—	—	—	89	89	20	nd–610	43	
Nonachlor, *trans-*	—	—	—	46	46	30	nd–190	nd	
Oxychlordane	—	—	—	4	4	nr	nr	nr	
PCA	—	—	—	14	14	nr	nr	nr	
Pentachloro-phenol	—	—	—	4	4	nr	nr	nr	
Photomirex	—	—	—	4	4	nr	nr	nr	

Table 2.1. Pesticides in bed sediment and aquatic biota from rivers and estuaries in the United States: National and multistate monitoring studies—*Continued*

Reference	Sam-pling Dates	Sampling Locations and Strategy	Sampling Media and Species	Number of Samples	Target Analytes	Bed Sediment % Detection	
						Sites	Samples
Clark and others, 1984	1980	12 sites in the Great Lakes Basin. Fish (adult) were collected as they began their autumn, upstream migration. Samples were composites of 5 fillets.	Fish (skin-on fillets). Species: coho salmon.	36	Chlordane, *cis-*	—	—
					Chlordane, *trans-*	—	—
					Chlorpyrifos	—	—
					Dacthal	—	—
					DDD, *p,p´-*	—	—
					DDE, *p,p´-*	—	—
					DDT, *p,p´-*	—	—
					DDT, total	—	—
					Diazinon	—	—
					Dieldrin	—	—
					Endrin	—	—
					HCH, α-	—	—
					Heptachlor	—	—
					Heptachlor epoxide	—	—
					Hexachlorobenzene	—	—
					Lindane	—	—
					Mirex	—	—
					Nonachlor, *cis-*	—	—
					Nonachlor, *trans-*	—	—
					Photomirex	—	—
					Toxaphene†	—	—
					Trifluralin	—	—

Table 2.1. Pesticides in bed sediment and aquatic biota from rivers and estuaries in the United States: National and multistate monitoring studies—*Continued*

Target Analytes	Bed Sediment (µg/kg dry weight)			Aquatic Biota (whole fish unless noted)					Comments
				% Detection		(µg/kg wet weight)			
	DL	Range	Median	Sites	Samples	DL	Range	Median	
Chlordane, *cis-*	—	—	—	100*	100*	5	trace–70	40	* Skin-on fish fillets.
Chlordane, *trans-*	—	—	—	100*	100*	5	trace–50	20	† "Apparent toxaphene," mixture of chlorinated camphenes. Objective:
Chlorpyrifos	—	—	—	8*	8*	5	nd–trace	nd	to evaluate contami-
Dacthal	—	—	—	67*	67*	5	nd–40	trace	nants in fillets of
DDD, *p,p´-*	—	—	—	92*	92*	5	nd–110	45	autumn run coho
DDE, *p,p´-*	—	—	—	100*	100*	5	20–980	370	salmon in the Great
DDT, *p,p´-*	—	—	—	92*	92*	5	nd–60	20	Lakes Basin. 30 pest-
DDT, total	—	—	—	100*	100*	5	20–1,030	440	icides and 36 industrial
Diazinon	—	—	—	25*	17*	5	nd–trace	nd	chemicals were ident-
Dieldrin	—	—	—	100*	100	5	trace–110	40	ified. Total DDT was
Endrin	—	—	—	100*	100	5	trace–10	trace	≥85% DDE, although
HCH, α-	—	—	—	100*	100	5	trace–10	trace	the DDE/DDT ratio
Heptachlor	—	—	—	17*	17*	5	nd–trace	nd	varied among sites.
Heptachlor epoxide	—	—	—	58*	56*	5	nd–trace	trace	Several presently used pesticides (dacthal,
Hexachloro-benzene	—	—	—	92*	92	5	nd–20	trace	trifluralin, diazinon, and chlorpyrifos) were
Lindane	—	—	—	33*	33*	5	nd–trace	nd	detected. Total DDT
Mirex	—	—	—	8*	8*	5	nd–160	nd	residues were lowest in
Nonachlor, *cis-*	—	—	—	58*	58*	5	nd–40	13	Lake Superior and highest in Lake
Nonachlor, *trans-*	—	—	—	100*	100*	5	trace–100	50	Michigan. "Apparent toxaphene" levels were
Photomirex	—	—	—	8*	8*	5	nd–90	nd	lowest in Lake Erie and
Toxaphene†	—	—	—	92*	92*	250	nd–1,700	800	highest in Lakes Huron
Trifluralin	—	—	—	8*	8*	5	nd–trace	nd	and Michigan. Mirex was detected only in Lake Ontario.

Table 2.1. Pesticides in bed sediment and aquatic biota from rivers and estuaries in the United States: National and multistate monitoring studies—*Continued*

Reference	Sam-pling Dates	Sampling Locations and Strategy	Sampling Media and Species	Number of Samples	Target Analytes	Bed Sediment % Detection	
						Sites	Samples
Martin and Hartman, 1985	1980–1982	Riverine and pothole wetlands at 9 (fish) or 17 (sediment) locations in the north central United States: 15 national wildlife refuges and 2 waterfowl production areas. Sediment was sampled in depositional areas. Fish were composites in size range of potential food for fish-eating birds.	Sediment (surficial), fish (whole). Species: 16 species, including black bullhead, common carp, fathead minnow, and white sucker.	117 66	Aldrin	6	1
					DDD	6	1
					DDE	6	1
					DDT	0	0
					Dieldrin	6	1
					Endrin	0	0
					HCH, α-	0	0
					HCH, β-	0	0
					Heptachlor	0	0
					Heptachlor epoxide	0	0
					Hexachlorobenzene	0	0
					Lindane	0	0
					Nonachlor, *trans-*	—	—
					Oxychlordane	0	0
Leiker and others, 1991	1987	17 sites on the Mississippi River and its tributaries in 7 states	Fish (whole). Species: blue catfish, black bullhead, channel catfish, and flathead catfish.	nr	Chlordane	—	—
					Dacthal	—	—
					DDD, *p,p´-*	—	—
					DDE, *p,p´-*	—	—
					DDT, *p,p´-*	—	—
					Dieldrin	—	—
					Hexachlorobenzene	—	—

Table 2.1. Pesticides in bed sediment and aquatic biota from rivers and estuaries in the United States: National and multistate monitoring studies—*Continued*

Target Analytes	Bed Sediment (µg/kg dry weight)			Aquatic Biota (whole fish unless noted)					Comments
				% Detection		(µg/kg wet weight)			
	DL	Range	Median	Sites	Samples	DL	Range	Median	
Aldrin	5	nd–70	nd	0	0	5	nd	nd	* Range shown is for
DDD	5	nd–17	nd	22	14	5	nd–60	nd	sum of dieldrin and
DDE	5	nd–19	nd	67	51	5	nd–512	nd	*trans*-nonachlor.
DDT	5	nd	nd	11	3	5	nd–20	nd	Objective: to determine
Dieldrin	5	nd–170	nd	22	6	5	nd–41*	nd	residues in wetlands
Endrin	5	nd	nd	0	0	5	nd	nd	managed by the FWS
HCH, α-	5	nd	nd	67	36	5	nd–27	nd	in the north central
HCH, β-	5	nd	nd	—	—	—	—	—	United States. Concen-
Heptachlor	5	nd	nd	0	0	5	nd	nd	trations and prevalence
Heptachlor epoxide	5	nd	nd	0	0	5	nd	nd	of organochlorines were highest in fish
Hexachloro-benzene	5	nd	nd	—	—	—	—	—	from the upper Rio Grande drainage
Lindane	5	nd	nd	0	0	5	nd	nd	(Alamosa and Monte
Nonachlor, *trans*-	—	—	—	11	1	5	nd–41*	nd	Vista National Wildlife Refuges). Sediment
Oxychlordane	5	nd	nd	—	—	—	—	—	and fish from this survey had low organo-chlorine levels compared with sam-ples from other parts of the United States (Schmitt and others, 1981).
Chlordane	—	—	—	100	nr	10	11–170*	54*	* Range or median of
Dacthal	—	—	—	24	nr	1	nd–9*	nd*	average values for the
DDD, *p,p´-*	—	—	—	94	nr	5	nd–373*	8*	17 sites. Objective: to
DDE, *p,p´-*	—	—	—	100	nr	1	16–269*	62*	investigate types and
DDT, *p,p´-*	—	—	—	71	nr	nr	nd–179*	14*	levels of halogenated
Dieldrin	—	—	—	35	nr	1	nd–75*	nd*	contaminants in catfish
Hexachloro-benzene	—	—	—	71	nr	1	nr	nr	from the Mississippi River and its tributaries.

Table 2.1. Pesticides in bed sediment and aquatic biota from rivers and estuaries in the United States: National and multistate monitoring studies—*Continued*

Reference	Sampling Dates	Sampling Locations and Strategy	Sampling Media and Species	Number of Samples	Target Analytes	Bed Sediment % Detection	
						Sites	Samples
Phillips and Birchard, 1991	1978–1987	Unspecified number of sites throughout the United States. Sites were classified into one of 9 regions (that were based on U.S. Census divisions).	Sediment; fish (tissue type not reported); groundwater. Species: not reported.	nr, nr	Chlordane	nr	nr
					DDE, *p,p′-*	nr	nr
					DDT	nr	nr
					Dieldrin	nr	nr
					Heptachlor epoxide	nr	nr
					Hexachlorobenzene	nr	nr
					Lindane	nr	nr
					Mirex	nr	nr

Table 2.1. Pesticides in bed sediment and aquatic biota from rivers and estuaries in the United States: National and multistate monitoring studies—*Continued*

| Target Analytes | Bed Sediment (µg/kg dry weight) | | | Aquatic Biota (whole fish unless noted) | | | | | Comments |
| | DL | Range | Median | % Detection | | (µg/kg wet weight) | | | |
				Sites	Samples	DL	Range	Median	
Chlordane	nr	nr	nr	nr	nr	nr	nr	nr	Objective: to evaluate
DDE, *p,p´-*	nr	nr	nr	nr	nr	nr	nr	nr	variation in regional
DDT	nr	nr	nr	nr	nr	nr	nr	nr	accumulation of toxic
Dieldrin	nr	nr	nr	nr	nr	nr	nr	nr	contaminants in the
Heptachlor epoxide	nr	nr	nr	nr	nr	nr	nr	nr	United States. Existing data were retrieved
Hexachloro-benzene	nr	nr	nr	nr	nr	nr	nr	nr	from STORET database. Ranks were
Lindane	nr	nr	nr	nr	nr	nr	nr	nr	assigned to each region
Mirex	nr	nr	nr	nr	nr	nr	nr	nr	according to the mean
									concentration of each contaminant in each medium. The west north-central region clearly outranked other regions for pesticides in sediment. Fish tissue rankings for pesticides were highest in the west south-central and west north-central regions.

Appendix B—Table 2.2. Pesticides in bed sediment and aquatic biota from rivers in the United States: State and local monitoring studies

[Studies are listed chronologically. Explanation of footnotes for each study is given in that study's Comments column. Sampling media and species, lists all sampling media analyzed by the study and, for aquatic biota, some or all species sampled. In this column, "Species include" indicates that the following species list is a partial list for that study; "sediment" is designated as either surficial (for surficial bed sediment), core (for bed sediment core), or suspended (for suspended sediment). Number of samples, number listed applies to the sampling medium in the same row in the preceding column (Sampling media and species). Target analytes, the complete list of target analytes in bed sediment or aquatic biota for that study; data in the next ten columns apply to the target analyte listed in the same row. Bed sediment, the five columns beneath this heading contain data that pertain to bed sediment; unless otherwise noted with a footnote, bed sediment concentrations are in µg/kg dry weight. Aquatic Biota, the five columns beneath this heading contain data that pertain to aquatic biota; unless otherwise noted with a footnote, data apply to whole fish and concentrations are in µg/kg dry weight. Abbreviations and symbols: BAF, bioaccumulation factor; BofCF, Bureau of Commercial Fisheries; BSAF, biota-sediment accumulation factor; C, channel; CB, chlorobenzene; cm, centimeter; DCA, 3,4-dichloroaniline; EDL, elevated data level; EP, Extraction Procedure; F, F test statistic in analysis of variance; FCL, Four County Landfill; FDA, Food and Drug Administration; ft, foot; FWS, Fish and Wildlife Service; GC/MS, gas chroma-tography with mass spectrometric detection; h, hour; HCB, hexachlorobenzene; HCBD, hexachlorobutadiene; in., inch; K_{ow}, n-octanol-water partitioning coefficient; kg, kilogram; km, kilometer; L, liter; lb, pound; mg, milligram; MIA, Miami International Airport; misc., miscellaneous; mm, millimeter; NAS/NAE, National Academy of Sciences and National Academy of Engineering; NCBP, National Contaminant Biomonitoring Program; nd, not detected; NOAA, National Oceanic and Atmospheric Administration; NOAEL, no observed adverse effect level; NoL, Niagara-on-the-Lake; nr, not reported; NWR, National Wildlife Refuge; O, overbank; OC, organochlorine; OP, organophosphate; PAH, polynuclear aromatic hydrocarbons; PCA, pentachloroanisole; PCB, polychlorinated-biphenyl; pg, picogram; R., River; RCRA, Resource Conservation and Recovery Act; r.m., river mile; sp., species; TOC, total organic carbon; U.S., United States; USEPA, U.S. Environmental Protection Agency; USGS, U.S. Geological Survey; VOC, volatile organic compound; WMA; wildlife management area; wt., weight; µg, micrograms; µm, micrometer; ~, approximately; —, not applicable; %, percent; >>, is much greater than]

[Table 2.2 begins on next page]

Table 2.2. Pesticides in bed sediment and aquatic biota from rivers in the United States: State and local monitoring studies—*Continued*

Reference	Sampling dates	Sampling locations and strategy	Sampling media and species	Number of samples	Target analytes	Bed sediment % Detection Sites	Bed sediment % Detection Samples
Mack and others, 1964	1963	16 (fish), 4 (water), and 2 (sediment) sites in waters of New York state. Sites include lakes and rivers.	Fish (whole, muscle, fat, testes, eggs, spleen, gills) Species include rainbow trout, largemouth bass.	20 6, 4 4, 12 1, 3	DDT, total	—	—
Bedford and others, 1968	1966	12 sites (water, sediment) and 2 sites (mussels) on the Red Cedar River, Michigan. Resident mussels were from 2 reference (upstream) sites. Mussels were transplanted to 1 reference site and 5 sites near municipal and residential sources.	Water (filtered), sediment (cores, suspended), mussels. Species include *Lasmigona ventricosa, Anodonta grandis, Eliptio dilatus.*	nr nr nr nr	DDT, total Methoxychlor	nr nr	nr nr
Kleinert and others, 1968	1965– 1967	109 inland lakes and streams in Wisconsin, the Mississippi River, and Wisconsin's coastal waters of Lakes Michigan and Superior. Most samples were composites of 3–10 fish of the same species.	Fish (whole). 35 species, including sucker, carp, northern pike, coho salmon, crappie, walleye, brook trout, rainbow trout, bluegill.	561	DDT Dieldrin	— —	— —
Lyman and others, 1968	1965– 1967	83 sampling sites located in rivers and 10 sites on ponds throughout the state of Massachusetts. A minimum of 5 individual fish were analyzed from each site.	Fish (whole). Species include yellow perch, white sucker, creek chub, small-mouth bass, bluegill.	1,310	DDD DDE DDT DDT, total	— — — —	— — — —
Modin, 1969	1966– 1968	21 sites in estuaries throughout California. Two of these sites were in the freshwater parts	Clams, mussels, oysters, crab (ova, whole),	18 18 120 6 1	Aldrin DDD DDE DDT Dieldrin	— — — — —	— — — — —

Table 2.2. Pesticides in bed sediment and aquatic biota from rivers in the United States: State and local monitoring studies—*Continued*

Bed sediment (μg/kg dry weight)			Aquatic biota (whole fish unless noted)					Comments
			% Detection		(μg/kg wet weight)			
DL	Range	Median	Sites	Samples	DL	Range	Median	
—	—	—	100	100	30	100–6,200	450	Objective: to do a preliminary survey of DDT in representative fish from New York lakes and streams. DDT levels in water ranged from 0.015–0.33 μg/L. Tissues such as visceral fat, gills, eggs, and testes contained up to 40 mg/kg DDT.
nr	<100–10,000	nr	100*	100*	nr	15.3–198*	nr*	* Data are for mussels. Objective: to assess background residues in mussels used as biomonitors in concurrent transplantation experiments. Transplantation experiments are described in Table 2.3. Total DDT in filtered water ranged from trace to 60 μg/L. Concentrations in suspended sediment ranged from 1,000–50,000 μg/kg dry wt., but there was relatively little suspended material.
nr	nd–trace	nr	0*	0*	nr	nd*	nd*	
—	—	—	100	100	nr	21–1,620	845*	* Mean values. Objective: to survey for dieldrin and DDT in Wisconsin fish. Highest DDT levels occurred in Lake Michigan and southeastern Wisconsin. Many streams had higher residues in downstream than upstream reaches. Dieldrin was elevated in fish from Pike and Milwaukee rivers, which pass through industrial regions. Results indicate widespread contamination.
—	—	—	84	70	nr	nd–12,500	158*	
—	—	—	100	nr	nr	0–12,500*	nr	*Range is of mean residue in individual fish from each site and sampling date. Objective: to characterize ambient DDT residues in fish in Massachusetts waterways. Pesticide concentrations increased at most sites over the 3 year monitoring period.
—	—	—	100	nr	nr	160–13,180*	nr	
—	—	—	100	nr	nr	0–1,230*	nr	
—	—	—	100	nr	nr	190–28,880*	nr	
—	—	—	0*	0*	10	nd*	nd*	* Data are for clam tissue from freshwater portions of estuaries (2 sites, 18 samples). Objective: to monitor pesticide residues in biota of California
—	—	—	100*	100*	10	93–490*	255*	
—	—	—	100*	100*	10	140–690*	325*	
—	—	—	100*	100*	10	130–1,100*	310*	
—	—	—	100*	94*	10	nd–28*	17.5*	

Table 2.2. Pesticides in bed sediment and aquatic biota from rivers in the United States: State and local monitoring studies—*Continued*

Reference	Sampling dates	Sampling locations and strategy	Sampling media and species	Number of samples	Target analytes	Bed sediment % Detection Sites	Bed sediment % Detection Samples
		of the Sacramento– San Joaquin River Delta.	fish (ova), prawn (ova). Species include Asiatic clam, bay mussel, king crab, halibut.	4 4	Endrin Heptachlor Heptachlor epoxide Lindane Methoxychlor	— — — — —	— — — — —
Anderson and Fenderson, 1970	1967	Unspecified number of sites in Jordan River, a spawning tributary to Sebago Lake, Maine. Landlocked salmon were targeted. Only males were sampled.	Fish (whole, brain, whole minus brain). Species: Atlantic salmon.	59 29, 29	DDD DDE DDT Dieldrin	— — — —	— — — —
Claeys and others, 1970	1961– 1969	12 general locations in inland and coastal Oregon. Inland sites include fish hatcheries. No information was provided on sampling strategy or methods.	Fish (flesh, liver, not reported). Species: albacore tuna (10 samples) and not reported (24 samples).	3 8 24	DDE DDT	— —	— —
Hartung and Klingler, 1970	nr	Unspecified number of sites on the lower Detroit River (Michigan).	Sediment, water (filtered, unfiltered).	30	DDD, *p,p'-* DDE, *p,p'-* DDT, *p,p'-* Dieldrin	nr nr nr nr	nr nr nr nr

Table 2.2. Pesticides in bed sediment and aquatic biota from rivers in the United States: State and local monitoring studies—*Continued*

Bed sediment (μg/kg dry weight)			Aquatic biota (whole fish unless noted)					Comments
			% Detection		(μg/kg wet weight)			
DL	Range	Median	Sites	Sam-ples	DL	Range	Median	
—	—	—	50*	17*	10	nd–10*	nd*	estuaries as part of nationwide monitoring by the U.S. Bureau of Commercial Fisheries. Results indicate that biota in estuaries receiving agricultural or urban runoff are more likely to have elevated residues than biota in isolated estuaries.
—	—	—	0*	0*	10	nd*	nd*	
—	—	—	0*	0*	10	nd*	nd*	
—	—	—	0*	0*	10	nd*	nd*	
—	—	—	0*	0*	10	nd*	nd*	
—	—	—	nr	100	nr	250–2,720	1,230*	* Value shown is mean value. Objective: to identify causes of variation in insecticide residues in landlocked salmon. Residues in brain samples were lower and more variable than in whole fish. DDE and DDD concentrations increased with increasing age (F<0.01). DDD and dieldrin concentrations increased with increasing fat content (F<0.01). Means and variances were reported for samples collected in 1962–1967. Authors recommend stratified random method of selection of fish for monitoring (considering factors such as age, length, sex, fat content).
—	—	—	nr	100	nr	4,300–14,600	9,380*	
—	—	—	nr	100	nr	250–3,610	800*	
—	—	—	nr	100	nr	11–37	19*	
—	—	—	67*	59*	nr	nd–160*	16*	* Data shown are for all fish, regardless of tissue type. Objective: to report DDT residues in Oregon fish. In inland fish (species and tissue type not reported), *p,p'*- DDT values were generally higher than *p,p'*- DDE, indicating recent exposure. Mean levels in coastal albacore flesh were 24 (*p,p'*-DDE) and 12 (*p,p'*- DDT) μg/kg.
—	—	—	67*	62*	nr	nd–560*	10*	
nr	nr	nr	—	—	—	—	—	* Estimated from figure. Objective: to evaluate partitioning of DDT into sedimented oils (polluting oils settling out as sediments). DDT concentration in Detroit R. water averaged 0.04 (unfiltered) and 0.007 (filtered) μg/L. DDT concentration in sediment was correlated with oil in sediment, indicating sedimented oils have a strong influence on DDT transfer between water and sediment.
nr	nr	nr	—	—	—	—	—	
nr	~10–1,300*	nr	—	—	—	—	—	
nr	nr	nr	—	—	—	—	—	

Table 2.2. Pesticides in bed sediment and aquatic biota from rivers in the United States: State and local monitoring studies—*Continued*

Reference	Sampling dates	Sampling locations and strategy	Sampling media and species	Number of samples	Target analytes	Bed sediment % Detection	
						Sites	Samples
Johnson and Lew, 1970	1965–1969	9 sites in the lower Colorado River Basin, Arizona.	Fish (whole, muscle, fat, skin, gills, liver, intestine, blood, kidney, ovaries, testes. Species include carp, channel catfish, Gila sucker.	4 45	DDD DDE DDT Dieldrin Toxaphene	— — — — —	— — — — —
Smith and Cole, 1970	1966–1967	The Weweantic River Estuary (Massachusetts), plus 4 other coastal or estuarine locations. The Weweantic River Estuary is a small tidal tributary of Buzzards Bay. Fish were classed into 4 age groups. Only juveniles (present year-round) were used for seasonal analysis.	Fish (muscle, ovaries). Species: winter flounder.	65 10	DDT DDE Heptachlor Heptachlor epoxide Dieldrin	— — — — —	— — — — —

Table 2.2. Pesticides in bed sediment and aquatic biota from rivers in the United States: State and local monitoring studies—*Continued*

Bed sediment			Aquatic biota (whole fish unless noted)						
(µg/kg dry weight)			% Detection		(µg/kg wet weight)				
DL	Range	Median	Sites	Sam-ples	DL	Range	Median	Comments	
								Detroit R. sediment containing 1% sedimented oils reached DDT levels near 1 mg/kg. *p,p'-* DDE, *p,p'-* DDD, and dieldrin also were detected in sediment.	
—	—	—	100	100	nr	20–1,500	160	Objective: to provide reference	
—	—	—	100	100	nr	30–7,250	310	data for assessing the effect of	
—	—	—	100	100	nr	30–6,750	500	DDT ban on DDT residues in	
—	—	—	25	100	nr	trace	—	fish; to evaluate the effect of	
—	—	—	100	100	nr	1,050–25,000		species and organ on tissue levels. In 1966, fishes in study area commonly exceeded the FDA action level for total DDT (5 mg/kg in edible flesh). DDT and metabolites appear to concentrate in the liver.	
—	—	—	100*	nr	10	nd–310*†	60*†	*Data are for juvenile fish fillets.	
—	—	—	100*	nr	10	30–1,070*†	40*†	†Range and median are of mean	
—	—	—	100*	nr	10	nd–1,100*†	80*†	residues of fish grouped by date. Objective: to evaluate insecti-	
—	—	—	100*	nr	10	nd–440*†	60*†	cide residues in flounder in the	
—	—	—	100*	nr	10	nd–50*†	nd*†	Weweantic R. in relation to use in the drainage basin, season, adult migration, and spawning. Maximum DDT residues oc-curred in association with spring runoff, not with any known DDT application in the basin. Maximum heptachlor residues occurred in winter (midwinter thaw with subsequent runoff); residues of its epoxide metabo-lite peaked <2 months later. Adult flounder appeared in the estuary from October to April. They had reduced DDT exposure relative to juveniles (which lived year-round in the estuary) but comparable heptachlor exposure during fall and winter. Adults had significantly lower DDT levels and similar heptachlor levels than juveniles (in winter). Based on chromato-graphic pattern of residues in flounder from other estuaries and in mussels from the Weweantic R. estuary, authors suggested local watershed was major	

Table 2.2. Pesticides in bed sediment and aquatic biota from rivers in the United States: State and local monitoring studies—*Continued*

Reference	Sampling dates	Sampling locations and strategy	Sampling media and species	Number of samples	Target analytes	Bed sediment % Detection Sites	Bed sediment % Detection Samples
Stucky, 1970	1964	18 sites from major drainage systems in Nebraska. 10 fish were collected per site. Fish of 25–35 cm length were targeted for sampling.	Fish (fat, blood). Species: channel catfish.	178	DDD, *p,p'*- DDE DDT, *o,p'*- DDT, *p,p'*- DDT, total Dieldrin	— — — — — —	— — — — — —
Kolipinski and others, 1971	1966–1968	10 sites (sediment) and 7 sites (biota) in Everglades National Park and Loxahatchee National Wildlife Refuge, Florida. Sites were located where water flows into and out of the park and the refuge.	Water, sediment (submerged soil), fish (whole), algae, plants, invertebrates. Species include mosquitofish, crayfish, gastropod, oyster, pond snail, crab, sawgrass.	19 11	Aldrin DDD DDE DDT Dieldrin Endrin Heptachlor Heptachlor epoxide Lindane	0 90 80 60 0 0 0 0 0	0 74 16 47 0 0 0 0 0
McPherson, 1971	1969	8 sites (sediment) and 6 sites (biota) in 3 areas in south Florida: near the Miami International Airport; wetlands in the Everglades; and near a jetport under construction.	Sediment, fish (tissue not reported), aquatic plants, water. Fish species: mosquitofish, sunfish, sailfin molly, flagfish.	8 5 6 6	DDD DDE DDT Dieldrin Heptachlor epoxide Lindane	38 50 50 38 13 0	38 50 50 38 13 0
Mick and McDonald, 1971	1968–1971	5 sites in Iowa: the Iowa River above and below Coralville Flood Control Reservoir, Lake MacBride and the Cedar River.	Sediment, fish (tissue not reported), periphyton, water. Species include carp, bluegill, black bull-	nr nr nr nr	Aldrin DDD, *p,p'*- DDE, *p,p'*- DDT, *p,p'*- DDT, total Dieldrin HCH, total Heptachlor Heptachlor epoxide	nr — — — nr nr nr nr —	nr — — — nr nr nr nr —

Table 2.2. Pesticides in bed sediment and aquatic biota from rivers in the United States: State and local monitoring studies—*Continued*

Bed sediment (µg/kg dry weight)			Aquatic biota (whole fish unless noted)					Comments
			% Detection		(µg/kg wet weight)			
DL	Range	Median	Sites	Samples	DL	Range	Median	
								pesticide source.
—	—	—	100*	nr	10	nr	nr	* Data are for fish fat samples.
—	—	—	100*	nr	10	nr	nr	Objective: to determine residues
—	—	—	100*	nr	10	nr	nr	in channel catfish in Nebraska
—	—	—	100*	nr	10	nr	nr	watersheds. Organochlorine
—	—	—	100*	100*	10	280–259,000*	nr	residues were lower in blood (trace to 160 µg/kg DDT, trace
—	—	—	100*	10*	10	nd–12,900*	nr	to 70 µg/kg dieldrin) than in fat samples.
0.1	nd	nd	0	0	0.1	nd	nd	Objective: to determine
0.1	nd–30	2.8	86	91	0.1	nd–120	48	residues in various levels in the
0.1	nd–21	1.8	86	91	0.1	nd–300	114	ecosystem. Concentrations in
0.1	nd–45	nd	86	91	0.1	nd–470	140	submerged soils, algal mats,
0.1	nd	nd	14	9	0.1	nd–1.1	nd	crustaceans were as much as
0.1	nd	nd	0	0	0.1	nd–0.9	nd	1,000-fold higher than water.
0.1	nd	nd	0	0	0.1	nd	nd	Concentrations in marsh fish
0.1	nd	nd	0	0	0.1	nd	nd	were 10,000-fold higher than in
0.1	nd	nd	14	9	0.1	nd	nd	water. The highest DDT and metabolite concentrations were in higher carnivores and omnivores. Residues at different times varied in one location. Residues in mosquitofish were higher in February than in October.
nr	nd–18.2	nd	67*	40*	0.01	nd–297.5*	nd*	* Data shown are for fish (tissue
nr	nd–27	nd/0.4	100*	60*	0.01	nd–358.7*	0.18*	type not reported).
nr	nd–40	nd/1.2	33*	20*	0.01	nd–189*	nd*	Objective: to assess baseline
nr	nd–2.5	nd	67*	40*	0.01	nd–1.2*	nd*	water quality near a jetport under
nr	nd–0.5	nd	67*	40*	0.01	nd–19*	nd*	construction near the Everglades
nr	nd	nd	0*	0*	0.01	nd*	nd*	National Park. Total DDT residues were highest in sunfish from a canal near the new jetport and mosquitofish from a canal near Miami International Airport (MIA). DDTs and dieldrin in sediment were highest near the MIA. DDD and DDE were detected in plants or algae, and traces of lindane and heptachlor epoxide in water.
nr	nd–28	nr	nr	nr	nr	nd–210*	nr	* Data shown are for fish (tissue
—	—	—	nr	nr	nr	nd–213.4*	nr	type not reported).
—	—	—	nr	nr	nr	nd–234.9*	nr	Objective: to determine pesticide
—	—	—	nr	nr	nr	nd–373.1*	nr	residues in water, mud, plants,
nr	nd–54	nr	—	—	—	—	—	and fish in Iowa; to identify factors affecting pesticide residues.
nr	nd–11	nr	nr	nr	nr	nd–859*	nr	tors affecting pesticide residues.
nr	nd–3	nr	—	—	—	—	—	Highest pesticide levels occurred
nr	nd	nr	nr	nr	nr	nd–84.1*	nr	during July–October. Residues:
—	—	—	nr	nr	nr	nd–200.8*	nr	fish > periphyton >> mud, water.

Table 2.2. Pesticides in bed sediment and aquatic biota from rivers in the United States: State and local monitoring studies—*Continued*

Reference	Sampling dates	Sampling locations and strategy	Sampling media and species	Number of samples	Target analytes	Bed sediment % Detection Sites	Bed sediment % Detection Samples
			head, black crappie, channel catfish, yellow bass.		Oxychlordane	—	—
Miles and Harris, 1971	1970	4 sites (sediment) and 2 sites (fish) in Big Creek, which flows into Lake Erie, and in an agricultural drainage system in Ontario, Canada. Water was sampled weekly, bottom sediment monthly.	Water (whole), sediment (surficial), fish (whole). Species: chub, suckers, catfish.	53 27 5	Aldrin DDD, *p,p'*- DDE, *p,p'*- DDT, *o,p'*- DDT, *p,p'*- DDT, total Dieldrin Endosulfan Endrin Heptachlor Lindane	0 100 100 100 100 100 100 50 0 0 0	0 96 100 67 100 100 33 37 0 0 0
Morris and Johnson, 1971	1970	35 sites in 17 rivers and 2 lakes in Iowa. Rivers having high dieldrin in earlier surface water study were targeted. Composites were made of fillets from fish of approximately the same size from each location.	Fish (muscle). Species: bigmouth buffalo, black bullhead, black crappie, bluegill, carp, carpsucker, catfish, largemouth bass, northern pike, walleye, white bass, white crappie.	76	Dieldrin	—	—
Wallace and Brady, 1971	1970	3 sites on Rocky River, South Carolina. Sites were upstream and downstream of a tributary receiving effluent from a woolen mill that uses dieldrin in the dyeing process.	Insects, snails. Species: mayflies, stoneflies, caddisflies, true flies, beetles, hell-grammites,	20	Dieldrin	—	—

Table 2.2. Pesticides in bed sediment and aquatic biota from rivers in the United States: State and local monitoring studies—*Continued*

Bed sediment (µg/kg dry weight)			Aquatic biota (whole fish unless noted)					Comments
			% Detection		(µg/kg wet weight)			
DL	Range	Median	Sites	Samples	DL	Range	Median	
—	—	—	nr	nr	nr	nd–78.1*	nr	Residues were affected by type of fish, fat content, and drainage area. Bottom-feeders had higher residues than predators. The highest residues were in fish with the highest fat content.
nr	nd	nd	—	—	—	—	—	Objective: to determine insecticide content of water systems draining areas containing contaminated farms. Residues in sediment were 2–4 orders of magnitude higher than in water. Residues in fish were 4 orders of magnitude higher than in water.
2	nd–680	55	100	100	nr	114–310	208	
3	3–200	34	100	100	nr	239–800	385	
3	nd–190	10	100	40	3	nd–92	nd	
nr	2–842	70	100	60	3	nd–409	60	
nr	8–1,730	250	100	100	nr	645–1,012	769	
1	nd–50	nd	100	100	nr	11–189	36	
1	nd–62	nd	—	—	—	—	—	
3	nd	nd	100	80	nr	nd–18	11	
nr	nd	nd	—	—	—	—	—	
nr	nd	nd	—	—	—	—	—	
—	—	—	100*	100*	nr	11–1,600*	225*	* Data are for fish fillets. Objective: to survey dieldrin in edible fish from Iowa streams. Species effect: catfish> rough fish (buffalo, carp, carpsuckers)> pan and game fish (white bass, crappie, bluegill, walleye). Traces of other organochlorines were detected in many samples. The dieldrin/aldrin ratio was generally about 10:1, although one sample contained 910 µg/kg aldrin. Authors stated that residues increased as row crop-draining streams proceed southwards due to accumulated siltation; streams in northeastern Iowa draining non-row crop areas and with less siltload did not have catfish with high dieldrin residues. Residues in catfish <15 in. are below FDA action level (300 µg/kg) in most locations.
—	—	—	100*	95*	nr	nd–103,000*	1,480*	* Data are for invertebrates. Objective: to quantify bottom fauna upstream and downstream of tributary containing effluent from woolen mill and to relate the results to dieldrin residues. Dieldrin residues were higher in downstream fauna than upstream (by up to 1100-fold). The number of invertebrate species

Table 2.2. Pesticides in bed sediment and aquatic biota from rivers in the United States: State and local monitoring studies—*Continued*

Reference	Sampling dates	Sampling locations and strategy	Sampling media and species	Number of samples	Target analytes	Bed sediment % Detection Sites	Bed sediment % Detection Samples
			dragonflies.				
Zabik and others, 1971	1966–1968	12 sites on Red Cedar River, Michigan. Land use is wetland and pasture upstream; and agriculture, urban, and industry downstream.	Sediment (surficial, suspended), water (dissolved)	771	DDT DDT metabolites	100 100	nr nr
Baetcke and others, 1972	1970	Unspecified number of sites from Oktibbeha and Lowndes Counties, Mississippi. All samples were taken from areas treated with mirex during 1962–1969. Some fish samples were composites.	Fish (muscle, adipose tissue), birds, deer, arthropods, earthworms, beef, cow's milk, crops. Species include catfish, sunfish.	5 9	DDD, *p,p'-* DDE, *p,p'-* DDT, *o,p'-* DDT, *p,p'-* DDT, total Mirex	— — — — — —	— — — — — —
Bevenue and others, 1972	1970–1971	5 sites (sediment) and 1 site (biota) on 2 canals on Oahu, Hawaii. Rain, drinking water, and nonpotable water also were sampled on Oahu, Kauai, and Maui.	Water (whole), sewage, rain, sediment (surficial), fish (muscle, whole), algae. Species include molly, guppy, tarbon, milkfish.	5 2 1 1	Chlordane DDD, *p,p'-* DDE, *p,p'-* DDT, *p,p'-* Dieldrin Lindane	100 80 80 100 100 0	100 80 80 100 100 0

Table 2.2. Pesticides in bed sediment and aquatic biota from rivers in the United States: State and local monitoring studies—*Continued*

Bed sediment (µg/kg dry weight)			Aquatic biota (whole fish unless noted)					Comments
			% Detection		(µg/kg wet weight)			
DL	Range	Median	Sites	Samples	DL	Range	Median	
								downstream has been reduced and community structure changed.
0.5	280–6,480	nr	—	—	—	—	—	* Approximate values, estimated from figure.
0.5	300–4,300*	nr						Objective: to quantify various sources (runoff, municipal waste effluents) to DDT levels in river. River became progressively more contaminated downstream. In sediment, DDTs increased during April–May (spring runoff), then decreased during summer. In water, DDT peaked in June and metabolites in April. Monthly variation in DDT levels in suspended particulates followed that in bottom sediment. DDTs in suspended matter were high at 2 sites downstream of 2 waste treatment facilities relative to upstream sites.
—	—	—	nr*	60*	1	nd–327*	118*	* Data are for fish muscle.
—	—	—	nr*	100*	1	16–251*	105*	Objective: to analyze mirex and
—	—	—	nr*	80*	1	nd–289*	37*	DDT residues in wildlife in
—	—	—	nr*	80*	1	nd–449*	142*	counties treated with mirex to
—	—	—	nr*	100*	1	24–1,316*	448*	control fire ants. Residues in
—	—	—	nr*	40*	2	nd–79*	nd*	fish were higher in adipose tissue than in muscle. All samples containing mirex were from areas treated recently with mirex.
nr	125–720	255	0	0*	nr	nd*	nd*	* Data aggregated for 1 whole
nr	nd–220	95	100	100*	nr	149–581*	210*	fish and 2 fish muscle samples.
nr	nd–100	10	100	100*	nr	80–298*	182*	Objective: to determine organo-
nr	30–170	50	100	100*	nr	101—170*	159*	chlorine contamination of water,
nr	0.1–370	100	100	100*	nr	141–486*	340*	sediment, algae, and fish in the
nr	nd	nd	0	0*	nr	nd*	nd*	state of Hawaii. The only algae sample contained *p,p'*- DDE (10 µg/kg wet weight), *p,p'*- DDD (30), *p,p'*- DDT (45), and dieldrin (45). Authors suggest that chlordane and dieldrin found in waters and sediment and pentachlorophenol found in sewage originated in urban areas.

Table 2.2. Pesticides in bed sediment and aquatic biota from rivers in the United States: State and local monitoring studies—*Continued*

Reference	Sampling dates	Sampling locations and strategy	Sampling media and species	Number of samples	Target analytes	Bed sediment % Detection	
						Sites	Samples
Bonderman and Slach, 1972	1966–1970	50 farms in eastern Iowa.	Soil, crops, fish (whole). Species include sunfish, crappie, bluegill, catfish, carp.	1,500 400 75	Chlordene, 1-hydroxy Heptachlor Heptachlor epoxide	— — —	— — —
Durant and Reimold, 1972	1971	4 sites (sediment) and 1 site (mollusks) on Terry Creek (Georgia), near a toxaphene plant outfall. Sediment cores were taken at 3 sites prior to dredging (in June). Surficial sediment was sampled following dredging at 1 site (dredge spoil area). Composite oyster samples (each of 12 oysters) were collected from April to Oct.	Sediment (cores, surficial), oysters. Species: *Crassostrea virginica*.	3 1 16	Toxaphene	100*	100*
Freiberger and McPherson, 1972	1971–1972	4 sites (sediment and water) and 3 sites (fish) in canals and ditches representing major drainage systems near Miami International Airport, Florida.	Water (whole), sediment, fish (whole). Species: Florida gar, bluegill, mosquitofish, spotted sunfish, freshwater glass-minnow.	16 16 12	Aldrin Chlordane D, 2,4- DDD DDE DDT Dieldrin Endrin Heptachlor Heptachlor epoxide Lindane Silvex T, 2,4,5- Toxaphene	0 0 0 100 100 100 75 0 0 — 0 50 0 —	0 0 0 83 82 33 60 0 0 — 0 29 0 —
Kramer and Plapp, 1972	nr *	Unspecified number of sites on the Brazos and Navasota rivers and Somerville Reservoir (Texas). Some samples were composites.	Fish (muscle). Species include carpsucker, freshwater drum, channel catfish, flathead catfish, carp,	53	DDD DDE DDT DDT, total	— — — —	— — — —

Table 2.2. Pesticides in bed sediment and aquatic biota from rivers in the United States: State and local monitoring studies—*Continued*

Bed sediment (µg/kg dry weight)			Aquatic biota (whole fish unless noted)					Comments
			% Detection		(µg/kg wet weight)			
DL	Range	Median	Sites	Samples	DL	Range	Median	
—	—	—	—	17	nr	13–78	16	Objective: to determine a hepta-chlor metabolite in soil, crops, and fish in an agricultural area with a documented history of heptachlor applications. Detections of 1-hydroxychlordene in fish was related to the presence of heptachlor and its epoxide.
—	—	—	—	9	nr	0–35	3	
—	—	—	—	15	nr	0–95	20	
250	*32,800–1,858,000	73,700*	100	100†	nr	710–5,700†	2,000†	* Data are for 1 surficial sample and top 10 cm of the 3 cores. † Data are for oysters; concentrations are estimated from figure. Objective: to determine the effect of dredging on toxaphene residues in oysters and sediment and to assess any ecological damage from dredging. Residues in sediment and oysters were extremely high. Top 10 cm sediment core samples contained 35,500–1,858,000 µg/kg toxaphene, and dredge spoil 32,800 µg/kg. Analyses of oysters revealed no increase in toxaphene residues resulting from dredging and spoil runoff; residues were highest in May, lowest in Sept.
0.1	nd	nd	0	0	0.1	nd	nd	Objective: to provide baseline data on surface-water quality at the commercial airport. DDTs, dieldrin, 2,4-D, parathion, diazinon, and methyl parathion were detected in some water samples. Organochlorine concentrations: fish> sediment> water, but lower than in fish. Organochlorines were generally higher at sites near the airport than in relatively undisturbed environments in southern Florida (data not presented).
1	nd	nd	33	20	1	nd–260	nd	
0.1	nd	nd	—	—	—	—	—	
0.1	nd–93	2.9	100	100	0.1	15–150	61	
0.1	nd–40	2.5	100	100	0.1	41–120	63	
0.1	nd–23	nd	100	100	0.1	5.8–79	22	
0.1	nd–7.8	0.3	100	92	0.1	nd–2,400	8.5	
0.1	nd	nd	0	0	0.1	nd	nd	
0.1	nd	nd	0	0	0.1	nd	nd	
—	—	—	100	67	0.1	nd–0.9	0.6	
0.1	nd	nd	0	0	0.1	nd	nd	
0.1	nd–7.2	nd	—	—	—	—	—	
0.1	nd	nd	—	—	—	—	—	
—	—	—	0	0	1	nd	nd	
—	—	—	100†	nr	100	nd–137†§	nr	* Data probably were collected between 1967 (date reservoir filled) and 1970 (date published). † Fish muscle tissue. § Range is of 42 samples from the 2 rivers only (not reservoir). Objective: to determine DDT residues in muscle tissue from the Brazos River Basin. Dieldrin
—	—	—	100†	nr	100	nd–624†§	nr	
—	—	—	67†	nr	100	nd–169†§	nr	
—	—	—	100†	nr	100	nd–930†§	nr	

Table 2.2. Pesticides in bed sediment and aquatic biota from rivers in the United States: State and local monitoring studies—*Continued*

Reference	Sampling dates	Sampling locations and strategy	Sampling media and species	Number of samples	Target analytes	Bed sediment % Detection Sites	Bed sediment % Detection Samples
			longnose gar, small-mouth buffalo, spotted gar, white crappie, alligator gar, black bullhead, blue catfish, bluegill.				
Morris and others, 1972	1968–1971	8 streams in Iowa. Land use in the stream basins was primarily agricultural with the exception of one targeted reference site. Catfish were 15–18 inches long.	Water, sediment (surficial), fish (muscle). Species: catfish, bass, walleye, northern pike, crappie, bluegill	nr 6 nr	Aldrin Chlordane DDD DDE DDT Dieldrin Heptachlor epoxide	17 nr nr nr nr 83 nr	17 nr nr nr nr 83 nr
Routh, 1972	1971	8 sites in the Salinas River (California). Samples were collected weekly.	Sediment (surficial, suspended).	32	DDT, total	100	nr
Klaassen and Kadoum, 1973	1967–1969	5 sites on the Smoky Hill River (Kansas). 25 fish per species were collected at each site 1–2 times during study period. Study area was dryland farming area with some irrigation	Fish (whole, muscle, gonads). Species: 24, including red shiner, green sunfish (whole), carp, channel catfish (muscle,	184 100 97	Aldrin DDD, *p,p'*- DDE, *p,p'*- DDT, *o,p'*- DDT, *p,p'*- Dieldrin Endrin Heptachlor Heptachlor epoxide	— — — — — — — — —	— — — — — — — — —

Table 2.2. Pesticides in bed sediment and aquatic biota from rivers in the United States: State and local monitoring studies—*Continued*

Bed sediment (µg/kg dry weight)			Aquatic biota (whole fish unless noted)					Comments
			% Detection		(µg/kg wet weight)			
DL	Range	Median	Sites	Samples	DL	Range	Median	
								and toxaphene were detected in fat samples from fish collected at same sites. DDT residues in fish were highest for Brazos R. and lowest for Somerville Reservoir fish. Brazos R. flows through agricultural land, Navasota R. through rangeland, and the reservoir (filled in 1967) occupied land that was formerly used in agriculture.
6	1	17	56	56	nr	nd–280*	46*	* Data are for catfish flesh. Objective: to evaluate distribution of pesticides in the aquatic environment. Dieldrin was present in catfish, water, and sediment samples. High dieldrin levels were not seen in bass, wall-eye, northern pike, crappie, and bluegill from the same area. Highest residues in fish and water may be related to high suspended sediment concentrations.
nr	nr	nr	33	33	nr	nd–200*	nd*	
nr	nr	nr	56	56	nr	nd–95*	19*	
nr	nr	nr	56	56	nr	nd–180*	12*	
nr	nr	nr	56	56	nr	nd–86*	15*	
1	nd–170	16	100	100	nr	34–1,600*	525*	
nr	nr	nr	22	22	nr	nd–270*	nd*	
nr	nr	nr	—	—	—	—	—	Objective: to determine the distribution of DDT residues in the Salinas River, and their mode and rate of translocation. Total DDT levels in sediment were <10 µg/kg except at 1 site. This site (Davis R.) had lower water flow and finer-grained sediment than other sites. At another site, suspended sediment sampled during high flow averaged 76 µg/kg, which was about 3 times the residues in bottom sediment at the same time. The main component of total DDT was *p,p'*- DDT.
—	—	—	0	0	10	nd	nr	Objective: to measure baseline pesticide residues in natural fish populations. Species diversity varied widely at 1 site and was fairly constant at 4 sites. No trend in residues with respect to trophic level was evident. Fish were clean relative to national averages from the FWS's NCBP.
—	—	—	40	2	10	nd–20	nr	
—	—	—	100	82	10	nd–90	nr	
—	—	—	20	1	10	nd–10	nr	
—	—	—	80	2	10	nd–20	nr	
—	—	—	80	15	10	nd–40	nr	
—	—	—	0	0	10	nd	nr	
—	—	—	0	0	10	nd	nr	
—	—	—	40	1	10	nd–trace	nr	

Table 2.2. Pesticides in bed sediment and aquatic biota from rivers in the United States: State and local monitoring studies—*Continued*

Reference	Sampling dates	Sampling locations and strategy	Sampling media and species	Number of samples	Target analytes	Bed sediment % Detection Sites	Bed sediment % Detection Samples
		developing.	gonad).				
McPherson, 1973	1970–1972	13 (sediment), 11 (water), and 7 (fish) sites in the conservation areas of Central and Southern Florida Flood Control District. Sites included canals and reservoirs.	Sediment, fish (tissue not reported), water. Species include lake chubsucker, golden shiner, bluegill, spotted sunfish, American eel, largemouth bass.	40 49 29	Aldrin Chlordane D, 2,4- DDD DDE DDT DDT, total Dieldrin Endrin Heptachlor Lindane Silvex T, 2,4,5- Toxaphene	0 8 14 100 100 54 100 62 0 0 0 29 0 —	0 6 8 81 81 21 84 21 0 0 0 23 0 —
Bulkley and others, 1974	1971–1973	A single site on the Des Moines River, Iowa, about 265 miles upstream from its mouth. Much of the drainage basin is cropland.	Water, sediment (surficial, suspended), fish (whole, fillet), invertebrates. Species include channel catfish, spotfin shiner, sand shiner, bluntnose minnow, carpsucker, crayfish, aquatic insects.	17 70 103 107	Dieldrin	100	65
Ginn and Fisher, 1974	nr	Unspecified number of canals associated with a marshland-ricefield ecosystem on the Texas coast.	Fish (whole, liver+muscle, eggs), Species: spotted gar, bluegill, mullet, menhaden, Altantic croaker, pipefish.	15 12 2	DDD DDE DDT	— — —	— — —

Table 2.2. Pesticides in bed sediment and aquatic biota from rivers in the United States: State and local monitoring studies—*Continued*

Bed sediment (µg/kg dry weight)			Aquatic biota (whole fish unless noted)					Comments
			% Detection		(µg/kg wet weight)			
DL	Range	Median	Sites	Samples	DL	Range	Median	
0.1	nd	nd	0*	0*	nr	nd*	nd*	* Data are for fish (tissue type not reported). Objective: to provide background data on water quality. Conservation areas were bounded by agricultural land (north), undeveloped wetlands (south), and partly drained land that is becoming increasingly urban and agricultural (east). Pesticides and suspended solids were higher in the north than south (e.g., total DDT in sediment averaged 192 µg/kg in the north and 13.8 µg/kg in the south).
1	nd–50	nd	0*	0*	nr	nd*	nd*	
0.1	nd–0.2	nd	—	—	—	—	—	
0.1	nd–870	6.2	—	—	—	—	—	
0.1	nd–740	2.9	—	—	—	—	—	
0.1	nd–8.1	nd	—	—	—	—	—	
0.1	nd–1,618	7.7	100*	100*	nr	5.7–>805*	100*	
0.1	nd–17	nd	100*	57*	nr	nd–130*	1*	
0.1	nd	nd	0*	0*	nr	nd*	nd*	
0.1	nd	nd	0*	0*	nr	nd*	nd*	
0.1	nd–78	nd	—	—	—	—	—	
0.1	nd	nd	—	—	—	—	—	
—	—	—	100*	93*	nr	nd–5,000*	60*	
0.1	nd–5.1	2.2	100	100	1	5–170	37.5	Objective: to determine dieldrin contamination in channel catfish in part of the Des Moines R.; to define the pathways of exposure. Dieldrin was detected in water, suspended sediment, bed sediment, crayfish, aquatic insects, catfish muscle tissue, and minnows. Dieldrin levels in catfish tissue varied with size and age of individuals but not with fat content. Seasonal trends in dieldrin concentration were observed in water, invertebrates, minnows, and catfish. The observed trends could not be directly related to dieldrin levels in river water. Average bioaccumulation factors of dieldrin in tissue relative to river water were 4,620 for aquatic insects, 7,540 for minnows and 6,920 for catfish.
—	—	—	nr	13	nr	nd–17.4	nd	Objective: to investigate organochlorine residues in a marshland-estuarine-ricefield ecosystem. Aldrin and dieldrin data are shown in Table 2.3 because aldrin was applied during the study. Residues of DDTs in whole-body bluegill, menhaden, coaker, pipefish, mullet (shown in this table) were much smaller than residues in 1:1 mixture of
—	—	—	nr	80	nr	nd–48.6	16.9	
—	—	—	nr	7	nr	nd–13.3	nd	

Table 2.2. Pesticides in bed sediment and aquatic biota from rivers in the United States: State and local monitoring studies—*Continued*

Reference	Sampling dates	Sampling locations and strategy	Sampling media and species	Number of samples	Target analytes	Bed sediment % Detection	
						Sites	Samples
Goerlitz and Law, 1974	nr	6 sites in rivers in California and Arizona. Samples were analyzed by particle-sized fractions: gravel (>2 mm), sand (2–0.062 mm), silt (0.062–0.004 mm), and clay (<0.004 mm).	Sediment (surficial).	6	DDD DDE DDT Lindane Total pesticides	83 83 83 17 83	83 83 83 17 83
Irwin and Lemons, 1974	1971–1972	5 sites (sediment) and 13 sites (water) in the Eel, Salinas, and Santa Ana rivers (California). One composite sediment sample was collected from each site. Water samples were taken 3–6 times between Oct. 1971 and July 1972.	Water (whole), sediment (surficial).	63 5	Aldrin Chlordane D, 2,4- DDD DDE DDT Dieldrin Endrin Heptachlor Heptachlor epoxide Lindane Silvex T, 2,4,5-	nr 60 0 ≥40 ≥40 nr nr nr nr nr nr 0 0	nr 60 0 ≥40 ≥40 nr nr nr nr nr nr 0 0
Johnson and Morris, 1974	1971	5 sites in the East Nishnabotna, West Nishnabotna, Mississippi, and Iowa rivers (Iowa). Samples were taken from individual fish.	Fish (eggs). Species: channel catfish, largemouth bass, walleye, northern pike.	13	Aldrin Chlordane, α- + γ- DDD DDE DDT Dieldrin Heptachlor epoxide	— — — — — — —	— — — — — — —

Table 2.2. Pesticides in bed sediment and aquatic biota from rivers in the United States: State and local monitoring studies—*Continued*

Bed sediment (µg/kg dry weight)			Aquatic biota (whole fish unless noted)					Comments
			% Detection		(µg/kg wet weight)			
DL	Range	Median	Sites	Samples	DL	Range	Median	
								liver and muscle tissue of spotted gar. For example, DDE residues in spotted gar liver-muscle were 524 (maximum) and 2,110 (median) µg/kg.
nr	nd–3,390*	157*	—	—	—	—	—	* Data are for clay fraction only. Objective: to determine the distribution of OCs by particle class in bed materials. The highest concentration was not always in the finer material. In 1 sample, the gravel fraction (mostly clam shells) contained the highest concentrations. Coarser fractions contained the highest percent of OC load for 4 of the 6 samples. Distribution of OCs may be controlled by organic matter and associated organisms.
nr	nd–1,380*	96*	—	—	—	—	—	
nr	nd–5,020*	141*	—	—	—	—	—	
nr	nd–221*	nd	—	—	—	—	—	
nr	nd–10,000*	444*	—	—	—	—	—	
0.1	nr	nr	—	—	—	—	—	Objective: to assess water quality in the named rivers. Pesticides were virtually undetected in water and sediment samples from the Eel R. Pesticides in water were detected frequently in the Santa Ana R. and in the Salinas R. at Spreckels. Sediment from the Salinas R. at Spreckels had detectable OC pesticides, principally chlordane. Chlordane, DDD, and DDE were detected in sediment from 2 of 3 sites on the Santa Ana R. Concentrations of many water-quality variables were quite different within rivers, especially the Salinas and Santa Ana rivers.
0.1	nd–47	nr	—	—	—	—	—	
0.1	nd	nd	—	—	—	—	—	
0.1	nr	nr	—	—	—	—	—	
0.1	nr	nr	—	—	—	—	—	
0.1	nr	nr	—	—	—	—	—	
0.1	nr	nr	—	—	—	—	—	
0.1	nr	nr	—	—	—	—	—	
0.1	nr	nr	—	—	—	—	—	
0.1	nr	nr	—	—	—	—	—	
0.1	nr	nr	—	—	—	—	—	
0.1	nd	nd	—	—	—	—	—	
0.1	nd	nd	—	—	—	—	—	
—	—	—	40*	31*	nr	nd–175*	nd*	* Data are for fish eggs. Objective: to investigate dieldrin residues in fish eggs from Iowa rivers. The highest residues in eggs occurred at sites where fish flesh residues were highest in previous studies. Dieldrin level was correlated with lipid content at 3 sites where multiple species were sampled.
—	—	—	60*	54*	nr	nd–350*	24*	
—	—	—	100*	100*	nr	30–180*	98*	
—	—	—	100*	100*	nr	57–360*	181*	
—	—	—	100*	100*	nr	15–175*	49*	
—	—	—	100*	100*	nr	37–950*	455*	
—	—	—	100*	100*	nr	5–93*	51*	

Table 2.2. Pesticides in bed sediment and aquatic biota from rivers in the United States: State and local monitoring studies—*Continued*

Reference	Sam-pling dates	Sampling locations and strategy	Sampling media and species	Number of samples	Target analytes	Bed sediment % Detection	
						Sites	Sam-ples
Kuhr and others, 1974	1972	1 site on a stream in an apple orchard in New York on the south shore of Lake Ontario. Fish were fingerlings.	Water, sediment, soil (cores), fish, snails, tadpoles, frogs, worms, slugs. Species: not reported.	nr nr 48 2 2 2	DDD DDE DDT DDT, total	100 100 100 100	nr nr nr nr
Law and Goerlitz, 1974	1972	39 sites on 26 tributary streams to the San Francisco Bay (California). Sediment samples were wet-sieved (2 mm).	Sediment (surficial).	39	Chlordane DDD DDE DDT, *o,p'*- DDT, *p,p'*-	92 97 95 100 95	92 97 95 100 95
Miller and Gomes, 1974	1971–1973	2 sites (sediment) and 4 sites (fish) in the lower Rio Grande Valley (Texas). From 1971 to 1973, water and sediment were sampled monthly at 2 sites and fish (menhaden) at 1 site. Fish were sampled at 4 sites during 1973–1974. Air was sampled at 2 sites during 1972–1973.	Sediment (surficial), fish (whole, flesh, liver, ovaries, testes, skin, viscera), water (whole), air. Species include menhaden (whole), mullet, sea trout (other organs).	72 35 94 92 71 19 1 36	Dacthal	50	1
Schacht, 1974	1970–1972	41 sites (sediment) and 15 sites (water) in Lake Michigan and tributary streams (Illinois). Fish were collected from Lake Michigan near	Sediment, water, fish (edible). Species: yellow perch, coho salmon, chubs,	50 45 54	Aldrin DDD, *o,p'*- DDD, *p,p'*- DDD, total DDE, *o,p'*- DDE, *p,p'*- DDT, *o,p'*- DDT, *p,p'*-	— 93* 100* — 87* 100* 93* 100*	— 94* 100* — 89* 100* 95* 100*

Table 2.2. Pesticides in bed sediment and aquatic biota from rivers in the United States: State and local monitoring studies—*Continued*

Bed sediment (µg/kg dry weight)			Aquatic biota (whole fish unless noted)					Comments
			% Detection		(µg/kg wet weight)			
DL	Range	Median	Sites	Samples	DL	Range	Median	
nr	nr	120*	100	nr	nr	nr	500*	* Mean values. Objective: to measure DDT residues in soil of apple orchards and in a stream running through an orchard. DDT was applied in study area from 1947 to 1960. Soil residues were higher under trees than in aisles and decreased with increasing soil depth. DDT and DDD were found in water samples, but not DDE. DDT, DDE, and DDD were found in all biota species. Residues in soil biota exceeded residues in aquatic biota. In soil, 70–90% of total DDT was DDT. In biota, 30–50% was DDT.
nr	nr	60*	100	nr	nr	nr	600*	
nr	nr	80*	100	nr	nr	nr	500*	
nr	nr	nr	100	nr	nr	nr	1,600*	
0.1	nd–800	45	—	—	—	—	—	Objective: to assess OC contamination in San Francisco Bay. OC pesticide contamination was widespread in tributary sediments. There was no significant different between streams discharging into the Bay north and south of San Francisco.
0.1	nd–160	8.8	—	—	—	—	—	
0.1	nd–61	5.5	—	—	—	—	—	
0.1	0.57–200	7.1	—	—	—	—	—	
0.1	nd–89	2.4	—	—	—	—	—	
nr	nd–42	nd	100	34	nr	nd–8,150	nd	Objective: to investigate DCPA contamination in the study area. DCPA was detected in 50% of monthly water samples during 1971–1972, and in 100% in 1973. The maximum level detected was 10 µg/L. DCPA was not detected in monthly menhaden samples from Jan. 1971 to March 1972. After that, it was detected in 11 of the next 20 monthly samples. The amount of DCPA in menhaden (sampled monthly at 1 site) varied widely. DCPA residues in trout were generally higher in liver than in flesh or gonads.
—	—	—	0†	0†	100	nd†	nd†	* For sites in tributaries only. (excluding offshore lake sites). † Data are for edible fish tissue. Objective: to determine present levels of pesticides in Lake Michigan water, sediment, and fish. DDT compounds and dieldrin were found consistently
nr	nd–62*	8*	100†	26†	100	nd–200†	nd†	
nr	0.08–353*	30.6*	100†	77†	100	nd–1,100†	100†	
—	—	—	100†	81†	100	nd–1,300†	100†	
nr	nd–17*	0.45*	100†	58†	100	nd–300†	100†	
nr	0.17–67*	6.4*	100†	94†	100	nd–2,000†	600†	
nr	nd–83*	4.54*	100†	59†	100	nd–600†	100†	
nr	0.11–375*	18*	100†	74†	100	nd–1,700†	200†	

Table 2.2. Pesticides in bed sediment and aquatic biota from rivers in the United States: State and local monitoring studies—*Continued*

Reference	Sampling dates	Sampling locations and strategy	Sampling media and species	Number of samples	Target analytes	Bed sediment % Detection	
						Sites	Samples
		Chicago and Waukegan. Fish samples were composites of about 5 individuals (by size and sex).	brown trout, carp, alewife.		DDT, total	100*	100*
					Dieldrin	100*	100*
					Endrin	—	—
					Heptachlor	85*	85*
					Heptachlor epoxide	100*	100*
					Lindane	31*	31*
					Methoxychlor	100*	100*
Claeys and others, 1975	1970–1973	19 sites in estuaries along the Oregon–Washington coast and in the Rogue River. Estuarine bivalves were collected quarterly from 5 estuaries (Oregon). Estuarine fish were collected from Coos Bay and Columbia River estuaries, and summer-run steelhead from the Rogue River.	Fish (whole), shrimp, euphausiids, mollusks. Species include Asiatic clam, cockle clam, bay mussel, starry flounder, sand sole, staghorn sculpin, finescale sucker, cod, steelhead.	44 32 29 41	Chlordane	—	—
					DDE, *p,p'-*	—	—
					DDD, *p,p'-*	—	—
					DDT, *p,p'-*	—	—
					DDT, total	—	—
					Dieldrin	—	—
					Endosulfan	—	—
Mattraw, 1975	1968–1972	Unspecified number of sites in southern Florida. Sites were located in agricultural, urban, and undeveloped wetland areas.	Sediment (surficial), water (whole).	287 368	Aldrin	nr	2*
					Chlordane	nr	32*
					DDD	nr	78*
					DDE	nr	78*
					DDT	nr	37*
					Dieldrin	nr	52*
					Endrin	nr	0*
					Heptachlor	nr	0*
					Heptachlor epoxide	nr	0*
					Lindane	nr	1*
					Toxaphene	nr	3*
Shampine, 1975	1972	13 sites in the White River Basin (Indiana), in areas affected by agricultural and urban land use.	Sediment (surficial)	13	Aldrin	8	8
					Chlordane	77	77
					DDD	69	69
					DDE	0	0
					DDT	0	0
					Dieldrin	100	100
					Endrin	0	0

Table 2.2. Pesticides in bed sediment and aquatic biota from rivers in the United States: State and local monitoring studies—*Continued*

Bed sediment (µg/kg dry weight)			Aquatic biota (whole fish unless noted)					Comments
DL	Range	Median	% Detection Sites	Samples	DL	Range (µg/kg wet weight)	Median	
nr	0.42–942*	79.35*	100†	100†	100	100–4,600†	1,450†	in fish, but below FDA action levels. Tributary sediments had higher pesticide levels than offshore lake sediments.
nr	0.01–31*	1.37*	100†	63†	100	nd–500†	100†	
—	—	—	0†	0†	100	nd†	nd†	
nr	nd–0.24*	0.05*	0†	0†	100	nd†	nd†	
nr	0.02–57*	2.6*	100†	13†	100	nd–100†	nd†	
nr	nd–0.15*	nd*	33†	3†	100	nd–trace†	nd†	
nr	0.19–175*	6.9*	33†	7†	100	nd–100†	nd†	
—	—	—	33*	7*	nr	nd–6*	nd*	* Data are for fish from Rogue R., Coos Bay, and Columbia R. † Data are for 1 site (Rogue R.) Objective: to collect baseline data on contaminants in the North Pacific Oregon estuaries. Columbia R. fish contained much higher OC levels than those from Coos Bay. Summer-run steelhead from Rogue R. contained the highest DDE and dieldrin levels of all species sampled. DDT levels were higher in mollusks from Columbia R. estuary than from other Oregon estuaries. Residues in mollusks were ≤4 µg/kg.
—	—	—	100*	93*	nr	nd–140*	40*	
—	—	—	33*	21*	nr	nd–62*	nd*	
—	—	—	33*	14*	nr	nd–143*	nd*	
—	—	—	100*	93*	nr	nd–143*	58*	
—	—	—	100*†	83*†	nr	nd–29*†	20*†	
—	—	—	33*	7*	nr	nd–6*	nd*	
0.05	nr	nr	—	—	—	—	—	* Estimated from figure. Objective: to determine OC insecticides in water and sediment in different land use areas in southern Florida. DDT, DDE, DDD, dieldrin, and lindane were detected in whole water. DDD levels in sediment were higher in agricultural and urban areas than in undeveloped wetlands. Dieldrin in sediment was more frequently detected and at higher concentrations in the urban area (perhaps due to use against termites) than in other land-use areas. Maximum residues in natural (undeveloped) areas were 6 µg/kg DDD, 9 µg/kg DDE, and <1 µg/kg dieldrin.
0.05	nr	nr	—	—	—	—	—	
0.05	nr	nr	—	—	—	—	—	
0.05	nr	nr	—	—	—	—	—	
0.05	nr	nr	—	—	—	—	—	
0.05	nr	nr	—	—	—	—	—	
0.05	nr	nr	—	—	—	—	—	
0.05	nr	nr	—	—	—	—	—	
0.05	nr	nr	—	—	—	—	—	
0.05	nr	nr	—	—	—	—	—	
0.05	nr	nr	—	—	—	—	—	
nr	nd–2.1	nd	—	—	—	—	—	* Detection limit for chlordane was reported as 0. Objective: to evaluate water quality in the upper White R. using mostly available data. Water quality is degraded where the river flows through urban
nr*	nd–20	7	—	—	—	—	—	
nr	nd–10	2	—	—	—	—	—	
nr	nd	nd	—	—	—	—	—	
nr	nd	nd	—	—	—	—	—	
nr	0.2–4.6	1.2	—	—	—	—	—	
nr	nd	nd	—	—	—	—	—	

Table 2.2. Pesticides in bed sediment and aquatic biota from rivers in the United States: State and local monitoring studies—*Continued*

Reference	Sampling dates	Sampling locations and strategy	Sampling media and species	Number of samples	Target analytes	Bed sediment % Detection Sites	Bed sediment % Detection Samples
					Heptachlor	0	0
					Lindane	0	0
Truhlar and Reed, 1975	1968–1971	4 sites (sediment) and 3 sites (fish) in Pennsylvania streams draining different land uses: forested, general farming, residential, and orchards. Sediment was collected from depositional areas 4–7 times during the study period. Fish were collected once during the study period and were composited by species and age.	Sediment (surficial), water (whole), fish (whole), soil. Species: blacknose dace, northern creek chub, white sucker.	21 83 12	Aldrin Chlordane DDD DDE DDT Dieldrin Endrin Lindane	— 25 100 100 100 75 25 0	— 5 71 62 67 33 10 0
Bulkley and others, 1976	1971–1973	1 site on the Des Moines River near Boone, Iowa. Samples were composited by length or analyzed individually.	Fish (muscle). Species: channel catfish.	nr	Dieldrin	—	—

Table 2.2. Pesticides in bed sediment and aquatic biota from rivers in the United States: State and local monitoring studies—*Continued*

Bed sediment (µg/kg dry weight)			Aquatic biota (whole fish unless noted)					Comments
			% Detection		(µg/kg wet weight)			
DL	Range	Median	Sites	Sam-ples	DL	Range	Median	
nr	nd	nd	—	—	—	—	—	areas, with the most severe
nr	nd	nd	—	—	—	—	—	degradation occurring
								downstream of Indianapolis.
—	—	—	0	8	0.1	nd–trace	nd	Objective: to determine the
0.1	nd–250	nd	—	—	—	—	—	relative pesticide contamination
0.1	nd–230	0.8	67	50	0.1	nd–110	nd/4.6	in four small drainage basins
0.1	nd–60	0.6	100	100	nr	15–350	27	representing different land uses.
0.1	nd–5900	0.5	33	8	0.1	nd–370	nd	Total DDT residues in soil were
0.1	nd–10	nd	67	42	0.1	nd–9.0	nd	highest in the orchard area
0.1	nd–120	nd	—	—	—	—	—	(40,000 µg/kg in the top 0.5
0.1	nd	nd	0	0	0.1	nd	nd	inch) and lowest in the forested
								area. In sediment, total DDT
								was highest in streams
								from residential and orchard
								areas, and lowest from forested
								and general farming areas. Very
								low pesticide levels were ob-
								served in water from 2 of 4
								streams during base flow and
								storm-runoff. In the other 2
								streams, pesticide detection
								frequencies and concentrations
								were higher in storm-runoff
								than baseflow samples. In these
								2 streams, total DDT levels in
								whole water correlated fairly
								well with suspended sediment
								levels (but with considerable
								scatter).
—	—	—	100	nr	nr	5–940*†	nr	* Data are for muscle tissue.
								† Range is of mean values for
								fish grouped by month and
								length.
								Objective: to determine residues
								in relation to age, length, sex
								and season. Dieldrin residues in-
								creased with increasing age and
								length, except in fish <400 mm.
								Residues in males and females
								were not significantly different.
								Fat content was correlated with
								age and length, but not with diel-
								drin residues. Seasonal trends
								varied for 200–299 and 300–399
								mm groups; peak residues oc-
								curred in Oct. and July,
								respectively.

Table 2.2. Pesticides in bed sediment and aquatic biota from rivers in the United States: State and local monitoring studies—*Continued*

Reference	Sampling dates	Sampling locations and strategy	Sampling media and species	Number of samples	Target analytes	Bed sediment % Detection Sites	Bed sediment % Detection Samples
Demas, 1976	1975	164 sites (sediment) and 123 sites (water) in selected reaches of 12 major navigable waterways in Louisiana, including the Mississippi River. Samples were collected twice at each site. Elutriate tests were performed for each site and sampling date.	Water (whole), sediment (surficial), elutriates.	272 384 268	Aldrin	7*	3.5*
					Chlordane	55*	46*
					DDD	83*	71*
					DDE	30*	19*
					DDT	72*	55*
					Diazinon	1.3*	0.66*
					Dieldrin	75*	61*
					Endrin	34*	20*
					Ethion	0*	0*
					Heptachlor	1.2*	0.58*
					Heptachlor epoxide	17*	11*
					Lindane	0*	0*
					Malathion	0*	0*
					Methyl parathion	0*	0*
					Methyl trithion	0*	0*
					Parathion	0*	0*
					Toxaphene	0*	0*
					Trithion	0*	0*
Iwatsubo and others, 1976	1974	10 sites in the Redwood Creek drainage basin in Redwood National Park, California.	Sediment (surficial).	10	Aldrin	0	0
					Chlordane	0	0
					D, 2,4-	0	0
					DDD	0	0
					DDE	0	0
					DDT	0	0
					Diazinon	0	0
					Dieldrin	0	0
					Endrin	0	0
					Ethion	0	0
					Heptachlor	0	0
					Heptachlor epoxide	0	0
					Lindane	0	0
					Malathion	0	0
					Methyl parathion	0	0
					Methyl trithion	0	0
					Parathion	0	0
					Silvex	0	0
					T, 2,4,5-	0	0
					Toxaphene	0	0
					Trithion	0	0
Kellogg and Bulkley, 1976	1971–1973	1 site on the Des Moines River, Iowa. Catfish and potential catfish food organisms were sampled. Sediment was sampled monthly. Biota were sampled between April and November.	Water (dissolved), sediment (surficial, suspended), fish (muscle, whole), insects. Species include chan-nel catfish	nr nr nr	Dieldrin	100	nr

Table 2.2. Pesticides in bed sediment and aquatic biota from rivers in the United States: State and local monitoring studies—*Continued*

Bed sediment (µg/kg dry weight)			Aquatic biota (whole fish unless noted)					Comments
			% Detection		(µg/kg wet weight)			
DL	Range	Median	Sites	Samples	DL	Range	Median	
0.1	nd–1.3	nd	—	—	—	—	—	* Data are for Mississippi R. samples only (171 samples at 83 sites). Objective: to evaluate possible effects of proposed dredging activities in major Louisiana waterways. Reports contains raw data, but no summary statistics or interpretation. An elutriate is the supernatant from shaking 1 part bottom sediment from a site with 4 parts native water from that site, followed by settling, centrifugation, and filtration.
1	nd–32	nd	—	—	—	—	—	
0.1	nd–20	nr	—	—	—	—	—	
0.1	nd–6.4	nd	—	—	—	—	—	
0.1	nd–14	nr	—	—	—	—	—	
0.1	nd–13	nd	—	—	—	—	—	
0.1	nd–8	nr	—	—	—	—	—	
0.1	nd–3	nd	—	—	—	—	—	
0.1	nd	nd	—	—	—	—	—	
0.1	nd–0.3	nd	—	—	—	—	—	
0.1	nd–3	nd	—	—	—	—	—	
0.1	nd	nd	—	—	—	—	—	
0.1	nd	nd	—	—	—	—	—	
0.1	nd	nd	—	—	—	—	—	
0.1	nd	nd	—	—	—	—	—	
0.1	nd	nd	—	—	—	—	—	
1	nd	nd	—	—	—	—	—	
0.1	nd	nd	—	—	—	—	—	
nr	nd	nd	—	—	—	—	—	Objective: to describe pesticide residues in bottom sediment in the Redwood Creek Basin as part of a larger study of Redwood National Park ecosystems, erosion and sedimentation processes, and the effects of road construction and timber harvest. No pesticides were detected in bottom sediment.
nr	nd	nd	—	—	—	—	—	
nr	nd	nd	—	—	—	—	—	
nr	nd	nd	—	—	—	—	—	
nr	nd	nd	—	—	—	—	—	
nr	nd	nd	—	—	—	—	—	
nr	nd	nd	—	—	—	—	—	
nr	nd	nd	—	—	—	—	—	
nr	nd	nd	—	—	—	—	—	
nr	nd	nd	—	—	—	—	—	
nr	nd	nd	—	—	—	—	—	
nr	nd	nd	—	—	—	—	—	
nr	nd	nd	—	—	—	—	—	
nr	nd	nd	—	—	—	—	—	
nr	nd	nd	—	—	—	—	—	
nr	nd	nd	—	—	—	—	—	
nr	nd	nd	—	—	—	—	—	
nr	nd	nd	—	—	—	—	—	
nr	nd	nd	—	—	—	—	—	
nr	nd	nd	—	—	—	—	—	
nr	<0.1–4 *	nr	100†	nr	nr	nd–207†	45§	* Range in monthly mean values. † Catfish muscle. § Mean value in catfish muscle. Objective: to determine seasonal patterns in dieldrin levels in channel catfish, and catfish-food organisms. Dieldrin levels in water were highest in June (after application). Residues were highest in July

Table 2.2. Pesticides in bed sediment and aquatic biota from rivers in the United States: State and local monitoring studies—*Continued*

Reference	Sampling dates	Sampling locations and strategy	Sampling media and species	Number of samples	Target analytes	Bed sediment % Detection	
						Sites	Samples
			(muscle), minnows, carpsucker (whole), crayfish, mayfly naiads.				
Laska and others, 1976	1975	Water and sediment were collected at 29 sites on the Mississippi River (Louisiana) and at 7 other sites not on the main river channel. Fish were collected at 3 sites and crayfish at 4 sites.	Water, sediment (surficial), fish, crayfish. Species: mosquitofish, *Procambarus clarki*)	36 36 3 4	Hexachlorobenzene Hexachloro- butadiene	61 28	61 28
Nelson, 1976	1974	Sediment was collected at 5 sites along the length of the Lower Santee River, South Carolina.	Water, sediment (surficial).	15	DDD DDE DDT Dieldrin Endrin	100 100 20 80 20	73 100 7 40 7
Truhlar and Reed, 1976	1969– 1971	Four small drainage basins in central Pennsylvania, each representing a land-use area: forested, general farming, residential, and orchards.	Water, soil, sediment (surficial), fish. Species: blacknose dace, white sucker, northern creek chub.	>80 6 21 12	Aldrin Chlordane DDD DDE DDT Dieldrin Endrin Lindane	0 25 100 100 100 75 25 0	0 5 71 62 67 33 10 0
Archer and Turk, 1977	1976	9 streams in Westchester County, New York. Streams with known pollution problems and reference streams were sampled.	Sediment (surficial).	20	Aldrin Chlordane D, 2,4- DDD DDE DDT Diazinon Dieldrin Endrin Ethion Heptachlor Heptachlor epoxide Lindane Malathion	0 94 0 88 59 82 6 76 0 0 0 53 0 0	0 95 0 90 55 80 5 75 0 0 0 45 0 0

Table 2.2. Pesticides in bed sediment and aquatic biota from rivers in the United States: State and local monitoring studies—*Continued*

Bed sediment (µg/kg dry weight)			Aquatic biota (whole fish unless noted)					Comments
			% Detection		(µg/kg wet weight)			
DL	Range	Median	Sites	Samples	DL	Range	Median	
								(sediment), June–July (water, insects, forage fish), July–August (catfish). Catfish of 200–299 mm length had the highest dieldrin residues.
0.7	nd–~900	nr	100	100	0.7	72–380	137	Objective: to determine the extent of HCB and HCBD contamination in the lower Mississippi R. Concentrations in water were generally <2 µg/L. Highest levels were downstream of an industrialized area. Fish residues at that site represent bioaccumulation factors of 172 for HCB and 435 for HCBD.
0.7	nd–~800	nr	100	100	0.7	113–827	197	
nr	nd–3.7	0.4	—	—	—	—	—	Objective: to determine baseline water quality data for use in future trend analysis. The report also surveys surveys aquatic plants and small mammals in the study area.
nr	0.1–2.5	0.2	—	—	—	—	—	
nr	nd–2.6	nd	—	—	—	—	—	
nr	nd–1.2	nd	—	—	—	—	—	
nr	nd–0.2	nd	—	—	—	—	—	
nr	nd	nd	33	8	nr	nd–trace	nd	Objective: to describe the extent of contamination of streams in basins characterized by different land uses. DDTs were detected in sediment of all watersheds, and dieldrin in all except forested. Endrin was only detected in sediment from the orchard use area. Total DDT was highest in sediment and fish of the residential use basin. Maximum total DDT in water generally occurred in storm runoff.
nr	nd–250	nd	0	0	nr	nd	nd	
nr	nd–230	0.8	67	50	nr	nd–110	2.3	
nr	nd–60	0.6	100	100	nr	15–350	27.5	
nr	nd–5,900	0.5	33	8	nr	nd–370	nd	
nr	nd–10	nd	67	42	nr	nd–9	nd	
nr	nd–120	nd	0	0	nr	nd	nd	
nr	nd	nd	0	0	nr	nd	nd	
nr	nd	nd	—	—	—	—	—	Objective: to obtain baseline data on water quality in selected streams in Westchester County at low flow. Pesticides also were analyzed in 6 bottom material samples from 6 lakes as part of eutrophication study. (Results not reported in this table.) Data were presented without interpretation of results. No information on sampling methods was provided.
nr	nd–530	14	—	—	—	—	—	
nr	nd	nd	—	—	—	—	—	
nr	nd–200	29.5	—	—	—	—	—	
nr	nd–7.6	0.25	—	—	—	—	—	
nr	nd–120	3.8	—	—	—	—	—	
nr	nd–6.3	nd	—	—	—	—	—	
nr	nd–37	0.55	—	—	—	—	—	
nr	nd	nd	—	—	—	—	—	
nr	nd	nd	—	—	—	—	—	
nr	nd	nd	—	—	—	—	—	
nr	nd–16	nd	—	—	—	—	—	
nr	nd	nd	—	—	—	—	—	
nr	nd	nd	—	—	—	—	—	

Table 2.2. Pesticides in bed sediment and aquatic biota from rivers in the United States: State and local monitoring studies—*Continued*

Reference	Sampling dates	Sampling locations and strategy	Sampling media and species	Number of samples	Target analytes	Bed sediment % Detection Sites	Bed sediment % Detection Samples
					Methyl parathion	0	0
					Methyl trithion	0	0
					Parathion	0	0
					Silvex	0	0
					T, 2,4,5-	0	0
					Toxaphene	0	0
					Trithion	0	0
Demas, 1977	1976	14 sites on the Atchafalaya River and 5 sites in Atchafalaya Bay (Louisiana). Sediment samples were cores (50 ft.) on the river bank at all river sites. The entire core was composited. At 4 bay sites, surficial sediment (top 6 in.) was sampled.	Water (whole), sediment (cores, surficial), elutriates.	19 15 4	Aldrin	0*	0*
					Chlordane	0*	0*
					DDD	42*	42*
					DDE	21*	21*
					DDT	0*	0*
					Diazinon	0*	0*
					Dieldrin	16*	16*
					Endrin	0*	0*
					Ethion	0*	0*
					Heptachlor	0*	0*
					Heptachlor epoxide	0*	0*
					Lindane	0*	0*
					Malathion	0*	0*
					Methyl parathion	0*	0*
					Methyl trithion	0*	0*
					Mirex	0*	0*
					Parathion	0*	0*
Green and others, 1977	1972– 1973	15 sites (sediment) in 2 estuaries (Walker and Kaiaka Bays) and their influent streams on Oahu, Hawaii. Sediment was collected on 3 sampling dates 6–7 months apart. Water was sampled at 9 sites, both near the bottom and near the surface. Soil and runoff samples were taken in a field dissipation study (results shown in Table 2.3).	Water (whole), sediment (surficial), soil.	18 38	Aldrin	0	
					Ametryn	20	16
					Atrazine	0	0
					Chlordane, α-	100	
					Chlordane, γ-	100	
					DDD	100	
					DDE	100	
					DDT	100	
					3,4-Dichloroaniline	60	47
					Dieldrin	67	
					Diuron	100	97
					Heptachlor	0	
					Heptachlor epoxide	0	
					Lindane	0	

Table 2.2. Pesticides in bed sediment and aquatic biota from rivers in the United States: State and local monitoring studies—*Continued*

Bed sediment (µg/kg dry weight)			Aquatic biota (whole fish unless noted)					Comments
			% Detection		(µg/kg wet weight)			
DL	Range	Median	Sites	Samples	DL	Range	Median	
nr	nd	nd	—	—	—	—	—	
nr	nd	nd	—	—	—	—	—	
nr	nd	nd	—	—	—	—	—	
nr	nd	nd	—	—	—	—	—	
nr	nd	nd	—	—	—	—	—	
nr	nd	nd	—	—	—	—	—	
nr	nd	nd	—	—	—	—	—	
0.1	nd*	nd*	—	—	—	—	—	* Data are for cores and surficial
1	nd*	nd*	—	—	—	—	—	samples, aggregated.
0.1	nd–5.8*	nd*	—	—	—	—	—	Objective: to evaluate possible
0.1	nd–3.6*	nd*	—	—	—	—	—	effects of proposed
0.1	nd*	nd*	—	—	—	—	—	channel-enlargement project.
0.1	nd*	nd*	—	—	—	—	—	Report contains raw data, but no
0.1	nd–0.6*	nd*	—	—	—	—	—	summary statistics or
0.1	nd*	nd*	—	—	—	—	—	interpretation. An elutriate is the
0.1	nd*	nd*	—	—	—	—	—	supernatant from shaking 1 part
0.1	nd*	nd*	—	—	—	—	—	bottom sediment from a site
0.1	nd*	nd*	—	—	—	—	—	with 4 parts native water from
0.1	nd*	nd*	—	—	—	—	—	that site, followed by settling,
0.1	nd*	nd*	—	—	—	—	—	centrifugation, and filtration.
0.1	nd*	nd*	—	—	—	—	—	
0.1	nd*	nd*	—	—	—	—	—	
0.1	nd*	nd*	—	—	—	—	—	
0.1	nd*	nd*	—	—	—	—	—	
nr	nd	nd	—	—	—	—	—	Objective: to determine if herbi-
nr	nd–4,000	nd	—	—	—	—	—	cides used in plantation crops
nr	nd	nd	—	—	—	—	—	were reaching estuaries. Ame-
nr	3.2–6.0	4.1	—	—	—	—	—	tryn was detected only in Walk-
nr	2.7–3.7	3.7	—	—	—	—	—	er Bay sediment. Diuron and its
nr	4.9–9.5	5	—	—	—	—	—	degradation product, 3,4-di-
nr	1.7–1.9	1.8	—	—	—	—	—	chloroaniline (DCA), were
nr	5.1–20.8	15.3	—	—	—	—	—	found in both estuaries. Diuron
20	nd–16,860	nd	—	—	—	—	—	levels varied up to 20-fold be-
nr	nd–2.4	1.3	—	—	—	—	—	tween sampling dates. There was
2	nd–147,786	201	—	—	—	—	—	no correlation between levels in
nr	nd	nd	—	—	—	—	—	sediments from Kaiaka Bay and
nr	nd	nd	—	—	—	—	—	its influent streams. Very high
nr	nd	nd	—	—	—	—	—	DCA and diuron levels in one

sample probably were from point source contamination (a nearby airport). Presence of OC insecticides (used on pineapple but not on sugarcane) was used to distinguish the probable source of diuron residues. A few OCs were present at low levels; the absence of lindane and heptachlor (heavily used in pineapple) suggested that soil erosion from pineapple fields

Table 2.2. Pesticides in bed sediment and aquatic biota from rivers in the United States: State and local monitoring studies—*Continued*

Reference	Sampling dates	Sampling locations and strategy	Sampling media and species	Number of samples	Target analytes	Bed sediment % Detection Sites	Bed sediment % Detection Samples
Johnson and others, 1977	1975	4 sites (sediment, water) in American Falls Reservoir, an impoundment in the Snake River (Idaho). Fish collections were not restricted to these sites.	Sediment, water (whole), fish (whole, muscle). Species: Utah chub, yellow perch, black crappie, black bullhead, carp, and Utah sucker.	12 ≥4 179 20	Aldrin	0	0
					DDD	100	100
					DDE	100	100
					DDT	0	0
					Dieldrin	0	0
					Endrin	0	0
					Heptachlor	0	0
					Heptachlor epoxide	0	0
					Lindane	0	0
Lake and Morrison, 1977	1972–1977	Unspecified number of sites in the Maumee River Basin (Ohio, Indiana, Michigan). Sampling was at low flow.	Sediment, water (whole), fish. Species: not reported.	14 7 18	Alachlor	nr	0
					Atrazine	nr	0
					Carbofuran	nr	0
					DDE	nr	0
					Dieldrin	nr	0
					Malathion	nr	0
					T, 2,4,5-	nr	0
McKenzie, 1977	1977	2 sites in the Willamette River, Portland Harbor (Oregon). Sites were potential dredging sites.	Sediment (surficial).	2	Aldrin	100	100
					Chlordane	100	100
					DDD	100	100
					DDE	100	100
					DDT	100	100
					Diazinon	0	0
					Dieldrin	100	100
					Endosulfan	0	0
					Endrin	0	0
					Ethion	0	0
					Heptachlor	0	0
					Heptachlor epoxide	0	0
					Lindane	50	50
					Malathion	0	0
					Methoxychlor	100	100
					Methyl parathion	0	0
					Methyl trithion	0	0
					Parathion	0	0
					Trithion	0	0
Rinella and MacKenzie, 1977	1977	1 site in the Willamette River (Oregon), the site of proposed dredging. Sediment was composited from 15	Sediment, water (<0.45 µm filtered), elutriate.	1 2 2	Aldrin	0	0
					Chlordane	100	100
					DDD	100	100
					DDE	100	100
					DDT	100	100
					Dieldrin	100	1
					Endosulfan	0	0

Table 2.2. Pesticides in bed sediment and aquatic biota from rivers in the United States: State and local monitoring studies—*Continued*

Bed sediment (µg/kg dry weight)			Aquatic biota (whole fish unless noted)					Comments
			% Detection		(µg/kg wet weight)			
DL	Range	Median	Sites	Sam-ples	DL	Range	Median	
								was not a major contributor of diuron to estuaries.
nr	nd	nd	nr	0	nr	nd*	nd*§	* Fish muscle.
nr	0.53–3.60†	1.60†	nr	100	nr	0.9–781.7*	25.8*§	† Data are in µg/kg wet wt.
nr	0.45–3.80†	2.16†	nr	100	nr	1.0–104.8*	24.6*§	§ Median of the mean values for the 6 fish species sampled. Objective: to determine OCs in
nr	nd	nd	nr	0	nr	nd*	nd*§	water, sediment, game and rough
nr	nd	nd	nr	nr	nr	nd–160.4*	nd*§	fish in the American Falls
nr	nd	nd	nr	0	nr	nd*	nd*§	Reservoir. OC residues were not
nr	nd	nd	nr	0	nr	nd*	nd*§	detectable in water and were low
nr	nd	nd	nr	0	nr	nd*	nd*§	in sediment. OC residues in
nr	nd	nd	nr	0	nr	nd*	nd*§	fish varied by species. Utah sucker had the highest OC residues, especially DDD. Dieldrin was detected only in yellow perch and black bullhead.
nr	nd	nd	nr	0*	nr	nd*	nd*	* Tissue type not reported.
nr	nd	nd	nr	0*	nr	nd*	nd*	† Mean residue.
nr	nd	nd	nr	0*	nr	nd*	nd*	Objective: to study agricultural
nr	nd	nd	nr	100*	nr	7–31*	16*†	pollution in the basin. Overall,
nr	nd	nd	nr	100*	nr	4–86*	22*†	level of pesticide pollution was
nr	nd	nd	nr	0*	nr	nd*	nd*	low. Traces of 2,4,5-T were
nr	nd	nd	nr	0*	nr	nd*	nd*	detected occasionally in water.
nr	2–7	4.5	—	—	—	—	—	Objective: to provide
nr	8–10	9	—	—	—	—	—	information on sediment
nr	4.6–6.7	5.7	—	—	—	—	—	characteristics and contaminants
nr	3.7–7.5	5.6	—	—	—	—	—	from potential dredging sites in
nr	1.4–2.7	2.1	—	—	—	—	—	the Willamette River. The
nr	nd	nd	—	—	—	—	—	proposed disposal site was the
nr	1.0–2.1	1.6	—	—	—	—	—	Columbia River. Results were
nr	nd	nd	—	—	—	—	—	presented without interpretation.
nr	nd	nd	—	—	—	—	—	
nr	nd	nd	—	—	—	—	—	
nr	nd	nd	—	—	—	—	—	
nr	nd	nd	—	—	—	—	—	
nr	nd–0.8	nd/0.8	—	—	—	—	—	
nr	nd	nd	—	—	—	—	—	
nr	6.1–9.0	7.6	—	—	—	—	—	
nr	nd	nd	—	—	—	—	—	
nr	nd	nd	—	—	—	—	—	
nr	nd	nd	—	—	—	—	—	
nr	nd	nd	—	—	—	—	—	
100	nd	nd	—	—	—	—	—	Objective: to conduct elutriate
nr	15	15	—	—	—	—	—	tests on sediment from a
nr	16	16	—	—	—	—	—	proposed dredging site on the
nr	9	9	—	—	—	—	—	Willamette R. to determine
nr	1.1	1.1	—	—	—	—	—	chemicals in the sediment that
nr	0.5	0.5	—	—	—	—	—	would dissolve into water during
100	nd	nd	—	—	—	—	—	dredging operations and possibly

Table 2.2. Pesticides in bed sediment and aquatic biota from rivers in the United States: State and local monitoring studies—*Continued*

Reference	Sampling dates	Sampling locations and strategy	Sampling media and species	Number of samples	Target analytes	Bed sediment % Detection Sites	Bed sediment % Detection Samples
		midchannel grabs. Water samples for elutriate tests were collected from the proposed dredging and disposal sites.			Endrin	0	0
					Heptachlor	0	0
					Heptachlor epoxide	0	0
					Lindane	0	0
					Methoxychlor	0	0
					Perthane	0	0
					Toxaphene	0	0
Sullivan and Atchison, 1977	1974	3 sites along the Rouge River, Michigan.	Water, sediment (suspended, surficial), snails. Species: *Physa gyrina*	~10 ~10 ~10 ~10	Methoxychlor	0	0
Ayers, 1978	1975–1976	3 sites in the Middle Fork Anderson River watershed, Indiana. Two sites were sampled once and 1 site 4 times during the study.	Sediment.	6	Aldrin	0	0
					Dieldrin	67	83
					Chlordane	33	17
					DDD	33	17
					DDE	67	33
					DDT	67	33
					Endrin	0	0
					Heptachlor	0	0
					Heptachlor epoxide	0	0
					Lindane	0	0
					Toxaphene	0	0
Borsetti and Roach, 1978	nr	1 site on the James River south of Hopewell (Virginia), where kepone was manufactured. Fish samples were composites of 30 fish. Soil was collected near the manufacturing plant.	Fish (fillets), soil. Species: mullet.	1 1	9-Chloro-homolog of kepone Kepone	— —	— —
Bowers and Irwin, 1978	1974–1976	4 sites in the Santa Clara River Basin, California.	Sediment (surficial), water (whole).	4	Aldrin	0	0
					Chlordane	25	25
					DDD	0	0
					DDE	50	50
					DDT	50	50
					Diazinon	0	0
					Dieldrin	25	25
					Lindane	0	0
					Malathion	0	0

Table 2.2. Pesticides in bed sediment and aquatic biota from rivers in the United States: State and local monitoring studies—*Continued*

Bed sediment (µg/kg dry weight)			Aquatic biota (whole fish unless noted)					Comments
			% Detection		(µg/kg wet weight)			
DL	Range	Median	Sites	Samples	DL	Range	Median	
100	nd	nd	—	—	—	—	—	be toxic to aquatic life. Low
100	nd	nd	—	—	—	—	—	levels of chlordane, dieldrin, and
100	nd	nd	—	—	—	—	—	DDT metabolites were detected
100	nd	nd	—	—	—	—	—	in bottom sediment, but not in
100	nd	nd	—	—	—	—	—	the 2 native waters tested or in
100	nd	nd	—	—	—	—	—	the elutriate samples.
100	nd	nd	—	—	—	—	—	
nr	—	nd	0	0	nr	—	nd	* Snails. Objective: To determine the extent of methoxychlor contamination in an urban river in an area of current use to control Dutch Elm disease. Methoxychlor was not detected in water, sediment, or snails.
nr	nd	nd	—	—	—	—	—	Objective: to define variation in
nr	nd–1.4	1	—	—	—	—	—	OC levels in bed materials. No
nr	nd–2	nd	—	—	—	—	—	OCs were detected at the most
nr	nd–1.2	nd	—	—	—	—	—	upstream site, possibly because
nr	nd–0.6	nd	—	—	—	—	—	OCs were not present in the
nr	nd–1.2	nd	—	—	—	—	—	largely forested drainage, and(or)
nr	nd	nd	—	—	—	—	—	the flood-retarding structure there
nr	nd	nd	—	—	—	—	—	may prevent migration of any
nr	nd	nd	—	—	—	—	—	contaminated sediment. Aldrin
nr	nd	nd	—	—	—	—	—	found at 1 site (adjacent to crop-
nr	nd	nd	—	—	—	—	—	land) suggests the bed material contained recent sediments.
—	—	—	100	100	nr	40	—	Objective: to identify kepone degradation products in
—	—	—	100	100	nr	560	—	contaminated soil and fish near a kepone production facility. Kepone and its 8- and 9-chloro homologs were detected in soil at 60,000, 10, and 1,000 µg/kg, respectively.
nr	nd	nd	—	—	—	—	—	Objective: to do reconnaissance
nr	nd–2	nd	—	—	—	—	—	survey of water quality in the
nr	nd	nd	—	—	—	—	—	study basin. Aldrin, chlordane,
nr	nd–2.9	nd/2	—	—	—	—	—	DDD, DDE, DDT, diazinon,
nr	nd–3.7	nd/3.2	—	—	—	—	—	dieldrin, lindane, and malathion
nr	nd	nd	—	—	—	—	—	were detected in water samples
nr	nd–0.1	nd	—	—	—	—	—	collected at 1 or more of 6 sites.
nr	nd	nd	—	—	—	—	—	
nr	nd	nd	—	—	—	—	—	

Table 2.2. Pesticides in bed sediment and aquatic biota from rivers in the United States: State and local monitoring studies—*Continued*

Reference	Sampling dates	Sampling locations and strategy	Sampling media and species	Number of samples	Target analytes	Bed sediment % Detection	
						Sites	Samples
Bulkley, 1978	1971	1 site on the Des Moines River, Iowa. 32 samples consisting of individual fish were collected in June, and one composite sample was collected once each month in April, May, and July through October.	Fish (muscle). Species: catfish.	38	DDD DDE DDT DDT, total	— — — —	— — — —
Eikenberry, 1978	1975–1976	4 sites in the Busseron Creek watershed, Indiana.	Water, sediment (surficial).	6	Aldrin Chlordane DDD DDE DDT Dieldrin Endrin Heptachlor Heptachlor epoxide Lindane Toxaphene	50 25 0 50 0 100 0 0 25 0 0	33 50 0 33 0 83 0 0 17 0 0
Frank and others, 1978	1968–1976	3 sites in Lake Superior and 4 sites in Lake Huron and its tributaries (Canada). Individual fish were analyzed, except that small fish were prepared as composites of similar-sized fish.	Fish (body with head and viscera removed). Species: 15 species, including bloater, white sucker, lake whitefish, lake trout, cisco, coho salmon.	724	Chlordane DDD DDE DDT DDT, total Dieldrin HCB Heptachlor epoxide	— — — — — — — —	— — — — — — — —

Table 2.2. Pesticides in bed sediment and aquatic biota from rivers in the United States: State and local monitoring studies—*Continued*

Bed sediment (µg/kg dry weight)			Aquatic biota (whole fish unless noted)					
			% Detection		(µg/kg wet weight)			
DL	Range	Median	Sites	Samples	DL	Range	Median	Comments
—	—	—	100	nr	nr	nr	222*	* Mean of individual fish. Objectives: to determine the effect of length, age, sex, and fat content on total DDT levels in catfish muscle. Mean total DDT increased with increasing body length (p<0.01) and with increasing age (p<0.01). Regression lines relating total DDT to body length were not statistically different for males and females. Mean total DDT was 1,221 µg/kg in males and 799 µg/kg in females. Fat content was correlated with both age and length (p<0.01), but not with total DDT (at 0.05 level). Length, age, and fat together accounted for 45% of variation in total DDT levels.
—	—	—	100	nr	nr	nr	1,675*	
—	—	—	100	nr	nr	nr	333*	
—	—	—	100	100	nr	78–6,336	1,006*	
0.1	nd–0.4	nd	—	—	—	—	—	Objective: to describe water quality and identify potential sources of contaminants. OC insecticides were detected and probably originated from agricultural lands in the basin.
0.1	nd–13	2	—	—	—	—	—	
0.1	nd	nd	—	—	—	—	—	
0.1	nd–0.3	nd	—	—	—	—	—	
0.1	nd	nd	—	—	—	—	—	
0.1	nd–9.8	1.2	—	—	—	—	—	
0.1	nd	nd	—	—	—	—	—	
0.1	nd	nd	—	—	—	—	—	
0.1	nd–1.0	nd	—	—	—	—	—	
0.1	nd	nd	—	—	—	—	—	
0.1	nd	nd	—	—	—	—	—	
—	—	—	nr*	nr*	10	nd–39*	nr	* Fish without head and viscera. Objective: to measure OC contaminants in Great Lakes fish, and to determine whether use restrictions were affecting residues. Mean total DDT residues during any year were highest for lake trout. Total DDT declined from 1968–71 and 1975–76 in ≥2 fish species from Lakes Superior and Huron. By 1975, mean total DDT levels had declined in lake trout and were highest in bloaters from both lakes. The DDT/total DDT ratio decreased for ≥2 and 3 species from Lakes Superior and Huron, suggesting lower intake of parent and(or) breakdown to metabolites.
—	—	—	nr*	nr*	10	nd–2,020*	nr	
—	—	—	nr*	nr*	10	nd–7,110*	nr	
—	—	—	nr*	nr*	10	nd–5,700*	nr	
—	—	—	nr*	nr*	10	nd–14,100*	nr	
—	—	—	nr*	nr*	10	nd–600*	nr	
—	—	—	nr*	nr*	10	nd–30*	nr	
—	—	—	nr*	nr*	10	nd–100*	nr	

Table 2.2. Pesticides in bed sediment and aquatic biota from rivers in the United States: State and local monitoring studies—*Continued*

Reference	Sampling dates	Sampling locations and strategy	Sampling media and species	Number of samples	Target analytes	Bed sediment % Detection	
						Sites	Samples
Hardy, 1978	1975–1976	3 sites (sediment) in the Muddy Fork Silver Creek watershed, Indiana.	Water, sediment (surficial).	7	Aldrin	67	86
					Chlordane	67	71
					DDD	67	71
					DDE	67	86
					DDT	67	71
					DDT, total	67	86
					Dieldrin	100	100
					Endrin	0	0
					Heptachlor	33	14
					Heptachlor epoxide	33	57
					Lindane	0	0
					Toxaphene	0	0
Lamb, 1978a	1976–1977	6 sites in the Village Creek watershed, Arizona. 3 sites were on Village Creek and 3 sites were on major tributaries.	Water, sediment.	6 12	Aldrin	33	17
					Chlordane	83	67
					DDD	100	100
					DDE	100	100
					DDT	83	67
					Diazinon	0	0
					Dieldrin	83	75
					Endrin	17	8
					Ethion	0	0
					Heptachlor	17	17
					Heptachlor epoxide	17	8
					Lindane	50	25
					Malathion	0	0
					Methoxychlor	0	0
					Methyl parathion	0	0
					Methyl trithion	0	0
					Parathion	0	0
					Toxaphene	0	0
					Trithion	0	0
Lamb, 1978b	1973	5 sites on the Tyronza River, Arizona.	Water, sediment.	5 10	Aldrin	20	10
					Chlordane	80	40
					DDD	100	100
					DDE	100	100
					DDT	100	60
					Diazinon	0	0
					Dieldrin	80	70
					Endrin	20	10
					Ethion	0	0
					Heptachlor	0	0
					Heptachlor epoxide	20	10
					Lindane	0	0
					Malathion	0	0

Table 2.2. Pesticides in bed sediment and aquatic biota from rivers in the United States: State and local monitoring studies—*Continued*

Bed sediment (µg/kg dry weight)			Aquatic biota (whole fish unless noted)						Comments
			% Detection		(µg/kg wet weight)				
DL	Range	Median	Sites	Sam-ples	DL	Range	Median		
									Dieldrin levels increased during the study period in 6 species from Lake Huron. Dieldrin residues exceeded the FDA action level (300 µg/kg) in some fish from Lake Huron.
0.1	nd–5.1	1.1	—	—	—	—		—	Objective: to evaluate water
0.1	nd–14	3	—	—	—	—		—	quality and potential problems
0.1	nd–1.8	0.6	—	—	—	—		—	related to pollutant inputs in the
0.1	nd–1.9	0.7	—	—	—	—		—	study area. Organochlorine
0.1	nd–19	1.3	—	—	—	—		—	compounds were detected in
0.1	nd–22	4.1	—	—	—	—		—	bottom sediment and could
0.1	nd–6.8	2.5	—	—	—	—		—	enter food chains.
0.1	nd	nd	—	—	—	—		—	
0.1	nd–0.1	nd	—	—	—	—		—	
0.1	nd–1.0	0.2	—	—	—	—		—	
0.1	nd	nd	—	—	—	—		—	
0.1	nd	nd	—	—	—	—		—	
0.1	nd–0.4	nd	—	—	—	—		—	Objective: to assess
0.1–1	nd–17	1	—	—	—	—		—	water-quality conditions in study
0.1	0.3–48	6.2	—	—	—	—		—	area before implementation of
0.1	0.2–40	6.8	—	—	—	—		—	Soil Conservation Service
0.1	nd–38	2.1	—	—	—	—		—	programs. Pesticide
0.1	nd	nd	—	—	—	—		—	concentrations were in the range
0.1	nd–4.2	0.3	—	—	—	—		—	of those expected in an
0.1	nd–0.2	nd	—	—	—	—		—	intensively farmed watershed.
0.1	nd	nd	—	—	—	—		—	Water samples were analyzed for
0.1	nd–2.5	nd	—	—	—	—		—	3 phenoxy herbicides, which
0.1	nd–0.2	nd	—	—	—	—		—	were detected at low
0.1	nd–2.8	nd	—	—	—	—		—	concentrations. Erosion control
0.1	nd	nd	—	—	—	—		—	may help reduce pesticide
0.1	nd	nd	—	—	—	—		—	residues in streams.
0.1	nd	nd	—	—	—	—		—	
0.1	nd	nd	—	—	—	—		—	
0.1	nd	nd	—	—	—	—		—	
0.1	nd	nd	—	—	—	—		—	
0.1	nd	nd	—	—	—	—		—	
0.1	nd–17	nd	—	—	—	—		—	Objective: to assess
0.1	nd–13	nd	—	—	—	—		—	water-quality conditions in study
0.1	0.5–12	1	—	—	—	—		—	area before implementation of
0.1	0.3–10	1.1	—	—	—	—		—	Soil Conservation Service
0.1	nd–12	0.7	—	—	—	—		—	programs. Pesticide
0.1	nd	nd	—	—	—	—		—	concentrations were in the range
0.1	nd–1.3	0.1	—	—	—	—		—	of those expected in an
0.1	nd–0.5	nd	—	—	—	—		—	intensively farmed watershed.
0.1	nd	nd	—	—	—	—		—	Water samples were analyzed for
0.1	nd	nd	—	—	—	—		—	3 phenoxy herbicides, one of
0.1	nd–0.3	nd	—	—	—	—		—	which (2,4,5-T) was detected at
0.1	nd	nd	—	—	—	—		—	low concentrations. Erosion
0.1	nd	nd	—	—	—	—		—	control may help reduce

Table 2.2. Pesticides in bed sediment and aquatic biota from rivers in the United States: State and local monitoring studies—*Continued*

Reference	Sampling dates	Sampling locations and strategy	Sampling media and species	Number of samples	Target analytes	Bed sediment % Detection	
						Sites	Samples
					Methoxychlor	0	0
					Methyl parathion	0	0
					Methyl trithion	0	0
					Parathion	0	0
					Toxaphene	0	0
					Trithion	0	0
Leone and Dupuy, 1978	1977	Proposed dredging sites in Yazoo River, Mississippi. Surficial sediment was sampled at 1 site. Surface (receiving) water was analyzed at 4 sites. River bank cores (20 ft) were sampled at 4 sites.	Water, (dissolved whole), sediment (surficial, core), elutriate (dissolved).	1 4	Aldrin	0	0
					Chlordane	0	0
					DDD	100	100
					DDE	100	100
					DDT	100	100
					Diazinon	0	0
					Dieldrin	0	0
					Endrin	0	0
					Ethion	0	0
					Heptachlor	0	0
					Heptachlor epoxide	0	0
					Lindane	0	0
					Malathion	0	0
					Methyl parathion	0	0
					Methyl trithion	0	0
					Mirex	0	0
					Parathion	0	0
					Toxaphene	100	100
					Trithion	0	0
Moccia, 1978	1976–1978	5 tributary streams to the Great Lakes (New York and Ontario, Canada); 1 site in Lake Ontario; 1 site (fish hatchery) on the Platte River, Michigan. All fish (except Lake Ontario) were adults. 17 of 25 samples were composites of 140 fish. Males and females were pooled separately.	Fish (muscle). Species: coho salmon, chinook salmon.	25	Chlordane	—	—
					DDD, *p,p'*-	—	—
					DDE, *p,p'*-	—	—
					DDT, *p,p'*-	—	—
					Dieldrin	—	—
					Mirex	—	—
					Photomirex	—	—
Silvey and Wheeler, 1978	1975–1976	16 sites in 4 watersheds in South Rockingham County, New Hampshire.	Sediment, water, ground water.	16 16 17	Aldrin	0	0
					Chlordane	6	6
					DDD	25	25
					DDE	38	38
					DDT	6	6
					DDT, total	50	50
					Dieldrin	19	19
					Endrin	0	0
					Heptachlor	0	0

Table 2.2. Pesticides in bed sediment and aquatic biota from rivers in the United States: State and local monitoring studies—*Continued*

Bed sediment (µg/kg dry weight)			Aquatic biota (whole fish unless noted)						Comments
			% Detection		(µg/kg wet weight)				
DL	Range	Median	Sites	Sam-ples	DL	Range	Median		
0.1	nd	nd	—	—	—	—	—		pesticide residues in streams.
0.1	nd	nd	—	—	—	—	—		
0.1	nd	nd	—	—	—	—	—		
0.1	nd	nd	—	—	—	—	—		
0.1	nd	nd	—	—	—	—	—		
0.1	nd	nd	—	—	—	—	—		
0.1	nd	nd	—	—	—	—	—		Objective: to sample river bank
1	nd	nd	—	—	—	—	—		(core) or bottom at proposed
nr	1.2	1.2	—	—	—	—	—		dredging sites. Pesticides
nr	0.9	0.9	—	—	—	—	—		detected in 1 or more cores were
nr	1	1	—	—	—	—	—		aldrin, endrin, methyl parathion,
0.1	nd	nd	—	—	—	—	—		dieldrin, DDT, DDD, and DDE.
0.1	nd	nd	—	—	—	—	—		Elutriate test with bottom
0.1	nd	nd	—	—	—	—	—		sediment contained diazinon,
0.1	nd	nd	—	—	—	—	—		DDT, dieldrin, and lindane.
0.1	nd	nd	—	—	—	—	—		Elutriate tests with 1 or more
0.1	nd	nd	—	—	—	—	—		cores contained diazinon,
0.1	nd	nd	—	—	—	—	—		dieldrin, heptachlor, 2,4-D,
0.1	nd	nd	—	—	—	—	—		2,4,5-T, and lindane. Receiving
0.1	nd	nd	—	—	—	—	—		water used in 1 or more core
0.1	nd	nd	—	—	—	—	—		elutriate tests contained
0.1	nd	nd	—	—	—	—	—		dieldrin, lindane, 2,4-D, and
0.1	nd	nd	—	—	—	—	—		2,4,5-T, but not diazinon (<0.01
1	4	4	—	—	—	—	—		µg/L).
0.1	nd	nd	—	—	—	—	—		
—	—	—	78	71	nr	nd–170	40		Objective: to determine if goiter
—	—	—	100	100	nr	70–670	210		(severe thyroid hyperplasia) in
—	—	—	100	100	nr	200–3,620	1,850		Great Lakes salmon populations
—	—	—	100	100	nr	60–1,300	290		was related to OC levels in fish.
—	—	—	78	76	nr	nd–140	70		No goiters were found in chi-
—	—	—	56	53	nr	nd–810	140		nook salmon. Goiter frequency
—	—	—	56	53	nr	nd–390	50		in coho salmon was 3–80% at
									different sites, but was not cor-
									related with OC residues.
									The OCs detected here do not
									appear to be solely responsible
									for goiter induction and do not
									explain interspecies and
									interlake differences in goiter
									frequency.
nr	nd	nd	—	—	—	—	—		Objective: to describe the study
nr	nd–36	nd	—	—	—	—	—		area; to show relative water
nr	nd–9.9	nd	—	—	—	—	—		quality in 4 watersheds. Some
nr	nd–4.4	nd	—	—	—	—	—		maximum pesticide
nr	nd–0.3	nd	—	—	—	—	—		concentrations cited in text did
nr	nd–9.9	nd/0.5	—	—	—	—	—		not appear to match those listed
nr	nd–19	nd	—	—	—	—	—		in data tables. Pesticide levels
nr	nd	nd	—	—	—	—	—		in Spicket R. watershed were
nr	nd	nd	—	—	—	—	—		high relative to the other

Table 2.2. Pesticides in bed sediment and aquatic biota from rivers in the United States: State and local monitoring studies—*Continued*

Reference	Sampling dates	Sampling locations and strategy	Sampling media and species	Number of samples	Target analytes	Bed sediment % Detection	
						Sites	Samples
					Heptachlor epoxide	6	6
					Lindane	0	0
					Methoxychlor	0	0
					Toxaphene	0	0
Van Hove Holdrinet and others, 1978	1968, 1976	229 sites in Lake Ontario were sampled in 1968. In 1976, transects across regions of high mirex were sampled, including Oswego and Niagara rivers (New York and Ontario, Canada). The top 3 cm were analyzed in 1968, the top 1–3 cm in 1976. 1976 sampling also included four 24-cm cores, subsampled at 1 cm increments, and 12 samples of Oswego River surficial sediment.	Sediment (surficial, cores, suspended).	271 4 2	Mirex (1968)* Mirex (1976)†	33* 88†	33* 92†
Wall and others, 1978	1973	6 sites located in the Maumee River Basin, Ohio.	Water, sediment (surficial).	6	Atrazine D, 2,4- DDE Endosulfan	nr nr nr nr	nr nr nr nr
Wiemeyer and others, 1978	1966, 1974	11 locations in Maine. Sampling locations included sites in eastern Maine and along the Kennebec River (where reproductive success of bald eagles had been good and poor, respectively). Fish were composited by site, collection date,	Fish (whole), birds (carcass). Fish species: 14 species, including yellow perch, white sucker, chain pickerel, carp, American	29	DDD, *p,p'-* DDE, *p,p'-* DDT, *p,p'-* DDT, total	— — — —	— — — —

Table 2.2. Pesticides in bed sediment and aquatic biota from rivers in the United States: State and local monitoring studies—*Continued*

Bed sediment (µg/kg dry weight)			Aquatic biota (whole fish unless noted)						Comments
			% Detection		(µg/kg wet weight)				
DL	Range	Median	Sites	Sam-ples	DL	Range	Median		
nr	nd–1.9	nd	—	—	—	—	—		watersheds. Water quality in the Powwow R. were excellent relative to the other watersheds.
nr	nd	nd	—	—	—	—	—		
nr	nd	nd	—	—	—	—	—		
nr	nd	nd	—	—	—	—	—		
nr	nd–40*	nd*	—	—	—	—	—		*Data from 1968 lake survey. †Data from 1976 Oswego R. surficial sediment samples only. Objective: to investigate sources and distribution of mirex in Lake Ontario sediment. Two zones of high contamination were identified in regions of Oswego and Niagara rivers. No mirex was detected in suspended sediment from St. Lawrence, Humber, Credit, and Oswego rivers; Niagara R. sample contained 12 µg/L, indicating recently active source. Oswego R. sediment sampling revealed original source was probably Armstrong Cork, 14 km upstream. Core samples in lake and Oswego R. showed increasing levels down the core, with maximum residues deposited about 8–13 years before.
nr	nd–1,666†	nr†	—	—	—	—	—		
10	nd–30	nr	—	—	—	—	—		Objective: to characterize bottom sediments that contribute to the suspended sediment load of the Maumee R. 34% of clay-sized particles were aggregated with courser fractions. Atrazine levels were higher in water than in sediment. DDE and endosulfan were found in all water samples, and 2,4-D in none.
10	nd–90	nr	—	—	—	—	—		
0.5	nd–4.8	nr	—	—	—	—	—		
1	nd–4,530	nr	—	—	—	—	—		
—	—	—	67	38	nr	nd–270	nd		Objective: to determine OC levels in potential bald eagle prey in areas with different rates of eagle reproduction; to assess changes over time. In 1966, OC residue patterns in fish were similar in the same species from different areas. In 1974, American eels and white suckers from Lincoln County (LC, with poor eagle reproduction) had higher DDT and PCB residues than those from Washington County
—	—	—	100	79	nr	nd–1,100	110		
—	—	—	58	41	nr	nd–600	nd		
—	—	—	100	79	nr	nd–1,940	160		

Table 2.2. Pesticides in bed sediment and aquatic biota from rivers in the United States: State and local monitoring studies—*Continued*

Reference	Sam-pling dates	Sampling locations and strategy	Sampling media and species	Number of samples	Target analytes	Bed sediment % Detection	
						Sites	Sam-ples
		and species. Some commercial fish samples also were obtained.	eel, landlocked salmon, sea herring.				
Boileau and others, 1979	1974–1975	The first 10 km of the Richilieu River flowing in Canada (Quebec). The Richilieu River is the outlet to Lake Champlain.	Fish (whole). Species: northern pike.	149	DDD, *p,p'-* DDE, *p,p'-* DDT, *p,p'-*	— — —	— — —
Bryant and others, 1979	1978	The L'Anguille River Basin, Arkansas. Sediment was collected at 1 site from 1971 to 1978. Fish were collected at 1 site in 1978. 5–7 fish thought to be representive of fish found in the basin were composited for each species.	Sediment, fish (edible, carcass), water (whole). Species: carp, buffalo.	15 2 2	Aldrin DDD DDD, *o,p'-* DDD, *p,p'-* DDE DDE, *o,p'-* DDE, *p,p'-* DDT DDT, *o,p'-* DDT, *p,p'-* Dieldrin Endrin Toxaphene	100 100 — — 100 — — 100 — — 100 100 100	57 93 — — 100 — — 87 — — 67 33 20
Dawson and others, 1979	nr	47 sites in the James River (Virginia), downstream of kepone production facilities in Hopewell.	Sediment (surficial, cores), soil, sewer wastes.	47 7	Kepone	94	94
Dupuy and Couvillion, 1979	1977–1978	163 total sites (136 with sediment data) in selected water-ways in southern Louisiana,	Sediment, water, elutriates.	181 189 117	Aldrin Chlordane DDD DDE DDT	7 29 64 37 14	6 33 63 33 12

Table 2.2. Pesticides in bed sediment and aquatic biota from rivers in the United States: State and local monitoring studies—*Continued*

| Bed sediment (µg/kg dry weight) | | | Aquatic biota (whole fish unless noted) | | | | | Comments |
DL	Range	Median	% Detection Sites	Samples	DL	Range (µg/kg wet weight)	Median	
								(WC, with higher eagle reproduction). From 1966 to 1974, DDT in some fish species decreased by >50%, but PCB levels were twice as high. DDT residues in herring gulls (1966) were higher than in fish (1966 or 1974). Gulls also contained dieldrin, heptachlor epoxide, mirex, and oxychlordane, which were rarely detected in fish.
—	—	—	100	100	1	1,000–31,000	nr	Objective: to determine residues in relation to season and age of fish. DDT levels increased with increasing age and size. Lowest total DDT residues occurred during Aug.–Nov., when water levels in lake were the lowest.
—	—	—	100	100	1	110,000–140,000	nr	
—	—	—	100	100	1	3,100–4,600	nr	
0.2	nd–2.1	0.7	0*	0*	nr	nd*	nd*	* Edible fish tissue only. Objective: to document types, sources, and occurrence of contaminants in the basin. High toxaphene levels were found in sediment and edible fish. In water, 2,4-D and 2,4,5-T occurred basinwide. Low concentrations of aldrin, chlordane, DDT, DDD, DDE, dieldrin, lindane, methyl parathion, and silvex were found in water-sediment mixtures.
nr	nd–55	30	—	—	—	—	—	
—	—	—	0*	0*	nr	nd*	nd*	
—	—	—	100*	100*	nr	150–320	235	
nr	3.2–39	21	—	—	—	—	—	
—	—	—	0*	0*	nr	nd*	nd*	
—	—	—	100*	100*	nr	240–630	435	
0.2	nd–110	6.4	—	—	—	—	—	
—	—	—	0*	0*	nr	nd*	nd*	
—	—	—	0*	0*	nr	nd*	nd*	
0.2	nd–7.7	2	0*	0*	nr	nd*	nd*	
0.2	nd–1	nd	0*	0*	nr	nd*	nd*	
nr	nd–45	nd	100*	100*	nr	1,400–1,900	1,650	
20	nd–66,100	307	—	—	—	—	—	Objective: to determine fate and distribution of kepone below known source. Transport downstream appears to be related to desorption from sediment. In sediment cores, kepone levels were highest at about 6 inches depth. Kepone was detected throughout a 70 mile reach from Richmond to the mouth on Chesapeake Bay. Kepone in the bottom sediment was estimated to be 78,000 to 103,000 lb.
0.1	nd–1.8	nd	—	—	—	—	—	Objective: to determine water quality and conduct elutriate studies to evaluate possible environmental effects of proposed dredging activities in
1	nd–560	nd	—	—	—	—	—	
0.1	nd–37	2.4	—	—	—	—	—	
0.1	nd–53	nd	—	—	—	—	—	
0.1	nd–8.2	nd	—	—	—	—	—	

Table 2.2. Pesticides in bed sediment and aquatic biota from rivers in the United States: State and local monitoring studies—*Continued*

Reference	Sampling dates	Sampling locations and strategy	Sampling media and species	Number of samples	Target analytes	Bed sediment % Detection	
						Sites	Samples
		including the Mississippi River, Tiger Pass, Bayou Long, Bayou Segnette Waterway, Lake Pontchartrain, Houma Navigation Canal, Atchafalaya River Channel and Bay, Mermentau River, Red River, and Calcasieu Ship Channel.			Diazinon	3	3
					Dieldrin	30	30
					Endrin	7	7
					Ethion	0	0
					Heptachlor	2	2
					Heptachlor epoxide	4	3
					Lindane	4	4
					Malathion	0	0
					Methyl parathion	0	0
					Methyl trithion	0	0
					Mirex	0	0
					Parathion	2	2
					Toxaphene	0	0
					Trithion	0	0
Eccles, 1979	1977	17 agricultural drains in the southeastern desert area of California. The top 6 in. were sampled.	Water (whole), sediment (surficial), tile drainage effluents, irrigation tailwater.	119 19 26 2	Aldrin	6	5
					Chlordane	0	0
					D, 2,4-	6	5
					DDD	94	84
					DDE	100	100
					DDT	76	68
					Diazinon	0	0
					Dieldrin	71	74
					DP, 2,4-	0	0
					Endosulfan	0	0
					Endrin	47	42
					Ethion	0	0
					Heptachlor	0	0
					Heptachlor epoxide	0	0
					Lindane	24	21
					Malathion	0	0
					Methyl parathion	0	0
					Parathion	35	32
					Silvex	12	11
					T, 2,4,5-	6	5
					Toxaphene	53	47
Evans and Tobin, 1979	1976	3 sites in Rattlesnake Creek watershed, Ohio.	Sediment, benthic survey.	3	Aldrin	0	0
					Chlordane	100	100
					D, 2,4-	0	0
					DDD	100	100
					DDE	100	100
					DDT	33	33
					Diazinon	0	0
					Dieldrin	67	67
					Endrin	0	0
					Ethion	0	0
					Heptachlor	0	0
					Heptachlor epoxide	0	0
					Lindane	0	0
					Malathion	0	0

Table 2.2. Pesticides in bed sediment and aquatic biota from rivers in the United States: State and local monitoring studies—*Continued*

Bed sediment (µg/kg dry weight)			Aquatic biota (whole fish unless noted)					Comments
			% Detection		(µg/kg wet weight)			
DL	Range	Median	Sites	Samples	DL	Range	Median	
0.1	nd–0.9	nd	—	—	—	—	—	southern Louisiana waterways. Data were presented without interpretation.
0.1	nd–16	nd	—	—	—	—	—	
0.1	nd–0.3	nd	—	—	—	—	—	
0.1	nd	nd	—	—	—	—	—	
0.1	nd–1.7	nd	—	—	—	—	—	
0.1	nd–0.1	nd	—	—	—	—	—	
0.1	nd–1.6	nd	—	—	—	—	—	
0.1	nd	nd	—	—	—	—	—	
0.1	nd	nd	—	—	—	—	—	
0.1	nd	nd	—	—	—	—	—	
0.1	nd	nd	—	—	—	—	—	
0.1	nd–1.1	nd	—	—	—	—	—	
1	nd	nd	—	—	—	—	—	
0.1	nd	nd	—	—	—	—	—	
0.1	nd–0.1	nd	—	—	—	—	—	Objective: to determine the occurrence, distribution, and sources of pesticides in the agricultural drains for 3/4 million irrigated acres in the southeastern desert area of California. Concentrations in the drains fluctuated widely. Frequency of detection and concentrations were highest in drains receiving surface runoff, such as drains in the Imperial Valley. Drains received negligible pesticides from tile drainage effluents. The probable sources of pesticides in drain water were slight concentrations from bed sediment in the drains, and high concentrations from aerial drift and irrigation tailwater.
0.1	nd	nd	—	—	—	—	—	
0.1	nd–8	nd	—	—	—	—	—	
0.1	nd–28	1.5	—	—	—	—	—	
0.1	0.2–110	9.8	—	—	—	—	—	
0.1	nd–63	1.6	—	—	—	—	—	
0.1	nd	nd	—	—	—	—	—	
0.1	nd–4.4	0.4	—	—	—	—	—	
0.1	nd	nd	—	—	—	—	—	
0.1	nd	nd	—	—	—	—	—	
0.1	nd–1.3	nd	—	—	—	—	—	
0.1	nd	nd	—	—	—	—	—	
0.1	nd	nd	—	—	—	—	—	
0.1	nd	nd	—	—	—	—	—	
0.1	nd–0.4	nd	—	—	—	—	—	
0.1	nd	nd	—	—	—	—	—	
0.1	nd	nd	—	—	—	—	—	
0.1	nd–4.6	nd	—	—	—	—	—	
0.1	nd–5	nd	—	—	—	—	—	
0.1	nd–3	nd	—	—	—	—	—	
0.1	nd–72	nd	—	—	—	—	—	
0.1	nd	nd	—	—	—	—	—	Objective: to evaluate water quality in the Rattlesnake Creek Basin. There appeared to be a downstream increase in concentrations of chlordane and dieldrin.
nr	3–12	10	—	—	—	—	—	
0.1	nd	nd	—	—	—	—	—	
0.1	0.4–1.5	0.4	—	—	—	—	—	
0.1	0.2–1.0	0.5	—	—	—	—	—	
0.1	nd–0.3	nd	—	—	—	—	—	
0.1	nd	nd	—	—	—	—	—	
0.1	nd–3.2	1.6	—	—	—	—	—	
0.1	nd	nd	—	—	—	—	—	
0.1	nd	nd	—	—	—	—	—	
0.1	nd	nd	—	—	—	—	—	
0.1	nd	nd	—	—	—	—	—	
0.1	nd	nd	—	—	—	—	—	
0.1	nd	nd	—	—	—	—	—	

Table 2.2. Pesticides in bed sediment and aquatic biota from rivers in the United States: State and local monitoring studies—*Continued*

Reference	Sampling dates	Sampling locations and strategy	Sampling media and species	Number of samples	Target analytes	Bed sediment % Detection Sites	Bed sediment % Detection Samples
					Methyl parathion	0	0
					Methyl trithion	0	0
					Parathion	0	0
					Silvex	0	0
					T, 2,4,5-	0	0
					Toxaphene	0	0
					Trithion	0	0
Hall and others, 1979	1977–1978	8 sites in Everglades National Park (Florida). Samples consisted of eggs that remained in nests after hatching. Whole egg except shell and shell membrane was analyzed.	Crocodile (eggs). Species: American crocodile.	23	Chlordane, *cis*-	—	—
					DDD, *p,p'*-	—	—
					DDE, *p,p'*-	—	—
					DDT, *p,p'*-	—	—
					Dieldrin	—	—
					Endrin	—	—
					HCB	—	—
					Heptachlor epoxide	—	—
					Mirex	—	—
					Nonachlor, *cis*-	—	—
					Nonachlor, *trans*-	—	—
					Oxychlordane	—	—
					Toxaphene	—	—
New York Department of Environmental Conservation, 1979	1977–1979	106 locations throughout the state of New York. Many fish samples were composites. Includes 61 sites (1977–1978) from New York's statewide Toxic Substances Monitoring Program (TSMP) and 8 other surveys and programs in the state.	Fish (tissue type not reported), reptiles, birds, mammals, amphibians. Fish species include largemouth bass, white sucker, carp, brown bullhead, lake trout, Atlantic salmon.	225 31 26 5	DDT, total	—	—
					Dieldrin	—	—
					Endrin	—	—
					Heptachlor, total	—	—
					HCH, total	—	—
					Mirex	—	—
Niimi, 1979	nr	3 locations in Ontario, Canada: eastern Lake Ontario; Ganaraska River (during spring spawning migration); and Credit River (during fall spawning migration).	Fish (whole). Species: lake trout, rainbow trout, coho salmon.	49	HCB	—	—

Table 2.2. Pesticides in bed sediment and aquatic biota from rivers in the United States: State and local monitoring studies—*Continued*

Bed sediment (µg/kg dry weight)			Aquatic biota (whole fish unless noted)					Comments
			% Detection		(µg/kg wet weight)			
DL	Range	Median	Sites	Samples	DL	Range	Median	
0.1	nd	nd	—	—	—	—	—	
0.1	nd	nd	—	—	—	—	—	
0.1	nd	nd	—	—	—	—	—	
0.1	nd	nd	—	—	—	—	—	
0.1	nd	nd	—	—	—	—	—	
0.1	nd	nd	—	—	—	—	—	
0.1	nd	nd	—	—	—	—	—	
—	—	—	38*	nr*	10	nd–10*†	nd*†	* Crocodile eggs.
—	—	—	88*	nr*	10	nd–70*†	25*†	† Range and median are of mean
—	—	—	100*	nr*	10	370–2,900*†	1,000*†	residues for the 8 sites.
—	—	—	100*	nr*	10	20–230*†	45*†	Objective: to examine OCs in eggs of the endangered American
—	—	—	88*	nr*	10	nd–30*†	20*†	crocodile from Everglades
—	—	—	0*	0*	10	nd*†	nd*†	National Park. No eggs
—	—	—	0*	0*	10	nd*†	nd*†	contained embryos. Eggs in the
—	—	—	75*	nr*	10	nd–40*†	10*†	same clutch had similar types
—	—	—	50*	nr*	10	nd–20*†	nd/10*†	and levels of contaminants.
—	—	—	63*	nr*	10	nd–30*†	15*†	DDT and DDD levels were
—	—	—	63*	nr*	10	nd–40*†	15*†	lower than in a 1972 study.
—	—	—	75*	nr*	10	nd–70*†	10*†	Levels in 1972 were lower than
—	—	—	0*	0*	10	nd*†	nd*†	residues in bald eagle eggs from the same area, and lower than levels reported to affect reproductive success in brown pelicans.
—	—	—	72*	nr	10	nd–3,720	nr	*Data are for fish in 1977–1978.
—	—	—	54*	nr	10	nd–260	nr	TSMP (tissue type not reported).
—	—	—	3*	nr	10	nd–20	nr	Objective: to evaluate the extent
—	—	—	2*	nr	10	nd–20	nr	of contamination in New York.
—	—	—	11*	nr	10	nd–60	nr	Methods of collection and
—	—	—	5*	nr	10	nd–200	nr	analysis for the different surveys were not included in report. Also reported were the results of laboratory studies on mirex elimination in brook trout and the effect of trimming and cooking on OC concentrations in edible portions of fish.
—	—	—	100	100	nr	16–125	57*	* Mean residue. Objective: to examine HCB levels in salmonids from Lake Ontario. HCB levels increased with body weight for all species. For fish of same weight, HCB levels were: lake trout > rainbow trout > coho salmon. Average body fat also were highest in lake trout.

Table 2.2. Pesticides in bed sediment and aquatic biota from rivers in the United States: State and local monitoring studies—*Continued*

Reference	Sampling date	Sampling location and strategy	Sampling media and species	Number of samples	Target analytes	Bed sediment % Detection	
						Sites	Samples
Perry, 1979	1976	12 sites (sediment, biota) on the Upper Snake River (Idaho) following the collapse of the Teton Dam. Sediment was collected from the top 3–4 cm. Plankton and drift were collected with a plankton net (153 μm mesh).	Sediment (surficial), plankton-drift, fish (whole), birds (whole). Fish species include Utah sucker, rainbow trout, white-fish, cut-throat trout, Utah chub.	13 3 31	Carbaryl D, 2,4- DDT, total Dieldrin HCB HCH Heptachlor epoxide Lindane OPs (not listed individually)	— — 50 8 — — — — —	— — 66 11 — — — — —
Punzo and others, 1979	1974	2 sites (toads, frogs, snakes) in Iowa, both in agricultural (corn, soybeans) areas. 1 site was not specified, the other was in the Wapsipinicon River floodplain. Lizards also were collected at 3 terrestrial sites (Arizona, New Mexico, Texas).	Toads (viscera), frogs (viscera), snakes (fat, eggs), turtles (fat, eggs), lizards (viscera, carcass). Species include northern water snake, snapping turtle.	 21 7 23 11 2 1 35 35	DDD, *p,p'-* Dieldrin Heptachlor epoxide	— — —	— — —
Wang and others, 1979	1977	6 sites in St. Lucie River and Estuary (Florida). Sampling was done once before and twice after water from Lake Okeechobee was discharged through the St. Lucie River. Sediment samples were composites of 3 subsamples.	Sediment (core), water (whole).	17 18	DDD, *o,p'-* DDD, *p,p'-* DDE, *o,p-* DDE, *p,p-* DDT, *o,p'-* DDT, *p,p'-* DDT, total	33* 83* 67* 83* 17* 17* 83*	18* 71* 29* 71* 6* 6* 82*
Bednar, 1980	1976	2 sites in the Pearl River (Mississippi). Data were collected at low stream flow.	Sediment, water.	2 2	Aldrin Chlordane D, 2,4- DDD DDE	50 50 0 0 0	50 50 0 0 0

Table 2.2. Pesticides in bed sediment and aquatic biota from rivers in the United States: State and local monitoring studies—*Continued*

Bed sediment (µg/kg dry weight)			Biota (whole fish unless noted) % Detection		Biota (whole fish unless noted) (µg/kg wet weight)			Comments
DL	Range	Median	Sites	Samples	DL	Range	Median	
—	—	—	nr	13	5–50	nd–50	nd	Objective: to evaluate pesticide contamination in the Upper Snake R. resulting from the Teton flood in 1976. Total DDT was low or not detected in sediment; up to 200 µg/kg in trout, 600 µg/kg in Utah sucker, and 1,200 µg/kg in Rocky Mountain whitefish. Theoretically, the levels observed in trout were sufficient to have adverse effects on the population. Sediment from Blackfoot area had the highest residue levels, probably because of irrigation return flows there.
—	—	—	nr	8	1	nd–6	nd	
20	nd–68*	nr	100	100	nr	40–1,897	nr	
1–5	nd–5	nr	67	58–64	1–5	nd–69	nr	
—	—	—	nr	0	5	nd	nd	
—	—	—	nr	57	5	nd–170	nr	
—	—	—	nr	81	1	nd–51	nr	
—	—	—	nr	3	5	nd–9	nd	
—	—	—	nr	0	nr	nd	nd	
—	—	—	100*	nr *	1	nd–625*	nr *	* Data shown are for snake fat. Objective: to collect baseline information on pesticide residues in reptiles and amphibians in agricultural areas in Iowa. DDE and dieldrin were detected in viscera of toads, and DDE, dieldrin, and heptachlor epoxide in fat and eggs of reptiles. There was no evidence of PCBs. No pesticides were detected in frog viscera. The highest residues were found in snakes. Only low or nondetectable residues were found in lizards.
—	—	—	100*	nr *	1	nd–193	nr *	
—	—	—	100*	nr *	1	nd–101	nr *	
1	nd–0.3*	nd*	—	—	—	—	—	* Data are for sediment cores. Objective: to survey water and sediment in the St. Lucie R. estuary for DDT and PCBs. No residues were detected in water samples (<0.01 µg/L). Total DDT levels were higher after canal water was discharged through the St. Lucie R.
1	nd–1.8*	0.22*	—	—	—	—	—	
1	nd–0.92*	nd*	—	—	—	—	—	
1	nd–0.94*	0.2*	—	—	—	—	—	
1	nd–1.1*	nd*	—	—	—	—	—	
1	nd–2.3*	nd*	—	—	—	—	—	
1	nd–6.15*	0.68*	—	—	—	—	—	
0.1	nd–8	nd/8	—	—	—	—	—	Objective: to obtain hydrologic data useful in developing water-resource management plan for the study area. Data were from 36-h intensive survey.
1	nd–24	nd/24	—	—	—	—	—	
1	nd	nd	—	—	—	—	—	
0.1	nd	nd	—	—	—	—	—	
0.1	nd	nd	—	—	—	—	—	

Table 2.2. Pesticides in bed sediment and aquatic biota from rivers in the United States: State and local monitoring studies—*Continued*

Reference	Sampling date	Sampling location and strategy	Sampling media and species	Number of samples	Target analytes	Bed sediment % Detection Sites	Bed sediment % Detection Samples
					DDT	0	0
					Diazinon	0	0
					Dieldrin	50	50
					Endrin	0	0
					Ethion	0	0
					Heptachlor	0	0
					Heptachlor epoxide	0	0
					Lindane	0	0
					Malathion	0	0
					Methyl parathion	0	0
					Methyl trithion	0	0
					Parathion	0	0
					Silvex	0	0
					T, 2,4,5-	0	0
					Toxaphene	0	0
					Trithion	0	0
Bobo and Peters, 1980	1979	8 sites in the Cypress Creek watershed, Indiana.	Water, sediment (surficial, core).	8 1	Aldrin	0	0
					Chlordane	100	79
					DDD	63	29
					DDE	63	50
					DDT	63	64
					Dieldrin	88	79
					Endosulfan	0	0
					Endrin	0	0
					Heptachlor	0	0
					Heptachlor epoxide	50	43
					Lindane	0	0
					Mirex	0	0
					Perthane	0	0
					Toxaphene	0	0
Bobo and Renn, 1980	1978	Sediment was collected at 10 sites in the Kankakee River Basin, Indiana.	Water, sediment (surficial, core).	19 2	Aldrin	20	16
					Chlordane	70	58
					DDD	70	68
					DDE	60	52
					DDT	60	63
					DDT, total	80	74
					Dieldrin	100	95
					Endosulfan	0	0
					Endrin	0	0
					Heptachlor	0	0
					Heptachlor epoxide	10	5
					Lindane	0	0
					Mirex	0	0
					Perthane	0	0
					Toxaphene	0	0
Brightbill and Treadway, 1980	1979	Proposed dredging sites in the Upper Yazoo River (Mississippi). 1 surficial sediment	Water (whole, dissolved), sediment (surficial,	1	Aldrin	0	0
					Chlordane	100	100
					DDD	100	100
					DDE	100	100
					DDT	100	100

Table 2.2. Pesticides in bed sediment and aquatic biota from rivers in the United States: State and local monitoring studies—*Continued*

| Bed sediment (µg/kg dry weight) | | | Biota (whole fish unless noted) | | | | | Comments |
DL	Range	Median	% Detection Sites	% Detection Samples	(µg/kg wet weight) DL	Range	Median	
0.1	nd	nd	—	—	—	—	—	Trace amounts of diazinon were
0.1	nd	nd	—	—	—	—	—	found in water samples at the
0.1	nd–5.3	nd/5.3	—	—	—	—	—	two sites sampled; silvex was
0.1	nd	nd	—	—	—	—	—	found in water at 1 site.
0.1	nd	nd	—	—	—	—	—	
0.1	nd	nd	—	—	—	—	—	
0.1	nd	nd	—	—	—	—	—	
0.1	nd	nd	—	—	—	—	—	
0.1	nd	nd	—	—	—	—	—	
0.1	nd	nd	—	—	—	—	—	
0.1	nd	nd	—	—	—	—	—	
0.1	nd	nd	—	—	—	—	—	
0.1	nd	nd	—	—	—	—	—	
1	nd	nd	—	—	—	—	—	
1	nd	nd	—	—	—	—	—	
0.1	nd	nd	—	—	—	—	—	
nr	nd	nd	—	—	—	—	—	Objective: to assess water quality
nr	nd–130	29.5	—	—	—	—	—	to aid in developing
nr	nd–5.5	nd	—	—	—	—	—	management plans for creeks
nr	nd–11	0.2	—	—	—	—	—	receiving acid-mine drainage
nr	nd–28	2.1	—	—	—	—	—	from coal-mine waste slurry and
nr	nd–17	4.7	—	—	—	—	—	runoff from mined or reclaimed
nr	nd–1.9	nd	—	—	—	—	—	lands. Organochlorine pesticides
nr	nd	nd	—	—	—	—	—	and PCBs were detected in
nr	nd	nd	—	—	—	—	—	Cyprus Creek and its tributaries.
nr	nd	nd	—	—	—	—	—	Polychlorinated naphthalenes
nr	nd	nd	—	—	—	—	—	were not detected.
nr	nd	nd	—	—	—	—	—	
nr	nd	nd	—	—	—	—	—	
nr	nd	nd	—	—	—	—	—	
nr	nd–0.9	nd	—	—	—	—	—	Objective: to identify areas
nr	nd–36	0.6	—	—	—	—	—	and possible sources of
nr	nd–27	0.8	—	—	—	—	—	water-quality problems in the
nr	nd–7.5	0.3	—	—	—	—	—	Kankakee River watershed. OC
nr	nd–16	0.3	—	—	—	—	—	compounds were detected in
nr	nd–43	2.6	—	—	—	—	—	surficial bed sediment. The site
nr	nd–74	3.8	—	—	—	—	—	with the highest dieldrin level
nr	nd	nd	—	—	—	—	—	was adjacent to cornfields. Two
nr	nd	nd	—	—	—	—	—	core sediment samples (2 ft
nr	nd	nd	—	—	—	—	—	depth) were collected at one site;
nr	nd–0.5	nd	—	—	—	—	—	one from the stream bank
nr	nd	nd	—	—	—	—	—	contained no detectable OCs and
nr	nd	nd	—	—	—	—	—	the core from the stream channel
nr	nd	nd	—	—	—	—	—	contained detectable dieldrin (4.6
nr	nd	nd	—	—	—	—	—	µg/kg).
100	nd	nd	—	—	—	—	—	Objective: to obtain data at
1,000	1,000	1,000	—	—	—	—	—	proposed dredging sites to
nr	5,100	5,100	—	—	—	—	—	ascertain influence of dredging
nr	10,000	10,000	—	—	—	—	—	on water quality. Core samples
nr	3,300	3,300	—	—	—	—	—	contained chlordane, DDT,

Table 2.2. Pesticides in bed sediment and aquatic biota from rivers in the United States: State and local monitoring studies—*Continued*

Reference	Sam-pling date	Sampling location and strategy	Sampling media and species	Number of samples	Target analytes	Bed sediment % Detection Sites	Bed sediment % Detection Sam-ples
		sample was taken at 1 site. Bank material (31–37 ft cores) was also sampled at 4 sites.	core), elutriates (dissolved).	4	Diazinon	0	0
					Dieldrin	0	0
					Endosulfan	0	0
					Endrin	0	0
					Ethion	0	0
					Heptachlor	0	0
					Heptachlor epoxide	0	0
					Lindane	0	0
					Malathion	0	0
					Methyl parathion	0	0
					Methyl trithion	0	0
					Mirex	0	0
					Parathion	0	0
					Perthane	0	0
					Toxaphene	0	0
					Trithion	0	0
Cartwright and Ziarno, 1980	1974	7 sites at community water supplies (New York State). Bed sediment was collected in 1974 only.	Water, sediment (surficial).	14 14	Aldrin	0	0
					Chlordane	17	10
					D, 2,4-	0	0
					DDT, total	57	64
					Diazinon	0	0
					Dieldrin	57	46
					Endrin	0	0
					Ethion	0	0
					Heptachlor	0	0
					Heptachlor epoxide	0	0
					Lindane	0	0
					Malathion	0	0
					Methyl parathion	0	0
					Methyl trithion	0	0
					Parathion	0	0
					Silvex	0	0
					Toxaphene	0	0
					Trithion	0	0
Demcheck and Dupuy, 1980	1978	5 sites near Sicily Island (Louisiana), including Fool River, Bayou Falcon, Bayou Louis, Haha Bayou. Sediment cores (1–20 ft) from each site were composited. Receiving waters (which will receive dredged material) and elutriates from the 5 sites also were analyzed.	Sediment (cores), water (whole, dissolved), elutriates (dissolved).	5 5 5 5	Aldrin	0*	0*
					Chlordane	0*	0*
					DDD	80*	80*
					DDE	80*	80*
					DDT	60*	60*
					Diazinon	0*	0*
					Dieldrin	0*	0*
					Endosulfan	20*	20*
					Endrin	0*	0*
					Ethion	0*	0*
					Heptachlor	0*	0*
					Heptachlor epoxide	0*	0*
					Lindane	0*	0*
					Malathion	0*	0*
					Methyl parathion	0*	0*
					Methyl trithion	0*	0*

Table 2.2. Pesticides in bed sediment and aquatic biota from rivers in the United States: State and local monitoring studies—*Continued*

Bed sediment			Biota (whole fish unless noted)					Comments
(µg/kg dry weight)			% Detection		(µg/kg wet weight)			
DL	Range	Median	Sites	Sam-ples	DL	Range	Median	
100	nd	nd	—	—	—	—	—	DDE, DDD, and dieldrin at 1 or
100	nd	nd	—	—	—	—	—	more sites. Surface water
100	nd	nd	—	—	—	—	—	samples were collected at 5
100	nd	nd	—	—	—	—	—	sites. Elutriates from all 4 core
100	nd	nd	—	—	—	—	—	sites contained diazinon, 2,4-D,
100	nd	nd	—	—	—	—	—	and 2,4,5-T; these compounds
100	nd	nd	—	—	—	—	—	were present in receiving water
100	nd	nd	—	—	—	—	—	at 3 (phenoxy compounds) or 4
100	nd	nd	—	—	—	—	—	(diazinon) of these sites. Data
100	nd	nd	—	—	—	—	—	presented without interpretation
100	nd	nd	—	—	—	—	—	of results.
100	nd	nd	—	—	—	—	—	
100	nd	nd	—	—	—	—	—	
100	nd	nd	—	—	—	—	—	
1,000	nd	nd	—	—	—	—	—	
100	nd	nd	—	—	—	—	—	
0.1	nd	nd	—	—	—	—	—	Objective: to compile data
0.1	nd–3	nd	—	—	—	—	—	collected over a 5 year period
0.1	nd	nd	—	—	—	—	—	by the USGS and New York
0.1	nd–70	4.2	—	—	—	—	—	State Department of Health at
0.1	nd	nd	—	—	—	—	—	sources of community water
0.1	nd–2.6	nd	—	—	—	—	—	supplies.
0.1	nd	nd	—	—	—	—	—	
0.1	nd	nd	—	—	—	—	—	
0.1	nd	nd	—	—	—	—	—	
0.1	nd	nd	—	—	—	—	—	
0.1	nd	nd	—	—	—	—	—	
0.1	nd	nd	—	—	—	—	—	
0.1	nd	nd	—	—	—	—	—	
0.1	nd	nd	—	—	—	—	—	
0.1	nd	nd	—	—	—	—	—	
0.1	nd	nd	—	—	—	—	—	
0.1	nd	nd	—	—	—	—	—	
0.1	nd	nd	—	—	—	—	—	
0.1	nd*	nd*	—	—	—	—	—	* Data are for 1–20 ft sediment
0.1	nd*	nd*	—	—	—	—	—	core samples.
0.1	nd–1.2*	0.3*	—	—	—	—	—	Objective: to investigate
0.1	nd–0.7*	0.4*	—	—	—	—	—	possible environmental effects
0.1	nd–1.7*	0.4*	—	—	—	—	—	of the Sicily Island area levee
0.1	nd*	nd*	—	—	—	—	—	(flood control) project, which
0.1	nd*	nd*	—	—	—	—	—	included levees, channel
0.1	nd–3*	nd*	—	—	—	—	—	excavation, and drainage
0.1	nd*	nd*	—	—	—	—	—	structures. Report contains raw
0.1	nd*	nd*	—	—	—	—	—	data, but no summary statistics
0.1	nd*	nd*	—	—	—	—	—	or interpretation. An elutriate is
0.1	nd*	nd*	—	—	—	—	—	the supernatant from shaking 1
0.1	nd*	nd*	—	—	—	—	—	part bottom sediment from a site
0.1	nd*	nd*	—	—	—	—	—	with 4 parts native water from
0.1	nd*	nd*	—	—	—	—	—	that site, followed by settling,
0.1	nd*	nd*	—	—	—	—	—	centrifugation, and filtration.

Table 2.2. Pesticides in bed sediment and aquatic biota from rivers in the United States: State and local monitoring studies—*Continued*

Reference	Sampling date	Sampling location and strategy	Sampling media and species	Number of samples	Target analytes	Bed sediment % Detection	
						Sites	Samples
					Mirex	20*	20*
					Parathion	0*	0*
					Perthane	0*	0*
					Toxaphene	0*	0*
					Trithion	0*	0*
Dudley and Karr, 1980	1977	Unspecified number of sites in the Black Creek drainage, Indiana.	Sediment (surficial), fish (whole), water (whole). Species: creek chub.	14 18 13	Alachlor	nr	0
					Atrazine	nr	0
					Carbofuran	nr	0
					DDE	nr	0
					Dieldrin	nr	0
					Malathion	nr	0
					T, 2,4,5-	nr	0
Faye, 1980	1977–1978	2 sites, 1 each on the Pascagoula and Escatawpa rivers (Mississippi).	Sediment, water.	3	Aldrin	0	0
					Chlordane	50	33
					D, 2,4-	0	0
					DDD	50	33
					DDE	50	33
					DDT	50	33
					Diazinon	0	0
					Dieldrin	0	0
					Endrin	0	0
					Ethion	0	0
					Heptachlor	0	0
					Heptachlor epoxide	0	0
					Lindane	0	0
					Malathion	0	0
					Methoxychlor	0	0
					Methyl trithion	0	0
					Parathion	0	0
					Silvex	0	0
					T, 2,4,5-	0	0
					Toxaphene	0	0
					Trithion	0	0
Kuehl and others, 1980	nr	Unspecified number of sites on the Wabash River (Indiana) and Ashtabula River (Ohio). Unknown number of fish were composited to form 1 sample from each river.	Fish (whole). Species: not reported.	2	Chlordane, *cis-*	—	—
					Chlordane, *trans-*	—	—
					Chlordene	—	—
					DDT	—	—
					Endrin	—	—
					HCB	—	—
					Heptachlor	—	—
					Heptachlor epoxide	—	—
					Nonachlor, *cis-*	—	—
					Nonachlor, *trans-*	—	—
					Oxychlordane	—	—
					Pentachlorophenol	—	—
Leard and others, 1980	1972–1973	12 sites in 5 major river basins in Mississippi.	Mollusks, water (whole). Species: 7 species of	54 26	Chlordane	—	—
					DDD, *p,p'-*	—	—
					DDE, *p,p'-*	—	—
					DDT, *o,p'-*	—	—
					DDT, *p,p'-*	—	—

Table 2.2. Pesticides in bed sediment and aquatic biota from rivers in the United States: State and local monitoring studies—*Continued*

Bed sediment			Biota (whole fish unless noted)						Comments
(μg/kg dry weight)			% Detection		(μg/kg wet weight)				
DL	Range	Median	Sites	Sam-ples	DL	Range	Median		
0.1	nd–0.1*	nd*	—	—	—	—	—		
0.1	nd*	nd*	—	—	—	—	—		
0.1	nd*	nd*	—	—	—	—	—		
0.1	nd*	nd*	—	—	—	—	—		
0.1	nd*	nd*	—	—	—	—	—		
100	nd	nd	nr	0	100	nd	nd		* Mean value.
100	nd	nd	nr	0	100	nd	nd		Objective: to assess contamina-
100	nd	nd	nr	0	100	nd	nd		tion at low stream discharge as
0.2	nd	nd	nr	100	0.2	7–31	16*		part of an effort to identify and
0.2	nd	nd	nr	100	0.2	4–86	23*		reduce agricultural sources of
1	nd	nd	nr	0	1	nd	nd		nonpoint pollution. 2,4,5-T
0.2	nd	nd	nr	0	0.2	nd	nd		was detected in water samples.
0.1	nd	nd	—	—	—	—	—		Objective: to provide
1	nd–39	nd	—	—	—	—	—		water-quality data for the
1	nd	nd	—	—	—	—	—		Pascagoula and Escatawpa
0.1	nd–8	nd	—	—	—	—	—		rivers for optimimal utilization
0.1	nd–7	nd	—	—	—	—	—		of surface water in southeastern
0.1	nd–3.2	nd	—	—	—	—	—		Mississippi. Diazinon and DDT
0.1	nd	nd	—	—	—	—	—		were detected in water from the
0.1	nd	nd	—	—	—	—	—		Pascagoula R., and DDT from
0.1	nd	nd	—	—	—	—	—		the Escatawpa R. Pesticide
0.1	nd	nd	—	—	—	—	—		concentrations in water and
0.1	nd	nd	—	—	—	—	—		sediment were extremely low or
0.1	nd	nd	—	—	—	—	—		below detection limits.
0.1	nd	nd	—	—	—	—	—		Storm-water quality was affected
0.1	nd	nd	—	—	—	—	—		by wastewater discharges, river
0.1	nd	nd	—	—	—	—	—		discharge, and tidal intrusion,
0.1	nd	nd	—	—	—	—	—		
0.1	nd	nd	—	—	—	—	—		
0.1	nd	nd	—	—	—	—	—		
1	nd	nd	—	—	—	—	—		
1	nd	nd	—	—	—	—	—		
0.1	nd	nd	—	—	—	—	—		
—	—	—	50	50	0.5	nd–13.0	nd/13.0		Objective: to identify
—	—	—	50	50	0.5	nd–8.54	nd/8.54		contaminants in fish from 2
—	—	—	50	50	nr	nr	nr		rivers using multiple analytical
—	—	—	100	100	0.5	130–310	220		procedures. Several chlorinated
—	—	—	50	50	nr	nr	nr		norbornene intermediates (and
—	—	—	100	100	0.5	30–3,140	1,585		their metabolites) in the
—	—	—	0	0	0.5	nd	nd		production of endrin were
—	—	—	50	50	0.5	nd–trace	nd/trace		identified in Wabash R. fish.
—	—	—	50	50	0.5	nd–4.89	nd/4.89		
—	—	—	50	50	0.5	nd–20.0	nd/20.0		
—	—	—	50	50	0.5	nd–trace	nd/trace		
—	—	—	50	50	nr	nr	nr		
—	—	—	8*	6*	10	nd–10*	nd*		* Data are for mollusks.
—	—	—	75*	76*	10	nd–490*	trace*		Objective: to investigate accu-
—	—	—	75*	78*	10	nd–390*	20*		mulation and elimination of pes-
—	—	—	33*	35*	10	nd–1,000*	nd*		ticides by clams. Purging was
—	—	—	67*	33*	10	nd–540*	nd*		expected in 1973 (relative to

Table 2.2. Pesticides in bed sediment and aquatic biota from rivers in the United States: State and local monitoring studies—*Continued*

Reference	Sampling date	Sampling location and strategy	Sampling media and species	Number of samples	Target analytes	Bed sediment % Detection	
						Sites	Samples
			freshwater clams, including *Corbicula manilensis.*		DDT, total Endrin Toxaphene	— — —	— — —
Pastel and others, 1980	1977	2 sites on the Hudson River (New York). Samples were composites of 12–13 female fish.	Fish (*). Species: American shad	2	DDE	—	—
Peterman and others, 1980	1976–1977	Unspecified number of sites in the lower Fox River (Wisconsin). Includes wastewaters from pulp or paper mills and municipal sewage treatment plants.	Sediment, fish (fillets), mollusks, seston, water, snowmelt, wastewater. Species: not reported.	About 250 samples total, mostly wastewater	Dieldrin Pentachloroanisole Pentachlorophenol	nr nr nr	nr nr nr
Qasim and others, 1980	nr	13 sites along the length of the Trinity River (Texas). Sites were selected based on their close proximity to contaminant sources such as municipal/industrial discharges and urban/rural runoff.	Water, sediment (surficial), elutriate.	13 13	Chlordane DDT Dieldrin Endrin Heptachlor Lindane	46 85 85 92 100 80	46 85 85 92 100 80
Simmons and Aldridge, 1980	1976–1978	4 sites in the Chicod Creek Basin (North Carolina). Analyses were made during periods of low base flow and storm runoff.	Sediment, water (whole).	8 8	DDD DDE DDT Diazinon Dieldrin Endrin	100 100 75 0 100 50	100 88 63 0 88 25

Table 2.2. Pesticides in bed sediment and aquatic biota from rivers in the United States: State and local monitoring studies—*Continued*

Bed sediment (µg/kg dry weight)			Biota (whole fish unless noted)						
			% Detection		(µg/kg wet weight)				
DL	Range	Median	Sites	Sam-ples	DL	Range	Median	Comments	
—	—	—	75*	78*	10	nd–1,780*	40*	1972) because of DDT ban in	
—	—	—	8*	2*	10	nd–trace*	nd*	1972 and extensive flooding	
—	—	—	67*	61*	100	nd–7,680*	trace*	during spring 1973. Highest residues in clams were at sites in agricultural basins. *o,p´-* and *p,p´-* DDT residues decreased from 1972 to 1973, but metabolite residues were variable.	
—	—	—	50*	50*	nr	nr	nr	* Data are for scaled, deboned, definned, beheaded, gutted fish. Objective: to identify pollutants in American shad from the Hudson R.	
nr	nr	nr	nr	nr	nr	8–22	nr	Objective: to assess sources and	
nr	nr	nr	nr	nr	nr	5–60	nr	distribution of organic	
nr	220–280	nr	nr	nr	nr	nr	nr	compounds in the lower Fox R., one of the most heavily developed industrial river basins in the world. Paper focused on identification of industrial compounds.	
0.3	nd–64	nd	—	—	—	—	—	Objective: to evaluate potential	
0.5	nd–53.6	5.4	—	—	—	—	—	mobility of sediment contami-	
0.3	nd–176	5	—	—	—	—	—	nants during proposed dredging	
0.3	nd–20.0	7	—	—	—	—	—	operations. The upper reach of	
nr	0.7–9.3	4.9	—	—	—	—	—	the Trinity R. was found to be	
0.2	nd–0.7	0.2	—	—	—	—	—	grossly polluted due to discharge of wastewater treatment plants in the Dallas–Fort Worth area. The quality of water and bottom sediment gradually improved in the lower reaches.	
nr	1.5–56	9.3	—	—	—	—	—	Objective: to collect baseline	
nr	nd–30	6.6	—	—	—	—	—	water-quality data to determine	
nr	nd–15	1.2	—	—	—	—	—	effects of planned channel	
nr	nd	nd	—	—	—	—	—	excavation and modification.	
nr	nd–5.6	2.5	—	—	—	—	—	OCs were frequently detected in	
nr	nd–1.0	nd	—	—	—	—	—	sediment, with levels generally higher under storm runoff than low flow conditions. OCs appeared to be associated with stream-bed materials that became waterborne during high flows and(or) with sediment or other materials recently washed into the stream by storm runoff.	

Table 2.2. Pesticides in bed sediment and aquatic biota from rivers in the United States: State and local monitoring studies—*Continued*

Reference	Sam- pling date	Sampling location and strategy	Sampling media and species	Number of samples	Target analytes	Bed sediment % Detection	
						Sites	Sam- ples
Water and Air Research, Inc., 1980	1979– 1980	6 sites (sediment) and 32 sites (fish) on the Tennessee River and 4 tributaries in Alabama. Most sites were below the Huntsville Spring Branch, which had received waste from a DDT- manufacturing facility. Whole body fish and fillets were analyzed for 3 species at 3 locations.	Water, sediment (cores), elutriates, fish (whole, fillet), invertebrates. Fish species: channel catfish, smallmouth- buffalo, largemouth bass, bluegill.	230 nr nr nr	DDT, total	80*	nr*
Zahnow and Riggleman, 1980	1977– 1978	22 sites in the Chesapeake Bay, including tributary streams in Maryland. Sites were downstream of agricultural lands where linuron use was heavy, low, or absent.	Water sediment (surficial, 7–8 cm deep).	79	Linuron	0	0
Bulkley and others, 1981	1977– 1978	3 sites on the Des Moines River, Iowa.	Fish (whole). Species: gizzard shad, river carpsucker, carp, channel catfish, white crappie, largemouth bass, walleye.	173	Chlordane DDT, total Dieldrin Heptachlor epoxide	— — — —	— — — —

Table 2.2. Pesticides in bed sediment and aquatic biota from rivers in the United States: State and local monitoring studies—*Continued*

Bed sediment (µg/kg dry weight)			Biota (whole fish unless noted)					Comments
			% Detection		(µg/kg wet weight)			
DL	Range	Median	Sites	Samples	DL	Range	Median	
nr*	nd–3.02x10⁷*	nr*	100†	100†	nr†	50–627,000†	nr †	* Data are from the top 0–6 in depth of 24 in cores. † Data are for fish fillets. Objective: to assess DDT contamination in water, sediment, and biota downstream of historic manufacturing waste inputs. About 99.9% of total DDT was in sediment, the rest in water or biota. Sediment from Huntsville Spring Branch and Indian Creek contained about 837 tons of total DDT. Annually, 0.04 to 0.2% of the sediment load is transported to the Tennessee R., evenly divided between suspended and coarser size fractions. Tennessee R. fish were significantly contaminated, especially catfish. Total DDT levels in birds, mammals, and reptiles were highest in the Huntsville Spring Branch sub-basin. Whole fish residues were 2–35 times higher than fillets.
10	nd	nd	—	—	—	—	—	Objective: to determine if the herbicide linuron was transported from its point of application on agricultural lands to adjacent water bodies or accumulating in Chesapeake Bay sediment. Linuron was not detected in the bay, or in water and sediment samples from tributaries draining high or low usage areas.
—	—	—	0	0	~10	nd	nd	* Range of median values for individual species.
—	—	—	100	100	~10	6–329	42–76*	Objective: to measure OC resi-
—	—	—	100	100	~10	7–301	28–114*	dues in fish of the Des Moines
—	—	—	nr	77–85*	~10	0–68	2–16*	R., compare differences among species, and relate those differences to trophic level and fat content. Species differences were significant for 3 analytes. Residues were significantly correlated to fat content for some species–analyte combinations. However, normalizing concentrations to fat content did not explain differences between species. When

Table 2.2. Pesticides in bed sediment and aquatic biota from rivers in the United States: State and local monitoring studies—*Continued*

Reference	Sampling date	Sampling location and strategy	Sampling media and species	Number of samples	Target analytes	Bed sediment % Detection Sites	Bed sediment % Detection Samples
Cutshall and others, 1981	1978	20 sites in the James River and Estuary, downstream of Hopewell, Virginia (former kepone production facility). Sediment cores (60 cm) were analyzed for kepone and radionuclides (^{137}Cesium and ^{60}Cobalt) as indicators of sedimentation areas and rates.	Sediment (cores).	24	Kepone	nr	nr
Frank, 1981	1976–1977	2 sites at the mouths of the Grand and Saugeen rivers, which drain into Lakes Erie and Huron, respectively (Ontario, Canada).	Water (whole), sediment (suspended, surficial).	84 22 8	Chlordane DDT, total Dieldrin Endrin HCH, α- Heptachlor epoxide Lindane	nr nr nr nr nr nr nr	nr nr nr nr nr nr nr
Hubert and Ricci, 1981	1980	4 sites on the Des Moines River (Iowa), at 2 impounded and 2 riverine locations. Five composite samples were collected at each site in June and	Fish (muscle). Species: carp.	40	DDT, total Dieldrin	— —	— —

Table 2.2. Pesticides in bed sediment and aquatic biota from rivers in the United States: State and local monitoring studies—*Continued*

Bed sediment (µg/kg dry weight)			Biota (whole fish unless noted)					Comments
			% Detection		(µg/kg wet weight)			
DL	Range	Median	Sites	Samples	DL	Range	Median	
								normalized for fat content, concentrations were greater for piscivorous fish than for forage fish; the relation was statistically different only for dieldrin (p<0.05).
nr	nr –740*	nr	—	—	—	—	—	* Range is for individual segments of sediment cores. Objective: to investigate sedimentation areas and rates in association with kepone distribution in the James R. estuary. Estimated sedimentation rates ranged from <1 to >19 cm/year. Highest kepone residues were at >55 cm segments at the 2 most upstream sites. Rapid burial has occurred at the most contaminated sites. Barring mixing by hurricanes or dredging, surface sediment residues are expected to decline.
nr	nd–17	0.2–6*	—	—	—	—	—	* Range of mean values for the Saugeen (lower value) and Grand (upper value) rivers. Objective: to integrate crop distribution and pesticide use information with pesticide residues in 2 agricultural basins in the Great Lakes region. Authors estimated pesticide use in the 2 basins, and pesticide loadings to the 2 rivers and at the mouths of the rivers. Loadings from the Grand R. to Lake Erie were greater than those from the Saugeen R. to Lake Huron. The relative importance of water and suspended solid phases to pesticide loadings depended on the year, river, and pesticide.
nr	nd–60	0.5–21*	—	—	—	—	—	
nr	nd–5	nd–2*	—	—	—	—	—	
nr	nd–2	nd–1*	—	—	—	—	—	
nr	nd–10	nd–3*	—	—	—	—	—	
nr	nd–3	nd–1*	—	—	—	—	—	
nr	nd–5	0.1–2*	—	—	—	—	—	
—	—	—	100*	100*	nr	18–191*	55*	* Data are for fish muscle. Objective: to assess influence of sampling month, fish age, sampling location, and lipid content on residues in carp muscle; to assess trends from 1977 to 1980. Normalizing on a fat basis did not reduce data variability. Residues fluctuated over time,
—	—	—	100*	100*	nr	10–128*	42*	

Table 2.2. Pesticides in bed sediment and aquatic biota from rivers in the United States: State and local monitoring studies—*Continued*

Reference	Sampling date	Sampling location and strategy	Sampling media and species	Number of samples	Target analytes	Bed sediment % Detection	
						Sites	Samples
		Sept. Composites contained equal weights of muscle from 10 fish 2–4 years in age.					
Kalkhoff, 1981	1980	1 site in Okatoma Creek (Mississippi).	Sediment, water.	1 1	Chlordane DDD DDE DDT Dieldrin	100 100 100 100 100	100 100 100 100 100
Leung and others, 1981a	1977– 1978	3 sites in the Saylorville Reservoir and Des Moines River (Iowa). Sites were located above, within, and below the reservoir, which was a new impoundment in the river. Fish were composited by species, site, collection date, and body length. Most fish were subadult. Gates to reservoir were closed in April 1977; the study was conducted from Oct. 1977 to Oct. 1978.	Sediment, fish (whole), water (dissolved). Species: river carpsucker, gizzard shad, common carp, channel catfish, white crappie, largemouth bass, walleye.	42 173 nr	Alachlor Atrazine Cyanazine DDT, total Dieldrin Heptachlor epoxide	— — — nr nr nr	— — — most 100 10
Leung and others, 1981b	1978	1 site in the Des Moines River (Iowa). Fish were composited by collection date and body length. Water samples were collected every 1–2 weeks. 1971–1973 fish and water data were available for same site from a previous study. Comparisons with	Fish (muscle), sediment (suspended), water (filtered). Species: channel catfish.	1* 29 29	Dieldrin	—	—

Table 2.2. Pesticides in bed sediment and aquatic biota from rivers in the United States: State and local monitoring studies—*Continued*

Bed sediment (µg/kg dry weight)			Biota (whole fish unless noted)					Comments
			% Detection		(µg/kg wet weight)			
DL	Range	Median	Sites	Sam-ples	DL	Range	Median	
								but there was no consistent seasonal pattern. Fish from impoundments had higher dieldrin than those from riverine sites, but not always higher total DDT. Residues were higher in 4-year than in 3-year olds.
nr	10	10	—	—	—	—	—	Objective: to provide water-quality data for determining waste-assimilation capacity of streams in study area. Reference did not list pesticides that were analyzed for but not detected.
nr	8.8	8.8	—	—	—	—	—	
nr	0.3	0.3	—	—	—	—	—	
nr	1.4	1.4	—	—	—	—	—	
nr	0.1	0.1	—	—	—	—	—	
—	—	—	0	0	10	nd	nd	* Range estimated from month-ly means and ranges provided. Objective: to determine effect of impoundment on pesticide accu-mulation in fish. Reference did not provide details of methods for sediment and water data. Pes-ticide concentration was related to body length for 1 species on-ly (largemouth bass). Levels of dieldrin and heptachlor epoxide were higher in river carpsucker from the reservoir than from the river. Other species showed no significant differences related to location. Seasonal differences in dieldrin levels in carp were significant; residues in and below the reservoir were highest in July 1978. Heptachlor epoxide residues in all species tended to be highest in July 1978.
—	—	—	0	0	10	nd	nd	
—	—	—	0	0	10	nd	nd	
0.1	nd–17.1	nr	100	100	10	6–329	nr	
nr	0.6–12.0	nr	100	100	10	7–240	nr	
nr	nd–2.0	nd	100	76–83*	10	nd–68	nr	
—	—	—	100	100*	10	46*	46*	* Data shown here assume that the results in the reference were for a single composite of 8 fish; it is possible that the authors meant to report the mean value for 8 samples. Objective: to compare dieldrin levels in water and fish before and after aldrin was banned (1975). July residues in fish decreased from 1973 (75 µg/kg) to 1978 (46 µg/kg), but this de-crease was less than differences among monthly samples in

Table 2.2. Pesticides in bed sediment and aquatic biota from rivers in the United States: State and local monitoring studies—*Continued*

Reference	Sampling date	Sampling location and strategy	Sampling media and species	Number of samples	Target analytes	Bed sediment % Detection Sites	Bed sediment % Detection Samples
		1971–1973 data were made for similar-length fish (201–300 mm) collected during same season of year (June–Sept.)					
Petersen, 1981	1980	3 sites in Larkin Creek and its tributaries (Arkansas). The top 1–2 in of sediment were sampled.	Sediment (surficial), water.	3 6	Aldrin	0	0
					Chlordane	0	0
					DDD	33	33
					DDE	0	0
					DDT	100	100
					Diazinon	0	0
					Dieldrin	100	100
					Endrin	67	67
					Ethion	0	0
					Heptachlor	0	0
					Heptachlor epoxide	0	0
					Lindane	33	33
					Malathion	0	0
					Methoxychlor	0	0
					Methyl parathion	0	0
					Methyl trithion	0	0
					Mirex	0	0
					Parathion	0	0
					Perthane	0	0
					Toxaphene	100	100
					Trithion	0	0
Roberts, 1981	1978	2 sites in the lower James River and 5 sites in the Chesapeake Bay (Virginia). Crabs collected were mature.	Crabs (backfin muscle, gonad, heptatopancreas). Species: Blue crab.	166 162 169	Kepone	—	—

Table 2.2. Pesticides in bed sediment and aquatic biota from rivers in the United States: State and local monitoring studies—*Continued*

Bed sediment (µg/kg dry weight)			Biota (whole fish unless noted)						Comments
			% Detection		(µg/kg wet weight)				
DL	Range	Median	Sites	Sam-ples	DL	Range	Median		
									1973. Dieldrin residues in whole water were highest in June–July (aldrin was applied April–May). Percent of dieldrin in dissolved phase was 65% in 1973 and 43% in 1978. Dissolved concentrations were comparable or slightly lower in 1978 than in 1973. Data suggest that dieldrin is still being washed into the river from croplands.
0.1	nd	nd	—	—	—	—	—		* Interferences raised detection limit in 2 samples. Objective: to assess water quality in the Larkin Creek watershed prior to land- and water-improvement measures. Several pesticides were present in water or sediment. Dieldrin in water exceeded USEPA aquatic life criterion. Toxaphene residues were high at all sites. Endrin, lindane, DDT, DDD, DDE, silvex, 2,4-D, and 2,4,5-T also were detected in water or sediment samples.
0.1	nd	nd	—	—	—	—	—		
<5–38*	nd–5.3	nd	—	—	—	—	—		
0.1–24*	nd	nd	—	—	—	—	—		
0.1	1.9	1.9	—	—	—	—	—		
0.1	nd	nd	—	—	—	—	—		
0.1	0.2–1.6	1	—	—	—	—	—		
0.1	nd–0.6	0.2	—	—	—	—	—		
0.1	nd	nd	—	—	—	—	—		
0.1	nd	nd	—	—	—	—	—		
0.1	nd	nd	—	—	—	—	—		
0.1	nd–0.1	nd	—	—	—	—	—		
0.1	nd	nd	—	—	—	—	—		
0.1	nd	nd	—	—	—	—	—		
0.1	nd	nd	—	—	—	—	—		
0.1	nd	nd	—	—	—	—	—		
0.1	nd	nd	—	—	—	—	—		
0.1	nd	nd	—	—	—	—	—		
0.1	nd	nd	—	—	—	—	—		
0.1	20–170	30	—	—	—	—	—		
0.1	nd	nd	—	—	—	—	—		
—	—	—	nr	nr	nr	10–1450*†	70*†		* Crabs from James R. only. † Range and median are of median values for 3 groups (males, ovigerous females, nonovigerous females) in each of 3 mos. Objective: to determine the distribution of kepone within blue crabs from the study area. Crabs from lower James R. were more frequently contaminated than those from the Bay. Median kepone levels in backfin muscle were higher in males than in females during all months. Females from the James R. usually had higher residues in the ovaries than in backfin muscle.

Table 2.2. Pesticides in bed sediment and aquatic biota from rivers in the United States: State and local monitoring studies—*Continued*

Reference	Sampling date	Sampling location and strategy	Sampling media and species	Number of samples	Target analytes	Bed sediment % Detection	
						Sites	Samples
Sloan, 1981a	1978–1980	8 studies in this report measured pesticides in aquatic biota in New York rivers: (1) 34 sites in the Statewide Toxic Substances Monitoring Program (1978–1979); (2) 15 sites in Lakes Ontario and Champlain (1978–1980); (3) 10 urban sites statewide (1979–1980); (4) 5 sites in Grass River drainage (1980); (5) 2 sites sampled as part of the Wetlands Toxic Substance Inventory (1979–1980); (6) 2 sites on Genesee River (1980); (7) 11 sites on 5 rivers (1979); (8) 4 sites on miscellaneous lakes and rivers sampled as part of various special fish collections. Results here are presented separately for the 8 studies.	(1) Fish (flesh). Species include white sucker, lake trout, channel catfish.	107	Chlordane	—	—
					DDT, total	—	—
					Dieldrin	—	—
					Endrin	—	—
					Heptachlor	—	—
					Lindane	—	—
					Mirex	—	—
			(2) Fish (whole, fillet, eggs, not reported). Species include coho salmon, white perch, brown trout.	119 33, 42 359	DDT, total	—	—
					Dieldrin	—	—
					Endrin	—	—
					HCB	—	—
					Mirex	—	—
			(3) Fish (edible, whole). Species include carp, largemouth bass, brown bullhead, white perch, northern pike.	14 1	Chlordane	—	—
					DDT	—	—
					Dieldrin	—	—
					Endrin	—	—
					HCB	—	—
					HCH, α-	—	—
					Heptachlor	—	—
					Heptachlor epoxide	—	—
					Lindane	—	—
					Mirex	—	—
					Parathion	—	—
			(4) Fish (*). Species include rock bass.	38	DDT, total	—	—
			(5) Fish (*), frogs, turtles, soil. Fish species: yellow bullhead, carp.	5	DDT, total	—	—
					Dieldrin	—	—
			(6) Fish (*). Species include white sucker, sculpin, smallmouth bass, brown trout.	6	Alachlor	—	—
					D, 2,4-	—	—
					DCPA	—	—
					DDT, total	—	—
					Dicofol	—	—
					Disulfoton	—	—
					Methoxychlor	—	—
			(7) Fish (*), sediment. Species include fallfish.	19 8	DDT, total	0	0
					Methoxychlor	0	0

Table 2.2. Pesticides in bed sediment and aquatic biota from rivers in the United States: State and local monitoring studies—*Continued*

Bed sediment (µg/kg dry weight)			Biota (whole fish unless noted)					
			% Detection		(µg/kg wet weight)			
DL	Range	Median	Sites	Samples	DL	Range	Median	Comments
—	—	—	100*	90*	10	nd–150*	20*	(1) * Data are for fish flesh. Objective: to monitor trends in OCs in fish flesh statewide.
—	—	—	84*	86*	10	nd–4,740*	80*	
—	—	—	64*	31*	10–20	nd–150*	nd*	
—	—	—	18*	9*	10	nd–30*	nd*	
—	—	—	6*	3*	10	nd–20*	nd*	
—	—	—	0*	0*	10	nd*	nd*	
—	—	—	12*	6*	10	nd–120*	nd*	
—	—	—	100*	100*	nr	10–5,000*	nr*	(2) * Data are for whole fish and not reported (i.e., excluding eggs and fillets). Objective: to update data on contaminant distribution and trends in Lake Ontario; to determine if contaminants in Lake Champlain pose a problem for implementing the management plan.
—	—	—	67*	nr*	10	nd–560*	nr*	
—	—	—	100*	100*	10	10–40*	nr*	
—	—	—	83*	nr*	10	nd–160*	nr*	
—	—	—	100*	100*	10	nd–1,810*	nr*	
—	—	—	22*	31*	10	nd–590*	nd*	(3) * Data are for edible fish tissue. Objective: to assess contaminant levels in fish from selected urban waterways.
—	—	—	100*	100*	nr	10–320*	40*	
—	—	—	0*	0*	10	nd*	nd*	
—	—	—	0*	0*	10	nd*	nd*	
—	—	—	0*	0*	10	nd*	nd*	
—	—	—	0*	0*	10	nd*	nd*	
—	—	—	0*	0*	10	nd*	nd*	
—	—	—	0*	0*	10	nd*	nd*	
—	—	—	0*	0*	10	nd*	nd*	
—	—	—	0*	0*	10	nd*	nd*	
—	—	—	11*	8*	10	nd–280*	nd*	
—	—	—	100*	84*	10	nd–110*	nr*	(4) * Tissue type not reported. Objective: to determine potential PCB sources in the Grass and St. Lawrence rivers.
—	—	—	100*	100*	nr	1–80*	20*	(5) * Tissue type not reported. Objective: to identify contaminants in selected wetlands, assess significance, and provide baseline data.
—	—	—	100*	75*	1	nd–20*	6*	
—	—	—	0*	0*	10	nd*	nd*	(6) * Tissue type not reported. Objective: to determine pesticide residues following a fire and clean-up activities.
—	—	—	0*	0*	10	nd*	nd*	
—	—	—	0*	0*	11	nd*	nd*	
—	—	—	100*	100*	nr	10–30*	20*	
—	—	—	0*	0*	12	nd*	nd*	
—	—	—	0*	0*	13	nd*	nd*	
—	—	—	0*	0*	14	nd*	nd*	
10	nd	nd	71*	84*	10	nd–390*	10*	(7) * Tissue type not reported. Objective: to determine effects of methoxychlor (larvicide) on nontarget organisms.
5	nd	nd	0*	0*	50	nd*	nd*	

Table 2.2. Pesticides in bed sediment and aquatic biota from rivers in the United States: State and local monitoring studies—*Continued*

Reference	Sam-pling date	Sampling location and strategy	Sampling media and species	Number of samples	Target analytes	Bed sediment % Detection Sites	Bed sediment % Detection Sam-ples
			(8) Fish (fat, muscle, brain, testes, kidney, liver, not reported). Species include large-mouth bass.	1 5 3, 2 2, 5 10	DDT, total Diazinon Dieldrin Endrin Ethion HCB Heptachor epoxide Lindane	— — — — — — — —	— — — — — — — —
Sloan, 1981b	1979–1981	3 studies in this report measured pesticides in sediment or aquatic biota in New York rivers: (1) 54 sites in the Statewide Toxic Substances Monitoring Program (1979–1980); (2) 2 sites in Lake Ontario (1980–1981); (3) 6 tributaries to Seneca Lake (1981). Results here are presented separately for the 3 studies.	(1) Fish (*). Species include lake trout, striped bass, goldfish, pumpkinseed.	202	Chlordane DDT, total Dieldrin Endrin Heptachlor, total Lindane Mirex	— — — — — — —	— — — — — — —
			(2) Fish (*). Species: rainbow trout.	68	DDT, total Dieldrin HCB Mirex Photomirex	— — — — —	— — — — —
			(3) Sediment.	24	DDT, total	100	71
U.S. Fish and Wildlife Service, 1981	1980	7 sites (sediment) and 4 sites (fish) in streams and ditches near an abandoned hazardous waste storage facility (Indiana). Fish samples were com-posites. Also, 1 dead fish was collected from the East Fork White River.	Sediment, fish (whole). Species: redhorse species, carpsucker.	7 5	Aldrin DDE, p,p'- Dieldrin Endrin HCH, α- Heptachlor epoxide Lindane	14 — 14 14 14 14 14	14 — 14 14 14 14 14
Water and Air Research, Inc., 1981	1981	11 sites (sediment) and 7 sites (biota) in the Savannah River and tributaries (Georgia, South Carolina). The study area was the site of a planned impoundment (Lake Russell). Sediment samples were composites of	Sediment (surficial), fish (skin-off fillets), invertebrates. Species include silver redhorse sucker, redbreast sunfish,	11 21	Aldrin Chlordane DDD DDD, o,p'- DDD, p,p'- DDE DDE, o,p'- DDE, p,p'- DDT DDT, o,p'- DDT, p,p'- Dieldrin	0 0 0 — — 0 — — 0 — — 0	0 0 0 — — 0 — — 0 — — 0

Table 2.2. Pesticides in bed sediment and aquatic biota from rivers in the United States: State and local monitoring studies—*Continued*

| Bed sediment (µg/kg dry weight) | | | Biota (whole fish unless noted) | | | | | Comments |
| | | | % Detection | | (µg/kg wet weight) | | | |
DL	Range	Median	Sites	Samples	DL	Range	Median	
—	—	—	50*	11*	30–150	nd–680*	nd*	(8) * Tissue type not reported. Objective: various, depending on site and study.
—	—	—	0*	0*	30	nd*	nd*	
—	—	—	0*	0*	10–50	nd*	nd*	
—	—	—	0*	0*	10–50	nd*	nd*	
—	—	—	0*	0*	200	nd*	nd*	
—	—	—	0*	0*	10	nd*	nd*	
—	—	—	0*	0*	10	nd*	nd*	
—	—	—	0*	0*	20	nd*	nd*	
—	—	—	39*	45*	10	nd–440*	nr*	(1) * Tissue type not reported, but was probably fish flesh based on statewide program description in Sloan (1981a). Objective: to monitor contaminants in 2 species of fish per site statewide.
—	—	—	98*	94*	10	nd–13,400*	nr*	
—	—	—	37*	40*	10	nd–130*	nr*	
—	—	—	11*	6*	10	nd–20	nr*	
—	—	—	0*	0*	10	nd*	nr*	
—	—	—	19*	nr*	10	nd–80*	nr*	
—	—	—	11*	14*	10	nd–200*	nr*	
—	—	—	100*	100*	nr	30–290*	nr*	(2) * Tissue type not reported. Objective: to monitor trends in Lake Ontario salmonoids.
—	—	—	50*	nr*	nr	10–220*	nr*	
—	—	—	100*	nr*	nr	10–220*	nr*	
—	—	—	100*	100*	nr	10–220*	nr*	
—	—	—	100*	100*	nr	10–300*	nr*	
0.4	nd–16.67	nr	—	—	—	—	—	(3) Objective: to determine source of DDT in Seneca Lake by monitoring its tributaries. Results suggested inputs may have occurred via Glen Eldridge, Sawmill, and Breakneck creeks.
nr	nd–30	nd	—	—	—	—	—	Objective: to investigate cause of fish kills at or downstream from confluence of waste site drainage ditches and the East Fork White River. Pesticides analyzed for but not found were not listed. The specific cause of fish death was unknown, but several priority pollutants were detected in sediment and fish near the waste site.
—	—	—	40	40	nr	nd–20	nd	
nr	nd–20	nd	20	20	nr	nd–70	nd	
nr	nd–10	nd	—	—	—	—	—	
nr	nd–30	nd	—	—	—	—	—	
nr	nd–20	nd	—	—	—	—	—	
nr	nd–50	nd	—	—	—	—	—	
1	nd	nd	—	—	—	—	—	* Data are for fish fillets. Objective: to determine preimpoundment water quality within the future area of Lake Russell. DDT compounds were high in biota from Savannah R. *p,p'*-DDE levels were 12–100 µg/kg in surface fish and 75–910 µg/kg in bottom-feeders. In the 2 tributaries, HCH, chlordane, heptachlor, and DDTs were occasionally detected in hellgrammites,
1	nd	nd	100*	65*	1	nd–150*	7*	
1	nd	nd	—	—	—	—	—	
—	—	—	0*	0*	1	nd*	nd*	
—	—	—	100*	61*	1	nd–170*	6*	
1	nd	nd	—	—	—	—	—	
—	—	—	50*	40*	1	nd–12*	nd*	
—	—	—	100*	94*	1	nd–910*	33*	
1	nd	nd	—	—	—	—	—	
—	—	—	0*	0*	1	nd*	nd*	
—	—	—	57*	24*	1	nd–400*	nd*	
1	nd	nd	0*	0*	1	nd*		

Table 2.2. Pesticides in bed sediment and aquatic biota from rivers in the United States: State and local monitoring studies—*Continued*

Reference	Sampling date	Sampling location and strategy	Sampling media and species	Number of samples	Target analytes	Bed sediment % Detection Sites	Bed sediment % Detection Samples
		subsamples. 2 invertebrate species (crayfish and insect) and 2 fish species (surface- and bottom-feeders) were sampled at each site.	bluegill, white bass, bullhead, crayfish, caddisfly larvae, cranefly larvae, hellgrammites.		Endrin	0	0
					HCH, α-	0	0
					HCH, β-	0	0
					Heptachlor	0	0
					Lindane	0	0
					Methoxychlor	—	—
					Mirex	0	0
					Toxaphene	0	0
Watson and others, 1981	1976–1977	6 sites in the Blind River (Louisiana). Most samples were collected monthly, except fishes were collected quarterly.	Sediment (surficial), fish (viscera), plants, water (whole). Species: gizzard shad, paddlefish, channel catfish, blue catfish, bowfin, largemouth bass, spotted gar, duckweed, arrow arum.	nr (~19) 35 23 50	Chlordane	100	100
					DDD, *p,p'*-	83–100	nr
					DDE, *p,p'*-	100	nr
					DDT, *p,p'*-	17	nr
					Dieldrin	100	nr
					HCB	100	nr
					Heptachlor epoxide	100	nr
Dick, 1982	1972–1977	176 sites (sediment) and 129 sites (water) were sampled as part of the Texas Statewide Monitoring Network. Of these sites, 110 (sediment) and 101 (water) were freshwater. The remaining sites were in bays, estuaries, and tidal portions of rivers. Sites were near urban or agricultural areas, or	(1) Sediment (surficial), water (whole), fish (whole, edible, other organs), oysters. Fish species include blue catfish, gizzard shad, channel catfish.	174* 147* 18 12 1	Aldrin	0	0
					Chlordane	4	2
					DDD	13	7
					DDE	32	20
					DDT	12	6
					Diazinon	0	0
					Dieldrin	12	5
					Endrin	0	0
					Heptachlor	1	1
					Heptachlor epoxide	0	0
					Lindane	0	0
					Malathion	0	0
					Methoxychlor	1	1
					Methyl parathion	0	0
					Parathion	0	0
					Toxaphene	0	0

Table 2.2. Pesticides in bed sediment and aquatic biota from rivers in the United States: State and local monitoring studies—*Continued*

Bed sediment			Biota (whole fish unless noted)					Comments
(µg/kg dry weight)			% Detection		(µg/kg wet weight)			
DL	Range	Median	Sites	Samples	DL	Range	Median	
1	nd	nd	—	—	—	—	—	crayfish, and surface fish, but at lower levels than in the Savannah R. PCBs were found in the Savannah R. but not in its tributaries. The authors concluded that low residues of OC pesticides in fish from tributaries were due to agricultural runoff (not to exposure by migration into the Savannah R.) No OC pesticides were detected in sediment. However, sediment generally was coarse sand or exposed bedrock.
1	nd	nd	86*	38*	1	nd–22*	nd*	
1	nd	nd	0*	0*	1	nd*	nd*	
1	nd	nd	100*	62*	1	nd*	nd*	
1	nd	nd	29*	14*	1	nd–2*	nd*	
—	—	—	0*	0*	10	nd*	nd*	
10	nd	nd	0*	0*	10	nd*	nd*	
25	nd	nd	0*	0*	25	nd*	nd*	
nr	4–107	nr	nr	nr	nr	nd–186*	nr	* Data are for fish viscera. Objective: to evaluate occurrence and distribution of OCs in the Blind R. No pesticides were detected in water. Mean chlordane residues in sediment increased gradually downstream. Ratio of *p,p'*- DDE: *p,p'* -DDD (E/D ratio) in sediment was about 2:1 at the 1st 3 (upstream) sites, 3:1 at site 5, and 5:1 at site 6, suggesting that upstream sediments (before a diversion canal) were more anaerobic. The E/D ratio was 3:1 in fish viscera. Ratio of *p,p'* -DDT: *p,p'*- DDE:*p,p'*- DDD was 26:1:20 in arrow-arum and 2:2:1 in duckweed. Residues in fish viscera were higher than in sediment or plants.
nr	nr	nr	nr	nr	nr	nd–68*	nr	
nr	nd–17	nr	nr	nr	nr	nd–161*	nr	
nr	nr	nr	0*	0*	nr	nd*	nd*	
nr	nd–2	nr	nr	nr	nr	nd–13*	nr	
nr	nd–2	nr	—	—	—	—	—	
nr	nd–2	nr	nr	nr	nr	nd–27*	nr	
0.2–1	nd*	nd*	—	—	—	—	—	*Data are for freshwater samples. Objective: to monitor pesticides in water, sediment, and fish statewide to identify and correct problems associated with use or misuse of pesticides. Of the pesticide detections in water, most were in streams and bayous draining Houston, Dallas, and San Antonio. For 12 of 18 pesticides, >50% of detections were at sites in these populated areas. In sediment, these same urbanized basins show significant pesticide contamination (especially
1–20	nd–46.9*	nd*	—	—	—	—	—	
0.1–3	nd–180*	nd*	73	78	nr	nd–3,150	61	
0.1–2	nd–190*	nd*	93	94	nr	nd–20,000	276	
0.1–5	nd–10.5*	nd*	80	83	nr	nd–2,300	105	
0.1–5	nd*	nd*	—	—	—	—	—	
0.1–3	nd–38*	nd*	—	—	—	—	—	
0.1–3	nd*	nd*	—	—	—	—	—	
0.2–1	nd–16.6*	nd*	—	—	—	—	—	
0.2–1	nd*	nd*	—	—	—	—	—	
0.2–3	nd*	nd*	—	—	—	—	—	
2–5	nd*	nd*	—	—	—	—	—	
0.2–20	nd–0.18*	nd*	—	—	—	—	—	
0.1–5	nd*	nd*	—	—	—	—	—	
0.1–5	nd*	nd*	—	—	—	—	—	
0.2–50	nd*	nd*	—	—	—	—	—	

Table 2.2. Pesticides in bed sediment and aquatic biota from rivers in the United States: State and local monitoring studies—*Continued*

Reference	Sampling date	Sampling location and strategy	Sampling media and species	Number of samples	Target analytes	Bed sediment % Detection Sites	Bed sediment % Detection Samples
		in systems receiving water and sediment from major rivers. Data are presented here in 2 parts: collected by (1) Texas Department of Water Resources and (2) USGS (with lower detection limits). In part (1), fish were sampled at 7 urban runoff sites throughout the state.	(2) Sediment (surficial), water (whole).	351* 568*	Aldrin	nr	2*
					Chlordane	nr	38*
					DDD	nr	45*
					DDE	nr	53*
					DDT	nr	33*
					Diazinon	0*	0*
					Dieldrin	nr	39*
					Endrin	0*	0*
					Heptachlor	nr	2*
					Heptachlor epoxide	nr	6*
					Lindane	nr	2*
					Malathion	0*	0*
					Methoxychlor	0*	0*
					Methyl parathion	0*	0*
					Parathion	0*	0*
					Toxaphene	nr	2*
Garrison, 1982	1981	Rivers and wetlands at 6 sites in Jean Lafitte National Park, Louisiana. Sediment was sampled at 3 sites.	Water, sediment (surficial).	3	Aldrin	0	0
					Chlordane	100	100
					DDD	67	67
					DDE	33	33
					DDT	33	33
					Diazinon	0	0
					Dieldrin	0	0
					Endosulfan	0	0
					Endrin	0	0
					Ethion	0	0
					Heptachlor	0	0
					Heptachlor epoxide	0	0
					Lindane	0	0
					Malathion	0	0
					Methoxychlor	0	0
					Methyl parathion	0	0
					Methyl trithion	0	0
					Mirex	0	0
					Parathion	0	0
					Perthane	0	0
					Toxaphene	0	0
					Trithion	0	0
Graves and others, 1982	1978–1979	Unknown number of sites in 3 watersheds in Louisiana that had high, medium, and low contamination in an earlier (1973) study. Fish samples were composites of 1–5 fish.	Fish (whole). Species: catfish, crappie, shad.	90	Chlordane	—	—
					DDD	—	—
					DDE	—	—
					DDT	—	—
					Dieldrin	—	—
					Endrin	—	—
					Toxaphene	—	—

Table 2.2. Pesticides in bed sediment and aquatic biota from rivers in the United States: State and local monitoring studies—*Continued*

Bed sediment (µg/kg dry weight)			Biota (whole fish unless noted)					Comments
			% Detection		(µg/kg wet weight)			
DL	Range	Median	Sites	Samples	DL	Range	Median	
0.1	nr	nr	—	—	—	—	—	chlordane and dieldrin).
10	nr	nr	—	—	—	—	—	However, several other basins
0.1	nr	nr	—	—	—	—	—	show substantial contamination
0.1	nr	nr	—	—	—	—	—	with 1 or more pesticides. For
0.1	nr	nr	—	—	—	—	—	example, high DDT levels were
0.1	nd*	nd*	—	—	—	—	—	found in sediment and fish from
0.1	nr	nr	—	—	—	—	—	the Colorado R. Basin (Pecan
0.1	nd*	nd*	—	—	—	—	—	Bayou) and Rio Grande R., and
0.1	nr	nr	—	—	—	—	—	in sediment from the San
0.1	nr	nr	—	—	—	—	—	Antonio R. and Leon Creek.
0.1	nr	nr	—	—	—	—	—	
0.1	nd*	nd*	—	—	—	—	—	
0.1	nd*	nd*	—	—	—	—	—	
0.1	nd*	nd*	—	—	—	—	—	
0.1	nd*	nd*	—	—	—	—	—	
0.1	nr	nr	—	—	—	—	—	
0.1	nd	nd	—	—	—	—	—	Objective: to obtain baseline
nr	8.0–13	8	—	—	—	—	—	data for waterways in Jean
0.1	nd–2.7	1.5	—	—	—	—	—	Lafitte National Historical Park.
0.1	nd–0.2	nd	—	—	—	—	—	No information provided on
0.1	nd–1.2	nd	—	—	—	—	—	sampling methods. Surface
0.1	nd	nd	—	—	—	—	—	water was sampled monthly at 6
0.1	nd	nd	—	—	—	—	—	sites from April 1981 to Jan.
0.1	nd	nd	—	—	—	—	—	1982. Data are presented without
0.1	nd	nd	—	—	—	—	—	interpretation of results.
0.1	nd	nd	—	—	—	—	—	
0.1	nd	nd	—	—	—	—	—	
0.1	nd	nd	—	—	—	—	—	
0.1	nd	nd	—	—	—	—	—	
0.1	nd	nd	—	—	—	—	—	
0.1	nd	nd	—	—	—	—	—	
0.1	nd	nd	—	—	—	—	—	
0.1	nd	nd	—	—	—	—	—	
0.1	nd	nd	—	—	—	—	—	
1	nd	nd	—	—	—	—	—	
0.1	nd	nd	—	—	—	—	—	
—	—	—	100*	nr	50	nd–240†	nd†	* Percent of the 3 watersheds.
—	—	—	100*	nr	50	nd–110†	nd†	† Range and median are of
—	—	—	100*	nr	50	nd–130†	10†	mean values for each species and
—	—	—	0*	0*	50	nd†	nd†	year (counting nondetects as 0).
—	—	—	100*	nr	50	nd–90†	nd†	Objective: to determine OC resi-
—	—	—	0*	0*	50	nd†	nd†	dues in fish from 3 watersheds
—	—	—	100*	nr	100	nd–460†	nd†	with high, medium, and low

contamination in an earlier (1973) study. DDE and DDD residues decreased from 1973 to 1978–

Table 2.2. Pesticides in bed sediment and aquatic biota from rivers in the United States: State and local monitoring studies—*Continued*

Reference	Sam-pling date	Sampling location and strategy	Sampling media and species	Number of samples	Target analytes	Bed sediment % Detection Sites	Bed sediment % Detection Sam-ples
Hochreiter, 1982	1980–1981	10 sites in the Delaware River estuary and adjacent tributaries (Pennsylvania, New Jersey)	Water (whole), sediment (surficial).	11	Aldrin	0	0
					Chlordane	70	73
					DDD, total	100	100
					DDE, total	100	100
					DDT, total	70	64
					Diazinon	10	10
					Dieldrin	50	55
					Endosulfan	20	18
					Endrin	0	0
					Ethion	0	0
					Heptachlor	0	0
					Heptachlor epoxide	20	18
					Lindane	0	0
					Malathion	0	0
					Methoxychlor	0	0
					Methyl parathion	0	0
					Methyl trithion	0	0
					Mirex	0	0
					Parathion	0	0
					Perthane	10	9
					Toxaphene	0	0
					Trithion	0	0
Loesch and others, 1982	1977–1978	6 rivers in Virginia near the western shore of the Chesapeake Bay. Samples were collected from river reaches used as nursery habitat for Chesapeake Bay fisheries. Fish sampled were less than one year old. Samples were composites.	Fish (whole). Species: alewife, American shad, blueback herring, striped bass.	124	Kepone	—	—

Table 2.2. Pesticides in bed sediment and aquatic biota from rivers in the United States: State and local monitoring studies—*Continued*

Bed sediment (µg/kg dry weight)			Biota (whole fish unless noted) % Detection		Biota (whole fish unless noted) (µg/kg wet weight)			Comments
DL	Range	Median	Sites	Samples	DL	Range	Median	
								1979. DDT was prohibited on cotton in 1972. Toxaphene was barely detectable in 1978–1979. In general, shad had higher levels than catfish or crappie.
0.1	nd	nd	—	—	—	—	—	Objective: to assess occurrence and distribution of contaminants in study area. Surface water was collected at 13 sites and analyzed for VOCs. Bed sediment at one site also contained traces (to 4.5 µg/kg) of chlorinated benzenes and PAHs.
1	nd–21	4	—	—	—	—	—	
nr	1.9–4,000	12	—	—	—	—	—	
nr	2.4–300	10	—	—	—	—	—	
0.1	nd–150	0.5	—	—	—	—	—	
0.1	nd–1.6	nd	—	—	—	—	—	
0.1	nd–10	nd	—	—	—	—	—	
0.1	nd–0.2	nd	—	—	—	—	—	
0.1	nd	nd	—	—	—	—	—	
0.1	nd	nd	—	—	—	—	—	
0.1	nd	nd	—	—	—	—	—	
0.1	nd–1.6	nd	—	—	—	—	—	
0.1	nd	nd	—	—	—	—	—	
0.1	nd	nd	—	—	—	—	—	
0.1	nd	nd	—	—	—	—	—	
0.1	nd	nd	—	—	—	—	—	
0.1	nd	nd	—	—	—	—	—	
0.1	nd	nd	—	—	—	—	—	
0.1	nd	nd	—	—	—	—	—	
0.1	nd–510	nd	—	—	—	—	—	
1	nd	nd	—	—	—	—	—	
10	nd	nd	—	—	—	—	—	
—	—	—	67	na	20	nd–2,800	na	Objective: to evaluate the occurrence and distribution of kepone in fish of Cheseapeake Bay's eastern tributaries. All 4 species collected in the James R. and alewife and blueback herring from the lower Chickahominy R. (a tributary of the James) had kepone residues exceeding the action level (300 µg/kg). Kepone was also detected in fish from the Pamunkey and Mattaponi rivers, for which there is no known instream or local source. A possible source is atmospheric transport from the James R. Basin. No kepone was detected in fish from Rappahannock or Potomac rivers.

Table 2.2. Pesticides in bed sediment and aquatic biota from rivers in the United States: State and local monitoring studies—*Continued*

Reference	Sampling date	Sampling location and strategy	Sampling media and species	Number of samples	Target analytes	Bed sediment % Detection Sites	Bed sediment % Detection Samples
McHenry and others, 1982	1979	3 sites in Wolf Lake, lower Yazoo River Basin (Mississippi). Sediment cores (50–100 cm) were sectioned into 10-cm increments. Like-depth samples from 6–8 cores per site were composited. Water was sampled at the surface and bottom of the channel.	Sediment (cores), water (whole).	3	Chlordane DDD DDE DDT Dieldrin Endrin Heptachlor Heptachlor epoxide Toxaphene	0* 100* 100* 100* 33* 33* 0* 0* 0*	0* 100* 100* 100* 33* 33* 0* 0* 0*
Rohrer and others, 1982	1980	7 tributaries to the Great Lakes (Michigan). Samples were collected during the fall spawning migration.	Fish (skin-on fillets, skin-off fillets). Species: coho salmon, chinook salmon.	113 10	Aldrin Chlordane DDT, total Dieldrin Heptachlor Heptachlor epoxide Toxaphene	— — — — — — —	— — — — — — —
Scrudato and DelPrete, 1982	1979	About 23 sites in the Oswego River, Oswego Harbor and eastern basin of Lake Ontario (New York). Oswego River has a known industrial source of mirex (Armstrong Cork Company). Surficial samples were taken at 11 river sites, and surficial or core samples at 12	Sediment (surficial, cores).	~18 5	Mirex	55*	55*

Table 2.2. Pesticides in bed sediment and aquatic biota from rivers in the United States: State and local monitoring studies—*Continued*

Bed sediment			Biota (whole fish unless noted)					Comments
(µg/kg dry weight)			% Detection		(µg/kg wet weight)			
DL	Range	Median	Sites	Sam-ples	DL	Range	Median	
0.01	nd*	nd*	—	—	—	—	—	* Data are for 0–10 cm depth.
0.01	18.2–328*	72.1*	—	—	—	—	—	Objective: to characterize the
0.01	15.5–146*	17.4*	—	—	—	—	—	sedimentation regime in Wolf
0.01	4.3–67*	4.9*	—	—	—	—	—	Lake, Lower Yazoo R. Basin.
0.01	nd–2.9*	nd*	—	—	—	—	—	Sediment profiles show DDT
0.01	nd–1.55*	nd*	—	—	—	—	—	levels were greatest after 1964
0.01	nd*	nd*	—	—	—	—	—	but have declined in recent years.
0.01	nd*	nd*	—	—	—	—	—	However, sediments are a
0.01	nd*	nd*	—	—	—	—	—	considerable reservoir of DDT.
								Heptachlor epoxide was detected
								at 50–60 cm depth at one site.
								especially at the surface. Sedi-
								ment entering the lake is still
								contaminated with pesticides.
—	—	—	0*	0*	10–200	nd*	nd*	* Data are skin-on fish fillets.
—	—	—	0*	0*	100–500	nd*	nd*	Objective: to assess levels and
—	—	—	100*	100*	50–200	60–3,400*	nr	trends of contaminants in Great
—	—	—	100*	nr	10–100	nd–170	nr	Lakes salmon. Chinook salmon
—	—	—	0*	0*	10–200	nd*	nd*	generally had higher DDT and
—	—	—	0*	0*	30–200	nd*	nd*	dieldrin residues than coho
—	—	—	0*	0*	2,000–5,000	nd*	nd*	salmon collected at the same
								time and site. This may be due
								to age differences; salmon were
								about 5 (chinook) and 3 (coho)
								years old. Almost all total DDT
								was *p,p'*- DDE. The highest
								total DDT levels were from the
								southernmost Lake Michigan
								tributary (St. Joseph R.), and
								the lowest levels from Lake Erie
								tributaries. OC residues in 1980
								were significantly lower than in
								salmon sampled in 1971.
								Skinless fillets had lower DDTs
								and PCBs than skin-on fillets.
nr	nd–1,834	7.3	—	—	—	—	—	* Data are for river sites only.
								Objective: to investigate mirex-
								sediment relations in Lake Onta-
								rio and the Oswego R. The
								highest mirex levels in the river
								were just downstream of the
								industrial discharge source. In
								harbor and offshore samples, the
								highest mirex residues were as-
								sociated with fine-grained sedi-
								ment. Organic carbon content
								and mirex in offshore samples
								were highly correlated. Mirex-
								contaminated sediment was
								estimated to be accumulating at

Table 2.2. Pesticides in bed sediment and aquatic biota from rivers in the United States: State and local monitoring studies—*Continued*

Reference	Sampling date	Sampling location and strategy	Sampling media and species	Number of samples	Target analytes	Bed sediment % Detection Sites	Bed sediment % Detection Samples
		harbor or offshore sites.					
Sloan, 1982	1979–1981	3 studies in this report measured pesticides in aquatic biota from New York: (1) 3 sites in Lake Ontario and Spring Brook; (2) 8 sites in Patroons Creek, Plattekill Brook, Belmont Lake, Hannacroix Creek, and Cayuga Creek; (3) 7 estuarine or marine sites in New York and New Jersey. Some samples were composites. Results here are presented separately for the 3 studies.	(1) Fish (flesh*). Species: rainbow trout, lake trout, brown trout, coho salmon.	62	Chlordane Dieldrin Endrin HCB HCH, α- Mirex OPs (not specified) Pentachlorophenyl methyl ether Photomirex Toxaphene	— — — — — — — — — —	— — — — — — — — — —
			(2) Fish (whole, flesh*), invertebrates, amphibians. Species include carp, bluegill, white sucker.	4 52 8	Chlordane DDE, p,p´- DDT, total Dieldrin Endosulfan HCB HCH, total Mirex	— — — — — — — —	— — — — — — — —
			(3) Fish (flesh*), invertebrates. Species include clams, striped bass.	nr nr	Chlordane DDD+DDE Dieldrin HCB Pentachlorophenol	— — — — —	— — — — —
Waller, 1982	1969–1977	12 sites in and near Everglades National Park, Florida. Sites were freshwater or estuarine, including rivers, ponds, canals, bays, and wetlands.	Water, sediment.	1,719 nr	Aldrin Chlordane D, 2,4- DDD DDE DDT DDT, total Diazinon Dieldrin Endrin Ethion Heptachlor	0 50 0 nr nr nr 92 0 92 0 0 0	0 16 0 nr nr nr 78 0 27 0 0 0

Table 2.2. Pesticides in bed sediment and aquatic biota from rivers in the United States: State and local monitoring studies—*Continued*

| Bed sediment (µg/kg dry weight) | | | Biota (whole fish unless noted) | | | | | Comments |
| DL | Range | Median | % Detection | | (µg/kg wet weight) | | | |
			Sites	Samples	DL	Range	Median	
								2.2–7.0 mm/year in deeper wa–ters. One offshore site used for several years for disposal of dredged Oswego Harbor sediment had sandy sediment and sporadic mirex residues, suggesting that fine-grained contaminated sediments have been redistributed elsewhere in the lake. Oswego R. was still a source of mirex to the lake, although inputs appeared to be declining.
—	—	—	100*	100*	nr	140–200*	180*§	(1) * Data are probably for fish
—	—	—	33*	*nr	0	nd–30*	nd*†	flesh. Tissue type was not
—	—	—	*nr	*nr	nr	nd–trace*	nr*	reported, but some analyses were
—	—	—	100*	*nr	10	nd–70*	30*†	done by FDA, which generally
—	—	—	*nr	*nr	nr	nd–trace*	nr*	analyzes edible portions.
—	—	—	100*	*nr	10	nd–580*	150*†	† Median of mean values for
—	—	—	0*	0*	nr	nd*	nd*	each site/species.
—	—	—	*nr	*nr	nr	nd–trace*	nr*	§ Mean of 3 composites. Objective: to measure
—	—	—	100*	*nr	10	nd–490*	130*†	contaminants in salmonoid
—	—	—	100*	100*	nr	500–1,000*	770*§	species from Lake Ontario.
—	—	—	100*	100*	nr	350–5,200*	2,780*	(2) * Data are probably mostly
—	—	—	80*	nr*	10	nd–220*	25*†	for fish flesh (4 samples were
—	—	—	100*	100*	nr	100–1,630*	330*	noted in report as whole fish).
—	—	—	100*	75*	10	nd–60*	20*	† Median of mean values for
—	—	—	0*	0*	40	nd*	nd*	each site/species.
—	—	—	60*	nr*	10	nd–280*	10*†	Objective: Special collections
—	—	—	100*	nr*	10–130	nd–3,750*	100*†	each had its own objectives
—	—	—	100*	nr*	10–30	nd–170	nd*†	relating to monitoring specific contaminants in the basin.
—	—	—	80	nr	nr	nd–480*	trace*	(3) * Data are probably for fish
—	—	—	60	nr	nr	nd–60*	trace*	flesh. Tissue type was not
—	—	—	40	nr	nr	nd–30*	nd*	reported, but some analyses were
—	—	—	20	nr	nr	nd–trace*	nd*	done by FDA, which generally
—	—	—	20	nr	nr	nd–100*	nd*	analyzes edible portions.
nr	nd	nd	—	—	—	—	—	* Mean value.
nr	nd–40	1.8*	—	—	—	—	—	Objective: to summarize
nr	nd	nd	—	—	—	—	—	water-quality characteristics and
nr	nd–90	3.5*	—	—	—	—	—	trends in Everglades National
nr	nd–41	4.2*	—	—	—	—	—	Park. The single silvex
nr	nd–35	0.9*	—	—	—	—	—	detection in sediment probably
nr	nr	nr	—	—	—	—	—	was result of direct aquatic
nr	nd	nd	—	—	—	—	—	application for weed control.
nr	nd–9.3	0.2*	—	—	—	—	—	DDD, DDE, DDT, dieldrin,
nr	nd	nd	—	—	—	—	—	diazinon, and 2,4-D were the
nr	nd	nd	—	—	—	—	—	only pesticides detected in
nr	nd	nd	—	—	—	—	—	surface water samples

Table 2.2. Pesticides in bed sediment and aquatic biota from rivers in the United States: State and local monitoring studies—*Continued*

Reference	Sampling date	Sampling location and strategy	Sampling media and species	Number of samples	Target analytes	Bed sediment % Detection Sites	Bed sediment % Detection Samples
					Heptachlor epoxide	0	0
					Lindane	0	0
					Malathion	0	0
					Methyl parathion	0	0
					Methyl trithion	0	0
					Parathion	0	0
					Silvex	8	1.3
					T, 2,4,5-	0	0
					Toxaphene	0	0
					Trithion	0	0
Water and Air Research, Inc., 1982	1989	19 sites (sediment) and 1 site (mollusks) on Lake Seminole and its major tributaries, the Chattahoochee and Flint rivers, and its outfall, the Apalachicola River (Georgia, Florida, Alabama). Mollusks were collected from 4 locations equally spaced along cross-section.	Sediment (surficial), mollusks, water (whole, dissolved). Species: *Corbicula* species.	19 1	Aldrin	0	0
					Chlordane	0	0
					D, 2,4-	0	0
					DDD, *p,p'*-	0	0
					DDE, *p,p'*-	32	32
					DDT, *o,p'*-	0	0
					DDT, *p,p'*-	0	0
					Dieldrin	0	0
					Endosulfan sulfate	—	—
					Endothal	0	0
					Endrin	0	0
					Endrin aldehyde	0	0
					Glyphosate	0	0
					HCH, α-	0	0
					HCH, β-	0	0
					Heptachlor	0	0
					Heptachlor epoxide	0	0
					Lindane	0	0
					Methoxychlor	0	0
					Mirex	0	0
					Pentachlorophenol	0	0
					Toxaphene	0	0
Fox and others, 1983	1981	1 site (surficial sediment, biota) and 2 sites (water, suspended sediment) on the Niagara River and 4 sites (surficial sediment) in western Lake Ontario. Benthic invertebrates present in sediment samples were surveyed. The fish sample was a composite of 1-year olds.	Water (whole), sediment (surficial, suspended) fish (whole), mysids, oligochaetes, amphipods. Fish species: lake trout.	2 9 4 1 1 9 10	HCB	100	100

Table 2.2. Pesticides in bed sediment and aquatic biota from rivers in the United States: State and local monitoring studies—*Continued*

Bed sediment (µg/kg dry weight)			Biota (whole fish unless noted)					Comments
			% Detection		(µg/kg wet weight)			
DL	Range	Median	Sites	Samples	DL	Range	Median	
nr	nd	nd	—	—	—	—	—	(collected from 1966 to 1977)
nr	nd	nd	—	—	—	—	—	within the park.
nr	nd	nd	—	—	—	—	—	
nr	nd	nd	—	—	—	—	—	
nr	nd	nd	—	—	—	—	—	
nr	nd	nd	—	—	—	—	—	
nr	nd–160	nd	—	—	—	—	—	
nr	nd	nd	—	—	—	—	—	
nr	nd	nd	—	—	—	—	—	
nr	nd	nd	—	—	—	—	—	
0.1	nd	nd	0	0	0.2	nd	nd	* Data are for *Corbicula*.
0.5	nd	nd	0	0	5	nd	nd	Objective: to determine baseline
2	nd	nd	—	—	—	—	—	conditions and identify water-
0.2	nd	nd	0	0	0.5	nd	nd	quality problems. *Corbicula*
0.1	nd–5.4	nd	—	—	—	—	—	were targeted for sampling at 4
0.2	nd	nd	0	0	0.5	nd	nd	sites, but found at only 1 site.
0.2	nd	nd	0	0	0.5	nd	nd	In an earlier sampling effort in
0.1	nd	nd	0	0	0.5	nd	nd	the same study area (Phase I),
—	—	—	0	0	5	nd	nd	2,4-D was found in sediment at
0.5	nd	nd	—	—	—	—	—	4 sites, and *p,p'*- DDE was below
0.1	nd	nd	—	—	—	—	—	detection at all sites; in
0.1	nd	nd	—	—	—	—	—	*Corbicula,* endosulfan sulfate
1	nd	nd	—	—	—	—	—	and chlordane were found during
0.1	nd	nd	0	0	0.2	nd	nd	Phase I at 1 site (the same site
0.1	nd	nd	0	0	0.2	nd	nd	sampled in Phase II).
0.1	nd	nd	0	0	0.2	nd	nd	
0.1	nd	nd	0	0	0.2	nd	nd	
0.1	nd	nd	0	0	0.1	nd	nd	
0.2	nd	nd	0	0	5	nd	nd	
0.1	nd	nd	0	0	5	nd	nd	
0.5	nd	nd	0	0	5	nd	nd	
0.5	nd	nd	0	0	5	nd	nd	
nr	62–840*	120*	100	100	nr	83*	83*	* µg/kg dry weight. Objective: to examine relations between chlorobenzene levels in Niagara R. water, suspended solids, surficial sediment, and biota. Chlorobenzene residues tended to be higher in large size fractions of suspended solids. Residues in water and suspended solids upstream were much lower than at Niagara-on-the-Lake, indicating sources on the river. HCB residues in sediment were correlated with those in oligochaetes (which live in and ingest sediment), but not amphipods (which live at the sediment–water interface and

Table 2.2. Pesticides in bed sediment and aquatic biota from rivers in the United States: State and local monitoring studies—*Continued*

Reference	Sampling date	Sampling location and strategy	Sampling media and species	Number of samples	Target analytes	Bed sediment % Detection Sites	Bed sediment % Detection Samples
Fuhrer and Rinella, 1983	1980	33 sites(*) in 14 rivers and estuaries in Oregon and Washington. Sites were associated with dredging operations. Elutriate tests were performed.	Sediment (surficial), water (dissolved), elutriates (dissolved).	33*	Aldrin	15	7
					Chlordane	31	16
					D, 2,4-	0	0
					DDD	54	48
					DDE	62	42
					DDT	31	19
					Dieldrin	38	19
					DP, 2,4-	0	0
					Endosulfan	0	0
					Endrin	0	0
					Heptachlor	8	3
					Heptachlor epoxide	0	0
					Lindane	8	3
					Methoxychlor	0	0
					Mirex	0	0
					Perthane	0	0
					Silvex	0	0
					T, 2,4,5-	0	0
					Toxaphene	0	0
Gaydos, 1983	1979– 1982	8 sites (sediment) and 12 sites (water) in rivers in western Tennessee.	Sediment, water.	23 108	Aldrin	0	0
					Chlordane, total	25	17
					DDD	50	35
					DDE	38	22
					DDT	50	26
					Dieldrin	25	13
					Endosulfan	0	0
					Endrin	0	0
					Heptachlor	0	0
					Heptachlor epoxide	0	0
					Lindane	0	0
					Methoxychlor	0	0
					Mirex	0	0
					Perthane	0	0
					Toxaphene	0	0
Kauss, 1983	1979	24 sites (sediment) on the Niagara River and 1 control site on Lake Erie (New York and Ontario, Canada). Sediment sites were located near potential sources and(or) in depositional areas. Samples were composites of 3	Sediment (surficial, suspended), water (whole).	24 7 nr	Aldrin	0	0
					Chlordane, α-	50	50
					Chlordane, γ-	41	41
					DDD, p,p'-	35	35
					DDE, p,p'-	100	100
					DDT, o,p'-	9	9
					DDT, p,p'-	9	9
					DDT, total	100	100
					Dieldrin	45	45
					Endosulfan I	30	30
					Endosulfan II	26	26
					Endrin	14	14

Table 2.2. Pesticides in bed sediment and aquatic biota from rivers in the United States: State and local monitoring studies—*Continued*

Bed sediment (µg/kg dry weight)			Biota (whole fish unless noted) % Detection		(µg/kg wet weight)			Comments
DL	Range	Median	Sites	Samples	DL	Range	Median	
								ingest detritus). Accumulation factor for HCB was higher than for trichlorobenzene (which had a lower K_{ow}).
0.1	nd–1.5	nd	—	—	—	—	—	* Data shown are for 31 sites in
1	nd–4.0	nd	—	—	—	—	—	13 rivers and estuaries.
0.1	nd	nd	—	—	—	—	—	Objective: to provide
0.1	nd–5.9	nd	—	—	—	—	—	reconnaissance data to
0.1	nd–6.8	nd	—	—	—	—	—	determine short-term
0.1	nd–0.5	nd	—	—	—	—	—	water-quality conditions
0.1	nd–0.5	nd	—	—	—	—	—	associated with dredging
0.1	nd	nd	—	—	—	—	—	operations in rivers and
0.1	nd	nd	—	—	—	—	—	estuaries. Methods and raw data
0.1	nd	nd	—	—	—	—	—	were reported without
0.1	nd–0.2	nd	—	—	—	—	—	interpretation.
0.1	nd	nd	—	—	—	—	—	
0.1	nd–0.4	nd	—	—	—	—	—	
0.1	nd	nd	—	—	—	—	—	
0.1	nd	nd	—	—	—	—	—	
1	nd	nd	—	—	—	—	—	
0.1	nd	nd	—	—	—	—	—	
0.1	nd	nd	—	—	—	—	—	
1	nd	nd	—	—	—	—	—	
0.1	nd	nd	—	—	—	—	—	Objective: to help determine
0.1–1	nd–11	nd	—	—	—	—	—	whether problems caused by
0.1	nd–4.4	nd	—	—	—	—	—	pesticide use could be identified.
0.1	nd–2.7	nd	—	—	—	—	—	Traces of commonly used
0.1	nd–4.6	nd	—	—	—	—	—	herbicides were found in water
0.1	nd–0.8	nd	—	—	—	—	—	from 10 of 12 sites. Low levels
0.1	nd	nd	—	—	—	—	—	of DDT were found in water
0.1	nd	nd	—	—	—	—	—	from 2 sites. Sources of
0.1	nd	nd	—	—	—	—	—	pesticides in sediment and water
0.1	nd	nd	—	—	—	—	—	were not identified.
0.1	nd	nd	—	—	—	—	—	
0.1	nd	nd	—	—	—	—	—	
0.1	nd	nd	—	—	—	—	—	
0.1	nd	nd	—	—	—	—	—	
1	nd	nd	—	—	—	—	—	
nr	nd	nd	—	—	—	—	—	Objective: to determine spatial
nr	nd–293	nd/1	—	—	—	—	—	and temporal distribution of
nr	nd–70	nd	—	—	—	—	—	contaminants. There was no
nr	nd–65	nd	—	—	—	—	—	significant correlation between
nr	1–36	5	—	—	—	—	—	sediment contaminant
nr	nd–21	nd	—	—	—	—	—	concentration and particle size.
nr	nd–74	nd	—	—	—	—	—	DDT, dieldrin, chlordanes,
nr	1–190	7	—	—	—	—	—	endosulfans, and mirex levels
nr	nd–26	nd	—	—	—	—	—	were higher in the lower river
nr	nd–15	nd	—	—	—	—	—	than in the upper river and Lake
nr	nd–45	nd	—	—	—	—	—	Ontario. *p,p'*- DDT comprised
nr	nd–13	nd	—	—	—	—	—	about 35% of total DDT in the

Table 2.2. Pesticides in bed sediment and aquatic biota from rivers in the United States: State and local monitoring studies—*Continued*

Reference	Sampling date	Sampling location and strategy	Sampling media and species	Number of samples	Target analytes	Bed sediment % Detection Sites	Bed sediment % Detection Samples
		grabs per site. Water samples were collected during 1979–1980.			HCB	39	39
					HCH, α-	22	22
					HCH, β-	0	0
					Heptachlor	0	0
					Heptachlor epoxide	30	30
					Lindane	5	5
					Mirex	61	61
Lurry, 1983	1979–1981	38 sites on southern Louisiana waterways and coastal areas. Sites included river, estuary, and ocean locations.	Water, sediment (surficial), elutriates.	38	Aldrin	0	0
					Chlordane	39	39
					DDD	53	53
					DDE	45	45
					DDT	0	0
					Diazinon	5	5
					Dieldrin	34	34
					Endosulfan	3	3
					Endrin	3	3
					Ethion	0	0
					Heptachlor	0	0
					Heptachlor epoxide	0	0
					Lindane	0	0
					Malathion	0	0
					Methyl parathion	0	0
					Methyl trithion	0	0
					Methoxychlor	0	0
					Mirex	0	0
					Parathion	0	0
					Toxaphene	0	0
					Trithion	0	0
Ray and others, 1983	1980	8 (sediment) and 2 (clams) sites in the Portland (Maine) area, including the Fore River. Sediment samples were collected at low tide.	Sediment (surficial), mollusks. Species: clams (*Neanthes virens*).	8 2	Chlordane	75	75
					DDT, total	88	88
					HCB	75	75
					HCH, α-	50	50
					Pentachlorophenol	100	100
Suns and others, 1983	1975–1981	7 sites on the Niagara River (New York), including Niagara-on-the-Lake (NoL). Annual collections were made at same time each year (not specified). Samples were composites.	Fish (whole). Species: spottail shiner.	112	Aldrin	—	—
					Chlordane	—	—
					DDT, total	—	—
					Dieldrin	—	—
					Endosulfan	—	—
					Endrin	—	—
					HCB	—	—
					HCH	—	—
					Heptachlor	—	—
					Heptachlor epoxide	—	—
					Mirex	—	—
					TCDD, 2,3,7,8-	—	—

Table 2.2. Pesticides in bed sediment and aquatic biota from rivers in the United States: State and local monitoring studies—*Continued*

Bed sediment (µg/kg dry weight)			Biota (whole fish unless noted)					
			% Detection		(µg/kg wet weight)			
DL	Range	Median	Sites	Sam-ples	DL	Range	Median	Comments
nr	nd–250	nd	—	—	—	—	—	lower river, 0% in the upper
nr	nd–110	nd	—	—	—	—	—	river, and 14% in Lake Erie
nr	nd	nd	—	—	—	—	—	sediment. o,p'- and p,p'-DDT
nr	nd	nd	—	—	—	—	—	were not detected in suspended
nr	nd–36	nd	—	—	—	—	—	solids collected at lower river
nr	nd–20	nd	—	—	—	—	—	sites at the same time.
nr	nd–640	5	—	—	—	—	—	
0.1	nd	nd	—	—	—	—	—	Objective: to obtain water and
0.1	nd–79	nd	—	—	—	—	—	sediment quality data for use in
0.1	nd–56	25	—	—	—	—	—	evaluating the environmental
0.1	nd–16	nd	—	—	—	—	—	effects of dredging in the
0.1	nd	nd	—	—	—	—	—	Calcasieu R. Analysis of
0.1	nd–0.2	nd	—	—	—	—	—	organic constituents was done
0.1	nd–1.6	nd	—	—	—	—	—	on both water and bed sediment
0.1	nd–0.1	nd	—	—	—	—	—	at all sites and on sediment
0.1	nd–0.4	nd	—	—	—	—	—	elutriate at 22 of the 38 sites.
0.1	nd	nd	—	—	—	—	—	
0.1	nd	nd	—	—	—	—	—	
0.1	nd	nd	—	—	—	—	—	
0.1	nd	nd	—	—	—	—	—	
0.1	nd	nd	—	—	—	—	—	
0.1	nd	nd	—	—	—	—	—	
0.1	nd	nd	—	—	—	—	—	
0.1	nd	nd	—	—	—	—	—	
0.1	nd	nd	—	—	—	—	—	
0.1	nd	nd	—	—	—	—	—	
0.1	nd	nd	—	—	—	—	—	
0.1	nd	nd	—	—	—	—	—	
0.03	nd–9.8	0.3	50*	50*	0.03	nd–1.8*	nd/1.8*	* Data are for clams.
0.03	nd–42	2.1	100*	100*	nr	4.0–11*	7.5*	Objective: to characterize con-
0.03	nd–0.37	0.12	50*	50*	0.1	nd–0.64*	nd/0.64*	taminants in sediment and biota
0.03	nd–0.7	nd/0.06	100*	100*	nr	0.22–0.23*	0.23*	in Portland Harbor area. Al-
nr	0.01–2.4	0.48	100*	100*	nr	1.7–3.4*	2.6*	though most industrial activity
								is in Fore R., contaminant le-
								vels were lower there than in
								Back Cove (poorly flushed).
								Tissue/sediment accumulation
								factors were calculated.
—	—	—	0	nr	1	nd*	nd*	* Range or median of the mean
—	—	—	100	nr	2	nd–50*	14*	values for all sites and years.
—	—	—	100	nr	5	9–244*	37*	Objective: to determine spatial
—	—	—	71	nr	2	nd–6*	trace*	distribution (1980–1981) and
—	—	—	43	nr	4	nd–trace*	nd*	(at NoL) trends (1975–1981).
—	—	—	86	nr	4	nd–7*	trace*	Concentrations of HCH and
—	—	—	100	nr	nr	trace–7*	5.5*	chlordane were high relative to
—	—	—	86	nr	1	nd–27*	4*	similar data for Lakes Erie and
—	—	—	0	nr	1	nd*	nd*	Huron. Significant mirex
—	—	—	57	nr	1	nd–3*	nd*	reductions in NoL samples
—	—	—	86	nr	5	nd–29*	9.5*	since 1978 suggest reduced in-
—	—	—	100	100	0.002	0.003–	13*	puts to Lake Ontario. Total DDT

Table 2.2. Pesticides in bed sediment and aquatic biota from rivers in the United States: State and local monitoring studies—*Continued*

Reference	Sampling date	Sampling location and strategy	Sampling media and species	Number of samples	Target analytes	Bed sediment % Detection Sites	Bed sediment % Detection Samples
Wangsness, 1983	1980–1982	13 sites in Eagle Creek watershed (Indiana). 11 sites were sampled in 1980; 2 of these sites in the most contaminated area plus 2 new sites downstream of these were sampled in 1982. The 1982 survey was done under extreme low flow conditions. Sediment samples were sieved (<63 um).	Sediment (surficial), water (whole).	15	Aldrin	8	7
					Chlordane	92	93
					D, 2,4-	0	0
					DDD	46	47
					DDE	23	20
					DDT	31	27
					Diazinon	15	13
					Dieldrin	92	93
					DP, 2,4-	0	0
					Endosulfan	0	0
					Endrin	0	0
					Ethion	0	0
					HCB	0	0
					Heptachlor	0	0
					Heptachlor epoxide	0	0
					Lindane	0	0
					Malathion	0	0
					Methoxychlor	0	0
					Methyl parathion	0	0
					Methyl trithion	0	0
					Mirex	0	0
					Parathion	0	0
					Pentachlorophenol	0	0
					Perthane	0	0
					Silvex	0	0
					T, 2,4,5-	0	0
					TCDD, 2,3,7,8-	0	0
					Toxaphene	0	0
					Trithion	0	0
White and others, 1983	1976–1979	23 sites in the lower Rio Grande Valley (Texas). Location was selected because of high DDE residues in FWS's NCBP samples from the Rio Grande River. Two composite fish samples, each of a different species and consisting of 2–10	Fish (whole, fillets). Species include channel catfish, blue catfish, gizzard shad, white bass.	49 2	Dacthal	—	—
					DDE	—	—
					Dieldrin	—	—
					Endrin	—	—
					Toxaphene	—	—

Table 2.2. Pesticides in bed sediment and aquatic biota from rivers in the United States: State and local monitoring studies—*Continued*

Bed sediment (µg/kg dry weight)			Biota (whole fish unless noted)						Comments
			% Detection		(µg/kg wet weight)				
DL	Range	Median	Sites	Samples	DL	Range	Median		
							0.060*		increased significantly from 1980 to 1981 at 2 sites on the Ontario shoreline (including NoL) and 1 on the U.S. side. At NoL, *p,p'*- DDT content of the total DDT increased from 0% in 1979 to 22% in 1980, implying recent inputs to the river.
0.1	nd–1.8	nd	—	—	—	—	—		Objective: The 1980 survey was to define the water quality of the Eagle Creek watershed. The 1982 survey was to determine whether the type and concentrations of substances in Eagle Creek had changed since 1980. Data are presented without interpretation.
1	nd–29	9	—	—	—	—	—		
0.1	nd	nd	—	—	—	—	—		
0.1	nd–3.3	nd	—	—	—	—	—		
0.1	nd–1.5	nd	—	—	—	—	—		
0.1	nd–2.1	nd	—	—	—	—	—		
0.1	nd–2.4	nd	—	—	—	—	—		
0.1	nd–19	5.3	—	—	—	—	—		
0.1	nd	nd	—	—	—	—	—		
0.1	nd	nd	—	—	—	—	—		
0.1	nd	nd	—	—	—	—	—		
0.1	nd	nd	—	—	—	—	—		
20	nd	nd	—	—	—	—	—		
0.1	nd	nd	—	—	—	—	—		
0.1	nd	nd	—	—	—	—	—		
0.1	nd	nd	—	—	—	—	—		
0.1	nd	nd	—	—	—	—	—		
0.1	nd	nd	—	—	—	—	—		
0.1	nd	nd	—	—	—	—	—		
0.1	nd	nd	—	—	—	—	—		
0.1	nd	nd	—	—	—	—	—		
0.1	nd	nd	—	—	—	—	—		
20	nd	nd	—	—	—	—	—		
1	nd	nd	—	—	—	—	—		
0.1	nd	nd	—	—	—	—	—		
0.1	nd	nd	—	—	—	—	—		
20	nd	nd	—	—	—	—	—		
10	nd	nd	—	—	—	—	—		
0.1	nd	nd	—	—	—	—	—		
—	—	—	>0	>0	nr	nd–13,400	nr		* Toxaphene reported for Arroyo Colorado sites (8) only. Objective: to determine the extent of DDT contamination in fishes and wildlife of the Rio Grande Valley and possible sources of contamination. DDE and toxaphene were elevated in the Arroyo Colorado. Fillets contained 25% and 50% of the DDD and toxaphene (respectively) in whole fish for blue catfish. Channel catfish, gizzard
—	—	—	100	94	100	nd–31,500	900		
—	—	—	>0	>0	nr	nr	nr		
—	—	—	>0	>0	nr	nr	nr		
—	—	—	88*	85*	nr	nd– 31,500*	18,500*		

Table 2.2. Pesticides in bed sediment and aquatic biota from rivers in the United States: State and local monitoring studies—*Continued*

Reference	Sampling date	Sampling location and strategy	Sampling media and species	Number of samples	Target analytes	Bed sediment % Detection Sites	Bed sediment % Detection Samples
		individuals, were collected at each site.					
Allen, 1984	1979–1980	4 sites on Spring Creek, Illinois.	Sediment (surficial).	4	Aldrin	50	50
					D, 2,4-	0	0
					DDD	50	50
					DDE	100	100
					DDT	50	50
					Diazinon	0	0
					Dieldrin	100	100
					Ethion	0	0
					Heptachlor	0	0
					Heptachlor epoxide	0	0
					Lindane	0	0
					Malathion	0	0
					Mirex	0	0
					Silvex	0	0
					T, 2,4,5-	0	0
					Toxaphene	0	0
Barker, 1984	1980	4 sites on the Schuylkill River (Pennsylvania).	Fish (whole, fillets), snails, macrophytes, green algae, periphyton. Fish species include minnows, black crappie, American eel, white sucker.	4 82 9 5 9 1	Aldrin Chlordane DDT, total Dieldrin Endosulfan Endrin Heptachlor Heptachlor epoxide Lindane Methoxychlor Mirex Perthane Toxaphene	— — — — — — — — — — — — —	— — — — — — — — — — — — —
DeVault and Weishaar, 1984	1982	9 tributary streams to Lakes Huron, Michigan, Erie, and Ontario (Illinois, Indiana, Michigan, Ohio, Pennsylvania, New York). Composite fish samples (5 individuals each) were collected during the fall coho salmon run. Each of the 27 samples	Fish (skin on fillets). Species: coho salmon.	27	Chlordane, total Chlordane, *cis-* Chlordane, *trans-* Dacthal DDD DDE DDT DDT, total Dieldrin Endrin HCB HCH, α- Heptachlor epoxide Lindane	— — — — — — — — — — — — — —	— — — — — — — — — — — — — —

Table 2.2. Pesticides in bed sediment and aquatic biota from rivers in the United States: State and local monitoring studies—*Continued*

Bed sediment (µg/kg dry weight)			Biota (whole fish unless noted)					Comments
			% Detection		(µg/kg wet weight)			
DL	Range	Median	Sites	Samples	DL	Range	Median	
								shad, and white bass from the study area should not be used as food. DDE also was elevated in birds. Primary sources of DDE and toxaphene may be runoff from agricultural fields or a pesticide manufacturing plant.
nr	nd–0.9	nd/0.9	—	—	—	—	—	Objective: to establish baseline sediment and water-quality conditions before urban development in the basin. Trace amounts of organochlorine compounds were detected.
nr	nd	nd	—	—	—	—	—	
nr	nd–0.5	nd/0.5	—	—	—	—	—	
nr	0.2–0.3	0.25	—	—	—	—	—	
nr	nd–1.2	nd/1.2	—	—	—	—	—	
nr	nd	nd	—	—	—	—	—	
nr	6–15	10.5	—	—	—	—	—	
nr	nd	nd	—	—	—	—	—	
nr	nd	nd	—	—	—	—	—	
nr	nd	nd	—	—	—	—	—	
nr	nd	nd	—	—	—	—	—	
nr	nd	nd	—	—	—	—	—	
nr	nd	nd	—	—	—	—	—	
nr	nd	nd	—	—	—	—	—	
nr	nd	nd	—	—	—	—	—	
nr	nd	nd	—	—	—	—	—	
—	—	—	0*	0*	0.1	nd*	nd*	* Data are for fish fillets. Objective: to evaluate potential OC pesticide bioaccumulation in 4 trophic levels. Chlordane and dieldrin were higher at downstream sites. Highest residues of all analytes were in American eels. In general, residues were low or nondetectable in water, higher in sediment and primary producers, still higher in primary consumers, and highest in secondary and tertiary consumers.
—	—	—	100*	nr*	0.1	nr–1,100*	nr	
—	—	—	100*	nr*	0.1	nr–500*	nr	
—	—	—	100*	nr*	0.1	nr–180*	nr	
—	—	—	0*	0*	0.1	nd*	nd*	
—	—	—	100*	nr*	0.1	nd–19*	nr	
—	—	—	0*	0*	0.1	nd*	nd*	
—	—	—	0*	0*	0.1	nd*	nd*	
—	—	—	0*	0*	0.1	nd*	nd*	
—	—	—	0*	0*	0.1	nd*	nd*	
—	—	—	0*	0*	0.1	nd*	nd*	
—	—	—	0*	0*	1	nd*	nd*	
—	—	—	0*	0*	10	nd*	nd*	
—	—	—	100	100	5	12–140	63	Objective: to assess occurrence and distribution of OC contaminants in coho salmon from the Great Lakes during the 1982 fall run. Most OC levels were below FDA action levels, except mirex and PCB in salmon from Lake Ontario. Mirex exceeded 100 µg/kg and PCBs exceeded 2,000 µg/kg in these fish. Comparison of 1982 data with earlier survey (1980) shows a significant decline for DDT, chlordane, and PCB in
—	—	—	89	93	5	nd–30	20	
—	—	—	89	74	5	nd–20	10	
—	—	—	11	4	50	nd–50	nd	
—	—	—	89	78	5	nd–50	20	
—	—	—	100	100	5	20–500	130	
—	—	—	67	59	5	nd–60	10	
—	—	—	100	100	5	42–600	153	
—	—	—	100	85	5–50	10–30	20	
—	—	—	0	0	5	nd	nd	
—	—	—	0	0	50	nd	nd	
—	—	—	11	4	50	nd–50	nd	
—	—	—	0	0	5	nd	nd	
—	—	—	0	0	50	nd	nd	

Table 2.2. Pesticides in bed sediment and aquatic biota from rivers in the United States: State and local monitoring studies—*Continued*

Reference	Sampling date	Sampling location and strategy	Sampling media and species	Number of samples	Target analytes	Bed sediment % Detection Sites	Bed sediment % Detection Samples
		analyzed was a composite of 5 fish.			Mirex	—	—
					Nonachlor, *cis-*	—	—
					Nonachlor, *trans-*	—	—
					Octachlor epoxide	—	—
					Pentachlorophenol methyl ether	—	—
					Photomirex	—	—
					Toxaphene	—	—
					Trifluralin	—	—
Elder and Mattraw, 1984	1979–1980	5 sites on the Apalachicola River (Florida). Surficial sediment was collected downstream and leeward of islands or point bars (depositional areas), composited and sieved (250 µm). Clams were composited (2–4 cm size range).	Sediment (surficial), bed-load, detritus, mollusks. Species: Asiatic clam.	12 9	Aldrin	0	0
					Chlordane	40	17
					D, 2,4-	0	0
					DDD	nr	67
					DDE	nr	92
					DDT	nr	25
					Diazinon	0	0
					Dieldrin	nr	25
					DP, 2,4-	0	0
					Endosulfan	nr	0
					Endrin	nr	0
					Ethion	0	0
					Heptachlor	0	0
					Heptachlor epoxide	nr	0
					Lindane	0	0
					Malathion	0	0
					Methoxychlor	0	0
					Methyl parathion	0	0
					Methyl trithion	0	0
					Mirex	0	0
					Parathion	0	0
					Perthane	0	0
					Strobane	nr	0
					T, 2,4,5-	0	0
					Toxaphene	nr	0
					Trithion	0	0
Fuhrer, 1984	1982	3 sites on Chetco and Rogue rivers (Oregon). Potential dredging sites were targeted.	Water (filtered), sediment (surficial), elutriate (filtered).	3	Aldrin	0	0
					Chlordane	67	67
					DDD	67	67
					DDE	33	33
					DDT	33	33
					Dieldrin	67	67
					Endosulfan	0	0
					Endrin	0	0
					Heptachlor	0	0
					Heptachlor epoxide	0	0
					Lindane	0	0
					Methoxychlor	67	67
					Mirex	0	0
					Perthane	33	33
					Toxaphene	0	0

Table 2.2. Pesticides in bed sediment and aquatic biota from rivers in the United States: State and local monitoring studies—*Continued*

Bed sediment (µg/kg dry weight)			Biota (whole fish unless noted)					Comments
			% Detection		(µg/kg wet weight)			
DL	Range	Median	Sites	Samples	DL	Range	Median	
—	—	—	11	100	5	150–170	160	salmon from Lakes Michigan
—	—	—	67	48	5	nd–30	10	and Huron and for PCB in
—	—	—	100	96	5	nd–50	20	salmon from Lake Erie.
—	—	—	33	22	5	nd–20	nd	Although the comparison
—	—	—	0	0	50	nd	nd	between years was limited to 3-
								year old fish, variation due to
—	—	—	11	100	5	60–80	70	size, sex, and lipid content was
—	—	—	78	78	250	nd–1,200	400	not factored into the analysis.
—	—	—	0	0	50	nd	nd	
0.1	nd	nd	0	0	0.1	nd*	nd*	* µg/kg dry weight.
1	nd–3	nd	100	100	1	16–68*	21*	Objective: to determine contami-
0.1	nd	nd	—	—	—	—	—	nant accumulation in bed
0.1	nd–25	0.4	nr	100	0.1	2.8–44*	8*	sediment and benthic organisms
0.1	nd–3.5	1.3	nr	100	0.1	9.1–46*	18*	in the study area. Seasonal
0.1	nd–0.9	nd	nr	89	0.1	nd–16*	3*	variability: concentrations were
0.1	nd	nd	—	—	—	—	—	lower in Feb. 1979 than in Aug.
0.1	nd–0.1	nd	nr	100	0.1	0.5–2.9*	2*	1979 or May 1980. But seasonal
0.1	nd	nd	—	—	—	—	—	variability was generally less
0.1	nd	nd	nr	11	0.1	nd–0.2*	nd*	than variability among sites on
0.1	nd	nd	nr	33	0.1	nd–0.4*	nd*	many sample dates. There was
0.1	nd	nd	—	—	—	—	—	no evidence of downstream
0.1	nd	nd	0	0	0.1	nd*	nd*	trends. River was dredged
0.1	nd	nd	nr	89	0.1	nd–0.6*	0.3*	shortly after first sampling. For
0.1	nd	nd	0	0	0.1	nd*	nd*	chlordane, DDT, and dieldrin,
0.1	nd	nd	—	—	—	—	—	dry weight concentrations were
0.1	nd	nd	0	0	0.1	nd*	nd*	as follows: clam> detritus> fine
0.1	nd	nd	—	—	—	—	—	sediment.
0.1	nd	nd	—	—	—	—	—	
0.1	nd	nd	0	0	0.1	nd*	nd*	
0.1	nd	nd	—	—	—	—	—	
0.1	nd	nd	0	0	0.1	nd*	nd*	
1	nd	nd	nr	33	1	nd–58*	nd*	
0.1	nd	nd	—	—	—	—	—	
10	nd	nd	nr	33	10	nd–300*	nd*	
0.1	nd	nd	—	—	—	—	—	
0.1	nd	nd	—	—	—	—	—	Objectives: to collect
1	nd–2	1	—	—	—	—	—	reconnaissance data on elutriates,
0.1	nd–1.0	0.1	—	—	—	—	—	native water, and bed sediment at
0.1	nd–2.3	0.1	—	—	—	—	—	potential dredging sites. DDT,
0.1	nd–1.9	nd	—	—	—	—	—	perthane, methoxychlor in bed
0.1	nd–0.1	0.1	—	—	—	—	—	sediment exceeded maximum
0.1	nd	nd	—	—	—	—	—	concentrations from a previous
0.1	nd	nd	—	—	—	—	—	(1980) reconnaissance study.
0.1	nd	nd	—	—	—	—	—	
0.1	nd	nd	—	—	—	—	—	
0.1	nd	nd	—	—	—	—	—	
0.1	nd–1.5	0.7	—	—	—	—	—	
0.1	nd	nd	—	—	—	—	—	
1	nd–1	nd	—	—	—	—	—	
10	nd	nd	—	—	—	—	—	

Table 2.2. Pesticides in bed sediment and aquatic biota from rivers in the United States: State and local monitoring studies—*Continued*

Reference	Sampling date	Sampling location and strategy	Sampling media and species	Number of samples	Target analytes	Bed sediment % Detection Sites	Bed sediment % Detection Samples
Granstrom and others, 1984	1979–1980	Unspecified number of sites on the Delaware and Raritan Canal (New Jersey). Sediment was drained and analyzed wet.	Sediment, water.	27 37	Chlordane, γ- DDD, *p,p'*- DDE, *p,p'*- HCH, β- Lindane	nr nr nr nr nr	22 11 52 74 0
Hardy, 1984	1978–1980	40 sites on 4 streams in the Indiana Dunes National Lakeshore (Indiana). Specific land uses, sites known to be contaminated, and tributary confluences were targeted for sampling.	Water (filtered), sediment (unsieved), ecological surveys.	58	Aldrin Chlordane DDD DDE DDT Dieldrin Endosulfan Endrin Heptachlor Heptachlor epoxide Lindane Mirex Perthane Toxaphene	0 78 78 73 68 78 0 3 3 20 0 0 0 0	0 72 74 59 62 66 0 2 2 14 0 0 0 0
Harmon Engineering and Testing, 1984	1978–1979	19 sites (sediment) and 6 sites (mollusks) along 108 mi. of the Chattahoochee River (Alabama), including two reservoirs. Sediment samples were composites of 3 cross-sectional subsamples in rivers, and 4 subsamples in lakes.	Sediment (surficial), mollusks. Species: *Corbicula* species.	19 6	Aldrin DDD, *p,p'*- DDE, *p,p'*- DDT, *o,p'*- DDT, *p,p'*- Endrin Endrin aldehyde HCH, α- HCH, β- Heptachlor Heptachlor epoxide Lindane Methoxychlor Mirex Pentachlorophenol	0 0 42 0 0 0 0 5 0 0 0 5 0 0 0	0 0 42 0 0 0 0 5 0 0 0 5 0 0 0
Kizlauskas and others, 1984	1981	20 sites on New York tributaries to Lake Ontario: 13 on the Genesee River, 3 on Wine Creek, and 4 on Eighteen Mile Creek. Sites were selected to find worst-case conditions (e.g., in depositional areas near point sources). Some samples were	Sediment (surficial, 20-in cores).	24	Aldrin Chlordane, γ- Chlorobenzilate D, 2,4 (isopropyl) DCPA DDD, *o,p'*- DDD, *p,p'*- DDE, *o,p'*- DDE, *p,p'*- DDT, *o,p'*- DDT, *p,p'*- DDT, total Dieldrin Endosulfan I	15 55 ≥1 ≥1 40 30 25 70 60 40 75 80 40 5	13 50 ≥1 ≥1 38 25 29 67 63 33 71 79 33 8

Table 2.2. Pesticides in bed sediment and aquatic biota from rivers in the United States: State and local monitoring studies—*Continued*

Bed sediment (µg/kg dry weight)			Biota (whole fish unless noted)						
			% Detection		(µg/kg wet weight)				
DL	Range	Median	Sites	Samples	DL	Range	Median	Comments	
nr	nr	nd	—	—	—	—	—	Objective: to describe relations	
nr	nr	nd	—	—	—	—	—	between pollutants in sediment	
nr	nr	nr	—	—	—	—	—	and overlying waters of the	
nr	nd–2,800	nr	—	—	—	—	—	Delaware and Raritan Canal.	
nr	nd	nd	—	—	—	—	—	Lindane, β-HCH, *o,p'*- DDT, and *p,p'*- DDD were detected in some water samples.	
nr	nd	nd	—	—	—	—	—	Objectives: to assess water	
nr	nd–480	3	—	—	—	—	—	quality in 4 drainage basins; to	
nr	nd–260	1.7	—	—	—	—	—	determine potential causes of	
nr	nd–260	0.8	—	—	—	—	—	water-quality variation. Stream	
nr	nd–180	0.7	—	—	—	—	—	water quality is affected by	
nr	nd–33	0.4	—	—	—	—	—	agricultural, urban, and	
nr	nd	nd	—	—	—	—	—	residential land uses. For	
nr	nd–1.3	nd	—	—	—	—	—	example, DDT and dieldrin enter	
nr	nd–0.6	nd	—	—	—	—	—	the Little Calumet R. from	
nr	nd–2.0	nd	—	—	—	—	—	agricultural areas, whereas the	
nr	nd	nd	—	—	—	—	—	primary source of chlordane,	
nr	nd	nd	—	—	—	—	—	DDTs, and PCBs to the Grand	
nr	nd	nd	—	—	—	—	—	Calumet R. lagoons is discharge	
nr	nd	nd	—	—	—	—	—	from storm sewers.	
nr	nd	nd	0*	0*	nr	nd*	nd*	* Data are for mollusks.	
nr	nd	nd	0*	0*	nr	nd*	nd*	Objective: to establish baseline	
nr	nd–5.14	nd	nr*	nr*	nr	nr–11*	nd*	conditions and identify	
nr	nd	nd	17*	17*	nr	nd–40*	nd*	water-quality problems in the	
nr	nd	nd	0*	0*	nr	nd*	nd*	study area. Mollusks were	
nr	nd	nd	0*	0*	nr	nd*	nd*	analyzed for pesticides only in	
nr	nd	nd	0*	0*	nr	nd*	nd*	the Walter F. George Reservoir.	
nr	nd–0.53	nd	0*	0*	nr	nd*	nd*	There were low but detectable	
nr	nd	nd	0*	0*	nr	nd*	nd*	levels of p,p'-DDE in sediment	
nr	nd	nd	0*	0*	nr	nd*	nd*	from the Chattahoochee R. and	
nr	nd	nd	0*	0*	nr	nd*	nd*	in sediment and mollusks from	
nr	nd–1.15	nd	0*	0*	nr	nd*	nd*	the Walter F. George Reservoir.	
nr	nd	nd	0*	0*	nr	nd*	nd*		
nr	nd	nd	0*	0*	nr	nd*	nd*		
nr	nd	nd	0*	0*	nr	nd*	nd*		
1	nd–16	nd	—	—	—	—	—	Objective: to determine the level	
1	nd–23	nd/1	—	—	—	—	—	of toxic substances in selected	
270	nr	nr	—	—	—	—	—	Great Lakes river sediments.	
670	nr	nr	—	—	—	—	—	Certain compounds (high	
1	nd–13	nd	—	—	—	—	—	detection limits) were identified	
1	nd–88	nd	—	—	—	—	—	in selected samples by GC/MS,	
1	nd–49	nd	—	—	—	—	—	but not quantified. Pesticides	
1	nd–33	1.5	—	—	—	—	—	were present at trace to low	
1	nd–19	1	—	—	—	—	—	levels. Levels were highest at	
1	nd–23	nd	—	—	—	—	—	the Riverview Yacht Basin in	
1	nd–49	nd	—	—	—	—	—	the Genesee R. and in a swampy	
1	nd–214	9	—	—	—	—	—	area on Wine Creek (near a	
1	nd–6	nd	—	—	—	—	—	former hazardous waste facility).	
1	nd–6	nd	—	—	—	—	—	Pesticides were not detected in	

Table 2.2. Pesticides in bed sediment and aquatic biota from rivers in the United States: State and local monitoring studies—*Continued*

Reference	Sampling date	Sampling location and strategy	Sampling media and species	Number of samples	Target analytes	Bed sediment % Detection Sites	Bed sediment % Detection Samples
		composites of multiple grabs.			Endosulfan II	45	46
					Endrin	5	4
					HCB	35	33
					HCH, α-	≥1	≥1
					HCH, β-	40	42
					Heptachlor	0	0
					Heptachlor epoxide	10	13
					Isodrin	≥1	≥1
					Isophorone	≥1	≥1
					Kepone	≥1	≥1
					Lindane	5	4
					Methoxychlor	20	17
					Mirex	≥1	≥1
					Oxychlordane	10	8
					Pentachlorophenol	≥1	≥1
					Tetradifon	≥1	≥1
					Trifluralin	15	13
					Zytron	70	67
Malins and others, 1984	1979–1982	About 40 sites in Puget Sound (Washington). Sites were in urban and non-urban embayments and adjacent waterways. Fish were bottom-feeders.	Fish (liver, muscle), sediment. Species: English sole, rock sole, Pacific staghorn sculpin.	nr nr nr	Aldrin	nr	nr
					Chlordane, α-	nr	nr
					DDT, total	nr	nr
					HCB	nr	nr
					Heptachlor	nr	nr
					Lindane	nr	nr
					Nonachlor, *trans*-	nr	nr
Pariso and others, 1984	1979–1981	3 surveys of Lakes Superior and Michigan and various Wisconsin tributaries: (1) 63 sites in 1979; (2) 17 sites in 1980; (3) 27 sites in 1981. Sampling locations began at river mouth and extended 1–2 miles upstream. Fish	(1) Fish (*). Species include burbot, walleye, lake trout.	284	Aldrin	—	—
					Chlordane	—	—
					DDT	—	—
					Dieldrin	—	—
					Endrin	—	—
					HCB	—	—
					HCH, α-	—	—
					Lindane	—	—
					Methoxychlor	—	—
					Pentachlorophenol	—	—
			(2) Fish (*), sediment,	123 66	Chlordane	nr	8
					Dacthal	—	—

Table 2.2. Pesticides in bed sediment and aquatic biota from rivers in the United States: State and local monitoring studies—*Continued*

Bed sediment (μg/kg dry weight)			Biota (whole fish unless noted)					Comments
			% Detection		(μg/kg wet weight)			
DL	Range	Median	Sites	Sam-ples	DL	Range	Median	Comments
1	nd–13	nd	—	—	—	—	—	samples from Eighteen Mile
1	nd–4	nd	—	—	—	—	—	Creek, except (at low levels) in
1	nd–12	nd	—	—	—	—	—	samples from a floating
4,060	nr	nr	—	—	—	—	—	mud-clay-oil slick in the harbor
1	nd–36	nd	—	—	—	—	—	at the river mouth.
1	nd	nd	—	—	—	—	—	
1	nd–4	nd	—	—	—	—	—	
600	nr	nr	—	—	—	—	—	
40	nr	nr	—	—	—	—	—	
970	nr	nr	—	—	—	—	—	
1	nd–trace	nd	—	—	—	—	—	
1	nd–40	nd	—	—	—	—	—	
500	nr	nr	—	—	—	—	—	
1	nd–10	nd	—	—	—	—	—	
800	nr	nr	—	—	—	—	—	
1,230	nr	nr	—	—	—	—	—	
1	nd–13	nd	—	—	—	—	—	
1	nd–67	4	—	—	—	—	—	
nr	<1–<2*	nr	nr	nr	nr	nr	nr	* Range is of mean residue in
nr	<1–<2*	nr	nr	nr	nr	nr	nr	nonurban embayments to mean
nr	<1–<10*	nr	nr	nr	nr	nr	nr	residue for urban embayments.
nr	0.1–70*	nr	nr	nr	nr	nr–170*†	nr	† Data are for English sole
nr	<1–<2*	nr	nr	nr	nr	nr	nr	muscle tissue, in μg/kg dry wt.
nr	<1–<2*	nr	nr	nr	nr	nr	nr	Objective: to examine relations
nr	<1–<2*	nr	nr	nr	nr	nr	nr	between contaminants, fish
								diseases, and urban land use in
								Puget Sound. Metabolically
								resistant compounds (e.g., HCB)
								were found at higher levels in
								tissues than sediment; metabo-
								lically labile compounds (e.g.,
								PAHs) were higher in sediment
								than tissues. Xenobiotics were
								detected at substantially higher
								levels in urban than nonurban
								embayments and were associated
								with hepatic diseases in English
								sole and other species.
—	—	—	nr*	0*	20	nd*	nd*†	* Tissue type not reported.
—	—	—	nr*	42*	50	nd–1,050*	100*†	† Mean of values above
—	—	—	nr*	93*	50	nd–11,370*	880*†	detection limit.
—	—	—	nr*	49*	20	nd–500*	100*†	Objectives: to survey toxic
—	—	—	nr*	5*	20	nd–60*	nd*†	substances in fish in Wisconsin
—	—	—	nr*	12*	10	nd–50*	10*†	coastal zone, identify problem
—	—	—	nr*	24*	10	nd–90*	20*†	drainages and point sources,
—	—	—	nr*	10*	10	nd–50*	20*†	determine spatial distribution,
—	—	—	nr*	1*	50	nd–80*	70*†	and evaluate trends in
—	—	—	nr*	2*	50	nd–90*	70*†	contaminant levels. DDT,
10	nd–70	30†	nr*	15*	50	nd–460*	130*†	chlordane, and dieldrin were
—	—	—	50*	50*	nr	nr*	nr*	identified as problem

Table 2.2. Pesticides in bed sediment and aquatic biota from rivers in the United States: State and local monitoring studies—*Continued*

Reference	Sampling date	Sampling location and strategy	Sampling media and species	Number of samples	Target analytes	Bed sediment % Detection	
						Sites	Samples
		samples were composites. Results here are presented separately for the three surveys.	effluents. Species include carp, walleye.	50	DDT	nr	52
					Dieldrin	nr	5
					HCH	—	—
					Toxaphene	—	—
			(3) Fish (*), sediment, effluents. Species include carp.	52	Chlordane	nr	14
				22	DDT	nr	77
				42	Dieldrin	nr	14
Radtke and others, 1984	1978–1979	10 sites (sediment) and 2 sites (fish) in the upper Chattahoochee River and its impoundment, West Point Lake (Alabama, Georgia). Sediment was sampled in the reservoir, its tributaries, and the Chattahoochee River downstream of the dam. Fish samples were composites of 3–5 young-of-the-year from 2 tributaries.	Sediment (surficial), fish (whole, fillets). Species: brown bullhead, largemouth bass.	20 6 4	Aldrin	0	0
					Chlordane	90	75
					DDD	80	10
					DDE	70	35
					DDT	40	30
					Dieldrin	80	70
					Endosulfan	0	0
					Endrin	0	0
					HCH, total	0	0
					Heptachlor	10	5
					Heptachlor epoxide	0	0
					Perthane	0	0
					Toxaphene	0	0
Rogers, 1984	1981	13 sites in the Saw Mill River (New York). Sediment was collected in the streambed at the center of flow where possible; otherwise, depositional areas were sampled.	Sediment (surficial), water (whole).	15 8	Aldrin	0	0
					Chlordane	92	93
					DDD	67	71
					DDE	67	57
					DDT	83	86
					Diazinon	0	0
					Dieldrin	92	93
					Endosulfan	0	0
					Endrin	0	0
					Ethion	0	0
					HCB	0	0
					Heptachlor	0	0
					Heptachlor epoxide	8	7
					Lindane	0	0
					Malathion	0	0
					Methoxychlor	0	0
					Methyl parathion	0	0
					Methyl trithion	0	0
					Mirex	0	0
					Parathion	0	0
					Pentachlorophenol	0	0
					Perthane	0	0
					TCDD, 2,3,7,8-	0	0

Table 2.2. Pesticides in bed sediment and aquatic biota from rivers in the United States: State and local monitoring studies—*Continued*

Bed sediment (µg/kg dry weight)			Biota (whole fish unless noted)					Comments
			% Detection		(µg/kg wet weight)			
DL	Range	Median	Sites	Samples	DL	Range	Median	
10	nd–2,280	160†	nr*	76*	50	nd–9,710*	600*†	contaminants in 1979 survey.
10	nd–90	40†	nr*	30*	20	nd–1,600*	110*†	Tributary streams to southern
—	—	—	25*	25*	50	nr*	nr*	Lake Michigan appeared to be
—	—	—	40*	50*	2,000	nr*	nr*	sources of various contaminants.
10	nd–340	140†	nr*	8*	50	nd–200*	150*†	Yearly mean levels of DDT,
10	nd–650	150†	nr*	73*	50	nd–3,160*	550*†	dieldrin, and chlordane decreased
10	nd–50	40†	nr*	21*	20	nd–300*	70*†	from 1979 to 1981.
0.1	nd	nd	0	0	0.1	nd	nd	Objective: to document post-
1	1–210	18	100	100	0.1	7–280	116	impoundment water-quality con-
0.1	nd–15	1.5	100	83	0.1	nd–34	14	ditions in the West Point Lake
0.1	nd–11	nd	100	100	0.1	6.5–49	12	reservoir; to establish baseline
0.1	nd–32	nd	50	33	0.1	nd–8.3	nd	conditions; to identify problems.
0.1	nd–3.2	0.4	100	83	0.1	nd–13	1.9	The highest concentrations of
0.1	nd	nd	0	0	0.1	nd	nd	OCs in sediment occurred
0.1	nd	nd	50	17	0.1	nd–1.0	nd	where clay and silt particles
0.1	nd	nd	0	0	0.1	nd	nd	constituted most of the sample.
0.1	nd–0.1	nd	50	17	0.1	nd–1.6	nd	A downstream gradation from
0.1	nd	nd	50	33	0.1	nd–1.3	nd	sand to silt plus clay between
0.1	nd	nd	0	0	0.1	nd	nd	the headwaters of the reservoir
0.1	nd	nd	0	0	0.1	nd	nd	and the dam pool was indicative of decreasing velocities through the reservoir. OC pesticides were detected in both fillet and whole-body fish.
0.1	nd	nd	—	—	—	—	—	Objective: to identify and
1	nd–91	35	—	—	—	—	—	quantitate contaminants in the
0.1	nd–11	1.8	—	—	—	—	—	Saw Mill R., describe their
0.1	nd–12	0.6	—	—	—	—	—	spatial distributions, and relate
0.1	nd–14	1.9	—	—	—	—	—	them to land use and possible
0.1	nd	nd	—	—	—	—	—	sources. None of the pesticides
0.1	nd–1.8	0.8	—	—	—	—	—	detected in sediment exhibited
0.1	nd	nd	—	—	—	—	—	a spatial pattern within the
0.1	nd	nd	—	—	—	—	—	basin. The greatest number of
0.1	nd	nd	—	—	—	—	—	organic contaminants was found
20	nd	nd	—	—	—	—	—	below river mile (r.m.) 2.0.
0.1	nd	nd	—	—	—	—	—	PCBs, heavy metals, and PAHs
0.1	nd–0.4	nd	—	—	—	—	—	increased downstream and were
0.1	nd	nd	—	—	—	—	—	highest in the urbanized reach
0.1	nd	nd	—	—	—	—	—	(below r.m 3.7.) Urban runoff
0.1	nd	nd	—	—	—	—	—	was considered the major cause
0.1	nd	nd	—	—	—	—	—	of downstream river
0.1	nd	nd	—	—	—	—	—	deterioration.
0.1	nd	nd	—	—	—	—	—	
0.1	nd	nd	—	—	—	—	—	
20	nd	nd	—	—	—	—	—	
0.1	nd	nd	—	—	—	—	—	
20	nd	nd	—	—	—	—	—	

Table 2.2. Pesticides in bed sediment and aquatic biota from rivers in the United States: State and local monitoring studies—*Continued*

Reference	Sampling date	Sampling location and strategy	Sampling media and species	Number of samples	Target analytes	Bed sediment % Detection Sites	Bed sediment % Detection Samples
					Toxaphene	0	0
					Trithion	0	0
Rompala and others, 1984	1981–1982	48 sites on Pennsylvania rivers. Waters with known or suspected contamination were given priority for sampling. Fish samples were composites.	Fish (whole), birds, mammals. Fish species include carp, white sucker, walleye, brown trout, largemouth bass.	48	Chlordane, α-	—	—
					Chlordane, γ-	—	—
					DDD	—	—
					DDE	—	—
					DDT	—	—
					Dieldrin	—	—
					Endrin	—	—
					HCB	—	—
					HCH, α-	—	—
					HCH, γ-	—	—
					Heptachlor	—	—
					Heptachlor epoxide	—	—
					Kepone	—	—
					Mirex	—	—
					Nonachlor, β-	—	—
					Nonachlor, *cis-*	—	—
					Nonachlor, *trans-*	—	—
					Oxychlordane	—	—
					Toxaphene	—	—
Sabourin and others, 1984	1977–1979	3 sites along the Mississippi River (Louisiana). Snakes were collected from borrow pits between the levee and the river.	Water snakes (whole, liver, muscle, fat bodies, embryos). Species: *Nerodia rhombifera* and *N. cyclopion.*	29 28 28 21 4	Aldrin	—	—
					DDD	—	—
					DDE	—	—
					DDT	—	—
					DDT, total	—	—
					Dieldrin	—	—
					HCB	—	—
					Heptachlor	—	—
					Heptachlor epoxide	—	—
					Lindane	—	—
Stephens, 1984	1980–1982	5 sites on the Jordan River (Utah) and 3 sites at mouths of major tributaries.	Water, sediment (surficial).	nr	D, 2,4-	nr	>0
					DDD	nr	>0
					DDE	nr	>0
					DDT	nr	>0
					Dieldrin	nr	>0
					Methoxychlor	nr	>0

Table 2.2. Pesticides in bed sediment and aquatic biota from rivers in the United States: State and local monitoring studies—*Continued*

| Bed sediment (µg/kg dry weight) | | | Biota (whole fish unless noted) | | | | | Comments |
DL	Range	Median	% Detection Sites	% Detection Samples	DL	Range (µg/kg wet weight)	Median	
1	nd	nd	—	—	—	—	—	
0.1	nd	nd	—	—	—	—	—	
—	—	—	48	48	50*	nd–1,090	nd	* This is the reported detection
—	—	—	25	24	50*	nd–970	nd	limit, although levels as low as
—	—	—	50	50	50*	nd–1,290	nd/10	10 µg/kg were reported).
—	—	—	88	86	50*	nd–1,460	70	Objective: to investigate
—	—	—	44	44	50*	nd–1,500	nd	contaminant levels in
—	—	—	38	37	50*	nd–70	nd	Pennsylvania fish in order to
—	—	—	0	0	50*	nd	nd	evaluate potential threats to fish
—	—	—	4	4	50*	nd–20	nd	and wildlife resources. 25
—	—	—	8	8	50*	nd–40	nd	streams had OC pesticide or
—	—	—	2	2	50*	nd–70	nd	PCB residues exceeding
—	—	—	6	6	50*	nd–40	nd	NAS/NAE criteria to protect
—	—	—	8	8	50*	nd–30	nd	piscivorous wildlife.
—	—	—	100	100	50*	5–150	nr	
—	—	—	63	63	50*	nd–980	10	
—	—	—	6	6	50*	nd–10	nd	
—	—	—	38	38	50*	nd–90	nd	
—	—	—	67	66	50*	nd–1,550	20	
—	—	—	21	20	50*	nd–30	nd	
—	—	—	2	2	50*	nd–200	nd	
—	—	—	67*	28*	nr	nd–10*†	nr*	* Data are for whole-body
—	—	—	33*	14*	nr	nd–70*†	nr*	water snakes (both species).
—	—	—	100*	100*	nr	100–220*†	nr*	† Range is of mean values for 5
—	—	—	67*	31*	nr	nd–110*†	nr*	species-site groups.
—	—	—	100*	100*	nr	250–580*†	nr*	Objective: to determine the use-
—	—	—	100*	93*	nr	nd–80*†	nr*	fulness of water snakes in pollu-
—	—	—	100*	90*	nr	<10–200*†	nr*	tion monitoring. Fishes consti-
—	—	—	0*	10*	nr	nd–<10*†	nr*	tute 95–98% of the total
—	—	—	100*	86*	nr	30–220*†	nr*	volume of food consumed by the
—	—	—	33*	10*	nr	nd–100*†	nr*	2 snake species. OC levels: fat>
								liver > muscle. No correlation was observed between OC load and sex, length, or weight. DDT itself was found only in fat (57% of samples). OCs in em- bryos followed those in adults.
nr	nr–320	nr	—	—	—	—	—	Objectives: to determine pesti-
nr	nr	nr	—	—	—	—	—	cides in water and sediment of
nr	nr	nr	—	—	—	—	—	Jordan R. and tributaries. Sam-
nr	nr	nr	—	—	—	—	—	pling and analytical methods
nr	nr	nr	—	—	—	—	—	were not reported. 8 of 17 pesti-
nr	nr	nr	—	—	—	—	—	cides (not specified) were detected in at least 1 sediment sample. DDTs, dieldrin, and methoxy- chlor were detected in sediment at most sites downstream of Jordan Narrows. OC levels generally were <15 µg/kg in sediment.

Table 2.2. Pesticides in bed sediment and aquatic biota from rivers in the United States: State and local monitoring studies—*Continued*

Reference	Sam-pling date	Sampling location and strategy	Sampling media and species	Number of samples	Target analytes	Bed sediment % Detection	
						Sites	Sam-ples
Thompson, 1984	1981	8 sites on the Jordan River and 3 major tributaries (Utah).	Water, sediment (surficial).	16	Aldrin	0	0
					D, 2,4-	43	27
					DDD	88	71
					DDE	88	71
					DDT	38	29
					Dieldrin	75	64
					DP, 2,4-	0	0
					Endosulfan	0	0
					Endrin	0	0
					Heptachlor	25	15
					Heptachlor epoxide	0	0
					Lindane	0	0
					Methoxychlor	100	100
					Perthane	0	0
					Silvex	0	0
					T, 2,4,5-	0	0
					Toxaphene	0	0
Watkins and Simmons, 1984	1976–1980	4 sites in the Chicod Creek Basin (North Carolina). Sampling was conducted before and during channel modifications over a broad range of flow conditions.	Sediment, water (dissolved).	24 31	Chlordane	75	21
					DDD	100	79
					DDE	100	83
					DDT	100	67
					Diazinon	0	0
					Dieldrin	100	79
					Endrin	50	13
					Heptachlor	25	4
Winger and others, 1984	1978	8 sites on the Apalachicola River (Florida). Samples were composited by size and for upper and lower reaches of the river (except bird samples, which were analyzed individually).	Fish (whole, eggs), clams, insects, snakes, birds. Species include large-mouth bass, channel cat-fish, Asiatic clam, may-fly, water snake.	21 6,3 3 3 7	Chlordane, *cis-*	—	—
					Chlordane, *trans-*	—	—
					DDD, *p,p'-*	—	—
					DDE, *p,p'-*	—	—
					DDT, total	—	—
					Dieldrin	—	—
					Endrin	—	—
					HCH, α-	—	—
					Heptachlor epoxide	—	—
					Nonachlor, *cis-*	—	—
					Nonachlor, *trans-*	—	—
					Oxychlordane	—	—
					Toxaphene	—	—
Boellstorff and others, 1985	1981	4 sites total in the Lower Klamath National Wildlife Refuge (NWR) and the Tule Lake NWR (California). Fish were collected monthly. From each collection, composites of 10–15 cm and of	Fish (whole), birds (eggs, brains). Fish species: tui chub, blue chub.	60 57 40	Chlordane, *cis-*	—	—
					DDD, *p,p'-*	—	—
					DDE, *p,p'-*	—	—
					DDT, *p,p'-*	—	—
					Dieldrin	—	—
					Endrin	—	—
					Heptachlor epoxide	—	—
					Nonachlor, *trans-*	—	—
					Oxychlordane	—	—
					Toxaphene	—	—

Table 2.2. Pesticides in bed sediment and aquatic biota from rivers in the United States: State and local monitoring studies—*Continued*

Bed sediment (µg/kg dry weight)			Biota (whole fish unless noted)					Comments
			% Detection		(µg/kg wet weight)			
DL	Range	Median	Sites	Samples	DL	Range	Median	
nr	—	nd	—	—	—	—	—	Objective: to conduct a
nr	nd–320	nd	—	—	—	—	—	reconnaissance study to
nr	nd–35	0.35	—	—	—	—	—	determine the occurrence of
nr	nd–14	0.45	—	—	—	—	—	selected toxic compounds in the
nr	nd–1.4	nd	—	—	—	—	—	Jordan R. near Salt Lake City.
nr	nd–1.8	0.15	—	—	—	—	—	11 of the 20 compounds on the
nr	—	nd	—	—	—	—	—	analytical schedule were
nr	—	nd	—	—	—	—	—	detected.
nr	—	nd	—	—	—	—	—	
nr	nd–.3	nd	—	—	—	—	—	
nr	—	nd	—	—	—	—	—	
nr	—	nd	—	—	—	—	—	
nr	1.1–80	7.5	—	—	—	—	—	
nr	—	nd	—	—	—	—	—	
nr	—	nd	—	—	—	—	—	
nr	—	nd	—	—	—	—	—	
nr	—	nd	—	—	—	—	—	
1	nd–80	nd	—	—	—	—	—	Objective: to assess water
0.1	nd–58	3.8	—	—	—	—	—	quality in the Chicod Creek
0.1	nd–37	2.6	—	—	—	—	—	Basin before and during channel
0.1	nd–20	0.7	—	—	—	—	—	modifications. No significant
0.1	nd	nd	—	—	—	—	—	differences were detected, except
0.1	nd–11	1	—	—	—	—	—	possibly during floods. DDT,
0.1	nd–1.0	nd	—	—	—	—	—	dieldrin, and diazinon were
0.1	nd–8.1	nd	—	—	—	—	—	detected (10–20 µg/L) in the
								dissolved phase during a flood.
—	—	—	100	87*	10	nd–180	nr	* Percent of all biological
—	—	—	0	38*	10	nd–50	nr	samples with detects.
—	—	—	100	nr	10	nd–400	nr	Objective: to determine residues
—	—	—	100	nr	10	nd–2,660	nr	in lower and upper reaches for
—	—	—	100	100*	nr	30–2,820	nr	use as baseline information. OC
—	—	—	100	86*	10	nd–30	nr	residues tended to be higher in
—	—	—	50	39*	10	nd–20	nr	samples from upper, than from
—	—	—	0	2*	10	nd	nr	lower, reaches (except female
—	—	—	0	35*	10	nd–20	nr	catfish, for which reverse was
—	—	—	100	90*	10	nd–130	nr	true). Compositing by size
—	—	—	100	90*	10	nd–140	nr	contributed to high variability.
—	—	—	50	28*	10	nd–20	nr	DDE was the principal
—	—	—	100	96*	100	nd–1,400	nr	component of total DDT.
—	—	—	0	0	106	nd	nd	Objective: to determine OCs in
—	—	—	0	0	101	nd	nd	white pelicans and western
—	—	—	0	0	100	nd	nd	grebes in the Klamath Basin; to
—	—	—	0	0	102	nd	nd	ascertain potential effects and ex-
—	—	—	0	0	103	nd	nd	posure source. DDD, DDT and
—	—	—	0	0	109	nd	nd	dieldrin in pelican eggs from
—	—	—	0	0	104	nd	nd	Lower Klamath NWR declined
—	—	—	0	0	107	nd	nd	from 1969 to 1981, but DDE and
—	—	—	0	0	105	nd	nd	PCB did not. No detectable resi-
—	—	—	0	0	108	nd	nd	dues in fish gill-netted from
								refuges or regurgitated from

Table 2.2. Pesticides in bed sediment and aquatic biota from rivers in the United States: State and local monitoring studies—*Continued*

Reference	Sampling date	Sampling location and strategy	Sampling media and species	Number of samples	Target analytes	Bed sediment % Detection	
						Sites	Samples
		23–29 cm fish were analyzed.					
Carleton, 1985	1985	3 sites in the Nashua impoundment of the Cedar River, Iowa. Sediment cores (33 in.) were taken at 3–5 points along each of 3 transects. Cores from each transect were composited into top (top 8–10 in.) and bottom (remainder) samples for each transect and site.	Sediment* (cores: top 8–10 in, bottom 23–25 in.)	8 total: 4 4	D, 2,4- Endrin Lindane Methoxychlor Toxaphene TP, 2,4,5 (silvex)	0* 0* 0* 0* 0* 0*	0* 0* 0* 0* 0* 0*
DeVault, 1985	1980–1981	9 sites in Great Lakes harbors and tributaries: Astabula and Black rivers (Ohio), Sheboygan, Milwaukee, Menominee, Kinnickinnic, Fox, and Wolf rivers (Wisconsin), Chequamegon Bay (Lake Superior). Most samples were composites of 3–5 fish.	Fish (whole). Species: carp, northern pike, bluegill, brown bullhead, yellow bullhead, redhorse sucker, black crappie, walleye, rock bass, white sucker.	22	Aldrin Chlordane, γ- Dacthal DDD, o,p'- DDD, p,p'- DDE, o,p'- DDE, p,p'- DDT, o,p'- DDT, p,p'- DDT, total Dieldrin Endosulfan II Endrin HCB HCH, β- Heptachlor Heptachlor epoxide Lindane Methoxychlor Oxychlordane Pentachlorophenol Trifluralin	— —	— —
Dowd and others, 1985	1978–1979	3 sites in 3 Louisiana watersheds.	Fish, invertebrates, amphibians, birds, reptiles, mammals. Species	24 6 12	Chlordane, α- Chlordane, γ- Compound E DDD DDE DDT Dieldrin	— — — — — — —	— — — — — — —

Table 2.2. Pesticides in bed sediment and aquatic biota from rivers in the United States: State and local monitoring studies—*Continued*

Bed sediment (µg/kg dry weight)			Biota (whole fish unless noted)					Comments
			% Detection		(µg/kg wet weight)			
DL	Range	Median	Sites	Samples	DL	Range	Median	
								pelicans suggests there was minimal exposure to OCs at the Klamath Basin in 1981. Pelican deaths due to OCs increased in 1981; birds probably are exposed on wintering grounds or during northward migration.
1,000*†	nd †	nd †	—	—	—	—	—	* Data shown are for EP Toxicity extracts of sediment cores.
2*†	nd †	nd †	—	—	—	—	—	† Concentrations are in µg/L.
40*†	nd †	nd †	—	—	—	—	—	Objective: to characterize
1,000*†	nd †	nd †	—	—	—	—	—	sediment contaminants; to
50*†	nd †	nd †	—	—	—	—	—	determine whether sediments
100*†	nd †	nd †	—	—	—	—	—	are hazardous waste as defined under the RCRA. EP Toxicity extraction test simulated physical processes occurring in a landfill and was used to identify wastes likely to leach toxic constituents into groundwater.
—	—	—	0	0	3	nd	nd	Objective: to locate source areas of contaminants to the Great Lakes and to identify previously unknown contaminants. Severe fish contamination occurred in the Sheboygan, Black, and Ashtabula rivers. Somewhat less severe contamination occurred in the Fox, Milwaukee, and Kinnickinnic rivers.
—	—	—	20	9	2	nd–40	nd	
—	—	—	90	73	2	nd–120	3.5	
—	—	—	80	55	5	nd–320	7.5	
—	—	—	40	23	5	nd–340	nd	
—	—	—	100	100	nr	10–360	75	
—	—	—	100	91	2	nd–1,100	120	
—	—	—	100	91	20	nd–260	35	
—	—	—	80	50	2	nd–150	2	
—	—	—	100	100	nr	23–1,703	262	
—	—	—	90	73	2	nd–90	10	
—	—	—	100	100	nr	4–40	17.5	
—	—	—	40	18	2	nd–10	nd	
—	—	—	90	91	2	nd–3,470	16.5	
—	—	—	80	55	2	nd–900	60	
—	—	—	10	5	2	nd–300	nd	
—	—	—	80	77	5	nd–480	10	
—	—	—	60	41	2	nd–120	nd	
—	—	—	30	14	10	nd–120	nd	
—	—	—	70	64	2	nd470	20	
—	—	—	10	5	80–3,000	nd–4,520	nd	
—	—	—	10	5	2	nd–7	nd	
—	—	—	100	22	50	nd–76	nd	Objective: to characterize ambient levels of OC residues in biota. DDE and chlordane levels declined from 1978 to 1979. Shad and herons appear to be the species most likely to accumulate OC compounds and
—	—	—	100	22	50	nd–82	nd	
—	—	—	33	0	50	nd	nd	
—	—	—	100	22	50	nd–106	nd	
—	—	—	100	67	50	nd–238	11	
—	—	—	0	0	50	nd	nd	
—	—	—	100	22	50	nd–92	nd	

Table 2.2. Pesticides in bed sediment and aquatic biota from rivers in the United States: State and local monitoring studies—*Continued*

Reference	Sampling date	Sampling location and strategy	Sampling media and species	Number of samples	Target analytes	Bed sediment % Detection	
						Sites	Samples
			include shad, channel catfish, crappie, frog (*Rana* sp.), water snakes (*Natrix* sp.)		Endrin HCH, α- HCH, β- Heptachlor Nonachlor, *cis-* Nonachlor, *trans-* Toxaphene	— — — — — — —	— — — — — — —
Eisenberg and Topping, 1985	1976–1980	Unspecified number of sites in the Maryland portion of the Chesapeake Bay and its tributaries.	Fish (edible, roe, gonads). Species include striped bass, American shad, white perch, yellow perch.	291 4 16	Aldrin Chlordane Dacthal DDD DDE DDT Dieldrin Endrin HCB HCH, α- Heptachlor Heptachlor epoxide Kepone Lindane Methoxychlor Mirex Toxaphene	— — — — — — — — — — — — — — — — —	— — — — — — — — — — — — — — — — —
Fallon and Horvath, 1985	1982	31 sites in the Detroit River (Michigan and Canada). Two 50-cm cores were taken in depositional areas at each site.	Sediment (cores).	62	Heptachlor Other OC pesticides	3* 0*	2* 0*
Hamdy and Post, 1985	1980	59 sites on the Detroit River, in both Canadian and American (Michigan) waters. Sites were located within 15–50 m from shore and midchannel. Each sample is a composite of 3 grabs.	Sediment (surficial).	59	Aldrin Chlordane, *cis-* Chlordane, *trans-* DDD, *p,p'-* DDE, *p,p'-* DDT, *o,p'-* DDT, *p,p'-* Dieldrin Endosulfan I Endosulfan II Endrin HCB HCH, α- HCH, β- Heptachlor	0 58 27 51 15 19 15 22 3 3 17 59 52 29 0	0 58 27 51 15 19 15 22 3 3 17 59 52 29 0

Table 2.2. Pesticides in bed sediment and aquatic biota from rivers in the United States: State and local monitoring studies—*Continued*

Bed sediment (µg/kg dry weight)			Biota (whole fish unless noted)					
			% Detection		(µg/kg wet weight)			
DL	Range	Median	Sites	Samples	DL	Range	Median	Comments
—	—	—	33	6	50	nd–22	nd	could serve as indicator species
—	—	—	33	6	50	nd–12	nd	in future monitoring.
—	—	—	0	0	50	nd	nd	
—	—	—	100	22	50	nd–600	nd	
—	—	—	33	0	50	nd	nd	
—	—	—	100	22	50	nd–44	nd	
—	—	—	100	28	100	nd–704	nd	
—	—	—	nr	nr	1	nr	nr	* Data are for edible fish.
—	—	—	nr	nr	1	nd–700*	50–90*††	† Range of 5 yearly median
—	—	—	nr	nr	1	nd–trace*	nr	values.
—	—	—	nr	nr	1	nd–490*	10–40*†	Objective: to establish baseline
—	—	—	nr	nr	1	nd–1,080*	10–60*†	data on OC residues in Maryland
—	—	—	nr	nr	1	nd–70*	nd*	finfish; to study temporal and
—	—	—	nr	nr	1	nd–50*	2–10*†	spatial variations; to ascertain
—	—	—	nr	nr	1	nd–trace*	nr	suitability for human consump-
—	—	—	nr	nr	1	nd–trace*	nr	tion; and to compare bioaccumu-
—	—	—	nr	nr	1	nd–340*	nd–3*†	lation in different tissue types.
—	—	—	nr	nr	1	nr	nr	There was little year-to-year var-
—	—	—	nr	nr	1	nr	nr	iation in flesh-tissue levels. Le-
—	—	—	nr	nr	1	nd–trace*	nr	vels of chlordane, DDD, and
—	—	—	nr	nr	1	nd–trace*	nr	DDE were greater in roe than in
—	—	—	nr	nr	1	nr	nr	flesh of yellow and white perch
—	—	—	nr	nr	1	nr	nr	and striped bass. OCs in striped
—	—	—	nr	nr	1	nr	nr	bass: gonads > flesh. OCs in
								shad: flesh ≥ gonads. Authors
								noted that shad flesh tends to
								have higher lipids than bass.
10	nd–12*	nd*	—	—	—	—	—	* Data are for sediment cores.
nr	nd*	nd*	—	—	—	—	—	Objective: to characterize priori-
								ty pollutants in Detroit R. sedi-
								ments. Metals and industrial or-
								ganics were widespread, with
								higher levels on the U.S. side
								downstream of the Rouge R. and
								in the Trenton Channel.
1	nd	nr	—	—	—	—	—	Objective: to determine spatial
1	nd–90	nr	—	—	—	—	—	distribution of organics in the
1	nd–30	nr	—	—	—	—	—	Detroit R. DDT compounds
5	nd–5,820	nr	—	—	—	—	—	were detected downstream of
1	nd–1,870	nr	—	—	—	—	—	Little R. (10 µg/kg), along the
5	nd–86	nr	—	—	—	—	—	Windsor waterfront (6–49
5	nd–8,240	nr	—	—	—	—	—	µg/kg), and downstream of
1	nd–35	nr	—	—	—	—	—	Rivière aux Canards, which
1	nd–10	nr	—	—	—	—	—	drains an agricultural area (5–7
1	nd–10	nr	—	—	—	—	—	µg/kg). The highest total DDT
1	nd–43	nr	—	—	—	—	—	levels were along the U.S.
1	nd–360	nr	—	—	—	—	—	shoreline near Fort Wayne
1	nd–65	nr	—	—	—	—	—	(16,016 µg/kg), Zug Island (140
1	nd–35	nr	—	—	—	—	—	µg/kg), and Trenton (186
1	nd	nr	—	—	—	—	—	µg/kg).

Table 2.2. Pesticides in bed sediment and aquatic biota from rivers in the United States: State and local monitoring studies—*Continued*

Reference	Sampling date	Sampling location and strategy	Sampling media and species	Number of samples	Target analytes	Bed sediment % Detection	
						Sites	Samples
					Heptachlor epoxide	8	8
					Lindane	27	27
					Mirex	0	0
Hopkins and others, 1985	1978–1984	10 sites (sediment) and 32 sites (biota) in Washington State. Sediment was sampled in 1984 and biota during 1978–1984. Fish of 2 trophic levels (grazer, predator) were collected at each site. Sediment samples were composites of 5 subsamples collected near each fish collection site.	Sediment, fish (whole, skin-on fillet, liver, viscera), mollusks. Species include long-nose sucker, mountain whitefish, northern squawfish, largemouth bass, bay mussel.	10 67 30 22 9 11	Aldrin	0	0
					Chlordane	0	0
					DDD, *o,p'*-	—	—
					DDD, *p,p'*-	40	40
					DDE, *o,p'*-	—	—
					DDE, *p,p'*-	80	80
					DDT, *o,p'*-	—	—
					DDT, *p,p'*-	40	40
					DDT, total	—	—
					Dieldrin	0	0
					HCB	0	0
					HCH, α-	40	40
					Methoxychlor	0	0
					Pentachlorophenol	0	0
Jaffe and others, 1985	1983	16 sites in embayments and tributaries of Lakes Huron and Superior (Michigan). Fish of similar size were composited. Composite samples of the largest fish at each site were analyzed.	Fish (whole). Species: carp, hog sucker, northern pike, white sucker.	21	Aldrin	—	—
					Chlordane, α-	—	—
					Chlordane, γ-	—	—
					Chlordane, misc.	—	—
					Chlordene	—	—
					Dacthal	—	—
					DDD, *p,p'*-	—	—
					DDE, *p,p'*-	—	—
					DDT, *p,p'*-	—	—
					Dieldrin	—	—
					Endotetrasulfuran I	—	—
					Endrin aldehyde	—	—
					Endrin ketone	—	—
					HCB	—	—
					HCH, α-	—	—
					HCH, β-	—	—
					HCH, δ-	—	—
					Heptachlor	—	—
					Heptachlor epoxide	—	—
					1-Hydroxychlordene	—	—
					Kepone	—	—
					Mirex	—	—
					Nonachlors, total	—	—
					Oxychlordane	—	—
					PCA	—	—
Kaiser and others, 1985	1983	20 sites on the Detroit River (Michigan and Ontario, Canada). Sediment samples were pressed through 5 µm filter	Sediment (surficial, suspended), water (dissolved), pore water.	13 11 20	OCs, total	69	69

Table 2.2. Pesticides in bed sediment and aquatic biota from rivers in the United States: State and local monitoring studies—*Continued*

| Bed sediment (µg/kg dry weight) | | | Biota (whole fish unless noted) | | | | | Comments |
| | | | % Detection | | (µg/kg wet weight) | | | |
DL	Range	Median	Sites	Samples	DL	Range	Median	
1	nd–10	nr	—	—	—	—	—	
1	nd–65	nr	—	—	—	—	—	
5	nd	nr	—	—	—	—	—	
1	nd	nd	0*	0*	1	nd*	nd*	* Data are for fish fillets from
1	nd	nd	0*	0*	1	nd*	nd*	1984 only (n=21).
—	—	—	20*	14*	1	nd–45*	nd*	Objective: to identify potential
1	nd–53	nd	100*	100*	1	nd–780*	58*	problem areas; to evaluate
—	—	—	90*	71*	1	nd–120*	23*	long-term water-quality changes
1	nd–51	6	100*	100*	1	nd–2,400*	370*	statewide. Report focused on
—	—	—	80*	86*	1	nd–220*	32*	1984 data; fish and shellfish data
1	nd–35	nd	100*	100*	1	nd–420*	64*	for 1978–1983 (mostly whole fish
—	—	—	100*	100*	1	nd–3,200*	1,000*	and viscera) were reported in an
1	nd	nd	0*	0*	1	nd*	nd*	appendix. Some edible fish
10	nd	nd	—	—	—	—	—	samples from the Yakima and
1	nd–13	nd	100*	100*	1	nd–172*	6*	and Okanogan rivers approached
1	nd	nd	0*	0*	1	nd*	nd*	the FDA action level for
20	nd	nd	80*	76*	1	nd–55	1.5	total DDT (5 mg/kg).
—	—	—	0	0	nr	nd*	nd*	* Concentrations are on a fat
—	—	—	100	100	nr	nd–170*	39*	weight basis (µg/kg-fat).
—	—	—	94	90	nr	nd–120*	28*	Objective: to identify
—	—	—	94	90	nr	nd–76*	18*	contaminants in fish tissue and
—	—	—	0	0	nr	nd*	nd*	to compare industrial and
—	—	—	31	90	nr	nd–15*	5.6*	nonindustrial areas. Correlation
—	—	—	69	62	nr	nd–2,900*	370*	was good between urban and
—	—	—	100	100	nr	37–6,100*	1,200*	industrial areas and the
—	—	—	6	5	nr	nd–3,200*	nd*	abundance and variety of
—	—	—	100	100	nr	nd–180*	43*	contaminants. Pesticide residues
—	—	—	0	0	nr	nd*	nd*	were generally higher in
—	—	—	0	0	nr	nd*	nd*	tributaries of Lake Huron than
—	—	—	0	0	nr	nd*	nd*	in those of Lake Superior.
—	—	—	100	100	nr	1.6–525*	37*	Total DDT levels were highest
—	—	—	31	24	nr	nd–15*	nd*	in Chippewa and Pine rivers.
—	—	—	0	0	nr	nd*	nd*	Dacthal was detected in all Lake
—	—	—	0	0	nr	nd*	nd*	Huron fish and many Lake
—	—	—	0	0	nr	nd*	nd*	Superior fish.
—	—	—	19	19	nr	nd–41*	nd*	
—	—	—	0	0	nr	nd*	nd*	
—	—	—	0	0	nr	nd*	nd*	
—	—	—	0	0	nr	nd*	nd*	
—	—	—	100	100	nr	nd–360*	140*	
—	—	—	25	24	nr	nd–23*	nd*	
—	—	—	100	100	nr	nd–33*	1.9*	
nr	nd–270*	50*	—	—	—	—	—	* Estimated from figure. Objective: to investigate con- taminant sources and distribu- tion in a large river. No data were provided for individual con- taminants. Detroit Sewage

Table 2.2. Pesticides in bed sediment and aquatic biota from rivers in the United States: State and local monitoring studies—*Continued*

Reference	Sam-pling date	Sampling location and strategy	Sampling media and species	Number of samples	Target analytes	Bed sediment % Detection Sites	Bed sediment % Detection Sam-ples
		to separate pore water.					
Mischke and others, 1985	1984–1985	Two studies in this report measured pesticides in sediment or soil in California. (1) Unspecified number of sites in the Blanco Drain near Salinas. (2) Soil samples were analyzed from 32 counties statewide.	Sediment, soil.	(1) nr (1) nr; (2) 99	DDT, total	100	100
Putnam and others, 1985	1984	1 site on the lower Lackawanna River (Pennsylvania). The watershed includes the Lackawanna Refuse site (an industrial waste site). Fish sample was a com-posite of 5 fish.	Fish (whole), rabbits, mice. Fish species: white sucker.	1 2 4	Dieldrin	—	—
Robbins and others, 1985	1984	3 sites on inflow tributary streams (Tennessee, Kentucky) to Reelfoot Lake in Tennessee. Sediment was collected at low flow.	Sediment.	3	Aldrin Chlordane DDD DDE DDT Dieldrin Endosulfan Endrin Heptachlor	0 0 67 0 0 0 0 0 0	0 0 67 0 0 0 0 0 0

Table 2.2. Pesticides in bed sediment and aquatic biota from rivers in the United States: State and local monitoring studies—*Continued*

Bed sediment (µg/kg dry weight)			Biota (whole fish unless noted)					Comments
DL	Range	Median	% Detection		(µg/kg wet weight)			
			Sites	Sam-ples	DL	Range	Median	
								Treatment Plant is a principal source of dieldrin, heptachlor and epoxide, endosulfan, and lindane. Trenton Channel has active inputs of lindane, dieldrin, endosulfan, heptachlor, and chlordane. *p,p'-* DDT and DDE were detected in surficial and suspended sediments, while *o,p'-* DDE was primarily associated with surface microlayer. Dieldrin heptachlor and epoxide were detected solely in water phase. Chlordane was associated with subsurface water and pore water.
nr	220–6,300	2,100*	—	—	—	—	—	* Mean concentration. Objective: to investigate sources of total DDT in environmental samples from California. (1) Total DDT contained an average of 51% DDT in sediment and 70% in soil. Total DDT in soils adjacent to the Blanco Drain averaged 2,600 µg/kg. *p,p'-* and *o,p'-* DDT have long lifetimes in Salinas clay soils. Soil erosion is the likely source of total DDT in the Blanco Drain. (Also see Agee, 1986.) (2) Statewide soil survey confirmed findings of the Blanco Drain study.
—	—	—	100	100	nr	6.43	6.43	Objective: to determine if contaminants were entering the food chain in the study area. Report did not list target analytes that were not detected. DDE, dieldrin and various industrial compounds were detected in mouse and(or) rabbit samples. Sampling design did not distinguish sources of contamination.
0.1	nd	nd	—	—	—	—	—	Objective: to collect hydrologic data and prepare a water budget for Reelfoot Lake. Eutrophication of the lake has been accelerated by land-use practices within the basin. Data are presented without interpretation.
1	nd	nd	—	—	—	—	—	
0.1	nd–12	0.4	—	—	—	—	—	
0.1	nd	nd	—	—	—	—	—	
0.1	nd	nd	—	—	—	—	—	
0.1	nd	nd	—	—	—	—	—	
0.1	nd	nd	—	—	—	—	—	
0.1	nd	nd	—	—	—	—	—	
0.1	nd	nd	—	—	—	—	—	

Table 2.2. Pesticides in bed sediment and aquatic biota from rivers in the United States: State and local monitoring studies—*Continued*

Reference	Sampling date	Sampling location and strategy	Sampling media and species	Number of samples	Target analytes	Bed sediment % Detection Sites	Bed sediment % Detection Samples
					Heptachlor epoxide	0	0
					Lindane	0	0
					Methoxychlor	0	0
					Mirex	0	0
					Perthane	0	0
Stamer and others, 1985	1978–1979	69 sites on the Schuylkill River and its tributaries (Pennsylvania).	Water, sediment (surficial).	92	Chlordane	—	—
					DDD	—	—
					DDE	—	—
					DDT	—	—
					Dieldrin	—	—
Suns and others, 1985	1978–1985	A total of 9 sites in Great Lakes (Michigan and Ontario, Canada): 6 sites in the Detroit River, 1 in Lake St. Clair, and 2 in Lake Erie. Annual collections were made at same time each year (not specified.) Samples were composites.	Fish (whole). Species: spottail shiners.	175	Aldrin	—	—
					Chlordane, total	—	—
					DDT, total	—	—
					HCB	—	—
					HCH	—	—
					Heptachlor	—	—
					Mirex	—	—
					Octachlorostyrene	—	—
Winger and others, 1985	1980–1983	5 sites (sediment) and 9 sites (fish) in the Yazoo National Wildlife Refuge (NWR) in in Mississippi. 4 sites were sampled in 1980; 2 of these plus 5 new sites were sampled during 1982–1983. Fish samples were composites of 3–5 fish (by species).	Sediment, fish (whole). Species: yellow bullhead, channel catfish, black crappie, white crappie, common carp, bluegill, bowfin,	15 75	DDD, *p,p'-*	100	nr
					DDE, *p,p'-*	100	nr
					DDMU	100	nr
					DDT, *o,p'-*	—	—
					DDT, *p,p'-*	100	nr
					DDT, total	100	nr
					Dieldrin	0	nr
					Endrin	0	nr
					HCH	0	nr
					Toxaphene	80	nr

Table 2.2. Pesticides in bed sediment and aquatic biota from rivers in the United States: State and local monitoring studies—*Continued*

Bed sediment (μg/kg dry weight)			Biota (whole fish unless noted)					Comments
			% Detection		(μg/kg wet weight)			
DL	Range	Median	Sites	Samples	DL	Range	Median	
0.1	nd	nd	—	—	—	—	—	
0.1	nd	nd	—	—	—	—	—	
0.1	nd	nd	—	—	—	—	—	
0.1	nd	nd	—	—	—	—	—	
1	nd	nd	—	—	—	—	—	
1	nd–140	23*	—	—	—	—	—	* Mean of average concentrations for tributary and main stem sites in each of two study years. Objective: to assess the distribution of trace substances in the Schuylkill R. and its tributaries; to estimate their average annual transport. Concentrations of chlordane, dieldrin, and total DDT increased in a downstream direction. Transport of OC insecticides appears to be small, but total load calculations were not made because of the large number of samples below detection limits.
0.1	nd–56	4.6*	—	—	—	—	—	
0.1	nd–43	4.3*	—	—	—	—	—	
0.1	nd–62	32.5*	—	—	—	—	—	
0.1	nd–11	2*	—	—	—	—	—	
—	—	—	nr	nr	nr	nd–trace*	nr	* Range or median of the mean values for all sites and years. Objective: to assess spatial distribution and trends. Total DDT was fairly uniform throughout study area (1982–1983). Total chlordane was higher at urban Detroit R. sites than 2 of 3 lake sites. Distribution within Detroit R. was fairly uniform. Trends data (1978–1983) showed a decline in total DDT at all 3 lake sites, especially during the late 1970s, and in total chlordane at only 1 lake site.
—	—	—	100	nr	2	nd–27*	8*	
—	—	—	89	nr	1	nd–141*	22*	
—	—	—	nr	nr	nr	nd–trace*	nr	
—	—	—	nr	nr	nr	nd–trace*	nr	
—	—	—	nr	nr	nr	nd–trace*	nr	
—	—	—	0	0	nr	nd*	nd*	
—	—	—	100	nr	1	nd–10*	5*	
nr	40–150*	nr	100	100	nr	30–10,300*	nr	* Range of mean residues for samples grouped by site, date, and (for fish) species. Objective: to determine OC pesticides in fish from the Yazoo NWR. OC pesticide levels were highest in areas receiving direct inflow of waters draining agricultural land and lowest in lakes receiving only local runoff or overflow. Spatial distribution of total DDT in sediment was similar to that in fish, but concentrations were much lower.
nr	50–200*	nr	100	100	nr	50–16,700*	nr	
nr	10–40*	nr	100	nr	nr	10–1,400*	nr	
—	—	—	100	nr	nr	50–720*	nr	
nr	10*	nr	100	nr	nr	nd–2,900*	nr	
nr	120–380*	nr	100	100	nr	120–23,490*	nr	
nr	nd*	nr	67	81	nr	nd–5,680*	nr	
nr	nd*	nr	56	73	nr	nd–2,060*	nr	
nr	nd*	nr	56	nr	nr	nd–160*	nr	
nr	nd–70*	nr	100	97	nr	nd–280,330*	nr	

Table 2.2. Pesticides in bed sediment and aquatic biota from rivers in the United States: State and local monitoring studies—*Continued*

Reference	Sam-pling date	Sampling location and strategy	Sampling media and species	Number of samples	Target analytes	Bed sediment % Detection	
						Sites	Sam-ples
			bigmouth buffalo, gizzard shad.				
Yorke and others, 1985	1978–1980	Sites on the Schuylkill River (Pennsylvania). Sediment was collected from depositional areas above low level dams at 6 sites, and at 2 mile intervals along the length of the lower river. Surficial sediment was analyzed by size fractions.	Water, sediment (surficial, cores, suspended).	25 3	Chlordane DDD DDE DDT Dieldrin	100* 100* 100* 100* 100*	100* 100* 100* 100* 100*
Agee, 1986	1984	Sediment was collected at 23 sites along the Blanco Drain and 2 sites in the Salinas River (California). Fish were collected at 1 site in the drain and 2 sites in the Salinas River. Soil was collected from fields adjacent to the drain.	Sediment (surficial, suspended), fish (whole), soil. Species: suckers.	38 4 6 23	DDT, o,p'- DDT, p,p'- DDE, o,p'- DDE, p,p'- DDD, o,p'- DDD, p,p'- DDT, total	100 100 100 100 100 100 100	95 100 95 100 95 100 100

Table 2.2. Pesticides in bed sediment and aquatic biota from rivers in the United States: State and local monitoring studies—*Continued*

Bed sediment (µg/kg dry weight)			Biota (whole fish unless noted)					
			% Detection		(µg/kg wet weight)			
DL	Range	Median	Sites	Samples	DL	Range	Median	Comments
								High proportion of DDE in total DDT in biota and sediment indicated that total DDT in the NWR is from past inputs.
nr	5–84*	44*	—	—	—	—	—	* Data are for 6 dam sites only.
nr	0.7–8.4*	5.3*	—	—	—	—	—	Objective: to describe the effect
nr	0.6–12*	8.4*	—	—	—	—	—	of low dams on the distribution
nr	0.3–4.6*	2.8*	—	—	—	—	—	of sediment-born contaminants
nr	0.2–4.5*	3*	—	—	—	—	—	on the lower Schuylkill R. Differences in OC levels between dam locations appear to be a result of different local land use patterns. OC levels in sediment were higher in the <62-µm size fraction than the <16-µm fraction. From analysis of cores, OC pesticides and PCBs were higher in pre-1972 sediments than in more recent deposits.
nr	nd–807	133	67	50	10	nd–32	nd/18	Objective: to determine probable
nr	29–3,752	522	100	100	10	10–220	104	source of DDT in the Blanco
nr	nd–201	20	0	0	10	nd	nd	Drain, a major DDT source to
nr	59–1,252	421	100	100	5	25–470	280	the Salinas R. (Note: Clam
nr	nd–1,312	51	67	67	10	nd–40	27	transplantation study results are
nr	28–2,453	160	67	67	10	nd–240	140	in Table 2.3.) Soils from nearby
nr	145–6,264	1,450	100	100	10	36–978	560	fields contained up to 5,000 µg/kg DDT, of which 66–80% was technical DDT, 1–6% DDD, and 15–20% *o,p'*- DDT (very similar to technical DDT). DDT levels in drain sediment were 200–400 times those in the Salinas R. above the drain outfall. Sediment samples were either: ≥60% technical DDT and <10% DDD (recently mobilized soils) or <45% technical DDT and >20% DDD (more typical of sediment). No "hot spots" were identified, and there was no evidence of recent use or leaky depositories of DDT. The probable source of DDT was soil erosion from adjoining fields. DDT may be very persistent in soils in study area, but appeared to break down slowly in sediment.

Table 2.2. Pesticides in bed sediment and aquatic biota from rivers in the United States: State and local monitoring studies—*Continued*

Reference	Sampling date	Sampling location and strategy	Sampling media and species	Number of samples	Target analytes	Bed sediment % Detection	
						Sites	Samples
Briggs and Feiffer, 1986	1978–1983	6 sites on 4 major streams in Rhode Island. Sediment was sampled yearly during low flow.	Water (filtered), sediment (surficial, suspended).	36	Aldrin	17	nr
					Chlordane, total	100	nr
					DDD, total	100	nr
					DDE, total	100	nr
					DDT, total	100	nr
					Dieldrin	83	nr
					Endosulfan, total	17	nr
					Endrin	17	nr
					Heptachlor	67	nr
					Heptachlor epoxide	0	nr
					Lindane	0	nr
					Mirex	17	nr
					Methoxychlor	0	nr
					Perthane	0	nr
					Toxaphene	0	nr
Bush and others, 1986	1981–1984	2 sites in Lake Oconee, a hydro-electric lake in Oconee River (Georgia). The lake is adjacent to a seed orchard (loblolly pine) treated with various insecticides. Storm runoff was measured from 2 treated watersheds and 1 control watershed. Fish populations were sampled from areas of the lake that received runoff from treated and untreated portions of the orchard. Fish were collected monthly during 1981–1983 and quarterly in 1984. Offsite drift was measured during 1981 treatment.	Water (storm runoff), fish (whole), air (drift). Species: carp, sucker, bowfin, channel catfish, spotted gar, largemouth bass, shad, bluegill.	nr 100 21	Aldrin	—	—
					Azinphosmethyl	—	—
					Carbofuran	—	—
					Carbophenothion	—	—
					Chlordane	—	—
					Chlorpyrifos	—	—
					DDT, total	—	—
					Diazinon	—	—
					Dicofol	—	—
					Dieldrin	—	—
					Dimethoate	—	—
					Endrin	—	—
					EPN	—	—
					Fenvalerate	—	—
					Heptachlor	—	—
					Heptachlor epoxide	—	—
					3-Hydroxy-carbofuran	—	—
					Lindane/HCH	—	—
					Malathion	—	—
					Methoxychlor	—	—
					Methyl parathion	—	—
					Mirex	—	—
					Parathion	—	—
					Phosdrin	—	—
					Tetrachlorvinphos	—	—
					Toxaphene	—	—
					Trichlorfon	—	—
Graczyk, 1986	1981	Unspecified number of sites above and below cranberry bogs that drain into the Namekagon River, Wisconsin.	Water (whole), sediment (surficial).	nr nr	Aldrin	0	0
					Chlordane	0	0
					D, 2,4-	0	0
					DDD	0	0
					DDE	0	0
					DDT	0	0
					Diazinon	0	0

Table 2.2. Pesticides in bed sediment and aquatic biota from rivers in the United States: State and local monitoring studies—*Continued*

Bed sediment (µg/kg dry weight)			Biota (whole fish unless noted) % Detection		(µg/kg wet weight)			Comments
DL	Range	Median	Sites	Samples	DL	Range	Median	
0.1	nd–1.0	nd*	—	—	—	—	—	* Median of site means.
0.1	nd–110	16.2*	—	—	—	—	—	Objectives: to determine
0.1	nd–160	11.6*	—	—	—	—	—	physical, chemical, and
0.1	nd–22	0.62*	—	—	—	—	—	biological quality of major
0.1	nd–19	1.50*	—	—	—	—	—	streams in Rhode Island.
0.1	nd–440	2.96*	—	—	—	—	—	
0.1	nd–11	nd*	—	—	—	—	—	
0.1	nd–7.5	nd	—	—	—	—	—	
0.1	nd–0.5	0.04	—	—	—	—	—	
0.1	nd	nd*	—	—	—	—	—	
0.1	nd	nd*	—	—	—	—	—	
0.1	nd–310	nd*	—	—	—	—	—	
0.1	nd	nd*	—	—	—	—	—	
0.1	nd	nd*	—	—	—	—	—	
0.1	nd	nd*	—	—	—	—	—	
—	—	—	0	0	3	nd	nd	* Residues could not be
—	—	—	0	0	300	nd	nd	confirmed by GC/MS due to
—	—	—	50	2	20–200	nd–560*	nd	analytical difficulties and may
—	—	—	0	0	50	nd	nd	represent interferences.
—	—	—	0	0	50	nd	nd	Objective: to measure insecticide
—	—	—	0	0	10	nd	nd	residues in fish near a seed
—	—	—	100	84	1–50	nd–550	nr	orchard. Application rates of
—	—	—	0	0	100	nd	nd	azinphosmethyl, carbofuran, and
—	—	—	0	0	10	nd	nd	fenvalerate were known, so
—	—	—	100	23	10	nd	nd	study also is listed in Table 2.3
—	—	—	0	0	41	nd	nd	for these compounds.
—	—	—	0	0	10	nd	nd	Toxaphene, once used on cotton
—	—	—	0	0	200	nd	nd	in the Oconee R. Basin, was
—	—	—	0	0	nr	nd	nd	detected at high levels in fish.
—	—	—	0	0	3	nd	nd	Its occurrence in fish increased
—	—	—	0	0	10	nd	nd	from 1981 to 1984 as rainfall
—	—	—	100	6	20–200	nd–1,490*	nd	and storm runoff increased during this period.
—	—	—	100	26	10–100	nd–20	nd	
—	—	—	0	0	50	nd	nd	
—	—	—	0	0	30	nd	nd	
—	—	—	0	0	50	nd	nd	
—	—	—	0	0	50	nd	nd	
—	—	—	0	0	30	nd	nd	
—	—	—	0	0	20	nd	nd	
—	—	—	0	0	20	nd	nd	
—	—	—	100	70	5–100	nd–1,070	nr	
—	—	—	0	0	80	nd	nd	
nr	nd	nd	—	—	—	—	—	Objectives: to assess water
nr	nd	nd	—	—	—	—	—	quality of the St. Croix Scenic
nr	nd	nd	—	—	—	—	—	Waterway (Wisconsin,
nr	nd	nd	—	—	—	—	—	Minnesota). None of the
nr	nd	nd	—	—	—	—	—	pesticides on the analytical
nr	nd	nd	—	—	—	—	—	schedule were detected in
nr	nd	nd	—	—	—	—	—	sediment.

Table 2.2. Pesticides in bed sediment and aquatic biota from rivers in the United States: State and local monitoring studies—*Continued*

Reference	Sampling date	Sampling location and strategy	Sampling media and species	Number of samples	Target analytes	Bed sediment % Detection	
						Sites	Samples
					Dieldrin	0	0
					DP, 2,4-	0	0
					Endosulfan	0	0
					Endrin	0	0
					Ethion	0	0
					Heptachlor	0	0
					Heptachlor epoxide	0	0
					Lindane	0	0
					Malathion	0	0
					Methoxychlor	0	0
					Methyl parathion	0	0
					Methyl trithion	0	0
					Mirex	0	0
					Parathion	0	0
					Perthane	0	0
					Silvex	0	0
					T, 2,4,5-	0	0
					Toxaphene	0	0
					Trithion	0	0
Jaffe and Hites, 1986	1984	15 sites at the mouths of tributaries to Lake Ontario, and in the Niagara River and its tributaries (New York). Fish were composited by site, species, and body size.	Fish (whole). Species: common carp, goldfish, sucker.	18	Aldrin	—	—
					Chlordane, α-	—	—
					Chlordane, γ-	—	—
					Chlordane, misc.	—	—
					Chlordene	—	—
					Dacthal	—	—
					DDD, *o,p'*-	—	—
					DDD, *p,p'*-	—	—
					DDE, *o,p'*-	—	—
					DDE, *p,p'*-	—	—
					Dieldrin	—	—
					Endosulfan I	—	—
					HCB	—	—
					HCH, α-	—	—
					HCH, δ-	—	—
					Heptachlor	—	—
					Heptachlor epoxide	—	—
					1-Hydroxychlordene	—	—
					Kepone	—	—
					Lindane + β-HCH	—	—
					Mirex	—	—
					Nonachlor, misc.	—	—
					Nonachlor, *trans*-	—	—
					Oxychlordane	—	—
					PCA	—	—
					Photomirex	—	—
Johnson and others, 1986	1985	11 sites (sediment) and 5 sites (fish) on the Yakima River (Washington). Most biota samples	Sediment (surficial), fish (whole, muscle, eggs),	13 13 19 10	Aldrin	9	8
					Chlordane	0	0
					DDD, *o,p'*-	0	0
					DDD, *p,p'*-	91	92
					DDE, *o,p'*-	0	0

Table 2.2. Pesticides in bed sediment and aquatic biota from rivers in the United States: State and local monitoring studies—*Continued*

Bed sediment (µg/kg dry weight)			Biota (whole fish unless noted)					Comments
			% Detection		(µg/kg wet weight)			
DL	Range	Median	Sites	Samples	DL	Range	Median	
nr	nd	nd	—	—	—	—	—	
nr	nd	nd	—	—	—	—	—	
nr	nd	nd	—	—	—	—	—	
nr	nd	nd	—	—	—	—	—	
nr	nd	nd	—	—	—	—	—	
nr	nd	nd	—	—	—	—	—	
nr	nd	nd	—	—	—	—	—	
nr	nd	nd	—	—	—	—	—	
nr	nd	nd	—	—	—	—	—	
nr	nd	nd	—	—	—	—	—	
nr	nd	nd	—	—	—	—	—	
nr	nd	nd	—	—	—	—	—	
nr	nd	nd	—	—	—	—	—	
nr	nd	nd	—	—	—	—	—	
nr	nd	nd	—	—	—	—	—	
nr	nd	nd	—	—	—	—	—	
nr	nd	nd	—	—	—	—	—	
nr	nd	nd	—	—	—	—	—	
nr	nd	nd	—	—	—	—	—	
—	—	—	0	0	nr	nd*	nd*	* Data are in µg/kg-fat.
—	—	—	100	100	nr	38–970*	170*	Objective: to demonstrate
—	—	—	100	100	nr	27–550*	96*	long-range transport and
—	—	—	100	100	nr	22–270*	54*	bioavailability of certain organic
—	—	—	0	0	nr	nd*	nd*	pollutants in Lake Ontario. Fish
—	—	—	100	100	nr	29–2,300*	150*	from Ellicott Creek and
—	—	—	80	78	nr	nd–420*	125*	Tonawanda Creek (Niagara R.
—	—	—	93	94	nr	nd–2,000*	885*	tributaries) showed highest
—	—	—	60	56	nr	nd–76*	21*	residues of many pesticides.
—	—	—	100	100	nr	510–7,200*	3,500*	HCHs appear to have point
—	—	—	100	100	nr	78–560*	190*	sources in the Love Canal area;
—	—	—	100	100	nr	3–285*	34*	HCB also was found at highest
—	—	—	100	100	nr	37–1,350*	130*	levels in this area. Dacthal
—	—	—	100	100	nr	9–610*	51*	appeared to have point source in
—	—	—	27	22	nr	nd–410*	nd*	the Ellicott and Tonawanda area.
—	—	—	0	0	nr	nd*	nd*	Photomirex was not detected in
—	—	—	100	100	nr	5–120*	34*	samples above Niagara Falls,
—	—	—	0	0	nr	nd*	nd*	but was prevalent in samples
—	—	—	0	0	nr	nd*	nd*	below the falls. Mirex also was
—	—	—	100	100	nr	12–260*	46*	lower above the falls. Genesee
—	—	—	100	100	nr	12–1,400*	393*	R. fish had high PCA residues.
—	—	—	100	100	nr	32–330*	105*	
—	—	—	100	100	nr	80–940*	310*	
—	—	—	100	100	nr	9–98*	43*	
—	—	—	87	83	nr	nd–160*	24*	
—	—	—	53	56	nr	nd–400*	136*	
0.1	nd–1,065	nd	0	0	10	nd	nd	Objective: to investigate DDT
0.1	nd	nd	0	0	100	nd	nd	contamination of Yakima R.
0.1	nd	nd	0	0	40	nd	nd	Basin; to evaluate hazards and
0.1	nd–37	2.2	20	15	40	nd–140	nd	identify sources. For fish species
0.1	nd	nd	20	8	20	nd–30	nd	present at multiple sites, levels

Table 2.2. Pesticides in bed sediment and aquatic biota from rivers in the United States: State and local monitoring studies—*Continued*

Reference	Sampling date	Sampling location and strategy	Sampling media and species	Number of samples	Target analytes	Bed sediment % Detection	
						Sites	Samples
		were composites of 2–16 individuals.	mussels, crayfish, water (whole). Species: rainbow trout, mountain whitefish, largescale sucker, bridgelip sucker, northern squawfish, smallmouth bass, channel catfish, *Margaritifera falcata, Pascifastacus* sp.	2 2 50	DDE, *p,p'-* DDT, *o,p'-* DDT, *p,p'-* DDT, total Dieldrin Endosulfan Endosulfan sulfate Endrin Endrin aldehyde HCH, α- HCH, β- HCH, δ- Heptachlor Heptachlor epoxide Lindane Methoxychlor Toxaphene	100 91 91 100 73 9 0 0 0 0 0 0 0 0 0 0 0	100 92 92 100 69 8 0 0 0 0 0 0 0 0 0 0 0
Kepner, 1986	1985	17 locations on the Gila River (Arizona) and one station below the rinse area of an agricultural air service facility. Samples consisted of triplicate composites of individual organisms or sediment samples from each site.	Sediment, fish, birds, reptiles. Aquatic species include carp, channel catfish, spiny softshell turtle.	nr nr, nr nr (208 total samples)	DDE, *p,p'-* Dicofol Toxaphene	nr nr nr	nr nr nr
Merna, 1986	1977– 1978	2 streams (biota) and 3 streams (sediment) tributary to Lake Michigan (Michigan). Two of the streams were accessible to anadromous fish and one was	Sediment (surficial), fish (fillets, whole, eggs), invertebrates. Species: sculpins, brown trout,	28 128 18 9 4	DDT Dieldrin	0 0	0 0

Table 2.2. Pesticides in bed sediment and aquatic biota from rivers in the United States: State and local monitoring studies—*Continued*

Bed sediment (µg/kg dry weight)			Biota (whole fish unless noted)						Comments
			% Detection		(µg/kg wet weight)				
DL	Range	Median	Sites	Samples	DL	Range	Median		Comments
0.1	0.1–129	7	100	100	20	50–2,900	520		increased by a factor of 2–6 at
0.1	nd–8.4	0.71	0	0	40	nd	nd		each successive downstream
0.1	nd–59	3.4	80	54	40	nd–120	50		site. DDT levels were lower in
0.1	0.1–234	11.6	100	100	20–40	50–3,000	570		muscle than in whole fish. Me-
0.1	nd–14.9	2.4	60	38	20	nd–240	nd		tabolites endrin aldehyde and en-
0.1	nd–4.1	nd	20	23	20	nd–170	nd		dosulfan sulfate were detected in
0.1	nd	nd	0	0	200	nd	nd		1 (each) of 10 fish egg samples.
0.1	nd	nd	20	15	20	nd–50	nd		When sediment concentrations
0.1	nd	nd	0	0	200	nd	nd		were normalized for TOC con-
0.1	nd	nd	0	0	100	nd	nd		tent, there was a downstream
0.1	nd	nd	0	0	100	nd	nd		trend in total DDT residues.
0.1	nd	nd	0	0	100	nd	nd		From water samples, the major
0.1	nd	nd	0	0	10	nd	nd		sources among the 11 tributaries
0.1	nd	nd	0	0	100	nd	nd		sampled were Sulphur Creek
0.1	nd	nd	0	0	100	nd	nd		(DDTs, dieldrin) and Granger
0.1	nd	nd	0	0	100	nd	nd		Drain (dieldrin). Historical data
4.6	nd	nd	0	0	2000	nd	nd		shows decreasing trend in total
									DDT and dieldrin levels since
									1970s, and that contamination
									occurs primarily during irriga-
									tion season. DDE comprised
									84% (on average) of total DDT
									in fish, which is consistent with
									a historical source. In most wa-
									ter and sediment samples, DDT≥
									DDE. DDT seems to have a long
									lifetime in Yakima Basin soils.
10	10–460	nr	nr	nr	10	40–23,000	nr		Objective: to describe
10	nr	nd	nr	nr	10	nd–560	nr		distribution of contaminants in
10	nr	nd	nr	nr	10	nd–8,400	nr		the Gila R. DDE was detected in
									all sample matrices. DDE was
									5–23 times higher at stations
									above Painted Rock than below.
									Flood control and agricultural
									diversion dams appear to contain
									contaminated sediment in the
									upper reaches of the study area.
									Although toxaphene was banned
									for use in Arizona in 1969,
									residues still present a threat to
									fish and wildlife in the region.
nr	nd	nd	100*	nr	nr	100–1,070*†	360*†		* Fish fillets.
									† Range and median of average
nr	nd	nd	0*	0*	nr	nd*	nd*		residues for fish grouped by age,
									species, and site.
									Objective: to evaluate role of
									salmon in transport of OC com-
									pounds from the Great Lakes to
									tributary streams. Consumption
									of eggs from Great Lakes

Table 2.2. Pesticides in bed sediment and aquatic biota from rivers in the United States: State and local monitoring studies—*Continued*

Reference	Sampling date	Sampling location and strategy	Sampling media and species	Number of samples	Target analytes	Bed sediment % Detection Sites	Bed sediment % Detection Samples
		blocked to fish migration.	rainbow trout, coho salmon, chinook salmon, crayfish.				
Oliver and Pugsley, 1986	1984–1985	45 sites (1984) and 20 sites (1985) on the St. Clair River, between Michigan and Canada. 1984 samples (surficial) were collected throughout the river. 1985 samples (cores collected on a transect) were from a 7 km (industrial) section of the river.	Sediment (surficial, cores).	45 65	HCB	96*	96*
Reich and others, 1986	1983	13 sites above and downstream of a historic contaminant source on the Huntsville Spring Branch near the Tennessee River (Alabama). At least 3 sediment samples were taken at each site along a cross-section, more than 3 if the cross-section was wider than 60 m.	Water, sediment (surficial), invertebrates. Species: chironomids, tubificid worms, water boatmen.	>39 5	DDD, o,p'- DDD, p,p'- DDE, o,p'- DDE, p,p'- DDT, o,p'- DDT, p,p'-	100 * 100 * 100 * 100 * 100 * 100 *	nr nr nr nr nr nr

Table 2.2. Pesticides in bed sediment and aquatic biota from rivers in the United States: State and local monitoring studies—*Continued*

Bed sediment (µg/kg dry weight)			Biota (whole fish unless noted)						
			% Detection		(µg/kg wet weight)				
DL	Range	Median	Sites	Samples	DL	Range	Median	Comments	
								salmon is a source of OC contamination to the food chain. Mean levels in salmon eggs were 660 µg/kg DDT and 2,590 µg/kg PCB (coho) and 1,070 µg/kg DDT and 4,020 µg/kg PCB (chinook). Eggs had higher mean levels than fillets, whole sculpins, and crayfish. Salmon eggs constituted up to 87% of stomach contents of trout sampled during salmon spawning season.	
nr	nd– 24,000*	nr	—	—	—	—	—	* Data are for surficial (1984) samples only. Objective: to investigate OCs in bottom sediment of the St. Clair R. Contaminants are high in the Sarnia industrial area, and remain high all the way to Lake St. Clair. Contaminated sediment is confined to the Canadian shoreline. 1985 data showed hot spots, including an industrial sewer discharge to the river, which also affects downstream concentrations.	
nr	nr	nr	100†	100†	nr	319– 30,680†	3,650†	* Estimated from figure. † Data are for invertebrates. Objective: to measure DDT residues in the study area. Levels in sediment were highly variable within transects closest to the original contamination source. Despite variability, concentrations showed a rapid, exponential decline in water and sediment within the first few miles downstream. No change was apparent when the sediment data (1983) were compared to a 1979 study (U.S. Army Corps of Engineers, 1987). Residues in invertebrates were highest at the first sampled transect below the old DDT point source and were an order of magnitude lower at sites further downstream. For total DDT, bioaccumulation factors (BAF) ranged from 2,300 to >99,000 and biota–sediment	
nr	nr	nr	100†	100†	nr	705– 38,520†	9,010†		
nr	nr	nr	100†	100†	nr	512– 18,875†	3,258†		
nr	nr	nr	100†	100†	nr	485– 124,242†	6,470†		
nr	nr	nr	80†	86†	nr	244– 87,133†	2,126†		
nr	nr	nr	100†	100†	nr	1,091– 440,670†	6,450†		

Table 2.2. Pesticides in bed sediment and aquatic biota from rivers in the United States: State and local monitoring studies—*Continued*

Reference	Sampling date	Sampling location and strategy	Sampling media and species	Number of samples	Target analytes	Bed sediment % Detection	
						Sites	Samples
Saiki and Schmitt, 1986	1981	8 sites on the San Joaquin River and 2 tributaries (Merced River and Salt Slough) in California. Samples were composites of 3–5 fish. 2–3 replicate samples were taken per site.	Fish (whole). Species: bluegill, common carp.	About 28	Aldrin Chlordane, *trans-* Chlordane * DCPA DDD, *p,p'-* DDE, *p,p'-* DDT, *o,p'-* DDT, *p,p'-* DDT, total Dieldrin Endrin HCB HCH, α- Heptachlor Heptachlor epoxide Lindane Methoxychlor Mirex Toxaphene	— — — — — — — — — — — — — — — — — — —	— — — — — — — — — — — — — — — — — — —
Sewell and Knight, 1986	1978–1981	Unspecified number of sites on the Wolf and Loosahatchie river basins in the Mississippi River Delta (Tennessee, Mississippi). Results were reported by trophic level: bottom feeders, herbivores, lesser carnivores, and top carnivores.	Fish (fillets). Species include common carp, bigmouth buffalo, gizzard shad, white crappie, channel catfish, bowfin, largemouth bass.	nr	DDD DDE DDT DDT, total	— — — —	— — — —

Table 2.2. Pesticides in bed sediment and aquatic biota from rivers in the United States: State and local monitoring studies—*Continued*

Bed sediment (µg/kg dry weight)			Biota (whole fish unless noted)					Comments
			% Detection		(µg/kg wet weight)			
DL	Range	Median	Sites	Sam-ples	DL	Range	Median	
								accumulation factors (units were not defined) ranged from 0.018 to 5.1. BAFs for invertebrates closest to the contaminant source were much higher than at other sites.
—	—	—	0	0	2	nd	nr	* Sum of *cis-* chlordane and
—	—	—	0	0	4	nd	nr	*trans-* nonachlor.
—	—	—	100†	nr	4	0–273 §	nr	† Percent of sites with mean
—	—	—	80†	nr	4	0–54 §	nr	concentration >0.
—	—	—	88†	nr	10	0–345 §	nr	§ Range is of mean values for
—	—	—	100†	100	6	13–1,866 §	nr	fish grouped by species and site.
—	—	—	13†	nr	10	0–2 §	nr	Objective: to determine OC con-
—	—	—	63†	nr	10	0–92 §	nr	taminants in fish from the San
—	—	—	100†	100	4–10	13–2,213 §	nr	Joaquin R. and tributaries. OCs
—	—	—	50†	nr	10	0–67 §	nr	were detected less often and at
—	—	—	0	0	10	nd	nr	lower levels in bluegills than
—	—	—	0	0	2	nd	nr	carp. Lipid content also was
—	—	—	63†	nr	2	0–3 §	nr	lower in bluegills. α-HCH and
—	—	—	0	0	2	nd	nr	toxaphene were found only in
—	—	—	0	0	4	nd	nr	carp and *p,p'-* DDT only in blue-
—	—	—	0	0	2	nd	nr	gills. OCs were generally higher
—	—	—	0	0	40	nd	nr	at downstream than upstream
—	—	—	0	0	20	nd	nr	sites. Levels of most OCs were
—	—	—	13†	nr	40	0–3,123 §	nr	significantly correlated with water quality, especially turbidity. Turbidity, conduc- tivity, and total alkalinity in- creased at downstream sites most heavily affected by irrigation return flows and agriculture.
—	—	—	nr*	nr*	10	63–412*	85*	* Data are for fillets; range and
—	—	—	nr*	nr*	10	338–1,085*	540*	median are of 4 trophic levels.
—	—	—	nr*	nr*	10	nd–85*	nd*	Objective: to determine persis-
—	—	—	nr*	nr*	10	400–1,500*	670*	tence of DDTs in aquatic food chains. Top carnivores had high- er residues than lesser carni- vores. Highest total DDT resi- dues were in herbivores (shad); most fish sampled were >7 in. long, so would be too large to be prey for secondary consumers. Bottom-feeders had higher DDT levels than DDD; DDT may be concentrated in top 2 cm of sediment, where bottom feeders tend to feed.

Table 2.2. Pesticides in bed sediment and aquatic biota from rivers in the United States: State and local monitoring studies—*Continued*

Reference	Sampling date	Sampling location and strategy	Sampling media and species	Number of samples	Target analytes	Bed sediment % Detection Sites	Bed sediment % Detection Samples
U.S. Fish and Wildlife Service, 1986	1985	6 sites (sediment) in streams and ditches near an abandoned hazardous waste storage facility (Seymour Recycling Corporation) in Indiana. Fine silt and clay sediments were collected.	Sediment (surficial, 1-ft cores, 2-ft cores).	6 3 3	Aldrin	0	0
					Chlordane	0	0
					DDD, *p,p'*-	0	0
					DDE, *p,p'*-	0	0
					DDT, *p,p'*-	0	0
					Dieldrin	17	0
					Endosulfan I	0	0
					Endosulfan II	0	0
					Endosulfan sulfate	0	0
					Endrin	17	17
					Endrin aldehyde	0	0
					Endrin ketone	0	0
					HCH, α-	0	0
					HCH, β-	0	0
					HCH, δ-	0	0
					Heptachlor	0	0
					Heptachlor epoxide	0	0
					Lindane	0	0
					Methoxychlor	0	0
					Toxaphene	0	0
Arruda and others, 1987	1986	13 sites on Kansas River, Kansas. Sites were targeted on basis of land use. Samples were composites.	Fish (muscle). Species: carp, river carpsucker.	25	Chlordane, technical	—	—
Butler, 1987	1976–1978	20 sites in Wyoming streams. Basins sampled include Bighorn, Powder, North Platte, Green, Wind, Bear, and Snake rivers. Sampling was done twice yearly, in spring/early summer and in the autumn. Sediment was collected from depositional areas.	Water (whole), sediment (surficial).	93 89	Aldrin	5	1
					Chlordane	15	7
					D, 2,4-	5	2
					DDD	45	20
					DDE	75	39
					DDT	40	7
					Dicamba	20	9
					Dieldrin	45	21
					DP, 2,4-	5	2
					Endosulfan	0	0
					Endrin	10	3
					Heptachlor	5	1
					Heptachlor epoxide	20	6
					Lindane	20	4
					Mirex	0	0
					Perthane	0	0
					Picloram	5	2
					Silvex	0	0
					T, 2,4,5-	0	0
					Toxaphene	5	1

Table 2.2. Pesticides in bed sediment and aquatic biota from rivers in the United States: State and local monitoring studies—*Continued*

| Bed sediment (µg/kg dry weight) | | | Biota (whole fish unless noted) | | | | | Comments |
DL	Range	Median	% Detection Sites	Sam-ples	DL	Range (µg/kg wet weight)	Median	
8–480	nd	nd	—	—	—	—	—	Objective: to evaluate off-site movement of contaminants from an abandoned hazardous waste storage facility into the East Fork of the White R.; to identify and quantitiate contaminants in off-site soil and sediment. Previous collections of sediment and biota analyzed for trace elements, and volatile and semivolatile organic compounds also were presented in this report. The number of compounds migrating from the site appeared to have decreased since 1980 (See U.S. Fish and Wildlife Service, 1981), probably due to USEPA clean-up efforts.
80–4,800	nd	nd	—	—	—	—	—	
16–960	nd	nd	—	—	—	—	—	
16–960	nd	nd	—	—	—	—	—	
16–960	nd	nd	—	—	—	—	—	
nr	nd–64	nd	—	—	—	—	—	
8–480	nd	nd	—	—	—	—	—	
16–960	nd	nd	—	—	—	—	—	
16–960	nd	nd	—	—	—	—	—	
nr	nd–11	nd	—	—	—	—	—	
16–960	nd	nd	—	—	—	—	—	
16–960	nd	nd	—	—	—	—	—	
8–480	nd	nd	—	—	—	—	—	
8–480	nd	nd	—	—	—	—	—	
8–480	nd	nd	—	—	—	—	—	
8–480	nd	nd	—	—	—	—	—	
8–480	nd	nd	—	—	—	—	—	
8–480	nd	nd	—	—	—	—	—	
80–4,800	nd	nd	—	—	—	—	—	
160–9,600	nd	nd	—	—	—	—	—	
—	—	—	92	68	30	nd–2,100	50	Objective: to assess the effect of land use on occurrence of chlordane. Mean chlordane concentration increased at or below each major urban area, then decreased further downstream.
0.1	nd–0.1	nd	—	—	—	—	—	Objective: to determine pesticide concentrations in major drainage basins in Wyoming. DDD and DDE residues were highest in the North Platte and Shoshone rivers. Dieldrin residues were highest in the Greybull and Bighorn rivers in the Bighorn River Basin and in the North Platte River. A single sample from the North Platte contained the highest residues of DDE, dieldrin, and toxaphene.
1	nd	nd	—	—	—	—	—	
0.1	nd	nd	—	—	—	—	—	
0.1	nd	nd	—	—	—	—	—	
0.1	nd	nd	—	—	—	—	—	
0.1	nd	nd	—	—	—	—	—	
0.1	nd	nd	—	—	—	—	—	
0.1	nd	nd	—	—	—	—	—	
0.1	nd	nd	—	—	—	—	—	
0.1	nd	nd	—	—	—	—	—	
0.1	nd	nd	—	—	—	—	—	
0.1	nd	nd	—	—	—	—	—	
0.1	nd	nd	—	—	—	—	—	
0.1	nd	nd	—	—	—	—	—	
0.1	nd	nd	—	—	—	—	—	
0.1	nd	nd	—	—	—	—	—	
0.1	nd	nd	—	—	—	—	—	
0.1	nd	nd	—	—	—	—	—	
0.1	nd	nd	—	—	—	—	—	
1	nd	nd	—	—	—	—	—	

Table 2.2. Pesticides in bed sediment and aquatic biota from rivers in the United States: State and local monitoring studies—*Continued*

Reference	Sampling date	Sampling location and strategy	Sampling media and species	Number of samples	Target analytes	Bed sediment % Detection	
						Sites	Samples
Camanzo and others, 1987	1983	14 sites from Lake Michigan tributaries aand embayments (Michigan). Fish were collected in the autumn (before migration). Sport fish and bottom feeders were targeted. 2 species were collected per site. Samples were composited by site, species, and size. Only the sample with the largest fish was analyzed.	Fish (whole). Species: common carp, smallmouth bass, largemouth bass, channel catfish, pumpkinseed, bowfin, northern pike, rock bass, lake trout.	28	Chlordane, *cis-* Chlordane, *trans-* Dacthal DDD, *p,p'-* * DDE, *p,p'-* DDT, *p,p'-* Dieldrin Endosulfan sulfate I Endrin HCH, α- HCH, β- Heptachlor Heptachlor epoxide Lindane Methoxychlor Mirex Nonachlor, *cis-* Nonachlor, *trans-* Oxychlordane Toxaphene Trifluralin Xytron	— —	— —
Crayton, 1987	1984– 1985	2 sites in Womens Bay (sediment, mussels), 1 site on the Tuluksak River (fish), and 2 on Kelly and Scenic lakes (fish, clams, leeches), Alaska. Clam and leech samples were composites.	Sediment, mollusks, fish (whole), leeches. Species include round whitefish, blue mussel, clams (*Anodonta* sp.)	2 3 9 1	Chlordane, *cis-* DDD, *p,p'-* DDE, *p,p'-* DDT, *p,p'-* Dieldrin Endrin Heptachlor epoxide Nonachlor, *cis-* Nonachlor, *trans-* Oxychlordane Toxaphene	0 0 0 0 0 0 0 0 0 0 0	0 0 0 0 0 0 0 0 0 0 0
Funk and others, 1987	1984– 1986	7 sites (sediment), 3 sites (fish), and 9 sites (water) in 4 drainage basins in North Cascades National Park (Washington). Principal water bodies include Chelan, Ross, and Diablo Lakes, the Stehekin and Skagit rivers. Fish were composited by site and species.	Sediment, fish (whole, fillets), water (whole). Species: brook trout, rainbow trout, northern squawfish, longnose suckers, Dolly Varden.	7 14 total fish, 9	Aldrin Chlordane D, 2,4- DDD † DDE DDT, *p,p'-* Dieldrin Endrin HCH, α- Heptachlor Lindane Methoxychlor Mirex Silvex T, 2,4,5- Toxaphene	14 — 0 57 71 29 14 0 29 0 14 0 0 14 29 —	14 — 0 57 71 29 14 0 29 0 14 0 0 14 29 —

Table 2.2. Pesticides in bed sediment and aquatic biota from rivers in the United States: State and local monitoring studies—*Continued*

Bed sediment (µg/kg dry weight)			Biota (whole fish unless noted)					Comments
			% Detection		(µg/kg wet weight)			
DL	Range	Median	Sites	Samples	DL	Range	Median	
—	—	—	100	96	<5	nd–211	7.5	* Residues also contain *o,p'*- DDT. Objective: to analyze nearshore fish to detect problems in resident fish populations. OC residues were positively correlated with fat content. Highest OC levels (DDT, toxaphene, chlordane, dieldrin, methoxychlor, PCBs) occurred in common carp from St. Joseph's R. Bottom-feeders had much higher OC residues than predator species. Bottom-feeders generally were older and higher in fat content. Some Lake Michigan tributaries have elevated levels of PCBs, DDT, toxaphene, chlordane components, and dieldrin, and constitute potential sources to the Great Lakes.
—	—	—	100	96	<5	nd–25	1	
—	—	—	0	0	<5	nd	nd	
—	—	—	100	100	<5	2–1,084	77	
—	—	—	100	100	<5	13–9,015	702	
—	—	—	100	89	<5	nd–343	22	
—	—	—	100	86	<5	nd–286	20	
—	—	—	0	0	<5	nd	nd	
—	—	—	0	0	<5	nd	nd	
—	—	—	100	100	<5	3–97	18	
—	—	—	0	0	<5	nd	nd	
—	—	—	93	68	<5	nd–8	1	
—	—	—	100	79	<5	nd–62	3	
—	—	—	86	64	<5	nd–26	1	
—	—	—	43	21	<5	nd–118	nd	
—	—	—	0	0	<5	nd	nd	
—	—	—	100	96	<5	nd–156	17	
—	—	—	100	96	<5	nd–406	44	
—	—	—	100	100	<5	nd–144	14.5	
—	—	—	71	54	200	nd–3,460	825	
—	—	—	100	96	<5	nd–126	11	
—	—	—	64	39	<5	nd–69	nd	
100	nd	nd	0*	0*	100	nd*	nd*	*Data are whole fish from (Tuluksak R. and Scenic L.). Objectives: Report describes several projects assessing pesticides in Alaska: (1) Womens Bay; (2) the Tuluksak R., especially effects of mining; quality and fish; (3) Kenai Wildlife Refuge (near Swanson R. Oil Field).
100	nd	nd	0*	0*	100	nd*	nd*	
100	nd	nd	0*	0*	100	nd*	nd*	
100	nd	nd	0*	0*	100	nd*	nd*	
100	nd	nd	0*	0*	100	nd*	nd*	
100	nd	nd	0*	0*	100	nd*	nd*	
100	nd	nd	0*	0*	100	nd*	nd*	
100	nd	nd	0*	0*	100	nd*	nd*	
100	nd	nd	0*	0*	100	nd*	nd*	
100	nd	nd	0*	0*	100	nd*	nd*	
0.1	nd–0.4	nd	0*	0*	0.1	nd*	nd*	* Reference did not report whether data were for whole fish or fillets. † It is ambiguous whether this is *o,p'*- DDD or total DDD. Objective: to document water-quality conditions and identify pollutants in North Cascades National Park complex. Residues of OCs in sediment and fish were very low. OC, organophosphate, and carbamate insecticides and various herbicides were not detected in water.
—	—	—	0*	0*	1	nd*	nd*	
0.1	nd	nd	0*	0*	0.1	nd*	nd*	
0.1	nd–3.7	trace	0*	0*	0.1	nd*	nd*	
0.1	nd–2.5	trace	100*	50*	0.1	nd–0.4*	trace*	
0.1	nd–7.8	nd	67*	29*	0.1	nd–0.12*	nd*	
0.1	nd–0.8	nd	0*	0*	0.1	nd*	nd*	
0.1	nd	nd	0*	0*	0.1	nd*	nd*	
0.1	nd–trace	nd	0*	0*	0.1	nd*	nd*	
0.1	nd	nd	0*	0*	0.1	nd*	nd*	
0.1	nd–trace	nd	0*	0*	0.1	nd*	nd*	
0.1	nd	nd	0*	0*	0.1	nd*	nd*	
0.1	nd	nd	0*	0*	0.1	nd*	nd*	
0.1	nd–trace	nd	0*	0*	0.1	nd*	nd*	
0.1	nd–3.4	nd	0*	0*	0.1	nd*	nd*	
—	—	—	0*	0*	10	nd*	nd*	

Table 2.2. Pesticides in bed sediment and aquatic biota from rivers in the United States: State and local monitoring studies—*Continued*

Reference	Sampling date	Sampling location and strategy	Sampling media and species	Number of samples	Target analytes	Bed sediment % Detection Sites	Bed sediment % Detection Samples
Prancke-vicius, 1987	1982	28 sites on the Detroit River and its 3 major U.S. tributaries, the Rouge, Ecorse, and Huron rivers (Michigan). Sites were selected to find worst-case conditions. Sediment was collected from depositional areas.	Sediment (surficial).	28	Aldrin	0	0
					Chlordane, γ-	68	68
					Chlorobenzilate	0	0
					D, 2,4-, isopropyl	0	0
					DCPA	4	4
					DDD, *o,p'*-	79	79
					DDD, *p,p'*-	86	86
					DDE, *o,p'*-	61	61
					DDE, *p,p'*-	86	86
					DDT, *o,p'*-	39	39
					DDT, *p,p'*-	50	50
					DDT, total	93	93
					Dieldrin	11	11
					Endosulfan I	0	0
					Endosulfan II	29	29
					Endrin	0	0
					HCB	86	86
					HCH, α-	0	0
					HCH, β-	32	32
					HCH, γ-	0	0
					Heptachlor	0	0
					Heptachlor epoxide	11	11
					Isodrin	0	0
					Kepone	0	0
					Methoxychlor	0	0
					Mirex	0	0
					Oxychlordane	11	11
					Tetradifon	0	0
					Trifluralin	11	11
					Zytron	0	0
Rice, 1987	1986	6 sites on 5 small streams in Centre County, Pennsylvania.	Fish. Species: brown trout, white sucker.	6	Chlordane, *cis*-	—	—
					Chlordane, *trans*-	—	—
					DDD, *p,p'*-	—	—
					DDE, *p,p'*-	—	—
					DDT, *p,p'*-	—	—
					Dieldrin	—	—
					Endrin	—	—
					Heptachlor epoxide	—	—
					Kepone	—	—
					Mirex	—	—
					Nonachlor, *cis*-	—	—
					Nonachlor, *trans*-	—	—
					Oxychlordane	—	—
					Toxaphene	—	—
Sloan, 1987	1980–1986	5 studies in this report measured pesticides in aquatic biota in New York rivers: (1) 213 sites in the	(1) Fish (fillets). Species include lake trout, carp, rainbow	1,126	Chlordane	—	—
					DDT, total	—	—
					Dieldrin	—	—
					Endrin	—	—
					HCB	—	—
					Heptachlor, total	—	—

Table 2.2. Pesticides in bed sediment and aquatic biota from rivers in the United States: State and local monitoring studies—*Continued*

Bed sediment (µg/kg dry weight)			Biota (whole fish unless noted)					Comments
			% Detection		(µg/kg wet weight)			
DL	Range	Median	Sites	Sam-ples	DL	Range	Median	
nr	nd	nd	—	—	—	—	—	Objective: to evaluate sediment
nr	nd–149	16	—	—	—	—	—	chemistry data from the Detroit
300	nd	nd	—	—	—	—	—	R. and its 3 major U.S.
400	nd	nd	—	—	—	—	—	tributaries; to identify "hot
nr	nd–11	nd	—	—	—	—	—	spots;" to contribute to
nr	nd–309	nd/15	—	—	—	—	—	environmental information base
nr	nd–1,255	63	—	—	—	—	—	for the Great Lakes. The highest
nr	nd–292	6	—	—	—	—	—	levels of total DDT were found
nr	nd–262	28	—	—	—	—	—	at Belle Isle, and were dominated
nr	nd–431	nd	—	—	—	—	—	by DDD. High levels of
nr	nd–180	nd/4	—	—	—	—	—	unmetabolized DDT were
nr	nd–2,265	231	—	—	—	—	—	detected at the Rouge R. mouth
nr	nd–14	nd	—	—	—	—	—	and in Trenton Channel. β-HCH
nr	nd	nd	—	—	—	—	—	levels were high at Belle Isle,
nr	nd–14	nd	—	—	—	—	—	above the Ecorse R. and below
nr	nd	nd	—	—	—	—	—	the Ecorse R. Chlordane was
nr	nd–106	14	—	—	—	—	—	found throughout study area,
330	nd	nd	—	—	—	—	—	with high levels at Connors
nr	nd–195	nd	—	—	—	—	—	Creek and in the Ecorse R.
nr	nd	nd	—	—	—	—	—	Residues of other pesticides did
nr	nd	nd	—	—	—	—	—	not show spatial patterns.
nr	nd–106	nd	—	—	—	—	—	
410	nd	nd	—	—	—	—	—	
950	nd	nd	—	—	—	—	—	
nr	nd	nd	—	—	—	—	—	
nr	nd	nd	—	—	—	—	—	
nr	nd–87	nd	—	—	—	—	—	
2,730	nd	nd	—	—	—	—	—	
nr	nd–25	nd	—	—	—	—	—	
nr	nd	nd	—	—	—	—	—	
—	—	—	17	14	10	nd–110	nd	Objective: to evaluate levels of
—	—	—	17	14	10	nd–58	nd	OC contamination in fish from
—	—	—	33	29	10	nd–300	nd	the study area. Previous studies
—	—	—	100	86	10	nd–610	130	had detected elevated levels of
—	—	—	33	29	10	nd–97	nd	PCBs. Fish from all three
—	—	—	17	14	10	nd–120	nd	watersheds sampled contained
—	—	—	0	0	10	nd	nd	elevated PCB levels (up to 5,100
—	—	—	0	0	10	nd	nd	µg/kg wet wt.). DDE levels at
—	—	—	17	14	10	nd–140	nd	one site were high in brown
—	—	—	33	29	10	nd–470	nd	trout (610 µg/kg) but
—	—	—	17	14	10	nd–68	nd	nondetectable in white suckers.
—	—	—	17	14	10	nd–150	nd	
—	—	—	0	0	10	nd	nd	
—	—	—	0	0	50	nd	nd	
—	—	—	86*	nr	10	nd–920*	nr	(1) * Data are for fish fillets.
—	—	—	99*	nr	10	nd–19,750*	nr	Objective: to identify
—	—	—	34*	nr	10	nd–120*	nr	contaminant problem areas and
—	—	—	4*	nr	10	nd–30*	nr	trouble-free areas in New York
—	—	—	18*	nr	10–100	nd–100*	nr	State.
—	—	—	17*	nr	10	nd–350*	nr	

Table 2.2. Pesticides in bed sediment and aquatic biota from rivers in the United States: State and local monitoring studies—*Continued*

Reference	Sampling date	Sampling location and strategy	Sampling media and species	Number of samples	Target analytes	Bed sediment % Detection Sites	Bed sediment % Detection Samples
		Statewide Toxic Substances Monitoring Program (1980–1986); (2) 1 site in the Hudson River (1982); (3) an unspecified number of sites in the Great Lakes and St. Lawrence River (1980–1986); (4) 53 sites in lakes and rivers on Long Island (1982–1985); (5) 3 sites on various lakes and rivers sampled as part of various special fish collections (1986).	trout, American eel. (2) Fish (fillets). Species: striped bass.	26	Lindane Mirex Aldrin Chlordane, total DDT, total Dieldrin Endosulfan I Endosulfan II Endosulfan sulfate Endrin Endrin aldehyde HCB HCH, α- HCH, β- HCH, δ- Heptachlor Heptachlor epoxide Lindane Toxaphene	— — — — — — — — — — — — — — — — — — —	— — — — — — — — — — — — — — — — — — —
		Results here are presented separately for the 5 studies.	(3) Fish (roe, (fillets, liver). Species include lake sturgeon, American eel, channel catfish.	1 100, 1	Chlordane DDT, total Dieldrin HCB Mirex Photomirex	— — — — — —	— — — — — —
			(4) Fish (*). Species include largemouth bass, white perch, brown bullhead.	nr	Chlordane, total DDT, total Dieldrin HCB Heptachlor epoxide Lindane	— — — — — —	— — — — — —
			(5) Fish (*). Species: rainbow trout, brown bullhead.	24	Chlordane, total DDT, total Dieldrin Endrin Mirex	— — — — —	— — — — —
Smith and others, 1987	nr	7 sites in the upper Rockaway River, New Jersey. Includes 2 forested (reference sites and 5 sites in a residential–commercial–industrial area containing several Superfund sites. Sediment	Sediment (surficial).	7	Chlordane DDD DDE DDT Dieldrin Heptachlor epoxide Mirex	86 100 100 71 100 57 71	86 100 100 71 100 57 71

Table 2.2. Pesticides in bed sediment and aquatic biota from rivers in the United States: State and local monitoring studies—*Continued*

Bed sediment (µg/kg dry weight)			Biota (whole fish unless noted)					Comments
			% Detection		(µg/kg wet weight)			
DL	Range	Median	Sites	Sam-ples	DL	Range	Median	
—	—	—	29*	nr	10	nd–110*	nr	
—	—	—	9*	nr	10	nd–270*	nr	
—	—	—	0*	0*	5	nd*	nd*	(2) * Data are for fish fillets.
—	—	—	100*	100*	50	50–500*	160*†	† Mean value.
—	—	—	100*	100*	nr	90–1,050*	340*†	Objective: to evaluate
—	—	—	100*	nr	5	nd–40*	10*†	contaminants in Hudson River
—	—	—	0*	0*	5	nd*	nd*	striped bass.
—	—	—	0*	0*	5	nd*	nd*	
—	—	—	0*	0*	5	nd*	nd*	
—	—	—	0*	0*	5	nd*	nd*	
—	—	—	0*	0*	5	nd*	nd*	
—	—	—	100*	nr	5	nd–10*	10*†	
—	—	—	0*	0*	5	nd*	nd*	
—	—	—	0*	0*	5	nd*	nd*	
—	—	—	0*	0*	5	nd*	nd*	
—	—	—	0*	0*	5	nd*	nd*	
—	—	—	100*	nr	5	nd–10*	nd*†	
—	—	—	0*	0*	5	nd*	nd*	
—	—	—	0*	0*	500	nd*	nd*	
—	—	—	nr	100*	10	20–370*	nr*	(3)* Data are for fish fillets
—	—	—	nr	100*	10	10–1,520*	nr*	from the St. Lawrence R. only.
—	—	—	nr	nr*	10	nd–160*	nr*	Objective: to evaluate
—	—	—	nr	nr*	10	nd–80*	nr*	contaminant conditions in the
—	—	—	nr	nr*	10	nd–540*	nr*	New York portion of the Great
—	—	—	nr	nr*	10	nd–160*	nr*	Lakes and its connecting waterways.
—	—	—	89*	nr*	10	nd–6,750*	nr*	(4) * Tissue type not reported.
—	—	—	100*	nr*	10	nd–7,040*	nr*	Objective: to survey chlordane
—	—	—	54*	nr*	10	nd–90*	nr*	residues in fish in ponds and
—	—	—	4*	nr*	10	nd–20*	nr*	waterways on Long Island.
—	—	—	45*	nr*	10	nd–140*	nr*	
—	—	—	33*	nr*	10	nd–20*	nr*	
—	—	—	33*	nr*	20	nd–110*	nr*	(5) * Tissue type not reported.
—	—	—	100*	nr*	20	nd–1,130*	nr*	Objective: various, depending on
—	—	—	33*	nr*	10	nd–90*	nr*	site and study.
—	—	—	33*	nr*	10	nd–20*	nr*	
—	—	—	33*	nr*	10	nd–50*	nr*	
1	nd–510	140	—	—	—	—	—	Objective: to measure contami-
0.1	1.9–74	50	—	—	—	—	—	nants in sediment of the upper
0.1	1.4–24	9.2	—	—	—	—	—	Rockaway R.; to determine the
0.1	nd–14	5.4	—	—	—	—	—	relation between contaminant
0.1	0.1–5.2	0.6	—	—	—	—	—	distribution and land use in the
0.1	nd–10	0.7	—	—	—	—	—	basin. Sediment at 5 developed
0.1	nd–80	23	—	—	—	—	—	sites was highly contaminated (with OCs and trace elements) relative to the 2 upstream forested sites. This relation held

Table 2.2. Pesticides in bed sediment and aquatic biota from rivers in the United States: State and local monitoring studies—*Continued*

Reference	Sampling date	Sampling location and strategy	Sampling media and species	Number of samples	Target analytes	Bed sediment % Detection Sites	Bed sediment % Detection Samples
		samples were sieved (63 μm).					
U.S. Army Corps of Engineers, 1987	1979–1983	19 sites (sediment) and 12 sites (biota) on the Black Warrior and Tombigbee rivers (Alabama), along 600 miles of the river potentially affected by dredging activities associated with channel maintenance.	Sediment (surficial), mollusks. Species: clam (*Corbicula* sp.)	19 / 21	Aldrin	0	0
					Chlordane	0	0
					DDD	0	0
					DDE	0	0
					DDT	0	0
					DDT, *o,p'*-	—	—
					Dieldrin	0	0
					Endosulfan I	0	0
					Endosulfan II	0	0
					Endosulfan sulfate	0	0
					Endrin	0	0
					Endrin aldehyde	0	0
					HCH, α-	0	0
					HCH, β-	0	0
					HCH, δ-	0	0
					Heptachlor	0	0
					Heptachlor epoxide	0	0
					Lindane	0	0
					Methoxychlor	—	—
					Mirex	—	—
					Toxaphene	0	0
Ward, 1987	1977–1979	7 sites (sub-basins) in Pequea Creek Basin, Pennsylvania.	Water, sediment (surficial, suspended).	35	Aldrin	0	nr
					Chlordane	100	nr
					D, 2,4-	0	nr
					DDD	86	nr
					DDE	100	nr
					DDT	86	nr
					Diazinon	0	nr
					Dieldrin	100	nr
					DP, 2,4-	0	nr
					Endosulfan	0	nr
					Endrin	0	nr
					Ethion	0	nr
					Heptachlor	0	nr
					Heptachlor epoxide	71	nr
					Lindane	0	nr
					Malathion	0	nr
					Methyl parathion	0	nr
					Methyl trithion	0	nr
					Mirex	0	nr
					Parathion	0	nr
					Perthane	0	nr
					Silvex	0	nr
					T, 2,4,5-	0	nr
					Toxaphene	0	nr
					Trithion	0	nr

Table 2.2. Pesticides in bed sediment and aquatic biota from rivers in the United States: State and local monitoring studies—*Continued*

Bed sediment (µg/kg dry weight)			Biota (whole fish unless noted)						Comments
DL	Range	Median	% Detection		(µg/kg wet weight)				
			Sites	Samples	DL	Range	Median		
									when concentrations were normalized for TOC.
10	nd	nd	56*	64*	nr	nd–18*	3*		* Data are for clam tissue.
50	nd	nd	100*	100*	nr	19.6–532*	160*		Objective: to do an
10	nd	nd	100*	81*	nr	nd–100*	27*		environmental impact statement
10	nd	nd	100*	100*	nr	8.9–111*	32*		for the continuation of
10	nd	nd	100*	76*	nr	nd–132*	14*		maintenance dredging of the
—	—	—	83*	81*	nr	nd–122*	3.6*		subject rivers. Bed sediment
10	nd	nd	75*	48*	nr	nd–7.54*	nd*		contamination and
20	nd	nd	83*	48*	nr	nd–22.3*	nd*		bioaccumulation in *Corbicula*
10	nd	nd	83*	48*	nr	nd–0.75*	nd*		was characterized to evaluate
50	nd	nd	0*	—	—	—	—		past effects of dredging.
10	nd	nd	0*	—	—	—	—		Concentrations of organic
50	nd	nd	0*	—	—	—	—		chemicals in bed sediment were
10	nd	nd	92*	100*	nr	2.34–64*	6.4*		below detection limits.
10	nd	nd	100*	100*	nr	2.98–19.5*	7.4*		Bioaccumulation in *Corbicula*
10	nd	nd	0*	—	—	—	—		was below FDA action levels
10	nd	nd	83*	100*	nr	1.35–37.8*	5.6*		for analytes in the upper
10	nd	nd	100*	100*	nr	6.1–24*	9.9*		reaches of the study area where
10	nd	nd	83*	57*	nr	nd–27.3*	3*		sufficient tissue samples for
—	—	—	83*	48*	nr	nd–4.05*	3.3*		analysis were collected.
—	—	—	83*	52*	nr	nd–85.2*	6.2*		
500	nd	nd	0*	0*	nr	nd*	nd*		
nr	nd	nd	—	—	—	—	—		* Value is median of the 7 site
0.1	nd–57	4*	—	—	—	—	—		medians reported.
nr	nd	nd	—	—	—	—	—		Objectives: to determine total
0.01	nd–12	0.8*	—	—	—	—	—		discharges of nutrients, sedi-
0.01	nd–45	3.2*	—	—	—	—	—		ment, and contaminants from
0.01	nd–97	3.4*	—	—	—	—	—		the basin; to determine factors
nr	nd	nd	—	—	—	—	—		affecting discharges. Median pes-
0.01	nd–8.0	1.0*	—	—	—	—	—		ticide levels in sediment were
nr	nd	nd	—	—	—	—	—		higher in the upper Pequea
nr	nd	nd	—	—	—	—	—		Creek Basin. Upper Basin
nr	nd	nd	—	—	—	—	—		samples also had higher TOC
nr	nd	nd	—	—	—	—	—		and higher silt-plus-clay
0.01	nd–2.9	nd	—	—	—	—	—		percentages. Pesticides detected
nr	nd	nd	—	—	—	—	—		in water were present at low
nr	nd	nd	—	—	—	—	—		levels during baseflows and
nr	nd	nd	—	—	—	—	—		storms. Triazine discharges from
nr	nd	nd	—	—	—	—	—		basin may temporarily increase
nr	nd	nd	—	—	—	—	—		load carried by Susquehanna R.
nr	nd	nd	—	—	—	—	—		to the Upper Chesapeake Bay by
nr	nd	nd	—	—	—	—	—		25%. Mean levels of chlordane
nr	nd	nd	—	—	—	—	—		and DDT in water exceeded USEP
nr	nd	nd	—	—	—	—	—		freshwater aquatic-life criteria.
nr	nd	nd	—	—	—	—	—		Lindane, heptachlor, and PCB
nr	nd	nd	—	—	—	—	—		slightly exceeded these criteria
nr	nd	nd	—	—	—	—	—		during storms.

Table 2.2. Pesticides in bed sediment and aquatic biota from rivers in the United States: State and local monitoring studies—*Continued*

Reference	Sampling date	Sampling location and strategy	Sampling media and species	Number of samples	Target analytes	Bed sediment % Detection	
						Sites	Samples
Arruda and others, 1988	1985	2 sites in Tuttle Creek Lake, a 21 km-long impound-ment in the Big Blue River (Kansas). Sites were located at the upper and lower ends of the impoundment. Fish samples were composites of 2–6 fish. Samples were taken in Apr., July, and Oct. (fish) and monthly from Apr. to Oct. (water).	Fish (whole, fillets), water (whole). Species: carp (whole), white bass (fillets).	8 12 14	Alachlor Atrazine Chlordane DDD, *p,p'*- DDE, *p,p'*- DDT, *o,p'*- Dieldrin HCH, α- Heptachlor epoxide Metolachlor	— — — — — — — — — —	— — — — — — — — — —
Foley and others, 1988	1982–1984	13 watersheds of New York State. Mink and otter were collected from 8 ecozones in New York. Fish were collected from the same areas, as part of New York's Toxic Substances Monitoring Program.	Fish (whole), mink (fat, liver), otter (fat, liver). Species include carp, white sucker, American eel, large-mouth bass, brown bull-head, chain pickerel, white perch.	394 66 total mink, 45 total otter	DDT, total	—	—
Illinois Environ-mental Protection Agency, 1988a	1983–1984	38 sites in the DesPlaines River Basin (Illinois). Sediment samples were sieved (<62 μm).	Sediment (surficial), ecological survey.	38	Aldrin Chlordane DDT, total Dieldrin Endrin HCB HCH, α- Heptachlor Heptachlor epoxide Lindane Methoxychlor	0 58 55 82 0 0 0 0 16 0 0	0 58 55 82 0 0 0 0 16 0 0
Illinois Environ-mental Protection Agency, 1988b	1985	5 sites in the Elkhorn River Basin (Illinois). Sediment samples were sieved (<62 μm).	Sediment (surficial), ecological survey.	5	Aldrin Chlordane DDT, total Dieldrin Endrin HCB	0 0 0 100 0 0	0 0 0 100 0 0

Table 2.2. Pesticides in bed sediment and aquatic biota from rivers in the United States: State and local monitoring studies—*Continued*

Bed sediment (µg/kg dry weight)			Biota (whole fish unless noted)					Comments
			% Detection		(µg/kg wet weight)			
DL	Range	Median	Sites	Samples	DL	Range	Median	
—	—	—	0	0	nr	nd–nr	nd*	* Mean value.
—	—	—	0	0	300–8,000	nd–nr	nd*	Objective: to investigate occurrence of pesticides in fish and
—	—	—	100	100	nr	nr–170	100*	water from Tuttle Creek Lake.
—	—	—	100	88	nr	nd–nr	8*	Whole carp had slightly higher
—	—	—	100	100	nr	nd–nr	62*	α-HCH levels than white bass
—	—	—	0	0	nr	nd–nr	nd*	fillets. The other OCs had no
—	—	—	100	100	nr	nr–170	69*	significant difference between
—	—	—	100	63	nr	nd–nr	3*	species. Levels of chlordane and
—	—	—	100	88	nr	nd–nr	8*	*p,p'*- DDD in fish were higher at
—	—	—	0	0	nr	nd–nr	nd*	upper than lower lake site; dieldrin in white bass also was higher at the upper lake site. Sampling date significantly affected tissue levels of heptachlor epoxide, dieldrin, and *p,p'*- DDE.
—	—	—	100*	nr	5–10	nr	1,500–10,000†	* Percentage of watersheds with detectable residues. † Range (13 watersheds) of geometric mean values (µg/kg-lipid). Objective: to correlate residues of DDTs and PCB in wild mink and otter with those in fish from the same areas. *p,p'*- DDE was detected in all mink and otter. OCs were highest in mink and otter near the Hudson Valley and within 5 mi of Lake Ontario. OC residues in mink and otter were positively correlated with those in fish when collection areas were 10–20 km apart.
1	nd	nd	—	—	—	—	—	Objective: to assess water and
5	nd–100	6	—	—	—	—	—	sediment quality and survey
10	nd–790	16	—	—	—	—	—	macroinvertebrates, fish, and
1	nd–45	3.9	—	—	—	—	—	habitat in the DesPlaines R.
1	nd	nd	—	—	—	—	—	Basin. Sediment from Cook
1	nd	nd	—	—	—	—	—	County (urban, industrial land
1	nd	nd	—	—	—	—	—	use) generally had higher OCs
1	nd	nd	—	—	—	—	—	than 2 other counties. Overall
1	nd–4.6	nd	—	—	—	—	—	evaluation indicated full aquatic
1	nd	nd	—	—	—	—	—	use at 6 sites, partial/minor support at 8, partial/moderate support at 27, and nonsupport at 10.
5	nd	nd	—	—	—	—	—	
1	nd	nd	—	—	—	—	—	Objective: to assess water and
5	nd	nd	—	—	—	—	—	sediment quality and survey
10	nd	nd	—	—	—	—	—	macroinvertebrates, fish, and
1	1.6–4.0	2.2	—	—	—	—	—	habitat in the Elkhorn R. Basin.
1	nd	nd	—	—	—	—	—	Overall evaluation indicated
1	nd	nd	—	—	—	—	—	partial/minor support at 4 of 5

Table 2.2. Pesticides in bed sediment and aquatic biota from rivers in the United States: State and local monitoring studies—*Continued*

Reference	Sampling date	Sampling location and strategy	Sampling media and species	Number of samples	Target analytes	Bed sediment % Detection Sites	Bed sediment % Detection Samples
					HCH, α-	0	0
					Heptachlor	0	0
					Heptachlor epoxide	0	0
					Lindane	0	0
					Methoxychlor	0	0
Illinois Environmental Protection Agency, 1988c	1984	6 sites in the Kyte River Basin (Illinois). Sediment samples were sieved (<62 μm).	Sediment (surficial), ecological survey.	6	Aldrin	0	0
					Chlordane	0	0
					DDT, total	17	17
					Dieldrin	67	67
					Endrin	0	0
					HCB	0	0
					HCH, α-	0	0
					Heptachlor	0	0
					Heptachlor epoxide	0	0
					Lindane	0	0
					Methoxychlor	0	0
Illinois Environmental Protection Agency, 1988d	1984	23 sites in the Pecatonica River and its tributaries (Illinois). Sediment samples were sieved (<62 μm).	Sediment (surficial), ecological survey.	23	Aldrin	0	0
					Chlordane	4	4
					DDT, total	0	0
					Dieldrin	57	57
					Endrin	0	0
					HCB	0	0
					HCH, α-	0	0
					Heptachlor	0	0
					Heptachlor epoxide	0	0
					Lindane	0	0
					Methoxychlor	0	0
Irwin, 1988	1985	27 sites along a 250 mile reach of the Trinity River (Texas), above and below sewage treatment plant discharges, industrial discharges, and nonpoint source inputs from the Dallas–Fort Worth metropolitan area. Mosquito fish were collected at all sites. Additional species were collected at 2 sites. Samples were composites.	Fish (whole), mollusks, invertebrates, reptiles. Species include carp, mosquitofish, channel catfish, longnose gar, redfin shiners, blue catfish, freshwater drum, crayfish, Asiatic clams, unionid clams, snapping turtle, spiny	64 (total for all biota)	Chlordane, *cis*-	—	—
					Chlordane, *trans*-	—	—
					Dacthal	—	—
					DDD, *o,p'*-	—	—
					DDD, *p,p'*-	—	—
					DDE, *o,p'*-	—	—
					DDE, *p,p'*-	—	—
					DDT, *o,p'*-	—	—
					DDT, *p,p'*-	—	—
					Dieldrin	—	—
					Endosulfan I	—	—
					Endosulfan II	—	—
					Endosulfan sulfate	—	—
					Endrin	—	—
					HCB	—	—
					HCH, α-	—	—
					HCH, β-	—	—
					HCH, δ-	—	—
					Heptachlor epoxide	—	—
					Lindane	—	—
					Mirex	—	—
					Nonachlor, *cis*-	—	—
					Nonachlor, *trans*	—	—

Table 2.2. Pesticides in bed sediment and aquatic biota from rivers in the United States: State and local monitoring studies—*Continued*

| Bed sediment (µg/kg dry weight) | | | Biota (whole fish unless noted) | | | | | |
| DL | Range | Median | % Detection | | (µg/kg wet weight) | | | Comments |
			Sites	Samples	DL	Range	Median	
1	nd	nd	—	—	—	—	—	sites and partial/moderate
1	nd	nd	—	—	—	—	—	support at 1 site.
1	nd	nd	—	—	—	—	—	
1	nd	nd	—	—	—	—	—	
5	nd	nd	—	—	—	—	—	
1	nd	nd	—	—	—	—	—	Objective: to assess water and
5	nd	nd	—	—	—	—	—	sediment quality and survey
10	nd–13	nd	—	—	—	—	—	macroinvertebrates, fish, and
1	nd–7.5	1.75	—	—	—	—	—	habitat in the Kyte R. Basin.
1	nd	nd	—	—	—	—	—	Overall evaluation indicated full
1	nd	nd	—	—	—	—	—	aquatic use at all 6 sites.
1	nd	nd	—	—	—	—	—	
1	nd	nd	—	—	—	—	—	
1	nd	nd	—	—	—	—	—	
1	nd	nd	—	—	—	—	—	
5	nd	nd	—	—	—	—	—	
1	nd	nd	—	—	—	—	—	Objective: to assess water and
5	nd–9.7	nd	—	—	—	—	—	sediment quality and survey
10	nd	nd	—	—	—	—	—	macroinvertebrates, fish, and
1	nd–4.1	1.2	—	—	—	—	—	habitat in the Pecatonica R.
1	nd	nd	—	—	—	—	—	Basin. Overall evaluation of 22
1	nd	nd	—	—	—	—	—	sites indicated full aquatic use at
1	nd	nd	—	—	—	—	—	13 sites, partial/minor support
1	nd	nd	—	—	—	—	—	at 7, and partial/moderate
1	nd	nd	—	—	—	—	—	support at 2. Dieldrin residues
1	nd	nd	—	—	—	—	—	were higher in tributary than
5	nd	nd	—	—	—	—	—	in mainstem sediments.
—	—	—	—	67	10	nr	nr	* Data are for whole body
—	—	—	—	66	10	nr	nr	tissues of all species sampled.
—	—	—	—	nr	nr	nr	nr	Objective: to determine occur-
—	—	—	—	nr	nr	nr	nr	rence and distribution of contam-
—	—	—	—	42	10	nd–170	nr	inants in biota of the study area.
—	—	—	—	nr	nr	nr	nr	PCBs, chlordane, lead, and mer-
—	—	—	—	95	10	nd–850	nr	cury were the most consistently
—	—	—	—	nr	nr	nr	nr	elevated contaminants in fish
—	—	—	—	nr	nr	nr	nr	and wildlife of the upper Trinity
—	—	—	—	77	10	nr	nr	R. Residues in about half the
—	—	—	—	nr	nr	nr	nr	samples were high enough to
—	—	—	—	nr	nr	nr	nr	justify concern for predatory
—	—	—	—	nr	nr	nr	nr	species. High levels of dieldrin,
—	—	—	—	nr	nr	nr	nr	PAHs, and chromium were
—	—	—	—	nr	nr	nr	nr	frequently detected at sites just
—	—	—	—	nr	nr	nr	nr	downstream of Dallas. Lindane
—	—	—	—	2	10	nd–10	nd	was detected downstream of
—	—	—	—	nr	nr	nr	nr	sewage treatment plants. Lead
—	—	—	—	8	10	nr	nr	and PCBs were significantly
—	—	—	—	11	10	nr	nr	higher at sites receiving runoff
—	—	—	—	5	10	nd–10	nd	from industrial areas, and
—	—	—	—	34	10	nr	nr	dieldrin and chlordane at sites
—	—	—	—	34	10	nr	nr	receiving runoff from residential

Table 2.2. Pesticides in bed sediment and aquatic biota from rivers in the United States: State and local monitoring studies—*Continued*

Reference	Sampling date	Sampling location and strategy	Sampling media and species	Number of samples	Target analytes	Bed sediment % Detection	
						Sites	Samples
			softshell turtle, Mississippi map turtle.		Oxychlordane Toxaphene	— —	— —
Johnson and others, 1988b	1985	11 sites (sediment) and 4 sites (fish) on Yakima River, Washington. Irrigation drainage areas were targeted for sampling. Fish samples were composites.	Water (whole), sediment (fine surficial), fish (whole). Species: mountain whitefish, bridgelip sucker, northern squawfish, largescale sucker.	13 11	DDD, *o,p'-* DDD, *p,p'-* DDE, *o,p'-* DDE, *p,p'-* DDT, *o,p'-* DDT, *p,p'-* DDT, total	0 91 0 100 91 91 100	0 92 0 100 92 92 100
Knapton and others, 1988	1986	3 sites (sediment) and 6 sites (fish) in rivers and lakes in Sun River drainage area, Montana. Study area is irrigation drainage area within wildlife refuge area.	Water, sediment (surficial), fish (whole), birds. Fish species include: brown trout, carp, white sucker, yellow perch.	3 9	Chlordane, *cis-* Chlordane, *trans-* D, 2,4- DDD, *p,p'-* DDE, *p,p'-* DDT, *p,p'-* Dicamba Dieldrin DP, 2,4- Endrin Heptachlor epoxide Nonachlor, *cis-* Nonachlor, *trans-* Oxychlordane Picloram Silvex T, 2,4,5-	— — 0 — — — 0 — 0 — — — — — 0 0 0	— — 0 — — — 0 — 0 — — — — — 0 0 0

Table 2.2. Pesticides in bed sediment and aquatic biota from rivers in the United States: State and local monitoring studies—*Continued*

Bed sediment (µg/kg dry weight)			Biota (whole fish unless noted)					Comments
			% Detection		(µg/kg wet weight)			
DL	Range	Median	Sites	Samples	DL	Range	Median	
—	—	—	—	33	10	nr	nr	areas. Sites dominated by
—	—	—	—	nr	nr	nr	nr	pollution-tolerant species and low species diversity correlated with urban runoff inputs.
0.1	nd	nd	0	0	40	nd	nd	Objective: to show persistence,
0.1	nd–37	2.2	25	18	40	nd–140	nd	trends, and sources of DDT, and
0.1	nd	nd	25	9	20	nd–30	nd	the effect of irrigation on DDT.
0.1	0.1–129	7	100	100	20	50–2,900	540	High DDE levels detected are
0.1	nd–8.4	0.71	0	0	40	nd	nd	consistent with historical use in
0.1	nd–59.2	3.4	75	55	40	nd–120	50	the study area. In some
0.1	0.1–234	11.6	100	100	40	50–3,000	560	tributaries, *p,p'-* DDT levels were greater than *p,p'-* DDE. Also, *p,p'* / *o,p'* DDT ratio was similar to that in technical DDT (5:1). These observations suggest *o,p'-* and *p,p'-* DDT degrade very slowly in Yakima soils. DDTs were routinely detected in whole water from several tributaries during irrigation (June–Aug.) but not after irrigation (Oct.). There was an increasing downstream trend for 3 species of fish, but not for sediment. Variability in fish residues cannot be explained by differences in lipid content. No relation was evident between estimated DDT concentrations in suspended solids (June–Aug.) and measured DDT concentrations in fine surficial sediment (Sept.).
—	—	—	0	0	nr	nd	nd	Objectives: to describe
—	—	—	0	0	nr	nd	nd	concentrations of constituents in
0.1	nd	nd	—	—	—	—	—	water, bed sediment, biota for
—	—	—	33	33	nr	nd–17	nd	comparison to guidelines and
—	—	—	33	33	nr	nd–25	nd	baseline information; to
—	—	—	67	67	nr	nd–27	nd	determine whether irrigation has
0.1	nd	nd	—	—	—	—	—	caused or may cause harmful
—	—	—	0	0	nr	nd	nd	effects on humans, fish,
0.1	nd	nd	—	—	—	—	—	wildlife, or beneficial uses of
—	—	—	0	0	nr	nd	nd	water. All bird eggs have
—	—	—	0	0	nr	nd	nd	detectable *p,p'-* DDE
—	—	—	0	0	nr	nd	nd	concentrations. Focus of study
—	—	—	0	0	nr	nd	nd	was on trace elements,
—	—	—	0	0	nr	nd	nd	especially selenium.
0.1	nd	nd	—	—	—	—	—	
0.1	nd	nd	—	—	—	—	—	
0.1	nd	nd	—	—	—	—	—	

Table 2.2. Pesticides in bed sediment and aquatic biota from rivers in the United States: State and local monitoring studies—*Continued*

Reference	Sampling date	Sampling location and strategy	Sampling media and species	Number of samples	Target analytes	Bed sediment % Detection Sites	Bed sediment % Detection Samples
Lambing and others, 1988	1986	1 (fish, sediment), 3 (water), and 4 (birds) sites in Bowdoin National Wildlife Refuge and adjacent areas of Milk River Basin (Montana). Sediment was sampled in Aug., when water levels had receded.	Sediment (surficial), fish (whole), birds (eggs, liver), water (whole). Fish species: white sucker, walleye.	1 2 4 3 5	Chlordane, *cis-* Chlordane, *trans-* D, 2,4- DDD, *p,p'-* DDE, *p,p'-* DDT, *p,p'-* Dicamba Dieldrin DP, 2,4- Endrin Nonachlor, *cis-* Nonachlor, *trans-* Oxychlordane Picloram Silvex T, 2,4,5-	— — 0 — — — 0 — 0 — — — — 0 0 0	— — 0 — — — 0 — 0 — — — — 0 0 0
Lewis and Makarewicz, 1988	1983–1984	3 sites in tributaries to Lake Ontario (New York). 1 site was accessible to migrating fish from the lake and 2 control sites were inaccessible because of a dam. Fish samples were individuals or composites. Tissues analyzed were fillets and sex products (salmon) or whole-body (all other species).	Fish (whole, fillets, eggs, milt). Species: coho salmon, chinook salmon, creek chub, smallmouth bass, bluntnose minnow, white sucker, brown bullhead.	40 4 1 1	Mirex	—	—
Matson and Hite, 1988	1986	3 sites in the Eagle Creek watershed (Illinois). Sieved (<63 µm) and unsieved sediments were analyzed.	Sediment (surficial).	3	Aldrin Chlordane, *cis-* Chlordane, total Chlordane, *trans-* DDD, *o,p'-* DDD, *p,p'-* DDE, *o,p'-* DDE, *p,p'-* DDT, *o,p'-* DDT, total Dieldrin	0 100 100 100 0 0 0 0 0 0 100	0 100 100 100 0 0 0 0 0 0 100

Table 2.2. Pesticides in bed sediment and aquatic biota from rivers in the United States: State and local monitoring studies—*Continued*

Bed sediment (µg/kg dry weight)			Biota (whole fish unless noted)					Comments
			% Detection		(µg/kg wet weight)			
DL	Range	Median	Sites	Sam-ples	DL	Range	Median	
—	—	—	0	0	nr	nd	nd	Objective: to describe
—	—	—	0	0	nr	nd	nd	concentrations of contaminants
0.1	nd	nd	—	—	—	—	—	in water, sediment, and biota of
—	—	—	0	0	nr	nd	nd	Bowdoin National Wildlife
—	—	—	0	0	nr	nd	nd	Refuge; to determine whether
—	—	—	0	0	nr	nd	nd	irrigation drainage has adversely
0.1	nd	nd	—	—	—	—	—	affected humans, fish, wildlife,
—	—	—	0	0	nr	nd	nd	or the suitability of water for
0.1	nd	nd	—	—	—	—	—	beneficial use. Low
—	—	—	0	0	nr	nd	nd	concentrations of 2,4-D,
—	—	—	0	0	nr	nd	nd	dicamba, and picloram were
—	—	—	0	0	nr	nd	nd	detected in water samples. No
—	—	—	0	0	nr	nd	nd	pesticides were detected in fish,
0.1	nd	nd	—	—	—	—	—	sediment, or bird livers.
0.1	nd	nd	—	—	—	—	—	*p,p'*- DDE was detected in 2 bird
0.1	nd	nd	—	—	—	—	—	egg samples.
—	—	—	33	30	nr	nd–25	nd	Objective: to estimate mirex loading to Marsh Creek (undammed tributary to Lake Ontario) and contamination of resident fish due to spawning migration of contaminated salmon. All contaminated fish were from Marsh Creek (52% contained detectable mirex); no fish collected at control sites had detectable mirex. Estimated mirex loading to Marsh Creek was 69 mg/km stream (based on residues in salmon carcasses found on Marsh Creek). Estimated mirex transported to all Lake Ontario tributaries from fall 1983 spawning salmon was 53–121 g, or 0.02% of total mirex load in Lake Ontario. From angler harvest, about 26–60 g/yr mirex would be removed from the Lake Ontario ecosystem.
nr	nr	nr	—	—	—	—	—	* Detected at low levels in both
nr	nr*	nr	—	—	—	—	—	sieved and unsieved samples.
nr	nr*	nr	—	—	—	—	—	Objective: to assess water
nr	nr*	nr	—	—	—	—	—	quality in the Eagle Creek
nr	nr	nr	—	—	—	—	—	watershed. Study included fish
nr	nr	nr	—	—	—	—	—	and macroinvertebrate surveys
nr	nr	nr	—	—	—	—	—	and biological stream
nr	nr	nr	—	—	—	—	—	characterization. Aquatic life
nr	nr	nr	—	—	—	—	—	support ranged from full support
nr	nr	nr	—	—	—	—	—	to partial support with moderate
nr	nr*	nr	—	—	—	—	—	impairment.

Table 2.2. Pesticides in bed sediment and aquatic biota from rivers in the United States: State and local monitoring studies—*Continued*

Reference	Sampling date	Sampling location and strategy	Sampling media and species	Number of samples	Target analytes	Bed sediment % Detection	
						Sites	Samples
					Endrin	0	0
					HCB	0	0
					HCH, α-	0	0
					Heptachlor	0	0
					Heptachlor epoxide	100	100
					Lindane	0	0
					Methoxychlor	0	0
					Nonachlor, *cis-*	0	0
					Nonachlor, *trans-*	0	0
					Pentachlorophenol	0	0
Pereira and others, 1988	nr	3 sites (catfish) and 1 site (other biota, water, surficial and suspended sediment) in the Calcasieu River estuary (Louisiana). The 1st site was near a chemical plant. Concentrations were normalized to lipid content (biota) or TOC (sediment).	Sediment (surficial, suspended), water (whole), fish (whole), shellfish. Species: blue catfish, Atlantic croaker, spotted sea trout, blue crab.	1 1 1 5 1	HCB	100	100
Peterson and others, 1988	1986– 1987	3 sites (sediment) and 5 sites (fish) in the Kendrick Reclamation Area (Wyoming). Sites are inflow and irrigation drainage sites.	Water (whole, dissolved), sediment (surficial), fish (whole), birds (eggs). Fish species: rainbow trout, carp, fathead minnow, white sucker, longnose sucker.	3 nr	Chlordane	—	—
					D, 2,4-	0	0
					DDD	—	—
					DDE	—	—
					DDT	—	—
					Diazinon	0	0
					Dicamba	0	0
					Dieldrin	—	—
					DP, 2,4-	0	0
					Endrin	—	—
					Ethion	0	0
					Malathion	0	0
					Methyl parathion	0	0
					Methyl trithion	0	0
					Nonachlor	—	—
					Parathion	0	0
					Picloram	0	0

Table 2.2. Pesticides in bed sediment and aquatic biota from rivers in the United States: State and local monitoring studies—*Continued*

Bed sediment (µg/kg dry weight)			Biota (whole fish unless noted)						Comments
			% Detection		(µg/kg wet weight)				
DL	Range	Median	Sites	Sam-ples	DL	Range	Median		Comments
nr	nr	nr	—	—	—	—	—		
nr	nr	nr	—	—	—	—	—		
nr	nr	nr	—	—	—	—	—		
nr	nr	nr	—	—	—	—	—		
nr	nr*	nr	—	—	—	—	—		
nr	nr	nr	—	—	—	—	—		
nr	nr	nr	—	—	—	—	—		
nr	nr	nr	—	—	—	—	—		
nr	nr	nr	—	—	—	—	—		
nr	nr	nr	—	—	—	—	—		
nr	7.5×10^6*	7.5×10^6*	100	100	nr	610–27,000†	7,700†		* µg/kg–sediment organic carbon † µg/kg–lipid. Objective: to evaluate chlorobenzene (CB) residues in various environmental compartments; to assess impacts of system dynamics on CB distribution. CB residues: surficial sediment>> suspended sediment> water. Lipid-based BAFs (catfish/water) were in agreement with literature values. Biota–sediment factors (BSAFs) were calculated from normalized catfish and sediment data. BSAFs were <1, suggesting that residues in biota are below equilibrium and desorption from sediment is slow and rate-limiting. Distributions between water, suspended sediment, and biota are much closer to equilibrium.
—	—	—	0	0	nr	nd	nd		Objective: to assess occurrence of trace elements and organic contaminants to evaluate impact of irrigation drainage on water quality. No pesticides were detected in surface water at 10 sites or in groundwater at 5 sites. DDE was the only organochlorine compound detected in 5 bird eggs collected from Rasmus Lee and Goose Lake. Its concentration ranged from 0.23 to 2.8 mg/kg wet weight.
0.1	nd	nd	—	—	—	—	—		
—	—	—	0	0	nr	nd	nd		
—	—	—	0	0	nr	nd	nd		
—	—	—	0	0	nr	nd	nd		
0.1	nd	nd	—	—	—	—	—		
0.1	nd	nd	—	—	—	—	—		
—	—	—	0	0	nr	nd	nd		
0.1	nd	nd	—	—	—	—	—		
—	—	—	0	0	nr	nd	nd		
0.1	nd	nd	—	—	—	—	—		
0.1	nd	nd	—	—	—	—	—		
0.1	nd	nd	—	—	—	—	—		
0.1	nd	nd	—	—	—	—	—		
—	—	—	0	0	nr	nd	nd		
0.1	nd	nd	—	—	—	—	—		
0.1	nd	nd	—	—	—	—	—		

Table 2.2. Pesticides in bed sediment and aquatic biota from rivers in the United States: State and local monitoring studies—*Continued*

Reference	Sampling date	Sampling location and strategy	Sampling media and species	Number of samples	Target analytes	Bed sediment % Detection Sites	Bed sediment % Detection Samples
					Silvex	0	0
					T, 2,4,5-	0	0
					Trithion	0	0
PTI Environmental Services and Tetra Tech Inc., 1988	1986	57 sites (sediment) and 11 sites (biota) in Everett Harbor, including Port Gardner and lower Snohomish River (Washington). These sites include 4 reference sites, 3 in Puget Sound (for sediment) and 1 in Port Susan (for biota). Crabs were male.	Sediment (surficial), fish (skin-off muscle), crab. Species: English sole, Dungeness crab.	66 55 11	Aldrin	0	0
					Chlordane	0	0
					DDD, *p,p'-*	0	0
					DDE, *p,p'-*	0	0
					DDT, *p,p'-*	2	2
					Dieldrin	0	0
					Endosulfan I	0	0
					Endosulfan II	0	0
					Endosulfan sulfate	0	0
					Endrin	0	0
					Endrin ketone	0	0
					HCB	0	0
					HCH, α-	0	0
					HCH, β-	0	0
					HCH, δ-	0	0
					Heptachlor	0	0
					Heptachlor epoxide	0	0
					Lindane	2	2
					Methoxychlor	0	0
					Pentachlorophenol	37	33
					Toxaphene	0	0
					12 OC pesticides*	—	—
Radtke and others, 1988	1986	8 sites (sediment) and 10 sites (biota) in the lower Colorado River valley (Arizona, California, Nevada). Sediment samples were unsieved composites of 6 or more subsamples. Triplicate biota samples, each a composite of 3–5 individuals, were collected.	Water, sediment (surficial), fish (whole), mollusks, birds (whole). Aquatic species: carp, spiny naiad.	8 31 27 9	Aldrin	0	0
					Chlordane	25	20
					Chlordane, *cis-*	—	—
					Chlordane, *trans-*	—	—
					DDD	75	60
					DDE	100	100
					DDT	38	30
					Dieldrin	0	0
					Endosulfan	0	0
					Endrin	0	0
					Heptachlor	0	0
					Heptachlor epoxide	0	0
					Lindane	0	0
					Methoxychlor	0	0
					Mirex	0	0
					Nonachlor, *cis-*	—	—
					Nonachlor, *trans-*	—	—
					Oxychlordane	—	—
					Perthane	0	0
					Toxaphene	0	0
Schroeder and others, 1988	1986	9 sites in Kern National Wildlife Refuge (NWR), Pixley NWR, and Westfarmers	Water (whole), sediment (surficial), fish (whole),	12 nr	Aldrin	0	0
					Chlordane	0	0
					Chlordane, *cis-*	—	—
					Chlordane, *trans-*	—	—
					D, 2,4-	0	0

Table 2.2. Pesticides in bed sediment and aquatic biota from rivers in the United States: State and local monitoring studies—*Continued*

Bed sediment (µg/kg dry weight)			Biota (whole fish unless noted)					Comments
			% Detection		(µg/kg wet weight)			
DL	Range	Median	Sites	Samples	DL	Range	Median	
0.1	nd	nd	—	—	—	—	—	
0.1	nd	nd	—	—	—	—	—	
0.1	nd	nd	—	—	—	—	—	
0.1–5	nd	nd	—	—	—	—	—	* Not specified.
0.1–5	nd	nd	—	—	—	—	—	† Data are for muscle tissue
0.1–5	nd	nd	—	—	—	—	—	(skin-off).
0.1–5	nd	nd	—	—	—	—	—	Objective: to identify spatial
1–10	nd–23	nd	—	—	—	—	—	patterns on contaminant
0.1–5	nd	nd	—	—	—	—	—	distribution, sediment toxicity,
0.1–5	nd	nd	—	—	—	—	—	and biological reponses in the
0.1–5	nd	nd	—	—	—	—	—	study area. Problem areas were
0.1–5	nd	nd	—	—	—	—	—	identified based on occurrence
0.1–5	nd	nd	—	—	—	—	—	of sediment contamination or
0.1–5	nd	nd	—	—	—	—	—	toxicity, benthic community
0.1–5	nd	nd	—	—	—	—	—	structure, fish pathology, or
0.1–5	nd	nd	—	—	—	—	—	bioaccumulation. None of the
0.1–5	nd	nd	—	—	—	—	—	12 OC pesticides (which were
0.1–5	nd	nd	—	—	—	—	—	not specified) that were
0.1–5	nd	nd	—	—	—	—	—	analyzed for in fish and crab
0.5–5	nd–1.0	nd	—	—	—	—	—	were detected. Pesticides in
0.1–5	nd	nd	—	—	—	—	—	sediment and biota in the study
0.1–5	nd–460	nr	—	—	—	—	—	area were very low.
0.1–5	nd	nd	—	—	—	—	—	
—	—	—	0†	0†	0.1–0.8	nd†	nd†	
0.1*	nd	nd	—	—	—	—	—	* Data are in µg/kg wet weight.
1*	nd–1*	nd	—	—	—	—	—	Objective: to determine if
—	—	—	0	0	nr	nd	nd	elevated concentrations of
—	—	—	0	0	nr	nd	nd	contaminants exist in the
0.1*	nd–2.4*	0.2*	0	0	nr	nd	nd	study area and if they are related
nr	0.1–7.5*	2.3*	90	81	10	nd–380	82	to irrigation drainage. Irrigation
0.1*	nd–0.8*	nd	0	0	nr	nd	nd	drainage did not appear to
0.1*	nd	nd	0	0	nr	nd	nd	contribute to trace element
0.1*	nd	nd	—	—	—	—	—	contamination in the study
0.1*	nd	nd	0	0	nr	nd	nd	area. OC pesticides detected
0.1*	nd	nd	—	—	—	—	—	most likely result from the past
0.1*	nd	nd	0	0	nr	nd	nd	use. OC pesticides in fish were
0.1*	nd	nd	—	—	—	—	—	below levels of concern. DDE
0.1*	nd	nd	—	—	—	—	—	residues were 40-fold greater in
0.1*	nd	nd	—	—	—	—	—	cormorants than in carp and
—	—	—	0	0	nr	nd	nd	exceeded NAS/NAE guidelines
—	—	—	0	0	nr	nd	nd	for protection of wildlife. No
—	—	—	0	0	nr	nd	nd	OC pesticides were detected in
1*	nd	nd	—	—	—	—	—	water.
10*	nd	nd	—	—	—	—	—	
0.1	nd	nd	—	—	—	—	—	* Median of 7 geometric means
1	nd	nd	—	—	—	—	—	for each species/site group.
—	—	—	67	nr	nr	nd–16	nd*	† One sample had detection
—	—	—	67	nr	nr	nd–14	nd*	limit of <45 µg/kg.
0.1	nd	nd	—	—	—	—	—	Objective: to determine whether

Table 2.2. Pesticides in bed sediment and aquatic biota from rivers in the United States: State and local monitoring studies—*Continued*

Reference	Sampling date	Sampling location and strategy	Sampling media and species	Number of samples	Target analytes	Bed sediment % Detection	
						Sites	Samples
		evaporation pond in the Tulare Lake Bed area (California).	aquatic insects, aquatic plants, birds (eggs, liver). Aquatic species: mosquito-fish, carp, yellow bullhead, water boat-man, widgeon grass.	nr nr nr nr	DDD	50	44
					DDE	100	100
					DDT	13	11
					Diazinon	63	67
					Dicamba	0	0
					Dieldrin	38	44
					DP, 2,4-	0	0
					Endosulfan	0	0
					Endrin	0	0
					Ethion	0	0
					Heptachlor	13	11
					Heptachlor epoxide	13	11
					Lindane	0	0
					Malathion	0	0
					Methoxychlor	0	0
					Methyl parathion	0	0
					Methyl trithion	0	0
					Mirex	0	0
					Nonachlor, *cis-*	—	—
					Nonachlor, *trans-*	—	—
					Oxychlordane	—	—
					Parathion	0	0
					Perthane	0	0
					Picloram	0	0
					Silvex	0	0
					T, 2,4,5-	0	0
					Toxaphene	0	0
					Trithion	0	0
Short, 1988	1987	22 sites in the Mackinaw River Basin (Illinois). Sediment samples were sieved (<63 μm).	Sediment (surficial), ecological survey.	22	Aldrin	0	0
					Chlordane	5	5
					DDT, total	0	0
					Dieldrin	55	55
					Endrin	0	0
					HCB	0	0
					HCH, α-	0	0
					Heptachlor	0	0
					Heptachlor epoxide	9	9
					Lindane	0	0
					Methoxychlor	0	0
Steffeck, 1988	1987	5 sites near the Four County Landfill (FCL), a hazardous waste landfill in Indiana and 1 control site (Lake Maxintuckee). Sites were in wetland areas where surface water runoff had occurred from	Fish (whole), amphibians, mammals. Species include: black bullhead, brown bullhead, yellow bullhead,	10 7 6	Chlordane, α-	—	—
					Chlordane, γ-	—	—
					DDD, *o,p'-*	—	—
					DDD, *p,p'-*	—	—
					DDE, *o,p'-*	—	—
					DDE, *p,p'-*	—	—
					DDT, *o,p'-*	—	—
					DDT, *p,p'-*	—	—
					DDT, total	—	—
					Dieldrin	—	—
					Endrin	—	—

Table 2.2. Pesticides in bed sediment and aquatic biota from rivers in the United States: State and local monitoring studies—*Continued*

Bed sediment (µg/kg dry weight)			Biota (whole fish unless noted)					Comments
			% Detection		(µg/kg wet weight)			
DL	Range	Median	Sites	Sam-ples	DL	Range	Median	
0.1	nd–2.0	nd	100	nr	nr	nd–110	31*	chemicals in irrigation drainage
nr	0.2–8.5	0.8	100	nr	nr	43–370	140*	threaten wildlife on or near Kern
0.1	nd–0.2	nd	67	nr	nr	nd–610	nd*	NWR. The highest DDD and
0.1	nd–0.2	0.1	—	—	—	—	—	DDE levels were at the 2 sites
0.1	nd	nd	—	—	—	—	—	most likely to receive surface
0.1	nd–1.1	nd	100	nr	nr	nd–190	50*	runoff or tailwater from fields.
0.1	nd	nd	—	—	—	—	—	Diazinon, prometryne, atrazine,
0.1	nd	nd	—	—	—	—	—	and 2,4-D were found in water,
0.1	nd	nd	0	nr	nr	nd	nd	but at low concentrations.
0.1	nd	nd	—	—	—	—	—	Sediment samples were
0.1	nd–0.1	nd	—	—	—	—	—	composits of 10–15 subsamples
0.1	nd–0.2	nd	0	nr	nr	nd	nd	per site and sieved (<62 µm).
0.1	nd	nd	—	—	—	—	—	Fish were composited (5–50
0.1	nd	nd	—	—	—	—	—	individuals per sample,
0.1 †	nd	nd	—	—	—	—	—	depending on species).
0.1	nd	nd	—	—	—	—	—	
0.1	nd	nd	—	—	—	—	—	
0.1	nd	nd	—	—	—	—	—	
—	—	—	0	nr	nr	nd	nd	
—	—	—	33	nr	nr	nd–50	nd*	
—	—	—	0	nr	nr	nd	nd	
0.1	nd	nd	—	—	—	—	—	
1	nd	nd	—	—	—	—	—	
0.1	nd	nd	—	—	—	—	—	
0.1	nd	nd	—	—	—	—	—	
0.1	nd	nd	—	—	—	—	—	
10	nd	nd	—	—	—	—	—	
0.1	nd	nd	—	—	—	—	—	
1	nd	nd	—	—	—	—	—	Objective: to assess water and
5	nd–5.0	nd	—	—	—	—	—	sediment quality and survey
10	nd	nd	—	—	—	—	—	macroinvertebrates, fish, and
1	1.9–7.9	5.9	—	—	—	—	—	habitat in the Mackinaw R.
1	nd	nd	—	—	—	—	—	Basin. Overall evaluation of 22
1	nd	nd	—	—	—	—	—	sites indicated full aquatic use at
1	nd	nd	—	—	—	—	—	18, partial/minor support at 3,
1	nd	nd	—	—	—	—	—	and partial/moderate support at 1
1	nd–1.4	nd	—	—	—	—	—	site.
1	nd	nd	—	—	—	—	—	
5	nd	nd	—	—	—	—	—	
—	—	—	50	50	10	nd–30	nd/10	Objective: to determine if biota
—	—	—	0	0	10	nd	nd	near a hazardous waste landfill
—	—	—	0	0	10	nd	nd	(FCL) had elevated contaminant
—	—	—	50	50	10	nd–70	nd/20	levels relative to a control site;
—	—	—	0	0	10	nd	nd	to determine if surrounding areas
—	—	—	75	80	10	nd–160	10	were impacted by contaminants.
—	—	—	0	0	10	nd	nd	OC pesticides were at or below
—	—	—	0	0	10	nd	nd	detection limits at sites near
—	—	—	75	80	10	nd–220	20	FCL, indicating these
—	—	—	75	80	10	nd–30	10	compounds were not migrating
—	—	—	0	0	10	nd	nd	off-site. Fish from control site

Table 2.2. Pesticides in bed sediment and aquatic biota from rivers in the United States: State and local monitoring studies—*Continued*

Reference	Sampling date	Sampling location and strategy	Sampling media and species	Number of samples	Target analytes	Bed sediment % Detection Sites	Bed sediment % Detection Samples
		the FCL. Biota samples were composites of 5 or more individuals.	white crappie, black crappie, walleye, white bass, carp, central mudminnow, green frog.		HCB HCH (4 isomers) Heptachlor epoxide Mirex Nonachlor, *cis-* Nonachlor, *trans-* Oxychlordane Toxaphene	— — — — — — — —	— — — — — — — —
Wells and others, 1988	1986	4 sites (sediment), 9 sites (fish), and 15 sites (water) in the lower Rio Grande Valley, including Laguna Atascosa National Wildlife Refuge (NWR), Texas. One (reference) site was upstream of irrigation drainage influences. Sediment samples were composites of 3 dredge grabs.	Sediment (surficial), water (whole), fish (whole), crabs, turtles, plants, birds. Species include: gizzard shad, freshwater drum, channel catfish, blue catfish, sea catfish, carp, tilapia, sheepshead minnow, gulf killifish, largemouth bass, striped bass, alligator gar, blue crab, *Chara* sp.	4 18 22 5 7 2, 1	Aldrin Chlordane Chlordane, *cis-* Chlordane, *trans-* DDD DDD, *p,p'-* DDE DDE, *p,p'-* DDT DDT, *p,p'-* Diazinon Dieldrin Endosulfan Endrin Ethion Heptachlor Heptachlor epoxide Lindane Malathion Methoxychlor Methyl parathion Methyl trithion Mirex Nonachlor, *cis-* Nonachlor, *trans-* Oxychlordane Parathion Perthane Toxaphene Trithion	0 25 — — 50 — 100 — 25 — 0 25 0 0 0 0 0 0 0 0 0 0 0 — — — 0 0 0 0	0 25 — — 50 — 100 — 25 — 0 25 0 0 0 0 0 0 0 0 0 0 0 — — — 0 0 0 0
Bopp and Simpson, 1989	1975–1986	Unspecified number of sites in Hudson River and New York Harbor (New York). Sediment cores were sectioned at 2- to 4-cm intervals. Radionuclides were used to date	Sediment (cores).	>200	Chlordane DDD, *p,p'-*	nr nr	nr nr

Table 2.2. Pesticides in bed sediment and aquatic biota from rivers in the United States: State and local monitoring studies—*Continued*

| Bed sediment (µg/kg dry weight) | | | Biota (whole fish unless noted) | | | | | Comments |
| | | | % Detection | | (µg/kg wet weight) | | | |
DL	Range	Median	Sites	Samples	DL	Range	Median	
—	—	—	0	0	10	nd	nd	had low OC levels (near or
—	—	—	0	0	10	nd	nd	below national mean values
—	—	—	0	0	10	nd	nd	from the FWS's NCBP). This
—	—	—	0	0	10	nd	nd	study could not explain: (1) dead
—	—	—	50	50	10	nd–20	nd/10	trees and reduced number of
—	—	—	50	60	10	nd–50	15	animal species in drainageway
—	—	—	0	0	10	nd	nd	between FCL and King Lake; or
—	—	—	0	0	10	nd	nd	(2) reduction in fish species diversity and production in Tippecanoe R. downstream of FCL.
0.1	nd	nd	—	—	—	—	—	Objective: to collect
1	nd–4	nd	—	—	—	—	—	water-quality data for the Lower
—	—	—	0	0	10	nd	nd	Rio Grande Valley–Laguna
—	—	—	0	0	10	nd	nd	Atascosa NWR area to use in
0.1	nd–9.7	nd	—	—	—	—	—	evaluating effects of irrigation
—	—	—	100	95	10	nd–180	22	waters on human health, fish,
0.1	0.2–34	nd	—	—	—	—	—	wildlife, or other beneficial
—	—	—	100	100	10	36–9,900	380	water uses. Maximum residues
0.1	nd–0.2	nd	—	—	—	—	—	of DDD, DDE, and DDT in fish
—	—	—	67	50	10	nd–66	21	exceeded geometric mean values
0.1	nd	nd	—	—	—	—	—	from the FWS's NCBP; the
0.1	nd–0.2	nd	0	0	10	nd	nd	median DDE concentration also
0.1	nd	nd	—	—	—	—	—	exceeded the NCBP mean.
0.1	nd	nd	0	0	10	nd	nd	DDT was detected in 1 of 2
0.1	nd	nd	—	—	—	—	—	algal (*Chara* sp.) samples. DDE
0.1	nd	nd	—	—	—	—	—	and toxaphene also were
0.1	nd	nd	0	0	10	nd	nd	detected in softshell turtles,
0.1	nd	nd	—	—	—	—	—	DDE was detected in blue
0.1	nd	nd	—	—	—	—	—	crabs, and DDT, DDD, and
0.1	nd	nd	—	—	—	—	—	DDE in birds. DDE, 3 triazine
0.1	nd	nd	—	—	—	—	—	herbicides, 3 OP insecticides,
0.1	nd	nd	—	—	—	—	—	and the phenoxy herbicide
0.1	nd	nd	—	—	—	—	—	2,4-D were detected in water or
—	—	—	0	0	10	nd	nd	runoff samples. Report also
—	—	—	0	0	10	nd	nd	presents data from a 1985
—	—	—	0	0	10	nd	nd	survey of 95 sites in the NWR.
0.1	nd	nd	—	—	—	—	—	
1	nd	nd	—	—	—	—	—	
10	nd	nd	67	50	500	nd–5,100	980	
0.1	nd	nd	—	—	—	—	—	
nr	nr–48*	nr	—	—	—	—	—	* Data are for all 2–4 cm
nr	nr–100*	nr	—	—	—	—	—	intervals of sediment cores. Objective: to reconstruct history of contaminated sediment in Hudson R. using data from dated sediment cores. Maximum values of *p,p'*- DDD and γ-chlordane occurred from the 1960s to the early 1970s. Local New

Table 2.2. Pesticides in bed sediment and aquatic biota from rivers in the United States: State and local monitoring studies—*Continued*

Reference	Sampling date	Sampling location and strategy	Sampling media and species	Number of samples	Target analytes	Bed sediment % Detection Sites	Bed sediment % Detection Samples
		sediment.					
Crayton, 1989	1988	9 sites (sediment) at Cape Newenham Air Force Station within the Togiak National Wildlife Refuge (NWR), Alaska.	Sediment, soil.	27 12	Aldrin	0	0
					Chlordane, α-	0	0
					Chlordane, γ-	0	0
					DDD, *o,p'*-	0	0
					DDD, *p,p'*-	0	0
					DDE, *o,p'*-	0	0
					DDE, *p,p'*-	0	0
					DDT, *o,p'*-	0	0
					DDT, *p,p'*-	0	0
					DDT, total	0	0
					Dieldrin	0	0
					Endrin	0	0
					HCB	0	0
					HCH, α-	0	0
					HCH, β-	0	0
					HCH, δ-	0	0
					Heptachlor	0	0
					Heptachlor epoxide	0	0
					Mirex	0	0
					Nonachlor, *cis*-	0	0
					Nonachlor, *trans*-	0	0
					Oxychlordane	0	0
					Toxaphene	0	0
Davenport, 1989	1986–1987	3 (sediment) and 19 sites (water) in the Reedy Fork and Buffalo Creek basins (North Carolina). Water was collected at high and low flow.	Water, sediment.	99 nr	OCs, not specified	0	0
					OPs, not specified	0	0

Table 2.2. Pesticides in bed sediment and aquatic biota from rivers in the United States: State and local monitoring studies—*Continued*

Bed sediment (µg/kg dry weight)			Biota (whole fish unless noted)						Comments
			% Detection		(µg/kg wet weight)				
DL	Range	Median	Sites	Samples	DL	Range	Median		
									York metropolitan inputs were significant sources of chlordane and DDD to the estuary. Tidal Hudson sediment upstream of New York Harbor contained <5 µg/kg chlordane, while residues in the harbor peaked at ~50 µg/kg. *p,p'-* DDD also was higher in harbor sediment (~150 µg/kg) than in sediment from the tidal Hudson R. Farther upstream, DDD residues reached almost 100 mg/kg, indicating the importance of downstream transport as well as inputs from New York metropolitan area.
50	nd	nd	—	—	—	—	—		Objective: to determine what contaminants from the Air Force Station may have entered the NWR. HCB was found in soil at 2 of 4 sites, and PCBs in soil at all 4 sites. No OCs were found in any sediment samples.
50	nd	nd	—	—	—	—	—		
50	nd	nd	—	—	—	—	—		
50	nd	nd	—	—	—	—	—		
50	nd	nd	—	—	—	—	—		
50	nd	nd	—	—	—	—	—		
50	nd	nd	—	—	—	—	—		
50	nd	nd	—	—	—	—	—		
50	nd	nd	—	—	—	—	—		
50	nd	nd	—	—	—	—	—		
50	nd	nd	—	—	—	—	—		
50	nd	nd	—	—	—	—	—		
50	nd	nd	—	—	—	—	—		
50	nd	nd	—	—	—	—	—		
50	nd	nd	—	—	—	—	—		
50	nd	nd	—	—	—	—	—		
50	nd	nd	—	—	—	—	—		
50	nd	nd	—	—	—	—	—		
50	nd	nd	—	—	—	—	—		
50	nd	nd	—	—	—	—	—		
50	nd	nd	—	—	—	—	—		
50	nd	nd	—	—	—	—	—		
nr	nd	nd	—	—	—	—	—		Objective: to define water quality of streams in the Greensboro (North Carolina) area. OP and OC analytes in sediment were not specified in the report; none were detected. 10 OCs and 7 OPs were detected in water samples.
nr	nd	nd	—	—	—	—	—		

Table 2.2. Pesticides in bed sediment and aquatic biota from rivers in the United States: State and local monitoring studies—*Continued*

Reference	Sampling date	Sampling location and strategy	Sampling media and species	Number of samples	Target analytes	Bed sediment % Detection	
						Sites	Samples
Fuhrer, 1989	1983	10 sites in Portland Harbor, Willamette River (Oregon). Samples were grab (1 site) or 1-m core (9 sites) samples.	Sediment (surficial, core).	1 9	Aldrin	0	0
					Chlordane	100	100
					D, 2,4-	0	0
					DDD	100	100
					DDE	80	80
					DDT	50	50
					Dieldrin	90	90
					DP, 2,4-	0	0
					Endosulfan	0	0
					Endrin	0	0
					Heptachlor	90	90
					Heptachlor epoxide	0	0
					Isophorone	0	0
					Lindane	0	0
					Methoxychlor	0	0
					Mirex	0	0
					Pentachlorophenol	0	0
					Silvex	0	0
					T, 2,4,5-	0	0
					TCDD, 2,3,7,8-	0	0
Fuhrer and Horowitz, 1989	1984	4 sites (sediment) and 1 site (water) in the lower Columbia River Navigation Channel (Washington). Sites were potential dredge sites. Cores were analyzed in segments. Top segments were 0.8–1.9 m in depth.	Sediment (cores), water (native), elutriates.	4 1 4	Aldrin	0*	0*
					Chlordane	25*	25*
					DDD	75*	75*
					DDE	75*	75*
					DDT	25*	25*
					Dieldrin	25*	25*
					Endosulfan	0*	0*
					Endrin	0*	0*
					Heptachlor	0*	0*
					Heptachlor epoxide	25*	25*
					Isophorone	0*	0*
					Lindane	0*	0*
					Methoxychlor	0*	0*
					Mirex	0*	0*
					Pentachlorophenol	0*	0*
					Perthane	0*	0*
					Toxaphene	0*	0*
Isaza and Dreisig, 1989	1986	5 sites on the Connecticut River (New Hampshire, Vermont). Sites were downstream of municipal and industrial development. Fish were composited by species.	Fish (skin-off fillets, carcasses). Species include smallmouth bass, white perch, walleye, white sucker.	17 16	DDD	—	—
					DDE	—	—
					DDT	—	—

Table 2.2. Pesticides in bed sediment and aquatic biota from rivers in the United States: State and local monitoring studies—*Continued*

Bed sediment (µg/kg dry weight)			Biota (whole fish unless noted)					Comments
			% Detection		(µg/kg wet weight)			
DL	Range	Median	Sites	Samples	DL	Range	Median	
0.1	nd	nd	—	—	—	—	—	Objective: to describe occurrence
nr	1–10	5	—	—	—	—	—	and spatial distribution of
0.1	nd	nd	—	—	—	—	—	chemicals in bottom material of
nr	0.7–14	5	—	—	—	—	—	Portland Harbor and to assess
0.1	nd–4.8	2.4	—	—	—	—	—	the disposal of dredged sediment
0.1	nd–1.3	nd–0.2	—	—	—	—	—	in the Columbia R. The Doane
0.1	nd–0.4	0.2	—	—	—	—	—	L. outlet to Portland Harbor
0.1	nd	nd	—	—	—	—	—	appeared to be highly
0.1	nd	nd	—	—	—	—	—	contaminated with DDT (2,700
0.1	nd	nd	—	—	—	—	—	µg/kg). High DDT to DDE and
0.1	0.2–0.4	0.3	—	—	—	—	—	DDD levels suggested recent
0.1	nd	nd	—	—	—	—	—	movement of sediment into
20–50	nd	nd	—	—	—	—	—	harbor.
0.1	nd	nd	—	—	—	—	—	
1	nd	nd	—	—	—	—	—	
0.1	nd	nd	—	—	—	—	—	
20–180	nd	nd	—	—	—	—	—	
0.1	nd	nd	—	—	—	—	—	
0.1	nd	nd	—	—	—	—	—	
200	nd	nd	—	—	—	—	—	
0.1	nd*	nd*	—	—	—	—	—	* Data are for surficial segments
1	nd–2.0*	nd*	—	—	—	—	—	of sediment cores only.
0.1	nd–23*	5.1*	—	—	—	—	—	Objective: to assess vertical
0.1	nd–5.0*	0.7*	—	—	—	—	—	distribution of contaminants in
0.1	nd–0.2*	nd*	—	—	—	—	—	sediment cores at proposed
0.1	nd–0.2*	nd*	—	—	—	—	—	dredging sites in the lower
0.1	nd*	nd*	—	—	—	—	—	Columbia R.; to determine
0.1	nd*	nd*	—	—	—	—	—	potential contribution of
0.1	nd*	nd*	—	—	—	—	—	chemicals from dredged
0.1	nd–0.1*	nd*	—	—	—	—	—	sediment to native water.
9–13	nd*	nd*	—	—	—	—	—	Elutriate tests were performed
0.1	nd*	nd*	—	—	—	—	—	with sediment samples. OCs in
0.1	nd*	nd*	—	—	—	—	—	the upper (to 0.9 m) and middle
0.1	nd*	nd*	—	—	—	—	—	(0.9–4.1 m) intervals of the
67–173	nd*	nd*	—	—	—	—	—	Skipanon R. core represented the
1	nd*	nd*	—	—	—	—	—	maxima for all cores. In 1 core
10	nd*	nd*	—	—	—	—	—	that was analyzed in detail,
								higher OC residues were
								associated with high silt-clay
								fraction and high sediment TOC.
—	—	—	0*	0*	1–20	nd*	nd*	* Data are for skin-off fillets.
—	—	—	100*	87*	1–10	nd–260*	20*	Objective: to identify problem
—	—	—	0*	0*	1–20	nd*	nd*	areas in the Connecticut R.
								Whole-body residues calculated
								from weighted fillet and carcass
								data. Whole-body residues did
								not exceed NAS/NAE maximum
								recommended tissue
								concentrations for total DDT.

Table 2.2. Pesticides in bed sediment and aquatic biota from rivers in the United States: State and local monitoring studies—*Continued*

Reference	Sampling date	Sampling location and strategy	Sampling media and species	Number of samples	Target analytes	Bed sediment % Detection	
						Sites	Samples
Kovats and Ciborowski, 1989	1987	8 sites on Detroit and St. Clair rivers (Great Lakes), Ontario, Canada. Reference sites and contaminated sites were targeted for sampling.	Adult aquatic insects. Species: caddisflies (*Hydropsyche* sp.), burrowing mayflies (*Hexagenia* sp.), true flies, beetles.	24 total	Aldrin DDE, *p,p'*- DDT, *p,p'*- Dieldrin HCB HCH, α- Heptachlor Heptachlor epoxide Lindane	— — — — — — — — —	— — — — — — — — —
Lau and others, 1989	1986	3 transects across the St. Clair River and 2 transects across the Detroit River (Michigan and Ontario, Canada).	Sediment (surficial, suspended), water (dissolved).	5* 5 5	HCB	100*	100*
Madden and others, 1989	1986–1987	7 sites in Louisiana: 5 in farm ponds and 2 in the Atchafalaya River Basin. Samples were collected 3 times per year, early (Dec.–Feb.), middle (Mar.–April), and late (May–July) in the crayfish production season.	Sediment, crayfish (hepatopancreas, muscle), water (filtered). Species: *Procambarus* sp.	27 81 81 27	DDD, *o,p'*- DDD, *p,p'*- DDE, *o,p'*- DDE, *p,p'*- Dieldrin Endrin Heptachlor epoxide Heptachlor Mirex	0 0 0 0 0 0 0 0 0	0 0 0 0 0 0 0 0 0
Murray and Beck, 1989	1984–1986	7 sites in the Calcasieu River and Lake system, Louisiana, which is a commercial harvesting area for shrimp.	Shellfish. Species: shrimp (*Penaeus* sp.)	31	HCB Hexachlorobutadiene	— —	— —

Table 2.2. Pesticides in bed sediment and aquatic biota from rivers in the United States: State and local monitoring studies—*Continued*

Bed sediment (µg/kg dry weight)			Biota (whole fish unless noted)						Comments
			% Detection		(µg/kg wet weight)				
DL	Range	Median	Sites	Sam-ples	DL	Range	Median		Comments
—	—	—	25*	nr	nr	nr	nd*†		* Data are for insects.
—	—	—	100*	nr	nr	nr	31.6*†		† Median of mean concentra-
—	—	—	50*	nr	nr	nr	nd/1*†		tions (dry wt.) at the 8 sites.
—	—	—	100*	nr	nr	nr	18.3*†		Objective: to compare contami-
—	—	—	100*	nr	nr	nr	11.3*†		nant burdens in reference and
—	—	—	100*	nr	nr	nr	7.4*†		contaminated areas; to assess
—	—	—	63*	nr	nr	nr	0.16*†		use of adult aquatic insects as
—	—	—	88*	nr	nr	nr	8.2*†		biomonitors. Concentrations of
—	—	—	100*	nr	nr	nr	4.7*†		most OCs (except dieldrin at one site) were higher at sites on Detroit and St. Clair rivers than at reference sites.
nr	6.1–300*	153*	—	—	—	—	—		* Data are for 2 transects of the St. Clair R. only. Objective: to determine relative importance of water, suspended sediment, and bed sediment in the transport of selected contaminants in the St. Clair and Detroit rivers. Suspended sedi-ments transported the largest amount of contaminants because of high suspended load (Detroit R.) or because industrial dischar-ges arrived in particulate form (St. Clair R.). The amount transported by bed sediment in these rivers was negligible.
nr	nd	nd	43*	6*	nr	nd–40*	nd*		* Data are for crayfish
nr	nd	nd	29*	5*	nr	nd–311*†	nd*		abdominal muscle tissue.
nr	nd	nd	0*	0*	nr	nd*	nd*		† The highest DDD value may
nr	nd	nd	14*	2*	nr	nd–11*	nd*		be a co-eluting artifact.
nr	nd	nd	0*	0*	nr	nd*	nd*		Objective: to determine OC
nr	nd	nd	0*	0*	nr	nd*	nd*		pesticides in edible tissues of
nr	nd	nd	0*	0*	nr	nd*	nd*		crayfish, the source of residues,
nr	nd	nd	0*	0*	nr	nd*	nd*		and spatial and seasonal distribu-
nr	nd	nd	0*	0*	nr	nd*	nd*		tion. OCs were not detected in water or sediment, and were de-tected occasionally in crayfish at low concentrations. Residues in heptatopancreas were comparable or higher than in muscle. OCs were detected in all 3 seasons.
—	—	—	57	16	0.01	nd–20.9	nd		Objective: to determine OC
—	—	—	57	19	0.01	61.1	nd		residues in shrimp in the study area. Other chlorinated benzenes were also measured. The highest OC levels were found in the upper Calcasieu Lake, which is nearest to urban, industrial, and

Table 2.2. Pesticides in bed sediment and aquatic biota from rivers in the United States: State and local monitoring studies—*Continued*

Reference	Sam-pling date	Sampling location and strategy	Sampling media and species	Number of samples	Target analytes	Bed sediment % Detection Sites	Bed sediment % Detection Sam-ples
Niimi and Oliver, 1989	nr	3 sites in Lake Ontario and its tributary, the Credit River (Ontario, Canada). Tissue samples were taken that would be representative of residues in whole fish and muscle.	Fish (whole, muscle). Species: brown trout, rainbow trout, lake trout, coho salmon.	59 59	Chlordane, γ- DDD, *p,p'*- DDE, *p,p'*- DDT, *p,p'*- HCB HCH, α- Lindane Mirex Photomirex	— — — — — — — — —	— — — — — — — — —
Ott and others, 1989	1988	24 sites in the Susquehanna River Basin (Pennsylvania). Samples were collected from depositional areas during summer low flow period. Sediment for organic analysis was sieved (2 mm).	Sediment (surficial).	24	Aldrin Dieldrin Chlordane DDT, *p,p'*- DDE, *p,p'*- DDD, *p,p'*- Endosulfan I Endosulfan II Endosulfan sulfate Endrin Endrin aldehyde Heptachlor Heptachlor epoxide HCH, α- HCH, β- HCH, δ- Lindane Toxaphene	0 0 0 0 0 0 0 0 0 0 0 0 0 0 0 0 0 0	0 0 0 0 0 0 0 0 0 0 0 0 0 0 0 0 0 0

Table 2.2. Pesticides in bed sediment and aquatic biota from rivers in the United States: State and local monitoring studies—*Continued*

Bed sediment			Biota (whole fish unless noted)					
(μg/kg dry weight)			% Detection		(μg/kg wet weight)			Comments
DL	Range	Median	Sites	Samples	DL	Range	Median	
								agricultural areas. The site with the highest levels of hexachloro-butadiene was in a depositional zone with minimal mixing and transport of sediment out of the zone.
—	—	—	nr	nr	nr	3–48*	nr	* Range of mean values for
—	—	—	nr	nr	nr	25–218*	nr	brown trout, lake trout, small
—	—		nr	nr	nr	257–1,982*	nr	and large rainbow trout, small
—	—	—	nr	nr	nr	35–160*	nr	and large coho salmon.
—	—	—	nr	nr	nr	20–90*	nr	Objective: to quantitate selected
—	—	—	nr	nr	nr	4–25*	nr	OCs in 4 species of Lake Onta-
—	—	—	nr	nr	nr	nd–5*	nr	rio fish; to compare residues in
—	—		nr	nr	nr	45–430*	nr	whole fish and muscle. PCB and
—	—		nr	nr	nr	25–196*	nr	pesticide residues were highest in lake trout. Percent lipid increased with weight in whole lake trout and small coho salmon and muscle from lake trout only. Paired t-tests between whole fish and muscle showed significantly higher percent chlorine content in muscle than whole fish in all species except lake trout.
nr	nd	nd	—	—	—	—	—	Objective: to determine the dis-
nr	nd	nd	—	—	—	—	—	tribution of pollutants in bed
nr	nd	nd	—	—	—	—	—	sediment in the Susquehanna
nr	nd	nd	—	—	—	—	—	R. system. In a previous survey
nr	nd	nd	—	—	—	—	—	(1974), OC pesticides were de-
nr	nd	nd	—	—	—	—	—	tected in 16 of 18 sites sampled.
nr	nd	nd	—	—	—	—	—	Chlordane, DDT, DDD, DDE,
nr	nd	nd	—	—	—	—	—	dieldrin, endrin, and heptachlor
nr	nd	nd	—	—	—	—	—	epoxide were detected in 1974. If
nr	nd	nd	—	—	—	—	—	sampling methods and analytical
nr	nd	nd	—	—	—	—	—	detection limits (not specified in
nr	nd	nd	—	—	—	—	—	reference) for the 1974 and 1988
nr	nd	nd	—	—	—	—	—	surveys were comparable, this
nr	nd	nd	—	—	—	—	—	suggests that OC residues
nr	nd	nd	—	—	—	—	—	declined during this time period.
nr	nd	nd	—	—	—	—	—	
nr	nd	nd	—	—	—	—	—	
nr	nd	nd	—	—	—	—	—	

Table 2.2. Pesticides in bed sediment and aquatic biota from rivers in the United States: State and local monitoring studies—*Continued*

Reference	Sampling date	Sampling locations and strategy	Sampling media and species	Number of samples	Target analytes	Bed sediment % Detection	
						Sites	Samples
Ryan, 1989	1988	3 sites on Threemile and Sixmile creeks (New York). One site on each creek was located downstream of Griffiss Air Force Base. The third (reference) site was upstream of the base on Sixmile Creek. Sediment samples were composites of 5 cores from depositional areas at the site. Fish samples were composites of the 5 largest fish of each species.	Sediment (cores), fish (whole). Species: white sucker, brook trout.	2 5	Chlordane, α-	0*	0*
					Chlordane, γ-	0*	0*
					DDD, *o,p'-*	0*	0*
					DDD, *p,p'-*	0*	0*
					DDE, *o,p'-*	0*	0*
					DDE, *p,p'-*	0*	0*
					DDT, *o,p'-*	0*	0*
					DDT, *p,p'-*	0*	0*
					DDT, total	0*	0*
					Dieldrin	0*	0*
					Endrin	0*	0*
					HCB	0*	0*
					HCH, α-	0*	0*
					HCH, β-	0*	0*
					HCH, δ-	0*	0*
					Heptachlor epoxide	0*	0*
					Lindane	0*	0*
					Mirex	0*	0*
					Nonachlor, *cis-*	0*	0*
					Nonachlor, *trans-*	0*	0*
					Oxychlordane	0*	0*
					Toxaphene	0*	0*
Webber and others, 1989	1983–1984	6 sites on Huntsville Spring Branch and Indian Creek, 2 tributaries to Wheeler Reservoir (Alabama). The study area is near Redstone Arsenal, a former DDT-manufacturing site. A reference site was upstream of the arsenal. Samples were taken quarterly within the channel and along the flooded margins. Sediment samples were composites from about 25 dredge grabs. Macroinvertebrates were pooled by trophic level (detritivores, herbivores, and predators).	Sediment (surficial), macroinvertebrates. Species include crayfish, snails, clams, mussels, burrowing mayflies, mayflies, chironomids, leeches, dragonflies.	24 (C)* 16 (O)* 111(C)* 84 (O)*	DDT, total	100†	100†

Table 2.2. Pesticides in bed sediment and aquatic biota from rivers in the United States: State and local monitoring studies—*Continued*

Bed sediment (µg/kg dry weight)			Aquatic biota (whole fish unless noted)					Comments
			% Detection		(µg/kg wet weight)			
DL	Range	Median	Sites	Sam-ples	DL	Range	Median	
10	nd*	nd*	0	0	10	nd	nd	* Data are for sediment cores.
10	nd*	nd*	0	0	10	nd	nd	Objective: to determine the
10	nd*	nd*	0	0	10	nd	nd	extent of contamination of fish
10	nd*	nd*	100	100	10	10–60	20	and sediment in Threemile and
10	nd*	nd*	0	0	10	nd	nd	Sixmile creeks, downstream of
10	nd*	nd*	100	80	10	nd–80	10	Griffiss Air Force Base. Several
10	nd*	nd*	0	0	10	nd	nd	OCs showed slight increases at
10	nd*	nd*	67	40	10	nd–20	nd	downstream sites; however, all
10	nd*	nd*	100	100	10	10–110	50	OC pesticide residues were close
10	nd*	nd*	100	80	10	nd–80	10	to the detection limit.
10	nd*	nd*	0	0	10	nd	nd	
10	nd*	nd*	0	0	10	nd	nd	
10	nd*	nd*	0	0	10	nd	nd	
10	nd*	nd*	0	0	10	nd	nd	
10	nd*	nd*	0	0	10	nd	nd	
10	nd*	nd*	0	0	10	nd	nd	
10	nd*	nd*	0	0	10	nd	nd	
10	nd*	nd*	0	0	10	nd	nd	
10	nd*	nd*	0	0	10	nd	nd	
10	nd*	nd*	67	40	10	nd–10	nd	
10	nd*	nd*	0	0	10	nd	nd	
10	nd*	nd*	0	0	10	nd	nd	
nr	300–4.44x10⁶†	34,100†	100§	100§	nr	900–157,900§	41,900§	* C= channel; O= overbank. † Data are for channel samples. § Data are for channel samples of macroinvertebrates. Range and median are of annual means for invertebrates grouped by site and trophic level. Objective: to analyze DDT in macroinvertebrates that inhabit contaminated sediment for evidence of biomagnification. Total DDT residues at the control site were <1 mg/kg. Other sites showed a decreasing gradient downstream. Overbank concentrations were more variable than those in the channel. Invertebrates from the reference site had lower total DDT than those from other sites. Except for the reference site, detritivores had the highest total DDT levels. Levels in herbivores and carnivores were correlated with those in associated sediment; detritivore levels were not.

Table 2.2. Pesticides in bed sediment and aquatic biota from rivers in the United States: State and local monitoring studies—*Continued*

Reference	Sampling date	Sampling locations and strategy	Sampling media and species	Number of samples	Target analytes	Bed sediment % Detection Sites	Bed sediment % Detection Samples
Crayton, 1990	1987	29 sites (soil), 9 sites (sediment), 7 sites (fish), and 1 site (sludge) in or near abandoned military bases on national wildlife refuge (NWR) lands in the Aleutian Islands, Alaska: Great Sitkin, Agattu, Adak, and Kiska Islands.	Soil, sediment (surficial), fish (*), sludge. Species: threespine stickleback, Dolly varden.	255 total samples (all media)	Aldrin	0	0
					Chlordane, α-	0	0
					Chlordane, γ-	0	0
					DDD, *o,p´-*	0	0
					DDD, *p,p´-*	0	0
					DDE, *o,p´-*	0	0
					DDE, *p,p´-*	0	0
					DDT, *o,p´-*	0	0
					DDT, *p,p´-*	0	0
					Dieldrin	0	0
					HCB	0	0
					HCH, α-	0	0
					HCH, β-	0	0
					HCH, δ-	0	0
					Heptachlor	0	0
					Heptachlor epoxide	0	0
					Lindane	0	0
					Mirex	0	0
					Nonachlor, *cis-*	0	0
					Nonachlor, *trans-*	0	0
					Oxychlordane	0	0
					Toxaphene	0	0
Fuhrer and Evans, 1990	1980–1983	Unspecified number of sites in 17 rivers and estuaries in Oregon and Washington. Dredging sites and point sources were targeted for sampling.	Sediment (surficial), elutriates.	39	Aldrin	nr	nr
					Chlordane	nr	nr
					D, 2,4-	nr	nr
					DDD	nr	nr
					DDE	nr	nr
					DDT	nr	nr
					Dieldrin	nr	nr
					DP, 2,4-	nr	nr
					Endosulfan	nr	nr
					Endrin	nr	nr
					Heptachlor	nr	nr
					Lindane	nr	nr
					Methoxychlor	nr	nr
					Mirex	nr	nr
					Perthane	nr	nr
					Silvex	nr	nr
					T, 2,4,5-	nr	nr
					Toxaphene	nr	nr
Gilliom and Clifton, 1990	1985	24 sites on the San Joaquin River (California). Samples were collected during low flow conditions. Several subsamples from a single depositional area within 10 m	Sediment (surficial).	24	Chlordane	17	17
					DDD	83	83
					DDE	100	100
					DDT	33	33
					Dieldrin	58	58
					Endosulfan	17	17
					Endrin	0	0
					Heptachlor	0	0
					Heptachlor epoxide	0	0
					Lindane	0	0

Table 2.2. Pesticides in bed sediment and aquatic biota from rivers in the United States: State and local monitoring studies—*Continued*

Bed sediment (µg/kg dry weight)			Aquatic biota (whole fish unless noted)					Comments
			% Detection		(µg/kg wet weight)			
DL	Range	Median	Sites	Samples	DL	Range	Median	
20–50	nd	nd	0*	0*	20–50	nd*	nd*	* Fish tissue type not reported.
20–50	nd	nd	0*	0*	20–50	nd*	nd*	Objective: to evaluate the
20–50	nd	nd	0*	0*	20–50	nd*	nd*	distribution and magnitude of
20–50	nd	nd	0*	0*	20–50	nd*	nd*	contamination at abandoned
20–50	nd	nd	0*	0*	20–50	nd*	nd*	military bases on NWR lands in
20–50	nd	nd	0*	0*	20–50	nd*	nd*	the Aleutian Islands. Pesticide
20–50	nd	nd	0*	0*	20–50	nd*	nd*	residues were detected in soil
20–50	nd	nd	0*	0*	20–50	nd*	nd*	samples at 3 sites on Great
20–50	nd	nd	0*	0*	20–50	nd*	nd*	Sitkin Island (up to 140 µg/kg
20–50	nd	nd	0*	0*	20–50	nd*	nd*	α-chlordane, 60 µg/kg
20–50	nd	nd	0*	0*	20–50	nd*	nd*	*cis-* nonachlor, 50 µg/kg dieldrin,
20–50	nd	nd	0*	0*	20–50	nd*	nd*	85 µg/kg oxychlordane) but no
20–50	nd	nd	0*	0*	20–50	nd*	nd*	pesticide residues were
20–50	nd	nd	0*	0*	20–50	nd*	nd*	detected in sediment or fish.
20–50	nd	nd	0*	0*	20–50	nd*	nd*	Environmentally significant
20–50	nd	nd	0*	0*	20–50	nd*	nd*	concentrations of petroleum-
20–50	nd	nd	0*	0*	20–50	nd*	nd*	based compounds were detected
20–50	nd	nd	0*	0*	20–50	nd*	nd*	in soil samples near old storage
20–50	nd	nd	0*	0*	20–50	nd*	nd*	facilities.
20–50	nd	nd	0*	0*	20–50	nd*	nd*	
20–50	nd	nd	0*	0*	20–50	nd*	nd*	
20–50	nd	nd	0*	0*	20–50	nd*	nd*	
20–50	nd	nd	0*	0*	20–50	nd*	nd*	
0.1	nd–0.1	nd	—	—	—	—	—	Objectives: to relate chemical
0.1	nd–10	nd	—	—	—	—	—	concentrations in bottom
0.1	nd–0.1	nd	—	—	—	—	—	sediment with those in elutriate
0.1	nd–14	nd	—	—	—	—	—	test filtrate; to identify sites
0.1	nd–6.8	0.1	—	—	—	—	—	where elutriate test filtrates
0.1	nd–1.3	nd	—	—	—	—	—	exceed native water
0.1	nd–0.5	nd	—	—	—	—	—	concentrations and(or) USEPA
0.1	nd	nd	—	—	—	—	—	aquatic-life guidelines; and to
0.1	nd–0.1	nd	—	—	—	—	—	describe limitations and factors
0.1	nd–0.1	nd	—	—	—	—	—	causing variability in elutriate
0.1	nd–0.4	nd	—	—	—	—	—	tests. Most of maximum OC
0.1	nd–0.4	nd	—	—	—	—	—	concentrations occurred in the
0.1	nd–1.5	nd	—	—	—	—	—	Willamette R. Sampling and
1	nd	nd	—	—	—	—	—	analytical methods were
1	nd	nd	—	—	—	—	—	reported elsewhere.
0.1	nd–0.1	nd	—	—	—	—	—	
0.1	nd–0.1	nd	—	—	—	—	—	
10	nd	nd	—	—	—	—	—	
1	nd–3	nd	—	—	—	—	—	Objective: to determine spatial
0.1	nd–260	1.6	—	—	—	—	—	distribution of OC pesticides in
0.1	0.1–430	3.9	—	—	—	—	—	San Joaquin R. sediment. Con-
0.1	nd–420	nd	—	—	—	—	—	centrations of DDT, DDD, DDE
0.1	nd–8.9	nd	—	—	—	—	—	and dieldrin were significantly
0.1	nd–87	nd	—	—	—	—	—	correlated. DDE accounted for
0.1	nd	nd	—	—	—	—	—	~60% of total DDT. DDE levels
0.1	nd	nd	—	—	—	—	—	were significantly correlated
0.1	nd	nd	—	—	—	—	—	with sediment TOC, and with
0.1	nd	nd	—	—	—	—	—	the <62 µm size fraction. After

Table 2.2. Pesticides in bed sediment and aquatic biota from rivers in the United States: State and local monitoring studies—*Continued*

Reference	Sampling date	Sampling locations and strategy	Sampling media and species	Number of samples	Target analytes	Bed sediment % Detection	
						Sites	Samples
		of the site location were combined into a composite sample for that site.			Methoxychlor	0	0
					Mirex	8	8
					Perthane	0	0
					Toxaphene	4	4
Greene and others, 1990	1988	5 sites (sediment), 2 sites (fish), and 6 sites (water) in the the Cheyenne River and tributaries (Angostura Reclamation Unit, South Dakota). Sediment samples were composites of 9 or more cross-sectional sub-samples. Samples were sieved (<62 μm).	Sediment (surficial), fish (whole), birds (eggs), water (whole). Species: carp, channel catfish, sauger, shorthead redhorse.	5 6 1 20	Atrazine	0	0
					Carbofuran	0	0
					Chlordane	—	—
					DDD, *o,p´-*	—	—
					DDD, *p,p´-*	—	—
					DDE, *o,p´-*	—	—
					DDE, *p,p´-*	—	—
					DDT, *o,p´-*	—	—
					DDT, *p,p´-*	—	—
					Dieldrin	—	—
					Endrin	—	—
					HCB	—	—
					Heptachlor epoxide	—	—
					Mirex	—	—
					Nonachlor, *cis-*	—	—
					Nonachlor, *trans-*	—	—
					Oxychlordane	—	—
					Toxaphene	—	—
Hickey and others, 1990	1984–1987	1 contaminated site (Love Canal, New York) and 1 reference site (Black Creek, Ontario, Canada) on the Niagara River. Results here are shown separately for 3 parts of the Niagara River Environmental Contaminants Study: (1) residues in young-of-the-year spottail shiners from Love Canal and reference sites (1984–1987); (2) residues in brown bullhead fillets from Love Canal and reference sites (1985);	(1) Fish (whole). Species: spottail shiner.	nr	DDT	—	—
					HCB	—	—
					HCH, total	—	—
					Mirex	—	—
			(2) Fish (fillets). Species: brown bullhead.	4	Chlordane, α-	—	—
					Chlordane, γ-	—	—
					DDD, *o,p´-*	—	—
					DDD, *p,p´-*	—	—
					DDE, *o,p´-*	—	—
					DDE, *p,p´-*	—	—
					DDT, *o,p´-*	—	—
					DDT, *p,p´-*	—	—
					Dieldrin	—	—
					Endrin	—	—
					HCB	—	—
					HCH, α-	—	—
					HCH, β-	—	—
					Heptachlor epoxide	—	—

Table 2.2. Pesticides in bed sediment and aquatic biota from rivers in the United States: State and local monitoring studies—*Continued*

Bed sediment (µg/kg dry weight)			Aquatic biota (whole fish unless noted)					Comments
			% Detection		(µg/kg wet weight)			
DL	Range	Median	Sites	Samples	DL	Range	Median	
0.1	nd	nd	—	—	—	—	—	normalizing residues by TOC,
0.1	nd–0.4	nd	—	—	—	—	—	only sites on westward tribu-
0.1	nd	nd	—	—	—	—	—	taries (which drain agricultural
10	nd–250	nd	—	—	—	—	—	land) had residues that clearly
								exceeded those at other sites.
								Pesticide loads for tributaries
								were estimated from streamflow
								and suspended sediment data.
100	nd	nd	—	—	—	—	—	Objective: to determine if
100	nd	nd	—	—	—	—	—	contaminants were present at
—	—	—	0	0	20	nd	nd	concentrations high enough to
—	—	—	0	0	10	nd	nd	cause harmful effects on human
—	—	—	50	67	10	nd–30	10	health or fish and wildlife
—	—	—	0	0	10	nd	nd	within or downstream of the
—	—	—	50	67	10	nd–50	15	Angostura Reclamation Unit.
—	—	—	0	0	10	nd	nd	Pesticide concentrations in
—	—	—	0	0	10	nd	nd	sediment and biota samples
—	—	—	50	17	10	nd–10	nd	from the study area were low.
—	—	—	0	0	10	nd	nd	Some triazine herbicides were
—	—	—	0	0	10	nd	nd	detected in some water
—	—	—	0	0	10	nd	nd	samples, but at fairly low
—	—	—	0	0	10	nd	nd	levels; the highest
—	—	—	0	0	10	nd	nd	concentration was 1 µg/L
—	—	—	50	33	10	nd–20	nd	prometon, which was measured
—	—	—	0	0	10	nd	nd	in the Cheyenne R. midway
—	—	—	0	0	50	nd	nd	through the unit.
—	—	—	100	nr	nr	2–23*	nr	Objective: to study histological
—	—	—	100	nr	nr	nd–36*	nr	indicators of pollution in fish at
—	—	—	50	nr	nr	nd–22*	nr	1 contaminated site (Love
—	—	—	50	nr	nr	nd–7*	nr	Canal) and 1 reference site on
								the Niagara R.
								(1) * Range of mean residues
								for all site–year combinations.
								OC residues at the Love Canal
								site were higher than at the refer-
								ence site, except for total DDT.
—	—	—	0*	0*	nr	nd*	nd*	(2) * Data are for fish fillets.
—	—	—	0*	0*	nr	nd*	nd*	Few OC pesticides (3 of 20
—	—	—	0*	0*	nr	nd*	nd*	analyzed) were detected in
—	—	—	50*	25*	nr	nd–20*	nd*	samples from either site.
—	—	—	0*	0*	nr	nd*	nd*	Residues of *trans-* nonachlor,
—	—	—	100*	100*	nr	10–70*	25*	*p,p´-* DDE, *p,p´-* DDD, and
—	—	—	0*	0*	nr	nd*	nd*	PCBs appeared to be higher at
—	—	—	0*	0*	nr	nd*	nd*	the Love Canal site than at the
—	—	—	0*	0*	nr	nd*	nd*	reference site.
—	—	—	0*	0*	nr	nd*	nd*	
—	—	—	0*	0*	nr	nd*	nd*	
—	—	—	0*	0*	nr	nd*	nd*	
—	—	—	0*	0*	nr	nd*	nd*	
—	—	—	0*	0*	nr	nd*	nd*	

Table 2.2. Pesticides in bed sediment and aquatic biota from rivers in the United States: State and local monitoring studies—*Continued*

Reference	Sampling date	Sampling locations and strategy	Sampling media and species	Number of samples	Target analytes	Bed sediment % Detection Sites	Bed sediment % Detection Samples
		(3) residues in whole brown bullhead from Love Canal (1986).			Lindane	—	—
					Mirex	—	—
					Nonachlor, *cis-*	—	—
					Nonachlor, *trans-*	—	—
					Oxychlordane	—	—
					Toxaphene	—	—
			(3) Fish (whole). Species: brown bullhead.	9	Aldrin	—	—
					Chlordane, *cis-*	—	—
					Chlordane, *trans-*	—	—
					Dacthal	—	—
					DDD, *o,p'-*	—	—
					DDD, *p,p'-*	—	—
					DDE, *o,p'-*	—	—
					DDE, *p,p'-*	—	—
					DDT, *o,p'-*	—	—
					DDT, *p,p'-*	—	—
					Dicofol	—	—
					Dieldrin	—	—
					Endosulfan I	—	—
					Endosulfan II	—	—
					Endosulfan sulfate	—	—
					Endrin	—	—
					HCB	—	—
					HCH, α-	—	—
					HCH, β-	—	—
					HCH, δ-	—	—
					Heptachlor	—	—
					Heptachlor epoxide	—	—
					Lindane	—	—
					Methoxychlor	—	—
					Mirex	—	—
					Nonachlor, *cis-*	—	—
					Nonachlor, *trans-*	—	—
					Oxychlordane	—	—
					Tetradifon	—	—
Hoffman and others, 1990	1986– 1987	18 sites (sediment) located above and below agricultural lands and irrigation drainage near and within Stillwater Wildlife Management Area, Nevada. Biota were sampled at an unspecified number of sites in 1986.	Water, sediment (surficial), fish (whole), birds (liver), aquatic insects, aquatic plants. Species: not reported.	18 181 total biota samples	Aldrin	6	6
					Chlordane	22	22
					DDD	28	28
					DDE	71	71
					DDT	12	12
					Dieldrin	17	17
					Endosulfan	0	0
					Endrin	0	0
					Heptachlor	0	0
					Heptachlor epoxide	11	11
					Lindane	17	17
					Methoxychlor	6	6
					Mirex	0	0
					Nonachlor	—	—

Table 2.2. Pesticides in bed sediment and aquatic biota from rivers in the United States: State and local monitoring studies—*Continued*

Bed sediment (µg/kg dry weight)			Aquatic biota (whole fish unless noted)					Comments
			% Detection		(µg/kg wet weight)			
DL	Range	Median	Sites	Samples	DL	Range	Median	
—	—	—	0*	0*	nr	nd*	nd*	
—	—	—	0*	0*	nr	nd*	nd*	
—	—	—	0*	0*	nr	nd*	nd*	
—	—	—	50*	50*	nr	nd–20*	nd/10*	
—	—	—	0*	0*	nr	nd*	nd*	
—	—	—	0*	0*	nr	nd*	nd*	
—	—	—	100	14	1	nd–3	nd	(3) Contaminant data indicated
—	—	—	100	100	1–2	13–31	27	that biological uptake of OC
—	—	—	100	100	1–2	2–14	6	contaminants was occurring at
—	—	—	100	100	4–6	39–150	140	the Love Canal site.
—	—	—	100	100	2	55–210	88	Histological examinatin of
—	—	—	100	100	1	30–270	52	brown bullheads from the Love
—	—	—	0	0	2–3	nd	nd	Canal and reference sites
—	—	—	100	100	1–2	9–430	46	showed increased incidence of
—	—	—	100	100	2–3	8–28	15	total and nonparasitic lesions.
—	—	—	100	57	2–3	nd–140	38	The most common lesions
—	—	—	0	0	1	nd	nd	observed at the Love Canal site
—	—	—	100	100	2	10–45	29	were inflammation of the gills,
—	—	—	0	0	2	nd	nd	kidney, and liver. The elevated
—	—	—	0	0	2	nd	nd	lesion incidence appeared to be
—	—	—	0	0	2	nd	nd	a response to contaminant-
—	—	—	0	0	2–3	nd	nd	induced stress and served as a
—	—	—	100	100	2	5–70	7	biological indicator of pollution
—	—	—	100	100	2–3	7–70	12	at the site.
—	—	—	0	0	2–3	nd	nd	
—	—	—	100	57	2–3	nd–22	14	
—	—	—	100	29	1–2	nd–3	nd	
—	—	—	100	100	1–2	5–10	7	
—	—	—	100	100	5–7	5–49	35	
—	—	—	100	100	43–60	21–130	46	
—	—	—	100	100	1	9–75	18	
—	—	—	100	100	1–2	4–1	8	
—	—	—	100	100	1	5–28	13	
—	—	—	100	100	1–2	11–27	23	
—	—	—	0	0	4–5	nd	nd	
0.1	nd–0.3	nd	—	—	—	—	—	Objective: to assess the
1	nd–45	nd	0	0	10	nd	nd	occurrence of contaminants and
0.1	nd–3.2	nd	0	0	10	nd	nd	evaluate the impact of irrigation
0.1	nd–2.1	0.3	0	0	10	nd	nd	drainage. Analytes included
0.1	nd–0.2	nd	0	0	10	nd	nd	inorganic constituents as well
0.1	nd–4.6	nd	0	0	10	nd	nd	as OC compounds. No OCs
0.1	—	nd	—	—	—	—	—	were detected in biological
0.1	—	nd	0	0	10	nd	nd	samples, so OC analysis was
0.1	—	nd	—	—	—	—	—	discontinued. TOC-normalized
0.1	nd–0.5	nd	0	0	10	nd	nd	concentration of lindane in the
0.1	nd–4.7	nd	—	—	—	—	—	sediment at 2 sites exceeded
0.1	nd–1	nd	—	—	—	—	—	the mean USEPA interim
0.1	—	nd	—	—	—	—	—	sediment criteria of 160 µg/kg.
—	—	—	0	0	10	nd	nd	Other pesticides were low or

Table 2.2. Pesticides in bed sediment and aquatic biota from rivers in the United States: State and local monitoring studies—*Continued*

Reference	Sam-pling date	Sampling locations and strategy	Sampling media and species	Number of samples	Target analytes	Bed sediment % Detection	
						Sites	Sam-ples
Kaiser and others, 1990	1985	6 and 7 sites (surficial and suspended sediment) in 3 riverine lakes of the St. Lawrence River (New York, Canada). Surficial sediment was sectioned into 3 1-cm sections.	Sediment (surficial, suspended).	18*	Chlordane, γ- DDD DDE DDT HCB HCH, α- Lindane Mirex	50* 83* 100* 17* 83* 0* 0* 75*	50* 83* 100* 17* 83* 0* 0* 75*
Marcus and Renfrow, 1990	nr	16 sites in major estuarine systems of South Carolina. Sediment and oysters were collected in Mar. from mid-intertidal oyster reefs for 2 and 3 years, respectively. Crabs were collected in July of 1 year.	Sediment (surficial), mollusks, crabs (somatic muscle). Species: American oyster, blue crab.	32 44 16	Aldrin Azinphos-methyl Chlordane DDD DDE DDT Diazinon Dieldrin Endrin Ethion Fenchlorphos HCB HCH, α- HCH, β- Heptachlor Heptachlor epoxide Lindane Malathion Methoxychlor Methyl parathion Mirex Parathion Toxaphene	0 0 0 0 0 0 0 0 0 0 0 0 0 0 0 0 0 0 0 0 0 0 0	0 0 0 0 0 0 0 0 0 0 0 0 0 0 0 0 0 0 0 0 0 0 0
Mason and others, 1990	1976–1980	4 sites in the Chicod Creek Basin (North Carolina). Samples were collected before and during channel modification.	Sediment, water.	24 31	Chlordane DDD DDE DDT Dieldrin Endrin Heptachlor	75 100 100 100 100 50 25	17 79 83 67 79 13 4

Table 2.2. Pesticides in bed sediment and aquatic biota from rivers in the United States: State and local monitoring studies—*Continued*

| Bed sediment (µg/kg dry weight) | | | Aquatic biota (whole fish unless noted) | | | | | Comments |
| | | | % Detection | | (µg/kg wet weight) | | | |
DL	Range	Median	Sites	Samples	DL	Range	Median	
nr	nd–0.91*	nd/0.2*	—	—	—	—	—	* Data are for the top 1-cm section only (6 samples). Objective: to investigate OC contaminants in surficial and suspended sediments in the St. Lawrence R. Concentrations in settling particulates generally reflected those in surficial sediment.
nr	nd–9.2*	1.85*	—	—	—	—	—	
nr	0.51–7.7*	2.65*	—	—	—	—	—	
nr	nd–1.8*	nd*	—	—	—	—	—	
nr	nd–3.70*	2*	—	—	—	—	—	
nr	nd*	nd*	—	—	—	—	—	
nr	nd*	nd*	—	—	—	—	—	
0.01	nd–0.95*	0.59*	—	—	—	—	—	
2	nd	nd	0*	0*	5	nd*	nd*	* Data are for oysters. Objective: baseline monitoring of selected pesticides in oysters, crabs, and sediment in South Carolina estuarine systems. Data showed only sporadic occurrence of pesticides in the study area: spatially in all media and temporally in oysters. DDT was the only pesticide detected in crabs, but 75% of crab sites and samples contained detectable DDT; concentrations detected ranged from 7.26 to 177 µg/kg. The highest DDT residues detected were at May R. (crabs), Stono R. and Whale Branch (oysters). The highest DDE residues detected were in Coosaaw and Stono rivers (oysters).
4	nd	nd	0*	0*	250	nd*	nd*	
2	nd	nd	0*	0*	5	nd*	nd*	
2	nd	nd	0*	0*	5	nd*	nd*	
2	nd	nd	13*	5*	5	nd–7.36*	nd	
2	nd	nd	6*	2*	5	nd–15.6*	nd	
4	nd	nd	0*	0*	250	nd*	nd*	
2	nd	nd	0*	0*	5	nd*	nd*	
2	nd	nd	0*	0*	5	nd*	nd*	
4	nd	nd	0*	0*	250	nd*	nd*	
4	nd	nd	0*	0*	250	nd*	nd*	
2	nd	nd	0*	0*	5	nd*	nd*	
2	nd	nd	6*	2*	5	nd–7.0*	nd*	
2	nd	nd	0*	0*	5	nd*	nd*	
2	nd	nd	0*	0*	5	nd*	nd*	
2	nd	nd	0*	0*	5	nd*	nd*	
2	nd	nd	0*	0*	5	nd*	nd*	
4	nd	nd	0*	0*	250	nd*	nd*	
2	nd	nd	0*	0*	5	nd*	nd*	
4	nd	nd	0*	0*	250	nd*	nd*	
2	nd	nd	0*	0*	5	nd*	nd*	
4	nd	nd	0*	0*	250	nd*	nd*	
2	nd	nd	0*	0*	5	nd*	nd*	
1	n2	nd	—	—	—	—	—	Objective: to determine the effect of channel modifications (1978–1981) on the hydrology and water quality of Chicod R. Basin. Pesticides were not analyzed at a reference site outside the basin. DDT, dieldrin, and diazinon were detected occasionally in water. There were insufficient samples to determine the effect of channel modifications on pesticide levels in water or sediment. DDT in water was associated with higher flows.
0.1	nd–58	3.8	—	—	—	—	—	
0.1	nd–37	2.8	—	—	—	—	—	
0.1	nd–20	0.7	—	—	—	—	—	
0.1	nd–11	1	—	—	—	—	—	
0.1	nd–1	nd	—	—	—	—	—	
0.1	nd–8.1	nd	—	—	—	—	—	

Table 2.2. Pesticides in bed sediment and aquatic biota from rivers in the United States: State and local monitoring studies—*Continued*

Reference	Sampling date	Sampling locations and strategy	Sampling media and species	Number of samples	Target analytes	Bed sediment % Detection	
						Sites	Samples
Metcalfe and Charlton, 1990	1985	20 sites on the St. Lawrence River, Ottawa River, and riverine lakes (Great Lakes) in New York and Ontario, Canada.	Mussels. Species: *Elliptio complanata, Lampsilis radiata.*	38	Chlordane, γ- DDD, *p,p´-* DDE, *p,p´-* DDT, *p,p´-* HCB HCH, α- Lindane Mirex	— — — — — — — —	— — — — — — — —
Murray and Beck, 1990	1985–1986	7 sites in the Calcasieu River and Lake system (Louisiana).	Shrimp. Species: *Penaeus* sp.	31	Aldrin DDD Endosulfan II Endrin Endrin aldehyde HCH, β- HCH, δ- Heptachlor	— — — — — — — —	— — — — — — — —
Nair, 1990a	1988	6 sampling sites in the nontidal reaches of the Patauxent and Little Patauxent rivers (Maryland).	Sediment (surficial), fish (whole, muscle). Species: brown bullhead, white catfish.	6 12 6	Chlordane, *cis-* Chlordane, *trans-* DDD, *o,p´-* DDD, *p,p´-* DDE, *o,p´-* DDE, *p,p´-* DDT, *o,p´-* DDT, *p,p´-* Dieldrin Endrin Heptachlor epoxide Hexachlorobenzene Lindane Mirex Nonachlor, *cis-* Nonachlor, *trans-* Oxychlordane Toxaphene	0 0 0 0 0 0 0 0 0 0 0 0 0 0 0 0 0 0	0 0 0 0 0 0 0 0 0 0 0 0 0 0 0 0 0 0
Nair, 1990b	1988	4 sites on the James River near Hopewell, Virginia: one reference site and 1 site each in the Presquile National Wildlife Refuge (NWR), an	Sediment, fish (whole, fillets, carcasses). Species: striped bass, channel	4 16 8 8	Aldrin Chlordane, total DDT, total Dieldrin Endrin HCB HCH	0 0 25 0 0 0 0	0 0 25 0 0 0 0

Table 2.2. Pesticides in bed sediment and aquatic biota from rivers in the United States: State and local monitoring studies—*Continued*

Bed sediment (µg/kg dry weight)			Aquatic biota (whole fish unless noted)					Comments
			% Detection		(µg/kg wet weight)			
DL	Range	Median	Sites	Sam-ples	DL	Range	Median	
—	—	—	65*	66*	0.01	nd–0.3*	nr	* Data are for mussels.
—	—	—	65*	47*	0.01	nd–6.80*	nd*	Objective: to determine the
—	—	—	95*	97*	0.01	nd–24*	nr	distribution of contaminants in
—	—	—	20*	21*	0.01	nd–0.80*	nd*	native mussels. *p,p´-* DDE levels
—	—	—	85*	66*	0.01	nd–0.5*	nr	were highest at the inlet from
—	—	—	80*	74*	0.01	nd–0.20*	0.1*	Lake Ontario, at the mouth of
—	—	—	10*	5.3*	0.01	nd–trace*	nd*	Grass R., and in Lac St. Francois.
—	—	—	75*	74*	0.01	nd–1.5*	nr	Mirex levels decreased down-stream from Lake Ontario, im-plicating the lake as the source. Chlordane was highest in mus-sels from the 3 riverine lakes. Mirex was generally higher in *L. radiata* and HCB in *E. complanata* .
—	—	—	57*	23*	0.01	nd–0.12*	nd*	* Data are for shrimp.
—	—	—	29*	6.5*	0.01	nd–0.25*	nd*	Objective: to determine OC pes-
—	—	—	57*	23*	0.01	nd–2.57*	nd*	ticide contamination of shrimp
—	—	—	100*	68*	0.01	nd–9.47*	0.25*	in the study area. Compounds
—	—	—	100*	55*	0.01	nd–0.69*	0.01*	that were not detected were not
—	—	—	43*	13*	0.01	nd–0.56*	nd*	reported. Shrimp with OC resi-
—	—	—	86*	42*	0.01	nd–1.41*	nd*	dues were distributed throughout
—	—	—	71*	35*	0.01	nd–0.75*	nd*	the study area, but the highest levels were in shrimp from the upper reaches of the system.
10	nd	nd	100	92	10	nd–30	10	Objective: to determine the
10	nd	nd	100	92	10	nd–100	20	degree and extent of chlordane
10	nd	nd	0	0	10	nd	nd	contamination in fish in the
10	nd	nd	0	0	10	nd	nd	Patuxent R.; to compare data to
10	nd	nd	0	0	10	nd	nd	results of previous monitoring.
10	nd	nd	0	0	10	nd	nd	Chlordane residues detected in
10	nd	nd	0	0	10	nd	nd	this study are less than those
10	nd	nd	0	0	10	nd	nd	measured at the same sites
10	nd	nd	0	0	10	nd	nd	several years earlier. Data
10	nd	nd	33	17	10	nd–20	nd	collected by a state agency
10	nd	nd	0	0	10	nd	nd	during 1983–1984 appear to
10	nd	nd	0	0	10	nd	nd	represent the maximum in a
10	nd	nd	0	0	10	nd	nd	trend of increasing residue
10	nd	nd	17	8	10	nd–10	nd	levels which has reversed and is now decreasing. No
10	nd	nd	100	92	10	nd–30	10	statistically significant
10	nd	nd	0	0	10	nd	nd	geographic trends were
10	nd	nd	0	0	50	nd	nd	described by the data set.
50*	nd*	nd*	0	0	50	nd	nd	* µg/kg wet weight.
50*	nd*	nd*	100	94	50	nd–870	285	† Equivalent to nd–140 µg/kg
50*	nd–60*†	nd*	100	100	50	60–640	305	dry wt.
50*	nd*	nd*	0	0	50	nd	nd	Objective: to determine the
50*	nd*	nd*	0	0	50	nd	nd	impact of contaminants dis-
50*	nd*	nd*	0	0	50	nd	nd	charged in the Hopewell area on
50*	nd*	nd*	0	0	50	nd	nd	fish and wildlife in and near the

Table 2.2. Pesticides in bed sediment and aquatic biota from rivers in the United States: State and local monitoring studies—*Continued*

Reference	Sampling date	Sampling locations and strategy	Sampling media and species	Number of samples	Target analytes	Bed sediment % Detection	
						Sites	Samples
		effluent area, and downstream of the effluent area. Sediment samples were composites of 3 grabs per site. 4 fish samples of each species were taken at each site: two whole-fish composites of 3 fish each, plus fillet and carcass samples from a single fish.	catfish.		Lindane	0	0
					Mirex	0	0
					Toxaphene	0	0
Orazio and others, 1990	1984–1987	1 site on the Missouri River (Missouri). Fish samples were composites of 5 individuals of comparable size and maturity. Chlordane concentrations also were reported for about 96 sites in Missouri.	Fish (fillets). Species: carp, shovelnose sturgeon, channel catfish, river carpsucker.	12	Aldrin	—	—
					Chlordane	—	—
					DDD, *p,p´-*	—	—
					DDE, *p,p´-*	—	—
					DDT, *p,p´-*	—	—
					Dieldrin	—	—
					HCH	—	—
					Heptachlor epoxide	—	—
Petersen, 1990	1974–1985	12 sites (sediment) on streams in northeastern Arkansas.	Sediment, water.	nr	Aldrin	57*	nr
					Chlordane	71*	nr
					DDD	86*	nr
					DDE	86*	nr
					DDT	71*	nr
					Diazinon	0*	0*
					Dieldrin	86*	nr
					Endosulfan	0*	0*
					Endrin	71*	nr
					Ethion	0*	0*
					Heptachlor	14*	nr
					Heptachlor epoxide	17*	nr
					Lindane	0*	0*
					Malathion	0*	0*
					Methoxychlor	14*	nr
					Methyl parathion	0*	0*
					Methyl trithion	0*	0*
					Mirex	0*	0*
					Parathion	0*	0*
					Toxaphene	57*	nr
					Trithion	0*	0*

Table 2.2. Pesticides in bed sediment and aquatic biota from rivers in the United States: State and local monitoring studies—*Continued*

Bed sediment (µg/kg dry weight)			Aquatic biota (whole fish unless noted)					Comments
			% Detection		(µg/kg wet weight)			
DL	Range	Median	Sites	Samples	DL	Range	Median	
0.1	nd–0.3	nd/0.3	—	—	—	—	—	Objective: to evaluate the quantity and quality of storm runoff from Western Daytona Beach. DDD and heptachlor were the only pesticides detected in bottom sediment. Diazinon and 2,4-D were detected in water at low concentrations at both sites. Target analytes that were not detected were not listed.
0.1	nd–0.2	nd/0.2	—	—	—	—	—	
0.1–1.0	nd	nd	—	—	—	—	—	Objective: to determine if selected contaminants were present in water flowing to Utah Lake, or in the outlet from the lake, at concentrations that could be hazardous to wildlife. Those OP, OC, and carbamate insecticides analyzed in water samples were not detected. Low but measurable residues of chlordane, DDD, DDE, and dieldrin were detected in some sediment samples.
1–7	nd	nd	—	—	—	—	—	
0.1	nd–2.3	0.6	—	—	—	—	—	
0.1–6	nd–1.4	1.4	—	—	—	—	—	
0.1–1.0	nd	nd	—	—	—	—	—	
0.1–1.0	nd	nd	—	—	—	—	—	
0.1	nd–0.4	nd	—	—	—	—	—	
0.1	nd	nd	—	—	—	—	—	
0.1	nd	nd	—	—	—	—	—	
0.1	nd	nd	—	—	—	—	—	
0.1–1.0	nd	nd	—	—	—	—	—	
0.1	nd	nd	—	—	—	—	—	
0.1	nd	nd	—	—	—	—	—	
0.1–1.0	nd	nd	—	—	—	—	—	
0.1	nd	nd	—	—	—	—	—	
0.1–1.0	nd	nd	—	—	—	—	—	
0.1	nd	nd	—	—	—	—	—	
0.1–1.0	nd	nd	—	—	—	—	—	
0.1–1.0	nd	nd	—	—	—	—	—	
1	nd	nd	—	—	—	—	—	
10	nd	nd	—	—	—	—	—	
0.1	nd	nd	—	—	—	—	—	
0.1	nd	nd	—	—	—	—	—	Objective: to evaluate whether high concentrations of contaminants exist within the Bear River Migratory Bird Refuge or proposed land acquisition areas. Oxamyl was detected in water at 1 site (5 µg/L). Low levels of DDD and dieldrin (1–2 µg/L) and 10 µg/L malathion were detected at a 2nd site. The latter site had the highest total DDT residues in sediment samples (31 µg/kg).
1	nd–1.0	nd	—	—	—	—	—	
0.1	nd–13	1.2	—	—	—	—	—	
0.1	0.1–17	1.1	—	—	—	—	—	
0.1	nd–5.9	nd/0.2	—	—	—	—	—	
0.1	nd–0.1	nd	—	—	—	—	—	
0.1	nd–0.1	nd	—	—	—	—	—	
0.1	nd	nd	—	—	—	—	—	
0.1	nd	nd	—	—	—	—	—	
0.1	nd	nd	—	—	—	—	—	
0.1	nd	nd	—	—	—	—	—	
0.1	nd	nd	—	—	—	—	—	
0.1	nd	nd	—	—	—	—	—	
0.1	nd	nd	—	—	—	—	—	
0.1–0.7	nd	nd	—	—	—	—	—	
0.1	nd	nd	—	—	—	—	—	
0.1	nd	nd	—	—	—	—	—	
0.1	nd	nd	—	—	—	—	—	

Table 2.2. Pesticides in bed sediment and aquatic biota from rivers in the United States: State and local monitoring studies—*Continued*

Reference	Sampling date	Sampling locations and strategy	Sampling media and species	Number of samples	Target analytes	Bed sediment % Detection Sites	Bed sediment % Detection Samples
Arizona Department of Health Services, 1991	1982–1990	Unspecified number of sites in Painted Rocks Borrow Pit Lake, an impoundment in the Gila River (Arizona). Sediment was sampled in 1982, 1985, and 1986; whole fish during 1985–1990, and fish fillets during 1986–1989.	Sediment, water, fish (whole, fillets), turtles, lizards, birds. Fish species: channel catfish, black crappie, largemouth bass, carp.	10 2 19 22 4 3 8	Chlordane, α- DDE DDT, total Dicofol Dieldrin Nonachlor TCDD Toxaphene 12 other OCs	— 100 — — — — — — 0	— nr — — — — — — 0
Ashby and others, 1991	1990–1991	20 sites (surficial sediment) in the upper watershed of Steele Bayou (Mississippi). Sediment cores were taken at 6 sites.	Sediment (surficial, cores), water.	20 6	D, 2,4- DB, 2,4- DDD, *p,p´*- DDE, *p,p´*- DDT, *p,p´*- DP, 2,4- Endosulfan I Endosulfan II Endosulfan sulfate Endrin Endrin aldehyde HCH, δ- Heptachlor Heptachlor epoxide T, 2,4,5- TP, 2,4,5- Trifluralin	20 15 90 90 65 0 25 10 60 15 40 5 40 5 10 0 5	20 15 90 90 65 0 25 10 60 15 40 5 40 5 10 0 5
Burt and others, 1991	1985	70 sites on the St. Marys River (Michigan and Ontario, Canada). 3 replicate benthic samples (for community analysis) and 1 sediment sample were taken at each site. Sediment samples were composites of 3 or more grabs per site.	Sediment (surficial).	70	Endosulfan Endrin Heptachlor epoxide Other pesticides (not specified)	nr nr nr nr	nr nr nr nr

Table 2.2. Pesticides in bed sediment and aquatic biota from rivers in the United States: State and local monitoring studies—*Continued*

Bed sediment (µg/kg dry weight)			Aquatic biota (whole fish unless noted)					Comments
			% Detection		(µg/kg wet weight)			
DL	Range	Median	Sites	Samples	DL	Range	Median	
—	—	—	100	37	nr	nd–30	nr	Objective: to determine the level of chemical contamination in environmental media from the study area; to estimate the associated health risks. Residues in edible portions of fish generally were lower than in whole fish. Total DDT was detected in 100% of edible fish samples (70–2,630 µg/kg), α-chlordane in 55% (nd–61 µg/kg). Nonachlor, toxaphene, α-HCH, and dieldrin were detected in 9–18% of edible fish samples.
10	nd–50	nr	—	—	—	—	—	
—	—	—	100	100	nr	670–4,630	nr	
—	—	—	100	41	nr	nd–560	nr	
—	—	—	100	53	nr	nd–40	nr	
—	—	—	100	47	nr	nd–70	nr	
—	—	—	100	33	nr	nd–0.0005	nr	
—	—	—	100	63	nr	nd–2,500	nr	
10–50	nd	nd	—	—	—	—	—	
64–176	nd–99	nd	—	—	—	—	—	Objective: to describe the distribution of water and sediment quality in the study area. Lindane, δ-HCH, $p,p´$- DDE, $p,p´$- DDT, heptachlor, dieldrin, endrin, diazinon, malathion, and 2,4,5-T were detected in water. Pesticides in water were detected primarily in Main Canal in July and Oct. Sediment cores showed decreasing levels of DDT compounds with increasing depth at 2 sites, peak levels at lower depths at 2 sites, and low or undetectable residues in the remaining 2 sites. Pesticide residues in surface sediment was not related to particle size or TOC.
64–176	nd–126	nd	—	—	—	—	—	
0.2	nd–46	15	—	—	—	—	—	
0.2	nd–81	26	—	—	—	—	—	
0.2	nd–78	5.7	—	—	—	—	—	
64–176	nd	nd	—	—	—	—	—	
0.2	nd–4.8	nd	—	—	—	—	—	
0.2	nd–0.4	nd	—	—	—	—	—	
0.2	nd–8.3	0.8	—	—	—	—	—	
0.2	nd–3.4	nd	—	—	—	—	—	
0.2	nd–16	nd	—	—	—	—	—	
0.2	nd–0.8	nd	—	—	—	—	—	
0.2	nd–3.9	nd	—	—	—	—	—	
0.2	nd–0.5	nd	—	—	—	—	—	
64–176	nd–105	nd	—	—	—	—	—	
64–176	nd	nd	—	—	—	—	—	
2	nd–3.8	nd	—	—	—	—	—	
nr	nr	nr	—	—	—	—	—	Objective: to assess spatial and temporal trends in benthic community structure in relation to sediment contaminaton. Contaminant data were not reported. Cluster analysis identified 7 benthic communities, 3 of which were pollution-impacted. Taxa at polluted sites were dominated by tubificids and nematodes. Impacted sites were downstream of industrial/municipal sources, in depositional areas, and mostly on Canada shoreline. Unimpacted sites had more taxa, occurred upstream of point sources, on the U.S. shoreline, and in most areas of downstream lakes. Im-
nr	nr	nr	—	—	—	—	—	
nr	nr	nr	—	—	—	—	—	
nr	nr	nr	—	—	—	—	—	

Table 2.2. Pesticides in bed sediment and aquatic biota from rivers in the United States: State and local monitoring studies—*Continued*

Reference	Sampling date	Sampling locations and strategy	Sampling media and species	Number of samples	Target analytes	Bed sediment % Detection	
						Sites	Samples
Butler and others, 1991	1988	2 sites (sediment) and 7 sites (fish) that are potentially affected by irrigation drainage in the Gunison and Uncompahgre river basins, Colorado.	Water (whole, dissolved), sediment (surficial), fish (whole). Species: rainbow trout, carp, brown trout.	3 8	Aldrin	0	0
					Chlordane	50	33
					Chlordane, α-	—	—
					Chlordane, γ-	—	—
					Dacthal	—	—
					DDD	50	67
					DDE	100	100
					DDT	0	0
					Dieldrin	100	100
					Endosulfan	0	0
					Endrin	0	0
					HCB	—	—
					HCH, β-	—	—
					Heptachlor	0	0
					Heptachlor epoxide	0	0
					Lindane	0	0
					Mirex	0	0
					Nonachlor, *trans-*	—	—
					Oxychlordane	—	—
					Perthane	0	0
					Toxaphene	0	0
Christian-sen and others, 1991	1988	10 sites on the Missouri River (Iowa, Nebraska). Several industrial and municipal facilities discharge to this river reach. 3 composite fish samples were collected per site. Fish ranged from 270 to 520 mm in length (175–1,235 g).	Fish (fillets). Species: channel catfish.	30	Chlordane, *cis-*	—	—
					Chlordane, technical	—	—
					Chlordane, total	—	—
					Chlordane, *trans-*	—	—
					DDD	—	—
					DDE	—	—
					DDT	—	—
					Dieldrin	—	—
					Heptachlor epoxide	—	—
					Nonachlor, *cis-*	—	—
					Nonachlor, *trans-*	—	—
					Oxychlordane	—	—
					Trifluralin	—	—
Cooper, 1991	1983–1984	21 sites (sediment), 10 sites (fish), and 4 sites (water) from the Phillips Bayou/ Moon Lake water-shed (Mississippi). Sediment cores (300 mm) were collected from 4	Water (whole), sediment (cores), fish (whole, flesh, viscera), soil (cores). Species in-	255 65 34 33 33 69	DDT, total	nr	nr
					Fenvalerate	nr	0§
					Methyl parathion	nr	15§
					Permethrin	nr	15§
					Toxaphene	nr	nr

Table 2.2. Pesticides in bed sediment and aquatic biota from rivers in the United States: State and local monitoring studies—*Continued*

Bed sediment (µg/kg dry weight)			Aquatic biota (whole fish unless noted)					Comments
			% Detection		(µg/kg wet weight)			
DL	Range	Median	Sites	Samples	DL	Range	Median	
								pacted and nonimpacted communities were separated along particle size and contaminant gradients in river sediment.
0.1	nd	nd	—	—	—	—	—	Objective: to identify the nature
1	nd–1.0	nd	—	—	—	—	—	and extent of irrigation-related
—	—	—	14	13	10	nd–10	nd	water-quality problems in the
—	—	—	14	13	10	nd–20	nd	study area. Irrigation drainage
—	—	—	71	75	10	nd–1,300	250	contributes substantial quantities
variable	nd–0.5	0.3	57	50	10	nd–70	nd	of inorganic constituents,
0.1	0.2–2.8	0.8	100	100	10	10–1,000	190	including selenium, into the
0.1–0.3	nd	nd	43	38	10	nd–90	nd	Gunnison R. Levels of DDE
nr	0.1–0.3	0.3	57	63	10	nd–50	10	and toxaphene were high in
0.1	nd	nd	—	—	—	—	—	some biota samples. DDE levels
0.1	nd	nd	29	25	10	nd–20	nd	in adult killdeer were as high as
—	—	—	43	38	10	nd–10	nd	110,000 µg/kg. Toxaphene
—	—	—	14	13	10	nd–10	nd	residues in one fish sample
0.1	nd	nd	—	—	—	—	—	exceeded values that have been
0.1	nd	nd	43	38	10	nd–30	nd	reported to be potentially
0.1	nd	nd	—	—	—	—	—	hazardous to fish health. Levels
0.1	nd	nd	—	—	—	—	—	of DDE and toxaphene in
—	—	—	43	38	10	nd–30	nd	sediment were relatively low.
—	—	—	14	13	10	nd–10	nd	
1	nd	nd	—	—	—	—	—	
10	nd	nd	57	57	50	nd–920	300	
—	—	—	90*	nr *	nr	nd–59*	nd–44†	* Data are for fish fillets.
—	—	—	90*	nr *	nr	nd–470*	nd–363†	† Range of site-mean values for fish fillets at 10 sites.
—	—	—	100*	nr *	nr	nd–154*	nd–116†	Objective: to measure OC contaminants in channel catfish; to
—	—	—	90*	nr *	nr	nd–73*	nd–31†	evaluate human health risks of
—	—	—	10*	nr *	nr	nd–20*	nd–12†	consuming these fish. Mean
—	—	—	100*	nr *	nr	nd–60*	2–33†	wet-weight concentrations of
—	—	—	100*	nr *	nr	nd–46*	nd–12†	DDT, chlordane, components,
—	—	—	90*	nr *	nr	nd–110*	nd–64†	and trifluralin were significantly
—	—	—	70*	nr *	nr	nd–13*	nd–5.9†	different among sites. Highest
—	—	—	90*	nr *	nr	nd–14*	nd–13†	levels were at Missouri R. at
—	—	—	100*	nr *	nr	nd–35*	nd–28†	Bellevue, perhaps because of
—	—	—	80*	nr *	nr	nd–14.8*	nd–7.8†	urban runoff. At this site, mean
—	—	—	90*	nr *	nr	nd–74*	nd–63†	chlordane residues exceeded the FDA action level.
nr	nd–651	228*	100†	100†	nr	80–463†	244*†	* Mean value.
nr	nd	nd	nr	2.2¶	nr	nd–11¶¥	nr	† Data are for Moon Lake fish.
nr	nr	nd	nr	17¶	nr	nd–16¶¥	nr	§ Data are for 20 wetland
nr	nr	nd	nr	6.7¶	nr	nd–0.53¶¥	nr	samples only.
								¶ Data are for all fish samples
nr	nd–36	8.6*	—	—	—	—	—	(whole-body, flesh, and viscera).
								¥ Range of mean concentrations for all species.
								Objective: to document residues

Table 2.2. Pesticides in bed sediment and aquatic biota from rivers in the United States: State and local monitoring studies—*Continued*

Reference	Sam-pling date	Sampling locations and strategy	Sampling media and species	Number of samples	Target analytes	Bed sediment % Detection	
						Sites	Sam-ples
		Moon Lake and 17 other watershed sites. Fish were collected pre- and post-spray season. Water was collected biweekly. Compos-ite soil samples (200 mm cores) were collected from the dominant land use in each square mile of the watershed.	clude channel catfish, gizzard shad, common carp, freshwater drum.				
Crayton, 1991	1987–1988	7 sites (sediment) and 1 site (fish) in the Fowler Creek drainage within the Yukon Delta National Wildlife Refuge (NWR), Alaska. The Cape Romanzof Long Range Radar Site is located within the drainage. Sediment samples were composites.	Sediment (surficial), fish (fillets), voles, red fox. Fish species: Dolly Varden.	14 5	Aldrin	0	0
					Chlordane, α-	0	0
					Chlordane, γ-	0	0
					DDD, *o,p*´-	0	0
					DDD, *p,p*´-	0	0
					DDE, *o,p*´-	0	0
					DDE, *p,p*´-	0	0
					DDT, *o,p*´-	0	0
					DDT, *p,p*´-	0	0
					Dieldrin	0	0
					Endrin	0	0
					HCB	0	0
					HCH, α-	0	0
					HCH, β-	0	0
					HCH, δ-	0	0
					Heptachlor	0	0
					Heptachlor epoxide	0	0
					Lindane	0	0
					Mirex	0	0
					Nonachlor, *cis*-	0	0
					Nonachlor, *trans*-	0	0

Table 2.2. Pesticides in bed sediment and aquatic biota from rivers in the United States: State and local monitoring studies—*Continued*

Bed sediment (µg/kg dry weight)			Aquatic biota (whole fish unless noted)					Comments
			% Detection		(µg/kg wet weight)			
DL	Range	Median	Sites	Samples	DL	Range	Median	
								of historical and currently used insecticides in major watershed compartments. Total DDT was higher in lake than other wetland sediment. 75% of soil samples contained total DDT. Mean total DDT in water was 0.11 µg/L. Toxaphene was detected in soil and lake sediment, but not in other wetland sediments. Mean levels of total DDT and toxaphene by crop type were: cotton> soybeans> hardwoods. The top 10-cm layer of sediment cores contained the highest DDD and DDE residues. Total DDT residues were higher during wet (Dec.–May) than dry (June–Nov.) seasons. Currently used insecticides were not found in soil. Permethrin, methyl parathion, and fenvalerate were sporadically detected in water and fish. There were no significant differences among pesticide levels in whole-body, flesh, or liver samples. Methyl parathion was detected twice as often in fish and water as the pyrethroids.
20	nd	nd	0*	0*	10	nd*	nd*	* Data are for fish fillets.
20	nd	nd	100*	80*	10	nd–50*	35*	Objective: to determine if
20	nd	nd	100*	80*	10	nd–30*	15*	contaminants from the radar site
20	nd	nd	0*	0*	10	nd*	nd*	activities had entered the NWR
20	nd	nd	100*	80*	10	nd–60*	25*	and its trust resources. OC
20	nd	nd	0*	0*	10	nd*	nd*	residues were elevated in fish,
20	nd	nd	100*	100*	10	30–170*	80*	vole, and fox samples.
20	nd	nd	0*	0*	10	nd*	nd*	
20	nd	nd	0*	0*	10	nd*	nd*	
20	nd	nd	100*	80*	10	nd–20*	15*	
20	nd	nd	0*	0*	10	nd*	nd*	
20	nd	nd	0*	0*	10	nd*	nd*	
20	nd	nd	0*	0*	10	nd*	nd*	
20	nd	nd	0*	0*	10	nd*	nd*	
20	nd	nd	100*	80*	10	nd–30*	25*	
20	nd	nd	0*	0*	10	nd*	nd*	
20	nd	nd	0*	0*	10	nd*	nd*	
20	nd	nd	0*	0*	10	nd*	nd*	
20	nd	nd	0*	0*	10	nd*	nd*	
20	nd	nd	100*	60*	10	nd–20*	10*	
20	nd	nd	100*	80*	10	nd–60*	30*	

Table 2.2. Pesticides in bed sediment and aquatic biota from rivers in the United States: State and local monitoring studies—*Continued*

Reference	Sampling date	Sampling locations and strategy	Sampling media and species	Number of samples	Target analytes	Bed sediment % Detection Sites	Bed sediment % Detection Samples
Griffiths, 1991	1985	78 sites in the St. Clair River (between Michigan and Ontario, Canada). At each site, the top 3 cm of 3 sediment grabs were composited. Also, 3 samples of benthic fauna were collected at each site.	Sediment (surficial), benthic survey.	78 78	Aldrin	16*	16*
					DDD, *p,p´-*	6*	6*
					DDE, *p,p´-*	45*	45*
					DDT, *p,p´-*	4*	4*
					Dieldrin	63*	63*
					DMDT methoxy-chlor	12*	12*
					Endosulfan II	6*	6*
					Endosulfan sulfate	14*	14*
					HCB	90*	90*
					Heptachlor epoxide	10*	10*
Jackson, 1991	1988	1 site (sediment) on a creek near an oil well and an associated landfill in Becharof National Wildlife Refuge (NWR), Alaska. Trash was eroding from the banks of the creek. Soil was collected at 2 sites.	Sediment (surficial), soil, waste material.	3 4 1	Aldrin	0	0
					Chlordane, α-	0	0
					Chlordane, γ-	0	0
					DDD, *o,p´-*	0	0
					DDD, *p,p´-*	0	0
					DDE, *o,p´-*	0	0
					DDE, *p,p´-*	0	0
					DDT, *o,p´-*	0	0
					DDT, *p,p´-*	0	0
					DDT, total	0	0
					Dieldrin	0	0
					HCB	0	0
					HCH, α-	0	0
					HCH, β-	0	0
					HCH, δ-	0	0
					Heptachlor	0	0
					Heptachlor epoxide	0	0
					Lindane	0	0
					Mirex	0	0
					Nonachlor, *cis-*	0	0
					Nonachlor, *trans-*	0	0
					Oxychlordane	0	0
					Toxaphene	0	0

Table 2.2. Pesticides in bed sediment and aquatic biota from rivers in the United States: State and local monitoring studies—*Continued*

Bed sediment (µg/kg dry weight)			Aquatic biota (whole fish unless noted)					Comments
			% Detection		(µg/kg wet weight)			
DL	Range	Median	Sites	Samples	DL	Range	Median	
								detection frequencies, concentrations, tissue types, and sources of catfish were not reported.
—	—	—	0	0	nr	nd	nd	* OC pesticides analyzed in sediment were not reported. Objective: to assess bald eagles' exposure to environmental contaminants; to collect baseline contaminants data for the NWR and WMAs. *p,p´*- DDE, *p,p´*- DDD, dieldrin, α-chlordane, *trans*- nonachlor, oxychlordane, heptachlor epoxide, mirex, and HCB were detected in turtle fat. Bird eggs contained detectable residues of *p,p´*- DDE, *p,p´*- DDT, dieldrin, α- and γ-chlordane, *trans*- nonachlor, oxychlordane, heptachlor epoxide, and mirex.
—	—	—	0	0	nr	nd	nd	
—	—	—	67	50	nr	nd–200	nd/10	
—	—	—	100	100	nr	20–30	20	
—	—	—	0	0	nr	nd	nd	
—	—	—	0	0	nr	nd	nd	
—	—	—	0	0	nr	nd	nd	
—	—	—	0	0	nr	nd	nd	
—	—	—	0	0	nr	nd	nd	
—	—	—	0	0	nr	nd	nd	
—	—	—	0	0	nr	nd	nd	
nr	nd–0.00026	nd	—	—	—	—	—	
nr	nd	nd	—	—	—	—	—	
—	—	—	67	33	5	nd–14.8	nd	* 15 OC pesticides that were not detected were not specified. Objective: to assess concentration and distribution of organic pollutants in water, suspended solids, and fish in the Rainy R. near 2 bleached kraft mills prior to remedial action at the mills. DDD was detected in fish and suspended solids in effluents or downstream of the mills. α-HCH, lindane, and pentachlorophenol were detected in water and α-HCH and pentachlorophenol in suspended solids. Contaminants were rarely detected upstream of the mills; those detections were much lower than in mill effluents and usually lower than at downstream sites.
—	—	—	0	0	5	nd	nd	
—	—	—	0	0	5	nd	nd	
—	—	—	100*	100*	0.01	1.3–60*	22*	*Data are for crab muscle. Objective: to determine PAH, PCB, and OC pesticides in biota from a contaminated estuary. PAH, PCB, DDE, and chlordane were detected at all sites. Highest concentrations were from the
—	—	—	100*	100*	0.01	1.4–41*	21.5*	

Table 2.2. Pesticides in bed sediment and aquatic biota from rivers in the United States: State and local monitoring studies—*Continued*

Reference	Sampling date	Sampling locations and strategy	Sampling media and species	Number of samples	Target analytes	Bed sediment % Detection	
						Sites	Samples
		heavily urbanized industrial areas and in more rural areas.					
Mueller and others, 1991	1988	4 sites on the Arkansas River and 5 sites on off-channel reservoirs (Colorado, Kansas). Sediment was collected at the 5 reservoir sites, fish at 2 reservoir and 4 stream sites, and birds at 4 reservoir sites. Sites were associated with irrigation drainage. Fish were composited by species and site; each fish sample was a composite of 3 individuals of comparable length. Sediment was sieved at 2 mm.	Water, sediment (surficial), fish (whole), birds (eggs, livers). Fish species: common carp, gizzard shad.	5 8 2 14	Aldrin	20	20
					Chlordane	100	100
					Chlordane, α-	—	—
					Chlordane, γ-	—	—
					DDD, *o,p´-*	—	—
					DDD, *p,p´-*	20	20
					DDE, *o,p´-*	—	—
					DDE, *p,p´-*	100	100
					DDT, *o,p´-*	—	—
					DDT, *p,p´-*	0	0
					Dieldrin	40	40
					Endosulfan	0	0
					Endrin	0	0
					HCB	—	—
					HCH	—	—
					HCH, α-	—	—
					HCH, β-	—	—
					Heptachlor	0	0
					Heptachlor epoxide	0	0
					Lindane	0	0
					Methoxychlor	0	0
					Mirex	0	0
					Nonachlor, *cis-*	—	—
					Nonachlor, *trans-*	—	—
					Oxychlordane	—	—
					Perthane	0	0
					Toxaphene	0	0
Ong and others, 1991	1987–1988	12 sites (sediment) and 2 sites (whole fish) in the Middle Rio Grande Valley and Bosque Del Apache National Wildlife Refuge (NWR), New Mexico. Areas receiving	Water (whole), sediment (surficial), ground water, fish (whole), birds (whole, eggs).	19 12 2 2 3	Aldrin	8	8
					Chlordane	33	33
					DDD, *p,p´-*	—	—
					DDD	33	33
					DDE, *p,p´-*	—	—
					DDE	58	58
					DDT, *p,p´-*	—	—
					DDT	17	17
					Dieldrin	0	0

Table 2.2. Pesticides in bed sediment and aquatic biota from rivers in the United States: State and local monitoring studies—*Continued*

Bed sediment (µg/kg dry weight)			Aquatic biota (whole fish unless noted)					Comments
			% Detection		(µg/kg wet weight)			
DL	Range	Median	Sites	Samples	DL	Range	Median	
								Elizabeth R., which is the most urbanized area in the study and has documented industrial contamination. Lower levels were found in the nearby Nansemond R. and the more distant York R. Residues were higher in hepatopancreas (higher fat content) than in muscle tissue. Mean DDE and chlordane levels were 275 and 270 µg/kg in hepatopancreas and 27 and 20 µg/kg in muscle. Mean PAH levels in hepatopancreas were 5,238 µg/kg.
0.1	nd–0.1	nd	0	0	nr	nd	nd	Objective: to assess the
0.1	1.0–2.0	1	—	—	—	—	—	occurrence of contaminants in
—	—	—	0	0	nr	nd	nd	the study area; to evaluate the
—	—	—	0	0	nr	nd	nd	impact of irrigation drainage.
—	—	—	0	0	nr	nd	nd	The only OC pesticide detected
0.1	nd–0.3	nd	17	25	10	nd–10	nd	in fish from the 4 stream sites
—	—	—	0	0	nr	nd	nd	(common carp) was *p,p´*- DDE in
0.1	0.4–1.0	0.6	50	63	10	10–80	15	1 of 4 samples. The following 8
—	—	—	0	0	nr	nd	nd	OC pesticides were detected in
0.1	—	nd	0	0	10	—	nd	fish and birds (egg and liver)
0.1	nd–0.3	nd	17	13	10	nd–10	nd	from the 4 reservoir sites:
0.1	—	nd	—	—	—	—	—	β-HCH, oxychlordane,
0.1	—	nd	0	0	nr	nd	nd	heptachlor epoxide, *trans-*
—	—	—	0	0	nr	nd	nd	nonachlor, *p,p´*- DDE, dieldrin,
—	—	—	0	0	nr	nd	nd	*p,p´*- DDD, and *p,p´*- DDT. DDE
—	—	—	0	0	nr	nd	nd	residues reached 1,000–2,300
—	—	—	0	0	10	—	nd	µg/kg in killdeer liver samples
0.1	—	nd	0	0	nr	nd	nd	from 2 sites. Other OCs in biota
0.1	—	nd	0	0	10	—	nd	from reservoir sites generally
0.1	—	nd	0	0	nr	nd	nd	were <260 µg/kg.
0.1	—	nd	0	0	nr	nd	nd	
0.1	—	nd	—	—	—	—	—	
—	—	—	0	0	nr	nd	nd	
—	—	—	33	38	10	nd–10	nd	
—	—	—	0	0	10	—	nd	
1	—	nd	—	—	—	—	—	
10	—	nd	0	0	nr	nd	nd	
0.1	nd–0.1	nd	—	—	—	—	—	Objectives: to collect
1	nd–3.0	nd	—	—	—	—	—	reconnaissance data to determine
—	—	—	0	0	nr	nd	nd	if chemical constituents in
0.1	nd–1.7	nd	—	—	—	—	—	water, bed sediment, and biota
—	—	—	50	50	nr	nd–70	nd/70	associated with irrigation
0.1	nd–1.3	—	—	—	—	—	—	drainage are potentially harmful
—	—	—	0	0	nr	nd	nd	to human health, fish, wildlife,
0.1	nd–0.1	nd	50	50	nr	nd–70	nd/70	or other beneficial uses of water.
0.1	nd	nd	—	—	—	—	—	In 4 bird samples at 4 sites,

Table 2.2. Pesticides in bed sediment and aquatic biota from rivers in the United States: State and local monitoring studies—*Continued*

Reference	Sampling date	Sampling locations and strategy	Sampling media and species	Number of samples	Target analytes	Bed sediment % Detection Sites	Bed sediment % Detection Samples
		irrigation drainage were sampled. Sediment was sampled after irrigation (Nov. 1987) and during herbicide application (Feb. 1988). Biota were sampled in June–July 1988.	Fish species: brown bullhead, threadfin shad.		Endosulfan	0	0
					Endrin	0	0
					Heptachlor	0	0
					Heptachlor epoxide	0	0
					Lindane	0	0
					Methoxychlor	0	0
					Mirex	0	0
					Oxychlordane	—	—
					Perthane	0	0
					Toxaphene	0	0
Pennington and others, 1991	1990	6 sites in Bear Creek watershed, 36 sites on Yazoo River and tributaries, and 2 sites on oxbow lakes (Mississippi). Surface sediment samples were composites of three 5-cm cores. Core samples were 76-cm deep, divided into 10-cm sections; 3 cores were composited by depth.	Sediment (surficial, cores), water.	44 14	Aldrin	0	0
					D, 2,4-	5	5
					DDD, p,p'-	23	23
					DDE, p,p'-	18	18
					DDT, p,p'-	23	23
					Dieldrin	0	0
					Endosulfan I	5	5
					Endosulfan II	0	0
					Endosulfan sulfate	5	5
					Endrin	2	2
					Endrin aldehyde	2	2
					HCH, δ-	0	0
					Heptachlor	27	27
					Heptachlor epoxide	2	2
					Methoxychlor	2	2
Peterson and others, 1991	1988	8 sites (sediment), 7 sites (water), 6 sites (birds) and 2 sites (fish) in the Riverton Reclamation Project, Wyoming. Sites were associated with inflow and irrigation drainage.	Water, sediment (surficial), fish (whole, eggs), birds (liver, eggs). Fish species: carp, rainbow trout, white sucker.	7 8 13 3 18 8	Aldrin	0	0
					Chlordane	0	0
					Chlordane, α-	—	—
					Chlordane, γ-	—	—
					DDD	12.5	12.5
					DDD, o,p'-	—	—
					DDD, p,p'-	—	—
					DDE	50	50
					DDE, o,p'-	—	—
					DDE, p,p'-	—	—
					DDT	12.5	12.5
					DDT, o,p'-	—	—
					DDT, p,p'-	—	—
					Diazinon	0	0
					Dieldrin	0	0
					Endosulfan	0	0
					Endrin	0	0
					Ethion	0	0
					HCB	—	—
					HCH, α-	—	—
					HCH, β-	—	—
					HCH, δ-	—	—
					Heptachlor	0	0
					Heptachlor epoxide	0	0
					Lindane	0	0

Table 2.2. Pesticides in bed sediment and aquatic biota from rivers in the United States: State and local monitoring studies—*Continued*

Bed sediment (µg/kg dry weight)			Aquatic biota (whole fish unless noted)					Comments
			% Detection		(µg/kg wet weight)			
DL	Range	Median	Sites	Sam-ples	DL	Range	Median	
0.1	nd	nd	—	—	—	—	—	*p,p´*- DDE was detected in all
0.1	nd	nd	—	—	—	—	—	samples. In an earlier study
0.1	nd	nd	—	—	—	—	—	(1986), 25 OCs were analyzed in
0.1	nd	nd	—	—	—	—	—	plants (bullrush, curlyleaf pond-
0.1	nd	nd	—	—	—	—	—	weed, coontail, sedge) but none
0.1	nd	nd	—	—	—	—	—	were detected. 1986 fish and
0.1	nd	nd	—	—	—	—	—	bird samples contained *p,p´*-
—	—	—	0	0	nr	nd	nd	DDE (6 of 7 samples), DDD (1
1	nd	nd	—	—	—	—	—	of 7), DDT (1 of 7), and PCBs
10	nd	nd	—	—	—	—	—	(2 of 7 samples).
0.2	nd	nd	—	—	—	—	—	Objective: to assess water and
100–230	nd–302	nd	—	—	—	—	—	sediment quality in the Upper
0.2	nd–68	nd	—	—	—	—	—	Yazoo R. and tributaries, includ-
0.2	nd–44	nd	—	—	—	—	—	ing the heavily agricultural Bear
0.2	nd–55	nd	—	—	—	—	—	Creek watershed. Only those
0.2	nd	nd	—	—	—	—	—	analytes detected in sediment
0.2	nd–5.6	nd	—	—	—	—	—	were specified in the report, so
0.2	nd	nd	—	—	—	—	—	the list of compounds not detect-
0.2	nd–7	nd	—	—	—	—	—	ed that is shown here is probably
0.2	nd–0.9	nd	—	—	—	—	—	incomplete. Data from sediment
0.2	nd–1.9	nd	—	—	—	—	—	cores are not included here.
0.2	nd	nd	—	—	—	—	—	Comparison with historical data
0.2	nd–5.3	nd	—	—	—	—	—	from these watersheds suggested
0.2	nd–0.2	nd	—	—	—	—	—	that OC levels in sediment
0.2	nd	nd	—	—	—	—	—	were declining.
0.1	nd	nd	—	—	—	—	—	Objective: to assess the
0.1	nd	nd	—	—	—	—	—	occurrence of contaminants in
—	—	—	50	15	nr	nd–10	nd	the study area to evaluate the
—	—	—	50	15	nr	nd–10	nd	impact of irrigation drainage.
0.1	nd–0.4	nd	—	—	—	—	—	DDT and(or) metabolites were
—	—	—	0	0	nr	nd	nd	detected in sediment at 4 sites,
—	—	—	50	31	nr	nd–30	nd	all of which were associated
0.1	nd–0.2	nd	—	—	—	—	—	with irrigation drainage. In fish,
—	—	—	0	0	nr	nd	nd	fish eggs, bird livers, and bird
—	—	—	100	100	nr	10–170	20	eggs, all OC pesticides were
0.1	nd–0.6	nd	—	—	—	—	—	below detection limits except
—	—	—	0	0	nr	nd	nd	DDT metabolites. One rainbow
—	—	—	0	0	nr	nd	nd	trout egg sample contained 240
0.1	nd	nd	—	—	—	—	—	µg/kg total PCBs, which
0.1	nd	nd	0	0	nr	nd	nd	approaches the concentration
0.1	nd	nd	—	—	—	—	—	found to cause deformities in
0.1	nd	nd	0	0	nr	nd	nd	rainbow trout fry. 3 of 8 bird
0.1	nd	nd	—	—	—	—	—	egg samples contained detectable
—	—	—	0	0	nr	nd	nd	DDT and DDE. American coot
—	—	—	0	0	nr	nd	nd	livers contained detectable
—	—	—	0	0	nr	nd	nd	oxychlordane, HCB, DDE, and
—	—	—	0	0	nr	nd	nd	total DDT. Dicamba, 2,4-D, and
0.1	nd	nd	—	—	—	—	—	methyl parathion were detected
0.1	nd	nd	0	0	nr	nd	nd	at low concentrations in surface
0.1	nd	nd	0	0	nr	nd	nd	water samples. Parathion was

Table 2.2. Pesticides in bed sediment and aquatic biota from rivers in the United States: State and local monitoring studies—*Continued*

Reference	Sampling date	Sampling locations and strategy	Sampling media and species	Number of samples	Target analytes	Bed sediment % Detection Sites	Bed sediment % Detection Samples
		supplemental samples at contaminated sites.			Dicofol	—	—
					Dieldrin	—	—
					Endosulfan I	—	—
					Endosulfan II	—	—
					Endosulfan sulfate	—	—
					Endrin	—	—
					Ethion	—	—
					HCB	—	—
					HCH, α-	—	—
					HCH, β-	—	—
					HCH, δ-	—	—
					HCH, total	—	—
					Heptachlor	—	—
					Heptachlor epoxide	—	—
					Lindane	—	—
					Methoxychlor	—	—
					Methyl parathion	—	—
					Nonachlor, *cis-*	—	—
					Nonachlor, *trans-*	—	—
					Oxadiazon	—	—
					Oxychlordane	—	—
					Parathion	—	—
					Pentachlorophenol	—	—
					Tetradifon	—	—
					Toxaphene	—	—
Rice, 1991	1986	2 sites located on 2 small streams in Quehanna Wildlife Area in central Pennsylvania.	Fish (whole). Species: brook trout.	2	Chlordane, *cis-*	—	—
					Chlordane, *trans-*	—	—
					DDD, *o,p´-*	—	—
					DDD, *p,p´-*	—	—
					DDE, *o,p´-*	—	—
					DDE, *p,p´-*	—	—
					DDT, *o,p´-*	—	—
					DDT, *p,p´-*	—	—
					Dieldrin	—	—
					Endrin	—	—
					Heptachlor epoxide	—	—
					Hexachlorobenzene	—	—
					Lindane	—	—
					Mirex	—	—
					Nonachlor, *cis-*	—	—
					Nonachlor, *trans-*	—	—
					Oxychlordane	—	—
					Toxaphene	—	—
Roddy and others, 1991	1988–1989	10 sites (water), 7 sites (sediment) and 3 sites (fish) in the Belle Fourche Reclamation Area, South Dakota. Sites were	Water, sediment (surficial), fish (whole), bird (egg). Fish species: carp,	20 9 5 2	Atrazine	0	0
					Carbofuran	0	0
					Chlordanes, other	—	—
					DDD, *o,p´-*	—	—
					DDD, *p,p´-*	—	—
					DDE, *o,p´-*	—	—
					DDE, *p,p´-*	—	—

Table 2.2. Pesticides in bed sediment and aquatic biota from rivers in the United States: State and local monitoring studies—*Continued*

Bed sediment (µg/kg dry weight)			Aquatic biota (whole fish unless noted)					Comments
			% Detection		(µg/kg wet weight)			
DL	Range	Median	Sites	Sam-ples	DL	Range	Median	
—	—	—	nr	nr	100	nr	nr	for 1988–1989 were 373 and
—	—	—	67*†	34*	5	nd–620*	nd*	1,020 µg/kg for fillets and whole
—	—	—	67*†	27*	5	nd–687*	nd*	fish, respectively. Principal
—	—	—	nr	nr	70	nr	nr	components of chlordane were
—	—	—	nr	nr	85	nr	nr	*cis-* and *trans-* nonachlor and
—	—	—	22*†	4*	15	nd–30*	nd*	*trans-* chlordane. Diazinon was
—	—	—	nr	nr	20	nr	nr	detected in only 2 of over 600
—	—	—	67*†	18*	2	nd–53	nd*	samples analyzed during 1978–
—	—	—	nr	9*	2	nd–8.3*	nd*	1987, but was detected in 3 sam-
—	—	—	0*†	0*	10	nd*	nd*	ples during 1988–1989 at up to
—	—	—	0*†	0*	5	nd*	nd*	98 µg/kg (the highest level yet
—	—	—	78*†	16*	2–10	nr	nd*	found). The first detection of
—	—	—	0*†	0*	5	nd*	nd*	methoxychlor occurred in 1987
—	—	—	1*†	2*	5	nd–14*	nd*	(at 4 sites at up to 28 µg/kg); it
—	—	—	nr	15*	2	nd–15*	nd*	was detected in 4 samples during
—	—	—	22*†	3*	15	nd–44*	nd*	1988–1989 at up to 44 µg/kg. The
—	—	—	22*†	2*	10	nd–11*	nd*	occurrence of parathion and
—	—	—	nr	nr	5	nr	nr	methyl parathion also may be
—	—	—	nr	nr	5	nr	nr	increasing; these two pesticides
—	—	—	44*†	7*	5	nd–2,200*	nd*	were detected in 6 and 3
—	—	—	nr	nr	5	nr	nr	samples, respectively, during
—	—	—	44*†	4*	10	nd–38*	nd*	1988–1989, compared with 6 and
—	—	—	100*§	14*§	2	nd–3.9*§	nd*§	0 samples during 1978–1987.
—	—	—	nr	nr	10	nr	nr	
—	—	—	56*†	16*	100	nd–6,800*	nd*	
—	—	—	0	0	10	nd	nd	Objective: to collect and analyze
—	—	—	0	0	10	nd	nd	samples for a reconnaisance
—	—	—	0	0	10	nd	nd	screening of contaminent
—	—	—	0	0	10	nd	nd	residues in fish from the study
—	—	—	0	0	10	nd	nd	area. The study area currently is
—	—	—	50	50	10	nd–10	nd	forested wildlife habitat, but has
—	—	—	0	0	10	nd	nd	a history of industrial activities.
—	—	—	0	0	10	nd	nd	Very low levels of *p,p´-* DDE
—	—	—	0	0	10	nd	nd	were detected in one sample. No
—	—	—	0	0	10	nd	nd	other pesticides on the analyte
—	—	—	0	0	10	nd	nd	list or PCBs were detected.
—	—	—	0	0	10	nd	nd	
—	—	—	0	0	10	nd	nd	
—	—	—	0	0	10	nd	nd	
—	—	—	0	0	10	nd	nd	
—	—	—	0	0	10	nd	nd	
—	—	—	0	0	10	nd	nd	
10	nd	nd	—	—	—	—	—	Objective: to assess the
10	nd	nd	—	—	—	—	—	occurrence of constituents in the
—	—	—	0	0	20	nd	nd	study area to evaluate the impact
—	—	—	0	0	10	nd	nd	of irrigation drainage.
—	—	—	33	40	10	nd–10	nd	Concentrations of OCs and
—	—	—	0	0	10	nd	nd	PCBs in fish samples were not
—	—	—	67	80	10	nd–30	10	significantly elevated. OCs were

Table 2.2. Pesticides in bed sediment and aquatic biota from rivers in the United States: State and local monitoring studies—*Continued*

Reference	Sampling date	Sampling locations and strategy	Sampling media and species	Number of samples	Target analytes	Bed sediment % Detection Sites	Bed sediment % Detection Samples
Sloterdijk, 1991	1979–1981	98 sites in the St. Lawrence River at Lake St. Francis (New York; Ontario and Quebec, Canada).	Sediment (surficial).	98	Aldrin	nr	nr
					Chlordane	nr	nr
					DDD	nr	nr
					DDE	nr	nr
					DDT	nr	nr
					DDT, *o,p´-*	nr	nr
					Dieldrin	nr	nr
					HCB	nr	nr
					HCH, total	nr	nr
					Heptachlor	nr	nr
					Methoxychlor	nr	nr
					Mirex	nr	nr
					Pentachlorophenol	nr	0
Sorenson and Schwarz-bach, 1991	1988	12 sites (sediment), 9 sites (fish), 2 sites (birds), and 1 site (invertebrate) in the Upper Klamath Basin (California and Oregon). Sites were located above and below irrigation drain inputs.	Sediment (surficial), fish (whole), invertebrates, birds (eggs, carcass). Aquatic species: mussels, Klamath sucker, rainbow trout, Sacramento perch, tui chub.	12 9 2 20 3	Aldrin	0	0
					Chlordane	8	8
					DDD	42	42
					DDD, *o,p´-*	—	—
					DDD, *p,p´-*	—	—
					DDE	75	75
					DDE, *o,p´-*	—	—
					DDE, *p,p´-*	—	—
					DDT	8	8
					DDT, *o,p´-*	—	—
					DDT, *p,p´-*	—	—
					Dieldrin	0	0
					Endosulfan	0	0
					Endrin	0	0
					HCB	—	—
					HCH, α-	—	—
					HCH, β-	—	—
					HCH, δ-	—	—
					Heptachlor	0	0
					Heptachlor epoxide	0	0
					Lindane	0	0
					Methoxychlor	0	0
					Mirex	0	0
					Nonachlor, *cis-*	—	—
					Nonachlor, *trans-*	—	—
					Oxychlordane	—	—
					Toxaphene	0	0
Sowards and others, 1991	1987–1989	1 site (sediment) and 6 sites (fish) in rivers and lakes in South Dakota. Sites were selected where contamination was known or suspected to occur, or where no background	Sediment, fish (whole), plants. Species: white sucker, common carp, goldeye, channel catfish,	2 17 2	Aldrin	—	—
					Chlordane, *cis-*	0	0
					Chlordane, total	—	—
					Chlordane, *trans-*	0	0
					Dacthal	—	—
					DDD, *o,p´-*	0	0
					DDD, *p,p´-*	0	0
					DDE, *o,p´-*	0	0
					DDE, *p,p´-*	0	0
					DDT, *o,p´-*	0	0

Table 2.2. Pesticides in bed sediment and aquatic biota from rivers in the United States: State and local monitoring studies—*Continued*

Bed sediment (µg/kg dry weight)			Aquatic biota (whole fish unless noted)					Comments
			% Detection		(µg/kg wet weight)			
DL	Range	Median	Sites	Samples	DL	Range	Median	
0.1	nd–1.1	nd	—	—	—	—	—	Objective: to evaluate Lake St.
0.1	nd–5.1	0.7	—	—	—	—	—	Francis as a depositional zone
0.1	nd–5.8	nd	—	—	—	—	—	for toxic substances transported
0.1	nd–8.7	0.9	—	—	—	—	—	from Lake Ontario and the
0.1	nd–9.7	nd	—	—	—	—	—	international portion of the St.
0.1	nd–1.1	nd	—	—	—	—	—	Lawrence R. PCBs were detected
0.1	nd–2.3	nd	—	—	—	—	—	at high concentrations relative to
0.1	nd–13.0	0.6	—	—	—	—	—	other OC compounds. Local
0.1	nd–2.0	0.6	—	—	—	—	—	sources were more significant
0.1	nd–1.2	nd	—	—	—	—	—	than transport from Lake
0.1	nd–3.2	nd	—	—	—	—	—	Ontario.
0.1	nd–3.3	nd	—	—	—	—	—	
1	nd	nd	—	—	—	—	—	
0.1	nd	nd	—	—	—	—	—	Objective: to assess the
1	nd–13	nd	13	11	nr	nd–30	nd	occurrence of constituents in the
0.1	nd–2.7	nd	—	—	—	—	—	study area to evaluate the impact
—	—	—	13	11	nr	nd–20	nd	of irrigation drainage on water
—	—	—	13	11	nr	nd–20	nd	quality. Only DDE was
0.1	nd–6.6	0.73	—	—	—	—	—	frequently detected in sediment
—	—	—	0	0	nr	nd	nd	and fish tissue in the study area.
—	—	—	63	67	nr	nd–50	10	Bird eggs contained HCB,
0.1	nd–0.4	nd	—	—	—	—	—	β-HCH, oxychlordane,
—	—	—	0	0	nr	nd	nd	heptachlor epoxide, chlordane,
—	—	—	0	0	nr	nd	nd	and *trans-* nonachlor.
0.1	nd	nd	13	11	nr	nd–10	nd	Concentrations of all OC
0.1	nd	nd	—	—	—	—	—	pesticides detected in biota
0.1	nd	nd	0	0	nr	nd	nd	were relatively low (≤50 µg/kg).
—	—	—	0	0	nr	nd	nd	Concentrations of OC pesticides
—	—	—	0	0	nr	nd	nd	in bed sediment and waterfowl
—	—	—	0	0	nr	nd	nd	tissues were highest in the Tule
0.1	nd	nd	—	—	—	—	—	Lake area downstream of
0.1	nd	nd	0	0	nr	nd	nd	irrigation drainage sources.
0.1	nd	nd	0	0	nr	nd	nd	
0.1	nd	nd	—	—	—	—	—	
0.1	nd	nd	0	0	nr	nd	nd	
—	—	—	0	0	nr	nd	nd	
—	—	—	13	11	nr	nd–10	nd	
—	—	—	0	0	nr	nd	nd	
10	nd	nd	0	0	nr	nd	nd	
—	—	—	0	0	2	nd	nd	Objective: to determine
nr	nd	nd	100	100	2	2–5	4	contaminant residues in aquatic
—	—	—	17	40	2–20	nd–8	nd	environments in federal lands
nr	nd	nd	100	100	2	3	3	(refuges) in South Dakota.
—	—	—	100	83	6	nd–14	5	Report compiles data from all
nr	nd	nd	17	35	2–10	nd–36	nd	studies conducted from 1985 to
nr	nd	nd	50	71	2–10	nd–30	10	1989, of which 4 studies
nr	nd	nd	0	0	2–10	nd	nd	measured pesticides in
nr	nd	nd	67	82	2–10	nd–50	22	sediment and(or) aquatic biota.
nr	nd	nd	17	29	2–10	nd–10	nd	Specific objectives of

Table 2.2. Pesticides in bed sediment and aquatic biota from rivers in the United States: State and local monitoring studies—*Continued*

Reference	Sampling date	Sampling locations and strategy	Sampling media and species	Number of samples	Target analytes	Bed sediment % Detection	
						Sites	Samples
		information was available. Fish were analyzed individually. Sediment and cattail samples were composites.	sauger, shorthead redhorse, black bullhead, northern pike, walleye, yellow perch, cattails.		DDT, *p,p´-*	0	0
					Dicofol	—	—
					Dieldrin	0	0
					Endosulfan I	—	—
					Endosulfan II	—	—
					Endosulfan sulfate	—	—
					Endrin	0	0
					HCB	0	0
					HCH, α-	0	0
					HCH, β-	0	0
					HCH, δ-	0	0
					Heptachlor	—	—
					Heptachlor epoxide	0	0
					Lindane	0	0
					Methoxychlor	—	—
					Mirex	0	0
					Nonachlor, *cis-*	0	0
					Nonachlor, *trans-*	0	0
					Oxychlordane	0	0
					Tetradifon	—	—
					Toxaphene	0	0
Welsh and Olson, 1991	1989	2 sites in Fort Clark and Heart Butte Irrigation Units in the Missouri River Basin (North Dakota). Sediment was sampled at both sites and fish only from Heart Butte (Heart River). Fish samples were composites of mixed species of minnows (forage fish).	Sediment (surficial), water, fish (whole), frogs and toads, plants, birds (liver). Fish species: minnows. Plant species: coontail, sago pondweed. Frogs and toads: species unknown.	3 3 2 3 4 5	Aldrin	0	0
					Chlordane, α-	0	0
					Chlordane, γ-	0	0
					DDD, *o,p´-*	0	0
					DDD, *p,p´-*	0	0
					DDE, *o,p´-*	0	0
					DDE, *p,p´-*	0	0
					DDT, *o,p´-*	0	0
					DDT, *p,p´-*	0	0
					Dieldrin	0	0
					Endrin	0	0
					HCB	0	0
					HCH, α-	0	0
					HCH, β-	0	0
					HCH, δ-	0	0
					HCH, total	0	0
					Heptachlor	0	0
					Heptachlor epoxide	0	0
					Lindane	0	0
					Mirex	0	0
					Nonachlor, *cis-*	0	0
					Nonachlor, *trans-*	0	0
					Toxaphene	0	0

Table 2.2. Pesticides in bed sediment and aquatic biota from rivers in the United States: State and local monitoring studies—*Continued*

Bed sediment (µg/kg dry weight)			Aquatic biota (whole fish unless noted)					Comments
			% Detection		(µg/kg wet weight)			
DL	Range	Median	Sites	Samples	DL	Range	Median	
nr	nd	nd	17	24	2–10	nd–33	nd	individual studies were not
—	—	—	0	0	2	nd	nd	reported. Data were presented
nr	nd	nd	50	47	2–10	nd–15	10	without interpretation.
—	—	—	0	0	2	nd	nd	
—	—	—	0	0	2	nd	nd	
—	—	—	0	0	2	nd	nd	
nr	nd	nd	17	6	2–10	nd–3	nd	
nr	nd	nd	17	35	2–10	nd–10	nd	
nr	nd	nd	17	24	2–10	nd–6	nd	
nr	nd	nd	0	0	2–10	nd	nd	
nr	nd	nd	0	0	4–10	nd	nd	
—	—	—	100	33	2	nd–2	nd	
nr	nd	nd	17	35	2–10	nd–5	nd	
nr	nd	nd	0	0	2–10	nd	nd	
—	—	—	100	83	10–44	nd–25	11	
nr	nd	nd	17	18	2–10	nd–3	nd	
nr	nd	nd	17	35	2–10	nd–4	nd	
nr	nd	nd	33	53	2–10	nd–20	3	
nr	nd	nd	17	35	2–10	nd–4	nd	
—	—	—	0	0	2–6	nd	nd	
nr	nd	nd	0	0	8–50	nd	nd	
10	nd	nd	0	0	10	nd	nd	Objective: to determine if
10	nd	nd	0	0	10	nd	nd	contaminant concentrations in
10	nd	nd	0	0	10	nd	nd	sediment and biota in the Fort
10	nd	nd	0	0	10	nd	nd	Clark and Heart Butte
10	nd	nd	0	0	10	nd	nd	reclamation units warrant a
10	nd	nd	0	0	10	nd	nd	complete reconnaissance study.
10	nd	nd	0	0	10	nd	nd	There was no evidence of OC
10	nd	nd	0	0	10	nd	nd	contamination based on the
10	nd	nd	0	0	10	nd	nd	samples taken. Authors
10	nd	nd	0	0	10	nd	nd	suggested that predator fish and
10	nd	nd	0	0	10	nd	nd	piscivorous birds would have
10	nd	nd	0	0	10	nd	nd	been better indicators of mercury
10	nd	nd	0	0	10	nd	nd	contamination; the same may be
10	nd	nd	0	0	10	nd	nd	true of OC pesticides.
100	nd	nd	0	0	100	nd	nd	
10	nd	nd	0	0	10	nd	nd	
10	nd	nd	0	0	10	nd	nd	
10	nd	nd	0	0	10	nd	nd	
10	nd	nd	0	0	10	nd	nd	
10	nd	nd	0	0	10	nd	nd	
10	nd	nd	0	0	10	nd	nd	
100	nd	nd	0	0	100	nd	nd	

Table 2.2. Pesticides in bed sediment and aquatic biota from rivers in the United States: State and local monitoring studies—*Continued*

Reference	Sampling date	Sampling locations and strategy	Sampling media and species	Number of samples	Target analytes	Bed sediment % Detection Sites	Bed sediment % Detection Samples
Allen, 1992	1991	4 sites on the Missouri and Kansas rivers (Kansas). Sites were located primarily to establish background contaminant levels. Fish samples were composites.	Sediment (surficial), fish (*). Species: common carp.	4 4	Aldrin	0	0
					Chlordane, α-	0	0
					Chlordane, γ-	0	0
					DDD, *o,p´-*	0	0
					DDD, *p,p´-*	25	25
					DDE, *o,p´-*	0	0
					DDE, *p,p´-*	25	25
					DDT, *o,p´-*	0	0
					DDT, *p,p´-*	0	0
					Dieldrin	0	0
					Endrin	0	0
					HCH, α-	0	0
					HCH, β-	0	0
					HCH, δ-	0	0
					Heptachlor	0	0
					Heptachlor epoxide	0	0
					Hexachlorobenzene	0	0
					Lindane	0	0
					Mirex	0	0
					Nonachlor, *cis-*	0	0
					Nonachlor, *trans-*	0	0
					Oxychlordane	0	0
					Toxaphene	0	0
Allen and Nash, 1992	1991	16 sites (sediment) and 7 sites (fish) on Marais des Cygnes River (Kansas, Missouri) and nearby tributaries and wetlands.	Sediment (surficial), fish (*). Species: river carpsucker, common carp, smallmouth buffalo, largemouth bass, bigmouth buffalo.	16 11	Chlordane, α-	0	0
					Chlordane, γ-	0	0
					DDD, *o,p´-*	0	0
					DDD, *p,p´-*	0	0
					DDE, *o,p´-*	0	0
					DDE, *p,p´-*	0	0
					DDT, *o,p´-*	0	0
					DDT, p,p´-	0	0
					Dieldrin	0	0
					Endrin	0	0
					HCH, α-	0	0
					HCH, β-	0	0
					HCH, δ-	0	0
					Heptachlor epoxide	0	0
					Hexachlorobenzene	0	0
					Lindane	0	0
					Mirex	0	0
					Nonachlor, *cis-*	0	0
					Nonachlor, *trans-*	0	0
					Oxychlordane	0	0
Cashman and others, 1992	1987	1 site in the Sacramento–San Joaquin Delta (California). From the delta, moribund and healthy fish (adult and juvenile)	Fish (liver). Species: striped bass.	24	Molinate	—	—
					Thiobencarb	—	—
					Triazines	—	—

Table 2.2. Pesticides in bed sediment and aquatic biota from rivers in the United States: State and local monitoring studies—*Continued*

Bed sediment (µg/kg dry weight)			Aquatic biota (whole fish unless noted)					Comments
			% Detection		(µg/kg wet weight)			
DL	Range	Median	Sites	Samples	DL	Range	Median	
10	nd	nd	0*	0*	10	nd*	nd*	* Data are for fish (tissue type
10	nd	nd	100*	100*	10	10–120*	25*	not reported).
10	nd	nd	100*	100*	10	10–100*	25*	† µg/kg wet wt.
10	nd	nd	0*	0*	10	nd*	nd*	Objective: to conduct
10	nd–10†	nd	50*	50*	10	nd–10*	nd/10*	reconnaissance survey of water
10	nd	nd	0*	0*	10	nd*	nd*	quality problems and collect
10	nd–10†	nd	100*	100*	10	30–130*	35*	baseline data. Concentrations of
10	nd	nd	0*	0*	10	nd*	nd*	cyclodiene compounds in
10	nd	nd	25*	25*	10	nd–10*	nd*	common carp exceeded
10	nd	nd	100*	100*	10	30–40*	40*	NAS/NAE guidelines.
10	nd	nd	0*	0*	10	nd*	nd*	Concentrations of all other
10	nd	nd	0*	0*	10	nd*	nd*	analytes were below levels of
10	nd	nd	0*	0*	10	nd*	nd*	concern.
10	nd	nd	0*	0*	10	nd*	nd*	
10	nd	nd	0*	0*	10	nd*	nd*	
10	nd	nd	0*	0*	10	nd*	nd*	
10	nd	nd	0*	0*	10	nd*	nd*	
10	nd	nd	0*	0*	10	nd*	nd*	
10	nd	nd	0*	0*	10	nd*	nd*	
10	nd	nd	50*	50*	10	nd–40*	nd/10*	
10	nd	nd	100*	100*	10	10–120*	25*	
10	nd	nd	0*	0*	10	nd*	nd*	
10	nd	nd	0*	0*	10	nd*	nd*	
10	nd	nd	100*	100*	10	11–55*	16*	* Data are for fish (tissue type
10	nd	nd	100*	100*	10	12–40*	12*	not reported).
10	nd	nd	86*	73*	10	nd–15*	11*	Objective: to evaluate
10	nd	nd	86*	82*	10	nd–30*	13*	contamination on land being
10	nd	nd	100*	100*	10	12–18*	12*	considered for acquisition as a
10	nd	nd	100*	100*	10	12–65*	38*	national wildlife refuge. No
10	nd	nd	57*	45*	10	nd–16*	nd*	OC compounds were
10	nd	nd	43*	36*	10	nd–25*	nd*	detected in sediment. The
10	nd	nd	0*	27*	10	nd–52*	nd*	highest concentrations of OC
10	nd	nd	29*	0*	10	nd*	nd*	compounds in fish were of
10	nd	nd	100*	91*	10	11–13*	11*	chlordane and nonachlor which
10	nd	nd	71*	73*	10	12–15*	14*	exceeded NAS/NAE guidelines
10	nd	nd	0*	0*	10	nd*	nd*	at two sites. Concentrations of
10	nd	nd	14*	18*	10	nd–13*	nd*	other compounds were below
10	nd	nd	0*	0*	10	nd*	nd*	levels considered harmful.
10	nd	nd	100*	100*	10	11*	11*	
10	nd	nd	0*	0*	10	nd*	nd*	
10	nd	nd	43*	45*	10	nd–33*	nd*	
10	nd	nd	100*	100*	10	11–78*	14*	
10	nd	nd	29*	36*	10	10–22*	nd*	
—	—	—	100*	nr	nr	nr	42–79*†	* Data are for fish liver.
—	—	—	100*	nr	nr	nr	496–1,190*†	† Range of mean values for 3 groups (ocean control, delta
—	—	—	100*	nr	nr	nr	23–56*†	control, and delta moribund fish). Objective: to compare contaminants in healthy and moribund striped bass from the delta dur-

Table 2.2. Pesticides in bed sediment and aquatic biota from rivers in the United States: State and local monitoring studies—*Continued*

Reference	Sampling date	Sampling locations and strategy	Sampling media and species	Number of samples	Target analytes	Bed sediment % Detection Sites	Bed sediment % Detection Samples
					Nonachlor, *trans-*	—	—
					Oxychlordane	—	—
					Perthane	0	0
					Toxaphene	0	0
Edwards, 1992; Edwards and Curtiss, 1993	1988	9 sites in Johnson Creek Basin (Oregon). Sediment was sampled during low flow, and water during low flow and storm runoff. Sediment samples were composites of subsamples from 10–20 (random) points at each site. Fractions <63 and >63 μm were analyzed separately.	Sediment (surficial), water (whole).	10 59	Aldrin	22	10
					Chlordane	22	10
					DDD, *p,p´-*	78	60
					DDE, *p,p´-*	78	65
					DDT, *p,p´-*	67	55
					Dieldrin	11	5
					Endrin	33	15
					HCH, α-	44	20
					HCH, β-	22	10
					Heptachlor	11	5
					Heptachlor epoxide	11	10
					Lindane	22	10
					Methoxychlor, *p,p´-*	78	50
King and others, 1992	1987	9 sites in the San Pedro River Basin (Arizona). Sediment samples were composites of 5 subsamples at each site. Fish samples were composites, for each species, of individuals of nearly equal weight and nearly equal length	Sediment (surficial), fish (whole), lizards, frogs, crayfish, birds. Species include green sunfish, black bullhead, longfin dace, desert sucker, Sonoran sucker, bullfrog.	9 16 7 2 1 9	Chlordane, γ-	0	0
					DDD, *o,p´-*	0	0
					DDD, *p,p´-*	0	0
					DDE, *o,p´-*	0	0
					DDE, *p,p´-*	0	0
					DDT, *o,p´-*	0	0
					DDT, *p,p´-*	0	0
					Dieldrin	0	0
					Endrin	0	0
					HCB	0	0
					HCH	0	0
					Heptachlor epoxide	0	0
					Mirex	0	0
					Nonachlor, *cis-*	0	0
					Nonachlor, *trans-*	0	0
					Oxychlordane	0	0
					Toxaphene	0	0
Nash and Charbonneau, 1992	1987	4 sites (fish) at Squaw Creek National Wildlife Refuge (NWR), Missouri. Sites were located where Davis and Squaw creeks enter and	Fish (whole), turtle (liver). Species include black buffalo, channel catfish, gizzard shad, carp,	16 1	Chlordane, total	—	—
					DDT, total	—	—
					Dieldrin	—	—
					Endrin	—	—
					HCB	—	—
					HCH, β-	—	—
					Heptachlor epoxide	—	—
					Lindane	—	—

Table 2.2. Pesticides in bed sediment and aquatic biota from rivers in the United States: State and local monitoring studies—*Continued*

Bed sediment (µg/kg dry weight)			Aquatic biota (whole fish unless noted)					Comments
			% Detection		(µg/kg wet weight)			
DL	Range	Median	Sites	Samples	DL	Range	Median	
—	—	—	0	0	10	nd	nd	
—	—	—	0	0	10	nd	nd	
0.1	nd	nd	—	—	—	—	—	
10	nd	nd	0	0	5	nd	nd	
5	nd–8	nd	—	—	—	—	—	Objective: to assess water quality in the basin. Edwards (1992) presented raw data and methods (water and sediment), but no interpretation. Edwards and Curtiss (1993) presented pesticide data for sediment only, including interpretation. Highest chlordane, DDTs, and methoxychlor were at river mile 17.4. At that site, residues were consistently higher in the <63 than in the >63 µm fraction. At other sites, one fraction was not consistently higher than the other. OC pesticides levels in water appeared to be higher in storm runoff than at low flow.
50	nd–290	nd	—	—	—	—	—	
5	nd–568	8	—	—	—	—	—	
5	nd–388	8	—	—	—	—	—	
5	nd–226	7	—	—	—	—	—	
5	nd–9	nd	—	—	—	—	—	
5	nd–6	nd	—	—	—	—	—	
5	nd–11	nd	—	—	—	—	—	
5	nd–13	nd	—	—	—	—	—	
5	nd–7	nd	—	—	—	—	—	
5	nd–5	nd	—	—	—	—	—	
5	nd–6	nd	—	—	—	—	—	
5	nd–366	nd/11	—	—	—	—	—	
10	nd	nd	0	0	10	nd	nd	Objective: to evaluate the impacts of environmental pollutants in the San Pedro R. on fish and wildlife. This study followed up a 1986 sampling in which longfin dace contained 11–56 times more DDT than DDE. In this study, DDT was not detected in longfin dace and DDE levels were 10–30 µg/kg (below the geometric mean from the FWS's NCBP). Longfin dace had higher lipid levels and DDE detection frequency than other fish species had. The highest DDE level detected was in western whiptail lizards (90 µg/kg). Longfin dace and whiptail lizards also contained low levels of some chlordane components.
10	nd	nd	0	0	10	nd	nd	
10	nd	nd	0	0	10	nd	nd	
10	nd	nd	0	0	10	nd	nd	
10	nd	nd	88	63	10	nd–30	10	
10	nd	nd	0	0	10	nd	nd	
10	nd	nd	0	0	10	nd	nd	
10	nd	nd	0	0	10	nd	nd	
10	nd	nd	0	0	10	nd	nd	
10	nd	nd	0	0	10	nd	nd	
10	nd	nd	0	0	10	nd	nd	
10	nd	nd	0	0	10	nd	nd	
10	nd	nd	0	0	10	nd	nd	
10	nd	nd	25	13	10	nd–10	nd	
10	nd	nd	0	0	10	nd	nd	
10	nd	nd	0	0	10	nd	nd	
—	—	—	100	94	nr	nd–80	nr	Objective: to determine baseline data on residues in the Squaw Creek NWR. Dieldrin is elevated relative to the national geometric mean from the FWS's NCBP.
—	—	—	100	94	nr	nd–40	30	
—	—	—	100	94	nr	nd–180	60	
—	—	—	0	0	nr	nd	nd	
—	—	—	0	0	nr	nd	nd	
—	—	—	0	0	nr	nd	nd	
—	—	—	100	69	nr	nd–30	10	
—	—	—	0	0	nr	nd	nd	

Table 2.2. Pesticides in bed sediment and aquatic biota from rivers in the United States: State and local monitoring studies—*Continued*

Reference	Sam-pling date	Sampling locations and strategy	Sampling media and species	Number of samples	Target analytes	Bed sediment % Detection Sites	Bed sediment % Detection Samples
		principal diet of river otters were collected.	white sucker, crayfish.		Dieldrin	—	—
					Endrin	—	—
					HCB	—	—
					HCH, α-	—	—
					HCH, β-	—	—
					Heptachlor epoxide	—	—
					Lindane	—	—
					Mirex	—	—
					Nonachlor, *trans-*	—	—
					Oxychlordane	—	—
					Toxaphene	—	—
Rice, 1992b	1991	10 sites (sediment) along the Mahoning River and its relatively uncontaminated tributary, Hickory Run (Pennsylvania). Fish were collected at 2 sites on the Mahoning and Shenango rivers. Sediment samples (fine-grained) were taken from depositional areas. Fish samples were composites of 2–5 fish.	Sediment (surficial), fish (whole, fillet). Species: carp, white sucker, rock bass, pumpkin-seed.	10 4 6	Chlordane, α-	0	0
					Chlordane, γ-	0	0
					DDD, *o,p´-*	0	0
					DDD, *p,p'*	0	0
					DDE, *o,p´-*	0	0
					DDE, *p,p´-*	0	0
					DDT *o,p´-*	0	0
					DDT, *p,p´-*	0	0
					Dieldrin	10	10
					Endrin	0	0
					HCB	0	0
					HCH, α-	0	0
					HCH, β-	0	0
					HCH, δ-	0	0
					Heptachlor epoxide	0	0
					Lindane	0	0
					Mirex	0	0
					Mirex, (*cis*)5,10-dihydro	0	0
					Mirex, (*trans*)5,10-dihydro	0	0
					Mirex, 10-monohydro	0	0
					Mirex, 2,8-dihydro	0	0
					Mirex, 8-monohydro	0	0
					Nonachlor, *cis-*	0	0
					Nonachlor, *trans-*	0	0
					Oxychlordane	0	0
					Toxaphene	0	0
Rinella and others, 1992	1987–1991	28 sites (sediment) and 33 sites (biota) in the Yakima River Basin (Washington). Sediment samples were collected in late summer or autumn (at the end	Water (whole, filtered), sediment (surficial: <2 mm, <180 um), <62 um),	81 53 20 3 19	Aldrin	7*	7*
					Chlordane, *cis-*	—	—
					Chlordane, total	18*	19*
					Chlordane, *trans-*	—	—
					DDD, *o,p´-*	—	—
					DDD, *p,p´-*	75*	79*
					DDE, *o,p´-*	—	—
					DDE, *p,p´-*	86*	86*

Table 2.2. Pesticides in bed sediment and aquatic biota from rivers in the United States: State and local monitoring studies—*Continued*

Bed sediment (µg/kg dry weight)			Aquatic biota (whole fish unless noted)					Comments
			% Detection		(µg/kg wet weight)			
DL	Range	Median	Sites	Samples	DL	Range	Median	
—	—	—	0*	0*	10	nd*	nd*	all sites except the Clarion R.
—	—	—	0*	0*	10	nd*	nd*	near Ridgeway. Oxychlordane
—	—	—	20*	11*	10	nd–48*	nd*	and DDT levels at this site ex-
—	—	—	0*	0*	10	nd*	nd*	ceeded geometric means from the
—	—	—	0*	0*	10	nd	nd*	FWS's NCBP. PCBs were elevat-
—	—	—	0*	0*	10	nd	nd*	ed at this site and at Tionesta
—	—	—	0*	0*	10	nd	nd*	Creek. The author concluded that
—	—	—	0*	0*	10	nd	nd*	no otter releases should be made
—	—	—	0*	0*	10	nd	nd*	at Clarion R. near Ridgeway, and
—	—	—	20*	22*	10	nd–18*	nd*	that the reproductive success of
—	—	—	0*	0*	100	nd*	nd*	otters previously released in Tionesta Creek be monitored.
10	nd	nd	100	100	10	20–50	25	Objective: to assess the extent
10	nd	nd	100	83	10	nd–30	15	and degree of contamination of
10	nd	nd	0	0	10	nd	nd	sediment and fish in the
10	nd	nd	100	83	10	nd–50	15	Pennsylvania reach of the
10	nd	nd	0	0	10	nd	nd	Mahoning R. PCBs were the
10	nd	nd	100	100	10	20–140	30	most frequently detected of the
10	nd	nd	0	0	10	nd	nd	OC compounds analyzed, and
10	nd	nd	0	0	10	nd	nd	had the highest concentrations.
10	nd–20	nd	100	67	10	nd–40	20	All fish samples had PCB
10	nd	nd	0	0	10	nd	nd	concentrations that exceeded
10	nd	nd	0	0	10	nd	nd	the geometric mean from the
10	nd	nd	0	0	10	nd	nd	FWS's NCBP. Chlordane in
10	nd	nd	0	0	10	nd	nd	several fish samples was
10	nd	nd	0	0	10	nd	nd	slightly elevated relative to the
10	nd	nd	100	67	10	nd–30	15	geometric mean concentration
10	nd	nd	0	0	10	nd	nd	from the NCBP.
10	nd	nd	0	0	10	nd	nd	
10	nd	nd	0	0	10	nd	nd	
10	nd	nd	0	0	10	nd	nd	
10	nd	nd	0	0	10	nd	nd	
10	nd	nd	0	0	10	nd	nd	
10	nd	nd	0	0	10	nd	nd	
10	nd	nd	0	0	10	nd	nd	
10	nd	nd	100	100	10	20–50	35	
10	nd	nd	100	83	10	nd–20	10	
50	nd	nd	0	0	50	nd	nd	
0.1–10	nd–0.4*	nd*	—	—	—	—	—	* Data are for all sediment
—	—	—	42	43	10	nd–40	nd	samples (<2 mm, <180 µm,
1–10	nd–15*	nd*	—	—	—	—	—	<62 µm), pooled.
—	—	—	12	10	10	nd–10	nd	Objective: to present methods
—	—	—	45	47	10	nd–70	nd	and data for study designed to
0.1–10	nd–440*	3.7*	52	62	10	nd–420	20	describe spatial distribution of
—	—	—	36	22	10	nd–30	nd	contaminants in surface water,
0.1–10	nd–1,700*	13*	82	83	10	nd–3,400	170	stream sediment, and biota in

Table 2.2. Pesticides in bed sediment and aquatic biota from rivers in the United States: State and local monitoring studies—*Continued*

Reference	Sampling date	Sampling locations and strategy	Sampling media and species	Number of samples	Target analytes	Bed sediment % Detection	
						Sites	Samples
		of irrigation season). Samples were sieved and the <2 mm fraction was analyzed; for some samples, the <180 μm and <62 μm fractions also were analyzed. Most fish samples were composites of about 10 individual fish.	sediment (suspended), soil, fish (whole), mollusks, crayfish, aquatic plants. Species include rainbow trout, mountain whitefish, sculpin, chiselmouth, carp, large-scale sucker, Asiatic clam.	60, 12, 58, 14, 3, 14	DDT, *o,p´-*	—	—
					DDT, *p,p´-*	61*	64*
					Dicofol	—	—
					Dieldrin	46*	62*
					Endosulfan I	32*	36*
					Endrin	21*	24*
					HCB	0*	0*
					HCH, α-	—	—
					HCH, β-	—	—
					HCH, δ-	—	—
					Heptachlor	21*	14*
					Heptachlor epoxide	18*	19*
					Kepone	—	—
					Lindane	4*	5*
					Methoxychlor, *p,p´-*	14*	10*
					Mirex	0*	0*
					Nonachlor, *cis-*	—	—
					Nonachlor, *trans-*	—	—
					Oxychlordane	—	—
					Pentachlorophenol	0*	0*
					Perthane	4*	2*
					Toxaphene	0*	0*
Rinella and Schuler, 1992	1988–1989	5 sites (sediment) and 3 sites (biota) in the Malheur National Wildlife Refuge (NWR), Oregon. Sediment samples were collected in July after the irrigation season. 5–7 subsamples were composited and sieved (2 mm). Fish samples were composites.	Sediment (surficial), fish (whole), birds (eggs, carcass), water. Fish species: white crappie.	6, 3, 10, 5, 3	Aldrin	0	0
					Chlordane, *cis-*	—	—
					Chlordane, total	0	0
					Chlordane, *trans-*	—	—
					DDD	80	67
					DDD, *o,p´-*	—	—
					DDD, *p,p´-*	—	—
					DDE	80	83
					DDE, *o,p´-*	—	—
					DDE, *p,p´-*	—	—
					DDT	40	33
					DDT, *o,p´-*	—	—
					DDT, *p,p´-*	—	—
					Dieldrin	0	0
					Endosulfan	0	0
					Endrin	20	17
					HCB	—	—
					HCH, α-	—	—
					HCH, β-	—	—
					HCH, δ-	—	—
					Heptachlor	0	0
					Heptachlor epoxide	0	0
					Lindane	0	0
					Methoxychlor	0	0
					Mirex	0	0
					Nonachlor, *cis-*	—	—
					Nonachlor, *trans-*	—	—

Table 2.2. Pesticides in bed sediment and aquatic biota from rivers in the United States: State and local monitoring studies—*Continued*

Bed sediment (µg/kg dry weight)			Aquatic biota (whole fish unless noted)					Comments
			% Detection		(µg/kg wet weight)			
DL	Range	Median	Sites	Samples	DL	Range	Median	
—	—	—	15	9	10	nd–40	nd	the Yakima R. Basin and to
0.1–10	nd–370*	0.9*	48	57	10	nd–960	15	identify factors affecting
—	—	—	36	33	10	nd–300	nd	distribution of trace
0.1–10	nd–47*	1*	58	60	10	nd–170	10	contaminants. Data were
0.1–10	nd–71*	nd*	—	—	—	—	—	presented without interpretation.
0.1–10	nd–17*	nd*	0	0	10	nd	nd	Some aquatic plant samples
200–400	nd*	nd*	3	2	10	nd–10	nd	contained $p,p´$- DDE.
—	—	—	0	0	10	nd	nd	
—	—	—	0	0	10	nd	nd	
—	—	—	0	0	10	nd	nd	
0.1–10	nd–0.1*	nd*	—	—	—	—	—	
0.1–10	nd–2.5*	nd*	24	17	10	nd–20	nd	
—	—	—	0	0	10	nd	nd	
0.1–10	nd–0.8*	nd*	0	0	10	nd	nd	
0.1–10	nd–3.2*	nd*	—	—	—	—	—	
0.1–10	nd*	nd*	0	0	10	nd	nd	
—	—	—	0	0	10	nd	nd	
—	—	—	39	38	10	nd–40	nd	
—	—	—	18	14	10	nd–20	nd	
600–1,200	nd*	nd*	—	—	—	—	—	
1–10	nd–2*	nd*	—	—	—	—	—	
10–500	nd*	nd*	18	21	50	nd–1,200	nd	
0.1	nd	nd	0	0	50	nd	nd	Objective: to determine whether
—	—	—	0	0	50	nd	nd	irrigation drainage in the study
1	nd	nd	0	0	50	nd	nd	area adversely affects human
—	—	—	0	0	50	nd	nd	health, fish, and wildlife
0.1	nd–0.5	0.2	—	—	—	—	—	populations and other beneficial
—	—	—	0	0	50	nd	nd	uses of water. OC residues were
—	—	—	0	0	50	nd	nd	low in sediment and not
0.1	nd–1.2	0.35	—	—	—	—	—	detectable in fish. All bird egg
—	—	—	0	0	50	nd	nd	samples and 1 carcass contained
—	—	—	0	0	50	nd	nd	detectable OCs.
0.1	nd–0.4	nd	—	—	—	—	—	
—	—	—	0	0	50	nd	nd	
—	—	—	0	0	50	nd	nd	
0.1	nd	nd	0	0	50	nd	nd	
0.1	nd	nd	—	—	—	—	—	
0.1	nd–0.2	nd	0	0	50	nd	nd	
—	—	—	0	0	50	nd	nd	
—	—	—	0	0	50	nd	nd	
—	—	—	0	0	50	nd	nd	
—	—	—	0	0	50	nd	nd	
0.1	nd	nd	0	0	50	nd	nd	
0.1	nd	nd	0	0	50	nd	nd	
0.1	nd	nd	—	—	—	—	—	
0.1	nd	nd	0	0	50	nd	nd	
—	—	—	0	0	50	nd	nd	
—	—	—	0	0	50	nd	nd	

Table 2.2. Pesticides in bed sediment and aquatic biota from rivers in the United States: State and local monitoring studies—*Continued*

Reference	Sampling date	Sampling locations and strategy	Sampling media and species	Number of samples	Target analytes	Bed sediment % Detection Sites	Bed sediment % Detection Samples
					Oxychlordane	—	—
					Perthane	0	0
					Toxaphene	0	0
Ryan and others, 1992	1987	8 sites (sediment), 5 sites (fish), and 18 sites (water) in the Great Dismal Swamp National Wildlife Refuge (NWR), Virginia. One site was a control site; other sites were located near possible sources of contamination, including ditches receiving runoff from junkyards, agricultural fields, and a landfill. Soil and sediment were collected near mammal collection sites. Fish samples were composites. All samples were taken in 1987 except water (1989).	Sediment, soil, fish (whole), mammals, water (whole). Fish species: creek chubsucker, golden shiner, bluespotted sunfish, flier, yellow bullhead, chain pickerel, yellow perch, bowfin.	9 10 14 9 36	Aldrin	0	0
					Chlordane, *cis-*	100	89
					Chlordane, *trans-*	100	100
					Dacthal	0	0
					DDD, *o,p´-*	25	22
					DDD, *p,p´-*	100	89
					DDE, *o,p´-*	0	0
					DDE, *p,p´-*	88	89
					DDT, *o,p´-*	0	0
					DDT, *p,p´-*	63	67
					Dieldrin	0	0
					Endosulfan I	0	0
					Endosulfan II	0	0
					Endosulfan sulfate	0	0
					Endrin	0	0
					HCH, α-	0	0
					HCH, β-	0	0
					HCH, δ-	0	0
					Heptachlor	0	0
					Heptachlor epoxide	38	33
					Hexachlorobenzene	0	0
					Lindane	0	0
					Methoxychlor	50	44
					Mirex	0	0
					Nonachlor, *cis-*	38	33
					Nonachlor, *trans-*	63	56
					Oxychlordane	25	22
					Tetradifon	13	11
See and others, 1992	1988–1990	2 sites in the Kendrick Reclamation Project Area (Wyoming).	Fish (*). Species: carp.	7	Chlordane, α-	—	—
					Chlordane, γ-	—	—
					DDD, *o,p´-*	—	—
					DDD, *p,p´-*	—	—
					DDE, *o,p´-*	—	—
					DDE, *p,p´-*	—	—
					DDT, *o,p´-*	—	—
					DDT, *p,p´-*	—	—
					Dieldrin	—	—
					Endrin	—	—
					HCB	—	—
					HCH, α-	—	—
					HCH, β-	—	—
					HCH, δ-	—	—
					Heptachlor epoxide	—	—
					Lindane	—	—
					Mirex	—	—
					Nonachlor, *cis-*	—	—
					Nonachlor, *trans-*	—	—

Table 2.2. Pesticides in bed sediment and aquatic biota from rivers in the United States: State and local monitoring studies—*Continued*

Bed sediment (µg/kg dry weight)			Aquatic biota (whole fish unless noted)					Comments
			% Detection		(µg/kg wet weight)			
DL	Range	Median	Sites	Sam-ples	DL	Range	Median	
10	nd	nd	56	42	1	nd–34	nd	FWS's NCBP. St. Mary's R.
10	nd–20	nd	56	42	nr , <5	nd–16	nd	sediment and fish also con-
—	—	—	60	44	1	nd–17	nd	tained organic contaminants.
10	nd	nd	67	53	nr , <7	nd–29	7	Total chlordane residues in
—	—	—	0	0	3	nd	nd	several fish samples approached
10	nd	nd	33	16	1	nd–15	nd	the FDA action level or exceeded
10	nd	nd	0	0	1	nd	nd	a NOAEL for fish survival from
10	nd	nd	100	100	nr , <20	20–110	40	Eisler (1990). The presence of
10	nd	nd	56	47	10	nd–66	nd	*o,p´*- DDT or aldrin in fish may
—	—	—	0	0	1	nd	nd	indicate point sources. Authors
10	nd	nd	0	0	7	nd	nd	did an ecological risk assessment on birds from fish tissue OCs.
0.17–0.49	nd–31.2†	nd†	—	—	—	—	—	* Percent of impoundments with detectable residues.
0.17–0.49	nd–43.2†	nd†	—	—	—	—	—	† Data are for all 3 layers of sediment cores, pooled.
0.49	nd†	nd†	—	—	—	—	—	§ Data are for top layer of
0.17	nd†	nd†	—	—	—	—	—	sediment cores.
0.17–0.49	nd–34.4†	nd†	—	—	—	—	—	Objective: to assess occurrence of OCs in mosquito control impoundments; to assess effect of impoundment techniques on
0.17–0.49	nd–7.5§	nd§	—	—	—	—	—	mobility into lagoon. In most
0.17–0.49	nd–26.9§	nd§	—	—	—	—	—	cases, OC concentrations decreased with depth. Maximum
0.49	nd†	nd†	—	—	—	—	—	residues of all compounds
0.17	nd†	nd†	—	—	—	—	—	detected occurred in the top
0.17–0.49	nd–5.92§	nd§	—	—	—	—	—	layers. Pore water and overlying water contained no detectable OCs. DDE occurred in impoundments near military training areas.
—	—	—	0	0	10	nd	nd	Objective: to assess contaminant
—	—	—	50	50	10	nd–10	nd/10	levels in the shovelnose
—	—	—	100	100	10	10	10	sturgeon from the confluence of
—	—	—	0	0	10	nd	nd	the Yellowstone and Missouri
—	—	—	100	100	10	10	10	rivers. OC pesticide residues
—	—	—	0	0	10	nd	nd	were below the national
—	—	—	100	100	10	50–80	65	geometric means from the
—	—	—	100	100	10	10	10	FWS's NCBP.
—	—	—	0	0	10	nd	nd	
—	—	—	100	100	10	10	10	
—	—	—	0	0	10	nd	nd	
—	—	—	0	0	10	nd	nd	
—	—	—	50	50	10	nd–10	nd/10	
—	—	—	0	0	10	nd	nd	
—	—	—	0	0	10	nd	nd	

Table 2.2. Pesticides in bed sediment and aquatic biota from rivers in the United States: State and local monitoring studies—*Continued*

Reference	Sampling date	Sampling locations and strategy	Sampling media and species	Number of samples	Target analytes	Bed sediment % Detection Sites	Bed sediment % Detection Samples
					Heptachlor	—	—
					Heptachlor epoxide	—	—
					Lindane	—	—
					Mirex	—	—
					Nonachlor, *cis-*	—	—
					Nonachlor, *trans-*	—	—
					Oxychlordane	—	—
					Toxaphene	—	—
Welsh and Olson, 1992	1991	1 site on the Missouri River near Bismarck (North Dakota) Site was selected because of urban and industrial land use. All fish were male.	Fish (liver, skinless fillets, testes). Species: shovelnose sturgeon.	5 5 4	Chlordane, α-	—	—
					Chlordane, γ-	—	—
					Chlordane, total	—	—
					DDD, *o,p´-*	—	—
					DDD, *p,p´-*	—	—
					DDE, *o,p´-*	—	—
					DDE, *p,p´-*	—	—
					DDT, *o,p´-*	—	—
					DDT, *p,p´-*	—	—
					Dieldrin	—	—
					Endrin	—	—
					HCB	—	—
					HCH, α-	—	—
					HCH, β-	—	—
					HCH, δ-	—	—
					Heptachlor epoxide	—	—
					Lindane	—	—
					Mirex	—	—
					Nonachlor, *cis-*	—	—
					Nonachlor, *trans-*	—	—
					Oxychlordane	—	—
					Toxaphene	—	—
Blanchard and others, 1993	1990– 1991	9 sites (sediment) and 6 sites (fish) in the San Juan River area (New Mexico). Sediment and fish samples were composites.	Sediment (surficial), fish (whole), water (whole). Species: flannel- mouth sucker.	9 6	Aldrin	0	0
					Chlordane, α-	—	—
					Chlordane, γ-	—	—
					Chlordane, total	44	40
					DDD, *o,p´ + p,p´*	56	50
					DDD, *o,p´-*	—	—
					DDD, *p,p´-*	—	—
					DDE, *o,p´ + p,p´*	78	70
					DDE, *o,p´-*	—	—
					DDE, *p,p´-*	—	—
					DDT, *o,p´ + p,p´*	11	10
					DDT, *o,p´-*	—	—
					DDT, *p,p´-*	—	—
					Dieldrin	0	0
					Endosulfan	0	0
					Endrin	0	0
					HCB	—	—
					HCH, α-	—	—
					HCH, β-	—	—
					HCH, δ-	—	—

Table 2.2. Pesticides in bed sediment and aquatic biota from rivers in the United States: State and local monitoring studies—*Continued*

Bed sediment (µg/kg dry weight)			Aquatic biota (whole fish unless noted)					Comments
			% Detection		(µg/kg wet weight)			
DL	Range	Median	Sites	Samples	DL	Range	Median	
—	—	—	0	0	10	nd	nd	
—	—	—	0	0	10	nd	nd	
—	—	—	0	0	10	nd	nd	
—	—	—	0	0	10	nd	nd	
—	—	—	0	0	10	nd	nd	
—	—	—	100	100	10	10	10	
—	—	—	0	0	10	nd	nd	
—	—	—	0	0	10	nd	nd	
—	—	—	0*	0*	10	nd*	nd*	* Data are for skinless fillets.
—	—	—	0*	0*	10	nd*	nd*	† Mean value.
—	—	—	0*	0*	10	nd*	nd*	Objective: to determine
—	—	—	0*	0*	10	nd*	nd*	contaminants in shovelnose
—	—	—	100*	nr	10	nd–20*	10*†	sturgeon in the Missouri R.
—	—	—	0*	0*	10	nd*	nd*	This species is a surrogate for
—	—	—	100*	100*	10	20–90*	50*†	the pallidnose sturgeon, an
—	—	—	0*	0*	10	nd*	nd*	endangered species that has not
—	—	—	0*	0*	10	nd*	nd*	reproduced for 10 years or more.
—	—	—	100*	nr	10	nd–10*	nd*	*p,p′-* DDD and *p,p′-* DDE were
—	—	—	0*	0*	10	nd*	nd*	detectable in most samples;
—	—	—	0*	0*	10	nd*	nd*	testes had the highest levels.
—	—	—	0*	0*	10	nd*	nd*	One pallid sturgeon collected 43
—	—	—	0*	0*	10	nd*	nd*	miles downstream in 1983
—	—	—	0*	0*	10	nd*	nd*	contained the same OC
—	—	—	0*	0*	10	nd*	nd*	pesticides, but at higher levels;
—	—	—	0*	0*	10	nd*	nd*	the pallid sturgeon was older and
—	—	—	0*	0*	10	nd*	nd*	probably had greater exposure
—	—	—	0*	0*	10	nd*	nd*	than shovelnose sturgeon in the
—	—	—	100*	nr	10	nd–10*	10*†	present study.
—	—	—	0*	0*	10	nd*	nd*	
—	—	—	0*	0*	50	nd*	nd*	
0.1	nd	nd	—	—	—	—	—	Objective: to present
—	—	—	83	83	nr	nd–30	20	information for determining
—	—	—	83	83	nr	nd–10	10	whether irrigation drainage in
1	nd–2	nd	—	—	—	—	—	the study area has caused or may
0.1	nd–0.2	0.05	83	83	nr	nd–20	20	cause adverse effects to human
—	—	—	0	0	nr	nd	nd	health, fish, or wildlife. OC
—	—	—	83	83	nr	nd–20	20	residues were less than national
0.1	nd–.4	0.15	100	100	nr	20–80	50	geometric means from the
—	—	—	0	0	nr	nd	nd	FWS's NCBP.
—	—	—	100	100	nr	20–80	50	
0.1	nd–.1	nd	83	83	nr	nd–20	15	
—	—	—	0	0	nr	nd	nd	
—	—	—	83	83	nr	nd–20	15	
0.1	nd	nd	0	0	nr	nd	nd	
0.1	nd	nd	—	—	—	—	—	
0.1	nd	nd	0	0	nr	nd	nd	
—	—	—	0	0	nr	nd	nd	
—	—	—	0	0	nr	nd	nd	
—	—	—	0	0	nr	nd	nd	
—	—	—	0	0	nr	nd	nd	

Table 2.2. Pesticides in bed sediment and aquatic biota from rivers in the United States: State and local monitoring studies—*Continued*

Reference	Sampling date	Sampling locations and strategy	Sampling media and species	Number of samples	Target analytes	Bed sediment % Detection Sites	Bed sediment % Detection Samples
					Heptachlor	0	0
					Heptachlor epoxide	0	0
					Lindane	0	0
					Methoxychlor	0	0
					Mirex	0	0
					Nonachlor, *cis-*	—	—
					Nonachlor, *trans-*	—	—
					Oxychlordane	—	—
					Perthane	0	0
					Toxaphene	0	0
Butler and others, 1993	1988–1989	10 sites (sediment) and 6 sites (biota) in streams and reservoirs receiving irrigation drainage from the Pine River Project Area, Southern Ute Indian Reservation (Colorado and New Mexico). Major streams included Los Pinos and Florida rivers. Composite sediment samples (fine-grained) were collected from depositional areas. Fish samples were composites of 4 fish.	Sediment (surficial), fish (whole), birds (egg, whole-body). Species: carp, white sucker.	11 4 5 2	Aldrin	0	0
					Chlordane	18	18
					Chlordane, α-	—	—
					Chlordane, γ-	—	—
					DDD	18	18
					DDD, *o,p'-*	—	—
					DDD, *p,p'-*	—	—
					DDE	18	18
					DDE, *o,p'-*	—	—
					DDE, *p,p'-*	—	—
					DDT	9	9
					DDT, *o,p'-*	—	—
					DDT, *p,p'-*	—	—
					Dieldrin	0	0
					Endosulfan	0	0
					Endrin	0	0
					HCB	—	—
					HCH, α-	—	—
					HCH, β-	—	—
					HCH, δ-	—	—
					Heptachlor	0	0
					Heptachlor epoxide	0	0
					Lindane	0	0
					Mirex	0	0
					Nonachlor, *cis-*	—	—
					Nonachlor, *trans-*	—	—
					Oxychlordane	—	—
					Perthane	0	0
					Toxaphene	0	0
DeWeese and others, 1993	1988	3 sites on the South Platte River (Colorado). About 30 state wildlife management areas are in or near the study area. For each site, all samples were composites of subsamples collected within 200	Sediment (surficial), crayfish, fish (whole). Species: common carp, shiners, crayfish.	3 3 6	Acephate	0	0
					Aldicarb	0	0
					Aldrin	0	0
					Azinphos-methyl	0	0
					Carbaryl	0	0
					Carbofuran	0	0
					Chlordane, total *	0	0
					Chlorpyrifos	0	0
					Coumaphos	0	0
					DDD, *o,p'-*	0	0
					DDD, *p,p'-*	0	0

Table 2.2. Pesticides in bed sediment and aquatic biota from rivers in the United States: State and local monitoring studies—*Continued*

| Bed sediment (µg/kg dry weight) | | | Aquatic biota (whole fish unless noted) | | | | | Comments |
| | | | % Detection | | (µg/kg wet weight) | | | |
DL	Range	Median	Sites	Sam-ples	DL	Range	Median	
0.1	nd	nd	—	—	—	—	—	
0.1	nd	nd	0	0	nr	nd	nd	
0.1	nd	nd	0	0	nr	—	—	
0.1–1	nd	nd	—	—	—	—	—	
0.1	nd	nd	0	0	nr	nd	nd	
—	—	—	0	0	nr	nd	nd	
—	—	—	83	83	nr	nd–20	15	
—	—	—	0	0	nr	nd	nd	
1	nd	nd	—	—	—	—	—	
10	nd	nd	0	0	nr	nd	nd	
0.1	nd	nd	0	0	0.01–0.05	nd	nd	Objective: to describe
1	nd–1.0	nd	—	—	—	—	—	concentrations of contaminants
—	—	—	0	0	0.01–0.05	nd	nd	in streams and reservoirs that
—	—	—	0	0	0.01–0.05	nd	nd	receive irrigation drainage from
0.1	nd–0.3	nd	—	—	—	—	—	the Pine River Project Area. OC
—	—	—	0	0	0.01–0.05	nd	nd	pesticide levels were low or
—	—	—	0	0	0.01–0.05	nd	nd	nondetectable in sediment, fish,
0.1	nd–0.2	nd	—	—	—	—	—	and birds.
—	—	—	0	0	0.01–0.05	nd	nd	
—	—	—	50	50	0.01–0.05	nd–0.04	nd/0.03	
0.1	nd–0.1	nd	—	—	—	—	—	
—	—	—	0	0	0.01–0.05	nd	nd	
—	—	—	0	0	0.01–0.05	nd	nd	
0.1	nd	nd	0	0	0.01–0.05	nd	nd	
0.1	nd	nd	—	—	—	—	—	
0.1	nd	nd	0	0	0.01–0.05	nd	nd	
—	—	—	0	0	0.01–0.05	nd	nd	
—	—	—	0	0	0.01–0.05	nd	nd	
—	—	—	0	0	0.01–0.05	nd	nd	
—	—	—	0	0	0.01–0.05	nd	nd	
0.1	nd	nd	0	0	0.01–0.05	nd	nd	
0.1	nd	nd	0	0	0.01–0.05	nd	nd	
0.1	nd	nd	0	0	0.01–0.05	nd	nd	
—	—	—	0	0	0.01–0.05	nd	nd	
—	—	—	0	0	0.01–0.05	nd	nd	
—	—	—	0	0	0.01–0.05	nd	nd	
1	nd	nd	0	0	0.01–0.05	nd	nd	
10	nd	nd	0	0	0.1–0.5	nd	nd	
50	nd	nd	0	0	50	nd	nd	* Data are for the sum of α- and
50	nd	nd	0	0	50	nd	nd	γ- chlordane and *trans*-
10	nd	nd	0	0	10	nd	nd	nonachlor.
50	nd	nd	0	0	50	nd	nd	Objective: to determine if biota
50	nd	nd	0	0	50	nd	nd	and sediment in the study area
50	nd	nd	0	0	50	nd	nd	showed evidence of
10	nd	nd	33	17	10	nd–50	nd	contamination detrimental to
50	nd	nd	0	0	50	nd	nd	threatened and endangered
50	nd	nd	0	0	50	nd	nd	species and migratory birds.
10	nd	nd	0	0	10	nd	nd	Fish contained more OC
10	nd	nd	33	17	10	nd–20	nd	compounds than crayfish, which

Table 2.2. Pesticides in bed sediment and aquatic biota from rivers in the United States: State and local monitoring studies—*Continued*

Reference	Sam-pling date	Sampling locations and strategy	Sampling media and species	Number of samples	Target analytes	Bed sediment % Detection Sites	Bed sediment % Detection Sam-ples
		m of that site. Sediment was collected from depositional areas.			DDE, *o,p´-*	0	0
					DDE, *p,p´-*	0	0
					DDT, *o,p´-*	0	0
					DDT, *p,p´-*	0	0
					Demeton	0	0
					Diazinon	0	0
					Dichlorvos	0	0
					Dieldrin	0	0
					Dimethoate	0	0
					Disulfoton	0	0
					Endrin	0	0
					EPA	0	0
					Ethoprop	0	0
					Famphur	0	0
					Fensulfothion	0	0
					Fenthion	0	0
					HCB	0	0
					HCH, α-	0	0
					HCH, β-	0	0
					HCH, δ-	0	0
					Heptachlor	0	0
					Heptachlor epoxide	0	0
					Lindane	0	0
					Malathion	0	0
					Methamidophos	0	0
					Methiocarb	0	0
					Methomyl	0	0
					Methyl parathion	0	0
					Mevinphos	0	0
					Mirex	0	0
					Monocrotophos	0	0
					Nonachlor, *cis-*	0	0
					Oxamyl	0	0
					Oxychlordane	0	0
					Parathion	0	0
					Phorate	0	0
					Terbufos	0	0
					Toxaphene	0	0
					Trichlorfon	0	0
Rinella, 1993	1990	12 sites on Amazon Creek and a tributary (Oregon). Sediment samples were composites of 20–25 cross-sectional subsamples. The <63 μm fraction was analyzed for organics.	Water (whole), sediment (surficial).	8 12	Aldrin	0	0
					Chlordane	100	100
					DDD	100	100
					DDE	58	58
					DDT	17	17
					Dieldrin	92	92
					Endosulfan	0	0
					Endrin	8	8
					Heptachlor	0	0
					Heptachlor epoxide	42	42
					Lindane	0	0

Table 2.2. Pesticides in bed sediment and aquatic biota from rivers in the United States: State and local monitoring studies—*Continued*

Bed sediment (µg/kg dry weight)			Aquatic biota (whole fish unless noted)					Comments
			% Detection		(µg/kg wet weight)			
DL	Range	Median	Sites	Sam-ples	DL	Range	Median	
10	nd	nd	0	0	10	nd	nd	contained only *p,p´*- DDE (2 of 3
10	nd	nd	100	100	10	50–290	170	samples). OC levels in fish were
10	nd	nd	0	0	10	nd	nd	not believed high enough to
10	nd	nd	33	17	10	nd–10	nd	cause toxicity to fish or wildlife.
50	nd	nd	0	0	50	nd	nd	
50	nd	nd	0	0	50	nd	nd	
50	nd	nd	0	0	50	nd	nd	
10	nd	nd	33	17	10	nd–10	nd	
50	nd	nd	0	0	50	nd	nd	
50	nd	nd	0	0	50	nd	nd	
10	nd	nd	0	0	10	nd	nd	
50	nd	nd	0	0	50	nd	nd	
50	nd	nd	0	0	50	nd	nd	
50	nd	nd	0	0	50	nd	nd	
50	nd	nd	0	0	50	nd	nd	
50	nd	nd	0	0	50	nd	nd	
10	nd	nd	0	0	10	nd	nd	
10	nd	nd	0	0	10	nd	nd	
10	nd	nd	0	0	10	nd	nd	
10	nd	nd	0	0	10	nd	nd	
10	nd	nd	0	0	10	nd	nd	
10	nd	nd	0	0	10	nd	nd	
10	nd	nd	0	0	10	nd	nd	
50	nd	nd	0	0	50	nd	nd	
50	nd	nd	0	0	50	nd	nd	
50	nd	nd	0	0	50	nd	nd	
50	nd	nd	0	0	50	nd	nd	
50	nd	nd	0	0	50	nd	nd	
50	nd	nd	0	0	50	nd	nd	
10	nd	nd	0	0	10	nd	nd	
50	nd	nd	0	0	50	nd	nd	
10	nd	nd	0	0	10	nd	nd	
50	nd	nd	0	0	50	nd	nd	
10	nd	nd	0	0	10	nd	nd	
50	nd	nd	0	0	50	nd	nd	
50	nd	nd	0	0	50	nd	nd	
50	nd	nd	0	0	50	nd	nd	
10	nd	nd	0	0	10	nd	nd	
50	nd	nd	0	0	50	nd	nd	
0.1–1	nd	nd	—	—	—	—	—	Objective: to identify extent of
nr	9.0–140	60	—	—	—	—	—	trace element and organic
nr	3.5–120	9.5	—	—	—	—	—	compound contamination within
10–20	nd–11	3.2	—	—	—	—	—	the Amazon Creek
0.1–10	nd–3.0	nd	—	—	—	—	—	Basin. Whole-water samples
1	nd–10	3.2	—	—	—	—	—	from some sites contained
0.1–1	nd	nd	—	—	—	—	—	2,4-D, 2,4,5-T, silvex, picloram,
0.1	nd–0.1	nd	—	—	—	—	—	dicamba, diazinon,
0.1–1	nd	nd	—	—	—	—	—	pentachlorophenol, and other
0.1	nd–1.8	nd	—	—	—	—	—	priority pollutants. Sediment
0.1–1	nd	nd	—	—	—	—	—	samples contained several OC

Table 2.2. Pesticides in bed sediment and aquatic biota from rivers in the United States: State and local monitoring studies—*Continued*

Reference	Sampling date	Sampling locations and strategy	Sampling media and species	Number of samples	Target analytes	Bed sediment % Detection Sites	Bed sediment % Detection Samples
					Methoxychlor	0	0
					Mirex	0	0
					Pentachlorophenol	33	33
					Perthane	0	0
					Toxaphene	0	0
Secor and others, 1993	1991	7 sites in 3 rivers and 3 lakes in New York. Sites were in polluted areas. Mussels were composited; most were 1–1.8 cm in length.	Mussels. Species: zebra mussel.	11	DDE, *p,p´-*	—	—
Seiler and others, 1993	1988– 1990	2 sites (sediment) and 3 sites (fish) in and near Humboldt Wildlife Management Area (WMA), Nevada. The principal river was the Humboldt River. Sediment samples were composites of cross-sectional samples at a given site. Sediment samples were sieved (<2 mm) using native water. 1-lb fish were targeted for collection.	Water (whole), sediment (surficial), fish (whole). Species: channel catfish, walleye, white bass, carp.	6 4 15	Aldrin	0	0
					Chlordane, α-	—	—
					Chlordane, γ-	—	—
					Chlordane, total	0	0
					DDD, *o,p´ + p,p´-*	50	25
					DDD, *o,p´-*	—	—
					DDD, *p,p´-*	—	—
					DDE, *o,p´ + p,p´-*	100	75
					DDE, *o,p´-*	—	—
					DDE, *p,p´-*	—	—
					DDT, *o,p´ + p,p´-*	0	0
					DDT, *o,p´-*	—	—
					DDT, *p,p´-*	—	—
					Dieldrin	0	0
					Endosulfan I	—	—
					Endosulfan II	—	—
					Endosulfan, total	0	0
					Endrin	0	0
					HCB	—	—
					HCH, α-	—	—
					HCH, β-	—	—
					HCH, δ-	—	—
					Heptachlor	0	0
					Heptachlor epoxide	0	0
					Lindane	0	0
					Methoxychlor	0	0
					Mirex	0	0
					Nonachlor, *cis-*	—	—
					Nonachlor, *trans-*	—	—
					Oxychlordane	—	—
					Perthane	0	0
					Toxaphene	0	0
Setmire and others, 1993	1988– 1990	29 sites in the Salton Sea National Wildlife Refuge (NWR) and in the Imperial Valley	Fish (whole, fillet), clams, crayfish, pelagic in-	26 1 12 4 7	Aldrin	—	—
					Chlordane, α-	—	—
					Chlordane, γ-	—	—
					Dacthal	—	—
					DDT, total	—	—

Table 2.2. Pesticides in bed sediment and aquatic biota from rivers in the United States: State and local monitoring studies—*Continued*

| Bed sediment (µg/kg dry weight) | | | Aquatic biota (whole fish unless noted) | | | | | Comments |
| | | | % Detection | | (µg/kg wet weight) | | | |
DL	Range	Median	Sites	Samples	DL	Range	Median	
0.1–1	nd	nd	—	—	—	—	—	pesticides, PCBs,
0.1–1	nd	nd	—	—	—	—	—	pentachlorophenol, and 23 other
600	nd–770	nd	—	—	—	—	—	semivolatile priority pollutants
1–10	nd	nd	—	—	—	—	—	at some or all sites sampled.
10	nd	nd	—	—	—	—	—	
—	—	—	29*	27*	10*	nd–40*	nd*	* Data are for mussel tissue (µg/kg dry weight). Objective: to examine zebra mussels from industrialized areas for contaminant levels and histologic lesions. Significant lesions or infectious agents were not observed.
0.1	nd	nd	—	—	—	—	—	Objective: to determine
—	—	—	0	0	10	nd	nd	concentrations of contaminants
—	—	—	0	0	10	nd	nd	in water, sediment, and biota of
1	nd	nd	—	—	—	—	—	the WMA. The report also
0.1	nd–0.1	nd	—	—	—	—	—	includes aquatic and sediment
—	—	—	0	0	10	nd	nd	biotoxicity tests (trace elements
—	—	—	0	0	10	nd	nd	only) and results of benthic
0.1	nd–0.2	0.1	—	—	—	—	—	surveys. The few OC detections
—	—	—	33	13	10	nd–30	nd	in fish came from a reference
0.1	nd	nd	—	—	—	—	—	site (not influenced by irrigation
—	—	—	0	0	10	nd	nd	drainage). The source of OCs is
—	—	—	33	7	10	nd–10	nd	unknown. Water samples were
0.1	nd	nd	0	0	10	nd	nd	analyzed at selected sites for
—	—	—	0	0	10	nd	nd	carbamate pesticides and
—	—	—	0	0	10	nd	nd	herbicides. One water sample
0.1	nd	nd	—	—	—	—	—	contained 2,4-D.
0.1	nd	nd	0	0	10	nd	nd	
—	—	—	0	0	10	nd	nd	
—	—	—	0	0	10	nd	nd	
—	—	—	0	0	10	nd	nd	
0.1	nd	nd	—	—	—	—	—	
0.1	nd	nd	0	0	10	nd	nd	
0.1	nd	nd	0	0	10	nd	nd	
0.1–10	nd	nd	—	—	—	—	—	
0.1	nd	nd	0	0	10	nd	nd	
—	—	—	0	0	10	nd	nd	
—	—	—	0	0	10	nd	nd	
—	—	—	0	0	10	nd	nd	
1	nd	nd	—	—	—	—	—	
10	nd	nd	0	0	100	nd	nd	
—	—	—	nr	3*	10	nd–10*	nr	* Data are for all biota samples.
—	—	—	nr	14*	10	nd–240*	nr	† Data are for all fish samples.
—	—	—	nr	3*	10	nd–80*	nr	Objective: to investigate effects
—	—	—	nr	64*	10	nd–320*	nr	of irrigation drainage in the
—	—	—	nr	100†	100	80–5,820†	nr	Salton Sea area. The highest

Table 2.2. Pesticides in bed sediment and aquatic biota from rivers in the United States: State and local monitoring studies—*Continued*

Reference	Sam- pling date	Sampling locations and strategy	Sampling media and species	Number of samples	Target analytes	Bed sediment % Detection	
						Sites	Sam- ples
		(California). Sites were located in rivers, creeks, drainage ditches, and freshwater impoundments. Biota represented various trophic levels. Small fish were composited.	vertebrates, frogs, turtles (fat, egg), birds (egg, liver, muscle, carcass, other). Aquatic spe- cies include redfin shiner, sailfin mol- ly, Tilapia, mosquito- fish, Asiatic clam, *Pro - cambarus clarkii* , wa- ter boatman.	2 6 1 84 3 9 38 3	Dicofol Dieldrin Endosulfan I Endosulfan II Endosulfan sulfate Endrin HCB HCH, α- HCH, β- HCH, δ- Heptachlor Heptachlor epoxide Lindane Methoxychlor Mirex Nonachlor, *cis-* Nonachlor, *trans-* Oxychlordane Tetradifon Toxaphene	— — — — — — — — — — — — — — — — — — — —	— — — — — — — — — — — — — — — — — — — —
Giesy and others, 1994b	1990	7 sites in the Au Sable, Manistee, and Muskegon rivers (Michigan). Samples were collected above and below dams separating fish populations that had access to the Great Lakes (Lakes Michigan and Huron).	Fish (whole). Species: chinook salmon, pike, walleye, white sucker, brown trout, steelhead trout, carp, perch.	23	Chlordane, α- Chlordane, γ- DDD, *o,p ´-* DDD, *p,p ´-* DDE, *o,p ´-* DDE, *p,p ´-* DDT, *p,p ´-* DDT, total Dieldrin Endosulfan I Endrin Heptachlor Heptachlor epoxide Lindane Methoxychlor Nonachlor, *trans-* Oxychlordane TCDD-equivalent §	— — — — — — — — — — — — — — — — — —	— — — — — — — — — — — — — — — — — —

Table 2.2. Pesticides in bed sediment and aquatic biota from rivers in the United States: State and local monitoring studies—*Continued*

Bed sediment (µg/kg dry weight)			Aquatic biota (whole fish unless noted)					Comments
			% Detection		(µg/kg wet weight)			
DL	Range	Median	Sites	Samples	DL	Range	Median	
—	—	—	nr	0*	10	nd*	nr	DDT residues in fish were from
—	—	—	nr	60*	10	nd–850*	nr	river and drain sites. Mean DDE
—	—	—	nr	9*	10	nd–10*	nr	residue in mosquitofish was
—	—	—	nr	9*	10	nd–20*	nr	almost 3-times greater than
—	—	—	nr	9*	10	nd–20*	nr	national mean from the FWS's
—	—	—	nr	3*	10	nd–30*	nr	NCBP. The highest DDE
—	—	—	nr	45*	10	nd–2,900*	nr	concentration in fish (5,700
—	—	—	nr	0*	10	nd*	nr	µg/kg) was in a redfin shiner
—	—	—	nr	36*	10	nd–360*	nr	from the Whitewater R. Asiatic
—	—	—	nr	1*	10	nd–50*	nr	clams had the highest DDE
—	—	—	nr	0*	10	nd*	nr	residues of all invertebrates.
—	—	—	nr	19*	10	nd–60*	nr	Dieldrin was widespread in the
—	—	—	nr	0*	10	nd*	nr	Imperial Valley, but at low
—	—	—	nr	27*	10	nd–30*	nr	levels. DDE was accumulating
—	—	—	nr	2*	10	nd–40*	nr	in tissues of resident and
—	—	—	nr	3*	10	nd–110*	nr	migratory birds. No OCs were
—	—	—	nr	20*	10	nd–290*	nr	detected in fish at levels greater
—	—	—	nr	36*	10	nd–640*	nr	than the NAS/NAE maximum
—	—	—	nr	0*	10	nd*	nr	threshold to protect fish-eating
—	—	—	nr	8*	10	nd–7,000*	nr	birds and wildlife.
—	—	—	100	100	*	0.02–35.6	1.89	* Method detection limits were
—	—	—	100	100	*	0.10–23.0	0.77	reported as between 0.11 pg/kg
—	—	—	100	100	*	0.13–81.7	5.83	(for heptachlor) and 1.6 pg/kg
—	—	—	100	100	*	0.04–58.8	3.06	(for *p,p´*- DDT).
—	—	—	100	100	*	0.15–38.0	1.25	† µmol/kg.
—	—	—	100	100	*	3.54–738	66.2	§ As measured by the H4IIE
—	—	—	86	74	0.0016	nd–89.6	2.44	bioassay.
—	—	—	100	100	*	4.71–977†	82.1†	Objective: to determine whether
—	—	—	100	100	*	0.06–107	2.09	fish transport contaminants from
—	—	—	100	74	*	nd–1.67	0.04	Great Lakes into tributaries.
—	—	—	100	74	*	nd–8.03	0.08	Total DDT levels at downstream
—	—	—	57	30	0.00011	nd–1.00	nd	sites (Great Lakes-influenced)
—	—	—	100	100	*	0.07–14.5	0.47	were greater than at upstream
—	—	—	100	100	*	0.02–0.84	0.12	sites for Au Sable and Manistee
—	—	—	14	4	*	nd–1.40	nd	rivers, but not Muskegon R. The
—	—	—	100	100	*	0.10–17.6	0.98	DDE/DDT ratio ranged from 8
—	—	—	100	100	*	0.15–16.0	1.31	to 758, suggesting that fish
—	—	—	100	100	*	0.0024–0.071	0.0074	obtained DDE directly from

diet, rather than from recent exposure to parent DDT. The authors suggested that DDT occurrence in fish was due to long-range atmospheric transport. Mean concentrations of TCDD-equivalents were greater at downstream than upstream sites. A hazard assessment for mink was reported in a related paper (Giesy and others, 1994c).

Table 2.2. Pesticides in bed sediment and aquatic biota from rivers in the United States: State and local monitoring studies—*Continued*

Reference	Sampling date	Sampling locations and strategy	Sampling media and species	Number of samples	Target analytes	Bed sediment % Detection	
						Sites	Samples
Rinella and others, 1994	1990	14 sites (sediment) and 15 sites (fish) in the Owyhee and Vale projects (Oregon and Idaho). Principal rivers were the Malheur, Owyhee, and Snake rivers. 2 reference sites were not influenced by irrigation drainage from the projects, but were influenced by agricultural drainage from higher in the Snake River Basin. Sediment was collected during irrigation (July–Aug.). Sediment samples were sieved (<2 mm). Most fish samples were composites of 2–3 fish.	Water (whole), sediment (surficial), fish (whole), birds (eggs, carcasses). Fish species: channel catfish, yellow bullhead, brown bullhead, smallmouth bass.	22 14 16 9 5	Aldrin	43	43
					Chlordane, α-	—	—
					Chlordane, γ-	—	—
					Chlordane, total	79	79
					DDD, $o,p' + p,p'$-	100	100
					DDD, o,p'-	—	—
					DDD, p,p'-	—	—
					DDE, $o,p' + p,p'$-	100	100
					DDE, o,p'-	—	—
					DDE, p,p'-	—	—
					DDT, $o,p' + p,p'$-	93	93
					DDT, o,p'-	—	—
					DDT, p,p'-	—	—
					DDT, total	100	100
					Dieldrin	93	93
					Endosulfan	21	21
					Endrin	0	0
					HCB	—	—
					HCH, α-	—	—
					HCH, β-	—	—
					HCH, δ-	—	—
					Heptachlor	0	0
					Heptachlor epoxide	7	7
					Lindane	0	0
					Methoxychlor	0	0
					Mirex	0	0
					Nonachlor, *cis*-	—	—
					Nonachlor, *trans*-	—	—
					Oxychlordane	—	—
					Perthane	0	0
					Toxaphene	0	0

Table 2.2. Pesticides in bed sediment and aquatic biota from rivers in the United States: State and local monitoring studies—*Continued*

Bed sediment (μg/kg dry weight)			Aquatic biota (whole fish unless noted)					Comments
			% Detection		(μg/kg wet weight)			
DL	Range	Median	Sites	Samples	DL	Range	Median	
0.1	nd–1.5	nd	—	—	—	—	—	Objective: to determine whether
—	—	—	67	69	10	nd–40	20	irrigation drainage in the study
—	—	—	33	38	10	nd–20	nd	area may cause toxicity to hu-
1	nd–30	2	—	—	—	—	—	mans, fish, or wildlife. Total
0.1	0.1–20	5	—	—	—	—	—	DDT, dieldrin, and toxaphene
—	—	—	0	0	10	nd	nd	were elevated in some samples
—	—	—	87	88	10	nd–230	50	of fish or birds. All fish
0.1	0.7–67	16	—	—	—	—	—	concentrations were below
—	—	—	0	0	10	nd	nd	dietary concentrations believed
—	—	—	87	88	10	nd–1,500	750	to impair reproduction of avian
0.1	nd–86	4.3	—	—	—	—	—	predators. Toxaphene in some
—	—	—	20	19	10	nd–150	nd	fish samples exceeded levels
—	—	—	67	69	10	nd–440	70	reported to impair growth and
0.1	0.8–114	23.3	87	88	10	nd–2,700	935	reproduction in fish. OCs were
0.1	nd–43	1.5	60	63	10	nd–370	15	higher in biota from the Snake R.
0.1–1	nd–0.2	nd	—	—	—	—	—	and Owyhee Project Area than
0.1–1	nd	nd	0	0	10	nd	nd	from the Vale Project Area. Most
—	—	—	20	25	10	nd–20	nd	OCs, when normalized for sedi-
—	—	—	0	0	10	nd	nd	ment TOC, had highest levels in
—	—	—	0	0	10	nd	nd	drainwater and at sites below irri-
—	—	—	0	0	10	nd	nd	gated areas in Owyhee and Mal-
0.1	nd	nd	—	—	—	—	—	heur rivers. Calculated pore-water
0.1	nd–0.8	nd	67	69	10	nd–40	10	concentrations exceeded
0.1	nd	nd	7	6	10	nd–10	nd	"no-effect" (safe) criteria for
0.1	nd	nd	—	—	—	—	—	protection of benthic fauna for
0.1	nd	nd	0	0	10	nd	nd	chlordane, DDT, heptachlor
—	—	—	7	6	10	nd–10	nd	epoxide, endosulfan, and total
—	—	—	60	63	10	nd–30	10	dieldrin in 7–100% of samples.
—	—	—	47	50	10	nd–10	nd/10	OC and OP insecticides and
1–10	nd	nd	—	—	—	—	—	various herbicides were detected
10	nd	nd	40	44	50	nd–1,900	nd	in water .

Appendix C—Table 2.3. Pesticides in bed sediment and aquatic biota from rivers in the United States: Process and matrix distribution studies

[Studies are listed chronologically. Sampling media and species, lists all sampling media analyzed by the study and, for aquatic biota, some or all species sampled. In this column, "Species include" indicates that the following species list a partial list for that study; "sediment" is designated as either surficial (for surficial bed sediment) or core (for bed sediment core), or suspended (for suspended sediment). Target analytes, the complete list of pesticide analytes in bed sediment or aquatic biota for that study. Abbreviations: a.i., active ingredient; BAF, bioaccumulation factor; BCF, bioconcentration factor; C, Celsius; CP, chlorophenol; DCA, dichloroaniline; DMA, dimethylamine salt; DTA, dodecyl-tetradecylamine salt; g, gram; h, hour; ha, hectare; kg, kilogram; L, liter; m, meter; mg, milligram; nd, not detected; OC, organochlorine; OP, organophosphate; ppm, part per million; sp., species; SPMD, semipermeable membrane sampling device; wt, weight; µg, microgram. Common names of pesticides are defined in the Glossary in Appendix E]

Reference	Sampling date	Sampling Locations and Strategy	Sampling Media and Species	Target Analytes	Comments
Sparr and others, 1966	1964–1965	3 field studies: (1) rice fields near Stuttgart (Arkansas), which drain into Little LaGrue Bayou and White River; (2) cotton field at Kelso (Arkansas) and impounded drainage water; (3) Kankakee River (Indiana) and 2 drainage ditches draining cornfields.	Water (decanted or filtered), sediment (mud), fish (whole, muscle), soil. Species: catfish, perch, and bluegill.	Aldrin, Dieldrin, Endrin	Objective: to determine whether treatment of (1) rice seed with aldrin, (2) cotton with endrin, and (3) corn with aldrin contribute to water contamination. (1) Rice seed was treated with aldrin prior to seeding. The highest concentrations in water (0.9 µg/L dieldrin and 1.1 µg/L aldrin) occurred after seeding. Residues in mud from the paddy remained the same (about 70 µg/kg total dieldrin) throughout the experiment (14 weeks). Residues in Little LaGrue Bayou and the White River in July ranged from 0.01–0.04 µg/L in water and 10–210 µg/kg in catfish, declining by harvest time to <0.01–0.02 µg/L in water and <10–50 µg/kg in fish. (2) Analyses of water collected from irrigation and drainage ditches before and after spraying showed that endrin was washed from the field by irrigation water. More soil particles containing endrin were dislodged from the first irrigation than the second; filtration removed 83% and 10% of endrin from water samples collected from the first 2 irrigations. Cotton field soil samples did not contain detectable residues from the previous season's treatment; following treatment, soil residues declined from 40–20 µg/kg. No endrin was detected in mud from beneath water impounded by a dam. Fish from the drainage ditch contained ≤20 µg/kg endrin, although this may have been due to interference from toxaphene (used to collect fish). (3) Water collected from the Kankakee River contained <0.6 µg/L total dieldrin. Residues in water from 2 ditches draining cornfields were even lower. Water residues were mostly dieldrin. Only 1 mud sample contained detectable aldrin (30–35 µg/kg), from 1 drainage ditch.

Table 2.3. Pesticides in bed sediment and aquatic biota from rivers in the United States: Process and matrix distribution studies—*Continued*

Reference	Sampling date	Sampling Locations and Strategy	Sampling Media and Species	Target Analytes	Comments
Terriere and others, 1966	1962–1964	2 mountain lakes (Oregon) treated with toxaphene for fish and lamprey control. Control samples were taken from nearby untreated bodies of water.	Water, sediment, fish (tissue not reported), invertebrates, aquatic plants. Fish species: rainbow trout, brook trout, and Atlantic salmon. Invertebrate and plant species: not reported.	Toxaphene	Objective: to determine the persistence of toxaphene in lakes of different types following treatment for rough fish control. Toxaphene residues in water declined slowly during the 1st year after treatment. Residues in water of Miller Lake (deep, small surface area, sparse in biological life) fluctuated seasonally, with higher levels in early spring than late summer, probably because bottom sediment was resuspended during spring turnover. Maximum water concentrations were 3.1 µg/L in Miller Lake and 0.9 µg/kg in Davis Lake, a wide, shallow, eutrophic lake. Plants accumulated appreciable toxaphene residues (up to 15,500 µg/kg). Roots contained higher residues than foliar portions. In Davis Lake, BAFs were about 500 for plant/water, 1,000–2,000 for invertebrates/water, and 10,000–20,000 for fish/water. In <1 year, Davis Lake was restocked; toxaphene deposits in fish reached a plateau in 30–50 days. Healthy trout contained up to 10,000 µg/kg toxaphene. Miller Lake detoxified more slowly, remaining toxic to fish for up to 5 years. Maximum residues in stocked fish in Miller Lake were 24,800 µg/kg 6 years after treatment.

Table 2.3. Pesticides in bed sediment and aquatic biota from rivers in the United States: Process and matrix distribution studies—*Continued*

Reference	Sam-pling date	Sampling Locations and Strategy	Sampling Media and Species	Target Analytes	Comments
Cole and others, 1967	1965	Bed sediment was sampled at 4 sites and stream biota was sampled at 3 sites before and following an application of DDT to surrounding forest lands (Pennsylvania). Samples were collected 32, 122, and 380 days after application.	Water, bed sediment, soil, fish (tissue type not reported), invertebrates. Species: brook trout, white sucker, and crayfish.	DDD DDE DDT, o,p'-DDT, p,p'-Dieldrin	Objective: to monitor DDT and metabolites in soil, water, sediment and aquatic organisms following aerial application of DDT to forest lands. DDE, p,p'-DDT, and dieldrin were detected in soil, stream sediment, and fish in the isolated study area prior to any known application of these compounds. Such background concentrations may be due to atmospheric transport into the study area. Pretreatment concentrations of DDE and dieldrin were much higher in fish than soil or stream sediment. 32 days after treatment, DDT levels in fish increased substantially. Maximum levels were 10,600 µg/kg in brook trout and 6,900 µg/kg in white suckers. Crayfish contained up to 1,600 µg/kg DDD. Concentrations in fish had returned to background levels when sampled 122 and 380 days after treatment. Residues in stream sediment remained at pretreatment levels throughout the monitoring period. DDT remained elevated in forest litter and surface soil 380 days after application. Background levels of dieldrin in fish were lower after DDT application.

Table 2.3. Pesticides in bed sediment and aquatic biota from rivers in the United States: Process and matrix distribution studies—*Continued*

Reference	Sam-pling date	Sampling Locations and Strategy	Sampling Media and Species	Target Analytes	Comments
Bedford and others, 1968	1966	Freshwater mussels were transplanted to the Red Cedar River (Michigan). Sites included an upstream reference site (site I) and several sites (II–VI) downstream of residential and municipal sources. Residues were analyzed in control mussels. Two-part experiment, with 2 species introduced for overlapping time periods.	Mussels. Species: *Lampsilis siliquoidea, Anodonta grandis.*	Aldrin DDD DDE DDT Methoxychlor	Objective: to measure pesticide residues in (1) *Lampsilis* mussels transplanted from the Cass River and (2) *Anodonta* mussels collected upstream of the reference site and transplanted. Mussels were analyzed after 2, 6, and 10 weeks. Control mussels contained only total DDT (background pesticide residues in resident species are described in Table 2.2). Total DDT levels in transplanted mussels increased significantly in a downstream direction. Levels increased significantly with time before reaching a plateau. No mussels survived after 2 weeks at the most downstream site. Residues in transplanted mussels at sites II–VI were 80–250 µg/kg DDT, 50–360 µg/kg DDD, and 5–80 µg/kg DDE. During this time, total DDT concentrations ranged from trace to 60 µg/L in filtered water and 1–50 mg/kg (dry wt.) in suspended sediment. Methoxychlor was detected at site III and downstream, reaching 80–100 µg/kg after 10 weeks. *Lampsilis* accumulated higher residues of total DDT and methoxychlor than *Anodonta*. Aldrin was detected in *Lampsilis, Anodonta,* and in water samples during one discrete time period, probably indicating a discrete input.

Table 2.3. Pesticides in bed sediment and aquatic biota from rivers in the United States: Process and matrix distribution studies—*Continued*

Reference	Sampling date	Sampling Locations and Strategy	Sampling Media and Species	Target Analytes	Comments
Dimond and others, 1968	1966–1967	21 watersheds in northern Maine. Watersheds were unsprayed, or sprayed 1 or 3 times during 1958–1967, with DDT. Most were completely forested. Samples were composites, except that 5 crayfish from 2 streams were analyzed individually to assess variability.	Crayfish. Species: *Cambarus bartoni*.	DDT, total	Objective: to investigate persistence of DDT residues following treatment for control of spruce budworm. The highest DDT residues were in streams sprayed during year of sampling (500–2,700 µg/kg). Lower residues were observed in larger watercourses, probably because pilots were instructed to avoid spraying within 150 m and to apply half the normal dosage within a second 150 m. Such precautions were not taken prior to 1967. After 3 years of treatment, residues declined to about 100 µg/kg, but remained at this level up to 9 years following treatment. Residues in untreated streams were <20 µg/kg. In 1967-treated streams, the bulk of the residue was DDT, with DDE averaging 27%. In earlier-treated and unsprayed streams, DDE comprised >50% of total DDT (average 60%).

Table 2.3. Pesticides in bed sediment and aquatic biota from rivers in the United States: Process and matrix distribution studies—*Continued*

Reference	Sampling date	Sampling Locations and Strategy	Sampling Media and Species	Target Analytes	Comments
Fahey and others, 1968	1963–1964	13 sites in ponds, creeks, and the Kalamazoo River (Michigan). Samples were collected before and after areas were treated with dieldrin.	Water (whole), silt, soil (turf and cultivated land).	Dieldrin	Objective: to survey occurrence and distribution of dieldrin in environmental samples before and after treatment with dieldrin to control Japanese beetles. Dieldrin was not detected in water on the dates sampled after treatment. Dieldrin residues in streambed and pond silt were generally low, not detectable, or inconclusive because of interferences. Substantial, but not uniform residues of chlordane, DDE, DDT, HCH, and heptachlor and its epoxide were present in almost all soil samples taken from turf or cultivated areas in Battle Creek. Only 3 pretreatment soil samples contained measurable dieldrin residues; posttreatment soil samples contained an average of 1.39 mg/kg dieldrin 2 months after treatment ended.
Moubry and others, 1968	1966	3 sites adjacent to an orchard and one upstream control site (Wisconsin).	Water, sediment (surficial), organic debris, invertebrates, fish. Species include caddisfly larvae, alderfly larvae, freshwater shrimp brook trout, northern creek chub, and blacknose dace.	DDE DDD DDT Dieldrin Endrin	Objective: to assess contamination in a stream receiving runoff from an orchard with a history of organochlorine pesticide use. The extent of contamination appeared to be relatively low. No detections were made in water samples despite a reported detection limit of 1 ppt (although not defined in text, appears to be part per trillion). Contaminant levels in benthic invertebrates were higher than concentrations in bed sediment, but similar to concentrations in stream bed organic debris. No endrin residues were detected in fish, but dieldrin residues were the highest of any other matrix. DDT compounds and dieldrin residues were higher on a lipid weight basis (average DDT of 2,310 ppm) than on a whole weight basis (average DDT of 90 ppm).

Table 2.3. Pesticides in bed sediment and aquatic biota from rivers in the United States: Process and matrix distribution studies—*Continued*

Reference	Sampling date	Sampling Locations and Strategy	Sampling Media and Species	Target Analytes	Comments
Yule and Tomlin, 1970	1967	A tributary of the Miramichi River, New Brunswick, Canada. Samples were collected before, during, and after DDT treatment to the surrounding forest. Loose sediment was targeted for sampling; stream bottom was mostly rocky beds.	Water (whole), sediment (surficial).	DDT, p,p'-DDT, total	Objective: to resolve the role of DDT in fish population declines and to provide ecological perspective on DDT residues in forest and stream environments. Only during and for a few hours after spraying did DDT concentration in stream water exceed 0.9 μg/L p,p'-DDT. Maximum residues in water occurred during spraying (17 μg/L p,p'-DDT at the surface and 1.69 μg/L at 12–18 inches depth). DDT residues in sediment were variable throughout the river system. A dilution effect from headwaters to estuary was apparent. Sediment residues averaged about 12.5% of DDT in soils of surrounding forests. Maximum residues in sediment were 420 μg/kg total DDT. From 43 to 81% of DDT in stream sediment had broken down to DDD and DDE.

Table 2.3. Pesticides in bed sediment and aquatic biota from rivers in the United States: Process and matrix distribution studies—*Continued*

Reference	Sampling date	Sampling Locations and Strategy	Sampling Media and Species	Target Analytes	Comments
Dimond and others, 1971	1967–1969	68 streams in small watersheds (Maine) that received 0–3 annual applications of DDT between 1958–1967. Of the 68 streams, 44 were sprayed once, 7 twice, and 7 three times; 10 streams were controls (no application).	Sediment (surficial), fish, mussels, insects, plants, birds. Species include trout.	DDD DDE DDT	Objective: to monitor DDT in streams 0–10 years after application. Following a single application, residues dropped rapidly for 1–2 years, but further degradation was difficult to demonstrate. Even 10 years after application, residues were well above those in control streams. This led to cumulative residues in streams sprayed 2–3 times. Residues ranged from nd–1,900 µg/kg (dry wt) in sediment, 40–15,300 µg/kg (wet wt.) in trout, 10–210 µg/kg (wet wt.) in mussels, nd–6,400 µg/kg (wet wt.) in insects, nd–900 µg/kg (wet wt.) in plants, and 300–13,500 µg/kg (wet wt.) in birds. Within-treatment (number of years since a single application) variability was very high. DDE was the principal component in animal samples (60–85%). Degradation of DDT appeared lower in sediment and plant samples (about 45% DDE, 20% DDD). Bioaccumulation factors (between trophic levels) were discussed.
Sears and Meehan, 1971	1968	Nakwasina River watershed (Alaska). 2,4-D was applied to inhibit growth of broad-leaved plants. Noxon Creek watershed was the unsprayed control. Fish and insects were placed in live-boxes prior to spraying.	Water, fish (fry). Species: coho salmon.	2,4-D	Objective: to determine the impact of 2,4-D spray operations on the quality of the aquatic environment. 2,4-D was sprayed on clearcut forest lands in southeastern Alaska. Residues were measured in 4 water samples and 1 composite fish sample (from live-boxes) taken 3 days after spraying. Treatment with 2,4-D caused no significant mortality to aquatic organisms; some mortality in test organisms was attributed by authors to handling and confinement. Spray drifted outside the spray area, including tributary streams. Residues in 4 water samples ranged from <0.5–200 µg/L. Single fish sample contained 500 µg/kg 2,4-D.

Table 2.3. Pesticides in bed sediment and aquatic biota from rivers in the United States: Process and matrix distribution studies—*Continued*

Reference	Sam-pling date	Sampling Locations and Strategy	Sampling Media and Species	Target Analytes	Comments
Naqvi and de la Cruz, 1973	1972	5 sites (Mississippi) in areas that had a history of mirex applications.	Fish, shellfish, inverte-brates, amphibians. Species: green sunfish, mosquito-fish, catfish, frog tadpole, shrimp, snail, crayfish, clam, oyster, crab, and miscellan-eous insects.	Mirex	Objective: to evaluate the distribution of mirex in nontarget organisms living in areas that had been treated with the pesticide to control fire ants. Mirex was also detected in untreated areas, suggesting that it is mobile in the environment. Data were grouped for analysis by type of organism, habitat, mirex treatment history, and trophic level. Average residues were highest in annelids, followed by crustaceans, insects, fish, and mollusks (630, 440, 290, 260, and 150 µg/kg, respectively). Habitats listed in decreasing order were pond, creek, grassland, lake, and estuary (370, 310, 280, 270, and 200 µg/kg, respectively). Omnivores and carnivores had higher average residues than herbivores (350, 300, and 230 µg/kg, respectively). Biota in areas that had received aerial applications in 1972 and in 1969–1970 averaged 330 and 380 µg/kg, respectively. Biota in areas that had received localized treatment on fire ant mounds had average residues of 120 µg/kg. In areas where no treatments were made, residues in biota were 190 to 200 µg/kg.

Table 2.3. Pesticides in bed sediment and aquatic biota from rivers in the United States: Process and matrix distribution studies—*Continued*

Reference	Sam-pling date	Sampling Locations and Strategy	Sampling Media and Species	Target Analytes	Comments
Wolfe and Norment, 1973	1971–1972	Mattuby Creek watershed (Mississippi). Organisms were sampled from ponds and terrestrial habitats before and after mirex treatment for fire ant control. Stream organisms were sampled from treated and adjacent untreated streams. Some pond fish were caged.	Fish (whole), inverte-brates, mammals (whole, brain, liver, and muscle). Aquatic species include green sunfish, largemouth bass, channel catfish, top minnow, crayfish, and mosquito larvae.	Mirex	Objective: to determine the extent of residues in a freshwater ecosystem following aerial application of mirex. In crayfish, no mirex was detected in pretreatment samples, but posttreatment residues averaged 70 μg/kg. In insects, there was no clear increase in residues during posttreatment (1 year) period. Caged channel catfish (pond) contained no detectable mirex before treatment and 10–20 μg/kg afterwards. Resident sunfish from the pond contained 20 μg/kg mirex before treatment and 0–10 μg/kg afterwards. Fish from treated streams had 2- to 10-fold higher residues than those in untreated streams, although this varied by species. Residues had not changed substantially 5 months after treatment. Residues in rat and mouse species were 0–10 μg/kg before treatment, and 0–210 μg/kg afterwards. Shrews (sampled only posttreatment) contained up to 1,890 μg/kg mirex. Comparison of mirex residues in various mammal tissues indicated that whole body residues approximate muscle residues, with brain levels slightly lower and liver levels much higher.

Table 2.3. Pesticides in bed sediment and aquatic biota from rivers in the United States: Process and matrix distribution studies—*Continued*

Reference	Sam- pling date	Sampling Locations and Strategy	Sampling Media and Species	Target Analytes	Comments
Dimond and others, 1974	1971– 1972	19 sites on Webster Brook (Maine), from its headwaters to its mouth, were sampled 7 years after a DDT application to the basin.	Sediment (surficial), fish, invertebrate. Species: brook trout, and crayfish.	DDD DDE DDT DDT, total	Objective: to determine if DDT loads in a stream resulting from a uniform application in its watershed are related to stream size. Residue data in sediment and biota did not indicate stream size was a factor in DDT concentrations. Concentrations in sediment were significantly higher in ponded reaches of the stream than in the faster flowing sections and were related to differences in organic matter content. There were no significant differences in concentrations in trout or crayfish along the length of the stream or between fast and slow water habitats. There were no significant differences in residue levels between age classes of trout. The smallest (slowest growing) trout within the same age class had the highest residue levels. Crayfish residue levels were 10–25% lower than trout. DDD was the predominant residue in sediment (30–60%). DDE was the predominant form in trout and crayfish (60–80%). Equal levels of DDT and DDD were seen in trout, but only traces of DDD were detected in crayfish.
Ginn and Fisher, 1974	nr	Canals associated with a marshland-ricefield ecosystem on the Gulf coast (Texas).	Water (whole), soil, sediment, fish (liver plus muscle), crustaceans. Species: spotted gar, bluegill, menhaden, blue crab, and grass shrimp.	Aldrin Dieldrin	Objective: to investigate the dynamics and distribution of aldrin and related compounds in a marshland-estuarine-ricefield ecosystem. Aldrin was applied as dressing on seed rice (applied aerially). Ricefields were flooded, then flood waters were discharged 24–48 h later into drainage canals. Dieldrin was primary residue detected in canal water, canal bottom sediment, and biota. Both aldrin and dieldrin accumulated in soil, but aldrin residues declined rapidly (nondetectable by 8 weeks). Dieldrin peaked at 1–2 weeks after discharge in grass shrimp and blue crab, and at 4–5 weeks in spotted gar and bluegill. Aldrin was detected for 4 weeks in grass shrimp and bluegill, 1 week in blue crab, and 7 weeks in spotted gar. Menhaden contained the highest residues found in any biota sampled. (Also see Table 2.2.)

Table 2.3. Pesticides in bed sediment and aquatic biota from rivers in the United States: Process and matrix distribution studies—*Continued*

Reference	Sampling date	Sampling Locations and Strategy	Sampling Media and Species	Target Analytes	Comments
Jackson and others, 1974	1967–1972	3 plots of Fraser fir in Mount Mitchell State Park (North Carolina) were sampled in conjunction with an HCH spray operation in 1967.	Water (whole), sediment (surficial), forest litter, small mammals (fat).	HCH, α-HCH, β-Lindane	Objective: to study the persistence of HCH in typical Fraser fir stand, including deposition and persistence in surface litter and soil, residues in water and sediment of streams within and below the treated area, and residues in small mammals in treated and untreated areas. Area was last sprayed in 1963; litter and soil residues before 1967 application contained 67 and 17 mg/kg HCH, respectively. After 1967 application, forest litter averaged 585 mg/kg, decreasing over 5 years to 27 mg/kg. Residues in water were generally below detection (0.06 µg/L). Sediment from the stream within the treated area contained 0.49–3.17 mg/kg HCH. Total residues in animal fat ranged from 0.1 to 176 mg/kg; there was no relation between residue and trapping location (within or outside treated area).
Markin and others, 1974b	1971–1972	2 farm ponds and 1 stream within a block of pasture and forest land (Louisiana) treated with mirex. Residues were monitored before treatment and for 1 year afterward.	Invertebrates. 25 groups of terrestrial and aquatic invertebrates.	Mirex	Objective: to determine mirex residues in invertebrate populations following a single application of bait to control fire ants. Mirex residues were detected in all but one group sampled (white fringed beetle larvae). Concentrations usually corresponded with feeding habit: general scavengers (that would feed directly on bait) contained the highest residues; predaceous insects showed slow, progressive accumulation for 20–90 days after treatment; smallest levels were seen in herbivorous animals. Most residues began to decline after 90 days. Mirex was detected in 8 of 25 invertebrates sampled 1 year following treatment.

Table 2.3. Pesticides in bed sediment and aquatic biota from rivers in the United States: Process and matrix distribution studies—*Continued*

Reference	Sampling date	Sampling Locations and Strategy	Sampling Media and Species	Target Analytes	Comments
Reimold and Durant, 1974	1972	5 sites were monitored weekly before, during, and after dredging in a contaminated channel of Terry Creek near Brunswick (Georgia).	Water, sediment, fish, shellfish, plants. Species: mummichog, anchovy, shrimp, marsh grass, and star drum.	Toxaphene	Objective: to evaluate the redistribution of toxaphene from contaminated sediment during a dredging operation. Dredging spoils were confined to diked areas to minimize exposure of wildlife to toxaphene. Toxaphene levels in dredge spoils ranged from 810 to over 812,000 µg/kg during active dredging (43 days). Toxaphene residues in march grass (*Spartina alterniflora*) increased about an order of magnitude during dredging (average of replicate samples ranged from 645 to 6,795 µg/kg). The highest residues in mummichog (174,000 µg/kg) during the period of dredging were also much higher than predredging samples (9,710 µg/kg). In contrast, residues in oysters did not change significantly during the monitoring period (<1,750 µg/kg). Although residues in biota increased during dredging, the highest values were less than background concentrations measured in previous years. Lower residues were attributed to reduced input from upstream toxaphene manufacturing facilities.
Schultz and Harman, 1974	1971	9 treatment ponds and 2 control ponds (Florida, Georgia, Missouri). In some ponds, fish were stocked prior to 2,4-D (methyl ester) treatment experiments.	Water (whole), sediment (mud), fish. Species: largemouth bass, channel catfish, bluegill.	2,4-D, methyl ester	Objective: to determine residue levels and dissipation rate of 2,4-D methyl ester in fish, bottom sediment, and water of ponds treated for aquatic plant control. The highest level of 2,4-D methyl ester in water (690 µg/L) occurred on the 3rd day following treatment in 5 of the 9 ponds. Residues declined to <5 µg/L within 14–56 days. Highest residues in mud (up to 170 µg/kg) occurred in the ponds that received the heaviest applications. Trace residues persisted in mud from 1 pond up to 112 days after treatment. Highest residues in fish (up to 1,075 µg/kg) were observed in caged fish collected on the 1st day. In some Fla. and Mo. ponds, fish contained detectable residues only after 14–28 days following treatment (perhaps due to release by decaying vegetation). Fish from control ponds (no treatment) did not contain detectable residues.

Table 2.3. Pesticides in bed sediment and aquatic biota from rivers in the United States: Process and matrix distribution studies—*Continued*

Reference	Sam-pling date	Sampling Locations and Strategy	Sampling Media and Species	Target Analytes	Comments
Schultz and Whitney, 1974	1971	Hillsboro Canal and Loxahatchee National Wildlife Refuge (Florida). Samples were taken before and up to 4 months after spraying.	Fish (muscle), sediment (hydrosol), water (whole), birds (breast muscle, liver). Fish species: gar, redear sunfish, largemouth bass, brown bullhead, chubsuckers, and bluegill.	2,4-D	Objective: to obtain residue data from fish, water, and sediment from the study area after application of 2,4-D DTA or 2,4-D DMA to control water hyacinth (to support pesticide registration). The highest residue in water was 37 µg/L, found 1 day after treatment. Most residues in water were <10 µg/L, decreasing to <1 µg/L within 30–56 days. Residues in hydrosol were not detectable (<1–5 µg/kg). Of 60 fish samples, 19 samples contained detectable 2,4-D residues. 3 samples had residues >100 µg/kg. With one exception, all fish samples containing 2,4-D residues were from a site that had been sprayed within the preceding 3 days. Residues in birds dropped to nondetectable levels 4 days after treatment.
Spence, and Markin, 1974	1971–1972	2 sites in ponds (Louisiana, Mississippi) were sampled following aerial application of mirex to surrounding farm and grasslands.	Water, sediment (surficial), plants. Species: Bahia grass.	Mirex	Objective: to monitor the movement of mirex to soil, water, sediment, and plant tissue following a single aerial application of mirex bait (1.74 g/acre a.i.) to control fire ants. The highest concentrations in water were detected in the first day after application (0.53 µg/L). The highest concentrations in sediment were detected 30–40 days following application (1.1 µg/kg). Concentrations in soil and bahia grass were highly variable (0.3–10.4 µg/kg in soil and 1.8–26 µg/kg in grass).

Table 2.3. Pesticides in bed sediment and aquatic biota from rivers in the United States: Process and matrix distribution studies—*Continued*

Reference	Sampling date	Sampling Locations and Strategy	Sampling Media and Species	Target Analytes	Comments
Spacie, 1975	1973–1975	15 sites on the Wabash River (Indiana), upstream and downstream of point source.	Water, fish (fat), algae, invertebrates. Species: sauger, shorthead redhorse, golden redhorse, carp, river carpsucker, mussels, crayfish, and aquatic insect larvae.	Trifluralin	Objective: to assess bioaccumulation of trifluralin downstream of manufacturing plant discharge to the Wabash River. Samples were collected before and after plant modifications were made to reduce trifluralin levels in effluent. Rates of uptake and elimination were determined in lab experiments. Biota downstream of the plant contained trifluralin: sauger had a mean of 145 mg/kg in fat (10 mg/kg on whole body wt. basis); aquatic insects had 1–2.5 mg/kg; mussels, algae, and crayfish had 550, 300, and 20 µg/kg, respectively. Mean levels did not decline from 3–32 km downstream. Models showed that the trifluralin level in fish was related to the concentration in water. After plant improvements were made that decreased trifluralin discharge to the stream by 99%, residues in biota also decreased 40–83%.

Table 2.3. Pesticides in bed sediment and aquatic biota from rivers in the United States: Process and matrix distribution studies—*Continued*

Reference	Sam-pling date	Sampling Locations and Strategy	Sampling Media and Species	Target Analytes	Comments
Wojcik and others, 1975	1971–1972	3 areas (southwest Georgia) where mirex was applied for fire ant control. Aquatic and terrestrial organisms were sampled for 1 year after application. Aquatic vertebrates were collected from farm ponds within each test area.	Fish, toads, frogs, reptiles, spiders, insects, worms, isopods, birds, mammals. Species: mosquito-fish, largemouth bass, bluegill, and green sunfish.	Mirex	Objective: to monitor mirex residues in nontarget organisms following applications of a standard bait formulation and 3 experimental formulations. Mirex residues were found in 20% of pretreatment samples (all from 1 test area). One year after treatment, residues in 6 of 28 species had returned to pretreatment levels; in the others, residues were fairly low (62% of samples had <50 µg/kg, and 92% had <500 µg/kg). After application, 72% of samples from the 3 test areas contained detectable mirex. Levels reached a maximum of 2,930 µg/kg in fish and 1,080 µg/kg in amphibians. Residues in all aquatic animals after 1 year were <90 µg/kg. No appreciable differences in mirex residues were noted as a result of different formulations.
Apperson and others, 1976	1975	A single site in Clear Lake (California) was sampled before and after (1–15 days) three applications of methyl parathion to the lake surface.	Water, sediment (surficial), fish, planktonic inverte-brates. Species: bluegill and crustaceans.	Methyl parathion	Objective: to monitor methyl parathion residues before and after applications of methyl parathion to control the Clear Lake gnat. Applications (at 3.3 µg/L) temporarily reduced zooplankton populations. Recovery of populations to pretreatment levels occurred within a few days or weeks. Residues in fish held in cages open to lake water circulation varied from 11–110 µg/kg. The residues accumulating in fish tissue were dependent on the concentration of methyl parathion in the water. No residues were detected in bottom sediment.

Table 2.3. Pesticides in bed sediment and aquatic biota from rivers in the United States: Process and matrix distribution studies—*Continued*

Reference	Sampling date	Sampling Locations and Strategy	Sampling Media and Species	Target Analytes	Comments
Reeves and others, 1977	1971	20 sites in North Carolina. Each site was a tobacco field with farm pond. Ten sites were inside and 10 outside a pest management program area. Samples were collected in spring and fall (before and after pesticide applications).	Sediment (surficial), water (whole), fish (whole), turtles, frogs, invertebrates, tobacco leaves, field soil. Species: bluegill, bullfrog, tiger beetles, snapping turtle, musk turtle, painted turtle, and yellow-bellied turtle.	Aldrin Carbaryl Carbofuran Chlordane DDD, o,p'- DDD, p,p'- DDE, o,p'- DDE, p,p'- DDT, o,p'- DDT, p,p'- Dieldrin Endosulfan Endrin Ethion Heptachlor Heptachlor epoxide Lindane Malathion Methomyl Methyl parathion Methyl trithion Mirex Parathion Strobane Toxaphene	Objective: to monitor pesticide residues in biotic and abiotic compartments of tobacco field ecosystems before and after pesticide application. Sites were located inside or outside a pest management program area in North Carolina. Pesticides were applied to farms inside and outside the program area; the specific pesticides applied (listed) varied among sites inside and outside the program area. Sediment samples contained o,p'-DDD (10–350 μg/kg) and p,p'-DDD (20–1,150 μg/kg); other organochlorines, organophosphates, carbaryl, carbofuran, and methomyl were not detected. Soil and sediment inside the program area generally had slightly higher pesticide residues than outside the program area. Inside and outside the program area, biota contained dieldrin, endrin, and all 6 DDT isomers. No significant differences were found between residues outside and inside the program area. No OPs (<10 μg/kg) were detected in biota. Results of water analyses were not reported. All reported residues were corrected for percent recovery.

Table 2.3. Pesticides in bed sediment and aquatic biota from rivers in the United States: Process and matrix distribution studies—*Continued*

Reference	Sampling date	Sampling Locations and Strategy	Sampling Media and Species	Target Analytes	Comments
Prest and others, 1992	1990	Clams were transplanted and semi-permeable membrane sampling devices (SPMD) deployed at 3 sites on the Sacramento and San Joaquin rivers (California). Clams and SPMDs were exposed for about 2 months.	Clams, SPMD. Species: freshwater clam (*Corbicula fluminea*).	Chlordane, *cis*-Chlordane, total Chlordane, *trans*-DDD, *p,p'*-DDE, *o,p'*-DDE, *p,p'*-DDT, *o,p'*-DDT, *p,p'*-DDT, total HCB HCH, α-HCH, β-HCH, δ-Heptachlor epoxide Lindane Nonachlor, *cis*-Nonachlor, *trans*-Oxychlordane TCDD, 2,3,7,8-	Objective: to investigate organochlorine accumulation by clams and SPMDs in the Sacramento–San Joaquin Delta, as part of study on impacts of industrial facilities in this area. A number of clams were held as controls and 5 SPMDs were analyzed as blanks. SPMDs contained triolein lipid (95%). On lipid-normalized basis, all OC concentrations were 10–100 times higher in clams than in SPMDs. Authors suggested this reflected nonequilibrium status of SPMDs. HCB levels in clams and SPMDs were highest in clams from Antioch, near the confluence of the San Joaquin and Sacramento rivers. Total DDT and total chlordanes were highest in clams from the San Joaquin River, intermediate in clams from the Sacramento River, and lowest in clams from Antioch. In contrast, SPMDs showed: Sacramento> Antioch> San Joaquin. SPMDs from San Joaquin River became biofouled, which probably suppressed contaminant uptake by these devices. Also, waters in the San Joaquin were stagnant with high suspended solids. Particle-bound contaminants there may have been available to clams, but not SPMDs.

APPENDIX D. Common names and taxonomic classifications of aquatic organisms sampled in the studies reviewed (listed in Tables 2.1, 2.2, and 2.3) or mentioned in the text

[Types: A, amphibian; F, fish; I, invertebrate; O, other; P, plant; R, reptile. Some common names are listed more than once because they correspoond to multiple species or taxonomic groups that were sampled by the studies reviewed (listed in Tables 2.1, 2.2, and 2.3); repeated taxonomic classifications indicate either multiple common names or different life stages. Abbreviation: sp., species]

Type	Common Name	Taxonomic Classification
F	Albacore tuna	*Thunnus alalunga*
I	Alderflies	family Sialidae, order Megaloptera
I	Alderfly larvae	family Sialidae, order Megaloptera
F	Alewife	*Alosa pseudoharengus*
P	Alga	*Chlorella*
P	Alga	*Dunaliella salina*
P	Algae	*Oedogonium* sp.
F	Alligator gar	*Lepidosteus spatula*
R	American alligator	*Alligator mississippiensis*
R	American crocodile	*Crocodilus aculatus*
F	American eel	*Anguilla rostrata*
I	American oyster	*Ostrea virginica*
F	American shad	*Alosa sapidissima*
I	Amphipod	*Pontoporeia affinis*
I	Amphipod	*Pontoporeia hoya*
F	Anchovy	family Engraulidae
F	Arctic grayling	*Thymus arcticus*
P	Arrow arum	*Peltandra* sp.
I	Asiatic clam	*Corbicula fluminea*
F	Atlantic croaker	*Micropogon undulatus*
F	Atlantic needlefish	*Strongylura marina*
F	Atlantic salmon	*Salmo salar*
F	Atlantic silverside	*Menidia menidia*
F	Ayu sweetfish	*Plecoglossus altivelis*
P	Bahia grass	*Paspalum notatum*
F	Barred sand bass	*Paralabrax nebulifer*
F	Bass	family Centrarchidae or Percichthyidae
I	Bay mussel	*Mytilus edulis*
I	Beetles	order Coleoptera
F	Bigmouth buffalo	*Ictiobus cyprinellus*
I	Bivalve mollusk	*Margaritifera falcata*
I	Bivalve	*Yoldia limatula*
F	Black bass	*Micropterus salmoides*
F	Black buffalo	*Ictiobus niger*
F	Black bullhead	*Ictalurus melas*
F	Black crappie	*Pomoxis migromaculatus*
F	Black croaker	*Cheilotrema saturnum*
F	Black grouper	*Mysteroperca bonaci*
F	Blacknose dace	*Rhinichthys atratulus*
F	Bleak	*Alburnus alburnus*
F	Bloater	*Coregonus hoyi*
F	Blue catfish	*Ictalurus furcatus*
F	Blue chub	*Gila coerulea*

APPENDIX D. Common names and taxonomic classifications of aquatic organisms sampled in the studies reviewed (listed in Tables 2.1, 2.2, and 2.3) or mentioned in the text—*Continued*

Type	Common Name	Taxonomic Classification
F	Sand sole	*Psettichthys melanostictus*
I	Sandworm	*Nereis virens*
F	Sauger	*Stizostedion canadense*
P	Sawgrass	*Cladium* sp.
I	Scallop	family Pectinidae
F	Sculpins	*Myoxocephalus scorpius*
F	Sculpins	family Cottidae
F	Scup	*Stenotomus chrysops*
F	Sea catfish	family Ariidae
F	Sea herring	*Clupea harengus*
F	Sea trout	*Cynoscion* sp.
F	Shad	*Alosa* or *Dorosoma* sp.
F	Shark	class Elasmobranchiomorphi
F	Sheepshead minnow	*Cyprinodon vanegatus*
F	Shiner	family Cyprinidae
F	Shorthead redhorse	*Moxostoma macrolepidotum*
F	Shortnose gar	*Lepidosteus platostomus*
F	Shovelnose sturgeon	*Scaphirhynchus platorynchus*
I	Shrimp	*Crangon septemspinosa*
I	Shrimp	*Penaeus* sp.
I	Shrimp	order Decapoda
F	Silver perch	*Bairdiella chrysura*
F	Silver redhorse	*Moxostoma anisurum*
F	Slimy sculpin	*Cottus cognatus*
F	Smallmouth bass	*Micropterus dolomieu*
F	Smallmouth buffalo	*Ictiobus bubalus*
F	Smelt	family Osmeridae
R	Smooth green snake	*Leopeltis vernalis*
I	Snail	*Physa gyrina*
I	Snail	*Physa* sp.
R	Snapping turtle	*Chelydra serpentina*
I	Softshell clam	order Pelecypoda
F	Sonora sucker	*Catostomus insignis*
F	Southern flounder	*Paralichthys lethostigma*
F	Spanish mackerel	*Scomberomorus maculatus*
F	Speckled Dace	*Rhinichthys osculus*
F	Spiny dogfish	*Squalus acanthias*
I	Spiny naiad	*Najas marina*
R	Spiny softshell turtle	*Trionyx ferox*
F	Splake	hybrid between brook trout and lake trout
F	Spot	*Leiostomos xanthurus*
F .	Spotfin shiner	*Cyprinella spiloptera*
F	Spottail shiner	*Notropis hudsonius*
F	Spotted gar	*Lepidosteus oculatus*
F	Spotted seatrout	*Cynoscion nebulosus*
F	Spotted sunfish	*Lepomis punctatus*
R	Spotted turtle	*Clemmys guttata*

APPENDIX D. Common names and taxonomic classifications of aquatic organisms sampled in the studies reviewed (listed in Tables 2.1, 2.2, and 2.3) or mentioned in the text—*Continued*

Type	Common Name	Taxonomic Classification
F	Sprat	*Clupea sprattus*
I	Squid	*Loligo* sp. or *Ommastrephes* sp.
F	Staghorn sculpin	*Leptcottus armatus*
F	Star drum	*Stellifer lanceolatus*
F	Starry flounder	*Paltichthys stellatus*
F	Steelhead	*Oncorhynchus mykiss*
F	Steelhead trout	*Oncorhynchus mykiss*
F	Sting ray	*Dasyatidae sabrina*
F	Sting ray	*Dasyatidae sayi*
I	Stoneflies	*Acroneuria abnormis*
I	Stoneflies	*Acroneuria* sp.
I	Stoneflies	*Pteronarcys* sp.
I	Stoneflies	order Plecoptera
P	Stonewort	*Chara* sp.
F	Striped bass	*Morone saxatilis*
F	Striped mullet	*Mugil cephalus*
F	Striped seaperch	*Embiotoca lateralis*
F	Suckers	*Catostomus* sp.
F	Summer flounder	*Paralichthys dentatus*
F	Sunfish	*Lepomis gibbosus*
F	Sunfish	family Centarchidae
I	Tanner crab	*Chionoecetes bairdi*
F	Tarbon	*Elops hawaiensis*
F	Threadfin shad	*Dorosoma petenense*
F	Threespine stickleback	*Gasterosteus aculeatus*
I	Tiger beetles	family Cicindela
F	Tilapia	*Tilapia* sp.
F	Tomcod	*Microgadus* sp.
F	Top minnow	*Fundulus* sp.
I	True flies	order Diptera
I	Tubificid worms	family Tubificidae
F	Tui chub	*Gila bicolor*
F	Tuna	*Thunnus thynnus*
F	Tuna	family Scombridae
I	Unionid clams	order Pelycopoda
F	Utah chub	*Gila atraria*
F	Utah sucker	*Catostomus ardens*
F	Vimba	*Vimba vimba* L.
F	Walleye	*Stizostedion vitreum*
F	Warmouth	*Lepomis gulosus*
I	Water boatman	family Corixidae
P	Water arum	*Calla palustris*
I	Water bug	order Hemiptera
R	Water snake	*Natrix* sp.
R	Water snake	*Nerodia cyclopion*
R	Water snake	*Nerodia rhombifera*
R	Water snake	*Nerodia* sp.

Appendix E. Glossary Of Common And Chemical Names Of Pesticides And Related Compounds Given In Text—*Continued*

Common Name	Chemical Class	Use	CAS No.	Chemical Nomenclature
Chlorthiamid	amide	H	1918-13-4	2,6-dichlorothiobenzamide
Cinerin I	pyrethroid	I	25402-06-6	[1R-[1α[S*(Z)],3β]]-3-(2-butenyl)-2-methyl-4-oxo-2-cyclopenten-1-yl 2,2-dimethyl-3-(2-methyl-1-propenyl)cyclopropanecarboxylate
Cinerin II	pyrethroid	I	121-20-0	[1R-[1α[S*(Z)],3β(E)]]-3-(2-butenyl)-2-methyl-4-oxo-2-cyclopenten-1-yl 3-(3-methoxy-2-methyl-3-oxo-1-propenyl)-2,2-dimethylcyclopropanecarboxylate
Clomazone	miscellaneous N	H	81777-89-1	2-[(2-chlorophenyl)methyl]-4,4-dimethyl-3-isoxazolidinone
Copper sulfate	inorganic	Fn,H	7758-99-8	copper sulfate
Coumaphos	organophosphorus	I,N	56-72-4	O,O-diethyl O-(3-chloro-4-methyl-2-oxo-2H-1-benzo-pyran-7-yl) phosphorothioate
p-Cresol	phenol	Ind	106-44-5	4-methylphenol
Crotoxyphos	organophosphorus	I	7700-17-6	α-methylbenzyl 3-hydroxy-cis-crotonate, dimethyl phosphate
Crufomate	organophosphorus	I	299-86-5	4-tert-butyl-2-chlorophenyl N-methyl O-methylphosphoramidate
Cryolite	inorganic	I	15096-52-3	sodium fluoaluminate
Cyanazine	triazine	H	21725-46-2	2-[[4-chloro-6-(ethylamino)-1,3,5-triazin-2-yl]amino]-2-methylpropionitrile
Cycloate	thiocarbamate	H	1134-23-2	S-ethyl N-ethyl N-cyclohexylthiocarbamate
Cyclodienes, total	organochlorine	I		mixture of aldrin, chlordane, dieldrin, endosulfan, endrin, heptachlor, isodrin, toxaphene
Cyfluthrin	pyrethroid	I	68359-37-5	cyano(4-fluoro-3-phenoxyphenyl)methyl 3-(2,2-dichloroethenyl)-2,2-dimethylcyclopropanecarboxylate

Appendix E. Glossary Of Common And Chemical Names Of Pesticides And Related Compounds Given In Text—*Continued*

Common Name	Chemical Class	Use	CAS No.	Chemical Nomenclature
Cyhalothrin	pyrethroid	I	91465-08-6	(RS)-α-cyano-3-phenoxybenzyl (Z)-(1R,3R)-3-(2-chloro-3,3,3-trifluoroprop-1-enyl)-2,2-dimethylcyclopropanecarboxylate
λ-Cyhalothrin	see Cyhalothrin			
Cypermethrin	pyrethroid	I	52315-07-8	cyano(3-phenoxyphenyl)methyl 3-(2,2-dichloroethenyl)-2,2-dimethylcyclopropanecarboxylate
Cyromazine	triazine	I	66215-27-8	N-cyclopropyl-1,3,5-triazine-2,4,6-triamine
Cythioate	organophosphorus	I	115-93-5	O-[4-(aminosulfonyl)phenyl] O,O-dimethyl phosphorothioate
1,3-D	see Telone II			
2,4-D	chlorophenoxy acid	H	94-75-7	(2,4-dichlorophenoxy)acetic acid
2,4-D, isopropyl ester	chlorophenoxy acid ester	H	94-11-1	isopropyl (2,4-dichlorophenoxy)acetate
2,4-D, methyl ester	chlorophenoxy acid ester	H	1928-38-7	methyl (2,4-dichlorophenoxy)acetate
Dacthal	chlorobenzoic acid	H	1861-32-1	dimethyl 2,3,5,6-tetrachloro-1,4-benzenedicarboxylate
Dalapon	miscellaneous	H	75-99-0	2,2-dichloropropanoic acid
2,4-DB	chlorophenoxy acid	H	94-82-6	4-(2,4-dichlorophenoxy)butanoic acid
DBCP	see 1,2-Dibromo-3-chloropropane			
DBP	see Dibutyl phthalate			
DCA	pesticide degradate		608-27-5	2,3-dichlorobenzeneamine
			554-00-7	2,4-dichlorobenzeneamine
			95-76-1	3,4-dichlorobenzeneamine
			626-43-7	3,5-dichlorobenzeneamine
			95-76-1	4,5-dichlorobenzeneamine

Appendix E. Glossary Of Common And Chemical Names Of Pesticides And Related Compounds Given In Text—*Continued*

Common Name	Chemical Class	Use	CAS No.	Chemical Nomenclature
DCNA	nitroaniline	Fn	99-30-9	2,6-dichloro-4-nitroaniline
DCPA	see Dacthal			
D-D	VOC	Fm,N	8003-19-8	mixture of 1,2-dichloropropane and 1,3-dichloropropene
DDA	DDT degradate		83-05-6	4-chloro-α-4-(chlorophenyl)-benzene acetic acid
DDD	organochlorine, DDT degradate	I	72-54-8	1,1-dichloro-2,2-bis(chlorophenyl)ethane; mixture of *o,p'*-DDD and *p,p'*-DDD
o,p'-DDD	*o,p'*-DDT degradate		53-19-0	1-chloro-2-[2,2-dichloro-1(4 chlorophenyl)ethyl]benzene
p,p'-DDD	organochlorine, *p,p'*-DDT degradate	I	72-54-8	1,1-dichloro-2,2-bis(*p*-chlorophenyl)ethane
DDE	organochlorine, DDT degradate		72-55-9	1,1-dichloro-2,2-bis(chlorophenyl)ethene; mixture of *o,p'*-DDE and *p,p'*-DDE
o,p'-DDE	*o,p'*-DDT degradate		3424-82-6	1-chloro-2-[(2,2-dichloro-1-(4-chlorophenyl)ethenyl]benzene
p,p'-DDE	*p,p'*-DDT degradate	I	72-55-9	1,1-dichloro-2,2-bis(*p*-chlorophenyl)ethene
p,p'-DDMS	*p,p'*-DDT degradate		2642-80-0	2-chloro-1,1-bis(*p*-chlorophenyl)ethane
DDMU	DDT degradate		1022-22-6	2-chloro-1,1-bis(*p*-chlorophenyl)ethylene
p,p'-DDMU	*p,p'*-DDT degradate		1022-22-6	2-chloro-1,1-bis(*p*-chlorophenyl)ethylene
DDT	organochlorine	I	50-29-3	1,1,1-trichloro-2,2-bis(chlorophenyl)ethane; mixture of *o,p'*-DDT and *p,p'*-DDT
DDT, technical	organochlorine	I		mixture of *p,p'*-DDT (about 80 percent) and *o,p'*-DDT (about 20 percent)

Appendix E. Glossary Of Common And Chemical Names Of Pesticides And Related Compounds Given In Text—*Continued*

Common Name	Chemical Class	Use	CAS No.	Chemical Nomenclature
DDT, total	organochlorine	I		mixture of o,p'-DDT, p,p'-DDT, o,p'-DDD, p,p'-DDD, o,p'-DDE, and p,p'-DDE
o,p'-DDT	organochlorine	I	789-02-6	1,1,1-trichloro-2-(p-chlorophenyl)-2-(o-chlorophenyl)ethane
p,p'-DDT	organochlorine	I	50-29-3	1,1,1-trichloro-2,2-bis(p-chlorophenyl)ethane
DDVP	see Dichlorvos			
Decanol	alcohol	Ind	36729-58-5	1-decanol
Deet	see Diethyltoluamide			
DEF	see Tribufos			
DEHP	phthalate	Ind	117-81-7	bis(2-ethylhexyl)phthalate
Demeton	organophosphorus	I,Ac	298-03-3	O,O-diethyl O-[2-(ethylthio)ethyl] phosphorothioate
DES	see Diethylstilbesterol			
Desmethyl fenitrooxon	fenitrothion degradate		15930-84-4	phosphoric acid, monomethyl mono(3-methyl-4-nitrophenyl) ester
Desmethyl fenitrothion	fenitrothion degradate		4321-64-6	phosphorothioic acid, O-(3-methyl-4-nitrophenyl) ester
Dextrin	miscellaneous	Ad	9004-53-9	hydrolyzed starch
Diazinon	organophosphorus	I,N	333-41-5	O,O-diethyl O-[6-methyl-2-(1-methylethyl)-4-pyrimidinyl] phosphorothioate
Dibromochloropropane	see 1,2-Dibromo-3-chloropropane			
1,2-Dibromo-3-chloropropane	VOC	Fm	96-12-8	1,2-dibromo-3-chloropropane
Dibutyl phthalate	miscellaneous	IR	84-74-2	dibutyl 1,2-benzenedicarboxylate

Appendix E. Glossary Of Common And Chemical Names Of Pesticides And Related Compounds Given In Text—*Continued*

Common Name	Chemical Class	Use	CAS No.	Chemical Nomenclature
Dicamba	chlorobenzoic acid, Vel-4207 degradate	H	1918-00-9	3,6-dichloro-2-methoxybenzoic acid
Dicarboxylic acid	carboxylic acid	Ind	1403-44-7	α,β-dicarboxylic acid
Dichlobenil	benzonitrile	H	1194-65-6	2,6-dichlorobenzonitrile
Dichlone	miscellaneous	Fn	117-80-6	2,3-dichloro-1,4-naphthoquinone
Dichloroaniline	see DCA			
p-Dichlorobenzene	see 1,4-Dichlorobenzene			
1,2-Dichlorobenzene	organochlorine	Fm,H, I,Ad	95-50-1	1,2-dichlorobenzene
1,3-Dichlorobenzene	organochlorine	U	541-73-1	1,3-dichlorobenzene
1,4-Dichlorobenzene	organochlorine	Fm,Fn, I,R	106-46-7	1,4-dichlorobenzene
4,4′-Dichlorobenzophenone	DDT degradate		90-98-2	bis(4-chlorophenyl)methanone
Dichlorobiphenyl, total	organochlorine	Ind	25512-42-9	dichloro-1,1′-biphenyl (all isomers)
4,4′-Dichlorodiphenylmethane	DDT degradate		101-76-8	bis(*p*-chlorophenyl)methane
Dichlorophen	miscellaneous	Fn,Ad	97-23-4	bis(5-chloro-2-hydroxyphenyl)methane
Dichlorophene	see Dichlorophen			
2,4-Dichlorophenol	phenol	Ind	120-83-2	2,4-dichlorophenol
Dichlorvos	organophosphorus	Fm,I, Ad	62-73-7	*O*,*O*-dimethyl *O*-(2,2-dichlorovinyl) phosphate

Appendix E.   Glossary Of Common And Chemical Names Of Pesticides And Related Compounds Given In Text—*Continued*

Common Name	Chemical Class	Use	CAS No.	Chemical Nomenclature
Diclofop	chlorophenoxy acid	H	40843-25-2	(*RS*)-2-[4-(2,4-dichlorophenoxy)phenoxy]propionic acid
Dicloran	benzene derivative	Fn	99-30-9	2,6-dichloro-4-nitrobenzenamine
Dicofol	organochlorine	I	115-32-2	4-chloro-α-(4-chlorophenyl)-α-(trichloromethyl) benzenemethanol
Dicrotophos	organophosphorus	I	141-66-2	(*E*)-2-dimethylcarbamoyl-1-methylvinyl dimethyl phosphate
Dieldrin	organochlorine	I	60-57-1	1,2,3,4,10,10-hexachloro-6,7-epoxy-1,4,4a,5,6,7,8,8a-octahydro (endo,exo) 1,4:5,8-dimethanonaphthalene
Dienochlor	organochlorine	Mi	2227-17-0	decachlorobis(2,4-cyclopentadien-1-yl)
Diethatyl ethyl	amino acid derivative	H	58727-55-8	*N*-(chloroacetyl)-*N*-(2,6-diethylphenyl)glycine ethyl ester
Diethyl aniline	benzene derivative	Ind	99-66-7	*N*,*N*-diethylbenzenamine
Diethylfumarate	malathion degradate		623-91-6	(*E*)-2-butenedioic acid, diethyl ester
Diethyl phosphate	organophosphorus insecticide degradate		598-02-7	phosphoric acid, diethyl ester
Diethyl phosphorothionate	organophosphorus insecticide degradate		2465-65-8	phosphorothioic acid, *O*,*O*-diethyl ester
Diethylstilbesterol	synthetic estrogen	Pha	56-53-1	(*E*)-4,4′-(1,2-diethyl-1,2-ethenediyl)bisphenol
Diethyltoluamide	amide	IR	134-62-3	*N*,*N*-diethyl-*m*-toluamide
Diflubenzuron	urea	I	35367-38-5	1-(4-chlorophenyl)-3-(2,6-difluorobenzoyl)urea
Difolitan	see Captafol			
5,6-Dihydro-dihydroxycarbaryl	carbaryl degradate		5375-49-5	5-(methylcarbamate)-1,2-dihydro-1,2,5-naphthalenetriol

Appendix E. Glossary Of Common And Chemical Names Of Pesticides And Related Compounds Given In Text—*Continued*

Common Name	Chemical Class	Use	CAS No.	Chemical Nomenclature
2,8-Dihydromirex	mirex degradate		57096-48-7	1,1a,2,2,3,3a,4,5,5,5a-decachloro-octahydro-1,3,4-metheno-1*H*-cyclobuta[*cd*]pentalene
5,10-Dihydromirex	mirex degradate		53207-72-0	1,1a,2,3,3a,4,5,5a,5b,6-decachloro-octahydro-1,3,4-metheno-1*H*-cyclobuta[*cd*]pentalene
Dimethoate	organophosphorus	I	60-51-5	*O,O*-dimethyl *S*-methylcarbamoylmethyl phosphorodithioate
Dimethyl phosphate	organophosphorus insecticide degradate		813-78-5	phosphoric acid, dimethyl ester
Dimethyl phosphorodithioic acid	organophosphorus insecticide degradate		756-80-9	*O,O*-dimethylphosphorodithioic acid
Dimethyl phosphorothionate	organophosphorus insecticide degradate		1112-38-5	phosphorothioic acid, *O,O*-dimethyl ester
Dinitramine	miscellaneous N	H	29091-05-2	N^4,N^4-diethyl-α,α,α-trifluoro-3,5-dinitrotoluene-2,4-diamine
Dinocap	miscellaneous N	Ac,Fn	39300-45-3	2,4-dinitro-6-octylphenylcrotonate
Dioxathion	organophosphorus	I,Ac	78-34-2	*S,S*'-(1,4-dioxane-2,3-diyl) *O,O,O',O'*-tetraethyl bis(phosphorodithioate)
Diphacinone	miscellaneous	I,R	82-66-6	2-(diphenylacetyl)-1,3-indandione
Diphenamid	amide	H	957-51-7	*N,N*-dimethyl-2,2-diphenylacetamide
Diquat	see Diquat dibromide			
Diquat dibromide	miscellaneous N	H	85-00-7	1,1'-ethylene-2,2'-bipyridylium dibromide, monohydrate
Disulfoton	organophosphorus	I	298-04-4	*O,O*-diethyl *S*-[2-(ethylthio)ethyl]phosphorodithioate
Dithane M-45	see Mancozeb			

Appendix E. Glossary Of Common And Chemical Names Of Pesticides And Related Compounds Given In Text—*Continued*

Common Name	Chemical Class	Use	CAS No.	Chemical Nomenclature
Diuron	urea	H	330-54-1	3-(3,4-dichlorophenyl)-1,1-dimethylurea
DMDT	see *p,p'*-Methoxychlor			
Dodecanol	alcohol	Ind	27342-88-7	1-dodecanol
Dodine	amine	Fn	2439-10-3	1-dodecylguanidine acetate
2,4-DP	chlorophenoxy acid derivative	H	120-36-5	(±)-2-(2,4-dichlorophenoxy)propanoic acid
DSMA	organic arsenical	H	144-21-8	disodium methanearsonate
Dyfonate	see Fonofos			
EBDC fungicides	ethylene bis-dithio-carbamates	Fn	111-54-6	
EDB	see Ethylene dibromide			
Endosulfan	organochlorine	I	115-29-7	6,7,8,9,10,10-hexachloro-1,5,5a,6,9,9a-hexahydro-6,9-methano-2,4,3-benzodioxathiepin-3-oxide
Endosulfan I	organochlorine	I	959-98-8	3α,5aβ,6α,9α,9aβ-6,7,8,9,10,10-hexachloro-1,5,5a,6,9,9a-hexahydro-6,9-methano-2,4,3-benzodioxathiepin-3-oxide
Endosulfan II	organochlorine	I	33213-65-9	3α,5aα,6β,9β,9aβ-6,7,8,9,10,10-hexachloro-1,5,5a,6,9,9a-hexahydro-6,9-methano-2,4,3-benzodioxathiepin-3-oxide
Endosulfan sulfate	endosulfan degradate		1031-07-8	3α,5aα,6β,9β,9aβ-6,7,8,9,10,10-hexachloro-1,5,5a,6,9,9a-hexahydro-6,9-methano-2,4,3-benzodioxathiepin-3,3-dioxide
Endothal	see Endothall			
Endothall	miscellaneous	H	129-67-9	7-oxabicyclo[2.2.1]heptane-2,3-dicarboxylic acid

Appendix E. Glossary Of Common And Chemical Names Of Pesticides And Related Compounds Given In Text—*Continued*

Common Name	Chemical Class	Use	CAS No.	Chemical Nomenclature
Endrin	organochlorine	I	72-20-8	1,2,3,4,10,10-hexachloro-6,7-epoxy-1,4,4a,5,6,7,8,8a-octahydro-(endo,endo)-1,4:5,8-dimethanonaphthalene
Endrin aldehyde	endrin degradate		7421-93-4	2,2a,3,3,4,7-hexachlorodecahydro-1,2,4-methenocyclopenta-(*cd*)-pentalene-*r*-carboxaldehyde
Endrin ketone	endrin degradate		53494-70-5	3b,4,5,6,6a-hexachlorodecahydro-(2α,3aβ,3bβ,4β,5β,6aβ,7α,7aβ,8R*)-2,5,7-metheno-3H-cyclopenta(a)pentalen-3-one
EPN	organophosphorus	I,Ac	2104-64-5	O-ethyl O-4-nitrophenyl phenylphosphonothioate
Eptam	thiocarbamate	H	759-94-4	S-ethyl dipropylthiocarbamate
EPTC	see Eptam			
Esfenvalerate	pyrethroid	I	66230-04-4	(S)-α-cyano-3-phenoxybenzyl (S)-2-(4-chlorophenyl)-3-methylbutyrate
Estradiol	estrogen	NO	50-28-2	(17B)-Estra-1,3,5(10)-triene-3,17-diol
Ethalfluralin	dinitroaniline	H	55283-68-6	N-ethyl-N-(2-methyl-2-propenyl)-2,6-dinitro-4-(trifluoromethyl)benzenamine
Ethion	organophosphorus	I,Ac	563-12-2	S,S'-methylene bis(O,O-diethyl phosphorodithioate)
Ethoprop	organophosphorus	I,N	13194-48-4	O-ethyl S,S-dipropyl phosphorodithioate
Ethylene dibromide	VOC	Fm,I,N	106-93-4	1,2-dibromoethane
Ethyl parathion	see Parathion			
Etridiazole	miscellaneous N	Fn	2593-15-9	5-ethoxy-3-trichloromethyl-1,2,4-thiadiazole
Famphur	organophosphorus	I	52-85-7	O-4-dimethylsulphamoylphenyl O,O-dimethyl phosphorothioate

Appendix E. Glossary Of Common And Chemical Names Of Pesticides And Related Compounds Given In Text—*Continued*

Common Name	Chemical Class	Use	CAS No.	Chemical Nomenclature
Fenamiphos	organophosphorus	N	22224-92-6	O-ethyl 3-methyl-4-(methylthio)phenyl 1-methylethyl phosphoramidate
Fenarimol	miscellaneous N	Fn	60168-88-9	α-(2-chlorophenyl)-α-(4-chlorophenyl)-5-pyrimidine methanol
Fenbutatin oxide	organotin	I,Ac	13356-08-6	bis[tris(2-methyl-2-phenylpropyl)tin] oxide
Fenchlorphos	see Ronnel			
Fenitrooxon	fenitrothion degradate		2255-17-6	phosphoric acid, dimethyl 3-methyl-4-nitrophenyl ester
Fenitrothion	organophosphorus	I,Ac	122-14-5	O,O-dimethyl O-(3-methyl-4-nitrophenyl)phosphorothioate
Fenoprop	see 2,4,5-TP			
Fenoxycarb	carbamate	IGR	79127-80-3	ethyl [2-(4-phenoxyphenoxy)ethyl]carbamate
Fensulfothion	organophosphorus	I	115-90-2	O,O-diethyl O-[4-(methylsulfinyl)phenyl] phosphorothioate
Fenthion	organophosphorus	I	55-38-9	O,O-dimethyl O-[3-methyl-4-(methylthio)phenyl] phosphorothioate
Fenvalerate	pyrethroid	I	51630-58-1	cyano-(3-phenoxyphenyl)methyl-4-chloro-(1-methylethyl)-benzeneacetate
Ferbam	thiocarbamate	Fn	14484-64-1	ferric dimethyldithiocarbamate
Fluazifop	miscellaneous N	H	69335-91-7	(RS)-2-[[[4-(5-trifluoromethyl)-2-pyridinyl]oxy]phenoxy]propanoic acid
Fluazifop-butyl	miscellaneous N	H	69806-50-4	butyl (RS)-2-[4-[[5-(trifluoromethyl)-2-pyridinyl]oxy]phenoxy]propanoate
Fluometuron	urea	H	2164-17-2	1,1-dimethyl-3-(α,α,α-trifluoro-m-tolyl)urea
Fluorodifen	miscellaneous N	U	15457-05-3	4-nitrophenyl α,α,α-trifluoro-2-nitro-p-tolyl ether

Appendix E. Glossary Of Common And Chemical Names Of Pesticides And Related Compounds Given In Text—*Continued*

Common Name	Chemical Class	Use	CAS No.	Chemical Nomenclature
Fluridone	miscellaneous N	H	59756-60-4	1-methyl-3-phenyl-5-[3-(trifluoromethyl)phenyl]-4(1*H*)-pyridinone
Folex	organophosphorus	Df	150-50-5	*S,S,S*-tributyl phosphorotrithioate
Folpet	imide	Fn	133-07-3	*N*-[(trichloromethyl)thio]phthalimide
Fonofos	organophosphorus	I	944-22-9	*O*-ethyl *S*-phenyl ethylphosphonodithioate
Formetanate HCl	carbamate	I,Ac	23422-53-9	3-dimethylaminomethyleneaminophenyl methylcarbamate hydrochloride
Fosetyl-Al	see Fosetyl aluminum			
Fosetyl aluminum	organophosphorus	Fn	39148-24-8	aluminum tris(*O*-ethyl phosphonate)
Glyphosate	amino acid derivative	H	1071-83-6	*N*-(phosphonomethyl)glycine, isopropylamine salt
Guthion	see Azinphos-methyl			
HCB	see Hexachlorobenzene			
α-HCH	organochlorine	I	319-84-6	1α,2α,3β,4α,5β,6β-hexachlorocyclohexane
β-HCH	organochlorine	I	319-85-7	1α,2β,3α,4β,5α,6β-hexachlorocyclohexane
γ-HCH	organochlorine	I	58-89-9	1α,2α,3β,4α,5α,6β-hexachlorocyclohexane
δ-HCH	organochlorine	I	319-86-8	1α,2α,3α,4β,5α,6β-hexachlorocyclohexane
HCH, technical	organochlorine	I	608-73-1	1,2,3,4,5,6-hexachlorocyclohexane (mixture of isomers)
Heptachlor	organochlorine	I	76-44-8	1,4,5,6,7,8,8-heptachloro-3a,4,7,7a-tetrahydro-4,7-methano-1*H*-indene
Heptachlor epoxide	heptachlor degradate		1024-57-3	2,3,4,5,6,7,8-heptachloro-1a,1b,5,5a,6,6a-hexahydro-2,5-methano-2*H*-indeno(1,2b)oxirene

Appendix E. Glossary Of Common And Chemical Names Of Pesticides And Related Compounds Given In Text—*Continued*

Common Name	Chemical Class	Use	CAS No.	Chemical Nomenclature
Hexachlorobenzene	organochlorine	Fn	118-74-1	hexachlorobenzene
2,4,5,2',4',5'-Hexachlorobiphenyl	organochlorine	Ind	35065-27-1	2,2',4,4',5,5'-hexachloro-1,1'-biphenyl
2,4,6,2',4',6'-Hexachlorobiphenyl	organochlorine	Ind	33979-03-2	2,2',4,4',6,6'-hexachloro-1,1'-biphenyl
Hexachlorobiphenyl, total	organochlorine	Ind	26601-64-9	hexachloro-1,1'-biphenyl (all isomers)
Hexachlorobutadiene	organochlorine	Ind	87-68-3	1,1,2,3,4,4-hexachloro-1,3-butadiene
Hexazinone	triazine	H	51235-04-2	3-cyclohexyl-6-(dimethylamino)-1-methyl-1,3,5-triazine-2,4(1H,3H)-dione
Hydramethylnon	miscellaneous N	I	67485-29-4	tetrahydro-5,5-dimethyl-2(1H)-pyrimidinone [3-[4-(trifluoromethyl)phenyl]-1-[2-[4-(trifluoromethyl) phenyl]ethenyl]-2-propenylidene]hydrazone
3-Hydroxycarbofuran	carbofuran degradate		16655-82-6	2,3-dihydro-2,2-dimethyl-7-methylcarbamate, 3,7-benzofurandiol
1-Hydroxychlordene	chlordane degradate		24009-05-0	4,5,6,7,8,8-hexachloro-3a,4,7,7a-tetrahydro-(endo,exo)-4,7-methanoinden-1-ol
syn-12-Hydroxydieldrin	dieldrin degradate		26946-01-0	3,4,5,6,9,9-hexachloro-1a,2,2a,3,6,6a,7,7a-octahydro-2,7;3,6-dimethanonaphth[2.3-*b*] oxiren-8-ol, stereoisomer
N-Hydroxymethyl carbaryl	carbaryl degradate		15386-08-0	(hydroxymethyl)methylcarbamic acid, 1-naphthalenyl ester
Imazaquin	imidazolinone	H	81335-37-7	(±)-2-[4,5-dihydro-4-methyl-4-(1-methylethyl)-5-oxo-1H-imidazol-2-yl]-3-quinolinecarboxylic acid
Iprodione	amide	Fn	36734-19-7	3-(3,5-dichlorophenyl)-*N*-(1-methylethyl)-2,4-dioxo-1-imidazolidinecarboxamide
Isobornyl thiocyanoacetate	miscellaneous N	I	115-31-1	thiocyanato-1,7,7-trimethylbicyclo-(2,2,1)hept-2-yl-exo-acetate

Appendix E. Glossary Of Common And Chemical Names Of Pesticides And Related Compounds Given In Text—*Continued*

Common Name	Chemical Class	Use	CAS No.	Chemical Nomenclature
Isodrin	organochlorine, aldrin isomer	I	465-73-6	1,2,3,4,10,10-hexachloro-1,4,4a,5,8,8a-hexahydro-1,4:5,8-(endo,endo)-dimethanonaphthalene
Isofenphos	organophosphorus	I	25311-71-1	1-methylethyl 2-[[ethoxy[(1-methylethyl)amino] phosphinothioyl] oxy]benzoate
Isophenfos	see Isofenphos			
Isophenphos	see Isofenphos			
Isophorone	miscellaneous	Ind	78-59-1	3,5,5-trimethyl-2-cyclohexene-1-one
Isopropalin	dinitroaniline	H	33820-53-0	2,6-dinitro-*N,N*-dipropylcumidine
Jasmolin I	pyrethroid	I	4466-14-2	[1*R*-[1α{*S***(Z)*},3β]]-2-methyl-4-oxo-3-(2-pentenyl)-2-cyclopenten-1-yl 2,2-dimethyl-3-(2-methyl-1-propenyl)cyclopropanecarboxylate
Jasmolin II	pyrethroid	I	1172-63-0	[1*R*-[1α{*S***(Z)*},3β(*E*)]]-2-methyl-4-oxo-3-(2-pentenyl)-2-cyclopenten-1-enyl 3-(3-methoxy-2-methyl-3-oxo-1-propenyl)-2,2-dimethylcyclopropanecarboxylate
Karathane	see Dinocap			
Kelthane	see Dicofol			
Kepone	see Chlordecone			
3-Keto carbofuran	carbofuran degradate		16709-30-1	2,2-dimethyl-7-[[(methylamino)carbonyl]oxy]-3(2*H*)-benzofuranone
Lambdacyhalothrin	see Cyhalothrin			
Leptophos	organophosphorus	I	21609-90-5	*O*-(4-bromo-2,5-dichlorophenyl) *O*-methylphenyl phosphonothioate
Limonene	hydrocarbon	Ad	138-86-3	1-methyl-4-(1-methylethenyl)cyclohexene
Lindane	see γ-HCH			

Appendix E. Glossary Of Common And Chemical Names Of Pesticides And Related Compounds Given In Text—*Continued*

Common Name	Chemical Class	Use	CAS No.	Chemical Nomenclature
Linuron	urea	H	330-55-2	*N′*-(3,4-dichlorophenyl)-*N*-methoxy-*N*-methylurea
Malathion	organophosphorus	I	121-75-5	*O,O*-dimethyl *S*-[1,2-bis(ethoxycarbonyl)ethyl] dithiophosphate
Maleic hydrazide	hydrazide	PGR	51542-52-0	1,2-dihydro-3,6-pyridazinedione
Mancozeb	ethylene bisdithiocarbamate	Fn	8018-01-7	coordination product of zinc ion and manganese ethylene bisdithiocarbamate
Maneb	ethylene bisdithiocarbamate	Fn	12427-38-2	manganese ethylenebisdithiocarbamate
MCPA	chlorophenoxy acid	H	94-74-6	(4-chloro-2-methyl)phenoxyacetic acid
MCPP	chlorophenoxy acid salt	H	1929-86-8	potassium (*RS*)-2-(4-chloro-2-methylphenoxy)propanoate
Mecoprop	chlorophenoxy acid	H	7085-19-0	(*RS*)-2-(4-chloro-2-methylphenoxy)propanoic acid
Metalaxyl	amino acid derivative	Fn	57837-19-1	*N*-(2,6-dimethylphenyl)-*N*-(methoxyacetyl)-DL-alanine methyl ester
Metaldehyde	miscellaneous	Mo	108-62-3	2,4,6,8-tetramethyl-1,3,5,7-tetroxocane
Metam-sodium	thiocarbamate	Fm,Fn, H,I,N	137-42-8	sodium *N*-methyldithiocarbamate
Methamidophos	organophosphorus	I	10265-92-6	*O,S*-dimethyl phosphoramidothioate
Methidathion	organophosphorus	I,Ac	950-37-8	[(5-methoxy-2-oxo-1,3,4-thiadiazol-3(2*H*)-yl)methyl] *O,O*-dimethylphosphorodithioate
Methiocarb	carbamate	I,Ac, Mo	2032-65-7	3,5-dimethyl-4-(methylthio)phenyl methylcarbamate
Methomyl	carbamate, thiodicarb degradate	I	16752-77-5	methyl *N*-[[(methylamino)carbonyl]oxy] ethanimidothioate

Appendix E. Glossary Of Common and Chemical Names Of Pesticides And Related Compounds Given In Text—*Continued*

Common Name	Chemical Class	Use	CAS No.	Chemical Nomenclature
Methoprene	miscellaneous	IGR	40596-69-8	isopropyl (2E,4E,7S)-11-methoxy-3,7,11-trimethyldodeca-2,4-dienoate
Methotrexate	miscellaneous	Pha	59-05-2	N-[4-[[(2,4-diamino-6-pteridinyl)methyl]methylamino]benzoyl]-L-glutamic acid
Methoxychlor	*p,p'*-Methoxychlor			
o,p'-Methoxychlor	organochlorine	I	30667-99-3	2,2-bis(2-methoxyphenyl)-1,1,1-trichloroethane
p,p'-Methoxychlor	organochlorine	I	72-43-5	2,2-bis(4-methoxyphenyl)-1,1,1-trichloroethane
Methylarsonic acid, mono-sodium salt	organic arsenical	H	2163-80-6	monosodium methanearsonate
Methyl bromide	VOC	Fm,Ad	74-83-9	bromomethane
Methylcarbamic acid	carbaryl degradate		6414-57-9	carbamic acid, methyl ester
3-Methyl-4-nitrophenol	fenitrothion degradate			3-methyl-4-nitrophenol
Methyl parathion	organophosphorus	I	298-00-0	O,O-dimethyl O-(4-nitrophenyl) phosphorothioate
Methyl trithion	organophosphorus	I,Ac	953-17-3	S-[[(4-chlorophenyl)thio]methyl] O,O-dimethyl phosphorodithioate
Metiram	thiocarbamate	Fn	9006-42-2	tris[ammine-[ethylen bis(dithiocarbamato)zinc(II)] [tetrahydro-1,2,4,7-dithiadiazocine-3,8-dithione] polymer
Metolachlor	acetanilide	H	51218-45-2	2-chloro-N-(2-ethyl-6-methylphenyl)-N-(2-methoxy-1-methylethyl)-acetamide
Metribuzin	triazine	H	21087-64-9	4-amino-6-(1,1-dimethylethyl)-3-(methylthio)-1,2,4-triazin-5(4H)-one
Mevinphos	organophosphorus	I,Ac	7786-34-7	methyl 3-[(dimethoxyphosphinyl)oxy]-2-butenoate

Appendix E. Glossary Of Common And Chemical Names Of Pesticides And Related Compounds Given In Text—*Continued*

Common Name	Chemical Class	Use	CAS No.	Chemical Nomenclature
MGK 264	imide	IS	113-48-4	*N*-(2-ethylhexyl)bicyclo(2.2.1)-hept-5-ene-2,3-dicarboximide
Mirex	organochlorine	I	2385-85-5	1,1a,2,2,3,3a,4,5,5,5a,5b,6-dodecachlorooctahydro-1,3,4-metheno-1*H*-cyclobuta[*cd*]pentalene
Molinate	thiocarbamate	H	2212-67-1	*S*-ethyl hexahydro-1*H*-azepine-1-carbothioate
Monocrotophos	organophosphorus	I,Ac	6923-22-4	dimethyl (*E*)-[1-methyl-3-(methylamino)-3-oxo-1-propenyl] phosphate
8-Monohydromirex	mirex degradate		39801-14-4	1,1a,2,2,3,3a,4,5,5,5a,5b-undecachlorooctahydro-1,3,4-metheno-1*H*-cyclobuta[*cd*]pentalene
10-Monohydromirex	mirex degradate		845-66-9	1,1a,2,2,3,3a,4,5,5,5a,5b,6-undecachlorooctahydro-1,3,4-metheno-1*H*-cyclobuta[*cd*]pentalene
Monuron	urea	H	150-68-5	*N'*-(4-chlorophenyl)-*N,N*-dimethylurea
MSMA	see Methylarsonic acid, monosodium salt			
Myclobutanil	triazole	Fn	88671-89-0	α-butyl-α-(4-chlorophenyl)-1*H*-1,2,4-triazole-1-propanenitrile
Nabam	thiocarbamate	Fn	142-59-6	disodium 1,2-ethanediylbis(carbamodithioate)
Naled	organophosphorus	I	300-76-5	1,2-dibromo-2,2-dichloroethyl dimethyl phosphate
1-Naphthol	carbaryl degradate		90-15-3	1-naphthalenol
Napropamide	amide	H	15299-99-7	(*RS*)-*N,N*-diethyl-2-(1-naphthyloxy)propionamide
Naptalam	amine	H	132-66-1	sodium 2-[(1-naphthalenylamino)carbonyl]benzoate
NCPP	see Nonachloro-2-phenoxyphenol			
Nitralin	dinitroaniline	H	4726-14-1	4-methylsulfonyl-2,6-dinitro-*N,N*-dipropylaniline

Appendix E. Glossary Of Common And Chemical Names Of Pesticides And Related Compounds Given In Text—*Continued*

Common Name	Chemical Class	Use	CAS No.	Chemical Nomenclature
Nitrobenzene	benzene derivative	Ind	98-95-3	nitrobenzene
Nitrofen	miscellaneous	H	1836-75-5	2,4-dichlorophenyl 4-nitrophenyl ether
4-Nitrophenol	nitrophenol, methyl parathion degradate	Fn,Ad	100-02-7	4-nitrophenol
Nonachlor	organochlorine	I	3734-49-4	1,2,3,4,5,6,7,8,8-nonachloro-2,3,3a,4,7,7a-hexahydro-4,7-methano-1*H*-indene; mixture of *cis* and *trans* isomers
cis-Nonachlor	organochlorine	I	5103-73-1	1,2,3,4,5,6,7,8,8-nonachloro-2,3,3a,4,7,7a-hexahydro-4,7-methano-1*H*-indene (combined nomenclature for *cis* and *trans* isomers)
trans-Nonachlor	organochlorine	I	39765-80-5	*trans*-nonachlor
Nonachloro-2-phenoxyphenol	impurity in technical pentachlorophenol		35245-80-8	2,3,4,5-tetrachloro-6-(pentachlorophenoxy)phenol
Norea	urea	H	18530-56-8	3-(hexahydro-4,7-methanoindan-5-yl)-1,1-dimethylurea
Norflurazon	amine	H	27314-13-2	4-chloro-5-methylamino-2-(α,α,α-trifluoro-*m*-tolyl) pyridazin-3(2*H*)-one
Noruron	see Norea			
Octachlor epoxide	see Oxychlordane			
Octachlorobiphenyl, total	organochlorine	Ind	31472-83-0	octachloro-1,1'-biphenyl (all isomers)
Octachlorostyrene	organochlorine	Ind	29082-74-4	pentachloro(trichloroethenyl)-benzene
Octylphenol	alkylphenol	Ind	27193-28-8	(1,1,3,3-tetramethylbutyl)phenol
Omite	see Propargite			

Appendix E. Glossary Of Common And Chemical Names Of Pesticides And Related Compounds Given In Text—*Continued*

Common Name	Chemical Class	Use	CAS No.	Chemical Nomenclature
Oryzalin	dinitroaniline	H	19044-88-3	3,5-dinitro-N^4,N^4-dipropylsulfanilamide
Oxadiazon	oxadiazol	H	19666-30-9	2-*tert*-butyl-4-(2,4-dichloro-5-isopropoxyphenyl)-Δ^2-1,3,4-oxadiazolin-5-one
Oxamyl	carbamate	I,N,Ac	23135-22-0	S-methyl N',N'-dimethyl-N-(methylcarbamoyloxy)-1-thio-oxaminidate
Oxychlordane	chlordane degradate		27304-13-8	2,3,4,5,6,6a,7,7-octachloro-1a,1b,5,5a,6,6-hexahydro-2,5-methano-2H-indeno(1,2b)oxirene
Oxydemeton-methyl	organophosphorus	I	301-12-2	S-[2-(ethylsulfinyl)ethyl] O,O-dimethyl phosphorothioate
Oxyfluorfen	diphenyl ether	H	42874-03-3	2-chloro-1-(3-ethoxy-4-nitrophenoxy)-4-(trifluoromethyl benzene)
Oxytetracycline	amide	An	79-57-2	4-(dimethylamino)-1,4,4a,5,5α,6,11,12a-octahydro-3,5,6,10,12,12a-hexahydroxy-6-methyl-1,11-dioxo-2-naphthacenecarboxamide
Oxythioquinox	dithiocarbonate	Ac,Fm, Fn	2439-01-2	6-methyl-1,3-dithiolo[4,5-b]quinoxalin-2-one
Paradichlorobenzene	see 1,4-Dichlorobenzene			
Paraoxon	organophosphorus	I	311-45-5	O,O-diethyl O-(4-nitrophenyl) phosphate
Paraquat	miscellaneous N	H	1910-42-5	1,1'-dimethyl-4,4'-bipyridinium ion, dichloride salt
Parathion	organophosphorus	I	56-38-2	O,O-diethyl O-(4-nitrophenyl) phosphorothioate
PBB	see Polybrominated biphenyl			
PCA	see Pentachloroanisole			
PCB	see Polychlorinated biphenyl			

Appendix E. Glossary Of Common And Chemical Names Of Pesticides And Related Compounds Given In Text—*Continued*

Common Name	Chemical Class	Use	CAS No.	Chemical Nomenclature
PCB-18	see 2,5,2'-Trichlorobiphenyl			
PCB-40	see 2,3,2',3'-Tetrachlorobiphenyl			
PCB-52	see 2,5,2',5'-Tetrachlorobiphenyl			
PCB-101	see 2,4,5,2',5'-Pentachlorobiphenyl			
PCB-153	see 2,4,5,2',4',5'-Hexachlorobiphenyl			
PCB-155	see 2,4,6,2',4',6'-Hexachlorobiphenyl			
PCNB	see Pentachloronitrobenzene	Fn		
PCP	see Pentachlorophenol			
Pebulate	thiocarbamate	H	1114-71-2	S-propyl butyl(ethyl)thiocarbamate
Pendimethalin	dinitroaniline	H	40487-42-1	N-(1-ethylpropyl)-3,4-dimethyl-2,6-dinitrobenzeneamine
Pentachloroanisole	pentachlorophenol degradate		1825-21-4	pentachloromethoxybenzene
Pentachlorobenzene	organochlorine	Ind	608-93-5	pentachlorobenzene
2,4,5,2',5'-Pentachlorobiphenyl	organochlorine	Ind	37680-73-2	2,2',4,5,5'-pentachloro-1,1'-biphenyl
Pentachlorobiphenyl, total	organochlorine	Ind	25429-29-2	pentachloro-1,1'-biphenyl (all isomers)

Appendix E. Glossary Of Common And Chemical Names Of Pesticides And Related Compounds Given In Text—*Continued*

Common Name	Chemical Class	Use	CAS No.	Chemical Nomenclature
Pentachloronitrobenzene	chlorinated nitroaromatic	Fn	82-68-8	pentachloronitrobenzene
Pentachlorophenol	organochlorine	Fn,Mo, Ad	87-86-5	pentachlorophenol
Pentachlorotoluene	organochlorine	Ind	877-11-2	2,3,4,5,6-pentachloro-1-methylbenzene
Permethrin	pyrethroid	I	52645-53-1	(3-phenoxyphenyl)methyl 3-(2,2-dichloroethenyl)-2,2-dimethylcyclopropanecarboxylate
cis-Permethrin	pyrethroid	I	61949-76-6	(3-phenoxyphenyl)methyl *cis*-3-(2,2-dichloroethenyl)-2,2-dimethylcyclopropanecarboxylate
trans-Permethrin	pyrethroid	I	61949-77-7	(3-phenoxyphenyl)methyl *trans*-3-(2,2-dichloroethenyl)-2,2-dimethylcyclopropanecarboxylate
Perthane	organochlorine	I	72-56-0	1,1-dichloro-2,2-bis(4-ethylphenyl)ethane
Phenol red	phenol	Pha	143-74-8	4,4′-(1,1-dioxido-3*H*-2,1-benzoxathiol-3-ylidene)bis-phenol
Phorate	organophosphorus	I	298-02-2	*O,O*-diethyl *S*-ethylthiomethyl phosphorodithioate
Phosalone	organophosphorus	Ac.I	2310-17-0	*S*-[(6-chloro-2-oxo-3(2*H*)-benzoxazolyl)methyl] *O,O*-diethyl phosphorodithioate
Phosdrin	see Mevinphos			
Phosmet	organophosphorus	I	732-11-6	*N*-(mercaptomethyl)phthalimide-*S*-(*O,O*-dimethylphosphorodithioate)
Photomirex	mirex degradtate			8-hydroxy-1,1a,2,2,3,3a,4,5,5,5a,5b,6-didecacgkiriictagtdri-1m3m4-metheno-1-*H*-cyclobuta[cd]pentalene
Picloram	amine	H	1918-02-1	4-amino-3,5,6-trichloropicolinic acid
Pindone	inandione	I,R	83-26-1	2-(2,2-dimethyl-1-oxopropyl)-1*H*-indene-1,3(2*H*)-dione
Piperonyl butoxide	miscellaneous	IS	51-03-6	5-[[2-(2-butoxyethoxy)ethoxy]methyl]-6-propyl-1,3-benzodioxole

Appendix E. Glossary Of Common And Chemical Names Of Pesticides And Related Compounds Given In Text—*Continued*

Common Name	Chemical Class	Use	CAS No.	Chemical Nomenclature
Polybrominated biphenyl	organohalogen	Ind		1,1-biphenyl, bromo derivatives
Polychlorinated biphenyl	organochlorine	Ind	1336-36-3	1,1-biphenyl, chloro derivatives
Procymidone	miscellaneous	Fn	32809-16-8	3-(3,5-dichlorophenyl)-1,2-dimethyl-3-azabicyclo[3.1.0]hexane-2,4-dione
Profenofos	organophosphorus	Ac,I	41198-08-7	*O*-4-bromo-2-chlorophenyl *O*-ethyl *S*-propyl phosphorothioate
Profluralin	dinitroaniline	H	26399-36-0	2,6-dinitro-*N*-cyclopropylmethyl-*N*-propyl-4-(trifluoromethyl)benzenamine
Prometon	triazine	H	1610-18-0	6-methoxy-*N*,*N'*-bis(1-methylethyl)-1,3,5-triazine-2,4-diamine
Prometryn	triazine	H	7287-19-6	*N*,*N'*-bis(1-methylethyl)-6-(methylthio)-1,3,5-triazine-2,4-diamine
Propachlor	acetanilide	H	1918-16-7	2-chloro-*N*-(1-methylethyl)-*N*-phenylacetanilide
Propanil	amide	H	709-98-8	*N*-(3,4-dichlorophenyl)propanamide
Propargite	sulfite ester	Ac	2312-35-8	2-[4-(1,1-dimethylethyl)phenoxy]cyclohexyl-2-propynyl sulfite
Propazine	triazine	H	139-40-2	6-chloro-*N*,*N'*-bis(1-methylethyl)-1,3,5-triazine-2,4-diamine
Propham	carbamate	H,PGR	122-42-9	1-methylethylphenyl carbamate
Propiconazole	triazole	Fn	60207-90-1	1-[[2-(2,4-dichlorophenyl)-4-propyl-1,3-dioxolan-2-yl] methyl]-1*H*-1,2,4-triazole
Propoxur	see Baygon			
Pyrethrin I	pyrethroid	I	121-21-1	[1*R*-[1α[*S**(*Z*)],3β]]-2-methyl-4-oxo-3-(2,4-pentadienyl)cyclopenten-1-yl 2,2-dimethyl-3-(2-methyl-1-propenyl)cyclopropanecarboxylate

Appendix E. Glossary Of Common And Chemical Names Of Pesticides And Related Compounds Given In Text—*Continued*

Common Name	Chemical Class	Use	CAS No.	Chemical Nomenclature
Pyrethrin II	pyrethroid	I	121-29-9	[1R-[1α[S*(Z)],3β(E)]]-2-methyl-4-oxo-3-(2,4-pentadienyl)-2-cyclopenten-1-yl 3-(3-methoxy-2-methyl-3-oxo-1-propenyl)-2,2-dimethylcyclopropanecarboxylate
Pyrethrins	pyrethroid	I	8003-34-7	six insecticidal constituents in extract of the flowers *Pyrethrum cineriaefolium* and other species; see Pyethrin I, Cinerin I, Jasmolin I, Pyethrin II, Cinerin II, Jasmolin II
Resmethrin	pyrethroid	I	10453-86-8	[5-(phenylmethyl)-3-furanyl]methyl 2,2-dimethyl-3-(2-methyl-1-propenyl)cyclopropanecarboxylate
Ronalin	see Ronilan			
Ronilan	miscellaneous	Fn	50471-44-8	3-(3,5-dichlorophenyl)-5-methyl-5-vinyl-1,3-oxazolidine-2,4-dione
Ronnel	organophosphorus	I	299-84-3	O,O-dimethyl O-(2,4,5-trichlorophenyl)phosphorothioate
Rotenone	miscellaneous	I	83-79-4	1,2,12,12a-tetrahydro-8,9-dimethoxy-2-(1-methylethenyl)-[1]benzopyrano[3,4-b]furo[2,3-h][1]-benzopyran-6(6H)-one
Sethoxydim	miscellaneous N	H	74051-80-2	2-[1-(ethoxyimino)butyl]-5-[2-(ethylthio)propyl]-3-hydroxy-2-cyclohexen-1-one
Silvex	see 2,4,5-TP			
Simazine	triazine	H	122-34-9	2-chloro-4,6-bis(ethylamino)-s-triazine
Sodium borate	inorganic	H,L	1303-96-4	Sodium tetraborate decahydrate
Sodium methyldithiocarbamate	see Metam-sodium			
Sodium thiocyanate	miscellaneous N	H	540-72-7	sodium thiocyanate
Streptomycin	amine	An	57-92-1	O-2-deoxy-2-(methylamino)-α-L-glucopyranosyl-(1→2)-O-5-deoxy-3-C-formyl-α-L-lyxofuranosyl-(1→4)-N,N'-bis(aminoiminomethyl)-D-streptamine
Strobane	organochlorine	I	8001-50-1	mixture of polychlorinated camphene, pinene and related terpenes

Appendix E. Glossary Of Common And Chemical Names Of Pesticides And Related Compounds Given In Text—*Continued*

Common Name	Chemical Class	Use	CAS No.	Chemical Nomenclature
Sulfometuron	sulfonylurea	H	74223-56-6	2-[[[[(4,6-dimethyl-2-pyrimidinyl)amino]carbonyl]amino] sulfonyl]benzoic acid
Sulfometuron-methyl	sulfonylurea	H	74222-97-2	methyl 2-[[[[(4,6-dimethyl-2-pyrimidinyl)amino]carbonyl]amino] sulfonyl]benzoate
Sulfuric acid	inorganic	H,Ds	7664-93-9	sulfuric acid
Sulprofos	organophosphorus	I	35400-43-2	O-ethyl O-[(4-methylthio)phenyl] S-propyl phosphorodithioate
Sumithrin	pyrethroid	I	26046-85-5	(3-phenoxyphenyl)methyl 2,2-dimethyl-3-(2-methyl-1-propenyl)cyclopropanecarboxylate
2,4,5-T	chlorophenoxy acid	H	93-76-5	2,4,5-trichlorophenoxyacetic acid
2,3,6-TBA	see 2,3,6-Trichloro-benzoic acid			
TCA	see Trichloroacetic acid			
TCDD	organochlorine	Ind, BP	1746-01-6	tetrachlorodibenzo-p-dioxin
2,3,7,8-TCDD	impurity in 2,4,5-T and other organochlorines		1746-01-6	2,3,7,8-tetrachlorodibenzo-p-dioxin
TDE	see DDD			
Tebuthiuron	urea	H	34014-18-1	N-[5-(1,1-dimethylethyl)-1,3,4-thiadiazol-2-yl]-N,N′-dimethylurea
Tefluthrin	pyrethroid	I	79538-32-2	[1α,3α(Z)]-(±)-(2,3,5,6-tetrafluoro-4-methylphenyl)methyl 3-(2-chloro-3,3,3-trifluoro-1-propenyl)-2,2-dimethylcyclopropane-carboxylate
Telone II	VOC	Fm,N	542-75-6	mixture of cis- and trans-1,3-dichloropropene

Appendix E. Glossary Of Common And Chemical Names Of Pesticides And Related Compounds Given In Text—*Continued*

Common Name	Chemical Class	Use	CAS No.	Chemical Nomenclature
Terbufos	organophosphorus	I,N	13071-79-9	S-[[(1,1-dimethylethyl)thio]methyl] O,O-diethyl phosphorodithioate
Terbutryn	triazine	H	886-50-0	2-(tert-butylamino)-4-(ethylamino)-6-(methylthio)-s-triazine
1,2,3,4-Tetrachlorobenzene	organochlorine	Ind	634-66-2	1,2,3,4-tetrachlorobenzene
1,2,4,5-Tetrachlorobenzene	organochlorine	Ind	95-94-3	1,2,4,5-tetrachlorobenzene
2,3,2′,3′-Tetrachlorobiphenyl	organochlorine	Ind	38444-93-8	2,2′,3,3′-tetrachloro-1,1′-biphenyl
2,5,2′,5′-Tetrachlorobiphenyl	organochlorine	Ind	35693-99-3	2,2′,5,5′-tetrachloro-1,1′-biphenyl
Tetrachlorobiphenyl, total	organochlorine	Ind	26914-33-0	tetrachloro-1,1′-biphenyl (all isomers)
Tetrachlorvinphos	organophosphorus	I	22248-79-9	(Z)-2-chloro-1-(2,4,5-trichlorophenyl)vinyl dimethyl phosphate
Tetradifon	chlorinated bridged diphenyl	Ac	116-29-0	1,2,4-trichloro-5-[(4-chlorophenyl)sulfonyl]benzene
Tetramethrin	pyrethroid	I	7696-12-0	(1,3,4,5,6,7-hexahydro-1,3-dioxo-2H-isoindol-2-yl)methyl 2,2-dimethyl-3-(2-methyl-1-propenyl)cyclopropanecarboxylate
Thiabendazole	imidazole	Fn	148-79-8	2-(4′-thiazolyl)-benzimidazole
Thiobencarb	thiocarbamate	H	28249-77-6	S-4-chlorobenzyl diethylthiocarbamate
Thiodan	see Endosulfan			
Thiodicarb	carbamate	I	59669-26-0	dimethyl N,N′-[thiobis[(methylimino)carbonyloxy]]bis(ethanimidothioate)
Thiophanate-methyl	carbamate	Fn	23564-05-8	dimethyl 4,4′-o-phenylenebis(3-thioallophanate)
Thiram	thiocarbamate	Fn	137-26-8	bis(dimethylthiocarbamyl) disulfide
Thymol	miscellaneous	Fn	89-83-8	5-methyl-2-(1-methylethyl)phenol

Appendix E. Glossary Of Common And Chemical Names Of Pesticides And Related Compounds Given In Text—*Continued*

Common Name	Chemical Class	Use	CAS No.	Chemical Nomenclature
Toxaphene	organochlorine	I	8001-35-2	polychlorinated camphene
2,4,5-TP	chlorophenoxy acid	H	93-72-1	(±)-2-(2,4,5-trichlorophenoxy)propanoic acid
Tralomethrin	pyrethroid	I	66841-25-6	cyano-(3-phenoxyphenyl)methyl 2,2-dimethyl-3-(1,2,2,2-tetrabromoethyl)cyclopropanecarboxylate
Triadimefon	triazole	Fn	43121-43-3	1-(4-chlorophenoxy)-3,3-dimethyl-1-(1*H*-1,2,4-triazol-1-yl)-2-butanone
Triallate	thiocarbamate	H	2303-17-5	S-(2,3,3-trichloro-2-propenyl) bis(1-methylethyl) thiocarbamate
Tribufos	organophosphorus, merphos degradate	Df	78-48-8	S,S,S-tributyl phosphorotrithioate
Trichlorfon	organophosphorus	I	52-68-6	dimethyl (2,2,2-trichloro-1-hydroxyethyl) phosphonate
Trichloroacetic acid	chlorinated acid	H	76-03-9	trichloroacetic acid
1,2,3-Trichlorobenzene	organochlorine	Ind	87-61-6	1,2,3-trichlorobenzene
1,2,4-Trichlorobenzene	organochlorine	Ind	120-82-1	1,2,4-trichlorobenzene
1,3,5-Trichlorobenzene	organochlorine	Ind	108-70-3	1,3,5-trichlorobenzene
2,3,6-Trichlorobenzoic acid	chlorobenzoic acid	H	50-31-7	2,3,6-trichlorobenzoic acid
2,5,2′-Trichlorobiphenyl	organochlorine	Ind	37680-65-2	2,2′,5-trichloro-1,1′-biphenyl
2,5,4′-Trichlorobiphenyl	organochlorine	Ind	16606-02-3	2,4′,5-trichloro-1,1′-biphenyl
Trichlorobiphenyl, total	organochlorine	Ind	25323-68-6	trichloro-1,1′-biphenyl (all isomers)
2,4,5-Trichlorophenol	phenol	Ind	95-95-4	2,4,5-trichlorophenol
Triclopyr	miscellaneous	H	55335-06-3	(3,5,6-trichloro-2-pyridinyloxy)acetic acid

Appendix E. Glossary Of Common And Chemical Names Of Pesticides And Related Compounds Given In Text—*Continued*

Common Name	Chemical Class	Use	CAS No.	Chemical Nomenclature
Tricosene	hydrocarbon	I	27519-02-4	*cis*-tricos-9-ene
Trifluralin	dinitroaniline	H	1582-09-8	2,6-dinitro-*N,N*-dipropyl-4-(trifluoromethyl) benzenamine
Triforine	amide	Fn	26644-46-2	*N,N'*-[1,4-piperazinediylbis(2,2,2-trichloroethylidene)] bis(formamide)
Trimethacarb	carbamate	I	12407-86-2	3,4,5- (or 2,3,5-)trimethylphenyl methylcarbamate
Triphenyltin hydroxide	organo tin	Fn	76-87-9	hydroxytriphenylstannane
Trithion	organophosphorus	I,Ac	786-19-6	*S*-[[(4-chlorophenyl)thio]methyl] *O,O*-diethyl phosphorodithioate
Vernolate	thiocarbamate	H	1929-77-7	*S*-propyl dipropylthiocarbamate
Vinclozolin	see Ronilan			
Warfarin	miscellaneous	R	81-81-2	3-(α-acetonylbenzyl)-4-hydroxycoumarin
Xytron	see Zytron			
Zearalenone	estrogen, resorcylic acid lactone	NO	17924-92-4	[*s*-(*E*)]-3,4,5,6,9,10-hexahydro-14,6-dihydroxy-3-methyl-1*H*-2-benzoxacyclotetradecin-1,7(8*H*)-dione
Zineb	thiocarbamate	Fn	12122-67-7	[[1,2-ethanediylbis[carbamodithioato]](-2-)]zinc complex
Ziram	thiocarbamate	Fn	137-30-4	zinc bis(dimethyldithiocarbamate)
Zytron	organophosphorus	I	299-85-4	*O*-2,4-dichlorophenyl *O*-methyl isopropylphosphoroamidothioate

REFERENCES

Ackerman, L.B., 1980, Overview of human exposure to dieldrin residues in the environment and trends of residue levels in tissue: *J. Pest. Monitor.*, v. 14, no. 2, pp. 64–69.

Adams, W.J., 1984 [1987], Bioavailability of neutral lipophilic organic chemicals contained on sediments: A review, *in* Dickson, K.L., Maki, A.W., and Brungs, W.A., eds.: *Fate and effects of sediment-bound chemicals in aquatic systems: Proceedings of the Sixth Pellson Workshop, Florisant, Colorado, August 12–17, 1984*: Pergamon Press, New York, pp. 219–242.

Adams, W.J., Kimerle, R.A., and Barnett, J.W., Jr., 1992, Sediment quality and aquatic life assessment: *Environ. Sci. Technol.*, v. 26, no. 10, pp. 1864–1875.

Adams, W.J., Kimerle, R.A., and Mosher, R.G., 1985, Aquatic safety assessment of chemicals sorbed to sediments, *in* Cardwell, R.D., Purdy, R., and Bahner, R.C., eds., *Aquatic toxicology and hazard assessment: Seventh symposium*: American Society for Testing and Materials, Philadelphia, Penn., Special Technical Publication series, v. 854, pp. 429–453.

Addison, R.F., 1976, Organochlorine compounds in aquatic organisms: Their distribution, transport, and physiological significance, *in* Lockwood, A.P.M., ed., *Effects of pollutants on aquatic organisms*: Cambridge University Press, Cambridge, Society for Experimental Biology seminar series, v. 2, p. 127–143.

Addison, R.F., and Smith, T.G., 1974, Organochlorine residue levels in arctic ringed seals: Variation with age and sex: *Oikos*, v. 25, no. 3, pp. 335–377.

Addison, R.F., and Zinck, M.E., 1986, PCBs have declined more than DDT-group residues in arctic ringed seals (*Phoca hispida*) between 1972 and 1981: *Environ. Sci. Technol.*, v. 20, no. 3, pp. 253–256.

Agee, B.A., 1986, DDT in the Salinas Valley: A special report on the probable source of technical grade DDT found in the Blanco Drain near Salinas, California: California Water Resources Control Board, Division of Water Quality, Water Quality Monitoring Report 86-2, 49 p.

Agricultural Research Service, 1997, The pesticide properties database: U.S. Department of Agriculture, from URL http://www.arsusda.gov, HTML format.

Aguilar, A., 1984, Relationship of DDE/Sum-DDT in marine mammals to the chronology of DDT input into the ecosystem: *Can. J. Fish. Aquat. Sci.*, v. 21, pp. 840–844.

Ahmad, N., Harsas, W., Marolt, R.S., Morton, M., and Pollack, J.K., 1988, Total DDT and dieldrin content of human adipose tissue: *Bull. Environ. Contam. Toxicol.*, v. 41, pp. 802–808.

Albaiges, J., Farran, A., Soler, M., Gallifa, A., and Martin, P., 1987, Accumulation and distribution of biogenic and pollutant hydrocarbons, PCBs and DDT in tissues of western fishes: *Mar. Environ. Res.*, v. 22, no. 1, pp. 1–18.

Alexander, M., 1981, Biodegradation of chemicals of environmental concern: *Science*, v. 211, no. 4478, pp. 132–138.

Ali, S.M., Bowes, G.W., and Cohen, D.B., 1984, Endosulfan (Thiodan): California State Water Resources Control Board, Toxic Substances Control Program, Water Quality and Pesticides series, v. 5, Special Projects Report 84-7SP, 131 p.

Allan, R.J., 1988, Toxic chemical pollution of the St. Lawrence River (Canada) and its upper estuary: *Water Sci. Technol.*, v. 20, no. 6/7, pp. 77–88.

———1989, Factors affecting sources and fate of persistent toxic organic chemicals: Examples from the Laurentian Great Lakes, *in* Boudou, A., and Ribeyre, F., eds., *Aquatic ecotoxicology: Fundamental concepts and methodologies*: CRC Press, Boca Raton, Fla., v. 1, pp. 219–248.

Allan, R.J., and Ball, A.J., 1990, An overview of toxic contaminants in water and sediments of the Great Lakes: *Water Pollut. Res. J. Can.*, v. 25, no. 4, pp. 387–680.

Allen, G.T., 1991a, Background contaminants evaluation of Flint Hills National Wildlife Refuge, 1989: U.S. Fish and Wildlife Service Contaminant Report R/502M/91, 40 p.

——1991b, Petroleum hydrocarbons, chlorinated hydrocarbons, and metals in soils and sediments of Quivira National Wildlife Refuge: U.S. Fish and Wildlife Service, 10 p.

——1992, Metals and organic compounds in fish and sediments from the Missouri and lower Kansas rivers in Kansas in 1991: U.S. Fish and Wildlife Service Contaminant Report R6/509M/92, 17 p.

Allen, G.T., and Nash, T., 1992, Contaminants survey of the proposed Marais des Cygnes National Wildlife Refuge: U.S. Fish and Wildlife Service Contaminant Report R6/508M/92, 19 p., appendix.

Allen, H.E., Jr., and Gray, J.R., 1984, Runoff, sediment transport, and water quality in a northern Illinois agricultural watershed before urban development, 1979–81: U.S. Geological Survey Water-Resources Investigations Report 82-4073, 55 p.

Allen, J.L., Sills, J.B., Dawson, V.K., and Amel, R.T., 1981, Residues of isobornyl thiocyanoacetate (Thanite) and a metabolite in fish and treated ponds: *J. Agric. Food Chem.*, v. 29, no. 3, pp. 634–636.

Aly, O.M., and El-Dib, M.A., 1971, Photodecomposition of some carbamate insecticides in aquatic environments, *in* Faust, S.D., and Hunter, J.V., eds., *Organic compounds in aquatic environments*: Marcel Dekker, New York, pp. 469–494.

Aly, O.M., and Faust, S.D., 1964, Studies on the fate of 2,4-D and ester derivatives in natural surface waters: *J. Agr. Food Chem.*, v. 12, no. 6, pp. 541–546.

American Chemical Society, 1998, Chemical Abstracts Service registry handbook—Common names: American Chemical Society, Columbus, Ohio.

Andelman, J.B., and Suess, M.J., 1971, The photodecomposition of 3,4-benzpyrene sorbed on calcium carbonate, *in* Faust, S.J., Hunter, J.V., eds., *Organic compounds in aquatic environments*: Marcel Dekker, New York, pp. 439–468.

Andersen, M.E., 1981, A physiologically based toxicokinetic description of the metabolism of inhaled gases and vapors: *Toxicol. Appl. Pharmacol.*, v. 60, no. 3, pp. 509–526.

Anderson, D.W., 1982, Residues of o,p'-DDT in southern California coastal sediments in 1971: *Bull. Environ. Contam. Toxicol.*, v. 29, no. 4, pp. 429–433.

Anderson, R.B., and Everhart, W.H., 1966, Concentrations of DDT in landlocked salmon (*Salmo salar*) at Sebago Lake, Maine: *Trans. Am. Fish. Soc.*, v. 95, pp. 160–164.

Anderson, R.B., and Fenderson, O.C., 1970, An analysis of variation of insecticide residues in landlocked Atlantic salmon (*Salmo salar*): *Can. J. Fish. Aquat. Sci.*, v. 27, no. 1, pp. 1–11.

Andersson, T., and Koivusaari, U., 1986, Oxidative and conjugative metabolism of xenobiotics in isolated liver cells from thermally acclimated rainbow trout: *Aquat. Toxicol.*, v. 8, pp. 85–92.

Andreasen, J.K., 1989, Environmental monitoring programs of the U.S. Fish and Wildlife Service, *in* Weigmann, D.L., ed., *Pesticides in terrestrial and aquatic environment: Proceedings of a national research conference*: Virginia Polytechnic Institute and State University, Virginia Water Resources Research Center, Blacksburg, Va., pp. 292–298.

Andrilenas, P.A., 1974, Farmers' use of pesticides in 1971—Quantities: U.S. Department of Agriculture, Economic Research Service, Agricultural Economic Report 252, 56 p.

Ankley, G.T., Cook, P.M., Carlson, A.R., Call, D.J., Swenson, J.A., Corcoran, H.F., and Hoke, R.A., 1992b, Bioaccumulation of PCBs from sediments by oligochaetes and fishes: Comparison of laboratory and field studies: *Can. J. Fish. Aquat. Sci.*, v. 49, pp. 2080–2085.

Ankley, G.T., Lodge, K., Call, D.J., Balcar, M.D., Brooke, L.T., Cook, P.M., Kreis, R.G., Jr., Carlson, A.R., Johnson, R.D., and Niemi, J.P., Hoke, R.A., West, C.W., Giesy, J.P., Jones, P.D., and Fuying, Z.C., 1992a, Integrated assessment of contaminated sediments in the lower Fox River and Green Bay, Wisconsin: *Ecotoxicol. Environ. Safety*, v. 23, no. 1, pp. 46–63.

Anonymous, 1992, Is our fish fit to eat?: *Consumer Reports*, v. 57, no. 2, p. 103–114.

Apperson, C.S., Elston, R., and Castle, W., 1976, Biological effects and persistence of methyl parathion in Clear Lake, California: *Environ. Entomol.*, v. 5, no. 6, pp. 1116–1120.

Archer, R.J., and Turk, J.T., 1977, Discharge and water-quality data for selected streams at low flow including some bottom-material analyses, and limnological studies of six lakes, Westchester County, New York: U.S. Geological Survey Open-File Report 77-781, 107 p.

Arizona Department of Health Services, 1991, Risk assessment for recreational usage of the Painted Rocks Borrow Pit Lake at Gila Bend, Arizona: Arizona Department of Health Services, Division of Disease Prevention, Office of Risk Assessment and Investigation report prepared for Arizona Department of Environmental Quality, 67 p., 3 appendices.

Armstrong, D.E., and Chesters, G.W., 1968, Adsorption catalyzed chemical hydrolysis of atrazine: *Environ. Sci. Technol.*, v. 2, no. 9, pp. 683–689.

Armstrong, D.E., and Konrad, J.G., 1974, Nonbiological degradation of pesticides, *in* Guenzi, W.D., ed., *Pesticides in soil and water*: Soil Science Society of America, Madison, Wis., pp. 123–130.

Armstrong, R.W., and Sloan, R.J., 1980, Trends in levels of several known chemical contaminants in fish from New York State waters: New York State Department of Environmental Conservation, Technical Report 80-2, 77 p.

Arnold, D.L., Nera, E.A., Stapley, R., Tolnai, G., Claman, P., Hayward, S., Tryphonas, H., and Bryce, F., 1996, Prevalence of endometriosis in rhesus (*Macaca mulatta*) monkeys ingesting PCB (Aroclor 1254): Review and evaluation: *Fundam. Appl. Toxicol.*, v. 31, no. 1, pp. 42–55.

Arruda, J.A., Cringan, M.S., Gilliland, D., Haslouer, S.G., Fry, J.E., Broxterman, R., and Brunson, K.L., 1987, Correspondence between urban areas and the concentrations of chlordane in fish from the Kansas River: *Bull. Environ. Contam. Toxicol.*, v. 39, no. 4, pp. 563–570.

Arruda, J.A., Cringan, M.S., Layher, W.G., Kersh, G., and Bever, C., 1988, Pesticides in fish tissue and water from Tuttle Creek Lake, Kansas: *Bull. Environ. Contam. Toxicol.*, v. 41, no. 4, pp. 617–624.

Ashby, S.L., Sturgis, T.C., Price, C.B., Brannon, J.M., and Pennington, J.C., 1991, Water quality studies in the upper watershed of Steele Bayou, Mississippi: U.S. Army Corps of Engineers Miscellaneous Paper EL-91-23, 48 p., 2 appendices.

Aspelin, A.L., 1994, Pesticides industry sales and usage: 1992 and 1993 market estimates: U.S. Environmental Protection Agency, Office of Pesticide Programs, Biological and Economic Analysis Division, Economic Analysis Branch Report 733-K-94-001, 33 p.

——1997, Pesticides industry sales and usage, 1994 and 1995 market estimates: U.S. Environmental Protection Agency, Office of Pesticide Programs, Biological and Economic Analysis Division Report 733-R-97-002, 35 p.

Aspelin, A.L., Grube, A.H., and Torla, R., 1992, Pesticides industry sales and usage: 1990 and 1991 market estimates: U.S. Environmental Protection Agency, Office of Pesticide Programs, Biological and Economic Analysis Division, Economic Analysis Branch Report 733-K-92-001, 37 p.

Ayers, M.A., 1978, Water-quality assessment of the Middle Fork Anderson River watershed, Crawford and Perry Counties, Indiana: U.S. Geological Survey Open-File Report 78-71, 31 p.

Baetcke, K.P., Cain, J.D., and Poe, W.E., 1972, Mirex and DDT residues in wildlife and miscellaneous samples in Mississippi—1970: *J. Pest. Monitor.*, v. 6, no. 1, pp. 14–22.

Bahner, L.H., and Oglesby, J.L., 1982, Models for predicting bioaccumulation and ecosystem effects of kepone and other materials, *in* Conway, R.A., ed., *Environmental risk analysis for chemicals*: Van Nostrand-Reinhold Company, New York, pp. 461–473.

Bahner, L.H., Wilson, A.J., Jr., Sheppard, J.M., Patrick James M., J., Goodman, L.R., and Walsh, G.E., 1977, Kepone bioconcentration, accumulation, loss, and transfer through estuarine food chains: *Chesapeake Sci.*, v. 18, no. 3, pp. 299–308.

Baker, J.E., Eisenreich, S.J., Johnson, T.C., and Haffman, B.M., 1985, Chlorinated hydrocarbon cycling in the benthic nepheloid layer of Lake Superior: *Environ. Sci. Technol.*, v. 19, no. 9, pp. 854–861.

Barbash, J., and Resek, E.A., 1996, *Pesticides in ground water: Distribution, trends, and governing factors*: Ann Arbor Press, Chelsea, Mich., Pesticides in the Hydrologic System series, v. 2, 588 p.

Barber, M.C., Suarez, L.A., and Lassiter, R.R., 1988, Modeling bioconcentration of nonpolar organic pollutants by fish: *Environ. Toxicol. Chem.*, v. 7, no. 7, pp. 545–558.

——1991, Modeling bioconcentration of organic pollutants in fish with an application to PCBs in Lake Ontario salmonids: *Can. J. Fish. Aquat. Sci.*, v. 48, no. 2, pp. 318–337.

Barker, J.L., 1984, Organochlorine pesticide and polychlorinated biphenyl residues at four trophic levels in the Schuylkill River, Pennsylvania, *in* Meyer, E.L., Selected Papers in the Hydrologic Sciences 1984: U.S. Geological Survey Water-Supply Paper 2262, pp. 25–31.

Barrick, R., Becker, S., Brown, L., Beller, H., and Pastorok, R., 1988, Sediment quality values refinement: 1988 update and evaluation of Puget Sound AET: PTI Environmental Services report prepared for the Puget Sound Estuary Program, Office of Puget Sound, v. 1, variously paged.

Barron, M.G., 1990, Bioconcentration: *Environ. Sci. Technol.*, v. 24, no. 11, pp. 1612–1618.

Barron, M.G., Gedutis, C., and James, M.O., 1988, Pharmacokinetics of sulphadimethoxine in the lobster, *Homarus americanus*, following intrapericardial administration: *Xenobiotica*, v. 18, no. 3, pp. 269–276.

Barron, M.G., Mayes, M.A., Murphy, P.G., and Nolan, R.J., 1990a, Pharmacokinetics and metabolism of triclopyr butoxyethyl ester in coho salmon: *Aquat. Toxicol.*, v. 16, no. 1, pp. 19–32.

Barron, M.G., Stehly, G.R., and Hayton, W.L., 1990b, Pharmacokinetic modeling in aquatic animals. I. Models and concepts—Review: *Aquat. Toxicol.*, v. 17, no. 3, pp. 187–212.

Barron, M.G., Tarr, B.D., and Hayton, W.L., 1987, Temperature dependence of di-2-ethylhexylphthalate (DEHP) pharmacokinetics in rainbow trout: *Toxicol. Appl. Pharmacol.*, v. 88, no. 3, pp. 305–312.

Barthel, W.F., Hawthorne, J.C., Ford, J.H., Bolton, G.C., McDowell, L.L., Grissinger, E.H., and Parsons, D.A., 1969, Pesticide residues in sediments of the lower Mississippi River and its tribuaries: *J. Pest. Monitor.*, v. 3, no. 1, pp. 8–34.

Baughman, G.L., and Lassiter, R.R., 1978, Prediction of environmental pollutant concentration, *in* Cairns, J., Jr., Dickson, K.L., and Maki, A.W., eds., *Estimating the hazard of chemical substances to aquatic life*: American Society for Testing and Materials, Philadelphia, Penn., Special Technical Publication series, v. 657, pp. 35–54.

Baughman, G.L., and Paris, D.F., 1981, Microbial bioconcentration of organic pollutants from aquatic systems—A critical review: *CRC Crit. Rev. Microbiol.*, v. 7, no. 1, pp. 205–207.

Baumann, P.C., and Whittle, D.M., 1988, The status of selected organics in the Laurentian Great Lakes—An overview of DDT, PCBs, dioxins, furans, and aromatic hydrocarbons: *Aquat. Toxicol.*, v. 11, no. 3–4, pp. 241–257.

Bedford, J.W., Roelofs, E.W., and Zabik, M.J., 1968, The freshwater mussel as a biological monitor of pesticide concentrations in a lotic environment: *Limnol. Ocean.*, v. 13, no. 1, pp. 118–126.

Bednar, G.A., 1980, Quality of water in the Pearl River, Jackson to Byram, Mississippi, September 21–22, 1976: U.S. Geological Survey Open-File Report 80-575, 35 p.

Begley, S., and Glick, D., 1994, The estrogen complex: *Newsweek*, v. 123, no. 12, pp. 76–77.

Beliaeff, B., O'Connor, T.P., Daskalakis, D.K., and Smith, P.J., 1997, U.S. Mussel Watch data from 1986 to 1994: Temporal trend detection at large spatial scales: *Environ. Sci. Technol.*, v. 31, no. 5, pp. 1411–1415.

Bell, G.R., 1956, On the photochemical degradation of 2,4-dichlorophenoxyacetic acid and structurally related compounds in the presence and absence of riboflavin: *Bot. Gaz.*, v. 118, pp. 133–136.

Bender, M.E., 1969, The toxicity of the hydrolysis and breakdown products of malathion to the fathead minnow (*Pimephales promelas*, Rafinesque): *Wat. Res.*, v. 3, no. 7, pp. 571–582.

Bennett, I.L., 1967, Forward: *J. Pest. Monitor.*, v. 1, no. 1, pp. i–ii.

Bennetts, H., Underwood, E., and Shier, F., 1946, A specific breeding problem of sheep on subterranean clover pastures in western Australia: *Aust. Vet. J.*, v. 22, no. 1, pp. 2–12.

Bero, A.S., and Gibbs, R.J., 1990, Mechanisms of pollutant transport in the Hudson estuary: *Sci. Total Environ.*, v. 97/98, pp. 9–22.

Bevans, H.E., Goodbred, S., Miesner, J.F., Watkins, S.A., Gross, T.S., Denslow, N.D., and Schoeb, T., 1996, Synthetic organic compounds and carp endocrinology and histology in Las Vegas Wash and Las Vegas and Callville bays of Lake Mead, Nevada, 1992 and 1995: U.S. Geological Survey Water-Resources Investigation 96-4266, 12 p.

Bevenue, A., 1976, The "bioconcentration" aspects of DDT in the environment: *Residue Rev.*, v. 61, pp. 37–112.

Bevenue, A., Hylin, J.W., Kawano, Y., and Kelley, T.W., 1972, Organochlorine pesticide residues in water, sediment, algae, and fish, Hawaii, 1970–71: *J. Pest. Monitor.*, v. 6, no. 1, pp. 56–64.

Biddinger, G.R., and Gloss, S.P., 1984, The importance of trophic transfer in the bioaccumulation of chemical contaminants in aquatic ecosystems: *Residue Rev.*, v. 91, pp. 103–145.

Bidleman, T.F., and Olney, C.E., 1974, High-volume collection of atmospheric polychlorinated biphenyls: *Bull. Environ. Contam. Toxicol.*, v. 11, no. 5, pp. 442–450.

Bidleman, T.F., Patton, G.W., Hinckley, D.A., Walla, M.D., Cotham, W.E., and Hargrave, B.T., 1990, Chlorinated pesticides and polychlorinated biphenyls in the atmosphere of the Canadian arctic, *in* Kurtz, D.A., ed., *Long range transport of pesticides*: Lewis Publishers, Chelsea, Mich., pp. 237–372.

Bidleman, T.F., Patton, G.W., Walla, M.D., Hargrave, B.T., Vass, W.P., Erickson, P., Fowler, B., Scott, V., and Gregor, D.J., 1989, Toxaphene and other organochlorines in Arctic Ocean fauna: Evidence for atmospheric delivery: *Arctic*, v. 42, no. 4, pp. 307–313.

Bidleman, T.F., Zaranski, M.T., and Walla, M.D., 1988, Toxaphene: Usage, aerial transport, and deposition, *in* Schmidtke, N.W., ed., *Chronic effects of toxic contaminants in large lakes*, v. 1 of *Toxic contamination in large lakes*: Lewis Publishers, Chelsea, Mich., pp. 257–284.

Bierman, V.J., Jr., 1990, Equilibrium partitioning and biomagnification of organic chemicals in benthic animals: *Environ. Sci. Technol.*, v. 24, no. 9, pp. 1407–1412.

Binder, R.L., Melancon, M.J., and Lech, J.J., 1984, Factors influencing the persistence and metabolism of chemicals in fish: *Drug Metab. Rev.*, v. 15, pp. 697–724.

Bishop, C.A., Brooks, R.J., Carey, J.H., Ng, P., Norstrom, R.J., and Lean, D.R.S., 1991, The case for a cause–effect linkage between environmental contamination and development in the eggs of the common snapping turtle (*Chelydra s. serpentina*) from Ontario, Canada: *J. Toxicol. Environ. Health*, v. 33, pp. 521–547.

Black, M.C., and McCarthy, J.F., 1988, Dissolved organic macromolecules reduce the uptake of hydrophobic organic contaminants by the gills of rainbow trout (*Salmo gairdnieri*): *Environ. Toxicol. Chem.*, v. 7, pp. 593–600.

Blanchard, P.J., Roy, R.R., and O'Brien, T.F., 1993, Reconnaissance investigation of water quality, bottom sediment, and biota associated with irrigation drainage in the San Juan River area, San Juan County, northwestern New Mexico, 1990–91: U.S. Geological Survey Water-Resources Investigations Report 93-4065, 141 p.

Bobo, L.L., and Peters, C. A., 1980, Water-quality assessment of the Cypress Creek watershed, Warrick County, Indiana: U.S. Geological Survey Water-Resources Investigations [Report] 80-35, 67 p.

Bobo, L.L., and Renn, D.E., 1980, Water-quality assessment of the Porter County watershed, Kankakee River Basin, Porter County, Indiana: U.S. Geological Survey Open-File Report 80-331, 54 p.

Boehm, P.D., and Quinn, J.G., 1976, Effect of dissolved organic matter in seawater on the uptake of mixed individual hydrocarbons and No. 2 fuel oil by a marine filter feeding bivalve (*Mercenaria mercenaria*): *Est. Coast Mar. Sci.*, v. 4, pp. 93–105.

———1977, The persistence of chronically accumulated hydrocarbons in the hard shell clam, *Mercenaria*: *Mar. Biol.*, v. 44, no. 3, pp. 227–233.

Boellstorff, D.E., Ohlendorf, H.M., Anderson, D.W., O'Neill, E.J., Keith, J.O., and Prouty, R.M., 1985, Organochlorine chemical residues in white pelicans and western grebes from the Klamath Basin, California: *Arch. Environ. Contam. Toxicol.*, v. 14, no. 4, pp. 485–493.

Boese, B.L., Lee, H., II, Specht, D.T., Pelletier, J., and Randall, R., 1996, Evaluation of PCB and hexachlorobenzene biota–sediment accumulation factors based on ingested sediment in a deposit-feeding clam: *Environ. Toxicol. Chem.*, v. 15, no. 9, pp 1584–1589.

Boese, B.L., Winsor, M., Lee, H., II, Echols, S., Pelletier, J., and Randall, R., 1995, PCB congeners and hexachlorobenzene biota–sediment accumulation factors for *Macoma nasuta* exposed to sediments with different total organic carbon contents: *Environ. Toxicol. Chem.*, v. 14, no. 2, pp. 303–310.

Boileau, S., Baril, M., and Alary, J.G., 1979, DDT in northern pike (*Esox lucius*) from the Richilieu River, Quebec, Canada, 1974–75: *J. Pest. Monitor.*, v. 13, no. 3, pp. 109–114.

Bollag, J.-M., and Liu, S.-Y., 1990, Biological transformation processes of pesticides, *in* Cheng, H.H., ed., *Pesticides in the soil environment: Processes, impacts, and modeling*: Soil Science Society of America, Madison, Wis., pp. 169–211.

Bonderman, D.P., and Slach, E., 1972, Appearance of 1-hydroxychlordene in soil, crops, and fish: *J. Agr. Food Chem.*, v. 20, no. 2, pp. 328–331.

Bopp, R.F., and Simpson, H.J., 1989, Contamination of the Hudson River: The sediment record, *in* National Research Council, *Contaminated marine sediments: Assessment and remediation*: National Academy Press, Washington, D.C., pp. 401–416.

Borgmann, U., and Whittle, D.M., 1992, Bioenergetics and PCB, DDE, and mercury dynamics in Lake lake trout (*Salvelinus namaycush*): A model based on data: *Can. J. Fish. Aquat. Sci.*, v. 49, no. 6, pp. 1086–1096.

Borsetti, A.P., and Roach, J.A.G., 1978, Identification of kepone alteration products in soil and mullet: *Bull. Environ. Contam. Toxicol.*, v. 20, no. 2, pp. 241–247.

Bowers, J.C., and Irwin, G.A., 1978, Water-quality investigation, upper Santa Clara River Basin, California: U.S. Geological Survey Water-Resources Investigations [Report] 77-99, 43 p.

Boyer, J.M., and Chapra, S.C., 1991, Fate of environmental pollutants: *Water Environ. Res.*, v. 63, no. 4, pp. 607–619.

Bradbury, R., and White, D., 1954, Estrogens and related substances in plants: *Vitam. Hormones*, v. 12, pp. 207–233.

Bradbury, S.P., 1983, The toxicity and toxicokinetics of fenvalerate in fish: A preliminary assessment: U.S Environmental Protection Agency, Environmental Research Laboratory, In-house report, cited in McKim and others (1985).

Bradbury, S.P., and Coats, J.R., 1989, Toxicokinetics and toxicodynamics of pyrethroid insecticides in fish: *Environ. Toxicol. Chem.*, v. 8, no. 5, pp. 373–380.

Bradshaw, J.S., Loveridge, E.L., Rippee, K.P., Peterson, J.L., White, D.A., Barton, J.R., and Fuhriman, D.K., 1972, Seasonal variations in residues of chlorinated hydrocarbon pesticides in the water of the Utah Lake drainage system—1970 and 1971: *J. Pest. Monitor.*, v. 6, no. 3, pp. 166–170.

Branson, D.R., Blau, G.E., Alexander, H.C., and Neely, W.B., 1975, Bioconcentration of 2,2',4,4'-tetrachlorobiphenyl in rainbow trout as measured by an accelerated test: *Trans. Am. Fish. Soc.*, v. 104, pp. 785–792.

Briggs, G.G., 1981, Theoretical and experimental relationships between soil adsorption, octanol-water partition coefficients, water solubilities, bioconcentration factors, and the parachor: *J. Agric. Food Chem.*, v. 29, no. 5, pp. 1050–1059.

Briggs, J.C., and Feiffer, J.S., 1986, Water quality of Rhode Island streams: U.S. Geological Survey Water-Resources Investigations Report 84-4367, 51 p.

Brightbill, D.B., and Treadway, J.B., Jr., 1980, Analysis of water, bank material, bottom material, and elutriate samples collected near Belzoni, Mississippi (upper Yazoo projects): U.S. Geological Survey Open-File Report 80-758, 17 p.

Bro, K.M., Sonzogni, W.C., and Hanson, M.E., 1987, Relative cancer risks of chemical contaminants in the Great Lakes: *Environ. Manag.*, v. 11, no. 4, pp. 495–505.

Brookes, D.N., Dobbs, A.J., and Williams, N., 1986, Octanol:water partition coefficients (P): Measurement, estimation and interporetation, particularly for chemicals with P>10[5]: *Ecotoxicol. Environ. Safety*, v. 11, no. 3, pp. 251–260.

Brown, D.A., Gossett, R.W., Hershelman, P., Schaefer, H.A., Jenkins, K.D., and Perkins, E.M., 1983, Bioaccumulation and detoxification of contaminants in marine organisms from southern California coastal waters, *in* Soule, D.F., and Walsh, D., eds., *Waste disposal in the oceans: Minimizing impact, maximizing benefits*: Westview Press, Boulder, Colo., pp. 171–193.

Brown, D.P., 1987, Mortality of workers exposed to polychlorinated biphenyls—An update: *Arch. Environ. Health*, v. 42, pp. 333–339.

Bruggeman, W.A., Martron, L.B.J.M., Kooiman, D., and Hutzinger, O., 1981, Accumulation and elimination kinetics of di-, tri-, and tetra-chlorobiphenyls by goldfish after dietary and aqueous exposure: *Chemosphere*, v. 10, no. 8, pp. 811–832.

Bruggeman, W.A., Opperhuizen, A., Wijbenga, A., and Hutzinger, O., 1984, Bioaccumulation of superlipophilic chemicals in fish: *Toxicol. Environ. Chem.*, v. 7, pp. 173–189.

Brungs, W.A., and Mount, D.I., 1978, Introduction to a discussion of the use of toxicity tests for evaluation of the effects of toxic substances, *in* Cairns, J., Jr., Dickson, K.L., and Maki, A.W., eds., *Estimating the hazard of chemical substances to aquatic life*: American Society for Testing and Materials, Philadelphia, Penn., Special Technical Publication series, v. 657, pp. 15–26.

Bryant, C.T., Morris, E.E., and Terry, J.E., 1979, Water-quality assessment of the L'Anguille River Basin, Arkansas: Progress report: U.S. Geological Survey Open-File Report 79-1482, 139 p.

Buchman, M.F., 1989, A review and summary of trace contaminants data for coastal and estuarine Oregon: National Oceanic and Atmospheric Administration Technical Memorandum NOS OMA 42, 115 p.

Buhler, D.R., Rasmusson, M.E., and Nakaue, H.S., 1973, Occurrence of hexachlorophene and pentachlorophenol in sewage and water: *Environ. Sci. Technol.*, v. 7, no. 10, pp. 929–934.

Bulkley, R.V., 1978, Variations in DDT concentration in muscle tissue of channel catfish, *Ictalurus punctatus*, from the Des Moines River, 1971: *J. Pest. Monitor.*, v. 11, no. 4, pp. 165–169.

Bulkley, R.V., Kellogg, R.L., and Shannon, L.R., 1976, Size-related factors associated with dieldrin concentrations in muscle tissue of channel crayfish, *Ictalurus punctatus*: *Trans. Am. Fish. Soc.*, v. 105, no. 2, pp. 301–307.

Bulkley, R.V., Leung, S.-Y.T., and Richard, J.J., 1981, Organochlorine insecticide concentrations in fish of the Des Moines River, 1977–78: *J. Pest. Monitor.*, v. 15, no. 2, pp. 86–89.

Bulkley, R.V., Shannon, L.R., and Kellogg, R.L., 1974, Contamination of channel catfish with dieldrin from agricultural runoff: Completion report, project no. A-042-1A, duration July 1971–June 1974: Iowa State Water Resources Research Institute, Ames, Iowa, ISWRRI series, v. 62, 144 p.

Bungay, P.M., Dedrick, R.L., and Guarino, A.M., 1976, Pharmacokinetic modeling of the dogfish shark (*Squalus acanthias*): Distribution and urinary and biliary excretion of phenol red and its glucuronide: *J. Pharmacol. Biopharm.*, v. 4, pp. 377–388.

Burke, A.B., Huckle, K.R., and Millburn, P., 1988, The metabolism of 3-phenoxybenzoic acid and benzoic acid in rat and rainbow trout: *Biochem. Soc. Trans.*, v. 16, pp. 25–26.

Burkhard, N., and Guth, J.A., 1981, Chemical hydrolysis of 2-chloro-4,6-bis-(alkylamino)-1,3,5-triazine herbicides and their breakdown in soil under the influence of adsorption: *Pest. Sci.*, v. 12, no. 1, pp. 45–52.

Burns, L.A., 1983, Fate of chemicals in aquatic systems: Process models and computer codes, *in* Swann, R.L., and Eschenroeder, A., eds., *Fate of chemicals in the environment: Compartmental and multimedia models for predictions*: American Chemical Society Symposium Series, no. 225, pp. 25–40.

Burt, A.J., McKee, P.M., Hart, D.R., and Kauss, P.B., 1991, Effects of pollution on benthic invertebrate communities of the St. Mary's River, 1985: *Hydrobiologia*, v. 219, pp. 63–81.

Bush, P.B., Neary, D.G., Taylor, J.W., Jr., and Nutter, W.L., 1986, Effects of pesticide use in a pine seed orchard on pesticide levels in fish: *Water Resour. Bull.*, v. 22, no. 5, pp. 817–827.

Bustos, S., Denegri, J.C., Diaz, F., and Tchernitchin, A.N., 1988, *p,p'*-DDT is an estrogenic compound: *Bull. Environ. Contam. Toxicol.*, v. 41, pp. 496–501.

Butler, D.L., 1987, Pesticide data for selected Wyoming streams, 1976–78: U.S. Geological Survey Water-Resources Investigations Report 83-4127, 41 p.

Butler, D.L., Krueger, R.P., Osmundson, B.C., Thompson, A.L., and McCall, S.K., 1991, Reconnaissance investigation of water-quality, bottom sediment, and biota associated with irritgation drainage in the Gunnison and Uncompahgre river basins and at Sweitzer Lake, west-central Colorado, 1988–89: U.S. Geological Survey Water-Resources Investigations Report 91-4103, 99 p.

Butler, D.L., Krueger, R.P., Thompson, A.L., Formea, J.J., and Wickman, D.W., 1993, Reconnaissance investigation of water quality, bottom sediment, and biota associated with irrigation drainage in the Pine River Project area, Southern Ute Indian Reservation, southwestern Colorado and northwestern New Mexico, 1988–89: U.S. Geological Survey Water-Resources Investigations Report 92-4188, 105 p.

Butler, P.A., 1964, DDT residues in oyster gametes (abs.), *in* National Shellfisheries Association, *Proceedings of the National Shellfisheries Association*: MPG Communications, Plymouth, Mass., v. 55.

——1966, The problem of pesticides in estuaries: *Trans. Am. Fish. Soc.*, v. 95, no. 4, supp., pp. 110–115.

——1969a, Monitoring pesticide pollution: *BioScience*, v. 19, pp. 889–891.

——1969b, The significance of DDT residues in estuarine fauna, *in* Miller, M.W., and Berg, G.G., eds., *Chemical fallout: Current research on persistent pesticides*: Charles C. Thomas Publishers, Springfield, Ill. pp. 205–220.

——1973a, Trends in pesticide residues in shellfish, *in* National Shellfisheries Association, *Proceedings of the National Shellfisheries Association*: MPG Communications, Plymouth, Mass., v. 64, pp. 77–81.

——1973b, Organochlorine residues in estuarine mollusks 1965–1972, National Pesticides Monitoring Program: *J. Pest. Monitor.*, v. 6, no. 4, pp. 238–362.

——1974, Biological problems in estuarine monitoring, *in* Verner, S.S., ed., *Proceedings of seminar on methodology for monitoring the marine environment*: U.S. Environmental Protection Agency, [Washington, D.C.], pp. 126–138.

——1975, National Estuarine Monitoring Program, in *Estuarine pollution control and assessment: Proceedings of a conference*: U.S. Environmental Protection Agency, [Washington, D.C.], v. 3, pp. 519–521.

——1978, EPA-NOAA Cooperative Estuarine Monitoring Program: U.S Environmental Protection Agency, Gulf Breeze, Florida, Final Report, 8 p., data sheets.

Butler, P.A., Childress, R., and Wilson, A.J., 1972, The association of DDT residues with losses in marine productivity, *in* Ruivo, M., ed., *Marine pollution and sea life*: Fishing (Books), London, England, pp. 262–266.

Butler, P.A., Kennedy, C.D., and Schutzmann, R., 1978, Pesticide residues in estuarine mollusks, 1977 versus 1982, National Pesticide Monitoring Program: *J. Pest. Monitor.*, v. 12, no. 3, pp. 99–101.

Butler, P.A., and Schutzmann, R.L., 1978, Residues of pesticides and PCBs in estuarine fish, 1972–76 National Pesticide Monitoring Program: *J. Pest. Monitor.*, v. 12, no. 2, p. 51–59.

——1979, Bioaccumulation of DDT and PCB in tissues of marine fishes, *in* Marking, L.L., and Kimerle, R.A., eds., *Aquatic toxicology: Proceedings of the Second Annual Symposium on Aquatic Toxicology*: American Society for Testing and Materials, Philadelphia, Penn., Special Technical Publication series, v. 667, pp. 212–220.

Cade, T.J., White, C.M., and Haugh, J.R., 1968, Peregrines and pesticides in Alaska: *The Condor*, v. 70, no. 2, pp. 170–178.

Calabrese, E.J., 1982, Human breast milk contamination in the United States and Canada by chlorinated hydrocarbon insecticides and industrial pollutants: Current status: *J. Am. College Toxicol.*, v. 1, no. 3, pp. 91–98.

Camanzo, J., Rice, C.P., Jude, D.J., and Rossman, R., 1987, Organic priority pollutants in nearshore fish from 14 Lake Michigan tributaries and embayments, 1983: *J. Great Lakes Res.*, v. 13, no. 3, pp. 296–309.

Canadian Council of Ministers of the Environment, 1995, Protocol for the derivation of Canadian sediment quality guidelines for the protection of aquatic life: Canadian Council of Ministers of the Environment Report CCME EPC-98E, 38 p.

——1998, Canadian sediment quality guidelines for the protection of aquatic life: Introduction and summary tables, in *Canadian environmental quality guidelines*: Canadian Council of Ministers of the Environment, Winnipeg, Canada, accessed December 21, 1998, at URL http://www.ec.gc.ca/ceqg-rcqe.

Capel, P.D., 1993, Organic chemical concepts, *in* Alley, W.M., ed., *Regional ground-water quality*: Van Nostrand Reinhold, N.Y., pp. 155–179.

Capel, P.D., and Eisenreich, S.J., 1985, PCBs in Lake Superior, 1978–1980: *J. Great Lakes Res.*, v. 11, no. 4, pp. 447–461.

———1990, Relationship between chlorinated hydrocarbons and organic carbon in sediment and porewater: *J. Great Lakes Res.*, v. 16, no. 2, pp. 245–257.

Capel, P.D., Giger, W., Reichert, P., and Wanner, O., 1988, Accidental input of pesticides into the Rhine River: *Environ. Sci. Technol.*, v. 22, no. 9, pp. 992–997.

Carey, A.E., and Kutz, F.W., 1985, Trends in ambient concentrations of agrochemicals in humans and the environment of the United States: *Environ. Monitor. Assess.*, v. 5, pp. 155–163.

Carleton, J.E., 1985, *Nashua impoundment sediment sampling, June 1985*: Iowa Department of Water, Air, and Waste Management, Des Moines, Iowa, [32] p.

Carlsen, E., Giwercman, A., Keiding, N., and Skakkeback, N.E., 1992, Evidence for decreasing quality of semen during past 50 years: *Br. Med. J.*, v. 304, pp. 609–613.

Cartwright, R.H., and Ziarno, J.A., compilers, 1980, Chemical quality of water from community systems in New York, November 1970 to May 1975: U.S. Geological Survey Water-Resources Investigation 80-77, 444 p.

Cashman, J.R., Maltby, D.A., Nishioka, R.S., Howard, A., Gee, S.J., and Hammock, B.D., 1992, Chemical contamination and the annual summer die-off of striped bass (*Morone saxatilis*) in the Sacramento–San Joaquin Delta: *Chem. Res. Toxicol.*, v. 5, no. 1, pp. 100–105.

Casper, V.L., Hammerstrom, R.J., Robertson, E.A., Jr., Bugg, J.C., Jr., and Gaines, J.L., 1969, *Study of chlorinated pesticides in oysters and estuarine environment of the Mobile Bay area*: U.S. Consumer Protection and Environmental Health Service, Bureau of Water Hygiene, Cincinnati, Ohio, 47 p.

Castro, O., Ferreira, A.M., and Vale, C., 1990, Organochlorine compounds in the Portuguese oyster: Importance of seasonal variations: *Mar. Pollut. Bull.*, v. 21, no. 11, pp. 545–547.

Cattani, O., Corni, M.G., Crisetig, G., and Serrazanetti, G.P., 1981, Chlorinated hydrocarbon residues in zooplankton from the northern Adriatic Sea: *J. Etud. Pollut. Mar. Mediterr.*, v. 5, pp. 353–360.

Chacko, C.I., and Lockwood, J.L., 1967, Accumulation of DDT and dieldrin by microorganisms: *Can. J. Microbiol.*, v. 13, p. 1123.

Chadwick, G.G., and Brocksen, R.W., 1969, Accumulation of dieldrin by fish and selected fish-food organisms: *J. Wildlife Manag.*, v. 33, no. 3, pp. 693–700.

Chapra, S.C., and Boyer, J.M., 1992, Fate of environmental pollutants: *Water Environ. Res.*, v. 64, no. 4, pp. 581–593.

Charles, M.J., and Hites, R.A., 1987, Sediments as archives of environmental pollution trends *in* Hites, R.A., and Eisenreich, S.J., eds., *Sources and fates of aquatic pollutants*: American Chemical Society Advances in Chemistry series, no. 216, pp. 365–389.

Check, R.M., and Canario, M.T., 1972, Residues of chlorinated hydrocarbon pesticides in the northern quahog (hard-shell clam), *Mercenaria mercenaria*—1968 and 1969: *J. Pest. Monitor.*, v. 6, no. 3, pp. 229–230.

Cheer, A.Y., Ogami, Y., and Sanderson, S.L., 1993, Fluid dynamics in the oral cavity of suspension feeding fishes, in *International Symposium on Nonlinear Theory and its Applications*: Research Society of Nonlinear Theory and its Applications, Honolulu, Hawaii, December 5–10, 1993, pp.1101–1104. .

Chiou, C.T., 1985, Partition coefficients of organic compounds in lipid-water systems and correlations with fish bioconcentration factors: *Environ. Sci. Technol.*, v. 19, no. 1, pp. 57–62.

———1998, Soil sorption of organic pollutants and pesticides, *in* Meyers, R.A., ed., *Encyclopedia of environmental analysis and remediation*: Wiley, New York, v. 7, pp 4517–4554.

Chiou, C.T., Freed, V.H., Schmedding, D.W., and Kohnert, R.L., 1977, Partition coefficient and bioaccumulation of selected organic chemicals: *Environ. Sci. Technol.*, v. 11, no. 5, pp. 475–478.

Chiou, C.T., Peters, L.J., and Freed, V.H., 1979, A physical concept of soil-water equilibria for nonionic compounds: *Science*, v. 206, pp. 831–832.

Christiansen, C.C., Hesse, L.W., and Littell, B., 1991, Contamination of the channel catfish (*Ictalurus punctatus*) by organochlorine pesticides and polychlorinated biphenyls in the Missouri River: *Trans. Nebraska Acad. Sci*, v. 18, pp. 93–98.

Claeys, R.R., Caldwell, R.S., Cutshall, N.H., and Holton, R., 1975, Chlorinated pesticides and ploychlorinated biphenyls in marine species, Oregon/Washington coast, 1972: *J. Pest. Monitor.*, v. 9, no. 1, pp. 2–10.

Claeys, R.R., Every, R.W., Capizzi, J., Kiigenaji, U., Shumway, D.L., and Norris, L.A., 1970, Report of Task Force on Persistent Pesticides in Oregon, Part I: *Environmental Sciences Center Newsletter, Environmental Research and Reports*, v. 2, no. 1, pp. 1–10.

Clark, D.R., Jr., and Krynitsky, A.J., 1983, DDT: Recent contamination in New Mexico and Arizona?: *Environment*, v. 25, no. 5, pp. 27–31.

Clark, J.R., DeVault, D., Bowden, R.J., and Weishar, J.A., 1984, Contaminant analysis of fillets from Great Lakes Coho salmon, 1980: *J. Great Lakes Res.*, v. 10, no. 1, pp. 38–47.

Clark, K.E., Gobas, F.A.P.C., and Mackay, D., 1990, Model of organic chemical uptake and clearance by fish from food and water: *Environ. Sci. Technol.*, v. 24, no. 8, pp. 1203–1213.

Clark, K.E., and Mackay, D., 1991, Dietary uptake and biomagnification of four chlorinated hydrocarbons by guppies: *Environ. Toxicol. Chem.*, v. 10, no. 9, pp. 1205–1217.

Clark, T., Clark, K., Paterson, S., Mackay, D., and Norstrom, R.J., 1988, Wildlife monitoring, modeling, and fugacity: *Environ. Sci. Technol.*, v. 22, no. 2, pp. 120–127.

Clarke, J.U., McFarland, V.A., and Dorkin, J., 1988, Evaluating bioavailability of neutral organic chemicals in sediments—A confined disposal facility case study, *in* Willey, R.G., ed., *Water quality '88: Seminar proceedings, 23–25 February 1988, Charleston, South Carolina*: U.S. Army Corps of Engineers, Committee on Water Quality, Washington, D.C., pp. 251–268.

Clayton, J.R., Jr., Pavlou, S.P., and Breitner, N.F., 1977, Polychlorinated biphenyls in coastal marine zooplankton: Bioaccumulation by equilibrium partitioning: *Environ. Sci. Technol.*, v. 11, no. 7, pp. 676–682.

Clegg, D.E., 1974, Chlorinated hydrocarbon residues in oysters (*Crassostrea commercialis*) in Moreton Bay, Queensland, Australia, 1970–72: *J. Pest. Monitor.*, v. 8, no. 3, pp. 162–166.

Clement Associates Inc., 1984, Review and evaluation of the evidence for cancer associated with air pollution. Final Report: U.S. Environmental Protection Agency, Office of Air Quality Planning and Standards, Pollutant Assessment Branch, EPA 450/5-83-006R, variously paged.

Clement, C.R., and Colburn, T., 1992, Herbicides and fungicides: A perspective on potential human exposure, *in* Colborn, T., and Clement, C.R., eds., *Chemically induced alterations in sexual and functional development: The wildlife/human connection*: Princeton Scientific Publishing, Princeton, N.J., Advances in Modern Environmental Toxicology series, v. 21, pp. 347–364.

Cohn, W.J., Boylan, J.J., Blanke, R.V., Farris, M.W., Howell, J.R., and Guzelian, P.S., 1978, Treatment of chlordecone (kepone) toxicity with cholestyramine: Results of a controlled clinical trial: *N. Engl. J. Med.*, v. 298, pp. 243–248.

Colborn, T., and Clement, C.R., eds., 1992a, Statement from the work session on chemically induced alterations in sexual development: The wildlife/human connection, *in* Colborn, T., and Clement, C.R., eds., *Chemically induced alterations in sexual and functional development: The wildlife/human connection*: Princeton Scientific Publishing, Princeton, N.J., Advances in Modern Environmental Toxicology series, v. 21, pp. 1–6.

——1992b, *Chemically induced alterations in sexual and functional development: The wildlife/ human connection*: Princeton Scientific Publishing, Princeton, N.J., Advances in Modern Environmental Toxicology series, v. 21, 403 p.

Colborn, T., Dumanoski, D., and Myers, J.P., 1995, *Our stolen future: Are we threatening our fertility, intelligence, and survival?: A scientific detective story*: Dutton, N.Y., 306 p.

Colborn, T., Vom Saal, F.S., and Soto, A.M., 1993, Developmental effects of endocrine-disrupting chemicals in wildlife and humans: *Environ. Health Perspect.*, v. 101, no. 5, pp. 378–384.

Cole, G.A., 1983, *Textbook of limnology*: Waveland Press, Prospect Heights, Ill., 401 p.

Cole, H., Jr., Barry, D., Frear, D.E.H., and Bradford, A., 1967, DDT levels in fish, streams, stream sediments, and soil after DDT aerial spray application for fall canker-worm in Pennsylvania: *Bull. Environ. Contam. Toxicol.*, v. 2, no. 3, pp. 127–146.

Cole, R.H., Frederick, R.E., Healy, R.P., and Rolan, R.G., 1983, *NURP (Nationwide Urban Runoff Program) Priority Pollutant Monitoring Project: Summary of findings*: Dalton-Dalton-Newport, Cleveland, Ohio, 149 p.

——, 1984, Preliminary findings of the Priority Pollutant Monitoring Project of the Nationwide Urban Runoff Program: *J. Water Poll Control Fed.*, v. 56, no. 7, pp. 898–908.

Connell, D.W., 1978, A kerosene-like taint in the sea mullet *Mugil cephalus* (Linneaus). II. Some aspects of the deposition and metabolism of hydrocarbons in muscle tissue: *Bull. Environ. Contam. Toxicol.*, v. 20, no. 3, pp. 492–498.

——1988, Bioaccumulation behavior of persistent organic chemicals with aquatic organisms: *Rev. Environ. Contam. Toxicol.*, v. 101, pp. 117–154.

Connell, D.W., and Schuurmann, G., 1988, Evaluation of various molecular parameters as predictors of bioconcentration in fish: *Ecotoxicol. Environ. Safety*, v. 15, no. 3, pp. 324–335.

Connolly, J.P., and Pedersen, C.J., 1988, A thermodynamic-based evaluation of organic chemical accumulation in aquatic organisms: *Environ. Sci. Technol.*, v. 22, no. 1, pp. 99–103.

Connolly, J.P., and Tonelli, R., 1985, Modelling kepone in the striped bass food chain of the James River estuary: *Estuar. Coastal Shelf Sci.*, v. 20, pp. 349–366.

Connor, M.S., 1984a, Fish/sediment concentration ratios for organic compounds: *Environ. Sci. Technol.*, v. 18, no. 1, pp. 31–35.

——1984b, Comparison of carcinogenic risks from fish vs. groundwater contamination by organic compounds: *Environ. Sci. Technol.*, v. 18, no. 8, pp. 628–631.

Conte, F.S., and Parker, J.C., 1975, Effect of aerially-applied malathion on juvenile brown and white shrimp *Penaeus aztecus* and *P. setiferus*: *Trans. Amer. Fish. Soc.*, v. 104, no. 4, pp. 793–799.

Cook, G.H., and Moore, J.C., 1976, Determination of malathion, malaoxon, and mono- and dicarboxylic acids of malathion in fish, oyster, and shrimp tissue: *J. Agric. Food Chem.*, v. 24, no. 3, pp. 631–634.

Cooper, C.M., 1991, Persistent organochlorine and current use insecticide concentrations in major watershed components of Moon Lake, Mississippi, USA: *Arch. fuer Hydrobiol.*, v. 121, no. 1, pp. 103–113.

Cope, O.B., 1961, Effects of DDT spraying for spruce budworm in the Yellowstone river system: *Trans. Am. Fish. Soc.*, v. 90, no. 3, pp. 239–251.

Cordle, F., Locke, R., and Springer, J., 1982, Risk assessment in a federal regulatory agency: An assessment of risk associated with the human consumption of some species of fish contaminated with polychlorinated biphenyls (PCBs): *Environ. Health Perspect.*, v. 45, pp. 171–182.

Courtney, W.A.M., and Langston, W.J., 1978, Uptake of polychlorinated biphenyl (Aroclor 1254) from sediment and from seawater in two intertidal polychaetes: *Environ. Pollut.*, v. 15, no. 4, pp. 303–309.

Cowgill, U.M., Williams, D.M., and Esquivel, J.B., 1984, Effects of maternal nutrition on fat content and longevity of neonates of *Daphnia magna*: *J. Crustacean Biol.*, v. 4, no. 2, pp. 173–190.

Cox, J.L., 1970a, DDT residues in marine phytoplankton: Increase from 1955 to 1969: *Science*, v. 170, no. 3953, pp. 71–73.

——1970b, Low ambient level uptake of [^{14}C]-DDT by three species of phytoplankton: *Bull. Environ. Contam. Toxicol.*, v. 5, pp. 218–221.

——1971, DDT residues in seawater and particulate matter in the California current system: *Fish. Bull.*, v. 69, p. 443.

——1972, DDT residues in marine phytoplankton: *Residue Rev.*, v. 44, pp. 23–38.

Crane, D.B., and Younghans-Haug, C., 1992, Oxadiazon residue concentrations in sediment, fish, and shellfish from a combined residential/agricultural area in southern California: *Bull. Environ. Contam. Toxicol.*, v. 48, no. 4, pp. 608–615.

Crawford, J.K., and Luoma, S.N., 1993, Guidelines for studies of contaminants in biological tissues for the National Water-Quality Assessment Program: U.S. Geological Survey Open-File Report 92-494, 69 p.

Crayton, W.M., 1987, *Summary of contaminant investigations, fiscal year 1985*: U.S. Department of the Interior, Fish and Wildlife Service, 49 p.

——1989, *Cape Newenham military cleanup, Togiak National Refuge: Partial report of findings 1*: U.S. Department of the Interior, Fish and Wildlife Service, 13 p.

——1990, *Aleutian Islands military contaminants, fiscal year 1987 data: Adak Island, Agattu Island, Great Sitkin Island, Kiska Island*: U.S. Department of the Interior, Fish and Wildlife Service, 36 p., appendix.

——1991, *Contaminant study of the environment surrounding the Cape Romanzof long range radar site: Report of findings*: U.S. Department of the Interior, Fish and Wildlife Service, 22 p., 3 appendices.

Crocker, R.A., and Wilson, A.J., 1965, Kinetics and effects of DDT in intidal marsh ditch: *Trans. Am. Fish. Soc.*, v. 94, pp.152–159.

Crockett, A.B., Weirsma, G.B., Tai, H., Mitchell, W.G., Sand, P.F., and Carey, A.E., 1974, Pesticide residue levels in soils and crops, FY-70—National Soils Monitoring Program (II): *J. Pest. Monitor.*, v. 8, no. 2, pp. 69–97.

Crosby, D.G., 1994, Photochemical aspects of bioavailability, *in* Hamelink, J.L., Landrum, P.F., Bergman, H.L., and Benson, W.H., eds., *Bioavailability: Physical, chemical, and biological interactions*: Lewis Publishers, Boca Raton, Fla., pp. 109–118.

——1998, Environmental toxicology and chemistry: Oxford University Press, New York, 336 p.

Crosby, D.G., and Tucker, R.K., 1971, Accumulation of DDT by *Daphnia magna*: *Environ. Sci. Technol.*, v. 5, no. 8, pp. 714–716.

Crosby, D.G., and Tutass, H.O., 1966, Photodecomposition of 2,4-dichlorophenoxyacetic acid: *J. Agric. Food Chem.*, v. 14, no. 6, pp. 596–599.

Crossland, N.O., Bennett, D., and Wolff, C.J.M., 1987, Fate of 2,5,4′-trichlorophenol in outdoor ponds and its uptake via the food chain compared with direct uptake via the gills in grass carp and rainbow trout: *Ecotoxicol. Environ. Safety*, v. 13, no. 2, pp. 225–238.

Crouch, E.A.C., and Wilson, R., 1982, *Risk/benefit analysis*: Ballinger Publishing Co., Cambridge, Mass., 218 p.

Crouch, E.A.C., Wilson, R., and Zeise, L., 1983, The risks of drinking water: *Water Resour. Res.*, v. 19, pp. 1359–1375.

Crouter, R.A., and Vernon, E.H., 1959, Effects of black-headed budworm on salmon and trout in British Columbia: *Can. Fish. Cult.*, v. 24, pp. 23–40.

Cutshall, N.H., Larsen, I.L., and Nichols, M.M., 1981, Man-made radionuclides confirm rapid burial of kepone in sediments: *Science*, v. 213, no. 4506, pp. 440–442.

Czuczwa, J.M., and Hites, R.H., 1985, Dioxins and dibenzofurans in air, soil and water, *in* Kamrin, M.A., and Rodgers, P.W., eds., *Dioxins in the environment*: Hemisphere Publishing Corp., Washington, DC, pp. 85–99.

D'Itri, F.M., 1988, Contaminants in selected fishes from the Upper Great Lakes, *in* Schmidke, N.W., ed., *Impact of toxic contaminants on fisheries management*, v. 2 of *Toxic contamination in large lakes*: Lewis Publishers, Chelsea, Mich., pp. 51–84.

Dagley, S., 1983, Biodegradation and biotransformation of pesticides in the earth's carbon cycle: *Residue Rev., v.* 85, pp. 127–137.

Davar, P., and Wightman, J.P., 1981, Interaction of pesticides with chitosan, *in* Tewari, P.H., ed., *Adsorption from aqueous solutions*: Plenum Press, N.Y., pp. 163–177.

Davenport, M.S., 1989, Water quality in Reedy Fork and Buffalo Creek basins in the Greensboro area, North Carolina, 1986–87: U.S. Geological Survey Water-Resources Investigations Report 88-4210, 81 p.

Davies, K., 1990, Human exposure pathways to selected organochlorines and PCBs in Toronto and southern Ontario, *in* Nriagu, J.O., and Simmons, M.S., eds., *Food contamination from environmental sources*: Wiley-Interscience, New York, Advances in Environmental Science and Technology series, v. 23, pp. 525–540.

Davies, R.P., and Dobbs, A.J., 1984, The prediction of bioconcentration in fish: *Water Res.*, v. 18, no. 10, pp. 1253–1262.

Davis, D.L., and Bradlow, H.L., 1995, Can environmental estrogens cause breast cancer?: *Sci. Am.*, v. 10, pp. 166–172.

Davis, W.S., and Denbow, T.J., 1988, Aquatic sediments: *J. Water Poll. Control Fed.*, v. 60, no. 6, pp. 1077–1088.

Davis, W.S., Denbow, T.J., and Lathrop, J.E., 1987, Aquatic sediments: *J. Water Poll. Control Fed.*, v. 59, no. 6, pp. 586–597.

Dawson, G.W., Weimer, W.C., and Shupe, S.J., 1979, Kepone—a case study of a persistent material, *in* Alleman, J.E., and Bennett, G.F., eds., *Water—1978*: American Institute of Chemical Engineers Symposium series, v. 75, no. 190, pp. 366–374.

Day, K.E., 1990, Pesticide residues in freshwater and marine zooplankton: A review: *Environ. Pollut.*, v. 67, no. 3, pp. 205–222.

——1991, Pesticide transformation products in surface waters: Effects on aquatic biota, *in* Somasundaram, L., and Coats, J.R., eds., *Pesticide transformation products: Fate and significance in the environment*: American Chemical Society Symposium series, v. 459, pp. 217–241.

Day, K., and Kaushik, N.K.K., 1987, The adsorption of fenvalerate to laboratory glassware and the alga, *Chlamydomonas reinhardii*, and its effect on uptake of the pesticide by *Damphnia galeata mendotae*: *Aquat. Toxicol.*, v. 10, no.2–3, pp. 131–142.

de Boer, J., 1988, Chlorobiphenyls in bound and non-bound lipids of fishes: Comparison of different extraction methods: *Chemosphere*, v. 17, no. 9, pp. 1803–1810.

Demas, C.R., 1976, Analysis of native water, bed material, and elutriate samples of major Louisiana waterways, 1975: United States Geological Survey Open-File Report 76-853, 304 p.

——1977, Analysis of native water, core material, and elutriate samples collected from the Atchafalaya River and Atchafalaya Bay: U.S. Geological Survey Open-File Report 77-769, 17 p.

Demcheck, D.K., and Dupuy, A.J., 1980, Analyses of water, core material, and elutriate samples collected near Sicily Island, Louisiana (Sicily Island Area Levee Project): U.S. Geological Survey Open-File Report 80-434, 17 p.

Denison, M.S., Chambers, J.E., and Yarbrough, J.D., 1981, Persistent vitellogenin-like protein and binding of DDT in the serum of insecticide-resistant mosquitofish (*Gambusia affinis*): *Compar. Biochem. Physiol.*, v. 69, no. 1, pp. 109–112.

——1985, Short-term interactions between DDT and endrin accumulation and elimination in mosquitofish (*Gambusia affinis*): *Arch. Environ. Contam. Toxicol.*, v. 14, no. 3, pp. 315–320.

Derr, S.K., and Zabik, M.J., 1974, Bioactive compounds in the aquatic environment: Studies on the mode of uptake of DDE by the aquatic midge, *Chironomus tentans* (Diptera: Chironomidae): *Arch. Environ. Contam. Toxicol.*, v. 2, no. 2, pp. 152–164.

DeVault, D.S., 1985, Contaminants in fish from Great Lakes harbors and tributary mouths: *Arch. Environ. Contam. Toxicol.*, v. 14, no. 5, pp. 587–594.

DeVault, D.S., and Weishaar, J.A., 1984, Contaminant analysis of 1982 fall run coho salmon (*Onchoryhnchus kisutch*): U.S. Environmental Protection Agency, Great Lakes National Program Office, EPA-905/3-84-004, 16 p.

Dewailly, E., Dodin, S., Verrault, R., Ayotte, P., Sauve, L., Morin, J., 1994, High organochlorine body burden in women with estrogen receptor-positive breast cancer: *J. Natl. Cancer Inst.*, v. 86, no. 3, pp. 232–234.

Dewailly, E., Nantel, A., Weber, J.-P., and Meyer, F., 1989, High levels of PCBs in breast milk of Innuit women from arctic Quebec: *Bull. Environ. Contam. Toxicol.*, v. 43, no. 5, pp. 641–646.

DeWeese, L.R., Smykaj, A.M., Miesner, J.F., and Archuleta, A.S., 1993, Environmental contaminants survey of the South Platte River in northeastern Colorado, 1988: U.S. Fish and Wildlife Service, Fish and Wildlife Enhancement, Colorado State Office, Contaminant Report series, R6/306G/93, 75 p.

Di Toro, D.M., Zarba, C.S., Hansen, D.J., Berry, W.J., Swartz, R.C., Cowan, C.E., Pavlou, S.P., Allen, H.E., Thomas, N.A., and Paquin, P.R., 1991, Technical basis for establishing sediment quality criteria for nonionic organic chemicals by using equilibrium partitioning: *Environ. Toxicol. Chem.*, v. 10, no. 12, pp. 1541–1583.

Dick, M., 1982, Pesticide and PCB concentrations in Texas—water sediment, and fish tissue: Texas Department of Water Resources, Austin, Tex., Report 264, 77 p.

Dierberg, F.E., and Pfeuffer, R.J., 1983, Fate of ethion in canals draining a Florida citrus grove: *J. Agric. Food Chem.*, v. 31, no. 4, pp. 704–708.

Dileanis, P.D., Sorenson, S.K., Schwarzbach, S.E., and Maurer, T.C., 1992, Reconnaissance investigation of water quality, bottom sediment, and biota associated with irrigation drainage in the Sacramento National Wildlife Refuge Complex, California, 1988–89: U.S. Geological Survey Water-Resources Investigations Report 92-4036, 79 p.

Dimond, J.B., Getchel, A.S., and Blease, J.A., 1971, Accumulation and persistence of DDT in a lotic ecosystem: *Can. J. Fish. Aquat. Sci.*, v. 28, no. 12, pp. 1877–1882.

Dimond, J.B., Kadunce, R.E., Getchell, A.S., and John, A., 1968, Persistence of DDT in crayfish in a natural environment: *Ecology*, v. 49, no. 4, pp. 759–762.

Dimond, J.B., Owen, R.B., Jr., and Getchell, A.S., 1974, Distribution of DDTR in a uniformly treated stream: *Bull. Environ. Contam. Toxicol.*, v. 12, no. 5, pp. 522–528.

Doguchi, M., 1973, Chlorinated hydrocarbons in the environment in the Kanto Plain and Tokyo Bay, as reflected in fishes, birds, and man, *in* Coulston, F., Korte, F., and Goto, M., eds., *New methods in environmental chemistry and toxicology: Collection of papers presented at the Research Conference on New Methodology in Ecological Chemistry, Susono, Japan, November 23, 24, and 25, 1973*: International Academic Printing Company, Tokyo, Japan, pp. 269–289.

Dowd, P.F., Mayfield, G.U., Coulon, D.P., Graves, J.B., and Newsom, J.D., 1985, Organochlorine residues in animals from three Louisiana watersheds in 1978 and 1979: *Bull. Environ. Contam. Toxicol.*, v. 34, no. 6, pp. 832–841.

Dudley, D.R., and Karr, J.R., 1980, Pesticides and PVB residues in the Black Creek watershed, Allen County, Indiana, 1977–78: *J. Pest. Monitor.*, v. 13, no. 4, pp. 155–157.

Duffy, J.R., and O'Connell, D., 1968, DDT residues and metabolites in Canadian Atlantic coast fish: *Can. J. Fish. Aquat. Sci.*, v. 25, no. 1, pp. 189–195.

Duggan, R.E., Lipscomb, G.Q., Cox, E.L., Heatwole, R.E., and Kling, R.C., 1971, Pesticide residue levels in foods in the United States from July 1, 1963 to June 30, 1969: *J. Pest. Monitor.*, v. 5, no. 2, pp. 73–80, 88, 168–176.

Duke, T.W., Lowe, J.I., and Wilson, A.J., Jr., 1970, A polychlorinated biphenyl (Aroclor 1254) in the water, sediment, and biota of Escambia Bay, Florida: *Bull. Environ. Contam. Toxicol.*, v. 5, pp. 171–180.

Duke, T.W., and Wilson, A.J., Jr., 1971, Chlorinated hydrocarbons in livers of fishes from the northeastern Pacific Ocean: *J. Pest. Monitor.*, v. 5, no. 2, pp. 228–232.

Dupuy, A.J., and Couvillion, N.P., 1979, Analyses of native water, bottom material, and elutriate samples of southern Louisiana waterways, 1977–78: U.S. Geological Survey Open-File Report 79-1484, 414 p.

Durant, C.J., and Reimold, R.J., 1972, Effects of estuarine dredging of toxaphene-contaminated sediments in Terry Creek, Brunswick, Georgia, 1971: *J. Pest. Monitor.*, v. 6, no. 2, pp. 94–96.

Eadie, B.J., Morehead, N.R., and Landrum, P.F., 1990, Three-phase partitioning of hydrophobic organic compounds in Great Lakes waters: *Chemosphere*, v. 20, no. 1/2, pp. 161–178.

Eadie, B.J., Nalepa, T.F., and Landrum, P.F., 1988, Toxic contaminants and benthic organisms in the Great Lakes. Cycling, fate, and effects, *in* Schmidke, N.W., ed., *Chronic effects of toxic contaminants in large lakes*, v. 1 of *Toxic contamination in large lakes*: Lewis Publishers, Chelsea, Mich., pp. 161–177.

Eadie, B.J., and Robbins, J.A., 1987, The role of particulate matter in the movement of contaminants in the Great Lakes, *in* Hites, R.A., and Eisenreich, S.J., eds., *Sources and fates of aquatic pollutants*: American Chemical Society Advances in Chemistry series, v. 216, pp. 319–364.

Eaton, L., and Carr, K., 1991, Contaminant levels in the Sudbury River, Massachusetts: U.S. Fish and Wildlife Service, RY91-NEFO-2-EC.

Eccles, L.W., 1979, Pesticide residues in agricultural drains, southeastern desert area, California: U.S. Geological Survey Water-Resources Investigations 79-16.

Edgren, M., Olsson, M., and Renberg, L., 1979, Preliminary results on uptake and elimination at different temperatures of *p,p'*-DDT and two chlorobiphenyls in perch from brackish water: *Ambio*, v. 8, no. 6, pp. 270–272.

Edgren, M., Olsson, M., and Reutergardh, L., 1981, A one-year study of the seasonal variations of DDT and PCB levels in fish from heated and unheated areas near a nuclear power plant: *Chemosphere*, v. 10, no. 5, pp. 447–452.

Edwards, R., Millburn, P., and Hutson, D.H., 1987a, The toxicity and metabolism of the pyrethroids *cis-* and *trans*-permethrin in rainbow trout, *Salmo gairdnieri*: *Xenobiotica*, v. 17, no. 10, pp. 1175–1193.

——1987b, Factors influencing the selective toxicity of *cis-* and *trans*-permethrin in rainbow trout, frog, mouse and quail: Biotransformation in liver, plasma, brain and intestine: *Pest. Sci.*, v. 21, no. 1, pp. 1–21.

Edwards, T.K., 1992, Water-quality and flow data for the Johnson Creek Basin, Oregon, April 1988 to January 1990: U.S. Geological Survey Open-File Report 92-73, 29 p.

Edwards, T.K., and Curtiss, D.A., 1993, Preliminary evaluation of water-quality conditions of Johnson Creek, Oregon: U.S. Geological Survey Water-Resources Investigations Report 92-4136, 15 p.

Eichers, T.R., Andrilenas, P.A., Blake, H., Jenkins, R., and Fox, A., 1970, Quantities of pesticides used by farmers in 1966: U.S. Department of Agriculture, Economic Research Service, Agricultural Economic Report series, no. 179, 61 p.

Eichers, T.R., Andrilenas, P.A., Jenkins, R., and Fox, A., 1968, Quantities of pesticides used by farmers in 1964: U.S. Department of Agriculture, Economic Research Service, Agricultural Economic Report series, no. 131, 37 p.

Eichers, T.R., Andrilenas, P.A., and Anderson, T.W., 1978, Farmers' use of pesticides in 1976: U.S. Department of Agriculture, Economics, Statistics, and Cooperative Service, Agricultural Economic Report series, no. 418, 58 p.

Eidt, D.C., Hollebone, J.E., Lockhart, W.L., Kingsbury, P.D., Gadsby, M.C., and Ernst, W.R., 1984, Pesticides in forestry and agriculture: Effects on aquatic habitats.: *Adv. Environ. Sci. Technol.*, v. 14, pp. 245–284.

Eikenberry, S.E., 1978, A water-quality assessment of the Busseron Creek watershed, Sullivan, Vigo, Greene, and Clay counties, Indiana: U.S. Geological Survey Open-File Report 78-13, 36 p.

Eisenberg, M., and Topping, J.J., 1984, Organochlorine residues in shellfish from Maryland waters, 1976–1980: *J. Environ. Sci. Health*, v. B19, no. 7, pp. 673–688.

——1985, Organochlorine residues in finfish from Maryland waters, 1976–1980: *J. Environ. Sci. Health*, v. B20, no. 6, pp. 729–742.

Eisenreich, S.J., Capel, P.D., Robbins, J.A., and Bourbonniere, R., 1989, Accumulation and diagenesis of chlorinated hydrocarbons in lacustrine sediments: *Environ. Sci. Technol.*, v. 23, no. 9, pp. 1116–1126.

Eisenreich, S.J., Hollod, G.J., Johnson, T.C., and Evans, J., 1980, Polychlorinated biphenyl and other microcontaminant–sediment interactions in Lake Superior, *in* Baker, R.A., ed., Contaminants and sediments. *Fate and transport, case studies, modeling, toxicity*: Ann Arbor Science, Ann Arbor, Mich., v. 1, pp. 67–94.

Eisenreich, S.J., Looney, B.B., and Thornton, J.D., 1981, Airborne organic contaminants in the Great Lakes ecosystem: *Environ. Sci. Technol.*, v. 15, no. 1, pp. 30–38.

Eisler, R., 1985, Mirex hazards to fish, wildlife, and invertebrates: A synoptic review: U.S. Fish and Wildlife Service, Biological Report series, no. 85(1.1), Contaminant Hazard Reviews Report 1, 42 p.

——1990, Chlordane hazards to fish, wildlife, and invertebrates: A synoptic review: U.S. Fish and Wildlife Service, Biological Report series, no. 85(1.21), Contaminant Hazard Reviews Report 21, 49 p.

——1992, Fenvalerate hazards to fish, wildlife, and invertebrates: A synoptic review: U.S. Fish and Wildlife Service, Biological Report series, no. 2, Contaminant Hazard Reviews Report 24, 43 p.

Eisler, R., and Jacknow, J., 1985, Toxaphene hazards to fish, wildlife, and invertebrates: A synoptic review: U.S. Fish and Wildlife Service, Biological Report series, no. 85(1.4), Contaminant Hazard Reviews Report 4, 26 p.

Eitzer, B.D., 1993, Comparison of point and nonpoint sources of polychlorinated dibenzo-*p*-dioxins and polychlorinated dibenzofurans to sediments of the Housatonic River: *Environ. Sci. Technol.*, v. 27, no. 8, pp. 1632–1637.

Elder, J.F., and Mattraw, H.C., Jr., 1984, Accumulation of trace elements, pesticides, and polychlorinated biphenyls in sediments and the clam *Corbicula manilensis* of the Apalachicola River, Florida: *Arch. Environ. Contam. Toxicol.*, v. 13, no. 4, pp. 453–469.

Ellgehausen, H., Guth, J.A., and Esser, H.O., 1980, Factors determining the bioaccumulation potential of pesticides in the individual compartments of aquatic food chains: *Ecotoxicol. Environ. Safety*, v. 4, no. 2, pp. 134–157.

Elzerman, A.W., and Coates, J.T., 1987, Hydrophobic organic compounds on sediments: Equilibria and kinetics of sorption, *in* Hites, R.A., and Eisenreich, S.J., eds., *Sources and fates of aquatic pollutants*: American Chemical Society Advances in Chemistry series, v. 216, pp. 263–317.

Engler, R., 1992, List of chemicals evaluated for carcinogenic potential, personal communication (memorandum) to the U.S. Environmental Protection Agency Health Effects Division Branch chiefs, October 14, U.S. Environmental Protection Agency, Office of Pesticide Programs.

Environment Canada, 1995, *Interim sediment quality guidelines*: Environment Canada, Ecosystem Conservation Directorate, Evaluation and Interpretation Branch, Ottawa, Canada, 9 p., 2 appendices.

Erickson, R.J., and McKim, J.M., 1990, A simple flow-limited model for exchange of organic chemicals at fish gills: *Environ. Toxicol. Chem.*, v. 9, no. 2, pp. 159–165.

Ernst, W., 1977, Determination of the bioconcentration potential of marine organisms—A steady-state approach. I. Bioconcentration data for seven chlorinated pesticides in mussels (*Mytilus edulis*) and their relation to solubility data: *Chemosphere*, v. 6, no. 11, pp. 731–740.

Esser, H.O., 1986, A review of the correlation between physicochemical properties and bioaccumulation: *Pest. Sci.*, v. 17, no. 3, pp. 265–276.

Esser, H.O., and Moser, P., 1982, An appraisal of problems related to the measurement and evaluation of bioaccumulation: *Ecotoxicol. Environ. Safety*, v. 6, no. 2, pp. 131–148.

Esworthy, R.F., 1987, Incremental benefit analysis: Restricted use of all pesticides registered for subterranean termite control: U.S. Environmental Protection Agency, Benefits and Use Division, Economic Analysis Branch.

Evans, K.F., and Tobin, R.L., 1979, Water-quality assessment of Rattlesnake Creek watershed, Ohio: U.S. Geological Survey Water-Resources Investigations [Report] 79-17, 28 p.

Fahey, J.E., Butcher, J.W., and Turner, M.E., 1968, Monitoring the effects of the 1963–64 Japanese beetle control program on soil, water, and silt in the Battle Creek area of Michigan: *J. Pest. Monitor.*, v. 1, no. 4, pp. 30–35.

Fallon, M.E., and Horvath, F.J., 1985, Preliminary assessment of contaminants in soft sediments of the Detroit River: *J. Great Lakes Res.*, v. 11, no. 3, pp. 373–378.

Farmer, G.J., and Beamish, F.S.H., 1969, Oxygen consumption of *Tilapia nitotica* in relation to swimming speed and salinity: *Can. J. Fish. Aquat. Sci.*, v. 26, no. 11, pp. 2807–2821.

Farrington, J.W., 1989, Bioaccumulation of hydrophobic organic pollutant compounds, *in* Levin, S.A., Harwell, M.A., Kelly, J.R., and Kimball, D., eds., *Ecotoxicology: Problems and approaches*: Springer-Verlag, New York, pp. 279–313.

Farrington, J.W., Davis, A.C., Tripp, B.W., Phelps, D.K., and Galloway, W.B., 1987, "Mussel Watch"— Measurements of chemical pollutants in bivalves as one indicator of coastal environmental quality, in *New approaches to monitoring aquatic ecosystems: A symposium ASTM E-47*: American Society for Testing and Materials, p. 125–139.

Farrington, J.W., Goldberg, E.D., Risebrough, R.W., Martin, J.H., and Bowen, V.T., 1983, U.S. "Mussel Watch" 1976–1978: An overview of the trace metal, DDE, PCB, hydrocarbon, and artificial radionuclide data: *Environ. Sci. Technol.*, v. 17, no. 8, pp. 490–496.

Faust, S.D., 1977, Chemical mechanisms affecting the fate of organic pollutants in natural aquatic environments, *in* Suffet, I.H., ed., *Chemical and biological fate of pollutants in the environment*, pt. 2 of *Fate of pollutants in the air and water environments*: John Wiley, New York, Advances in Environmental Science and Technology series, v. 8, pp. 317–365.

Faye, R.E., 1980, Hydrologic reconnaissance of the Pascagoula and Escatawpa rivers, Jackson County, Mississippi—May 1974 to July 1978: U.S. Geological Survey Open-File Report 80-727, 109 p.

Fehringer, N.V., Walters, S.M., Ayers, R.J., Kozara, R.J., Ogger, J.D., and Schneider, L.F., 1985, A survey of 2,3,7,8-TCDD residues in fish from the Great Lakes and selected Michigan rivers: *Chemosphere*, v. 14, no. 6–7, pp. 909–912.

Fehringer, N.V., Walters, S.M., Kozara, R.J., and Schneider, L.F., 1985, A survey of 2,3,7,8-TCDD residues in fish from the Great Lakes and selected Michigan rivers: *J. Agric. Food Chem.*, v. 14, no. 33, pp. 626–630.

Fein, G.G.., Jacobson, J.L., Jacobson, S.W., Schwartz, P.M., and Dowler, J.K., 1984, Prenatal exposure to polychlorinated biphenyls: Effects on birth size and gestational age: *J. Pediatr.*, v. 105, no. 2, pp. 315–320.

Feng, J.C., Thompson, D.G., and Reynolds, P.E., 1990, Fate of glyphosate in a Canadian forest watershed. 1. Aquatic residues and off-target deposit assessment: *J. Agric. Food Chem.*, v. 38, no. 4, pp. 1110–1118.

Ferguson, D.E., and Bingham, C.R., 1966, Endrin resistance in the yellow bullhead: *Trans. Am. Fish. Soc.*, v. 95, no. 3, pp. 325–326.

Ferguson, D.E., Ludke, J.L., and Murphy, G.G., 1966, Dynamics of endrin uptake and release by resistant and susceptible strains of mosquitofish: *Trans. Am. Fish. Soc.*, v. 95, no. 4, pp. 335–344.

Ferraro, S.P., Lee, H., II, Ozretich, R.J., and Specht, D.T., 1990, Predicting bioaccumulation potential: A test of a fugacity-based model: *Arch. Environ. Contam. Toxicol.*, v. 19, no. 3, pp. 386–394.

Ferraro, S.P., Lee, H., II, Smith, L.M., Ozretich, R.J., and Specht, D.T., 1991, Accumulation factors for eleven polychlorinated biphenyl congeners: *Bull. Environ. Contam. Toxicol.*, v. 46, no. 2, pp. 276–283.

Fisher, N.S., Graham, L.B., Carpenter, E.J., and Wurster, C.F., 1973, Geographic differences in phytoplankton sensitivity to PCBs: *Nature*, v. 241, no. 5391, pp. 548–549.

Fletcher, C.L., and McKay, W.A., 1993, Polychlorinated dibenzo-p-dioxins (PCDD's) and dibenzofurans (PCDF's) in the aquatic environment — A literature review: *Chemosphere*, v. 26, no. 6, pp. 1041–1069.

Flynn, G.L., and Yalkowski, S.H., 1972, Correlation and prediction of mass transport across membranes. I. Influence of alkyl chain length on flux-determining properties of barrier and diffusant: *J. Pharm. Sci.*, v. 61, pp. 838–851.

Foehrenbach, J., 1972, Chlorinated pesticides in estuarine organisms: *J. Water Pollut. Control Fed.*, v. 44, no. 4, pp. 619–624.

Foley, R.E., Jackling, S.J., Sloan, R.J., and Brown, M.K., 1988, Organochlorine and mercury residues in wild mink and otter: Comparison with fish: *Environ. Toxicol. Chem.*, v. 7, no. 5, pp. 363–374.

Folmar, L.C., 1993, Effects of chemical contaminants on blood chemistry of teleost fish: A bibliography and synopsis of selected effects—Review: *Environ. Toxicol. Chem.*, v. 12, no. 2, pp. 337–375.

Food and Drug Administration, 1980, Compliance Program report of findings, FY75 Pesticide Program and FY76 Pesticide and Metals Program: Department of Health and Human Services, Food and Drug Administration, Final Report FDA/BF-82/86, 19 p., 2 attachments.

——1981, Compliance Program report of findings, FY77 Pesticides and Metals Program: Department of Health and Human Services, Food and Drug Administration, Final Report FDA/BF-82/87, 28 p.

——1988, Food and Drug Administration Pesticide Program, residues in foods, 1987: *J. Assoc. Off. Anal. Chem.*, v. 71, no. 6, pp. 156A–174A.

——1989a, Food and Drug Administration Pesticide Program, residues in foods, 1988: *J. Assoc. Off. Anal. Chem.*, v. 72, no. 5, pp. 133A–152A.

——1989b, Action levels for unavoidable pesticide residues in food or feed commodities, attachment B *of* Pesticide residues in food or feed—Enforcement criteria: Department of Health and Human Services, U.S. Food and Drug Administration Compliance Policy Guide 7141.01, variously paged.

——1990a, Food and Drug Administration Pesticide Program, residues in foods, 1989: *J. Assoc. Off. Anal. Chem.*, v. 73, no. 5, p. 127A–146A.

——1990b, Action levels for residues of certain pesticides in food or feed: *Fed. Reg.*, v. 55, no. 74, pp. 14359–14363.

——1991, Food and Drug Administration Pesticide Program, residues in foods, 1990: *J. Assoc. Off. Anal. Chem.*, v. 74, no. 5, pp. 121A–141A.

——1992, Residue monioring 1991: *J. Assoc. Off. Anal. Chem.*, v. 75, no. 5, pp. 135A–157A.

Ford, W.M., and Hill, E.P., 1991, Organochlorine pesticides in soil sediments and aquatic animals in the upper Steele Bayou watershed of Mississippi: *Arch. Environ. Contam. Toxicol.*, v. 20, no. 2, pp. 161–167.

Forlin, L., 1980, Effects of clophen A50, 3-methylcholanthrene, pregnenolone-16a-carbonitrile and phenobarbital on the hepatic microsomal cytochrome P-450 dependent monooxygenase system in rainbow trout, *Salmo gairdnieri*, of different age and sex: *Toxicol. Appl. Pharmacol.*, v. 54, no. 3, pp. 420–430.

Forlin, L., and Hansson, T., 1982, Effects of oestradil-17 and hypophysectomy on hepatic mixed function oxidases in rainbow trout: *J. Endocrin.*, v. 95, pp. 245–252.

Foster, G.D., Baksi, S.M., and Means, J.M., 1987, Bioaccumulation of trace organic contaminants from sediment by Baltic clams (*Macoma balthica*) and soft-shell clams (*Mya arenaria*): *Environ. Toxicol. Chem.*, v. 6, no. 12, pp. 969–976.

Fowler, S.W., 1990, Critical review of selected heavy metal and chlorinated hydrocarbon concentrations in the marine environment: *Mar. Environ. Res.*, v. 29, no. 1, pp. 1–64.

Fowler, S.W., Polikarpov, G.G., Elder, D.L., Parsi, P., and Villeneuve, J.-P., 1978, Polychlorinated biphenyls: Accumulation from contaminated sediments and water by the polychaete *Nereis diversicolor*: *Mar. Biol.*, v. 48, no. 4, pp. 303–309.

Fox, G.A., 1992, Epidemiological amd pathological evidence of contaminant-induced alterations in sexual development in free-living wildlife, *in* Colborn, T., and Clement, C.R., eds., *Chemically induced alterations in sexual and functional development: The wildlife/human connection*: Princeton Scientific Publishing, Princeton, N.J., Advances in Modern Environmental Toxicology series, v. 21, pp. 147–158.

Fox, G.A., Collins, B., Hayakawa, E., Weseloh, D.V., Ludwig, J.P., Kubiak, T.J., and Erdman, T.C., 1991, Reproductive outcomes in colonial fish-eating birds: A biomarker for developmental toxicants in Great Lakes food chains. II. Spatial variation in the occurrence and prevalance of bill defects in young double-crested cormorants in the Great Lakes: *J. Great Lakes Res.*, v. 17, no. 2, pp. 158–167.

Fox, M.E., Carey, J.H., and Oliver, B.G., 1983, Compartmental distribution of organochlorine contaminants in the Niagara River and the western basin of Lake Ontario: *J. Great Lakes Res.*, v. 9, no. 2, pp. 287–294.

Frank, R., 1981, Pesticides and PCB in the Grand and Saugeen river basins: *J. Great Lakes Res.*, v. 7, no. 4, pp. 440–454.

Frank, R., Armstrong, A.E., Boelens, R.G., Braun, H.E., and Douglas, C.W., 1974, Organochlorine insecticide residues in sediment and fish tissues, Ontario, Canada: *J. Pest. Monitor.*, v. 7, no. 3/4, pp. 165–180.

Frank, R., Holdrinet, M., Braun, H.E., Dodge, D.P., and Sprangler, G.E., 1978, Residues of organochlorine insecticides and polychlorinated biphenyls in fish from Lakes Huron and Superior, Canada—1968–76: *J. Pest. Monitor.*, v. 12, no. 2, pp. 60–68.

Franklin, R.B., Elcombe, C.R., Vodicinik, M.J., and Lech, J.J., 1980, Comparative aspects of the disposition and metabolism of xenobiotics in fish and mammals: *Fed. Proc. Fed. Am. Soc. Exp. Biol.*, v. 39, pp. 3144–3149.

Freed, V.H., 1984, Chemical and biological agents in forest pest management: Historical overview, *in* Garner, W.Y., and Harvey, J., Jr., eds., *Chemical and biological controls in foresty: Based on a symposium sponsored by the Division of Pesticide Chemistry at the 185th Meeting of the American Chemical Society, Seattle, Washington, March 20–25, 1993*: American Chemical Society Symposium series, v. 238, pp. 1–10.

Freiberger, H.J., and McPherson, B.F., 1972, Water quality at Miami International Airport, Miami, Florida, 1971–72: U.S. Geological Survey Open-File Report 72-023. 50 p.

Freitag, D., and Klein, W., 1982, Pesticide residues in fish with emphasis on the application of nuclear techniques, *in Agrochemicals: Fate in food and the environment: Proceedings of an International Symposium on Agrochemicals*: International Atomic Energy Agency, Vienna, Austria, pp. 143–166.

Friedman, H.I., and Nylund, B., 1980, Intestinal fat digestion, absorption and transport: *Am. J. Clin. Nutr.*, v. 33, pp. 1108–1139.

Fries, C.R., and Lee, R.F., 1984, Pollutant effects on the mixed function oxygenase (MFO) and reproductive systems of the marine polychaete *Nereis virens*: *Mar. Biol.*, v. 79, no. 2, pp. 187–193.

Fromm, P.O., and Hunter, R.C., 1969, Uptake of dieldrin by isolated perfused gills of rainbow trout: *Can. J. Fish. Aquat. Sci.*, v. 26, no. 7, pp. 1939–1942.

Fry, D.M., and Toone, C.K., 1981, DDT-induced feminization of gull embryos: *Science*, v. 213, no. 4510, pp. 922–924.

Fry, D.M., Toone, C.K., Speich, S.M., and Peard, R.J., 1987, Sex ratio skew and breeding patterns of gulls: Demographic and toxicological considerations: *Stud. Avian Biol.*, v. 10, pp. 26–43.

Fuhrer, G.J., 1984, Chemical analyses of elutriates, native water, and bottom material from the Chetco, Rogue, and Columbia rivers in western Oregon: U.S. Geological Survey Open-File Report 84-133, 57 p.

——1989, Quality of bottom material and elutriates in the lower Willamette River, Portland Harbor, Oregon: U.S. Geological Survey Water-Resources Investigations Report 89-4005, 30 p.

Fuhrer, G.J., and Evans, D., 1990, Use of elutriate tests and bottom-material analyses in simulating dredging effects on water quality of selected rivers and estuaries in Oregon and Washington: U.S. Geological Survey Water-Resources Investigations Report 89-4051, 54 p.

Fuhrer, G.J., and Horowitz, A.J., 1989, The vertical distribution of selected trace metals and organic compounds in bottom materials of the proposed lower Columbia River Export Channel, Oregon, 1984: U.S. Geological Survey Water-Resources Investigations Report 88-4099 40 p.

Fuhrer, G.J., and Rinella, F.A., 1983, Analyses of elutriates, native water, and bottom material in selected rivers and estuaries in western Oregon and Washington: U.S. Geological Survey Open-File Report 82-922, 147 p.

Fukami, J.-I., Shishido, T., Fukunaga, K., and Casida, J.E., 1969, Oxidative metabolism of rotenone in mammals, fish, and insects and its relation to selective toxicity: *J. Agric. Food Chem.*, v. 17, no. 6, pp. 1217–1226.

Funk, K., Zabik, M.E., and Smith, S.L., 1971, Dieldrin residues in sausage patties cooked by three methods: *J. Food Sci.*, v. 36, no. 4, pp. 616–618.

Funk, W.H., Hindin, E., Moore, B.C., Wasem, C.R., Larsen, C.P., McKarns, T.C., Juul, S.T., Trout, C.K., Porter, J.P., and Lafer, J.E., 1987, Water quality benchmarks in the North Cascades: State of Washington Water Research Center Report 68, 83 p.

Gakstatter, J.H., and Weiss, C.M., 1967, The elimination of DDT-^{14}C, dieldrin-^{14}C, and lindane-^{14}C following a single sublethal exposure in aquaria: *Trans. Am. Fish. Soc.*, v. 96, no. 3, pp. 301–307.

Galassi, S., Gandolfi, G., and Pacchetti, G., 1981, Chlorinated hydrocarbons in fish from the River Po (Italy): *Sci. Total Environ.*, v. 20, no. 3, pp. 231–240.

Garrison, C.R., 1982, Water quality of the Barataria Unit, Jean Lafitte National Historical Park, Louisiana (April 1981–March 1982): U.S. Geological Survey Open-File Report 82-691, 34 p.

Gaydos, M.W., 1983, Results of pesticide sampling in Tennessee: U.S. Geological Survey Open-File Report 83-148, 22 p.

George, J.L., and Frear, E.H., 1966, Pesticides in Antarctica: *J. Appl. Ecol.*, v. 3, no. 3, pp. 155–176.

Geyer, H., Scheunert, I., and Korte, F., 1986, Bioconcentration potential of organic environmental chemicals in humans: *Reg. Toxicol. Pharmacol.*, v. 6, no. 4, pp. 313–347.

Giam, C.S., Hanks, A.R., Richardson, R.L., Sackett, W.M., and Wong, M.K., 1972, DDT, DDE, and polychlorinated biphenyls in biota from the Gulf of Mexico and Caribbean Sea, 1971: *J. Pest. Monitor.*, v. 6, no. 3, pp. 139–143.

Gianessi, L.P., and Puffer, C.A., 1991, *Herbicide use in the United States*: Resources for the Future, Quality of the Environment Division, Washington, D.C., 128 p.

——1992a, *Fungicide use in U.S. crop production*: Resources for the Future, Quality of the Environment Division, Washington, D.C., variously paged.

——1992b, *Insecticide use in U.S. crop production*: Resources for the Future, Quality of the Environment Division, Washington, D.C., variously paged.

Giesy, J.P., Ludwig, J.P., and Tillitt, D.E., 1994a, Deformities in birds of the Great Lakes region. Assigning causality: *Environ. Sci. Technol.*, v. 28, no. 3, pp. 128A–135A.

Giesy, J.P., Verbrugge, D.A., Othout, R.A., Bowerman, W.W., Mora, M.A., Jones, P.D., Newsted, J.L., Vandervoort, C., Heaton, S.N., Aulerich, R.J., Bursian, S.J., Ludwig, J.P., Ludwig, M., Dawson, G.A., Kubiak, T.J., Best, D.A., and Tillitt, D.E., 1994b, Contaminants in fishes from Great Lakes-influenced sections and above dams of three Michigan rivers. I: Concentrations of organochlorine insecticides, polychlorinated biphenyls, dioxin equivalents, and mercury: *Arch. Environ. Contam. Toxicol.*, v. 27, no. 2, pp. 202–212.

——1994c, Contaminants in fishes from Great Lakes-influenced sections and above dams of three Michigan rivers. II: Implications for health of mink: *Arch. Environ. Contam. Toxicol.*, v. 27, no. 2, pp. 213–223.

Gilliom, R.J., Alexander, R.B., and Smith, R.A., 1985, Pesticides in the Nation's rivers, 1975–1980, and implications for future monitoring: U.S. Geological Survey Water-Supply Paper 2271, 26 p.

Gilliom, R.J., Alley, W.M., and Gurtz, M.E., 1995, Design of the National Water-Quality Monitoring Program: Occurrence and distribution of water-quality conditions: U.S. Geological Survey Circular 1112, 33 p.

Gilliom, R.J., and Clifton, D.G., 1990, Organochlorine pesticide residues in bed sediments of the San Joaquin River, California: *Water Resour. Bull.*, v. 26, no. 1, pp. 11–24.

Ginn, T.M., and Fisher, F.M., 1974, Studies on the distribution and flux of pesticides in waterways associated with a ricefield-marshland ecosystem: *J. Pest. Monitor.*, v. 8, no. 1, pp. 23–32.

Glickman, A.H., and Lech, J.J., 1981, Hydrolysis of permethrin, a pyrethroid insecticide, by rainbow trout and mouse tissues *in vitro*: *Toxicol. Appl. Pharmacol.*, v. 60, pp. 186–192.

Gobas, F.A.P.C., Bedard, D.C., Ciborowski, J.J., and Haffner, G.D., 1989a, Bioaccumulation of chlorinated hydrocarbons by the mayfly (*Hexagenia limbata*) in Lake St. Clair: *J. Great Lakes Res.*, v. 15, no. 4, pp. 581–588.

Gobas, F.A.P.C., Clark, K.C., Shiu, W.Y., and Mackay, D., 1989b, Bioconcentration of polybrominated benzenes and biphenyls and related superhydrophobic chemicals in fish: Role of bioavailability and elimination into the feces: *Environ. Toxicol. Chem.*, v. 8, no. 3, pp. 231–245.

Gobas, F.A.P.C., and Mackay, D., 1987, Dynamics of hydrophobic organic chemical bioconcentration in fish: *Environ. Toxicol. Chem.*, v. 6, no. 7, pp. 495–504.

Gobas, F.A.P.C., McCorquodale, J.R., and Haffner, G.D., 1993a, Intestinal absorption and biomagnification of organochlorines: *Environ. Toxicol. Chem.*, v. 12, no. 3, pp. 567–576.

Gobas, F.A.P.C., McNeil, E.J., Lovett-Doust, L., and Haffner, G.D., 1991, Bioconcentration of chlorinated aromatic hydrocarbons in aquatic macrophytes: *Environ. Sci. Technol.*, v. 25, no. 5, pp. 924–929.

Gobas, F.A.P.C., Muir, D.C.G., and Mackay, D., 1988, Dynamics of dietary bioaccumulation and fecal elimination of hydrophobic organic chemicals in fish: *Chemosphere*, v. 17, no. 5, pp. 943–962.

Gobas, F.A.P.C., Opperhuizen, A., and Hutzinger, O., 1986, Bioconcentration of hydrophobic chemicals in fish: Relationship with membrane permeation: *Environ. Toxicol. Chem.*, v. 5, no. 7, pp. 637–646.

Gobas, F.A.P.C., and Russell, R.W., 1991, Bioavailability of organochlorines in fish: *Comp. Biochem. Physiol.*, v. 100C, no. 1/2, pp. 17–20.

Gobas, F.A.P.C., Zhang, X., and Wells, R., 1993b, Gastrointestinal magnification: The mechanism of biomagnification and food chain accumulation of organic chemicals: *Environ. Sci. Technol.*, v. 27, no. 13, pp. 2855–2863.

Godsil, R.J., and Johnson, W.C., 1968, Pesticide monitoring of the aquatic biota at the Tule Lake National Wildlife Refuge: *J. Pest. Monitor.*, v. 1, no. 4, pp. 21–26.

Goerke, H., Eder, G., Wober, K., and Ernst, W., 1979, Patterns of organochlorine residues in animals of different trophic levels from the Weber Estuary: *Mar. Pollut. Bull.*, v. 10, pp. 127–133.

Goerlitz, D.F., and Law, L.M., 1974, Distribution of chlorinated hydrocarbons in stream-bottom material: *J. Res. U.S. Geol. Sur.*, v. 2, no. 5, pp. 541–543.

Gohre, K., and Miller, G.C., 1986, Singlet oxygen reactions on irradiated soil surfaces: *J. Agric. Food Chem.*, v. 34, no. 4, pp. 709–713.

Goksoyr, A., 1985, Purification of hepatic microsomal cytochromes P-450 from B-naphthoflavone-treated Atlantic cod (*Gadus morhua*), a marine teleost fish: *Biochim. Biophys. Acta.*, v. 840, pp. 409–417.

Goksoyr, A., Andersson, T., Hansson, T., Klingsoyr, J., Zhang, Y., and Forlin, L., 1987, Species characteristics of the hepatic xenobiotic and steroid biotansformation systems of two teleost fish, Atlantic cod (*Gadus morhua*) and rainbow trout (*Salmo gairdnieri*): *Toxicol. Appl. Pharmacol.*, v. 89, no. 3, pp. 347–360.

Goldbach, R.W., Van Generen, H., and Leewangh, P., 1976, Hexachlorobutadiene residues in aquatic fauna from surface water fed by the River Rhine: *Sci. Total Environ.*, v. 6, pp. 31–40.

Goodbred, S., Gilliom, R., and Gross, T., 1995, Reconnaissance of 17-estradiol and 11-ketotestosterone levels in fish of United States streams, *in Abstract book, Second SETAC World Congress: 16th annual meeting: Global environmental protection: Science, politics, and common sense: 5–9 November 1995, Vancouver Trade and Convention Centre, Vancouver, B.C., Canada*: SETAC Press, Pensacola, Fla., no. 322, p. 60.

——1996, Reconnaissance of sex steroid hormones, vitellogenin and gonad histopathology in common carp from United States streams [abs.]: Society of Environmental Toxicology and Chemistry (SETAC) Annual Meeting, 17th, Wasington, D.C., 1996, Abstract Book, no. 94, p. 18.

Goodbred, S., Gilliom, R., Gross, T., Denslow, N.P., Bryant, W.L., and Schoeb, T.R., 1997, Reconnaissance of 17β-estradiol, 11-ketotestosterone, vitellogenin, and gonad histopathology in common carp of United States streams: Potential for contaminant-induced endocrine disruption: U.S. Geological Survey Open-File Report 96-627, 47 p.

Gossett, R., Wikholm, G., Ljubenkov, J., and Steinman, D., 1989, Human serum DDT levels related to consumption of fish from the coastal waters of Los Angeles: *Environ. Toxicol. Chem.*, v. 8, no. 10, pp. 951–955.

Graczyk, D.J., 1986, Water quality in the St. Croix National Scenic Riverway, Wisconsin: U.S. Geological Survey Water-Resources Investigations Report 85-4319, 48 p.

Granstrom, M.L., Ahlert, R.C., and Wiesenfeld, J., 1984, The relationships between the pollutants in the sediments and in the water of the Delaware and Raritan Canal: *Water Sci. Technol.*, v. 16, no. 5–7, pp. 375–380.

Grant, B.F., 1976, Endrin toxicity and distribution in freshwater: A review: *Bull. Environ. Contam. Toxicol.*, v. 15, no. 3, pp. 283–290.

Grassle, J.F., Caswell, H., Farrington, J.W., Stegman, J.J., and Grassle, J.P., 1986, Contaminant levels and relative sensitivities to contamination in the deep-ocean communities: National Oceanic and Atmospheric Administration Technical Memorandum NOS OMA 26, variously paged.

Graves, J.B., Mayfield, G.U., and Newsom, J.D., 1982, Residues of 'persistent' pesticides decreasing: *Louisiana Agri.*, v. 25, no. 2, pp. 8–9.

Gray, L.E. Jr., 1992, Chemical induced alterations of sexual differentiation: A review of effects in humans and rodents, *in* Colborn, T., and Clement, C.R., eds., *Chemically induced alterations in sexual and functional development: The wildlife/human connection*: Princeton Scientific Publishing, Princeton, N.J., Advances in Modern Environmental Toxicology series, v. 21, pp. 203–230.

Gray, L.E. Jr., Ostby, J.S., and Kelce, W.R., 1994, Developmental effects of an environmental androgen: The fungicide vinclozolin alters sex differentiation of the male rat: *Toxicol. Appl. Pharmacol.*, v. 129, no. 1, pp. 46–52.

Green, R.E., Goswami, K.P., Mukhtar, M., and Young, H.Y., 1977, Herbicides from cropped watersheds in stream and estuarine sediments in Hawaii: *J. Environ. Qual.*, v. 6, no. 2, pp. 145–154.

Greene, E.A., Sowards, C.L., and Hansmann, E.W., 1990, Reconnaissance investigation of water quality, bottom sediment, and biota associated with irrigation drainage in the Angostura Reclamation Unit, southwestern South Dakota, 1988–89: U.S. Geological Survey Water-Resources Investigations Report 90-4152, 75 p.

Gregor, D.J., 1990, Deposition and accumulation of selected agricultural pesticides in Canadian arctic snow, *in* Kurtz, D.A., ed., *Long range transport of pesticides*: Lewis Publishers, Chelsea, Mich., pp. 373–386.

Greve, P.A., and Wit, S.L., 1971, Endosulfan in the Rhine River: *J. Water Poll. Control Fed.*, v. 43, no. 12, pp. 2338–2348.

Griffiths, R.W., 1991, Environmental quality assessment of the St. Clair River as reflected by the distribution of benthic macroinvertebrates in 1985: *Hydrobiologia*, v. 219, pp. 143–164.

Gross, D.A., Gross, T.S., Johnson, B., and Folmar, L., 1995, Characterization of endocrine disruption and clinical manifestations in large-mouth bass from Florida lakes [abs.]: 16th annual meeting of the Society of Environmental Toxicology and Chemistry, Vancouver, B.C., Canada, November 5–9, 1995, Abstract, p. 185.

Gross, T.S., and Guillette, L.J., Jr., 1995, Pesticide induction of developmental abnormalities of the reproductive system of alligators (*Alligator missisiippiensis*) and turtles (*Trachemys scripta*) [abs.], *in* McLachlan, J.A. and Korach, K.S., eds., *Estrogens in the Environment, III: Global helath implications: January 9–10, 1994, Washington, DC*: National Institutes of Health, National Institute of Environmental Health Sciences, [Bethesda, Md.].

Gruger, E.H., Jr., Karrick, N.L., Davidson, A.I., and Hruby, T., 1975, Accumulation of 3,4,3′,4′-tetrachlorobiphenyl and 2,4,5,2′,4′,5′-and 2,4,6,2′,4′,6′-hexachlorobiphenyl in juvenile coho salmon: *Environ. Sci. Technol.*, v. 9, no. 2, pp. 121–127.

Grzenda, A.R., Paris, D.F., and Taylor, W.J., 1970, The uptake, metabolism, and elimination of chlorinated residues by goldfish (*Carassius auratus*) fed a ^{14}C-DDT contaminated diet: *Trans. Amer. Fish. Soc.*, v. 99, no. 2, pp. 385–396.

Grzenda, A.R., Taylor, W.J., and Paris, D.F., 1971, The uptake and distribution of chlorinated residues by goldfish (*Carassius auratus*) fed a ^{14}C-dieldrin contaminated diet: *Trans. Am. Fish. Soc.*, v. 100, no. 2, pp. 215–221.

Guarino, A.M., and Lech, J.J., 1986, Metabolism, disposition, and toxicity of drugs and other xenobiotics in aquatic species: *Vet. Hum. Toxicol.*, v. 28, supp. 1, pp. 38–44.

Guenzi, W.D., and Beard, W.E., 1967, Anaerobic biodegradation of DDT to DDD in soil: *Science*, v. 156, p. 1116.

Guillette, L.J., Jr., Gross, T.S., Masson, G.R., Matter, J.M., Percival, H.F., and Woodward, A.R., 1994, Developmental abnormalities of the gonad and abnormal sex hormone concentrations in juvenile alligators from contaminated and control lakes in Florida: *Environ. Health Perspect.*, v. 102, no. 8, pp. 680–688.

Halfon, E., 1987, Modeling of mirex loadings to the bottom sediments of Lake Ontario within the Niagara River plume: *J. Great Lakes Res.*, v. 13, no. 1, pp. 18–23.

Hall, R.J., Kaiser, T.E., Robertson, W.B., and Patty, P.C., 1979, Organochlorine residues in eggs of the endangered American crocodile (*Crocodylus acutus*): *Bull. Environ. Contam. Toxicol.*, v. 23, no. 1–2, pp. 87–90.

Hall, W.S., Dickson, K.L., Saleh, F.Y., and Rodgers, J.H., 1986, Effects of suspended solids on the bioavailability of chlordane to *Daphnia magna*: *Arch. Environ. Contam. Toxicol.*, v. 15, no. 5, pp. 529–534.

Hallett, D.J., 1985, Dioxin in the Great Lakes ecosystem. A Canadian perspective: *Chemosphere*, v. 14, no. 6–7, pp. 745–753.

Hallett, D.J., and Brooksbank, M.G., 1986, Trends of TCDD and related compounds in the Great Lakes: The Lake Ontario ecosystem: *Chemosphere*, v. 15, no. 9–12, pp. 1405–1416.

Hamdy, Y.S., and Post, L., 1985, Distribution of mercury, trace organics, and other heavy metals in Detroit River sediments: *J. Great Lakes Res.*, v. 11, no. 3, pp. 353–365.

Hamelink, J.L., Landrum, P.F., Bergman, H.L., and Benson, W.H., 1994, *Bioavailability: Physical, chemical, and biological interactions*: Lewis Publishers, Boca Raton, Fla., 239 p.

Hamelink, J.L., and Spacie, A., 1977, Fish and chemicals: The process of bioaccumulation: *Ann. Rev. Pharmacol. Toxicol.*, v. 17, pp. 167–177.

Hamelink, J.L., and Waybrant, R.C., 1976, DDE and lindane in a large-scale model lentic ecosystem: *Trans. Am. Fish. Soc.*, v. 105, no. 1, pp. 124–134.

Hamelink, J.L., Waybrant, R.C., and Ball, R.C., 1971, A proposal: Exchange equilibria control the degree chlorinated hydrocarbons are biologically magnified in lentic environments: *Trans. Am. Fish. Soc.*, v. 100, no. 2, pp. 207–214.

Hamelink, J.L., Waybrant, R.C., and Yant, P.R., 1977, Mechanisms of bioaccumulation of mercury and chlorinated hydrocarbon pesticides by fish in lentic ecosystems, *in* Suffet, I.H., ed., *Chemical and biological fate of pollutants in the environment*, pt. 2 *of Fate of pollutants in the air and water environments*: John Wiley, New York, Advances in Environmental Science and Techology series, v. 8, pp. 261–282.

Hanninen, O., Koivusaari, U., and Lindstrom-Seppa, P., 1984, Biotransformation, especially glucuronidation and their control by temperature in fish, *in* Matern, S., Bock, K.W., and Gerok, W., eds., *Advances in glucuronide conjugation: Proceedings of the 40th Falk Symposium held at Titisee, West Germany, May 31st to June 2nd, 1994*: MTP press, Lancaster, pp. 101–107.

Hannon, M.R., Greichus, Y.A., Applegate, R.L., and Fox, A.C., 1970, Ecological distribution of pesticides in Lake Poinsett, South Dakota: *Trans. Am. Fish. Soc.*, v. 99, no. 3, pp. 496–500.

Hansch, C., 1969, A quantitative approach to biochemical structure-activity relationships: *Acc. Chem. Res.*, v. 2, pp. 232–240.

Hansch, C., and Clayton, J.M., 1973, Lipophilic character and biological activity of drugs: *J. Pharm. Sci.*, v. 62, pp. 1–21.

Hansen, D.J., and Wilson, A.J., 1970, Significance of DDT residues from the estuary near Pensacola, Florida: *J. Pest. Monitor.*, v. 4, no. 2, pp. 51–56.

Hansen, P.-D., 1980, Uptake and transfer of the chlorinated hydrocarbon lindane (γ-BHC) in laboratory freshwater food chain: *Environ. Pollut.*, v. A21, no. 2, pp. 97–108.

Hanson, P.J., Evans, D.W., Fortner, A.R., and Siewicki, T.C., 1989, Contaminant assessment for the southeast Atlantic and Gulf of Mexico coasts: 1984 (cycle I) and 1985 (cycle 2), Results of the National Benthic Surveillance Project: U.S. Department of Commerce, National Oceanic and Atmospheric Administration, National Marine Fisheries Service, Southeast Fisheries Science Center, Beaufort Laboratory, North Carolina, variously paged.

Harding, G.C.H., and Vass, W.P., 1979, Uptake from seawater of DDT by marine planktonic crustacea: *Can. J. Fish. Aquat. Sci.*, v. 36, no. 3, pp. 247–254.

Harding, L.W., Jr., and Phillips, J.H., Jr., 1978, Polychlorinated biphenyl (PCB) uptake by marine phytoplankton: *Mar. Biol.*, v. 49, no 2, pp. 103–111.

Hardy, M.A., 1978, Water-quality assessment of the Muddy Fork Silver Creek watershed, Clark, Floyd, and Washinton counties, Indiana: U.S. Geological Survey Open-File Report 78-202, 41 p.

——1984, Chemical and biological quality of streams at the Indiana Dunes National Lakeshore, Indiana, 1978–80: U.S. Geological Survey Water-Resources Investigations [Report] 83-4208 95 p.

Hargrave, B.T., Vass, W.P., Erickson, P.E., and Fowler, B.R., 1988, Atmospheric transport of organochlorines to the Arctic Ocean: *Tellus*, v. B40, pp. 480–493.

Harless, R.L., Oswald, E.O., Lewis, R.G., Dupuy, A.E., Jr., McDaniel, D.D., and Tai, H., 1982, Determination of 2,3,7,8-tetrachlorodibenzo-*p*-dioxin in fresh water fish: *Chemosphere*, v. 11, no. 2, pp. 193–198.

Harmon Engineering and Testing, 1984, Water quality management study of the Walter F. George and George W. Andrews Reservoirs, Chattahoochee River, April 1978 through December 1979: U.S. Department of the Army, Corps of Engineers, Mobile, Alabama, HE&T Project 362-03, v. 1, variously paged.

Harris, C.R., and Miles, J.R.W., 1975, Pesticide residues in the Great Lakes region of Canada: *Residue Rev.*, v. 57, pp. 27–79.

Harrison, H.L., Loucks, O.L., Mitchell, J.W., Parkhurst, D.F., Tracey, C.R., Watts, D.G., and Yannaconne, V.J., Jr., 1970, Systems studies of DDT transport: *Science*, v. 170, no. 3957, pp. 503–508.

Harrison, S.A., Watschke, T.L., Mumma, R.O., Jarrett, A.R., and Hamilton, G.W., Jr., 1993, Nutrient and pesticide concentrations in water from chemically treated turfgrass, *in* Racke, K.D., and Leslie, A.R., eds., *Pesticides in urban environments: Fate and significance*: American Chemical Society Symposium Series, v. 522, pp. 191–207.

Hartung, R., and Klingler, G.W., 1970, Concentration of DDT by sedimented polluting oils: *Environ. Sci. Technol.*, v. 4, no. 5, pp. 407–410.

Harvey, G.R., Miklas, H.P., Bowen, V.T., and Steinhauer, W.G., 1974, Observations on the distribution of chlorinated hydrocarbons in Atlantic Ocean organisms: *J. Mar. Res.*, v. 32, pp. 103–118.

Hattula, M.L., Janatuinen, J., Sarkka, J., and Paasivirta, F., 1978, A five-year monitoring study of the chlorinated hydrocarbons in the fish of a Finnish lake ecosystem: *Environ. Pollut.*, v. 15, no, 2, pp. 121–139.

Hawker, D.W., and Connell, D.W., 1985, Relationships between partition coefficient, uptake rate constant, clearance rate constant and time to equilibrium for bioaccumulation: *Chemosphere*, v. 14, no. 9, pp. 1205–1219.

——1986, Bioconcentration of lipophilic compounds by some aquatic organisms: *Ecotoxicol. Environ. Safety*, v. 11, no. 2, pp. 184–197.

Hayton, W.L., and Barron, M.G., 1990, Rate-limiting barriers to xenobiotic uptake by the gill: *Environ. Toxicol. Chem.*, v. 9, no. 2, pp. 151–157.

Hebert, C.E., and Keensleyside, K.A., 1995, To normalize or not to normalize? Fat is the question: *Environ. Toxicol. Chem.*, v. 14, no. 5, p. 801–807.

Hedlund, R.T., and Youngson, C.R., 1972, The rates of photodecomposition of picloram in aqueous solution, *in* Faust, S.D., ed., *Fate of organic pesticides in aquatic environments: A symposium sponsored by the Division of Pesticide Chemistry at the 161st meeting of the American Chemical Society, Los Angeles, California, March 29–31, 1971*: American Chemical Society Advances in Chemistry series, v. 111, pp. 159–172.

Hektoen, H., Ingebribsten, K., Brevik, E.M., and Oehme, M., 1992, Interspecies differences in tissue distribution of 2,3,7,8-tetrachlorodibenzo-*p*-dioxin between cod (*Gadus morhua*) and rainbow trout (*Oncorhynchus mykiss*): *Chemosphere*, v. 24, no. 5, pp. 581–587.

Hellawell, J.M., 1988, Toxic substances in rivers and streams: *Environ. Pollut.*, v. 50, no. 1–2, pp. 61–85.

Henderson, C., Inglis, A., and Johnson, W.L., 1971, Organochlorine insecticide residues in fish—Fall, 1969. National Pesticide Monitoring Program: *J. Pest. Monitor.*, v. 5, no. 1, pp. 1–11.

Henderson, C., Johnson, W.L., and Inglis, A., 1969, Organochlorine insecticide residues in fish (National Pesticide Monitoring Program): *J. Pest. Monitor.*, v. 3, no. 3, pp. 145–171.

Henderson, R.J., and Tocher, D.R., 1987, The lipid composition and biochemistry of freshwater fish: *Prog. Lipid Res.*, v. 26, pp. 281–347.

Hickey, J.T., Bennett, R.O., Reimschuessel, R., and Merckel, C., 1990, Biological indicators of environmental contaminants in the Niagara River: Histological evaluations of the tissues from brown bullheads at the Love Canal-102nd Street dump site compared to the Black Creek, reference site: U.S. Fish and Wildlife Service, New York Field Office, Cortland, N.Y., 37 p.

Hidaka, H., Tanabe, S., and Tatsukawa, R., 1983, DDT compounds and PCB isomers and congeners in Weddell seals and their fate in the Antartctic marine ecosystem: *Agric. Biol. Chem.*, v. 47, no. 9, pp. 2009–2017.

Hileman, B., 1993, Concerns broaden over chlorine and chlorinated hydrocarbons: *Chem. Eng. News*, v.71, no. 16, pp. 11–20.

———1994, Environmental estrogens linked to reproductive abnormalities, cancer: *Chem. Eng. News*, v. 72, no. 5, pp. 19–23.

Hirsch, R.M., Alley, W.M., and Wilber, W.G., 1988, Concepts for a national water-quality assessment program: U.S. Geological Survey Circular 1021, 42 p.

Hitchcock, M., and Murphy, S.D., 1967, Enzymatic reduction of *O,O*-(4-nitrophenyl) phosphorothioate, *O,O*-diethyl *O*-(4-nitrophenyl) phosphate, and *O*-ethyl *O*-(4-nitrophenyl) benzene thiophosphonate by tissues from mammals, birds, and fishes: *Biochem. Pharmacol.*, v. 16, pp. 1801–1811.

Hitt, K.J., 1994, Refining 1970's land-use data with 1990 population data to indicate new residential development: U.S. Geological Survey Water-Resources Investigations Report 94-4250, 15 p.

Hochreiter, J.J., Jr., 1982, Chemical-quality reconnaissance of the water and surficial bed material in the Delaware River estuary and adjacent New Jersey tributaries, 1980-81: U.S. Geological Survey Water-Resources Investigations Report 82-36, 41 p.

Hodge, J.E., 1993, Pesticide trends in the professional and consumer markets, *in* Racke, K.D., and Leslie, A.R., eds., *Pesticides in urban environments: Fate and significance*: American Chemical Society Symposium series, v. 522, pp. 11–17.

Hoffman, R.J., Hallock, R.J., Rowe, T.G., Lico, M.S., Burge, H.L., and Thompson, S.P., 1990, Reconnaissance investigation of water quality, bottom sediment, and biota associated with irritgation drainage in and near Stillwater Wildlife Management Area, Churchill County, Nevada, 1986–1987: U.S. Geological Survey Water-Resources Investigations Report 89-4105, 150 p.

Hoigne, J., Faust, B.C., Haag, W.R., Scully, F.E., Jr., and Zepp, R.G., 1989, Aquatic humic substances as sources and sinks of photochemically produced transient reactants, *in* Suffet, I.H., and MacCarthy, P., eds., *Aquatic humic substances: Influence on fate and treatment of pollutants*: American Chemical Society Advances in Chemistry series, v. 219, pp. 363–381.

Holden, A.V., 1962, A study of the absorption of ^{14}C-labeled DDT from water by fish: *Ann. Appl. Biol.*, v. 50, pp. 467–477.

——1970, International cooperative study of organochlorine pesticide residues in terrestrial and aquatic wildlife, 1967/1968: *J. Pest. Monitor.*, v. 4, no. 3, pp. 117–135.

——1973, International cooperative study of organochlorine and mercury residues in wildlife, 1969–71: *J. Pest. Monitor.*, v. 7, no. 1, pp. 37–52.

Hopkins, B.S., Clark, D.K., Schlender, M., and Stinson, M., 1985, Basic water monitoring program: Fish tissue and sediment sampling for 1984: Washington State Department of Ecology Report 85–7, 43 p.

Hora, M.E., 1981, Reduction of polychlorinated biphenyl (PCB) concentrations in carp (*Cyprinus carpio*) fillets through skin removal: *Bull. Environ. Contam. Toxicol.*, v. 26, no. 3, pp. 364–366.

House, L.B., 1990, Data on polychlorinated biphenyls, dieldrin, lead, and cadmium in Wisconsin and upper Michigan tributaries to Green Bay, July 1987 through April 1988: U.S. Geological Survey Open-File Report 89-52, 7 p.

Howard, P.H., ed., 1991, *Pesticides*, v. 3 of *Handbook of environmental fate and exposure data for organic chemicals*: Lewis Publishers, Chelsea, Mich., 684 p.

Howard, P.H., Boethling, R.S., Jarvis, W.F., Meylan, W.M., and Michalenko, E.M., 1991, *Handbook of environmental degradation rates*: Lewis Publishers, Chelsea, Mich., 725 p.

Howard, P.H., and Neal, M., 1992, *Dictionary of chemical names and synonyms*: Lewis Publishers, Chelsea, Mich., variously paged.

Hubert, W.H., and Ricci, E.D., 1981, Factors influencing dieldrin and DDT residues in carp from the Des Moines River, Iowa, 1977–80: *J. Pest. Monitor.*, v. 15, no. 3, pp. 111–116.

Huckins, J.N., Schwartz, T.R., Petty, J.D., and Smith, L.M., 1988, Determination, fate, and potential significance of PCBs in fish and sediment samples with emphasis on selected AHH-inducing congeners: *Chemosphere*, v. 17, no. 10, pp. 1995–2016.

Huckle, K.R., and Millburn, P., 1990, Metabolism, bioconcentration, and toxicity of pesticides in fish, *in* Hutson, D.H., and Roberts, T.R., eds., *Environmental fate of pesticides*: Wiley, Chichester, England, Progress in Pesticide Biochemistry and Toxicology series, v. 7, pp. 175–243.

Huggett, R.J., 1989, Kepone and the James River, *in* National Research Council, *Contaminated marine sediments: Assessment and remediation*: National Academy Press, Washington, D.C., pp. 417–424.

Huggett, R.J., and Bender, M.E., 1980, Kepone in the James River: *Environ. Sci. Technol.*, v. 14, no. 8, pp. 918–923.

Hughes, C., 1988, Phytochemical mimicry of reproductive hormones and modulation of herbivore fertility by phytoestrogens: *Environ. Health Perspect.*, v. 78, pp. 171–175.

Humphrey, H.E.B., 1976, Evaluation of changes of the level of polychlorinated biphenyls (PCB) in human tissues: Final report prepared for the Food and Drug Administration by the Michigan Department of Public Health, 86 p.

——1983a, Evaluation of humans exposed to water borne chemicals in the Great Lakes: Final report prepared for the U.S. Environmental Protection Agency by the Michigan Department of Public Health, 198 p., appendices.

——1983b, Population studies of PCBs in Michigan residents, *in* D'Itri, F.M., and Kamrin, M.A., eds., *PCBs, human and environmental hazards*: Butterworth Publishers, Boston, Mass., pp. 299–310.

——1987, The human population—An ultimate receptor for aquatic contaminants: *Hydrobiologia*, v. 149, pp. 75–80.

Hunt, E.G., 1966, Biological magnification of pesticides, in *Scientific aspects of pest control*: National Academy of Sciences—National Research Council, Washington, D.C., National Research Council Publication series, no. 1402, pp. 251–262.

Hunt, E.G., and Bischoff, A., 1960, Inimical effects on wildlife of periodic DDD applications to Clear Lake: *California Fish and Game*, v. 46, pp. 91–106.

Hunt, E.G., and Linn, J.D., 1970, Fish kills by pesticides, *in* Gillett, J.W., ed., *The biological impact of pesticides in the environment*: Oregon State University Press, Corvallis, Oreg., pp. 97–103.

Hurlbert, S.H., 1975, Secondary effects of pesticides on aquatic ecosystems: *Residue Rev.*, v. 57, pp. 81–148.

Hutson, D.H., and Roberts, T.R., 1985, Insecticides, *in* Hutson, D.H., and Roberts, T.R., eds., *Insecticides*: Wiley, Chichester, England, Progress in Pesticide Biochemistry and Toxicology series, v. 5, pp. 1–34.

Illinois Environmental Protection Agency, 1988a, An intensive survey of the Des Plaines River Basin from the Wisconsin state line to Joliet, Illinois, 1983–1984: Illinois Environmental Protection Agency, Division of Water Pollution Control Staff Report IEPA/WPC 88-014, 95 p.

——1988b, An intensive survey of the Elkhorn Creek Basin, 1985: Staff report: Illinois Environmental Protection Agency, Division of Water Pollution Control Staff Report IEPA/WPC 88-017, 30 p.

——1988c, An intensive survey of the Kyte River Basin, 1984: Illinois Environmental Protection Agency, Division of Water Pollution Control Staff Report IEPA/WPC/88-013, 32 p.

——1988d, An intensive survey of the Pecatonica River Basin, 1984–1985: Illinois Environmental Protection Agency, Division of Water Pollution Control Staff Report IEPA/WPC/88-012, 56 p.

Imanaka, M., Hino, S., Matsunaga, K., and Ishida, T., 1985, Oxadiazon residues in surface water and crucian carps (*Carassius cuvieri*) of Lake Kojima: *J. Pest. Sci.*, v. 10, no. 1, pp. 125–134.

Imanaka, M., Matsunaga, K., Shigeta, A., and Ishida, T., 1981, Oxadiazon residues in fish and shellfish: *J. Pest. Sci.*, v. 6, no. 4, pp. 413–417.

Immerman, F.W., and Drummond, D.J., 1984, National Urban Pesticide Applicator Survey: Final report—overview and results: U.S. Environmental Protection Agency, Office of Pesticides Program, Economic Analysis Branch, variously paged.

Insalaco, S.E., Makarewicz, J.C., and McNamara, J.N., 1982, The influence of sex, size, and season on mirex levels within selected tissues of Lake Ontario salmon: *J. Great Lakes Res.*, v. 8, no. 4, pp. 660–665.

Irwin, G.A., and Lemons, M., 1974, Reconnaissance study of selected nutrients, pesticides, and trace elements in the Eel, Salinas, and Santa Ana rivers, California, October 1971 through July 1972: U.S. Geological Survey Water-Resources Investigation Report 73-16, 55 p.

Irwin, R.J., 1988, Impacts of toxic chemicals on Trinity River fish and wildlife: U.S. Fish and Wildlife Service, Contaminants Report, Fort Worth Field Office, Fort Worth, Tex., 82 p.

Isaza, J., and Dreisig, J., 1989, Metals and organics survey of fish from the Connecticut River in New Hampshire: U.S. Fish and Wildlife Service and New Hampshire Division of Public Health Services Joint Report PHS/FWS/89-2.

Iwatsubo, R.T., Nolan, K.M., Harden, D.R., and Glysson, G.D., 1976, Redwood National Park studies, data release number 2, Redwood Creek, Humboldt County, and Mill Creek, Del Norte County, California, April 11, 1974–September 30, 1975: U.S. Geological Survey Open-File Report 76-678, 247 p.

Jackson, M.D., Sheets, T.J., and Moffett, C.L., 1974, Persistence and movement of BHC in a watershed, Mount Mitchell State Park, North Carolina, 1967–72: *J. Pest. Monitor.*, v. 8, no. 3, pp. 202–208.

Jackson, R., 1990, Alaska Maritime National Wildlife Refuge, Womens Bay Contaminant Study: U.S. Fish and Wildlife Service Report of Findings, Anchorage, Alaska, 17 p., 7 appendices.

——1991, Becharof National Wildlife Refuge Contaminants Study: U.S. Fish and Wildlife Service Report of Findings, Anchorage, Alaska, 12 p., 6 appendices.

Jacobson, J.L., Fein, G., G,, Jacobson, S.W., Schwartz, P.M., and Dowler, J.K., 1984, The transfer of polychlorinated biphenyls (PCBs) and polybrominated biphenyls (PBBs) across the human placenta and into maternal milk: *Am. J. Public Health*, v. 74, no. 4, pp. 378–379.

Jacobson, J.L., Humphrey, H.E.B., Jacobson, S.W., Schantz, S.L., Mullin, M.D., and Welch, R.W., 1989, Determinants of polychlorinated biphenyls (PCBs), polybrominated biphenyls (PBBs), and dichlorodiphenyl trichloroethane (DDT) levels in the sera of young children: *Am. J. Public Health*, v. 79, no. 10, pp. 1401–1404.

Jacobson, J.L., Jacobson, S.W., and Humphrey, H.E.B., 1990a, Effects of in utero exposure to polychlorinated biphenyls and related contaminants on cognitive functioning in young children: *J. Pediatr.*, v. 116, no. 1, pp. 38–45.

——1990b, Effects of exposure to PCBs and related compounds on growth and activity in children: *Neurotoxicol. Teratol.*, v. 12, no. 4, pp. 319–326.

Jacobson, S.W., Fein, G.G., Jacobson, J.L., Schwartz, P.M., and Dowler, J.K., 1985, The effect of intrauterine PCB exposure on visual recognition memory: *Child Dev.*, v. 56, pp. 853–860.

Jaffe, R., and Hites, R.A., 1986, Anthropogenic, polyhalogenated, organic compounds in non-migratory fish from the Niagara River area and tributaries to Lake Ontario: *J. Great Lakes Res.*, v. 12, no. 1, pp. 63–71.

Jaffe, R., Stemmler, E.A., Eitzer, B.D., and Hites, R.A., 1985, Anthropogenic, polyhalogenated, organic compounds in sedentary fish from Lake Huron and Lake Superior tributaries and embayments: *J. Great Lakes Res.*, v. 11, no. 2, pp. 156–162.

Jager, G., 1983, Herbicides, *in* Buchel, K.H., ed., *Chemistry of pesticides*: Wiley, New York, pp. 322–392.

James, M.O., 1978, Taurine conjugation of carboxylic acids in some marine species, *in* Aitio, A., ed., *Conjugation reactions in drug biotransformation: Proceedings of the Symposium on Conjugation Reactions in Drug Biotransformation held in Turku, Finland, July 23–26, 1978*: Elsevier/North-Holland Biomedical Press, Amsterdam, The Netherlands, pp. 121–129.

——1987, Conjugation of organic pollutants in aquatic species: *Environ. Health Perspect.*, v. 71, pp. 97–103.

James, M.O., Fouts, J.R., and Bend, J.R., 1977, Xenobiotic metabolizing enzymes in marine fish, *in* Khan, M.A.Q., ed., *Pesticides in aquatic environments*: Plenum Press, New York, pp. 171–189.

Jarvinen, A.W., Hoffman, M.J., and Thorslund, T.W., 1977, Long-term toxic effect of DDT food and water exposure on fathead minnows (*Pimephales promelas*): *Can. J. Fish. Aquat. Sci.*, v. 34, no, 11, pp. 2089–2103.

Jensen, A.A., and Slorach, S.A., 1991, *Chemical contaminants in human milk*: CRC Press, Boca Raton, Fla., 298 p.

Jensen, A.L., Spigarelli, S.A., and Thommes, M.M., 1982, PCB uptake by species of fish in Lake Michigan, Green Bay of Lake Michigan, and Cayuga, New York: *Can. J. Fish. Aquat. Sci.*, v. 39, no. 5, pp. 700–709.

Jensen, S., Gothe, R., and Kindstedt, M.-O., 1972, Bis-(*p*-chlorophenyl)-acetonitrile (DDN), a new DDT derivative formed in anaerobic digested sludge and bed sediment: *Nature*, v. 240, no. 5381, pp. 421–422.

Jensen, S., Johnels, A.G., Odsjo, T., Olsson, M., and Otterlind, G., 1970, PCB—Occurrence in Swedish wildlife, in *Proceedings of the PCB Conference, Wenner-Gren Centre, Stockholm:* Research Secretariat, Statens Naturvardsverk, Solna, Sweden.

Jensen, S., Johnels, A.G., Olsson, M., and Otterlind, G., 1969, DDT and PCB in marine animals from Swedish waters: *Nature*, v. 224, no. 5216, pp. 247–250.

Johnson, A., Norton, D.E., and Yake, W.E., 1986, Occurrence and significance of DDT compounds and other contaminants in fish, water, and sediment from the Yakima River Basin: Washington (State) Department of Ecology, Water-Quality Investigations Section Report 86-5, 89 p.

——1988b, Persistence of DDT in the Yakima River drainage, Washington: *Arch. Environ. Contam. Toxicol.*, v. 17, no. 3, pp. 289–297.

Johnson, D.W., 1968, Pesticides and fishes: A review of selected literature: *Trans. Am. Fish. Soc.*, v. 97, no. 4, pp. 398–424.

——1973, Pesticide residues in fish, in Edwards, C.A., ed., *Environmental pollution by pesticides*: Plenum Press, London, England, Environmental Science Research series, v. 3, pp. 181–212.

Johnson, D.W., Kent, J.C., and Campbell, D.K., 1977, *Availability and concentration of pollutants from American Falls Reservoir sediments to forage and predaceous fishes: Technical completion report*: Moscow, Idaho, University of Idaho, Idaho Water Resources Research Institute, 95 p.

Johnson, D.W., and Lew, S., 1970, Chlorinated hydrocarbon pesticides in representative fishes of southern Arizona: *J. Pest. Monitor.*, v. 4, no. 2, pp. 57–61.

Johnson, H.E., 1978, Toxic organic residues in fish, in Mount, D.I., ed., Proceedings of the first and second USA-USSR symposia on the effects of pollutants upon aquatic ecosystems: U. S. Environmental Protection Agency, Office of Research and Development, EPA-600/3-78-076, v. 1, pp. 115–120.

Johnson, H.E., and Ball, R.C., 1972, Organic pesticide pollution in an aquatic environment, in Faust, S.D., ed., *Fate of organic pesticides in the aquatic environment*: American Chemical Society Advances in Chemistry series, no. 111, pp. 1–10.

Johnson, J.L., Stalling, D.L., and Hogan, J.W., 1974, Hexachlorobenzene (HCB) residues in fish: *Bull. Environ. Contam. Toxicol.*, v. 11, no. 5, pp. 393–398.

Johnson, L.G., and Morris, R.L., 1974, Chlorinated insecticide residues in the eggs of some freshwater fish: *Bull. Environ. Contam. Toxicol.*, v. 11, no. 6, pp. 503–510.

Johnson, L.L., Casillas, E., Collier, T.K., McCain, B.B., and Varanasi, U., 1988a, Contaminant effects on ovarian development in English sole (*Parophrys vetulus*) from Puget Sound, Washington: *Can. J. Fish. Aquat. Sci.*, v. 45, no. 12, pp. 2133–2146.

Johnson, L.L., Stehr, C.M., Olson, O.P., Myers, M.S., Pierce, S.M., McCain, B.B., and Varanasi, U., 1992a, National Benthic Surveillance Project: Northeast coast: Fish histopathology and relationships between lesions and chemical contaminants (1987–89): U.S. Department of Commerce, National Oceanic and Atmospheric Administration, National Marine Fisheries Service, Northwest Fisheries Science Center, NOAA Technical Memorandum NMFS-NWFSC series, no. 4, 96 p.

Johnson, L.L., Stehr, C.M., Olson, O.P., Myers, M.S., Pierce, S.M., Wigren, C.A., McCain, B.B., and Varanasi, U., 1993, Chemical contaminants and hepatic lesions in winter flounder (*Pleuronectes americanus*) from the northeast coast of the United States: *Environ. Sci. Technol.*, v. 27, no. 13, pp. 2759–2771.

Johnson, L.L., Stein, J.E., Collier, T.K., Casillas, E., and Varanasi, U., 1992b, Bioindicators of contaminant exposure, liver pathology, and reproductive development in prespawning female winter flounder (*Pleuronectes americanus*) from urban and nonurban estuaries on the northeast Atlantic coast: U.S. Department of Commerce, National Oceanic and Atmospheric Administration, National Marine Fisheries Service, Northwest Fisheries Science Center, NOAA Technical Memorandum NMFS-NWFSC series, no. 1, 76 p.

——1994, Indicators of reproductive development in prespawning female winter flounder (*Pleuronectes americanus*) from urban and non-urban estuaries in the northeast United States: *Sci. Total Environ.*, v. 141, no. 1–3, pp. 241–260.

Johnston, J.J., and Corbett, M.D., 1986, The uptake and in vivo metabolism of the organophosphate insecticide fenitrothion by the blue crab, *Callinectes sapidus*: *Toxicol. Appl. Pharmacol.*, v. 85, pp. 181–188.

Jones, P.D., Giesy, J.P., Newsted, J.L., Verbrugge, D.A., Beaver, D.L., Ankley, G.T., Tillitt, D.E., Lodge, K.B., and Niemi, G.J., 1993, 2,3,7,8-Tetrachlorodibenzo-*p*-dioxin equivalents in tissues of birds at Green Bay, Wisconsin, USA: *Arch. Environ. Contam. Toxicol.*, v. 24, no. 3, pp. 345–354.

Jones, W.E., Palawski, D.U., and Malloy, J.C., 1990, Red Rock Lakes National Wildlife Refuge Contaminant Survey: U.S. Fish and Wildlife Service, Region 6, Helena, Mont., 26 p.

Jordan, P.R., and Stamer, J.K., 1991, Surface water-quality assessment of the lower Kansas River Basin, Kansas and Nebraska: Analysis of available water-quality data through 1986: U.S. Geological Survey Open-File Report 91-75, 172 p.

Jury, W.A., Focht, D.D., and Farmer, W.J., 1987, Evaluation of pesticide groundwater pollution potential from standard indices of soil-chemical sorption and biodegradation: *J. Environ. Qual.*, v. 16, no. 4, pp. 422–428.

Kaczmar, P., Zabik, M.J., and D'Itri, F.M., 1985, Part per trillion residues of 2,3,7,8-tetrachlorodibenzo-*p*-dioxin in Michigan fish, *in* Keith, L.H., Rappe, C., and Choudhary, G., eds., *Chlorinated dioxins and dibenzofurans in the total environment II*: Butterworth Publishers, Boston, Mass., Ann Arbor Science Book series, pp. 103–110.

Kaiser, K.L.E., 1974, Mirex: An unrecognized contaminant of fishes from Lake Ontario: *Science*, v. 185, no. 4150, pp. 523–525.

——1978, The rise and fall of mirex: *Environ. Sci. Technol.*, v. 12, no. 5, pp. 520–528.

Kaiser, K.L.E., Comba, M.E., Hunter, H., Maguire, R.J., Tkacz, R.J., and Platford, R.F., 1985, Trace organic contaminants in the Detroit River: *J. Great Lakes Res.*, v. 11, no. 3, pp. 386–399.

Kaiser, K.L.E., Oliver, B.G., Charlton, M.N., Nicol, K.D., and Comba, M.E., 1990, Polychlorinated biphenyls in St. Lawrence River sediments: *Sci. Total Environ.*, v. 97/98, pp. 495–506.

Kalkhoff, S.J., 1981, Quality of water and time of travel in Goodwater and Okatoma Creeks near Magee, Mississippi: U.S. Geological Survey Open-File Report 81-1012, 34 p.

Kammann, U., Knickmeyer, R., and Steinhart, H., 1990, Distribution of polychlorobiphenyls and hexachlorobenzene in different tissues of the dab (*Limanda limanda* L.) in relation to lipid polarity: *Bull. Environ. Contam. Toxicol.*, v. 45, no. 4. pp. 552–559.

Kanazawa, J., 1982, Relationship between the molecular weights of pesticides and their bioconcentration factors by fish: *Experientia*, v. 38, no. 9, pp. 1045–1046.

Kannan, K., Tanabe, S., Hoang, T.Q., Nguyen, D.H., and Tatsukawa, R., 1992, Residue pattern and dietary intake of persistent organochlorine compounds in foodstuffs from Vietnam: *Arch. Environ. Contam. Toxicol.*, v. 22, no. 4, p. 367–374.

Kapoor, I.P., Metcalf, R.L., Hirwe, A.S., Coats, J.R., and Khalsa, M.S., 1973, Structure activity correlations of biodegradability of DDT analogs: *J. Agric. Food Chem.*, v. 21, no. 2, pp. 310–315.

Karickhoff, S.W., 1981, Semi-empirical estimation of sorption of hydrophobic pollutants on natural sediments and soils: *Chemosphere*, v. 10, p. 833–846.

——1984, Organic pollutant sorption in aquatic systems: *J. Hydraulic Eng.*, v. 110, pp. 707–735.

Karickhoff, S.W., Brown, D.S., and Scott, T.A., 1979, Sorption of hydrophobic pollutants on natural sediments: *Water Res.*, v. 13, no. 3, pp. 241–248.

Kauss, P.B., 1983, Studies of trace contaminants, nutrients, and bacteria levels in the Niagara River: *J. Great Lakes Res.*, v. 9, no. 2, pp. 249–273.

——1991, Biota of the St. Marys River: Habitat evaluation and environmental assessment: *Hydrobiologia*, v. 219, pp. 1–35.

Kauss, P.B., Suns, K., and Johnson, A.F., 1983, Monitoring of PCBs in water, sediments and biota of the Great Lakes—Some recent examples, *in* Mackay, D., Paterson, S., Eisenreich, S.J., Simmons, M.S., eds., *Physical behavior of PCBs in the Great Lakes*: Ann Arbor Science, Ann Arbor, Mich., pp. 385–409.

Kay, S.H., 1984, Potential for biomagnification of contaminants within marine and freshwater food webs: US Army Corps of Engineers, Technical Report D-84-7, 166 p.

Kelce, W.R., Stone, C.R., Laws, S.C., Gray, L.E., Kemppainen, J.A., and Wilson, E.M., 1995, Persistent DDT metabolite *p,p'*-DDE is a potent androgen receptor antagonist: *Nature*, v. 375, no. 6532, pp. 581–585.

Kellogg, R.L., and Bulkley, R.V., 1976, Seasonal concentrations of dieldrin in water, channel catfish, and catfish-food organisms, Des Moines River, Iowa, 1971–73: *J. Pest. Monitor.*, v. 9, no. 4, pp. 186–194.

Kelso, J.R.M., and Frank, R., 1974, Organochlorine residues, mercury, copper, and cadmium in yellow perch, while bass and smallmouth bass, Long Point Bay, Lake Erie: *Trans. Am. Fish. Soc.*, v. 103, no. 3, pp. 577–581.

Kelso, J.R.M., MacCrimmon, H.R., and Ecobichon, D.J., 1970, Seasonal insecticide residue changes in tissues of fish from the Grand River, Ontario: *Trans. Am. Fish. Soc.*, v. 99, no. 2, pp. 423–426.

Kemp, M.V., and Wightman, J.P., 1981, Interaction of 2,4-D and dicamba with chitan and chitosan: *Virginia J. Sci.*, v. 32, no. 2, pp. 34–37.

Kenaga, E.E., 1972, Factors related to bioconcentration of pesticides, *in* Matsumura, F., Boush, G.M., and Misato, T., eds., *Environmental Toxicology of Pesticides*: Academic Press, New York, pp. 193–228.

——1980a, Correlation of bioconcentration factors of chemicals in terrestrial organisms with their physical and chemical properties: *Environ. Sci. Technol.*, v. 14, no. 5, pp. 553–556.

——1980b, Predicted bioconcentration factors and soil sorption coefficients of pesticides and other chemicals: *Ecotoxicol. Environ. Safety*, v. 4, no. 1, pp. 26–38.

Kenaga, E.E., and Goring, C.A.I., 1980, Relationship between water solubility, soil sorption, octanol-water partitioning, and concentration of chemicals in biota, *in* Eaton, J.G., Parrish, P.R., and Hendricks, A.C., eds., *Aquatic toxicology*: American Society for Testing and Materials, Philadelphia, Penn., Special Technical Publication series, v. 707, pp. 78–115.

Kennicutt, M.C., II, Brooks, J.M., Atlas, E.L., and Giam, C.S., 1988, Organic compounds of environmental concern in the Gulf of Mexico: A review: *Aquat. Toxicol.*, v. 11, no. 1–2, pp. 191–212.

Kepner, W.G., 1986, Lower Gila River contaminant study, *in* Summers, J.B., and Anderson, S.S., eds., *Toxic substances in agricultural water supply and drainage: Defining the problems: Proceedings from the 1986 regional meetings sponsored by the U.S. Committee on Irrigation and Drainage*: U.S. Committee on Irrigation and Drainage, Denver, Colo., pp. 251–258.

Kerr, S.R., and Vass, W.P., 1973, Pesticide residues in aquatic invertebrates, *in* Edwards, C.A., ed., *Environmental pollution by pesticides*: Plenum Press, London, England, Environmental Science Research series, v. 3, pp. 134–180.

Kerwill, C.J., and Edwards, H.E., 1967, Fish losses after forest spraying with insecticides in New Brunswick, 1957–62, as shown by caged specimens and other observations: *J. Fish. Res. Board Can.*, v. 24, no. 4, pp. 708–729.

Kevern, N.R., 1966, Feeding rate of carp estimated by a radioisotopic method: *Trans. Am. Fish. Soc.*, v. 95, no. 4, pp. 363–371.

Kezic, N., Britvic, S., Protic, M., Rijavec, M., Zahn, R.K., Kurelee, B., and Simmons, J.E., 1983, Activity of benzo(*a*)pyrene monooxygenase in fish from the Sava River, Yugoslavia: Correlation with pollution: *Sci. Total Environ.*, v. 27, no. 1, pp. 59–69.

Khan, S.U., 1980, *Pesticides in the soil environment*: Elsevier Scientific Publishing Company, Amsterdam, The Netherlands, Fundamental Aspects of Pollution Control and Environmental Science, v. 5, 240 p.

Kim, N.K., and Stone, D.W., 1981, *Organic chemicals and drinking water* (rev. ed.): New York State Department of Health, Albany, N.Y., 140 pp.

King, K.A., Baker, D.L., and Kepner, W.G., 1992, Organochlorine and trace element concentrations in the San Pedro River Basin, Arizona: U.S. Fish and Wildlife Sevice, Phoenix, Ariz., 17 p.

King, K.A., Baker, D.L., Kepner, W.G., and Krausmann, J.D., 1991, Contaminants in prey of bald eagles nesting in Arizona: U.S. Fish and Wildlife Sevice Phoenix, Ariz., 16 p.

Kingsbury, P.D., and Kreutzweiser, D.P., 1987, Permethrin treatments in Canadian forests, Part 1: Impact on stream fish: *Pest. Sci.*, v. 19, no. 1, pp. 35–48.

Kizlauskas, A.G., Rockwell, D.C., and Claff, R.E., 1984, Great Lakes National Program Office Harbor Sediment Program, Lake Ontario 1981: Rochester, New York, Oswego, New York, Olcott, New York: U.S. Environmental Protection Agency, Great Lakes National Program Office, EPA-905/4-84-002, 46 p.

Klaassen, C.D., Amdur, M.D., and Doull, J., eds., 1996, *Casarett and Doull's Toxicology: The basic science of poisons* (5th ed.): McGraw-Hill, Health Professions Division, New York, 1111 p.

Klaassen, H.E., and Kadoum, A.M., 1973, Pesticide residues in natural fish populations of the Smoky Hill River of eastern Kansas, 1967–69: *J. Pest. Monitor.*, v. 7, no. 1, pp. 53–61.

Klaseus, T.G., Buzicky, G.C., and Schneider, E.C., 1988, Pesticides and groundwater: Surveys of selected Minnesota wells: Minnesota Department of Health and Minnesota Department of Agriculture report prepared for the Legislative Commission on Minnesota Resources, 95 p.

Kleeman, J.M., Olson, J.R., Chen, S.M., and Peterson, R.E., 1988, Metabolism and disposition of 2,3,7,8-tetrachlorodibenzo-*p*-dioxin in rainbow trout: *Toxicol. Appl. Pharmacol.*, v. 83, p. 391–401.

Kleinert, S.J., Degurse, P.E., and Wirth, T.L., 1968, Occurrence and significance of DDT and dieldrin residues in Wisconsin fish: Wisconsin Department of Natural Resources Technical Bulletin 41, 43 p.

Kleinow, K.M., Melancon, M.J., and Lech, J.J., 1987, Biotransformation and induction: Implications for toxicity, bioaccumulation and monitoring of environmental xenobiotics in fish: *Environ. Health Perspect.*, v. 71, pp. 105–119.

Kleopfer, R.D., Bunn, W.W., Yue, K.T., and Harris, D.J., 1983, Occurrence of tetrachlorodibenzo-*p*-dioxins in environmental samples from southwest Missouri, *in* Choudhary, G., Keith, L.H., and Rappe, C., eds., *Chlorinated dioxins and dibenzofurans in the total environment*: Butterworth Publishers, Boston, Mass., Ann Arbor Science Book series, pp. 193–201.

Knapton, J.R., Jones, W.E., and Sutphin, J.W., 1988, Reconnaissance investigation of water quality, bottom sediment, and biota associated with irrigation drainage in the Sun River area, west-central Montana, 1986–87: U.S. Geological Survey Water-Resources Investigations Report 87-4244, 78 p.

Knezovich, J.P., Harrison, F.L., and Wilhelm, R.G., 1987, The bioavailability of sediment-sorbed organic chemicals—A review: *Water Air Soil Poll.*, v. 32, pp. 233–245.

Knickmeyer, R., and Steinhart, H., 1990, Seasonal variations and sex related differences of organochlorines in whelks (*Buccinum undatum*) from the German Bight: *Chemosphere*, v. 20, no. 1–2, pp. 109–122.

Kobylinski, G.J., and Livingston, R.J., 1975, Movement of mirex from sediment and uptake by the hogchoker, *Trinectes maculatus*: *Bull. Environ. Contam. Toxicol.*, v. 14, no. 6, pp. 692–698.

Koenig, C.C., 1977, The effects of mirex alone and in combination on the reproduction of a salt marsh cyprinodont fish, *Adinia xenica*, *in* Vernberg, F.J., Calabrese, A., Thurberg, F.P., and Vernberg, W.B., eds., *Physiological reponses of marine biota to pollutants*: Academic Press, New York, pp. 357–376.

Koivusaari, U., Harri, H., and Hanninen, O., 1981, Seasonal variation of hepatic transformation in female and male rainbow trout (*Salmo gairdnieri*): *Comp. Biochem. Physiol.*, v. 70C, no. 2, pp. 149–157.

Kolipinski, M.C., Higer, A.L., and Yates, M.L., 1971, Organochlorine insecticide residues in Everglades National Park and Loxahatchee National Wildlife Refuge, Florida: *J. Pest. Monitor.*, v. 5, no. 3, pp. 281–288.

Köneman, H., and van Leeuwen, K., 1980, Toxicokinetics in fish: Accumulation and elimination by guppies: *Chemosphere*, v. 9, no. 1, pp. 3–19.

Kovats, Z.E., and Ciborowski, J.J.H., 1989, Aquatic insect adults as indicators of organochlorine contamination: *J. Great Lakes Res.*, v. 15, no. 4, pp. 623–634.

Kramer, R.E., and Plapp, F.W., Jr., 1972, DDT residues in fish from the Brazos River Basin in central Texas: *Environ. Entomol.*, v. 1, no. 4, pp. 406–409.

Kreis, R.G., and Rice, C.P., 1985, Status of organic contaminants in Lake Huron: Atmosphere, water, algae, fish, herring gull eggs, and sediment: University of Michigan, Great Lakes Research Division, Special Report 114, 169 p., 1 appendix.

Kreutzweiser, D.P., and Wood, G.A., 1991, Permethrin treatments in Canadian forests, Part 3: Fate and distribution in streams: *Pest. Sci.*, v. 33, no. 1, pp. 35–46.

Krieger, N., Wolff, M.S., Hiatt, R.A., Rivera, M, Vogelman, J., and Orentreich, N., 1994, Breast cancer and serum organochlorines: A prospective study among white, black, and Asian women: *J. Natl. Cancer Inst.*, v. 86, no. 8, pp. 589–599.

Kubiak, T.J., Harris, H.J., Smith, L.M., Schwartz, T.R., Stalling, D.L., Trick, J.A., Sileo, L., Docherty, D.E., and Erdman, T.C., 1989, Microcontaminants and reproductive impairment of the Forster's tern on Green Bay, Lake Michigan—1983: *Arch. Environ. Contam. Toxicol.*, v. 18, pp. 706–727.

Kuehl, DW., Butterworth, B.C., McBride, A., Kroner, S., and Bahnick, D., 1989, Contamination of fish by 2,3,7,8-tetrachlorodibenzo-*p*-dioxin: A survey of fish from major watersheds in the United States: *Chemosphere*, v. 18, no. 9/10, pp. 1997–2014.

Kuehl, D.W., Cook, P.M., Batterman, A.R., Lothenbach, D., and Butterworth, B.C., 1987, Bioavailability of polychlorinated dibenzo-*p*-dioxins and dibenzofurans from contaminated Wisconsin River sediment: *Chemosphere*, v. 16, no. 4, pp. 667–679.

Kuehl, D.W., Leonard, E.N., Butterworth, B.C., and Johnson, K.L., 1983, Polychlorinated chemical residues in fish from major watersheds near the Great Lakes, 1979: *Environ. Int.*, v. 9, no. 4, pp. 293–299.

Kuehl, D.W., Leonard, E.N., Welch, K.J., and Veith, G.D., 1980, Identification of hazardous organic chemicals in fish from the Ashtabula River, Ohio, and Wabash River, Indiana: *J. Assoc. Off. Anal. Chem.*, v. 63, no. 6, pp. 1238–1244.

Kuhr, R.J., Davis, A.C., and Bourke, J.B., 1974, DDT residues in soil, water, and fauna from New York apple orchards: *J. Pest. Monitor.*, v. 7, no. 3/4, pp. 200–204.

Kuivila, K.M., and Foe, C.G., 1995, Concentrations, transport and biological effects of dormant spray pesticides in the San Francisco Estuary, California: *Environ. Toxicol. Chem.*, v. 14, no. 7, pp. 1141–1150.

Kuwabara, K., Yakushji, T., Watanabe, I., Yoshida, S., Yoyama, K., and Kunita, N., 1979, Increase in the human blood PCB levels promptly following ingestion of fish containing PCBs: *Bull. Environ. Contam. Toxicol.*, v. 21, no. 1–2, pp. 273–278.

Laitinen, M., Nieminen, M., and Heitanen, E., 1982, The effect of pH changes of water on the hepatic metabolism of xenobiotics in fish: *Acta Pharmacol. Toxicol.*, v. 51, pp. 24–29.

Lake, J., and Morrison, J., 1977, Environmental impact of land use on water quality: Final report on the Black Creek Project (technical report): U.S. Environmental Protection Agency, Great Lakes National Program Office, EPA-905/9-77-007-B, 280 p.

Lake, J.L., Rubinstein, N.I., Lee, H., II, Lake, C.A., Heltshe, J., and Pavignano, S., 1990, Equilibrium partitioning and bioaccumulation of sediment-associated contaminants by infaunal organisms: *Environ. Toxicol. Chem.*, v. 9, no. 8, pp. 1095–1106.

Lake, J.L., Rubinstein, N.I., and Pavignano, S., 1987, Predicting bioaccumulation: Development of a simple partitioning model for use as a screening tool for regulating ocean disposal of wastes, *in* Dickson, K.L., Maki, A.W., and Brungs, W.A., eds., *Fate and effects of sediment-bound chemicals in aquatic systems: Proceedings of the Sixth Pellston Workshop, Florissant, Colorado, August 12–17, 1984*, Society of Environmental Toxicology and Chemistry Special Publications series: Permagon Press, New York, pp. 151–167.

Lamb, T.E., 1978a, Water-quality data for the Village Creek watershed, northeast Arkansas: U.S. Geological Survey Open-File Report 78-497, 46 p.

——1978b, Water-quality investigation of the Tyronza River watershed, Arkansas: U.S. Geological Survey Open-File Report 78-175, 32 p.

Lambing, J.H., Jones, W.E., and Sutphin, J.W., 1988, Reconnaissance investigation of water quality, bottom sediment, and biota associated with irrigation drainage in Bowdoin National Wildlife Refuge and adjacent areas of the Milk River Basin, northeastern Montana, 1986–87: U.S. Geological Survey Water-Resources Investigations Report 87-4243, 71 p.

Landrum, P.F., Lee, H., II, and Lydy, M.J., 1992, Toxicokinetics in aquatic systems: Model comparisons and use in hazard assessment: *Environ. Toxicol. Chem.*, v. 11, no. 12, pp. 1709–1725.

Landrum, P.F., Nihart, S.R., Eadie, B.J., and Herche, L.R., 1987, Reduction in bioavailability of organic contaminants to the amphipod *Pontoporeia hoyi* by disslved organic matter of sediment interstitial waters: *Environ. Toxicol. Chem.*, v. 6, no. 1, pp. 11–20.

Landrum, P.F., Reinhold, M.D., Nihart, S.R., and Eadie, B.J., 1985, Predicting the bioavailability of organic xenobiotics to *Pontoporeia hoyi* in the presence of humic and fulvic materials and natural dissolved organic matter: *Environ. Toxicol. Chem.*, v. 4, no. 4, pp. 459–467.

Landrum, P.F., and Scavia, D., 1983, Influence of sediment on anthracene uptake, depuration, and biotransformation by the amphipod *Hyalella azteca*: *Can. J. Fish. Aquat. Sci.*, v. 40, no. 3, pp. 298–305.

Landrum, P.F., and Stubblefield, C.R., 1991, Role of respiration in the accumulation of organic xenobiotics by the amphipod *Diporeia* sp.: *Environ. Toxicol. Chem.*, v. 10, no. 8, pp. 1019–1028.

Langston, W.J., 1978, Persistence of polychlorinated biphenyls in marine bivalves: *Mar. Biol.*, v. 46, no.1, pp. 35–40.

Larson, R.A., and Weber, E.J., 1994, *Reaction mechanisms in environmental organic chemistry*: Lewis Publishers, Boca Raton, Fla., 433 p.

Larson, S.J., Capel, P.D., and Majewski, M.S., 1997, *Pesticides in surface waters: Distribution, trends, and governing factors*: Ann Arbor Press, Chelsea, Mich., Pesticides in the Hydrologic System series, v. 3, 373 p.

Larsson, P., 1986, Zooplankton and fish accumulate chlorinated hydrocarbons from contaminated sediments: *Can. J. Fish. Aquat. Sci.*, v. 43, no. 7, pp. 1463–1466.

Larsson, P., Hamrin, S., and Okla, L., 1991, Factors determining the uptake of persistent pollutants in an eel population (*Anguilla anguilla* L.): *Environ. Pollut.*, v. 69, no. 1, pp. 39–50.

Laska, A.L., Bartell, C.K., and Laseter, J.L., 1976, Distribution of hexachlorobenzene and hexachlorobutadiene in water, soil, and selected aquatic organisms along the lower Mississippi River, Louisiana: *Bull. Environ. Contam. Toxicol.*, v. 15, no. 5, pp. 535–542.

Lathrop, J.E., and Davis, W.S., 1985, Aquatic sediments: *J. Water Poll. Control Fed.*, v. 57, no. 6, pp. 712–724.

——1986, Aquatic sediments: *J. Water Poll. Control Fed.*, v. 58, no. 6, pp. 684–699.

Lau, Y.L., Oliver, B.G., and Krishnappan, B.G., 1989, Transport of some chlorinated contaminants by the water, suspended sediments, and bed sediments in the St. Clair and Detroit rivers: *Environ. Toxicol. Chem.*, v. 8, no. 4, pp. 293–301.

Lauenstein, G.G., Cantillo, A.Y., and Dolvin, S.S., 1993, Overview and summary of methods, v. 1 *of* Sampling and analytical methods of the National Status and Trends Program, National Benthic Surveillance and Mussel Watch Projects, 1984–1992: U.S. Department of Commerce, National Oceanic and Atmospheric Administration Technical Memorandum NOS ORCA 71, 117 p.

Law, L.M., and Goerlitz, D.F., 1974, Selected chlorinated hydrocarbons in bottom material from streams tributary to San Francisco Bay: *J. Pest. Monitor.*, v. 8, no. 1, pp. 33–36.

Lawson, L., and Machado, M.L., 1991, Report on the Arizona Priority Pollutant Program—1989: Arizona Department of Environmental Quality, Phoenix, Ariz., 22 p., 3 appendices.

Lay, M.M., and Menn, J.J., 1979, Mercapturic acid occurrence in fish bile: A terminal product of metabolism of the herbicide molinate: *Xenobiotica*, v. 9, pp. 669–673.

Leard, R.L., Grantham, B.J., and Pessoney, G.F., 1980, Use of selected freshwater bivalves for monitoring organochlorine pesticide residues in major Mississippi stream systems, 1972–73: *J. Pest. Monitor.*, v. 14, no. 2, pp. 47–52.

Leatherland, J.F., 1992, Endocrine and reproductive function in Great Lakes salmon, *in* Colborn, T., and Clement, C.R., eds., *Chemically induced alterations in sexual and functional development: The wildlife/human connection*: Princeton Scientific Publishing, Princeton, N.J., Advances in Modern Environmental Toxicology series, v. 21, pp. 129–145.

Lech, J.J., 1973, Isolation and identification of 3-fluoromethyl-4-nitrophenyl glucuronide from bile of rainbow trout exposed to 3-fluoromethyl-4-nitrophenol: *Toxicol. Appl. Pharmacol.*, v. 24, no. 1, pp. 114–124.

Leiker, T.J., Rostad, C.E., Barnes, C.R., and Wilfred, E., 1991, A reconnaissance study of halogenated organic compounds in from the lower Mississippi River and its major tributaries: *Chemosphere*, v. 23, no. 7, pp. 817–829.

Lemonick, M.D., 1994, Not so fertile ground: *Time*, v. 144, no. 12 (September 19, 1994), pp. 68–70.

Lenon, H.L., 1968, Accumulation of dietary polychlorinated biphenyls (Aroclor 1254) by rainbow trout (*Salmo gairdnieri*): Michigan State University, East Lansing, Mich., Ph.D. dissertation, 85 p.

Leonard, R.A., 1990, Movement of pesticides into surface waters, *in* Cheng, H.H., ed., *Pesticides in the soil environment: Processes, impacts, and modeling*: Soil Science Society of America, Madison, Wis., pp. 303–349.

Leone, H.L., Jr., and Dupuy, A.J., 1978, Analyses of water, core material, and elutriate samples near Yazoo City, Mississippi (Yazoo Headwater Project): U.S. Geological Survey Open-File Report 78-792, 10 p, 1 pl.

Leopold, L.B., Wolman, M.G., and Miller, J.P., 1964, *Fluvial processes in geomorphology*: W.H. Freeman and Company, San Francisco, 522 p.

Leung, S.-Y.T., Bulkley, R.V., and Richard, J.J., 1981a, Influence of a new impoundment on pesticide concentrations in warmwater fish, Saylorville Reservoir, Des Moines River, Iowa, 1977–78: *J. Pest. Monitor.*, v. 15, no. 3, pp. 117–122.

——1981b, Persistence of dieldrin in water and channel catfish from the Des Moines River, Iowa, 1971–73 and 1978: *J. Pest. Monitor.*, v. 15, no. 2, pp. 98–102.

Levin, S.A., Kimball, K.D., McDowell, W.H., and Kimball, S.F., 1985, New perspectives in ecotoxicology: *Environ. Manag.*, v. 8, pp. 375–442.

Levy, H., II, 1990, Regional and global transport and distribution of trace species released at the earth's surface, *in* Kurtz, D.A., ed., *Long range transport of pesticides*: Lewis Publishers, Chelsea, Mich., pp. 83–95.

Lewis, T.W., and Makarewicz, J.C., 1988, Exchange of mirex between Lake Ontario and its tributaries: *J. Great Lakes Res.*, v. 14, no. 4, pp. 388–393.

Li, M.Y., 1977, Pollution in the nation's estuaries originating from the agricultural use of pesticides, *in Estuarine Pollution Control and Assessment: Proceedings of a conference*: U. S. Environmental Protection Agency, Office of Water Planning and Standards, Washington, D. C, pp. 451–466.

Lichtenstein, E.P., Fuhremann, T.W., and Schulz, K.R., 1971, Persistence and vertical distribution of DDT, lndane, and aldrin residues, 10 and 15 years after a single soil application: *J. Agr. Food Chem.*, v. 19, no. 4, pp. 718–720.

Lick, W., 1984, The transport of sediments in aquatic systems, *in* Dickson, K.L., Maki, A.W., and Brungs, W.A., eds., *Fate and effects of sediment-bound chemicals in aquatic systems: Proceedings of the Sixth Pellston Workshop, Florissant, Colorado, August 12–17, 1984,* Society of Environmental Toxicology and Chemistry Special Publications series: Permagon Press, New York, pp. 61–74.

Lieb, A.J., Bills, D.D., and Sinnuber, R.O., 1974, Accumulation of dietary polychlorinated biphenyls (Aroclor 1254) by rainbow trout (*Salmo gairdnieri*): *J. Agric. Food Chem.*, v. 22, no. 4, pp. 638–642.

Liska, B.J., Stemp, A.R., and Stadelman, W.J., 1967, Effect of method of cooking on chlorinated insecticide residues in edible chicken tissues: *Food Technol.*, v. 21, pp. 117.

Liu, L.C., Fernandez-Horta, D., and Santiago-Cordova, 1985, Diuron and ametryn runoff from a plantain field: *J. Agric. Univ. Puerto Rico*, v. 69, no. 2, pp. 177–183.

Livingston, R.J., Thompson, N.P., and Meeter, D.A., 1978, Long-term variation of organochlorine residues and assemblages of epibenthic organisms in a shallow North Florida (USA) estuary: *Mar. Biol.*, v. 46, no. 4, pp. 355–372.

Lockhart, W.L., Metner, D.A., and Solomon, J., 1977, Methoxychlor residue studies in caged and wild fish from the Athabasca River, Alberta, following a single application of blackfly larvicide: *Can. J. Fish. Aquat. Sci.*, v. 34, no. 5, pp. 626–632.

Loesch, J.G., Huggett, R.J., and Foell, E.J., 1982, Kepone concentration in juvenile anadromous fishes: *Estuaries*, v. 5, no. 3, pp. 175–181.

Logan, T.J., 1987, Diffuse (non-point) source loading of chemicals to Lake Erie: *J. Great Lakes Res.*, v. 13, no. 4, pp. 649–658.

Loganathan, B.G., and Kannan, K., 1991, Time perspectives of organochlorine contamination in the global environment: *Mar. Pollut. Bull.*, v. 22, no. 12, pp. 582–584.

Long, E.R., MacDonald, D.D., and Cairncross, C., 1991, Status and trends in toxicants and the potential for their biological effects in Tampa Bay, Florida: U.S. Department of Commerce, National Oceanic and Atmospheric Administration Technical Memorandum NOS OMA 58, 77 p.

Long, E.R., MacDonald, D.D., Matta, M.B., VanNess, K., Buchman, M., and Harris, H., 1988, Status and trends in concentrations of contaminants and measures of biological stress in San Francisco Bay: National Oceanic and Atmospheric Administration Technical Memorandum NOS OMA 41, 268 p.

Long, E.R., MacDonald, D.D., Smith, S.L., and Calder, F.D., 1995, Incidence of adverse biological effects within ranges of chemical concentrations in marine and estuarine sediments: *Environ. Manag.*, v. 19, no. 1, pp. 81–97.

Long, E.R., and Morgan, L.G., 1991, The potential for biological effects of sediment-sorbed contaminants tested in the National Status and Trends Program: U.S. Department of Commerce, National Oceanic and Atmospheric Administration Technical Memorandum NOS OMA 52, variously paged.

Lord, K.A., Briggs, G.G., Neale, M.C., and Manlove, R., 1980, Uptake of pesticides from water and soil by earthworms: *Pest. Sci.*, v. 11, no. 4, pp. 401–408.

Low, P.A., 1983, Upstream transport of the Lake Ontario contaminant, mirex, by Pacific salmon (*Onchorhynchus* spp.) in tributaries of the Salmon River, New York: State University of New York College of Environmental Science and Forestry, Syracuse, N.Y., M.S. thesis.

Lowe, J.A., Farrow, D.R.G., Pait, A.S., Arenstam, S.J., and Lavan, E.F., 1991, Fish kills in coastal waters 1980–1989: National Oceanic and Atmospheric Administration, National Ocean Service, Office of Ocean Resources Conservation and Assessment, Strategic Environmental Assessments Division, 69 p.

Lu, P.-Y., 1973, Model aquatic ecosystem studies of the environmental fate and biodegradability of industrial compounds: University of Illinois, Urbana-Champaign, Ill., Ph.D dissertation.

Lu, P.-Y., and Metcalf, R., 1975, Environmental fate and biodegradability of benzene derivatives as studied in a model aquatic ecosystem: *Environ. Health Perspect.*, v. 10, pp. 269–284.

Ludke, J.L., and Schmitt, C.J., 1980, Monitoring contaminant residues in freshwater fishes in the United States: National Pesticide Monitoring Program, *in* Swain, W.R., and Shannon, V.R., eds., *Proceedings of the Third USA-USSR Symposium on the Effects of Pollutants Upon Aquatic Ecosystems:Theoretical aspects of aquatic toxicology, July 2–6, 1979, Borok, Jaroslav Oblast, USSR*: U.S. Environmental Protection Agency, Office of Research and Development, Environmental Research Laboratory (Duluth, Minn.), Research Reporting series, no. 9, Miscellaneous Report EPA-600/9-80-034, pp. 97–110.

Lum, K.R., Kaiser, K.L.E., and Comba, M.E., 1987, Export of mirex from Lake Ontario to the St. Lawrence estuary: *Sci. Total Environ.*, v. 67, pp. 41–51.

Lunsford, C.A., and Blem, C.R., 1982, Annual cycle of kepone residue and lipid content of the estuarine clam, *Rangia cuneata*: *Estuaries*, v. 5, no. 2, pp. 121–130.

Lunsford, C.A., Weinstein, M.P., and Scott, L., 1987, Uptake of kepone by the estuarine bivalve *Rangia cuneata*, dredging of contaminated sediments in the James River: *Water Res.*, v. 21, no. 4, pp. 411–416.

Lurry, D.L., 1983, Analyses of native water, bottom material, elutriate samples, and dredged material from selected southern Louisiana waterways and selected areas in the Gulf of Mexico, 1979–81: U.S. Geological Survey Open-File Report 82-690, 105 p.

Lydy, M.J., Oris, J.T., Baumann, P.C., and Fisher, S.W., 1992, Effects of sediment organic carbon content on the elimination rates of neutral lipophilic compounds in the midge (*Chironomus riparius*): *Environ. Toxicol. Chem.*, v. 11, no. 3, pp. 347–356.

Lyman, L.D., Tompkins, W.A., and McCann, J.A., 1968, Massachusetts Pesticide Monitoring Study: *J. Pest. Monitor.*, v. 2, no. 3, pp. 109–122.

Lyman, W.J., 1990, Adsorption coefficient for soils and sediments, in Lyman, W.J., Reehl, W.F., and Rosenblatt, D.H., eds., *Handbook of chemical property estimation methods: Environmental behavior of organic compounds*: American Chemical Society, Washington, D.C., pp. 4–1 to 4–33.

Lyman, W.J., Reehl, W.F., and Rosenblatt, D.H., eds., 1990, *Handbook of chemical property estimation methods: Environmental behavior of organic compounds*: American Chemical Society, Washington, D.C., variously paged.

Lynch, T.R., and Johnson, H.E., 1982, Availability of a hexachlorobiphenyl isomer to benthic amphipods from experimentally contaminated natural sediments, *in* Pearson, J.G., Foster, R.B., and Bishop, W.E., eds., *Aquatic toxicology and hazard assessment*: Proceedings of the Fifth Annual Symposium on Aquatic Toxicology: American Society for Testing and Materials, Philadelphia, Penn., Special Technical Publication series, v. 766, pp. 273–287.

Mabey, M., and Mill, T., 1978, Critical review of hydrolysis of organic compounds in water under environmental conditions: *J. Phys. Chem. Ref. Data*, v. 7, no. 2, pp. 383–423.

Macalady, D.L., Tratnyek, P.G., and Grundl, T.J., 1986, Abiotic reduction reactions of anthropogenic organic chemicals in anaerobic systems: A critical review: *J. Contam.. Hydrol.*, v. 1, no. 1, pp. 1–28.

MacDonald, D.A., 1991, Status and trends in concentrations of selected contaminants in Boston Harbor sediments and biota: U.S. Department of Commerce, National Oceanic and Atmospheric Administration Technical Memorandum NOS OMA 56, variously paged.

MacDonald, D.D., 1994, Approach to the assessment of sediment quality in Florida coastal water, Vol. 1, Development and evaluation of sediment quality assessment guidelines: MacDonald Environmental Services (Ladysmith, B.C.) report prepared for Florida Department of Environmental Protection, Office of Water Policy, Tallahassee, Florida, 123 p.

Macek, K.J., and Korn, S., 1970, Significance of the food chain in DDT accumulation by fish: *Can. J. Fish. Aquat. Sci.*, v. 27, no. 8, pp. 1496–1498.

Macek, K.J., Petrocelli, S.R., and Sleight, B.H., III, 1979, Considerations in assessing the potential for, and significance of, biomagnification of chemical residues in aquatic food chains, *in* Marking, L.L., and Kimerle, R.A., eds., *Aquatic toxicology: Proceedings of the Second Annual Symposium on Aquatic Toxicology*: American Society for Testing and Materials, Philadelphia, Penn., Special Technical Publication series, v. 667, pp. 251–268.

Macek, K.J., Rodgers, C.R., Stalling, D.L., and Korn, S., 1970, The uptake, distribution and elimination of dietary ^{14}C-DDT and ^{14}C-dieldrin in rainbow trout: *Trans. Am. Fish. Soc.*, v. 99, no. 4, pp. 689–695.

Mack, G.L., Corcoran, S.M., Gibbs, S.D., Gutenmann, W.H., Reckahn, J.A., and Lisk, D.J., 1964, The DDT content of some fishes and surface waters of New York State: *New York Fish Game J.*, v. 11, pp. 148–153.

Mackay, D., 1979, Finding fugacity feasible: *Environ. Sci. Technol.*, v. 13, no. 10, pp. 1218–1223.

——1982, Correlation of bioconcentration factors: *Environ. Sci. Technol.*, v. 16, no. 5, pp. 274–278.

Mackay, D., Bobra, A., Shiu, W.Y., and Yalkowsky, S.M., 1980, Relationships between aqueous solubility and octanol-water partition coefficients: *Chemosphere*, v. 9, no. 11, pp. 701–711.

Mackay, D., and Hughes, A.I., 1984, Three parameter equation describing the uptake of organic compounds by fish: *Environ. Sci. Technol.*, v. 18, no. 6, pp. 439–444.

Mackay, D., and Paterson, S., 1981, Calculating fugacity: *Environ. Sci. Technol.*, v. 15, no. 9, pp. 1006–1014.

——1991, Evaluating the multimedia fate of organic chemicals: A level III fugacity model: *Environ. Sci. Technol.*, v. 25, no. 3, pp. 427–436.

Mackenthun, K.M., and Keup, L.E., 1972, Water pollution: Freshwater macroinvertebrates: *J. Water Poll. Control Fed.*, v. 44, no. 6, pp. 1137–1150.

Madden, J.D., Finerty, M.W., and Grodner, R.M., 1989, Survey of persistent pesticide residues in the edible tissues of wild and pond-raised Louisiana crayfish and their habitat: *Bull. Environ. Contam. Toxicol.*, v. 43, no. 5, pp. 779–784.

Madden, J.D., Grodner, R.M., Feagley, S.E., Finerty, M.W., and Andrews, L.S., 1991, Minerals and xenobiotic residues in the edible tissues of wild and pond-raised Louisiana crayfish: *J. Food Safety*, v. 12, no. 1, pp. 1–15.

Madhun, Y.A., and Freed, V.H., 1990, Impact of pesticides on the environment, *in* Cheng, H.H., ed., *Pesticides in the soil environment: Processes, impacts, and modeling*: Soil Science Society of America, Madison, Wisc., pp. 429–466.

Majewski, M.S., and Capel, P.D., 1995, *Pesticides in the atmosphere: Distribution, trends, and governing factors*: Ann Arbor Press, Chelsea, Mich., Pesticides in the Hydrologic System series, v. 1, 214 p.

Malins, D.C., McCain, B.B., Brown, D.W., Chan, S.-L., Myers, M.S., Landahl, J.T., Prohaska, P.G., Friedman, A.J., Rhodes, L.D., Burrows, D.G., Gronlund, W.D., and Hodgins, H.O., 1984, Chemical pollutants in sediments and diseases of bottom-dwelling fish in Puget Sound, Washington: *Environ. Sci. Technol.*, v. 18, no. 9, pp. 705–713.

Marchand, M., Vas, D., and Duursma, E.K., 1976, Levels of PCBs and DDTs in mussels from the N.W. Mediterranean: *Mar. Pollut. Bull.*, v. 7, no. 4, pp. 65–69.

Marcus, J.M., and Renfrow, R.T., 1990, Pesticides and PCBs in South Carolina estuaries: *Mar. Pollut. Bull.*, v. 21, no. 2, pp. 96–99.

Maren, T.H., Embry, R., and broder, L.E., 1968, The excretion of drugs across the gill of the dogfish, *Squalus acanthus*: *Comp. Biochem. Physiol.*, v. 26, no. 3, pp. 853–864.

Markin, G.P., Collins, H.L., and Davis, J., 1974b, Residues of the insecticide mirex in terrestrial and aquatic invertebrates following a single aerial application of mirex bait, Louisiana, 1971–72: *J. Pest. Monitor.*, v. 8, no. 2, pp. 131–134.

Markin, G.P., Hawthorne, J.C., Collins, H.L., and Ford, J.H., 1974a, Levels of mirex and some other organochlorine residues in seafood from Atlantic and Gulf coastal states: *J. Pest. Monitor.*, v. 7, no. 3–4, pp. 139–143.

Marthinsen, I., Staveland, G., Skaare, J.U., Ugland, K.I., and Haugen, A., 1991, Levels of environmental pollutants in male and female flounder (*Platichthys flesus* L.) and cod (*Gadus morhua* L.) caught during the year 1988 near or in the waterway of Glomma, the largest river of Norway. I. Polychlorinated biphenyls: *Arch. Environ. Contam. Toxicol.*, v. 20, no. 3, pp. 353–360.

Martin, D.B., and Hartman, W.A., 1985, Organochlorine pesticides and polychlorinated biphenyls in sediment and fish from wetlands in the north central United States: *J. Assoc. Off. Anal. Chem.*, v. 68, no. 4, pp. 712–717.

Maslova, O.V., 1981, Accumulation of DDT as a function of lipid content of tissues of fish from estuaries: *Hydrobiol. J.*, v. 17, no. 4, pp. 56–59.

Mason, R.R., Simmons, C.E., and Watkins, S.A., 1990, Effects of channel modifications on the hydrology of Chicod Creek Basin, North Carolina, 1975–87: U.S. Geological Survey Water-Resources Investigations Report 90-4031, 83 p.

Matson, M.R., and Hite, R.L., 1988, An intensive survey of the Eagle Creek Basin, Saline and Gallatin counties, Illinois, 1986–1987: Illinois Environmental Protection Agency, Division of Water Pollution Report IEPA/WPC 88-019, 42 p.

Matsumura, F., 1977, Absorption, accumulation, and elimination of pesticides by aquatic organisms, *in* Khan, M.A.Q., ed., *Pesticides in aquatic environments*: Plenum Press, New York, pp. 77–105.

Matsuo, M., 1979, The i/o* (inorganic/organic) characters to describe ecological magnification of some organophosphorous insecticides in fish: *Chemosphere*, v. 7, no. 8, pp. 477–485.

——1980a, The i/o* characters to correlate bioaccumulation of some chlorobenzenes in guppies with their chemical structures: *Chemosphere*, v. 9, no. 7/8, pp. 409–413.

——1980b, A thermodynamic interpretation of bioaccumulation of aroclor 1254 (PCB) in fish: *Chemosphere*, v. 9, no. 10, pp. 671–675.

Mattraw, H.C., Jr., 1975, Occurrence of chlorinated hydrocarbon insecticides, South Florida, 1968–72: *J. Pest. Monitor.*, v. 9, no. 2, pp. 106–114.

Maul, R.E., Funk, K., Zabik, M.E., and Zabik, M.J., 1971, Dieldrin residues and cooking losses in pork loins: *J. Am. Dietetic Assoc.*, v. 59, pp. 481.

Mayer, F.L., Jr., and Ellersieck, M.R., 1986, Manual of acute toxicity: Interpretation and data base for 410 chemicals and 66 species of freshwater animals: U.S. Department of the Interior, Fish and Wildlife Service Resource Publication 160, variously paged.

Mayer, F.L., Jr., Street, J.C., and Neuhold, J.M., 1970, Organochlorine insecticide interactions affecting residue storage in rainbow trout: *Bull. Environ. Contam. Toxicol.*, v. 5, no. 4, pp. 300–310.

McBride, A.C., 1987, The USEPA National Dioxin Study—Tiers 3, 5, 6 and 7: *Chemosphere*, v. 16, no. 8, pp. 2169–2173.

McCain, B.B., Brown, D.W., Krahn, M.M., Myers, M.S., Clark, R.C., Jr., Chan, S.-L., and Malins, D.C., 1988, Marine pollution problems, North American west coast: *Aquat. Toxicol.*, v. 11, no. 1/2, pp. 143–162.

McCarthy, J.F., 1983, Role of particulate organic matter in decreasing accumulation of polynuclear aromatic hydrocarbons by *Daphnia magna*: *Arch. Environ. Contam. Toxicol.*, v. 12, no. 5, pp. 559–568.

McCarthy, J.F., Jimenez, B.D., and Barbee, T., 1985, Effect of dissolved humic material on accumulation of polycyclic aromatic hydrocarbons: Structure-activity relationships: *Aquat. Toxicol.*, v. 7, no. 1–2, pp. 15–24.

McCartney, T., 1991, Environmental quality assessment of the Iroquois National Wildlife Refuge, Alabama, New York, 1988–1990: U.S. Fish and Wildlife Service, New York Field Office, Cortland, N.Y., 26 p., 2 appendices.

McCaskey, T.A., Stemp, A.R., Liska, B.J., and Stadelman, W.J., 1968, Residue in egg yolks and raw and cooked tissues from laying hens administered chlorinated hydrocarbon insecticide: *Poultry Sci.*, v. 47, no. 2, pp. 564–569.

McConnell, L.L., Cotham, W.E., and Bidleman, T.F., 1993, Gas exchange of hexachlorocyclohexane in the Great Lakes: *Environ. Sci. Technol.*, v. 27, no. 7, pp. 1304–1311.

McDonald, D.B., 1973, Some chemical and biological characteristics of the Mississippi River bordering Iowa.: American Institute of Chemical Engineers (AIChE) Symposium Series, v. 129, no. 69, pp. 380–382.

McElroy, A.E., and Means, J.C., 1988, Factors affecting the bioavailability of hexachlorobiphenyls to benthic organisms, *in* Adams, W.J., Chapman, G.A., and Landis, W.G., eds., *Aquatic toxicology and hazard assessment: 10th volume*: American Society for Testing and Materials, Philadelphia, Penn., Special Technical Publication series, v. 971, pp. 149–158.

McFarland, V.A., 1984, Activity-based evaluation of potential bioaccumulation from sediments, *in* Montgomery, R.L., and Leach, J.W., eds., *Dredging and dredged material disposal*, v. 1 of *Proceedings of the Conference Dredging '84*: American Society of Civil Engineers, New York, pp. 461–466.

McFarland, V.A., Feldhaus, J., Ace, L.N., and Brannon, J.M., 1994, Measuring the sediment/organism accumulation factor of PCB-52 using a kinetic model: *Bull. Environ. Contam. Toxicol.*, v. 52, no. 5, pp. 699–705.

McHenry, J.R., Cooper, C.M., and Ritchie, J.C., 1982, Sedimentation in Wolf Lake, lower Yazoo River Basin, Mississippi: *J. Freshwater Ecol.*, v. 1, no. 5, pp. 547–558.

McKenzie, S.W., 1977, Analysis of bottom material from the Willamette River, Portland Harbor, Oregon: U.S. Geological Survey Open-File Report 77-740, 10 p.

McKim, J.M., 1994, Physiological and biochemical mechanisms that regulate the accumulation and toxicity of environmental chemicals in fish, *in* Hamelink, J.L., Landrum, P.F., Bergman, H.L., and Benson, W.H., eds., *Bioavailability: Physical, chemical, and biological interactions*: Lewis Publishers, Boca Raton, Fla., pp. 179–201.

McKim, J.M., and Goeden, H.M., 1982, A direct measure of the uptake efficiency of a xenobiotic chemical across the gills of a brook trout (*Salvelinus fontinalis*) under normoxic and hypoxic conditions: *Comp. Biochem. Physiol.*, v. 72C, no. 1, pp. 65–74.

McKim, J.M., Schmieder, P.K., and Erickson, R.J., 1986, Toxicokinetic modeling of [^{14}C] pentachlorophenol in the rainbow trout: *Aquat. Toxicol.*, v. 9, no. 1, pp. 59–80.

McKim, J.M., Schmeider, P.K, and Veith, G., 1985, Absorption dynamics of organic chemical transport across trout gills as related to octanol-water partition coefficient: *Toxicol. Appl. Pharmacol.*, v. 77, no. 1, pp. 1–10.

McKinney, J.D., and Waller, C.L., 1994, Polychlorinated biphenyls as hormonally active structural analogues: *Environ. Health Perspect.*, v. 102, no. 3, pp. 290–297.

McPherson, B.F., 1971, Water quality at the Dade-Collier Training and Transition Airport, Miami International Airport, and Cottonmouth Camp-Everglades National Park, Florida, November, 1969: U.S. Geological Survey Open-File Report [71-200], 29 p.

——1973, Water quality in the conservation areas of the Central and Southern Florida Flood Control District, 1970–72: U.S. Geological Survey Open-File Report 73-0174, 44 p.

Means, J.C., and Wijayaratne, R., 1982, Role of natural colloids in the transport of hydrophobic pollutants: *Science*, v. 215, pp. 968–970.

Mearns, A.J., Matta, M.B., Simecek-Beatty, D., Buchmen, M.F., Shigenaka, G., and Wert, W., 1988, PCB and chlorinated pesticide contamination in U.S. fish and shellfish: A historical assessment report: U.S. Department of Commerce, National Oceanic and Atmospheric Administration Technical Memorandum NOS OMA 39, 140 p.

Meeks, R.C., 1968, The accumulation of ^{36}Cl ring-labeled DDT in a freshwater marsh: *J. Wildl. Manag.*, v. 32, no. 2, pp. 376–398.

Mehrle, P.M., Buckler, D.R., Little, E.E., Smith, L.M., Petty, J.D., Petersman, P.H., Stalling, D.L., DeGraeve, G.E., Coyle, J.J., and Adams, W.J., 1988, Toxicity and bioconcentration of 2,3,7,8-tetrachlorodibenzo-*p*-dioxin and 2,3,7,8-tetrachlorodibenzofuran in rainbow trout: *Environ. Toxicol. Chem.*, v. 7, no. 1, pp. 47–62.

Meister PublishingCo., 1970, *1970 Farm chemicals handbook*: Meister Publishing, Willoughby, Ohio, 496 p.

——1995, *Farm chemicals handbook '95*: Meister Publishing , Willoughby, Ohio, variously paged.

——1996, *Farm chemicals handbook '96*: Meister Publishing, Willoughby, Ohio, variously paged.

Melancon, M.J., Yeo, S.E., and Lech, J.J., 1987, Induction of microsomal hepatic monooxygenase activity in fish by exposure to river water: *Environ. Toxicol. Chem.*, v. 6, no. 2, pp. 127–135.

Merna, J.W., 1986, Contamination of stream fishes with chlorinated hydrocarbons from eggs of Great Lakes salmon: *Trans. Am. Fish. Soc.*, v. 115, no. 1, pp. 69–74.

Merriman, J.C., Anthony, D.H.J., Kraft, J.A., and Wilkinson, R.J., 1991, Rainy river water quality in the vicinity of bleached kraft mills: *Chemosphere*, v. 23, no. 11–12, pp. 1605–1615.

Mes, J., Davies, D.J., Turton, D., and Sun, W.F., 1986, Levels and trends of chlorinated hydrocarbon contaminants in the breast milk of Canadian women: *Food Addit. Contam.*, v. 3, no. 4, pp. 313–322.

Mes, J., Doyle, J.A., Barrett, R.A., Davies, D.J., and Turton, D., 1984, Polychlorinated biphenyls and organochlorine pesticides in milk and blood of Canadian women during lactation: *Arch. Environ. Contam. Toxicol.*, v. 13, no. 2, pp. 217–223.

Mes, J., and Malcolm, S., 1992, Comparison of chlorinated hydrocarbon residues in human populations from the Great Lakes and other regions of Canada: *Chemosphere*, v. 25, no. 3, pp. 417–424.

Metcalf, R.L., 1977, Biological fate of pollutants in the environment, *in* Suffet, I.H., ed., *Chemical and biological fate of pollutants in the environment*, pt. 2 of *Fate of pollutants in the air and water environments*: John Wiley, New York, Advances in Environmental Science and Techology series, v. 8, pp. 195–221.

Metcalf, R.L., Kapoor, I.P., Lu, P.-Y., Schuth, C.K., and Sherman, P., 1973, Model ecosystem studies of the environmental fate of six organochlorine pesticides: *Environ. Health Perspect.*, v. 4, p. 35–44.

Metcalfe, J.L., and Charlton, M.N., 1990, Freshwater mussels as biomonitors for organic industrial contaminants and pesticides in the St. Lawrence River: *Sci. Total Environ.*, v. 97/98, pp. 595–615.

Metcalfe, J.L., and Hayton, A., 1989, Comparison of leeches and mussels as biomonitors for pollution: *J. Great Lakes Res.*, v. 15, no. 4, pp. 654–68.

Mick, D.L., and McDonald, D.B., 1971, Limnological factors affecting pesticide residues in surface waters: Completion report: Iowa Water Resources Research Institute Report 36, 10 p.

Miles, J.R.W., and Harris, C.R., 1971, Insecticide residues in a stream and a controlled drainage system in agricultural areas of southwestern Ontario, 1970: *J. Pest. Monitor.*, v. 5, no. 3, pp. 289–294.

Miller, F.M., and Gomes, E.D., 1974, Detection of DCPA residues in environmental samples: *J. Pest. Monitor.*, v. 8, no. 1, pp. 53–58.

Miller, G.C., and Donaldson, S.G., 1994, Factors affecting photolysis of organic compounds on soils, *in* Helz, G.R., Zepp, R.G., and Crosby, D.G., eds., *Aquatic and surface photochemistry*: Lewis Publishers, Boca Raton, Fla., pp. 97–109.

Miller, G.C., and Zepp, R.G., 1979a, Photoreactivity of aquatic pollutants sorbed on suspended sediments: *Environ. Sci. Technol.*, v. 13, pp. 860–863.

——1979b, Effects of suspended sediment on photolysis rates of dissolved pollutants: *Water Res.*, v. 13, pp. 453–459.

Miller, T.J., and Jude, D.J., 1984, Organochlorine pesticides, PBBs, and mercury in roundwhite fish fillets from Saginaw Bay, Lake Huron, 1977–1978: *J. Great Lakes Res.*, v. 10, no. 2, pp. 215–220.

Milne, G.W.A., ed., 1995, *CRC handbook of pesticides*: CRC Press, Boca Raton, Fla., 402 p.

Mingelgrin, U., and Gerstl, Z., 1983, Reevaluation of partitioning as a mechanism of nonionic chemicals adsorption in soils: *J. Environ. Qual.*, v. 12, no. 1, pp 1–11.

Mischke, T., Brunetti, K., Acosta, V., Weaver, D., and Brown, M., 1985, Agricultural sources of DDT residues in California's environment: California Department of Food and Agriculture, Environmental Hazards Assessment Program report prepared in response to House Resolution no. 53 (1984), 42 p.

Moccia, R.D., 1978, Are goiter frequencies in Great Lakes salmon correlated with organochlorine residues?: *Chemosphere*, v. 7, no. 8, pp. 649–652.

Modin, J.C., 1969, Chlorinated hydrocarbon pesticides in California bays and estuaries: *J. Pest. Monitor.*, v. 3, no. 1, pp. 1–7.

Montgomery, J.H., ed., 1993, *Agrochemicals desk reference: Environmental data*: Lewis Publishers, Chelsea, Mich., 625 p.

Moore, D.E., Hall, J.D., and Hug, W.L., 1974, Endrin in forest streams after aerial seeding with endrin-coated Douglas-fir seed: U.S. Department of Agriculture, Pacific Northwest Forest and Range Experiment Station, Forest Service Research Note PNW-219, 14 p.

Moore, R., Toro, E., Stanton, M., and Khan, M.A.Q., 1977, Absorption and elimination of [14]C-*alpha*- and *gamma*-chlordane by a freshwater alga, daphnid, and goldfish: *Arch. Environ. Contam. Toxicol.*, v. 6, no. 4, pp. 411–420.

Moriarty, F., 1985, Bioaccumulation in terrestrial food chains, in Sheehan, P., Korte, F., Klein, W., and Bourdeau, P., eds., *Appraisal of tests to predict the environmental behavior of chemicals*: Wiley, New York, pp. 257–284.

Moriarty, F., and Walker, C.H., 1987, Bioaccumulation in food chains: A rational approach: *Ecotoxicol. Environ. Safety*, v. 13, no. 2, pp. 208–215.

Morris, R.L., and Johnson, L.G., 1971, Dieldrin levels in fish from Iowa streams: *J. Pest. Monitor.*, v. 5, no. 1, pp. 12–16.

Morris, R.L., Johnson, L.G., and Ebert, D.W., 1972, Pesticides and heavy metals in the aquatic environment: *Health Lab. Sci.*, v. 9, no. 2, pp. 145–151.

Morris, S.C., Fischer, H., Moskowitz, P.D., Rybicka, K., and Thode, H.C., Jr., 1984, Risk assessment: Cancer: *Mech. Eng.*, v. 106, no. 11, pp. 40–45.

Mosher, L., 1985, U.S. farms are losing cropland soil at rate of three billion tons a year: *Ambio*, v. 14, no. 6, pp. 357–358.

Mosser, J.L., Fisher, N.S., and Wurster, C.F., 1972, Polychlorinated biphenyls and DDT alter species composition in mixed cultures of algae: *Science*, v. 176, no. 4034, pp. 533–535.

Mothershead, R.F., Hale, R.C., and Greaves, J.G., 1991, Xenobiotic compounds in blue crabs from a highly contaminated urban estuary: *Environ. Toxicol. Chem.*, v. 10, no. 3, pp. 1341–1349.

Moubry, R.J., Helm, J.M., and Myrdal, G.R., 1968, Chlorinated pesticide residues in an aquatic environment located adjacent to a commercial orchard: *J. Pest. Monitor.*, v. 1, no. 4, pp. 27–29.

Mount, D.I., and Putnicki, G.J., 1966, Summary report of the 1963 Mississippi fish kill: *Trans. N. Am. Wildl. Conf.*, v. 31, pp. 177–184.

Mudambi, A.R., Hassett, J.P., McDowell, W.H., and Scrudato, R.J., 1992, Mirex-photomirex relationships in Lake Ontario: *J. Great Lakes Res.*, v. 18, no. 3, p. 405–414.

Mueller, D.K., DeWeese, L.R., Garner, A.J., and Sprull, T.B., 1991, Reconnaissance investigation of water quality, bottom sediment, and biota associated with irrigation drainage in the middle Arkansas River Basin, Colorado and Kansas, 1988–89: U.S. Geological Survey Water-Resources Investigations Report 91-4060, 84 p.

Muir, D.C.G., Grift, N.P., and Strachan, W.M.J., 1990, Herbicides in rainfall in northwest Ontario, 1989: Presented at the 18th Workshop on the Chemistry and Biochemistry of Pesticides, Regina, Saskatchewan, Canada, April 24–25, 1990.

Muir, D.C.G., Lawrence, S., Holoka, M., Fairchild, W.L., Segstro, M.D., Webster, G.R.B., and Servos, M.R., 1992, Partitioning of polychlorinated dioxins and furans between water, sediments and biota in lake mesocosms: *Chemosphere*, v. 25, no. 1–2, pp. 119–124.

Muir, D.C.G., and Yarechewski, A.L., 1988, Dietary accumulation of four chlorinated dioxin congeners by rainbow trout and fathead minnows: *Environ. Toxicol. Chem.*, v. 7, no. 3, pp. 227–236.

Muir, D.C.G., Yarechewski, A.L., and Knoll, A., 1986, Bioconcentration and disposition of 1,3,6,8-tetrachlorodibenzo-*p*-dioxin and octachlorodibenzo-*p*-dioxin by rainbow trout and fathead minnows: *Environ. Toxicol. Chem.*, v. 5, no. 3, pp. 261–272.

Muir, D.C.G., Yarechewski, A.L., and Webster, G.R.B., 1985, Bioconcentration of four chlorinated dioxins by rainbow trout and fathead minnows, *in* Bahner, R.C., and Hansen, D.J., eds., *Aquatic toxicology and hazard assessment: Eighth symposium*: American Society for Testing and Materials, Philadelphia, Penn., Special Technical Publication series, v. 891, pp. 440–454.

Munkittrick, K.R., McMaster, M.E., Portt, C.B., Van Der Kraak, G.J., Smith, I.R., and Dixon, D.G., 1992a, Changes in maturity, plasma sex steroid levels, hepatic mixed-function oxygenase activity, and the presence of external lesions in lake whitefish exposed to bleached kraft mill effluent: *Can. J. Fish. Aquat. Sci.*, v. 49, no. 8, pp. 1560–1569.

Munkittrick, K.R., Van Der Kraak, G.J., McMaster, M.E., and Portt, C.B., 1992b, Response of hepatic MFO activity and plasma sex steroids to secondary treatment of bleached kraft pulp mill effluent and mill shutdown: *Environ. Toxicol. Chem.*, v. 11, no. 10, pp. 1427–1439.

Munkittrick, K.R., Van Der Kraak, G.J., McMaster, M.E., Portt, C.B., Van den Hevvel, M.R., and Servos, M.R., 1994, Survey of receiving-water environmental impacts associated with dishcarges from pulp mills. 2. Gonad size, liver size, hepatic EROD activity and plasma sex steroid levels in white sucker: *Environ. Toxicol. Chem.*, v. 13, no. 7, pp. 1089–1101.

Murphy, D.L., 1990, Contaminant levels in oysters and clams from the Chesapeake Bay 1981–1985: Maryland Department of the Environment, Water Management Administration Technical Report 102, 112 p.

Murphy, P.G., 1970, Effects of salinity on uptake of DDT, DDE and DDD by fish: *Bull. Environ. Contam. Toxicol.*, v. 5, no. 5, pp. 404–407.

——1971, Effect of size on the uptake of DDT from water by fish: *Bull. Environ. Contam. Toxicol.*, v. 6, no. 1, pp. 20–23.

Murphy, P.G., and Murphy, J.V., 1971, Correlations between respiration and direct uptake of DDT in the mosquito fish: *Bull. Environ. Contam. Toxicol.*, v. 6, no. 6, pp. 581–588.

Murray, H.E., and Beck, J.N., 1989, Halogenated organic compounds found in shrimp from the Calcasieu estuary: *Chemosphere*, v. 19, no. 8–9, pp. 1367–1374.

——1990, Concentrations of selected chlorinated pesticides in shrimp collected from the Calcasieu River/Lake Complex, Louisiana: *Bull. Environ. Contam. Toxicol.*, v. 44, no. 5, pp. 798–804.

Murty, A.S., 1986a, *Toxicity of pesticides to fish*: CRC Press, Boca Raton, Fla., v. 1, 178 p.

——1986b, *Toxicity of pesticides to fish*: CRC Press, Boca Raton, Fla., v. 2, 143 p.

Myers, M.S., Stehr, C.M., Olson, O.P., Johnson, L.L., McCain, B.B., Chan, S.-L., and Varanasi, U., 1993, National Benthic Surveillance Project: Pacific Coast: Fish histopathology and relationships between toxicopathic lesions and exposure to chemical contaminants for cycles I to V (1984–88): U.S. Department of Commerce, National Oceanic and Atmospheric Administration Technical Memorandum MFS NWFSC 6, 160 p.

——1994, Relationships between toxicopathic hepatic lesions and exposure to chemical contaminants in English sole (*Pleuronectes vetulus*), starry flounder (*Platichthys stellatus*), and white croaker (*Genyonemus lineatus*) from selected marine sites on the Pacific Coast, USA: *Environ. Health Perspect.*, v. 102, no. 2, pp. 200–215.

Nagel, R., 1983, Species differences, influence of dose and application on biotransformation of phenol in fish: *Xenobiotica*, v. 13, pp. 101–106.

Nair, A., and Pillai, M.K.K., 1989, Monitoring of hexachlorobenzene residues in Delhi and Faridabad, India: *Bull. Environ. Contam. Toxicol.*, v. 42, no. 5, pp. 682–686.

——1992, Trends in ambient levels of DDT and HCH residues in humans and the environment of Delhi, India: *Sci. Total Environ.*, v. 121, pp. 145–157.

Nair, P.S., 1990a, Chlordane residue levels and geographic distribution of chlordane in fish from the non-tidal portion of the Patuxent River: U.S. Fish and Wildlife Service Report AFO-C90-04, 26 p.

——1990b, Potential contaminant threats to Presquile National Wildlife Refuge: U.S. Fish and Wildlife Service, Environmental Contaminants Division, Annapolis (Maryland) Field Office, 126 p.

Najdek, M., and Bazulic, D., 1988, Chlorinated hydrocarbons in mussels and some benthic organisms from the northern Adriatic Sea: *Mar. Pollut. Bull.*, v. 19, no. 1, pp. 37–38.

Nalepa, T.F., and Landrum, P.F., 1988, Benthic invertebrates and contaminants levels in the Great lakes: Effects, fates, and role in cycling, *in* Evans, M.S., and Gannon, J.E., eds., *Toxic contaminants and ecosystem health: A Great Lakes focus*: Wiley, New York, Advances in Environmental Science and Technology series, v. 21, pp. 77–102.

Naqvi, S.M., and de la Cruz, A.A., 1973, Mirex incorporation in the environment: Residues in nontarget organisms: *J. Pest. Monitor.*, v. 7, no. 2, pp. 104–111.

Nash, T., and Charbonneau, C., 1992, Squaw Creek National Wildlife Refuge Contaminant Survey results: U.S. Fish and Wildlife Service, Columbia (Mo.) Field Office, 15 p., 9 appendices.

National Academy of Sciences and National Academy of Engineering, 1973, Water quality criteria, 1972: U.S. Environmental Protection Agency Report EPA-R3-033, 594 p.

National Oceanic and Atmospheric Administration, 1987, A summary of selected data on chemical contaminants in tissues collected during 1984, 1985, and 1986.: U.S. Department of Commerce, National Oceanic and Atmospheric Administration Technical Memorandum NOS OMA 38, variously paged.

——1988, A summary of selected data on chemical contaminants in sediments collected during 1984, 1985, 1986, and 1987: U.S. Department of Commerce, National Oceanic and Atmospheric Administration Technical Memorandum NOS OMA 44, variously paged.

——1989, A summary of data on tissue contaminantion [sic] from the first three years (1986–1988) of the Mussel Watch Project: U.S. Department of Commerce, National Oceanic and Atmospheric Administration Technical Memorandum NOS OMA 49, variously paged.

——1991, Second summary of data on chemical contaminants in sediments from the National Status and Trends Program: U.S. Deprtment of Commerce, National Oceanic and Atmospheric Administration Technical Memorandum NOS OMA 59, variously paged.

National Research Council, 1979, Polychlorinated biphenyls: A report: National Academy of Sciences, Washington, D.C., 182 p.

——1980, The international mussel watch: Report of a workshop: National Academy of Sciences, Washington, DC, 248 p.

——1985, *Oil in the sea: Inputs, fates, and effects*: National Academy Press, Washington, D.C. 601 p.

National Research Council of Canada, 1974, Chlordane: Its effects on Canadian ecosystems and its chemistry: National Research Council of Canada, Asssociate Committee on Scientific Criteria for Environmental Quality, Publication 14094 of the Environmental Secretariat, 189 p.

Nations, B.K., and Hallberg, G.R., 1992, Pesticides in Iowa precipitation: *J. Environ. Qual.*, v. 21, no. 3, pp. 486–492.

Neary, D.G., Bush, P.B., and Douglass, J.E., 1983, Off-site movement of hexazinone in stormflow and baseflow from forested watersheds: *Weed Sci.*, v. 31, no. 4, pp. 543–551.

Neary, D.G., and Michael, J.L., 1989, Effect of sulfometuron methyl on ground water and stream quality in coastal plain forest watersheds: *Water Resour. Bull.*, v. 25, no. 3, p. 617–623.

Neely, W.B., 1979, Estimating rate constants for the uptake and clearance of chemicals by fish: *Environ. Sci. Tech.*, v. 13, no. 12, pp. 1506–1510.

Neely, W.B., Branson, D.R., and Blau, G.E., 1974, Partition coefficient to measure bioconcentration potential of organic chemicals in fish: *Environ. Sci. Technol.*, v. 8, no. 13, pp. 1113–1115.

Neff, J.M., 1984, Bioaccumulation of organic micropollutants from sediments and suspended particulates by aquatic animals: *Fresenius Z. Anal. Chem.*, v. 319, pp. 132–136.

Neff, J.M., Bean, D.J., Cornaby, B.W., Vaga, R.M., Gulbransen, T.C., and Scanlon, J.A., 1986, Sediment quality criteria methodology validation: Calculation of screening level concentrations from field data: Final report: Battell Washington Environmental Program Office report prepared for the U.S. Environmental Protection Agency, Criteria and Standards Division, Office of Water Regulation and Standards, variously paged.

Nelson, F.P., 1976, Lower Santee River environmental quality study: An assessment of selected biological and physical parameters: South Carolina Water Resources Commission, State Water Plan Estuarine Studies series, no. 122, 60 p.

Nettleton, J.A., Allen, W.H., Jr., Klatt, L.V., Ratnayake, W.M.N., and Ackman, R.G., 1990, Nutrients and chemical residues in one- to two-pound farm-raised channel catfish (*Ictalurus punctatus*): *J. Food Sci.*, v. 55, no. 4, pp. 954–958.

New York State Department of Environmental Conservation, 1979, Toxic substances in fish and wildlife, 1978 annual report: New York State Department of Environmental Conservation, Division of Fish and Wildlife, Bureau of Environmental Protection Technical Report 79-1, v. 2.

Newell, A.J., Johnson, D.W., and Allen, L.K., 1987, Niagara River biota contamination project: Fish flesh criteria for piscivorous wildlife: New York State Department of Environmental Conservation, Division of Fish and Wildlife, Bureau of Environmental Protection Technical Report 87-3, 182 p.

Nichols, M.M., 1990, Sedimentologic fate and cycling of kepone in an estuarine system: Example from the James River estuary: *Sci. Total Environ.*, v. 97/98, pp. 407–440.

Nichols, M.M., and Cutshall, N.H., 1981, Tracing kepone contamination in James estuary sediments: *Rapp. P.-V. Reun. Cons. Int. Explor. Mer.*, v. 181, pp. 102–110.

Niimi, A.J., 1979, Hexachlorobenzene (HCB) levels in Lake Ontario salmonids: *Bull. Environ. Contam. Toxicol.*, v. 23, no. 1–2, pp. 20–24.

Niimi, A.J., and Beamish, F.W.H., 1974, Bioenergetics and growth of largemouth bass (*Micropterus salmonides*) in relation to body weight and temperature: *Can. J. Zool.*, v. 52, pp. 447–456.

Niimi, A.J., and Oliver, B.G., 1983, Biological half-lives of polychlorinated biphenyl (PCB) congeners in whole fish and muscle of rainbow trout (*Salmo gairdnieri*): *Can. J. Fish. Aquat. Sci.*, v. 40, no. 9, pp. 1388–1394.

——1989, Distribution of polychlorinated biphenyl congeners and other halocarbons in whole fish and muscle among Lake Ontario salmonids: *Environ. Sci. Technol.*, v. 23, no. 1, pp. 83–88.

Nilsson, S., 1986, Control of gill blood flow, *in* Nilsson, S., and Holmgren, S., eds., *Fish physiology: Recent advances*: Croom Helm, London, U.K., pp. 86–101.

Noegrohati, S., Sardjoko, Untung, K., and Hammers, W.E., 1992, Impact of DDT spraying on the residue levels in soil, chicken, fish-pond water, carp, and human milk samples from malaria-infested villages in central Java: *Toxicol. Environ. Chem.*, v. 34, no. 2–4, pp. 237–251.

Norris, L.A., 1976, Forests and rangelands as sources of chemical pollutants, *in Non-Point Sources of Water Pollution*: Oregon State University, Water Resources Research Institute (Corvallis) report prepared for the Office of Water Research and Technology, Washington, D.C., pp. 17–35.

——1981, Phenoxy herbicides and TCDD in forests: *Residue Rev.*, v. 80, pp. 65–135.

Norris, L.A., Montgomery, M.L., Warren, L.E., and Mosher, W.D., 1982, Brush control with herbicides on hill pasture sites in southern Oregon: *J. Range Manag.*, v. 35, no. 1, pp. 75–80.

Norstrom, R.J., McKinnon, R.E., and DeFreitas, A.S.W., 1976, A bioenergetic-based model for pollutant accumulation by fish: Simulation of PCB and methylmercury residue levels in Ottawa River yellow perch (*Perca flavescens*): *Can. J. Fish. Aquat. Sci.*, v. 33, no. 2, pp. 248–267.

Norstrom, R.J., and Muir, D.C.G., 1988, Long-range transport of organochlorines in the Arctic and sub-Arctic: Evidence from analysis of marine mammals and fish, *in* Schmidtke, N.W., ed., *Toxic contamination in large lakes*: Lewis Publishers, Chelsea, Mich., pp. 83–112.

Nowak, B., 1990, Residues of endosulfan in the livers of wild catfish from a cotton growing area: *Environ. Monitor. Assess.*, v. 14, pp. 347–351.

Nowak, B., and Julli, M., 1991, Residues of endosulfan in wild fish from cotton growing areas in New South Wales, Australia: *Toxicol. Environ. Chem.*, v. 33, no. 3–4, pp. 151–167.

Nowell, L.H., and Resek, E.A., 1994, National standards and guidelines for pesticides in water, sediment, and aquatic organisms: Application to water-quality assessments: *Rev. Environ. Contam. Toxicol.*, v. 140, p. 1–164.

Nygren, M., Rappe, C., Lindstrom, G., Hansson, M., Berqvist, P.-A., Marklund, S., Domellof, L., Hardell, L., and Olsson, M., 1987, Identification of 2,3,7,8-substituted polychlorinated dioxins and dibenzofurans in environmental and human samples, *in* Exner, J.H., ed., *Solving hazardous waste problems: Learning from dioxins*: American Chemical Society Syposium series, v. 338, pp. 20–33.

O'Connor, J.M., and Huggett, R.J., 1988, Aquatic pollution problems, North Atlantic coast, including Chesapeake Bay: *Aquat. Toxicol.*, v. 11, no. 1–2, pp. 163–190.

O'Connor, J.M., Kneip, T.J., and Lee, C.C., 1980, Biological monitoring of PCBs in Hudson River biota: Progress report for 1979–80: New York State Department of Environmental Conservation, Office of Water Research, 37 p.

O'Connor, J.M., and Pizza, J.C., 1987, Pharmacokinetic model for the accumulation of PCBs in marine fish, *in* Capuzzo, J.M., Kester, D.R., eds., *Biological processes and wastes in the ocean*: Krieger Publishing, Malabar, Fla., Oceanic Process in Marine Pollution series, v. 1, pp. 119–129.

O'Connor, T.P., 1991, Concentrations of organic contaminants in mollusks and sediments at NOAA national status and trends sites in the coastal and estuarine United States: *Environ. Health Perspect.*, v. 90, no. 1, pp. 69–73.

———1992, Recent trends in coastal environmental quality: Results from the first five years of the NOAA Mussel Watch Program: U.S. Department of Commerce, National Oceanic and Atmospheric Administration, Rockville, Md., 46 p.

O'Connor, T.P., and Ehler, C.N., 1991, Results from the NOAA National Status and Trends Program on distribution and effects of chemical contamination in the coastal and estuarine United States: *Environ. Monit. Assess.*, v. 17, pp. 33–49.

O'Keefe, P., Hilker, D., Meyer, C., Aldous, K., Shane, L., and Donnelly, R., 1984, Tetrachlorodibenzo-*p*-dioxins and tetrachlorodibenzofurans in Atlantic Coast striped bass and in selected Hudson River fish, waterfowl and sediments: *Chemosphere*, v. 13, no. 8, pp. 849–860.

O'Keefe, P., Meyer, C., Hilker, D., Aldous, K., Jelus-Tyror, B., and Dillon, K., 1983, Analysis of 2,3,7,8-tetrachlorodibenzo-*p*-dioxin in Great Lakes fish: *Chemosphere*, v. 12, no. 3, pp. 325–332.

O'Shea, T.J., Ludke, J.L., and Hines, B., 1979, Monitoring fish and wildlife for environmental pollution: U.S. Department of the Interior, Fish and Wildlife Service monograph, 12 p.

Oladimeji, A.A., and Leduc, G.V., 1975, Effects of dietary methoxychlor on the food maintenance requirements of brook trout: *Water Sci. Technol.*, v. 7, no. 3–4, pp. 587–597.

Oliver, B.G., 1984, Distribution and pathways of some chlorinated benzenes in the Niagara River and Lake Ontario: *Water Poll. Res. J. Canada*, v. 19, no. 1, pp. 47–58.

———1987, The fate of some chlorobenzenes from the Niagara River in Lake Ontario, *in* Hites, R.A., and Eisenreich, S.J., eds., *Sources and fates of aquatic pollutants*: American Chemical Society Advances in Chemistry series, no 216, pp. 471–489.

Oliver, B.G., and Nicol, K.D., 1984, Chlorinated contaminants in the Niagara River, 1981–1983: *Sci. Total Environ.*, v. 39, no. 1–2, pp. 57–70.

Oliver, B.G., and Niimi, A.J., 1983, Bioconcentration of chlorobenzenes from water by rainbow trout: Correlations with partition coefficients and environmental residues: *Environ. Sci. Technol.*, v. 17, no. 5, pp. 287–291.

——1985, Bioconcentration factors of some halogenated organics for rainbow trout: Limitations in their use for prediction of environmental residues: *Environ. Sci. Technol.*, v. 19, pp. 842–849.

——1988, Trophodynamic analysis of polychlorinated biphenyl congeners and other chlorinated hydrocarbons in the Lake Ontario ecosystem: *Environ. Sci. Technol.*, v. 22, no. 4, pp. 388–397.

Oliver, B.G., and Pugsley, C.W., 1986, Chlorinated contaminants in St. Clair River sediments: *Water Pollut. Res. J. Can.*, v. 21, no. 3, pp. 368–379.

Olsson, M., Jensen, S., and Reutergardh, L., 1978, Seasonal variation of PCB levels in fish—An inportant factor in planning aquatic monitoring programs: *Ambio*, v. 8, p. 270.

Ong, K., O'Brien, T.F., and Rucker, M.D., 1991, Reconnaissance investigation of water quality, bottom sediment, and biota associated with irrigation drainage in the Middle Rio Grande Valley and Bosque del Apache National Wildlife Refuge, New Mexico, 1988–89: U.S. Geological Survey Water-Resources Investigations Report 91-4036, 113 p.

Opperhuizen, A., Gobas, F.A.P.C., and Hutzinger, O., 1984, Unmetabolized compounds, their properties and implications, *in* Caldwell, J., and Paulson, G.D., eds., *Foreign compound metabolism*: Taylor and Francis, London, U.K., pp. 109–117.

Opperhuizen, A., and Schrap, S.M., 1987, Relationships between aqueous oxygen concentration and uptake and elimination rates during bioconcentration of hydrophobic chemicals in fish: *Environ. Toxicol. Chem.*, v. 6, no. 5, pp. 335–342.

Opperhuizen, A., and Sijm, D.T.H.M., 1990, Bioaccumulation and biotransformation of polychlorinated dibenzo-*p*-dioxins and dibenzofurans in fish: *Environ. Toxicol. Chem*, v. 9, no. 2, pp. 175–186.

Opperhuizen, A., and Stokkel, R.C.A.M., 1988, Influence of contaminated particles on the bioaccumulation of hydrophobic organic micropollutants in fish: *Environ. Pollut.*, v. 51, no. 3, pp. 165–177.

Opperhuizen, A., Velde, E.W.v.d., Gobas, F.A.P.C., Liem, D.A.K., Steen, J.M.D.v.d., and Hutzinger, O., 1985, Relationship between bioconcentration in fish and steric factors of hydrophobic chemicals: *Chemosphere*, v. 14, no. 11/12, pp. 1871–1896.

Orazio, C.E., Kapila, S., Puri, R.K., Meadows, J., and Yanders, A.F., 1990, Field and laboratory studies on sources and persistence of chlordane contamination in the Missouri aquatic environment: *Chemosphere*, v. 20, no. 10–12, pp. 1581–1588.

Ott, A.N., Takita, C.S., and Bollinger, S.W., 1989, Estimate of the occurrence and distribution of toxicants in streambed sediments of the Susquehanna River system: Susquehanna River Basin Commission Publication Number 123, 20 p.

Parejko, R., and Wu, C.-L.J., 1977, Chlorohydrocarbons in Marquette Fish Hatchery lake trout (*Salvelinus namaycush*): *Bull. Environ. Contam. Toxicol.*, v. 17, no. 1, pp. 90–97.

Paris, D.F., Lewis, D.L., and Barnett, J.T., 1977, Bioconcentration of toxaphene by microorganisms: *Bull. Environ. Contam. Toxicol.*, v. 17, no. 5, pp. 564–572.

Paris, D.F., Steen, W.C., and Baughman, G.L., 1978, Rate of physico-chemical properties of aroclors 1016 and 1242 in determining their fate and transport in aquatic environments: *Chemosphere*, v. 7, no. 4, pp. 319–325.

Pariso, M.E., St. Amant, J.R., and Sheffy, T.B., 1984, Microcontaminants in Wisconsin's coastal zone, *in* Nriagu, J.O., and Simmons, M.S., eds., *Toxic contaminants in the Great Lakes*: Wiley, New York, Advances in Environmental Science and Technology series, v. 14, pp. 265–285.

Parsons, A.M., and Moore, D.J., 1966, Some reactions of dieldrin and the proton magnetic resonance spectra of the products: *J. Chem. Soc.*, v 22C, pp. 2026–2031.

Pastel, M., Bush, B., and Kim, J.S., 1980, Accumulation of polychlorinated biphenyls in American shad during their migration in the Hudson River, spring 1977: *J. Pest. Monitor.*, v. 14, no. 1, pp. 11–22.

Patton, G.W., Hinckley, D.A., Walla, M.D., and Bidleman, T.F., 1989, Airborne organochlorines in the Canadian high arctic: *Tellus*, v. 41B, no. 3, pp. 243–255.

Pavlou, S.P., and Dexter, R.N., 1979, Distribution of polychlorinated biphenyls (PCB) in estuarine ecosystems: Testing the concept of equilibrium partitioning in the marine environment: *Environ. Sci. Technol.*, v. 13, no. 1, pp. 65–70.

Pedersen, M.G., Herschberger, W.K., Zachariah, P.K., and Juchau, M.R., 1976, Hepatic biotransformation of environmental xenobiotics in six strains of rainbow trout (*Salmo gairdnieri*): *Can. J. Fish. Aquat. Sci.*, v. 33, no. 4, pp. 666–675.

Pennington, J.C., Bunch, B.W., Ruiz, C.E., Brandon, D.L., Sturgis, T.C., Price, C.B., and Brannon, J.M., 1991, Water quality studies in the Upper Yazoo Project area, Mississippi: U.S. Army Engineer Waterways Experiment Station Miscellaneous Paper EL-91-22, 157 p.

Pereira, W.E., and Rostad, C.E., 1990, Occurrence, distributions, and transport of herbicides and degradation products in the lower Mississippi River and its tributaries: *Environ. Sci. Technol.*, v. 24, no. 9, pp. 1400–1406.

Pereira, W.E., Rostad, C.E., Chiou, C.T., Brinton, T.I., Barber, L.B., II, Demcheck, D.K., and Demas, C.R.,1988, Contamination of estuarine water, biota, and sediment by halogenated organic compounds: A field study: *Environ. Sci. Technol.*, v. 22, no.7, pp. 772–778.

Perry, J.A., 1979, Pesticide and PCB residues in the upper Snake River ecosystems, southeastern Idaho, following the collapse of the Teton Dam, 1976: *Arch. Environ. Contam. Toxicol.*, v. 8, no. 2, pp. 139–159.

Persaud, D., Jaagumagi, R., and Hayton, A., 1993, Guidelines for the protection and management of aquatic sediment quality in Ontario: Report: Ontario Ministry of the Environment, Water Resources Branch, [Toronto], Ontario, Canada, 23 p.

Pesonen, M., Andersson, T., and Forlin, L., 1985, Characterization and induction of cytochrome P-450 in the rainbow trout kidney: *Mar. Environ. Res.*, v. 17, pp. 106–108.

Peterle, T.J., 1969, DDT in Antarctic snow: *Nature*, v. 224, no. 5219, p. 620.

Peterman, P.H., Delfino, J.J., Dube, D.J., Gibson, T.A., and Priznar, F.J., 1980, Chloro-organic compounds in the lower Fox River, Wisconsin: *Environ. Sci. Res.*, v. 16, pp. 145–160.

Peters, L.S., and O'Connor, J.M., 1982, Factors affecting short-term PCB and DDT accumulation by zooplankton and fish from the Hudson estuary, *in* Mayer, G.F., ed., *Ecological stress and the New York Bight: Science and management*: Estuarine Research Foundation, Columbia, S.C, pp. 451–465.

Petersen, J.C., 1981, Water-quality reconnaissance of the Larkin Creek watershed, Lee and St. Francis counties, Arkansas: U.S. Geological Survey Open-File Report 81-819, 20 p.

——1990, Trends and comparison of water quality and bottom material of northeastern Arkansas streams, 1974–85, and effects of planned diversions: U.S. Geological Survey Water-Resources Investigations Report 90-4017, 215 p.

Peterson, D.A., Harms, T.F., Ramirez, J., Pedro, Allen, G.T., and Christenson, A.H., 1991, Reconnaissance investigation of water quality, bottom sediment, and biota associated with irritgation drainage in the Riverton Reclamation Project, Wyoming, 1988–89: U.S. Geological Survey Water-Resources Investigations Report 90-4187, 84 p.

Peterson, D.A., Jones, W.E., and Morton, A.G., 1988, Reconnaissance investigation of water quality, bottom sediment, and biota associated with irritgation drainage in the Kendrick Reclamation Project area, Wyoming, 1986–87: U.S. Geological Survey Water-Resources Investigations Report 87-4255, 57 p.

Petrocelli, S.R., Anderson, J.W., and Hanks, A.R., 1975, Seasonal fluctuations of dieldrin residues in the tissues of marsh clam, *Rangia cuneata*, from a Texas estuary: *Tex. J. Sci.*, v. 26, no. 3–4, pp. 443–448.

Phillips, D.J.H., 1978, Use of biological indicator organisms to quantitate organochlorine pollutants in aquatic environments—A review: *Environ. Pollut.*, v. 16, no. 3, pp. 167–229.

——1980, Quantitative aquatic biological indicators: Their use to monitor trace metal and organochlorine pollution, *in* Mellanby, K., ed., *Pollution Monitoring Series*: Elsevier Applied Science, London, U.K., 488 p.

Phillips, D.J.H., and Spies, R.B., 1988, Chlorinated hydrocarbons in the San Francisco estuarine ecosystem: *Mar. Pollut. Bull.*, v. 19, no. 9, pp. 445–453.

Phillips, L.J., and Birchard, G.F., 1991, Regional variations in human toxics exposure in the USA: An analysis based on the National Human Adipose Tissue Survey: *Arch. Environ. Contam. Toxicol.*, v. 21, no. 2, pp. 159–168.

——1991, Use of STORET data to evaluate variations in environmental contamination by census division: *Chemosphere*, v. 22, no. 9–10, pp. 835–848.

Phillips, P.T., 1988, California State mussel watch, ten year data summary, 1977–1987: California State Water Resources Control Board, Water Quality Monitoring Report 87-3, variously paged.

Pierce, R.H., Jr., Brent, C.R., Williams, H.P., and Reeves, 1977, Pentachlorophenol distribution in fresh water ecosystem: *Bull. Environ. Contam. Toxicol.*, v. 18, no. 2, pp. 251–258.

Pierce, R.H., Jr., Olney, C.E., and Felbeck, G.T., Jr., 1971, Pesticide adsorption in soils and sediments: *Environ. Let.*, v. 1, no. 2, pp. 157–172.

Pinza, M.R., Word, J.Q., Barrows, E.S., Mayhew, H.L., and Clark, D.R., 1991, Snake and Columbia River Sediment Sampling Project: U.S. Army Corps of Engineers, Walla Walla District, Wash., variously paged.

Pionke, H.B., and Chesters, G., 1973, Pesticide-sediment-water interactions: *J. Environ. Qual.*, v. 2, no. 1, pp. 29–45.

Pizza, J.C., and O'Connor, J.M., 1983, PCB dynamics in Hudson River striped bass. II. Accumulation from dietary sources: *Aquat. Toxicol.*, v. 3, pp. 313–327.

Plant, A.L., Pownall, H.J., and Smith, L.C., 1983, Transfer of polycyclic aromatic hydrocarbons between model membranes: Relation to carcinogenicity: *Chem.-Biol. Interact.*, v. 44, no. 3, pp. 237–246.

Porte, C., and Albaiges, J., 1994, Bioaccumulation patterns of hydrocarbons and polychlorinated biphenyls in bivalves, crustaceans, and fishes: *Arch. Environ. Contam. Toxicol.*, v. 26, no. 3, pp. 273–281.

Porte, C., Barcelo, D., and Albaiges, J., 1992, Monitoring of organophosphorus and organochlorinated compounds in a rice crop field (Ebro Delta, Spain) using the mosquitofish *Gambusia affinis* as indicator organism: *Chemosphere*, v. 24, no. 6, pp. 735–743.

Pranckevicius, P.E., 1987, 1982 Detroit, Michigan area sediment survey: U.S. Environmenal Protection Agency, Great Lakes Nation Program Office Report GLNPO 87-11, variously paged.

Premdas, F.H., and Anderson, J.M., 1963, The uptake and detoxification of C^{14}-labeled DDT in Atlantic salmon, *Salmo salar*: *Can. J. Fish. Aquat. Sci.*, v. 20, p. 827.

Prest, H.F., Jarman, W.M., Burns, S.A., Weismuller, T., Martin, M., and Huckins, J.N., 1992, Passive water sampling via semipermeable membrane devices (SPMDs) in concert with bivalves in the Sacramento/San Joaquin River Delta: *Chemosphere*, v. 25, no. 12, pp. 1811–1823.

Pritchard, J.B., Cotton, C.U., James, M.O., Giguere, D., and Koschier, F.J., 1978, Role of metabolism and transport in the excretion of phenylacetic acid and 2,4-dichlorophenoxyacetic acid by marine fish: *Bull. Mt. Desert Isl. Biol. Lab.*, v. 18, pp. 58–60.

Pritchard, J.B., and James, M.O., 1979, Determinants of the renal handling of 2,4-dichlorophenoxyacetic acid by winter flounder: *J. Pharmacol. Exp. Ther.*, v. 208, pp. 280–286.

Pritchard, J.B., Karnaky, K.J., Jr., Guarino, A.M., and Kinter, W.B., 1977, Renal handling of the polar DDT metabolite DDA (2,2-bis[p-chlorophenyl] acetic acid) by marine fish: *Am. J. Physiol.*, v. 233, pp. F126–F132.

Pruell, R.J., Rubinstein, N.I., Taplin, B.K., LiVolsi, J.A., and Bowen, R.D., 1993, Accumulation of polychlorinated organic contaminants from sediment by three benthic marine species: *Arch. Environ. Contam. Toxicol.*, v. 24, no. 3, pp. 290–297.

PTI Environmental Services and Tetra Tech Inc., 1988, Everett Harbor Action Program: Analysis of toxic problem areas. Puget Sound Estuary Program: Final report and appendices: Prepared for the U.S. Environmental Protection Agency, Seattle, Wash., EPA 9-88-241, 286 p., 10 appendices.

Punzo, F., Laveglia, J., Lohr, D., and Dahm, P.A., 1979, Organochlorine insecticide residues in amphibians and reptiles from Iowa and lizards from the southwestern United States: *Bull. Environ. Contam. Toxicol.*, v. 21, no. 6, pp. 842–848.

Purdom, C.E., Hardiman, P.A., Bye, V.J., Eno, N.C., Tyler, C.R., and Sumpter, J.P., 1994, Estrogenic effects of effluents from sewage treatment works: *Chem. Ecol.*, v. 8, no. 4, pp. 275–285.

Puri, R.K., Orazio, C.E., Kapila, S., Clevenger, T.E., Yanders, A.F., McGrath, K.E., Buchanan, A.C., Czarnezki, J., and Bush, J., 1990, Studies on the transport and fate of chlordane in the environment, *in* Kurtz, D.A., ed., *Long range transport of pesticides*: Lewis Publishers, Chelsea, Mich., pp. 271–289.

Putnam, D.J., Plewa, F.R., Gustafson, R.D., Rice, C.L., and Rutkosky, F.W., 1985, Lackawanna Refuse Site. A National Priority List (NPL) site. Lakawanna County, Pennsylvania: U.S. Fish and Wildlife Service Resource Contaminant Assessment Report 84-3, 16 p., 3 appendices.

Qasim, S.R., Armstrong, A.T., Corn, J., and Jordan, B.L., 1980, Quality of water and bottom sediments in the Trinity River: *Water Resour. Bull.*, v. 16, no. 3, pp. 522–531.

Racke, K.D., 1993, Urban pest control scenarios and chemicals, *in* Racke, K.D., and Leslie, A.R., eds., *Pesticides in urban environments: Fate and significance*: American Chemical Society Symposium series, v. 522, pp. 2–9.

Radtke, D.B., Buell, G.R., and Perlman, H.A., 1984, Water quality management studies. West Point Lake, Chattahoochee River, Alabama-Georgia, April 1978–December 1979: U.S. Army Corps of Engineers Report COESAM/PDEE-84/004, 527 p.

Radtke, D.B., Kepner, W.G., and Effertz, R.J., 1988, Reconnaissance investigation of water quality, bottom sediment, and biota associated with irrigation drainage in the Lower-Colorado River Valley, Arizona, California, and Nevada: U.S. Geological Survey Water-Resources Investigation Report 88-4002, 77 p.

Raloff, J., 1994a, The gender benders: Are environmental hormones emasculating wildlife?: *Sci. News*, v. 145, no. 2, pp. 24–27.

——1994b, That feminine touch: Are men suffering from prenatal or childhood exposure to "hormonal" toxicants?: *Sci. News*, v. 145, no. 4, pp. 56–58.

——1994c, Beyond estrogens, Why unmasking hormone-mimicking pollutants proves so challenging: *Sci. News*, v. 148, no. 3, pp. 44–46.

Ramade, F., 1989, The pollution of the hydrosphere by global contaminants and its effects on aquatic ecosystems, *in* Boudou, A., and Ribeyre, F., eds., *Aquatic ecotoxicology: Fundamental concepts and methodologies*: CRC Press, Fla., v. 1, pp. 151–183.

Randall, R.C., Lee, H.L., II, Ozretich, R.J., Lake, J.L., and Pruell, R.J., 1991, Evaluation of selected lipid methods for normalizing pollutant bioaccumulation: *Environ. Technol. Chem.*, v. 10, no. 11, pp. 1431–1436.

Rapaport, R.A., and Eisenreich, S.J., 1986, Atmospheric deposition of toxaphene to eastern North America derived from peat accumulation: *Atmospheric Environ.*, v. 20, no. 12, pp. 2367–2379.

——1988, Historical inputs of high molecular weight chlorinated hydrocarbons to eastern North America: *Environ. Sci. Technol.*, v. 22, no. 8, pp. 931–941.

Rapaport, R.A., Urban, N.R., Capel, P.D., Baker, J.E., Looney, B.B., and Eisenreich, S.J., 1985, "New" DDT inputs to North America—Atmospheric deposition: *Chemosphere*, v. 14, no. 9, pp. 1167–1173.

Rasmussen, D., 1992, Toxic substances monitoring program, 1990 data report: California Environmental Protection Agency, Water Resources Control Board, Water Quality Monitoring Report 92-1WQ, variously paged.

Rasmussen, D., and Blethrow, H., 1990, Toxic substances monitoring program, ten year summary report, 1978–1987: California Environmental Protection Agency, Water Resources Control Board, Water Quality Monitoring Report 90-1WQ, variously paged.

——1991, Toxic substances monitoring program, 1988–89 data report: California Environmental Protection Agency, Water Resources Control Board, Water Quality Monitoring Report 91-1WQ, variously paged.

Rasmussen, J.B., Rowan, D.J., Lean, D.R.S., and Carey, J.H., 1990, Food chain structure in Ontario lakes determines PCB levels in lake trout (*Salvelinus namaycush*) and other pelagic fish: *Can. J. Fish. Aquat. Sci.*, v. 47, no. 10, pp. 2030–2038.

Rathbun, R.E., 1998, Transport, behavior, and fate of volatile organic compounds in streams: U.S. Geological Survey Professional Paper 1589, 151 pp.

Ray, L.E., Murray, H.E., and Giam, C.S., 1983, Organic pollutants in marine samples from Portland, Maine: *Chemosphere*, v. 12, no. 7/8, pp. 1031–1038.

Reed, R.J., 1966, Some effects of DDT on the ecology of salmon streams in southeastern Alaska: U.S. Fish and Wildlife Service, Bureau of Commercial Fisheries, Special Scientific Report— Fisheries series, no. 542, 15 p.

Reeves, R.G., Woodham, D.W., Ganyard, M.C., and Bond, C.A., 1977, Preliminary monitoring of agricultural pesticides in a cooperative tobacco pest management project in North Carolina, 1971: First-year study: *J. Pest. Monitor.*, v. 11, no. 2, pp. 99–106.

Reich, A.R., Perkins, J.L., and Cutter, G., 1986, DDT contamination of a north Alabama aquatic ecosystem: *Environ. Toxicol. Chem.*, v. 5, no. 8, pp. 725–736.

Reijnders, P.J.H., 1986, Reproductive failure in common seals feeding on fish from polluted coastal waters: *Nature*, v. 324, pp. 456–457.

Reijnders, P.J.H., and Brasseur, S.M.J.M., 1992, Xenobiotic induced hormonal and associated developmental disorders in marine organisms and related effects in humans; An overview, *in* Colborn, T., and Clement, C.R., eds., *Chemically induced alterations in sexual and functional development: The wildlife/human connection*: Princeton Scientific Publishing, Princeton, N.J., Advances in Modern Environmental Toxicology series, v. 21, pp. 159–174.

Reimold, R.J., and Durant, C.J., 1972, Survey of toxaphene levels in Georgia estuaries: Georgia Marine Science Center, Skidaway Island, Georgia, Technical Report series, no. 72-2, 51 p.

——1974, Toxaphene content of estuarine fauna and flora before, during, and after dredging toxaphene–contaminated sediments: *J. Pest. Monitor.*, v. 8, no. 1, pp. 44–49.

Reinbold, K.A., Kapoor, I.P., Childers, W.F., Bruce, W.N., and Metcalf, R.L., 1971, Comparative uptake and biodegradability of DDT and methoxychlor to aquatic organisms: *Bull. Ill. Nat. Hist. Surv.*, v. 30, no. 6, pp. 405–417.

Reinbold, K.A., and Metcalf, R.L., 1976, Effects of the synergist piperonyl butoxide on metabolism of pesticides in green sunfish: *Pest. Biochem. Physiol.*, v. 6, no. 5, pp. 401–412.

Reinert, R.E., 1967, The accumulation of dieldrin in an algal (*Scenedesmus obliquus*), daphnia (*Daphnia magna*), guppy (*Lebistus reticulatus*) food chain: *Diss. Abstr.*, v. 28, pp. 2210B.

——1970, Pesticide concentrations in Great Lakes fish: *J. Pest. Monitor.*, v. 3, no. 4, pp. 233–240.

——1972, Accumulation of dieldrin in an alga (*Scenedesmus obliquus*), *Daphnia magna*, and the guppy (*Pecilia reticulata*): *Can. J. Fish. Aquat. Sci.*, v. 29, no. 10, pp. 1413–1418.

Reinert, R.E., and Bergman, H.L., 1974, Residues of DDT in lake trout (*Salvelinus namaycush*) and Coho salmon (*Oncorhynchus kisutch*) from the Great Lakes: *Can. J. Fish. Aquat. Sci.*, v. 31, no. 2, pp. 191–199.

Reinert, R.E., Knuth, B.A., Kamrin, M.A., and Stober, Q.J., 1991, Risk assessment, risk management, and fish consumption advisories in the United States: *Fisheries,* v. 16, no. 6, pp. 5–12.

Reinert, R.E., Stone, L.J., and Bergman, H.L., 1974a, Dieldrin and DDT: accumulation from water and food by lake trout (*Salvelinus namaycush*) in the laboratory, in *Proceedings, Seventeenth Conference on Great Lakes Research*: International Association for Great Lakes Research, pt. 1, pp. 52–58.

Reinert, R.E., Stone, L.J., and Willford, W.A., 1974b, The effect of temperature on accumulation of methylmercuric chloride and *p,p'*-DDT by rainbow trout (*Salmo gairdnieri*): *Can. J. Fish. Aquat. Sci.*, v. 31, no. 10, pp. 1649–1652.

Reish, D.J., Geesey, G.G., Wilkes, F.G., Oshida, P.S., A.J.S, M., Rossi, S.S., and Ginn, T.C., 1982, Marine and estuarine pollution: *J. Water Poll. Control Fed.*, v. 54, no. 6, pp. 786–812.

Ribick, M.A., Dubay, G.R., Petty, J.D., Stalling, D.L., and Schmitt, C.J., 1982, Toxaphene residues in fish: Identification, quantification and confirmation at part per billion levels: *Environ. Sci. Technol.*, v. 16, no. 6, pp. 310–318.

Rice, C.L., 1987, Follow-up survey of contaminants in fish from four streams in Centre County, Pennsylvania: U.S. Fish and Wildlife Service Environmental Contaminant Report 88-1, 63 p.

——1990, Chemical analysis of fish and wood duck eggs from the Erie National Wildlife Refuge, Crawford County, Pennsylvania, 1986: U.S. Fish and Wildlife Service Special Project Report 90-4, 32 p.

——1991, Chemical analysis of two whole brook trout samples from the Quehanna Wildlife Area, Pennsylvania: U.S. Fish and Wildlife Service Special Project Report 91-3, 9 p., 2 appendices.

——1992a, Concentrations of mercury and organochlorine compounds in river otter prey organisms in Pennsylvania: U.S. Fish and Wildlife Service Special Project Report 93-3, 8 p.

——1992b, Chemical analysis of sediments and fish from the Mahoning River, Lawrence County, Pennsylvania: U.S. Fish and Wildlife Service Special Project Report 93-2, 15 p., 2 appendices.

Rice, C.P., Samson, P.J., and Noguchi, G.E., 1986, Atmospheric transport of toxaphene to lake Michigan: *Environ. Sci. Technol.*, v. 20, no. 11, pp. 1109–1116.

Rinella, F.A., 1993, Evaluation of organic compounds and trace elements in Amazon Creek Basin, Oregon, September 1990: U.S. Geological Survey Water-Resources Investigations Report 93-4041, 41 p.

Rinella, F.A., Mullins, W.H., and Schuler, C.A., 1994, Reconnaissance investigation of water quality, bottom sediment, and biota associated with irrigation drainage in the Owyhee and Vale Projects, Oregon and Idaho, 1990–91: U.S. Geological Survey Water-Resources Investigations Report 93-4156, 101 p.

Rinella, F.A., and Schuler, C.A., 1992, Reconnaissance investigation of water quality, bottom sediment, and biota associated with irrigation drainage in the Malheur National Wildlife Refuge, Harney County, Oregon, 1988–89: U.S. Geological Survey Water-Resources Investigations Report 91-4085, 106 p.

Rinella, J.F., Hamilton, P.A., and McKenzie, S.W., 1993, Persistence of the DDT pesticide in the Yakima River Basin, Washington: U.S. Geological Survey Circular 1090, 24 p.

Rinella, J.F., and McKenzie, S.W., 1977, Elutriation study of Willamette River bottom material and Willamette-Columbia River water: U.S. Geological Survey Open-File Report 78-28, 8 p.

Rinella, J.F., McKenzie, S.W., Crawford, J.K., Foreman, W.T., Gates, P.M., Fuhrer, G.J., 1992, Surface-water-quality assessment of the Yakima River Basin, Washington: Pesticide and other trace-organic-compound data for water, sediment, soil, and aquatic biota, 1987–91: U.S. Geological Survey Open-File Report 92-644, 154 p.

Risebrough, R.W., 1990, Beyond long-range transport: A model of a global gas chromatographic system, *in* Kurtz, D.A., ed., *Long range transport of pesticides*: Lewis Publishers, Chelsea, Mich., pp. 417–426.

Ritchey, S.J., Young, R.W., and Essary, E.O., 1967, The effects of cooking on chlorinated hydrocarbon pesticide residues in chicken tissues: *J. Food Sci.*, v. 32, no. 2, pp. 238–240.

——1969, Cooking methods and heating effects of DDT residues in chicken tissues: *J. Food Sci.*, v. 34, no. 6, pp. 569–571.

Robbins, C.H., Garrett, J.W., and Mulderink, D.M., 1985, May 1984–April 1985 water budget of Reelfoot Lake with estimates of sediment inflow and concentrations of pesticides in bottom material in tributary streams: Basic data report: U.S. Geological Survey Open-File Report 85-498, 37 p.

Roberts, J.R., De Freitas, A.S.W., and Gidley, M.A.J., 1977, Influence of lipid pool size on bioaccumulation of the insecticide chlordane by northern redhorse suckers: *Can. J. Fish. Aquat. Sci.*, v. 34, no. 1, pp. 89–97.

Roberts, M.H., Jr., 1981, Kepone distribution in selected tissues of blue crabs, *Callinectes sapidus*, collected from the James River and lower Chesapeake Bay: *Estuaries*, v. 4, no. 4, pp. 313–320.

Robinson, J., Richardson, A., Bush, B., and Elgar, K.E., 1966, A photoisomerisation product of dieldrin: *Bull. Environ. Contam. Toxicol.*, v. 1, no. 4, pp. 127–132.

Robinson, J., Richardson, A., Crabtree, A.N., Coulson, J.C., and Potts, G.R., 1967, Organochlorine residues in marine organisms: *Nature*, v. 214, no. 5095, pp. 1307–1311.

Roddy, W.R., Greene, E.A., and Sowards, C.L., 1991, Reconnaissance investigation of water quality, bottom sediment, and biota associated with irrigation drainage in the Belle Fourche

Reclamation Project, western South Dakota, 1988–89: U.S. Geological Survey Water-Resources Investigations Report 90-4192, 113 p.

Roesijadi, G., Anderson, J.W., and Blaylock, J.W., 1978, Uptake of hydrocarbons from marine sediments contaminated with Prudhoe Bay crude oil: Influence of feeding type of test species and availability of polycyclic aromatic hydrocarbons: *Can. J. Fish. Aquat. Sci.*, v. 35, no. 5, pp. 608–614.

Rogers, R.J., 1984, Chemical quality of the Saw Mill River, Westchester County, New York, 1981–83: U.S. Geological Survey Water-Resources Investigations Report 84-4225, 51 p.

Rohrer, T., Forey, J.C., and Hartig, J.H., 1982, Organochlorine and heavy metal residues in standard fillets of coho and chinook salmon of the Great Lakes—1980: *J. Great Lakes Res.*, v. 8, no. 4, pp. 623–634.

Rompala, J.M., Rutkosky, F.W., and Putman, D.J., 1984, Concentrations of environmental contaminants in fish from selected waters in Pennsylvania: U.S. Fish and Wildlife Service, 113 p.

Rosen, J.D., and Strusz, R.F., 1968, The nature and toxicity of the photoconversion products of aldrin: *J. Agric. Food Chem.*, v. 16, no. 2, pp. 568–570.

Rosen, J.D., and Sutherland, D.J., 1967, Photolysis of 3-(*p*-bromophenyl)-1-methoxy-1-methylurea: *Bull. Environ. Contam. Toxicol.*, v. 2, no. 1, pp. 1–9.

Rosenberg, D.M., 1975, Food chain concentration of chlorinated hydrocarbon pesticides in invertebrate communities: A reevaluation: *Quaest. Entomol.*, v. 11, pp. 97–110.

Rossman, R., 1986, Lake Huron 1980 intensive surveillance: Management and summary: University of Michigan, Great Lakes Research Division, Special Report No. 118, variously paged.

Routh, J.D., 1972, DDT residues in Salinas River sediments: *Bull. Environ. Contam. Toxicol.*, v. 7, no. 2/3, pp. 168–176.

Rowan, D.J., and Rasmussen, J.B., 1992, Why don't Great Lakes fish reflect environmental concentrations of organic contaminants?—An analysis of between-lake variability in the ecological partitioning of PCBs and DDT: *J. Great Lakes Res.*, v. 18, no. 4, pp. 724–741.

Rowe, T.G., Lico, M.S., Hallock, R.J., Maest, A.S., and Hoffman, R.J., 1991, Physical, chemical, and biological data for detailed study of irrigation drainage in and near Stillwater, Fernley, and Humboldt wildlife management areas, and Carson Lake, west-central Nevada, 1987–89: U.S. Geological Survey Open-File Report 91-185, 199 p.

Rubinstein, N., Gilliam, W.T., and Gregory, N.R., 1984, Dietary accumulation of PCBs from a contaminated sediment source by a demersal fish (*Leiostomus xanthurus*): *Aquat. Toxicol.*, v. 5, pp. 331–342.

Rubinstein, N.I., Lake, J.L., Pruell, R.J., Lee, H., II, Taplin, B., Heltshe, J., Bowen, R., and Pavignano, S., 1987, Predicting bioaccumulation of sediment-associated organic contaminants: Development of a regulatory tool for dredged material evaluation: U.S. Environmental Protection Agency, Environmental Research Laboratory, Narragansett, R.I.

Rudd, R.L., 1964, *Pesticides and the living landscape*: University of Wisconsin Press, Madison, Wis., 320 p.

Ruelle, R., 1991, A pesticide and toxicity evelution of wetland waters and sediments on national wildlife refuges in South Dakota: U.S. Fish and Wildlife Service, Fish and Wildlife Enhancement, South Dakota State Office Contaminant Report R6/804P/91, variously paged.

Ruiz, X., and Llorente, G.A., 1991, Seasonal variation of DDT and PCB accumulation in muscle of carp (*Cyprinus carpio*) and eels (*Anguilla anguilla*) from the Ebro Delta, Spain: *Vie Milieu*, v. 41, no. 2/3, p. 133–140.

Russell, R.W., Lazar, R., and Haffner, G.D., 1995, Biomagnification of organochlorines in Lake Erie white bass: *Environ. Toxicol. Chem.*, v. 14, no. 4, pp. 719–724.

Ryan, D.A., 1989, Contaminants in fish and sediment from Sixmile Creek and Threemile Creek in the vicinity of Griffiss Air Force Base, Oneida County, New York: U.S. Fish and Wildlife Service, Cortland Field Office, N.Y., 18 p., 5 appendices.

Ryan, J.M., Stilwell, D.A., Kane, D., Rice, S.O., and Morse, N.J., 1992, A survey of contaminants in the Great Dismal Swamp National Wildlife Refuge, Virginia: U.S. Fish and Wildlife Service, Virginia Field Office, White Marsh, Va., variously paged.

Saarikoski, J., Lindstrom, R., Tyynela, M., and Viluksela, M., 1986, Factors affecting the absorption of phenolics and carboxylic acids in the guppy (*Poecillia reticulata*): *Ecotoxicol. Environ. Safety*, v. 11, no. 2, pp. 158–173.

Sabljic, A., 1987, The prediction of fish bioconcentration factors of organic pollutants from the molecular connectivity model: *Z. Hygiene Grenzgeb.*, v. 33, no. 10, pp. 493–496.

Sabljic, A., and Protic, M., 1982, Molecular connectivity: A novel method for prediction of bioconcentration factor of hazardous chemicals: *Chem.-Biol. Interact.*, v. 42, no. 3, pp. 301–310.

Sabourin, T.D., Stickle, W.B., Michot, T.C., Villars, C.E., Garton, D.W., and Mushinsky, H.R., 1984, Organochlorine residue levels in Mississippi River water snakes in southern Louisiana: *Bull. Environ. Contam. Toxicol.*, v. 32, no. 4, pp. 460–468.

Safe, S.H., 1995, Environmental and dietary estrogens and human health: Is there a problem?: *Environ. Health Perspect.*, v. 103, pp. 346–351.

Safe, S.H., and Gaido, K., 1998, Phytoestrogens and anthropogenic estrogenic compounds: *Environ. Toxicol. Chem.*, v. 17, no. 1, pp. 119–126.

Saiki, M.K., and Schmitt, C.J., 1986, Organochlorine chemical residues in bluegills and common carp from the irrigated San Joaquin Valley floor, California: *Arch. Environ. Contam. Toxicol.*, v. 15, no. 4, pp. 357–366.

Sallenave, R.M., Day, K.E., and Kreutzweiser, D.P., 1994, The role of grazers and shredders in the retention and downstream transport of a PCB in lotic environments: *Environ. Toxicol. Chem.*, v. 13, no. 11, pp. 1843–1847.

Sanderson, S. L., Cech, J.J., Jr., and Petterson, M.R., 1991, Fluid dynamics in suspension-feeding blackfish: *Science*, v. 251, pp. 1346–1348.

Sanderson, S.L., and Cheer, A.Y., 1993, Fish as filters: An empirical and mathematical analysis: *Contemp. Math.*, v. 141, pp. 135–160.

Saunders, J.W., 1969, Mass mortalities and behavior of brook trout and juvenile atlantic salmon in a stream polluted by agricultural pesticides: *J. Fish. Res. Board Can.*, v. 26, no. 3, pp. 695–699.

Schacht, R.A., 1974, Pesticides in the Illinois waters of Lake Michigan: U.S. Environmental Protection Agency, Office of Research and Development, Ecological Research series, EPA-660/3-74/002, 55 p.

Schecter, A., Fuerst, P., Krueger, C., Meemken, H.A., Groebel, W., and Constable, J.D., 1989, Levels of polychlorinated dibenzofurans, dibenzodioxins, PCBS, DDT, DDE, hexachlorobenzene, dieldrin, hexachlorocyclohexanes, and oxychlordane in human breast milk from the United States, Thailand, Vietnam and Germany: *Chemosphere*, v. 18, no. 1/6, pp. 445–454.

Schell, J.D., Jr., Campbell, D.M., and Lowe, E., 1993, Bioaccumulation of 2,3,7,8-tetrachlorodibenzo-*p*-dioxin in feral fish collected from a bleach-kraft mill receiving stream: *Environ. Toxicol. Chem.*, v. 12, no. 11, pp. 2077–2082.

Schmitt, C.J., Ludke, J.L., and Walsh, D.F., 1981, Organochlorine residues in fish: National Pesticide Monitoring Program, 1970–74: *J. Pest. Monitor.*, v. 14, no. 4, pp. 136–206.

Schmitt, C.J., Ribick, M.A., Ludke, J.L., and May, T.W., 1983, National Pesticide Monitoring Program: Organochlorine residues in freshwater fish, 1976–79: U.S. Department of the Interior, Fish and Wildlife Service Resource Publication 152, 62 p.

Schmitt, C.J., Zajicek, J.L., and Peterman, P.H., 1990, National Contaminant Biomonitoring Program: Residues of organochlorine chemicals in U.S. freshwater fish, 1976–1984: *Arch. Environ. Contam. Toxicol.*, v. 19, no. 5, pp. 748–781.

Schmitt, C.J., Zajicek, J.L., and Ribick, M.A., 1985, National Pesticide Monitoring Program: Residues of organochlorine chemicals in freshwater fish, 1980–81: *Arch. Environ. Contam. Toxicol.*, v. 14, no. 2, pp. 225–260.

Schneider, R., 1982, Polychlorinated biphenyls (PCBs) in cod tissues from the western Baltic: Significance of equilibrium partitioning and lipid composition in the bioaccumulation of lipophilic pollutants in gill-breathing animals: *Meeresforschung*, v. 29, pp. 69–79.

Schnoor, J.L., 1981, Fate and transport of dieldrin in Coralville Reservoir: Residues in fish and water following a pesticide ban: *Science*, v. 211, no. 4484, pp. 840–842.

Schottler, S.P., Eisenreich, S.J., and Capel, P.D., 1994, Atrazine, alachlor, and cyanazine in a large agricultural river system: *Environ. Sci. Technol.*, v. 28, no. 6, pp. 1079–1089.

Schroeder, R.A., Palawski, D.U., and Skorupa, J.P., 1988, Reconnaisance investigation of water quality, bottom sediment, and biota associated with irrigation drainage in the Tulare Lake bed area, southern San Joaquin Valley, California, 1986–87: U.S. Geological Survey Water-Resources Investigations Report 88-4001, 86 p.

Schultz, D.P., and Harman, P.D., 1974, Residues of 2,4-D in pond waters, mud and fish, 1971: *J. Pest. Monitor.*, v. 8, no. 3, pp. 173–179.

Schultz, D.P., and Whitney, E.W., 1974, Monitoring 2,4-D residues at Loxahatchee National Wildlife Refuge: *J. Pest. Monitor.*, v. 7, no. 3/4, pp. 146–152.

Schuurmann, G., and Klein, W., 1988, Advances in bioconcentration prediction: *Chemosphere*, v. 17, no. 8, pp. 1551–1574.

Schwarzenbach, R.P., Gschwend, P.M., and Imboden, D.M., 1993, *Environmental organic chemistry*: Wiley and Sons, New York, 681 p.

Schwarzenbach, R.P., and Westfall, J., 1981, Transport of nonpolar organic compounds from surface water to groundwater: Laboratory sorption studies: *Environ. Sci. Technol.*, v. 15, no. 11, pp. 1360–1367.

Scow, K.M., 1990, Biodegradation, *in* Lyman, W.F., Reehl, W.F., and Rosenblatt, D.H., eds., *Handbook of chemical property estimation methods: Environmental behavior of organic compounds*: American Chemical Society, Washington, D.C.

Scrudato, R.J., and DelPrete, A., 1982, Lake Ontario sediment-mirex relationships: *J. Great Lakes Res.*, v. 8, no. 4, pp. 695–699.

Sears, H.S., and Meehan, W.R., 1971, Short-term effects of 2,4-D on aquatic organisms in the Nakwasina River watershed, southeastern Alaska: *J. Pest. Monitor.*, v. 5, no. 2, pp. 213–217.

Secor, C.L., Mills, E.L., Harshbarger, J., Kuntz, H.T., Gutenmann, W.H., and Lisk, D.J., 1993, Bioaccumulation of toxicants, element and nutrient composition, and soft tissue histology of zebra mussels (*Dreissena polymorpha*) from New York State waters: *Chemosphere*, v. 26, no. 8, pp. 1559–1575.

See, R.B., Peterson, D.A., and Ramirez, P., Jr., 1992, Physical, chemical, and biological data for detailed study of irrigation drainage in the Kendrick Reclamation Project area, Wyoming, 1988–90: U.S. Geological Survey Open-File Report 91-533, 272 p.

Segar, D.A., and Davis, P.G., 1984, Contamination of populated estuaries and adjacent coastal ocean—A global review: National Oceanic and Atmospheric Administration, National Ocean Service, NOAA Technical Memorandum NOS OMA 11.

Seiber, J.N., Madden, S.C., McChesney, M.M., and Winterlin, W.L., 1979, Toxaphene dissipation from treated cotton field environments: Component residual behavior on leaves and in air, soil, and sediments determined by capillary gas chromatography: *J. Agric. Food Chem.*, v. 27, no. 2, pp. 284–291.

Seiler, R.L., Ekechukwu, G.A., and Hallock, R.J., 1993, Reconnaissance investigation of water quality, bottom sediment, and biota associated with irrigation drainage in and near Humboldt Wildlife Management Area, Churchill and Pershing counties, Nevada, 1990–91: U.S. Geological Survey Water-Resources Investigations Report 93-4072, 115 p.

Sergeant, D.B., and Onuska, F.I., 1989, Analysis of toxaphene in environmental samples, *in* Afghan, B.K., and Chau, A.S.Y., eds., *Analysis of trace organics in the aquatic environment*: CRC Press, Boca Raton, Fla., pp. 69–118.

Sericano, J.L., Atlas, E.L., Wade, T.L., and Brooks, J.M., 1990b, NOAA's Status and Trends Mussel Watch Program: Chlorinated pesticides and PCBs in oysters (*Crassostrea virginica*) and sediments from the Gulf of Mexico, 1986–1987: *Mar. Environ. Res.*, v. 29, no. 3, pp. 161–203.

Sericano, J.L., Wade, T.L., Atlas, E.L., and Brooks, J.M., 1990a, Historical perspective on the environmental bioavailability of DDT and its derivatives to Gulf of Mexico oysters: *Environ. Sci. Technol.*, v. 24, no. 10, pp. 1541–1548.

Setchell, K., 1985, Naturally occurring non-steroidal estrogens of dietary origin, *in* McLachlan, J.A., ed.,*Estrogens in the environment II: Influences on development: Proceedings of the symposium, Estrogens in the environment—Influences on development, Raleigh, North Carolina, United States, April 10–12, 1995*: Elsevier, New York, 435 p.

Setmire, J.G., Schroeder, R.A., Densmore, J.N., Goodbred, S.L., Audet, D.J., and Radke, W.R., 1993, Detailed study of water quality, bottom sediment, and biota associated with irrigation drainage in the Salton Sea area, California, 1988–90: U.S. Geological Survey Water-Resources Investigations Report 93-4014, 102 p.

Setmire, J.G., Wolfe, J.C., and Stroud, R.K., 1990, Reconnaissance investigation of water quality, bottom sediment, and biota associated with irrigation drainage in the Salton Sea area, California, 1986–87: U.S. Geological Survey Water-Resources Investigations Report 89-4102, 68 p.

Sewell, S.A., and Knight, L.A., Jr., 1986, Continued DDT persistence in Mississippi River delta streams: A case study: *Proc. Ark. Acad. Sci.*, v. 40, pp. 56–58.

Shampine, W.J., 1975, A river-quality assessment of the upper White River, Indiana: U.S. Geological Survey Water-Resources Investigations [Report] 10-75, 68 p.

Shaw, G.R., and Connell, D.W., 1980, Relationships between steric factors and bioconcentration of polychlorinated biphenyls (PCB's) by the sea mullet (*Mugil cephalus* Linnaeus): *Chemosphere*, v. 9, no. 12, pp. 731–743.

——1982, Factors influencing concentrations of polychlorinated biphenyls in organisms from an estuarine ecosystem: *Aust. J. Mar. Freshwater Res.*, v. 33, no. 6, pp. 1057–1070.

——1984, Physicochemical properties controlling polychlorinated biphenyl (PCB) concentrations in aquatic organisms: *Environ. Sci. Technol.*, v. 18, no. 1, pp. 18–23.

——1987, Comparative kinetics for bioaccumulation of polychlorinated biphenyls in organisms from an estuarine ecosystem: *Ecotoxicol. Environ. Safety*, v. 13, pp. 84–91.

Shaw, T.L., Karwowski, K., and Mann-Kleger, D.P., 1992, Chemical analysis of sediments from the St. Lawrence River: U.S. Fish and Wildlife Service, New York Field Office, Cortland, N.Y, 18 p., 7 appendices.

Sherwood, M.J., 1982, Fin erosion, liver condition, and trace contaminant exposure in fishes from three coastal regions, *in* Mayer, G.F., ed., *Ecological stress and the New York Bight: Science and management*: Estuarine Research Foundation, Columbia, S.C, pp. 359–377.

Shigenaka, G., 1990, Chlordane in the marine environment of the United States: Review and results from the National Status and Trends Program: U.S. Department of Commerce, National Oceanic and Atmospheric Administration Technical Memorandum NOS OMA 55, 230 p., 3 appendices.

Short, M., 1988, An intensive survey of the Mackinaw River Basin, 1987: Illinois Environmental Protection Agency, Division of Water Pollution Control Report IEPA/WPC/88-034, 51 p.

Sijm, D.T.H.M., and Opperhuizen, A., 1988, Biotransformation, bioaccumulation and lethality of 2,8-dichlorodibenzo-*p*-dioxin: A proposal to explain the biotic fate and toxicity of PCDDs and PCDFs: *Chemosphere*, v. 17, no. 1, pp. 83–99.

Sijm, D.T.H.M., Seinen, W., and Opperhuizen, A., 1992, Life-cycle biomagnification study in fish: *Environ. Sci. Technol.*, v. 26, no. 11, pp. 2162–2174.

Sikka, H.C., 1977, Fate of 2,4-D in fish and blue crabs: Department of Defense, Department of the Army, Corps of Engineers, Waterways Experiment Station, Contract Report A-77-2, 12 p.

Silvey, W.D., and Wheeler, R.L., 1978, Water-quality conditions in southern Rockingham County, New Hampshire: U.S. Geological Survey Open-File Report 78-224, 51 p.

Simmons, C.E., and Aldridge, M.C., 1980, Hydrology of the Chicod Creek Basin, North Carolina, prior to channel improvements: U.S. Geological Survey Open-File Report 80-680, 32 p.

Singh, S., and Singh, T.P., 1987, Evaluation of toxicity limits and sex hormone production in response to cythion and BHC in the vitellogenic catfish *Clarias batrachus*: *Environ. Res.*, v. 42, pp. 482–488.

Skaar, D.B., Johnson, B.T., Jones, J.R., and Huckins, J.N., 1981, Fate of kepone and mirex in model aquatic environment: Sediment, fish, and diet: *Can. J. Fish. Aquat. Sci.*, v. 38, no. 8, pp. 931–938.

Skea, J.C., Simonin, H.J., Jackling, S., and Symula, J., 1981, Accumulation and retention of mirex by brook trout fed a contaminated diet: *Bull. Environ. Contam. Toxicol.*, v. 27, no. 1, pp. 79–83.

Skinner, L.C., 1992, Chemical contaminants in wildlife from the Mohawk Nation at Akwesasne and the vicinity of the General Motors Corporation/Central Foundry Division, Massena, New York plant: New York State Department of Environmental Conservation, Bureau of Environmental Protection Technical Report 92-4, 113 p.

——1993, Dioxins and furans in fish below Love Canal, New York: Concentration reduction following remediation: New York State Department of Environmental Conservation, Albany, N.Y., 52 p.

Sladen, W.J.L., Menzie, C.M., and Reichel, W.L., 1966, DDT residues in Adelie penguins and a crabeater seal from Antarctica: *Nature*, v. 210, no. 5037, pp. 670–673.

Sloan, R., 1981a, Toxic substances in fish and wildlife, 1979 and 1980 annual reports, v. 4, no.1: New York State Department of Environmental Conservation, Division of Fish and Wildlife, Bureau of Environmental Protection Technical Report 81-1.

——1981b, Toxic substances in fish and wildlife, May 1 to November 1, 1981, v. 4, no. 2: New York State Department of Environmental Conservation, Division of Fish and Wildlife, Bureau of Environmental Protection Technical Report 82-1.

——1982, Toxic substances in fish and wildlife, November 1, 1981 to April 30, 1982, v. 5, no. 1: New York State Department of Environmental Conservation, Division of Fish and Wildlife, Bureau of Environmental Protection Technical Report 82-2.

——1987, Toxic substances in fish and wildlife, analyses since May 1, 1982, v. 6: New York State Department of Environmental Conservation, Bureau of Environmental Protection Technical Report 87-4, 182 p.

Sloan, R.J., and Jock, K., 1990, Chemical contaminants in fish from the St. Lawrence River Drainage on lands of the Mohawk Nation at Akwesasne and near the General Motors Corporation/Central Foundry Division, Massena, New York plant: New York State Department of Environmental Conservation, Bureau of Environmental Protection Technical Report 90-1, 96 p.

Slooff, W., and DeZwart, D., 1983, The growth, fecundity, and mortality of bream (*Abramis brama*) from polluted and less polluted surface waters in the Netherlands: *Sci. Total Environ.*, v. 27, pp. 149–162.

Sloterdijk, H.H., 1991, Mercury and organochlorinated hydrocarbons in surficial sediments of the St. Lawrence River (Lake St. Francis): *Water Pollut. Res. J. Can.*, v. 26, no. 1, pp. 41–60.

Smith, B.J., 1984, PCB levels in human fluids: Sheboygan case study: University of Wisconsin Sea Grant Institute Technical Report WIS-SG-83-240, 104 p.

Smith, J.A., Harte, P.T., and Hardy, M.A., 1987, Trace-metal and organochlorine residues in sediments of the upper Rockaway River, New Jersey: *Bull. Environ. Contam. Toxicol.*, v. 39, no. 3, pp. 465–473.

Smith, J.A., Witkowski, P.J., and Chiou, C.T., 1988, Partition of nonionic organic compounds in aquatic systems: *Rev. Environ. Contam. Toxicol.*, v. 103, pp. 127–151.

Smith, L.M., Schwartz, T.R., Feltz, K., and Kubiak, T.J., 1990, Determination and occurrence of AHH-active polychlorinated biphenyls, 2,3,7,8-tetrachloro-*p*-dioxin and 2,3,7,8-tetrachloro-dibenzofuran in Lake Michigan sediment and biota. The question of their relative toxicological significance: *Chemosphere*, v. 21, no. 9, pp. 1063–1085.

Smith, R.A., Alexander, R.B., and Lanfear, K.J., 1993, Stream water quality in the conterminous United States. Status and trends of selected indicators during the 1980's, in National Water Summary 1990–91, Hydrologic events and stream water quality: U.S. Geological Survey Water-Supply Paper 2400, pp. 111–140.

Smith, R.M., and Cole, C.F., 1970, Chlorinated hydrocarbon insecticide residues in winter flounder, *Pseudopleuronectes americanus*, from the Weweantic River estuary, Massachusetts: *Can. J. Fish. Aquat. Sci.*, v. 27, no. 12, pp. 2374–2380.

Smith, R.M., O'Keefe, P.W., Aldous, K.M., Hilker, D.R., and O'Brien, J.E., 1983, 2,3,7,8-Tetrachorodibenzo-*p*-dioxin in sediment samples from Love Canal storm sewers and creeks: *Environ. Sci. Technol.*, v. 17, no. 1, pp. 6–10.

Smith, S.B., and Sorenson, S.K., 1995, Contaminant impacts to the endocrine system in largemouth bass in northeast U.S. rivers, in *Abstract book, Second SETAC World Congress: 16th annual meeting: Global environmental protection: Science, politics, and common sense: 5–9 November 1995, Vancouver Trade and Convention Centre, Vancouver, B.C., Canada*: SETAC Press, Pensacola, Fla., p. 60.

——1997, Endocrine disruption biomarkers of common carp and largemouth bass in relation to sediment and tissue residue from the Connecticut River and Hudson River and Potomac River watersheds [abs.]: Annual Meeting of the Society of Environmental Toxicology and Chemistry, 18th, San Francisco, California, 1997, Abstract, p. 255.

Sobiech, S.A., and Sparks, D.W., 1992, Contaminant study of the sediments and fish of the St. Mary's, St. Joseph, and Maumee rivers of the Maumee River watershed, Allen County,

Indiana: U.S. Fish and Wildlife Service, Biological Report, Bloomington Field Office, Bloomington, Ind.,131 p.

Sodergren, A., 1968, Uptake and accumulation of C^{14}-DDT by *Chlorella* sp. (Chlorophyceae): *Oikos*, v. 19, no. 1, pp. 126–138.

Sodergren, A., Svensson, B., and Ulfstrand, S., 1972, DDT and PCB in south Swedish streams: *Environ. Pollut.*, v. 3, no. 1, pp. 25–36.

Sol, S.Y., Lomas, D.P., Jaconson, J.C., Sommers, F.C., Analacion, B.F., and Johnson, L.L., 1995, Effects of chronic contaminant exposure in reproductive development of English sole (*Pleuronectes vetulus*), in *Abstract book, Second SETAC World Congress: 16th annual meeting: Global environmental protection: Science, politics, and common sense: 5–9 November 1995, Vancouver Trade and Convention Centre, Vancouver, B.C., Canada*: SETAC Press, Pensacola, Fla., p. 60.

Sorenson, S.K., and Schwarzbach, S.E., 1991, Reconnaissance investigation of water quality, bottom sediment, and biota associated with irrigation drainage in the Klamath Basin, California and Oregon, 1988–89: U.S. Geological Survey Water-Resources Investigations Report 90-4203, 64 p.

Southward, G.R., Beauchamp, J.J., and Schnieder, P.K., 1978, Bioaccumulation potential of polycyclic aromatic hydrocarbons in *Daphnia pulex*: *Water Res.*, v. 12, no. 11, pp. 973–977.

Sowards, C., Maxwell, S., and Ruelle, R., 1991, A compendium of environmental contaminants in South Dakota fish, wildlife, and habitats: U.S. Fish and Wildlife Service Contaminant Report R6/812P/91, 93 p.

Spacie, A., 1975, The bioconcentration of trifluralin from a manufacturing effluent by fish in the Wabash River: Purdue University, West Lafayette, Ill., Ph.D dissertation, 136 p.

Spacie, A., and Hamelink, J.L., 1979, Dynamics of trifluralin accumulation in river fishes: *Environ. Sci. Technol.*, v. 13, no. 7, pp. 817–822.

——1982, Alternative models for describing the bioconcentration of organics in fish: *Environ. Toxicol. Chem.*, v. 1, no. 4, pp. 309–320.

——1985, Bioaccumulation, in Rand, G.M., and Petrocelli, S.R., eds., *Fundamentals of aquatic toxicology: Methods and applications*: Hemisphere Publishing Corporation, Washington, pp. 495–525.

Sparr, B.I., Appleby, W.G., De Vries, D.M., Osmun, J.V., McBride, J.M., and Foster, G.L., 1966, Insecticide residues in waterways from agricultural use, in Rosen, A.A., and Kraybill, H.F., eds., *Organic pesticides in the environment: A symposium*: American Chemical Society Advances in Chemistry series, v. 60, pp. 146–162.

Spence, J.H., and Markin, G.P., 1974, Mirex residues in the physical environment following a single bait application, 1971–72: *J. Pest. Monitor.*, v. 8, no. 2, pp. 135–139.

Spooner, J., Wyatt, L., Berryhill, W.S., Lanier, A.L., Brichford, S.L., Smolen, M.D., Coffey, S.W., and Bennett, T.B., 1989, Fate and effect of pollutants: Nonpoint sources: *J. Water Pollut. Control Fed.*, v. 16, no. 6, pp. 911–924.

Sprague, J.B., Elson, P.F., and Duffy, J.R., 1971, Decrease in DDT residues in young salmon after forest spraying in New Brunswick: *Environ. Pollut.*, v. 1, no 3, pp. 191–202.

Squillace, P.J., and Thurman, E.M., 1992, Herbicide transport in rivers: Importance of hydrology and geochemistry in nonpoint-source contamination: *Environ. Sci. Technol.*, v. 26, no. 3, pp. 538–545.

Squillace, P.J., Thurman, E.M., and Furlong, E.T., 1993, Groundwater as a nonpoint source of atrazine and deethylatrazine in a river during base flow conditions: *Water Resour. Res.*, v. 29, no. 6, pp. 1719–1729.

Stalling, D.L., Smith, L.M., Petty, J.D., Hogan, J.W., Johnson, J.L., Rappe, C., and Buser, H.R., 1983, Residues of polychlorinated dibenzo-*p*-dioxins and dibenzofurans in Laurentian Great Lakes fish, *in* Tucker, R.E., and Young, A.L., eds., *Human and environmental risks of chlorinated dioxins and related compounds*: Plenum Press, New York, Environmental Science Research series, v. 26, pp. 221–240.

Stamer, J.K., Yorke, T.H., and Pederson, G.L., 1985, Distribution and transport of trace substances in the Schuylkill River Basin from Berne to Philadelphia, Pennsylvania: U.S. Geological Survey Water-Supply Paper 2256-A, 45 p.

Statham, C.N., Pepple, S.K., and Lech, J.J., 1975, Biliary excretion products of 1-[1-^{14}C]naphthyl-*N*-methylcarbamate (carbaryl) in rainbow trout (*Salmo gairdnieri*): *Drug Metab. Dispos.*, v. 3, pp. 400–406.

Steffeck, D.W., 1988, A survey for contaminants in selected biota near the Four County Landfill, Fulton County, Indiana: U.S. Fish and Wildlife Service, Bloomington Field Office, Bloomington, Ind., 16 p., 2 appendices.

Steffeck, D.W., and Striegl, R.G., 1989, An inventory and evaluation of biological investigations that relate to stream-water quality in the upper Illinois River Basin of Illinois, Indiana, and Wisconsin: U.S. Geological Survey Water-Resources Investigations Report 89-4041, 54 p.

Stegeman, J.J., 1981, Polynuclear aromatic hydrocarbons and their metabolites, *in* Gelboin, H.V., Ts'O, P.O.P., and Andrews, L.S., eds., *Polycyclic hydrocarbons and cancer*: Academic Press, New York, v. 3, pp. 1–60.

Stegeman, J.J., and Chevion, M., 1980, Sex differences in cytochrome P-450 and mixed function oxygenase activity in gonadally mature trout: *Biochem. Pharmacol.*, v. 29, pp. 553–558.

Stegeman, J.J., and Teal, J.M., 1973, Accumulation, release, and retention of petroleum hydrocarbons by the oyster *Crasseostrea virginica*: *Mar. Biol.*, v. 22, no. 1, pp. 37–44.

Stehly, G.R., and Hayton, W.L., 1989, Disposition of pentachlorophenol in rainbow trout (*Salmo gairdneri*): Effect of inhibition of metabolism: *Aquat. Toxicol.*, v. 14, no. 2, pp. 131–148.

Stephan, C.R., Mount, D.I., Hansen, D.J., Gentile, J.H., Chapman, G.A., and Brungs, W.A., 1985, Guidelines for deriving numerical national water quality criteria for the protection of aquatic organisms and their uses: U.S. Environmental Protection Agency, Office of Water Regulations and Standards, Criteria and Standards Division, PB85-227049, 98 p.

Stephens, D.W., 1984, Water-quality investigations of the Jordan River, Salt Lake County, Utah, 1980–82: U.S. Geological Survey Water-Resources Investigations Report 84-4298, 45 p.

Stephenson, M.D., Martin, M., and Tjeerdema, R.S., 1995, Long-term trends in DDT, polychlorinated biphenyls, and chlordane in California mussels: *Arch. Environ. Contam. Toxicol.*, v. 28, no. 4, pp. 443–450.

Stephenson, M.D., Smith, D., Ichikawa, G., Goetzl, J., and Martin, M., 1986, State Mussel Watch preliminary data report, 1985–86: A report to the State Water Resources Control Board: California Department of Fish and Game, 31 p.

Stoll, M.F., 1990, 1988 Montezuma National Wildlife Refuge Contaminant Study: U.S. Fish and Wildlife Service, New York Field Office, Cortland, N.Y., 46 p., 5 appendices.

Stott, G.G., Haensly, W., Neff, J., and Sharp, J., 1983, Histopathologic surveys of ovaries of plaice, *Pleuronectes platessa* L., from Aberwrac'h and Aber Benoit, Brittany, France, oil spills: *J. Fish Dis.*, v. 6, no. 5, pp. 429–437.

Stout, V.F., 1968, Pesticide levels of fish of the northeast Pacific: *Bull. Environ. Contam. Toxicol.*, v. 3, no. 4, pp. 240–246.

——1980, Organochlorine residues in fish from the northwest Atlantic Ocean and the Gulf of Mexico: *Fish. Bull.*, v. 78, no. 1, pp. 51–58.

Stout, V.F., and Beezhold, F.L., 1981, Chlorinated hydrocarbon levels in fishes and shellfishes of the northeastern Pacific Ocean, including the Hawaiian Islands: *Mar. Fish. Rev.*, v. 43, no. 1, pp. 1–13.

Strachan, W.M.J., and Edwards, C.J., 1984, Organic pollutants in Lake Ontario, *in* Nriagu, J.O., Simmons, M.S., eds., *Toxic contaminants in the Great Lakes*: Wiley, N.Y., pp. 239–264.

Strachan, W.M.J., and Eisenreich, S.J., 1990, Mass balance accounting of chemicals in the Great Lakes, *in* Kurtz, D.A., ed., *Long range transport of pesticides*: Lewis Publishers, Chelsea, Mich., pp. 291–301.

Strachan, W.M.J., and Glass, G.E., 1978, Organochlorine substances in Lake Superior: *J. Great Lakes Res.*, v. 4, no. 3–4, pp. 389–397.

Stratton, G.W., 1984, Effects of the herbicide atrizine and its degradation products, alone and in combination, on phototrophic microorganisms: *Arch. Environ. Contam. Toxicol.*, v. 13, no. 1, p. 35.

Stucky, N.P., 1970, Pesticide residues in channel catfish from Nebraska: *J. Pest. Monitor.*, v. 4, no. 2, pp. 62–66.

Stumm, W., Schwarzenbach, R., and Sigg, L., 1983, From environmental analytical chemistry to ecotoxicology: A plea for more concepts and less monitoring and testing: *Angew. Chem. Int. Ed. Engl.*, v. 22, no. 5, pp. 380–389.

Sudershan, P., and Khan, M.A.Q., 1979, Metabolic and elimination products of [^{14}C]photodieldrin from bluegill fish: *Pest. Biochem. Physiol.*, v. 12, pp. 216–223.

——1980a, Metabolic fate of [^{14}C]endrin in bluegill fish: *Pest. Biochem. Physiol.*, v. 14, pp. 5–12.

——1980b, Metabolism of [^{14}C]chlordane and cis-[^{14}C]photochlordane in bluegill fish: *J. Agric. Food Chem.*, v. 28, no. 2, pp. 291–296.

——1980c, Metabolic fate of [^{14}C]photodieldrin in goldfish: *Pest. Biochem. Physiol.*, v. 13, pp. 148–157.

——1981, Metabolism of [^{14}C]dieldrin in bluegill fish: *Pest. Biochem. Physiol.*, v. 15, pp. 192–199.

Suffet, I.H., Jafvert, C.T., Kukkonen, J., Servos, M.R., Spacie, A., Williams, L.L., and Noblet, J.A., 1994, Synopsis of discussion session: Influences of particulate and dissolved material on the bioavailability of organic compounds, *in* Hamelink, J.L., Landrum, P.F., Bergman, H.L., and Benson, W.H., eds., *Bioavailability: Physical, chemical, and biological interactions*: Lewis Publishers, Boca Raton, Fla., pp. 93–108.

Sugiura, K., Washino, T., Hattori, M., Sato, E., and Goto, M., 1979, Ecological chemistry. XVI. Accumulation of organochlorine compounds in fishes. Difference of accumulation factors by fishes: *Chemosphere*, v. 8, no. 6, pp. 359–364.

Sullivan, J.F., and Atchison, G.J., 1977, Impact of an urban methoxychlor spraying program on the Rouge River, Michigan: *Bull. Environ. Contam. Toxicol.*, v. 17, no. 1, pp. 121–126.

Sullivan, J.R., and Armstrong, D.E., 1985, Toxaphene status in the Great Lakes: University of Wisconsin Sea Grant Institute, Madison, Wisconsin, Priority Pollutants Status Report series, no. 2, WIS-SG-85-241, 39 p.

Sundaram, K.M.S., 1987, Persistence characteristics of operationally sprayed fenitrothion in nearby unsprayed areas of a conifer forest ecosystem in New Brunswick: *J. Environ. Sci. Health*, v. B22, no. 4, pp. 413–438.

——1991, Fate and short-term persistence of permethrin insecticide in a northern Ontario (Canada) headwater stream: *Pest. Sci.*, v. 31, no. 3, pp. 281–294.

Sundaram, K.M.S., Holmes, S.B., Kreutzweiser, D.P., Sundaram, A., and Kingsbury, P.D., 1991, Environmental persistence and impact of diflubenzuron in a forest aquatic environment following aerial application: *Arch. Environ. Contam. Toxicol.*, v. 20, no 3, pp. 313–324.

Sundaram, K.M.S., and Szeto, S.Y., 1987, Distribution and persistence of carbaryl in some terrestrial and aquatic components of a forest environment: *J. Environ. Sci. Health*, v. B22, no. 5, pp. 579–599.

Suns, K., Craig, G.R., Crawford, G., Rees, G.A., Tosine, H., and Osborne, J., 1983, Organochlorine contaminant residues in spottail shiners (*Notropis hudsonius*) from the Niagara River: *J. Great Lakes Res.*, v. 9, no. 2, pp. 335–340.

Suns, K., Crawford, G., and Russell, D., 1985, Organochlorine and mercury residues in young-of-the-year spottail shiners from the Detroit River, Lake St. Clair, and Lake Erie: *J. Great Lakes Res.*, v. 11, no. 3, pp. 347–352.

Suntio, L.R., Shie, W.Y., Mackay, D., Seiber, J.N., and Glotfelty, D.E., 1988, Critical review of Henry's law constants for pesticides: *Rev. Environ. Contam. Toxicol.*, v. 103, pp. 1–59.

Suta, B.E., 1977, Human population exposures to mirex and kepone: U.S. Environmental Protection Agency, Office of Research and Development, Environmental Health Effects Research series, EPA-600/1-78-045, 139 p.

Swackhamer, D.L., and Skoglund, R.S., 1991, The role of phytoplankton in the partitioning of hydrophobic organic contaminants in water, *in* Processes and analytical, v. 2 *of* Baker, R.A., ed., *Organic substances and sediments in water*: Lewis Publishers, Chelsea, Mich., pp. 91–105.

Swain, W.R., 1978, Chlorinated organic residues in fish, water, and precipitation from the vicinity of Isle Royale, Lake Superior: *J. Great Lakes Res.*, v. 4, no. 3–4, pp. 398–407.

——1991, Effects of organochlorine chemicals on the reproductive outcome of humans who consumed contaminated Great Lakes fish: An epidemiologic consideration: *J. Toxicol. Environ. Health*, v. 33, no. 4, pp 587–639.

Szeto, S., and Sundaram, K.M.S., 1981, Residues of chlorpyrifos-methyl in balsam fir foliage, forest litter, soil, stream water, sediment and fish tissue after double aerial application of reldan: *J. Environ. Sci. Health*, v. B16, no. 6, pp. 743–766.

Takimoto, Y., Oshima, M., and Miyamoto, J., 1987, Comparative metabolism of fenitrothion in aquatic organisms. 1, Metabolism in the euryhaline fish *Oryzias latipes* and *Mugil cephalus*: *Ecotoxicol. Environ. Safety*, v. 13, no. 1, pp. 104–117.

Tanabe, S., Hidaka, H., and Tatsukawa, R., 1984, Polychlorinated biphenyls, DDT, and hexachlorocyclohexane isomers in the western North Pacific ecosystem: *Arch. Environ. Contam. Toxicol.*, v. 13, no. 6, pp. 731–738.

Tanabe, S., Tatsukawa, R., Kawano, M., and Hidaka, H., 1982, Global distribution and atmospheric transport of chlorinated hydrocarbons: HCH (BHC) isomers and DDT compounds in the western Pacific, eastern Indian, and Antarctic oceans: *J. Oceanograph. Soc. Jpn.*, v. 38, pp. 137–148.

Tanita, R., Johnsom, J.M., Chun, M., and Maciolek, J., 1976, Organochlorine pesticides in the Hawaii Kai Marina, 1970–74: *J. Pest. Monitor.*, v. 10, no. 1, pp. 24–29.

Taylor, G.F., 1990, Quantity and quality of stormwater runoff from western Daytona Beach, Florida, and adjacent areas: U.S. Geological Survey Water-Resources Investigations Report 90-4002, 88 p.

Teal, J.M., 1977, Food chain transfer of hydrocarbons, *in* Wolfe, D.A., and Anderson, J.W., eds., *Fate and effects of petroleum hydrocarbons in marine ecosystems and organisms*: Pergamon Press, Oxford, England, pp. 71–77.

ten Hulscher, Th.E.M., and Cornelissen, G., 1996, Effect of temperature on sorption equilibrium and sorption kinetics of organic micropollutants: *Chemosphere*, v. 32, no. 4, pp. 609–626.

Terriere, L.C., Kiigemagi, U., Gerlach, A.R., and Borovicka, R.L., 1966, The persistence of toxaphene in lake water and its uptake by aquatic plants and animals.: *J. Agric. Food Chem.*, v. 14, no. 1, pp. 66–69.

Terry, R.D., and Hughes, G.M., 1976, Pollution effects on surface waters and waters: *J. Water Pollut. Control Fed.*, v. 48, no. 6, pp. 1420–1433.

Tetra Tech Inc., 1986, Development of sediment quality values for Puget Sound, volume 1: U.S. Army Corps of Engineers, Seattle District, Puget Sound Dredged Disposal Analysis report, 129 p.

Thomann, R.V., 1981, Equilibrium model of fate of microcontaminants in diverse aquatic food chains: *Can. J. Fish. Aquat. Sci.*, v. 38, no. 3, pp. 280–296.

——1989, Bioaccumulation model of organic chemical distribution in aquatic food chains: *Environ. Sci. Technol.*, v. 23, no. 6, pp. 699–707.

Thomann, R.V., and Connolly, J.P., 1984, Model of PCB in the Lake Michigan lake trout food chain: *Environ. Sci. Technol.*, v. 18, no. 2, pp. 65–71.

Thomas, K.B., and Colborn, T., 1992, Organochlorine endocrine disruptors in human tissue, *in* Colborn, T., and Clement, C.R., eds., *Chemically induced alterations in sexual and functional development: The wildlife/human connection*: Princeton Scientific Publishing, Princeton, N.J., Advances in Modern Environmental Toxicology series, v. 21, pp. 365–394.

Thomas, R.L., Gannon, J.E., Hartig, J.H., Williams, D.J., and Whittle, D.M., 1988, Contaminants in Lake Ontario: A case study, *in* Schmidke, N.W., ed., *Sources, fate, and controls of toxic contaminants*, v. 3 *of Toxic contamination in large lakes*: Lewis Publishers, Chelsea, Mich., pp. 327–387.

Thompson, K.R., 1984, Reconnaissance of toxic substances in the Jordan River, Salt Lake County, Utah: U.S. Geological Survey Water-Resources Investigations Report 84-4155, 31 p.

Thomson, A.B.R., and Dietschy, J.M., 1981, Intestinal lipid absorption: Major extracellular and intracellular events, *in* Johnson, L.R., ed., *Physiology of the gastrointestinal tract*: Raven Press, New York, v. 2, pp. 1147–1220.

Thomson, V.E., Jones, A., Haemisegger, E., and Steigerwald, B., 1985, The air toxics problem in the United States: An analysis of cancer risks posed by selected air pollutants: *J. Air Pollut. Contr. Fed.*, v. 35, pp. 535–544.

Thybaud, E., and Caquet, T., 1991, Uptake and elimination of lindane by *Lymnaea palustris* (Mollusca: Gastropoda): A pharmacokinetic approach: *Ecotoxicol. Environ. Safety*, v. 21, no. 3, pp. 365–376.

Tillitt, D.E., Ankley, G.T., Giesy, J.P., Ludwig, J.P., Kurita-Matsuba, H., and Weseloh, D.V., 1992, Polychlorinated biphenyl residues and egg mortality in double-crested cormorants from the Great Lakes: *Environ. Toxicol. Chem.*, v. 11, no. 9, pp. 1281–1288.

Tinsley, I.J., 1979, *Chemical concepts in pollutant behavior*: Wiley, New York, Environmental Science and Technology series, 265 p.

Tomlin, C., ed., 1994, *The pesticide manual: Incorporating the agrochemicals handbook* (10th ed.): Crop Protection Publications, Farnham, U.K, 1341 p.

Tracey, G.A., and Hansen, D.J., 1996, Use of biota–sediment accumulation factors to assess similarity of nonionic organic chemical exposure to benthically-coupled organisms of differing trophic mode: *Arch. Environ. Contam. Toxicol.*, v. 30, pp. 467–475.

Trim, A.H., and Marcus, J.M., 1990, Integration of long-term fish kill data with ambient water quality monitoring data and application to water quality management: *Environ. Manag.*, v. 14, no. 3, pp. 389–396.

Trotter, W.J., Corneliussen, P.E., Laski, R.B., and Vannelli, J.J., 1989, Levels of polychlorinated biphenyls and pesticides in bluefish before and after cooking: *J. Assoc. Off. Anal. Chem.*, v. 72, no. 3, pp. 501–503.

Truhlar, J.F., and Reed, L.A., 1975, Occurrence of pesticide residues in four streams draining different land-use areas in Pennsylvania: U.S. Geological Survey Water-Resources Investigations [Report] 6-75, 23 p.

——1976, Occurrence of pesticide residues in four streams draining different land use areas in Pennsylvania, 1969–71: *J. Pest. Monitor.*, v. 10, no. 3, pp. 101–110.

Tsao, R., and Eto, M., 1994, Effects of some natural photosensitizers on photolysis of some pesticides, *in* Helz, G.R., Zepp, R.G., and Crosby, D.G., eds., *Aquatic and surface photochemistry*: Lewis Publishers, Ann Arbor, Mich., pp. 163–171.

Tsuda, T., Aoki, S., Kojima, M., and Harada, H., 1990, Accumulation and excretion of oxadiazon, CNP, and chlomethoxynil by willow shiner: *Comp. Biochem. Physiol.*, v. 96C, no. 2, pp. 373–375.

——1991, Pesticides in water and fish from rivers flowing into Lake Biwa: *Toxicol. Environ. Chem.*, v. 34, no. 1, pp. 39–55.

Tulp, M.T.M., Haya, K., Carson, W.G., Zitko, V., and Hutzinger, O., 1979, Effect of salinity on uptake of ^{14}C-2,2′,4,5,5′-pentachlorobiphenyl by juvenile Atlantic salmon: *Chemosphere*, v. 8, no. 4, pp. 243–249.

Tulp, M.T.M., and Hutzinger, O., 1978, Some thoughts of aqueous solubilities and partition coefficients of PCB, and the mathematical correlation between bioaccumulation and physicochemical properties: *Chemosphere*, v. 7, no. 10, pp. 849–860.

U.S. Army Corps of Engineers, 1987, Final supplement to the final environmental impact statement: Black Warrior and Tombigbee rivers, Alabama (Maintenance): U.S. Army Corps of Engineers, Mobile (Ala.) District, Final environmental impact statement COESAM/PDEI-87/01, variously paged.

U.S. Department of Commerce, 1995, 1987 Census of Agriculture: Washington, D.C. [CD-ROM, dBase format].

U.S. Environmental Protection Agency, 1975, DDT, a review of scientific and economic aspects of the decision to ban its use as a pesticide: U.S. Environmental Protection Agency, EPA-540/1-75-022, 300 p.

——1976, National study to determine levels of chlorinated hydrocarbon insecticides in human milk: U.S. Environmental Protection Agency, EPA/540/9-78/005, v. 1, 151 p.

——1980a, Ambient water quality criteria for aldrin/dieldrin: U.S. Environmental Protection Agency, Office of Water Regulations and Standards, Criteria and Standards Division, Ambient Water Quality Criteria Publications series, no. 4, EPA 440/5-80-019, variously paged.

——1980b, Ambient water quality criteria for chlordane: U.S. Environmental Protection Agency, Office of Water Regulations and Standards, Criteria and Standards Division, Ambient Water Quality Criteria Publications series, no. 13, EPA 440/5-80-027, variously paged.

——1980c, Ambient water quality criteria for chlorinated benzenes: U.S. Environmental Protection Agency, Office of Water Regulations and Standards, Criteria and Standards Division, Ambient Water Quality Criteria Publications series, no. 14, EPA 440/5-80-028, variously paged.

——1980d, Ambient water quality criteria for DDT: U.S. Environmental Protection Agency, Office of Water Regulations and Standards, Criteria and Standards Division, Ambient Water Quality Criteria Publications series, EPA 440/5-80-038, variously paged.

——1980e, Ambient water quality criteria for endosulfan: U.S. Environmental Protection Agency, Office of Water Regulations and Standards, Criteria and Standards Division, Ambient Water Quality Criteria Publications series, no. 31, EPA 440/5-80-046, variously paged.

——1980f, Ambient water quality criteria for endrin: U.S. Environmental Protection Agency, Office of Water Regulations and Stndards, Criteria and Standards Division, Ambient Water Quality Criteria Publications series, no. 32, EPA 440/5-80-047, variously paged.

——1980g, Ambient water quality criteria for heptachlor: U.S. Environmental Protection Agency, Office of Water Regulations and Standards, Criteria and Standards Division, Ambient Water Quality Criteria Publications series, no. 35, EPA 440/5-80-052, variously paged.

——1980h, Ambient water quality criteria for hexachlorocyclohexane: U.S. Environmental Protection Agency, Office of Water Regulations and Standards, Criteria and Standards Division, Ambient Water Quality Criteria Publications series, no. 37, EPA 440/5-80-054, variously paged.

——1980i, Ambient water quality criteria for toxaphene: U.S. Environmental Protection Agency, Office of Water Regulations and Standards, Criteria and Standards Division, Ambient Water Quality Criteria Publications series, no. 54, EPA 440/5-80-076, variously paged.

——1982, Toxaphene: Decision document: U.S. Environmental Protection Agency, Office of Pesticide Programs, EPA 540/9-82-027. 184 p.

——1983, Analysis of the risks and benefits of seven chemicals used for subterranean termite control: U.S. Environmental Protection Agency, Office of Pesticides and Toxic Substances, Office of Pesticide Programs, EPA 540/9-83-005, variously paged.

——1984, Dicofol special review position document 2/3: U.S. Environmental Protection Agency, Office of Pesticides and Toxic Substances, Office of Pesticide Programs, variously paged.

——1985, Work/quality assurance project plan for the bioaccumulation study: U.S. Environmental Protection Agency, Office of Water Regulations and Standards, Monitoring and Data Support Division, variously paged.

——1986a, Dicofol: Intent to cancel registration of pesticide products containing dicofol; denial of application for registration of pesticide products containing dicofol; conclusion of special review: U.S. Environmental Protection Agency, Office of Pesticides and Toxic Substances.

——1986b, Ambient aquatic life water quality criteria for chlorpyrifos: U.S. Environmental Protection Agency, Office of Water Regulations and Standards, Criteria and Standards Division, EPA 440/5-86-005, 64 p.

——1986c, Ambient aquatic life water quality criteria for parathion: U.S. Environmental Protection Agency, Office of Water Regulations and Standards, Criteria and Standards Division, EPA 440/5-86-007, 65 p.

——1986d, Ambient aquatic life water quality criteria for pentachlorophenol: U.S. Environmental Protection Agency, Office of Water Regulations and Standards, Criteria and Standards Division, EPA 440/5-86-009, 127 p.

——1986e, Ambient aquatic life water quality criteria for toxaphene: U.S. Environmental Protection Agency, Office of Water Regulations and Standards, Criteria and Standards Division, EPA 440/5-86-006, 85 p.

——1986f, Quality criteria for water 1986 ("Gold Book"): U.S. Environmental Protection Agency, Office of Water Regulations and Standards, EPA 440/5-86-001, loose-leaf.

——1986g, Work/quality assurance project plan for the Bioaccumulation Study: U.S. Environmental Protection Agency, Ofice of Water Regulations and Standards, Monitor and Data Support Division, variously paged.

——1986h, Environmental Protection Agency: Guidelines for the health assessment of suspect developmental toxicants: *Fed. Reg.*, v. 51, no. 185, pp. 34028-34040.

——1986i, Environmental Protection Agency: Guidelines for carcinogen risk assessment: *Fed. Reg.*, v. 51, no. 185, pp. 33992-34003.

——1986j, Environmental Protection Agency: Guidelines for the health risk assessment of chemical mixtures: *Fed. Reg.*, v. 51, no. 185, pp. 34014-34025.

——1987, FIFRA scientific advisory approval of proposed changes to inert ingredient lists 1 and 2: U.S. Environmental Protection Agency internal memorandum from F.S. Bishop, 5 p.

——1988, Draft guide to drinking water health advisories: U.S. Environmental Protection Agency, Office of Drinking Water, Criteria and Standards Division.

——1989, *Drinking water health advisory. Pesticides*: Lewis Publishers, Chelsea, Michigan, 819 p.

——1990a, National survey of pesticides in drinking water wells: Phase I report: U.S. Environmental Protection Agency, Office of Pesticides and Toxic Substances, EPA 570/9-90-015, 98 p.

——1990b, Suspended, cancelled, and restricted pesticides: U.S. Environmental Protection Agency, Office of Pesticides and Toxic Substances, 20T-4002, variously paged.

——1991, One liner data base (version 2.06), available from U.S. Environmental Protection Agency, Environmental Fate and Groundwater Branch.

——1992a, National study of chemical residues in fish: U.S. Environmental Protection Agency, Office of Science and Technology, Standards and Applied Science Division, v. 1, EPA-823-R-92-008a, 166 p., 2 appendices.

——1992b, National study of chemical residues in fish: U.S. Environmental Protection Agency, Office of Science and Technology, Standards and Applied Science Division, v. 2, EPA-823-R-92-008b, variously paged.

——1992c, Fish sampling and analysis: A guidance document for issuing fish advisories: U.S. Environmental Protection Agency, Office of Science and Technology, Fish Contamination Section, draft, variously paged.

——1992d, Water quality standards, establishment of numeric criteria for priority toxic pollutants: States' compliance, final rule: *Fed. Reg.*, v. 57, no. 246, pp. 60848-60923.

——1992e, Handbook of RCRA ground-water monitoring constituents: Chemical and physical properties (40 CFR part 264, appendix IX): U.S. Environmental Protection Agency, Office of Solid Waste, EPA 530-R-92-022, 269 p.

——1992f, MIXTOX [toxicological interactions data base], diskette for personal computers (version 1.5 ECAO) available from U.S. Environmental Protection Agency, Cincinnati, Ohio.

——1992g, List of chemicals evaluated for carcinogenic potential, Memorandum from Reto Engler to Health Effects Division Branch Chiefs (10/14/92): U.S. Environmental Protection Agency, Office of Pesticide Programs.

——1992h, Pesticides in ground water database: A compilation of monitoring studies, 1971–1991: U.S. Environmental Protection Agency, EPA 734-12-92-001, variously paged.

——1993a, Sediment quality criteria for the protection of benthic organisms: dieldrin: U.S. Environmental Protection Agency, Office of Water, Office of Research and Development, Office of Science and Technology, Health and Ecological Criteria Division, EPA-822-R-93-015, variously paged.

——1993b, Sediment quality criteria for the protection of benthic organisms: endrin: U.S. Environmental Protection Agency, Office of Water, Office of Research and Development, Office of Science and Technology, Health and Ecological Criteria Division, EPA-822-R-93-016, variously paged.

——1993c, Toxic substance spreadsheet: U.S. Environmental Protection Agency, Region IV, Water Management Division.

——1994a, Notice of availability and request for comment on sediment-quality criteria and support documents: *Fed. Reg.*, v. 59, no. 11, pp. 2652–2656.

——1994b, Guidance for assessing chemical contaminant data for use in fish advisories. Volume 2, Risk assessment and fish consumption limits: U.S. Environmental Protection Agency, Office of Water, Office of Science and Technology, EPA 823-R-94-004, variously paged.

——1995a, Guidance for assessing chemical contaminant data for use in fish advisories. Volume 1, Fish sampling and analysis (2nd ed.): U.S. Environmental Protection Agency, Office of Water, Office of Science and Technology, EPA 823-R-95-007, variously paged.

——1995b, Guidance for assessing chemical contaminant data for use in fish advisories. Volume 4: Risk communication: U.S. Environmental Protection Agency, Office of Water, Office of Science and Technology, EPA 823-R-95-001, variously paged.

——1995c, Public draft health effects test guidelines OPPTS 870-3800: Reproduction and fertility effects: U.S. Environmental Protection Agency, Office of Prevention, Pesticides and Toxic Substances 712-C-94-208.

——1996a, Integrated Risk Information system (IRIS) data base, available from U.S. Environmental Protection Agency, Office of Health and Environmental Assessment, Environmental Criteria and Assessment Office, Cincinnati, Ohio.

——1996b, Drinking water regulations and health advisories: U.S. Environmental Protection Agency, Office of Water, EPA 822-B-96-002, variously paged.

——1966c, Guidelines for reproductive toxicity risk assessment: *Fed. Reg.*,, v. 61, pp. 56274–56322

——1997a, National sediment quality survey, v. 1 of The incidence and severity of sediment contamination in surface waters of the United States: U.S. Environmental Protection Agency, Office of Science and Technology, EPA 823-R-97-006, variously paged.

——1997b, Special report on environmental endocrine disruption: An effects assessment and analysis: U.S. Environmental Protection Agency, Risk Assessment Forum, EPA/630/R-96/012, 116 p.

——1997c, Office of Pesticide Programs reference dose tracking report, February 19, 1997: U.S. Environmental Protection Agency, Office of Pesticide Programs, 79 p.

——1998, The Food Quality Protection Act (FQPA) of 1996: Office of Pesticide Programs (Accessed November 16, 1998 on the World Wide Web at URL http://www.epa.gov/oppfead1/fqpa; last update January 22, 1998).

U.S. Fish and Wildlife Service, 1981, Contaminants in sediments and fish in the vicinity of Seymour Recycling Corporation, Seymour, Indiana: U.S. Fish and Wildlife Service, Report to U.S. Environmental Protection Agency, Seymour, IN.

——1986, A report on off-site contaminant migration and distribution from Seymour Recycling Corporation, Seymour, Indiana: U.S. Environmental Protection Agency report prepared by the U.S. Fish and Wildlife Service.

——1992, U.S. Fish and Wildlife Service National Contaminant Biomonitoring Program fish data data file, 1969-1986 [ASCII and Lotus files (8/24/92) available from Fish and Wildlife Service National Fisheries Contaminant Research Center, 4200 New Haven Road, Columbia, Mo].

U.S. Forest Service, 1978, *Report of the Forest Service, fiscal year 1977*: U.S. Department of Agriculture—Forest Service, Washington, D.C., variously paged.

——1985, *Report of the Forest Service, fiscal year 1984*: U.S. Department of Agriculture—Forest Service, Washington, D.C., variously paged.

——1989, *Report of the Forest Service, fiscal year 1988*: U.S. Department of Agriculture—Forest Service, Washington, D.C., variously paged.

——1990, *Report of the Forest Service, fiscal year 1989*: U.S. Department of Agriculture— Forest Service, Washington, D.C., variously paged.

——1991, *Report of the Forest Service, fiscal year 1990*: U.S. Department of Agriculture— Forest Service, Washington, D.C., variously paged.

——1992, *Report of the Forest Service, fiscal year 1991*: U.S. Department of Agriculture— Forest Service, Washington, D.C., variously paged.

——1993, *Report of the Forest Service, fiscal year 1992*: U.S. Department of Agriculture— Forest Service, Washington, D.C., variously paged.

——1994, *Report of the Forest Service, fiscal year 1993*: U.S. Department of Agriculture— Forest Service, Washington, D.C., variously paged.

U.S. Geological Survey, 1996, National Water Information System (NWIS) Database, data retrieved January 1996, available from the California District Office, Reports Services Section, Sacramento.

——1970, The national atlas of the United States of America: U.S. Department of the Interior, U.S. Geological Survey, Washington, D.C., pp. 158–159.

Unger, M., Kiaer, H., and Blinchert-Toft, M., 1984, Organochlorine compounds in human breast fat from deceased with and without breast cancer and in a biopsy material from newly diagnosed patients undergoing breast surgery: *Environ. Res.*, v. 34, pp. 24–28.

Van Hove Holdrinet, M., Frank, R., Thomas, R.L., and Hetling, L.J., 1978, Mirex in the sediments of Lake Ontario: *J. Great Lakes Res.*, v. 4, no. 1, pp. 69–74.

Vanderford, M.J., and Hamelink, J.L., 1977, Influence of environmental factors on pesticide levels in sport fish: *J. Pest. Monitor.*, v. 11, no. 3, pp. 138–145.

Varanasi, U., Chan, S.-L., McCain, B.B., Landahl, J.T., Schiewe, M.H., Clark, R.C., Brown, D.W., Myers, M.S., Krahn, M.M., Gronlund, W.D., and Macleod, W.D., Jr., 1989, National Benthic Surveillance Project: Pacific Coast. Part II, Technical presentation, the results for cycles I to III (1984–86): U.S. Department of Commerce, National Oceanic and Atmospheric Administration Technical Memorandum NMFS F/NWC 170, variously paged.

Varanasi, U., Chan, S.-L., McCain, B.B., Schiewe, M.H., Clark, R.C., Brown, D.W., Myers, M.S., Landahl, J.T., Krahn, M.M., Gronlund, W. D., and Macleod, W.D., Jr.,1988, National Benthic Surveillance Project: Pacific Coast, Part I. Summary and overview of the results for cycles I to III (1984–86): U.S. Department of Commerce, National Oceanic and Atmospheric Administration Technical Memorandum NMFS F/NWC 156, variously paged.

Varanasi, U., Reichert, W.L., Stein, J.E., Brown, D.W., and Sanborn, H.R., 1985, Bioavailability and biotransformation of aromatic hydrocarbons in benthic organisms exposed to sediment from an urban estuary: *Environ. Sci. Technol.*, v. 19, no. 9, pp. 836–841.

Veith, G.D., DeFoe, D.L., and Bergstedt, B.V., 1979a, Measuring and estimating the bioconcentration factor of chemicals in fish: *Can. J. Fish. Aquat. Sci.*, v. 36, pp. 1040–1048.

Veith, G.D., Kuehl, D.W., Leonard, E.N., Puglisi, F.A., and Lemke, A.E., 1979b, Polychlorinated biphenyls and other organic chemical residues in fish from major watersheds of the United States, 1976: *J. Pest. Monitor.*, v. 13, no. 1, pp. 1–11.

Veith, G.D., Kuehl, D.W., Leonard, E.N., Welch, K., and Pratt, G., 1981, Polychlorinated biphenyls and other organic chemical residues in fish from major United States watersheds near the Great Lakes, 1978: *J. Pest. Monitor.*, v. 15, no. 1, pp. 1–8.

Vetter, R.D., Carey, M.C., and Patton, J.S., 1985, Coassimilation of dietary fat and benzo(*a*)pyrene in the small intestine: *J. Lipid Res.*, v. 26, pp. 428–434.

Virginia State Health Department, 1976–1985, Pesticide residue reports: Virginia State Health Department, Bureau of Shellfish Sanitation.

Vogel, T.M., Criddle, C.S., and McCarthy, P.L., 1987, Transformations of halogenated aliphatic compounds: *Environ. Sci. Technol.*, v. 21, no. 8, pp. 722–736.

Voice, T.C., and Weber, W.J., Jr., 1983, Sorption of hydrophobic compounds by sediments, soils, and suspended solids—I, Theory and background: *Water Res.*, v. 17, no. 10, pp. 1433–1441.

Vonier, P.M., Crain, D.A., McLachlan, J.A., Guillette, L.J., and Arnold, S.F., 1996, Interaction of environmental chemicals with the estrogen and progesterone receptors from the oviduct of the American alligator: *Environ. Health Perspect.*, v. 104, no. 12, pp. 1318–1322.

Waddell, B., and Coyner, J., 1990, Screening evaluation of biologically active elements and pesticides in the Utah Lake wetlands proposed for a National Wildlife Refuge: U.S. Fish and Wildlife Service, Utah State Office, 13 p.

Waddell, B., Dolling, J., Linner, S., Stephens, D., and Stephensen, S., 1990, A preliminary evaluation of contaminants at the Bear River Migratory Bird Refuge acquisitions: U.S. Fish and Wildlife Service, Utah State Office, 15 p., appendix.

Wagner, S.L., Durand, L.R., Inman, R.D., Kiigemagi, U., and Deinzer, M.L., 1991, Residues of pentachlorophenol and other chlorinated contaminants in human tissues: Analysis by electron capture gas chromatography and electron capture negative ion mass spectrometry: *Arch. Environ. Contam. Toxicol.*, v. 21, no. 4, pp. 596–606.

Wall, G.J., Wilding, L.P., and Smeck, N.E., 1978, Physical, chemical, and mineralogical properties of fluvial unconsolidated bottom sediments in northwestern Ohio: *J. Environ. Qual.*, v. 7, no. 3, pp. 319–325.

Wallace, J.B., and Brady, U.E., 1971, Residue levels of dieldrin in aquatic invertebrates and effect of prolonged exposure on populations: *J. Pest. Monitor.*, v. 5, no. 3, pp. 295–300.

Waller, B.G., 1982, Water-quality characteristics of Everglades National Park, 1959–77, with reference to the effects of water management: U.S. Geological Survey Water-Resources Investigations Report 82-34, 51 p.

Walsh, G.E., 1972, Insecticides, herbicides, and polychlorinated biphenyls in estuaries: *J. Wash. Acad. Sci.*, v. 62, no. 2, pp. 122–139.

Wang, T.C., Hoffman, M.E., David, J., and Parkinson, R., 1992, Chlorinated pesticide residue occurrence and distribution in mosquito control impoundments along the Florida Indian River Lagoon: *Bull. Environ. Contam. Toxicol.*, v. 49, no. 2, pp. 217–223.

Wang, T.C., Krivan, J.P., Jr., and Johnson, R.S., 1979, Residues of polychlorinated biphenyls and DDT in water and sediment of the St. Lucie estuary, Florida, 1977: *J. Pest. Monitor.*, v. 13, no. 2, pp. 69–71.

Wangsness, D.J., 1983, Water and streambed-material data, Eagle Creek watershed, August 1980 and October and December 1982: U.S. Geological Survey Open-File Report 83-215, 41 p.

Ward, J.R., 1987, Surface-water quality in Pequea Creek Basin, Pennsylvania, 1977–79: U.S. Geological Survey Water-Resources Investigations Report 85-4250, 66 p.

Ware, G.W., Estesen, B.J., Buck, N.A., and Cahill, W.P., 1978, DDT moratorium in Arizona: Agricultural residues after seven years: *J. Pest. Monitor.*, v. 12, no. 1, pp. 1–3.

Ware, G.W., Estesen, B.J., and Cahill, W.P., 1971, DDT moratorium in Arizona: Agricultural residues after 2 years: *J. Pest. Monitor.*, v. 5, no. 3, pp. 276–280.

——1974, DDT moratorium in Arizona: Agricultural residues after 4 years: *J. Pest. Monitor.*, v. 8, no. 2, pp. 98–101.

Ware, G.W., Estesen, B.J., Jahn, C.D., and Cahill, W.P., 1970, DDT moratorium in Arizona: Agricultural residues after 1 year: *J. Pest. Monitor.*, v. 4, no. 1, pp. 21–24.

Ware, G.W., and Roan, C.C., 1970, Interaction of pesticides with aquatic microorganisms and plankton: *Residue Rev.*, v. 33, pp. 15–45.

Warner, K., and Fenderson, O.C., 1962, Effects of DDT spraying for forest insects on Maine trout streams: *J. Wildl. Manag.*, v. 26, no. 1, pp. 86–93.

Water and Air Research Inc., 1980, Engineering and environmental study of DDT contamination of Huntsville Spring Branch, Indian Creek and adjacent lands and waters, Wheeler Reservoir, Alabama: Prepared for the U.S. Army Corps of Engineers (Mobile, Ala.) by Water and Air Research Inc., Gainesville, Fla., v. 1, 56 p.

——1981, Richard B. Russell preimpoundment water quality study: Water and Air Research, Inc., Gainesville, Fla., variously paged.

——1982, Water quality management studies: Lake Seminole—February–December 1979, phase II.: U.S. Army Corps of Engineers, Technical Publication ACF 80-11, variously paged.

Watkins, S.A., and Simmons, C.E., 1984, Hydrologic conditions in the Chicod Creek Basin, North Carolina, before and during channel modifications, 1975–81: U.S. Geological Survey Water-Resources Investigations 84-4025, 36 p.

Watson, M.B., Killebrew, C.J., Schurtz, M.H., and Landry, J.L., 1981, A preliminary survey of Blind River, Louisiana, *in* Krumholz, L.A., ed., *The Warmwater Streams Symposium: A national symposium on fisheries aspects of warmwater streams: Proceedings of a symposium held at the University of Tennessee, Knoxville, Tennessee, 9–11 March, 1980*: American Fisheries Society, Southern Division, Lawrence, Kan., pp. 303–319.

Wauchope, R.D., 1978, The pesticide content of surface water draining from agricultural fields: A review: *J. Environ. Qual.*, v. 7, no. 4, pp. 459–472.

Wauchope, R.D., Buttler, T.M., Hornsby, A.G., Augustijn-Beckers, P.W.M., and Burt, J.P., 1992, The SCS/ARS/CES pesticide properties database for environmental decision-making: *Rev. Environ. Contam. Toxicol.*, v. 123, pp. 1–153.

Webber, E.C., Bayne, D.R., and Seesock, W.C., 1989, DDT contamination of benthic macroinvertebrates and sediments from tributaries of Wheeler Reservoir, Alabama: *Arch. Environ. Contam. Toxicol.*, v. 18, no. 5, pp. 728–733.

Weber, J.B., 1972, Interaction of organic pesticides with particulate matter in aquatic and soil systems, in Faust, S.D., ed., Fate of organic pesticides in the aquatic environment: A symposium sponsored by the Division of Pesticide Chemistry at the 161st meeting of the American Chemical Society, Los Angeles, Calif., March 29–31, 1971: American Chemical Society Advances in Chemistry series, no. 111, pp. 55–120.

Wedemeyer, G., 1968, Role of intestinal microflora in the degradation of DDT by rainbow trout (*Salmo gairdnieri*): *Life Sci.*, v. 7, pp. 219–223.

Wehr, M.A., Mattson, J.A., Bofinger, R.W., and Sajdak, R.L., 1992, Ground-application trial of hexazinone on the Ottawa National Forest: U.S. Department of Agriculture, Forest Service, North Central Forest Experiment Station, Research Paper NC series, no. 308, 34 p.

Weininger, D., 1978, Accumulation of PCBs by lake trout in Lake Michigan: *Diss. Abstr.*, v. 39, p. 1323.

Wells, F.C., Jackson, G.A., and Rogers, W.J., 1988, Reconnaissance investigation of water quality, bottom sediment, and biota associated with irrigation drainage in the lower Rio Grande Valley and Laguna Atascosa National Wildlife Refuge, Texas, 1986–87: U.S. Geological Survey Water-Resources Investigations Report 87-4277, 89 p.

Welsh, D., 1992, Concentrations of inorganic and organic chemicals in fish and sediments from the confluence of the Missouri and Yellowstone rivers, North Dakota, 1988–90: U.S. Fish and Wildlife Service, Fish and Wildlife Enhancement, Contaminant Report R6/110K/92, 38 p.

Welsh, D., and Olson, M.M., 1991, Pre-reconnaissance investigation of trace elements and organic contaminants in biota and sediments at Fork Clark and Heart Butte Irrigation Units: U.S. Fish and Wildlife Service Contaminant Report R6/105K/91, 15 p.

——1992, Concentrations of potential contaminants in shovelnose sturgeon from the Missouri River at Bismarck, North Dakota, 1991: U.S. Fish and Wildlife Service, Contaminant Report R6/111K/92, 15 p.

Wennekens, M.P., 1983, Western Alaska Ecological Services Resource Contaminant Assessment Program, Annual progress report, F.Y. 83: U.S. Fish and Wildlife Service, Annual Progress Report, variously paged.

Wenning, R.J., Harris, M.A., Ungs, M.J., Paustenbach, D.J., and Bedbury, H., 1992, Chemometric comparisons of polychlorinated dibenzo-*p*-dioxin and dibenzofuran residues in surficial sediments from Newark Bay, New Jersey and other industrialized waterways: *Arch. Environ. Contam. Toxicol.*, v. 22, no. 4, pp. 397–413.

West, S.D., Burger, R.O., Poole, G.M., and Mowrey, D.H., 1983, Bioconcentration and field dissipation of the aquatic herbicide fluridone and its degradation products in aquatic environments: *J. Agric. Food Chem.*, v. 31, pp. 579–585.

Wetzel, R.G., 1983, *Limnology* (2nd ed.): Saunders College Publishing, Harcourt Brace Jovanovich College Publishers, New York, 767 p.

Wharfe, J.R., and van den Broek, W.L.F., 1978, Chlorinated hydrocarbons in macroinvertebrates and fish from the lower Medway Estuary, Kent: *Mar. Pollut. Bull.*, v. 9, no. 3, pp. 76–79.

Wheeler, W.B., Jouvenaz, D.P., Wojcik, D.P., Banks, W.A., VanMiddelem, C.H., Lofgren, C.S., Nesbitt, S., Williams, L., and Brown, R., 1977, Mirex residues in nontarget organisms after application of 10-5 bait for fire ant control, northeast Florida, 1972–74: *J. Pest. Monitor.*, v. 11, no. 3, pp. 146–156.

White, D.H., and Krynitsky, A.J., 1986, Wildlife in some areas of New Mexico and Texas accumulate elevated DDE residues, 1983: *Arch. Environ. Contam. Toxicol.*, v. 15, no. 2, pp. 149–157.

White, D.H., Mitchell, C.A., Kennedy, H.D., Krynitsky, A.J., and Ribick, M.A., 1983, Elevated DDE and toxaphene residues in fishes and birds reflect local contamination in the lower Rio Grande Valley, Texas: *Southwest. Nat.*, v. 28, no. 3, pp. 325–333.

White, R., Jobling, S., Hoare, S.A., Sumpter, J.P., and Parker, M.G., 1994, Environmentally persistent alkylphenolic compounds are estrogenic: *Endocrinology*, v. 135, no. 1, pp. 175–182.

Whitmore, R.W., Kelly, J.E., and Reading, P.L., 1992, Executive summary, results, and recommendations, v. 1 *of* National home and garden pesticide use survey: U.S. Environmental Protection Agency Report EPA RTI/5100/17-01F, 140 p.

Whitten, P., Lewis, C., and Naftolin, F., 1993, A phytoestrogen diet induces the premature anovulatory syndrome in lactationally exposed female rats: *Biol. Reprod.*, v. 49, no. 5, pp. 1117–1121.

Whitten, P., and Naftolin, F., 1992, Effects of a phytoestrogen diet on estrogen-dependent reproductive processes in immature female rats: *Steroids*, v. 57, pp. 56–61.

Whitten, P., Russell, E., and Naftolin, F., 1992, Effects of a normal human-concentration, phytoestrogen diet on rat uterine growth: *Steroids*, v. 57, pp. 98–106.

Whittle, D.M., and Fitzsimons, J.D., 1983, The influence of the Niagara River on contaminant burdens of Lake Ontario biota: *J. Great Lakes Res.*, v. 9, pp. 295–302.

Whittle, K.J., Hardy, R., Holden, A.V., Johnston, R., and Pentreath, R.J., 1977, Occurrence and fate of organic and inorganic contaminants in marine animals: *Ann. New York Acad. Sci.*, v. 298, pp. 47–79.

Wiemeyer, S.N., Belisle, A.A., and Gramlich, F.J., 1978, Organochlorine residues in potential food items of Maine bald eagles (*Haliaeetus leucocephalus*), 1966 and 1974: *Bull. Environ. Contam. Toxicol.*, v. 19, no. 1, pp. 64–72.

Wilkes, F.G., and Weiss, C.M., 1971, The accumulation of DDT by the dragonfly nymph, *Tetragoneuria*: *Trans. Am. Fish. Soc.*, v. 100, no. 2, pp. 222–236.

Willford, W.A., Sills, J.B., and Whealdon, E.W., 1969, Chlorinated hydrocarbons in the young of Lake Michigan coho salmon: *Prog. Fish-Cult.*, v. 31, no. 4, p. 220.

Williams, D.E., and Buhler, D.R., 1984, Benzo(*a*)pyrene hydroxylase catalyzed by purified isozymes of cytochrome P-450 from β-naphthoflavone-fed rainbow trout: *Biochem. Pharmacol.*, v. 33, pp. 3743–3753.

Williams, R., and Holden, A.V., 1973, Organochlorine residues from plankton: *Mar. Pollut. Bull.*, v. 4, no. 7, pp. 109–111.

Williams, R.T., and Millburn, P., 1975, Detoxication mechanisms: The biochemistry of foreign compounds, *in* Blaschko, H.K.F., ed., *Physiological and pharmacological biochemistry*: Butterworths, London, MTP International Review of Science, Biochemistry, series one, v. 12, pp. 211–266.

Windholz, M., ed., 1976, *The Merck Index* (9th ed.): Merck & Co., Rahway, N.J., variously paged.

Winger, P.V., Schultz, D.P., and Johnson, W.W., 1985, Organochlorine residues in fish from the Yazoo National Wildlife Refuge: Proceedings of the Annual Conference of the Southeastern Association of Fish and Wildlife Agencies, 39th, Lexington, Kentucky, pp. 125–131.

Winger, P.V., Sieckman, C., May, T.W., and Johnson, W.W., 1984, Residues of organochlorine insecticides, polychlorinated biphenyls, and heavy metals in biota from Apalachicola River, Florida, 1978: *J. Assoc. Off. Anal. Chem.*, v. 67, no. 2, pp. 325–333.

Wojcik, D.P., Banks, W.A., Wheeler, W.B., Jouvenaz, D.P., VanMiddelem, C.H., and Lofgren, C.S., 1975, Mirex residues in nontarget organisms after application of experimental baits for fire ant control, southwest Georgia, 1971–72: *J. Pest. Monitor.*, v. 9, no. 3, pp. 124–133.

Wolfe, J.L., and Norment, B.R., 1973, Accumulation of mirex residues in selected organisms after an aerial treatment, Mississippi, 1971–72: *J. Pest. Monitor.*, v. 7, no. 2, pp. 112–116.

Wolfe, N.L., and Macalady, D.L., 1992, New perspectives in aquatic redox chemistry: Abiotic transformations of pollutants in ground-water and sediments: *J. Contam. Hydrol.*, v. 9, no. 1–2, pp. 17–34.

Wolfe, N.L., Mingelgrin, U., and Miller, G.C., 1990, Abiotic transformations in water, sediments, and soil, *in* Cheng, H.H., ed., *Pesticides in the soil environment: Processes, impacts, and modeling*: Soil Science Society of America, Madison, Wis., pp. 103–168.

Wolff, M.S., Toniolo, P.G., and Lee, E.W., 1993, Blood levels of organochlorine residues and risk of breast cancer: *J. Natl. Cancer Inst.*, v. 85, no. 8, pp. 648–652.

Woodham, D.W., Robinson, H.F., Reeves, R.G., Bond, C.A., and Richardson, H., 1977, Monitoring agricultural insecticides in the cooperative cotton pest management program in Arizona, 1971: First-year study: *J. Pest. Monitor.*, v. 10, no. 4, pp. 159–167.

Woodwell, G.M., 1967, Toxic substances and ecological cycles. Radioactive substances and pesticides such as DDT that are released in the environment may enter meteorological and biological cycles that distribute them and can concentrate them to dangerous levels: *Sci. Am.*, v. 216, no. 3, pp. 24–31.

Woodwell, G.M., Craig, P.P., and Johnson, H.A., 1971, DDT in the biosphere: Where does it go?: *Science*, v. 174, no. 4014, pp. 1101–1107.

Woodwell, G.M., Wurster, C.F., Jr., and Isaacson, P.A., 1967, DDT residues in an east coast estuary: Case of biological concentration of a persistent insecticide: *Science*, v. 156, no. 3776, pp. 821–824.

Worthing, C.R., and Walker, S.B., eds., 1987, *The pesticide manual: A world compendium* (8th ed.): The British Crop Protection Council, Thornton Heath, United Kingdom, 1081 p.

Wu, S.C., and Gschwend, P.M., 1986, Sorption kinetics of hydrophobic organic compounds to natural sediments and soils: *Environ. Sci. Technol.*, v. 20, no. 7, pp. 717–725.

Yalkowsky, S.H., and Morozowich, W., 1980, A physical chemical basis for the design of orally active prodrugs, *in* Ariens, E.J., ed., *Drug design*: Academic Press, New York, Medical Chemistry series, v. 11, pp. 121–185.

Yang, C.F., and Sun, Y.P., 1977, Partition distribution of insecticides as a critical factor affecting their rates of absorption from water and relative toxicities to fish: *Arch. Environ. Contam. Toxicol.*, v. 6, no. 2/3, pp. 325–335.

Yess, N.J., Houston, M.G., and Gunderson, E.L., 1991a, Food and Drug Administration pesticide residue monitoring of foods: 1978–1982: *J. Assoc. Off. Anal. Chem.*, v. 74, no. 2, pp. 265–272.

——1991b, Food and Drug Administration pesticide residue monitoring of foods: 1983–1986: *J. Assoc. Off. Anal. Chem.*, v. 74, no. 2, pp. 273–280.

Yorke, T.H., Stamer, J., and Pederson, G.L., 1985, Effects of low-level dams on the distribution of sediment, trace metals, and organic substances in the lower Schuylkill River Basin, Pennsylvania: U.S. Geological Survey Water-Supply Paper 2256-B.

Young, A.L., Thalken, C.E., and Harrison, D.D., 1981, Persistence, bioaccumulation, and toxicology of TCDD in an ecosystem treated with massive quantities of 2,4,5-T herbicide: *Proceedings of the Western Society of Weed Science*, v. 35, pp. 70–77.

Young, D.R., Heesen, T.C., and McDermott, D.J., 1976, An offshore biomonitoring system for chlorinated hydrocarbons: *Mar. Pollut. Bull.*, v. 7, no. 8, pp. 156–159.

Young, R.G., St. John, L., and Lisk, D.J., 1971, Degradation of DDT by goldfish: *Bull. Environ. Contam. Toxicol.*, v. 6, no. 4, pp. 351–354.

Yule, W.N., and Tomlin, A.D., 1970, DDT in forest streams: *Bull. Environ. Contam. Toxicol.*, v. 5, no. 6, pp. 479–488.

Yurawecz, M.P., and Roach, J.A.G., 1978, Gas-liquid chromotographic determination of chlorinated norbornene derivatives in fish: *J. Assoc. Off. Anal. Chem.*, v. 61, no. 1, pp. 26–31.

Zabik, M.E., 1974, Polychlorinated biphenyl levels in raw and cooked chicken and chicken broth: *Poultry Sci.*, v. 53, no. 1, p. 1785–1790.

——1984, Environmental contaminants' potential reduction by the cooking of meat, poultry, and fish: *School Food Serv. Res. Rev.*, v. 8, no. 1, pp. 17–21.

Zabik, M.E., Merrill, C., and Zabik, M.J., 1982, PCBs and other xenobiotics in raw and cooked carp: *Bull. Environ. Contam. Toxicol.*, v. 28, no. 6, pp. 710–715.

Zabik, M.E., Olson, B., and Johnson, T.M., 1978, Dieldrin, DDT, PCBs, and mercury levels in freshwater mullet from the upper Great Lakes, 1975–76: *J. Pest. Monitor.*, v. 12, no. 1, pp. 36–39.

Zabik, M.J., Pape, B.E., and Bedford, J.W., 1971, Effect of urban and agricultural pesticide use on residue levels in the Red Cedar River: *J. Pest. Monitor.*, v. 5, no. 3, pp. 301–308.

Zaharko, D.S., Dedrick, R.L., and Oliverio, V.T., 1972, Prediction of the distribution of methotrexate in the sting rays *Dasyatidae sabrina* and *D. sayi* by use of a model developed in mice: *Comp. Biochem. Physiol.*, v. 42A, no. 1, pp. 183–194.

Zahnow, E.W., and Riggleman, J.D., 1980, Search for linuron residues in tributaries of the Chesapeake Bay: *J. Agric. Food Chem.*, v. 28, no. 5, pp. 974–978.

Zdanowicz, V.S., and Gadbois, D., 1990, Contaminants in sediment and fish tissue from estuarine and coastal sites of the northeastern United States: Data summary for the baseline phase of the National Status and Trends Program Benthic Surveillance Project, 1984–1986: U.S. Department of Commerce, National Oceanic and Atmospheric Administration Technical Memorandum NMFS-F/NEC 79. 138 p.

Zepp, R.G., Baughman, G.L., and Schlotzhauer, P.F., 1981, Comparison of photochemical behavior of various humic substances in water; I. Sunlight induced reactions of aquatic pollutants photosensitized by humic substances: *Chemosphere*, v. 10, no. 1, pp. 109–117.

Zepp, R.G., and Cline, D.M., 1977, Rates of direct photolysis in aquatic environment: Environ. Sci. Technol., v. 11, no. 4, pp. 359–366.

Zepp, R.G., and Schlotzhauer, P.F., 1983, Influence of algae on photolysis rates of chemicals in water: *Environ. Sci. Technol.*, v. 17, pp. 462–468.

Ziliukiene, V.R., 1989, Distribution of chlororganic pesticides in the organs and tissues of vimba (*Vimba vimba* L.) from selected water bodies of Lithuania: *Acta Hydrobiol.*, v. 31, no. 3/4, pp. 319–328.

Zitko, V., 1980, Metabolism and distribution by aquatic animals, *in* Hutzinger, O., ed.,*The handbook of environmental chemistry*: Springer-Verlag, Berlin, pp. 221–229.

Zitko, V., and Hutzinger, O., 1976, Uptake of chloro- and bromobiphenyls, hexachloro- and hexabromobenzene by fish: *Bull. Environ. Contam. Toxicol.*, v. 16, no. 6, pp. 665–673.

Index